Sedimentary Environments:

Processes, Facies and Stratigraphy

Sedimentary Environments:
Processes, Facies and Stratigraphy

Edited by H.G. Reading

Department of Earth Sciences, University of Oxford

THIRD EDITION

**Blackwell
Science**

© 1978, 1986, 1996 by
Blackwell Science Ltd
Editorial Offices:
Osney Mead, Oxford OX2 0EL
25 John Street, London WC1N 2BL
23 Ainslie Place, Edinburgh EH3 6AJ
238 Main Street, Cambridge
  Massachusetts 02142, USA
54 University Street, Carlton
  Victoria 3053, Australia

Other Editorial Offices:
Arnette Blackwell SA
  224, Boulevard Saint Germain
  75007 Paris, France

Blackwell Wissenschafts-Verlag GmbH
  Kurfürstendamm 57
  10707 Berlin, Germany

  Zehetnergasse 6
  A-1140 Wien
  Austria

Set by Setrite Typesetters, Hong Kong
Printed and bound in Great Britain
at the University Press, Cambridge

The Blackwell Science logo is a
trade mark of Blackwell Science Ltd,
registered at the United Kingdom
Trade Marks Registry

First published 1978
Second edition 1986
Third edition 1996

A catalogue record for this title
is available from the British Library

ISBN 0-632-03627-3

Library of Congress
Cataloging-in-Publication Data

Sedimentary environments:
processes, facies, and stratigraphy/
edited by H.G. Reading. — 3rd ed.
    p.   cm.
  Rev. ed. of: Sedimentary environments
and facies. 1978.
  Includes bibliographical references
and index.
  ISBN 0-632-03627-3
  1. Rocks, Sedimentary.   2. Facies
(Geology)   3. Sedimentation and
deposition.
  I. Reading, H.G.
  II. Sedimentary environments and
facies.
  QE471.S378   1996
  539.7′4—dc20   95-48457
                                      CIP

DISTRIBUTORS

Marston Book Services Ltd
PO Box 269
Abingdon Oxon OX14 4YN
(*Orders*: Tel: 01235 465500
              Fax: 01235 465555)

USA
Blackwell Science, Inc.
238 Main Street
Cambridge, MA 02142
(*Orders*: Tel: 800 215-1000
                  617 876-7000
              Fax: 617 492-5263)

Canada
Copp Clark, Ltd
2775 Matheson Blvd East
Mississauga, Ontario
Canada, L4W 4P7
(*Orders*: Tel: 800 263-4374
                  905 238-6074)

Australia
Blackwell Science Pty Ltd
54 University Street
Carlton, Victoria 3053
(*Orders*: Tel: 03 9347 0300
              Fax: 03 9349 3016)

Contents

6 Clastic coasts, 154
H.G. Reading & J.D. Collinson

7 Shallow clastic seas, 232
H.D. Johnson & C.T. Baldwin

Contents

8 Marine evaporites: arid shorelines and basins, 281

A.C. Kendall & G.M. Harwood

9 Shallow-water carbonate environments, 325

V.P. Wright & T.P. Burchette

10 Deep seas, 395

D.A.V. Stow, H.G. Reading & J.D. Collinson

Contents

13 Problems and perspectives, 568
H.G. Reading

Contributors

Philip A. Allen
Department of Earth Sciences, University of Oxford, Parks Road, Oxford OX1 3PR, UK

Christopher T. Baldwin
College of Arts and Sciences, Sam Houston State University, Box 2209, Huntsville, Texas 77341, USA

Trevor P. Burchette
BP Exploration Operating Company, Uxbridge One, 1 Harefield Road, Uxbridge, Middlesex UB8 1PD, UK

John D. Collinson
Barrow Cottage, Marchamley, Shrewsbury SY4 5LD, UK

Gillian M. Harwood
Died on 12th March 1996

Howard D. Johnson
Department of Geology, Imperial College, Prince Consort Road, London SW7 2BP, UK

Alan C. Kendall
Department of Environmental Sciences, University of East Anglia, Norwich NR4 7TJ, UK

Gary A. Kocurek
Department of Geological Sciences, University of Texas, PO Box 7909, Austin, Texas 78713, USA

Bruce K. Levell
Shell UK Exploration and Production, Shell-Mex House, Strand, London WC2R 0DX, UK

Julia M.G. Miller
Department of Geology, Vanderbilt University, Box 7-B, Nashville, Tennessee 37235, USA

Geoffrey J. Orton
Geology Department, McMaster University, 1280 Main Street West, Hamilton, Ontario L8S 4MI, Canada

Harold G. Reading
Department of Earth Sciences, University of Oxford, Parks Road, Oxford OX1 3PR, UK

Dorrik A.V. Stow
Department of Geology, University of Southamptom, Empress Dock, European Way, Southampton SO14 3ZH, UK

Michael R. Talbot
Geologisk Institutt, Universitet i Bergen, Allégt. 41, 5007 Bergen, Norway

V. Paul Wright
Postgraduate Research Institute for Sedimentology, University of Reading, Whiteknights, Reading RG6 2AB, UK

Preface

The first edition of this book was conceived in 1974 to provide a comprehensive text, covering modern and ancient environments, suitable for advanced university students, research workers and professional geologists. To cover all environments and facies with the authority of an active research worker, we formed a group of authors who knew each other well and shared a similar philosophical view. We could criticize, amend and integrate each other's contributions, while retaining individual styles and responsibility for each chapter.

The outlines of the second edition remained the same as the first edition, although each chapter was extensively rewritten, with many new figures, and a few changes of authors. The third edition has also been changed substantially. This is primarily because of the need to add sequence stratigraphic models to those of facies sedimentology. Two chapters have been dropped ('Pelagic Environments' and 'Sedimentation and Tectonics'); 'Volcanic Environments' has been added and the two chapters previously devoted to 'Deltas' and 'Siliciclastic Shorelines' have been combined under 'Clastic Coasts'.

As with all textbooks, our main problems have been the selection of material and the need to strike a balance between comprehensiveness and cost. In the first edition many chapters had to be reduced to half of their original length and references pruned. In this edition the problem has been exacerbated by the growth of sedimentological material and concepts in the past 20 years. Almost every chapter now has at least one textbook devoted to it, as well as several special publications.

In addition, we have tried to incorporate some of the concepts and ideas of sequence stratigraphy, yet retain the fundamentals of sedimentology, without which no sequence stratigraphic model can be validated.

No book is solely the product of its authors. In this book we have incorporated facts, ideas, philosophies and prejudices of many others; some are quoted and acknowledged; others have been absorbed by us over many years from teachers, colleagues, friends and students. It was the late Maurits de Raaf who taught many of us to combine careful facies analysis and an examination of every detail in a rock with an unceasing search for the processes which formed it, to doubt any hypothesis we may be defending and to be aware of becoming too dogmatic. This philosophy, we hope, will be followed by all who read this book, look at rocks or develop models.

We want first to thank those authors of previous editions without whose insights this edition would not have been possible, Marc M. Edwards, Trevor Elliott, Hugh Jenkyns, Andrew H.G. Mitchell, Nicholas A. Rupke, Charlotte Schreiber, Bruce W. Sellwood, Maurice E. Tucker and Roger Till. Individual chapters or parts of chapters were read by Benjamin F. Adams, Gail M. Ashley, Bernard M. Besly, Ronald C. Blakey, Michael D. Blum, Michael J. Branney, Stephen P. Carey, Ray A. Cas, John C. Crowell, Cristino J. Dabrio, Robert W. Dalrymple, José M.L. Dominguez, Marc B. Edwards, Richard V. Fisher, Peter Francis, Bibek Ghosh, Michael J. Hambrey, Karen G. Havholm, Markes E. Johnson, Nicholas Lancaster, Francesco Massari,

Tom Pierson, Guy Plint, George Postma, Greg Valentine, Jonathon E. Verlander. We wish to thank all those who typed individual chapters and in particular Mrs P. McNiff who typed many drafts and much of the final manuscript. Responsibility, however, for omissions, lack of balance or for errors must remain with us. We wish also to acknowledge the many societies and publishing houses who permitted illustrations from articles in their journals and books to be used as the basis for our figures; to the Geological Society of America, the Geological Society of London, the Society for Sedimentary Geology (SEPM), the International Association of Sedimentologists, the American Association of Petroleum Geologists, Canadian Society of Petroleum Geologists, International Glaciological Society, Norges Geologiske Undersøkelse, Scott Polar Research Institute, Elf Aquitaine, American Geophysical Union, US Geological Survey, American Association for the Advancement of Science, The Royal Society, Geological Association of Canada and Smithsonian Institution; to Academic Press, American Journal of Science, Annual Reviews Inc., A.A. Balkema Publishers, Cambridge University Press, Chapman and Hall, Economic Geology Publishing Co., Edward Arnold Publishers, Elsevier Science, Geological Society Publishing House, Graham and Trotman Ltd., John Wiley and Sons, Longmans Group, Reidel, Macmillan Magazines Ltd, Springer-Verlag, Oxford University Press, University of Chicago Press and Unwin Hyman.

The book could never have come to fruition without Blackwell Science, in particular Simon Rallison and Jonathan Rowley, always supported by Robert Campbell. Not only do more figures in this book emanate from the publications of Blackwell Science than any other publisher, but their constant encouragement has supported us, especially when we ran into difficulties.

Finally, no book can be completed without families and friends whose forebearance and help has sustained us over the years that this book has been in preparation.

Harold G. Reading
June, 1996 Oxford

Gill Harwood died on 12th March 1996 after a long struggle against cancer. We trust that her chapter with Alan Kendall will remain a permanent tribute to her devotion to science and research on evaporites.

Introduction

H.G. Reading

1.1 Development of sedimentology and sedimentary geology

Sedimentology is concerned with the composition and genesis of sediments and sedimentary rocks, and the creation of predictive models. It includes sedimentary petrology, which is the study of the nature and relationships of the constituent particles and their diagenesis. It differs from stratigraphy in that time is not of prime importance except in so far as it deals with sequences and the law of Superposition is fundamental. It overlaps with other geological disciplines such as geochemistry, mineralogy, palaeontology and tectonics. In addition, sedimentology takes from and contributes to chemistry, biology, physics, geomorphology, oceanography, soil science, civil engineering, climatology, glaciology and fluid dynamics. When sedimentology and stratigraphy are combined they become the science of sedimentary geology, a term that is generally used in a wider sense than sedimentology. However, since the genesis of sedimentary rocks cannot be understood without reference to the time framework within which they were deposited, sedimentology has always, and still does, embrace a large element of stratigraphy as well.

Modern sedimentology, characterized by the study of processes, can be said to have started with the publication of Kuenen and Migliorini's (1950) paper on turbidity currents as a cause of graded bedding (see Sect. 10.2.3). Before 1950, the sciences of stratigraphy, concerned primarily with correlation and broad palaeogeographic reconstructions, and sedimentary petrology, concerned primarily with the microscopic examination of sedimentary rocks, had evolved more or less independently, with the exception of a few notable contributions such as those of Sorby (1859, 1879).

The turbidite concept developed from Daly's (1936) hypothesis that turbidity currents might be the agent of erosion of submarine canyons and from the model flume experiments of Kuenen (1937, 1950). Under the impact of this concept, geologists, who for years had been working on 'flysch' began to realize that an actual mechanism of flow could be envisaged as the agent of transport and deposition of graded sand beds. Geologists could now look at sedimentary rocks as sediments that had modern analogues, some aspects of which could be simulated by experiment. Familiar rocks could be examined with new insight and such features as sole marks, previously largely undetected, because they were not understood, could be described and perhaps explained.

As data on the composition, texture and structures of sedimentary rocks have grown, models have been developed that lean on a comparison with processes observed in modern environments and in experiments. Although many of these comparative models are founded on observations that can be made in now-active environments, others are the result of a creative blend of experience and imagination. Matching process with the corresponding sedimentary product is often difficult. In present-day shallow-water environments, processes are readily studied and measured, but data on their products are difficult to collect. In the Ancient, composition, texture and sedimentary structures are normally easily observed, but the processes which produced the observed features cannot be directly measured. A prime aim of sedimentology is to narrow the gap between modern process and past product, in some cases aided by an understanding of diagenesis.

The many books on sedimentology which appeared in the 1960s and 1970s reflected the surge of new concepts. Sedimentary structures and their use in basinal reconstruction were emphasized by Potter and Pettijohn (1963). Physical processes of sedimentation and their importance in understanding sedimentary structures were first brought to the attention of geologists in the volume edited by Middleton (1965) and a deeper understanding of some of the processes has been developed by J.R.L. Allen (1968, 1984). A succinct description and explanation of the processes of formation of sedimentary structures is that of Collinson and Thompson (1982). In the carbonate field the first book to reflect the progress in matching process with product was that edited by Ham (1962). However, it scarcely mentioned diagenesis, understanding of which advanced rapidly in the 1960s to culminate in the most important book in the limestone field, that by Bathurst (1971, 1975). The importance of biological processes was for long underestimated by most sedimentologists and few textbooks mentioned organisms except as disturbers of sediment. German authors, however, such as Seilacher and the Tübingen school, and Schäfer (1972) cultivated the science of analysing faunas and the effect of their life and death patterns upon sediments. The genesis of sediments was stressed particularly by Blatt, Middleton and Murray (1972, 1980) who emphasized the mechanisms and processes of physical and chemical sedimentation.

In these early books, environmental analysis was not discussed at length. However, Reineck and Singh (1973) covered both physical and biological sedimentary processes and structures and also modern clastic sedimentary environments, with particular emphasis on the shallow-marine. J.L. Wilson (1975) did the same for carbonate facies, emphasizing the impact of organic evolution on carbonate build-ups.

These were followed by Reading (1978) and R.G. Walker (1979). The former started from a consideration of modern environments and moved through process to facies. R.G. Walker (1979, 1984) emphasized facies using modern environments as an aid to their interpretation. The latest edition (R.G. Walker & James, 1992), which is the best general introduction, expands to include sequence stratigraphic concepts, especially sea-level controls. Galloway and Hobday (1983) stressed terrigenous

clastic sediments, especially their economic aspects while the Association of American Petroleum Geologists published two magnificently illustrated volumes, one on sandstone depositional environments (Scholle and Spearing, 1982) and the other on carbonate depositional environments (Scholle, Bebout & Moore, 1983). The applied elements of sedimentology were also emphasized by Brenchley and Williams (1985). The Spanish equivalent to the present book is that edited by Arche (1989).

Meanwhile there was a growing number of special publications from societies and associations and of textbooks. The most important special publications are those published by the Society of Economic Paleontologists and Mineralogists (SEPM) and the International Association of Sedimentologists (IAS). Many of these deal with specific environments and are listed under Further Reading in the appropriate chapters. In addition, the economic importance, especially to the petroleum industry, of sedimentology has resulted in a number of special publications from the American Association of Petroleum Geologists (AAPG) and the Canadian Association of Petroleum Geologists (CAPG).

Throughout these decades some geologists kept a broader perspective alive by considering the relationship between sedimentation and tectonics, firstly through the concept of the geosyncline (Kay, 1951; Aubouin, 1965) and then through that of plate tectonics (Mitchell & Reading, 1969; Dewey & Bird, 1970; Dickinson, 1971a). Many books were devoted to tectonics and sedimentation (e.g. Burke & Drake, 1974; Dickinson, 1974a; Dott & Shaver, 1974). However, it was not until the late 1970s that two major developments took place. One was the consideration of sedimentary rocks on a broader basis than had been done by the rather narrowly focused sedimentologists of the 1960s and 1970s. This was to look at basins in their entirety, using both geophysical and stratigraphical methods and data acquired from extensive outcrops and the subsurface. This led to a number of books (e.g. Miall, 1984; P.A. Allen & J.R. Allen, 1990) on basin analysis. Meanwhile, petroleum geologists developed seismic stratigraphy, turning it into sequence stratigraphy (Sect. 2.2) (Payton, 1977; Wilgus, Hastings *et al.*, 1988), a philosophy rooted in the concepts of stratigraphy and sedimentation of Sloss (1950, 1963) (Sect. 2.2). Thus there has been a return, in part, to the stratigraphical concepts of the early part of this century that were put into abeyance by the process, facies sedimentologists of the 1960s and 1970s.

1.2　Scope and philosophy of this book

The prime purpose of this book is to show how ancient environments may be reconstructed by interpreting first the process or processes which gave rise to sedimentary rocks and then the environment in which these processes operated. To achieve this, an understanding of the factors, such as climate, tectonics and changing base level, that control the environment, both modern and ancient, is essential.

The reconstruction of environments requires the following.

1 A thorough description of the rocks, either in the field or in core, with additional laboratory data obtained from samples collected to answer specific questions. Since time is limited, rock description is inevitably selective, emphasizing some features, underplaying others and rejecting yet others as quite unimportant. The selection depends on the judgement, experience and purpose of the investigator. Judgement and experience take time to acquire and can only be gained by seeing lots of rocks. The absence of certain features is often as important as their presence. For example, the consistent absence of shallow-water features, rather than any positive evidence for great depth, leads sedimentologists to infer that most turbidites were deposited in deep water. The utilization of negative evidence requires a familiarity with a wide range of sedimentary rocks and environments.

2 An awareness of processes so that, simultaneously with rock description, the strength or direction of the current or the type of flow which carried and deposited each grain is being considered. Such questions as 'What was the oxidation state, salinity or pH of the water?' or 'What forms of life were extant?' can also be asked. We also have to consider the later alteration or diagenetic processes which may have changed not only the colour of the rocks but also their grain size and composition. Particular processes are seldom confined to one environment, though they may be absent from some, and therefore similar rocks may form in different environments.

3 A knowledge of present-day environments, the processes which operate within them and the factors that control them. We need to know how environments evolve both under stable conditions and as sea level, climate, tectonic activity or sediment supply change. Our understanding of environments is bounded not only by the limits on knowledge of the present day, whole regions still being virtually unexplored, but also by the uniqueness of the present. For example, the recent rise of sea level allows us readily to develop models of transgressive sedimentation in shallow seas but makes it difficult to develop models for periods of relatively stable or falling sea level. It is salutary to consider how difficult it would be to conceive a model of glacial sedimentation had the human race developed in an entirely non-glacial period. It would have taken a courageous scientist to postulate, from a limited knowledge of sea ice and snow falls, the hypothesis of large ice caps and glaciers which could erode and deposit large quantities of sediment.

Thus the emphasis of the book will be on:

1 *environments*, reviewing modern environments, with their associated processes and products;

2 *processes*, concentrating on those that occur in each environment and showing how they relate to the resultant sediment: they are not discussed for their own sake as there are already several good textbooks on processes (e.g. Pye, 1994) and the genesis of sedimentary rocks and structures (e.g. Collinson & Thompson, 1982; J.R.L. Allen, 1984, 1985);

3 *facies*, stressing field data, facies relationships, sequences and associations;

4 *sequence stratigraphy*, or facies analysis using a chronostratigraphic framework defined by extensive correlative surfaces;

5 *controls*, emphasizing the interplay of sediment supply, tectonics and sea-level changes; and

6 *geological applications*, illustrating how sedimentary rocks are related to their geological background and how the recognition of sedimentary processes and environments illuminates our understanding of past climates, the chemistry of the oceans and the land, the development of life and world tectonics.

Only a fraction of the material available has been incorporated in this book. Innumerable examples, ideas and alternative models have had to be eliminated to save space. We hope we have brought out the more important ones but judgement is subjective, and none of us can claim to have read everything, let alone understood all that has been published on the subjects covered in our chapter. The chapters are not comprehensive reviews of the subject. To achieve that would require at least one textbook for each chapter; in most cases, these have already been published. What we have tried to do is to produce a readable text, covering the essentials, which is a starting point for the subject, in particular for those either just entering sedimentology, or for those who want an introduction into those areas of sedimentology that they are not working on.

1.3   Organization of the book

There is no unique division of environments and there is no simple match between environment, processes and facies. An environment is a particular set of physical, chemical and biological variables; a facies is a body of rock with specified characteristics, and many processes operate in more than one environment.

Matching environment, process and facies is seldom easy, and frequently decisions have had to be made between dividing the book on a basis of environment, process or of facies. Division of most chapters was on environment, because that is the prime emphasis of the book and most major environments are dominated by a particular suite of processes, which are then covered in that chapter. Many facies, however, cut across several chapters. A particular difficulty is presented by evaporites which are now known to have occurred in almost as many environments as have sandstones, ranging from deserts to lakes, coastal flats and deep seas, and therefore the chapter on arid shorelines includes deep-water evaporites as well as those found in sabkhas. Inevitably, because there is a continuum from subaerial through coastal to deep seas, environments have had to be arbitrarily divided. Artificial boundaries have had to be placed between individual chapters. Although chapters have been selected mainly on geographical environment, in some cases emphasis has been given to the facies, as in the separation of 'Shallow clastic seas' from 'Shallow-water carbonate environments'. Here the processes which transport and deposit the sediments are essentially the same but, because the sediment is derived, in one case from the erosion of mainly extrabasinal sources and in the other case from biochemical intrabasinal sources, the facies types and facies patterns are very different, especially the relationship of sequences to relative sea-level changes. In the case of 'Desert aeolian systems' and 'Glacial sediments' climate, with consequent distinctive processes, is the prime factor in division. The same is true for 'Clastic coasts' and 'Marine evaporites: arid shorelines and basins' which have been separated on the basis of their distinctive climates and, therefore, facies. 'Volcanic environments', like 'Glacial sediments', have unique sources of sediment and embrace all other environments.

Within chapters, the organization of sections is no less difficult. Should the prime section headings be based on controlling factors, environment, process, facies or even grain size? How closely should modern and ancient be intertwined, especially when there is a complete gradation from a process and product that can be measured and observed today through those that we can observe within a generally known environment that was formed a few years ago, to the clearly ancient rock record where all the controlling processes and factors have to be inferred? In many chapters, the section headings and their organization are substantially different from previous editions, and during revisions they were changed more than once. No pattern is ideal, and the end is a compromise that must result in some overlap. Because of this overlap between chapters and

sections within chapters, the reader needs to make the links by reference to the Contents and the Index. Cross referencing between chapters has been kept to the minimum to avoid breaking the text.

Further reading

Arche A. (Ed.) (1989) *Sedimentología*, 1, 541 pp.; 2, 526 pp. Nuevas tendencias 11, 12, Consejo Superior de Investigaciones Científica, Madrid.

Brenchley P.J. & Williams B.P.J. (Eds) (1985) *Sedimentology: Recent Developments and Applied Aspects*, 342 pp. *Spec. Publ. geol. Soc. Lond.*, **18**.

Einsele G., Ricken W. & Seilacher A. (Eds) (1991) *Cycles and Events in Stratigraphy*, 955 pp. Springer-Verlag, Berlin.

Friedman G.M., Sanders J.E. & Kopaska-Merkel D.C. (1992) *Principles of Sedimentary Deposits: Stratigraphy and Sedimentology*, 717 pp. Macmillan, New York.

Galloway W.E. & Hobday D.K. (1983) *Terrigenous Clastic Depositional Systems*, 480 pp. Springer-Verlag, New York.

Leeder M.R. (1982) *Sedimentology: Process and Product*, 344 pp. George, Allen & Unwin, London.

Scholle P.A. & Spearing D. (Eds) (1982) *Sandstone Depositional Environments*, 410 pp. *Mem. Am. Ass. petrol. Geol.*, **31**, Tulsa.

Scholle P.A., Bebout D.G. & Moore C.H. (Eds) (1983) *Carbonate Depositional Environments*, 708 pp. *Mem. Am. Ass. petrol. Geol.*, **33**, Tulsa.

Walker R.G. & James N.P. (Eds) (1992) *Facies Models: Response to Sea Level Change*, 409 pp. Geol. Ass. Can., Waterloo, Ontario.

Controls on the sedimentary rock record

2

H.G. Reading & B.K. Levell

....

2.1 Controlling factors

Sedimentation results from the interaction of the *supply* of sediment, its *reworking* and modification by physical, chemical and biological processes, and *accommodation space* – that is, the space available for potential sediment accumulation. In many settings reworking allows only a small proportion of the delivered sediment to be preserved. Most is removed either almost immediately by an increase in physical energy, such as that produced by a storm or tidal current, by sediment instability as with deposition on a slope, by chemical dissolution, or, over a longer period of time, by environmental changes such as channel migration or shoreline advance or retreat. The accommodation space is controlled largely by external processes such as changes in sea level, climate, tectonic movements, volcanic activity, compaction and longer-term subsidence rates which together define a depositional base level.

2.1.1 Sediment supply

The supply of sediment varies in volume, composition and grain size, as well as in the mechanism and rate of delivery. These variations are the result of the climate, basinal water chemistry and the tectonics and bedrock geology of the source area. Where the supply of land-derived (terrigenous) sediment is abundant, siliciclastic sediments predominate. Where this is low or absent, physical erosion may be effective and chemical, biochemical and biological processes have a chance to produce or modify sediments. Evaporites, carbonates, diatomites, cherts, ironstones, phosphorites and carbonaceous sediments then predominate. There is thus a fundamental distinction between terrigenous *extrabasinal* particles derived ultimately, if not immediately, from outside the sedimentary basin, and biochemical *intrabasinal* particles generated within the depositional basin. Erupting volcanoes, however, upset this general rule and may supply large volumes of intrabasinal terrigenous volcaniclastic material of unusual compositions. In addition, substantial amounts of 'intrabasinal' terrigenous sediment may be supplied, for instance, from uplifted fault blocks or pre-rift thermal domes within large basins.

TERRIGENOUS SYSTEMS

A knowledge of the source or provenance of detrital sediment adds substantially to our understanding of depositional basins. The approximate grain size and composition of the lighter fraction, the heavy mineral population, and isotopic signatures can yield invaluable information on the nature of the bedrock and weathering processes in the source area.

Each depositional system has its own immediate source area. Deep-water systems such as submarine fans are supplied from an adjacent shelf or delta whose morphology, size, tectonics and climate they reflect. Shelves are supplied from coasts or coastal plains, which may be in part deltaic; deltas are supplied from alluvial systems which themselves reflect the features of the hinterland. Thus contemporaneous depositional systems are linked together as 'systems tracts' (Sect. 2.4). The downcurrent systems are directly controlled by the sediment supplied, or not supplied, by the upcurrent systems.

The rate of sediment supply is generally controlled more by the volume of sediment available in a given time interval than by transport capacities. Rates vary by many orders of magnitude, even within those systems dominated by terrigenous sediment.

The sizes and gradients of terrigenous depositional systems are related to the *grain size* or *calibre* of the sediment (Reading & Orton, 1991). Coarse-grained or gravel-rich alluvial, deltaic and deep-sea systems are all relatively small and steep with delta plain areas of 1–100 km² and gradients of >5 m km⁻¹ (Table 6.1). Deep-water coarse-grained submarine fans have radii of 1–50 km and slope gradients of 20–250 m km⁻¹ (1°–14°). Owing to the high competence required, sediment transport and deposition are largely by short-lived, but frequent, catastrophic events such as floods initiated by rain storms or slumping caused by seismic shocks.

Medium-grained or sand-rich systems tend to be intermediate in size, with moderate gradients. Delta plains have areas of 100–25 000 km² and gradients of 5.0–0.1 m km⁻¹. Deep-water fan systems have radii of 10–100 km and gradients of 18–6 m km⁻¹. The range of grain sizes available means that physical processes of transport and deposition operate over a wide range of energy levels. The resulting facies tend to be well differentiated, reflecting the full range of basin processes – river flow, tidal, storm, wave and wind energy of the basin as well as its morphological pattern. These systems are therefore particularly suitable both for modelling sedimentary dynamics and for the application of process models to environmental interpretation.

Although some fine-grained mud-rich systems can be small, the majority are very large, with low gradients. Delta plains have areas of 20 000–460 000 km² and gradients of 0.1–0.001 m km⁻¹. Submarine fans have radii of 100–3000 km and gradients of 5–1 m km⁻¹. The very low gradients of muddy shorelines make them very sensitive to sea-level changes caused either by tides or by longer-term rises and falls. Where sediment supply is high, the rapid deposition of mud and silt causes frequent slumping on delta slopes and submarine fans, in spite of the low gradients. On the other hand, slope aprons, where sediment accumulation is slow, are characterized by infrequent but very large slumps. The size of most mud-rich systems means that there is a considerable time lag between major changes in the controlling factors and the sedimentary response. Changes in one system may eventually affect patterns of sedimentation in the next system downcurrent, though with delay times of perhaps millions of years.

The pattern of sediment delivery to the basin is also important. Sediment may be supplied from a single point source, from multiple sources, from a linear source, from all around the basin, or from one end or the side of the basin. Such patterns may change with time, with important consequences. At one instant there may be a single point source and, in some situations, this may stay fixed for a long period. However, more commonly, sources change over time and thus over the longer, geological time periods in which rock successions accumulate, sediment sources are likely to have switched on one or more occasions. Modern siliciclastic depositional models, based on 'instantaneous' present-day examples, tend to emphasize single point sources. Ancient examples more commonly suggest multiple sources.

Volcanoes are also major contributors of sediment, yielding the whole range of grain sizes both below and above any water level and with variable composition (Sect. 2.1.10; Chapter 12).

BIOCHEMICAL AND CHEMICAL SYSTEMS

The production of biochemical sediment in lakes and the sea is controlled primarily by the nature and productivity of the biota which, in turn, depend on temperature, water chemistry, and the penetration of light into the water. In shallow water, although non-skeletal grains such as ooids, peloids and some lime mud can be important, carbonate production is primarily from the skeletal parts of animals and plants (algae) (Sect. 9.2). Thus the productivity of the 'carbonate factory' depends not only on the appropriate conditions of salinity, nutrients and temperature, but especially on light intensity. This is because some producers such as red and green algae are phototrophic. Others, such as mixotrophs (hermatypic corals and larger benthic forams), are light dependent because they use symbiont algae. Yet others, such as molluscs (especially bivalves), bryozoans, crinoids and brachiopods, are or were suspension feeders that ultimately depend on phytoplankton.

In deep basins only near surface waters penetrated by light produce significant sediment. Planktonic foraminifera and coccoliths yield calcareous material; Radiolaria, diatoms, silici-flagellates and some sponges yield siliceous material; upwelling phosphorus-rich waters yield phosphates. Productivity varies substantially (Sect. 10.2.1), with nutrient-rich waters in zones of oceanic upwelling producing 10 times as much pelagic sediment as nutrient-poor oceanic waters. However, of equal importance to productivity in determining sediment accumulation is the rate of dissolution of the particles as they descend through the water column. Calcareous particles are particularly susceptible to dissolution such that below a certain depth, the calcite compensation depth (CCD), few calcareous particles survive and the bottoms of deeper basins are covered by siliceous rather than calcareous oozes.

Evaporites are precipitated directly from sea or lake waters that have become concentrated to form brines. Their composition depends not only on the salinity of the brine but on its ionic make up, which, in lake waters, may vary considerably. Rates of deposition can be faster than for any other normal process of sedimentation, with vertical accretion up to 100 m ky^{-1} (Table 8.3).

2.1.2 Climate

The two main aspects of climate are temperature and precipitation, but, locally, wind regimes may also be significant. Not only are mean annual temperature and precipitation important, but also their fluctuations, both seasonal and non-seasonal, and the magnitude and frequency of extreme events.

The meteorological patterns of the Earth are primarily a consequence of the interaction of the Sun's radiation, the rotation of the Earth and the distribution of continents and oceans. If the Earth were to stop rotating, rising hot air at the equator would blow towards the poles where it would be cooled, become heavier, descend, and then return to the equator. However, this simple convection pattern (Fig. 2.1) can be considered to be disturbed by the Earth's rotation, which sets up a powerful deflecting effect, known as the *Coriolis force*. At the equator the Earth rotates at a velocity of 1600 km h^{-1} in an easterly direction; this velocity decreases away from the equator until it reaches zero at the poles. If a mass of air or water moves *away* from the equator it starts with the Earth's equatorial velocity and passes to places where the rotational velocity is

lower. The moving mass therefore tends to travel eastward faster than the Earth beneath it and the further north or south it goes the more it turns towards the east (Fig. 2.1). If a mass moves *towards* the equator, where the rotational velocity is higher, it tends to be left behind by the movement of the Earth and is deflected towards the west. Thus in the northern hemisphere the deflection is towards the right; in the southern hemisphere it is towards the left. This produces the anticlockwise rotation of water entering lakes and oceans such as the North Atlantic, in the northern hemisphere, and clockwise rotation in the southern hemisphere.

Hot air rising at the equator is responsible for the humid tropical rain belt. This is because, for precipitation to occur, an air mass must rise sufficiently for cooling by expansion to produce condensation of moisture. Although the hot air that rises at the equator and blows towards the poles is deflected towards the east, surface winds in the tropical belts either side of the equator blow towards the equator and westwards as the NE and SE trade winds. This is because the heated air from the tropics passes into latitudes that have a smaller circumference than at the equator, and the air becomes crowded, thus raising the pressure and forcing air downwards to give the high-pressure subtropical calm belt where deserts form. Secondary Hadley circulation cells known as the Hadley, Ferrel and Polar cells (Fig. 2.1) are created with winds blowing towards the equator deflected westwards by the Coriolis force and those blowing towards the poles deflected towards the east. In this way belts of trade winds, and of westerlies, develop either side of the high pressure belt. Meanwhile, the descending air at the high pressure arid polar regions is deflected westwards as cold surface

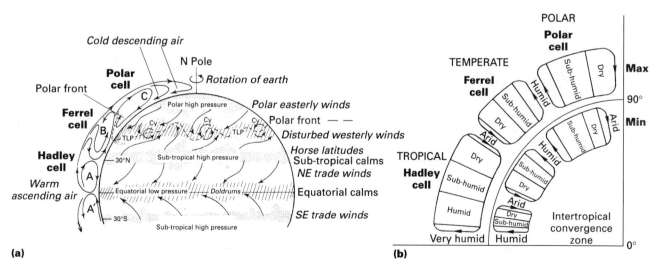

Figure 2.1 General planetary circulation of the lower atmosphere showing the Hadley circulation pattern of the three main cells in the northern hemisphere. (a) Present-day zonal pattern of high and low pressure belts and associated pattern of prevailing winds (after Duff, 1992). (b) Predicted position of circulation patterns and climatic belts at both climatic maximum and climatic minimum due to changes in the amount and distribution of heat received from the Sun (after Mathews & Perlmutter, 1994).

air wedges under the warmer westerlies along the Polar front. These moisture-laden winds flow up above the cold wedge to cause the relatively steady precipitation of the temperate regions. Such idealized cell patterns are perturbed by the distribution of areas of land and ocean. They also migrate on seasonal and longer-term cycles.

The climates of land areas have been divided in many different ways on the basis of average temperatures and rainfall, and their fluctuations and extremes, and there is no simple method of classification based on a single parameter. A practical zonation that relates continental geomorphic processes, environments and the supply of sediment to temperature and rainfall patterns, and one which is quoted in many text books (e.g. Chorley, Schumm & Sugden, 1984; Summerfield, 1991), is that based on Tricart and Cailleux (1972) (Fig. 2.2; Table 2.1). In that scheme eight morphoclimatic regions are defined – that is, large areas where distinctive associations of geomorphic processes occur. The classification is based on: (i) mean annual temperature; (ii) mean annual precipitation; (iii) mean number of wet (>50 mm precipitation) months; and (iv) mean temperature of the warmest month. In addition, there are azonal mountain

belts that do not fit the spatial pattern because they have a suite of climatic zones dependent on altitude and on the presence of orographic barriers.

At the extremes are the relatively simple non-seasonal climatic zones *polar glacial*, *arid* and *humid tropical*. In *polar glacial* zones frost weathering is at a maximum, other types of mechanical weathering are moderate and chemical weathering at a minimum. There is little mass wasting or river flow, but glacial scour and wind action are at a maximum, the latter removing most of the silt and leaving gravel, sand and boulder clay. In *arid* desert zones, mechanical weathering by salt and diurnal temperature fluctuations and wind action are at a maximum; chemical weathering is at a minimum. Sand and silt are the dominant mobile sediments. In *humid tropical* zones covered by rainforests, chemical weathering may give rise to a thick blanket (up to 100 m) of fine-grained debris and deep soil development. This prevents mechanical weathering and the consequent lack of coarse sediment limits mechanical erosion. Fine-grained clay-rich sediments are dominant.

The other zones have distinct seasons. The *periglacial* zone, characterized by permafrost and tundra vegetation, has a very

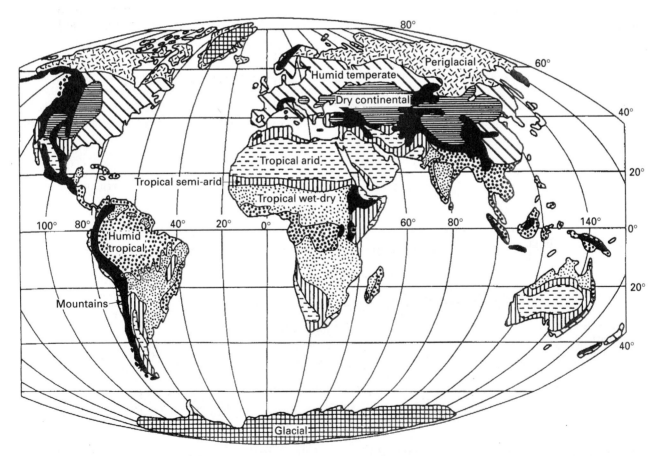

Figure 2.2 Present world distribution of climates and corresponding morphoclimatic zones (from Tricart & Cailleux, 1972; Summerfield, 1991). Notice latitudinal zonal pattern is modified by azonal factors, especially the distribution of land masses and oceans and the effects of prevailing winds on coasts.

Table 2.1 Principal features of morphoclimatic zones (based partly on data in Chorley, Schumm & Sugden, 1984; Summerfield, 1991).

	Glacial	Periglacial	Dry continental	Tropical arid	Tropical semi-arid	Tropical wet-dry	Humid temperate	Humid tropical
Frost weathering	High	High	Very seasonal	Low	Low	Nil	Low–high	Nil
Mechanical weathering	Moderate	High	Low–moderate	High	Low–moderate	Low–moderate	Low–moderate	High
Chemical weathering	Low	Low	Low–moderate	Low	Low–moderate	High	Moderate	High
Mass wasting	Low	High	Moderate	Low	Moderate, infrequent	Moderate–high	Moderate–high, creep	High, very episodic
Fluvial processes	Low	Moderate	Moderate–high episodic	Low, very episodic	High, episodic	Moderate–high, episodic, seasonal	Moderate	Moderate–low gradient
Glacial scour	High	Low	Nil	Nil	Nil	Nil	Nil except in mountains	Nil
Wind action	High	Moderate–high	Moderate	High	Moderate–high	Moderate–low	Low	Nil
Mean annual temp. (°C)	−25–10	−10–2	2–12	5–30	8–30	12–30	18–30	20–30
Mean annual rainfall (mm)	0–1000	0–1250	0–800	0–300	300–1000	1000–1800	800–1700	1800–2750
Mean no. of wet months (>50 mm)	—	0–7	0–5	<2	2–5	5–11	4–10	<11
Mean temp. of warmest month (°C)	<0	0–10	10–12	>22	>22	>24	>26	>22
Soils	None	Acid tundra	Grassland	None	Calcrete	Alcretes, ferricretes, silcretes	Pedalfers	Thick
Drainage density (km km^{-2})	—	—	5–10	0–5	1–25+	3–8	2–4	1–20
Infiltration	Low	High	—	Low	—	Moderate–high	High	High
Sediment grain size	Gravel and sand	Gravel, sand and clay	Gravel, sand and silt	Sand and silt	Gravel, sand and silt	Sand, silt and clay	Sand, silt and clay	Clay

wide seasonal temperature range and extensive frost, mechanical weathering and wind action. Chemical weathering and glacial scour are minimal but on slopes mass wasting by talus creep and solifluction are ubiquitous. Fluvial processes are very strongly seasonal with a short sharp peak discharge in the spring. Though dominated by gravel and sand, there is more clay than in the glacial zone. The *dry continental* zone, with its grassland steppes, also has a very wide seasonal temperature range. This leads to highly seasonal flooding and mass wasting processes. There is low to moderate mechanical (frost and wind) and chemical weathering. The *tropical semi-arid* zone, composed of dry, thorny savanna, marginal to hot deserts, is similar to the dry continental zone, except that it lacks the low temperatures. There is no frost weathering and both mechanical and chemical weathering are low to moderate. Fluvial processes are episodic and may be powerful where vegetation is sparse. Wind is locally important. The *tropical wet–dry* zone or wet savanna is a very broad type that embraces quite a range of climates, which includes both those that have extremely seasonal wet–dry periods and those that have long wet periods interspersed with short dry ones (e.g. Cecil, 1990). Mechanical weathering is low but mass wasting and fluvial processes are considerable. The dominant weathering process is chemical with penetration deep below the surface giving a range of surface duricrusts, bauxite (alcrete), laterite (ferricrete), and silcrete, depending on the degrees of temperature and rainfall, and the composition of the bedrock. The *humid temperate mid-latitude* zone, covered by well-developed soils and deciduous broadleaved trees, is also a very broad group. It includes the whole of subarctic Europe and the eastern half of North America. Most processes operate to a moderate extent. Frost action is very varied and mechanical weathering somewhat less important than chemical weathering. Erosion is modest in this zone.

Although climatic zones broadly reflect latitude, the distribution of continents and oceans, as well as mountains, substantially modifies this general pattern. Owing to the circulation of ocean currents, similar temperatures are found today on either side of the Atlantic at latitudes that differ by as much as 20°. Extremes of rainfall may occur at the same latitude. For example, in the subtropical trade wind belts the westerly flow of winds draws heavy moisture from the oceans on to eastward-facing coastlines, while the complementary west-facing coasts are where all the present-day coastal deserts lie. Southern Asia sees a transition from some of the highest rainfalls in the world in the east (Bangladesh) to the lowest (Arabia) (Fig. 2.2). In addition, it must be remembered that the local relief is crucial in governing rainfall amounts, and in a land of rugged relief, whether due to tectonic activity or volcanoes, arid basins may pass over distances of less than 50 km into mountains with substantial precipitation.

Climatic zones are never stationary and there is abundant evidence in the sedimentary record that climatic changes occur on a variety of scales. The thermal equator shifts seasonally, interacting with large landmasses to intensify summer heating and winter cooling of the atmosphere relative to the ocean. This seasonal differential heating leads to a significant poleward shift of the junction between the northern and southern hemispheric Hadley cells (the intertropical convergence zone). It creates monsoonal climates near the equator by drawing in monsoonal rains during the summer as the Asian continent heats up.

Longer-term climatic change is inevitable. This is in part due to plate tectonic movements resulting in the creation of new oceans and mountains, and variations in the amount of volcanic dust and CO_2 in the atmosphere. It is also due to extraterrestrial factors such as variations in insolation – that is, in the strength and distribution of incoming solar radiation as predicted by Milankovitch theory (Sect. 2.1.5).

Migration of the three Hadley circulation cells by as much as 30° latitude has been postulated (Perlmutter & Mathews, 1989; Mathews & Perlmutter, 1994). This results in a similar movement of latitudinal climatic belts, termed *cyclostratigraphic belts*, within which climatic and environmental end members can be defined (Fig. 2.1b). These are the extremes that can occur. More commonly the climatic shift is less. For example, at the present day, the subtropical high pressure zone, where deserts form, is between 20° and 30° latitude. However, during phases of climatic minimum (maximum cooling) it might have lain between 10° and 20° and at phases of climatic maxima between 30° and 40° (Fig. 2.1b). The high pressure zone also probably contracted during the climatic minimum with a consequent increase in wind power far beyond that of the present day. Oscillation between the two is over any of the time scales of Milankovitch cycles, but may be as little as half a precession cycle (less than 10 ky).

Each cyclostratigraphic belt has a different sequence of climatic change. In general, during a climatic minimum the Earth is expected to be cooler and drier than during a climatic maximum, and this is true for those areas affected by the Hadley and Polar cells. However, for the mid-latitude Ferrel cell, which moves towards the equator during a climatic minimum, belts between latitudes 35° and 50° become wetter (Fig. 2.1b).

Climate is a prime control on many sedimentary facies, and therefore they can be excellent palaeoclimatic indicators. The facies of lagoons and lakes, as well as soils, are particularly valuable as indicators of past climates. Temperature is a major control on the formation of evaporites, glacial tills, some oolites, palaeosols, vegetation, and many faunas. Rainfall affects evaporites, aeolian dunes, clay mineral provinces, palaeosols, vegetation, and fluvial, lagoonal and lacustrine morphology. In lakes, climatic changes are recorded not only in facies changes, but also in base-level and salinity changes which can affect their stratigraphy as much as tectonic movements.

Climate also governs sediment yield. After the evolution of land plants in Devonian times, equable climates, as in the present-day humid tropical belt and temperate humid mid-latitude regions, have promoted a good cover of vegetation, and erosion is kept to a minimum with clays being the dominant sediment. Inequable climates tend to yield greater volumes of sediment, especially of sands and gravels. In pre-Devonian times, the absence of land plants promoted highly variable discharges and the development of braided rivers even in regions of relatively constant and abundant rainfall.

2.1.3 Tectonic movements and subsidence

Tectonic movements affect sedimentation in a number of different ways and on many different scales. Isostatic movements are essentially an attempt by the lithosphere to balance spatially variable loads: for example, vertical adjustments in the lithosphere arising from its loading or unloading by water, sediment or ice, or by thermal or dynamic changes in the mantle. Sediment loading may enhance crustal subsidence by a factor of 3 when compared with an exactly similar basin that is sediment starved and loaded only with water. However, it is important to stress that the loading or unloading of one part of the crust also has substantial effects outside the immediate area as a result of the flexural rigidity of the lithosphere. Extended discussion of basin analysis can be found in Miall (1990) and P.A. Allen and Allen (1990).

On a global scale, the distribution and movements of lithospheric plates lead to the changing pattern of oceans and continents that controls the size and nature of the larger source areas, sediment transport paths and sedimentary depocentres. Continental collision zones, such as the Himalayas, produce the largest volumes of sediment. Many mountain belts, such as the American cordillera, the Pyrenees, the Alps, and Zagros mountains of Iran have an adjacent foreland basin, formed as a result of crustal loading by the nearby mountain belt. These are ready receptacles for sediments derived from the mountains. Inboard from subduction zones there is a discernible pattern of accretionary wedge, forearc, volcanic arc and backarc basin. Strike-slip belts are characterized by a linear series of rather small basins and highs of great local complexity. Extensional rifts also have relatively small scale basins and highs but these are generally of a larger size (50–70 km × 20–40 km) than those of strike-slip basins (20–40 km × 10–20 km). In both cases there is much seismic activity and small-scale but frequent mass flows are deposited close to fault scarps. Individual fault blocks may be important local sources of sediment within larger basins. In contrast, the broad passive continental margins show very large-scale asymmetric facies patterns that, although segmented by transform faults, may extend for thousands of kilometres along depositional strike. They have an early phase of rapid rift-related, fault-controlled subsidence, creating very large accommodation space, followed by slower subsidence due to thermal cooling of the lithosphere. On continents, large, gently subsiding sag basins give relatively slowly deposited, extensive, thin sedimentary sequences in which individual progradational and transgressive sequences migrate laterally over hundreds of kilometres, resulting in rather rapid changes in the vertical sequence and many hiatuses. The large stable cratons of the North American, Russian or Australian shield areas have little accommodation space and have been relatively starved of sediment. Deposition on their margins is consequently very responsive to relative sea-level changes whether eustatic or due to gentle tectonic warping. Overall sedimentation rates have been so slow that sedimentary processes were dominated by erosion and non-deposition and the formation of glauconite and phosphorite when the area was flooded. Unconformities, discontinuities and hiatuses abound.

On a smaller scale, movements along faults, the growth of folds, block tilting, differential subsidence and uplift on scales of 1–100s m provide a critical and delicate control on the type, thickness and distribution of sedimentary facies. The migration of rivers and deltas, the location of fluvial and deltaic depocentres and of carbonate banks, may all be governed by differential tectonic movements. These may be a consequence of differences in the underlying basement, in particular of heat distribution and buoyancy of crust, of active tectonic movements across faults, of differential compaction in the underlying sediment, or of lateral and vertical movements of mobile salt and shale.

2.1.4 Sea-level changes

Local changes in water depth (the distance between the water surface and water bottom) may be caused by changes in deposition: for example, in the terrigenous sediment budget, the production of biochemical sediment, or by variations in the rate of removal of sediment by marine processes.

Sea-level changes are of two types. *Eustatic sea-level change* is a function of sea-surface movement relative to some fixed point such as the centre of the Earth. *Relative sea-level change* is measured relative to some moving point in the underlying subsiding crust, or near the sea floor. It is therefore a function of both sea-surface movement and sea-floor movement in the fixed frame (Posamentier, Jervey & Vail, 1988; Posamentier & James, 1993).

Sea-level changes occur on a variety of scales. Short-term variations include those due to waves, tides, either daily (*diurnal*), or more commonly twice daily (*semi-diurnal*), so-called annual tides, more or less instantaneous storm surges, and tsunamis (giant sea waves produced by earthquake shocks and volcanic phenomena such as caldera collapse or entry of pyroclastic flows into the sea). Waves may be up to 20 m high. Semi-diurnal tides can also have ranges of 20 m, but most tidal ranges are much less, with a tidal range over 4 m considered to be macrotidal. No seas are completely tideless. The so-called tideless seas, such as the Mediterranean and the semi-enclosed Gulf of Mexico, have a tidal rise and fall of a few tens of centimetres. Even Lake Michigan has a tidal range of 10 cm.

Slightly longer-term, seasonal or annual tides also raise and lower sea level. Sea level may be lowered by a decrease in temperature and/or rise in salinity (which increases the density of sea water), by an increase in atmospheric pressure, or by an offshore wind. Sea level may be raised by an increase in temperature, a drop in salinity such as when melting of winter snow may add substantially to the water budget, by a decrease in atmospheric pressure, by oceanic currents that impinge on a coastline or by strong seasonal winds (e.g. the monsoons) that drive water on to the windward coasts of South East Asia, especially in the Bay of Bengal. Episodic catastrophic events that alter sea level include storm surges associated with onshore winds. These are often combined with sudden falls in atmospheric pressure during the passage of storms or hurricanes that raise sea level several metres above normal. Exceptional events such as tsunamis and/or major sedimentary slides into the ocean may cause sea level to rise instantaneously 27 m or more above the normal on the margin of the open ocean, and up to 500 m when amplified in enclosed fjords.

Longer-term variations in relative sea level arise through the interplay of changes in global sea level and basin floor subsidence and uplift. The most important effects on a regional scale are mountain building, volcanism, sediment compaction and isostatic movements due to sediment loading, ice loading and unloading (glacioisostasy), water loading and unloading (hydroisostasy) and thermal mechanisms. Glacioisostatic oscillations are a consequence of loading of crust by ice followed by crustal rebound as the ice melts and the load is removed. Rates and amounts of uplift are enormous, with rises of over 250 m in less than 10 000 years in northern Canada and Scandinavia (Sect. 11.5.2). Here the relative sea-level falls far outpace any rise of sea level due to the melting of the ice. Around this zone of subsidence and rebound is a peripheral flexural forebulge which operates in a reverse fashion – that is, it rises as ice loads the centre and sinks as the ice melts, so adding to the sea-level fall and rise in the peripheral region (Lambeck, Cloetingh & McQueen, 1987). Hydroisostasy is a consequence of any form of sea-level change. A fall in sea level reduces the water load on the sea floor or continental margin and causes uplift. A rise in sea level increases the water load and causes subsidence. The amount of movement reduces the effect of the sea-level change by about a third. This effect can be significant on continental margins and even more so during large desiccating and refilling events in restricted basins such as the Mediterranean during Messinian times when many hundreds of metres of sea-level movement may have taken place (Sect. 8.1).

Global (eustatic) sea-level changes may be brought about by changes in the volume of the ocean basins, changes in volumes of water in the world's oceans, changes to the hypsometric curve (the aerial distribution of global elevations) and in changes to the *geoid* (an equipotential surface of the Earth's gravitational field corresponding to mean sea level in the oceans). This latter surface today has a relief of 180 m relative to the centre of the

Earth and must have fluctuated in the past, perhaps by as much as 50–250 m on a My time scale and 60 m over the last 20 ky. These fluctuations are a result of shifting distributions of plates, ice or water masses, as well as changes in gravity forces such as those set up during Milankovitch effects (Sect. 2.1.5) (Mörner, 1994).

Changes in the volume of the ocean basins may have many causes (Donovan & Jones, 1979; Harrison, Brass *et al.*, 1981; Pitman & Golovchenko, 1983).

1 Changes in the total volume of mid-ocean ridges may be caused by subduction of existing ridges, by the creation of new ones, or by changes in spreading rates; an increase in spreading rates increases the volume of the ridges, due to thermal/convective buoyancy, and hence causes a rise of sea level; a decrease in spreading rate allows sea level to fall by perhaps as much as 350 m over 70 My.

2 Continental collision reduces the area of continent, increases that of the ocean and sea level drops in consequence.

3 Influx of terrigenous sediment to the oceans and the sharp increases in carbonate production especially since the mid-Cretaceous may raise sea level though this effect is normally reduced by isostatic depression beneath the sedimentary wedge.

4 Mid-plate, thermally induced (hotspot) uplift of oceanic floor may also decrease the volume of the oceans and cause eustatic rise (Schlanger, Jenkyns & Premoli-Silva, 1981).

5 Owing to thermal cooling and densification of the oceanic lithosphere, during periods in the past when the mean crustal age of the world ocean was relatively high, sea level would have been low and vice versa (Berger & Winterer, 1974).

All these changes are long lived, lasting for millions of years, and slow, only about 1 m My^{-1} (1.0 mm ky^{-1}).

Changes in the volume of water in the world's oceans may be caused by changes in mean ocean temperature (Donovan & Jones, 1979), by the waxing and waning of ice sheets, or by the sudden flooding or desiccation of isolated ocean basins such as the Miocene Mediterranean or the Cretaceous South Atlantic. The first mechanism may produce sea-level changes of as much as 10 m, but these are very slow. The second and third mechanisms, however, are several orders of magnitude faster. As the ice melted after the last Ice Age, sea level rose at a rate of 10 m ky^{-1} from about 15 000 to 6000 BP with a maximum rate of 2.4 m per century. Total rise due to melting of ice was about 100–130 m, but this figure varied around the world due to geoidal changes. If the remaining ice sheets in Antarctica and Greenland were to melt fully, the resulting rise should be about another 65–80 m. However, the compensatory effects of hydroisostasy on the ocean floor and continental margins would reduce the rise of sea level to about 40–50 m. If glacioeustatic falls have similar rates, we have to envisage sea-level rises and falls of up to 150 m at rates that average about 10 m ky^{-1} and are occasionally more rapid. However, as mentioned above, these rates may be counteracted on a more local scale by glacioisostatic effects.

Desiccation of small ocean basins and their reflooding, as has been documented for the Miocene of the Mediterranean, is another means by which substantial volumes of water could be added to the oceans during desiccation and lost during the reflooding. Although the amount of rise and fall is perhaps only a tenth of that due to glaciation (10–15 m if the whole present Mediterranean were desiccated) the rate would be very fast, perhaps as little as 1000 years if the present Mediterranean were suddenly cut off at the Straits of Gibraltar.

Thus it appears that relative sea-level changes that are both rapid and large scale can only be brought about by glacioisostasy and glacioeustasy, and, so far as we know, major build-ups of continental ice sheets that produced significant volumes of ice are confined to several relatively short periods of the Earth's history in the latest Precambrian, the Ordovician–Silurian, Carboniferous–Permian, and the mid-Cenozoic to the present day (Sect. 11.6). In addition there were several events in the earlier Precambrian. Probably for some 75% of Phanerozoic time there was no low altitude continental ice sheet. Yet there is evidence from stratigraphic palaeontology, outcrop geology and seismic stratigraphy that relatively rapid 1–10 My 'third order' cycles and even more rapid 100 000 year cycles of relative sea-level changes did occur during those periods, especially during the Mesozoic and early Cenozoic when there is no direct evidence of glaciation. In addition, frequent, very rapid but small-scale oscillations have been inferred to have taken place during the last 5.1 ky. During this time, on the east coast of Brazil, far away from any direct glacial effects, sea level has fallen 4–5 m overall, but within this period there have been very high frequency oscillations of about 5 m in a few hundred years (Sect. 6.7.1). Possible explanations for these changes are that they were caused by super-regional relative sea-level change such as: (i) the large-scale flexural behaviour of plates related to fluctuations in intraplate stress fields (Cloetingh, McQueen & Lambeck, 1985) that, according to models, may raise and lower large regions of the crust as much as 10–100 mm ky^{-1} – that is, rates 10 times more rapid than the longer-term mechanisms can manage, or truly global sea-level change; (ii) deformation of the geoid of as much as 5000 mm ky^{-1}; and (iii) differential rotation of the various layers of the Earth (Mörner, 1994).

The effects of relative sea-level change are felt primarily along shorelines and are reflected in transgressions (landward retreat of the shoreline) and regressions (seaward advance of the shoreline). Although, at times, rapid glacioeustatic sea-level changes have overwhelmed other effects, relative sea-level changes are generally the result of an interplay of many factors that do not necessarily work in unison. At any one place tectonic movements, compaction subsidence or changes in sediment supply may dominate shoreline behaviour.

Inland from shorelines, relative sea-level changes alter the gradient of the coastal plains and river systems. This is particularly true of fine-grained systems where rises and falls of a few metres significantly modify the relatively low gradients. As base level drops, erosion and the volume and calibre of the sediment transported increases. A relative sea-level rise causes reduction in gradient and thereby capacity, and sediment aggrades. On carbonate platforms the reverse may happen. A relative sea-level rise increases the production of carbonate and offshore areas receive an increased supply of sediments. As sea-level falls expose carbonate platforms, loose sediments are lithified, dissolved and possibly karstified. Areas of active carbonate production may be reduced to small fringing areas clinging to the platform slope and the supply of sediment offshore is reduced. As a consequence, mixed carbonate/siliciclastic basins commonly show reciprocal sedimentation

Table 2.2 Approximate magnitude, rates and durations of changes in sea level, eustatic, regional and local (after Donovan & Jones, 1979; Pitman & Golovchenko, 1983; Cloetingh, McQueen & Lambeck, 1985; Mörner, 1994).

Mechanism	Probable maximum magnitude (m)	Probable average maximum rate (mm ky^{-1}) (short term max.)	Time interval (My)
Ridge volume change	350	7.5	70
Orogeny	70	1.0	70
Sediment accumulation on the sea floor	60	1.1	70
Hotspot related sea-floor movements	100	Very slow	100
Intraplate stress	100	10–100	10
Flooding of ocean basin	15	Instantaneous	<0.013
Local tectonic movements	1000	10 000	<10
Glacioisostasy	250	10 000 (40 000)	<0.1
Glacioisostasy	150	10 000 (24 000)	<0.1
Tsunamis and landslides	500	Instantaneous	Hours
Differences in geoid relief	250	5000 (10 000)	<0.1

with carbonates during high stands and clastics during low stands.

In deep seas the effects of relative sea-level change are less direct. In general, a transgression reduces the clastic supply to the ocean by trapping sediment on shelves and coastal plains so that only the finest material enters deeper water. In contrast, an influx of coarse clastics may reflect a fall of relative sea level and regression. However, there are exceptions to this general rule.

2.1.5   Milankovitch processes and orbital forcing

The idea that astronomical forces have caused repeated climatic fluctuations on the Earth that may influence the character of sedimentary sequences, particularly during ice ages, goes back to the 19th century (Gilbert, 1895; Berger, 1988; Weedon, 1993). However, it is only in recent years that the Milankovitch theory – that cyclic changes in the nature of the Earth's orbit have caused repeated climatic variations – has become accepted as an important control on climatic variations outside the ice ages and on the nature of sedimentary sequences not only on pelagic and lacustrine sediments, but also on terrigenous clastic sediments (de Boer & Smith, 1994a). In addition, Mörner (1976, 1994) has argued that changes in astronomical forces also affect: (i) the gravity potential and hence sea level; (ii) differential rotation, and hence sea level, oceanic circulation, climate and ocean/atmosphere interaction; and (iii) palaeomagnetics, and hence the atmospheric shielding of the Earth.

There are three astronomical variables that are considered to influence the Earth's climate (de Boer & Smith, 1994b) (Fig. 2.3). *Precession* is the spinning of the Earth's axis due to the attraction of both the sun and the moon on the equatorial bulge of the Earth. The absolute period of orbital precession cycles is about 26 ky, but since the elliptical orbit of the Earth also rotates, the average mean periods are about 19 ky and 23 ky, and extremes of 14 ky and 28 ky can be reached. The main effect of precession cycles is to move the equinox as the Earth moves around its orbit. Since the equinox is the position of the Earth in its orbit around the sun at the moment when the sun is exactly above the equator, and day and night have equal length, precession cycles lead to regular and predictable changes in the insolation over the Earth and hence timing of seasons relative to the perihelion (closest point to the sun in each orbit or year), in particular changes in the contrast between summer and winter. The effect of precession is 180° out of phase between the northern and southern hemispheres (de Boer & Smith, 1994b).

The *obliquity* of the Earth's axis is its angle of tilt with respect to the perpendicular to the ecliptic, the plane in which the Earth rotates around the sun. Obliquity varies between 22° and 24.5° with a mean period of 41 ky. It determines the degree of seasonality, especially at high latitudes.

Eccentricity is the degree of ellipticity of the Earth's orbit around the sun. This varies from about 0.06 to close to zero over a period of about 100 ky, with major periods of 99 ky and 123 ky. Superimposed regular variations of eccentricity occur over 400 ky, 1300 ky and 2 My (de Boer & Smith, 1994b). Eccentricity itself has a relatively small impact on changes in insolation. However, it amplifies the changes of the precession cycles and hence controls the size of their effects. It also is the Milankovitch process that has the strongest signal in time series analysis of ancient successions. This is because the eccentricity cycle causes variations that are much more regular than those of precession cycles whose frequency varies by a factor of 2 (14–28 ky) (de Boer & Smith, 1994b).

As one goes back in time, conservation of angular momentum has dictated that the distance between the Earth and the moon was shorter, and so was the length of the day. In addition, the dynamical ellipticity of the Earth's orbit lengthens. Together these effects induce a shortening of the astronomical periods of precession cycles from 19 ky and 23 ky today to 16 ky and 19 ky in the Early Palaeozoic and of the obliquity cycle from about 41 ky and 54 ky today to 30 ky and 37 ky in the Early Palaeozoic (Berger & Loutre, 1994). Eccentricity cycles do not appear to have changed from the present 100 ky with time. One effect of these contrasting changes with time is that the ratio between precession/eccentricity and obliquity/eccentricity will have decreased with geological time and there should be fewer precession and obliquity cycles per eccentricity cycle at the present day than in the past.

Although cyclicity is apparent in nearly every sedimentary succession, the identification of Milankovitch cycles is not easy. The usual method is to count the number of stratigraphic cycles between levels that can be approximately dated and arrive at average periods. Absolute dating to sufficient accuracy is of course unreliable, particularly before the Tertiary. In addition, it is often difficult to define a cycle because variations in, for example, composition are often irregular or two or more regular cycles may be superimposed on each other. The method of *spectral analysis* is therefore used (Weedon, 1991, 1993). It

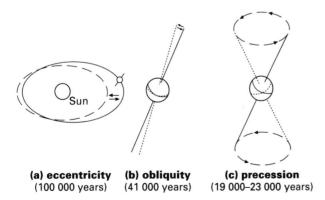

(a) eccentricity   **(b) obliquity**   **(c) precession**
(100 000 years)   (41 000 years)   (19 000–23 000 years)

Figure 2.3 The three orbital parameters that cause orbital cyclicity (from de Boer, 1983; de Boer & Smith, 1994b).

provides a more objective method of testing orbital–climatic control because regular cyclicity can be looked for, and super-imposed regular cycles can be distinguished from each other and from noise. It involves the measurement of one or more parameters taken at regularly spaced or well-defined intervals along a stratigraphic section. The parameters that may be used are bed thickness, grain size, fossil abundance, percentages of $CaCO_3$, organic carbon, Si/Al, total salt, ^{18}O and ^{13}C. The collection of values is called a *time series* (or stratigraphic depth series) even though the values obtained are a function of distance rather than time. The implicit assumptions of the above method are that deposition is maintained at a constant rate without breaks and that the measured parameter can be, indirectly, related to changes in insolation. Obviously, one or other of these criteria is frequently not met.

In applying the concept of 'orbital forcing' to sedimentary sequences, it must be remembered that orbital forcing only effects change in insolation as received at the top of the Earth's atmosphere and this varies with latitude. Although insolation is the major input into the atmosphere/ocean system, interaction of these systems is very complex and there may also be a time lag between a particular insolation change and its effect on the climate. Furthermore, there are many other mechanisms, tectonics, etc. by which cyclicity is caused (Algeo & Wilkinson, 1988) and there is a whole complex system of interaction and feedback through which the orbital signal must pass before it stands a chance of being encoded in the stratal record; the 'Stratigraphical Machine' of D.G. Smith (1994).

2.1.6   Intrinsic sedimentary processes

Processes intrinsic to the depositional setting itself may be responsible for changes in sedimentation patterns. These processes are frequently called 'autocyclic' and contrast with the 'allocyclic' processes described in the preceding sections (Beerbower, 1964). The progradation of distributaries of an elongate delta so reduces the gradient that the river eventually finds a steeper, shorter route to the sea and a new 'cycle' of deposition is initiated. High sinuosity rivers aggrade above the surrounding flood plain because deposition is largely confined to the channel and its levees. Thus, sooner or later they break through their banks to find a new course. On the slopes of deltas, sediment build-up may cause slumping to take place when the sediment load exceeds the strength of the sediment. Compaction of peat beneath a swamp may eventually exceed the organic production so that the swamp changes to a lake, allowing clastic infill and the re-establishment of a new swamp.

By the very nature of the sedimentary environment these changes are inevitable, though their exact timing is commonly governed by an unusual event such as an exceptionally violent flood, storm or seismic shock. This 'triggering' mechanism must be distinguished from the fundamental cause of inherent instability.

2.1.7   Physical processes

Physical processes are the main means by which terrigenous clastic sediments are eroded, transported and deposited. They are also important in the transport and deposition of many biochemical and chemical sediments. The principal physical agents are winds, river currents, wind-, wave-, tide- and storm-induced currents, gravity-induced mass flows, especially turbidity currents, slow moving oceanic and lacustrine currents, and glaciers.

The essential features of these processes are: (i) their ability (capacity) and competence to transport various grain sizes; and (ii) whether they are more or less steady or fluctuate in strength, and, if so, over what range and at what periodicity or episodicity, or whether they are catastrophic (Sect. 2.1.11). Based on competence, they fall into three groups: (i) wind; (ii) water currents; and (iii) mass flow. Because air has a low density and viscosity, wind is a very efficient sorter of the finer-grained sediment particles. Only particles finer than about 0.1 mm (very fine sand) can normally be picked up in suspension, while particles coarser than 1.00 mm can only be moved as bedload by exceptionally strong winds. Intermediate sizes move readily as bedload. Thus there are three clearly defined sediment populations: silt and clay moved in suspension; sand moved as bedload; and gravel that is seldom moved at all. Water currents can transport a whole range of sediment sizes up to cobbles and boulders, provided they are powerful enough, as in some high-gradient stream floods and during storms at the shoreline. However, tidal currents, many low gradient rivers and continuous oceanic currents all have upper limits to the grain sizes they can transport, often in the fine sand and silt grade. The ability of water to sort sediment varies greatly. The continuous reworking along shores and offshore by waves and tidal currents leads to a higher degree of sorting than can be achieved by river flow. Both gravity mass flow processes, and glacial ice, have an almost infinite capacity to transport large blocks not only cubic metres but cubic kilometres in size. Changes in sorting during gravity mass flow are invariably small or non-existent, not least because such flows commonly 'freeze' below some critical value of internal shear stress. Ice flow has many features similar to gravity mass flow, but some distinctions are apparent (Sect. 11.3).

Flow fluctuation gives rise to stratification in sediments. The most continuous currents are those due to oceanic and lacustrine circulation which commonly lead to the gravity settling of fine-grained suspended sediments. Yet invariably even these sediments show lamination that indicates some degree of fluctuation either of flow strength or sediment availability. Tidal currents are day by day 'normal' processes and they fluctuate in such a predictable fashion that diurnal and semi-diurnal tidal cycles can be recognized in the sedimentary record (see Sect. 6.2.7). Storms occur at less predictable intervals. River flows fluctuate, some randomly, especially in high-gradient flashy rivers, whereas others, such as some large tropical rivers and proglacial streams, do so on a periodic, seasonal basis.

Mass flow processes are generally episodic, caused by the build-up of unstable sediment and their triggering perhaps by sediment instability or earthquake shock. Yet a periodicity may develop in certain situations. For example, the annual spring flood in a small steep coarse-grained delta may result in an annual flow that triggers a slump, or, at the other extreme, a large mud flow may occur on average only once every 20 000 years in a large, fine-grained depositional system such as the present Madeira Abyssal Plain of northwest Africa (Sect. 10.3.6).

In the past, the magnitude and frequency of physical processes were generally very similar to those in equivalent climatic or oceanographic settings today. However, in the Precambrian, the smaller Earth–moon distance probably meant that tides were in general stronger and the lack of pre-Devonian land plants led to increased run-off rates and flood intensities.

2.1.8 Biological activity

Animals and plants not only produce large quantities of potential sediment; they trap it, stabilize it, mix it, erode it, and they govern a whole range of chemical and biological processes of sedimentation and diagenesis.

Their first role is as precipitators of mineral matter, mainly calcium carbonate which, in many instances, leads to the construction of major reef features and platforms. Corals, bryozoans, sponges, stromatoporoids, algae and rudists are or were the main organisms in this activity. Upward growth rate can be enormous with corals growing 6 m ky^{-1}, much faster than all mechanisms proposed for sea-level rise, except glacio-eustasy, and enormous volumes of carbonate sediment are produced. In the open ocean, micro-organisms such as Foraminifera, Radiolaria, algae and diatoms, which commonly live in near-surface waters, may provide a constant rain of pelagic sediment. In lakes, algae and diatoms are important sediment producers. On land, plants produce thick accumulations of peat and also inhibit erosion by promoting soil development and moderating the erosive effects of rainfall, run-off and wind. The absence of plant cover, whether on beaches or in deserts, allows wind to transport sediment. Bacteria are particularly important in the formation of soils, as weathering agents, in the oxidation and reduction of iron, and as reducers of sulphate.

The main mixers of sediment are: (i) plant roots that physically disturb a substrate and also concentrate solutions around them to form concretions; and (ii) burrowing animals that not only destroy sedimentary lamination and homogenize sediment, but also sort sediment. They too may nucleate concretions as a result of the deposition of mucus in the lining of burrows and microbial action in or around the burrow.

Bioerosion by borers, especially microborers, raspers, crushers and burrowers can remove carbonate sediments as fast as they are produced, and can degrade lithified limestones to fine

particles. It is thus also one of the principal sources of mud-grade carbonate sediment.

Organisms are closely associated with the chemistry of sea and lake waters, and of sediment pore waters. They have a strong effect on the acidity (pH) and oxidation–reduction potential (Eh) of water in their immediate vicinity; some use bicarbonate ions. During photosynthesis plants consume CO_2 and give off O_2, and this can lead to supersaturation of surface waters for calcite and aragonite during phases of intense biological activity such as occur during algal blooms. On the other hand, O_2 is lost and CO_2 produced by the respiration of animals and by oxidation of decaying organic material. Thus, when plant life is abundant and there are large masses of plant material to decay, the oxygen content is lowered and both near-bottom waters and pore waters become reducing.

Since organisms have evolved through geological time, the type, amount and the sites of biological activity have continually changed. An understanding of each contemporary biosphere is necessary if ancient and modern facies, or ancient facies from different ages, are to be compared. Although all environments may be affected to some degree, this factor is of greatest importance in pelagic environments and the carbonate build-ups of shallow seas, and in the development of coals and petroleum source rocks.

2.1.9 Water chemistry

The ionic concentration and composition of sea and lake waters are prime controls on the formation of evaporites and carbonates and other biochemical sediments. They are governed by temperature, organic activity (see above), the nature of inflowing waters and the degree of isolation from, or communication with, open oceans that restricted basins have. Sediments form when water composition or salinity changes as a consequence either of the movement of water from one place to another, or as a consequence of changing conditions in the basin. For example, oceanic circulation, resulting in upwelling of nutrient-rich waters, is responsible for the accumulation of some oozes, phosphates and diatomites. The ionic concentration of HCO_3 anions governs whether calcareous skeletons will be corroded, dissolved, or preserved, or whether carbonate deposits will be precipitated.

Although the composition of lake waters is primarily a function of catchment geology, with local volcanics playing a particularly important part, the temperature and salinity of sea and lake waters are largely the result of climate. Their chemistry therefore depends on climatic zonation and varies according to climatic fluctuations. Chemical variations in sedimentation frequently reflect climatic changes.

2.1.10 Volcanic activity

Volcanic activity essentially involves catastrophic processes which can upset completely many of the normal geological 'rules' that govern patterns of sedimentation and tectonics. Ash falls are the most reliable precise chronostratigraphic markers that exist in geology. They also preserve fossils, including footprints and soft bodies, in places where fossils are not normally preserved, such as the savannas of East Africa (Hay, 1986; Pickford, 1986). Volcanic activity can provide coarse clastic sediment from within the basin, as well as from outside, sometimes in enormous quantities over a short space of time. The sediment may be of very unusual composition and can give rise to very peculiar, sometimes unique, precipitates in lakes. The coarse clastic sediment is texturally and compositionally immature and often localized in distribution. In deep ocean basins, leaching of hot pillow lavas by sea water, formation of clay minerals by chemical exchange with sea water and associated hydrothermal discharge of metal-rich fluids have an important effect upon sedimentation. In addition, volcanoes not only have closely associated tectonic activity, simultaneous with and consequent upon the volcanic activity, but they create edifices on all scales. These range from those of the Pacific, such as Hawaii, with a relative relief greater than anywhere else in the world (including mountain ranges like the Himalayas), to lava flows and ash falls that form build-ups, both subaerial and subaqueous, that can transform the topography overnight and form both physiographic barriers which isolate depositional systems and temporary, local sources of sediment.

2.1.11 Normal vs. catastrophic sedimentation

There has been a long history of contention between those who considered the stratigraphical record to be dominantly catastrophic – that is, largely the result of floods, storms, earthquakes, etc. – and those who considered it to be explicable largely in terms of the relatively slowly acting phenomena that we can observe day by day. However, there is now an appreciation of the importance of both normal and catastrophic sedimentation, and of the necessity to distinguish between them in the rock record. The distinction may not always be easy, but must be attempted if correct interpretations of the processes of sedimentation and the controlling factors are to be made.

Normal or background sedimentary processes persist most of the time. Except for unusual processes such as algal blooms, biochemical processes are almost invariably steady and include pelagic settling, organic growth, evaporite precipitation, bioturbation and diagenesis. However, normal processes also include ice, wind, fluvial currents, tidal currents and waves, which, in some cases, have very high levels of physical energy. Some of these processes transport or deposit very slowly; others such as ice, wind, tidal and fluvial currents, deposit very quickly, but, because they often erode almost as fast as they deposit, they generally give rise to a low net sedimentation rate.

Catastrophic processes occur almost instantaneously. They frequently involve 'energy' levels several orders of magnitude greater than those operating during normal sedimentation. Typically they are the result of storms, heavy rain, small-scale earthquakes in seismically active basins, volcanic eruptions. They set off mass flows, such as turbidity currents, storm surges and flood currents, and may be locally very erosive. They typically deposit debrites or sharp-based, often graded, beds which can collectively be termed *event* beds. They may deposit a small proportion of the total succession and give rise to only an occasional bed, or they may deposit a large proportion of the total rock record and so become the dominant process of preserved deposition.

A final consideration is the *exceptional* event or process which produces a very small number of deposits of unique character. These exceptional events are the consequence of a process that either has a physical intensity at least an order of magnitude greater than a 'normal' catastrophic event or it is rare for a particular environment. They may be caused by an earthquake that sets off a major landslide up to thousands of cubic kilometres in volume involving a large part of the coastal plain or continental slope or be associated with a tsunami or giant sea wave caused by very large displacements of water (Sect. 2.1.4). Other exceptional processes are ice surges, extraterrestrial events, such as meteorite impacts, and the volcanic explosions that may affect even those environments that are far from active volcanoes.

The deposits that are produced by exceptional events are recognized by their uniqueness in the sedimentary record. They are so rare that they can be used as reference points to pinpoint particular instants of time in the stratigraphic record. The deposits are therefore *key beds* and may be seen as megabeds (i.e. unusually thick or slumped beds) in turbidite sequences (see Sect. 10.3.6), as unusual conglomerates in shoreline sequences, or as volcanic bentonite and ashfall horizons in otherwise non-volcanic sequences.

2.1.12 Rates of sedimentation and preservation potential

A constant concern in sedimentary analysis is the rate of sedimentation. Thickness is relatively easy to measure, though we have to allow for progressive compaction and reduction in thickness of sediments with time and burial, and thus calculate the original decompacted sediment thickness to make comparable measurements.

The duration of sedimentation is more difficult to measure. At one extreme we have chronostratigraphic ages, based on a mixture of biostratigraphy and isotopically determined 'absolute' ages. Normally the time units that can be calculated are in millions of years, and at best, with goniatites and ammonites, resolutions of about 200 ky can be estimated. Yet even these estimates are based on calibration with uncertain chronostratigraphic data. For example, in the Namurian, there are 60 biostratigraphic subzones, but two recent age calculations for the Namurian give either 11 My or 18 My for the duration of the stage. These ages yield either 185 ky or 300 ky per subzone. Yet that is not all, for the figures are averages, based on the unjustified assumption that goniatite-bearing marine bands recurred at similar intervals. Individual zones may have had half or double the average duration and thus could have lasted for, say, as little as 100 or as much as 500 ky.

Magnetostratigraphy, based on magnetic reversals, can be used for some geological periods to identify phases of normal or reversed magnetism lasting about 400–500 ky, with a magnetic reversal taking place over perhaps 10 ky. However, for other time periods, noticeably the Upper Permian and Upper Cretaceous, the method is virtually useless, as no reversals occurred. Relationships between sedimentary cyclicity and Milankovitch orbital forcing allow the potential for determining durations in the 100 ky or 20 ky range, although there is a clear danger of circular argument.

At the other extreme we can determine the occurrences of episodes of sedimentation such as turbidity currents and floods measured in hours and we can sometimes measure with precision the number of days a tidal package took to form (Sect. 6.2.7) or a coral took to grow.

Furthermore, we have a paradox, which can be illustrated by considering a hypothetical tidally-influenced shallow marine sequence. Using chronostratigraphy, we can calculate that a 100-m sequence was deposited in say 1000 ky at a rate of 100 mm ky^{-1}. Yet we can also calculate that each individual bed, 40-cm thick, was deposited during one lunar half cycle of 2 weeks and therefore the 100-m sequence should have formed in 10 years. Assuming no major breaks in the succession, that means 10 000 000 mm ky^{-1}, five orders of magnitude greater than the rate calculated using chronostratigraphy; or to put it another way, assuming tidal sedimentation persisted in the area for 1 My, for every tidal cycle preserved at a particular point, 50 000 were not recorded through erosion or non-deposition at that point.

The resolution of the paradox is twofold. First, most sediments are removed by physical or biological erosion soon after deposition, and in many, especially shallow-water and continental environments, few deposits survive. Thus *preservation* or *fossilization potential* (Goldring, 1965) is an important factor in any interpretation. The coarse channel base deposits in fluvial environments are more likely to be preserved than the topographically high levees (Sect. 3.3) and so are transgressive shoreface sands rather than those of barriers (Sect. 6.7.6). The chances of particular facies being preserved vary considerably, depending on the environment, subsidence rate and position of the deposit with respect to erosional base level. An assessment of preservation potential is necessary in order to interpret a facies sequence or to use the relative abundance of facies to deduce the importance of various processes.

Second, we have to beware of longer-term hiatuses in sedimentation and the fact that in most environments areas of non-deposition and erosion are both more widespread and persistent

than those of net deposition. Only locally, and for relatively short time periods (100s–1000s of years), as with some prograding deltas and in the deep sea, can we say that sedimentation is continuous and the whole record preserved. Even in the deep sea, where once we thought in terms of the unrelenting snowfall of pelagic sediments, we now recognize major breaks in deposition or even erosion. Preservation in most shallow-water environments is directly dependent on the generation of accommodation space by erosion, subsidence or base-level rise. Where subsidence or base-level rise is rapid, or recent erosion has created space, sediments can be preserved which would otherwise be reworked or removed. Where subsidence is slow and sediments are lithologically more mature, hiatuses in sedimentation and unconformities are widespread. For example, the preservation of thick piedmont fan deposits on the hanging wall of a fault contrasts with the thin veneer of gravel which is all that is preserved during fan retreat in a more stable area.

2.2 Facies and sequences

In interpreting the rock record, the factors that control sedimentary processes must be ever present in our minds. However, in order to reliably infer both processes and their controls we have to have a descriptive methodology, a requirement fulfilled in large measure by the concept of sedimentary facies. Ideas on sedimentary facies have come from several quarters, each developing its concepts on the basis of its own traditions, particular regional geology and nature of the data (outcrop, wireline log, core or seismic). Essentially three schools developed and their separate approaches are still evident, forming the basis for many of today's controversies. In particular, each school has used the word 'sequence' in a fundamentally different way. The realization of how and why these three schools developed assists us to understand the strengths and weaknesses of each approach and select the most appropriate to the solution of our problems.

First, there is the *British/Dutch/Shell School*, represented in the 1960s by J.R.L. Allen, Bouma, Collinson, de Raaf, Kruit, Kuenen, Middleton, Oomkens, Reading, van Straaten and Walker in Britain, Holland and Canada, by Bernard, Ginsburg, Hsü, Le Blanc, Visher, Wilson of Shell Oil in the USA, by Fischer, Klein and Van Houten from the eastern USA, and by many others such as Mutti and Ricci Lucchi in Italy. They paid close attention to the small-scale features of rocks, sedimentary processes and structures, and developed relatively local, static models centred around Walther's Law (Sect. 2.2.2). They underplayed stratigraphic changes, external controls and the large-scale linkage of depositional systems. The approach stemmed from the nature of inland outcrops within NW Europe and the eastern USA, where tectonic activity both during and after deposition made facies both laterally variable and difficult to trace. On the other hand, good coastal outcrops and stream sections allowed vertical sections to be measured. Emphasis was laid upon sedimentological, intrinsic or autocyclic causes, and external, allocyclic controls were only invoked when the scope for explaining facies relationships by intrinsic controls had been exhausted.

The second group, the *Gulf Coast School*, from Texas and Louisiana of south-central USA, is represented by Brown, Fisher, Fisk, Frazier, Galloway and McGowen. They were working over large areas where outcrop is almost non-existent but where there was an extensive coverage of rather poor-quality subsurface data. Cores and good-quality seismic lines were few, but electric logs abundant. This allowed regional isopach, sand thickness and sand percentage mapping which automatically lends itself to interpretation of regional scale, three-dimensional facies relationships. Tectonic deformation is not very significant and the present-day geomorphological pattern of the Gulf Coast provided a broadly similar model that could be relatively easily applied to the interpretation of Tertiary and Upper Palaeozoic rocks. Thus detailed process facies analysis on the lines of the British/Dutch/Shell School was neither possible nor necessary in order to interpret sedimentary environments. Instead the lateral coverage gave broad, large-scale facies patterns where thick 'facies sequences' could be easily linked together laterally as *genetic stratigraphic units* (Frazier, 1974; Galloway, 1989a). The emphasis was always on the large-scale systems, bounded by breaks in sedimentation (hiatal surfaces) of very slow or no deposition with minor emphasis on erosion. Their models were particularly suitable for large regions of substantial but even subsidence and a high influx of sediment where clastic sediment supply is the dominant control on facies patterns.

The third group is the *Cratonic/North Western/Exxon School*, developed initially by Sloss (1950, 1963), and revived many years later by the Exxon School of seismic interpreters and geologists such as Haq, Mitchum, Posamentier, Sangree, Sarg, Vail and Van Wagoner. It originated on the central, stable, cratonic shield region of the USA where similar lithofacies, separated by unconformities, could be traced almost horizontally for hundreds if not thousands of kilometres with little apparent facies change (Sloss, 1963). The ideas developed there could also be applied to the Carboniferous of central-eastern North America and the Cretaceous of the mid-west. Here too, laterally extensive and thin packets of sedimentary facies were separated by major hiatal bounding surfaces that could be traced for hundreds of kilometres. In all these regions there was tectonic stability and sedimentary supply was limited. Discussion on controls revolved round whether the episodic sedimentation was caused by tectonic movements or sea-level changes. These concepts were then extended by the Exxon Production Research Group (Vail, Mitchum & Thompson, 1977a) after examining many seismic sections across the world to show that large packets of seismic facies, bounded by unconformities, could also be traced laterally across passive continental margins where sediment thicknesses were much greater. At first the control that was considered almost exclusively was eustatic sea-level change, and

process sedimentology, even lithofacies, played only a small part in interpretation. Later, changing subsidence rates and sediment influx were included and the methodology and ideas were applied to onshore outcrop geology (Posamentier & Vail, 1988; Van Wagoner, Posamentier *et al.*, 1988). Facies patterns were fitted into models of sequence stratigraphy that were largely based on theoretical concepts, and had been generated to deal with large-scale geology (see Sects 2.2.2 & 2.4).

2.2.1 Rock facies definitions

The concept of facies has been used ever since geologists, engineers and miners recognized that features found in particular rock units were useful in correlation and in predicting the occurrence of coal, oil or mineral ores. The term itself was introduced by Gressly (1838) and has long been the subject of debate, well summarized by Middleton (1973).

A *rock facies* is a body of rock with specified characteristics. It may be a single bed, or a group of multiple beds. Ideally, it should be a distinctive rock that formed under certain conditions of sedimentation, reflecting a particular process, set of conditions, or environment. Facies may be subdivided into subfacies or grouped into facies associations (see Sect. 2.2.2). Where sedimentary rocks can be handled at outcrop or in cores, a facies may be defined on the basis of colour, bedding, composition, texture, fossils and sedimentary structures. A *biofacies* is one for which prime consideration is given to the biological content. If fossils are absent or of little consequence and emphasis is on the physical and chemical characteristics of the rock, then the term *lithofacies* is appropriate. Where definition depends on features seen in thin section, as is often the case with carbonates, the term *microfacies* is used.

Apart from these uses of the word facies, which are all primarily descriptive, it has also been used in more genetic senses – that is, for the products of a *process* by which a rock is thought to have formed, for example 'turbidite facies' for the inferred deposits of turbidity currents; or in an environmental sense for the *environment* in which a rock or suite of rocks is thought to have formed, for example 'fluvial facies' or 'shelf facies'; or as a *tectofacies*, for example 'postorogenic facies' or 'molasse facies'.

These uses of the term 'facies' are justified so long as we are aware of the sense in which the word is being used. For example, we can attempt to define objectively a sedimentary product, for example 'cross-bedded ooid grainstones', 'red, parallel-bedded sandstone facies'; or we can subjectively interpret a process, for example, 'turbidite facies', meaning that we *believe* it to have been deposited by turbidity currents, not that we can ever be sure that it *was* deposited by turbidity currents. If we use an environmental term such as 'basinal facies', 'platform facies', 'fluvial facies', it means that it has the characteristics compatible with such an environmental interpretation, but clearly, nevertheless, it is only an interpretation.

The selection of features to define facies and the weight attached to each of them are dependent on a subjective personal evaluation, based on the material to be examined, the type of outcrop or core, time available and research objective. Nevertheless, each unit must then be assigned to a facies based on objective observation and possibly measurement. It is very difficult to lay down universal rules for facies selection as each set of rocks is different and the facies criteria will vary accordingly. However, two schemes are now being widely applied, though with adjustments to local conditions. These are the turbidite scheme of Mutti and Ricchi Lucchi (1972) (see Fig. 10.29) and the fluvial scheme of Miall (1978) (Sect. 3.5).

2.2.2 Facies relationships, associations and sequences

Individual facies vary in their interpretative value. A rootlet bed and coal seam indicate that the depositional surface was very close to, or above, water level. However, while a current-rippled sandstone implies that deposition took place in the lower part of the lower flow regime from a current that flowed in a particular direction, it indicates little about depth, salinity or environment. Even a rootlet bed cannot be said to have formed in any one environment. It may have formed in a backswamp, on an alluvial fan, on a river levee or at a shoreline. We therefore have to recognize, at the outset, the interpretative limitations of individual facies, taken in isolation. A knowledge of the context of a facies – that is, the relationship of one facies to another – is essential before proposing an environmental interpretation. For example, a graded sandstone bed found interbedded with quiet-water pelagic limestones and mudstones would be interpreted differently from an identical graded sandstone interbedded with wave rippled sandstones clearly deposited in agitated shallow water.

Thus facies have to be interpreted, at the environmental level, by reference to their neighbours and are consequently grouped together as *facies associations* that are thought to be genetically or environmentally related (Collinson, 1969). The association provides additional evidence which makes environmental interpretation, particularly in the elimination of alternative interpretations, easier than treating each facies in isolation. Facies associations are thus the essential building blocks of facies analysis and, though the terminology may differ, 'model' groups of facies associations have been constructed for most environments. For example, Miall (1985a), following J.R.L. Allen (1983), suggests there are eight *architectural elements* that can be used for fluvial environments (Sect. 3.6.2) and Mutti and Normark (1991), following Mutti and Normark (1987), postulate five *primary elements* in deep-sea fans, including large-scale erosional features as one of their elements (Sect. 10.3.2).

In some successions the facies within an association are interbedded, so far as we can tell, randomly. In others, the facies may lie in a preferred order with vertical transitions from one facies into another occurring regularly or more often than would be expected in a random succession. There is thus a predictability about such successions that enables us to anticipate within known limits what we shall encounter as we move upwards or downwards through the succession. A *facies sequence* therefore is a series of facies which pass gradually from one into the other. R.G. Walker (1990) renamed such facies sequences 'facies successions' to avoid confusion with the term 'sequence' as used by seismic and sequence stratigraphers, which commonly involves inferences regarding relative sea-level change (Sect. 2.4). However, this is not necessary since ambiguity seldom arises in practice. Most facies sequences are bounded at top and bottom by a sharp or erosive junction, or by a hiatus in deposition indicated by a bioturbated horizon, a rootlet bed, hardground or early diagenesis. Some sequences may have no such sharp boundary and the junctions have to be taken at an arbitrary point within a progressively changing facies. A sequence may occur only once, or it may be repeated (i.e. cyclic). This concept of cyclic sedimentation, or the idea that patterns of facies repeat through a succession, tends to emphasize similarities between sequences. It can bring order out of apparent chaos and the recognition of cycles and rhythms may form the basis for the statistical analysis of successions. On the other hand, as with models (see Sect. 2.5), significant facts essential to sedimentological interpretation may be omitted, usually due to selection of features which help to establish an 'ideal' cycle. It is therefore preferable, after establishing an ideal, simplified cycle, to consider each individual sequence on its own. Then the small differences between sequences that may be crucial to environmental interpretation can be actively exploited rather than lost in the generality of establishing a simplified modal cycle.

The importance of facies sequences has long been recognized, at least since Walther's *Law of Facies* (1894) which states that 'The various deposits of the same facies area and, similarly, the sum of rocks of different facies areas were formed beside each other in space, but in a crustal profile we see them lying on top of each other … it is a basic statement of far-reaching significance that only those facies and facies areas can be superimposed, without a break, that can be observed beside each other at the present time' (translation from Blatt, Middleton & Murray, 1972, pp. 187–188). Walther's Law has been taken to indicate that facies occurring in vertical contact with each other must be the product of spatially neighbouring environments and that facies occurring in a sequence conformably above one another were formed in laterally adjacent environments. This principle has long been used, for example, to explain how a prograding delta yields a coarsening-upward sequence (Fig. 6.44).

It follows that the vertical succession of facies, laid on its side, may reflect the lateral juxtaposition of environments. Conversely, it may be possible to predict the sequence penetrated by a borehole through a prograding delta or coastal sabkha if the geographical distribution of their constituent depositional environments are known.

However, as Middleton (1973) has pointed out, Walther stressed that the law applies only to successions *without a major break*. A break in the succession, perhaps, though not necessarily, marked by an erosive contact, may represent the passage of any number of environments whose products were subsequently removed. This may occur, for example, in a prograding delta (Figs 6.52 & 6.54) where an advancing feeder channel may erode into the recently deposited delta top sediments and come to lie directly on delta-slope sediments, or in a transgressive sequence where the entire barrier/lagoonal system may be removed by shoreface processes so that nothing or very little is preserved (Fig. 6.87).

Walther's warning about sedimentary breaks has often been ignored by field geologists who have failed to describe the *contact* between facies or, even when they described them, failed to check or appreciate the significance of hiatuses in the record as shown particularly by biostratigraphic data. The three main types of contact are gradational, sharp and erosive, though sometimes one needs to differentiate those that are abruptly gradational, where a transition occurs over a few centimetres. Gradational contacts indicate that the facies immediately followed each other in time, probably by the migration of adjacent depositional environments. If contacts are sharp, even when erosion cannot be demonstrated, the facies may have formed in depositional environments which were widely separated in space, or in time. Thus it is essential to examine sharp contacts as closely as possible for the degree of burrowing, boring, penecontemporaneous deformation, diagenesis and hard ground formation of the underlying sediments, as well as biostratigraphical information on the duration of any hiatus, or of a jump in depositional environment.

When using Walther's Law to interpret evolving environments in a vertical sequence, one major assumption is made – this is that there has been no significant change in the external factors that control sedimentation, that rates of sea-level change, subsidence and sediment supply, as well as climate and tectonics, have remained essentially the same. For relatively small-scale sequences, this assumption is often sustainable. However, for larger sequences, those embracing longer periods of time, and volcaniclastic sequences, this is normally not justified.

Sequences may not only reflect changes brought about by natural sedimentary processes such as progradation and switching of a delta, the lateral migration and avulsion of a river channel or the build-up of a carbonate platform – that is, be *autocyclic* – but may also reflect changing external controls – that is, be *allocyclic*. Before the days of process sedimentology, most geologists emphasized allocyclic controls – that is, they interpreted sequences and cycles as the result of tectonics and base-level changes (e.g. Krumbein & Sloss, 1963) (Sect. 2.2). Process sedimentologists of the British/Dutch/Shell School, on the other hand, explored to the limit the ability of intrinsic properties of sedimentary settings to explain observed facies patterns. For example, the fining-upward sequence in alluvial sediments was interpreted as the consequence of the lateral migration of a meandering river (J.R.L. Allen, 1963a; Bernard & Major, 1963; Visher, 1965), the coarsening-upward sequence in deltaic cyclothems as the progradation of a distributary mouth bar (D. Moore, 1959; Scruton, 1960), coarsening/thickening-upward sequences in submarine fans as prograding lobes (Mutti & Ricci Lucchi, 1972), the sabkha cycle as a prograding coastline (Shearman, 1966) and many carbonate sequences as prograding platforms (J.L. Wilson, 1975). Although most process sedimentologists were well aware of allocyclic controls, in particular tectonic controls, some lost sight of them, in particular the effect on sedimentary patterns of changes in climate, sediment supply and sea level (Reading, 1987).

One of these factors, the stratigraphic effects of sea-level changes, was given a boost in the late 1970s from an unexpected quarter, seismic explorationists, in particular those working for Exxon. Seismic data quality had by then reached such a point that not only could seismic facies (see Sect. 2.3.1) be identified and attributed to particular depositional processes and environments, but apparent breaks in deposition could be traced laterally with confidence from one depositional environment to another. Thus began the science of *seismic stratigraphy* (Brown & Fisher, 1977; Mitchum, Vail & Sangree, 1977; Vail, Mitchum & Thompson, 1977a) that evolved into *sequence stratigraphy* (see Sect. 2.4) (Van Wagoner, Posamentier *et al.*, 1988; Galloway, 1989a).

2.3 Facies in the subsurface

In the past two decades such strides have been made in the acquisition, processing and interpretation of seismic data that it is now possible to define seismic facies from the reflection character and in many cases to determine the three-dimensional shape of seismic facies units and their relationships to other facies. Together with the greater use of cores and improvements in wireline logging, three-dimensional environmental models can now often be generated better from the subsurface than from surface geology. In the subsurface, not only can the overall framework of the sedimentary system be established but the underlying palaeogeology can commonly be seen and its relationship to the overlying facies pattern observed. This enables the effects of contemporaneous bathymetry and tectonic movements, sedimentary deformation and differential compaction of underlying sediments on the succeeding facies and environmental patterns to be determined.

The limitations of the subsurface are the widely separated scales of observation. At one extreme, seismic facies units have horizontal dimensions of tens of metres to kilometres, beyond the scale of most outcrops. At the other extreme, cores are usually widely spaced, are often limited to potential reservoir rocks, have widths of only a few centimetres and do not yield any of the information available at outcrop on medium-scale lateral variability. However, by careful calibration of cores to wireline logs, and of these to seismic data (3D) it is now possible to extrapolate facies analysis to uncored sections and to have a direct check on the sedimentary facies of seismically defined units.

2.3.1 Seismic facies

A seismic facies is defined on the basis of reflection configuration, continuity, amplitude, frequency and interval velocity together with the external form of the unit (Mitchum, Vail & Sangree, 1977) and may be mappable as a two- or three-dimensional unit.

The most obvious feature of a seismic facies is the reflection configuration (Fig. 2.4) which may yield information on bedding patterns, depositional and erosional processes, channel complexes and penecontemporaneous deformation. Given beds of similar velocity and density, reflector continuity broadly reflects lateral depositional continuity and amplitude reflects the vertical contrast in facies; high amplitude reflectors, for example, indicate large-scale interbedding of shales with thick units of sandstone or carbonate rocks; low amplitude reflectors are the result of monotonous lithological profiles. However, this picture can be substantially complicated by changes in reflection strength due to differences in velocity and density among different kinds of shale (clays vs. silts) or sands (cemented vs. non-cemented) as well as depth and acquisition-related changes in frequency.

Unlike outcrop facies analysis, where the shape of the facies unit is often difficult to ascertain, the two- or three-dimensional external form of the seismic facies unit is an essential element in subsurface analysis. The most common shapes are sheets, wedges, banks and lenses (Fig. 2.5). Topographic build-ups, due to organic growth or to clastic or volcanic deposition, are known as mounds. There are many forms of fill above discordances (channel, trough, basin and slope-front) which have a variety of internal reflection configurations (Fig. 2.6).

2.3.2 Seismic-stratigraphic units and seismic sequences

Seismic facies are grouped together into packages of *seismic-stratigraphic units* (Brown & Fisher, 1977) that are seismic reflection-bounded units composed of contemporaneous depositional systems (Brown & Fisher, 1977) or *seismic sequences* of internally concordant reflections bounded by 'unconformities and their correlative conformities' (Fig. 2.7; Mitchum, Vail & Thompson, 1977). These sequences are 10s–100s m thick and may include a wide range of depositional environments and facies and facies associations as understood by outcrop geologists. The integration of the concepts developed by seismic stratigraphers and sequence stratigraphers is addressed in Section 2.4.

2.3.3 Rocks from the subsurface

Cuttings and sidewall cores are of use in the identification of lithology and in palynology and micropalaeontology, but their value in clastic facies analysis is limited to indicating lithology and grain size. Cores, on the other hand, are vitally important

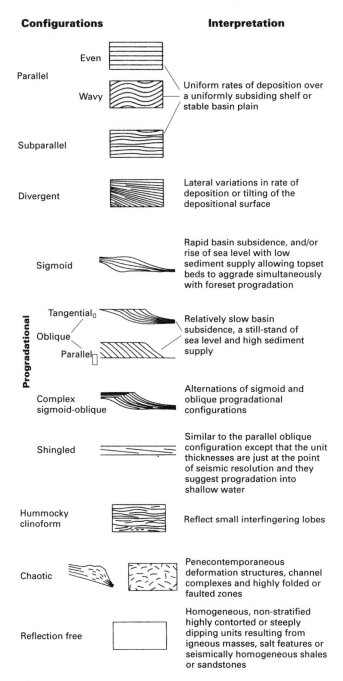

Figure 2.4 Seismic reflection configurations (from Mitchum, Vail & Sangree, 1977).

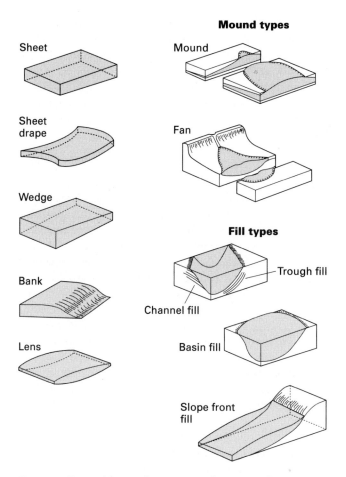

Mound types

Sheet

Sheet drape

Wedge

Bank

Lens

Mound

Fan

Fill types

Trough fill

Channel fill

Basin fill

Slope front fill

Figure 2.5 External forms of some seismic facies units (from Mitchum, Vail & Sangree, 1977).

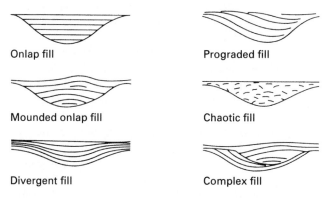

Onlap fill

Mounded onlap fill

Divergent fill

Prograded fill

Chaotic fill

Complex fill

Figure 2.6 Examples of reflection configurations that fill negative relief features in the underlying strata. Underlying reflections may be either truncated or concordant with the fill reflections (from Mitchum, Vail & Sangree, 1977).

in this regard, especially where they are continuous, since they provide the only means of directly observing and measuring facies. Facies analysis of cores is similar to that of outcrops and

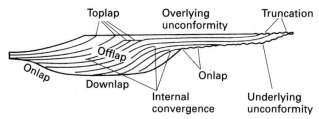

Toplap — Overlying unconformity — Truncation
Onlap — Offlap — Downlap — Internal convergence — Onlap — Underlying unconformity

Figure 2.7 Seismic stratigraphic reflection terminations within an idealized seismic sequence (from Mitchum, Vail & Sangree, 1977). *Top-discordant* relations include erosional truncation and toplap. Toplap is the termination of reflections, interpreted as strata, against an overlying surface as a result of non-deposition and minor erosion. Minor toplap boundaries are commonly included within depositional sequences. *Base-discordant* relations or baselap include onlap and downlap which may be difficult to distinguish in subsequently deformed sequences.

in terms of actual rock facies is the most reliable method of subsurface facies study. It is widely and effectively used in the coal and mineral industries. In the petroleum industry cores were historically relatively few in number, typically widely spaced and expensive to obtain, though recently their value has been increasingly appreciated. Nevertheless, coring is commonly only undertaken in hydrocarbon exploration when the borehole is already within the prospective reservoir unit and hence important information about its relationship with neighbouring units may be missed.

2.3.4 Wireline logs and log facies

Wireline logs measure the electrical, radioactive and acoustic properties of rocks, which are used to derive information on lithology, grain size, density, porosity and the pore fluids. Using these data, *log facies* can be defined and log sequences identified (see applications in Hurst, Lovell & Morton, 1990; Hurst, Griffiths & Worthington, 1992). Since logs are run continuously up a borehole they are particularly valuable as indicators of sequences on a scale of metres to hundreds of metres and can be used in sequential environmental analysis, provided the logs reflect the lithological properties and not fluid properties or other secondary features.

The main types of logs are neutron, density, sonic and gamma ray together with spectral gamma ray, spontaneous potential, resistivity logs and their derivative dipmeter and borehole imaging logs (Fig. 2.8). Some can be used individually to indicate lithology but are better used in concert and calibrated to cuttings or core. *Neutron* logs respond to hydrogen content and have negative values in porous rocks containing hydrogen in the form of water, oil or gas and in coal and organic-rich shales. Thus coal and organic-rich shales have high neutron porosities; tight reservoir rocks, very compacted shales, anhydrite and salt have low neutron porosities; porous sandstone, limestones and dolomites and moderately compacted shales have intermediate

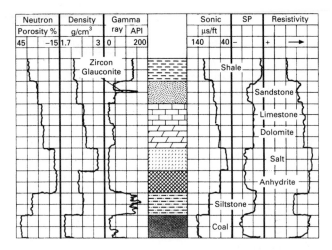

Figure 2.8 Typical wireline log responses to characteristic lithologies (by courtesy of G. Al-Murani and D.A.V. Stow, from various sources). Notice that the responses are not necessarily unique to particular lithologies. For example, since neutron, density and sonic logs are essentially porosity logs, the responses in sandstone, limestone and dolomite depend largely on whether the rock is porous or dense.

neutron porosities. *Density* logs record the electron density of rocks and therefore respond to both grain density and fluid density. They are normally used in combination with neutron logs and this combination is probably the most sensitive and reliable log for identification of lithology and calibration to core. The electron density is transformed into a bulk density equivalent. Salt and coal have low densities; anhydrite and tight reservoir rocks have high densities; porous sandstones, limestones and dolomites and moderately compacted shales have intermediate densities. *Gamma ray* logs measure natural gamma radiation emitted by the formation and indicate the integrated concentration of potassium, uranium and thorium. They most clearly reflect clay content but they are commonly taken as indirect indicators of mean grain size in clastic sequences because a high reading normally indicates clays. Caution is needed, however, because: (i) while illite, being rich in potassium, gives a high reading, kaolinitic shales and clays do not; and (ii) though the curve may be a measure of the argillaceous content of the rock (clay/non-clay ratios), variations in grain size of clay-free sandstones are not recorded. In addition, high concentrations of certain radioactive minerals in sandstones (e.g. mica, zircon, glauconite, K feldspar) or clay pebbles in a conglomerate yield a 'shale' response. Pure evaporites give either very low readings (e.g. anhydrite and salt) or very high readings (potassium salts). Coals and tuffs usually have rather low gamma ray readings; black shales, if they contain uranium, an exceptionally high reading. *Spectral gamma ray* logs allow discrimination of the radiation components due to potassium, thorium and uranium. Thus condensed marine intervals with high uranium radiation or heavy mineral concentrations rich in thorium, such as may

occur in shoreface sands due to wave winnowing or at the base of channel sands, may be detected. *Spontaneous potential* logs indicate permeability and are used as indicators of sand–shale ratio. However, tight sandstones, limestones and dolomites react in the same way as impermeable shales. The log is difficult to interpret quantitatively, has poor vertical resolution and is not easily run in offshore wells. *Sonic* logs measure the velocity of both compressional (P) and shear (S) sound waves passing through a formation and respond to both grains and fluids. They can be used to measure both porosity and lithology. They record the interval transit time – that is, the velocity of compressional sound waves passing through a formation. Sandstone, limestone and dolomite have low transit times (high velocities); coals and shales, particularly where undercompacted, have high transit times; anhydrite is low and salt intermediate. As well as allowing some determination of the lithology, sonic logs may also be used to determine shale velocity and hence give an indication of the burial and uplift history of a succession and allow the identification of concretions and cemented layers and detection of major unconformities. *Resistivity* records the resistance of rock formations to electric current flow. In general, shales and saltwater saturated and porous rocks have low resistivities; tight and hydrocarbon-bearing formations and coals have high resistivities. Shallow penetrating resistivity logs resolve thinner beds and allow the determination of patterns of interbedding down to the scale of 1 cm.

Dipmeter profiles may also be used in facies analysis. They are based on the angle and direction of dip constructed from correlated resistivity profiles measured around the borehole. Primarily they are used for the interpretation of structure, both tectonic and synsedimentary and these structural effects have first to be removed before information of sedimentological significance is apparent. Very low uniform dips suggest horizontal beds or lamination. High angle uniform dips suggest large-scale foresets as in aeolian dunes. Upward-decreasing dips may indicate convex upward bedforms or a decrease in cross-bed size as can be found in fining-upward channel-fills. However, in order to resolve smaller-scale cross-bedding and thereby extract palaeocurrent information from aqueous sandstones, it is necessary to process dipmeter data to smaller vertical intervals than is usual. Only then are foresets, rather than set boundaries or other surfaces, likely to be resolved (Cameron, 1992). Upward-increasing dips suggest prograding sequences and random profiles indicate no clearly defined bedding as in carbonate reef cores, debris flows and conglomerates. Recently, borehole imaging tools (CBIL, FMS) based either on the sonic or resistivity tool are beginning to give an image of the borehole wall with a resolution of approximately 1 cm. This allows very detailed structures such as cross-bedding styles and even burrows to be interpreted which can be used to orientate structures and to extrapolate from calibrated core to uncored intervals.

Patterns of sequences are apparent in all types of logs, in particular spontaneous potential, gamma ray, density, neutron and sonic logs (Fig. 2.9). They can be used to distinguish changing lithology and thereby discern vertical facies relationships and infer evolving environments. If related to clay content or to grain size, cylinder shapes on gamma ray or spontaneous potential logs indicate thick relatively homogeneous sediments bounded by argillaceous sediments such as channel-fills with sharp tops. Bell-shaped profiles indicate upward fining, possibly due to channel-fills. Funnel-shaped profiles indicate upward coarsening probably produced by prograding systems such as deltas, submarine fan lobes, regressive shallow marine bars, barrier islands or carbonate forereefs prograding over basin mudstones. Egg-shaped profiles might suggest fining-upwards channel-fills with basal shale clast conglomerates or breccias, progradational–regressional sequences or submarine fan channel–lobe systems. Linear profiles can indicate thick mudstone sequences, possibly with interbedded sandstones or siltstones, interfluvial deposits, marsh coals or shales.

Interpretations need to use all available logs, including dipmeter profiles, and an additional technique is to cross-plot different log values and check these against cores. In this way valid facies models can be erected and extrapolated out to uncored sequences in the same field (e.g. Rider & Laurier, 1979).

2.4 Sequence stratigraphy

Sequence stratigraphy is the analysis of genetically related depositional units within a *chronostratigraphic framework*. The roots of the ideas on sequence stratigraphy go back at least several decades to the two main schools of North American outcrop and subsurface geologists (Sect. 2.2) and it is from these two schools that the current models of sequence stratigraphy have come. The main difference between the two schools is in the nature of the boundaries between which the sequences are defined (Fig. 2.10). The EPR (Exxon Production Research) *depositional sequence model* of the Cratonic/North Western/Exxon school emphasizes unconformities and their correlative conformities that bound relatively conformable packages of genetically related strata. These unconformities may mark important hiatuses in areas marginal to basins and are generally well seen in seismic sections. The Galloway (1989a) *genetic stratigraphic sequence* model (Fig. 2.10) of the Gulf Coast school emphasizes marine flooding surfaces as the important boundaries because they mark major reorganizations of the sedimentary systems and are easier to correlate on logs and at outcrop.

The EPR depositional sequence model originated from the classic AAPG Memoir 26 on Seismic Stratigraphy (Payton, 1977). In this memoir, Exxon authors and others showed how seismic facies could be identified and integrated into large-scale seismic sequences that were separated by discordances. Of enormous value was the codification of seismic facies and of stratal relationships at their bounding surfaces. Although, initially, a depositional sequence was not defined as being formed in a cycle of sea-level change, a prime aim of the study had been to attempt worldwide correlations of continental margin sediments and an excessive emphasis was placed on eustasy as the cause of sequences and as the control on stratigraphic geometries and facies. Stratigraphic charts were produced in varying detail for Phanerozoic time (Vail, Mitchum & Thompson, 1977b). These earliest charts attempted to show the behaviour of global sea level through time – that is, to develop a *global sea-level model* (Carter, Abbott et al., 1991) derived primarily from the analysis of seismic profiles in terms of unconformity-bounded units termed sequences and the extent of their onlap. The earlier Mesozoic charts (Vail, Mitchum & Thompson, 1977b), as published, showed only absolute time, the main stratigraphic divisions and changes in sea level for 1st-, 2nd- and 3rd-order cycles. Since it was assumed that these cycles of sea-level change were eustatic, they were used as a means of inferring chronostratigraphic age in areas where other stratigraphic data are poor or absent. Later, the emphasis on eustasy alone was moderated (Van Wagoner, Posamentier et al., 1988), and relative sea-level changes, brought about by the interaction of subsidence with eustasy, were considered to be the main explanation of the sequence organization and formally introduced into the definition of a depositional sequence. At the same time, the methodology was extended to smaller-scale cycles or parasequences (see below). In addition, magnetochronostratigraphic and biochronostratigraphic data were inserted on the Mesozoic charts (Haq, Hardenbol & Vail, 1988) (as they were from the beginning for the Cenozoic) and the sequence stratigraphic data are shown as 'relative changes

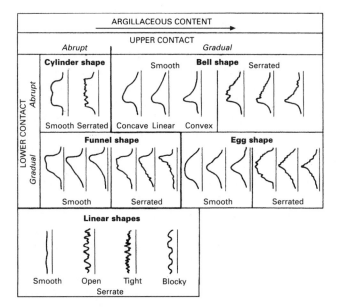

Figure 2.9 Some shapes of gamma ray/spontaneous potential log profiles, based on argillaceous content. Notice the nature of the upper and lower contacts (after Serra & Sulpice, 1975).

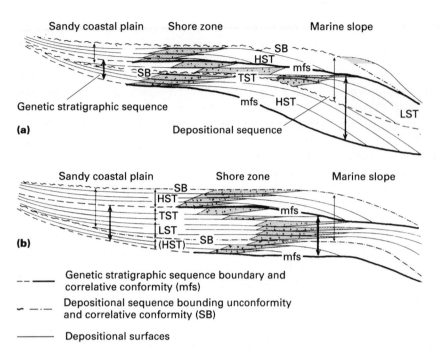

Figure 2.10 Contrast between EPR type 1(a) and type 2(b) depositional sequences and the Galloway genetic stratigraphic sequence (from Galloway, 1989a). In (a) EPR sequence boundary (SB) is a prominent subaerial unconformity. In (b) this unconformity is obscure and of limited extent. In contrast, in both examples, downlapped hiatal surfaces created by transgression of coastal plain (maximum flooding surfaces = mfs) are easily correlated. HST, highstand systems tract; LST, lowstand systems tract; TST, transgressive systems tract.

of coastal onlap' rather than of relative sea level (Figs 2.11, 2.12, 2.13 & 2.14). The stratigraphic information is probably the most useful data on the charts and is valuable because it emphasizes the desirability of comparing successions within a chronostratigraphic framework that is as refined as possible. It is for this reason they have been included in this book. However, despite acknowledgement of the local variation of relative sea level, the charts still show a eustatic curve directly related to the 'relative changes of coastal onlap', implying that there is a global sea-level curve that, if discovered anywhere, can be used directly as a stratigraphic predictor/correlation tool elsewhere. That this is partially misconceived has been pointed out by many authors in the last decade. Some have shown (Cloetingh, 1986, 1988; Hubbard, 1988; Underhill, 1991; Carter, Abbott *et al.*, 1991) that super-regional intraplate stresses and local tectonics may be as important as sea-level changes on the higher frequency 3rd-order cycles, and hence shed doubt on the basis of the curves at that level of resolution. Others (e.g. Miall, 1986, 1991b, 1992; Aubry, 1991) have been critical on the grounds that the field chrono- or biostratigraphic data, and the ties of bio- to chronostratigraphy, are often so imprecise that many of the 'type' sequences used to construct the curve and many of the sequences now being described by others can be placed at whatever position is convenient to fit the global sea-level model. It must be stressed, however, that the larger-scale 1st- and 2nd-order global relative sea-level cycles are probably valid and may well be the response to ocean-basin volume changes and/or thermo-tectonic subsidence (Sect. 2.2.4, Carter, Abbott *et al.*, 1991). Many studies suggest that 3rd-order cycles may well have at least local validity near their 'type'

sections. In addition glacioeustatic sea-level changes are an essential driving force for higher-order cycles during times of continental glaciation, provided allowance is made for the modifying effects of hydroisostasy, glacioisostasy and geoidal changes. It would be surprising if so detailed a model were to prove either entirely correct or, in its longer period trends, entirely incorrect, and, providing it is considered to be a useful conceptual framework for correlation and interpretation, and not a dogma, the global sea-level curves represent a valuable framework, particularly for those areas where only seismic data are available.

The second concept that emerged from the Exxon work was the *sequence stratigraphic model* (Carter, Abbott *et al.*, 1991) – that is, the idealized stratigraphic architecture deposited at a continental shelf margin during a single sea-level cycle. It involved the recognition of a *predictable lithologic succession* that resulted from variations of *relative* sea level, a function of both eustasy and tectonics, as well as sediment flux and physiography (Posamentier & James, 1993). Two-dimensional seismic sections were described in terms of *systems tracts*, which are a series of linked contemporaneous depositional systems (Brown & Fisher, 1977) such as a series of laterally inter-gradational coeval fluvial, deltaic and deepwater systems. Systems tracts are the equivalent of seismic-stratigraphic units (Sect. 2.3.2).

In the EPR sequence stratigraphic model, systems tracts are interpreted in terms of their position in the sea-level cycle, comprising a major sea-level fall, a lowstand, a sea-level rise, and a highstand (Haq, 1991) (Figs 2.15 & 2.16). Sea-level falls result in the creation of unconformities called sequence

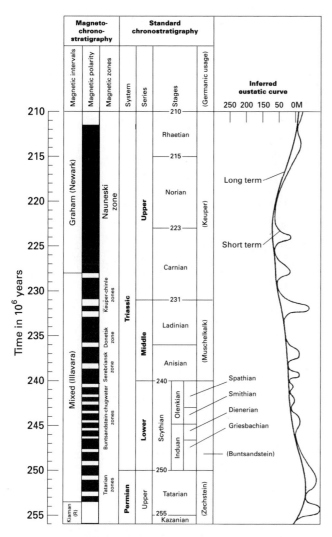

Figure 2.11 Chronostratigraphic chart and inferred eustatic curve for the Triassic (from Haq, Hardenbol & Vail, 1988).

boundaries and their correlative conformities in the EPR sense (Vail & Todd, 1981; Posamentier, Jervey & Vail, 1988) of which, originally, two types were distinguished and related to the inferred rate of relative sea-level fall. Both boundaries were characterized by subaerial exposure, a downward shift in coastal onlap, and onlap of overlying strata. A *type 1 sequence boundary* (Figs 2.10 & 2.15) has concurrent subaerial erosion associated with stream rejuvenation and a basinward shift of facies. Non-marine or very shallow marine rocks, such as braided-stream or estuarine sandstones may directly overlie deeper water marine rocks, such as lower shoreface sandstones or shelf mudstones across the sequence boundary. There are no intervening rocks deposited in intermediate depositional environments. A *type 2 sequence boundary* (Figs 2.10 & 2.15) lacks the extensive subaerial erosion associated with stream rejuvenation and the basinward shift in facies is less dramatic.

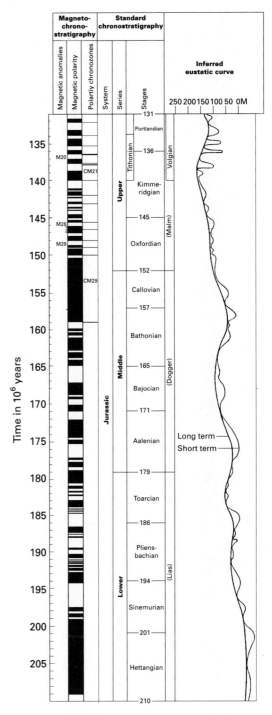

Figure 2.12 Chronostratigraphic chart and inferred eustatic curve for the Jurassic (from Haq, Hardenbol & Vail, 1988).

It forms where no relative sea-level fall occurs at the depositional shoreline break because the maximum rate of eustatic fall at the offlap break or break in the slope of the previous highstand (inflection point) never quite attains the rate of subsidence (Jervey, 1988). It is characterized by having a *shelf-margin*

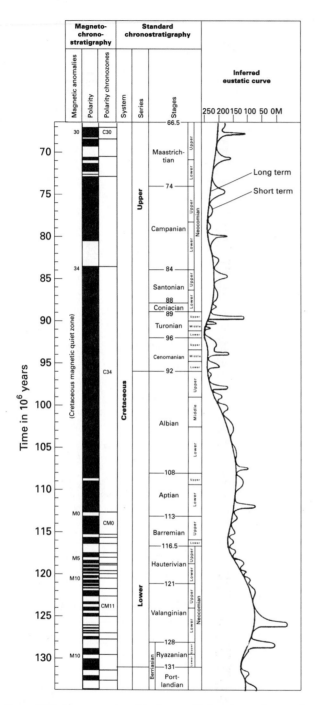

Figure 2.13 Chronostratigraphic chart and inferred eustatic curve for the Cretaceous (from Haq, Hardenbol & Vail, 1988).

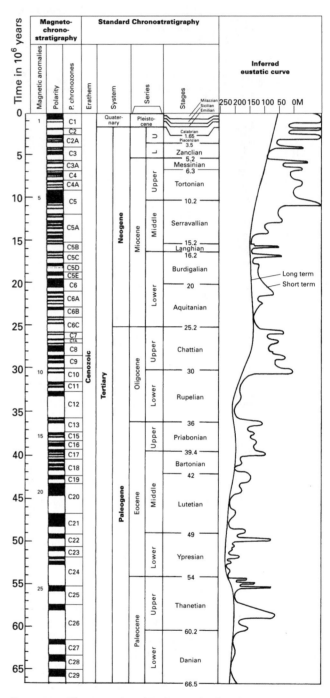

Figure 2.14 Chronostratigraphic chart and inferred eustatic curve for the Cenozoic (from Haq, Hardenbol & Vail, 1988).

systems tract below the transgressive and highstand system tracts. Such sequences are, however, no longer recognized by Exxon and a type 2 sequence boundary is considered synonymous with a parasequence set boundary and a type 2 sequence is considered synonymous with linked progradational and retrogradational parasequence sets (see below).

In the *initial sea-level fall* (Fig. 2.16a), the main effect is the formation of a type 1 sequence boundary, with substantial erosion of the alluvial/coastal system and the shelf to produce deeply incised valleys. Deposition is confined to basin floor fans fed by sediment bypassing. During formation of the *lowstand systems tract* (LST) (falling, stable low and slowly rising sea

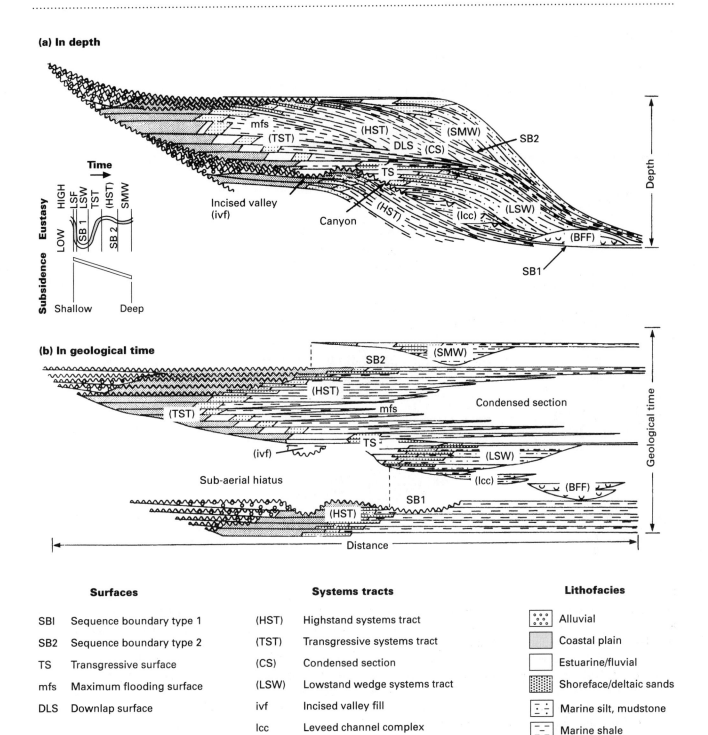

(a) In depth

mfs · (TST) · (HST) · (SMW) · SB2 · DLS · (CS) · TS · Incised valley (ivf) · Canyon · (HST) · (Icc) · (LSW) · (BFF) · SB1

Time · Eustasy · HIGH · LOW · LSF · LSW · TST · (HST) · SMW · SB 1 · SB 2 · Subsidence · Shallow · Deep · Depth

(b) In geological time

SB2 · (SMW) · (HST) · mfs · Condensed section · (TST) · TS · (LSW) · (ivf) · (Icc) · Sub-aerial hiatus · (BFF) · SB1 · (HST) · Geological time · Distance

Surfaces		**Systems tracts**		**Lithofacies**	
SBI	Sequence boundary type 1	(HST)	Highstand systems tract		Alluvial
SB2	Sequence boundary type 2	(TST)	Transgressive systems tract		Coastal plain
TS	Transgressive surface	(CS)	Condensed section		Estuarine/fluvial
mfs	Maximum flooding surface	(LSW)	Lowstand wedge systems tract		Shoreface/deltaic sands
DLS	Downlap surface	ivf	Incised valley fill		Marine silt, mudstone
		lcc	Leveed channel complex		Marine shale
		(SMW)	Shelf margin wedge systems tract		Deep water sands
		(BFF)	Basin floor fan		

Figure 2.15 Exxon Production Research (EPR) sequence stratigraphic depositional model showing siliciclastic systems tracts in depth (a) and time (b), their bounding surfaces and contained siliciclastic facies (from Haq, Hardenbol & Vail, 1988; Haq, 1991).

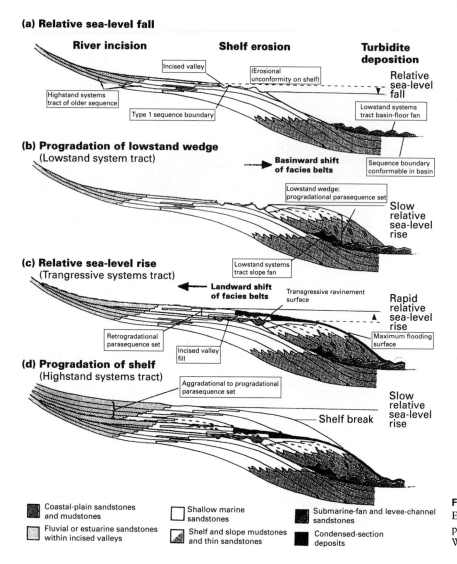

(a) Relative sea-level fall

River incision   Shelf erosion   **Turbidite deposition**

Incised valley

(Erosional unconformity on shelf)

Relative sea-level fall

Highstand systems tract of older sequence

Lowstand systems tract basin-floor fan

Type 1 sequence boundary

(b) Progradation of lowstand wedge
(Lowstand system tract)

Basinward shift of facies belts

Sequence boundary conformable in basin

Lowstand wedge: progradational parasequence set

Slow relative sea-level rise

(c) Relative sea-level rise
(Trangressive systems tract)

Lowstand systems tract slope fan

Landward shift of facies belts

Transgressive ravinement surface

Rapid relative sea-level rise

Retrogradational parasequence set

Incised valley fill

Maximum flooding surface

(d) Progradation of shelf
(Highstand systems tract)

Aggradational to progradational parasequence set

Slow relative sea-level rise

Shelf break

Coastal-plain sandstones and mudstones

Fluvial or estuarine sandstones within incised valleys

Shallow marine sandstones

Shelf and slope mudstones and thin sandstones

Submarine-fan and levee-channel sandstones

Condensed-section deposits

Figure 2.16 Step by step formation of a type 1 EPR sequence where sea level falls below the previous shelf break (modified from van Wagoner, Posamentier *et al.*, 1988).

level) (Fig. 2.16b) valley floor and canyon incision continues and a lowstand wedge of sediment is deposited, consisting of leveed channel complexes of slope fans and shelf-edge deltaic complexes. These prograding wedges of deltaic and shoreline deposits frequently become translated seaward and detached from the underlying shoreline with which their contact is sharp. Such *lowstand prograding wedges* have been interpreted as due to *forced regressions* – that is, a regression brought about by sea-level fall rather than an excess of sediment supply over accommodation space (Posamentier, Allen *et al.*, 1992). As sea level rises the *transgressive systems tract* (TST) (rapidly rising sea level) forms (Fig. 2.16c). Deposition is reduced in the basin and transgressive surfaces form on the shelf that extend on to the coastal plain. On the coastal plain of humid settings widespread peats aggrade while the incised valleys become filled with sediment. The transgressive surface of erosion at, or close to, the shoreline (initial ravinement surface or initial flooding surface) is succeeded by back-stepping retrogradational

parasequences. These are bounded at the top by a maximum flooding surface which is indicated by a condensed section due to sediment starvation. It is this maximum flooding surface that is used as the bounding surface in the Galloway genetic stratigraphic-sequence model. The *highstand systems tract* (HST) (stable high and slowly falling sea level) is marked by stacked parasequences that initially tend to aggrade, and then, as the accommodation space created by the sea-level rise diminishes, tend to prograde seaward.

Each systems tract is composed of *parasequences*, relatively conformable successions of genetically related beds or bedsets bounded by marine-flooding surfaces or their correlative surfaces (Van Wagoner, 1985; Van Wagoner, Posamentier *et al.*, 1988). Initially, the term was used to refer to sediments deposited during a *paracycle* – that is, they were part of a hierarchy of supercycles, cycles and paracycles that was meant to reflect relative changes of sea level of different orders of magnitude (Vail, Mitchum & Thompson, 1977b; Posamentier

& James, 1993). The duration of the paracycle was considered to be about 2 million years and was described as a 4th-order cycle of sea-level change. Thus a parasequence was considered to be part of the sequence stratigraphic model in that it should be interpreted in terms of a sea-level cycle. Subsequently, this view was changed (Van Wagoner, 1985) and the term has been used for shoaling-upwards sequences without any particular inferred genesis other than that they record shoreline progradation and are capped by a marine flooding surface (R.G. Walker, 1990). They may range in thickness from decimetres to tens of metres, and may be deposited over periods of a few days to millions of years. Many parasequences are similar to and of the same scale as certain facies sequences of the process sedimentologists (R.G. Walker, 1990) (Sect. 2.2.2). However, the concept of facies sequences embraces a much wider range of environments and facies associations than the shallowing-upward sequences that typify most parasequences.

A *parasequence set* is a succession of genetically related para-sequences which form a distinctive stacking pattern bounded, in many cases, by major marine flooding surfaces and their correlative surfaces. As with parasequences, they are independent of time and space inferences. Within sets, parasequences may be stacked to form progradational, aggradational or retrogradational patterns. Within the EPR sequence stratigraphic model, progradational parasequence sets are considered to form during late highstands and early falling stage, aggradational parasequence sets during early highstands and late lowstands (lowstand deltas) and retrogradational (backstepping) parasequence sets during transgressions (Posamentier & James, 1993).

Sequence stratigraphic models for carbonates were developed later than those for siliciclastic sediments and are rather more controversial. Earlier models were essentially modifications of the original EPR models for siliciclastic sediments (cf. Figs 2.15 & 2.17). The production, distribution and deposition of carbonates differ from siliciclastic sediments because: (i) most carbonate production, organic or inorganic, occurs in shallow water in tropical or subtropical areas; (ii) they can build wave-resistant structures with steep slopes; and (iii) they undergo extensive diagenetic alteration and induration soon after deposition, particularly where subaerially exposed.

Of particular importance is the restriction of high rates of carbonate production to the photic zone, with maximum production in the top 10–20 m of the water column. The presence of organic frameworks and the formation of steep, wave-resistant rims and slump-resistant slopes, allows the construction of sequence geometries which are foreign to siliciclastic depositional systems and which have the potential to build rapidly upwards, maintaining pace with rises in relative sea level. In addition, rates of sediment production may vary between the environments which make up a carbonate platform so that these may show varying responses to rises in relative sea level. It is particularly important to realize that major differences in response to relative sea-level changes are shown by carbonate platforms of different types, namely ramps (with very low angle slopes), rimmed shelves (which may have almost vertical slopes), and isolated buildups (Sect. 9.5).

According to the EPR model (Sarg, 1988) (Fig. 2.17), the *initial sea-level fall* is characterized by erosion and slumping along the platform slope to form very coarse debris flow aprons of limited basinward extent because the volume of sediment available for deposition in the deep basin is much less than in siliciclastic systems and induration of the platform margin leads to the formation of boulders. Subaerial exposure of the platform leads to the formation of karsts and meteoric diagenesis. Two types of *lowstand systems tract* form: (i) allochthonous deposits derived from erosion of the platform margin and slope and characterized by megabreccias; and (ii) autochthonous wedges deposited on the upper slope. More recent views (Handford & Loucks, 1993) suggest that in fact the amount of material that is resedimented during lowstands is small in most cases, the high solubility of carbonate minerals promoting solutional rather than physical erosion. Lowstand carbonate production is strongly dependent on the area of substrate which remains around the emergent platform and may develop extensively in basins with good circulation and gentle depositional slopes.

The *transgressive systems tract* (TST) in the EPR sequence stratigraphic model is composed of retrogradational (backstepping) parasequences organized into distinctive stacking patterns. It has a transgressive surface at its base which passes landward to onlap the unconformity which forms the sequence boundary. At the top of the tract is the maximum flooding surface, indicated by a condensed section of hemipelagic or pelagic sediment. In theory, individual cycles in any single section within the transgressive systems tract should become thicker and more open marine upwards towards the maximum flooding surface as the amount of accommodation space increases. In practice, it is now clear that carbonate platforms of different types react as differently at this stage in the relative sea-level cycle as at any other (Tucker, Wilson et al., 1990; Handford & Loucks, 1993) and that the exact response is determined by slope angle and the rate of any relative sea-level rise. Backstepping is common in platforms of all types, but is most characteristic of ramps and other low-productivity systems. Carbonate platforms with high productivity, or where the rate of relative sea-level rise is only slow or moderate, may accrete vertically, or even continue to prograde, or backstep on one side and prograde on the other (Chapter 9). Interpretation of the exact position of the maximum flooding surface within a sequence can be difficult in such cases and this has been responsible for much of the confusion that has hindered the emergence of viable sequence stratigraphic models for carbonate sediments.

The *highstand systems tract* (HST) is characterized by deceleration in the rate of relative sea-level rise. Carbonate production and accumulation can therefore catch up with and eventually exceed the amount required to maintain pace with any increase

(a) In depth

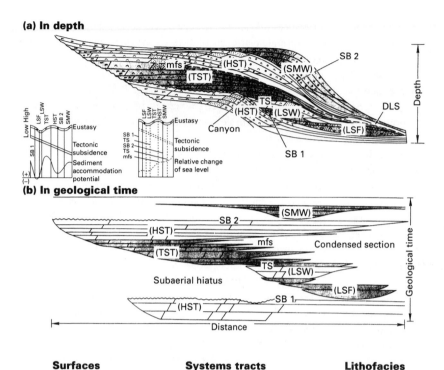

(b) In geological time

Surfaces		**Systems tracts**		**Lithofacies**
SB	Sequence boundaries	(HST)	Highstand systems tract	
SB 1	Type-1	(TST)	Transgressive systems tract	Supratidal
SB 2	Type-2	(LST)	Lowstand systems tract	Platform
DLS	Downlap surfaces	(LSF)	Lowstand fan	Platform-margin grainsupportstone/reefs
mfs	maximum flooding surface	(LSW)	Lowstand wedge	Megabreccias/sand
TS	TRANSGRESSIVE SURFACE (First flooding surface above maximum regression)	(SMW)	Shelf margin wedge systems tract	Foreslope
				Toe-of-slope/basin

Figure 2.17 Exxon Production Research (EPR) sequence stratigraphic depositional model showing carbonate depositional systems tracts in depth (a) and time (b), their bounding surfaces and contained carbonate facies (from Sarg, 1988).

in accommodation space, thus allowing the platform to expand. Volumetrically, the HST represents the bulk of most carbonate sequences and is the time when most sediment is shed from the platform to the basin (highstand shedding; Sect. 9.4.4). In this respect it differs from its siliciclastic counterpart where it is during the lowstand that most sediment is delivered to the basin, thus giving a pattern of reciprocal sedimentation. The carbonate HST is composed of parasequences that are initially aggradational, becoming progradational upward. In a platform interior, highstand parasequences become thinner upward, and the period of exposure or non-accumulation at each cycle top longer, as the accommodation space created by each flooding event reduces by increments to zero.

Although carbonate platforms show very high rates of sediment production, it is nevertheless possible for growth to be terminated by major, rapid relative sea-level rises that exceed their growth potential (Schlager, 1981). Commonly, some extrinsic factor such as an oceanographic or environmental change, or exposure, which hinders carbonate production appears to be responsible for platform demise. Drowning and burial of the platform beneath deeper-water sediments leads to the development of a drowning unconformity (Schlager, 1989), a term used to describe this stratigraphic morphology and may

or may not have sequence stratigraphic significance since it can occur with or without preceding subaerial exposure of the carbonate platform (Sect. 9.3.5).

Clearly, both the EPR siliciclastic and carbonate models, and the Galloway model, are simplifications of reality and should not be applied too uncritically. Sequence stratigraphy should be used as a way of looking at and ordering geological data rather than as an end in itself (Posamentier & James, 1993).

The model, sinusoidal relative sea-level curve (Figs 2.15 & 2.17) is a simplification that will rarely occur in nature, even when it is the result of glacioeustasy because there are bound to be both global and local perturbations and an interaction with subsidence that will affect the curve. In any case, the sedimentary response will be complex. In addition, the models were first erected to interpret the geology of cratonic areas and relatively passive continental margins where subsidence rates were predictable, moderate and regular, and sediment supply was relatively small. Consequently, the models work better where the systems tracts are laterally extensive than in tectonically active basins where relative sea-level changes are both rapid and irregular. Similarly, the models are more appropriate for gently sloping carbonate ramps than for steep, rimmed carbonate platforms.

There are two main general considerations. First, how far can the models and their constituent systems tracts be recognized in different geological situations (Figs 2.18 & 2.19), and secondly, where systems tracts have been identified, how valuable are the models in identifying the controls of sedimentary facies patterns and in predicting such patterns?

One of the most difficult tasks is the recognition of the EPR sequence boundaries. Of the suggested criteria (Vail, 1987), the most useful are onlap patterns, representing rising base levels after a base-level fall, and abrupt basinward shifts of facies. Onlap patterns may be observed on seismic, or inferred from well log correlation panels. In many cases, for instance with gently-dipping basin margins or high subsidence rates, they are difficult to pick. Abrupt basinward shifts of facies can, in principle, be spotted in vertical successions from sharp changes in depositional facies indicative of shallowing upwards, but these do not always coincide (Fig. 2.20). Furthermore, in many cases the resolution of depositional environments from logs, cores or fauna/flora is too poor to show clearly that a shift of depositional environment is more abrupt than would be expected from normal progradation under steady conditions. In many interpretations of prograding sequences it is often assumed that all coastal progradation is sea-level driven, and therefore any sharp contact within a progradational succession is a candidate for a sequence boundary. This viewpoint ignores evidence from modern coastlines (Sects 6.6.5 & 6.7.4) that progradation and transgression may be related to changes in sediment supply and basinal reworking, rather than changes in relative sea level. For

example, channelled surfaces overlain by coarse sands may be formed by autocyclic mechanisms such as channel switching. A fall in relative sea level might be distinguished by evidence of pronounced incision, but this will usually be laterally restricted and it might then prove difficult to decide which of several broadly equivalent interfluve palaeosols coincides with the incision (Sect. 6.7.6).

Another problem with identifying the EPR sequence boundary is that the unconformable surface is both difficult to determine and may itself be eroded by the subsequent transgression, in which case the transgressive ravinement surface becomes the sequence boundary for practical purposes. Transgressions may cut a ravinement surface by shoreface erosion leaving a typically very planar surface overlain by a pebble or shell lag, a thin unit of storm sands, and a condensed marine section. These surfaces may cut down through transgressive or lowstand deposits, or even parts of older sequences, and become the basal boundary of a sequence. This is particularly common in slowly subsiding areas. It is frequently these surfaces that are the detectable unconformities both at outcrop and in seismic sections. In addition, a marine erosion plane can be cut by waves during falling sea level in very shallow-water shelf or ramp settings with slow subsidence; a particular type of forced regression (Posamentier, Allen *et al.*, 1992). In other cases, unconformities can be formed by tectonic movements during times of rising relative sea level.

Because of the limitations on unconformities as sequence boundaries, it is commonly appropriate to use the Galloway

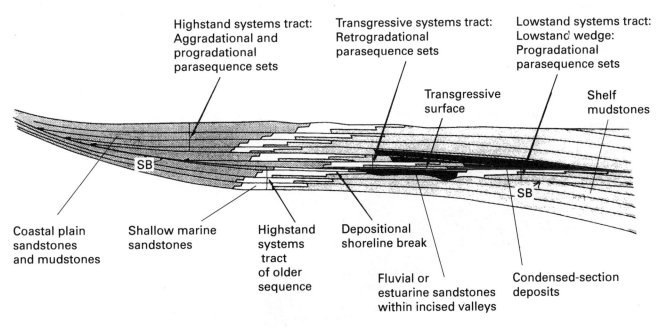

Figure 2.18 Modified sequence stratigraphic model for a wide cratonic basin with very slow subsidence rates. Highstand deposits are widely separated from lowstand deposits (from Van Wagoner, Posamentier *et al.*, 1988).

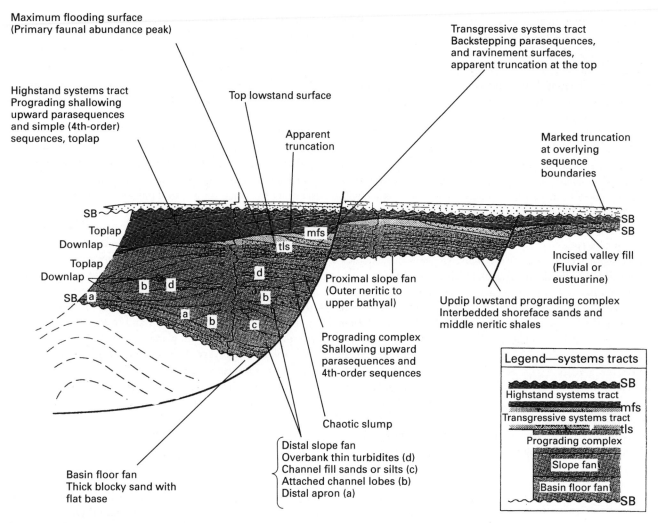

Maximum flooding surface
(Primary faunal abundance peak)

Highstand systems tract
Prograding shallowing
upward parasequences
and simple (4th-order)
sequences, toplap

Top lowstand surface

Apparent
truncation

Transgressive systems tract
Backstepping parasequences,
and ravinement surfaces,
apparent truncation at the top

Marked truncation
at overlying
sequence
boundaries

Incised valley fill
(Fluvial or
eustuarine)

Updip lowstand prograding complex
Interbedded shoreface sands and
middle neritic shales

Proximal slope fan
(Outer neritic to
upper bathyal)

Prograding complex
Shallowing upward
parasequences and
4th-order sequences

Chaotic slump

Distal slope fan
Overbank thin turbidites (d)
Channel fill sands or silts (c)
Attached channel lobes (b)
Distal apron (a)

Basin floor fan
Thick blocky sand with
flat base

SB
Toplap
Downlap
Toplap
Downlap
SB

mfs
tls

SB
SB

Legend—systems tracts

SB
Highstand systems tract
mfs
Transgressive systems tract
tls
Prograding complex

Slope fan

Basin floor fan
SB

Figure 2.19 Systems tracts in sequences with high subsidence rates related to growth faults (from Mitchum, Sangree *et al.*, 1990).

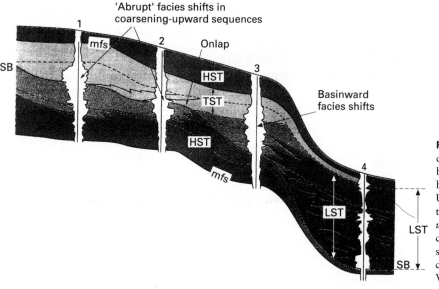

'Abrupt' facies shifts in
coarsening-upward sequences

Onlap

SB

mfs

HST

TST

HST

mfs

Basinward
facies shifts

LST

LST

SB

Figure 2.20 Diagram illustrating criteria and difficulties óf recognizing sequence boundaries. Onlap geometries and *abrupt* basinward shifts of facies are the main criteria. Using log data alone, it would not be possible to pick the sequence boundary in well 1 as it is *within* the sand towards the top of the coarsening-upward sequence. In well 3 the sequence boundary is at the *base* of the coarsening-upward sequence. (After Vail & Wornardt, 1990).

genetic stratigraphic sequence model as a complement, and, in some cases, a more practical alternative, to the EPR model. This has the advantage of using maximum flooding surfaces as sequence boundaries. These pass from the coast to the deep basin and, though they are diachronous, in outcrop and subsurface geology are the traditional marker horizons due to their distinctive character (fine grain size, richness in fauna and flora, sometimes high organic content and detectability on wireline logs). They also frequently represent more important reorganizations of the depositional setting than EPR sequence boundaries.

Sequences are often incomplete either vertically or laterally. Deeper water sediments such as the lowstand wedge are generally separated by a wide shelf in a ramp setting, or a fault in a rift setting, from the stacked highstand and transgressive systems tracts on the basin margin or on the upthrown block. Thus it is possible to find two sets of incomplete sequences, a HST and TST updip and a LST with pelagic drape downdip. In addition, two-dimensional sections across a shelf are certain to miss significant features. They tend to ignore the plan view and lateral changes parallel to the coast (e.g. Galloway, 1989b; Martinsen, 1993). For example, lowstand fans and deltas may be restricted to areas seaward of incised valleys and dip-lines across the shelf away from incised fluvial systems may lack the LST, which is therefore not visible on all dip sections. Similarly, structure may limit the location of system tracts to particular faulted sub-basins.

2.5 Models

In nature there are no models. Each environment and rock sequence is unique. We create models from present-day environments, and we also create them by interpreting facies in older rocks and by blending ideas from the modern and ancient together. However, situations never exactly repeat themselves. Each environment has its own unique set of internal processes and external conditions. It may have similarities to other examples, but it is never exactly the same.

Models are simplifications set up to aid our understanding of complex natural phenomena and processes. They serve several purposes and come in many forms. In this book we are primarily concerned with models of sedimentation, facies models and stratigraphic models, though climatic, tectonic and basin models are not completely neglected.

The aim of sedimentation models is to understand the processes by which individual beds – facies or subfacies – were formed. This is achieved primarily by reference to experiments, for example when sedimentary transport and deposition are generated in flumes under controlled conditions where it is posssible to determine how variations in grain size, sediment supply, flow rate and flow depth affect the kinds of sedimentary structure that are formed. Ideally, the knowledge gained by such experimental and theoretical research should be blended with

a study of the products (the rocks) and the processes observed in nature at the present day, as, for example, was done by Kuenen and Migliorini (1950) when they first postulated the turbidity current concept from flume experiments, from examination of outcropping graded beds and from sands discovered in deep oceans. However, experimental work on its own can lead to new insights and to the formulation of hypotheses that can later be tested in the field.

Facies models may be used both to interpret known facies distributions and to predict where as yet undiscovered facies may be found. They may be designed to show the effects of varying processes and external controls on measurable data and may be expressed as idealized sequences of facies, as cross-sections, as plan views or combined as block diagrams, as graphs, or as equations.

Mathematical and computer models simulate geological processes so that, for example, the results of a change in a sea level, subsidence rate or sediment supply can be predicted. Visual models of environments help us to see the relationship of environments to each other and to picture the processes and facies that we should anticipate. Visual models based on facies sequences and lateral facies relationships are used in reverse to determine the processes and environments in which we would expect deposition to have taken place. Here it is important to distinguish between a case history of a modern environment – a North Sea tidal shallow marine case – essentially a unique example, and an inductive model based on *inter alia* the North Sea, such as a 'tidally dominated shallow marine model'. The latter is a generalization inferred from several examples and its details and balance may be changed as our knowledge and understanding increases. However, the former, the North Sea case, will always remain valid, however much its perceived general applicability changes. Visual interpretations constructed from ancient sedimentary facies and facies sequences may also be either case histories, if they are simply interpretations of a particular sequence and local situation, or they may be models if they can extend our understanding of controlling processes by making generalizations (R.G. Walker, 1990). Although, ideally, we might like to work only with general process-based models, in practice we need to use individual cases as well – that is, examples of areas and environments that have been well studied and where the relationships of multiple processes to facies are well known – even though they probably represent a unique combination of processes.

Facies models are essentially autocyclic giving a snapshot of the lateral relationships of environments and facies over a limited period of time, under a particular set of steady, rather stable controls. What they do not show, without careful modification, are allocyclic changes, brought about by changes in rates of sea-level change, of tectonic movements, of basin subsidence and of climate. To convey these more complex controls we need dynamic stratigraphic models which are essentially evolutionary.

These are best illustrated by a series of palaeogeographic maps, or block diagrams. The latter are particularly valuable because they not only show a three-dimensional plan view, but the dimension of time.

Models have contributed greatly to sedimentology, particularly in improving the integration of depositional/biochemical processes and the sedimentary product. They are essential for instruction and learning, for the consolidation of disparate data, to promote new ideas and concepts and new interpretations of facies, environments and processes. They also enable us to search for new data that will support or disprove a model and to discover new facts by focusing attention on particular points in an obscure outcrop, core, well log or seismic section where the model suggests that geologically critical events may occur, such as marine fossils, an erosion surface, or a hiatus in deposition.

Yet we have to be careful how we use models (Anderton, 1985). All models are constructed on the basis of certain assumptions, the necessary suppression of some data as irrelevant, emphasis on other data, and the selection of some controlling factors at the expense of others. Too rigid an attachment to a particular model or the dogmatic assertion of its applicability may lead to the neglect of alternatives. For example, after the fining-upward sequence was so elegantly interpreted as a consequence of the lateral migration of a meandering river (J.R.L. Allen, 1963a), alternative hypotheses were conveniently forgotten, as also happened after the deep-sea fan became virtually the only model for turbidite sediments (Reading, 1987). In recent years many sequence stratigraphic interpreters have fallen into the same trap.

We have also to beware of simple models, whether visual or mathematical, particularly how they are used. Simple solutions may be easy to comprehend, but are never real. Because geologists always work with incomplete data and because a combination of different processes may have contributed to the final product, they have to use multiple working hypotheses and apply multiple models. Quantitative mathematical models, because of the necessity to make simplifying assumptions, often fail because our understanding of the rocks and their formative processes is inadequate. It is too easy to forget this when seeking to use the essential output of such models. Mathematical expressions or equations can sometimes be used to give a spurious authority to what is inherently unverifiable. On the other hand, they may provide a valuable and educational insight into consequences of different processes which might not have been intuitively obvious. They may thereby provide a theoretical framework for interpreting some more complex depositional systems. Where quantitative predictions are needed, as in reservoir modelling, it will commonly be most valid to express possible outcomes in terms of statistical probabilities, based, in many cases, on data collected from supposedly analogous successions elsewhere.

Some ancient environments and combinations of processes bear little similarity to any at the present day because some conditions in the past have no analogues today. Non-uniformitarian models have to be created because organisms have evolved, and there have been changes in the chemistry of the oceans, in earlier atmospheres, in the obliquity of the Earth's axis, and in the length of the day. Others are necessary because combinations of palaeogeographical conditions were quite different from any we experience today.

Further reading

Allen P.A. & Allen J.R. (1990) *Basin Analysis: Principles and Applications*, 451 pp. Blackwell Scientific Publications, Oxford.

Einsele G., Ricken W. & Seilacher A. (Eds) (1991) *Cycles and Events in Stratigraphy*, 955 pp. Springer-Verlag, Berlin.

Friedman G.M., Sanders J.E. & Kopaska-Merkel D.C. (1992) *Principles of Sedimentary Deposits*, 717 pp. Macmillan, New York.

Miall A.D. (1990) *Principles of Sedimentary Basin Analysis*, 2nd edn, 668 pp. Springer-Verlag, New York.

Scholle P.A., Bebout D.G. & Moore C.H. (Eds) (1983) *Carbonate Depositional Environments*, 708 pp. *Mem. Am. Ass. petrol. Geol.*, 33, Tulsa.

Scholle P.A. & Spearing D. (Eds) (1982) *Sandstone Depositional Environments*, 410 pp. *Mem. Am. Ass. petrol. Geol.*, 31, Tulsa.

Van Wagoner J.C., Mitchum R.M., Campion K.M. & Rahmanian V.D. (1990) *Siliciclastic Sequence Stratigraphy in Well Logs, Cores, and Outcrops: Concepts for High-resolution Correlation of Time and Facies. AAPG Methods in Exploration Series*, 7, 55 pp. Am. Ass. Petrol. Geol., Tulsa.

Walker R.G. & James N.P. (Eds) (1992) *Facies Models: Response to Sea Level Change*, 409 pp. Geol. Ass. Can., Waterloo, Ontario.

Wilgus C.K., Hastings B.S., Kendall C.G. St C., Posamentier H.W., Ross C.A. & Van Wagoner J.C. (Eds) (1988) *Sea-level Changes: An Integrated Approach*, 407 pp. *Spec. Publ. Soc. econ. Paleont. Miner.*, **42**.

Alluvial sediments

3

J.D. Collinson

3.1 Introduction

Rivers have long been recognized as the main means by which sediment is transferred across the land surface towards its eventual resting place. While temporary storage of sediment along the way is self evident, the recognition that alluvial sediments could form large parts of the long-term rock record is comparatively recent. Apart from alluvial 'fanglomerates', recognition of alluvial sediments in the rock record originally depended on the observation of channel forms, as in coal measure 'wash-outs'. Non-marine sequences lacking channels would commonly have been regarded as lake deposits.

Geomorphological and experimental hydrodynamic work in the 1950s provided important insights into sediment transport processes and the behaviour of different types of river channel (e.g. Sundborg, 1956; Leopold & Wolman, 1957; Simons & Richardson, 1961). Studies of ancient sequences in the early 1960s (e.g. Bersier, 1959; Bernard & Major, 1963; J.R.L. Allen, 1964) applied these insights and, in particular, recognized that lateral migration of channels generated fining-upwards sequences in tabular sandbodies. This early emphasis on lateral migration, particularly of point bars in meandering streams, led to other types of channel being largely ignored. Few attempts were made to offer depositional models for other types of river (Ore, 1963).

As other types of channels have been increasingly studied, it has become apparent that deposition in channels is very complex. A range of accretionary styles may occur across a spectrum of channel types so that no unique vertical facies sequence is diagnostic of a particular type of channel. Lateral, downstream and even upstream accretion on lateral and mid-channel bars can occur in any type of channel. In consequence, it is better to document ancient alluvial sequences in terms of 'alluvial architecture' (J.R.L. Allen, 1983; Miall, 1985a) – that is, the two- or, ideally, three-dimensional geometry of the constituent accretionary elements and their bounding surfaces. Establishment of the hierarchical status of erosional bounding surfaces is vital for understanding medium and long-term channel behaviour and hence for discriminating between internal autocyclic and external allocyclic controls.

Knowledge of soil-forming processes has increasingly aided the characterization and interpretation of ancient alluvium. Palaeosols, usually in floodplain sediment, give important information about the drainage state of the area and the climate. The maturity of palaeosols and their stratigraphic relationships allow insights into channel migration and incision which, when combined with inferences drawn directly from the channel sandstones, provide a basis for reconstruction of the whole alluvial system.

This chapter first deals with the processes of erosion, transport and deposition in alluvial settings and early post-depositional alteration and pedogenesis. Present-day river systems are discussed within a broad, grain-size classification while alluvial fans are dealt with in terms of their dominant transport processes. Overbank setttings, both close to an active channel and a long way removed from it, are discussed in terms of both depositional processes and soil types.

Ancient alluvial sediments are discussed first in terms of facies and then in terms of larger scale organization, in particular the role of architectural elements and their organization within an alluvial architecture. Emphasis is placed on the relationships between channel type, channel behaviour and the features of associated overbank sediment, especially palaeosols. Finally, the impact of external changes in base level and their role in the development of palaeovalleys are discussed.

3.2 Alluvial processes

Alluvial morphologies and deposits are products of complex interactions of erosion and deposition with the balance between them varying between different settings.

3.2.1 Erosional processes

Erosion occurs at a range of physical and temporal scales. At the smallest scale, floods erode loose sediment and cut small scours in cohesive fine sediment. At the larger scale, floods cut new channels that range in size from channel belts to thalweg channels and overbank/crevasse channels. Once channels are initiated, they may expand and shift position through a combination of vertical incision and lateral migration.

Incision is the vertical cutting of the substrate so that the channel deepens. It may occur during a single flood, when the bed is vertically scoured and then totally or partially filled as the flood wanes. On a longer time scale, incision may be the result of progressively increasing discharge due to climatic change or capture of some other channel. Deepening to accommodate increased discharge may be accompanied by widening. In addition, incision may be a response to lowering of the base level to which the stream flows. In that case, river banks are never overtopped by floods; floodplains become terraces, starved of sediment and susceptible to erosion and/or pedogenesis, and the initial channel may widen to become an incised valley (Sect. 3.6.9) within which flow from all but major floods is carried in a narrower channel. In settings where sediment is accumulating, incision is only important during initiation and relocation of channels and during short-lived scour and fill.

Migration of channels is associated with lateral erosion of bank material. Such erosion involves both removal of loose grains by fluid scour and gravity-driven mass movement when banks are oversteepened. The balance between scour and mass movement is largely determined by the nature of the bank

material and, in particular, its cohesive strength. For banks made up mainly of non-cohesive sand and gravel, fluid scour dominates and gravity collapse mainly results in sediment flows of loose grains. With cohesive, fine-grained bank material, gravity-driven movements may occur across discrete shear surfaces, subparallel to the bank (Fig. 3.1). The surfaces are commonly listric, often soling out at or below the channel floor at the foot of the cut bank. Blocks tend to rotate as they slide. Where slip surfaces extend beneath the channel floor, rotated blocks of bank material may either be emplaced beneath channel scour depth and thereby have enhanced preservation potential (Turnbull, Krinitzsky & Weaver, 1966; Laury, 1971) or become susceptible to scouring and fluting (Arnborg, 1957). Where cohesive material overlies non-cohesive sediment, it may develop an overhang prior to collapse and falling into the stream.

3.2.2 Transport and depositional processes

Four types of mechanism are responsible for the transport and deposition of sediment in alluvial settings: debris flows, bedload, suspended load and by the wind.

Debris flows are mobile, high-density sediment–water mixtures. By definition, they have a yield strength and 'freeze' once the applied shear (commonly the downslope component of gravity) no longer exceeds their strength (Fig. 3.2) (e.g. Nemec & Steel, 1984). Both cohesive and cohesionless debris flows are recognized. In cohesionless flows, excess pore water pressure and intergranular collisions allow grains to move relative to one another. Such flows freeze rather rapidly as pore water is lost, either across the upper surface of the flow or by infiltration into the bed.

Cohesive debris flows (mudflows) usually involve a fine-grained matrix component that increases the viscosity of the

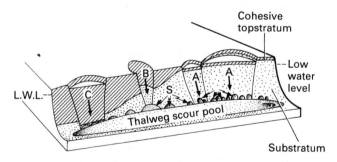

Figure 3.1 The influence of thickness of cohesive top stratum on channel bank failure. Scour at the foot of the bank is followed by failures in the overlying bank. Non-cohesive sands and gravels flow and shear subaqueously. With thin (A) or very thick (C) cohesive top stratum, subaqueous failures (S) are small, initiating shallow upper bank failures by shear. Intermediate thicknesses of top stratum (B) give rise to small to large subaqueous failures followed by deep, bowl-shaped upper-bank collapse (after Turnbull, Krinitzsky & Weaver, 1966).

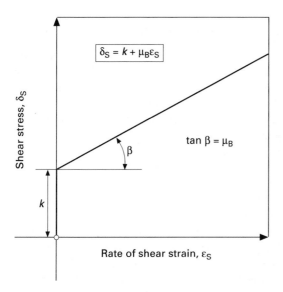

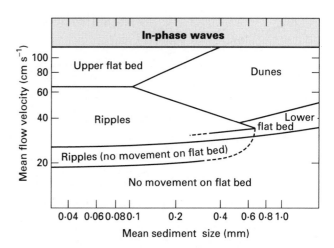

Figure 3.3 The equilibrium occurrence conditions of common sand bedforms plotted in the field of current velocity and sediment grain size. The occurrences of both ripples and dunes are limited by critical values of grain size (after Harms, Southard *et al.*, 1975).

Figure 3.2 Stress–strain relationship for a Bingham plastic, showing the critical value of shear (k) that must be exceeded before deformation occurs. This critical strength leads to such flows 'freezing' as the shear threshold is passed during deceleration (after Nemec & Steel, 1984).

sediment–water mixture which, along with buoyancy, inhibits settling of larger particles. Mudflows commonly undergo laminar deformation when the applied shear is close to the yield strength. Under higher shear or, more commonly, as a result of diminished viscosity through dilution, mudflows may become turbulent and are transitional to suspended sediment flows, the 'hyperconcentrated' or 'fluidal' flows that may be part of sheet floods (e.g. Pierson, 1981; Nemec & Steel, 1984). As shear stress falls through lateral expansion and thinning or through loss of gradient or as flows lose fluid, they 'freeze' to deposit a unit that is poorly sorted with a matrix-supported fabric. Debris flows are important agents of transport and deposition on certain types of alluvial fan. They also occur in both subaerial and subaqueous parts of channels.

Bedload transport is the movement of non-cohesive grains close to the bed by rolling or bouncing as a result of the shear of the overriding flow. Non-cohesive particles begin to move when the boundary shear stress exceeds some critical value determined by the size, shape and density of the particles. Movement of sand grains causes the bed to deform into a series of bedforms including ripples, dunes, plane beds and standing waves (Fig. 3.3). Some uncertainty surrounds the classification and description of dunes, which have variously been termed sandwaves, dunes and megaripples (e.g. Ashley, 1990). The consensus is that such larger forms are 'dunes', though the literature abounds with other terms and care must be take when comparing accounts of modern river beds. Dunes occur on river beds at a wide range of scales from around 2 m spacing and a few tens of centimetres height to spacing of hundreds of metres

and heights up to several metres. Forms vary between straight crested and strongly three-dimensional. Dunes are most active during floods and shift rapidly or are created and destroyed over short time intervals. Several scales of dune may coexist, especially where water depth changes rapidly, and superimposed relationships reflect disequilibrium (Fig. 3.4) (Neill, 1969; J.R.L. Allen & Collinson, 1974).

Dunes relate to specific ranges of hydrodynamic conditions and their size scales broadly with flow depth. They commonly cover the surface of larger compound features on the river bed which scale broadly with channel width. These larger forms respond primarily to geomorphic processes and are termed 'macroforms' or 'bars'. They are the main sites of longer-term bedload storage and are described in the context of particular channel types in Section 3.3.

While most sediment moved as bedload is sand and gravel, finer-grained material may also be transported in this way (Rust & Nanson, 1989). Deeply cracked floodplain soils (vertisols), dominated by clays, may be eroded as sand-size aggregates which then may be transported as bedload, forming bedforms similar to those of normal sands. In the rock record, compaction makes it difficult to recognize such bedforms but at least one example is documented (Ekes, 1993).

Suspended load is carried in the flowing water as a result of fluid turbulence. In many rivers, most sediment is transported in this way and sediment concentrations may be very high, particularly where the load is dominated by clays. However, silt and sand are also commonly carried in suspension. While most suspended load finds its way to the sea or lake at the end of the river's course, significant volumes are deposited both in channels, in overbank areas and on fan surfaces.

Wind may erode and transport both sand and finer-grained sediment as bedload and in suspension during periods when

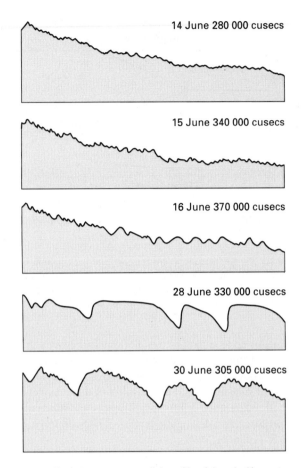

14 June 280 000 cusecs

15 June 340 000 cusecs

16 June 370 000 cusecs

28 June 330 000 cusecs

30 June 305 000 cusecs

Figure 3.4 Changing stream-parallel profile of dune bedforms in a reach of the Fraser River, British Columbia during a flood event. Dunes continue to grow after the flood peak and smaller forms become superimposed during falling stage. Length of surveyed reach is 670 m (after Pretious & Blench, 1951).

river bed and alluvial plain dry out. It is most effective in desert or semi-arid settings and in periglacial areas (see Sects 5.4 and 11.4.7). Sand, moved by the wind as a bedload, is likely to remain in or close to the river bed from which it was eroded. Wind ripples and small dunes form quickly in windy conditions but their preservation potential is low as the sediment is likely to be reworked by subsequent river flow. Sand that escapes the river channel to form dunes on the floodplain is more likely to be preserved, interbedded with fluvial overbank fines (e.g. Higgins, Ahmad & Brinkman, 1973). Fine-grained sediment may be thrown into suspension and be spread far afield as shown by thick deposits of loess that extend for great distances downwind of proglacial outwash areas. In addition, wind-blown fines contribute significantly to floodplains as in the Indus valley (Lambrick, 1967) and to the infilling of the pore space in framework gravels on exposed channel floors.

Sustained deflation of fines from alluvial surfaces, particularly on terraces, leads to the development of gravel pavements and, over time, to the erosion of larger clasts into *dreikanter*.

3.2.3 Postdepositional alteration and pedogenesis

When sediments deposited on alluvial fans, floodplains and, to a lesser extent, in river channels are subaerially exposed for long periods they become susceptible to alteration.

Physical modification results from both wind deflation and through the introduction of fine-grained aeolian material which may infiltrate the primary pore space as rain- and meltwater move downwards. In addition, products of *in situ* weathering, mainly clays, may be added to the pore fill (e.g. T.R. Walker, Waugh & Crone, 1978).

Soil-forming processes and early diagenesis modify both texture and mineralogy of host sediments while they are close to the surface as terraces, floodplains, fan surfaces or drained channel floors. While exposure renders near-surface sediments susceptible to soil-forming processes, the development of a recognizable soil demands that the processes operate over a considerable period of time before either sedimentation and burial or erosion and removal intervene. Immature soils characterized only by root disturbance and possibly animal bioturbation form on temporarily exposed surfaces in actively aggrading areas. These are known as 'alluvial soils'. When longer periods (hundreds–thousands of years) of uninterrupted alteration occur, soils develop vertical profiles of pedogenic features, keyed to the emergent surface. Soil formation involves a complex and variable suite of processes; physical, chemical and biological (Fig. 3.5). The nature and balance of processes are functions of the host sediment, the topography and drainage of the area, and the prevailing climate. In a geological perspective, the biological processes have changed with geological age, reflecting

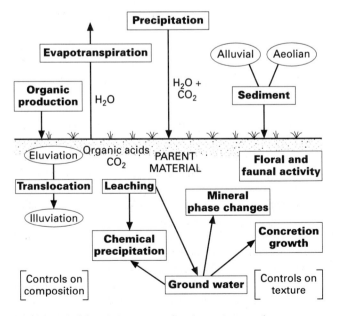

Figure 3.5 Soil-forming processes showing main controls on composition and texture.

changing floral, faunal and microbial constituents of soils and their surface coverings (Retallack, 1981, 1986a, 1990). Diversification of soil types through time can, however, be linked to a few major evolutionary 'events' such as the advent of vascular land plants in the Palaeozoic and the emergence of grasslands in the Tertiary (Retallack, 1981). In older geological periods soils probably formed under different atmospheric compositions.

A thorough review of soil-forming processes is beyond the scope of this account but several books are devoted to the topic (e.g. Bunting, 1967; Bridges, 1970; Hunt, 1972; Goudie, 1973; FitzPatrick, 1980; Duchaufour, 1982; Birkeland, 1984; Retallack, 1990). Not all the processes and products of modern soils are relevant to the geological perspective. Many of the more subtle textures, as well as features that depend on organic content, do not survive deep and sustained burial but are obliterated by compaction and diagenetic alteration.

Soil-forming processes of potential geological significance involve the vertical movement of water within the near-surface sediment and the action of organisms, including plant roots. These lead to a suite of alteration products involving both the textural and chemical properties of the sediment.

The movement of pore water through a developing soil is controlled by its drainage state, which in turn is commonly controlled by topography. With a low water table, near-surface layers may be well drained and a dominantly downward movement of water, run-off or precipitation, occurs. Such movement is likely to remove dissolved solutes and finer particles (*eluviation*), usually clays, from upper layers and carry them downwards to accumulate at lower levels (*illuviation*). Downward movement of water also dissolves material from the upper layers. This material may be reprecipitated within the soil as siderite, sesquioxides, calcite or sulphates, depending on the pore-water chemistry, or it may be removed from the system through the migration of groundwater. Both the eluviated clay and the dissolved ions may be weathering products of the host material.

In humid settings the net water movement is downwards, with progressive leaching of soluble ions from the upper levels. Plant colonization influences a developing soil in a number of ways. First, organic matter is added to the developing soil giving an organic-rich upper layer. Water passing through this layer becomes more acidic through the addition of CO_2 and organic acids accelerating weathering of unstable minerals and the dissolution of some alkali and alkali earth ions. In addition, iron, which is normally insoluble, may be mobilized by organic complexing. Second, plant roots tends to destroy depositional textures and structures and lead to local chemical and biochemical heterogeneities which influence mineral precipitation. They also abstract material in solution and return ions to the surface in organic tissue.

In arid or semi-arid settings, the lower volumes of water penetrating the soil tend not to escape into the groundwater but are lost through evaporation or transpiration. Plant growth is less prolific and oxidation more rapid so that accumulation of organic matter is reduced. This leads to alkaline soil waters and to precipitation of soluble minerals. The most common mineral is calcite but gypsum and silica are also important.

In order to account for the amount of mineral matter precipitated in some soils it is necessary that some new ions are delivered at the sediment surface in addition to those generated internally by weathering and dissolution of the host. Some additional material may arrive via leaf drop or drip (Goudie, 1973) but some must also come from mineral matter delivered as wind-blown dust (e.g. Reeves, 1970; Yaalon & Dan, 1974). Areas downwind of sites of gypsum precipitation have accelerated rates of calcite precipitation, presumably owing to the introduction of wind-blown gypsum and the operation of a common-ion effect (Lattman & Lauffenberger, 1974).

Iron mobilized in soil waters is precipitated in both ferrous or ferric form. With time ferrous oxides or hydroxides may transform to ferric oxides, depending on the Eh of the groundwater. In some developing soils, heterogeneous distribution of Eh leads to patchy transformation and colour-mottled textures involving red, green, brown and ochre, and bleached patches. The permeability of the host sediment also controls the distribution of ferric and ferrous pigments.

Red coloration in alluvial sediments is often general and pervasive and unconnected with soils. It occurs most commonly in sediments deposited under hot, semi-arid conditions. The colour results from diagenetic hematite coatings on detrital grains, commonly quartz and appears to result from the maturation of iron hydroxides (T.R. Walker, 1967; T.R. Walker, Waugh & Crone, 1978). These, in turn, are alteration products of unstable ferro-magnesian minerals (commonly hornblende and biotite) that may have already been at least partially altered prior to deposition. Chemical breakdown of these minerals in a regime of good drainage, intermittent wetting and high temperature causes the weathering products to be washed into the sediment. Permeable sediments become red more quickly compared with clay-rich layers. However, because of their lower permeability, reddened clay-rich layers resist subsequent reduction rather better than associated sandstones and conglomerates.

The rock record shows that red colours also develop in sediments laid down in more humid regimes. However, no examples are known from present-day humid settings where iron is present mainly as hydrated oxides (Van Houten, 1972; P. Turner, 1980). Lowered water tables and the resultant oxidizing conditions might allow these precursors to mature to red pigments (cf. Besly & Turner, 1983).

Where the host sediment is rich in swelling clays, wetting and drying resulting from changes in rainfall and flood inundation lead to physical disturbance. Consequent loss of original structure and texture is accompanied, in some cases, by the development of new textures including polished slickensided surfaces and pseudoanticlines (Fig. 3.6) (Gustavson, 1991).

Deep desiccation and shrinkage

Reworking of dry surface infilling of shrinkage cracks

Wetting — swelling of clays — compressive structures formed

Figure 3.6 Vertical mixing and development of compressional structures (thrusts, pseudoanticlines) resulting from the operation of vertic soils processes due to cyclical wetting and drying of a soil rich in swelling clays (after Duchaufour, 1982).

In settings where the water table is high, accumulating organic matter ensures that the pore waters are reducing. Oxidation of the organic-rich layer is inhibited and peat may accumulate. In addition, organic matter, including roots, in the lower parts of the soil layer is also preserved. Anaerobic decay leads to sulphate reduction and the precipitation of pyrite. Siderite, whose precipitation is also favoured by reducing conditions, is often localized on heterogeneities such as those created by roots. The acidic pore water also tends to dissolve cations, leading to a leached substrate.

3.3 Present-day alluvial settings

There are many ways of classifying alluvial settings. One way is to distinguish between alluvial plains, valleys and alluvial fans. On plains or in valleys, where flow is confined to channels except during floods, channels are distinguished from overbank settings. Channels are commonly described in terms of their plan-view pattern, in particular their sinuosity and degree of channel splitting (Sect. 3.3.1). Overbank or interchannel areas are differentiated by their proximity to an active channel. Near-channel levees and crevasse splays pass into more distal floodplains which are modified by soil-forming processes. A very high variation in discharge distinguishes ephemeral streams of arid and semi-arid regions from perennial rivers of more

humid settings. For alluvial fans, the discharge patterns and transport processes separate gravity-flow fans from fluvial fans (Sect. 3.3.8). On large alluvial fans with long-lived channels, channel morphology is important. Terminal fans are distinguished from alluvial fans by developing primarily as a response to loss of water discharge through evaporation and infiltration rather than being a response to a sharp loss of gradient where streams emerge from a mountain front.

General morphological descriptions provide useful starting points for interpreting ancient alluvium. Even though the distinctions between major settings are not always clear cut with certain morphological elements common to more than one setting, discussion based on the larger morphological elements, channels, overbank areas and fans, provides a basis for more detailed discussion. It should, however, be realized that there is a continuum of alluvial morphology and systems. Alluvial fans, in particular, act as sites for the operation of a range of different channel types.

3.3.1 River channel classification and controls

River channels may be classified on the basis of hydrology, grain size of bed material or channel pattern. In ancient alluvial sediments, grain size is the feature most easily diagnosed. It may be expressed in two ways:

1 the grain size of the bedload sediment forming the bedforms and bars of the channel floor will normally fall within the range pebble gravel to fine sand although both coarser and finer examples may be found; and

2 the ratio between bedload and suspended load.

Although there is scope for discussion about how any classification might be precisely formulated, here we broadly follow Schumm's (1972) scheme of 'bedload', 'mixed load' and 'suspended load' streams (Fig. 3.7) even though the nature and, still less, the amount of suspended sediment carried by a river may not be readily apparent from its channel deposits. The amount of suspended sediment has been suggested as a controlling factor on channel morphology through its influence on bank stability (e.g. Schumm, 1972; Orton & Reading, 1993). However, this factor, like most controls of complex systems, is far from perfect (Schumm, 1981). Exceptions, such as gravelly meandering streams and silty braided streams, are not uncommon.

There are two ways of characterizing plan view of channels. The first is sinuosity, the ratio of channel length to valley length, which reflects curvature of the channel thalweg. The second is the degree to which the flow is split (braided) by permanent or ephemeral bars and islands. Both may be expressed as numerical indices (e.g. Leopold & Wolman, 1957; Brice, 1964; Rust, 1978) but problems of consistent application, especially of braiding index, commonly make it more appropriate to use qualitative terms (e.g. highly sinuous, moderately braided, etc.).

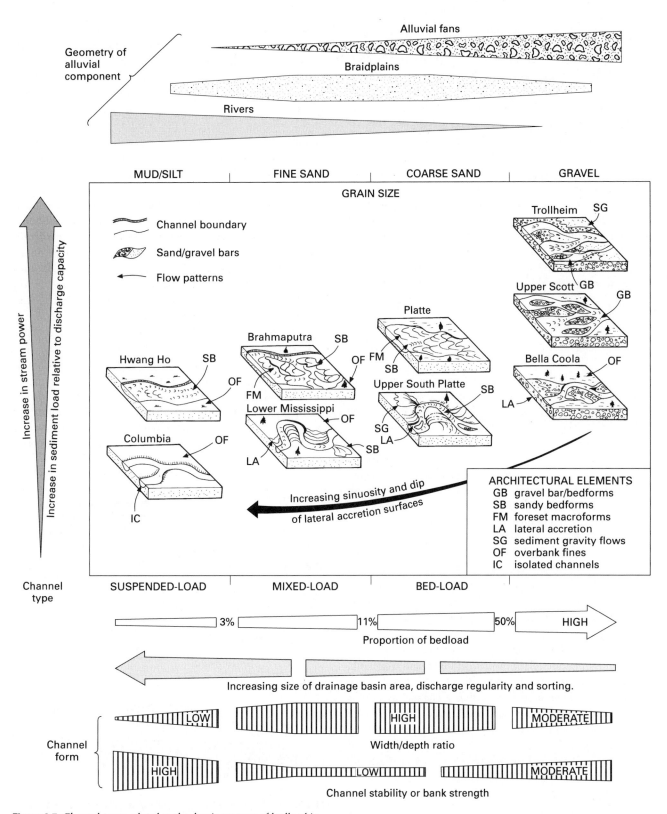

Figure 3.7 Channel types related to the dominant type of bedload in the stream and the relative stability of channel banks. The relationships are not absolute and many exceptions occur as a result of differences in gradient and discharge pattern (after Orton & Reading, 1993).

3.3.2 Coarse-grained bedload rivers

These streams carry both gravel (mainly pebbles and cobbles) and sand, in varying proportions. They also carry substantial volumes of suspended load, especially during floods. However, this is mostly carried through and beyond the gravel reach. Discharges and gradients are high though the largest clasts are commonly transported only during floods. Because bed material is non-cohesive, both the river bed and its banks are readily eroded and channels tend to be rather mobile. Coarse-grained bedload streams, therefore, tend to show low sinuosity and moderate or strong braiding. Braiding is also favoured by settings where streams are overloaded, generally as a result of a downstream reduction in gradient and lateral expansion which are features of alluvial fans (Sect. 3.3.8).

Coarse-grained bedload streams are common in proglacial settings where predictable patterns of seasonal discharge facilitate study. In semi-arid settings, similar channel patterns form in response to less predictable floods. In both settings, the lack of prolific vegetation contributes to bank erodibility.

Glacial outwash settings are well documented from the sandur plains of Iceland (e.g. Hjulström, 1952; Krigström, 1962; Bluck, 1974; Sect. 11.4.4) and from the subArctic and mountainous areas of North America (Williams & Rust, 1969; Boothroyd, 1972; Church, 1972, 1983; Rust, 1972; N.D. Smith, 1974; Boothroyd & Ashley, 1975; Southard, Smith & Kuhnle, 1984). Such outwash areas show complex patterns of rapidly changing channels and bars. Where gravel-bed rivers are confined by valley sides they commonly show alternating mobile reaches where sediment is being transported and more stable reaches where it is stored (Church, 1983). In mobile reaches the active river bed widens and braided patterns develop.

On more open fan-like settings, systems may be highly mobile and the whole depositional surface may be covered by active and abandoned channels and bars (Sect. 3.3.8). Overall, the channel pattern reflects an attempt by the river to adjust its frictional characteristics to prevailing flow but, because of the unsteady nature of the discharge, full equilibrium is never achieved.

Braiding occurs when sediment accumulates on the channel floor to split the flow, commonly at several coexisting scales, around *braid bars* that have more than one origin (Ashmore, 1991a,b). Experiments run with steady discharges show that: (i) *mid-channel bars* form within a channel as flow is progressively split; (ii) *point bars* on the insides of even quite gently curved channel bends may be detached through the erosion of chute channels; (iii) *transverse bars* may be converted to mid-channel bars; and (iv) *multiple bars* may be dissected. Most bars have complex histories of erosion and deposition so that morphology does not necessarily indicate origin.

Mid-channel bars, often termed longitudinal bars, may involve three concurrent depositional processes. On the bar top, clasts accrete by lodgement to develop an imbricated pebble framework and across the bar top there is commonly a downstream reduction in clast size. At the upstream end, the bar surface slopes gently into the adjacent channel without a break of slope or clast size. At the downstream ends of many bars, fine slip-face gravels contrast sharply with the coarse clasts of the adjacent channel floor over which they are advancing. Where coarser sediment is swept over a bar or at the channel confluence at the downstream end of a bar, a *riffle* may be present. This steeper sector of the channel floor is a site of rapid, shallow flow and a site of coarse gravel accumulation. Other flanks of a bar may, at the same time, be eroded by flow in adjacent channels (Boothroyd, 1972).

Many bars are emergent for long periods. As flow over a bar reduces, sand which during high stage is carried in rapid saltation or in suspension, becomes deposited on already static gravel. It infills pore spaces of gravel framework and accumulates on the surface. Bar flanks, which were depositional at high stage, may become erosional at low stage so that internal structures reflect repeated erosion and deposition. Fluctuations of discharge result in alternating layers of matrix-filled and openwork gravel within the bar (N.D. Smith, 1974) and tabular sets of cross-bedded gravel and sand wedges at the downstream end of a bar.

Bank-attached bars share many features with mid-channel bars, for example downstream diminution in clast size over the bar top and slip-faces or riffles at their downstream end. The riffle reach is probably the main site of gravel accumulation which leads, through channel migration, to the emergence of the bar (Fig. 3.8) (Bluck, 1974, 1976). Because they occur at

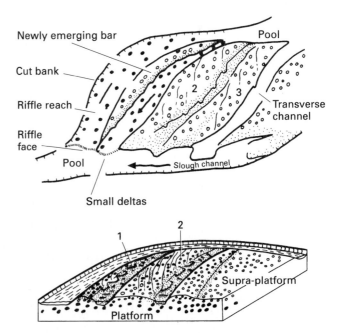

Figure 3.8 Morphology, terminology and structure of a typical bank-attached bar. (1), (2) and (3) are individual units of the bar, with (1) the most recently emergent. The slough channel will be reactivated during highest flood stages but accumulates fine-grained sediment at low stage (after Bluck, 1974).

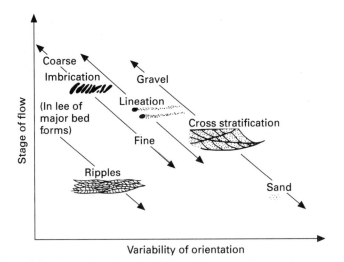

Figure 3.9 Likely relationships between flow stage and the type and variability of orientation of active structures on a mixed sand and gravel channel floor. Note that the ripples lie off the main trend as they tend to occur in major channel troughs which then determine their orientation (after Bluck, 1974).

channel bends, flow patterns over bank-attached bars are more predictable than for mid-channel bars. Bars on a convex bank are similar to point bars of meandering channels and flow over the bar top is likely to be diagonally towards the inner bank. This leads to dissection of the area into smaller bars separated by intermittently active chutes.

Some gravel bedload streams also have sectors where the flow is less divided. These typically occur in 'wandering' reaches where stable, single-channel sectors, with bank-attached bars, alternate with mobile, braided-channel sectors (Church, 1983; Desloges & Church, 1987).

Chutes commonly converge downstream into a channel close to the convex bank (Fig. 3.8) which is commonly cut off from significant flow during low discharges and becomes a quiet backwater, a *slough channel*. Small delta-like lobes form at the downstream ends of chute channels. As flow wanes, framework gravel, sandy framework infill, migrating sand bedforms and suspended load fines are successively deposited (Williams & Rust, 1969). Emergence leads to drying out and wind deflation.

In rivers that carry a wide range of grain sizes, different sizes are deposited over different ranges of discharge (Fig. 3.9). Directional variability increases as discharge wanes and flows collapse around emergent bars (Collinson, 1971). Structures in gravels, such as clast imbrication, are associated with high discharge and show unimodal trends with low variability within single bar or channel elements (Rust, 1972; Bluck, 1974). Cross-bedding at the downstream ends of bars tends to be more variable and bimodal either side of the downstream trend, reflecting skewed bar margins. Continued movement of sand at reduced discharge leads to small delta lobes at channel

confluences, to the migration of dunes and ripples and to sand ribbons on gravel surfaces giving further directional variability (Collinson, 1971; Bluck, 1974).

3.3.3 Sandy bedload rivers

While these river beds are dominated by sand, gravel may be present dispersed within the sand or as winnowed lags on scoured surfaces. The rivers, especially during floods, may have heavy loads of suspended sediment some of which may be deposited temporarily in overbank and channel settings. Accumulations of fine sediment are seldom thick enough to significantly resist channel migration and erosion. Highly erodible banks give rise to high width/depth ratios and to lateral movement both of the whole channel tract and of bars and islands within the tract. Thus sinuosity is rather low and braiding well developed though some meander with a less-divided channel. The availability of abundant sand is a major control on braided patterns (N.D. Smith & Smith, 1984) which result from dissection of both large repetitive dune forms and of larger compound bars. Most well-described examples are small- to medium-scale cases (e.g. Brice, 1964; N.D. Smith, 1970, 1971; Collinson, 1970a; Cant & Walker, 1978; Blodgett & Stanley, 1980). Several are in high latitudes where postglacial isostatic rebound has led to overall erosional regimes. While these may not offer as good analogues as rivers in sites of net accumulation, they have the advantage of displaying their deposits in banks cut into terraces. While much of our information on larger systems is very general and derived from descriptions prepared for non-geological purposes (e.g. NEDECO, 1959; Chien, 1961), some, such as the Brahmaputra, are becoming known in more geologically focused detail (e.g. Coleman, 1969; Bristow, 1987).

The dominant features of sandy bedload rivers are hierarchies of repetitive bedforms, ripples and dunes, often with more than one scale of coexisting dune. Dunes of different sizes and morphologies are superimposed upon one another due to a lag effect as they attempt to equilibrate to changing discharge prior to emergence (Fig. 3.4) (Pretious & Blench, 1951; Neill, 1969; J.R.L. Allen & Collinson, 1974).

Falling discharge also leads to partial emergence of the river bed (Fig. 3.10). High areas that emerge first split the flow at a variety of scales ranging from individual bedforms to larger compound 'bars' (see below). As a consequence, ripples change their orientation as flow paths become more tortuous. Dissection of dunes leads to small delta-lobes in front of normal slip-faces (Collinson, 1970a; N.D. Smith, 1971; Blodgett & Stanley, 1980). Some interdune troughs become cut off from significant flow and accumulate thin deposits of suspended sediment. Waves rework dune slip-faces to lower angles and concentrate heavy minerals.

Internal structures often reflect these stage-induced modifications (Figs 3.11 & 3.12) (Collinson, 1970a; N.D. Smith, 1971; Boothroyd, 1972; Boothroyd & Ashley, 1975; Blodgett

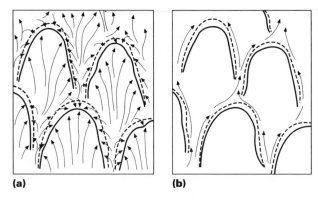

(a)   **(b)**

Figure 3.10 Flow patterns associated with (a) high and (b) low discharge over a bed dominated by low-relief dunes (linguoid bars). The concentration of flow between the dunes at low stage leads either to erosion or to lateral accretion on the flanks of the dunes (after Collinson, 1970a).

& Stanley, 1980). Straight-crested and low-relief linguoid dunes (linguoid bars), which produce tabular sets of cross-bedding, commonly show low-angle erosion surfaces (*reactivation surfaces*) which record slip-face reworking. However, similar structures are also produced by the migration and overtaking of individual forms in a bedform population (P.J. McCabe & Jones, 1977) and where dunes migrate down the lee side of a compound bar to give descending sets (cf. Banks, 1973a; Haszeldine, 1983).

The extent of modification depends upon the hydrology of the river. Rapid fall in water stage, as in the Tana River, favours abandonment of large bedforms and erosion of reactivation surfaces (Collinson, 1970a). Slower rates of fall, as in the Platte River, lead to bar dissection and the development of delta lobes and skewed crest lines (Blodgett & Stanley, 1980). These differences might provide a basis for reconstructing the discharge regime of an ancient river from structures in channel sands (cf. Jones, 1977).

Repetitive bedforms commonly construct higher-order components, 'bars', both mid-channel and bank-attached. Some are relatively ephemeral, perhaps surviving only a few flood cycles (e.g. Coleman, 1969; Bristow, 1987) while others may take on the role of semi-permanent islands, stabilized by vegetation (e.g. NEDECO, 1959; Collinson, 1970a). Where mid-channel bars divide the flow, smaller scale mid-channel and bank-attached bars may characterize the split reaches. The larger the river, the greater the extent of such hierarchies.

Understanding the development of bars, in particular their initiation and the ways in which dunes accrete, is vital for interpreting the internal organization of ancient channel sandbodies. Some bars originate from a chance accumulation of dunes. In other cases, individual large, low-relief dunes emerge during falling stage and develop highly skewed crest lines (Cant & Walker, 1978). Further dunes then accrete and build a mid-channel 'sandflat'. Once initiated, a bar influences local flow

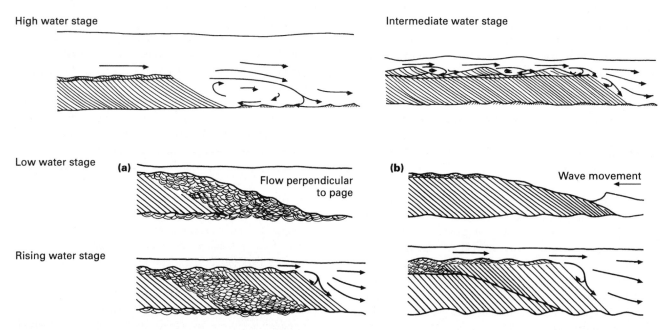

Figure 3.11 Modification of the internal lamination within tabular cross-bedded sets as a result of changing water-stage over linguoid bars (low relief dunes). Flow separation at high stage gives asymptotic foresets and counter-current ripples. At intermediate water-stage, the separation eddy is weaker and angular foresets develop through avalanching with burial of inactive ripples.

(a) Falling water stage promotes flow around the margin of the emergent bedform leading to the lateral accretion of ripple laminated sand. (b) Wave action during falling stage reduces the gradient of the lee face through erosion giving a reactivation surface when the bedform advances at the next high stage (after Collinson, 1970a).

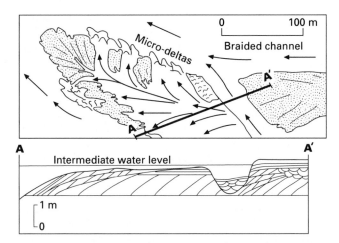

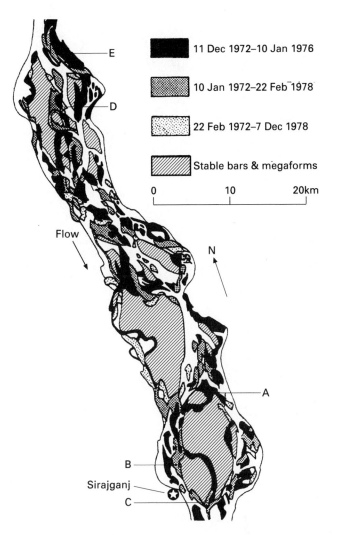

Figure 3.12 Falling-stage modifications to a low-relief dune (linguoid bar) in the Platte River, Nebraska. The top of the bar is dissected and lobate microdeltas form at the downstream ends of the dissecting subchannels. Deeper incision leads to braided channels and the dissection of earlier composite bars. Cross-section **A–A′** shows the internal structure resulting from alternating progradation and scour of the bar front and lobate microdeltas (after Blodgett & Stanley, 1980).

and this then controls further bed accretion. Flow across a bar may expand vertically at its lee-side causing sedimentation and downsteam accretion. Flow around a bar may develop helicoidal secondary flow and give rise to lateral accretion while dunes may accrete on the upstream side of a bar as they decelerate on meeting an upslope. In addition, bars may form at channel confluences giving in-channel deltas (Collinson, 1970a; Best, 1987; Bristow, Best & Roy, 1993). Once developed to near bank-full levels, bars may grow vertically through the fall-out of fine-grained sediment from suspension, perhaps aided by baffling effects of pioneering vegetation (NEDECO, 1959).

The fullest account of channel shifting and bar accretion is that for the Brahmaputra (Fig. 3.13) (Bristow, 1987) which carries a heavy load of suspended sediment during monsoonal floods and is transitional to a mixed-load stream. Its channel belt is around 15 km wide and expands and contracts along its length as the river alternates between divided and single-channel reaches. The whole belt shifts suddenly by avulsion, generally over distances of tens of kilometres on a frequency of a few hundred years, apparently associated with major floods. It also shifts gradually, possibly owing to tectonic tilting (Coleman, 1969). On a shorter time scale, individual channels within divided reaches shift rapidly, generally through lateral migration (Fig. 3.14). As the channels shift position, bars accrete, mostly as additions to the flanks of existing mid-channel bars, suggesting lateral accretion. Ridge and swale topography on the bar tops suggests analogies with point bars (Sect. 3.3.4). Considerable sediment is also added at the downstream ends of bars with minor amounts on their upstream sides. Bars are

Figure 3.13 A reach of the Brahmaputra showing the areas of deposition inferred from comparison of maps made between successive floods. A, upstream accretion; B, flank (lateral) accretion; C, downstream accretion; D, new mid-channel bar complex; E, large bank-attached point bar. Blank areas are zones of non-deposition or net erosion (after Bristow, 1987).

covered with dunes and accreted sediment will therefore be characterized by cross-bedding. Because bars are so large in the Brahmaputra, the gradients of accretion surfaces are small and it may therefore be difficult to detect them in preserved sediments.

At a much smaller scale, the Calamus River of Nebraska is a 15–20 m wide sandbed stream with low sinuosity and weakly developed braiding (Bridge, Smith *et al.*, 1986). Local reaches are either strongly curved or split by a mid-channel bar. Curved reaches migrate as point bars (Sect. 3.3.4) with associated neck and chute cut-offs but both point bars and mid-channel bars generate small-scale upwards-fining units dominated by low-angle cross-bedding (Fig. 3.15).

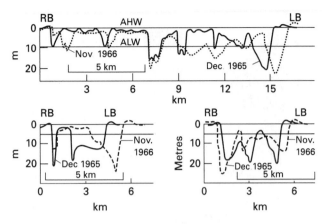

Figure 3.14 Transverse profiles across the Brahmaputra measured before and after a single monsoonal flood. Over this short time interval, huge volumes of sediment were eroded and deposited as the pattern of channels and bars was modified (compare with Fig. 3.13) (after Coleman, 1969).

Internally, accretionary bars are dominated by trough and tabular cross-bedded sands (Fig. 3.16) (Cant & Walker, 1978; Singh & Bhardwaj, 1991). These often overlie basal gravels, recording flows in the deepest parts of channels. Upwards, cross-bedded sands pass into finer sands and silts with smaller scale cross-bedding and ripple cross-lamination. Accretion surfaces, where detectable, may dip in any direction relative to the local current though surfaces dipping normal to flow are probably most common. Vertical facies sequences differ little from those of more sinuous channels, bearing out Brierley's (1989, 1991) contention that channel pattern is not readily diagnosed from channel facies. Most accretionary elements in braided channels are likely to be truncated by erosion, resulting from migration of channels within the channel belt. Preserved elements are therefore likely to have limited lateral continuity.

Where tabular cross-bedding is important, foreset dip azimuths may diverge considerably from the true downstream

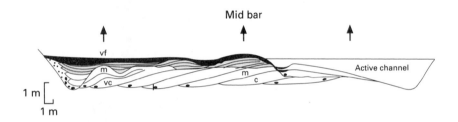

▪ Peat, with sand and silt	vc Very coarse-grained sand
▨ Vegetation-rich, small-scale cross-stratified sand	c Coarse
☐ Large-scale cross-stratified sand	m Medium
⬡ Unsorted sands and gravels	vf Very fine

Figure 3.15 Generalized facies model of the deposits of the Calamus River, Nebraska, a small, low sinuosity sandbed stream. Profile along a single bar shows both lateral accretion and the vertical channel filling associated with abandonment of a flanking channel sector (after Bridge, Smith et al., 1986).

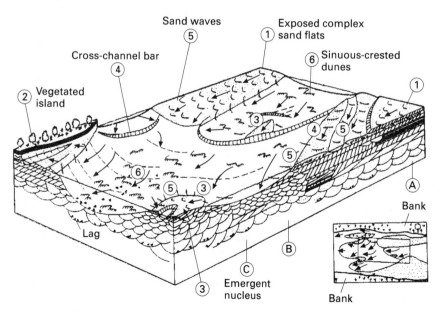

Figure 3.16 Composite facies model for a sandy braided river, based on the South Saskatchewan River. Significantly different sequences are predicted for mid-channel sandflats (A) and for channels (C); (B) is an intermediate case. The model suggests that morphological elements are somewhat static in position and accrete vertically, a situation that appears somewhat at variance with data from other sandbed braided streams. The inset shows the variations in local flow directions and the widespread of orientations of the slip-faces of the cross-channel bars (after Cant & Walker, 1978).

direction, reflecting highly skewed crestlines of straight-crested and linguoid dunes (Fig. 3.16) (Cant & Walker, 1978). This directional pattern may be more diagnostic of sandy braided stream deposits than their vertical facies sequence.

3.3.4 Mixed-load rivers

Mixed-load rivers transport fine, suspended sediment as well as significant bedload. Their bedload is dominantly sandy though there may be some gravel and sufficient fine-grained sediment accumulates to enhance bank stability (e.g. Bluck, 1971; Gustavson, 1978; Forbes, 1983). There is no clear demarcation from sandy bedload streams.

While some mixed-load streams have braided reaches with low sinuosity channels, the majority meander. Meandering channels occur in discrete belts on alluvial plains, on valley floors and between terraces. Where unconfined by terraces or valley walls, the active channel belt usually occupies only part of the available space (e.g. Bernard, Le Blanc & Major, 1962; Bernard & Major, 1963). The meander belt is a zone of active and recently abandoned channels and their adjacent overbank environments. Outside the meander belt lies the floodplain. While channels constantly shift position within meander belts, the belts themselves are relatively stable. Clay plugs generated by the infill of channel cut-offs help to stabilize banks and resist lateral erosion.

Overbank sedimentation is most rapid close to an active channel, leading to the growth of an alluvial ridge above the surrounding floodplain (Fisk, 1952). This increasingly unstable situation is periodically relieved by channel *avulsion*. During a flood, a channel bank is breached and a new channel course is established on the lowest part of the floodplain (e.g. Speight, 1965), the transfer of discharge commonly taking place over several years. For example, an avulsion of the meandering Saskatchewan River into adjacent marshland and shallow lakes began in 1873 (Fig. 3.17) (N.D. Smith, Cross *et al.*, 1989).

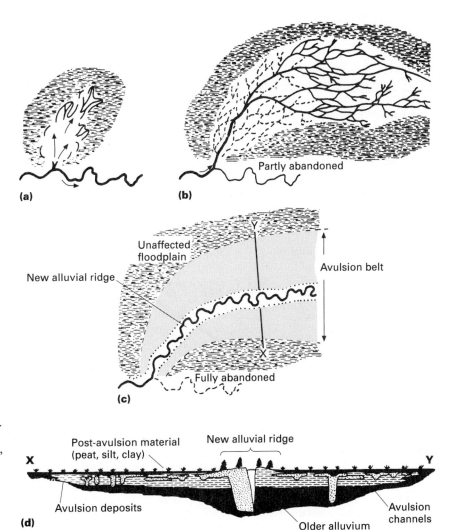

Figure 3.17 Evolution of a major channel avulsion from a crevasse event. (a) Following the initial crevassing, a splay expands rapidly into the overbank marsh. (b) Continued avulsion expands the splay and develops a network of anastomosing channels that gradually coalesces into fewer, larger channels. (c) In the final stage, one channel comes to dominate and incises the earlier splay deposits, initiating the growth of a new alluvial ridge. (d) Cross-section showing that migration of the new single channel may lead to erosion of the splay and of minor channel sediments associated with early stages of avulsion (after N.D. Smith, Cross *et al.*, 1989).

After more than 100 years it has only partially evolved towards a new meander belt. The initial stage was a large splay lobe crossed by rather unstable small channels. These channels gradually stabilized with an anastomosing pattern and some then grew at the expense of others until one became the locus of the new meander belt. Because this channel is deeper than precursor splay channels, it will, through migration, erode earlier splay deposits.

Within meandering, mixed-load channels *point bars* develop on the insides of meander bends and accrete laterally as the meander bend migrates with both downstream and transverse components of movement. While point bars are quite variable in detail, their behaviour conforms, to a greater or lesser degree, to the classical *point-bar model* which dominated the early days of fluvial sedimentology (Fig. 3.18) (e.g. Bernard & Major, 1963; J.R.L. Allen, 1964).

In this model, flow of water around bends leads to helicoidal secondary currents with surface components towards the outer bank and near-bed flows towards the inner bank (e.g. van Bendegom, 1947; Sundborg, 1956). Maximum velocities and depths occur close to the outer bank so that the channel has an asymmetric cross-section (Fig. 3.18) (e.g. Bridge & Jarvis, 1976, 1982; Geldof & de Vriend, 1983). As flow passes downstream through bends of alternating curvature, so the sense of rotation should alternate and the thalweg shift from side to side. Thus, outer banks of bends are predominantly sites of erosion and inner banks are predominantly sites of sediment accumulation through lateral accretion.

As banks retreat, both fluid scour and mass movement lead to erosion and large clasts may become buried in the deepest part of the channel as the point bar advances over them. The inner banks of channels accrete sediment on the *point bar*, which

is inclined at a gradient inversely related to the size of the channel. The surface is usually covered by dunes of various sizes (Fig. 3.19) (Sundborg, 1956; Jordan & Pryor, 1992) though ripples and plane beds occur on some higher, shallower areas. On the lower parts of point bars, high flow velocities, with their helicoidal component, sweep bedload particles along and up the surfaces. Where a range of grains sizes is present, finer particles are swept further up the surface while coarser particles lag behind, giving a systematic reduction of grain size upslope (van Bendegom, 1947; J.R.L. Allen, 1970a,b; Bridge, 1975, 1977). Point bars also show an upslope diminution in size of bedforms, reflecting diminishing depth and boundary shear stress.

In addition to transverse bedforms, point bar surfaces may also show elongate ridges of sand that broadly parallel the contours. These *scroll bars* originate low on the point bar and migrate upwards until they reach bank-full level. Here they become inactive and give rise to roughly concentric ridges and swales that characterize the upper surfaces of point bars (Fig. 3.20) and allow reconstruction of the migration of individual meander bends (Sundborg, 1956; Hickin, 1974).

Many rivers fit this rather idealized model quite well but others differ from it. Where the bedload is coarse grained, finer sediment is swept to the downstream end and an imbricated pebble surface may form at the upstream end (e.g. Bluck, 1971; Levey, 1978). In some cases the gravel may be swept to the upper levels of the point bar, contributing to a ridge and swale topography (Arche, 1983). In some meandering channels the helicoidal secondary flow may be out-of-phase with the geometry of the bends. Secondary rotation established around one bend does not switch over at the downstream inflection point but persists for some distance around the next bend. In the Wabash River, the pattern of flow at high stage over the upstream part of a point bar is inherited from the bend upstream (Jackson, 1975, 1976). Only over the downstream part of the point bar surface is the rotation as the classical model predicts. Consequently, distributions of grain size and of bedforms differ from the classical sequence in the upstream part and different vertical facies sequences can be produced within the same point bar.

In relatively low-energy streams with a substantial load of suspended sediment, mud may be laid down on point bar surfaces, especially their upper parts (Fig. 3.19) (Jackson, 1981; D.G. Smith, 1987; Jordan & Pryor, 1992). Owing to their cohesion, such layers are not eroded during later high-stage flows. They help to define *lateral accretion surfaces* or *epsilon cross-bedding* (Sect. 3.6.2). Where rivers are subjected to even minor tidal influence, such muddy layers are more common and more extensive (D.G. Smith, 1987). Many point bars, however, are dominated by trough cross-bedded sands without apparent muddy interbeds (e.g. Frazier & Osanik, 1961).

Point bar surfaces are commonly cut by subsidiary channels termed *chutes* (e.g. Harms, McKenzie & McCubbin, 1963; McGowen & Garner, 1970; Levey, 1978; Jordan & Pryor, 1992).

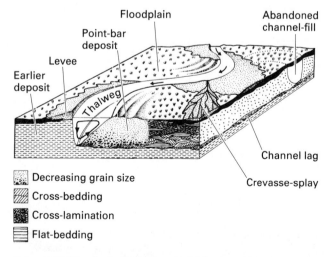

Figure 3.18 The classical point-bar model for deposition in a meandering stream channel (after J.R.L. Allen, 1964, 1970b).

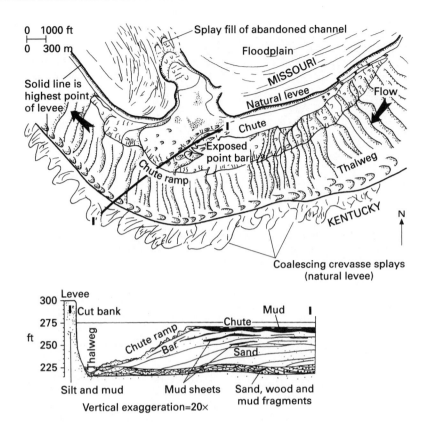

Figure 3.19 A large point bar in the Mississippi River showing a chute, its associated chute ramp, and an abandoned crevasse channel with the proximal part of its associated splay. Cross-section shows the relationships of the chute and chute ramp to the inferred internal structure. Note the muddy drape of the chute and the presence of mud sheets in the upper part of the point bar deposit (after Jordan & Pryor, 1992).

First-order transverse sand waves

Second-order transverse sand waves

Third-order barchanoid dunes

Fourth-order barchanoid dunes/transverse sand waves

Fifth-order ripples

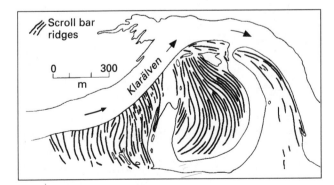

Figure 3.20 Scroll bar morphology on the upper surface of a point bar of the Klarälven, Sweden (after Sundborg, 1956).

These occur high on the point bar and are therefore most active at high discharge when they accommodate flow diverted across the point bar. A chute developed on a large point bar of the Mississippi (Jordan & Pryor, 1992) (Fig. 3.19) is some 5 metres deep and appears to have migrated along with the point bar. The floors of chutes are commonly covered by dunes though in coarser-grained examples there may be a coating of gravel, especially towards the upstream end. During falling water stages, chutes are abandoned and become draped by fine-grained sediment (Harms, McKenzie & McCubbin, 1963; Jordan & Pryor, 1992). Some chutes enlarge over several flood cycles until they capture discharge permanently and the original channel is abandoned.

At their downstream ends, chutes either shallow into a ramp that forms part of the main point bar surface (Fig. 3.19) (Jordan

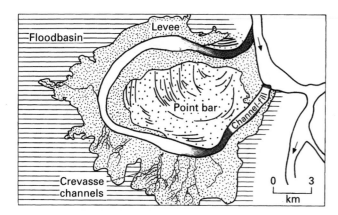

Figure 3.21 Levee, crevasse and crevasse splay topography preserved around an ox-bow lake caused by neck cut-off. False River cut-off channel, Mississippi River (after Fisk, 1947).

& Pryor, 1992) or feed a *chute bar* (McGowen & Garner, 1970). This is a lobe of sediment that protrudes downstream on to the point bar surface. Coarse-grained chute bars have avalanche slip-faces that are strongly convex downstream. Because they are only intermittently active, their internal cross-bedding is punctuated by reactivation surfaces (Levey, 1978).

A consequence of meandering is that sections of channel commonly become cut-off. Differing rates of lateral erosion on adjacent bends leads to the development of a narrow neck, which on breaching gives a *neck cut-off*. This tends to be a rather sudden process with the ends of the abandoned channel reach being rapidly plugged by channel sediment (Fig. 3.21) (Fisk, 1944, 1947). The central part of the abandoned channel is then an *ox-bow lake* which accumulates overbank fines and organic matter. *Chute cut-off* occurs through the gradual enlarge-

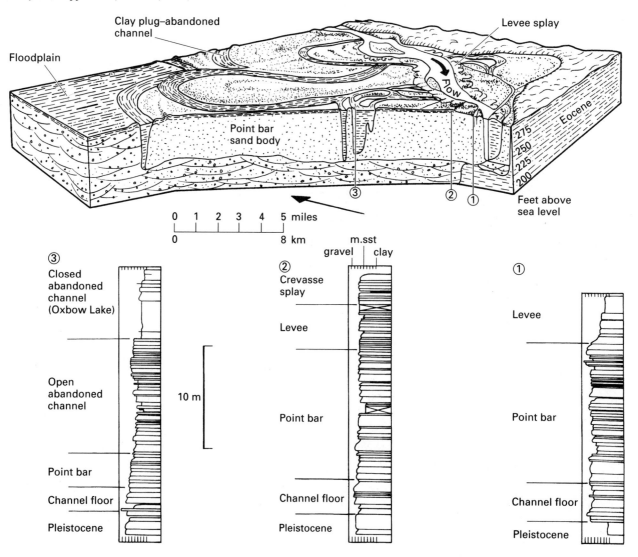

Figure 3.22 A point-bar complex from the Mississippi River showing the relationships between the deposits of actively migrating channels and the dominantly muddy infills of closed abandoned channels.

Vertical logs show the details of sections located within the complex (after Jordan & Pryor, 1992).

ment of an adjacent chute. An upwards-fining sequence is produced as increasing amounts of mud are laid down (Fig. 3.22) (Jordan & Pryor, 1992). The clay plugs that fill abandoned channels resist later erosion and help stabilize meander belts.

3.3.5 Suspended-load rivers

Rivers that carry a very high proportion of their load in suspension, and that flow on low gradients, deposit fine-grained sediment both on the floodplain and, to some degree, within the channels. The combination of low stream energy and abundant fine-grained cohesive sediment in channel banks means that channels, once established, tend not to migrate (e.g. D.G. Smith & Smith, 1980; Rust, 1981; D.G. Smith, 1983, 1986; Nadon, 1994). In plan view, such channels commonly have rather high sinuosities but are mainly characterized by a tendency to split and rejoin at a scale many times the channel width giving an *anastomosed* pattern (Fig. 3.23). Anastomosing channels are commonly characterized by prominent levees. Their associated floodplains may vary from swamps and marshes to arid desert plains, depending on the climatic setting and the position of the water table.

Anastomosing channel patterns occur in both small streams and major rivers. The low gradients that give rise to them may be features of distal, downstream settings or they may reflect local controls on base level such as damming of valley floors. In desert settings, anastomosed channel patterns result from radical reductions in river discharge caused by major climatic shifts (Rust & Legun, 1983).

The channels are mainly floored by sand which migrates as repetitive bedforms, possibly superimposed on alternating side bars (Rust & Legun, 1983). Individual channels stay active for long periods though some avulsion and abandonment takes place as mud-draped and mud-filled channels are features of the floodplains of such systems. The rapid sedimentation that

commonly occurs in these systems means that channel floor, levees and floodplain areas all accrete vertically (Fig. 3.23). The channels therefore give rise to elongate and often sinuous shoe-string sandbodies, of low width–thickness ratio, isolated in floodplain fines. The channel sandbodies also have 'wings' of levee sediment extending into the adjacent floodplain fines. Sequences developed under arid conditions are likely to have a much lower channel density than those developed in humid settings (Rust, 1981).

While most suspended-load streams have stable anastomosing channels, some have channels that migrate to form muddy point bars. Migrating muddy systems are commonly highly sinuous, best exemplified by tidal creeks (Sect. 6.6.1; Fig. 6.33), and this may lead to deposition of muddy benches on the outside of bends, possibly associated with erosion of the inner bank (G. Taylor & Woodyer, 1978; Woodyer, Taylor & Crook, 1979; Page & Nanson, 1982). Such bench deposits are restricted in their distribution and rest in erosive contact with point bar deposits. Their preservation potential is low and they might be confused with infills of abandoned channels.

3.3.6 Overbank areas

Overbank areas are associated with channels of all types. They can be divided into: (i) proximal areas close to active channels (levees and crevasse splays): and (ii) distal areas some distance from a channel (floodplain).

LEVEES AND CREVASSE SPLAYS

Levees are ridges built on either side of a channel but commonly better developed on the outer margins of bends (Fig. 3.24). They grow through the deposition of fine-grained, suspended load sediment during submergence by major floods. During lesser floods they may be the only dry land on the floodplain. As floodwaters overtop levees, turbulence diminishes and suspended sediment is deposited, commonly becoming finer grained away from the channel (Hughes & Lewin, 1982; Guccione, 1993). Banks are seldom uniformly overtopped and flows are commonly localized. Discrete lobes of silty or sandy sediment extend down the levee from shallow crevasses on the crest. These lobes spread on to the floodplain as *crevasse splays* with fingers of sand extending beyond the limits of the main lobe (Fig. 3.24) (O'Brien & Wells, 1986). In many cases, the levee itself is largely the product of laterally coalescing splay lobes (Coleman, 1969; Jordan & Pryor, 1992). Deeply incised crevasses may tap lower levels of the main channel and allow coarser sediment to escape on to the floodplain.

Levee deposits are mainly fine-grained sands and silts dominated by ripple cross-lamination and small-scale cross-bedding (e.g. Singh, 1972). However, lamination is commonly disturbed or destroyed by bioturbation. Further from the channel, isolated crevasse splays may be interbedded with floodplain fines as

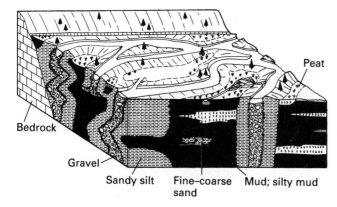

Figure 3.23 Three-dimensional model for the distribution of facies beneath an anastomosing channel system, based on closely spaced boreholes in the plains of the Alexandra and North Saskatchewan Rivers, Alberta (after D.G. Smith & Smith, 1980).

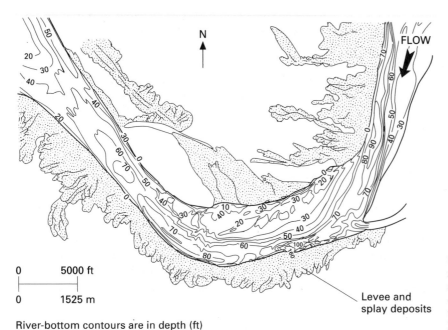

N

FLOW

0   5000 ft

0   1525 m

River-bottom contours are in depth (ft)

Levee and
splay deposits

Figure 3.24 Map of crevasse splays around a meander bend of the Mississippi River showing their coalescence into a levee on the outer bend and their more elongate nature on the surface of the point bar (after Jordan & Pryor, 1992).

discrete sandy beds (Farrell, 1987). Patterns of vertical facies change, such as upwards increasing sand layers probably reflect migration and avulsion of adjacent channels. Flow vectors derived from cross-lamination may diverge from those in the adjacent channel.

FLOODPLAINS

Floodplain sedimentation is relatively poorly documented in spite of extensive description of postdepositional alteration to form soils (Sect. 3.3.7). Many such areas are intensely cultivated at the present day and thus much disturbed. Apart from wind-blown dust, most sedimentation occurs during floods and is mostly from suspension. In some cases, floodwaters escape from through-flowing river channels. However, many floodplains are only rarely flooded by overtopping of river banks and are more commonly inundated through a rise of water table and the formation of floodplain lakes. This may result from intense precipitation on the floodplain or from increased discharge in a through-flowing river. Where flow is established on the flood-plain between lakes or close to flooding river channels, local erosion can lead to reworking of earlier floodplain sediments. Pedogenic nodules may be eroded and redeposited, both as the fills of shallow channels and as graded beds. Where flood flows emanate from major river channels, the grain size of floodplain sediments tends to diminish distally (cf. Pizzuto, 1987; Guccione, 1993).

Between floods, floodplains commonly dry out and various features of subaerial exposure develop, largely depending upon climate, water table and vegetation. In humid settings, or where the river flows close to its base level, the floodplain sediments

may stay wet with swamps and lakes as important elements in the landscape as in the 'Sud' of the Upper Nile (Rzoska, 1974), the present-day Atchafalaya River Basin (e.g. Coleman, 1966) and the Cumberland Marshes of Canada (N.D. Smith, Cross *et al.*, 1989). Where vegetation is abundant and the water table is high, peat accumulates, especially if suspended sediment is excluded through the baffling effect of plants around the margins of the swamps. Some interchannel areas of alluvial systems are raised swamps which are never inundated by river waters. Such areas derive their moisture from rainfall and from groundwater and may be the sites of accumulation of very thick, low-ash peat (e.g. Barber, 1981; McCabe, 1984).

Where plant growth is less prolific and more sediment can enter the floodplain, lakes and swamps develop. Floodplain lakes, which may arise through compaction of earlier fine-grained and organic-rich sediments, are commonly infilled by small deltas fed by crevasse flows from adjacent channels (Farrell, 1987). The resultant emergence allows new plant colonization as a marsh or swamp.

In semi-arid settings, vegetation is less prolific and less organic matter is incorporated into floodplain sediments, as it is mostly oxidized. Fine-grained floodplain sediments may be eroded by floods as pedogenic aggregates and transported as bedload (Rust & Nanson, 1989) while strong winds may deflate and redistribute sediment (Sect. 5.4) (Lambrick, 1967).

While local climate strongly influences the character of floodplain sediments, especially as reflected in their early diagenesis and pedogenesis, associated rivers (e.g. the Nile) may have a hydrology that reflects the climate of a distant catchment area. Accordingly, apparently contradictory evidence may be presented by associated floodplain and channel sediments.

3.3.7 Soils and their distribution

Soils have been classified by pedologists into several schemes with daunting nomenclature (see FitzPatrick, 1980, for illustration). Most of these schemes were not devised with the study of palaeosols in mind but were driven by biogeographic and agricultural needs. Many of the criteria for classification depend on properties that have little or no chance of preservation in palaeosols. It therefore can be difficult to directly compare modern soils to palaeosols. However, some features survive even deep burial so that parts of some schemes are useful to the geologist.

Soils develop in response to a suite of controlling factors among which the nature of the host material, the prevailing climate, drainage state and topography are the most important. Soil-forming processes primarily involve the vertical movement of mineral and organic matter within the active layer (Sect. 3.2.3). If these movements are sustained over a significant period of time, they lead to the development of profiles of vertically distinct layers each characterized by a combination of diagnostic or highly characteristic structures, many of them microscopic in scale. Not all of them survive deep burial and compaction but many are recognized in modified form in palaeosols. Present-day soils are characterized by networks of open voids that allow movement of fluids and mineral matter. These voids may be planar, as joint sets, tubular or more equidimensional.

Networks of surfaces enriched in illuviated mineral matter are referred to as *coatings* or *cutans*. The mineral matter may be clay, oxides, silica, carbonates or organic matter. The composition of cutans is an important indicator of the chemical conditions of soil formation. Some cutans, formed under conditions of stress, as in swelling soils, may develop slickensides. Bodies of host material which may be cut by coated surfaces are known as *peds*. These can occur at a variety of scales within the same soil and have a variety of shapes, including angular, prismatic, columnar, granular and platy types. Nodular or concretionary segregations of mineral matter within soils are known as *nodules* or *glaebules* that may be irregular or nearly spherical in shape. Glaebules have a variety of compositions: carbonates, oxides, silica and pyrite. In some cases, they may pseudomorph earlier structures such as burrows or root traces. Such biogenic structures constitute a further class of soil structures and are generally in the form of *tubules*.

Structures and compositions of soils vary vertically to define profiles or series of horizons whose character and distribution form the basis of most schemes of classification. An extensive account of classification is beyond the scope of this chapter. The great variability of present-day soils is illustrated in textbooks on soil science and on the Elsevier/ISRIC Wallchart (Lof, 1987). Of the schemes of classification available, the American (Soil Survey Staff, 1975), the FAO and the French schemes are among the most widely used (cf. Duchaufour, 1982). The American classification relies upon, *inter alia*, the organic content of the near-surface layer, the chemical characteristics of the deeper horizons and quantitative aspects of the climatic regime. Application of these criteria results in a complex hierarchical scheme. The criteria are either very susceptible to diagenetic alteration on burial or unknowable in the case of an ancient soil and are therefore inappropriate for describing palaeosols. The FAO scheme, which is simpler, recognizes a set of basic types, based on considerations of maturity, host material and, for the more mature soils, climate. It is more applicable to ancient examples. The French classification (Fig. 3.25) (Duchaufour, 1982) relies less on climatic criteria and more on the nature of soil-forming processes. As many of these processes find direct expression in the resulting soil profiles, this scheme is broadly followed here though not all the soil types defined and illustrated are relevant to alluvial depositional settings. Discussion here focuses on types that are particularly important where the main control on soil character is the drainage state of the material, usually floodplain sediment, in which they form.

Alluvial soils or *accumulative hydromorphic soils* are features of most vegetated floodplains. They are immature because of continual sedimentation and thus lack clearly developed profiles. They show a general mottling characterized by drab haloes and by preserved plant roots when the water table is high.

Hydromorphic gley soils develop where a surface is starved of sediment for long periods, where the water table is permanently high and where rather immobile pore waters are reducing in character (Duchaufour, 1982). Such conditions are most likely on the topographically lowest parts of the flood plain. In such soils, iron is mainly in a reduced state and is slightly mobilized to accumulate as a dark layer towards the base of the profile. The upper parts of the profile are characterized by grey colours and siderite nodules and cements may be present. Gleys also commonly contain preserved organic matter, some in the form of roots and rootlets. Highly organic gley soils may be transitional to peat.

Peat in alluvial settings is a special end-member of the soil spectrum and would be described as an *organic hydromorphic soil* by Duchaufour (1982). It records the sustained growth of plants on a soil made up largely of organic debris. A high water table and reducing pore waters ensure that the rate of accumulation of organic matter exceeds the rate at which it is oxidized.

Well-drained soils in humid cold or temperate settings may experience an active downwards movement of water often introduced by direct precipitation. This leads to the chemical and physical translocation of material within a profile to give a *podzol* or *Spodosol*. Organic matter in the near-surface layer provides organic acids that help to break down unstable minerals in underlying sediment so that clays and other weathering products are eluviated from the upper A horizon and become concentrated in a lower, illuviated B horizon. The depleted A horizon is effectively enriched in quartz while the B horizon is

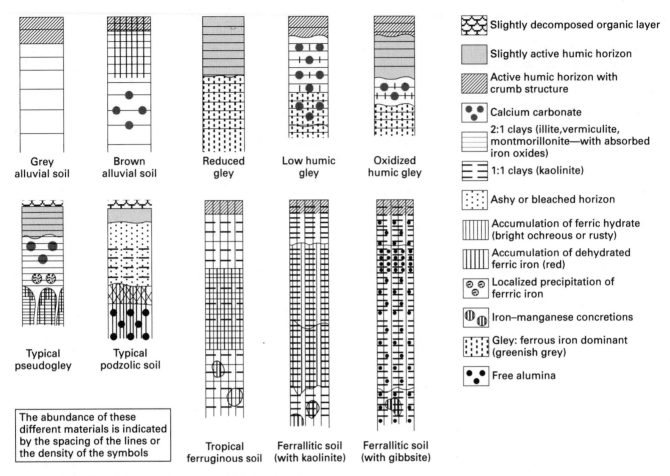

Figure 3.25 Profiles of representative recent soils likely to form in depositional alluvial settings (after Duchaufour, 1982).

enriched in clays, humus and sesquioxides. In alluvial settings this type of soil is most likely to form on terraces following periods of incision.

In warmer climates, iron oxides play a much more important role and many soils are characterized by red pigments. This is primarily because: (i) high temperatures lead to high rates of weathering reactions; (ii) accelerated rates of organic decay lead to less acidic and less reducing pore waters; and (iii) the iron oxides, once formed, are relatively immobile. In addition, crystallization of the iron oxides, goethite and hematite, is not inhibited by the presence of organic acids, as is the case in more temperate settings (Duchaufour, 1982). These two iron oxides reflect different moisture conditions; goethite crystallizes slowly under rather humid conditions while hematite forms rather rapidly in dry settings. However, goethite may mature to hematite over time, thus complicating the interpretation of red palaeosols.

In humid tropical conditions, the development of red soils is associated with weathering of virtually all minerals except quartz and with the lower parts of the profile being enriched in kaolinite. These are *ferrallitic soils* (Duchaufour, 1982), and

include *laterites*, which require a significant dry season and long periods of time to reach maturity. In some highly evolved examples, hematite locally segregates into indurated nodules, making a zone of plinthite or incipient ferricrete.

In drier conditions, such as occur in present-day savanna belts, weathering processes are less extreme and *ferruginous soils* develop. In these, weathering of unstable minerals is less complete. Upper parts of profiles are brownish-yellow while deeper, B, horizons show clay illuviation and intense red, brown or ochre colours.

In semi-arid settings, ferruginous soils are replaced by *fersiallitic soils* in which weathering reactions are less complete and in which calcite nodules and concretions occur if appropriate ions are available. Where present, calcite occurs as networks of veins, as elongate nodules and as more or less continuous layers as maturity (see below) increases (Machette, 1985). Such horizons are commonly referred to as *caliches* or *calcretes* (Goudie, 1973).

Soils developed under regimes of alternating wetting and drying in materials rich in swelling clays (montmorillonite) are termed *Vertisols*. These are characterized by the opening and closing of deep vertical cracks and by collapse of near-surface material into open cracks (Fig. 3.6). The result is homogenization of a sometimes deeply affected layer and the development of internal slickensided surfaces. In true vertisols organic and mineral components are fully homogenized. However, vertic processes may also occur during the formation of ferruginous and fersiallitic soils and calcretes. The expansion of clays during wetting may lead to the deformation of the soil surface as *gilgai*.

The spectrum of individual soil types outlined above forms in particular climatic and drainage conditions. These conditions may vary quite significantly at a local scale so that an association of soils may develop within a particular geomorphic setting. Such a soil association is termed a *catena* (Milne, 1935). Catenas are most clearly defined on hill slopes and valley sides where conditions change with height and gradient. However, catenas also occur in alluvial settings. In immature alluvial soils on an alluvial plain, for example, the elevated and better drained levee sediments may be coloured brown while the lower-lying flood-plain may be the site of an organic-rich hydromorphic soil. Even in semi-arid areas, gley soils may form in localized topographic depressions while calcrete forms in surrounding areas (e.g. Williams, 1969).

The soil types outlined above are defined on the basis of particular profiles of texture and composition. For such profiles to be clearly developed, a soil must develop over a considerable period of time under more or less stable conditions and without significant sedimentation. In alluvial settings, such stability may not occur (e.g. D.L. Johnson, Keller & Rockwell, 1990). Repeated deposition of sediment may prevent the development of a profile resulting in immature alluvial soils. Caliche soils, for example, have a succession of intergradational stages that reflect the extent to which profiles have developed (Machette, 1985).

If pedogenesis is sustained while an earlier buried soil is within range of broadly similar processes, there will be interference between the two profiles and a *composite* soil will develop (Morrison, 1978). If an intervening sediment is sufficiently thick, two soils may be distinct and separate, giving a *compound* soil (Figs 3.26 & 3.27) (Machette, 1978; Morrison, 1978). If erosion follows the formation of the first soil, a later profile may overprint the earlier one and penetrate to deeper levels, giving a *complex* or *reformed* profile. *Polygenetic* soils form where shifts in climate or drainage lead to overprinting of different types of profile.

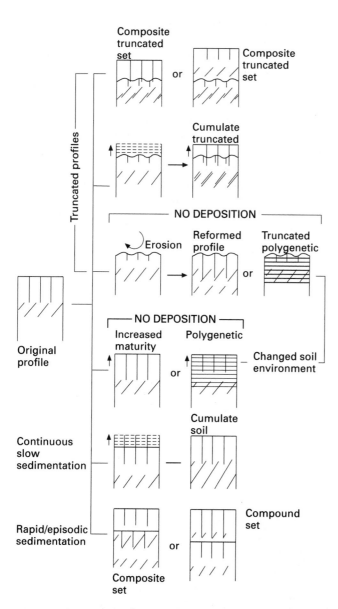

Figure 3.26 The development of different types of multiple soil profiles in response to changes in conditions through time, showing the terminology commonly used (after Marriott & Wright, 1993).

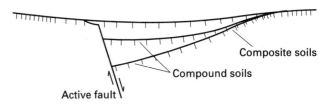

Figure 3.27 Stratigraphic relationships close to an active fault in a half-graben. The changes in stacking pattern may be accompanied by changes in soil character (after Machette, 1978).

3.3.8 Alluvial fans

Alluvial fans are localized areas of enhanced sedimentation downstream of points where laterally confined flows expand. Confinement is usually within a narrow valley or gorge cut into an area of high relief. Expansion is commonly on to a low-lying valley floor or a coastal plain. Flow expansion causes a reduction in depth and velocity with a consequent loss of competence leading to deposition. In many, but by no means all cases, the line across which topography changes is close to the trace of an active fault zone.

Alluvial fans vary enormously in scale from features a few tens of metres in radius to enormous cones extending more than 100 km. They also vary in terms of their active processes. Both scale and processes reflect a combination of source area lithology, catchment size and climate, especially as expressed through the pattern of discharge.

Fans typically approximate to a segment of a cone, commonly with channels radiating from the apex. In cases where a fan is formed within a valley, a true radial morphology may be constricted by valley walls or lateral terraces (e.g. Boothroyd, 1972; Boothroyd & Ashley, 1975). Where several catchments discharge along an escarpment, laterally adjacent fans coalesce to form a continuous ramp, a 'bajada'.

The gradient of most fans decreases downslope to give a concave-upwards radial profile. However, in detail, many fans reveal a profile that is a series of straighter segments rather than a smooth curve. These segments are related to incision of the upper part of the fan by the feeder channel. Since the degree of incision of the upper fan varies through time, the point at which flows expand laterally, the *intersection point* (Fig. 3.28) (Hooke, 1967), shifts up and down the fan surface. On small fans, during a major flood event, deposition may take place on the upper fan surface when the intersection point is close to the apex of the fan (Fig. 3.29). As discharge decreases, incision of the upper fan occurs and deposition shifts down the fan, partly reworking earlier flood peak deposits (cf. Blair, 1987a). On larger fans and on a longer time scale, a similar transition may follow a phase of tectonic uplift in the source area (cf. Harvey, 1987).

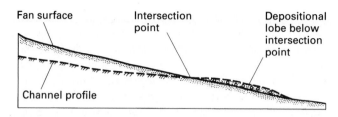

Figure 3.28 Radial profile of an alluvial fan showing the position of the intersection point (after Hooke, 1967).

The processes by which sediment is transported and deposited on fans range from debris flows to sustained stream flow. Two main classes of fan are recognized on the basis of dominant process; gravity-flow (semi-arid) fans and fluvial (humid) fans. The terms in parentheses are commonly used but they may mislead as processes are not simply equated with climate. Some workers regard the term 'alluvial fan' as only applicable to small, high-gradient features dominated by sediment gravity flows, the 'gravity-flow fans' of this account (Blair & McPherson, 1994). Such a usage seems unnecessarily restrictive even though there may be 'natural' breaks in the distributions of parameters such as radius and gradient between the types.

A further class of fans that do not entirely conform to the above generalizations are 'terminal fans'. These show less dependence on a point source and topographic discontinuity and result from an interaction of river discharge and the climate and hydrology of certain semi-arid settings.

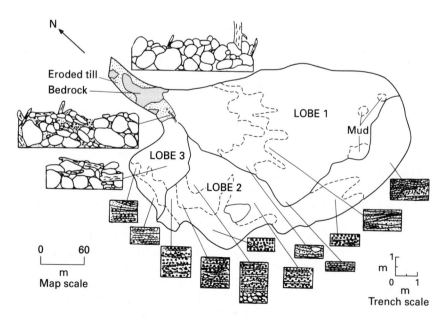

Figure 3.29 Map of a small alluvial fan developed in coarse gravels as a result of torrential flood events. Grain size diminished rapidly downfan. All but the coarsest deposits are horizontally stratified (after Blair, 1987a).

GRAVITY-FLOW FANS

Gravity-flow fans are the classic small to medium fans that commonly develop at tectonically active basin margins and that are best displayed in arid settings. However, they also occur in more humid climates, especially where very heavy rainfall is a recurring feature and where there is an abundance of fine-grained material in the source area. Volcanic terrains with abundant pyroclastic material are especially able to generate debris flows (e.g. Vessell & Davies, 1981) (Sect. 12.4.4). In semi-arid settings, debris flows are generated by rare but intense rainstorms which flush accumulated debris from the catchment valley or canyon. The resultant debris flows may be cohesive or non-cohesive, depending on the content of fine-grained sediment. Below the intersection point, flows decelerate on the fan surface and deposit elongate lobes as a result of active flows on the lobe surface being confined within levees of 'frozen' debris. Lobes may extend for several kilometres down larger fans, with sharp downslope ends (Sharp & Nobles, 1953; Hubert & Filipov, 1989; Beaty, 1990). Debris flow units are typically up to 2 m thick with matrix-supported fabrics and anomalously large 'floating' clasts, some of which may protrude from the top surface. Some debris flows show inverse grading in their lower parts as a result of the high shear in that zone.

While debris flows are important agents on fans of this type, they are not the only depositional process. As flow events subside residual *stream flows* may persist and upper surfaces of debris flow deposits are then reworked. Finer fractions are winnowed out and larger clasts concentrated at scoured surfaces. Where sand is abundant, stream processes produce ripples and dunes. A major debris flow may be followed by an interval during which little loose debris is available in the catchment, despite recurrent floods. Stream and sheetflood processes then dominate the fan surface (Beaty, 1990).

Sheet floods transport sediment both in suspension and as bedload and, on many fans they lose volume downfan through infiltration of water into the fan (Rahn, 1967). The resultant sheets and lobes are commonly dominated by laminated sediment with small scours, cross-lamination and cross-bedding. Where infiltration is rapid and the sediment load is deficient in fine-grained material, a discrete lobe of clast-supported framework gravel may result, a so-called *sieve deposit* (Hooke, 1967). Such lobes commonly have sharply defined downstream margins and tend to develop during the earliest, clearwater stages of a flood event (Wasson, 1974).

Deposition on gravity-flow fans is intermittent as a result of the episodic and localized nature of the processes, and the common incision of the upper fan. As a result, large areas of fans are starved of sediment for long periods during which *in situ* alteration may occur (e.g. Denny, 1967). This commonly takes the form of pedogenesis and the development of red hematite pigments through the weathering of ferro-magnesian minerals and the infiltration of weathering products (T.R. Walker, Waugh

& Crone, 1978). Soil development varies with extent of incision and with position on the fan surface (Fig. 3.30) (V.P. Wright & Alonso Zarza, 1990). On fans without incision, the least mature soils occur on the upper fan where depositional rates are highest. On incised fans, areas flanking the channel above the intersection point are likely to show the most mature soils. In addition, wind deflation winnows and redistributes sediment with aeolian dunes developing on the surfaces of some fans.

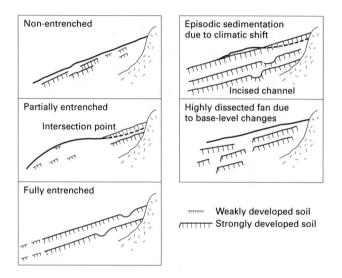

Figure 3.30 Model for the distribution of palaeosols of contrasting maturity associated with alluvial fans subjected to differing degrees of incision and different long-term controls (after V.P. Wright & Alonso Zarza, 1990).

FLUVIAL FANS

These are generally larger than gravity-flow fans and are characterized by the migration of a permanent or intermittent channelized stream. Channel shifting is largely random, related to crevassing and the development of channel bars (e.g. Knight, 1975). Where coarse material is available, as in proglacial outwash fans (Sect. 11.4.4), there is also commonly a downstream change in the style of channels and bars (Fig. 3.31). Proximal sheet bars with clasts up to boulder size pass progressively downstream into pebbly longitudinal braid bars and then into sandy transverse bedforms with associated compound bars (Boothroyd, 1972). Where sediment is very coarse, small, steep fans sometimes have convex-upwards profiles and both prograde and grow by lobe-shifting (e.g. Nemec & Postma, 1993). Overall deposition results from an interaction of channel scour and sheet-bar growth.

The largest examples are enormous and have a low gradient. Without extensive mapping, aerial photography or satellite images, they would not be recognized as fans but as alluvial plains. Of those that build out from the Himalayas into the

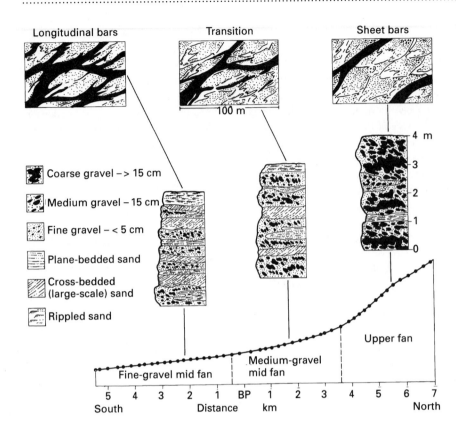

Longitudinal bars   Transition   Sheet bars

100 m

■ Coarse gravel – > 15 cm

▨ Medium gravel – 15 cm

▦ Fine gravel – < 5 cm

▤ Plane-bedded sand

▨ Cross-bedded (large-scale) sand

▨ Rippled sand

4 m
3
2
1
0

Upper fan

Medium-gravel mid fan

Fine-gravel mid fan

5   4   3   2   1   BP   1   2   3   4   5   6   7
South   Distance   km   North

Figure 3.31 Changes in gradient, bar type and internal structure along a fluvial braided outwash fan in front of the Scott Glacier, Alaska (after Boothroyd, 1972).

Ganges Valley, the best known is the Kosi fan (Fig. 3.32) (Gole & Chitale, 1966; N.A. Wells & Dorr, 1987; Gohain & Parkash, 1990) though the Gandak and Son fans to the west are equally large (Mohindra, Parkash & Prasad, 1992). The Kosi fan extends 160 km from its apex to the valley floor and is some 120 km wide. The active river channel has systematically shifted across the fan, the last sweep having occurred over a period of 230 years up to 1963. Engineering work has halted this migration, at least for the present. The migration appears to result from a combination of small avulsive steps and channel shifting. The continuous sweep from east to west may have been driven by neotectonic tilting, though sedimentation may also have played a role. The active channel, currently located close to the western margin of the fan, shows a systematic downfan change from a gravelly sand braided pattern through sandy braided and straight reaches into a meandering planform towards the low gradient toe (Gohain & Parkash, 1990). Out of context, the deposits are likely to be indistinguishable from those of floodplain streams. Abandoned channels are colonized by moisture-loving plants and, low on the fan, ox-bow lakes are common (N.A. Wells & Dorr, 1987). In addition to the main channel flow sourced from the Himalayas, smaller meandering channels, fed by springs, rework the lower fan. Outside the channels the fan surface is intensively cultivated but soil formation and intense bioturbation would dominate were the fan undisturbed.

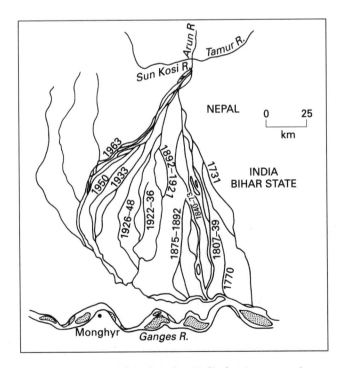

Figure 3.32 Kosi megafan of northern India showing progressive channel shifting over a period of 230 years (after Gole & Chitale, 1966).

TERMINAL FANS

A special case of fluvial fans is where water discharge progressively reduces downfan through evaporation and infiltration so that no water exits the surface of the fan. These *terminal fans* typically occur in semi-arid basins of inland drainage and are not necessarily associated with sharp topographic breaks. Discharge may be as as short-lived flood events or be more steady from a larger river (Mukherji, 1976; Parkash, Awasthi & Gohain, 1983; Abdullatif, 1989; Kelly & Olsen, 1993). On the surfaces of terminal fans, channels split into distributary networks. In-channel deposition due to loss of discharge induces splitting of flow and a pattern of subfans or lobes develops. At their distal ends, such fans end in playas, floodplains or in aeolian dune fields. Deposition on terminal fans may expand and contract in response either to short-lived flood events or to longer-term changes of discharge. Deposits appear to be mixtures of channel facies, dominated by cross-bedding, and sheet flood deposits of wider lateral extent dominated by parallel lamination and ripple cross-lamination and commonly showing grading. Overall, there is a distal decrease in the sand/mud ratio.

3.4 Ancient alluvial sediments

Alluvial sediments are commonly recognized by an absence of marine indicators, especially fossils, by relatively immature sediment textures and compositions, by channel geometries, by broadly unidirectional mean palaeocurrents, especially in channel facies and by evidence of subaerial emergence such as palaeosols and desiccation mudcracks. Red coloration is commonly used as a criterion but it must be applied with care as later secondary reddening can affect sediments of any facies, especially beneath unconformities, and early red pigments also form in aeolian and some marine sediments. Indeed, none of the features listed is an unambiguous criterion and diagnosis of an alluvial sequence must be based on a consideration of all aspects. In thick, Precambrian sandstone successions, where fossils are lacking and soils appear much less commonly, it may be difficult to distinguish fluvial, shallow marine and even aeolian elements. Where thick conglomerates are associated with active tectonics, the distinction between fluvial and deep-basinal deposits may be far from obvious (Harms, Tackenberg *et al.*, 1981; Stow, Bishop & Mills, 1982; Hein, 1984; Nemec & Steel, 1984).

Once an alluvial interpretation has been established, a sequence may be characterized in more detail. Fluvial sequences may be described in terms of channel stacking patterns, bounding surface hierarchies, facies organization, palaeocurrent populations and palaeosol type and maturity. These, in turn, may be used to infer accretion mechanisms, both in-channel and overbank, channel type and channel behaviour, floodplain drainage conditions and the role of external controls such as tectonics, base level and climate. Fan sequences may be characterized in terms of dominant processes of deposition and their vertical succession interpreted in terms of external controls operating at a range of time scales.

Because of the role of vegetation in modifying and stabilizing both source areas and alluvial sediments, the changing nature of plant cover on the Earth's surface through geological time means that present-day alluvial settings may not be fully representative of earlier regimes, especially those older than Devonian. For example, pre-Devonian alluvial sediments are less likely to contain deposits of meandering streams because stabilizing vegetation was then absent from overbank areas (Cotter, 1978). Similarly, changes in atmospheric composition, plant populations and soil organisms through geological time mean that soil types have become progressively more diverse (Retallack, 1986a, 1990).

The trend in recent years has been to try to interpret fluvial systems as a whole, giving appropriate weight to both channel units and overbank fines and to their mutual relationships. This is an important advance from earlier times when disproportionate emphasis was given to channel units. Successions are subdivided into genetic units which are organized into hierarchies and separated by bounding surfaces, again with their own hierarchy based on shape and extent. The genetic depositional units reflect hydrodynamics and styles of accretion over a wide range of scales and the bounding surfaces reflect erosion or non-deposition at different scales of time and extent.

3.5 Alluvial facies

Alluvial facies span a wide range of grain sizes and include some chemically and organically produced and modified sediments. The spectrum of lithologies may be subdivided and classified according to several facies schemes. Some sequences demand their own facies scheme because particular features are considered of special importance. For other sequences, a suite of more or less standard facies may be adequate. A scheme that has been widely applied is that suggested by Miall (1977). It recognizes three major grain-size classes – gravel, sand and fines (G, S, F) – each of which may be further subdivided according to texture and style of bedding and lamination (e.g. m, massive; t, trough cross-bedded; p, planar (i.e. tabular) cross-bedding; r, rippled; h, horizontal laminated, etc.). This scheme has the merit of being hierarchical and readily memorized and may be expanded and refined. It is a useful starting point and its application can aid preliminary working descriptions. Its uncritical application may, however, lead to pigeon-holing of successions with the risk that unusual or intermediate facies are forced into predetermined classes. For presentation of observations it is a poor substitute for full verbal description and/or well-constructed graphic logs. Facies are best defined mainly on the basis of internal textures and structures but a more complete understanding may only be possible when shapes, sizes, orientations and mutual relationships of the various facies units are also considered. Some workers (e.g. Ramos & Sopeña, 1983) have incorporated properties

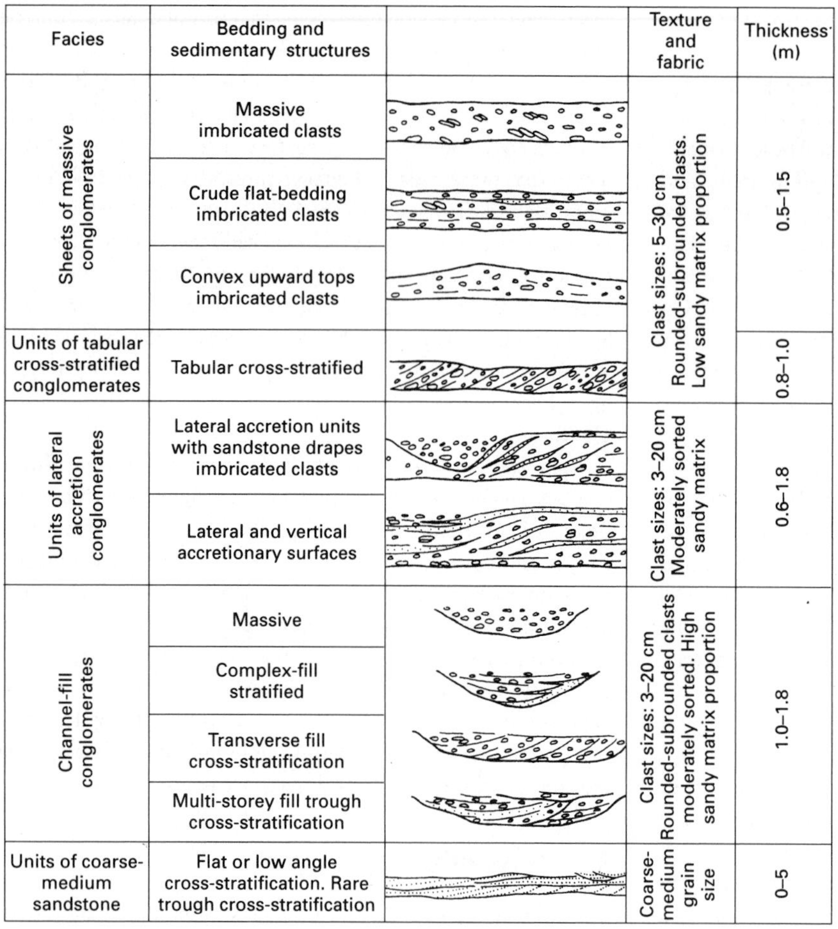

Facies	Bedding and sedimentary structures		Texture and fabric	Thickness (m)
Sheets of massive conglomerates	Massive imbricated clasts		Clast sizes: 5–30 cm. Rounded-subrounded clasts. Low sandy matrix proportion	0.5–1.5
Sheets of massive conglomerates	Crude flat-bedding imbricated clasts		Clast sizes: 5–30 cm. Rounded-subrounded clasts. Low sandy matrix proportion	0.5–1.5
Sheets of massive conglomerates	Convex upward tops imbricated clasts		Clast sizes: 5–30 cm. Rounded-subrounded clasts. Low sandy matrix proportion	0.5–1.5
Units of tabular cross-stratified conglomerates	Tabular cross-stratified			0.8–1.0
Units of lateral accretion conglomerates	Lateral accretion units with sandstone drapes imbricated clasts		Clast sizes: 3–20 cm. Moderately sorted sandy matrix	0.6–1.8
Units of lateral accretion conglomerates	Lateral and vertical accretionary surfaces		Clast sizes: 3–20 cm. Moderately sorted sandy matrix	0.6–1.8
Channel-fill conglomerates	Massive		Clast sizes: 3–20 cm. Rounded-subrounded clasts moderately sorted. High sandy matrix proportion	1.0–1.8
Channel-fill conglomerates	Complex-fill stratified		Clast sizes: 3–20 cm. Rounded-subrounded clasts moderately sorted. High sandy matrix proportion	1.0–1.8
Channel-fill conglomerates	Transverse fill cross-stratification		Clast sizes: 3–20 cm. Rounded-subrounded clasts moderately sorted. High sandy matrix proportion	1.0–1.8
Channel-fill conglomerates	Multi-storey fill trough cross-stratification		Clast sizes: 3–20 cm. Rounded-subrounded clasts moderately sorted. High sandy matrix proportion	1.0–1.8
Units of coarse-medium sandstone	Flat or low angle cross-stratification. Rare trough cross-stratification		Coarse-medium grain size	0–5

Figure 3.33 A scheme for conglomeratic alluvial facies. The facies are defined on the basis of texture, structures and the shape of the units and therefore constitute types of architectural element (after Ramos & Sopeña, 1983).

such as shape of the facies unit into their facies definitions (Fig. 3.33).

3.5.1 Conglomeratic facies

Conglomerates and gravel contain a significant proportion of pebble-size and larger clasts, usually more than 10%. Sediments with smaller proportions are best described as pebbly sandstones. Conglomerates can be divided into matrix-supported (paraconglomerates) and clast-supported (framework or ortho-conglomerates) textures. These classes are susceptible to further subdivision based on grain-size profile, clast orientation and bedding style.

Matrix-supported conglomerates commonly lack both internal bedding and clast imbrication though some may show weakly developed coarse-tail grading or inverse grading (e.g. Shultz, 1984). Large clasts may project above the top of the conglomerate layer (e.g. Tanner & Hubert, 1991). Such conglomerates are generally regarded as the products of *cohesive debris flows* (*sensu* Nemec & Steel, 1984). Grading suggests somewhat dilute and turbulent flows while inverse grading indicates that dispersive pressures operated, the density-modified grain flow of Lowe (1976). Inverse grading may also result from

growth of low-relief bars though a clast-supported fabric is then more likely (Todd, 1989; Nemec & Postma, 1993). Units with interbedded better- and less well-sorted components suggest flow surges (Fig. 3.34). Conglomerates in which larger clasts, floating in a sandy matrix, show imbrication suggest hyper-concentrated sheet floods, especially where units are tabular and interbedded with finer sediment (Laming, 1966; Nemec & Steel, 1984).

Clast-supported conglomerates and also some better-sorted pebbly sandstones with floating larger clasts record bedload deposition from *stream flows*. Such conglomerates commonly are characterized by lenticular bedding and erosion surfaces with conspicuous relief. Horizontal stratification and clast imbrication suggest deposition on near-horizontal pavements, either the tops of braid bars or as lags on channel floors (Fig. 3.35) (Nemec & Postma, 1993). Stratification defined by contrasting grain size and texture may record changing water stage over flood cycles (Fig. 3.36) (Steel & Thompson, 1983). Some conglomerates, especially those with more lenticular forms and erosive bases, show trough or tabular cross-bedding in which the foresets may be graded and which commonly show quite radical changes in grain size along the set, probably reflecting that stage fluctuations over the crest of the bar are responsible. Some sets may include reactivation surfaces or show gravel foresets interdigitating downdip with sandy toesets (Steel & Thompson, 1983). While most sets of cross-bedding in alluvial conglomerates have thicknesses of tens of centimetres or a few metres, very large sets also occur up to 20 m or more thick (e.g. Massari, 1983; Ori & Roveri, 1987; Collinson, Bevins & Clemmensen, 1989). Some may represent mid-channel or channel-confluence bars formed in deep sectors of a channel (Steel & Thompson, 1983) while others may compare with the large bedforms developed on the flanks of bedrock valleys as a result of catastrophic flooding (cf. Baker, 1973).

Sheet-like conglomerates with either clast-supported or matrix-supported textures are commonly inferred to be products of catastrophic flows (Fig. 3.34). This is supported by a positive linear correlation between maximum particle size and bed

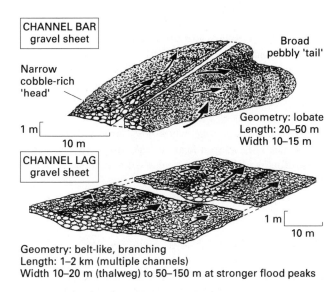

Figure 3.35 Interbedding of conglomeratic sheets interpreted as the products of lags and bar sheets on a fluvial fan, based on the Sfakia fan complex, Crete. Bar sheets are coarser and thicker while channel lags are finer and thinner (after Nemec & Postma, 1993).

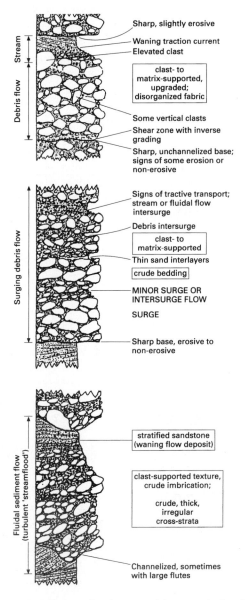

Figure 3.34 Typical textural and structural features of subaerial mass-flow deposits that characterize many debris-flow alluvial fans (after Nemec & Steel, 1984).

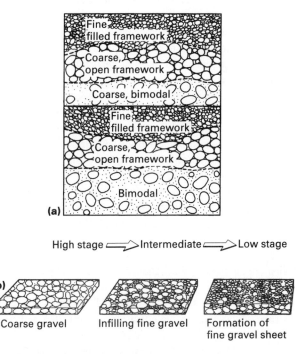

Figure 3.36 (a) Variable interbedded textures in bedded conglomerates from the Triassic of Cheshire, England. The variation is interpreted (b) in terms of varying transport populations during waning water stage (after Steel & Thompson, 1983).

thickness thought to reflect a relationship between competence and size of flow (e.g. Bluck, 1967a; Larsen & Steel, 1978; Nemec & Steel, 1984).

Units of low-angle, large-scale cross-stratification in conglomerates and pebbly sandstones are likely to have scoured bases with gravel lags. Imbrication in the gravel indicates transport at a high angle to the dip azimuth of the inclined surfaces (Ramos & Sopeña, 1983). Such units record either bar shifting within a braided system (cf. N.D. Smith, 1974) or migration of point bars in gravel-bed meandering streams (cf. Gustavson, 1978; Arche, 1983).

A less common facies is very angular, clast-supported breccia which occurs in alluvial associations close to steep palaeorelief (Collinson, Bevins & Clemmensen, 1989; Tanner & Hubert, 1991). The angular clasts may be very large and the usually sandy matrix shows stratification, suggesting later deposition. These deposits are interpreted as talus which accumulated directly at the foot of a steep slope.

3.5.2 Sandstone channel facies

The facies of alluvial channel sandstones are divided here into six facies on the basis of the type and scale of stratification.
1 *Conglomerates* are mostly confined to thin beds, commonly only a few clasts thick. These record winnowing by strong currents, either at erosion surfaces or, in some cases, on the tops of bars. Erosion surfaces may be major channel bases, when the conglomerate layer may be laterally extensive, or more local scoured surfaces related to the migration of bars or dunes.

Clasts may be of intraformational or extrabasinal origin. Intraformational clasts are more commonly associated with erosion surfaces related to channel migration, when overbank floodplain sediments have been eroded and reworked. Such clasts are most commonly blocks and flakes of mudstone, pebbles of nodular calcite (caliche) and reworked siderite concretions. In addition, large plant fragments, logs and blocks of peat/coal may also be present. Where a sequence is dominated by channel deposits, intraformational clasts give important clues as to the nature of contemporaneous interchannel settings, even though their deposits are poorly preserved.

Intraformational clasts sometimes occur as large blocks that must have slid or fallen from the cut bank and accumulated with little erosion or reworking (e.g. Laury, 1971; Gibling & Rust, 1984). Steep dips within blocks record rotation during sliding on listric failure surfaces (cf. Fig. 3.1) (Gersib & McCabe, 1981). Other conglomerates show chaotic textures with plastic deformation of clasts suggesting slumping of a cut-bank (e.g. Bøe, 1988). Some larger muddy clasts display flutes and linear scours on their surfaces recording corrasion by powerful, sediment-laden currents, probably in the deeper part of a channel (Gibling & Rust, 1984; Plint, 1986). Rill-like features on some blocks may have formed during brief periods of subaerial exposure at low-water stage. Intraformational conglomerates involving large fallen blocks often occur close to a preserved cut-bank.

While most extrabasinal clasts are concentrated in lags on scour surfaces or on bar tops, some pebbles and even coarser clasts, associated with plant debris and logs, form conglomerates in the upper parts of channel sandbodies (Liu & Gastaldo, 1992). Finer conglomerates form internal casts of hollow-stemmed plants which floated to the depositional site, probably as rafts. Much larger clasts, up to boulder size, occur outside the log casts and were probably transported enmeshed in roots or lodged in the raft. Grain sizes of such conglomerates bear no relationship to depositional hydrodynamics.

Very poorly sorted pebbly lithic sandstones in more or less structureless beds occur in channel sandbodies in volcanic basins where abundant volcaniclastic debris, hyperconcentrated flows and short transport distances prevented effective textural sorting (e.g. G.A. Smith, 1987a,b; Besly & Collinson, 1991; Haughton, 1993).
2 *Cross-bedded sandstones* are the most abundant facies in fluvial channel sandstones (Fig. 3.37). In some cases they account

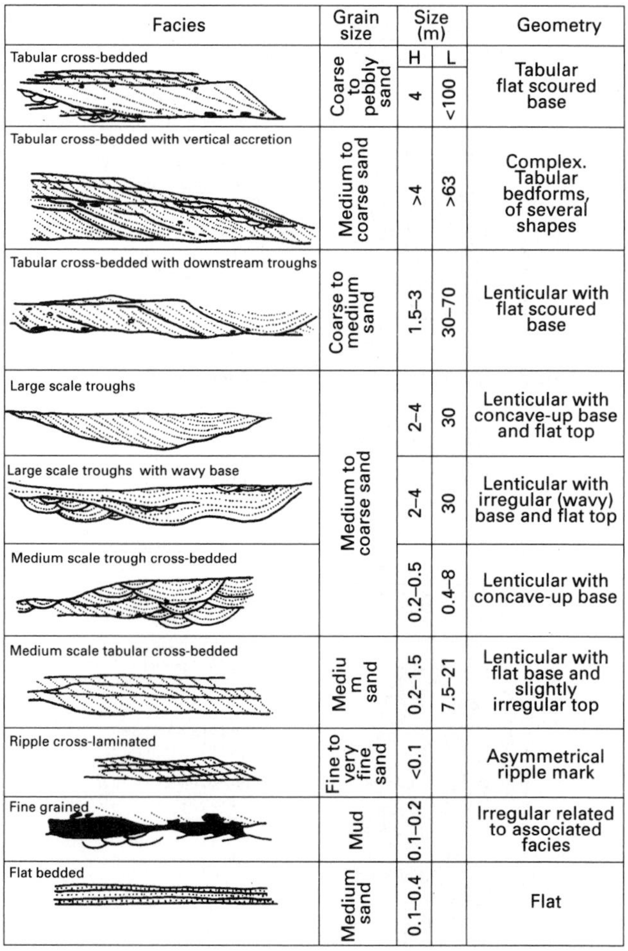

Facies	Grain size	Size (m) H	Size (m) L	Geometry
Tabular cross-bedded	Coarse to pebbly sand	4	<100	Tabular flat scoured base
Tabular cross-bedded with vertical accretion	Medium to coarse sand	>4	>63	Complex. Tabular bedforms, of several shapes
Tabular cross-bedded with downstream troughs	Coarse to medium sand	1.5–3	30–70	Lenticular with flat scoured base
Large scale troughs	Medium to coarse sand	2–4	30	Lenticular with concave-up base and flat top
Large scale troughs with wavy base	Medium to coarse sand	2–4	30	Lenticular with irregular (wavy) base and flat top
Medium scale trough cross-bedded	Medium to coarse sand	0.2–0.5	0.4–8	Lenticular with concave-up base
Medium scale tabular cross-bedded	Medium sand	0.2–1.5	7.5–21	Lenticular with flat base and slightly irregular top
Ripple cross-laminated	Fine to very fine sand	<0.1		Asymmetrical ripple mark
Fine grained	Mud	0.1–0.2		Irregular related to associated facies
Flat bedded	Medium sand	0.1–0.4		Flat

Figure 3.37 Typical channel sandstone depositional facies based on the Triassic of Central Spain (after Ramos, Sopeña & Perez-Arlucea, 1986).

for virtually all the channel sediments. Grain sizes range from coarse, pebbly sandstones to fine sandstone. Set thickness of cross-bedding varies from a few centimetres to several or many metres.

Trough sets are most abundant and are commonly a few tens of centimetres thick, though they may range up to 3 m or more. They are the product of migrating three-dimensional dunes.

Tabular cross-bedding is less abundant but still common. Set thicknesses are commonly of the order of 1 m but giant examples up to 40 m are known (Fig. 3.38) (Collinson, 1968; Conaghan & Jones, 1975; McCabe, 1977). Smaller tabular sets are produced by straight-crested dunes and, less commonly, scroll bars and chute bars. Some giant sets were probably formed by large transverse or linguoid bars which may have been bank attached or of mid-channel origin (McCabe, 1977), others as channel confluence deltas. No modern analogues are known at this scale.

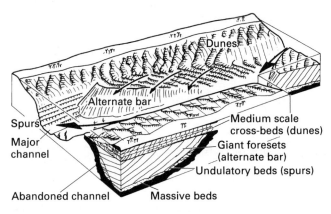

Figure 3.38 Model for the large-scale, fluvial channels of the Namurian Kinderscout Grit of northern England (after McCabe, 1977).

Some sets show sigmoidal foresets which pass upwards into horizontally laminated topsets (Fig. 3.39) (Røe, 1987). These record dunes or simple bars which were transitional to upper-phase plane bed. Foresets of both tabular and trough sets may be overturned (e.g. J.R.L. Allen & Banks, 1972; Hendry &

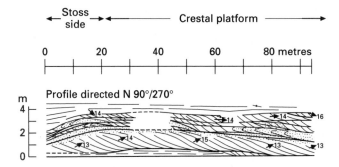

Figure 3.39 Sigmoidal foresets in a large tabular set of cross-bedding overlain by smaller sets which descend at the downstream end of the larger structure. Rapid deposition is indicated by overturned foresets and structureless sand. The overall geometry shows the evolution of a simple dune to a compound bar (after Røe, 1987).

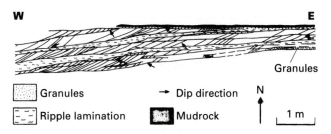

Figure 3.40 Flow-parallel section through descending sets of cross-bedding in a large channel sandbody, interpreted as the product of downstream accretion of a large compound bar. Upper Carboniferous Coal Measures, northeast England (after Haszeldine, 1983).

Stauffer, 1977; B.R. Turner & Munro, 1987) as a result of temporary loss of strength shortly after deposition, probably through rapid deposition. Some overturned foresets are associated with structureless sand, suggesting flow of liquefied sediment (Røe & Hermansen, 1993).

Evidence of water-stage fluctuations occurs as erosional reactivation surfaces (e.g. Rust & Jones, 1987) and as mud-draped foresets. Lee faces of dunes sometimes weather out in full relief to show gullies and small terraces (Banks, 1973b) and the superimposition of ripples (Stear, 1978).

Inclined bounding surfaces between cross-bedded sets that dip downstream record accretion on the lee-side of a compound bar (Fig. 3.40) (Banks, 1973a; McCabe & Jones, 1977; Haszeldine, 1983; Røe & Hermansen, 1993). Upstream dipping bounding surfaces record bedform climbing, probably on the upstream flank of a bar.

3 *Cross-laminated sandstones* make up significant volumes of channel sandstones, on occasion the whole channel unit. Ripple cross-laminated sandstones usually occur towards the tops of channel units and are finer grained and more micaceous and carbonaceous than underlying sandstones. They reflect relatively weak currents. Ripple-drift cross-lamination (climbing ripple lamination) records high rates of bed aggradation (e.g. Stear, 1985).

4 *Parallel laminated sandstones* are normally of minor significance but in some cases they form substantial thicknesses. They are usually fine-grained with well-defined horizontal lamination and parting lineation on bedding surfaces. While they may occur at any level in a channel sandstone, they tend to be more common towards the top. Where the sands are medium- to fine-grained and mica-free, the facies is interpreted as the product of upper-stage plane bed transport (upper flow regime) (e.g. Stear, 1985). Where sandstones are highly micaceous, flow conditions may have been gentle as platy particles inhibit ripple formation (Manz, 1978).

Lamination may show distinct undulations while retaining its essential parallelism. The undulations are gentle with some low-angle truncation of laminae (Cotter & Graham, 1991). In plan view, some examples show repetitive three-dimensional domes and the bedding surfaces carry parting lineations (e.g. Collinson, Bevins & Clemmensen, 1989; Rust & Gibling, 1990).

The structures are similar to hummocky cross-stratification (Sect. 7.7.2) but their context rules out a marine storm origin. Their fluvial setting suggests an upper-phase standing wave regime and they thus represent fortuitous preservation of highly ephemeral structures.

5 *Massive structureless sandstones* occur both as tabular beds and as more lenticular bodies within stratified sandstones. Where tabular, they commonly occur towards the base of channel sandbodies (e.g. Collinson, 1969; McCabe, 1977) and are attributed to rapid deposition from suspension during floods.

Where massive sands occur as lenticular channels within cross-stratified sandstones, the axes of the channels trend normal to palaeoflow. In the Carboniferous Fell Sandstone of north-eastern England, channels are narrow and steep-sided with sharp contacts to adjacent stratified sandstone. The sides are too steep ever to have stood unsupported in cohesionless sand (B.R. Turner & Munro, 1987). The margins of the massive sand units show narrow zones of weakly defined, diffuse lamination which parallels the margin while their cores appear structureless. In the Triassic Hawkesbury Sandstone of New South Wales, Australia, units of massive sandstone contain deformed clasts of cohesive overbank material (Jones & Rust, 1983; Rust & Jones, 1987). Both examples record bank collapse, liquefaction and flow of both in-channel bars and the channel bank itself (Fig. 3.41).

6 *Large-scale, low-angle cross-bedding* is a feature of many channel sandstones, the *epsilon cross-bedding* of J.R.L. Allen (1963b). The dip azimuth of the large sets is more or less normal to the palaeocurrents derived from smaller-scale structures within the set. The bedding characterizes an accretionary style and is best regarded as an architectural element (J.R.L. Allen, 1983; Miall, 1985a, 1988) to be discussed in detail in Section 3.6.2.

3.5.3 Fine-grained facies

These facies mostly comprise mudstones, siltstones and fine-grained sandstones although coarser intraclasts occur in some sandstones. Depositional facies may be modified by early diagenesis and pedogenesis so that formulation of a facies scheme presents a formidable challenge. Here depositional facies are separated, as far as possible, from the products of early alteration but the two may be combined in complex ways.

Mudstones and siltstones are commonly the most abundant facies. Some are thin bedded or laminated, with variable fissility, while others are homogeneous, often with blocky fracture patterns. Colours range from black through greys, greens, ochres, browns and reds.

Well-laminated examples are commonly rich in mica and/or organic matter if the sediment is unoxidized, when it is usually black or dark grey in colour. Some laminated mudstones carry a fauna of freshwater bivalves, ostracods and fish scales (e.g. Scott, 1978; Gersib & McCabe, 1981; Fielding, 1984a,b, 1986; Haszeldine, 1984). Others show fine, varve-like lamination and

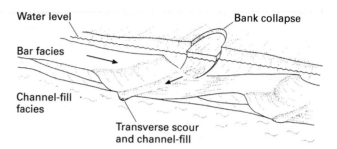

Figure 3.41 Model for the emplacement of channelized plugs of structureless sand oriented normal to palaeoflow by the collapse, liquefaction and flowage of sand from an adjacent sandy bank (after Turner & Munro, 1987).

carry both bedded and nodular siderite of early diagenetic origin (e.g. Boardman, 1989). Such facies were deposited from suspension either in extensive lakes on the floodplain or in abandoned channels such as ox-bow lakes.

Grey massive siltstones are the result either of rapid deposition by flood or of homogenization by bioturbation. The former contain abundant plant debris and may envelop tree trunks in growth position (e.g. Fielding, 1984a,b). The latter were bioturbated by plant roots or animals. Some examples show evidence of subaerial emergence in the form of mudcracks, rain pits and footprints (e.g. Thompson, 1970). Such facies were mainly laid down on floodplains submerged only during floods. Evidence for incremental deposition is commonly lost through homogenization though a major crevasse or avulsion may have led to sustained deposition of floodplain fines (cf. N.D. Smith, Cross et al., 1989).

Colour is a valuable indicator of the drainage state of the floodplain and the level of the water table. Red colours suggest good drainage and an oxidizing early diagenetic regime, while grey beds and the preservation of organic matter record water-logged conditions and reducing pore water. Early alteration of floodplain fines is dealt with in more detail in Section 3.5.4.

While much fine-grained sediment is deposited in overbank settings, it can also occur within channels, on distal terminal fans, as mud aggregates transported as bedload (Ekes, 1993), and as thick units of wind-lain siltstone (loess), particularly where structureless and homogeneous (cf. Lambrick, 1967) (Sect. 3.2.2). Thick intervals of fine-grained subaerial sediment may also accumulate as volcanic ash downwind of erupting volcanoes (Sect. 12.4.2).

Sharp-based sandstone beds are commonly interbedded with mudstone or siltstones. They are usually thin, less than 20-cm thick and seldom over 2 m. Examples over 2 m are more likely to be products of ephemeral streams. Many appear parallel-sided within the confines of small exposures (e.g. Tunbridge, 1981) but extensive exposures show many to be lenticular (e.g. Leeder, 1974; Mjøs, Walderhaug & Prestholm, 1993). Such sandstones may have minor erosive relief on their bases and they share many features with turbidites, including sharp bases, sole marks, graded bedding and a Bouma sequence of internal structures. Parallel lamination and ripple cross-lamination, particularly climbing ripple cross-lamination, are common (Fig. 3.42) (Steel & Aasheim, 1978; Tunbridge, 1981, 1984). Thicker beds may show small-scale cross-bedding.

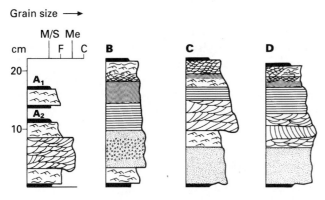

Figure 3.42 Sharp-based sandstone beds in a fine-grained alluvial succession resulting from catastrophic overbank flows on floodplains and from sheet floods on the more distal parts of alluvial fans (after Steel & Aasheim, 1978).

The similarities with turbidites reflect episodic, decelerating flows and climbing, ripple lamination indicates high rates of aggradation. Interbedded, finer-grained facies with palaeosols and other evidence of subaerial exposure indicate deposition by floods. Many sharp-based sandstone beds, particularly more lenticular and erosive-based examples, are commonly associated with channels formed as a result of the crevassing of a major river channel. They may show a distal transition to more sheet-like, finer-grained and commonly graded beds. Others, particularly more sheet-like beds, occur in successions lacking major channel sandstones and are most likely products of sheet floods, either on distal parts of alluvial fans (e.g. Steel & Aasheim, 1978; Tunbridge, 1981, 1984; Hubert & Hyde, 1983) or on alluvial plains (McKee, Crosby & Berryhill, 1967; Stear, 1985). On alluvial plains, flows may either emanate from channels or result from local heavy rainfall on the plain itself.

3.5.4 Palaeosols

Palaeosols are recognized on the basis of organic, textural and stratigraphic features (Retallack, 1988; V.P. Wright, 1992a; Kraus & Aslan, 1993). Root traces, which characterize many soils, are preserved either as carbonaceous films where conditions were reducing, or as casts, bleached zones or colour contrasts where organic matter has been oxidized. Where trace fossils are present, they belong to the *Scoyenia* ichnofacies (e.g. Blodgett, 1985). Characteristic textures include blocky, columnar and prismatic structures (peds), especially in fine-grained host sediment where structures may be picked out by clay-enriched surfaces. Such subtle and delicate features do not always survive deep burial, diagenesis and tectonic deformation. Concretions have better chances of preservation.

Textural and compositional features characterize the horizons that provide the basis for definition of soil profiles. Horizons tend to have gradational boundaries and to trend parallel with the upper surface of the palaeosol. This may be either an erosion surface or a sharp contact with unaltered overlying material. There is commonly a gradational downwards passage from palaeosol to unaltered sediment. In some cases, diagenesis accentuates original pedogenic differences between horizons making them appear more distinct than in analogous modern soils. In spite of diagenetic modification, palaeosols are commonly interpreted in terms of modern analogues (Fig. 3.25), with due allowance for differing biogenic components and activity. The classification suggested by Mack, James and Monger (1993) may provide a basis for classification in the future but is not followed strictly here.

Simple alluvial palaeosols are rooted and root-mottled sediments lacking any well-defined horizons or profile (Kraus & Aslan, 1993). This reflects rather continuous sedimentation, such as might be found in low-lying floodplains. Alluvial palaeosols are particularly common in coal-bearing sequences where they are characterized by rootlets and larger root systems (e.g. *Stigmaria*), by slickensided surfaces and by scattered siderite nodules (Besly & Fielding, 1989). This suggests a hydromorphic regime (Kraus & Aslan, 1993; cf. Duchaufour, 1982) though they may also occur in red-bed successions where root traces are associated with colour mottling.

Hydromorphic conditions are also suggested by cumulative *gley palaeosols* (Kraus & Aslan, 1993). These show a distinct profile that commonly has a bleached upper horizon and that may be overlain by coal. These are the typical 'seat earths' of coal-bearing successions (Raistrick & Marshall, 1939; Huddle & Patterson, 1961) though not all such palaeosols culminate in coal. Some palaeogleys show an upwards increase in siderite and well-preserved examples preserve micromorphological textures (e.g. Buurman, 1980; Kraus & Aslan, 1993). Similar palaeosols, but showing an upwards increase in small sphaero-siderite nodules, record conditions of slightly better drainage and are regarded as semi-gleys (Besly & Fielding, 1989; Besly & Collinson, 1991).

Where the host sediment is sandy, better drainage and downward movement of water leach the upper parts of soils to give albic A horizons from which clay and free iron have been removed and spodic B horizons with illuviation of clays and organic matter. These soils have been interpreted as *spodosols* (podzols) or *alfisols*, depending upon the degree of organic accumulation (Bown & Kraus, 1987). Similar highly siliceous palaeosols constitute the 'ganisters' of some coal-bearing sequences (Percival, 1986).

In alluvial red-bed successions, a fluctuating water table oxidized the pedogenic layer. *Ferruginous and ferralitic palaeosols* are characterized by strongly leached profiles with distinctly coloured horizons. Pale grey A horizons overlie colour-mottled B horizons with grey, red, purple and ochre pigments, mainly relating to differing contents of goethite and hematite (Besly & Fielding, 1989). Mature examples may contain hematite concretions that represent stages in the development of plinthite (laterite).

Other red-bed sequences show evidence of more arid or fluctuating climates with the development of textures comparable with modern *Aridisols/Vertisols* and *caliches*. Vertisol features include sand-filled cracks and pseudo-anticlinal slip surfaces attributed to alternating wetting and drying (e.g. J.R.L. Allen, 1973; V.P. Wright & Robinson, 1988; Tandon & Friend, 1989; Marriott & Wright, 1993). Commonly associated with such features are calcite nodules, which occur in a wide variety of sizes, forms and profiles, ranging from small isolated nodules with no orientation or subhorizontal elongation to vertical prismatic forms which may be closely enough spaced to constitute virtually the whole of the rock. Others follow pseudo-anticlinal slip surfaces. Calcite nodules commonly occur in profiles within which they increase upwards in abundance, in some cases culminating in a continuous and laminated limestone layer (Fig. 3.43) (J.R.L. Allen, 1974, 1986; Steel, 1974; Leeder, 1975;

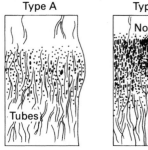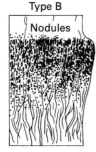

Figure 3.43 Three intergradational stages in the development of a caliche palaeosol profile. Types A–C show increasing maturity. Old Red Sandstone, South Wales (after J.R.L. Allen, 1974).

Blodgett, 1985; Tandon & Friend, 1989; Marriott & Wright, 1993). These calcite-bearing palaeosols compare with modern caliche or calcrete soils (e.g. Machette, 1985) and record arid to subhumid conditions where evaporative loss exceeded the supply of water to the surface by rainfall or flooding.

Polygenetic and composite palaeosols occur where a later soil profile has become superimposed upon an earlier one (Figs 3.26 & 3.44). Where conditions remain essentially unchanged, a composite profile results and horizons of the upper profile overprint the earlier profile with displacement of horizons (Fig. 3.26) (Marriott & Wright, 1993; cf. Morrison, 1978).

Where drainage conditions change, overprinting of features of contrasting profiles result. Palaeogleys have been overprinted with features of ferruginous palaeosols as a result of improved drainage (Fig. 3.44) while upper parts of ferruginous or ferrallitic palaeosols have been reduced when conditions became wetter (e.g. Besly & Fielding, 1989; Glover, Powell & Waters, 1993).

Early groundwater diagenesis may alter the appearance of alluvial sediments in ways which may be confused with pedogenic features. The development of red pigments and their precursors may take place away from the pedogenic zone and lead to colour mottling while mineral precipitation can lead to groundwater silcretes, calcretes and ferricretes (e.g. Spotl & Wright, 1992; V.P. Wright, Sloan *et al.*, 1992). In some cases, diagenesis may overprint palaeosol textures (e.g. Goldbery, 1982).

3.5.5 Biological and biochemical sediment

Coals and *peats* occur as both extensive sheets and as more local lenticular accumulations, both types varying in thickness from a few centimetres to many metres. Autochthonous peats and coals directly overlie palaeosols which show features of reducing and waterlogged conditions such as preserved organic matter, commonly as rootlets. Allochthonous coals lack an associated underlying palaeosol and accumulated in a body of still water as detrital organic matter or as a floating peat layer.

Coals vary in their content of mineral matter, which determines their ash content. Very low ash coals often accumulated as peats in raised mires (McCabe, 1984). They also vary in their maceral content, thus forming the basis for a scheme of coal facies. At its simplest, bright, humic coals are derived from macroscopic plant parts with high contents of vitrain indicating woody debris. Dull coals generally contain mainly smaller plant debris or are derived from degraded peats (McCabe, 1984). Vertical seam profiles commonly show systematic trends of maceral content related to the establishment and development of plant communities (e.g. Hacquebard & Donaldson, 1969; Scott, 1978). Sapropelic coals, in contrast, were deposited as subaqueous muds in reducing conditions and have a higher algal content. In alluvial settings, coals commonly show rather low sulphur contents compared with those of lower delta plain settings where either terrestrial sulphate is ponded and concentrated (Hunt & Hobday, 1984) or marine sulphate is introduced from overlying marine sediments deposited as a result of transgression (e.g. Shimoyama, 1984).

Freshwater limestones may be important in some alluvial sequences. They commonly occur as sheets of micritic limestone, associated with lacustrine mudstones and at the base of lake infill sequences. In some cases they provide the entire lake fill. They result from both inorganic precipitation and organic accumulation and will not be discussed in further detail here (see Sects 4.7.1, 4.12).

1) Persistent alluvial swamp sedimentation

Water table permanently at or above sediment surface

2) Progradation of well-drained swamp and alluvial plain—water table drop in well-drained areas induced by evaporation

3) Resumption of coal-forming alluvial swamp conditions after decrease in sediment supply and/or increase in subsidence rate

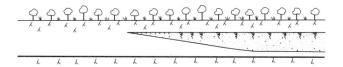

4) Renewed progradation of well-drained swamp and floodplain

Prolonged drop in water table allows oxidation of earlier organic rich sediment

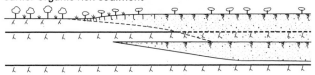

⊥ ⊥ ⊥ Coal and seat-earth (swamp palaeosol)

⊤ ⊤ ⊤ Coal and seat-earth: oxidized after burial

⊤ ⊤ ⊤ Well-drained swamp palaeosol

⊥ ⊥ ⊥ Mature palaeosol

Swamp flora

Floodplain flora

------- Water table

Reddened sediment

Figure 3.44 Model for the development of reddened units within alluvial swamp successions as a result of the local improvement of drainage state and shifts in the position of the water table (after Besly & Turner, 1983).

3.6 Larger-scale geometry, organization and controls

The facies outlined above do not, in isolation, permit alluvial style or external control to be interpreted. Such interpretations come from the geometries and mutual relationships of component facies at a wide range of scales. In some settings, vertical facies successions provide the key to understanding the system and the larger controls. In other settings, more complex two- and three-dimensional relationships, involving a systematic analysis of both bounding surfaces and facies associations, must be delineated. This commonly requires laterally extensive exposures, without which description and interpretation will always remain ambiguous.

3.6.1 Bounding surfaces

The subdivision of channel sandbodies into units separated by different orders of bounding surface was formally enunciated by J.R.L. Allen (1983) and developed by Miall (1985a, 1988) though the principles date back to Campbell (1976) and Brookfield (1977). Such subdivision helps in the discrimination of features related to internal controls (autocyclic) from those demanding an external explanation (allocyclic). Different suggested schemes were largely dictated by the nature of the successions studied. Miall (1985a, 1988), for example, had a sixfold hierarchy with rank 6 the largest scale. J.R.L. Allen (1983) made do with a three-rank scheme which did not extend to surfaces as large as Miall's 6th rank and which grouped some of Miall's lower-order surfaces. Published schemes may not, therefore, be comparable. Some channel sandbodies are intrinsically complex and a scheme devised for one need not apply to another. The terminology used here broadly follows the usage of Miall (1985a) (Table 3.1). Not all ranks need be present in any particular case and appropriate adaptations must be made (Røe & Hermansen, 1993).

The problem of establishing hierarchical relationships is particularly acute in studies of borehole core. Bounding surfaces between cross-bed sets and between cosets of contrasting character (1st and 2nd order of Miall, respectively) are readily recognized, as is the base of a channel unit overlying non-channel facies (6th rank). Problems arise in categorizing surfaces of intermediate rank which define channels at a range of sizes or bound the products of channel macroforms (bars) (cf. Bridge, 1993).

3.6.2 Architectural elements

Alluvial sequences that show laterally persistent vertical sequences of grain size and sedimentary structures may be adequately described on the basis of vertical sequence alone. However, the majority of sequences are laterally variable and many processes of bedform movement and bed accretion are

Table 3.1 Bounding surfaces.

Order (Miall, 1988)	Nature of surface	Significance and process
1st	Separates individual cross-bedded sets	Migration of dune bedforms under steady flow conditions. Surface a function of dune geometry
2nd	Separates cosets of contrasting set type	Change in hydrodynamic conditions through time, related to short-term unsteady flow or local non-uniformity
3rd	Inclined erosion surfaces within coset or group of cosets	Medium-term change in hydrodynamic conditions related to stage fluctuation or major shifting of flow across/around a bar form
4th	Separates units with discrete accretionary integrity	Shift of bar/subchannel pattern related to inherent channel-floor instability or to reorganization during a major flood
5th	Surfaces with a marked shift in grain size, bedform scale, etc. Laterally extensive with relief	Shifting and erosion of a channel floor. Isolated channels with relief reflect channel switching. Extensive surfaces within larger sandbodies record channel migration
6th	Separates major channel sandbodies from contrasting facies (fine-grained sediment or contrasting channel facies)	Major change of fluvial regime. May record shifts of base level or climatic or tectonic changes

common to a range of channel types. This has led to some disillusionment (Jackson, 1978; Bridge, 1985; Miall, 1985a) with vertical facies sequence models which sought to interpret alluvial style on the basis of vertical sequence alone (e.g. J.R.L. Allen, 1965a, 1970a; Miall, 1977, 1978). With laterally restricted outcrop or in boreholes, multistorey channel sandstones should be interpreted with circumspection as any one section may not be representative. Cant (1978) and Cant and Walker (1976) recognized this problem and suggested a solution that involved a composite vertical model based on many profiles which, though broadly similar, were different in detail (Fig. 3.45). Such analysis provides useful insights and is also valuable in closely integrating palaeocurrent data with the vertical sequence. However, it distils away valuable information on lateral variability (cf. Lawrence & Williams, 1987).

A more realistic solution is to recognize that alluvial successions are made up of building blocks termed 'architectural elements' (J.R.L. Allen, 1983; Miall, 1988) whose definition and recognition are easiest in extensive two- or three-dimensional exposure. Quite a small number of categories are adequate to describe the majority of successions (Fig. 3.46). The definition of architectural elements depends upon both their external shape, commonly defined by bounding surfaces, and the organization of their internal structures. They mainly reflect different styles of bed accretion and most may be related to morphological features at the scale of channels, bars or macroforms. These include lateral accretion and downstream accretion as well as cosets of cross-strata within which vertical accretion prevailed.

Channel elements are characterized by basal erosional relief and by fills which are either concordant or discordant with the base (cf. Bluck & Kelling, 1963). Some beds onlap an inclined erosion surface. Where preservation is incomplete, it may be difficult to distinguish a channel with inclined, asymmetrical fill

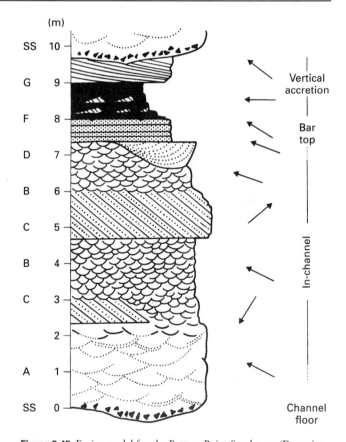

Figure 3.45 Facies model for the Battery Point Sandstone (Devonian, Quebec) based on analysis of vertical facies transitions through many channel units. The model presents an ideal that is seldom complete in nature and pays no explicit attention to lateral facies variations. The model presents the relationships between particular facies and their associated palaeocurrents. The sequence has been compared with similarly idealized models for sedimentation in the modern South Saskatchewan River (Fig. 3.16) (after Cant & Walker, 1976).

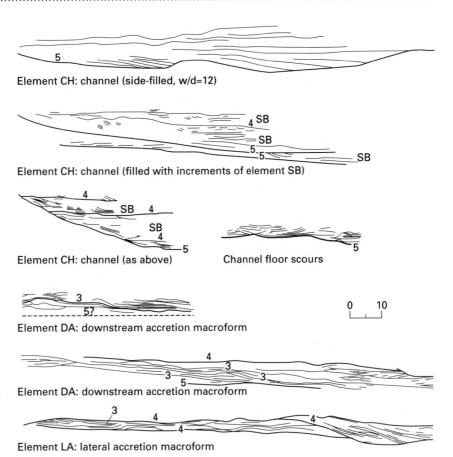

Figure 3.46 Examples of alluvial architectural elements recognized in the Westwater Canyon Member of the Morrison Formation (Jurassic, New Mexico). These elements are defined on the basis of accretionary style and the type and scale of associated bounding surfaces. These elements are common to many sandy fluvial channel sandbodies though may not fully reflect their variability (after Miall, 1988).

from lateral accretion associated with channel migration (Cowan, 1991). Recognition of *lateral accretion elements* depends upon showing that inclined accretionary layers dip roughly normal to the palaeoflow as measured from smaller-scale sedimentary structures within them. They are discussed in more detail in the context of tabular channel sandbodies. *Downstream accretion elements* have horizontal basal bounding surfaces (possibly 5th order) and upper bounding surfaces (usually 4th order) often with convex-upwards relief. Third-order bounding surfaces, inclined downcurrent, separate cross-bedded sets (Banks, 1973a; Haszeldine, 1983; Kirk, 1983; Røe & Hermansen, 1993; Miall, 1994). The simple pattern of descending sets may be complicated by reactivation surfaces both between sets and within individual sets. Such elements record the downstream accretion of dunes on larger compound bars (Crowley, 1983; Miall, 1994).

In addition to the elements shown in Fig. 3.46, there are:
1 *sheet-like elements* characterized by horizontal lamination with minor cross-stratification and primary current lineation (cf. Bromley, 1991) reflecting vertical bed accretion under upper flow regime flood conditions (e.g. McKee, Crosby & Berryhill, 1967);
2 minor units of *massive sandstone* occurring within cross-bedded units (e.g. B.R. Turner & Munro, 1987; Rust & Jones, 1987) and reflecting within-channel slumping and liquefaction

(Fig. 3.41) (Sect. 3.5.2); and
3 cosets of descending sets (downstream accretionary elements) that pass downcurrent into *single large tabular sets* recording large slip-face bars (Fig. 3.40) (Conaghan & Jones, 1975; McCabe, 1977; Rust & Jones, 1987).

Architectural elements occur at any size relative to the overall sediment body. Some sandbodies, for example, comprise entirely one element while others are three-dimensional mosaics of several types of element of different scales. There is no rigid relationship between architectural elements and particular orders of bounding surface, though more extensive bounding surfaces commonly encompass more complex assemblages of architectural element. Particular architectural elements are discussed below in the context of the types of alluvial successions in which they most commonly occur.

3.6.3 Fan conglomerates

Because the architectural elements of debris-flow fans are broadly parallel-sided beds related to discrete flow events, one-dimensional vertical profiles are commonly adequate to describe their deposits. Individual beds are highly stochastic in terms of grain size, facies and bed thickness. The magnitude and nature of debris flows reflect: (i) storm precipitation; (ii) the nature and abundance of available debris; and (iii) the time between flushing events. A positive linear correlation between maximum particle size and bed thickness means that upwards fining and coarsening coincides with upwards thinning and thickening respectively. If storms were frequent, insufficient debris may have accumulated for mobilization of debris flows, and stream flood deposits may dominate (Schumm, 1973; Heward, 1978).

Trends in facies, in bed thickness and in grain size (commonly maximum particle size) are commonly apparent across a range of scales (Steel, 1974, 1976; Heward, 1978a). Thick sequences (100s of metres) (megasequences *sensu* Heward, 1978a) probably record external controls such as climatic change and tectonic activity (e.g. Ethridge, Tyler & Burns, 1984; Blair, 1987b). Such effects are, however, not easily deciphered. Linked changes in climate and vegetation lead to facies and grain-size changes that complicate interpretation (e.g. Croft, 1962; Lustig, 1965). The evolution of drainage catchments following a period of tectonism at a basin margin can be complex with geomorphic controls modulating the more obvious climatic or tectonic signals (Fraser & DeCelles, 1992).

Thin sequences (10s of metres) are commonly of limited lateral extent and probably relate to intrinsic fan processes such as channel incision, filling and switching and lobe switching (Fig. 3.47) (DeCelles, Gray *et al.*, 1991). Some thickening-upwards sequences involving sheet-flood deposits culminate in an incised channel conglomerate in axial parts of fan lobes, giving an abrupt increase in bed thickness.

Deposits of fluvial fans have the same facies as other conglomeratic fluvial settings. Alternations of channel-floor lags and more stratified channel-bar sheets occur (Nemec & Postma, 1993). The large scale of many fluvial fans in relation to the size of typical study areas commonly precludes recognition of fan morphology and interpretation as braided stream deposits may be all that is possible. In fans where earlier deposits are reworked by later floods, interpretation is particularly difficult. Coarsening-upwards sequences may record either relaxation and adjustment following rapid tectonic uplift of the source area or periods of slow tectonic movement. Alternatively, they may record decreases in basinal subsidence or long-term hydrological changes (Heller & Paola, 1992; Paola, Heller & Angevine, 1992). The complexity of possible responses and the ambiguity of interpretation mean that controls on deposition will commonly be deduced from an integration of other regional information with that of the fan sequence.

3.6.4 Channel conglomerates

Conglomeratic channel deposits are composed of interbedded and interdigitating sandstones as well as conglomerate. Such deposits commonly occur as stacked multistorey sheets with little or no preserved overbank fines, a reflection of the lateral

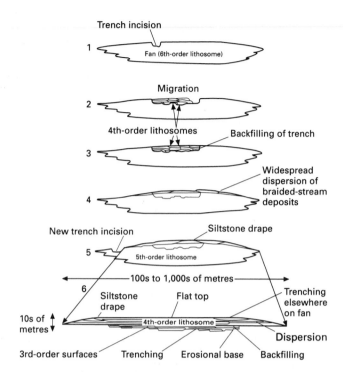

Figure 3.47 Cross-section through the proximal part of an alluvial fan, oriented normal to the fan axis. Stages 1–5 show the stages in the construction of a major element of fan sediment (5th-order lithosome) as a result of incision, migration and infilling of a channel and the subsequent overspilling to a braided sheet. Final abandonment occurs as a result of channel avulsion and is accompanied by pedogenesis of the earlier deposit. Section 6 shows some of the detail of the internal structure of the depositional unit (after DeCelles, Gray *et al.*, 1991).

mobility of high-energy bedload streams. However, in tropical settings, with abundant, readily weathered volcanic rocks, little sand-grade detritus is produced and conglomeratic channel fills occur as isolated lenses in a background of overbank fines (Tiercelin, Vincennes *et al.*, 1987). Similar restricted channel conglomerates occur in proximal parts of distributary alluvial systems (Alonso Zarza, Calvo & Garcia del Cura, 1993).

In the more common multistorey sheets, it can be difficult to establish a hierarchy of bounding surfaces, especially with laterally limited exposure. Widespread erosion surfaces (5th or 6th order of Miall, 1985a), associated with a marked increase in grain size, probably record shifts of a main channel while less extensive surfaces, defining more lenticular elements, record local shifting of bars and subchannels (e.g. S.A. Smith, 1990).

Most channel conglomerates comprise sheets of structureless or flat-bedded and commonly imbricated gravel and cross-bedded gravel, commonly in rather tabular sets (Fig. 3.33) (e.g. Ramos & Sopeña, 1983; Steel & Thompson, 1983; Ramos, Sopeña & Perez-Arlucea, 1986; Dawson & Bryant, 1987; S.A. Smith, 1990). In addition, lenticular channel-fill elements, and those showing lateral accretion surfaces, may be interbedded with sheet-like elements. Flat-bedded gravels formed by vertical accretion on longitudinal bars, and the narrow spread of palaeocurrents derived from clast imbrication, suggest deposition at high water stages (e.g. Rust, 1984). Such gravels may pass downstream into low-angle dipping beds (Nemec & Postma, 1993). Upwards coarsening within some sheets suggests downstream advance of a coarse bar head over a finer bar tail (cf. Bluck, 1982a) and some sheets of flat-bedded gravel form couplets with trough cross-bedded sandstone (S.A. Smith, 1990). Lenticular, channelized units occur laterally adjacent to sheet gravels and may also overlie them erosively. Laterally accreted gravels probably record late stages of a bar's development when vertical growth had reached some limiting height and flows were increasingly diverted around its flanks (Fig. 3.48) (Ramos, Sopeña & Perez-Arlucea, 1986). Lateral accretion units commonly terminate downdip in a plug of massive conglomerate.

Interbedded sandstones within channel conglomerates are commonly pebbly. Alternating foresets of sandstone and conglomerate within cross-bedded sets reflect discharge fluctuations (Steel & Thompson, 1983). Some channel conglomerate bodies fine upwards into sandstone, suggesting either gradual diminution of flow through diversion or systematic lateral migration of a curved channel reach (e.g. Campbell & Hendry, 1987). Where channel conglomerates contain very thick sets of cross-bedding and are interbedded with thick units of overbank fines, flows were probably confined within cohesive banks (Massari, 1983). The resultant deeper channels may have allowed the development of higher transverse bar forms.

At a regional scale, conglomeratic channel deposits may show systematic changes in grain size and facies parallel to palaeoflow. The Van Horn Sandstone of Texas, which is traced laterally for several tens of kilometres, shows proximal to distal facies changes across a stream-dominated alluvial fan (McGowen & Groat, 1971). Proximal facies include thick sheets of framework gravel with convex-upwards tops in sections normal to palaeocurrent. These formed as longitudinal gravel bars and are flanked by cross-bedded pebbly sandstone deposited in interbar channels at low discharge. Mid-fan facies contain less gravel and a higher proportion of beds have scoured bases. In distal areas gravels are confined to thin beds and lenses scattered within cross-bedded sandstones. Such changes compare with those of modern outwash fans (e.g. Boothroyd, 1972).

3.6.5 Channel sandbodies

Channel sandbodies range in shape and size from small-scale, isolated units, isolated in finer sediment to thick, laterally extensive sheets made up almost entirely of channel sediments. There is a gradational spectrum of both the shape of sandbodies and their degree of mutual erosion and connectivity. This spectrum is discussed initially under headings based on shape of the channel sandbodies. Larger-scale relationships between sandbodies and their associated overbank sediments are discussed in Section 3.6.7.

Ribbon sandbodies have low width/thickness ratios, commonly less than 15 (Fig. 3.49a), and are usually less than 10 m thick with widths of tens of metres isolated in fine-grained floodplain sediments (e.g. Hirst, 1991; Nadon, 1994). In cross-section, ribbon sandbodies are highly lenticular with steep concave-upwards lower bounding surfaces. 'Wings' of sand, representing levees, commonly extend laterally into the overbank fines (Friend, 1983; Friend, Hirst & Nichols, 1986; Hirst, 1991; Mjøs, Walderhaug & Prestholm, 1993). The sandbodies

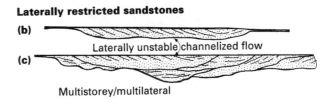

Ribbon sandstone

(a)

Wing   Laterally unstable flow

Laterally restricted sandstones

(b)

(c)   Laterally unstable channelized flow

Multistorey/multilateral

Multistorey/multilateral sheet sandstones

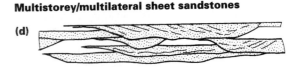

(d)

Figure 3.49 Range of contrasting channel sandbody cross-sections and internal bedding recognized in the Huesca System (Oligo-Miocene) of northern Spain (after Hirst, 1991).

Figure 3.48 Model for a pebbly braided river dominated by longitudinal and transverse bars. Some larger bars develop within-channel lateral accretion. Based on the Triassic of Central Spain (after Ramos, Sopeña & Perez-Arlucea, 1986).

differ little in shape from the original channels as they did not migrate laterally to any significant degree. In plan view, sandbodies may be either highly sinuous, as in the Tertiary of the Ebro Basin, northern Spain (Friend, Marzo *et al.*, 1981), or, in high-gradient settings, rather straight (Dreyer, 1993).

Most sandbodies comprise cross-bedded or cross-laminated sandstone but others may have heterolithic or even muddy fills (e.g. Kirschbaum & McCabe, 1992). Many of the fills fine upwards. They show a variety of fill styles, vertical accretion, concentric accretion and restricted lateral accretion (J.P. Turner, 1992; Kirschbaum & McCabe, 1992; Nadon, 1994). Where the channel sediments are fine-grained sands the rivers probably operated on very low gradients and were possibly anastomosing (cf. D.G. Smith, 1983; Rust & Legun, 1983). Where the fills are coarser grained and deposited by ephemeral floods surrounding fines show evidence of good drainage, straight incised channels (arroyos) may have been responsible (Dreyer, 1993).

Laterally restricted sandbodies with lateral accretion have essentially flat-lying erosional bases, commonly a 5th-order bounding surfaces (*sensu* Miall, 1988), and flat tops (Fig. 3.49b,c). They range up to around 15 m thick, though thicker examples are known (e.g. Mossop & Flach, 1983; Diemer & Belt, 1991). Width/thickness ratios are commonly in the range 50–100 (Collinson, 1978; Fielding & Crane, 1987) so that in reasonably extensive exposure, channel margins are commonly seen.

Single sandbodies commonly fine upwards from basal erosion surfaces that carry lag conglomerates. The sandbodies are domi-

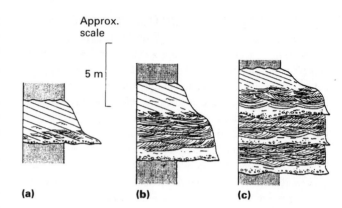

Figure 3.51 Examples of lateral accretion cross-bedding in small channel sandbodies in the Tertiary of the Pyrenees, Spain. (a) Lateral accretion bedding occupies the full thickness of the sandbody. (b) Lateral accretion bedding is confined to the upper part of the sandbody, a situation probably related to the vertical range of water-stage fluctuations. (c) Multistorey example of the type illustrated in (b) with lateral accretion bedding only apparent in the uppermost units (after Puigdefabregas & Van Vliet, 1978).

nated by inclined accretion surfaces or 'epsilon' cross-bedding (Figs 3.50 & 3.51) (J.R.L. Allen, 1963b), which is defined either by fluctuations in grain size or by interbedded sandstone and silt- or mudstone. Lateral accretion is unambiguously recognized where palaeocurrents, derived from cross-bedding and cross-lamination, trend roughly parallel with the strike of the inclined surfaces (e.g. Mossop & Flach, 1983). Where the sandbodies

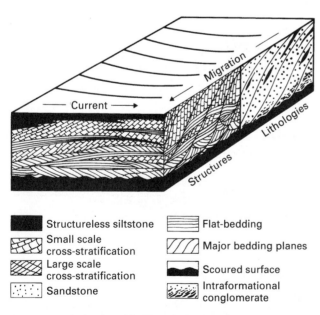

■ Structureless siltstone	▤ Flat-bedding
▨ Small scale cross-stratification	▨ Major bedding planes
▨ Large scale cross-stratification	▬ Scoured surface
⠂ Sandstone	▧ Intraformational conglomerate

Figure 3.50 Idealized model of lateral accretion (epsilon) cross-bedding based on examples in the Old Red Sandstone of the British Isles. Major bedding surfaces are inclined at between 4° and 14° depending on the thickness of the unit. Large vertical exaggeration (after J.R.L. Allen, 1965b).

Figure 3.52 Exposed scroll bars on the upper surface of a channel sandbody in the Scalby Formation (Middle Jurassic), Yorkshire, England.

have this simple organization, they are usually ascribed to deposition on the point bar of a meandering channel.

In many cases lateral accretion surfaces cross the entire thickness of the sandbody though they tend to be increasingly well defined upwards (e.g. Nami & Leeder, 1978; Puigdefabregas & Van Vliet, 1978; Stewart, 1983; D.G. Smith, 1987). In other cases, the lower part of the sandbody lacks accretion surfaces and is characterized by trough cross-bedding (Fig. 3.51b) (e.g. Puigdefabregas & Van Vliet, 1978; Gibling & Rust, 1987). Lateral accretion surfaces dip rather constantly in the direction of channel migration but the pattern may be broken by erosional discontinuities (3rd-order bounding surfaces of Miall, 1988) (e.g. J.R.L. Allen & Friend, 1968), some of which may be associated with scour steps on the base (Diemer & Belt, 1991) and ascribed to discharge fluctuations. The lack of accretion surfaces in the lower parts of some units may record a base-flow discharge which prevented fines accumulating in the deeper parts of the channel. In muddy examples, it may record accretion on a more complex, stepped channel bank with a lower platform and an upper accretionary wedge (Gibling & Rust, 1987).

Upper bedding surfaces of some laterally accreted sandbodies display roughly concentric ridges and swales with complex patterns of erosional cut-off (Fig. 3.52) (e.g. Puigdefabregas, 1973; Nami, 1976; Puigdefabregas & Van Vliet, 1978; B.M. Edwards, Eriksson & Kier, 1983; R.M.H. Smith, 1987; Alexander, 1992). These have been interpreted either as scroll bars like those of modern point bars (cf. Fig. 3.20), as counter point bars on the outer banks of channel bends (Alexander, 1992; cf. Taylor & Woodyer, 1978; Nanson & Page, 1983) or as chute-channel levees (Gibling & Rust, 1993).

Lateral accretion bedding commonly terminates downdip in a plug of fine-grained sediment bounded on the opposite side by a steep erosion surface (Fig. 3.53) (Nami & Leeder, 1978; Puigdefabregas & Van Vliet, 1978; R.M.H. Smith, 1987). This plug records cut-off of the active meander and establishment of an ox-bow lake. Lateral accretion surfaces provide evidence not just of accretionary style but also allow the width of the active channel to be estimated thus providing a starting point for palaeohydraulic reconstruction (e.g. Leeder, 1973; Bridge & Diemer, 1983).

Increasing tidal influence, in the form of salinity changes and short-term fluctuations in current strength, in the lower reaches of rivers leads to an increasing proportion of the channel body being made up of inclined sand–mud interbeds (D.G. Smith, 1987). Associated sediments should therefore be scrutinized for evidence of marine influence (e.g. Eschard, Ravenne *et al.*, 1991; Shanley, McCabe & Hettinger, 1992).

Lateral accretion surfaces are not exclusive to point bars of meandering channels but also occur through lateral shifting of compound bars in braided channels (e.g. J.R.L. Allen, 1983; cf. Bristow, 1987; Miall, 1994). This is dealt with in the following section.

Multistorey and multilateral sheet sandbodies are made up of several channel units and channel remnants (Fig. 3.49d). Many consist entirely of cosets of cross-strata with no discernible lateral or downstream accretion (e.g. Ramos, Sopeña & Perez-Arlucea, 1986). Cosets are often vertically stacked and separated by 5th-order bounding surfaces, suggesting that broadly vertical accretion prevailed. In the Lower Cretaceous of Spitzbergen, sheet-like channel sandstones show distinct patterns of internal lateral variability (Fig. 3.54). Restricted areas of trough cross-bedding are separated by wider zones of tabular sets, possibly reflecting areas of more persistent channel flow and areas of sandflat in the inferred braidplain (Nemec, 1992).

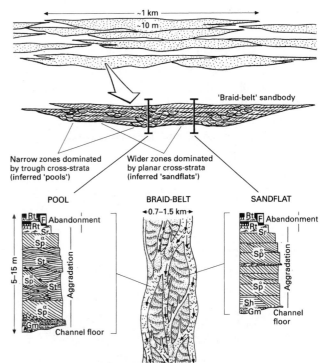

Figure 3.54 Geometry and internal organization of channel sandbodies of inferred low sinuosity river origin in the Helvetiafjellet Formation, Lower Cretaceous, Spitzbergen (cf. Fig. 3.16) (after Nemec, 1992).

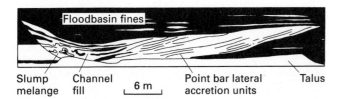

Figure 3.53 Channel sandbody of restricted width and characterized by lateral accretion surfaces. Lateral migration ended with channel abandonment and the infilling of the residual channel by bank slumping and deposition of fine-grained sediment. Scalby Formation, (Middle Jurassic) Yorkshire, England (after Nami & Leeder, 1978).

The lateral variability within these sandbodies means that interpretation of one-dimensional vertical facies sequences (Fig. 3.45) (e.g. Cant & Walker, 1976) is fraught with danger unless the status of erosion surfaces within the sequences is first established (Fig. 3.55) (Godin, 1991). Within many such sandbodies, various orders of surface may be present and elements showing vertical, downstream and lateral accretion may occur.

Most major sheet-like sandstones that have been analysed in this way appear to be the products of rather mobile, commonly braided river systems in which channel and bar shifting occurred at a variety of physical and temporal scales. Careful study of accretion directions reveals changing balances between lateral and downstream accretion (Fig. 3.56) (Godin, 1991; Miall, 1994), or may show preferred directions of channel shifting (e.g. Todd & Went, 1991). The erosional relief associated with some 5th-order bounding surfaces may reflect the importance of vertical scour and fill during flood events (Cowan, 1991).

Mudstone and siltstone units within the sandbodies are commonly restricted in extent and are usually remnants of overbank sediment laid down on floodplains or on the tops of mid-channel islands. Some may be fills of abandoned channels. The presence of palaeosols within such units may point towards an overbank setting while more laminated and possibly organic-rich sediment might suggest a channel plug (cf. H.D. Johnson & Stewart, 1985).

3.6.6 Overbank sequences and palaeosols

Overbank or floodplain sediments comprise floodplain fines, crevasse splay sands, thin channel sandbodies, palaeosols and, in some cases, coal. Overbank sequences are the product of both major floods on the alluvial plain and longer-term channel migration, avulsion and incision. They may also record tectonic or climatic changes.

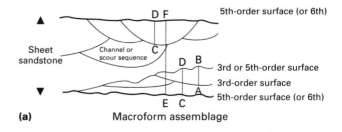

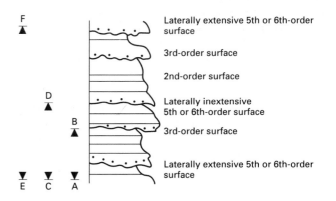

A–B   Macroform scale fining upward cycle

C–D   Channel/scour scale fining upward cycle

E–F   Sheet scale fining upward cycle

(b)

Figure 3.55 Vertical profiles through a complex, multistorey channel sandbody. (a) The two-dimensional distribution of different orders of bounding surface and their associated fining upwards cycles related to the migration of different scales of morphological element. (b) The difficulty of distinguishing different scales of cycle, and hence interpreting morphological elements, without independent evidence of the hierarchical significance of individual bounding surfaces (after Godin, 1991).

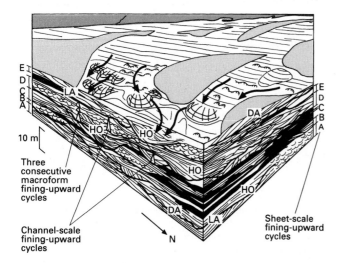

Figure 3.56 Model for the large-scale organization of the Westwater Canyon Member of the Morrison Formation, New Mexico. The model is drawn at waning water stage when large-scale scour pits develop at channel confluences to create the scour hollows (HO). Sand deposition is by filling of hollows (HO), migrating active channels (LA) and downstream accretion (DA). Longer-term shifting of channels by avulsion gives the major sheet-scale fining upwards units A–E (after Godin, 1991, based on Cowan, 1991).

Some sequences of interbedded floodplain muds and crevasse splay sands display systematic bed-thickness and grain-size variations in the form of 'bundling' whereby units, several metres thick, are alternately mud-rich or sand-rich. Sand-rich bundles occur both as the 'wings' lateral to ribbon channel sandbodies and in successions with tabular channel sandbodies of inferred meandering stream origin. In some examples, the sand-rich units have sharp bases and show an upwards diminution in thickness and abundance of sandbeds (e.g. Bridge, 1984). Such bundling is interpreted as a consequence of channel avulsion, each sharp base indicating a shift of the main channel towards the depositional site. The upwards thinning and fining probably records abandonment of a crevasse channel.

The number of avulsion-related bundles between successive channel sandstones may indirectly reflect floodplain width. The wider the floodplain the less likely an active channel belt is to return to a particular site and the greater the number of avulsion events may be preserved between channels (cf. Bridge & Leeder, 1979). Quantitative estimates based on such an idea must take account of tectonic tilting of the floodplain which tends to force channels into more rapidly subsiding areas of the floodplain (Bridge & Leeder, 1979; Alexander & Leeder, 1987).

Patterns of palaeosol stacking in the Willwood Formation (Eocene) of Wyoming (M.J. Kraus & Aslan, 1993) have been interpreted in terms of the avulsion mechanism documented for the Saskatchewan River (N.D. Smith, Cross *et al.*, 1989). Immature hydromorphic palaeosols, without profiles, result from rapid deposition associated with frequent splays during the earliest stages of avulsion. Mature palaeosols, showing much clearer profile development and evidence for better drainage, are associated with episodic overbank deposition following the development of a new alluvial channel belt and an associated alluvial ridge.

Patterns of major channel movement across a floodplain may also be inferred from the maturity of palaeosols where erosion is insignificant but where sedimentation rates vary. Palaeosols provide evidence of punctuated deposition and their maturities reflect the relative durations of periods of sediment starvation. Thicker, more mature palaeosols are related to infrequent and/or distant channel avulsion and to wide floodplains (e.g. J.R.L. Allen, 1974; Marriott & Wright, 1993; Willis & Behrensmeyer, 1994). Palaeosols that show exceptional maturity in comparison with others in the sequence may have resulted from palaeovalley incision and the development of terraces (e.g. J.R.L. Allen, 1974; Behrensmeyer & Tauxe, 1982; Retallack, 1986b; Leckie, Fox & Tarnocai, 1989).

The concept of *pedofacies* (Kraus & Bown, 1988) relates palaeosol maturity to distance from the active channel, lower maturities being inferred to occur in near-channel settings (Bown & Kraus, 1987). In the Eocene Willwood Formation, where the concept was enunciated, five 'stages' of soil formation are recognized. The least mature (stage 1) is characterized by bioturbation and weakly developed horizons while the most mature (stage 5) shows well-defined horizons which characterize spodosols (Bown & Kraus, 1987) (Fig. 3.57a). The very particular usage of the term 'pedofacies' means that it is not synonymous with

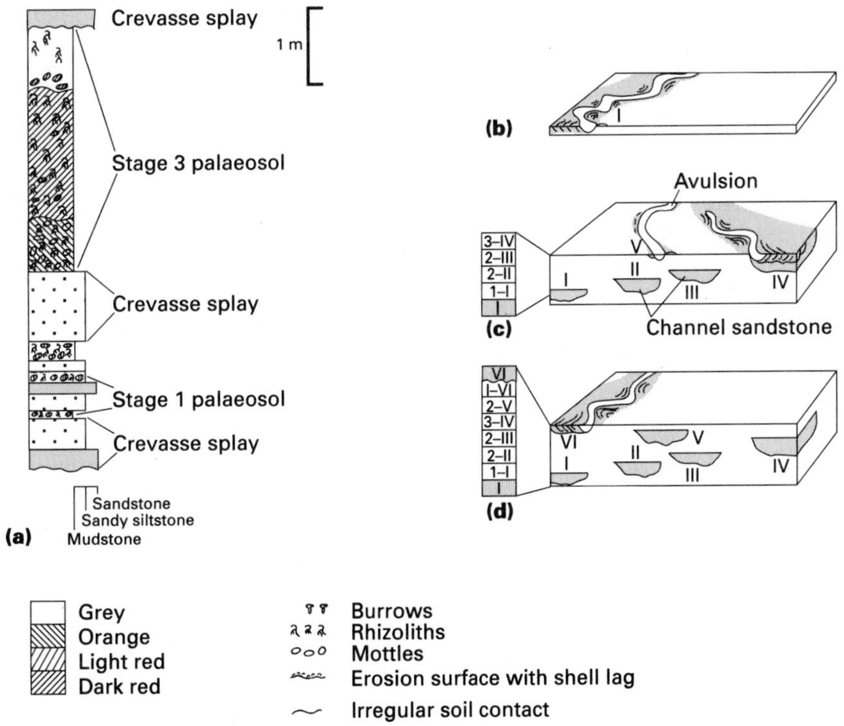

Figure 3.57 Pedofacies in the Eocene Willwood Formation, Wyoming. (a) Contrasting palaeosol maturities related to depositional rates. Stage 1 palaeosols are interbedded with crevasse splays reflecting a proximal setting while a stage 3 palaeosol records slower deposition between widely separated splays. (b–d) Application of the pedofacies concept to an idealized alluvial plain. Pedofacies stages are indicated by arabic numerals. Roman numerals indicate successive channel belts. The developing pedofacies sequence on the left reflects the proximity of successive channels (after Kraus, 1987).

'palaeosol type' or 'palaeosol facies', which express the character rather than the maturity of a palaeosol.

Successions of stacked palaeosols show various types of vertical organization which can be related to patterns of channel shifting (Kraus, 1987). Progressively increasing and then decreasing maturity through a sequence between channel sandstones may reflect successive avulsions away from and then towards the depositional site (Fig. 3.57b–d).

Changes in the character, as opposed to the maturity, of palaeosols reflect changes in the drainage state of the floodplain or an alluvial fan (e.g. Atkinson, 1986; R.M.H. Smith, 1990; Platt & Keller, 1992) and provide evidence for ancient catenas. Such differences may relate to purely alluvial controls or to differential subsidence, probably controlled by basinal tectonics.

Palaeosol profiles are either directly overlain by fine-grained sediments without erosion or by erosive lags of pedogenic material. In the former case, sedimentation resumed on to the floodplain as supply migrated back or as base level rose. In the latter case, a climatic change or intense rainstorms caused stripping of the floodplain as it was incised by shallow ephemeral floodplain channels (arroyos) (e.g. V.P. Wright & Robinson, 1988; Marriott & Wright, 1993).

3.6.7 Channel–overbank relationships

This aspect of 'alluvial architecture' compliments and extends the discussion of internal organization of individual channel sandbodies in terms of architectural elements (Sect. 3.6.2). The way in which channel sandbodies are stacked in relation to one another and to associated overbank sediments controls their degree of interconnectedness and reservoir character (e.g. H.D. Johnson & Stewart, 1985; MacDonald & Halland, 1993). Stacking pattern is primarily a function of the sandbody type, reflecting channel type and channel migration behaviour (Galloway, 1981; Friend, 1983). It also reflects the tectonic, climatic and base-level regime under which a sequence accumulated.

Ribbon sandbodies are most likely to be isolated in overbank fines, especially where overall sedimentation rates were high. Laterally restricted sandbodies, the product of avulsion and limited lateral migration of meanders, are the most readily modelled (e.g. J.R.L. Allen, 1978; Leeder, 1978; Bridge & Leeder, 1979). On an open floodplain, the interaction of avulsion frequency, floodplain width and overall subsidence rate determines the degree of mutual erosion and intercommunication of channel bodies. Tilting of the flood basin, as with axial flow in a half-graben, leads to: (i) preferential stacking of channel bodies in the hanging wall area while the footwall areas may lack channel units; and (ii) better drainage in footwall areas expressed by the character of floodplain palaeosols (Fig. 3.58) (e.g. Besly & Collinson, 1991; cf. Alexander & Leeder, 1987). Sheet-like, multi-storey channel sandbodies tend to have little preserved overbank

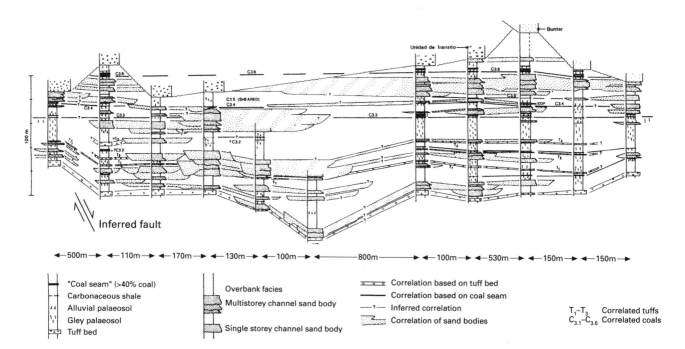

Figure 3.58 Correlated sections in a coal-bearing sequence from a small, asymmetrically subsiding basin in the Stephanian of the Spanish Pyrenees. The profile is normal to axial palaeoflow. Channel sandbodies are concentrated in the hanging wall of a syn-depositionally active fault and are associated with alluvial palaeosols. In footwall areas channels are less abundant and are associated with slightly better drained gley palaeosols (after Besly & Collinson, 1991).

sediment within them, largely as a result of the greater mobility of the active channels and the erodibility of bank material.

At the larger scale, major sheet sandstones may be interbedded with thick intervals of overbank fines as a result of major tectonic, base-level or climatic controls (e.g. Olsen & Larsen, 1993).

Thin channel sandbodies within floodplain successions may record ephemeral flows on the surface of the floodplain as a result of direct storm precipitation rather than overbank flooding of a major river. Such ephemeral streams may have ended on the floodplain through infiltration or they may have been tributaries to a major channel.

3.6.8 Coal in alluvial settings

In those settings where channels were small and actively avulsing over the floodplain, contemporaneous coals occur as elongate strips parallel with the channel trend reflecting local sedimentary

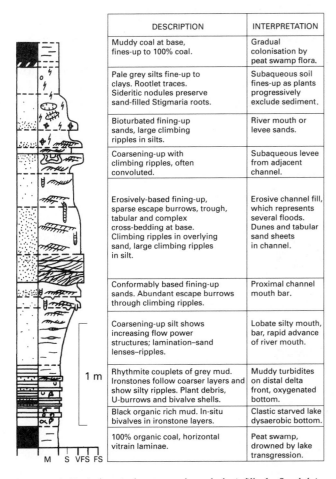

DESCRIPTION	INTERPRETATION
Muddy coal at base, fines-up to 100% coal.	Gradual colonisation by peat swamp flora.
Pale grey silts fine-up to clays. Rootlet traces. Sideritic nodules preserve sand-filled Stigmaria roots.	Subaqueous soil fines-up as plants progressively exclude sediment.
Bioturbated fining-up sands, large climbing ripples in silts.	River mouth or levee sands.
Coarsening-up with climbing ripples, often convoluted.	Subaqueous levee from adjacent channel.
Erosively-based fining-up, sparse escape burrows, trough, tabular and complex cross-bedding at base. Climbing ripples in overlying sand, large climbing ripples in silt.	Erosive channel fill, which represents several floods. Dunes and tabular sand sheets in channel.
Conformably based fining-up sands. Abundant escape burrows through climbing ripples.	Proximal channel mouth bar.
Coarsening-up silt shows increasing flow power structures; lamination–sand lenses–ripples.	Lobate silty mouth, bar, rapid advance of river mouth.
Rhythmite couplets of grey mud. Ironstones follow coarser layers and show silty ripples. Plant debris, U-burrows and bivalve shells.	Muddy turbidites on distal delta front, oxygenated bottom.
Black organic rich mud. In-situ bivalves in ironstone layers.	Clastic starved lake dysaerobic bottom.
100% organic coal, horizontal vitrain laminae.	Peat swamp, drowned by lake transgression.

Figure 3.59 Typical vertical sequence through the infill of a floodplain lake developed by the drowning of an active peat-generating mire. The progradation of a crevasse delta leads to quite rapid deposition in the central, more sandy part of the sequence. Fining upwards towards the top of the sequence is caused by growing plants excluding the ingress of coarse sediment (after Haszeldine, 1984).

(autocyclic) control. Laterally extensive coals, on the other hand, are more usually related to allocyclic regional reorganization of the drainage system or to changes in base level (Belt, Sakimoto & Rockwell, 1992).

Coals and palaeosols indicative of poor drainage formed in floodplains that were predominantly submerged, with abundant shallow lakes (e.g. Boardman, 1989). Such settings are most likely in the lower reaches of river systems often transitional to a delta plain. Coals most commonly occur at the top of small-scale upwards-coarsening sequences which result from progradational infill of lakes by crevasse deltas (Fig. 3.59) (e.g. Gersib & McCabe, 1981; Haszeldine, 1984, Fielding, 1984a,b, 1986). The colonization by vegetation of the tops of these deltas led to poorly drained soils and mires and to accumulation of peat. Where peat accumulated sufficiently rapidly to raise the surfaces of the mires above the flood level of rivers or when the mire reached great areal extent, low-ash coal resulted due to the exclusion of terrigenous material (Flores, 1981; Ethridge, Jackson & Youngbergh, 1981; McCabe, 1984). Where avulsion terminated peat growth, splay deposits directly overlie the coal. Where an associated river channel eroded such avulsion deposits, the depth of erosion is commonly restricted by the coal seam which, as a tough and elastic peat, was highly resistant to erosion. In such cases, coal may also occur as slabs and clasts within channel sandstones.

Channel sandbodies in coal-bearing successions vary greatly in type and size. Major ribbon-like channel sandstones cut across thick coal seams with highly sinuous and possibly anastomosed patterns (e.g. Eggert, 1984; Treworgy & Jacobson, 1985). These channels appear to have been active during at least some of the time of peat accumulation (McCabe, 1984). The stabilizing effect of peat adjacent to channels and the elevation of raised mires probably contributed to the lack of migration and to the localized channel stacking in multistorey complexes (Fig. 3.60) (McCabe, 1984).

In active braidplain regimes where overbank settings were short-lived, coals are likely to be thin and of limited lateral extent.

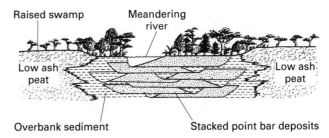

Figure 3.60 A model for fluvial deposition in an area of raised swamps where plant growth is sustained largely by precipitation. The elevated swamp inhibits overbank flooding and prevents avulsion, causing the channels to migrate within a fixed belt and give stacked, multistorey sandbodies (after McCabe, 1984).

Peat appears to have accumulated shortly after the abandonment of a channel belt while the area was elevated and starved of sediment. As adjacent areas aggraded, sediment eventually spilled into the abandoned channel and brought peat growth to an end (Nemec, 1992).

3.6.9 Incised palaeovalleys

Lowstands of base level lead to localized erosion of floodplains and to the development of incised valleys and flanking terraces. Incision is most likely during times of rapid and large fall of base level. Slower and smaller falls may be accommodated by a change of fluvial style, rather than incision (Schumm, 1993). Infill of incised valleys commonly commences as the rate of fall diminishes and base level begins to rise, though in some cases the lower reaches of the valley may experience sedimentation while the upstream parts are still undergoing erosion.

Evidence for incision and for palaeovalleys therefore comes from two sources: (i) the channel sandbodies and their distribution and stacking patterns; and (ii) overbank or floodplain sediments where terrace development might have given rise to mature palaeosols and to local erosion and reworking of palaeosols.

Most palaeovalleys are likely to be quite large and will only be detected through mapping and comparison of correlated sections (e.g. Andrews, Turner *et al.*, 1991). Some may be imaged

on (ideally three-dimensional) seismic data. Examples at outcrop are, as yet, poorly documented, especially in floodplain settings though some thick multistorey stacked units might be inferred to be of this origin (e.g. Okolo, 1983). In some cases, the erosional base of the palaeovalley will juxtapose alluvial valley-fill sediments above more distal, possible offshore facies.

Vertical changes in alluvial architecture may be related to changing base level, associated with filling of palaeovalleys (Fig. 3.61) (Shanley & McCabe, 1991). In some cases, the valley fill may be compound, recording multiple cycles of erosion and deposition (Dalrymple, Boyd & Zaitlin, 1994; Zaitlin, Dalrymple & Boyd, 1994; Sect. 6.7.6). In the Upper Cretaceous sediments of Utah, the early stages of base-level rise following an erosional sequence boundary (palaeovalley) gave rise to braided river deposits characterized by highly amalgamated channel sandbodies. With increasing accommodation space, larger volumes of overbank fines were preserved culminating in channel sandbodies isolated in fines, reflecting rapid vertical aggradation and low channel gradients (cf. Törnqvist, 1993). As base level rises and coastal gradients decrease, tidal currents may influence the channels. In the most landward parts of a coastal plain succession, beyond the extent of coastal transgression, tidal influence within fluvial channels may equate with a maximum flooding surface in more seaward areas (Shanley, McCabe & Hettinger, 1992). The change from braided to meandering styles compares with that of the post-

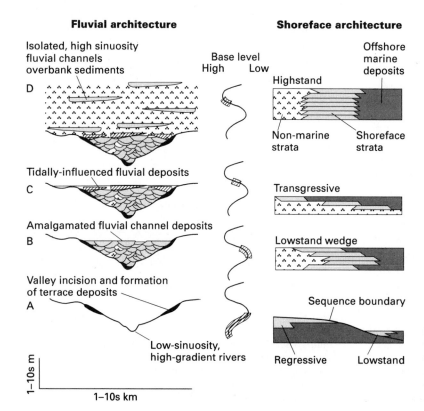

Figure 3.61 Model for the evolution of an incised valley and its subsequent filling. Coeval changes in nearshore stacking patterns are also shown. Note the incidence of tidal influence and the low channel/overbank ratio once the rim of the palaeovalley has been overtopped (after Shanley & McCabe, 1993).

Pleistocene of the Mississippi valley (Fisk, 1944) though this may be, to some extent, atypical on account of the higher river discharges associated with deglaciation and the subsequent switching of meltwater to the St Lawrence. With reduced fluvial supply of sediment, incised palaeovalleys may become estuaries during and following a rise of relative sea level and the fill may include large proportions of fine-grained sediment, especially in the middle reaches (G.P. Allen & Posamentier, 1993) (Sect. 6.7.5).

Evidence for base-level fall in interfluve areas centres on the identification of improved drainage or of particularly extensive and mature palaeosols. Such soils are most likely when a river became incised and was unable to deliver sediment to the neighbouring floodplain. In the Westphalian of eastern Canada, laterally confined channel sandstones appear to have been restricted by the presence of resistant duricrust palaeosols in the channel banks (Fig. 3.62) (Gibling & Rust, 1990). The improved floodplain drainage that gave rise to these palaeosols may also have resulted from lowering of base level such that both palaeosols and incision reflect a common control.

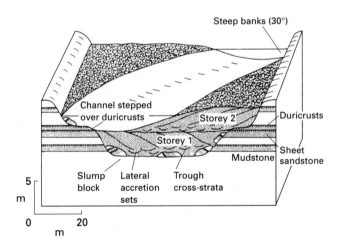

Figure 3.62 The suggested role of indurated palaeosols (duricrusts) in stabilizing the margins of a channel and allowing deep incision. Blocks of indurated fine sediment are preserved on the floor of the channel sandbody (after Gibling & Rust, 1990).

3.6.10 Ephemeral stream and terminal fan deposits

Ephemeral streams may have discharge cycles that equate with short-lived storms or with longer-term climatic changes. Short-term flood deposits, whether on fans or on alluvial plains, commonly reflect upper flow regime conditions commonly in the form of parallel lamination (Stear, 1985). In rare cases the combination of rapid flow and rapid sedimentation leads to the preservation of standing wave lamination (Hand, Wessel & Hayes, 1969; Collinson, Bevins & Clemmensen, 1989). Massive

sandstones, sometimes pebbly, were rapidly dumped from hyperconcentrated flows. Waning of flows through time commonly gives rise to upwards fining units within which there is evidence for waning flow regime. Other upwards-fining units may result from the shifting or filling of shallow channels. Interbedded sediments show evidence for quiet conditions and subaerial emergence. Desiccation cracks record drying out of the sediment surface between flood events.

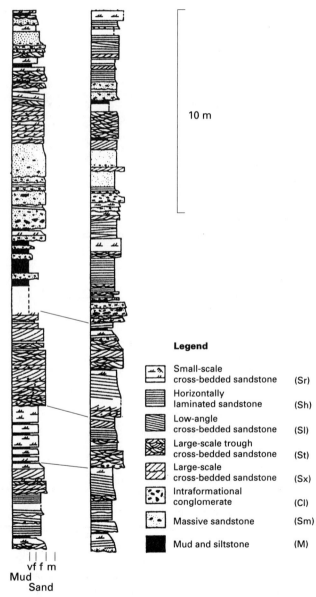

Legend

	Small-scale cross-bedded sandstone	(Sr)
	Horizontally laminated sandstone	(Sh)
	Low-angle cross-bedded sandstone	(Sl)
	Large-scale trough cross-bedded sandstone	(St)
	Large-scale cross-bedded sandstone	(Sx)
	Intraformational conglomerate	(Cl)
	Massive sandstone	(Sm)
	Mud and siltstone	(M)

Figure 3.63 Cored intervals of Bunter Sandstone (Triassic) in the Danish Sector of the North Sea showing the large-scale interbedding of sandy, channelized units and sheet-flood deposits and playa mudstones. The sequence is interpreted as the product of a low-gradient terminal fan upon which ephemeral streams shifted through time (after Olsen, 1987).

Ephemeral streams that expanded and contracted on a longer time scale gave rise to successions dominated by fining-upwards sequences. These result from various combinations of channel migration and waning discharge. Interbeds may be waterlain fines, often with evidence for emergence, palaeosols or may show evidence of aeolian winnowing or deposition during periods when the stream system migrated or dried up.

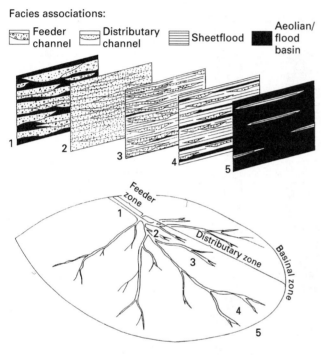

Figure 3.64 Idealized model of a terminal fan showing the downfan changes in sandbody shape and stacking (after Kelly & Olsen, 1993).

Sandstones in terminal fan successions range from thin sheets, some with locally erosive bases, the products of single flood events through simple channel units to more complex and extensive multistorey channel bodies (e.g. Tunbridge, 1981; Hubert & Hyde, 1983; Olsen, 1987, 1989; Kelly & Olsen, 1993). Where no sandstone beds exceed around 2 m in thickness it can be difficult to discriminate between channels and sheet-flood deposits (Fig. 3.63). A model based on modern analogues and examples from Devonian sequences in Ireland and Greenland (Kelly & Olsen, 1993) suggests that in more proximal settings, larger channel bodies occur in association with abundant overbank fines, reflecting the rather stable position and possible incision of the main feeder channel (Fig. 3.64). Distally, channels become more widely distributed but simpler in form, reflecting downfan channel bifurcation and increased mobility. In the most distal settings, sheet-like sandstones are interbedded with finer-grained sediments, of floodplain, playa or possible aeolian origin.

Further reading

Allen P.A. & Homewood P. (Eds) (1986) *Foreland Basins*, 453 pp. *Spec. Publ. int. Ass. Sediment.*, **8**. Blackwell Scientific Publications, Oxford.

Best J.L. & Bristow C.S. (Eds) (1993) *Braided Rivers*, 432 pp. *Spec. Publ. geol. Soc. Lond.*, **75**, Bristol.

Collinson J.D. & Lewin J. (Eds) (1983) *Modern and Ancient Fluvial Systems*, 575 pp. *Spec. Publ. int. Ass. Sediment.*, **6**. Blackwell Scientific Publications, Oxford.

Duchaufour P. (1982) *Pedology*, 449 pp. George Allen & Unwin, London.

Etheridge F.G. & Flores R.M. (Eds) (1981) *Recent and Ancient Non-marine Depositional Systems: Models for Exploration*, 389 pp. *Spec. Publ. Soc. econ. Paleont. Miner.*, **31**, Tulsa.

Ethridge F.G., Flores R.M. & Harvey M.D. (Eds) (1987) *Recent Developments in Fluvial Sedimentology*, 389 pp. *Spec. Publ. Soc. econ. Paleont. Miner.*, **39**, Tulsa.

Fielding C.R. (Ed.) (1993) *Current Research in Fluvial Sedimentology*, 656 pp. *Sediment. Geol.* (Special Issue), **85**.

Flint S.S. & Bryant I.D. (Eds) (1993) *The Geological Modelling of Hydrocarbon Reservoirs and Outcrop Analogues*, 269 pp. *Spec. Publ. Int. Ass. Sediment.*, **15**. Blackwell Scientific Publications, Oxford.

Frostick L.E. & Reid I. (Eds) (1987) *Desert Sediments: Ancient and Modern*, 401 pp. *Spec. Publ. geol. Soc. Lond.*, **35**, Bristol.

Goudie A. (1973) *Duricrusts in Tropical and Sub-tropical Landscapes*, 174 pp. Clarendon Press, Oxford.

Koster E.H. & Steel R.J. (Eds) (1984) *Sedimentology of Gravels and Conglomerates*, 441 pp. *Mem. Can. Soc. petrol. Geol.*, **10**, Calgary.

Marzo M. & Puigdefabregas C. (Eds) (1993) *Alluvial Sedimentation*, 586 pp. *Spec. Publ. int. Ass. Sediment.*, **17**. Blackwell Scientific Publications, Oxford.

Miall A.D. (Ed.) (1978) *Fluvial Sedimentology*, 859 pp. *Mem. Can. Soc. petrol. Geol.*, **5**, Calgary.

Miall A.D. (1996) *The Geology of Fluvial Deposits*, 586 pp. Springer Verlag, New York.

Miall A.D. & Tyler N. (Eds) (1991) *The Three-dimensional Facies Architecture of Terrigenous Clastic Sediments and its Implications for Hydrocarbon Discovery and Recovery*, 309 pp. *Concepts in Sedimentology and Paleontology*, **2**. Soc. econ. Paleont. Mines. (Soc. for Sedim. Geol.), Tulsa.

North C.P. & Prosser D.J. (Eds) (1993) *Characterisation of Fluvial and Aeolian Reservoirs*, 450 pp. *Spec. Publ. geol. Soc. Lond.*, **73**, Bristol.

Rachocki A.H. & Church M. (Eds) (1990) *Alluvial Fans: A Field Approach*, 391 pp. John Wiley, Chichester.

Rahmani R.A. & Flores, R.M. (Eds) (1984) *Sedimentology of Coal and Coal-bearing Sequences*, 412 pp. *Spec. Publ. int. Ass. Sediment.*, **7** Blackwell Scientific Publications, Oxford.

Reinhardt J. & Sigleo W.R. (Eds) (1988) *Paleosols and Weathering through Geologic Time*, 181 pp. *Spec. Pap. geol. Soc. Am.*, **216**, Boulder.

Retallack G.J. (1990) *Soils of the Past: An Introduction to Paleopedology*, 520 pp. Unwin Hyman, Boston.

Schumm S.A. (1977) *The Fluvial System*, 338 pp. Wiley, New York.

Wright V.P. (Ed.) (1986) *Paleosols: Their Recognition and Interpretation*, 315 pp. Blackwell Scientific Publications, Oxford.

Lakes

4

M.R. Talbot & P.A. Allen

....

4.1 Introduction

At the present day, lakes form only about 1% of the Earth's continental surface and contain less than 0.02% of the water in the hydrosphere, yet their geological significance is far greater than these meagre figures suggest. Many ideas on deltas, littoral processes and turbidity currents have developed from lake studies, and investigations in deep, stratified lakes have led to a better understanding of how some ocean basins may have become periodically anoxic. Recently, much attention has been directed towards the rift lakes of East Africa, as they probably provide the best modern analogues for sedimentary basins formed during the earliest stages of continental rifting. Because many of the processes operating in lakes are the same as those in other environments, some aspects of lake sedimentation are covered in other chapters.

A major stimulus to the study of lakes and lake deposits has undoubtedly been their economic importance (e.g. Tiercelin, 1991). Many modern lakes are a vital source of food and water, and the sediments of some are a valuable source of minerals. Ancient lacustrine basins contain extensive evaporite and oil shale deposits, which have provided sites for uranium fixation and serve as a source for hydrocarbons.

Two features of lakes stand out. The first is their sensitivity to climatic change – ancient lake deposits are among our best indicators of continental palaeoclimate. The second is the abrupt variation of sedimentary facies in vertical sequences as a result of shifting of the shorelines and biochemical fluctuations in lake waters. For this reason, many lake sequences need to be studied centimetre by centimetre in order to document the full range of sedimentary environments.

The very great diversity of lake basins and lake water presents something of a problem in classification. As the hydrological conditions determine the nature and arrangement of sedimentary facies, a fundamental distinction must be made between those lakes that have an outlet and are *hydrologically open* and those that lack an outlet and are *hydrologically closed*. However, individual lakes may pass through 'open' and 'closed' phases many times in their history; for example, *all* the major East African lakes (Lakes Tanganyika, Malawi, Victoria, Albert, Kivu, Turkana, etc.) have experienced one or more periods of hydrological closure during the last 20 000 years (Street-Perrott & Harrison, 1984; Talbot, 1988). It is probable that many ancient lake basins also oscillated between open and closed conditions. Such variations can be expected to have left an imprint upon the sedimentary facies of these basins. Other water bodies change between lacustrine and marine conditions, depending upon the height of relative sea level. The most spectacular example of this is the Black Sea, which becomes a huge lake during periods of glacially lowered sea level. Non-marine conditions have probably dominated the Neogene history of this basin (Hsü & Kelts, 1978).

4.2 Diversity of present-day lakes

Present-day lakes vary greatly in form, size and stability. Lakes of volcanic origin may be formed by lava damming (e.g. Lake Kivu) or by crater explosion and collapse. Many of the latter are small in area, but can be very deep (e.g. Crater Lake, Oregon). Lakes in glaciated regions (Sect. 11.4.5) may be proglacial (e.g. Lake Malaspina, Alaska; Gustavson, 1975) or formed by ice-damming, by barriers composed of moraine (e.g. Finger Lakes, New York State), by ice scour, freeze-thaw and by valley glaciation (overdeepened valleys and fjords) (e.g. Pitt Lake, British Columbia; Ashley, 1979). These glacial lakes are generally small, but notable exceptions are the large lakes bordering the Canadian Shield (Great Bear, Great Slave, Athabasca, Winnipeg and the Laurentian Great Lakes) which were formed by repeated glaciation. Lakes in fluvial environments (e.g. ox-bows: Sect. 3.3.4; Fig. 3.21) and those associated with shorelines (e.g. coastal lagoons: Sect. 6.7.4) are usually small and short-lived. Lakes may also develop as a result of wind, such as the impounding of rain and flood waters by dunes (Les Landes, southern France; Lake Faguibine, Mali) or by the formation of deflation basins. Other lake basins result from solution of rocks at depth, permafrost melting or even meteorite impact (e.g. Lake Bosumtwi, Ghana).

Larger lakes are primarily tectonic in origin and they fall into two groups.

1 Lakes that are formed in active tectonic areas either in extensional rift-valleys such as the East African and Baikal rifts or along strike-slip belts such as the Jordan Valley. Subsidence is rapid, sediment supply from nearby margins is typically substantial, and the sedimentary fill is thick – over 4 km in some East African lakes (Rosendahl, 1987); 2 km for the 5 My-old Lake Biwa, Japan (Ikebe & Yokoyama, 1976), and perhaps as much as 7.5 km for parts of the Baikal rift (Hutchinson, Golmshtok *et al.*, 1992).

2 Lakes, such as Lake Chad and Lake Eyre (Australia) that are formed on slowly subsiding sags in cratonic areas persist for long periods of geological time, margins fluctuate over hundreds of square kilometres in response to climatic changes and sediment influx is relatively low (e.g. Magee, Bowler *et al.*, 1995).

Sedimentation in lakes is affected by four principal factors:– the properties of the lake's water, the chemistry of these waters, fluctuations of the shoreline and the relative abundance of river-derived clastics and autochthonous sediment.

4.3 Properties of lake water

The regulation of the entire physical and chemical dynamics of lakes and their interactions is governed to a very great extent by differences in water density. Physical work is required to mix fluids of differing density. The density of water is a function of temperature and to a lesser extent of salinity and sediment concentration (Fig. 4.1). The temperature–density relationship of water is anomalous compared with other fluids, the greatest density being at 4°C. The rate of decrease of density increases with increasing temperature (summary in Ragotzkie, 1978) so that, for example, the amount of work required to mix two layered water masses at 29°C and 30°C is 40 times that required for two similar masses at 4°C and 5°C (Fig. 4.1). Tropical lakes therefore tend to become stratified more easily than temperate lakes. However, relatively minor cooling in a tropical lake can set up convection currents which, if prolonged, may eventually affect the entire waterbody, leading to mixing. In glacial lakes, the concentration of suspended sediment may be the fundamental control on density, differences in temperature being negligible by comparison (Gustavson, 1975). Density also increases with increasing concentrations of dissolved salts, but salinity-induced density stratification is important only in certain highly saline lakes or lakes influenced by deep-water hydrothermal springs (e.g. Lake Kivu; Degens, von Herzen *et al.*, 1973).

The greatest source of heat to lakes is solar radiation, although it has become apparent that heat flow from geothermal sources can also be of significance in some of the deep basins of major rift lakes (e.g. Malawi, Tanganyika, Kivu, Baikal). Loss of heat is the result of thermal radiation from the surface. The vertical temperature profile of a lake is a direct response to the penetration of solar radiation (Fig. 4.2). In thermally stratified lakes, an upper, warm, oxygenated and circulating layer termed the *epilimnion* overlies a lower, cooler and relatively undisturbed region, the *hypolimnion*. The hypolimnion may be anoxic, favouring preservation of organic matter on the lake floor. The intervening zone is termed the *metalimnion* and the zone where temperature decreases most rapidly with depth is called a *thermocline* (Figs 4.2 & 4.3). The extent of thermal density stratification and resistance to mixing (that is, the stability) of a lake is very strongly influenced by its size and morphology. A study of several lakes in Wisconsin and central Canada (Ragotzkie, 1978; Fig. 4.4) showed a simple relationship between the depth of the summer thermocline and the maximum fetch (the distance over which the wind is uninterrupted by land). Seasonal changes in air temperature, and turbulence induced by wind disturbance cause a breakdown of stratification in the upper layers and lowering of the thermocline. Lakes which circulate completely to the bottom at the time of winter cooling are termed *holomictic*, whereas those that undergo only a partial circulation, leaving a permanently stagnant bottom layer, are termed *meromictic*. Some lakes exhibit very peculiar

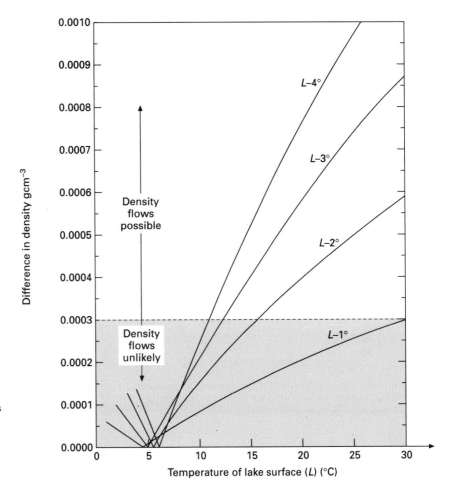

Figure 4.1 Difference in density between a lake surface (L) and inflow or bottom water that is 1, 2, 3 or 4°C cooler than surface waters. Note how the difference in density rises with increasing temperature, thus favouring density stratification and density flows in warmwater lakes. Density differences of as little as 0.0003 are sufficient to impede mixing and allow density underflows. Similar effects may result from high suspended sediment or dissolved salt contents (adapted from Axelsson, 1967).

stratification, including reversals of the conventional temperature gradient. Commonly, these paradoxes result from marked variations in salt concentration (Ruttner, 1952, p. 39).

Temperate and tropical lakes have rather different stratification behaviours. In temperate lakes immediately after the spring thaw, lake water is at about 4°C and only small amounts of wind energy are required to mix the entire water column. As spring progresses, heating of the surface layers causes a thermal stratification to develop. In the summer and autumn, mixing occurs in two phases. First, declining air temperatures cause the cooling and sinking of surface water, resulting in progressive erosion of the metalimnion. Second, there is usually a dramatic change from the final stages of weak summer stratification to autumnal circulation when overturn can occur in a few hours, especially if associated with high winds. As winter proceeds an ice cover may protect the lake from wind and a period of winter stagnation may follow. In tropical lakes, where seasonal changes in solar radiation are not so marked, thermal gradients are low but the large density contrasts that occur at tropical temperatures may nevertheless favour vertical stability of the water column, producing an immense, isolated reservoir of water at the bottom,

as in Lake Tanganyika (Fig. 4.3). Wind, rather than temperature change, is the principal agent of mixing in tropical lakes (Beadle, 1981) and, where orientation of the basin is favourable with respect to prevailing wind patterns, can cause periodic complete mixing, even in relatively deep lakes (e.g. Lake Albert).

4.4 Kinetics of lake water

Knowledge of water circulation patterns in lakes is an important component in the science of limnology and is ultimately a problem of fluid mechanics (see Csanady, 1972; 1978), the details of which need not concern us here. More extended discussions can be found in limnology textbooks such as Hutchinson (1957), Beadle (1981) and Tilzer and Serruya (1990). Wind and density flows are the most important physical inputs; barometric and gravity effects are minimal except in the very largest lakes.

4.4.1 Surface waves

Progressive surface waves are important in two respects. First,

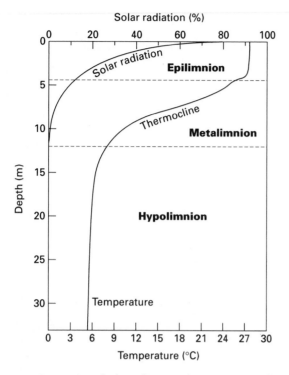

Figure 4.2 Penetration of solar radiation and temperature profile in Crooked Lake, Indiana, 18 July 1964. The thermocline marks the zone of maximum temperature change (based on Wetzel, 1983).

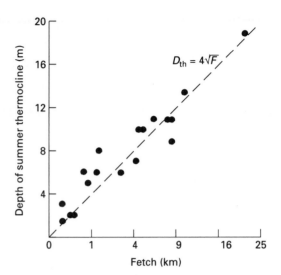

Figure 4.4 Relationship between depth of the summer thermocline and fetch based on temperate lakes in Wisconsin and central Canada (after Ragotzkie, 1978).

the orbital motion of water particles extends to a depth determined by surface wave characteristics. This wave energy is transferred in the form of turbulence to the metalimnion, but the hypolimnion is relatively unaffected because the density gradient in the metalimnion acts as a barrier to energy transfer. Second, in water shallow enough for the surface waves to 'feel' the bottom, the orbital velocities of water particles may cause sediment transport and thereby inhibit growth of aquatic plants.

In broad, shallow lakes like Lake Balaton, Hungary (mean depth of 3.3 m) waves affect the bottom over almost the entire lake area (Györke, 1973). In some calcareous hardwater lakes, wave turbulence may limit the aggradation of flat marl benches that extend out from the shoreline for considerable distances at a water depth roughly corresponding to wave base (Murphy & Wilkinson, 1980) (see also Sect. 4.6.3.). Waves have less effect in deep lakes, but the development of large waves during periods of intense wind can nevertheless cause considerable sedimentological effects at quite substantial depths. Storm waves prevent sediment accumulation offshore in Lake Superior wherever the lake is shallower than c. 100 m (T.C. Johnson, 1984). The dimensions of surface waves depend not only on wind strength and duration but also on fetch. This dependency on fetch has long been recognized (see Hutchinson, 1957, for summary) and,

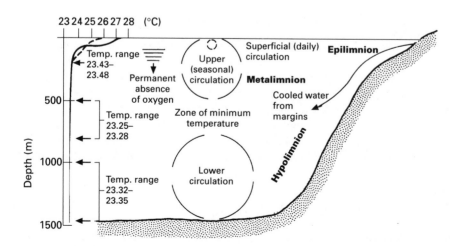

Figure 4.3 Temperature profiles and inferred circulation in different parts of the water column of Lake Tanganyika. The *epilimnion* extends down to about 50–80 m and is subject to daily circulation, the maximum depth of movement being largely determined by wind stress. The *metalimnion* extends to at least 200 m and is subject to seasonal circulation. Below this lies the anoxic *hypolimnion* (after Beadle, 1981, Fig. 6.2).

if the effective fetch is known, estimates can be made of the maximum sediment particle size that may be moved at different depths (T.C. Johnson, 1980). These relationships may also be used to estimate ancient lake sizes (Sect. 4.13.1).

4.4.2 Currents in lakes

Currents in lakes are of various kinds, the most important being the wind-driven circulation. However, currents set up by river inflows, by littoral warming or cooling (Fig. 4.3) and the hydrographic slope current from river inflows to outflowing spillways may also be significant.

Wind stress, mostly exerted during storms, sets up very complex patterns of water motion. In the nearshore zone, currents are strong (of the order of 0.30 m s^{-1} in the Great Lakes following storms) and are directed parallel to the coast, whereas deeper currents tend to be weaker and lacking a preferred direction. The typical circulation pattern is one of closed gyres lying to either side of the deepest axis of the lake basin. Wind-driven surface currents can be very effective in distributing suspended sediment, especially in lakes where there are minimal density contrasts between inflow and lake water (Weirich, 1986). Geostrophic effects (Coriolis force) are superimposed on these circulation patterns, causing a deflection to the right in the northern hemisphere and to the left in the southern hemisphere.

Inflowing rivers, coupled with geostrophic effects, can provide a major driving force for water circulation and consequently for sedimentation patterns. This effect has been demonstrated both in relatively small Swiss lakes (R.F. Wright & Nydegger, 1980) and in larger lakes such as Lake Ontario (Simons & Jordan, 1972). It is of major importance to the distribution of sediment in Lake Turkana, where the bulk of the suspended sediment is supplied by a single river, the Omo, which enters this elongate lake at one end of the basin (Yuretich, 1979).

Forel (1892), in his monograph on Lake Geneva, recognized that inflowing river water does not always mix vertically with lake water and introduced the concepts of *hyperpycnal* (under-) flow and *hypopycnal* (over-) flow later developed by Bates (1953) (Sect. 6.2.2; Fig. 6.5). In stratified lakes, sediment-laden river water may even flow along the thermocline as an *interflow* (Fig. 4.5). Movement of such currents is also greatly influenced by the Coriolis force (Giovanoli, 1990). Dispersion of river water commonly varies seasonally, sometimes forming an interflow or *underflow*, sometimes an *overflow*. For example, in the summer the waters of the Rhine and the Rhône enter the relatively warm water of Lakes Constance and Geneva respectively at densities greater than the surface water and form underflows, but this effect is not marked in winter months because of the similar temperatures of river and lake waters. Sediment-laden underflows are evidently common events, capable of rapidly introducing large volumes of suspended sediment to the bottom of many lakes (Lambert, Kelts & Marshall, 1976; Pharo & Carmack, 1979; Weirich, 1986; Fig. 4.5). Because of greater

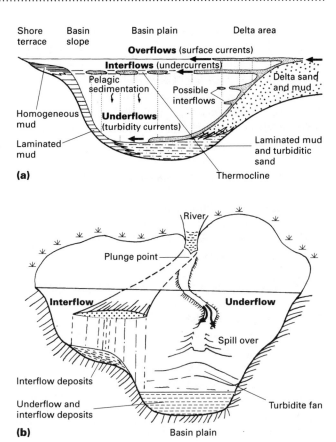

Figure 4.5 (a) Distribution mechanisms and resulting sediment types proposed for clastic sedimentation in oligotrophic lakes with annual thermal stratification. Based on Lake Brienz, Switzerland (after Sturm & Matter, 1978). (b) Density flows disappear beneath the lake surface at the *plunge point*. Spreading of interflows and their possible deflection by the Coriolis effect distributes suspended sediment to otherwise sediment-starved parts of the basin. Turbidity currents are initially confined within leveed channels, spreading out on the basin floor to form a fan (after Giovanoli, 1990).

density contrasts (Sect. 4.3) underflows may be of particular significance in tropical and subtropical lakes, providing an almost continual supply of sediment to the deep basins of some lakes (e.g. Tiercelin, Soreghan *et al.*, 1992). Spectacular evidence for the importance of this process can sometimes be observed in warm lakes after heavy rain, when relatively cool, muddy flood waters plunge abruptly beneath the clear lake waters within a very short distance of the river mouth (e.g. Tiercelin & Mondeguer, 1991, fig. 2).

4.4.3 Seiches

Wind drag on the water surface causes a piling up of water downwind. In wide, shallow lakes this piling up of water produces return currents that curve around the sides of the lake and converge in the windward half of the lake. In deep unstratified lakes there is a return current near the bed, a null-point being reached at some point in the water column. When the wind drops, the water flows back as an unopposed gradient current and a periodic rocking motion is initiated – such oscillations are termed *seiches*. The period of a seiche is determined by the shape and dimensions of the lake and varies between a few minutes to over 2 h. Vertical amplitudes of seiche waves may be considerable, the largest seiche observed in Europe being 1.87 m in amplitude in Lake Geneva in 1841. Motions are more complex when the waterbody is stratified. Motion in the surface layer sets up a corresponding motion in the opposite direction in the layer below. The thermocline (or uppermost thermocline in lakes with a multiple stratification) may be tilted. Internal seiches due to oscillation of the thermocline have a period and amplitude much greater than the ordinary seiche. Periods are of the order of days to weeks, with amplitudes up to 10–50 m (Beadle, 1981). Such waves are an important form of deep water movement in lakes as they can generate strong bottom currents. These may be particularly effective in causing local erosion, even at depths of several hundred metres, if the current is focused by topographic features (T.C. Johnson, 1984). Tilting of the thermocline in stratified lakes can cause nutrient-rich deep waters to be 'eroded' and mixed into the photic zone, triggering phytoplankton blooms which may provide a significant biogenic component to the local sediment (e.g. Haberyan & Hecky, 1987).

True lunar tides can be detected only in the largest lakes. Even in Lakes Baikal and Superior they have a maximum range of only 0.03 m, although larger ranges may occur when a tidal pulse coincides with a seiche, as is occasionally the case in Lake Huron. Generally, their influence on sedimentary processes is minimal.

4.5 Chemistry of lake waters

Non-marine waters are dominated by the four cations calcium, magnesium, sodium and potassium, and three anions, bicarbonate, sulphate and chloride. Silica can also be a significant dissolved component in alkaline lakes. The relative proportions of these and other ions are largely determined by the geology of the catchment, although it has recently been demonstrated that ions of marine origin, notably sodium, chloride and sulphate, transported as aerosols, probably provide a significant proportion of the salt budget in some lakes, even at a considerable distance from the coast (e.g. Chivas, Andrew *et al.*, 1991). Processes operating within the lake, both chemical and biological, will further influence the composition of the water.

The salinity is governed by contributions from drainage-basin run-off, aerosols and the balance between evaporation and precipitation. In closed basins, intense evaporation may give rise to very high salt concentrations, well in excess of sea water (Kelts, 1988), and the ionic composition of the water can be greatly modified by the precipitation of salts (see

Sect. 4.7.4). The salt concentration at which a lake can be called saline depends very much on the criteria used. Beadle (1981) has suggested that a good practical limit occurs at 5‰, which is the approximate upper limit for the viable existence of what are generally regarded as typical freshwater organisms. Such an evaluation is obviously almost impossible to apply to ancient lacustrine deposits, where fossils, if present, may be very different from modern lake organisms. Generally speaking, sedimentologists only regard facies as being of possible saline lake origin when evidence of evaporite mineral precipitation is present. However, even this criterion is not infallible. Some Neogene lakes in northeast Spain were fed by rivers draining areas of exposed Triassic gypsum. The lake waters had such high sulphate contents that gypsum formed together with deposits containing fossils, which suggests that the lake waters were not truly saline (Salvany, Muñoz & Pérez, 1994).

Lake waters show a very wide range of pH conditions from as low as 1.7 in some volcanic lakes to as high as 12.0 in some closed lakes such as the soda lakes of East Africa and South America. Nearly all waters with pH less than 4 occur in volcanic regions that receive strong mineral acids, particularly sulphuric acid produced by the weathering of sulphide minerals. Low pH values are also found in natural waters rich in dissolved organic matter, such as bog pools. The usual range of pH for open lakes is between 6 and 9, as most are strongly buffered by the CO_2–HCO_3^-–CO_3^{2-} system. It is a common misconception that soda lakes exhibiting extremely high pH conditions are restricted to regions of alkaline volcanicity. Similar water compositions can be achieved by evaporative concentration of waters derived from the chemical weathering of many common rock types (Eugster, 1970; Livingstone & Melack, 1984). Saline pools formed by seepage from Lake Chad into interdune depressions along the northern margin of the lake are accumulating an assemblage of evaporite minerals very similar to those found in the East African soda lakes. Dissolved ions in Lake Chad are derived almost entirely from the weathering of typical continental granitic rocks (Maglione, 1980).

In most lakes the water chemistry is determined by the composition of the surface inflow and ions supplied as aerosols. Rift and volcanic crater lakes, however, may also receive significant contributions from hydrothermal activity. Sublacustrine springs can supply large amounts of dissolved solids (including metals) and gases (Degens & Kulbicki, 1973; TANGANYDRO, 1992). The hypolimnion of Lake Kivu, for example, contains a vast reservoir of dissolved magmatic CO_2 (Schoell, Tietze & Schoberth, 1988); catastrophic outgassing of similar CO_2 from the volcanic crater Lake Nyos (Cameroon) in 1986 asphyxiated over 1700 people and 3000 cattle (Kling, Clark *et al.*, 1987).

Many chemical characteristics of lake waters are closely related to various sorts of biological activity, and vice versa. Both are in turn strongly influenced by physical processes operating within the waterbody. The dissolved oxygen content of a lake is thus of fundamental importance to organic productivity as oxygen is essential to the metabolism of all aerobic aquatic organisms. Oxygen concentrations are at the same time strongly dependent upon the rates of organic matter production and decomposition. Dissolved oxygen is supplied by chemical equilibration with the atmosphere and by photosynthesis, and consumed by aerobic respiration. Surface waters of low productivity (oligotrophic) lakes are nearly always close to saturation with respect to oxygen. However, in highly productive (eutrophic) lakes the waters may be supersaturated, due to high rates of photosynthesis, or subsaturated due to respiration and oxidation of organic matter. Dull, windless weather following an algal bloom may cause catastrophic deoxygenation because oxygen demand caused by the decay of organic matter exceeds the rate of renewal. Mass mortality of lake biota, especially fish, can result, as has been observed in Lake George, East Africa (Beadle, 1981). The development of stratification typically leads to an oxygen deficit in the hypolimnion, where sinking organic matter may consume all available oxygen, resulting in the formation of anoxic bottom waters. In lakes which undergo seasonal overturn, oxygen renewal will occur during the period of mixing, but in deep lakes the water must circulate for some weeks before saturation is reached. Meromictic lakes (Sect. 4.3) have a permanently anoxic hypolimnion; in some stratified lakes this anoxic water mass may constitute a large proportion of the total water volume. Mixing of such waters into the epilimnion, due to partial overturn or the effects of a seiche, can be another reason for mass fish deaths (Livingstone & Melack, 1984).

In general, the oxygen content of surface waters is sufficient to allow aerobic processes to function normally. Productivity is largely determined by the availability of key nutrients, notably phosphorus, nitrogen and silica. Phosphorus and nitrogen have probably always been vital nutrients and complete consumption of one or the other may severely limit organic growth, even when conditions are otherwise favourable. Silica is important today because diatoms form a major group of lacustrine phytoplankton organisms. Non-marine diatoms appeared in the Oligocene; before that time silica budgets in lakes are likely to have been rather different from today. Nutrients may be renewed by river inflow and rainfall, but in many large lakes the nutrient budget depends upon regeneration processes. During periods of stratification, the surface waters of such lakes can become depleted in phosphorus and nitrogen because of the incorporation of these elements in tissues of planktonic organisms which sink and accumulate below the thermocline. This removal of nutrients from surface waters profoundly influences their primary productivity, which then becomes dependent upon the recirculation of nutrients from below the thermocline. Thus, periods of partial or complete overturn may be accompanied by a marked burst of productivity, leading to phytoplankton blooms.

4.6 Clastic sedimentation

Most of the siliciclastic sediment deposited in lakes is transported there by rivers, either in suspension or as bedload. Wind-blown, ice-rafted and volcanic material may be locally important. The nature and size of the surrounding drainage basins exert a major influence on the input of sediment. The supply of sediment is often seasonally controlled; seasonal contrasts are particularly marked in high latitude lakes, fed wholly or in part by glacial meltwaters, and in seasonally dry regions of the warm climatic belts. In glacially fed lakes the extremely high load of the early summer contrasts with the small and almost clearwater discharge of the winter. Episodic run-off in seasonally dry regions may lead to the interlamination of clastic and chemical sediments.

4.6.1 Beaches and other nearshore zones

Siliciclastic sediments at lake margins are generally concentrated around river mouths. Beaches, spits and barriers may be formed by wave action, the processes and products differing little from their counterparts in low to intermediate wave-energy marine environments as described in Chapter 6 (Reid & Frostick, 1985; Renaut & Owen, 1991). The absence of tides means that wave attack may be limited to a narrow zone along the shores of hydrologically open lakes; in closed lakes, on the other hand, seasonal or longer-term variations in lake level may spread the effects of littoral processes over a large area. Abandoned beach ridges, preserved as staircase-like features around the margins of modern closed basins (e.g. Great Salt Lake, Dead Sea, Lakes Ziway–Shala (Ethiopia)), provide some of the most striking evidence for former high lake levels (Sect. 4.9).

One important contrast between marine and lacustrine environments is that in the latter, shorelines may be marked not by beaches but by stands of macrophytes. Today these are typically reeds and sedges, which can form extremely extensive swamps (Livingstone & Melack, 1984) and have a profound effect on lake-margin sedimentation, forming very effective wave baffles and traps for fine-grained sediment. Well-defined beaches also tend to be absent where the lake floor shoals very gently, particularly if frequent changes in lake level cause the shoreline to shift laterally over large distances. Marginal swamps may not form under these conditions, especially if the water has a tendency to be saline. Extensive *sand-* or *mudflats* (also known as *playa* or *inland sabkha*) are particularly characteristic of saline lakes. These are very low gradient features, which merge landward into alluvial fans or talus aprons along mountain fronts (e.g. in the intermontane basins of western USA, Iran and China) or, where relief is low, into alluvial or desert plains (e.g. central Australia; Magee, Bowler *et al.*, 1995), and lakeward into a saline or ephemeral waterbody. The flats are dry for much of the time, being covered by water only during periods of flood or high lake level. Most of the sediment is of fluvial origin, introduced by unconfined sheet floods, although windblown dust, trapped on the sticky hygroscopic surface, may also make a significant contribution (Smoot & Castens-Seidell, 1994). The northern margin of Lake Eyre is a vast playa covering c. 5000 km², which has formed from sediment carried to the basin by exceptional floods originating mainly in subtropical Queensland (Dulhunty, 1982; Magee, Bowler *et al.*, 1995).

Sheet floods typically produce flat, lenticular sandy or silty laminae, capped, as the flow dissipates, by a mud drape (Hardie, Smoot & Eugster, 1978, p. 20). Shallow water may cover the playa surface, due to ponding of floodwaters or expansion of the lake, and rework the surface sediments to form a wave-rippled cap. The deposits are thus initially horizontally- or wavy-laminated. However, they are frequently subject to prolonged periods of exposure, when the primary sedimentary textures may be disrupted by a variety of processes. Desiccation causes extensive cracking; salt crusts form at the surface and interstitial precipitation of evaporites from saline groundwaters or springs may deform the lamination. Bioturbation, due to the growth of halophyte vegetation and the activities of salt-tolerant insects such as ants (Hardie, Smoot & Eugster, 1978; Rosen, 1991), and multiple cycles of precipitation and dissolution of saline minerals can lead to a massive, apparently structureless mudstone or siltstone (Smoot & Olsen, 1988). Collapse due to salt dissolution can also result in the development of a characteristic *sand-patch* fabric (Smoot & Castens-Seidell, 1994). Desiccation, salt-heaving and the formation of efflorescent crusts make the exposed playa surface prone to deflation. Like a marine sabkha (Sect. 8.4), it is an equilibrium deflation surface, whose level is determined by the top of the capillary fringe. Below this, the sediments remain moist and cohesive; above the capillary fringe they dry out and are readily blown away. Playas can be the source not only of aeolian dust, but of abundant sand-sized clay pellets, which accumulate around the margins of some salt lakes as a distinctive aeolian sediment body (*clay dune* or *lunette* – Bowler, 1973).

Where lake margins are very steep, as may be the case where the shore coincides with an active border fault in rift or strike-slip basins, coarse colluvium, including blocks of basement rocks many metres in diameter, can be carried directly into the lake to form talus cones and megabreccias (Cohen, 1990; Renaut & Tiercelin, 1994). Such fault escarpments are normally found in areas of maximum subsidence, which limits the basinward progradation of these marginal fans. Thus the coarse deposits typically pass rapidly basinward into fine-grained deepwater facies (Baltzer, 1991).

4.6.2 Deltas

Many types of delta occur at lake margins (Chapter 6). The main differences between those that form in lakes and in the sea are determined by the following.

1 Water salinity. In most lakes, salinity contrasts between inflow and the standing body of water are minimal. This influences the way in which the two waters mix.
2 Tidal effects are negligible.
3 Wave modification of deltas. This tends to be modest, although it can be of significance in large, windy lakes.

Thus lacustrine deltas are limited to a range of coarse-grained deltas, particularly Gilbert-type, that are found mainly in tectonically active basins (Fig. 6.26) and a range of finer-grained fluvial and wave-modified deltas (e.g. Catatumbo delta in Lake Maracaibo; Hyne, Cooper & Dickey, 1979).

Gilbert-type deltas (Figs 6.22, 6.23) consist of extensive simple, steep foresets overlain by flat topsets (Sect. 6.5). They are found in relatively deep, freshwater lakes with rather steep-gradient inflowing rivers. Good modern examples are the Rhine River delta in Lake Constance (Müller, 1966; Förstner, Müller & Reineck, 1968) and several rivers entering Lake Malawi (Scholz, Johnson & McGill, 1993). The 'type' Gilbert deltas are of late Pleistocene age and were produced by rivers flowing into Lake Bonneville, which at the time was some 200 m deeper than the present Great Salt Lake (Gilbert, 1885). Remnants of these deltas, with foresets of coarse sand and gravel in excess of 70 m high, can still be observed along the western edge of the Wasatch Mountain range, Utah (D.G. Smith & Jol, 1992a; Fig. 4.6). Foresets in the late Pleistocene Kicking Horse River delta, British Columbia, are up to 100 m high (D.G. Smith, 1991).

In freshwater lakes, hypopycnal flow is much less likely than in marine coastal settings because river water is seldom less dense than lake water. In contrast, temperature differences and the high suspended loads in river waters (see Fig. 4.1) result in sediment-laden underflows (hyperpycnal flow) which may produce sublacustrine channels and levees, as in the Rhône delta in Lake Geneva (Forel, 1892; Lambert & Giovanoli, 1988; Fig. 4.5). Underflows may also play a part in determining the internal fabric of Gilbert-type foresets. In sections through exposed glacial and tropical Quaternary Gilbert deltas, pebble imbrication and climbing-ripple lamination indicate the frequent occurrence of strong downslope currents, probably during

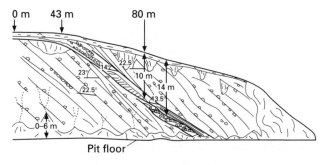

Figure 4.6 Giant foresets in a Gilbert-delta from the type area along the Bonneville (late Pleistocene) shoreline above the present Great Salt Lake. The unit with lower angle cross-bedding in the middle of the section was formed by wave reworking of the delta front (after D.G. Smith & Jol, 1992a).

periods of river flood. Basinward continuation of such flows is responsible for the formation of a distinctive lacustrine prodelta facies composed of a heterolithic assemblage of interlaminated muds, silts and very fine sands.

Large sediment loads, or a rapid fall in lake level, can cause lacustrine deltas to prograde rapidly over finer grained, prodelta deposits. Liquefaction and load structures, due to density instability, are thus relatively common. In some instances, quite large, intrusive bodies of fine-grained sediment may be produced, akin to those that form in some marine deltas (Sect. 6.6.4). Spectacular mud diapirs (*mud lumps*) with diameters of up to 300 m intrude the Truckee River delta (Pyramid Lake, Nevada; Born, 1972). Rapid accumulation and steep accretionary foresets also favour slope instability; for this reason, deltas provide the dominant source of slumps and major turbidity flows in many lake basins (e.g. Giovanoli, 1990). Wave attack can also be important in redistributing deltaic sediment. Gilbert-type deltas, which slope steeply into deep water, may be particularly prone to wave reworking, especially in large lakes with long fetch (Smith & Jol, 1992a). Wave-driven longshore drift in lakes with steep bottom slopes can produce spits with angle-of-repose foresets similar in scale to those typically associated with Gilbert-type deltas (Smith & Jol, 1992b).

4.6.3 Offshore zones

Sediment is supplied to offshore zones by pelagic fall-out, surge-like and semi-permanent density currents and mass flows. Accumulation rates are typically in the range of 0.1–0.5 kg m^{-2} year^{-1} (T.C. Johnson, 1984). These values represent less than half a millimetre of accumulation per year. Very much higher values occur in lakes fed by rivers draining areas of high relief. In Lake Turkana, for example, accumulation rates are of the order of 2 kg m^{-2} year^{-1} (Cerling, 1986).

The dispersion and sedimentation of fine-grained suspended matter are determined by lake circulation patterns (Sect. 4.4). When lakes are stratified, the distribution of suspended sediment may be greatly influenced by density contrasts within the water column. In Lake Brienz, Switzerland, during the period of summer stratification river water is less dense than the lake water and overflows result (Sturm & Matter, 1978). More frequently, incoming sediment-laden river water is denser than water in the epilimnion but less dense than the cold water of the hypolimnion and it therefore flows along or near the thermocline as an interflow. A rain of mostly silt-sized particles falls to the lake floor from these interflows, which can be particularly effective in transporting sediments to otherwise clastic-starved areas of the basin, where they accumulate as finely laminated silts (Fig. 4.5). Several laminae may be deposited in the space of a year (Giovanoli, 1990). The finest material is held in suspension and settles out only after turnover of the water column. This fine-grained suspended load forms a light-coloured, winter 'blanket' over the interflow silts or basin plain

underflow and turbidity current deposits. Even gentle circulation and upwelling can hinder settling of extremely fine particles. However, more rapid sedimentation is promoted by the presence of flocculating agents. Faecal pellet formation, due to the grazing activities of zooplankton, probably also enhanced the settling of suspended sediment in some lakes. It is estimated that copepod grazing may be responsible for as much as 30% of the total bulk sedimentation of clay- and silt-grade material in Lake Tanganyika (Haberyan, 1985).

Bottom-hugging density underflows are also effective in distributing fine-grained sediments over large areas of deep lake basins. Such flows may be permanent, or at least persist for long periods and can maintain their identity over very large distances, the best-known example probably being in Lake Mead (formed by damming the Colorado River), where underflows have been observed to traverse the length of the lake, a distance of approximately 150 km (Grover & Howard, 1938). These underflows, which apparently result from the combined effects of suspended sediment content and slightly elevated salinities in the Colorado River inflow, are the most important sediment-transporting agent within Lake Mead (Gould, 1960). Some of the large rivers entering Lake Tanganyika maintain more-or-less continuous underflows of relatively dense water (Cohen, 1990), which, in addition to supplying suspended sediment, can erode, rework and mould the basinal sediments into distinctive sediment bodies. Deep-water canyons off some deltas in Lake Malawi seem to be scoured by underflows (Scholz & Rosendahl, 1990; Scholz, Johnson & McGill, 1993) and wave-like, internally laminated bedforms with amplitudes of 2–3 m and wavelengths of 500–800 m occur in basinal muds at a depth of nearly 500 m at the southern end of Lake Tanganyika (Tiercelin, Soreghan et al., 1992). These structures have apparently been formed by density currents flowing down a nearby fault scarp (Fig. 4.7). Wave- and seiche-related currents can rework deepwater sediments; bottom currents similar to contour currents (Sect. 10.2.2) winnow fines at depths of up to 200 m in Lake Superior, producing planar and trough cross-laminated sands with sharp upper and lower boundaries (T.C. Johnson, Carlson & Evans, 1980).

Where an adequate clastic supply exists, typically provided by a river, surge-like, sporadic density currents (turbidity currents) can be a major source of sediment to the offshore areas of large lakes. Individual turbidite beds are commonly a result of floods, while exceptionally large turbidite units (examples metres thick are known from some Swiss lakes – Sturm & Matter, 1978; Kelts & Hsü, 1980; Lambert & Giovanoli, 1988) are caused by rare, catastrophic events. Turbidite fans with well-developed channels and levees are common in some large lakes and fine-grained turbidites blanket wide areas of some of the deep basins (e.g. Tiercelin, Mondeguer et al., 1988; Giovanoli, 1990, Johnson & Ng'ang'a, 1990; Scholz & Rosendahl, 1990; Scholz, Johnson & McGill, 1993). Rivers entering lakes at steep basin margins are probably a

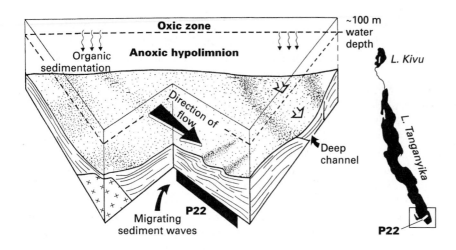

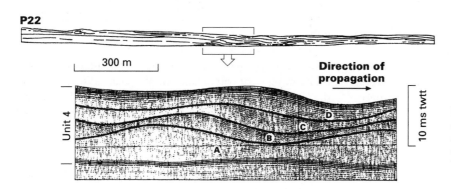

Figure 4.7 Seismic section and interpretation of low amplitude progradational and aggradational waves of muddy sediment at a depth of 400 m in the southern basin of Lake Tanganyika (after Tiercelin, Soreghan *et al.*, 1992).

particularly important source of turbidity flows, as the abrupt offshore increase in depth limits the development of a subaerial delta plain. Seismic profiles from such deltas in Lake Malawi indicate widespread chaotic bedding, suggestive of frequent slumping of the delta slope deposits (Scholz, Johnson & McGill, 1993). One further consequence of steep slopes and high rates of sediment supply is a tendency for mass movement by creep, sliding or slumping of large bodies of newly deposited sediment (Mondeguer, Tiercelin *et al.*, 1986; Eyles, Clark & Clague, 1987; Johnson & Ng'ang'a, 1990). Tectonic activity, undercompaction and abundant interstitial biogenic gas may further enhance sediment instability (Bouroullec, Thouin *et al.*, 1991; Tiercelin, Soreghan *et al.*, 1992).

4.7 Chemical and biochemical sedimentation

Where sedimentation is not overwhelmed by clastic material, chemical or biochemical deposits may dominate the basin. In dilute lakes these are typically carbonate, siliceous or iron mineral accumulations, while saline waterbodies can produce a wide range of carbonate, evaporite and silicate minerals.

Calcium carbonate precipitation is by far the most widespread chemical sedimentary process in lakes.

4.7.1 Lacustrine carbonates

Calcareous sediments are formed by one or more of four processes (Kelts & Hsü, 1978): (i) inorganic precipitation generally associated with the photosynthetic activities of plants, or, less commonly, induced by evaporation, changes in temperature, or mixing of water masses; (ii) production of calcareous shells, surface encrustations or skeletal elements of living organisms; (iii) clastic input of allochthonous carbonate particles derived from the drainage basin; and (iv) post-depositional or early diagenetic precipitation ((iii) and (iv) are not discussed further). Calcite is the commonest lacustrine carbonate mineral, but high-magnesian calcite and aragonite are also widespread. In addition, normally rare carbonate minerals, such as vaterite ($CaCO_3$), monohydrocalcite ($CaCO_3.H_2O$) and ikaite ($CaCO_3.6H_2O$), are recorded from lakes. There is also circumstantial evidence that primary dolomite may be forming in some shallow, saline waterbodies (De Deckker & Last, 1988).

INORGANIC PRECIPITATION

The most important control on primary carbonate precipitation is exercised by the $CO_2 - HCO_3^- - CO_3^{2-}$ system. Removal of CO_2, accomplished most effectively by photosynthesis, raises pH and promotes carbonate precipitation. Removal of CO_2 by degassing into the atmosphere appears to be a much less important and slower process (Kelts & Hsü, 1978; Dean, 1981). Carbonate precipitation can also be caused by the warming of lake waters, leading to carbonate supersaturation, or by the mixing of two waters of different composition (e.g. inflow and lake waters). These processes are of little significance in comparison with those related to photosynthesis.

Because of its association with photosynthesis, primary carbonate precipitation is restricted to shallow, marginal areas and the photic zone of deep, open lake waters. An annual cycle of calcite precipitation has been described from Lakes Zürich (Kelts & Hsü, 1978) and Greifensee, Switzerland (Weber, 1981), and other temperate-zone lakes (Dean, 1981; Koschel, Benndorf *et al.*, 1983). Late spring and summer diatom blooms abstract large quantities of dissolved CO_2, causing marked increases in pH and carbonate ion activity. This leads to supersaturation with respect to calcite. Photosynthetically induced precipitation in association with phyto- and picoplankton blooms is a major carbonate-producing process in the open waters of temperate, tropical, and even Antarctic lakes (Lawrence & Hendy, 1989). Simple chemical supersaturation is apparently not in itself sufficient to cause carbonate precipitation; phytoplankton or similar substrates also appear to be required to provide nuclei around which crystal formation can commence, even in lakes that are highly supersaturated with respect to all the common carbonate minerals (Galat & Jacobsen, 1985; Stabel, 1986). Once precipitation begins, it may proceed very rapidly and the concentration of suspended carbonate crystals can cause surface waters to become cloudy, events known as *whitings*.

Calcite, monohydrocalcite, Mg-calcite and aragonite may all be produced by direct precipitation. Which mineral precipitates is apparently determined by water chemistry, although the controlling factors are not known with certainty. In most temperate lakes, as well as Lakes Turkana and Malawi, the carbonate is calcite. In Lakes Tanganyika and Van (Turkey) it is aragonite, and, in Kivu, aragonite and monohydrocalcite. The Mg/Ca ratio of the lake water may be significant, calcite precipitating when the ratio is low, Mg-calcite or aragonite at higher ratios (Müller, Irion & Forstner, 1972). Experimental evidence, however, indicates that other ions, notably sulphate and phosphate, can also influence carbonate mineralogy. In general, calcite is characteristic of low-salinity lakes, while Mg-calcite and aragonite are precipitated in brackish and saline waterbodies. The minerals form silt-sized crystals that may have a variety of morphologies (Raidt &

Koschel, 1988). They settle as a rain of discrete grains or in larger aggregates of faecal origin. In some carbonate-precipitating lakes much of the calcareous mud may be deposited in pellet form (e.g. Spencer, Baedecker *et al.*, 1984; Kelts & Shahrabi, 1986). Because of its association with phytoplankton blooms, inorganic carbonate precipitation tends not to be a continuous process (despite permanent supersaturation), but occurs in seasonal bursts. During the intervening periods other sorts of sediment may accumulate. In these situations the carbonates typically occur as discrete laminae which alternate with other lithologies. Sediments consisting of carbonates interlaminated with clastic material, diatomaceous ooze, organic matter or even evaporites are a characteristic facies in the deep basins of a number of lakes (Eugster & Kelts, 1983).

Ooids are another form of inorganic carbonate. They are much less common than the carbonate silts described above, but all three of the principal mineralogies have been recorded and protodolomite ooids occur in the late Pleistocene sediments of Lake Albert (Stoffers & Singer, 1979). In some lakes, ooids of mixed mineralogy form (Popp & Wilkinson, 1983). Calcitic ooids commonly have a cortex in which the crystals have a radial fabric or no preferred orientation at all (e.g. Wilkinson, Pope & Owen, 1980). In common with marine ooids (Sect. 9.1.3), lacustrine ooids are restricted to shallow, wave-agitated marginal areas. In Lake Tanganyika and Great Salt Lake, littoral ooid shoals and beaches are widespread in areas starved of clastic sediments (Halley, 1977; Cohen, 1990). In the former lake, subsequent precipitation of calcite has cemented some of the sands into beachrock.

Inorganic carbonate may be precipitated where spring waters enter a lake or are ponded on a playa surface. Extensive carbonate precipitation can result from mixing or CO_2 outgassing, forming *travertine* or *tufa* mounds and chimneys. The most spectacular developments of this sort are found in some Pleistocene lakes in the western USA, where there are massive, cathedral-like carbonate towers and pinnacles tens of metres high (Scholl, 1960). Although algal textures can be recognized in these structures, discharge of calcium-rich springs along fractures into lakes rich in dissolved HCO_3^- was the primary cause of carbonate precipitation. The mineral framework is now calcite, but parts contain pseudomorphs of ikaite, a low-temperature calcium carbonate polymorph, suggesting that the tufa formed around springs entering the lakes during cold climatic episodes (Shearman, McGugan *et al.*, 1989). In spring-fed ponds, pisoids, concretions and dripstone overgrowths form (e.g. Risacher & Eugster, 1979; Jones & Renaut, 1994). If the pond is very still, sheets of calcite crystals, supported by surface tension, may even precipitate at the water surface (Chafetz, Rush & Utech, 1991). These subsequently sink and can be incorporated into the sediment as distinctive and diagnostic carbonate grains.

BIOGENIC CARBONATE PRODUCTION

Unlike the pelagic environment in the oceans, there are no common calcareous planktonic organisms in lakes. Some freshwater ostracods are free-swimming and one group of calcified planktonic green algae is known, but there are few records of these as major sediment producers (Müller & Oti, 1981; Koschel & Raidt, 1988). The bulk of biogenic carbonate production is therefore due to benthic organisms and is thus restricted to shallow, mainly marginal areas. The major sediment producers are molluscs, ostracods and green algae of the charophyte group. Carbonates also precipitate extensively in association with benthic microbial communities.

Gastropods are the most widespread lacustrine molluscs and include species that can tolerate salinities of up to 100‰. After death, gastropods tend to float because decomposition of the body tissue produces gas inside the shell. When this happens, shells may drift ashore and accumulate in large numbers to form littoral shell beds (De Deckker, 1988). Sublittoral accumulations can also occur in areas where gastropods are abundant. These may form because of clastic sediment starvation or by winnowing of finer sediment, perhaps due to a slight fall in lake level. In either case gastropod sand can result; some of the shallow platforms in Lake Tanganyika carry shell lags covering several hundred square kilometres (Cohen, 1989; Fig. 4.8).

Charophytes are a widespread group of aquatic algae which are commonest in freshwater lakes, although living specimens have also been recorded from saline waterbodies (Burne, Bauld & De Deckker, 1980). At times in the past they have been major carbonate sediment producers in lakes (Sects 4.12.1 & 4.15) and are still of importance today. Some charophytes (principally species of *Chara* and *Nitella*) produce calcified female reproductive cells (oogonia) and may have carbonate-coated stems; up to 60% of the dry weight of the plant can be calcite. Other macrophytes may also become encrusted with calcite. This calcification process seems to be the result of a particular metabolism employed by some plants that inhabit alkaline waters. Many carbonate-producing lakes are characterized by relatively high pH (8.5–9.5) and, where the pH exceeds 8.7, the amount of dissolved CO_2 available for photosynthesis is very limited (Wetzel, 1983). A number of aquatic macrophytes and cyanobacteria have the ability to metabolize HCO_3^-, which in alkaline waters is present in relatively much greater concentrations than CO_2. Under such conditions these plants can outcompete species that are unable to utilize HCO_3^-. Calcification seems to be an integral part of bicarbonate assimilation (McConnaughey, 1991); hence the widespread occurrence of calcite crusts on the aquatic macroflora of alkaline lakes. 'Meadows' of charophytes and plants such as *Potamogeton* can be a prolific source of calcitic sediment, yielding sands and muds with a gritty texture due to the presence of oogonia and calcified stem fragments. Similarly, abundant carbonate is being produced in the lakes of the Florida Everglades through the photosynthetic utilization of HCO_3^- by cyanobacteria (Merz, 1992).

Benthic communities of microphytic algae and cyanobacteria are relatively common in lakes and can form substantial organosedimentary bodies called *microbialites*. They may have internal structures that are stromatolitic (fine, planar or wavy lamination), oncolitic (concentrically laminated), thrombolitic (clotted texture) or cryptic (no obvious internal structure) (Burne & Moore, 1987). Although some of these strongly resemble marine stromatolites and oncoids (Sect. 9.4.5), a major difference is that in most lacustrine examples the trapping and binding of discrete sedimentary particles seems to have played little part in the development of the structures. They are formed mainly

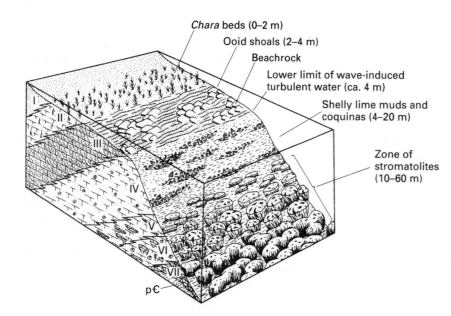

Chara beds (0–2 m)
Ooid shoals (2–4 m)
Beachrock
Lower limit of wave-induced turbulent water (ca. 4 m)
Shelly lime muds and coquinas (4–20 m)
Zone of stromatolites (10–60 m)

Figure 4.8 Schematic representation of the distribution of various carbonate facies along the margin of Lake Tanganyika, based upon observations from the northeast shore of the lake. Beachrock is formed by cementation of the ooid sands. Some of the coquinas and stromatolites are relict features from lowstands of the lake (after Cohen & Thouin, 1987).

by carbonate precipitation on and within the organically formed structure. It is still a matter of debate whether these are true biogenic deposits. Although microbialites may be produced by a variety of algae and cyanobacteria, none of the genera seems to be an obligate calcifier (Pentecost & Riding, 1986). Calcification apparently occurs mainly as a by-product of the organisms' photosynthetic activity (see above).

Microbialites typically develop on firm or lithified substrates. They may encrust boulders or cobbles along beaches and at the toes of alluvial fans (e.g. Osborne, Licari & Link, 1982), outcrops of basement rocks, lithified lacustrine sediments such as beachrock, or lacustrine deposits that were indurated due to exposure and pedogenesis. Some very substantial microbialite structures are known from modern lakes. In Fayetteville Green Lake, New York, a thrombolitic microbialite ledge 2–8 m thick extends up to 7 m into the lake from a shoreline outcrop of Palaeozoic rocks (Thompson & Ferris, 1990; Fig. 4.9). In Lake Tanganyika, thrombolitic lithoherm mounds up to 2 m in diameter have coalesced to form reef-like features at depths of 15–50 m and covering several 100 km² (Cohen & Thouin, 1987; Fig. 4.8). Studies of Quaternary lacustrine microbialites suggest that they may form over a depth range of 0–30 m (Casanova, 1991). Some spectacular examples are reported from Lake Van, where they are apparently forming at depths down to 100 m (Kempe, Kazmierczak et al., 1991). These match in scale the tufa pinnacles of the western USA and, like these, form where calcium-rich springs emerge into alkaline lake water. However, cyanobacteria seem to have played a much greater role in the development of the Lake Van structures, hence their designation as microbialites.

Where microbes coat grains such as skeletal debris, pebbles, wood fragments, or even reworked fragments of earlier microbialites, oncoids may form. These are concentrically laminated structures, commonly 0.01–0.05 m in diameter. Their size and shape initially reflects the form of the nucleus, but the addition of successive growth laminae can build up a substantial layer of carbonate, resulting in oncoids as much as 0.3 m in diameter (Schäfer & Stapf, 1978). As concentric growth requires that the clasts are periodically rolled, oncoids are characteristic of areas of wave or current activity and mainly form in shallow, marginal areas of lakes.

Microbialites occur in many types of lake, from the extremely dilute to the hypersaline and in all climatic settings from the arid tropics to Antarctica (Parker, Simmons et al., 1991). Calcite-, Mg-calcite- and aragonite-bearing varieties are known. Their common association with the littoral zone makes microbialite occurrences sensitive indicators of palaeoshorelines. They have been particularly valuable for tracing previous lake levels in a number of existing lake basins (e.g. Hillaire-Marcel & Casanova, 1987; Casanova & Hillaire-Marcel, 1992), while the alternation of dark and light growth laminae in some Quaternary stromatolitic varieties is interpreted as a seasonal signal and has been used to reconstruct variations in the relative intensity of dry and wet seasons (Casanova, 1986).

In hardwater lakes with little clastic input, extensive shallow-water carbonate deposits (commonly referred to as *marl*) can be formed from the accumulation of biogenic and inorganic carbonates. In Littlefield Lake, Michigan (Murphy & Wilkinson, 1980), a flat, shore-attached platform is covered with oncoids and bordered by a swamp accumulating peats. The platform passes lakeward into a marl bench slope, extensively colonized by *Chara*, and the site of algal mud accumulation. In this lake the slope facies grades into gastropodal and ostracodal muds in deeper water (c. 10 m) (Fig. 4.10), but in other waterbodies

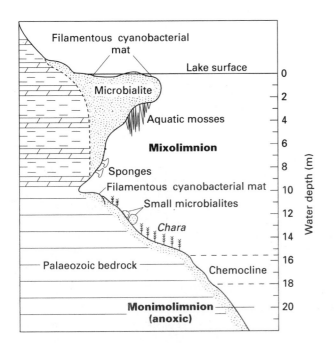

Figure 4.9 Section through a microbialite ledge currently forming at the margin of Green Lake, New York. The lake is stratified and anaerobic bacteria are the only life beneath the chemocline (after Thompson & Ferris, 1990).

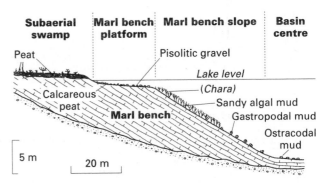

Figure 4.10 Typical cross-section through the margin of Littlefield Lake, Michigan, showing the morphology of the marl bench and the distribution of modern facies (after Murphy & Wilkinson, 1980).

there can be an abrupt change in gradient at the platform margin, causing large amounts of shallow-water carbonate material to be redeposited into deeper water by turbidity and mass flows (Dean, 1981). Marshes can form in shallows and on the deltas of inflowing rivers, such that carbonate sedimentation is associated with the accumulation of large quantities of organic matter that form peats (Treese & Wilkinson, 1982). Similar conditions exist in the Florida Everglades, where there are some very extensive swamps (Platt & Wright, 1992; Fig. 4.11). Here seasonal and longer-term periods of low water cause extensive desiccation of the sediments.

Significant carbonate accumulations also occur in lakes that receive abundant clastic sediment, typically where clastic sediment bypasses parts of the basin. In this way, gastropod coquinas, ooid sand shoals, *Chara* and *Potamogeton* 'meadows', which are the source of *Chara* – ostracod sands and silts, and microbial lithoherms, form extensive areas of varied carbonate sediment accumulation on some of the marginal platforms of Lake Tanganyika (Tiercelin, Soreghan *et al.*, 1992; Fig. 4.8).

4.7.2   Siliceous deposits

Large quantities of biogenic opaline silica are accumulating in some modern lakes. Diatoms, a major group of lacustrine phytoplanktonic organisms, are the principal source of this silica, although sponges may locally make a contribution. Diatoms comprise a diverse group of organisms which tolerate a very wide range of conditions, ranging from very dilute to hypersaline.

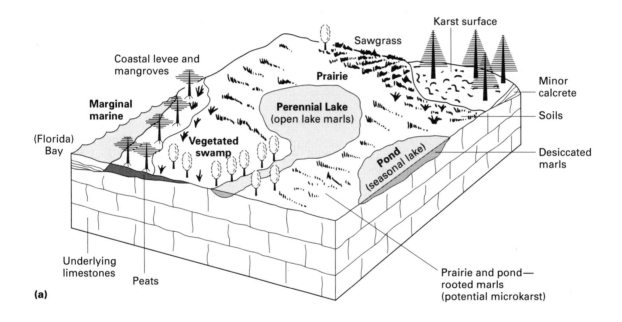

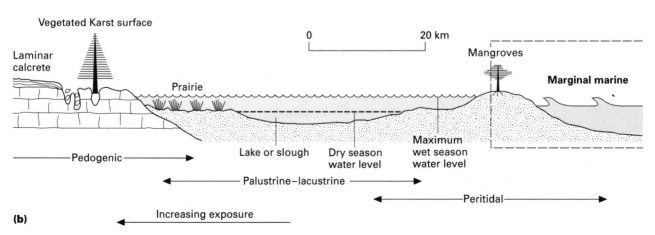

Figure 4.11 Simplified block diagram (a) and section (b) from the Everglades area of Florida, a modern example of an environment where palustrine carbonates form. The coastal levee and mangrove swamps are not essential parts of the palustrine environment; many ancient palustrine limestones formed in areas remote from any marine influence (after Platt & Wright, 1992).

Where silica supplies are not limiting, diatoms can outcompete other planktonic algal species to become extremely abundant and major sediment contributors. Lakes, or areas of lake floor starved of clastic sediment, may accumulate considerable thicknesses of diatomaceous ooze interrupted only by rare turbidites (e.g. Lake Malawi; Owen & Crossley, 1992) and seasonal plankton blooms can produce diatom-rich laminae within other types of sediments. A characteristic feature of some blooms is the tendency for diatoms, especially those with acicular frustules such as species of *Nitzschia*, to aggregate into 'mats' that sink to form distinctive laminae with a 'felted' fabric; faecal pellets rich in diatom debris cause muds to have a flaky texture (Tiercelin, Soreghan *et al.*, 1992). Lakes fed by silica-rich groundwaters are especially favourable sites for the accumulation of diatomaceous sediments. This is the case in flooded interdune depressions; the position of many former lakes of Holocene age in the southern Sahara, for example, is marked by extensive deposits of *diatomite*, the lithified equivalent of diatomaceous muds.

As the solubility of opaline silica increases sharply with rising pH, alkaline lake waters are agressive to diatom frustules and few traces of these organisms may be found in the sediments of such lakes, even if they were a dominant component of the phytoplankton flora.

In lakes associated with faults and active hydrothermal systems, alkaline hot springs can be an additional source of siliceous deposits. At Lake Bogoria, Kenya, extensive Holocene opaline chert deposition has occurred in association with saline hot springs. Silica precipitation resulted from a rapid drop in temperature, and possibly pH, as spring waters emerged at the lake floor (Renaut & Owen, 1988).

4.7.3 Iron-rich deposits

Although not widespread, sediments rich in iron minerals occur in some lakes. Iron-bearing ooids (up to 49% Fe_2O_3) are found at water depths of c. 1–3 m off the Chari Delta, Lake Chad (Lemoalle & Dupont, 1976). The iron is thought to be derived from regolith and transported in colloidal or adsorbed form as part of the suspended load of incoming rivers. In the lake it coprecipitates with silica (also a weathering product). In Lake Malawi, ferro-manganous ooids and pisoids occur at depths of 80–160 m (Williams & Owen, 1992). Growth of these is evidently relatively rapid; some sites where concretions up to 14 mm in diameter have formed became submerged by the last major transgression of the lake only 150–500 years ago. The concretions are of early diagenetic origin, but may be concentrated at the surface due to winnowing or low net accumulation rates.

4.7.4 Saline minerals

Saline minerals form *after* any Ca–Mg carbonates have precipitated from a waterbody. They are typically the result of evaporative concentration, but temperature change can also cause the precipitation of some common minerals. Saline minerals are found in three principal environments: (i) in perennial brine bodies or *saline lakes*; (ii) as efflorescent crusts on and around ephemeral *salt pans*; and (iii) as *cements* within sediment deposited in and around these waterbodies. Whereas perennial *saline lakes* receive inflow from at least one perennial river or spring, ephemeral *saline pans* are fed only from episodic run-off, although springs and groundwater may contribute significant inflow. The latter appear to have been both the commonest and most prolific sites for non-marine evaporite formation during the Holocene (Lowenstein & Hardie, 1985). Salts may come from normal surface or shallow groundwaters, from hydrothermal sources, from recycling of older evaporite deposits, or be transported to the basin as aeolian dust and aerosols. Concentration to the point where saline minerals begin to precipitate can occur entirely within the lake, as in the Great Salt Lake, a perennial saline lake which receives dilute river water direct from the surrounding mountains (Eugster & Hardie, 1978). In other basins, such as Death Valley and Bristol Dry Lake, California, and many central Australian playas (Arakel & Hongjun, 1994), much of the carbonate and gypsum is precipitated *before* the inflow reaches the saline pan, at springs or zones where groundwater seeps from the toes of alluvial fans.

Gypsum is commonly the first mineral to form after the Ca–Mg carbonates, but it precipitates only if the alkaline earth elements have not previously been depleted by carbonate production. Thereafter, the saline minerals that precipitate depend upon the initial composition of the inflow, which is in turn a function of the geology of the catchment. Typically the saline water or *brine* evolves along one of several pathways, precipitating a characteristic suite of minerals along the way (Fig. 4.12); the mineral assemblage in ancient saline lake deposits thus preserves a record of the composition of the inflow to the basin.

In lakes the modes of formation of saline minerals are similar to those in marine hypersaline lagoons and salinas (Smoot & Lowenstein, 1991; Sect. 8.5.2). Laminites of gypsum occur in some perennial saline lakes, but thick salt accumulations are more commonly of halite or sodium carbonate and sulphate minerals (Fig. 4.12). Drilling in connection with economic minerals exploration has demonstrated that most modern saline lakes have experienced periods of more dilute water composition, usually due to climatic variations. Ephemeral saline pans alternate between: (i) periods of flooding, when a temporary, brackish lake covers the pan; (ii) evaporative concentration which results in a saline lake; and (iii) desiccation (Fig. 4.13). The resulting deposits are cyclic and consist of alternating millimetre- to decimetre-scale layers of halite and mud (Lowenstein & Hardie, 1985), the muds being mainly introduced during brief periods of flooding. Thick deposits of saline minerals may accumulate in this way – several hundred metres

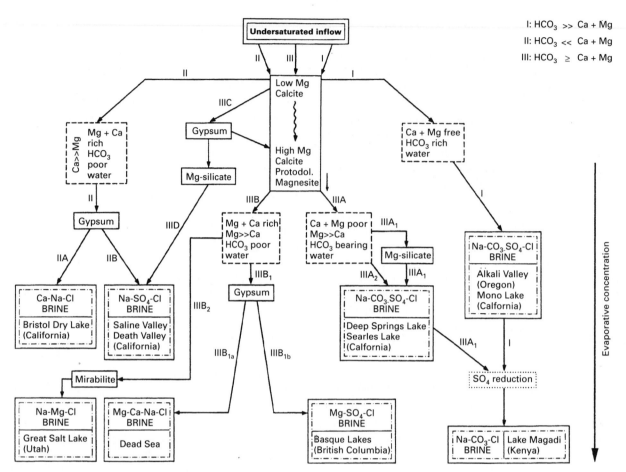

Figure 4.12 Flow diagram for continental brine evolution. Solid rectangles represent critical precipitates; rectangles with dashed borders are typical water compositions. Final brine types together with examples of salt lakes are surrounded with dash-dot rectangles. Three major geochemical paths are indicated. The first critical step is the precipitation of calcite. Waters initially enriched in HCO_3 compared to Ca + Mg travel along path I towards alkaline brines, such as in Alkali Valley, Oregon. Waters initially depleted in HCO_3 compared to Ca + Mg travel along path II toward Ca–Na sulphate–chloride brines, as in Bristol Dry Lake and Death Valley, California. Waters with intermediate HCO_3/(Ca + Mg) ratios (path III) initially precipitate low-Mg calcite followed by high-Mg calcite, then protodolomite and eventually magnesite. Depending on the relative abundance of alkaline earths and bicarbonate, further evaporation follows path IIIA (e.g. Deep Springs, California) or IIIB (e.g. Dead Sea) (after Eugster and Hardie, 1978).

of salt have been proven beneath some of the saline pans of the western USA (Lowenstein & Hardie, 1985). These commonly show a crude concentric zoning of minerals, with the most soluble being found in the centre, where they may become concentrated by repeated cycles of dissolution and reprecipitation (e.g. Deep Springs and Bristol Dry Lake, California; Jones, 1965; Rosen, 1991). The crystals of saline crusts grow both displacively and as void-fillings. Repeated precipitation ultimately leads to the formation of large upthrust ridges, as in Death Valley. Subsequent flooding may smooth out the ridges and if the water table is too deep for capillary evaporation to be effective, the old lake floor remains very flat. The Bonneville saltflats, along the western margin of the Great Salt Lake, are a good example of the operation of this process.

A significant proportion of the salts that ultimately accumulate in perennial saline lakes and ephemeral pans is of early diagenetic origin. Some of the most soluble salts, such as the potassium minerals, may be almost exclusively preserved as diagenetic phases (e.g. Casas, Lowenstein et al., 1992). There are two principal reasons for the importance of early diagenetic precipitates in saline mineral deposits. Primary precipitates commonly have significant intercrystalline porosity, due to uneven crystal growth or the loose packing of crystal cumulates. Furthermore, contact with undersaturated waters during floods or humid climatic intervals produces extensive dissolution cavities. All these voids provide sites for the subsequent precipitation of salts from saline pore fluids (Fig. 4.13). In soda lakes, diagenetic precipitation of minerals like nahcolite ($NaHCO_3$)

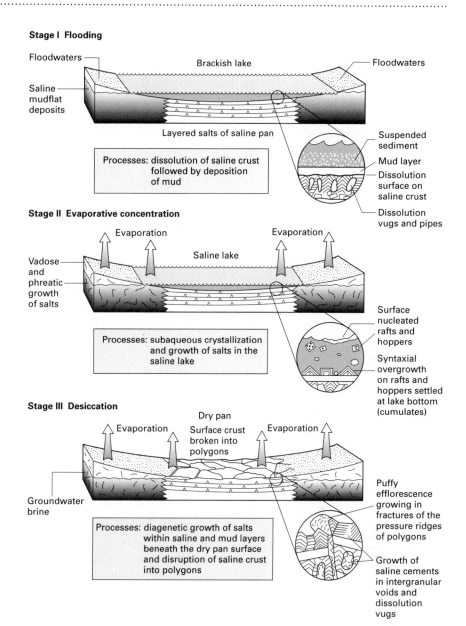

Stage I Flooding

Floodwaters

Brackish lake

Floodwaters

Saline mudflat deposits

Layered salts of saline pan

Processes: dissolution of saline crust followed by deposition of mud

Suspended sediment

Mud layer

Dissolution surface on saline crust

Dissolution vugs and pipes

Stage II Evaporative concentration

Evaporation

Evaporation

Saline lake

Vadose and phreatic growth of salts

Evaporation

Evaporation

Processes: subaqueous crystallization and growth of salts in the saline lake

Surface nucleated rafts and hoppers

Syntaxial overgrowth on rafts and hoppers settled at lake bottom (cumulates)

Stage III Desiccation

Dry pan

Evaporation

Surface crust broken into polygons

Evaporation

Groundwater brine

Processes: diagenetic growth of salts within saline and mud layers beneath the dry pan surface and disruption of saline crust into polygons

Puffy efflorescence growing in fractures of the pressure ridges of polygons

Growth of saline cements in intergranular voids and dissolution vugs

Figure 4.13 Summary of the cyclic development of saline pan deposits. Salts accumulate by settling of crystals from the water surface (crystal *cumulates*), growth at the sediment–water interface, and diagenetic precipitation in dissolution and intercrystal voids and in desiccation cracks (after Lowenstein & Hardie, 1985).

or trona (NaHCO$_3$. Na$_2$CO$_3$.2H$_2$O) is favoured by high concentrations of carbon dioxide produced by the degradation of organic matter (Eugster & Hardie, 1978). As sediments rich in organic matter typically accumulate during periods of dilute water (e.g. Renaut & Tiercelin, 1994), the presence within them of diagenetic saline minerals may give a misleading impression of the conditions under which the organic-rich unit accumulated.

Because precipitation of many saline minerals requires the evaporative concentration of surface waters, the formation of saline mineral deposits tends to be associated with hot, arid climatic conditions. While many do form in such environments, cooling a brine may also bring it to supersaturation with

respect to some common minerals. As low temperatures are a normal feature of subtropical and high latitude arid continental interiors, diurnal or seasonal cooling can be an important process in the formation of salt crusts in some basins. Thus, mirabilite (Na$_2$SO$_4$.10H$_2$O) precipitates extensively during cold periods in the Great Salt Lake and the lakes of central Australia and Canada. Warming may cause mirabilite to dehydrate to thenardite (Na$_2$SO$_4$) or simply to melt. In some saline lake basins, repeated cycles of crystallization and melting probably contribute to the extensive development of massive structureless muds.

Carbonate, sulphate and chloride minerals are the commonest saline minerals, reflecting the importance of these anions in continental waters. Authigenic silicate minerals, notably zeolites, are also relatively common in some saline lake deposits and are particularly characteristic of alkaline lakes in volcanic regions. Volcanic glass reacts readily with the alkaline brine to form zeolites or feldspar (Eugster & Hardie, 1978), but glass is not essential, as zeolites also occur in lakes remote from any volcanicity. In these settings siliceous microfossils such as diatoms have probably been the starting point for silicate mineral formation. In some soda lakes, one of the earliest silicates to form is magadiite ($NaSi_7O_{13}(OH)_3.3H_2O$) which occurs as nodules with a rubbery or gel-like consistency and a characteristic mammilated surface. Magadiite is relatively unstable and readily decomposes, especially on exposure to more dilute waters, to form chert. The chert inherits the morphology of the magadiite gel to form distinctive beds and nodules known as *magadi-type chert* (Eugster, 1969, 1980). These have a high preservation potential and are valuable indicators of alkaline waters in ancient lacustrine deposits (Sect. 4.13.1).

4.7.5 Organic matter

Lakes can be important sites for the accumulation of organic matter. Lacustrine deposits as a whole have contents of organic matter that are significantly above the average for sediments and sedimentary rocks in general. Contents of 1–5% (by weight) are common and in exceptional circumstances may be as high as 40–50%. Lacustrine peats can be composed of almost pure plant remains. The organic matter has three principal sources: terrestrial vegetation, marginal macrophyte swamps, and phytoplankton. During the closing phase of their history, many lakes become the sites of extensive swamp development, such that the sediment fill is capped by a bed of peat, mostly formed from *in situ* plant growth. Fluvial transport can also be an effective means of supplying organic matter to a lake. In Sucker Lake, Michigan, fluvially supplied allochthonous plant debris has formed a delta that is in many ways analogous to typical terrigenous lacustrine Gilbert deltas (Treese & Wilkinson, 1982).

Phytoplankton remains are most abundant in the open-water, offshore sediments of large lakes, in waterbodies (or parts of waterbodies) receiving little clastic supply, in the central areas of large, crustal sag basins (e.g. Lake Victoria), and where low slope gradients do not favour the widespread distribution of terrestrial material by density currents (Talbot, 1988). Where the lake margin is steep, density flows can carry terrestrial and marginal plant debris far offshore. The organic fraction of offshore muds in many large rift lakes is therefore a mixture of macrophytic and planktonic remains (e.g. Huc, Le Fournier *et al.*, 1990).

The preservation of large amounts of organic matter in lakes (and the ocean) is dependent upon rates of organic productivity, anoxic bottom conditions, or a combination of the two. The occurrence of anoxia is determined by the relative rates of oxygen supply and its consumption by decomposition of organic matter. When the supply of reactive organic matter exceeds the rate at which oxygen can be replenished, anoxic conditions will result. Periodic or permanent stratification is particularly effective at restricting oxygen renewal in the hypolimnion. However, in highly productive waterbodies, neither stratification nor permanently anoxic conditions are essential for the preservation of large amounts of organic matter (Talbot, 1988).

4.8 Rhythmites

Rhythmites are sequences of finely laminated, regular alternations of two or more contrasting sediment types. They are probably the most characteristic lacustrine facies and in their commonest development, as horizontal, millimetre- to submillimetre-thick laminae may also be called *laminites*. *Couplets* and *triplets* are rhythmites composed of either two or three lamina types respectively, and combinations of even more different laminae are known. The component sediments may be combinations of clastic, chemical or organic deposits rich in organic matter.

If individual rhythms can be shown to represent a single year's sediment accumulation, they are usually called *varves*. These are by far the best-known rhythmites, but non-annual rhythmites are by no means uncommon (see below). Classic varves, as originally defined by de Geer (1912), comprise couplets of clastic sediment which differ in colour and texture. They are the result of glacially influenced sedimentation (Sect. 11.4.5). The light, silt or sand layer forms from spring and summer meltwater and river run-off, while the dark, clay-rich layer is deposited during winter, when the lake is ice-covered. Critical to the development of rhythmites of this sort is the temporal separation of coarse- and fine-sediment deposition. In the case of classic varves this is achieved by winter freezing of rivers and the lake surface, cutting off supplies of coarser clastic sediment while providing a calm environment in which clay-grade material can settle from suspension. This separation may, in fact, be achieved in several different ways, by no means all of which involve glacial influence, and not all of which necessarily lead to the development of varves. In many stratified lakes suspended sediment carried by over- and interflows may be trapped at the thermocline (Sect. 4.6.3), to be released during the winter overturn. In unstratified lakes no such trapping can occur and grain-size separation is much less clear-cut. Sturm (1979) has reviewed the influence of various combinations of suspended sediment supply and stratification regime on clastic rhythmite formation. He suggests that the ideal clastic varve is formed in lakes typical of temperate, high-latitude and alpine regions that go through a seasonal cycle of stratification and mixing. Suspended sediment is introduced episodically during the period of stratification and released to the lake floor during the subsequent overturn event when it forms a clay blanket covering much of the lake floor (Fig. 4.14). In contrast to the clay layers, which tend to occur as laminae

Stratification	Suspended sediment supply		Depositional features

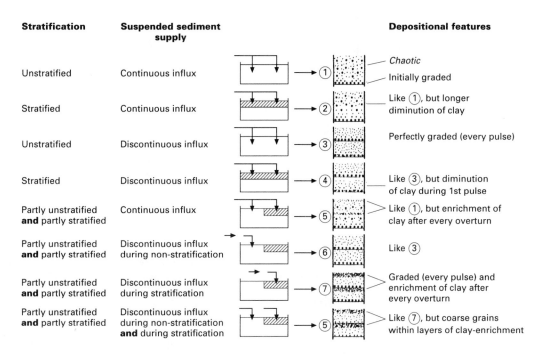

Figure 4.14 Lamination style in offshore areas of an oligotrophic lake depends upon stratification regime and suspended sediment supply. Case 7 represents the structure of an ideal clastic varve (heavy dots, sand; light dots, silt; dashes, clay) (after Sturm, 1979).

of uniform thickness and great lateral extent, the coarser units can be much more variable. Thickness and texture varies, depending upon distance from source; in proximal areas, close to river deltas, beds can be tens of centimetres thick and relatively coarse. They can also comprise several discrete depositional units, representing the sediments of individual density or turbidity flow events. In areas distant from inflowing rivers, the spring and summer silt layer is thinner and finer grained; the deposits of individual flow events may not be distinguishable (e.g. Ashley, 1975; Sturm & Matter, 1978).

Seasonal or episodic changes between a stratified and unstratified state can lead to the development of rhythmites which are very different from the classical varves. During the period of stratification, the hypolimnion may become enriched in ions that are stable only under anoxic conditions. During overturn, mixing with oxic waters can cause these ions to precipitate as insoluble phases. Iron and manganese are two elements that are particularly commonly involved in processes of this sort; the former, in particular, can be the source of a prominent iron oxide and hydroxide lamina following a mixing event (Fig. 4.15). A variant of this can occur in lakes where a zone of photosynthetic sulphide-oxidizing bacteria occurs at the chemocline. These anaerobic bacteria concentrate elemental sulphur in their cells. They are killed by overturn events and the introduction of large quantities of sulphur and organic matter to the sediment may form a dark, iron- sulphide- and organic-rich lamina (Dickman, 1985; Fig. 4.15).

Changes in solar radiation can lead to the development of yet other sorts of rhythmites. Primary biological production at mid- to high latitudes is largely controlled by the seasonal solar radiation cycle. The burst of productivity that occurs in spring and summer can produce a distinctive biogenic lamina that is rich in organic matter or diatom remains (Fig. 4.15). In hardwater lakes, this increased photosynthetic activity finds its most prominent expression in the formation of a spring–summer carbonate layer (Sect. 4.7.1), which can form couplets with clastic, organic or biogenic laminae (Fig. 4.15).

At low latitudes, seasonal variations in solar radiation are relatively small; the light reaching the surface of tropical and subtropical lakes is always more than adequate for photosynthesis. Thus other factors will normally determine whether or not there are regular seasonal (or longer term) variations in lake sedimentation. Upwelling of nutrient-rich waters, perhaps related to a windy season, may trigger phytoplankton blooms. Processes controlling the supply of terrigenous material may also contribute to the development of rhythmites. Many tropical rainfall regimes are highly seasonal in character and may provide an annual pulse of clastic sediment. However, bimodal seasons are also common, and in many tropical areas there is a tendency for much of the rain to fall in a few, intense storms. Not all rhythmites can therefore automatically be assumed to be varves.

The accumulation and preservation of thick rhythmite sequences requires a special combination of conditions. There must be both periodic variation in the nature of the sediment

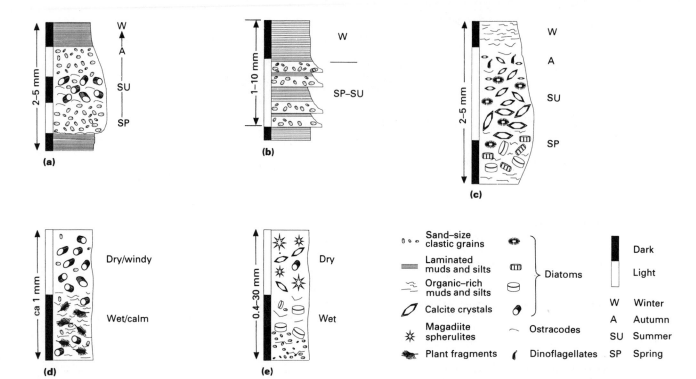

Figure 4.15 Some rhythmite types showing lithologies and colour changes and relationship between compositional variations and seasonal or longer-term climatic change. (a) Varve from a subarctic Swedish lake (after Renberg, 1981); (b) Varve from a glacial lake (after Smith & Ashley, 1985); (c) non-glacial varve from a temperate hardwater lake (Lake Zürich, after Kelts & Hsü, 1978); (d) non-glacial varve from tropical Lake Malawi (after Pilskaln & Johnson, 1991); (e) rhythmite from the Upper Pleistocene section in Lake Magadi, Kenya. Each couplet represents c. 2–3 years accumulation (after Damnati, Taieb & Williamson, 1992).

reaching a lake floor and favourable conditions for rhythmite preservation (Kelts & Hsü, 1978). These may be one or more of the following.

1 An absence of bottom-dwelling organisms that graze or churn the sediment. This is most likely to occur under oxygen-deficient conditions, although permanent anoxia is not essential.

2 High sedimentation rates even in oxic bottom conditions (e.g. Lake Turkana; Cerling, 1986).

3 Minimal bottom current activity that disturbs newly deposited, soupy lake muds.

4 A relatively extensive flat floor below wave base so that incoherent sediment will not readily creep or slide downslope.

5 Minimal biogenic gas bubble generation, so that the lamination is not disrupted by bubble migration once the rhythmites have become buried beneath the sediment–water interface.

4.9 Lake-level changes

One of the most characteristic features of lakes is the tendency for lake level to oscillate in response to variations in hydrological budget. Such oscillations have a major impact upon facies distribution within lacustrine basins and are one of the principal reasons why lakes can preserve outstanding records of past climate. Mapping the history of lake-level variations in tropical African lakes during the late Quaternary indicates that virtually all lakes were low and hydrologically closed during the late Pleistocene, very high during the early Holocene and have oscillated at somewhat lower levels since c. 4000 years BP (Street-Perrott & Harrison, 1984; Fig. 4.16). During the period of maximum lake levels (c. 10 000–5000 BP), innumerable perennial lakes, including some very large waterbodies, occupied interdunes and other depressions in what is now the southern Sahara desert. These changes have been essentially synchronous across the continent and have occurred in lakes of several different origins, indicating that the primary control has been climatic, principally variations in inflow versus water losses through evaporation. A similar pattern is apparent in the lakes of subtropical Australia (e.g. Lake Eyre and its precursor Lake Dieri; Dulhunty, 1982). Tropical lake levels have apparently varied in unison with astronomically forced changes in global climate (Kutzbach & Street-Perrott, 1985). Dry conditions, with low lake levels in the tropics and subtropics, are contemporaneous with cooler, glacial conditions at high latitudes;

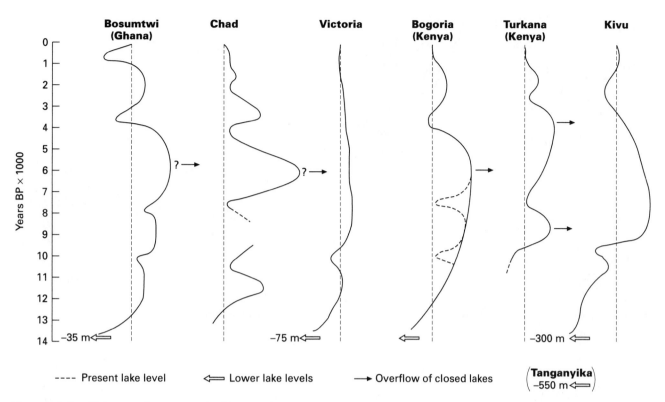

Figure 4.16 Late Pleistocene–Recent water-level history of various African lakes (from Talbot, 1988).

humid periods, with positive water balance and thus high lake levels, coincide with interglacials. Many lakes in present-day mid-latitude dry belts have also periodically been of much greater extent. Expanded ('pluvial') lakes existed in North America (e.g. Lake Lahontan, Great Salt Lake and precursor Lake Bonneville); South America (e.g. Titicaca and precursor Lake Ballivián) and Asia (Aral–Caspian Sea, Lake Van, Dead Sea and precursor Lake Lisan). In contrast to the African and Australian examples, these lakes appear to have expanded during glacial maxima, reflecting the combined effects of increased run-off and lowered evaporation rates due to depressed temperatures (Flint, 1971; G.I. Smith & Street-Perrott, 1983; Phillips, Campbell et al., 1992).

Open lakes, such as in the Great Lakes of North America, are characterized by relatively stable shorelines because inflow plus precipitation is balanced by outflow plus evaporation, the outflow acting as a buffer against extreme fluctuations in lake level. In tropical lakes, such as the large, hydrologically open East African lakes, direct evaporation may nevertheless be the main cause of water loss, inducing considerable seasonal or longer-term lake-level fluctuations (Beadle, 1981; Livingstone & Melack, 1984). Shoreline fluctuations may also result from such effects as glacioisostatic rebound following glaciations; the northern shore of Lake Superior is rising relative to the south by as much 0.46 m per 100 years (Kite, 1972) and the outlet of Lake Ontario is rising at about 0.37 m per 100 years

(Sly & Lewis, 1972). Other lakes, such as Lake Maracaibo, Venezuela, are directly connected to the sea which also serves as a control on lake level. An unusual situation exists in Pitt Lake, British Columbia (Ashley, 1979) where lake levels are controlled by tidal flows originating from the Fraser River estuary.

Closed lakes possess a net water budget in which loss by evaporation and infiltration commonly exceeds inflow plus precipitation. Subtle changes in water budget are reflected in substantial changes in lake levels (e.g. Spencer, Baedecker et al., 1984) and lake water compositions. In shallow, sag lakes, shoreline positions are very mobile, as for example in Lakes Chad (Servant & Servant, 1970), Victoria (Talbot & Livingstone, 1989) and Great Salt Lake (Benson, Currey et al., 1990), and the advance and retreat of facies belts may give rise to successive transgressive–regressive cycles in the sedimentary record. Sediments are generally a complex mixture of river-derived detritus, clastic material cannibalized from older, exposed lake beds and chemical and biochemical components. In rift-basin lakes lateral changes in shoreline position may be relatively subdued, because of the steeper lake basin morphology, but vertical changes can be dramatic. During the late Pleistocene, the levels of Lakes Malawi, Kivu and Tanganyika were possibly as much as 350 m lower than today (Haberyan & Hecky, 1987; Scholz & Rosendahl, 1990). These fluctuations can be extremely rapid; Lake Malawi, for example, is believed to have dropped as much as 140 m in the space of just 340 years (Owen, Crossley

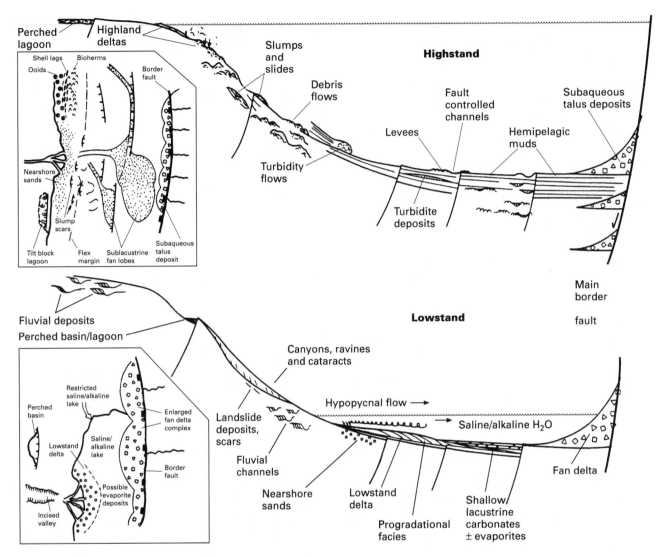

Figure 4.17 Sediment accumulation in an idealized, low-latitude rift (half-graben) lake during (a) humid, highstand conditions, and (b) arid, lowstand conditions (after Scholz, Rosendahl & Scott, 1990).

et al., 1990). Such large oscillations lead to deep incision of highstand deposits and the exposed sediments may be severely affected by subaerial weathering and soil formation. Abrupt vertical facies changes are another consequence; lowstand deltaic deposits, for example, may prograde rapidly over basinal sediments, which in turn may accumulate directly upon an exposed surface if the lake deepens abruptly (e.g. Scholz, Rosendahl & Scott, 1990; Fig. 4.17). One further consequence of large-scale changes in level is that some lakes are divided into two or more isolated waterbodies during lowstands. Lake Tanganyika, a single waterbody today, became three separate, closed lakes during periods of the Pleistocene (Fig. 4.18). In contrast, the Dead Sea and Lake Kinneret (Sea of Galilee) are today separated by c. 100 km of the Jordan Valley, but in the late Pleistocene were united within the very large Lake Lisan

(Horowitz, 1979). At times of separation each sub-basin may have a very different chemical, sedimentological and biological evolution. The present Lake Kinneret is a fresh, hydrologically open lake accumulating a mixture of carbonate and clastic sediments, while the Dead Sea is devoid of macrofauna, closed, hypersaline and dominated by evaporitic sedimentation (Sect. 8.6.2).

Given the rate and scale of climatically induced lake-level variations, it is not surprising to find that changes in lake level have had a major impact upon the sedimentary history of many ancient lakes.

4.10 Ancient lake deposits

Ancient lake sediments are preserved in a variety of tectonic

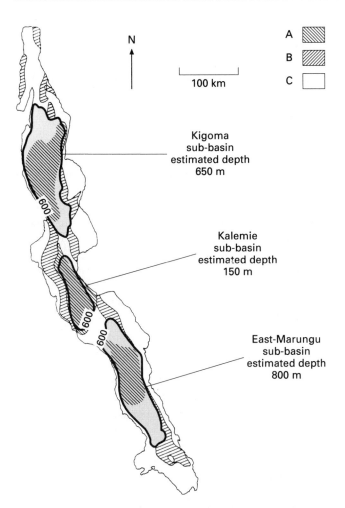

A
B
C

Kigoma
sub-basin
estimated depth
650 m

Kalemie
sub-basin
estimated depth
150 m

East-Marungu
sub-basin
estimated depth
800 m

N

100 km

600
600
600

Figure 4.18 Subdivision of Lake Tanganyika into three isolated sub-basins as a result of a Mid- to Late Pleistocene lowstand. Facies A, which appears as a highly continuous unit with parallel reflections on seismic sections, probably accumulated under open lake conditions. Over large areas of the lake it is truncated by seismic zone B, suggesting subaerial exposure and extensive erosion during a prolonged period of low water. The 600-m depth contour (C) shows the inferred extent of three separate lowstand lakes (after Tiercelin & Mondeguer, 1991).

settings: (i) in extensional rift basins, such as the Triassic basins of northeastern USA (Van Houten, 1964; Olsen, 1990), the Permo-Triassic (Karoo) of southern Africa (Yemane, 1993) and the Early Cretaceous of the South Atlantic margins; (ii) in strike-slip basins, such as the Pliocene Ridge Basin of California (Crowell, 1974a,b; Crowell & Link, 1982); (iii) in foreland basins like the intermontane Green River Formation lakes and the Oligocene lakes of the eastern Ebro Basin, Spain (Anadón, Cabrera *et al.*, 1989); (iv) in cratonic basins related to lithospheric stretching and thermal subsidence such as the Chad and Victoria basins and the middle Proterozoic of North Greenland (Collinson, 1983); and (v) interdune and

dune-dammed depressions, such as the Quaternary of the Sahara and some Mesozoic cratonic basins of the western USA.

4.10.1 Criteria for recognition of ancient lake deposits

Ancient lake deposits are directly identifed with the help of three groups of criteria: faunal, chemical and physical, or indirectly from their association with other demonstrably continental facies, particularly fluvial or aeolian deposits and palaeosols (see Feth, 1964; Picard & High, 1972, for reviews).

The first group includes an absence of a marine fauna represented by the stenohaline invertebrate groups such as corals, articulate brachiopods, echinoderms, cephalopods and byrozoans. Euryhaline faunas typical of lagoons are also absent. Instead, a non-marine flora such as some algae, and fauna such as certain gastropods, bivalves, ostracods and fish may be present (Heckel, 1972). Because lakes in closed basins are sensitive to frequent climatic changes, faunas may be under high stress and thus of low diversity. However, faunal diversity must be used with great caution as a criterion for distinguishing marine and lacustrine environments. Many of the existing large, ancient lake basins, such as Lake Baikal and the lakes of East Africa, contain extremely diverse faunas, including exceptionally rich species flocks of fish. It is probable that climatically and tectonically induced variations in these lakes, far from causing a stressed environment, have provided the rich and varying habitats that allowed such striking evolutionary diversity.

Physical processes recorded in ancient lake sediments are similar to those associated with marine environments. However, tidal currents are lacking, wave activity is reduced and subaerial emergence common, reflecting the frequent, even annual, oscillations in lake level and in the position of the shoreline.

The composition of saline deposits may be diagnositic because certain unusual minerals or mineral assemblages can only form from waters whose chemistry is very different from that of normal sea water. Fluctuations in chemical composition and salinity are also much more rapid and substantial than in the oceans. The most reliable criteria for confirming lacustrine deposits are geochemical. Continental waters have a much wider range of oxygen isotopic compositions than the ocean. This isotopic signature can be preserved in the oxygen contained in primary lacustrine carbonates such as molluscs, ostracods, microbialites and inorganic precipitates. Oxygen isotopic analysis of such carbonates can thus confirm whether they formed in marine or non-marine waters. The strontium isotopic composition of marine minerals such as carbonates and evaporites has varied in systematic fashion over at least the last 550 My (Smalley, Higgins *et al.*, 1994). The range of possible strontium isotopic compositions is, however, much greater in continental waters, which vary locally, depending upon the geology of the catchment. Thus, if a carbonate or evaporite mineral of known age has a primary strontium isotopic composition that falls outside the range of marine strontium for that period, it is likely to have formed in non-marine waters.

4.10.2 Ancient lacustrine facies

A wide range of facies has been identified in ancient lacustrine sequences, even within a single lake basin. A number of contrasting facies may develop. This is because water composition and lake level fluctuate, and there are frequent environmental changes. Lacustrine sequences of all ages also commonly show a cyclic arrangement of facies. Smaller-scale cycles may reflect expansion and contraction of lake margins. Larger-scale cycles may reflect changes in the entire hydrological status of the lake. Investigation of cyclic facies variations in ancient lacustrine basins has provided detailed information on past continental climate variations. Tectonic setting also has an important influence upon facies development. In strike-slip and rifted basins, facies may change rapidly in a lateral sense. In sag basins where the margins slope gently, facies can have a very wide lateral extent but may change rapidly in a vertical sense, as small changes in lake volume give rise to large changes in lake area and hence the position of the shoreline.

The sedimentary fill of ancient lake basins can be divided into four main facies associations: *clastic-dominated, carbonate-dominated, evaporite-dominated* and *organic-matter-dominated*. Because of variations in the state of the lake through time, facies typical of each association may occur within a single basin.

4.11　Ancient clastic-dominated basins

Many ancient lakes accumulated sediments that were almost entirely of siliciclastic origin. Sediment was mainly supplied by rivers, although glacial sources were important to some high latitude lakes (Sect. 11.4.5) and wind-blown sand and silt contributed to the fill of basins in arid climatic settings. Coarser clastic deposits (conglomerates and sandstones) occur in three main settings: (i) as cross-stratified units representing the mouth bar and foreset sediments deposited in association with fluvial inflows; (ii) relatively well-sorted, ripple and hummocky cross-laminated sandstones deposited in the littoral and nearshore zone; and (iii) as matrix-rich conglomerates and graded sandstones deposited at the foot of fault scarps and in offshore areas. Mudstones are commonly associated with these coarser deposits and may also constitute the major rock type in some basins. They may be of either shallow or deepwater origin, red, green or grey in colour, and rhythmically laminated or massive. Many dark, finely laminated mudstones are rich in organic matter.

Apart from climate and climatically induced variations in lake level (Sects 4.6 & 4.9), the factors that determine the nature and distribution of clastic facies in ancient lake basins are basin morphology and basin size. Basin morphology governs the topography of the source area, the slope of the lake floor and the absolute bathymetry of the lake. These, in turn, influence the overall character of clastic material delivered to the lake and the degree to which basinal deposits are influenced by waves. The size and orientation of the basin also affect wave-related processes through their influence on wind-fetch across the surface of the lake.

4.11.1　Lakes with steep margins

Lakes of this type tend to occur in regions of high topographic relief and thus receive abundant, coarse clastic sediment. Steep slopes can be a result of physical erosion, notably glaciation (Sect. 4.2), tectonic activity or volcanism. The slope favours rapid transfer of sediment to the basin floor and the marked bathymetric relief commonly results in abrupt lateral facies variations. In some basins collapse of the fault scarp has caused huge blocks of bedrock to slide into the lake where they are preserved as *olistoliths*, isolated within basinal mudstones (e.g. Anadón, Cabrera *et al.*, 1991; Tiercelin, Soreghan *et al.*, 1992; Renaut & Tiercelin, 1994).

THE RIDGE BASIN GROUP

The *Ridge Basin Group* (upper Miocene–Pliocene, California) accumulated in a small, wedge-shaped trough measuring only 15 km by 40 km. Despite its size, the Ridge basin contains over 9000 m of lacustrine and fluvial rocks overlying marine deposits (Crowell & Link, 1982). The basin formed at a bend in the strike-slip San Gabriel fault, which in Mio-Pliocene times was the locally active segment of the San Andreas fault system. The distribution of lacustrine facies reflects differing tectonic activity along the eastern and western margins of the basin; the western margin was a very active strike-slip zone (San Gabriel fault), with a dip-slip (listric) component, whereas the eastern margin was influenced by a series of high angle reverse faults (Crowell, 1974a,b; May, Ehman *et al.*, 1993; Fig. 4.19).

The fill of the Ridge basin is marked by the spectacular development of coarse, marginal clastic facies, while axial areas were dominated by the acccumulation of laminated muds alternating with sandy turbidites and cross-bedded and channelled sands. There were also minor developments of carbonate sediment. Facies relationships and geochemical analysis provide clear evidence for frequent, large-scale variations in the level of the Ridge basin lake.

Basin Margin Facies Associations differ along the eastern and western margins of the Ridge basin, reflecting the contrasting topographic settings of these two areas. *Poorly sorted conglomerates and breccias* (the Violin Breccia) characterize the western margin, where they have an apparent thickness in excess of 11 000 m. The deposits generally lack well-defined stratification and are devoid of fossils, apart from scattered microbial carbonate-encrusted clasts. Clasts are angular and up to 2 m in diameter; some horizons are inversely graded and matrix-supported textures are relatively common. The Violin

Breccia is restricted to a narrow zone (maximum width 1500 m) adjacent to the San Gabriel fault; in a basinward direction it interfingers with finer-grained, axial deposits. The breccia accumulated as scree, debris flow and landslide deposits, which formed small debris cones and sublacustrine fans along the active fault scarp (Fig. 4.19).

The generally lower slopes of the eastern margin are reflected in better sorting, organization and lateral persistence of the marginal facies. *Conglomerate and coarse-grained sandstone* facies dominate. These deposits formed on relatively large alluvial fans and deltas, which built into the basin from highlands associated with the zone of reverse faults (Fig. 4.19). *Well-sorted, ripple-, horizontally-laminated and low-angle, large-scale, cross-stratified sandstones* with layers of mudstone rip-up clasts occur; bioturbation is common locally, and thin *carbonate* units are also present. The latter are mixtures of ooïds, oncoids, fragments

of microbialite crusts and shell debris (molluscs and ostracods) or relatively massive, *in situ* microbialite lithoherms (Link, Osborne & Awramik, 1978). Some of the microbialite crust fragments bear the impression of a woody substrate, indicating that they formed on logs or standing trees (Talbot, 1994). The carbonates tend to occur at the tops of fining-upward clastic cycles and the substrate upon which they formed commonly carries evidence of some degree of pedogenesis. All these facies are interbedded with or may grade laterally into mudstones. The well-sorted sandstones are beach and nearshore bar deposits formed by wave reworking of fan-delta sediments, while the carbonates accumulated on clastic-starved portions of the fans following a rise in lake level; some of the microbialite crusts evidently forming on trees drowned by the transgression.

Axial facies associations are concentrated along the main depocentre of the Ridge basin, which seems generally to have been displaced towards the San Gabriel margin. Axial facies interfinger with marginal deposits originating along both the eastern and western sides of the basin. *Massive to horizontally laminated mudstones* dominate the axial zone. In detail these vary considerably in composition from clay shales to siltstones; some are dolomitic, sideritic, pyritic or contain analcime. Intervals of rhythmite lamination are present and bioturbation tends to be minor. Organic contents are up to 2.75% and fossils include plant, shell, fish and insect fragments. Interbedded with the mudstones are *sharp-based, thin- to thick-bedded graded sandstones*. The bases of individual beds are typically erosive with abundant sole marks or load structures; shale clasts are common. Some beds contain dish structures and slump folds. Close to the basin margin the tops of sandstone beds may be marked by desiccation features, oscillation ripples or rare vertebrate tracks. Both the sandstones and the associated mudstones accumulated under a variety of conditions. The mudcracked, wave-rippled and track-marked horizons are clearly of shallow-water origin, while the sharp-based sandstones and finely laminated, organic-rich mudstones accumulated under significantly deeper water conditions, perhaps in a stratified lake as much as 100 m deep. Dolomite- and analcime-bearing mudstones are thought to have formed when the lake was at intermediate depths and the waters relatively brackish due to evaporative concentration (Crowell & Link, 1982).

The abundant, fluvially supplied sediment indicates that in pluvial periods the Ridge basin received large amounts of inflowing water. Lake levels would have been high at these times. On the other hand, abrupt horizontal and vertical transitions between facies indicative of contrasting water depths suggest that lake level varied considerably and almost certainly included prolonged periods of low water (Talbot, 1994). Ultimately, as movement on the San Gabriel fault declined and then ceased, subsidence was no longer able to outpace sedimentation, the lake disappeared and the basin was filled with coarse fluvial deposits.

(a) Open lake basin

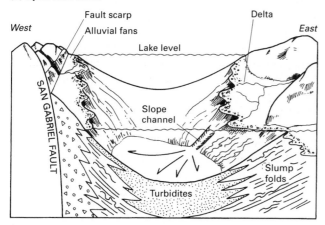

(b) Closed lake basin

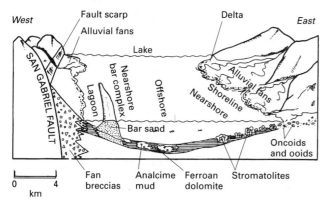

Figure 4.19 Palaeoenvironmental reconstruction of the Miocene Ridge basin lake, California, during (a) a deep-water lacustrine phase and (b) a shallow-water lacustrine phase (after Link & Osborne, 1978).

4.11.2 Wave-dominated lakes

Storms of the scale and intensity of those known from the oceans are very rare in inland waterbodies. Nevertheless, we have seen in Section 4.6 that wind-related processes can have a significant impact upon sedimentation in lakes. The commonest effects are wave reworking and transport of sediment away from the shore and river mouths. Storm deposits analogous to those seen in shallow marine sequences have been documented from a number of Pleistocene lake basins (e.g. Duke, 1984; Eyles & Clark, 1986) and an ancient example is provided by the *Kap Stewart Formation* (Rhaetian–Sinemurian) of east Greenland.

THE KAP STEWART FORMATION

The Kap Stewart Formation accumulated in a large lake (c. 80×250 km) which in Late Triassic–Early Jurassic time occupied a low-lying, fault-controlled basin along the eastern margin of Greenland. Due to its size this lake was strongly influenced by wave and storm processes; large changes in lake level also had a major impact upon facies distribution. The lacustrine succession is 185–300 m thick and contains 2 principal facies associations that formed in contrasting environments (Dam & Surlyk, 1992).

The *mudstone* facies association occurs as units 1–35 m thick and is dominated by dark grey to black, finely laminated mudstones containing up to 10% organic matter. Layers of siderite concretions and isolated sandstone beds are also present. The latter average 25 cm in thickness, are coarse to very coarse grained and laterally persistent. They commonly display large (ripple heights up to 20 cm), symmetrical wave-ripple forms. Associated fine-grained sandstones show hummocky cross-stratification.

Organic matter in the mudstones is a mixture of *Botryococcus*, other algae and higher plant remains. The well-preserved organic matter suggests accumulation under anoxic conditions, probably in a stratified lake. The rippled sandstones were probably derived from storm erosion of shallow-water sands and transported offshore by near-bottom basinward flow. The hummocky cross-bedded sandstones also suggest periodic storm activity.

The *sheet sandstone* facies association comprises variable proportions of mudstone, siltstone and sandstone. Sets of sandstone containing abundant wave-generated structures are interpreted as shoreface deposits. These are commonly overlain by medium- to coarse-grained sandstones with single foresets up to 5 m thick which are thought to have a terminal lobe origin. Medium-grained to very coarse, pebbly cross-bedded sandstones erosively overlie the lobe deposits. This subfacies tends to fine upwards, beginning with a conglomerate of logs, intraclasts and quartzite pebbles and is interpreted as the product of distibutary channel sedimentation. As a whole, the sheet sandstone facies represents a delta-front environment. It is capped by coal beds with associated rootlets or wave-dominated shoreface deposits indicative of delta abandonment. These abandonment facies are commonly abruptly overlain by open lake mudstones (Dam & Surlyk, 1993; Fig. 4.20).

Delta-front sandstones may have a gradational or erosive lower contact with deposits typical of more offshore conditions. Where the contact is sharp, upper shoreface, terminal lobe or distributary channel sandstones rest abruptly upon open-lake mudstones or fine-grained lower shoreface rocks. Contacts of this sort tend to occur in basin-margin areas, while a gradual transition from offshore to shallower water deposits is more typical of distal areas. In both cases the sandstones have a laterally persistent, sheet-like geometry.

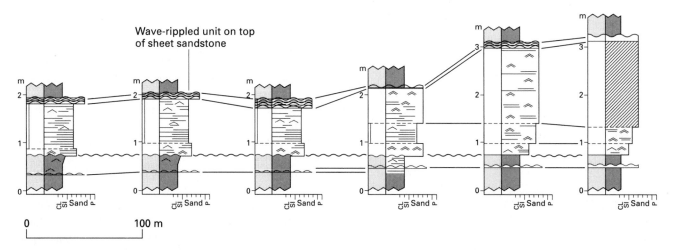

Figure 4.20 Strike section through a delta-front sheet sandstone that prograded into the Kap Stewart lake (Rhaetian–Sinemurian), East Greenland. The basal shoreface sandstone rests with sharp, erosive contact (wavy line) on the underlying open lacustrine mudstones, indicating that the delta formed during a period of falling water level in the lake. The top of the delta was subsequently reworked by waves and is abruptly overlain by open lake mudstones deposited following a rapid rise in lake level (after Dam & Surlyk, 1993).

There is a basic environmental incompatibility in the close spatial association between the two principal facies. The mudstones formed under calm, perhaps anoxic, conditions in a deep, periodically stratified lake. The coarse sediments present were reworked from shallow, marginal areas. The sheet sandstones, on the other hand, testify to turbulent, shallow-water conditions. Given the size of the Kap Stewart lake and the evidence for frequent windy conditions, it seems unlikely that the organic-rich mudstones can have formed at depths of much less than 100 m (Dam & Surlyk, 1993; cf. Sect. 4.6.3). The abrupt transition from mudstone to sheet sandstone therefore indicates a marked shallowing of the depositional environment. Dam and Surlyk suggest that this was caused by rapid, high amplitude fluctuations in lake level, the mudstones being deposited during periods of highstand, the sheet sandstones during periods of falling and low lake level, when marginal facies prograded far into the basin. The thin sequences that record an equally abrupt return to deep-water conditions above the sheet sandstones (Fig. 4.20) suggest rates of lake-level rise comparable to those caused by Late Quaternary climatic change.

4.11.3 Shallow, low-relief basins

THE MERCIA MUDSTONE GROUP

Many ancient lake deposits accumulated in basins that were neither especially deep nor particularly stormy. The fill of such basins tends to be predominantly fine grained. Typical examples are provided by the shallow lake facies that are commonly associated with floodplain and deltaic sequences (Sects 6.6.1 & 6.6.5). Others accumulated in much larger basins and under rather different conditions. An example is provided by the Triassic *Mercia Mudstone Group* of Great Britain (Ziegler, 1990), which is dominated by a characteristic assemblage of fine-grained facies, the most widespread of which is a *dolomitic red mudstone*. This is typically massive and unfossiliferous, with few macroscopic features apart from a sporadic, vague horizontal lamination and, at some horizons, irregularly distributed, centimetre-scale bodies of sandier sediment similar to the sand-patch fabric associated with some modern playas (Sect. 4.6.1). Desiccation structures are common and the rocks generally display a tendency to blocky fracturing. Associated with this facies are beds of nodular gypsum or celestine, gypsum-filled desiccation cracks and horizons with abundant halite pseudomorphs. Limestone and dolostone beds 10–30 cm thick are present locally. They commonly contain wave and current ripples, small scour depressions and brecciated horizons and are capped by one or more desiccation surfaces. Most of the limestones are micritic, but ooid grainstones and packstones containing freshwater algal remains (charophytes and *Botryococcus*) are also present (Talbot, Holm & Williams, 1994). At the margins of the basin these facies pass laterally into conglomerates and breccias that accumulated around the flanks of uplands of Palaeozoic rocks. Soil horizons typical of dry climates are common in these marginal sequences (V.P. Wright, 1992b). Gravel beaches and wave-reworked screes have also been recognized locally (Tucker, 1978). Boreholes in the central parts of several basins reveal extensive deposits of bedded halite (Arthurton, 1980).

The uniformity of the Mercia Mudstone Group over a very large area and the general scarcity of coarse-grained facies, other than those found close to the basin margins, indicates that the sediments accumulated in an area of low topographic relief. This geomorphic setting resembles the present semi-arid interior of Australia and it seems likely that the red Triassic mudstones accumulated mainly in playa lakes and on very wide alluvial plains, from ponded flood waters brought to the basin after exceptional rains (Fig. 4.21). Some of the silt-sized clastic grains in the mudstones are probably of aeolian origin. Early diagenetic

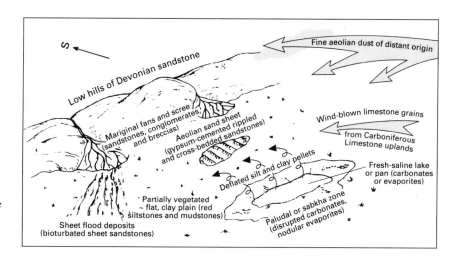

Figure 4.21 Depositional environments of the Mercia Mudstone Group (Triassic) of west Somerset, England (after Talbot, Holm & Williams, 1994).

dolomite forms in some Australian playa lakes today; much of the dolomite in the Mercia Mudstone Group is likely to have formed in a similar fashion. The limestones also have their counterpart in Australia and represent intervals of more humid climate when semipermanent, shallow, fresh to brackish water lakes developed in parts of the basin (Talbot, Holm & Williams, 1994). Accumulation of the bedded evaporites probably occurred during periods of intense aridity when the lakes retreated to the topographically lowest parts of the basins and were transformed to ephemeral saline pans.

4.12 Ancient carbonate-dominated basins

Carbonates are a major component of the fill in ancient lacustrine basins where supplies of clastic sediment have been limited or where bedrock geology has favoured a plentiful supply of dissolved calcium carbonate to the regional groundwater and fluvial systems. Both conditions are satisfied where the catchment is dominated by limestones and carbonate-rich lacustrine sequences are therefore particularly common in regions that previously had a long history of marine limestone accumulation.

Most well-studied ancient lacustrine carbonates are of shallow-water origin (Platt & Wright, 1991) because the bulk of carbonate production in lakes occurs in shallow areas, either through biological activity or the growth of ooids. Preferential production and accumulation of shallow-water carbonate resulted in the formation of various types of ramp or bench features around the margins of a number of ancient lakes (Fig. 4.22). In some lakes, microbial carbonate buildups were major basin margin rock-formers. The morphology, internal structure and facies associations of these marginal deposits were largely determined by the degree of exposure to wave attack (see Platt & Wright, 1991, for a full discussion). Relatively deep-water limestone deposits, similar to the lime muds currently accumulating in some Alpine lakes (Sect. 4.7.1) are known. A number of

lacustrine intervals in non-marine sections from beneath the Black Sea are of this type (Hsü & Kelts, 1978).

4.12.1 Low energy water bodies

In southern Europe the deposits of shallow, generally sheltered lakes are especially common. The combination of exposed Mesozoic marine platform limestones and active basin formation in Cretaceous and Tertiary times led to the widespread development of carbonate-rich lakes (Truc, 1978).

Much of the carbonate in these deposits is of biogenic origin; charophytes, ostracods and gastropods are common and the mud-grade material seems mainly to have originated from calcite encrustations on macrophytes and cyanobacterial mats (Fig. 4.22) (Platt & Wright, 1992). The lakes accumulated a characteristic suite of sediments, which tend to be arranged in a cyclic fashion. Well- to poorly laminated *charophyte–ostracod mudstones and wackestones* are commonly succeeded by bioturbated *wackestones* with intercalated *grainstone* beds. Horizons with desiccation cracks and other indicators of exposure, such as fenestral cavities, are also present. The grainstones are largely composed of angular to subrounded intraclasts of mudstone and wackestone. *Microbrecciated* and *peloidal mudstones* dominate the upper part of the cycle. These are generally poorly fossiliferous, although scattered charophyte and gastropod remains are present. Cracks and dissolutional cavities, many filled by geopetal sediment and sparry calcite, are abundant. Colour mottles and rootlet traces are also common. Some levels are marked by laminar micrites or rubbly horizons composed of a mixture of reddened, subangular clasts. The red colour also penetrates down cracks into the underlying microbrecciated beds. Clastic mudstones, coals (Sect. 4.15) and, more rarely, evaporite beds are associated with these limestones.

These facies evidently accumulated in shallow waterbodies, subject to frequent variations in lake level. Lake margins had low gradients, so extensive areas were exposed during low

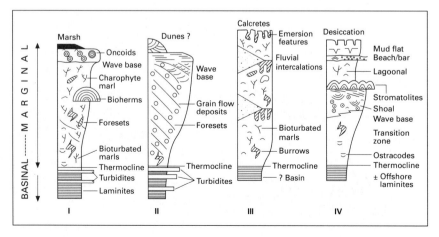

Figure 4.22 Idealized vertical sequences of carbonate and associated facies that accumulate in various sorts of lacustrine settings. The sequences are shallowing upwards, beginning with relatively deep-water deposits (rhythmites that possibly accumulated under anoxic conditions) and ending with deposits or features indicative of very shallow-water or subaerial conditions (after Platt & Wright, 1991).

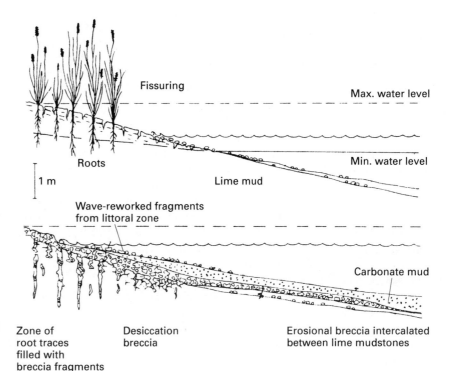

Figure 4.23 Typical features of palustrine limestones formed under low-energy conditions at the margin of a shallow lake subject to frequent changes in water level. Rootlet traces, synsedimentary brecciation and dissolution fissures are particularly characteristic of this environment (after Freytet & Plaziat, 1982).

water periods. The dominance of fine-grained facies and rarity of oncolite beds or shell lags indicate that wave energies were generally low, in part, perhaps, because of extensive stands of aquatic vegetation. The mudstones and wackestones accumulated under high water conditions, but the rarity of well-preserved lamination suggests that even then the lakes were rarely stratified. Indicators of subaerial conditions become increasingly common upward through the cycle and evidence of prolonged exposure is provided by the existence of pedogenic features such as brecciated or rubbly horizons, laminar micrites (calcrete) and colour mottling. Dissolutional cavities are of karstic origin

(Fig. 4.23). This association of shallow-water lacustrine carbonate deposits with fabrics and other features formed during exposure and pedogenesis has been termed the *palustrine* facies (Freytet & Plaziat, 1982; Platt & Wright, 1992). The relative abundance of aquatic facies compared with subaerial and evaporitic facies basins was largely controlled by climate (Platt & Wright, 1992). Exposure, brecciation and evaporite precipitation occurred most frequently under relatively arid conditions, karstic and calcrete formation in a slightly wetter climate and associated coals and fluvial channels under the most humid conditions (Fig. 4.24).

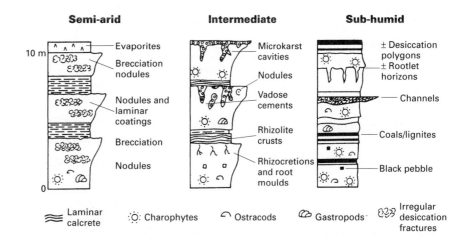

Figure 4.24 Simplified vertical facies sequences for palustrine carbonates that accumulated under different climates (after Platt & Wright, 1992).

4.12.2 High energy margins

THE GLENNS FERRY FORMATION

Wave processes also had a major influence upon facies development in some ancient carbonate-producing lakes (Sect. 4.6.1). In the *Shoofly Oolite* of the *Glenns Ferry Formation* (Pliocene, Idaho), ooids accumulated on sufficient scale to produce a nearshore *bench*, preserved as a laterally extensive oolitic limestone up to 12 m thick with large-scale cross-stratification dipping between 7° and 28° towards the ancient lake centre (Swirydczuk, Wilkinson & Smith, 1979, 1980; Fig. 4.25). Ooids were formed on a nearshore, wave-washed bench and deposited on the lakeward-dipping bench slope during periods of seasonally higher wave activity. The associated fish and molluscan fauna is clearly a freshwater assemblage. Crenulation of the low-Mg calcite laminae within the cortex of ooids is common (cf. Jones & Wilkinson, 1978) and external cortical laminae are patchily bored by endolithic algae. Individual grains therefore exhibit features typical of both ooids and algal oncoids. As the original aragonitic ooids grew in size they were transported less frequently, allowing algae an increased role in ooid formation. The large foresets in the oolite bench suggest accumulation by basinward progradation, probably during a period of relative stillstand.

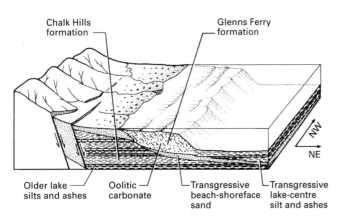

Figure 4.25 Schematic reconstruction of a lake-margin ooïd sand platform, Pliocene Glenns Ferry Formation, Snake River Plain (after Swirydczuk, Wilkinson & Smith, 1979).

4.12.3 Microbial build-ups

THE CHALK HILLS FORMATION

Rather different high-energy carbonate deposits occur below the Glenns Ferry Formation. In the Miocene *Chalk Hills Formation*, massive algal limestones which developed when a large lake formed over a fluvial floodplain predominate (Straccia, Wilkinson & Smith, 1990). Algal colonization of the firm floodplain surface produced cylindrical or vase-shaped structures 10–150 cm high and 10–40 cm in diameter. Both growth forms typically have a hollow central tube. The dominant carbonates are calcite micrite and radial fans of calcite crystals. Cylinders and vases are grouped into vertical masses or columns up to 7 m high and several tens of metres in diameter (Fig. 4.26). These formed reef-like structures rising from the floor of the lake and at times evidently had considerable topographic relief. Intercolumnar sediment was initially a mixture of coarse, generally well-sorted siliciclastic and carbonate sand. The latter includes ooids, shell debris and fragments broken from the algal columns, but is dominated by a gastropod typical of high-energy freshwater environments. Finer-grained ashy siltstone fills the upper parts of the intercolumn depressions and also caps the algal deposits.

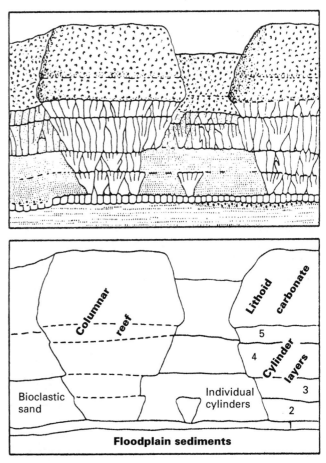

Figure 4.26 Sketch and interpretation of a typical microbialite column in the Hot Spring Limestone (Miocene, Idaho). The columns overlie floodplain deposits; growth was initiated on a transgressive strandline gravel deposited during initial flooding of the fluvial valley. Fully developed columnar reefs are c. 7 m high (after Straccia, Wilkinson & Smith, 1990).

The presence of carbonate deposits between clastic-dominated units suggests that the former developed during an initial period of lacustrine flooding. Column growth may have kept pace with the rising water level and certainly exceeded the general rate of sediment accumulation; some of the reefs may have reached as much as 6 m above the level of the surrounding lake floor. As the water deepened, wave-generated turbulence at the lake floor declined, coarse intercolumn sediment therefore gradually gave way to finer grained siliciclastic deposits, which eventually blanketed the whole reef complex, putting an end to carbonate sedimentation.

Massive microbialite deposits that developed under less turbulent conditions are excellently displayed by the uppermost limestones of the Ries Crater. The Ries basin of southern Germany is a shallow, circular depression 20–25 km in diameter formed in late Miocene time by a meteorite impact. The crater was rapidly filled by over 300 m of lacustrine sediments through which a borehole has been drilled (Füchtbauer, von der Brelie *et al.*, 1977). A central inner crater contains the thickest lacustrine sequence and is composed of bituminous and dolomitic clays and marls. Thick tufa deposits and shelly carbonate sands are restricted to the marginal zone. The Ries Crater evolved from a closed basin, saline playa-type lake to a perennial, saline lake and finally, when outlets had been established, to a freshwater system with a basinal marl facies and marginal algal build-ups.

The marginal carbonate facies of the Ries Crater contains bioherms up to 7 m high and 15 m across formed by green algae (Riding, 1979). The bioherms comprise compound cones and nodules which are themselves the result of amalgamation of smaller cones and nodules composed of *Cladophorites* tufts. Skeletal (ostracod, gastropod) and peloidal sands surround the bioherms, and laminated sinter, which veneers the cones and nodules, formed inorganically during periods of subaerial exposure at the lake margin.

4.13 Mixed clastic–carbonate basins

In several ancient lacustrine basins dominated by siliciclastic sedimentation, carbonate deposits nevertheless accumulated in areas that were locally starved of clastic sediment (cf. Sect. 4.6.1; Figs 4.8 & 4.17). Today, a variety of clastic and carbonate facies with complex vertical and horizontal relationships are preserved.

4.13.1 The Devonian Orcadian Basin of northeast Britain

The Middle Devonian Orcadian Basin was a land-locked, postorogenic extensional basin, stretching from northern Scotland, through the Orkney and Shetland Islands to perhaps as far as the west coast of Norway. Sedimentation generally kept pace with subsidence, allowing up to 3–4 km of continental and lacustrine sediment to accumulate. Lakes were generally confined to half-graben sub-basins, but during humid intervals these united to form one or more very large lakes, which perhaps on occasions overflowed the basin threshold to connect with the sea (Donovan, 1980; Hamilton & Trewin, 1988; Janaway & Parnell, 1989). Siliciclastic deposits predominate, reflecting accumulation in a variety of lake margin and central lake environments. There is, however, still some dispute as to the nature of the lake at some periods. Some workers consider that it held significant amounts of water throughout, while others (e.g. Rogers & Astin, 1991) suggest that the basins may have been dry for long periods of time, when aeolian deposits formed.

The most characteristic and widespread of the carbonate-bearing facies in the Orcadian Basin is the *carbonate–clastic–organic rhythmite facies* which comprises couplets and triplets averaging 5 mm in thickness in Orkney and Caithness, but up to 30 mm in southeast Shetland. The triplets are composed of silt-grade clastic material, organic-rich layers and micritic carbonate (calcite or dolomite), much of which is neomorphosed to microspar. Couplets lack either the clastic or carbonate lamina (Trewin, 1986; Janaway & Parnell, 1989). Organic matter came from the lake phytoplankton and the carbonate was probably precipitated at times of high primary productivity. Clastic laminae have a variety of origins; some were introduced by density flows, but wind may also have been important (Trewin, 1986). The rhythmites are interpreted as forming within the hypolimnion of a dilute, seasonally or permanently stratified lake (Donovan, 1980; Trewin, 1986). However, chert nodules within some rhythmite units contain pseudomorphs after probable evaporite minerals and have been compared with the Magadi-type cherts typical of saline–alkaline lakes (Sect. 4.7.4; Parnell, 1986), suggesting that some rhythmites may have formed in lakes with relatively elevated salinites. Particularly characteristic of the rhythmite sequences are *fish-beds* – intervals notably rich in well-preserved fish fossils. Some of the fish-beds can be traced over very wide areas and provide the most persuasive evidence for the periodic existence of a very large, deep lake within the Orcadian basin (Trewin, 1986). At its greatest extent, during deposition of the fish-bearing Achanarras Limestone, the Orcadian lake extended over an area of at least 50 000 km². When lakes were confined to individual sub-basins they were considerably smaller than this. Using the dimensions of wave ripple marks it is possible to reconstruct lake wave conditions (P.A. Allen, 1984). The period of waves which formed the wave ripple-marks in the Shetland lake is thought to be low. Since wave period is controlled primarily by fetch (Sect. 4.4), this suggests that the Shetland lake was relatively small and certainly less than 20 km wide (P.A. Allen, 1981).

The rhythmites are typical open lake deposits; other types of carbonate deposits accumulated at the lake margins and during periods of lowstand. Outcrops of basement rocks which formed cliffs and rocky shorelines around the lakes are locally covered by limestone and dolomite up to 3 m thick. Stromatolites and algal coatings are common, but the primary fabrics may be disrupted by fenestral structures suggesting periodic exposure. Rapid facies changes occur in these lake margin deposits. For example, birds-eye limestones pass within a few metres into carbonate-rich laminites (Fig. 4.27). Some siltstones show slump features, suggesting a steep local topographic slope at the lake margin. The penecontemporaneous nature of the dolomitization of the limestone is suggested by the absence of dolomitized carbonate in interstitial cavity fillings. Geochemical evidence indicates that the dolomite probably precipitated at times of low water when playa-like conditions existed in parts of the basin (Janaway & Parnell, 1989).

4.14 Evaporitic lake basins

Since evaporite minerals occur widely in areas of dry or arid climate, saline lake deposits are typically associated with aeolian and alluvial fan facies. However, as the precipitation of evaporites is primarily dependent upon water balance, saline minerals can form almost anywhere where hydrological (and temperature) conditions are suitable and local occurrences of ancient evaporites have been found together with a wide variety of continental facies. In central and western Canada, for example, saline lakes are widely developed upon the irregular topography left after the retreat of the last ice sheets, so here the saline deposits are closely associated with glaciogenic facies. One feature common to almost all ancient saline lake deposits is the evidence for

cyclic variations in lake hydrology, usually due to climatic change. This cyclicity leads to a characteristic spatial and vertical arrangement of facies in the fill of saline lake basins (Kendall, 1984; Fig. 4.28).

4.14.1 The Green River Formation (Palaeogene) of Utah, Wyoming and Colorado

The Eocene Green River Formation has been studied in great detail, not only because of its unusual and complex nature, but also because it contains some of the world's largest reserves of oil shale and trona and has sourced significant quantities of oil (Tuttle, 1991). The Green River Formation was deposited in several basins, each of which probably corresponded to a separate lake basin. The wide fluctuations in areal extent of the lake deposits and in the various lacustrine facies through time can best be explained by long-term climatic changes.

A large number of facies have been described from the different basins, including sandstones of deltaic and shoreface origin, limestones (particularly stromatolitic varieties) and volcanic ash horizons which are useful for correlation. Many ashes are altered to zeolites such as analcime ($NaAlSi_2O_6H_2O$), or to feldspars, authigenic minerals which compare closely with the reaction products of tuffs in present-day alkaline lakes (e.g. Hay, 1968; Eugster, 1969; Renaut, 1993). Most characteristic of the Green River Formation facies are, however, the *trona–halite facies* and the *oil shale facies* (Fig. 4.29).

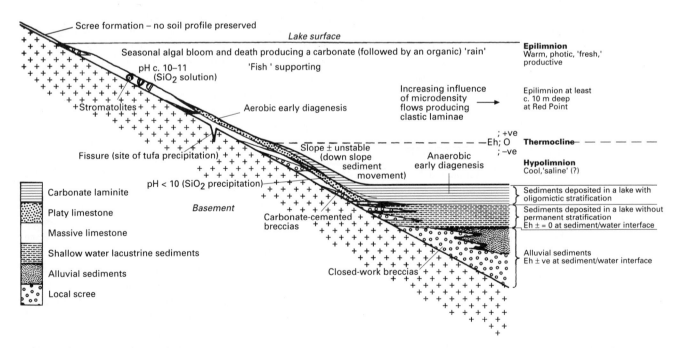

Figure 4.27 Depositional model for the margin of the Devonian Orcadian basin, Scotland, illustrating vertical and lateral facies relationships and processes that influence sedimentation at a *coincident* lake margin during lake transgression (after Donovan, 1975).

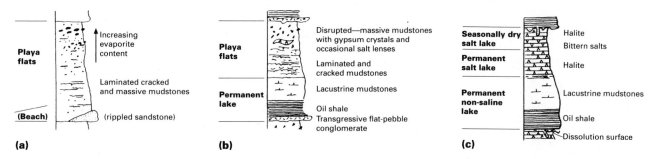

Figure 4.28 Hypothetical cycles from the margin (a) to centre (c) of a saline lake basin showing sedimentary response to increasing aridity and consequent shrinkage and evaporative concentration of the lake. The cycle ends with a new phase of flooding and a return to a relatively deep and dilute waterbody (after Kendall, 1984).

In the trona-halite facies, beds of trona are up to 11 m thick in some basins. Halite may be associated with the trona, either mixed with it or as separate beds. In general, the evaporites are best developed in the central part of the basin and commonly occur above a bed of oil shale. Conditions of accumulation have been compared with those in present-day soda lakes such as Lake Magadi (e.g. Eugster, 1970). Trona and halite deposits are thought to have precipitated from alkaline brines concentrated by evaporative reflux of groundwater and spring water and by the re-solution of efflorescent crusts. Precipitation of calcite and dolomite in more marginal areas led to high concentrations of alkali metal ions in the residual brines and to such increased pH values that trona could precipitate. Seasonal variations in rainfall and evaporation led to dry and flood periods in the lake with resultant precipitation and partial re-solution of trona (cf. Eugster, 1970). By analogy with Quaternary soda lakes, the stromatolitic limestones probably represent periods of fresher conditions in the lake.

In detail, the *oil shale facies* contains a wide variety of rock types but at its most characteristic comprises rhythmic alternations of mainly amorphous organic matter (*kerogen*) and slightly thicker laminae of clay or carbonate (commonly dolomitic). Organic matter contents may exceed 25% (Dean & Anders, 1991). Interbeds of algal boundstone, oolitic–pisolitic grain-stone, ostracod grainstone, sandstone and siltstone may also be present (Fig. 4.29). The thickest beds of rich oil shale tend to occur towards the centre basin, but some oil shale intervals can be traced over very large distances, indicating accumulation in a substantial waterbody. Towards the margins, the deposits change to marlstone, kerogen-rich marlstone and lean oil shale (Trudell, Beard & Smith, 1974; Dean & Anders, 1991). Some basin-centre oil shale accumulations are accompanied by dawsonite ($NaAl(CO_3)(OH)_2$), halite ($NaCl$) and nahcolite ($NaHCO_3$) (Dyni, 1974). Oil shale sequences may also contain mudcracked and brecciated horizons, but the accumulation of finely laminated, very organic rich muds over very large areas suggests that most oil shales probably accumulated in deep, stratified waterbodies (e.g. Bradley & Eugster, 1969; Desborough, 1978).

The apparently contrasting depositional environments of the closely juxtaposed oil shale and evaporite facies has been a source of considerable controversy. To satisfy the sedimentological and geochemical evidence, such as the highly dolomitic nature of many of the oil shales, some workers (e.g. Eugster & Surdam, 1973; Smoot, 1983) proposed a shallow, playa lake depositional model. Others (e.g. Bradley & Eugster, 1969; Desborough, 1978), impressed by the similarity between the oil shale and the organic-rich muds found in many modern stratified lakes, have suggested that the accumulation of the oil shales, at least, must have occurred in deep, meromictic lakes. The abundant dolomite is certainly no longer a hindrance to such an interpretation. Studies of modern lakes have

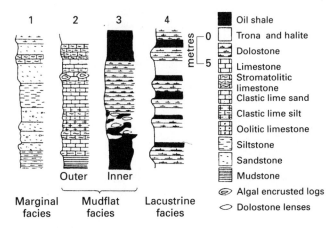

Figure 4.29 Schematic vertical sequences of lithofacies deposited in marginal, mudflat and open lake environments of the Green River Formation (after Surdam & Wolfbauer, 1975).

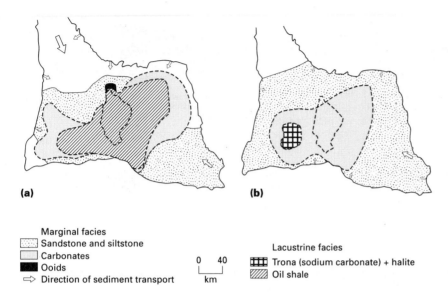

(a)   **(b)**

Marginal facies
- Sandstone and siltstone
- Carbonates
- Ooids
- ⇨ Direction of sediment transport

0   40
km

Lacustrine facies
- ⊞ Trona (sodium carbonate) + halite
- ▨ Oil shale

Figure 4.30 Typical lithofacies distribution during two contrasting phases of Green River Formation deposition. (a) Highstand conditions with extensive accumulation of oil shale. (b) Lowstand conditions when marginal facies extended towards the basin centre; there was no oil shale formation but abundant evaporite precipitation (after Surdam and Wolfbauer, 1975).

demonstrated that the formation of various carbonate minerals, particularly dolomite and siderite, in association with bacterial decomposition of organic matter is an important diagenetic process in organic-rich lacustrine muds that does not require evaporative conditions (Talbot & Kelts, 1990).

The very large areal extent of microlaminated oil shale deposits containing abundant oil-prone organic matter also argues in favour of the periodic existence of deep, probably meromictic lakes (Dean & Anders, 1991; Fig. 4.30). In fact playa lakes and stratified lakes can exist in association with each other. Shallow ephemeral lakes with extensive playa-flats can easily evolve into chemically stratified lakes as a result of increased freshwater inflow, but incomplete mixing, allowing the establishment of a density-stratified waterbody (*ectogenic meromixis*). For example, the thick oil shales of the Laney Member of the Bridger–Washakie basins or the Mahogany Bed of the Uinta–Piceance Creek basins may have accumulated under anoxic conditions on the floor of a permanently stratified lake with a surface layer of low salinity, oxygenated water overlying a saline hypolimnion (Boyer, 1982). Such an interpretation would explain why virtually the entire Green River fish fauna consists of *freshwater* taxa (Grande, 1980). Climatic changes, similar to those responsible for abrupt vertical facies changes in the soda lakes of East Africa (e.g. Renaut & Tiercelin, 1994), probably controlled the short-term variations in the water balance of the Green River lakes (Sect. 4.16.1).

4.14.2 The Ebro basin (Oligocene–Miocene), Spain

The Ebro basin is a foreland basin formed by uplift of the Pyrenean chain to the north and bounded also by the Catalan Coastal range to the southeast and Iberian range to the southwest. Within this major continental basin large volumes of evaporites have accumulated, notably during Oligocene and Miocene times (Fig. 4.31). A generally rather dry climate was one reason for the extensive formation of saline minerals, but in addition, the

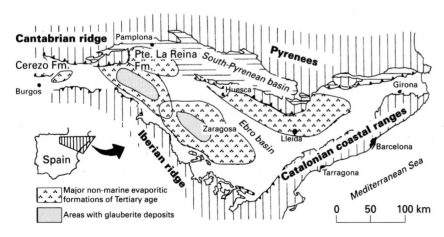

Figure 4.31 Map of the Ebro basin, Spain, showing zonal arrangement of Tertiary salt deposits (after Salvany & Orti, 1994).

mountains surrounding the basin contain both major Triassic evaporites and Mesozoic marine limestones. Surface and sub-surface run-off from these uplands have carried an abundant supply of dissolved ions to the Ebro basin. The area demonstrates the importance of groundwater in the formation of a major evaporite deposit.

The evaporite minerals show the concentric zonation typical of many evaporite basins, with the most soluble minerals located in central areas, having precipitated from residual brines that accumulated in the topographically lowest parts of the basin (Fig. 4.31). The Ebro basin is notable for the abundance of glauberite ($Na_2Ca(SO_4)_2$) which is largely confined to the centre of the basin. Gypsum, halite and polyhalite ($K_2MgCa_2(SO_4)_2.2H_2O$) are also common (Salvany & Ortí, 1994). Solute transport from the source areas was dominated by subsurface flow, notably through karst systems developed in the Mesozoic limestones, and significant brine concentration occurred as shallow ground-waters moved from the basin margin to the basin centre (Salvany, Muñoz & Pérez, 1994; Salvany & Ortí, 1994; Fig. 4.32).

Surface run-off was responsible for constructing a chain of alluvial fan systems along the southern margin. These fans graded northwards into marginal saline lakes. Extensive preci-pitation of calcium carbonate and gypsum occurred in both the fans and marginal lakes. During humid periods the lakes also accumulated substantial deposits of charophytic limestones and dolomites. Both primary and early diagenetic gypsum are present. A common type consists of microlenticular crystals 0.1–0.5 mm long which grew on the bottom of the lakes, at or just below the sediment–water interface. Other gypsum units contain current and oscillation ripples indicating that the gypsum crystals were reworked to form *gypsarenite*, a com-mon deposit in many other sulphate-rich saline lake sequences (e.g. Sanz, Rodríguez-Aranda *et al.*, 1994). In the Ebro basin, some gypsarenites are bioturbated, suggesting that the lakes here were periodically habitable despite their high sulphate content (Salvany, Muñoz & Pérez, 1994).

The basin floor sloped gently north from the marginal zone towards the central saline lake complex (Fig. 4.33). Marshes, where palustrine (Sect. 4.12.1) carbonate muds accumulated, formed in areas receiving dilute spring water or surface run-off, but much of this transition zone was characterized by vast saline mudflats where gypsum precipitated extensively, mainly just below the surface as early diagenetic nodules. The marginal lakes and mudflats evidently functioned as preconcentrators for shallow groundwaters draining towards the central saline lake (Salvany, Muñoz & Pérez, 1994). Brine evolution continued as the waters moved north; those that reached the central lake were already highly concentrated and led to the precipitation of thick beds of gypsum, halite and glauberite (Fig. 4.32). Here, too, the glauberite, polyhalite and some of the gypsum were precipitated interstitially, while the halite formed beds of hopper crystals interbedded with laminated, primary gypsum (Salvany, & Ortí, 1994).

The whole Oligocene–Miocene succession comprises six 100–200-m-thick cycles. Each cycle commences with saline lake deposits showing their greatest areal extent; glauberite is preferentially developed in this part of the cycle. The saline facies gradually become more restricted upward through the cycle, retreating towards the basin centre. This change is mainly due to progressive progradation of the alluvial fan systems and the top of some cycles is formed entirely of clastic deposits, mainly clays and fine-grained sandstones (Salvany & Ortí, 1994). A gradual shift in depocentre towards the northwest is apparent through the sequence as a whole, suggesting that the cycles may be primarily of tectonic origin.

4.15 Organic-matter-dominated basins

Organic matter is present in the sediments of most lakes, and in some it was a major component through long periods of the lake's history. Higher plants, phytoplankton and bacteria may all have made significant contributions. Such organic-rich deposits can be of considerable economic importance. Remains of higher plants in the form of *limnic* (lacustrine) coals and lignites compose an important primary source of fuel, and oil shales provide a source for migrated hydrocarbons. Some organic-rich lacustrine mudrocks are also notable for their exceptional fossil fauna and flora.

The preservation of significant quantities of organic matter is dependent upon the fortuitous combination of particular sedimentary conditions. The organic matter content of the average sedimentary rock is considerably less than 1%, so it is clear that preservation of organic matter is not a normal feature of most sedimentary environments. In typical, near-surface oxidizing environments reduced carbon is highly unstable. Two conditions, in particular, must be satisfied if organic-matter-rich sediments are to accumulate: (i) the rate of supply of clastic sediment should be relatively low in comparison with the rate of supply of organic matter; and (ii) the rate of accumulation of organic matter must exceed the rate of degradation due to inorganic and microbiological oxidation.

4.15.1 The Fort Union Formation

The combined effects of high rates of macrophyte production in extensive marginal swamps and locally reduced rates of clastic sediment accumulation are the apparent reasons for the formation of a number of limnic coal deposits. The *Fort Union Formation* (Palaeocene, Wyoming and Montana, USA) is a major coal-bearing unit which accumulated in the Powder River basin, one of several north–south elongate, intermontane basins which developed in early Tertiary time along the eastern margin of the Laramide fold belt (Yuretich, 1989). Coal seams up to 30 m thick may be correlated along strike for more than 60 km (Ayers & Kaiser, 1984). At the time of peat accumulation a large lake (Lake Lebo) supplied with abundant clastic sediment

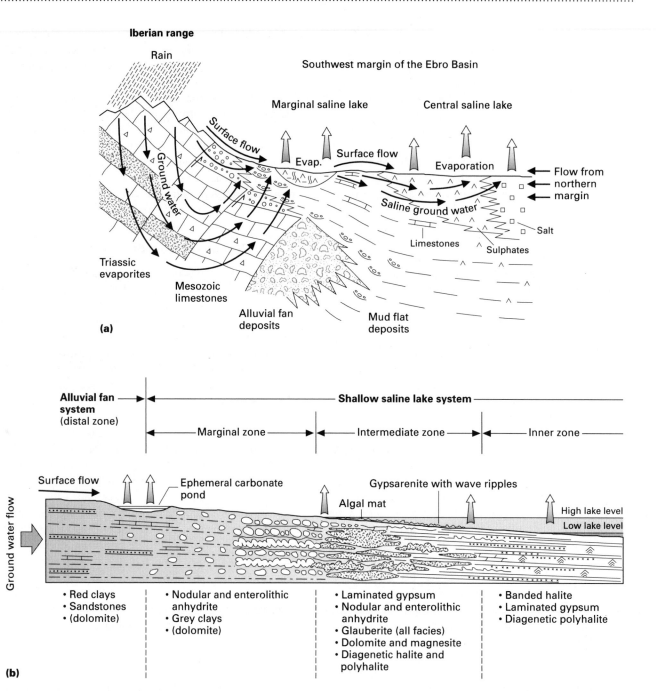

Figure 4.32 (a) Palaeohydrology of the southwestern margin of the Ebro basin during Miocene time, showing flowpaths for dissolved ion transport to the basin and environments of evaporite formation.

(b) Facies arrangement and sedimentological model for the principal evaporite deposits in the Ebro basin (after Salvany & Ortí, 1994; Salvany, Muñoz & Pérez, 1994).

occupied the basin. A number of river systems fed into Lake Lebo (Fig. 4.33), forming elongate, high constructive deltas. Extensive swamps developed between the deltas and persisted for long periods, allowing exceptionally thick peat deposits to accumulate. The factors responsible for peat formation are listed below (Ayers & Kaiser, 1984; Ayers, 1986).

1 *Stable lake level*, this was the result of a humid climate and continuous open lake conditions. In the absence of lowstands, there was no prolonged exposure and drainage of the swamps, which would have disrupted peat production by desiccation, oxidation and possibly fire. Abundant groundwater supply may also have helped maintain the high water table.

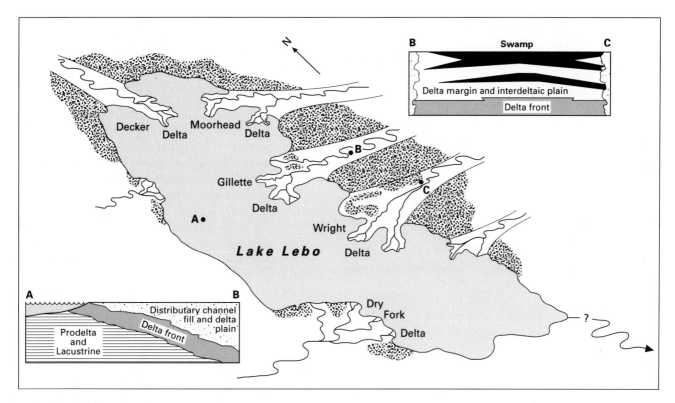

Figure 4.33 Depositional model for coal formation in the Powder River basin. The lake measured c. 90 × 300 km. Lacustrine mudstones dominate the centre of the basin, coal swamps were best developed in the very extensive low-lying areas between major deltas that built out from the eastern margin of the lake (after Ayers, 1986).

2 *Rapid basin subsidence*, which provided accommodation space for the peats and associated overbank deposits.

3 *The fluvial systems*, these exhibited little tendency for lateral migration, areas between the deltas were therefore starved of clastic sediment. Fluvial channel stability may in part have been due to rapid subsidence, but it is probable that fluvial access to the basin was constrained by regional structure or geomorphology. Vegetated levees may also have helped stabilize individual channels. Isolation from the main fluvial systems explains the low ash content of the coals.

4.15.2 The Calaf and Mequinenza basins

Rather different coal-bearing lacustrine sequences are found in the Oligocene deposits of northeast Spain. Here, in the *Calaf* and *Mequinenza* basins, peats formed in lakes that were of variable, generally shallow depth and accumulated abundant carbonate (Cabrera & Saez, 1987). The limestones are predominantly mudstones, wackestones and packstones with variable contents of charophytes, ostracods, gastropods and plant fragments. Many carbonate units, especially in marginal areas, display typical palustrine features, such as exposure surfaces and rootlet horizons. One further facies of note in the

Mequinenza basin is thin, lenticular deposits of nodular gypsum and anhydrite. Coal seams are widely distributed within both basins, but individual seams are thin (<1 m, commonly 0.2–0.4 m), laterally impersistent and have a tendency to split into thinner seams. Seat earths, rootlet or stem horizons are rarely associated with the coals.

Cabrera and Saez (1987) interpret the sedimentary setting to have been one of generally shallow lakes, fringed by small deltas and low angle fans (Fig. 4.34). Much of the clastic supply was evidently trapped in the deltas, allowing carbonate and organic matter sedimentation to dominate the open lake. Frequent fluctuations of lake level exposed the carbonate muds, prolonged periods of low water leading to the precipitation of evaporites. The lack of associated rootlet horizons or seat earths suggests that many of the coals are allochthonous.

4.15.3 Oil shales

The depositional settings in which *oil shales* and their precursors accumulate have been discussed in Sections 4.7.5 and 4.14.1. They are the products of sedimentation in basinal areas relatively remote from the main sources of clastic sediment supply. A persistent oxygen deficit, due to a stably stratified watermass

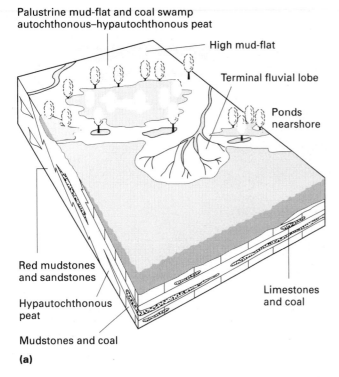

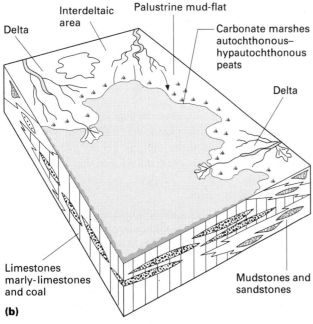

Figure 4.34 Depositional models for the development of coal deposits in the Calaf (a) and Mequinenza (b) basins, Spain. The coal seams are thin and discontinuous, much of the organic matter was apparently transported to shallow, offshore areas of the lakes (after Cabrera & Saez, 1987).

may also be important, but where rates of organic production are high, anoxic conditions are evidently not essential to the preservation of organic matter.

One other aspect of oil shales is that they not only have unusually high organic contents, but in some cases also an exceptional fossil content. Many oil shale basins are renowned for specimens of otherwise rare fossil groups such as leaves, flowers, insects, reptiles, birds and other vertebrates. Some of these were aquatic, but many were terrestrial organisms that fell or were washed or blown into the lake (e.g. M.V.H. Wilson, 1977, 1980). Here they sank beneath the oxicline, where the calm, anoxic waters provided a fossilization environment free from scavengers and physical disturbance (see, for example, Trewin, 1986, for a detailed discussion of fish preservation in the Orcadian rhythmites). Most noteworthy of all is probably the Messel Oil Shale (Eocene), which accumulated in a small, fault-bounded lake within the Rhine Graben. The oil shales have been subject to intense exploration for fossils which provide a unique glimpse into life in the early Tertiary. Beetles and spiders, some of the former still with iridescent wing cases, a 2-m long primitive python, feathered birds, rodents with clear impressions of a thick pelt and scaly tail, plus many other plants and animals, have been exquisitely preserved in the shales (Schaal & Ziegler, 1988).

4.16 Cycles in lake deposits

Cyclic alternations of different lithologies or facies are a characteristic feature of lake deposits of all ages. Cyclicity occurs at all scales from the submillimetre microlaminae of rhythmites (Sect. 4.8) to regional-scale cycles of lake basin filling. Study of these cycles can provide insight into the nature of the water-body, the sedimentary history of the lake and regional climatic variations.

4.16.1 The Green River Formation

The existence within the Green River Formation of well-developed cycles at a variety of scales has been apparent for many years. Bradley's (1929) detailed investigation of some of the oil shale rhythmites represents a pioneering step in the study of pre-Quaternary lacustrine deposits. By analogy with the Holocene sediments of Lake Zürich, Bradley identified the Green River rhythmites as non-glacial varves. Subsequent statistical analysis of rhythmite thickness variations has confirmed this interpretation and in addition demonstrated that the varves are bundled into a number of longer-term cycles, including ones with an approximately 3–5 year rhythm comparable to the El Niño–Southern Oscillation (ENSO) cycle, and an 11-year rhythm, which has been attributed to the sunspot cycle (Fischer & Roberts, 1991; Ripepe, Robert & Fischer, 1991). Within the Wilkins Peak Member, cycles up to 5 m thick can be traced for distances greater than 20 km without significant thickness change. In all there are about 50 cycles which comprise either oil shale–marlstone or oil shale–trona–marlstone alternations. These are interpreted as the results of climatically controlled transgressive–regressive events that caused fluctuations of lake level at a frequency of c. 20 000 years. This period represents one of the precessional rhythms of the Milankovitch astronomical cycles. The 100-ky eccentricity cycle has also been recognized (Fischer & Roberts, 1991).

4.16.2 The Rubielos de Mora basin

The Miocene *Rubielos de Mora* basin, northeast Spain, illustrates in a particularly striking manner the interplay between clastic and carbonate sedimentation in a relatively deep lake (Anadón, Cabrera & Julià, 1988; Anadón, Cabrera *et al.*, 1991). The basin fill is characterized by a pervasive cyclicity on several scales.

Basinal areas are characterized by a facies assemblage dominated by interbedded mudstones and microlaminated rhythmites. The mudstones are dark grey in colour and contain a limited fossil assemblage of leaves, insects and amphibian skeletons with rare laminae containing lacustrine molluscs and ostracods. Bioturbation is absent. Many of these mudstones are oil shales, containing up to 15% organic matter (Anadón, Cabrera *et al.*, 1991). The rhythmite units are up to 2.5 m thick and consist of submillimetre-thick couplets of clay and carbonate. Of particular note is the common presence of apparently primary aragonite, which has probably been protected from diagenesis or dissolution by the hydrophobic nature of the enclosing oil shales. This assemblage is interpreted as a suite of offshore deposits that accumulated in relatively deep water beneath a permanent chemocline.

At the western end of the basin a non-laminated facies assemblage is interbedded with the laminated mudstones and rhythmites. It is composed of white marls and grey, bioturbated mudstones. White marls are predominantly fine-grained limestones containing plant debris and ostracod shells. They grade upwards into grey and green, massive, blocky-fracturing mudstones (Fig. 4.35). Bioturbation is ubiquitous, including levels with abundant rootlet traces. Some horizons are also disrupted by microbrecciation. The association of the latter with bedding surfaces strewn with the carbonized seeds of *Potamogeton* suggests that the roots belonged to aquatic macrophytes. Cyclic alternation of laminated and non-laminated deposits indicates alternating anoxic and oxic bottom conditions, presumably because of periodic but minor variations in lake level. This part of the lake is thought to have been a ramp-like area which lay at such a depth that it was periodically above or below the oxicline. During highstands anoxic conditions prevailed and laminated mudstones accumulated; at times of lowstand, waters over the ramp were shallow and oxygenated, allowing a rich flora and fauna to flourish. Periodic exposure is indicated by the microbreccia fabric, suggesting incipient soil formation.

These oxic–anoxic cycles are just one of several scales of cyclicity apparent in the fill of the Rubielos de Mora basin. Rhythmite units within the anoxic cycles may record annual changes in the lake, while the oxic–anoxic cycles themselves are organized into larger cycles that probably reflect pulsatory deepening of the lake (Anadón, Cabrera *et al.*, 1991; Fig. 4.36).

4.16.3 The Newark Supergroup

The early Mesozoic *Newark Supergroup* of eastern North America demonstrates how lake sedimentation is controlled by regional-scale climatic variation. This is a suite of entirely continental sedimentary rocks and associated basalt flows which accumulated in elongate half-graben basins that form a NE–SW oriented belt parallel to the eastern seaboard of North America. Lacustrine deposits constitute a significant proportion of the sedimentary fill, covering almost all known lacustrine clastic rock types, which varied both between basins and within basins (Olsen, 1990; Smoot, 1991). Diversity between basins was because these were spread over more than 10° of latitude; in early Mesozoic time they spanned the humid to dry tropical climatic zones (c. 3°S–8°N) and were probably under

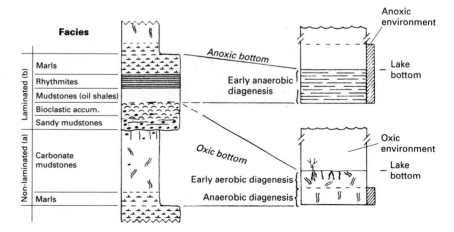

Figure 4.35 Idealized marginal cycles in the Rubielos de Mora basin, Spain. The cycles developed on a marginal platform and record variations in the depth of the chemocline. At times the platform lay below the chemocline and was subject to anoxic conditions, during other periods it was subject to shallow water, oxic conditions and may even have been periodically exposed. Cycles are typically 1–5 m thick (after Anadón, Cabrera & Julià, 1988).

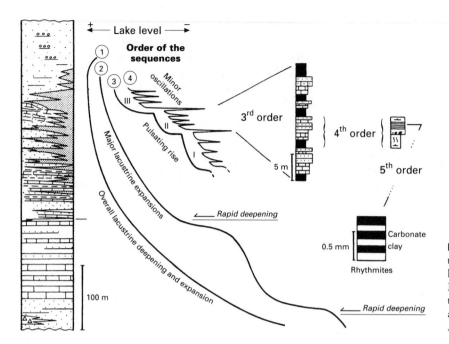

Figure 4.36 Diagrammatic representation of the various orders of cyclicity within the lacustrine sedimentary fill to the Rubielos de Mora basin. Low order cycles (1, 2, 3) are of tectonic origin. Fourth- and fifth-order cycles are largely climatically controlled (after Anadón, Cabrera *et al.*, 1991).

the influence of a monsoonal climate. Thus, the southerly basins were mainly located in relatively moist regions, where lakes were high for much of their history and coal beds could form. In the north the climate was much drier, lakes were generally low or ephemeral and frequently saline; aeolian dune sandstones are widespread in some of these basins.

Diversity of clastic facies within basins was because, in common with their Quaternary counterparts in the tropics and subtropics, the Newark Supergroup lakes were greatly influenced by climatically induced variations in water balance. All the basins underwent frequent oscillations in water level and sedimentation responded accordingly. Sequences of lacustrine and associated rocks from these basins thus display cyclicity on a number of scales. Some sequences have proved to be of critical importance in confirming the impact of cyclic climatic change upon continental sedimentation.

The cyclicity of the Newark Supergroup was first pointed out by Van Houten (1964) who, in a study of the Newark basin, noted that the rocks that constitute the 1200-m-thick Triassic Lockatong Formation (Carnian) can be divided into two principal types: (i) *detrital cycles* about 5 m thick occurring mainly in the lower part of the sequence; and (ii) *chemical cycles* about 3 m thick. The detrital cycles were formed in a hydrologically open lake during relatively humid periods, through-flow ensuring the maintenance of low ionic concentrations in the lake water. The chemical cycles, on the other hand, were formed during periods of closed drainage. Van Houten's pioneering work has subsequently been expanded in both scope and detail by Olsen (1984, 1986, 1990) who recognized that the lacustrine rocks of the Lockatong and overlying Passaic Formations are almost exclusively calcareous mudstones and siltstones that can be subdivided into seven

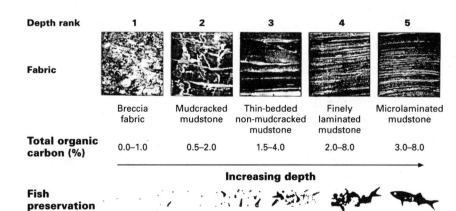

Figure 4.37 Ranks of sedimentary fabrics of mudstones and siltstones of the Lockatong Formation arranged in order of interpreted relative depth of water in which they were deposited and their relationship to organic carbon content and fossil fish preservation. Fabric 1 indicates prolonged periods of exposure and desiccation. Fabric 5 accumulated under calm, probably anoxic conditions in a deep lake. Rank 0 (not shown) is similar to Rank 1, but contains in addition traces of evaporite minerals (after Olsen, 1984).

facies, principally on the basis of differences in sediment fabric. These represent deposits that accumulated under conditions ranging from predominantly exposed mudflats to calm, periodically anoxic environments at depths exceeding 100 m in a stratified lake. By assigning a *depth rank* to each facies (Fig. 4.37) the nature of the cyclicity becomes apparent (Fig. 4.38). Radiometrically dated basalts and varve chronologies (from rhythmites) provide an absolute chronology for long sequences; time-series analysis of the vertical recurrence of the different depth ranks, reveals that cycles are present on a number of scales (Olsen, 1986, 1990). The periodicities of the cycles conform to those of the Milankovitch astronomical cycles (Sect. 2.1.5). Subsequent studies have shown that the cycles can be correlated to other basins of the Newark Group, where they are expressed by quite different combinations of facies.

4.17 Economic importance of lake deposits

The preservation of large amounts of organic matter in some ancient lake deposits has been noted in Section 4.15. Limnic coals and lignites provide an important fuel source, and lacustrine oil shales rich in algal and bacterial remains have functioned as source rocks in a number of major oil provinces that range in age from the Proterozoic to the Cenozoic (Fleet, Kelts & Talbot, 1988; M.A. Smith, 1990). For example, a significant proportion of China's current oil production is from reservoirs sourced by Palaeozoic–Cenozoic oil shales which accumulated either in intermontane basins or in large sag basins and much of the oil produced from fields along the African and Brazilian margins of the South Atlantic is sourced by oil shales deposited in large lakes formed during the initial rifting of Africa from South America (Katz, 1995).

Lacustrine mudstones rich in higher plant debris are favourable for humic and fulvic acid production and thereby uranium fixation. Tabular sandstone uranium deposits occur in association with lacustrine deposits such as the Triassic Lockatong Formation (Sect 4.16.3) and the Salt Wash Member of the Morrison Formation, southern Utah (Peterson, 1979; Turner-Peterson, 1979). A number of modern and ancient saline lake basins are important sources for a variety of evaporites, including otherwise rare minerals such as nitrates and borates. The deposits of saline–alkaline lakes can be rich in zeolites, which are today widely used in a number of industrial processes. Pure lacustrine diatomites also have a variety of industrial and domestic applications (Tiercelin, 1991).

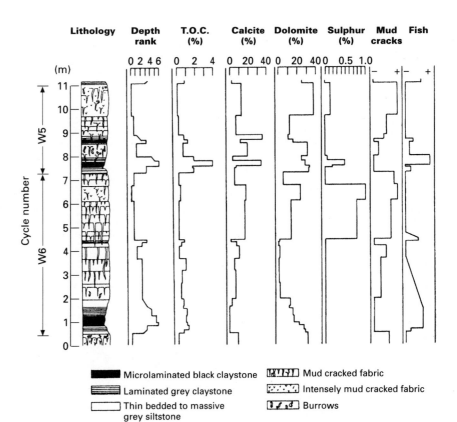

Figure 4.38 Representative section from the Lockatong Formation showing cyclic arrangement of depth ranks, organic carbon content, fish fossils (see Fig. 4.37 for details) and other features (after Olsen, 1986).

■ Microlaminated black claystone

≡ Laminated grey claystone

▭ Thin bedded to massive grey siltstone

▨ Mud cracked fabric

▨ Intensely mud cracked fabric

▨ Burrows

Further reading

Anadón P., Cabrera L. & Kelts K. (Eds) (1991) *Lacustrine Facies Analysis*, 318 pp. *Spec. Publ. int. Ass. Sediment.*, **13**, Blackwell Scientific Publications, Oxford.

Beadle L.C. (1981) *The Inland Waters of Tropical Africa*, 2nd edn, 365 pp. Longman, London.

Fleet A.J., Kelts K. & Talbot M.R. (Eds) (1988) *Lacustrine Petroleum Source Rocks*, 390 pp. *Spec. Publ. geol. Soc. Lond.*, **40**, Blackwell Scientific Publications, Oxford.

Frostick L.E., Renaut R.W., Reid I. & Tiercelin J.-J. (Eds) (1986) *Sedimentation in the African Rifts*, 382 pp. *Spec. Publ. geol. Soc. Lond.*, **25**, Blackwell Scientific Publications, Oxford.

Katz B.J. (Ed.) (1990) *Lacustrine Basin Exploration – Case Studies and Modern Analogs*, 340 pp. *Am. Ass. petrol. Geol. Mem.*, **50**, Tulsa.

Lerman A. (Ed.) (1978) *Lakes: Chemistry, Geology, Physics*, 363 pp. Springer-Verlag, Berlin.

Matter A. & Tucker M.E. (Eds) (1978) *Modern and Ancient Lake Sediments*, 290 pp. *Spec. Publ. int. Ass. Sediment.*, **2**, Blackwell Scientific Publications, Oxford.

Renaut R.W. & Last W.M. (Eds) (1994) *Sedimentology and Geochemistry of Modern and Ancient Saline Lakes*, 334 pp. *Soc. econ. Paleont. Miner. Spec. Publ.*, **50**, Tulsa.

Wetzel R.G. (1983) *Limnology*, 2nd edn, 767 pp. W.B. Saunders, Philadelphia.

Desert aeolian systems

5

G.A. Kocurek

5.1 Introduction

The wind is an important geomorphic agent wherever it is sufficient to surmount any surface stabilization. Aeolian processes are, therefore, important in environments as diverse as beaches, glacial outwash plains, and climatic deserts that range from those in cold climates to those in hot, arid climates (Sect. 2.1.2). Before vegetation, land surfaces on the Earth were generally dominated by aeolian processes far more extensively than today, and surfaces of other planets such as Mars have been shown to be strongly affected by the wind (e.g. Greeley & Iverson, 1985). The breadth of Earth systems affected by the wind is reflected in the range of disciplines that include aeolian studies. Geologists and geomorphologists have long been attracted to deserts and sand dunes, although the modern science was born with R.A. Bagnold and his classic work *The Physics of Blown Sand and Desert Dunes* (1941). Works by K.W. Glennie (1970) on the desert environment, E.D. McKee and his associates (e.g. McKee, 1966; McKee, Douglass & Rittenhouse, 1971; McKee & Moiola, 1975) on dune internal structure, and I.G. Wilson (e.g. 1971, 1973) on the dynamics of aeolian systems are also important mileposts. Because of their sensitivity to climatic change, aeolian accumulations have long been used to extract palaeoclimatic information, both for the Quaternary

(e.g. Lancaster, 1990) and for the more distant geological record (e.g. Parrish & Peterson, 1988). Palaeoclimatic inferences drawn from the aeolian record range from interpretation of climatic change such as with cyclic climatic shifts over the Sahara (e.g. Petit-Maire, 1989), to interpretations of global atmospheric circulation based upon the aeolian dust record from cores of deep-sea sediment (e.g. Sarnthein & Koopmann, 1980) and glacial ice (e.g. Grousset, Biscaye *et al.*, 1992). More recent work has drawn inferences about the interrelationships of climate, sediment supply, and relative sea-level change from aeolian units (e.g. Crabaugh & Kocurek, 1993; Clemmensen, Oxnevad & de Boer, 1994). *Desertification* has become prominent today both in science and in the public eye because of its impact on human life and as a barometer for changing global climatic conditions (e.g. Mainguet, 1991). Management of wind-blown sand is important for development in arid lands and along coasts (e.g. Pye & Tsoar, 1990, pp. 286–318). Many important insights for the transport of sediment by the wind are from soil scientists confronted by severe deflation of soil from the interior United States during the 1930s and today with increasing development in Third World countries (e.g. Chepil & Woodruff, 1963). The quantity of atmospheric dust derived from aeolian processes has a serious impact on populations in numerous regions (e.g. Nickling & Gillies, 1983). Aeolian sandstones are important aquifers and contain substantial hydrocarbon reserves as in the Permian Rotliegend of the North Sea area (e.g. Glennie & Provan, 1990), the Jurassic of the Gulf of Mexico (Mancini, Mink *et al.*, 1985), and the Jurassic of the western United States (e.g. Lindquist, 1988). A significant part of the recent approach to reservoir characterization and flow modelling is based upon aeolian rocks (e.g. Goggin, Chandler *et al.*, 1992). Because of the inaccessibility of many deserts, remote sensing has played a key role in aeolian studies, and these studies have in turn helped develop remote-sensing technology (e.g. Breed, Fryberger *et al.*, 1979).

A treatment of the total realm of aeolian systems is beyond the scope of this chapter, and this chapter is mostly about aeolian environments that occur in hot, arid deserts. The review by Koster (1988) is a good starting point for aeolian systems in cold climates. Emphasis here is on basic physical processes, with examples from present-day aeolian systems and the rock record. Inevitably, although our understanding of aeolian systems has rapidly evolved in recent years, the complexities of these systems have also become apparent and many basic questions remain unanswered.

5.2 The desert aeolian system

5.2.1 Setting

Figure 5.1 shows the location of present-day desert in the low and middle latitudes. A comparison with Fig. 2.2 shows that these deserts occur in arid, semi-arid and dry continental global climates. As discussed in Section 2.2.2 and illustrated by Fig.

2.1, one major concentration of deserts is within 30° belts centred on the Tropics of Cancer and Capricorn, where the climate is dominated by the *subtropical high-pressure cells* in the atmosphere. The *trade winds* diverging from these cells consist of air that is descending and stable, and therefore unlikely to produce rain. The position of the high-pressure cells over oceans causes the western sides of continents to be most affected, and deserts are confined to the west where upland barriers occur (e.g. Atacama, Namib), but are extensive where there are no orographic barriers (e.g. Sahara, Arabian, Australian deserts). A second concentration of deserts is within the interiors of large continents, remote from sources of oceanic moisture (e.g. Gobi, Turkestan, Taklamakan). Because mountain belts tend to rim continents, aridity is enhanced within interior *rain shadows* where mountains force air to rise, cool, and lose its moisture on the upwind side of the mountains (e.g. Patagonia, Mojave, Great Basin, Gobi). Aridity is enhanced along coasts where prevailing winds blow parallel to the shoreline (and not onshore) so that little moisture is carried inland (e.g. eastern Africa). Cold, upwelling currents adjacent to coasts also increase aridity because the currents cool the air, thereby favouring stability and suppressing convection needed to raise the air mass (e.g. Atacama, Namib).

The popular notion of deserts as vast areas of wind-blown sand is incorrect, and present-day deserts house a diverse array of environments that differ widely because of tectonic setting, coastal or inland location, local climate, and *antecedent conditions*. *Sandy aeolian systems* with *dune fields* and their larger counterparts, aeolian sand seas or *ergs*, cover only about

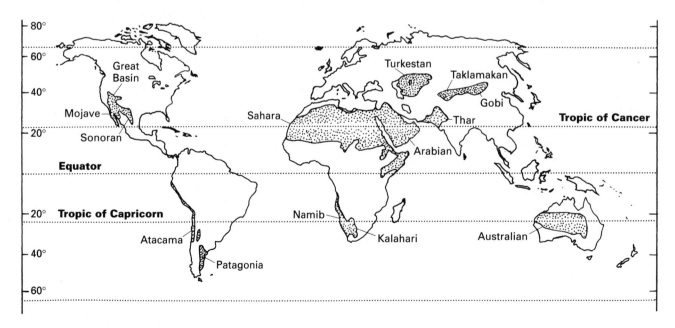

Figure 5.1 Low and middle latitude climatic deserts.

20% of modern deserts, ranging from the sand-poor deserts of North America to the relatively sand-rich deserts of central Asia. Where a paucity of sand exists, desert plains are floored by exposed bedrock or *desert pavements* of stones referred to as *reg, serir, hammada, gobi*, or *gibber plains* largely depending on the geographic area in which they occur. Bedrock or overlying consolidated sediments are wind-sculptured into streamlined hills called *yardangs*, whereas *demoiselles* or *hoodooes* refer to more irregular, peculiar shapes. Wind polishing is evident on faceted or grooved stones called *ventifacts* (e.g. Breed, McCauley & Whitney, 1989). Where precipitation is sufficient, fluvial features are present. *Arid alluvial fans* occur as aprons of sediment emanating from uplands, and ephemeral desert streams or *wadis* can reach into the basin (Sect. 3.3.8). Fine sediment deflated from aeolian provinces can accumulate as *loess*, and approximately 10% of the Earth's land surface is covered by loess or reworked loess, with the most spectacular examples being from China and Central Asia (Pye, 1987, 1995). *Sabkha* generally refers to any evaporite-encrusted flat, and can be inland or coastal (Sect. 8.4). *Playa* refers to a desert lake that dries, while *saline lake* refers to a more permanent lake with large fluctuations in water level and salinity (Chapter 4). The reader is referred to Cooke, Warren and Goudie (1993) for a full treatment of the diversity of the desert setting.

5.2.2 Overview of processes

Processes that operate within aeolian systems in deserts are complex and operate at a variety of spatial and temporal scales. In general, aeolian processes dominate in arid deserts because aridity limits vegetation and its stabilizing effect on the substrate. The relationship between wind and substrate susceptibility, however, is complex and depends not only on aridity (i.e. precipitation and evapotranspiration), but also wind energy and sand supply (e.g. Lancaster, 1988). For example, in many present-day deserts the 15-cm annual rainfall isohyet marks the division between active aeolian systems and those stabilized by vegetation (I.G. Wilson, 1973). However, some Australian ergs now vegetated and largely stabilized would be active under conditions of greater wind energy (Wasson, 1984), and coastal dune fields have been documented to form even under humid conditions where the sand supply is plentiful and the wind energy strong (Pye, 1983). In addition, features such as reg surfaces, a high water table, and evaporitic or pedogenic crusts can stabilize a substrate, while vegetation can cause the accumulation of aeolian sand and loess, and influence dune morphology (Thomas & Tsoar, 1990). In arid deserts, the degree of aeolian-sand development typically reflects sediment supply, and where alluvial sediment sources have been depleted, extensive reg development occurs. Without any external forcing factors, as sand supply wanes over a period of time, sandy aeolian systems contract, allowing adjacent environments to expand. As with other systems, the aeolian system is sensitive to global and regional changes in tectonism, sea level and climate.

Aeolian systems interact with adjacent environments such as fluvial, lacustrine and marine systems, but these interactions are varied. For example, the degree to which fluvial processes dominate in deserts depends on the slope, the nature of the surface and of run-off, and the extent of aeolian sand. Rainfall in deserts tends to be erratic and where the surfaces expose bedrock or a reg or crust of any sort, the substrate is relatively impermeable. Run-off can be substantial and the effects of flowing water are concentrated. In contrast, rainfall quickly soaks into aeolian sand and run-off is minimal. At the marine–aeolian interface, marine processes such as waves and tides, the extent of aeolian sand development, and wind energy and direction are factors that determine the nature of the interface. Where there is little or no freshwater recharge, a low gradient, and a high permeability of the sediments, marine waters can penetrate inland through a coastal desert aquifer for several or even tens of kilometres (Sects 8.4 & 8.5). The intense evaporation rates characteristic of deserts produce evaporite-encrusted surfaces where the surface is in contact with the water table. Landscape and surface evolution, weathering, pedogenesis, diagenesis, and groundwater dynamics are equally varied and commonly unique to arid systems (see overview in Cooke, Warren & Goudie, 1993).

On a larger scale, desert processes have an effect beyond the desert. The simple presence of deserts has an effect on the surface albedo of the Earth, which in turn exerts a control on global climate (Barron, Sloan & Harrison, 1980). The aeolian realm provides atmospheric dust that is an effective agent in inducing climatic change (e.g. Carlson & Benjamin, 1980). Dalrymple, Narbonne and Smith (1985) argued for long-term aeolian reworking of thin Cambro-Ordovician cratonic sequences in North America as an explanation for their paucity of clay and silt. Aeolian sands have been transported into oceanic basins as a result of marine reworking of aeolian accumulations during transgressions (Sarnthein & Diester-Haass, 1977).

It is important to understand that in interpreting geomorphic features in deserts, or using modern deserts to interpret the rock record, that the system may or may not be in *equilibrium* with current conditions. The system *response time* may be slow so that specific features are *relicts* (see *sensitivity* of systems in Thomas & Allison, 1993). Large alluvial fans or extensive wadi networks in deserts may be in equilibrium because they are the products of rare intense rainfall events, or they may represent a previous more humid time. Large dunes require time to respond to changed wind regimes because of the volume of sand needed to be mobilized in order to reshape the dune.

5.3 Aeolian processes and theory

5.3.1 Sediment transport

Since the seminal work of Bagnold (1941), tremendous development in the field of aeolian sediment transport has occurred. A treatment of the present thinking is beyond the scope of this chapter, but many of the salient points are summarized by McEwan and Willetts (1993). The aim here is to highlight a few points about sediment and wind that are basic to the understanding of bedforms, dune fields, and ergs.

The volume or mass of sand that is transported to a dune field or erg is a function of: (i) the generation of *sediment* supply of a suitable grain size at the *source area*; (ii) the *availability* of this sediment; and (iii) the *transport capacity* of the wind.

Typical source areas for sandy aeolian systems are the deflationary areas of alluvial accumulations, shoreline sands, and older sediments, including cannibalization of previously accumulated aeolian sediment. It has long been known that aeolian sands can be transported across the distances of continents (e.g. Fryberger & Ahlbrandt, 1979), so that identifying the source area for a particular dune field or erg is not necessarily easy. The range of grain sizes typically transported by the wind on Earth is more limited than in subaqueous transport because of the much lower density and viscosity of air compared with water. Clay and silt are transported by *suspension* in which the grains are held above the substrate by atmospheric turbulence. Grains smaller than medium silt can be suspended indefinitely and constitute atmospheric dust, whereas medium silt (the typical size of loess) can be transported up to 300 km in one wind storm, and grains approaching the limit of suspension (coarse silt) are likely to fall from suspension within 30 km (Pye & Tsoar, 1987). The common *bedload* is sand, but wind can transport pebbles up to 2 cm. Bedload transport occurs as a cloud of moving grains extending a few centimetres above the surface. Distinct grain motions can be seen within the cloud as end-members that together form a continuum. Most grains bounce or *saltate* with enough energy to bounce back into the flow and/or dislodge other grains (*ejecta*). Some grains are only weakly ejected and do not rebound or eject other grains upon impact with the bed (*reptation*), while others are affected by the turbulence in the air (*modified saltation*). Yet other grains move in short jerks along the bed when they are impacted by saltating grains, which is termed *surface creep*. Most steep dunes consist of grains between 0.1 and 0.3 mm in diameter (very fine–medium sand), the grain size that most easily saltates. Coarse sand and granules move by surface creep and typically only during episodes of high winds. By way of *selective transport*, the wind is extremely effective in *sorting* a sediment.

The generation of a suitable grain size for wind transport at the source area does not mean that this sediment is available for transport. Wind tunnel work is nearly always done using a substrate covered by dry, loose, well-sorted sand (i.e. sand that is perfectly *available* for transport by the wind). However, in nature surface dampness, encrustation, cementation, vegetation, reg surfaces, and poorly sorted grain populations all restrict the availability of sand for transport. Thus it is important to

consider the *available sediment supply* – that is, the amount of sediment on the surface at any given time and available for transport (Kocurek & Havholm, 1993).

As a first approximation, the time-averaged grain *transport rate*, q, is a cubic function of the *shear velocity*, u^*, above threshold values. The shear velocity is roughly a measure of the velocity gradient above the bed, and, therefore, proportional to the *shear stress* exerted on the bed. Although shear stress is believed to be the primary control on transport rate, the role of *turbulence* is increasingly recognized as important. Most of the quantification of sand transport is from wind tunnel studies over flat surfaces. Taking this work to nature where sloping surfaces, non-neutral atmospheric conditions, and changing surface roughness occur is not straightforward (e.g. Rasmussen, Sorensen & Willetts, 1985; Frank & Kocurek, 1994). In general, however, a given wind has a *potential transport rate*, q_p, and where the surface is covered by dry, loose, well-sorted grains, this potential is realized in a matter of seconds (R.S. Anderson & Haff, 1988). Where grains are restricted in availability for any reason or the surface is only partly covered by sand, the *actual transport rate*, q_a, may only slowly reach the potential transport rate, if at all. The *sediment-saturation* level of the flow is the ratio q_a/q_p. When this ratio is less than one, the flow is at least potentially *erosional*. A *steady-state* transport rate exists when the ratio is one and, on average, the surface is one of *bypassing*. When the flow decelerates, q_p decreases and *deposition* occurs until the ratio again reaches unity to adjust to the new, lower transport capacity.

When sand is transported by the wind usually *bedforms* or regular, asymmetric surface undulations form. Within minutes after transport begins *ripples* form. Requiring more time to form, from hours for the smallest (Kocurek, Townsley *et al*., 1992) to possibly thousands of years for the largest, are *dunes*. Dune fields and ergs can be thought of as collections of dunes, and they require only sand and wind to form. The exception to the formation of bedforms with bedload transport is a flat bed (analogous to upper flow regime in water) that occurs with sand at high wind speeds. Recently, ripples have been explored as self-organizing patterns of sediment transport (R.S. Anderson, 1990), and perhaps similar thinking can be applied to dunes. In this view, bedforms are nature's way of organizing and transporting sediment. There is, however, almost always a component of *throughgoing* sediment in dune fields and ergs that is distinct from the sediment moved with the dunes and consists of the suspended load and the bedload that bypasses the dunes.

5.3.2 Dunes and airflow

Airflow and sediment transport over dunes present a very challenging problem, but some general aspects can be outlined that are important in understanding dune strata in the rock record. Dunes are regularly spaced obstacles in the path of the general or *primary* wind. The wind causes the dunes to form, but the dunes also modify the wind in both speed and direction. This dune-modified airflow is the *secondary flow*.

For the dune shown in cross-section in Fig. 5.2(a), the windward or *stoss slope* of the dune is inclined at about 10–15°, and the *lee slope* is at the *angle of repose*, about 33°. From a consideration of *continuity*, the flow must *accelerate* over the inclined stoss slope, which causes an increase in the transport rate, q, and the stoss slope must be erosional. Recent work shows that the zone of accelerated flow occurs within a very thin internal boundary layer (Burkinshaw, Illenberger & Rust, 1993; Frank & Kocurek, 1996). In some cases between the highest part of the dune or the *crest*, and the break in slope at the *brink* is a flat or gently downwind-sloping surface. In other cases the crest and the brink coincide. The flow *decelerates* somewhat as it approaches the brink, in part because of *flow expansion* between the crest and brink, and in part because of a back pressure exerted on the flow by the *back eddy* or *separation cell*. *Flow separation* begins near the brinkline, and the flow *reattaches* at a distance of a few dune heights downwind as measured from the brink. The *Bernoulli* inverse relationship between velocity and pressure causes the fluid in the expanded flow field of the separation cell to flow back toward the higher velocity, lower pressure zone near the brink, only to be sheared by the overflow. On the lee slope, q is markedly lowered and

deposition occurs. Wind ripples that typically cover the stoss slope end abruptly at the brinkline and the lee face is dominated by gravity-driven processes (Fig. 5.2b). From the *reattachment* point, the near-surface airflow again accelerates within a thin growing boundary layer for a distance of several dune heights until *flow recovery* occurs. The portion of the interdune area over which the flow accelerates must be at least potentially erosional.

Important departures from the case in Fig. 5.2(a,b) occur primarily because of the nature of the *break in slope* between the stoss and the lee faces, and the *angle of incidence* at which the wind strikes the brinkline or crestline. The dune illustrated in Fig. 5.2(c), as with Fig. 5.2(a,b), has an incidence angle of 90° or the flow is *transverse* and can be described as *two-dimensional*, but in Fig. 5.2(c) the lee slope is gentle ($\approx 15°$ or less). Flow separation does not occur and the flow remains *attached*. Flow expansion and deposition still occur on the lee face, but wind ripples that migrate up the stoss slope can migrate down the lee slope, and the deposits on the lee slope are *tractional* in origin. In Fig. 5.2(d), the wind strikes the brinkline *obliquely*, and the flow is deflected *alongslope* on the lee face. In some cases, the alongslope lee airflow is *deflected* and *attached*, but in other cases an alongslope back eddy *vortex* forms (e.g. Sweet & Kocurek, 1990). In either case, the flow is *three-dimensional* and the lee-face flow direction can be at a

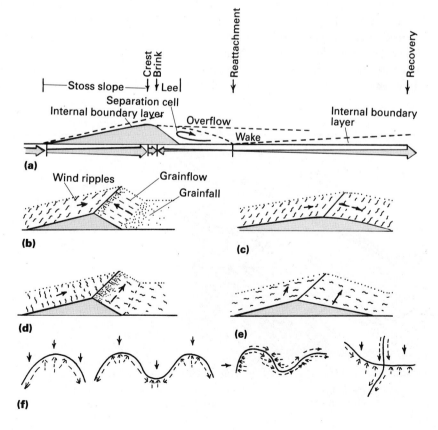

Figure 5.2 Airflow over dunes. (a) Dune in cross-section with transverse flow, showing secondary airflow zones and relative transport rates (large arrows). (b) Transverse flow over steep dune with back eddy showing surface processes. In (b)–(f), solid arrow is primary flow, dashed arrow is surface secondary flow. (c) Transverse flow over rounded dune. (d) Oblique flow. (e) Longitudinal flow. (f) Common dune crestlines in plan-view (solid lines) showing primary and secondary lee flow (modified from Sweet & Kocurek, 1990; Kocurek, 1991; Sweet, 1992).

high angle to the primary wind direction. Away from the lee face, the flow again recovers in speed and direction over several dune heights. Because the airflow on the lee face is alongslope, wind ripples that form are oriented with their crestlines parallel to the dune lee slope and they migrate alongslope. Gravity-driven depositional processes can still predominate on the lee face where alongslope winds are insufficient to rework the sand transported over the brinkline. On larger dunes that have smaller dunes superimposed on the lee face, the smaller dunes also migrate alongslope (e.g. Sweet, 1992). Tsoar (1983) and Sweet and Kocurek (1990) have shown that the wind speed on the lee slope is a cosine function of the incidence angle, such that the lee wind speed approaches zero at a 90° incidence angle (transverse), increases as the incidence angle becomes more oblique, and is approximately equal to the primary wind speed at a 0° incidence angle (*longitudinal*, Fig. 5.2e). Some dunes in nature can be represented by the single configurations shown in Fig. 5.2(a–e), but many dunes consist of transverse, oblique, and longitudinal *elements* for any given primary wind, and it is the configuration of these elements that determines the secondary flow and the nature of the lee-face surface processes (Fig. 5.2f).

5.3.3   Lee-face processes and stratification

From the previous section, deposition on dunes must be considered largely a lee-face process. The recognition of distinct lee-face processes and stratification dates from Bagnold (1941, pp. 236–246), with better definition by McKee, Douglass and Rittenhouse (1971), but Hunter (1977a) provided the first systematic description of *stratification types*. On lee faces where tractional processes dominate, wind ripples occur (Fig. 5.2c–e), and *wind-ripple laminae* form as the stratification type where deposition occurs. Where gravity-driven processes dominate,

two distinct deposits form with dry sand (Fig. 5.2b). *Grainfall* occurs when the haze of grains in saltation shoots past the brinkline to fall rather passively on to the lee face. The grains do not settle from suspension because they were never suspended by turbulence. Grainfall on the lee face results in a wedge of sediment that is thickest just downslope of the brinkline and thins downslope, possibly as a negative power function (Hunter, 1985a; Anderson, 1988). For 'small' dunes, the grainfall wedge can extend to the base of the lee face, and an *apron* of grainfall forms (Figs 5.2b & 5.3). For a given wind energy, as dune size increases, the grainfall wedge extends proportionally shorter distances downslope, so that on 'large' dunes the wedge is confined to the upper portion of the lee face. The steepest part of the wedge builds to the *angle of initial yield*, then fails or avalanches to give rise to the second gravity-driven process, *grainflow* or *sandflow*. The grainfall–grainflow process can be viewed as a repetitive redistribution of lee-face sediment to a static angle-of-repose slope. The portion of the lee face marked by grainflow is termed the *slipface*. Grainfall deposits on the lower portions of the lee face of small dunes are overridden by grainflow sediments, and the two stratification types occur interstratified. Grainflow deposits on small dunes usually occur in tongues encased in grainfall strata (Fig. 5.3). On larger dunes, where the grainfall wedge does not extend to the base of the slipface, grainflow deposits alone are represented on the lower portions of the slipface. With the introduction of moisture, the grainflow process becomes modified to reflect a variety of *cohesive-structure failures* such as *faults, slumps*, and *breccias* (McKee, Douglass & Rittenhouse, 1971). Hunter, Richmond and Alpha (1983) described '*wet*' *grainfall* during rain, and this and other types of *adhesion processes* that occur with damp sand result in the building of deposits beyond the static angle of repose for dry sand.

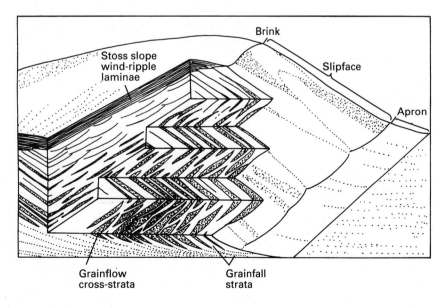

Figure 5.3 'Small' transverse dune, with slipface and basal apron, showing horizontal and cross-sections with stratification (from Hunter, 1977a).

In appearance the stratification types are distinct and usually can be recognized with confidence on the outcrop, and are sufficiently fine in scale that they can also be identified in core. *Grainflow cross-strata* are tabular, tapering-upward wedges at the angle of repose that exhibit loose packing and typically a two-fold grain-size segregation. The process of *shear sorting* during the flow causes coarser grains to rise to the surface, resulting in *inverse grading*. At the surface, some of the coarser grains outpace the finer grains and become concentrated at the toe of the flow. *Grainfall deposits* are typically indistinctly laminated and show an intermediate packing. Differences do exist between subaqueous and aeolian grainflow strata, and grainfall is less readily preserved in subaqueous cross-strata (Hunter, 1985b; Hunter & Kocurek, 1986). *Wind-ripple forms* are sometimes preserved, such as where they are buried by grainfall deposits, but *wind-ripple laminae* are more common. Without question, these are the most distinctive aeolian stratification type, and their recognition allows the identification of aeolian cross-strata in the rock record. Each lamina is the product of a single wind ripple. The laminae are thin (a few millimetres thick), closely packed and distinct. In some cases foresets can be seen, especially if the surface is shaved so that the thickness of individual laminae is exaggerated, but typically foresets are not apparent. The laminae are commonly inversely graded, reflecting grain segregation on wind ripples, in which the coarser grains are near the upper part of the lee slope.

Important inferences can be drawn from stratification patterns regarding the nature of lee-face processes and winds during deposition. Wind ripples on the lee face necessarily mean that lee-face winds were sufficient to cause traction transport. The presence of grainflow or grainfall deposits means that gravity-driven processes overwhelmed any tractional processes that may have existed. Because wind energy increases on the lee face as the angle of incidence decreases, a progression from grainfall/grainflow processes to wind ripples is expected with a decreasing incidence angle. Deposits shown in Fig. 5.4(a–c) are marked by grainfall and/or grainflow deposits, indicating that gravity processes dominated, a situation most typical of transverse flow. Grainfall deposits in Fig. 5.4(a) indicate either a 'smaller' dune (more likely) or higher winds than in Fig. 5.4(b) (see Hunter, 1981). The unusual occurrence of grainfall in Fig. 5.4(c) might indicate an abnormally high wind. Figure 5.4(d–f) shows grainflow and wind-ripple deposits that differ in their spatial arrangement. In Fig. 5.4(d), grainflow and wind-ripple migration occurred simultaneously, indicating the presence of both lee-face, along-slope winds that dominated the lower part of the lee face (where grainfall rates were lowest) and sufficient grainfall higher on the lee face to cause frequent avalanching that dominated here in spite of the lee-face winds. This configuration is indicative of an oblique flow condition. In contrast, Fig. 5.4(e) shows alternate periods of grainflow and wind-ripple deposition, with the wind-ripple wedges overlying an erosional surface. This scenario best represents alternate periods of transverse winds

causing slipface advance by grainflow and periods of oblique winds causing lee-face modification accompanying alongslope transport. In Fig. 5.4(f), grainflow and wind-ripple deposits are repeated, but in contrast to Fig. 5.4(e), no evidence is seen for any significant change in dune shape. Figure 5.4(e,f) illustrates patterns that are *cyclic*. Figure 5.4(g) shows deposits of only wind-ripple laminae, which can be explained by three interpretations. Firstly, the deposits could represent the same oblique configuration as Fig. 5.4(d), but more severe truncation has occurred. Secondly, the deposits could also represent an oblique or longitudinal lee face characterized by ripples only. Finally, the deposits could represent a low-relief transverse element, in which the ripples migrated down the lee face. Note that the slope of the laminae is not indicative of the migration direction of the ripples.

It is important to recognize that a dune can have several configurations of stratification types. As is evident from Fig. 5.2(f), a curving brinkline or multiple dune arms allow a full range of transverse to oblique to longitudinal local flow conditions to occur simultaneously on a single dune. It is erroneous to assume that the style of stratification seen in a two-dimensional outcrop represents the style throughout the entire dune deposit. Conversely, where the deposits of the total bedform can be seen, as in a horizontal section, then the stratification types can be used to reconstruct flow conditions.

5.3.4 Accumulation

Accumulation is the generation of a body of strata that may be *preserved* and incorporated into the rock record. Figure 5.5 shows migrating dunes that are leaving an accumulation. The *accumulation surface* is a plane joining the bedform *troughs* such that all sediment above the accumulation surface is in transport either as bedforms or as throughgoing sediment, and the sediment below the surface is not in transport and constitutes the accumulation. In Fig. 5.5 the accumulation surface is horizontal, but it can be inclined upwind or downwind. The behaviour of the accumulation surface over time ($\delta h/\delta t$) can be described by the *sediment conservation equation*:

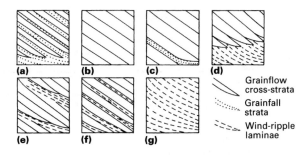

Figure 5.4 Common variations in lee-face stratification types (from Kocurek, 1991).

$$\left[\frac{\delta h}{\delta t} = -\left(\frac{\delta q}{\delta x} + \frac{\delta c}{\delta t} \right) \right]$$

where q (dimensions of $l^3/lt = l^2/t$) is the spatially averaged bulk volume *transport rate* per unit width of the flow, and c (dimensions of $l^3/l^2 = 1$) is the *concentration* of sediment in transport or the spatially averaged bulk volume of sediment above a unit area of the accumulation surface (Rubin & Hunter, 1982; Middleton & Southard, 1984). For accumulation to occur, a decrease in the downstream transport rate and/or a decrease in concentration over time must occur, regardless of whether accumulation is from the bedload or the suspended load (i.e. loess accumulation). Accumulation of dune deposits means that sediment deposited on the lee slope is not completely eroded on the stoss slope and in the *interdune trough* with migration. Similarly, interdune accumulation occurs where interdune deposits are not cannibalized before burial with the passage of the next dune.

Migrating dunes that leave an accumulation must move upward or *climb* with respect to the accumulation surface, and this *angle of climb* (θ) is measured with respect to this surface (Fig. 5.5). The accumulation from each migrating and climbing bedform is a *climbing translatent stratum* (Hunter, 1977b), which crosses accumulation surfaces and is time-transgressive. For dunes, this is a *set of cross-strata*, and for wind ripples a single lamina. The *vector of climb* (V) can be resolved into x and y components in which Δh is Vy or the *rate of accumulation*, and Vx is the *rate of bedform migration*. The tan $\theta = Vy/Vx$. For most natural situations, the vertical accumulation rate (Vy) is small in comparison to the downwind migration rate (Vx), so that the angle of climb is very low. Only the lower portions of foresets accumulate, and the set is just the record of the basal lee processes. This is termed a *subcritical* angle of climb (Hunter, 1977a,b). A *critical* angle of climb equals the steepest part of the stoss slope, and the set represents the entire bedform. At a *supercritical* angle of climb, lee and stoss accumulation occur, and laminae can be traced uninterrupted from set to set. Where dunes are good sand traps or conserve their sand, the bulk of the sand must either be in the dune or in the set of cross-strata.

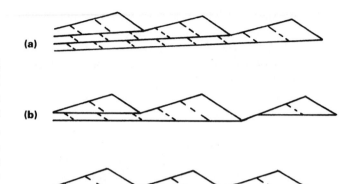

Figure 5.6 Set thickness as a function of scour depth. (a) Uniform depth of scour. (b) Spatial variation in scour depth where although the angle of climb is zero, a set is formed because of the greater scour depth of the middle dune. (c) Variation in scour depth with dune migration (here cyclic) causing wavy bounding surfaces (in part from Paola & Borgman, 1991).

A bedform can then generate a finite set, and set length decreases as the angle of climb increases. Figures 5.5 & 5.6(a) are over-simplified because they show a constant depth of scour and, therefore, a constant set thickness. In nature, there may be variation in the scour depth from dune to dune (Fig. 5.6b), or scour depth may change with migration of a single dune (Fig. 5.6c). The stochastic treatment of variation in the depth of scour by Paola and Borgman (1991) initiates a more realistic approach to climbing models.

5.3.5 Modelling of sets through space and time

In order to visualize the set of cross-strata that a present-day dune might generate, or to interpret a set of ancient cross-strata, it is necessary to link the *shape* of the dune to its *behaviour* over time. This is far from a simple task for most dunes. One approach is to trench modern dunes or to find natural exposures where the internal structure is revealed (e.g. McKee, 1966). There are limits to trenching, however, and while new technologies such as ground-penetrating radar (Schenk, Gautier *et al.*, 1993) are promising, there is always the danger of assuming that the internal structure of one dune is characteristic of all morphologically similar dunes without understanding why specific features of the cross-strata occur. The most creative approach to date is computer simulation (Rubin, 1987a), in which aspects of bedform shape and behaviour are fed into a program that uses sine curves to simulate the bedform and then derive a variety of information from the simulations (Fig. 5.7a,b). The program allows one to see why a particular aspect of the internal structure occurs, and sensitivity testing can be done in which specific parameters are allowed to vary. The program is, however, purely geometric; airflow dynamics are ignored and stratification types *per se* are not identified.

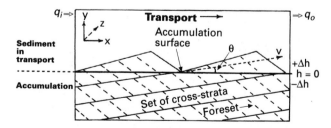

Figure 5.5 Generation of sets of cross-strata by migrating and climbing dunes shown in cross-section (x,y), where all transport is in the x direction and a unit width (z) is representative. Symbols given in text (modified from Kocurek & Havholm, 1993).

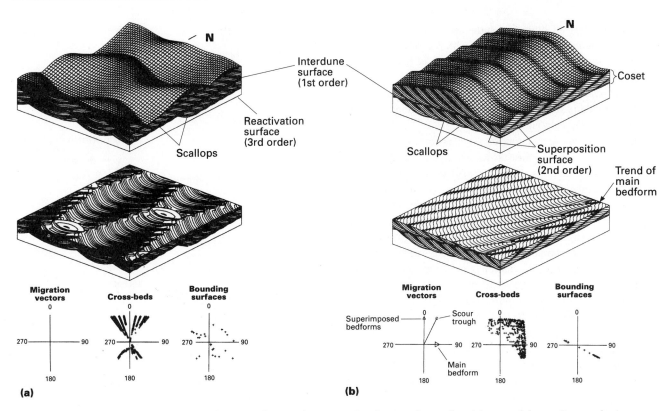

Figure 5.7 Computer simulation of migrating dunes with internal structure and plots of cross-strata and bounding surfaces. (a) Simple, sinuous dunes. (b) Straight-crested main dune with alongslope-migrating superimposed straight-crested dunes. Generated using program from Rubin (1987a).

Figure 5.7(a) shows *sinuous* dunes that migrate normal to their crestlines, in which the crestline curvature from dune to dune is *out-of-phase*. Migration speed has been varied with time to simulate cycles in which a period of advance is punctuated by a period of lee-face reworking and change in asymmetry. Migration of the stoss slope and interdune trough form scour depressions that are subsequently filled by the lee-face deposits. The dunes are subcritically climbing. The sets show classic *trough cross-strata* in which cross-strata symmetrically fill the trough-shaped scours, and the sets are symmetrically truncated by younger sets. Because the depth of scour is greatest in the lee of concave-downwind dune segments, only these filled troughs are accumulated and the entire bedform is not represented. Basal wedges accumulate when the dune profile is reworked to one that is nearly symmetrical, and the wedges are buried by steeper foresets when the dune becomes asymmetrical and advances most quickly. A plot of the foresets reflects both the span of the troughs and the low-angle foresets of the basal wedge. Similarly, the *bounding surfaces between sets* and the *internal bounding surfaces* plot differently; the latter are subparallel to the foresets, while the former reflect the angle of climb.

Figure 5.7(b) shows a dune in which the main bedform migrates normal to its crestline, while *superimposed* dunes migrate alongslope. The troughs of the two scales of *straight-crested* bedforms intersect to form a topographic depression that behaves like a scour depression. As seen in the horizontal section, migration of the larger bedform while the superimposed dunes migrate alongslope causes an oblique trace to the sets of the latter. In cross-section parallel to the crestline of the main bedform, the climbing aspect of the superimposed dunes is evident, and in the perpendicular cross-section *scallops* result. The scallops are formed because each successive trough cuts into the deposits of the previous bedform, truncating the foresets in the general migration direction.

It is evident from Fig. 5.7(a,b) that the simple measuring of foreset dip directions to interpret bedform type and migration, and to draw inferences about the palaeowind direction must be done with great care. Measurement of foreset distribution is not very meaningful unless the bedforms are very simple or the bedforms are reconstructed and it is understood why particular patterns result.

5.3.6 Generation of bounding surfaces

A bounding surface is an erosional surface within or between sets of cross-strata. A set of cross-strata is *simple* if it does not contain internal bounding surfaces, *compound* if it has internal bounding surfaces (e.g. Fig. 5.7a,b). A *coset* is composed of related sets, each separated by a surface (e.g. Fig. 5.7b). *Scalloped cross-strata* are compound sets where the bounding surfaces cyclically scoop into the previously deposited foresets within the set and sometimes into underlying sediment (Rubin & Hunter, 1983) as in Fig. 5.7(a,b).

The *reactivation* or *redefinition surfaces* formed within the set in Fig. 5.7(a) occur because the lee face is periodically eroded. In computer simulations, Rubin (1987a) formed this type of bounding surface by changes in dune migration direction, dune asymmetry, dune migration speed such that a period of advance is punctuated by a period of erosion, and dune height such that the dune grows by scour of the underlying sediment. These sorts of changes are common because natural flows are rarely steady, and if the period of flow fluctuation is regular, then the surfaces are cyclic.

Superposition surfaces form by migration of dunes or scour troughs superimposed on the lee face of the main bedform (e.g. Fig. 5.7b). Although superimposed dunes can migrate directly up or down the lee face of the main bedform, some component of alongslope migration is probably the most common scenario because of deflected, alongslope secondary airflow.

Interdune surfaces are formed between sets (Fig. 5.7a) or between the cosets (Fig. 5.7b), and separate accumulations of the different bedforms. These surfaces originate with erosion that begins on the stoss slope and progresses to the depth of scour defined by the passage of the interdune trough. Although the angle is not generally recognizable on a single outcrop because it is so low, over greater distances in the migration direction this surface is inclined at the angle of climb.

Identification of the type of bounding surface in the field is not a trivial exercise. For outcrops oriented parallel to the migration direction of the dune or the main bedform where superimposed dunes occur (e.g. Fig. 5.7a,b, respectively), scalloped surfaces occur with both lee-face reactivation and migration of superimposed dunes. Where an outcrop perpendicular to the migration direction also occurs, the two types of surfaces can be distinguished. As a general rule, reactivation surfaces can be distinguished from superposition surfaces because the former are subparallel to the foresets, while for the latter a difference in the mean dip direction occurs between the surfaces and the foresets (Rubin, 1987a,b). The exception here is where the superimposed dunes migrate directly up or down the lee face of the main bedform, so that foresets and bounding surfaces are subparallel. Using a stereonet to plot the mean attitude of the cross-strata as one plane, and the bounding surface mean orientation as a second plane, the line of intersection of the two planes defines the trend of the superimposed dunes (Fig. 5.8). Assuming that the superimposed dunes migrate perpendicular to this general crestline trend, then the line perpendicular to the line of intersection defines the migration direction of the superimposed bedforms (Rubin & Hunter,

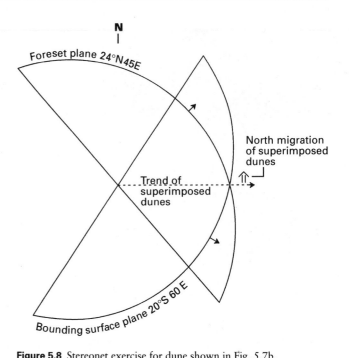

Figure 5.8 Stereonet exercise for dune shown in Fig. 5.7b, determining the trend and migration direction of superimposed dunes from plot of foresets and bounding surfaces, using method in Rubin and Hunter (1983).

1983). In a general sense, the migration direction of the main bedform is evident by the trend of the outcrop that shows the scallops (Fig. 5.7b). If a horizontal section is available, adjacent troughs migrate through the horizontal along a trend that approximately defines the main bedform trend (Fig. 5.7b).

In an earlier generation of thinking on bounding surfaces, Brookfield (1977) erected a hierarchy of bounding surfaces (i.e. *1st-*, *2nd-*, and *3rd-order surfaces*) that in general can be related to present thinking (Fig. 5.7a,b). However, much of the usefulness of this scheme has been lost with the advent of the computer' simulations, and this terminology should be abandoned. Examples, including the ones Brookfield first used, can be found in which all three surfaces are present, and the hierarchical level of truncation and extent of the surfaces that Brookfield envisaged are apparent. However, as already addressed for Fig. 5.7(a,b), the 3rd- and 2nd-order surfaces are very similar in sections parallel to bedform migration, and a hierarchical scheme does not then help in surface identification. Moreover, the surfaces do not break into universally distinct groups by extent or dip angle.

5.3.7 Generation of aeolian sequences

At a dune-field or erg scale, times of overall accumulation must end, necessitating yet a fourth type of bounding surface, a *super surface*, which caps the accumulation (Talbot, 1985; Kocurek, 1988). Considering the accumulation and its bounding super surface to define a *sequence*, with the surface being the *sequence boundary*, Kocurek and Havholm (1993) have suggested an approach to aeolian *sequence stratigraphy* based on the sediment conservation equation introduced in Section 5.3.4. Figure 5.5 can be used to represent an entire aeolian system or some portion thereof. Defining a *control volume* (x,y,z in Fig. 5.5) is a useful tool that allows a mass balance between the sediment *flux* of the system, measured as *influx* (q_i) and the *outflux* (q_o) across the yz planes of the control volume, and the sediment within the control volume that is present as either sediment in transport or as the accumulation. The *sediment budget* of the system, a concept from Mainguet and Chemin (1983), is the balance between influx and outflux, and can be *positive* ($q_i > q_o$), *neutral* ($q_i = q_o$), or *negative* ($q_i < q_o$). The influx is a function of the sediment availability from upwind of the control volume and the transport capacity of the flow. The outflux is a function of spatial changes that may occur within the control volume. From the sediment conservation equation, for accumulation to occur $\delta h/\delta t$ must be positive, and a super surface or sequence boundary occurs when $\delta h/\delta t$ changes from positive to zero or a negative value. Because accumulation can only occur when $q_i > q_o$, with the balance stored within the control volume, the sediment budget must be positive. Similarly, when $q_i = q_o$, a *bypass super surface* occurs ($\delta h/\delta t = 0$), and when $q_i < q_o$, an *erosional super surface* occurs ($\delta h/\delta t$ is negative). Changes in the elevation of the accumulation surface over time occur because of spatial change in the transport rate ($\delta q/\delta x$), and temporal change in the concentration of sediment in transport ($\delta c/\delta t$). Because concentration includes all sediment above the accumulation surface, concentration varies with average dune height.

In order to understand why accumulation and sequence boundaries occur in aeolian systems, it is necessary to understand what cause changes in the transport rate and concentration. Processes operating in aeolian systems define three end-members: dry, wet, and stabilizing systems (Kocurek & Havholm, 1993). In *dry aeolian systems*, the water table has no effect on the substrate and no stabilizing factors occur so that deposition, bypass, and erosion along the substrate are controlled by the aerodynamic configuration alone. *Wet aeolian systems* are those where the water table or its capillary fringe intersects the accumulation surface, so that both the aerodynamic configuration and the moisture content of the substrate determine whether the accumulation surface is depositional, bypassing or erosional. *Stabilizing aeolian systems* are those in which stabilizing factors such as vegetation or surface cementation play a significant role in determining the behaviour of the accumulation surface. A spectrum from subaqueous environments to sabkhas to wet aeolian systems to dry aeolian systems can be envisaged as a function of the available sediment supply over time or spatially (Fig. 5.9).

Interdune areas over which flow accelerates are at least potentially erosional (Sect. 5.3.2), so that in dry systems where aerodynamics alone are important, it is unlikely that accumulation

occurs until *interdune flats* have been eliminated by dune growth and only an *interdune depression* remains (Fig. 5.10). Accumulations in dry systems, therefore, should be characterized by an absence of interdune-flat accumulations, but the basal portions of sets may represent accumulation in interdune depressions. The simplest explanation for accumulation in a dry system is a downwind decrease in the transport rate caused by a downwind deceleration. A sequence boundary in a dry system could form when flow deceleration ceases, or the sediment-saturation level of the influx decreases to the point where the downstream deceleration is no longer sufficient to maintain the positive sediment budget.

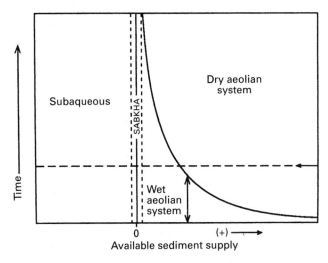

Figure 5.9 Environments based on available sediment supply while a relative rise in the water table occurs. Initially the depositional surface coincides with the capillary fringe. At any given time (t), the environmental tract reflects the available sediment supply laterally (modified from Kocurek & Havholm, 1993).

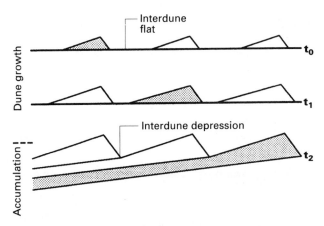

Figure 5.10 Period of dune growth through time (t_0–t_1) precedes onset of accumulation (by t_2). With dune growth, interdune flats reduce in size until only a depression remains between dunes.

In wet aeolian systems, the effect of the high water table greatly reduces erosion in interdune flats and the basal portion of dunes within the capillary fringe. With a relative rise in the water table, both dune and interdune deposits accumulate, the vertical accumulation rate (V_y from Sect. 5.3.4) equalling the rate of the water table rise (Fig. 5.11a). As the water table rises and accumulations form, the interdune flats grow at the expense of the dunes. In this example, because dune height decreases over time, a temporal decrease in concentration explains accumulation, and at any given time the transport rate does not decrease downstream because the dunes are all the same size and migrating at the same rate. If interdune flats parallel the accumulation surface, then even at very low angles of climb, the thicknesses of sets and interdune accumulations are proportional to their downwind extents along the depositional surface, but are very thin (Fig. 5.11b). If the interdune flats, however, show a central depression and truncation occurs at the level of the dune bases, only amalgamated interdune deposits accumulate (Fig. 5.11c). Where the net angle of climb is positive so that accumulations form, but on a shorter time scale, the angle of climb fluctuates from positive to zero and negative values, sets and interdune accumulations are likely to consist of amalgamated, discontinuous and dune lenses separated by surfaces representing periods of erosion and bypass (Fig. 5.11d). While a rising water table causes accumulation, a water table static relative to the accumulation promotes a bypass super surface, and a relative falling water table promotes an erosional super surface.

Stabilizing aeolian systems are potentially more varied than either wet or dry systems, because the nature of what causes accumulations or super surfaces can be so varied. For example, where vegetation baffles transport, the transport rate may be progressively reduced downwind. On the other hand, periodic introduction of stabilizing factors, such as with deposition of a subaqueous mud drape over interdune areas and the lower portions of dunes, affects the concentration in that average dune height is lowered. Super surfaces develop when the cause for accumulation in the system overall (e.g. a partial vegetative cover) is reduced so that the sediment budget switches from positive to neutral or negative, or conversely, when the effects

of the stabilizing agent are enhanced, such an increase in vegetation to the point where dunes are totally stabilized and the system becomes inactive.

Super surfaces, already classified as surfaces of erosion or bypass, can be further divided by the nature of the substrate: *stabilized* or *unstabilized*, and *dry* or *damp* (Fig. 5.12). Although super surfaces may initially reflect the type of aeolian system, the surfaces may continue to evolve so that the final configuration seen in the rock record does not reflect the initial or intermediate configurations. For example, an erosional, unstabilized, dry surface that initially cannibalizes the accumulation may progress until it reaches the water table and be represented in the rock record by a stabilized, damp surface of bypass. Because a super surface is defined as marking the end of aeolian accumulation, the special case exists where aeolian accumulation ends because of a change from aeolian to a non-aeolian environment without a gap in the stratigraphic record (Fig. 5.12). In practice, a super surface can be difficult to distinguish from an interdune surface (1st-order surface) at the outcrop or even greater scales. The following are general criteria that favour an interpretation of a

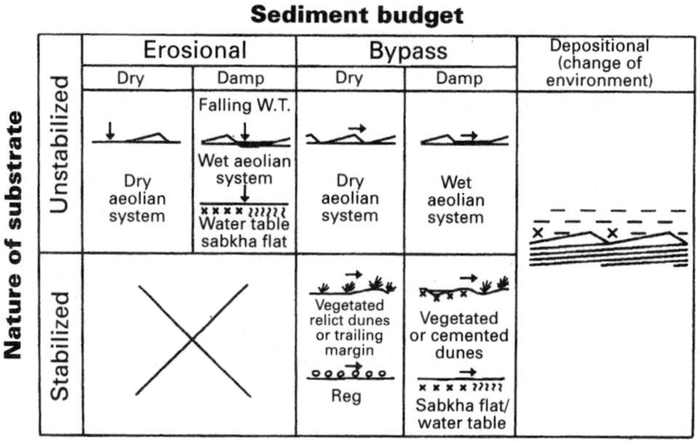

Sediment budget

	Erosional		Bypass		Depositional (change of environment)
	Dry	Damp	Dry	Damp	
Unstabilized	Dry aeolian system	Falling W.T. Wet aeolian system / Water table sabkha flat	Dry aeolian system	Wet aeolian system	
Stabilized			Vegetated relict dunes or trailing margin / Reg	Vegetated or cemented dunes / Sabkha flat/ water table	

Figure 5.12 Classification of super bounding surfaces based on sediment budget, then nature of the substrate. Downward arrows indicate erosion, horizontal arrows indicate bypass (from Kocurek & Havholm, 1993).

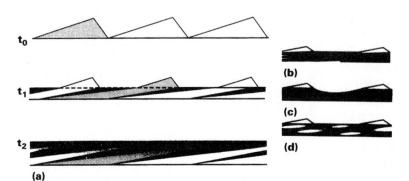

(a) **(b)** **(c)** **(d)**

Figure 5.11 Wet aeolian system. (a) Accumulation with a rising water table with time (t_0–t_2). One dune and its accumulation shaded for continuity. (b) Thin, continuous dune and interdune accumulations where interdune flats coincide with accumulation surface. (c) Amalgamation of interdune accumulations where interdune flats are concave and depth of scour is at or below dune base. (d) Lensoidal accumulation of dune and interdune strata because angle of climb fluctuates between positive and negative, or variation in scour depth occurs (modified from Kocurek & Havholm, 1993).

super surface: (i) different styles of cross-strata above and below a surface, suggesting accumulation within separate aeolian systems; (ii) distinct laterally extensive surfaces; (iii) surface features that are different from those of other surfaces suggesting a different origin; (iv) surfaces that bound accumulations of entire dune fields or ergs, as opposed to those that cap accumulations of single bedforms; (v) correlation of surfaces to basin-wide events such as marine transgressions; and (vi) surfaces that are horizontal and truncate climbing surfaces made by migrating bedforms.

The sequence approach by Kocurek and Havholm (1993) was developed for dune systems, but because the accumulation of loess from the suspended load must follow the same conservation principle (Sect. 5.3.4), a parallel system can be erected. Loess accumulates in areas of wind deceleration, on to a wet or damp substrate, or in vegetation (Pye & Tsoar, 1987; Tsoar & Pye, 1987); these respectively correspond to accumulation in dry, wet and stabilizing systems.

5.3.8   Preservation of aeolian sequences

Preservation of an aeolian sequence occurs when the sequence is placed below some regional baseline of erosion. Aeolian systems differ from some other types of systems, such as marine systems, in that an accumulation can build well above any baseline of erosion and have little *preservation potential*. In other words, *accumulation space* may not coincide with *preservation space* (Kocurek & Havholm, 1993). Factors that promote preservation are subsidence and a rise of the water table (Fig. 5.13). Subsidence occurs because of tectonism, loading and compaction. A rise in the water table can be relative with the accumulation subsiding through a static water table, or be absolute with a climatic change to more humid conditions or the inland response to a rise in sea level. An accumulation can also be stabilized by vegetation, a reg or some other factor, and while the accumulation is more resistant to erosion, it may still be above a regional baseline of erosion (Fig. 5.13). Preservation of sequences of dry systems can occur with subsidence, a rise of the water table through the dry accumulations, or by some combination of subsidence and a water table rise. Because accumulation in a wet aeolian system is determined by the water table, then this component of preservation space coincides with the accumulation space.

5.4   Present-day aeolian systems

5.4.1   Occurrence, accumulation and preservation

Dune fields and ergs require only available sand and wind to form (Sect. 5.3.1). Whether these leave accumulations (Sect. 5.3.7), and whether these accumulations are preserved in the rock record (Sect. 5.3.8) are very different matters. The Sahara (Fig. 5.14) shows a full range of aeolian systems from

Rise in relative sea level caused by subsidence or eustatic sea level rise

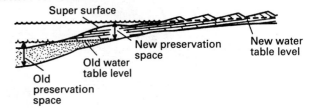

Rise in relative continental water table caused by subsidence or absolute water table rise

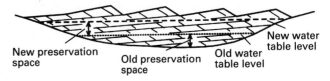

Subsidence of the accumulation below the baseline of erosion

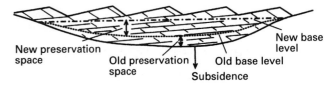

Stabilization of the accumulation surface above baseline of erosion

Figure 5.13 Modes of preservation of aeolian accumulations (from Kocurek & Havholm, 1993).

those with a positive to those with a negative sediment budget (Mainguet & Chemin, 1983). At present, most of the ergs of the Sahara are dry aeolian systems so that, given a supply of available sand, the sediment budget is determined by aerodynamic conditions (Sect. 5.3.7). A positive sediment budget can be expected where wind deceleration occurs, a negative sediment budget where acceleration occurs, and a neutral budget where the flow is uniform. Deceleration can occur because of large-scale atmospheric circulation patterns, or the interaction of the wind and the land surface. Examples of the latter include topographic basins where the wind decelerates because the flow expands vertically, and deceleration in the lee or upwind of highlands.

Most of the central and southern Sahara has a negative sediment budget in which the ergs are exporting sand and very little influx occurs because the sand sources, largely alluvial

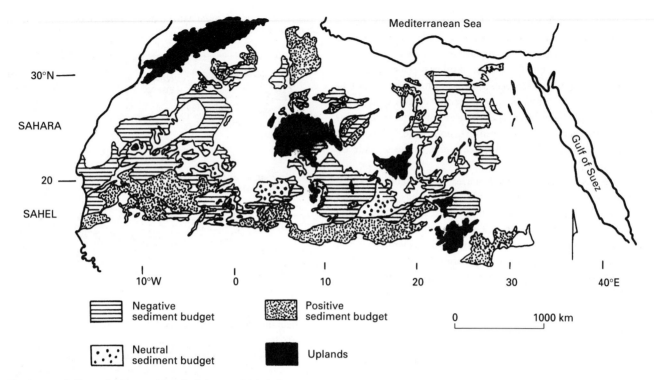

Figure 5.14 Sediment budget map for the Sahara and Sahel (from Mainguet & Chemin, 1983).

systems from previous Pleistocene–Holocene humid periods, have been exhausted (I.G. Wilson, 1973). Examples include the Erg Chech in Algeria where dunes are separated by wide interdune flats deflated to reg or bedrock (Mainguet & Chemin, 1983), and in Mauritania where the present generation of dunes is cannibalizing Pleistocene dunes (Kocurek, Deynoux *et al.*, 1991). Ultimately, areas characterized by a negative sediment budget deflate to regs or bedrock.

A positive sediment budget is found in some northern Saharan ergs and in the Sahel to the south. The Grand Ergs Occidental and Oriental in Algeria occupy topographic basins and have accumulated with an influx of sand from alluvial and lacustrine sources adjacent to the surrounding uplands. In the Sahel, the trade winds transport sand from the Sahara and decelerate as they approach the intertropical convergence zone (Sect. 2.1.2). Vegetation is also important here in trapping sand and promoting accumulation (Mainguet & Chemin, 1983), so that these ergs have to be considered in part as stabilizing systems (Sect. 5.3.7). Neutral systems (Fig. 5.14) occur between sand-exporting systems to the north and ergs in the Sahel with a positive sediment budget.

The Alexandria coastal dune field of South Africa is a dry system where a positive sediment budget occurs because winds decelerate onshore owing to increased frictional drag of the land surface (Illenberger & Rust, 1988). In Namibia, onshore winds produce a system of bypassing and erosional dunes that feed into the main erg system inland. These winds decelerate

inland and become more variable so that a positive sediment budget characterizes portions of the inland Namibian erg (Lancaster, 1989a; Corbett, 1993). Documented examples of wet systems are coastal or inland systems within basins where the water table is high. The dunes on Padre Island, Texas, are a microcosm of the control the water table exerts on the depositional surface. Although influx of new sediment is zero and the sediment budget is neutral because no sediment leaves the field, accumulation occurs in the winter when the water table rises and dune and interdune sediment accumulate. In the summer as the water table falls, the accumulation surface falls and the dunes grow larger by eroding the substrate (Kocurek, Townsley *et al.*, 1992). Accumulation of wet system deposits at Guerrero Negro in Baja, Mexico, and the Jafurah Erg of Saudi Arabia occurred with a relative rise of the water table (Fryberger, Al-Sari & Clisham, 1983; Fryberger, Schenk & Krystinik, 1988; Fryberger, Krystinik & Schenk, 1990).

Regardless of the sediment budget, preservation will not occur without subsidence or a relative rise of the water table so that preservation space is generated. Most of the Saharan ergs, even where an accumulation is building because of a positive sediment budget, are poor candidates for preservation because they occupy an inland stable cratonic setting. Ergs, however, in the Algerian–Tunisian foreland basin have a stratigraphic record where aeolian sediments are interbedded with alluvial and lacustrine sediments (Coque, 1962). During Pleistocene lowstands of sea level, ergs on the western African coast prograded far on to the now submerged continental shelf; where not totally reworked by marine processes, these erg accumulations have been incorporated into the stratigraphic record by a rise in sea level (Sarnthein, 1978).

5.4.2 Variations

A striking aspect of many present-day sandy aeolian systems is their variety of geomorphic elements. Different types of dunes (Sect. 5.4.3) occur not only in different ergs, but also within a single system. Interdune flats range in features, shape and extent (Sect. 5.4.5), and a variety of sand sheets occurs (Sect. 5.4.6). Loess occurs along the margins of some deserts, but is relatively

uncommon (Tsoar & Pye, 1987; Pye, 1995). Aeolian systems occur adjacent to a variety of coastal, alluvial and lacustrine environments, which are spatially arranged over an intricate topography of uplands, depressions and plains. Mapping from Landsat images by Breed, Fryberger et al. (1979) is still the best readily available source for documenting the complexity of major aeolian systems. Variation may well be expected in vast ergs like those of the Sahara, but even small dune fields such as the Algodones can be very complex (Fig. 5.15).

The spatial proximity of environments implies an interrelationship of processes. Sourcing of aeolian systems by alluvial or lacustrine systems may occur contemporaneously, or may follow in time, as with deflation of alluvial or lacustrine sediments only after a humid period has ended and an arid period has ensued (e.g. many Saharan ergs, Sect. 5.4.1). The Algodones, sourced by deflation of lacustrine and beach sand, has now largely lost its sand source with the drying of the lake and vegetating of the substrate. Aeolian and alluvial systems are simultaneously active at the Great Sands Dunes in Colorado, where fluvial–aeolian interactions include: (i) the control of fluvial-flood paths by aeolian landforms; (ii) interdune deposition caused by fluvial overbanking and an associated rise in the water table; and (iii) the recycling of sands through aeolian and subaqueous environments (Langford, 1989). At the aeolian–marine transition (Fig. 5.16), a complex array of interactions is common (Chan & Kocurek, 1988). The complexity of the accumulations in this transition is illustrated by the deposits at Guerrero Negro (Fig. 5.17) where tidal flooding of the dune field occurs (Fryberger, Krystinik & Schenk, 1990).

The cause for the spatial arrangement of geomorphic elements and environments presents the most difficult question. While the configuration and interaction of environments within some aeolian and associated environments can be understood in terms of present-day processes, our overall understanding of the dynamics of these systems is poor. Although wind regime and sediment availability are important factors (Sect. 5.4.3), it is unclear why different dune types evolve within an erg or differ from system to system. For example, at the Algodones, evidence indicates that the variety of dunes are contemporaneous; can sand supply, grain size, dune-field-wide secondary flow patterns,

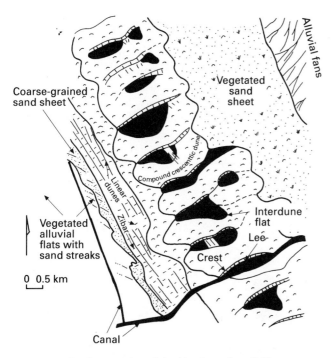

Figure 5.15 Southern portion of the Algodones dune field, southeastern California, drawn from aerial photo, illustrating the spatial variation across the dune field.

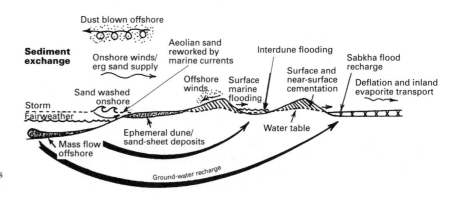

Figure 5.16 Common aeolian–marine process interactions (from Chan & Kocurek, 1988).

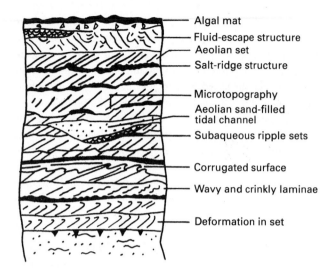

Figure 5.17 Schematic diagram illustrating coastal interdune sabkha and dune deposits at Guerrero Negro, Baja, Mexico (simplified from Fryberger, Krystinik & Schenk, 1990).

and perhaps yet other factors explain the variation in dune type (Fig. 5.15)? The present, however, is but a point in time, and as sandy aeolian systems evolve with changing conditions, not all features may be in equilibrium with current conditions. In the Gran Desierto of northwestern Mexico, a mosaic of dune types occurs (Fig. 5.18a) along with three distinct sand populations spatially (Fig. 5.18b). Stratigraphically, it has been demonstrated that three separate aeolian accumulations or sequences bounded by super surfaces occur, and these are exposed along the present surface (Blount & Lancaster, 1990; Lancaster, 1992, 1993) (Fig. 5.18c). The present-day configuration, therefore, is a composite, probably dating from the middle Pleistocene, in which periods of construction were separated by periods of relative geomorphic stability. The slow rate of accumulation with respect to the frequency of climatic change causes the separate sequences to be exposed; a more rapid rate of accumulation would foster their burial.

5.4.3 Classification of dunes

Thus far in this text, dunes have been treated 'generically' in order both to simplify the discussion and to express the view that because dunes occur in many complex shapes, any classification is somewhat arbitrary. Unfortunately the long history of work in present-day sandy aeolian systems has generated a correspondingly lengthy and conflicting terminology

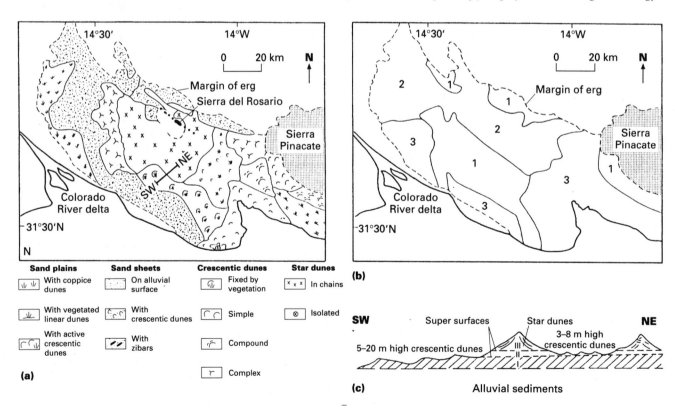

Figure 5.18 Gran Desierto, Sonora, Mexico. (a) Surface geomorphic elements. (b) Distribution of three distinct surface grain populations (1, 2, 3) by composition reflecting different generations of sand accumulation. (c) Generalized cross-section (SW–NE) indicated on (a), showing composite nature of surface features and corresponding stratigraphy (modified from Blount & Lancaster, 1990; Lancaster, 1992).

for classifying dunes (see table of synonyms assembled by Breed & Grow, 1979, pp. 284–296), and no one classification scheme is universally accepted, and the same term may differ in usage in the literature. The approach taken here is to classify dunes both descriptively or *morphologically* (e.g. Figs 5.15 & 5.18a), and *morphodynamically*. The morphological scheme (McKee, 1979) is based on a global view from space, a vantage point from which similarities between dunes are more apparent than differences (Fig. 5.19a). The morphodynamic scheme (Hunter, Richmond & Alpha, 1983) classifies dunes by crestline orientation with respect to the long-term *resultant transport vector* (Fig. 5.19b). Because morphologic and morphodynamic classifications are based upon different parameters, different sorts of information are needed to classify dunes in these dual schemes, and the classifications overlap (Fig. 5.19c).

The most common morphological dune types are linear and crescentic, with star dunes less common (Fig. 5.19a). The family of *crescentic* dunes includes *barchan* dunes, and *barchanoid* or *crescentic ridges*, and consists of dunes with a distinct asymmetry and common crestline sinuosity (e.g. Fig. 5.7a). *Linear* dunes are more symmetrical, with sinuous or straight crestlines, and typically have a gentle- to moderate-dipping *plinth* upon which the slip-face is perched. *Star* dunes are multi-armed dunes with distinct crestlines radiating from one or two central peaks. A variety of more specialized dune types also occurs. *Parabolic* dunes are U-shaped dunes associated with partial stabilization by vegetation in which the central portion migrates while the stabilized portions give rise to extended arms that point upwind. Low-relief, rounded dunes associated with a grain size too coarse for dunes with slipfaces are *zibars*. The term *dome* dune has been used to refer to barchan dunes lacking a slipface. Mounds of sand developed around and in the lee of clumps of vegetation are called *coppice* dunes. In addition, dunes can be *simple* (e.g. Fig. 5.7a) or they can have *superimposed* dunes (e.g. Fig. 5.7b). Where the same kinds of dunes are superimposed (e.g. crescentic dunes superimposed on a larger crescentic dune), the entire bedform is termed a *compound* dune. Where different kinds of dunes are superimposed (e.g. crescentic dunes superimposed on a larger linear dune), the entire bedform is a *complex* dune (Fig. 5.18a). *Draa* has been used to refer generically to large compound or complex dunes.

Morphodynamically, dunes can be classified as longitudinal, transverse and oblique (Fig. 5.19b). *Transverse* dunes are those with crestlines oriented perpendicular (±15°) to the resultant transport vector. *Longitudinal* dunes are those with crestlines oriented parallel (±15°) to the resultant transport vector. *Oblique* dunes have crestlines oriented roughly between 15° and 75° from the vector. The most satisfying explanation as to why some dunes are transverse, while others are oblique or longitudinal, arises from the concept of 'maximum gross bedform–normal transport' (Fig. 5.20a) by Rubin and Hunter (1987) and Rubin and Ikeda (1990). Bedforms tend towards equilibrium with their flow conditions, but there is a time lag in response that, all else

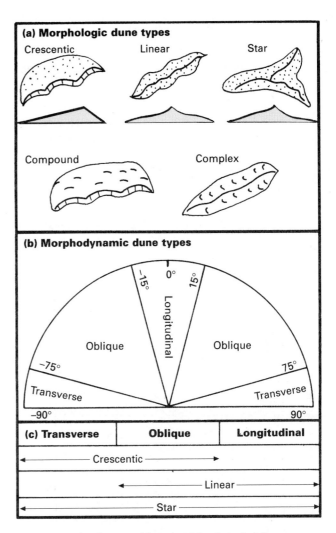

Figure 5.19 Classification of dunes. (a) Morphological dune types shown in plan view and cross-section for simple dunes, and plan view for compound and complex dunes. (b) Morphodynamic dune types based on orientation of crestline relative to resultant transport direction. (c) Probable range of morphological and morphodynamic dune types (modified from Hunter, Richmond & Alpha, 1983; Kocurek, 1991).

being equal, increases with bedform size. Where a period of flow fluctuation occurs and this is shorter than the bedform response time, but long enough that the bedforms can begin to respond (e.g. many larger dunes in a seasonally varying wind regime), the bedforms tend to orient themselves to be as transverse as possible to each flow direction. The bedform trend relative to the resultant transport direction (i.e. transverse, oblique, longitudinal) is always such that maximum gross transport occurs normal to the bedform crestlines. Using two flow directions, the range of morphodynamic bedforms was found experimentally to vary only with the angle between the two flow directions and the proportions of sand transported in the two directions (Fig. 5.20b).

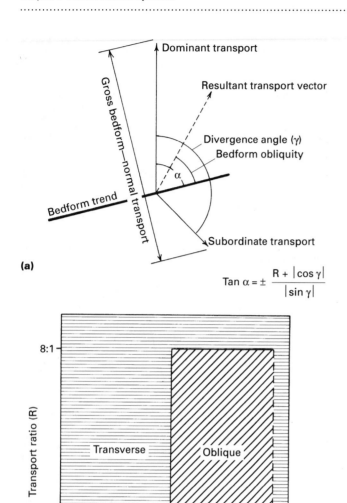

(a)

$$\text{Tan } \alpha = \pm \frac{R + |\cos \gamma|}{|\sin \gamma|}$$

(b)

Figure 5.20 Gross-bedform-normal transport where (a) the divergence angles (γ) and the transport ratio (R) determine bedform trend. Angles are defined in a clockwise direction, with a plus sign in the equation giving the correct value for divergence angles from 90 to 180° (modified from Rubin & Hunter, 1987; Rubin & Ikeda, 1990). (b) Generalized occurrence of morphodynamic dune types found experimentally as a function of the transport ratio and the divergence angle for two transport directions.

The cause of the different types of dunes has long been controversial, largely because the causes of dune formation are not well understood. The concept of the gross-bedform-normal argues that transverse, oblique and longitudinal bedforms are not distinctly different features, but rather are following the same rules of orientation (Rubin & Hunter, 1987). Morphologically, it can be argued that nature forms crescentic

dunes, and while these occur where a single wind direction predominates, other dune shapes occur with more complex wind regimes. Linear dunes occur especially in a bimodal wind regime where dune asymmetry reverses so that plinths evolve from stoss slopes, and develop with alongslope lee airflow. Star dunes are typical of multidirectional wind regimes. The role of the dune type and the volume of available sand for dune building is controversial. Dune size, however, is important in dune type at least to the extent that a combination of wind energy and dune size determine the reconstitution time of dunes, essentially the 'dune memory' of A. Warren and Kay (1987). As illustrated at Dumont, California, large star dunes represent too great a mass of sand to be completely reformed by seasonal wind directional change, and they maintain their star characteristics, while small crescentic dunes become star-dune like and then evolve into crescentic dunes of a new orientation that is transverse to the new wind direction (Nielson & Kocurek, 1987). While compound or complex dunes can occur wherever dune size reaches a point where smaller dunes can occur superimposed (Lancaster, 1988), it is clear that some compound or complex dunes, especially those of very complex shape, are multigenerational. The complex dunes of the Akchar Erg, for example, show modern crescentic dunes superimposed on Pleistocene linear features, and it is unknown whether this configuration represents an equilibrium condition (Kocurek, Deynoux *et al.*, 1991).

5.4.4 Dunes, airflow, stratification and cycles

For illustration of stratification types in dry sand in horizontal and various cross-sections, the definitive work of Hunter (1977a) from dunes on Padre Island and the Oregon coast remains a standard. The geometric arrangement of dry slip-face deposits of grainflow and grainfall on small dunes during transverse flow (Fig. 5.3) is readily seen on Padre Island. Strong winter-storm winds are responsible for the shape, orientation, and migration of the Oregon dunes. These winds are typically accompanied by rain, whereas dry, moderate summer winds only modify the dunes (Hunter, Richmond & Alpha, 1983). The occurrence of the most important winds during the wet season results in reduction of the actual transport rate (q_a, Sect. 5.3.1) to about a third of the calculated potential rate (q_p) had the sand been dry. Using this reduced transport rate to determine the resultant transport direction, these crescentic dunes are oblique. In addition, because much of the transport occurs during strong winds with rain, the stratification is dominated by adhesion structures, wet grainfall deposits, and sliding and slumping of cohesive masses of sand. The latter are also illustrated by Fryberger (1991). These winter deposits, plus less common dry grainflow strata, dominate the lee-face deposits, and override summer deposits consisting of a basal wedge of alongslope-migrating wind-ripple laminae that pinch out against reactivation surfaces (Fig. 5.21).

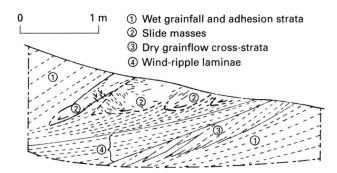

① Wet grainfall and adhesion strata
② Slide masses
③ Dry grainflow cross-strata
④ Wind-ripple laminae

Figure 5.21 Drawing from trench in the lower stoss slope of Oregon dune revealing cyclic winter (deposits 1–3) and summer (deposit 4) lee-face deposits (from Hunter, Richmond & Alpha, 1983).

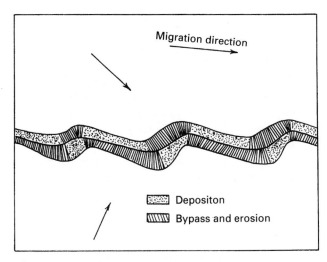

Figure 5.22 Composite areas of deposition, erosion and bypass on a sinuous linear dune that occur on the lee of the dune for the two primary wind directions (arrows) (from Tsoar, 1983). Winds from the upper portion of the diagram strike the dune obliquely, causing erosion on the entire upwind face of the dune, but alternating sites of deposition and erosion/bypass on the lee face according to the incidence angle. Oblique primary winds from the lower portion of the diagram largely reverse the process.

Winter–summer changes in stratification in the Oregon dunes result in *annual cycles* (Sect. 5.3.3). Both the Oregon and Padre Island dunes show *daily cycles* formed during the summer in which sea breezes, calm in the morning, progressively strengthen during the day with little directional change (Hunter & Richmond, 1988). Because the rate of sand transport increases as a cubic function of u^* (Sect. 5.3.1), while the migration rate of wind ripples increases linearly, the angle of climb of the ripples increases during the day, and ripple laminae may give way to grainfall deposits. When the angle of climb is low, the fine, dense, heavy minerals in the ripple troughs are preferentially enriched in the laminae, but as the angle of climb increases, the coarser, less-dense quartz grains on the upper part of the ripple lee slope become incorporated into the laminae.

The relationship of the incidence angle of the primary wind, secondary flow on the lee face, and lee surface processes are especially well illustrated on linear and star dunes (Sects 5.3.2 & 5.3.3). Winds striking linear dunes obliquely are deflected along the lee slope (Fig. 5.22) as shown by Tsoar (1982, 1983). Because of the dune sinuosity, the incidence angle varies along the dune from transverse (where grainflow and grainfall deposition occur) to highly oblique (where alongslope migrating ripples and erosion occur). Secondary flow on star dunes at Dumont maintains the arms of the dune as simultaneously active transverse, oblique, and longitudinal elements for any given wind (Nielson & Kocurek, 1987). Patterns of erosion and deposition on a star dune in the Gran Desierto in three primary wind directions (Fig. 5.23) show that seemingly complex flow conditions can be understood by analysis of individual dune segments with respect to the incidence angle of each primary wind (Lancaster, 1989b).

Monitoring of the dynamics of a compound crescentic dune over a 4-year period in the Algodones (Fig. 5.15) demonstrated that although these compound dunes are transverse to the resultant transport vector, seasonal components of the wind strike the crestline obliquely from alternate directions, producing a reversal in the migration direction of superimposed dunes.

Curvature of the main bedform apparently causes the lee face to vary from a simple slope with alongslope-migrating ripples (and grainflow during wind storms), to superimposed dunes migrating alongslope and obliquely down the lee face. Distinctly different styles of stratification, including a lateral change from simple to compound cross-strata occur, therefore, along the lee of a single bedform (Sweet & Kocurek, 1990; Sweet, 1992).

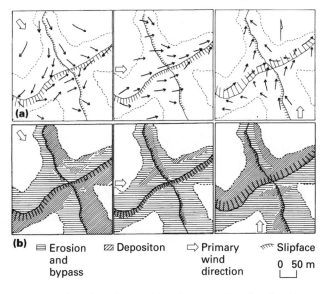

Figure 5.23 Star dune showing (a) surface secondary flow for three primary wind directions, and (b) resultant areas of erosion, bypass and deposition (modified from Lancaster, 1989b).

5.4.5   Interdune areas

Interdune areas are the troughs between dunes, ranging from interdune flats to interdune depressions (Sect. 5.3.7). In present-day aeolian systems there is a complete gradation from where the flats cover a greater portion of the depositional surface than the dunes, to where dunes and flats are equally represented, to where depressions only occur between the lee and stoss slopes of adjacent dunes. Interdune areas vary in shape by dune type: (i) ellipsoidal, commonly fully enclosed areas elongate with the dune trend for crescentic dunes; (ii) very long corridors paralleling linear dunes; and (iii) more varied, enclosed to interconnected, equidimensional to irregular areas for star dunes. Interdune areas also range from those that are deflationary and exist only as geomorphic features, to those where interdune strata accumulate. Deflationary interdune areas range from where bedrock, reg or older, unrelated deposits are exposed in some dry systems, to interdune flats floored by deflating cross-strata in wet systems. Particularly characteristic of damp or cemented surfaces in wet systems are *corrugated surfaces* developed by differential erosion of dune stratification types in which wind-ripple deposits typically form ridges and grainflow units form swales.

Processes and sedimentary structures in depositional interdune areas have been classified by the nature of the interdune depositional surface – *dry, damp* and *wet* (Kocurek, 1981; Ahlbrandt & Fryberger, 1981). This classification of interdune deposits largely corresponds to dry and wet aeolian systems (Sect. 5.3.7), but not entirely as, for example, wind ripples may form on a dry interdune substrate within a wet system. In dry systems where interdune accumulation is restricted to interdune depressions, wind ripples and grainfall, extending from dune plinths or aprons into the depressions, are very characteristic, as are extensions of superimposed lee dunes. Where there is a marked grain-size change between dunes and interdune areas, granule ripples and zibars characterize the interdune areas while fine sand is swept on to the dunes (e.g. linear dunes with zibars in the Algodones, Fig. 5.15). The greatest variety of interdune structures occurs in wet aeolian systems, with a major division being between interdune areas that contain evaporites and those that do not. Damp/wet interdune areas of the Oregon dune field (Hunter, Richmond & Alpha, 1983) and Padre Island (Hummel & Kocurek, 1984) lack evaporites and are dominated by *adhesion laminae* and *adhesion ripples* formed by wind-blown sand over wet surfaces (Hunter, 1973; Kocurek & Fielder, 1982) plus small-scale subaqueous features such as wrinkle marks, ripples, channels, and fluid-escape structures formed in ponds or during floods. The interdune flats on Padre Island are spatially and temporally heterogeneous, depending upon moisture content and sand availability (see surface maps in Hummel & Kocurek, 1984). Glennie (1970) was the first to emphasize the distinctly wavy and typically contorted and brecciated nature of interdune sabkha bedding. Much of this

fabric results from the nature of deposition over the irregular salt-ridge microtopography (Fryberger, Al-Sari & Clisham, 1983; Fryberger, Al-Sari *et al.*, 1984). Penecontemporaneous and subsurface deformation continues with evaporite dissolution and reprecipitation, collapse of the salt-ridge fabric, dune loading, and fluid escape. This overall wet-system interdune fabric, occurring with thin sets showing corrugated upper surfaces, is seen at Guerrero Negro (Fig. 5.17), the Jafurah area and at White Sands by Fryberger, Schenk and Krystinik (1988). Damp/wet surfaces are commonly sites where fine sediment, otherwise transported out of the system, collects. Bioturbation is also common in some interdune areas.

5.4.6   Sand sheets

Sand sheets are areas of aeolian sand where dunes with slip-faces are generally absent. The mere occurrence of sand sheets raises a basic question – why is the sand not organized into dunes? A comparison of six present-day sand sheets in North America led Kocurek and Nielson (1986) to suggest that sand sheets occur because dune development is inhibited by one or more factors such as: (i) a limit on the supply of available sand for dune building because of a high water table surface, cementation or binding; (ii) a limit on the time available for dunes to form, as when there is periodic flooding; (iii) a predominance of sediment too coarse for dune formation; and (iv) the presence of vegetation that not only limits the available sand supply by surface stabilization, but disrupts the airflow and breaks up migrating dunes.

The range of factors that promote sand sheets suggests their occurrence in diverse settings. Many sand sheets are transitional to dune fields or ergs (e.g. Figs 5.15 & 5.18). The sand sheets at Great Sand Dunes, Colorado, occur on the trailing upwind margin of the main dune mass where coarser sediment, vegetation, a high water table and surface cementation by trona are all factors that promote the sand-sheet development (Fryberger, Ahlbrandt & Andrews, 1979; Kocurek & Nielson, 1986). Similarly, the western coarse-grained sand sheets of the Algodones (Fig. 5.15) occur on the trailing margin of the dune field as a wind-reworked lag (Sweet, Nielson *et al.*, 1988). The sand sheets on the Colorado River delta pass transitionally into sabkha as a high water table and surface cementation by evaporites progressively decrease the available sediment supply (Kocurek & Nielson, 1986). The Selima Sand Sheet of southwestern Egypt and northern Sudan forms a vast area (on the order of 100 000 km²) of imperceptible slope and unbroken wind fetch across which a relatively thin veneer of coarse sediment migrates (Breed, McCauley & Davis, 1987).

The range of sedimentary structures in sand-sheet accumulations reflects the many processes that form sand sheets. Because, by definition, deposits of dune slip-faces are relatively rare, wind-ripple laminae are the primary structures of sand sheets. These range from sand ripples to granule ripples, such as those illustrated by Fryberger, Hesp and Hastings (1992) in Namibia. Ripples associated with bedforms such as zibars or coppice dunes occur in sets, while a more chaotic arrangement of laminae is characteristic of wind-rippled flats. The zibars at the Algodones migrate and leave an accumulation of relatively finer sediment deposited as low-angle (<15°) wind-ripple laminae on the zibar lee face where flow expansion occurs, and a coarse lag sediment in the interzibar troughs (Nielson & Kocurek, 1986). Bedforms on the Selima Sand Sheet are probably also a type of zibar (Breed, McCauley & Davis, 1987). These bedforms show relief up to 10 m with a spacing of just under 0.5 km. The internal structure consists of nearly horizontal laminae. Also within the Selima Sand Sheet, Landsat images show light and dark chevron-shaped patterns formed by discrete bedforms migrating at a rate up to 500 m year^{-1} (Maxwell & Haynes, 1989). These features are hardly discernible on the ground because of their low amplitude (10–30 cm) and long spacing (130–1200 m). Sand sheets associated with damp – wet surfaces or periodic flooding contain structures similar to wet interdune areas (Sect. 5.4.5), but chanelling is more common. Vegetated sand sheets have a complex internal structure because irregular topography associated with vegetation gives rise to laterally varying sites of deposition and erosion, and root-turbation, pedogenesis, and bioturbation are common. In some vegetated sand-sheet accumulations, the internal structure has been completely destroyed by vegetation.

5.4.7 Pleistocene–Holocene sequences

Recognition of the composite nature of some present-day aeolian systems such as the Gran Desierto (Sect. 5.4.2; Fig. 5.18), is important not only in understanding how systems evolved to their present configurations, but also in providing insights as to how aeolian sequences develop. The approach to aeolian sequence stratigraphy in Section 5.3.7 defined the response of aeolian systems in terms of first principles (i.e. sediment conservation principle), but our understanding of how a specific change (e.g. increased rainfall, tectonic uplift, rise in sea level) affects aeolian and related systems is not yet developed into a coherent process–response model. While the Pleistocene–Holocene is not representative of most geological time, it is a period of rapid and dramatic change in climate and sea level, which has directly or indirectly influenced aeolian systems. Particularly advantageous for the study of Pleistocene–Holocene sequences is the application of new dating methods such as luminescence, which measures for quartz or feldspar grains the time since the last exposure to light (e.g. Wintle, 1993), thereby allowing the establishment of a chronology of events in absolute time.

The Sahara–Sahel (Fig. 5.14) is one region where the evolution of aeolian systems is chronologically linked to climatic factors. The last glacial maximum (18 000 years BP) coincided with an aeolian *constructional* phase marked by both an expansion and

increased intensity of aeolian processes. The aeolian realm extended well into the Sahel where currently relict dunes are largely stabilized by vegetation (e.g. Sarnthein, 1978; Talbot, 1980). Offshore cores containing fine sand deposited from the aeolian suspended load occur up to 800 km offshore, arguing for intensified winds (Sarnthein & Koopmann, 1980). Beginning about 13 000 years BP, and well established between 11 000 and 4000 years BP, a widespread humid period occurred. This *destructional* phase of the aeolian systems coincided with development of a regional super surface (Talbot, 1985), with stabilization of dunes by vegetation as vegetative zones shifted northward (e.g. Lezine & Casanova, 1989). Lacustrine and marsh deposits from this period are now widely recognized over much of the Sahara (e.g. Petit-Maire, 1989), as are relict fluvial channels, including some of those spectacularly seen with Shuttle Imaging Radar (McCauley, Schaber *et al.*, 1982). Widespread hospitable conditions resulted in an influx of

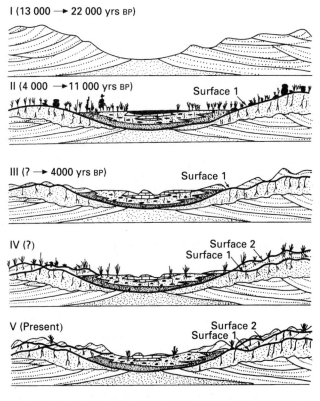

I (13 000 → 22 000 yrs BP)

II (4 000 → 11 000 yrs BP)   Surface 1

III (? → 4000 yrs BP)   Surface 1

IV (?)   Surface 2 / Surface 1

V (Present)   Surface 2 / Surface 1

Figure 5.24 Interpreted sequence of events leading to the present-day complex linear dunes and interdune accumulations in the Akchar Erg of Mauritania. Glacial maximum (I) coincides with dune constructional phase. Humid period (II) results in dune stabilization and pedogenesis with super-surface formation, and accumulation of interdune lacustrine and marsh deposits. For this area, a second, less significant period of dune activation (III), and super-surface formation (IV) pre-date the current reactivation stage (V) marked by continued deflation of interdune accumulations (from Kocurek, Deynoux *et al.*, 1991).

Neolithic peoples (Petit-Maire, 1989). Humid conditions ended about 4000 years BP, with the onset of hyperarid conditions in much of the Sahara by 2000 years BP. Aeolian systems were reactivated, and present-day aeolian offshore dust distributions more closely match glacial maximum than more humid periods (Sarthein & Koopmann, 1980), but, as cautioned by Tsoar and Pye (1987), modern dust accumulation rates are high largely because of the effects of human activities.

Climatic models (e.g. Kutzbach & Street-Perrott, 1985) suggest that the Saharan sequences have been caused by Milankovitch forcing in which the humid period resulted from a strong northward shift in the summer migration of the intertropical convergence zone, bringing monsoonal moisture into the Sahara during maximum solar insolation (about 10 000 years BP). The last Pleistocene aeolian constructional phase was induced by both aridity that reduced the vegetative cover and greater wind energy (Talbot, 1984). The subsequent humid periods induced the destructional phase, resulting in the sequence boundary, characterized by bioturbation and pedogenesis, and the superimposition of fluvial/lacustrine environments over the relict dune topography. Parts of the Sahara today show reactivated dunes as composite features stratigraphically adjacent to lacustrine or fluvial accumulations that are not contemporaneous (Fig. 5.24). Evidence exists for older arid–humid cycles in the Sahara, although these are less well documented. The interrelationship of depositional systems is evident in that the tremendous volume of sand emplaced during the last glacial period would not have occurred if this constructional phase had not been preceded by a long humid phase of alluvial accumulation, which then later sourced the aeolian systems by deflation. The last humid period apparently provided smaller volumes of source-area sands, and as discussed in Section 5.4.1, much of the Sahara today is characterized by a negative sediment budget.

5.5 Ancient aeolian systems

5.5.1 The record

Ancient aeolian systems provide the best record of the dynamics of these systems because their response to forcing factors over significant lengths of geological time can be evaluated. However, the difficulty is that the record is of the response, and the forcing factors typically have to be interpreted. This task, difficult even with Pleistocene–Holocene systems where much of the overall configuration of the setting may still be intact and absolute dating is possible, becomes increasingly hard with the progression of geologic time. Moreover, the selective potentials for accumulation and preservation of aeolian systems (Sect. 5.4.1) cause a bias in the rock record for those systems where accumulations occurred and were preserved, and do not necessarily represent aeolian systems in general.

The most extensive preserved and studied aeolian record is from the Late Palaeozoic and Mesozoic of the Western Interior of the United States (see syntheses in Blakey, Peterson & Kocurek, 1988). This thick, heterogeneous, widespread assemblage is the result of different types of dunes and aeolian systems, complex interactions between aeolian, fluvial, marine and lacustrine systems, and tectonic, eustatic and climatic forcing factors. The overall tectonic setting evolved from a passive margin during much of the Palaeozoic into a retroarc foreland basin in the west and craton in the east during the Mesozoic (Riggs & Blakey, 1993). While the overall trend in the basin-fill geometry can be understood in the regional tectonic setting, subtle tectonic elements also controlled the alignment and distribution of aeolian and other systems (F. Peterson, 1986; Blakey, 1988). Climatic forcing factors ranged from Pennsylvanian–Permian glacial cycles that controlled aeolian sequences (Loope, 1985), to monsoonal circulation from the Late Permian into the Mesozoic (Kutzbach & Gallimore, 1989), to the extreme aridity and high temperatures of portions of the Jurassic (Chandler, Rind & Ruedy, 1992). Aeolian transport patterns for the region hindcast from global climatic models conform fairly well with data from aeolian cross-strata (F. Peterson, 1988; Parrish & Peterson, 1988). Good examples of initial efforts at a wide-ranging view of the systems with respect to the whole configuration of North America, palaeoclimate, tectonic pulses, and fluvial and littoral systems that sourced the aeolian units are illustrated by Marzolf (1988) and Johansen (1988).

Other well-studied systems include the Permian Rotliegend and related units of northwestern Europe, the Proterozoic of India and northwestern Africa, and Lower Palaeozoic of the mid-continent of North America. In contrast to the tectonic setting of the Western Interior, the Rotliegend accumulated in a series of grabens as fluvial, aeolian, sabkha and lacustrine units. This continental facies tract ended with a very rapid Zechstein transgression (see overview in Glennie, 1983, 1986). The largely cratonic Proterozoic and Lower Palaeozoic examples illustrate aeolian systems before the advent of vegetation.

Major recent advances in the understanding of the ancient aeolian record have occurred because of: (i) the application of concepts of sequence stratigraphy (Sects 5.3.4, 5.3.7 & 5.3.8); and (ii) the use of stratification types, bounding surfaces, and computer simulations to reconstruct bedform shape and behaviour, and primary and secondary flow (Sects 5.3.2, 5.3.3, 5.3.5 & 5.3.6). These new methods require a high level of detailed study from the set to the formation scale, and some traditional methods of study such as columnar measured sections, or analysis of foreset attitude without bedform reconstruction have proven inadequate to address current questions.

5.5.2 Sequences

Recognition of sequences (i.e. accumulations with bounding super surfaces) has allowed the primary architectural units of aeolian formations to be identified and correlated over a region.

Within individual sequences, correlative system tracts can be reconstructed from the facies, including types of aeolian systems (Sect. 5.3.7). If the causes of accumulation, super-surface formation, and preservation (Sects 5.3.4, 5.3.7 & 5.3.8) can be determined, it is then possible to relate sequences to specific tectonic, eustatic, climatic and other forcing factors.

The Jurassic Page Sandstone of the Colorado Plateau is an example of a dry aeolian system (see Sect. 5.5.3) that has been mapped in detail to show that the formation consists of a three-dimensional mosaic of accumulations and their bounding super surfaces (Havholm, Blakey *et al.*, 1993; Havholm & Kocurek, 1994). The Page Sandstone developed in a coastal erg that lay along the Carmel inland coastal complex and was bounded to the east by the Monument structural bench (Fig. 5.25a). The Page has a Basal Unit that is a heterogeneous, discontinuous assemblage of aeolian and sabkha units (Fig. 5.25b,c). The main body of the formation consists of three units (Fig. 5.25b,c). The Lower Unit has a central depocentre and pinches out against the Monument Bench. It is bounded in stepwise fashion by super

surfaces that correlate with progressive eastward emplacement of tongues of the Carmel. The Middle Unit consists of several closely spaced sequences bounded by mutually truncating super surfaces. The Upper Unit shows a pronounced southward shift of the depocentre and is locally scoured by large troughs and an unconformity progressively downcuts through the Page eastward (Fig. 5.25c). Facies transitional to coastal sediments and local aeolian build-ups characterize the uppermost Page and lowermost Carmel (Fig. 5.25b,c). Super surfaces throughout the Page Sandstone are characterized by a lateral extent much greater than surfaces formed by migrating bedforms, polygonal fractures (Kocurek & Hunter, 1986), corrugated, erosional relief, and, commonly, overlying sabkha and subaqueous strata. With the recognition and mapping of sequences within the Page Sandstone, this formation that might otherwise be considered a single accumulation, can be viewed as a complex assemblage of pod-like, separate erg bodies. Erg spatial accumulation was, therefore, episodic and punctuated by the formation of super surfaces. The sequence surfaces in some cases can be directly

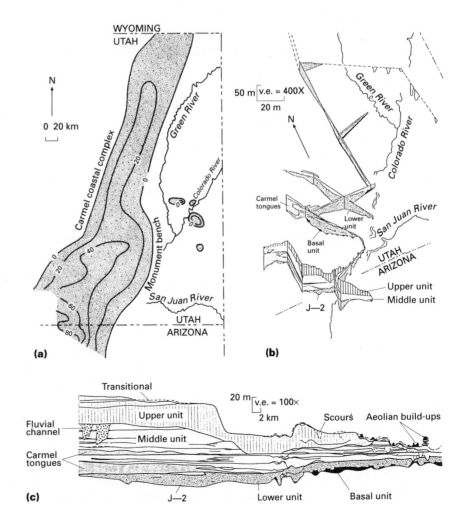

Figure 5.25 Jurassic Page Sandstone. (a) Palaeographic map showing Page isopachs (in metres). Note truncation to the south. (b) Fence diagram showing units discussed in text. (c) Detailed panel approximately along eastern half of the southwestern E–W fence in (b). Lines within major units and the continuous lines bounding units are super surfaces (simplified from Havholm, Blakey *et al.*, 1993).

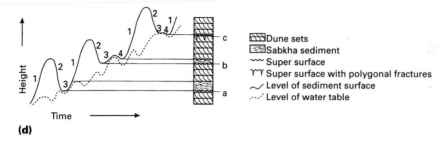

(d)

Figure 5.25 (*continued*) (d) Diagramme illustrating stratigraphy given different patterns of change in accumulation and deflation (solid line) and water table rise (dashed line). Surface *a* is overlain by sabkha sediments that were not eroded, surface *b* is overlain by eroded sabkha sediments, and surface *c* is an erosional surface with polygonal fractures. Numbers 1–4 on the sediment surface curve indicate type of activity: (1) dune accumulation: (2) deflation; (3) sabkha accumulation or formation of polygonal fractures; and (4) deflation of surface features (from Havholm & Kocurek, 1994).

traced to transgressive tongues of the Carmel Formation, and facies relationships show that the water table rose overall through Page time, culminating in transgression of Carmel coastal sediments. The effect, inland, of the water-table rise was to place a limit on the extent of deflation, manifested by super surfaces of bypass or erosion, thereby preserving the unit.

A detailed 2.7-km traverse by Crabaugh and Kocurek (1993) across outcrops of the Jurassic Entrada Sandstone illustrates an approach to the recognition of sequences that differs from the regional correlation of surfaces in the Page Sandstone. The Entrada Sandstone formed as a complex of coastal to inland aeolian and sabkha facies (Fig. 5.26a). Surveying of three-dimensional exposures of sets, 'flat' strata, and bounding surfaces allowed the attitudes of these to be determined (Fig. 5.26b,c). Surfaces interpreted as sequence boundaries are parallel to the lower formational contact and a bentonite drape mid-way in the unit. In some cases they are marked by polygonal fractures or they separate distinct facies. Sets and interdune surfaces and overlying interdune strata climb at angles of tenths of a degree in the transport direction. The traverse reconstruction (Fig. 5.26b) and an idealized version (Fig. 5.26c) show a lower and two upper sequences of climbing sets and interdune strata, each bounded by a super surface. A middle portion consists largely of sabkha strata with thin, discontinuous dune sets, which overall show two cycles of 'wetting-upward'. Subsequent study along eastern Utah has shown that the sequences recognized by the more limited traverse can be traced regionally (Crabaugh & Kocurek, 1996). The combined regional and field evidence indicates that the Entrada is a wet aeolian system (see Sect. 5.5.3) so that accumulation occurred during periods of a rising water table, with super surfaces forming during static or falling water-table levels. The available sediment supply during a period of rising water table and accumulation is a function of the influx of sediment relative to the water-table rise, and can be approximated by the proportion of dune and interdune sabkha along the accumulation surface (Sects 5.3.1 & 5.3.7; Figs 5.9 & 5.11). Either one can hold the rate of water-table rise constant (except during formation of super surfaces) and estimate the sediment influx to the system, or one can hold the influx of sediment constant and plot a relative sea-level curve from this coastal–aeolian unit (Fig. 5.26d). However, it is difficult to distinguish the effects of sediment influx from water-table controls on the stratigraphic record because both controls can yield similar results.

Viewing the Page Sandstone as a collection of sequences preserved by a water-table rise, or recognizing the effects of the balance between sediment influx and the water table on facies of the Entrada Sandstone, is mechanical in that the external driving factors are not identified. In glacial periods such as the Pennsylvanian, cyclically interbedded marine, fluvial and aeolian accumulations in the Hermosa Formation in Utah have been interpreted as driven by glacioeustasy superimposed upon longer-term cycles of tectonic subsidence (Atchley & Loope, 1993). In non-glacial periods, where glacioeustasy is not a factor, climatic change can have a profound effect on rates of weathering and erosion, sediment influx and conditions within the depositional basin, including water-table level and extent of vegetation. The Jurassic Wingate Sandstone of Utah and Permian sandstones from northwestern Europe (Clemmensen, Oxnevad & de Boer, 1994) are all interior continental systems, where eustasy was not a factor, tectonic controls are assumed unlikely to produce short-term cycles, and where the fluvial–alluvial sediment sources can be expected to be strongly affected by climatic change. The Wingate Sandstone consists of five sequences averaging about 23 m thick. Each sequence contains a basal sand sheet accumulation overlain by large sets of cross-strata capped by a super surface (Clemmensen & Blakey, 1989). Clemmensen, Oxnevad and de Boer (1994) infer that the sequences were caused by climatic fluctuations from relatively arid to relatively humid conditions caused by regional shifts in monsoonal circulation, possibly induced by orbital (Milankovitch) variations. Erg accumulation (i.e. a positive sediment budget) is inferred to have occurred during arid climates when sand was deflated from marginal fluvial systems, and sequence boundary formation (i.e. a negative sediment budget) occurred during continued aridity with depletion of

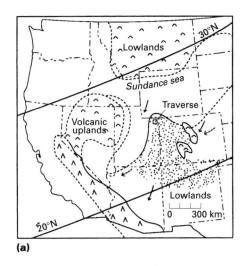

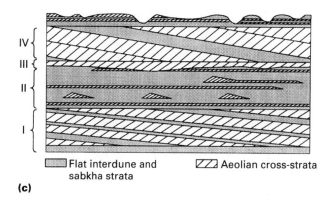

Flat interdune and sabkha strata   Aeolian cross-strata

(c)

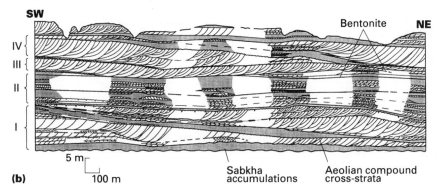

(b)

Figure 5.26 Jurassic Entrada Sandstone. (a) Entrada palaeogeography with box in north patterned area of Entrada indicating area of traverse in (b). Solid arrows represent summer winds, dashed arrow indicates winter wind. (b) Reconstruction of 2.7-km traverse of outcrop located near the general coastline. (c) Idealized version of (b) emphasizing lower (I) and two upper (III & IV) sequences of climbing sets and interdune–sabkha accumulations, separated by middle zone (II) of largely sabkha facies.

the source sands. Basal sand-sheet accumulation may then record the onset of more humid climates.

Intepreting the controls on sequence formation and their underlying forcing factors is often extremely difficult because of the lack of direct evidence. Given the poor time resolution for most aeolian sequences, our limited understanding of the rates of change in eustasy, basin tectonism and climate and their effects on erg dynamics, the possible long periods of time between the production of a source of sand and its delivery to a place where it may accumulate and be preserved, and a large variety of other unknowns, the danger is to replace documentation and a clear understanding of the mechanisms with speculation.

5.5.3   System reconstruction

Reconstruction of aeolian systems involves the interpretation of depositional systems tracts for the aeolian and adjacent accumulations within sequence boundaries. One goal in reconstructing a systems tract is to provide a map of the systems during accumulation, which can then be used in practical application such as hydrocarbon exploration. A systems reconstruction, however, also can be used to aid in interpreting the dynamics

of an entire sequence. For aeolian accumulations, 1st-order interpretations are identification of dune, sand sheet, and loess deposits, followed by the recognition of the type of aeolian system (Sect. 5.3.7). Lateral passage along the systems tract from aeolian into contemporaneous, non-aeolian systems may then allow interpretations of the interactions between the systems. The primary difficulty in reconstructing systems tracts is to determine whether adjacent facies are contemporaneous (i.e. within a sequence), and not merely adjacent stratigraphically and representing distinct events or conditions.

Sequences within the Jurassic Page Sandstone (see Sect. 5.5.2) largely consist of dune deposits, and because the sets of cross-strata are stacked without any intervening interdune-flat deposits, the Page is thought to represent a dry system (Havholm & Kocurek, 1994). However, the sequence-bounding surfaces with polygonal fractures, corrugated relief, and overlying sabkha accumulations, indicate a high water table associated with sequence boundaries. The overall dynamics of the system to emerge is all or part of a four-part cycle: (i) accumulation of aeolian dune deposits within a dry system; (ii) deflation of a portion of this accumulation down to a new, higher water table; (iii) formation of a polygonally fractured crust and/or accumulation

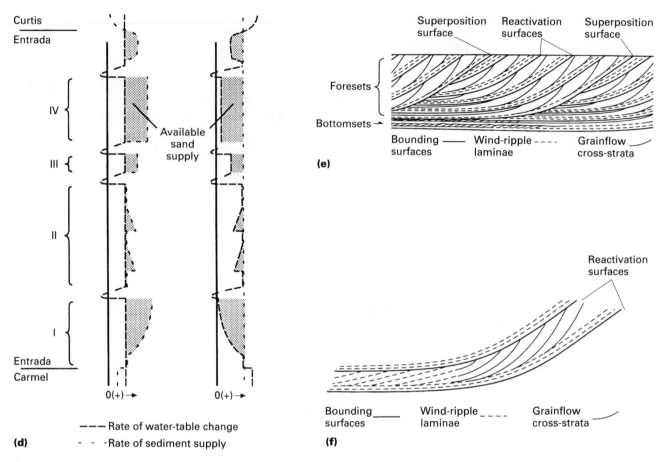

Figure 5.26 (*continued*) (d) Alternate plots of available sand supply for units I–IV, left diagram holds rate of water-table rise constant except at super surfaces separating sequences (I – IV); and right diagram holds rate of sediment supply constant. (e) Compound cross-strata showing two scales of scallops. (f) Details of cyclic stratification bounded by small scale scalloped bounding (reactivation) surfaces (from Crabaugh & Kocurek, 1993).

of sabkha sediments where the water table continued to rise; and (iv) fall of the water table and deflation of sabkha sediments or crust (Fig. 5.25d). Because there is no evidence within the aeolian accumulations that the water table played a role in accumulation, the aerodynamic and sediment supply conditions must have controlled accumulation. Regional mapping of the Page systems tracts show that accumulation occurred during regression of the Carmel sea whereas surface formation was associated with maximum flooding of the erg margin, thereby implying that sand was sourced to the system during regressions and the supply diminished during transgressions.

Within the Entrada Sandstone, however, features indicating a shallow water table occur throughout the sequences (see Sect. 5.5.2) and show the contemporaneous accumulation of dune, sabkha, and interdune sabkha deposits. Dune foresets can be traced into interdune sabkha bedding, and the bases of sets are loaded into underlying sabkha accumulations, while the upper set boundaries are corrugated. The sabkha accumu-

lations are characterized primarily by wavy bedding, but also contain subaqueous ripple deposits, contorted strata, breccias, collapse features, polygonal fractures, and ball-and-pillow structures. Thicker, dominantly sabkha accumulations contain foundered sets of cross-strata and trains of small dunes 'frozen' in place by vertically accreting sabkha strata (Crabaugh & Kocurek, 1993). The controls on Entrada accumulation are more apparent within the overall systems tract, in which the wet-aeolian facies of the Entrada passes into expansive sabkha beds (Fig. 5.26a). This lateral facies change suggests that the apparent water-table changes (Fig. 5.26d) may have been controlled by relative sea level such that the Entrada sequences record an inland relative sea-level curve.

Good examples of aeolian systems tracts on the Colorado Plateau that grade laterally into other continental systems include the aeolian Permian Cedar Mesa Sandstone of the Colorado Plateau into fluvial facies of the Cutler Formation (Langford & Chan, 1989), the Jurassic aeolian Navajo Sandstone into the

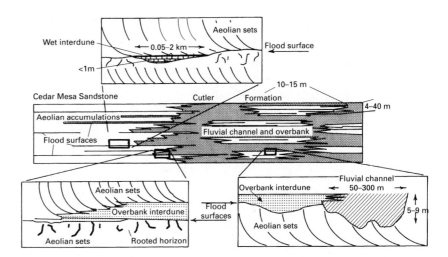

Figure 5.27 Schematic model of Permian Cutler/Cedar Mesa fluvial–aeolian intertonguing indicating geometry and scale of intertonguing (from Langford & Chan, 1989).

fluvial Kayenta Formation in Arizona (Middleton & Blakey, 1983), and the Jurassic Wingate Sandstone into the Moenave Formation (Clemmensen, Olsen & Blakey, 1989). While the systems tracts are fairly predictable, the vertical changes within the systems, and the scale of intertonguing of systems, have been interpreted to represent major climatic and other changes that produce sequence boundaries. Within a sequence, the Wingate–Moenave transition occurs over a 50–100 km belt showing dune to sand sheet to sabkha to fluvial and lacustrine facies. Within sequences, however, these marginal facies show drying–wetting-upward cycles that are thought to reflect Milankovitch climatic fluctuations between arid and humid periods (Clemmensen, Olsen & Blakey, 1989). Aeolian and fluvial strata in the Cedar Mesa Sandstone intertongue on a large scale, recording climatic change (Loope, 1985; Langford & Chan, 1993), but also intertongue on a small scale, which is thought to reflect common interactions between adjacent environments (Fig. 5.27). The Permian Rotliegend shows a systems tract of fluvial, aeolian, sabkha and lacustrine facies. Within the Southern Permian Basin of the North Sea, fluvial accumulations, derived from tectonically uplifted highlands give way to aeolian sands with sabkha/lacustrine facies to the north within the central basin (Glennie, 1986).

Proterozoic accumulations of India (Chakraborty, 1991; Chakraborty & Chaudhuri, 1993) and Cambro-Ordovician units of the northeastern Mississippi Valley of the United States (Dott, Byers *et al.*, 1986) represent continental system tracts before the advent of vegetation. Both show a central erg facies of sets of cross-strata that grade laterally into extensive sand plains of mixed aeolian and fluvial deposits. A distinctive characteristic of these accumulations is the very rich vertical and lateral assemblage of sedimentary structures, which is thought to reflect the sensitivity of the substrate to environmental shifts because of the lack of surface stabilization by vegetation (Fig. 5.28). Although distinct from later aeolian

accumulations, where the exclusion of vegetation is typically by aridity so that subaqueous and aeolian processes are not intermingled on such a fine scale, these accumulations compare well with Pleistocene and Holocene glacial sand plains (e.g. Koster, 1988).

A special case of interrelationships between aeolian and non-aeolian systems, especially marine systems, occurs at sequence boundaries where palaeotopographic patterns are created. These range from an inherited dune relief with minor reworking, to a reworked remnant dune topography, and finally to an erosional relief that is not representative of original dune topography (Fryberger, 1986; Eschner & Kocurek, 1988). The most pronounced relief in the rock record that best approaches inherited dune relief formed with the Zechstein transgression into the basins of the Rotliegende/Weissliegende complex, with 67 m measured by Blaszczyk (1981) in southwestern Poland. The equivalent Yellow Sands of England show tens of metres of relief from which Steele (1983) mapped a family of linear dunes. Glennie and Buller (1983) credited the inherited relief in the system to an extremely rapid transgression by a quiescent sea into the graben after breaching of a barrier. A far greater degree of reworking occurred with the high energy marine transgression of the Entrada Sandstone in northeastern Utah, where several metres of relief occur along the aeolian–marine contact, and mass-flow deposits within marine strata can be traced to remnant dune topographic highs from which they were shed (Eschner & Kocurek, 1986). Similar mass-flow units but containing large aeolian clasts are described from the reworked, channelized upper contact of the Late Proterozoic Bakoye 3 Formation in Mali (Deynoux, Kocurek & Proust, 1989). Huntoon and Chan (1987) interpreted 76 m relief on the Permian White Rim Sandstone as primarily the result of erosion during a major transgression. Tewes and Loope (1992), however, interpreted a much smaller scale of relief along the same contact as palaeoyardangs.

(a) Large-scale cross-stratified facies

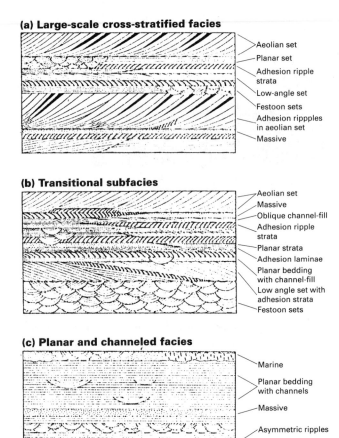

Aeolian set
Planar set
Adhesion ripple strata
Low-angle set
Festoon sets
Adhesion rippples in aeolian set
Massive

(b) Transitional subfacies

Aeolian set
Massive
Oblique channel-fill
Adhesion ripple strata
Planar strata
Adhesion laminae
Planar bedding with channel-fill
Low angle set with adhesion strata
Festoon sets

(c) Planar and channeled facies

Marine
Planar bedding with channels
Massive
Asymmetric ripples
Festoon sets

1 m
0   1 m

Figure 5.28 Vertical and lateral heterogeneous assemblage of sedimentary structures in Cambrian aeolian dune (a), transitional (b), and aeolian sand-sheet and fluvial facies (c) (from Dott, Byers *et al.*, 1986). Note the fine scale of interbedding of features indicating alternating subaqueous and aeolian processes.

5.5.4   Dune reconstruction

Reconstructing the shape and behaviour of dunes from sets of cross-strata typically means that the entire bedform and its dynamics must be inferred from its basal lee deposits because the angle of climb is usually small (Sect. 5.3.4). However, for numerous examples in the rock record, analysis of the type and distribution of stratification types, documentation of the geometries of bounding surfaces made by migrating bedforms, and comparisons of postulated models of dune shape and behaviour with computer simulations have led not only to reconstructed bedforms, but also to an interpretation of primary and secondary flow patterns.

Stratification types in ancient aeolian units were first used to distinguish aeolian from subaqueous cross-strata, but it

was quickly realized that they could also be used for dune reconstruction (Hunter, 1981; Kocurek & Dott, 1981). An early example of the definitive recognition of the aeolian origin of a controversial cross-stratified sandstone is the Permian Cedar Mesa Sandstone of Utah (Loope, 1984), and, more recently, aeolian strata have been identified in the Mississippian cross-stratified carbonate units of the North American mid-continent, which had been believed to represent marine shoals (Hunter, 1993). Wind-ripple laminae, abundant grainfall deposits, and the intertonguing of grainflow cross-strata into basal grainfall laminae are all now recognized as distinctive aeolian features. Although the mechanics remain unsolved, there is a general correlation between slip-face height and the thickness of individual grainflow cross-stratum (Hunter, 1981; Kocurek & Dott, 1981).

Stratification types, because they record surface processes during deposition, are valuable for the interpretation of secondary lee airflow, and, in conjunction with bedform reconstruction, can be used to infer primary palaeowinds (Sect. 5.3.3). For example, sets in the Proterozoic Dhandraul Quartzite of India consist of oppositely dipping foresets arranged in a zigzag pattern showing that this interpreted linear dune reversed asymmetry without net migration (Chakraborty, 1993). Wind-ripple laminae dominate the foresets, and can be shown to have moved alongslope, thereby indicating that the dune formed by two palaeowinds that struck the dune obliquely and were deflected alongslope.

Cyclic cross-strata caused by flow fluctuation (Sects 5.3.3, 5.3.5 & 5.3.6) are especially useful in inferring palaeowind conditions, as first recognized in the Navajo Sandstone in Utah by Hunter and Rubin (1983). In these Navajo outcrops, both concordant cross-strata (where grainfall and wind-ripple deposits alternate), and compound cross-strata occur. The compound cross-strata show reactivation-surface-bounded cycles of foresets of grainflow that toe into wind-ripple bottomsets, upon which a basal wedge of ripple strata rests. Judging by the implied dune advance rate of several tens of centimetres for each cycle, the cycles are interpreted as annual and corresponding to seasonal wind regimes. Additional examples from the Navajo and other units of sets formed under variable flow conditions, along with computer simulations, are shown in Rubin (1987a). Cyclic stratification in the Page Sandstone consists of packages of grainflow foresets passing into bottomsets of darker, coarser-grained, wind-ripple laminae, separated by thin deposits of lighter, finer-grained, ripple laminae that extend up the entire set (Kocurek, Knight & Havholm, 1991). In the Entrada Sandstone, cyclic scallops occur in which reactivation surfaces are overlain by thin wind-ripple laminae concordant with the surface, upon which grainflow cross-strata prograde and are contemporaneous with thick, steeply climbing ripple laminae, with both ripple and grainflow strata then truncated by the next reactivation surface (Fig 5.26e,f) (Crabaugh & Kocurek, 1993). All these Jurassic cycles show periods of dune migration

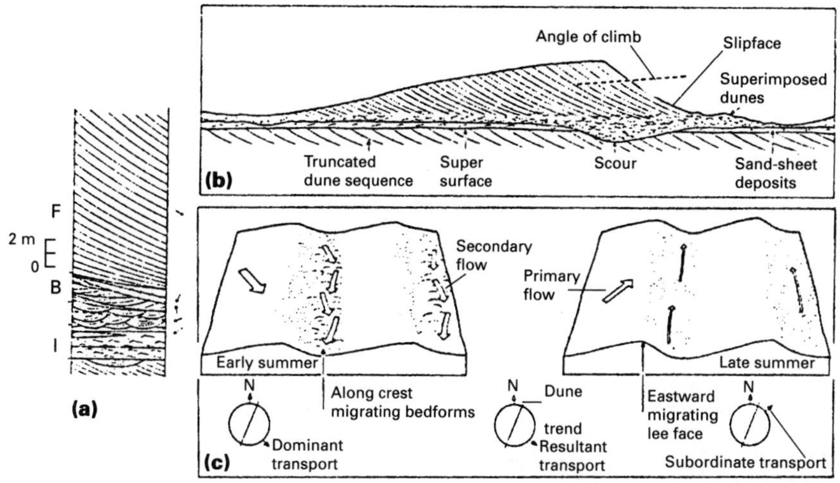

Figure 5.29 (a) Vertical column in Jurassic Wingate Sandstone showing foresets (F) overlying bottomsets (B) of small troughs, then interdune accumulations (I). Arrows indicate cross-strata dip directions. (b) Composite oblique dune model for origin of vertical section in (a). (c) Reconstructed dune and flow patterns (from Clemmensen & Blakey, 1989).

when slip-face progradation and some alongslope transport occurred, punctuated by periods of more pronounced alongslope transport with lesser lee deposition or even lee-face reworking. This pattern corresponds well to the palaeoclimatic reconstructions of Parrish and Peterson (1988) in which a bimodal wind regime was hindcast. In an example of pronounced flow variation, the Triassic Tumlin Sandstone of central Poland shows variation in trough scour depth such that multiple sets are removed (Gradzinski, 1992). These depressions are thought to have occurred randomly and may be an example of variation in scour depth by migrating bedforms (Sect. 5.3.4).

Cyclic compound cross-strata formed by the migration of superimposed dunes (Sects 5.3.5 & 5.3.6) have proved to be abundant in the rock record. The Navajo Sandstone especially shows a rich assemblage of cosets formed by superimposed dunes, as illustrated in Rubin and Hunter (1983) and Rubin (1987a), the latter of which also shows computer simulation of some of the bedforms. In addition to the compound cross-strata caused by annual cycles in the Entrada Sandstone, a yet larger scale of scallops occurs caused by the migration of superimposed dunes (Fig. 5.26e). Small satellite dunes also extend from the main dune on to the interdune flat and occur interbedded with wet-interdune strata. Similar compound cross-strata are shown by Clemmensen and Blakey (1989) in which bottomsets of small troughs overlain by foresets are interpreted as resulting from superimposed alongslope migrating dunes near the base of larger oblique dunes (Fig. 5.29). Permian linear dunes in the Yellow Sands of England mapped by Steele (1983) have been further computer modelled by Chrintz and Clemmensen (1993) to show that the outcrop pattern can be reproduced by superimposed reversing, sinuous linear dunes in which the sinuosity migrated along-crest.

Further reading

Brookfield M.E. & Alhbrandt T.S. (Eds) (1983) *Eolian Sediments and Processes*, 660 pp. *Developments in Sedimentology*, 38. Elsevier, Amsterdam.

Cooke R., Warren A. & Goudie A. (1993) *Desert Geomorphology*, 526 pp. University College London Press, London.

Frostick L.E. & Reid I. (Eds) (1987) *Desert Sediments: Ancient and Modern*, 401 pp. *Spec. Publ. geol. Soc. Lond.*, 35, Bristol.

Hesp P. & Fryberger S.G. (Eds) (1988) *Eolian Sediments*, pp. 1984. *Sediment. Geol.* (Special Issue), 55.

Kocurek G. (Ed.) (1988) *Late Paleozoic and Mesozoic Eolian Deposits of the Western Interior of the United States*, 413 pp. *Sediment. Geol.* (Special Issue), 56.

Pye K. (Ed.) (1993) *The Dynamics and Environmental Context of Aeolian Sedimentary Systems*, 332 pp. *Spec. Publ. geol. Soc. Lond.*, 72, Bristol.

Pye K. & Lancaster N. (Eds) (1993) *Aeolian Sediments Ancient and Modern*, 167 pp. *Spec. Publ. int. Ass. Sediment.*, 16. Blackwell Scientific Publications, Oxford.

Pye K. & Tsoar H. (1990) *Aeolian Sand and Sand Dunes*, 396 pp. Unwin Hyman, London.

Rubin D.M. (1987) *Cross-bedding, Bedforms, and Paleocurrents*, 187 pp. *Concepts in Sedimentology and Paleontology*, 1, Soc. Econ. Paleont. Miner., Tulsa.

Clastic coasts

H.G. Reading & J.D. Collinson*

6

6.1 Introduction

The *coast* is a broad zone that reaches from the landward limit of marine processes to the seaward limit of alluvial and shoreline processes. It includes not only deltas, beaches, barrier islands, tidal flats, tidal inlets, estuaries and cheniers but all those parts of the coastal plain affected by the proximity of a shoreline (Summerfield, 1991, p. 313) (Fig. 6.1).

The *shoreline*, on the other hand, is a line of demarcation between sea, lake, or lagoonal waters and an exposed beach (Komar, 1976, p. 13). It is therefore a local and transient feature.

Shorelines are obvious geomorphological boundaries and are the first to be drawn on many palaeogeographic maps. In ancient successions, coastal zones are relatively easy to identify using fossils because not only may they contain a brackish-water fauna but they also coincide with vertical and/or lateral changes from strata with dominantly marine fauna to strata with freshwater or no fauna. Hence landward migrations of the shoreline (transgressions) and seaward migrations of the shoreline (regressions) are comparatively easy to pin-point in both space and time. Identification of shoreline deposits may

be crucial in the documentation of sea-level changes and discussion in sequence stratigraphic analysis.

Coasts may be either dominantly erosional or dominantly depositional. Erosional coasts may be backed by cliffs of consolidated rock, or by loose or semi-consolidated sediment. Depositional coasts may be either devoid of terrigenous sediment, allowing biochemical sediments to form (Chapters 8 & 9), or receive a substantial supply of terrigenous sediment, derived either directly from the land via a river system, or from an adjacent coast or shelf by marine processes. As alluvial sediment reaches the shore, it is redistributed by basinal processes such as longshore drift, coastal current drift, waves, storms and tidal currents. Thus siliciclastic coastlines reflect the interplay of two competing suites of processes, fluvial currents and basinal energy, and two distinct sources of sediment. Where the coastline is fed directly from a contemporary river that supplies sediment more rapidly than basinal energy can redistribute, a discrete shoreline protuberance develops that has been called a *delta* (Sect. 6.6). Where sediment is derived largely from an adjacent coast or shelf, coastlines tend to be either linear, with strandplains, cheniers or barrier islands where wave and storm intensity are high, or indented with drowned river mouths or estuaries where tidal range is high (Sect. 6.7).

---

* Parts of this chapter are based on the second edition, written by T. Elliott.

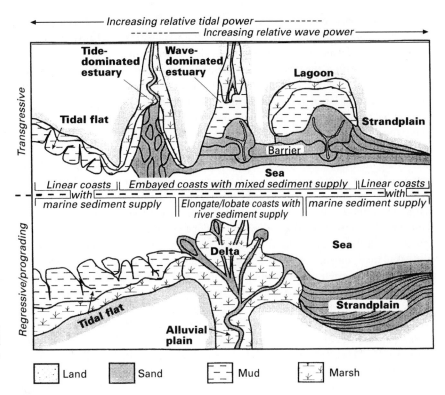

Figure 6.1 Plan views of transgressive and regressive/progradational coasts under varying conditions of tidal power and wave power and of marine or fluvial sediment supply (based on Heward, 1981; Boyd, Dalrymple & Zaitlin, 1992).

In the earlier editions of this book, 'siliciclastic shorelines' and 'deltas' were dealt with in separate chapters. There is, however, a continuum of settings, with many shared processes. Differences result mainly from the balance of fluvial and basinal processes, and in the manner and rates of sediment supply and reworking. For example, coastlines tend to straighten as: (i) fluvial input is reduced; (ii) coastal plains are supplied by longshore drift or numerous small, closely spaced rivers rather than by a single large river; and (iii) wave processes increase in intensity and frequency.

The association between ancient shorelines and some uncon-formities has long been recognized. During the 19th century and for much of the 20th century, the emphasis was on rocky shorelines, a type neglected in most recent textbooks, including earlier editions of this one (M.E. Johnson, 1992). Shorelines were generally recognized on the basis of: (i) an unconformity where strata had been stripped away and overlain by marine deposits, the only agent of denudation envisaged being marine peneplanation; (ii) a 'basal conglomerate', interpreted as a beach gravel; and (iii) a hardened surface, covered by encrusting organisms and riddled by organic borings, interpreted as a rocky 'beach-like' shore (e.g. Surlyk & Christensen, 1974). Ancient sea stacks could also be mapped along the unconformity with palaeogullies filled by boulder conglomerates (e.g. Whittard, 1932; Ricketts, Ballance *et al.*, 1989).

The first detailed description of a depositional shoreline facies was made by Gilbert (1885, 1890) in his studies of Pleistocene deltas in Lake Bonneville. Glacial streams transporting coarse sediment produced a series of fan-shaped lacustrine deltas now exposed and dissected by a fall in lake level. As they prograded, the deltas generated a threefold lateral and vertical sequence of bedding types (cf. Fig. 6.22). Subsequently, Barrell (1912, 1914) proposed criteria for the recognition of ancient deltaic deposits based on Gilbert's descriptions and applied these to the Devonian Catskill Formation. The terms 'topset', 'foreset' and 'bottomset' were used to describe the structure of the delta, and the bedding, texture, colour and fauna of each component were discussed, thus initiating the facies approach in the study of deltaic deposits. Barrell interpreted the Catskill cycles as a consequence of changes in sediment supply in the hinterland with a 'youthful' source giving delta progradation, and a reduced supply from a mature source causing marine planation of the delta. Although Barrell stressed that not all deltas exhibit a Gilbert-type structure, the concept conditioned thinking on deltas for several decades, and the presence of large-scale inclined foresets was considered an important criterion in the recognition of ancient deltaic successions.

In the 1940s, in both the USA and Europe, coarsening-upward cycles or cyclothems, which incorporated an upward transition from marine facies into terrestrial facies, were often attributed to delta progradation though without recourse to detailed comparison with modern deltas. In North America, however, sedimentological studies of modern deltas had begun with Johnston's (1921) account of the Frazer River delta and continued with that of Sykes (1937) of the Colorado River of California. Important though these studies were, the classic work on the Mississippi delta (Trowbridge, 1930; Russell & Russell, 1939; Fisk, 1944, 1947) provided the basis for the interpretation of most ancient deltaic deposits. This was largely because of the necessity to interpret the coals in the Carboniferous and oil- and gas-rich sandstones in the mid-West and Gulf Coast regions of the USA. This large, fine-grained fluvial-dominated delta became *the* model for the interpretation of ancient deltaic systems, and while it worked for some systems (e.g. R.C. Moore, 1959), it did not work for others. Furthermore, and perhaps of greater importance, shoreline facies sequences that did not fit the pattern of the coarsening-upward cyclothem were considered not to be deltaic. Meanwhile Van Andel and Curray (1960) recognized the need for critical comparison *between* deltas. While noting basic similarities between deltas, they stressed the striking variation in structure and lithology of the Rhône and Mississippi deltas, and asked for a comparative study of modern deltas. A response to this plea was given mainly by oil companies who initiated extensive borehole programmes in many of the world's major deltas, including the fluvial-dominated Mississippi (Fisk, McFarlan *et al.*, 1954; Fisk, 1955, 1961), the wave-dominated Rhône (Oomkens, 1970), the tide–wave interaction Niger (J.R.L. Allen, 1965c; Weber, 1971; Oomkens, 1974; Sects 6.6.1 & 6.6.5) and the mixed tide- and fluvial-dominated Mahakam (G.P. Allen, Laurier & Thouvenin, 1979). These studies showed that the variability between deltas was a consequence of the interplay between fluvial and basinal processes, and the significance of wave power and tidal strength at the coast was stressed (W.L. Fisher, Brown *et al.*, 1969; Wright & Coleman, 1973; Coleman & Wright, 1975; Galloway, 1975) (Fig. 6.2). They also demonstrated the wide variety of vertical facies sequences in deltaic deposits. Of particular importance was the recognition that facies sequences not only varied *between* deltas, but also *within* them so that delta type could not be diagnosed from single vertical sequences.

Meanwhile, the morphology of non-deltaic, linear coasts was also considered to reflect the effectiveness of wave and tidal processes. Coasts were divided into microtidal (<2 m tidal range), mesotidal (2–4 m), and macrotidal (>4 m) (J.L. Davies, 1964; Hayes, 1975, 1979) (Fig. 6.3).

These process-related models dominated interpretation of ancient coastlines, including deltas, up to the late 1970s. Since that time the importance of 'fan deltas' with their coarse grain size, steep slopes and short radii has been brought to everyone's attention in the major compilations by Koster and Steel (1984), Nemec and Steel (1988a), Colella and Prior (1990) and Dabrio, Zazo and Goy (1991). In addition, the grain size or calibre of the sediment has been recognized as a major component of facies models (Orton, 1988; Reading & Orton, 1991; Orton & Reading, 1993) as have water depth (Postma, 1990), the nature of the feeding system (Nemec & Steel, 1988b; McPherson, Shanmugam & Moiola, 1988), and tectonic–physiographic setting (Ethridge & Wescott, 1984; Leeder, Ord & Collier, 1988; Gawthorpe & Colella, 1990; Wescott & Ethridge, 1990).

Finally, the effects of sea-level changes, first documented by Curray (1964), have become increasingly appreciated as concepts of sequence stratigraphy (Sect. 2.4) developed (e.g. Posamentier & Vail, 1988; Swift, Phillips & Thorne, 1991b; Dalrymple, Zaitlin & Boyd, 1992; Posamentier, Allen *et al.*, 1992).

6.2 Shoreline processes

Shoreline morphology essentially reflects the interaction between supply of sediment and basinal reworking processes. The principal processes that move sediment at shorelines are fairweather waves and tides, episodically enhanced by storms. *Waves* give rise to a range of currents which may be directed offshore (*rip currents*), parallel to the shore (*longshore currents*), obliquely (*obliquely-directed currents*) and onshore (*onshore residual motions*). Tides primarily affect shorelines by raising and lowering sea level, thus not only shifting the site of wave action but also influencing the biology of the nearshore zone. They also generate currents as large volumes of water move between the sea and the coastal plain, lagoons and estuaries via inlets and delta distributaries. *Storms* interrupt these day-by-day processes by increasing their intensity and by giving rise to heightened turbulence and sudden movement of water and sediment both offshore and onshore.

6.2.1 Sediment supply

Sediment is primarily supplied to the oceans from the land, both from large continental regions, and from islands. Globally some 15×10^9 t year^{-1} are supplied by fluvial systems (Milliman & Meade, 1983). An additional 5–10% may come from erosion of cliffs, mostly of soft sediment but including every rock type. However, many coasts are supplied primarily from the sea, by longshore drift, by erosion at the shoreline, or from the shelf.

The sediment supplied from the land is delivered by a variety of alluvial systems, notably rivers, estuaries, delta distributaries, point-sourced alluvial fans and multiple sourced braid plains and by scree cones. The nature and volume of sediment is primarily determined by the alluvial catchment area, its size, relief, tectonics, climate, vegetation and the character of its bedrock and superficial deposits. These factors govern the amount, grain size and manner of delivery of the sediment to the shoreline. Larger catchment areas are associated with greater sediment volumes, generally finer-grained material (coarser material being stored within the alluvial system), more steady water and sediment discharge and lower coastal-plain gradients. Small catchment areas, especially tectonically active ones with high relief close to the shoreline, tend to supply coarser sediment, typically gravel and sand. However, many local catchment areas only reach a short distance inland, sometimes within the coastal plain, and may yield any type of sediment. In semi-arid regions sand is more abundant. In humid tropical regions silt and clay

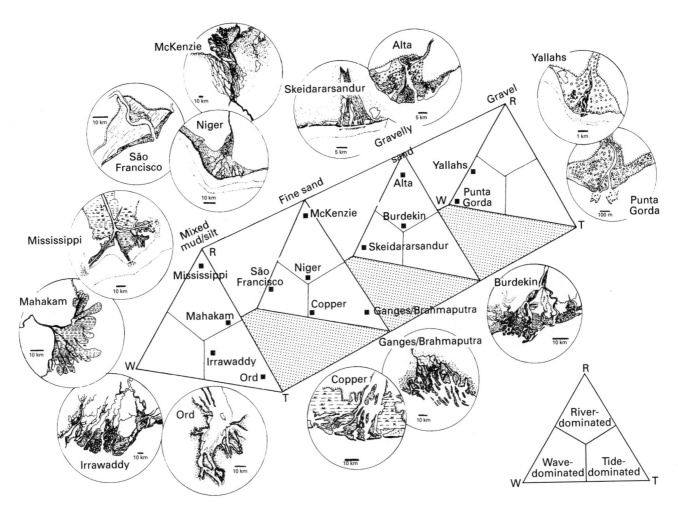

Figure 6.2 Classification of modern deltas based on the dominant process (fluvial, wave or tide) of sediment dispersal at the delta front (after Galloway, 1975) and on the prevailing grain size. Note the differences in scale (from Orton & Reading, 1993).

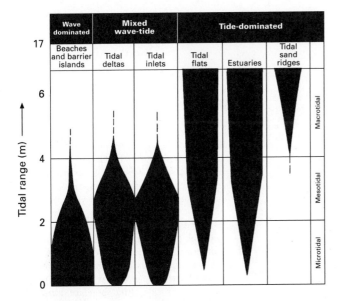

Figure 6.3 Types of coastline with respect to tidal range and subdivision into wave-dominated, mixed wave–tide and tide-dominated groups (after Hayes, 1975, 1979).

are nearly always dominant, except where a cratonic granitic basement (e.g. Brazil) provides abundant sands.

Terrigenous sediment, other than mud, does not easily pass out on to the shelf. This is because there is a *littoral energy fence* (J.R.L. Allen, 1970c, p. 169) at the shoreline where shoaling and breaking waves have the potential to move more sediment landward than seaward (Sect. 6.2.4; Fig. 6.7). This fence can be broken or 'bypassed' and sediment then escapes out on to the shelf by one of three methods: (i) *river mouth bypassing* (Fig. 6.4a) where deltas enter the sea and especially during the flood stages of rivers; (ii) *estuary mouth bypassing* (Sect. 6.7.5), mainly by ebb-tide enhanced currents; and (iii) *shoreface bypassing* (Fig. 6.4b) by which shoreface sediments are removed as the shoreface is eroded by wave/storm processes.

6.2.2 Sediment delivery to the basin

Where rivers flow into a basin, the effluent behaviour and consequent depositional patterns depend upon the relative dominance of: (i) the *inertia* of the inflowing water as it enters the basin, and its diffusive mixing with basin water; (ii) the *friction* of the inflow at and basinward of the river mouth; and

(iii) the *buoyancy* process at the river mouth (L.D. Wright, 1977). The factors that determine the role played by each of these processes include: (i) density contrast between river and basin waters; (ii) the concentration, grain size and suspension/total load ratio of the sediment; (iii) water depths at and basinward of the river mouth; (iv) the water discharge; and (v) the inflow velocity of the river.

The main consideration is the density contrast between river and basin waters. River waters may be equally dense (*homopycnal*), more dense (*hyperpycnal*) or less dense (*hypopycnal*) than the basin waters (Bates, 1953) (Fig. 6.5). *Homopycnal flow* generates intense local three-dimensional mixing at the river mouth causing appreciable sedimentation around this point, especially of bedload. It is the main form of flow in coarse-grained river mouths, especially where they enter fresh water, and is characteristic of lacustrine Gilbert-type deltas (Sect. 6.5.1). It may also occur in finer-grained marine deltas at times of brief, very high flood discharges. *Hyperpycnal flows* occur during floods and pass beneath the basin waters as density currents causing

sediment to bypass the shoreline and to be deposited on the lower delta front or on the prodelta. They thus inhibit delta progradation. They are characteristic of cold river waters, with heavy suspended sediment loads entering fresh water, especially warmwater, lakes (Sect. 4.4.2; Fig. 4.5) and of slope-type coarse-grained deltas building into deep water (Sect. 6.5.1). *Hypopycnal flow* extends into the basin as a buoyantly supported surface jet or plume. It is especially common where rivers enter the sea or a saline lake, because of the higher density of sea water, and is very effective at separating suspended load from bedload.

Differing combinations of inertial, frictional and buoyancy processes produce distinctive river mouth configurations where rivers debouch into basins with low wave and tidal energy (L.D. Wright, 1977) (Fig. 6.5).

1 *Inertia-dominated* river mouths form where there is a steep slope allowing expansion of the effluent in both the horizontal and vertical direction and where high velocity rivers enter fresh water and/or carry substantial quantities of coarse sediment (Sect. 6.5.1).

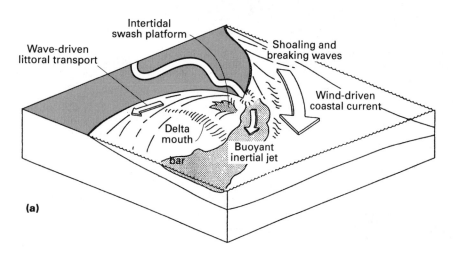

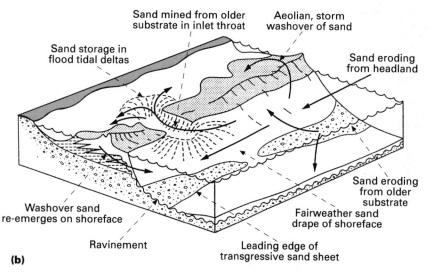

Figure 6.4 Two principal methods by which sediment is transported on to the shelf through the *littoral energy fence*. (a) *River mouth bypassing* – a river flood transports sediment on to a delta mouth bar and beyond. Sand is mostly stored in the mouth bar and slowly re-entrained in the littoral sand stream. Fine sand, silt and clay are carried as a buoyant half-jet and rained on the shelf floor. (b) *Shoreface bypassing* – storm washover sand is buried and eroded as it emerges on the shoreface. Erosion of the shoreface during its retreat allows transport alongshore and on to the shelf (from Swift & Thorne, 1991; Swift, Phillips & Thorne, 1991a).

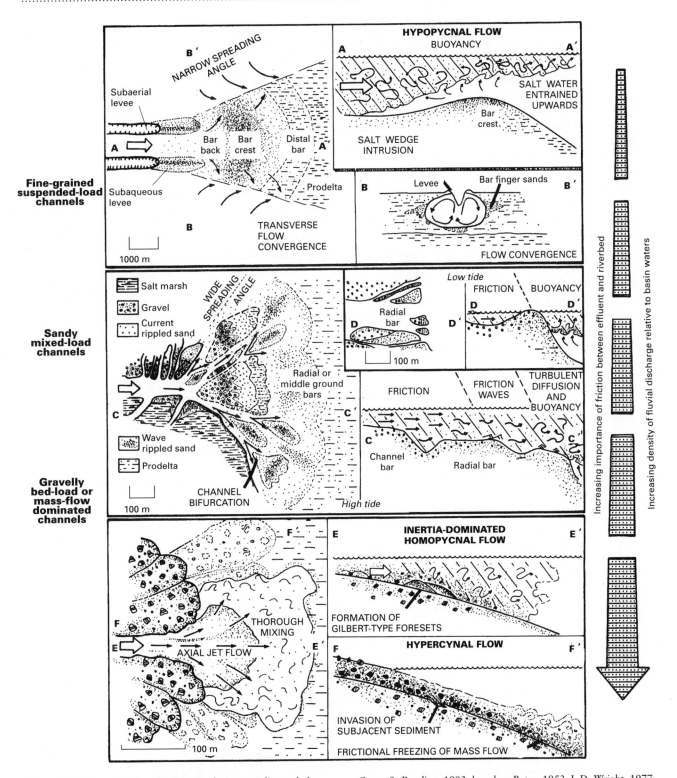

Figure 6.5 Different modes of interaction between sediment-laden river water and basin water, determined by the relative density of the waterbodies and the grain size of the incoming river sediment (from Orton & Reading, 1993, based on Bates, 1953; L.D. Wright, 1977; Kostaschuk, 1985).

2 *Friction-dominated* river mouths occur where rivers enter very shallow water so that effluent expansion occurs only in a horizontal direction. Frictional interaction between the flow and the sediment surface increases the lateral spreading and deceleration of the jet and produces a triangular 'middle ground bar' in the river mouth leading to channel bifurcation (Fig. 6.5). As progradation continues, new bars form at the mouths of the split channels.

3 *Buoyancy-dominated* river mouths form where the effluent extends as a plume into a relatively deep, commonly marine basin. The flow detaches from the bed and is unable to move bedload beyond the detachment point. A saltwater wedge intrudes the lower part of the channel especially at low water stage. Turbulent mixing is particularly intense near the river mouth as internal waves are generated between the plume and the underlying salt wedge. An appreciable amount of the suspended sediment load, particularly the coarser fraction, as well as any bedload, is deposited near the river mouth to form a bar, while the finer-grained sediment is transported further into the basin and deposited from suspension as the plume decelerates and mixes with basin waters. A dominance of buoyancy processes produces a mouth bar whose crest is considerably higher than the floor of the distributary channel. The leading slope projects a long way into the basin with a gentle gradient.

6.2.3 Zonation of the shoreline profile

Shoreline profiles have a number of zones each with its characteristic processes, morphology and facies (Fig. 6.6) (Komar, 1976, pp. 2–14; Bourgeois & Leithold, 1984). Although the precise terminology may vary between different studies, depending on whether emphasis is put upon process or morphology, the zones are differentiated primarily upon the positions of storm and fairweather wave base, and on mean high- and low-tide levels, and secondarily upon the nature of wave transformation (Sect. 6.2.4). They are introduced here in order to give a basis for the discussion of specific processes in later sections.

The *offshore-transition* zone extends from mean storm wave base to mean fairweather wave base and is therefore characterized by alternations of high and low energy conditions. During fairweather, fine-grained sediment settles from suspension and the bottom sediments are bioturbated. During storms the bottom is affected by oscillatory and shoaling waves, supplemented by storm-generated currents.

The *nearshore zone* extends from mean fairweather wave base to mean high water level. It comprises a *shoreface*, below mean low water level and a *foreshore* between mean low water and mean high water level.

The subtidal *shoreface* extends from mean fairweather wave base, where fairweather waves first feel the bottom, to mean low water level. It is the zone of maximum sediment movement (Howard & Reineck, 1981). During fairweather, oscillatory and shoaling wave processes operate in the lower part of the shoreface, and breaker/surf zone processes in the upper shoreface. Relatively weak rip currents and longshore currents may operate on the upper shoreface on sandy barred shorelines but otherwise are insignificant. During storms, shoaling waves, storm currents and enhanced rip currents erode the shoreface,

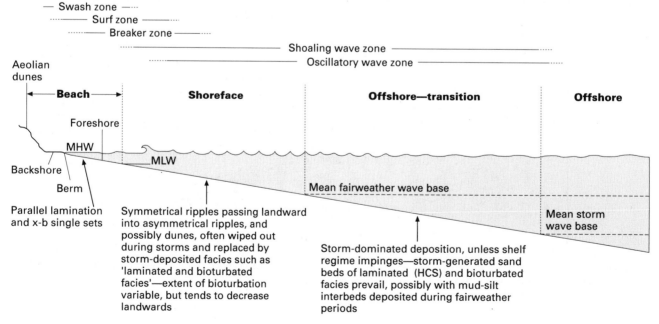

Figure 6.6 Generalized shoreline profile showing subenvironments, processes and facies.

particularly the upper shoreface. Sediment thus eroded is either carried landward on to the beach or swept offshore to be redeposited on the lower shoreface and on to the shelf.

The intertidal *foreshore* is regularly covered and exposed by most tides and thus is subjected to the daily swash of the waves. It is the littoral zone of biologists. To most geologists, however, the littoral zone also includes much of the shoreface, affected by surf and breaking waves, since they are concerned with all the physical processes that go to form a beach. These do not change sharply at low water level in the way that organic communities do. The foreshore generally has a steep sloping profile termed the *beach face* which is affected strongly by breaker, surf and swash zone processes, supplemented by longshore currents. Morphologies of foreshores vary considerably, depending on wave power, tidal range and available grain sizes. At its landward limit, the foreshore is generally separated by a *berm crest* from the *berm* or nearly horizontal portion of the beach profile.

The *backshore* is the relatively flat, unvegetated, supratidal area only inundated by rare very high tides and storms that carry sediment landward, to be winnowed and reworked. It frequently passes landward into aeolian dune fields formed of sand derived from the exposed beach. Backshores are particularly well developed on sandy coastlines, where the prevailing winds are onshore and where tidal range is high so that an enhanced supply of sand is provided by a wide beach at low tide.

6.2.4 Wave processes

Nearshore wave energy is the most significant marine process governing coastline development (L.D. Wright & Coleman, 1973; Davis & Hayes, 1984; Roy, Cowell *et al.*, 1994). Waves sort and redistribute the sediment delivered by rivers and mould it into shoreline and inshore features such as beaches, barriers and spits.

Two types of waves affect shorelines, fast travelling long-period *swell waves* that may be generated thousands of kilometres out in the ocean, and short-period *sea waves* generated by nearby storms or prevailing winds. In deep water the water particles move in almost circular orbits with the orbital diameter equalling wave height at the surface and decreasing progressively with depth (Sect. 7.4.1) (Fig. 6.7a). As the waves enter shallow water they 'feel' the sediment surface at a point termed 'wave base' which tends to occur where water depth is half the wave length ($D = L/2$ where D = water depth and L = wave length). Between this point and the point where $D = L/4$–6, the *oscillatory wave zone* (Fig. 6.6), the passage of each wave results in a symmetrical, straight-line, to-and-fro motion in the direction of wave propagation at the sediment surface. As waves enter the *shoaling wave zone* they are extensively modified and change from a symmetrical, sinusoidal form to an asymmetrical, solitary form: wave velocity and wave length decrease; wave height and

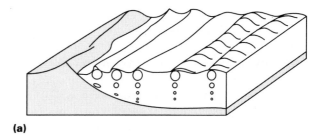

(a)

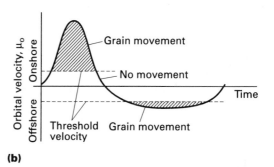

(b)

Figure 6.7 The littoral energy fence. (a) Wave transformation as a shoreline is approached. The orbital diameter decreases with depth, moving to-and-fro as it nears the bottom and frictional drag increases. (b) The effects on sediment movement during the passage of a shoaling wave. The onshore stroke of the wave as the crest passes carries more sediment than the offshore stroke associated with the passage of the trough (from Swift & Thorne, 1991).

steepness increase; only wave period remains constant. Wave motions at the sediment surface involve a brief landward-directed surge, and a rather longer, but weaker, seaward-directed return flow (Fig. 6.7b). Thus more sediment is carried landward than seaward. Progressive steepening of the wave as it approaches the shoreline eventually causes it to oversteepen and break in a landward direction, thus initiating the *breaker zone*. High energy conditions in the breaker zone cause fine sand to be suspended temporarily while coarser sand is concentrated at the bed. Breaking of waves, particularly as surging breakers, generates the *surf zone* in which a shallow, high velocity bore is directed up the shoreface on to the foreshore. Coarse sediment is transported landward while finer sand and silt are suspended briefly in bursting clouds. Finally, at the landward limit of wave penetration, in the *swash zone*, each wave produces a shallow, high velocity, landward-directed swash flow followed almost immediately by an even shallower, seaward-directed backwash which may disappear by infiltration into the bed. Plane bed or standing wave/antidune conditions predominate in this zone.

The general trend across the shoreface (Fig. 6.6) is, therefore, one in which oscillatory flow transforms into asymmetrical, landward-directed flow of increasing power. Bedforms reflect this change with symmetrical ripples of the oscillatory zone passing into asymmetrical wave ripples and possibly dunes in the shoaling wave zone, and finally an area of predominantly plane bed conditions in the breaker, surf and swash zones (Clifton, 1976).

Upon this simple model are superimposed many complicating factors. First, the position of these wave transformation zones varies daily and seasonally because of oscillation in tidal height and the alternation of storm and fairweather conditions. Hence, there is substantial overlap of the bedform zones. Second, there may be a considerable contrast between deepwater wave energy and that which reaches the shoreline. Wave energy at the shoreline depends not only on deepwater wave energy and the water depth of the basin, but also on the frictional attenuation that occurs on the shelf as wave power is transmitted from deeper water to the shoreline. Frictional attenuation is a function of the gradient of the sea floor which itself depends on the nature and width of the shelf and the rate and type of sediment supply to the nearshore zone.

Thus two types of wave-influenced shoreline are recognized (L.D. Wright, Chappell *et al.*, 1979; Short, 1984; L.D. Wright & Short, 1984; Fig. 6.8): *reflective* and *dissipative*.

Reflective shorelines have smooth, steep beach/shoreface profiles. Although offshore wave energy may not be high, most waves break directly against the beachface. Wave energy is reflected from the beachface and often refracted alongshore. Beach cusps form, but ridge (bar) and runnel (trough) topography and swash bars are absent. Reflective shorelines develop particulaly well where gravel is the dominant sediment (Sect. 6.5.2) because the swash percolates into the foreshore sediments and there is virtually no backwash. Thus landward-directed processes predominate. Apart from winnowing of the finer material, the sediment is not very mobile under most fairweather conditions, only being shifted during storms. Most of the sediment accumulates on the upper foreshore to form a high berm, while the coarsest gravels accumulate on the lower foreshore (Dabrio, 1990). Below the berm the foreshore is steep, occasionally reaching almost the angle of repose. However, deposition by waves and erosion by storms tend to smooth the foreshore profile. At the base of the foreshore there is a sharp change in slope as the relatively steep foreshore passes basinwards into a relatively gently inclined shoreface.

Dissipative shorelines have an irregular, gentler profile with high energy waves breaking a considerable distance from the shore. The width of the surf zone can range from 50 to 500 m depending on wave strength and the grain size. If sand is the predominant sediment then the surf zone develops a pattern of ridges (bars) separated by runnels (troughs). A complex three-dimensional topography occurs where longshore bars are cut by rip current channels. Percolation of swash is limited by the sediment texture and backwash carries the highly mobile sediment seaward across the gently dipping shoreface, thus dissipating wave energy and further reducing the wave power at the beach. When sand and gravel are available, a steep gravelly reflective foreshore may lie landward of a gentler finer-grained

more dissipative shoreface. Where mud is the predominant sediment and is transported in suspension, the soft unconsolidated bottom and subaqueous fluid mud shoals can completely attenuate wave energy and protect the coastline from erosion. Under such highly dissipative conditions most waves neither reach the shoreline, nor break. Continual landward motion of water carries mud particles landward, with no counterbalancing offshore movement so that muds are trapped in the nearshore zone.

6.2.5 Wave-induced nearshore currents

As waves approach the shoreline they not only transport sediment landward but generate two types of current: shore-parallel *longshore currents* and offshore-directed *rip currents* (Fig. 6.9).

Longshore currents are caused by variations in wave height parallel with the shore that induce the waves to approach the shore obliquely. They operate in the surf and breaker zones. As waves approach the shoreline they are refracted in a complex manner due to irregularities in depth of the shelf and the coastline. This produces zones of convergence and divergence (Fig. 6.9) even on a long straight beach. Here the *wave set-up*, or rise in the mean water level above stillwater level due to the persence of waves, produces a nearshore circulation cell with: (i) landward-directed mass transport where set-up is high, passing, at the breaker zone, into (ii) a longshore-directed current flow in the surf zone; and (iii) seaward-directed flow by rip currents that transfer water and sediment back to the sea. Where the general wave approach is orthogonal to the shoreline symmetrical circulation cells are set up. However, most shoreline waves approach obliquely and hence a more general littoral or longshore drift of sediment takes place as witnessed by the accumulation of sediment on the up-drift side of groynes. Estimates of longshore sediment drift of $100–700 \times 10^3$ m^3 year^{-1} have been made for coastlines facing oceans. Sediment can also be transported laterally by *beach drifting*, a sawtooth movement effected by the swash and backwash action, resulting from oblique wave approach.

Rip currents comprise narrow, high velocity storm-generated seaward-directed currents which start in the surf zone. Current velocities are typically 2–3 m s^{-1} but values of 10 m s^{-1} have been recorded. They erode shallow channels that pass seaward into rip current fans containing seaward-directed current ripples and dunes and massive, graded beds. Gruszczyński, Rudowski *et al.* (1993) have suggested that rip currents can transport coarse sediment several tens of kilometres seaward and deposit offshore storm-generated sandbeds.

6.2.6 Fairweather vs. storm conditions

During *fairweather*, waves are relatively low amplitude, long-period swells with a shallow wave base and shorelines tend to

DISSIPATIVE

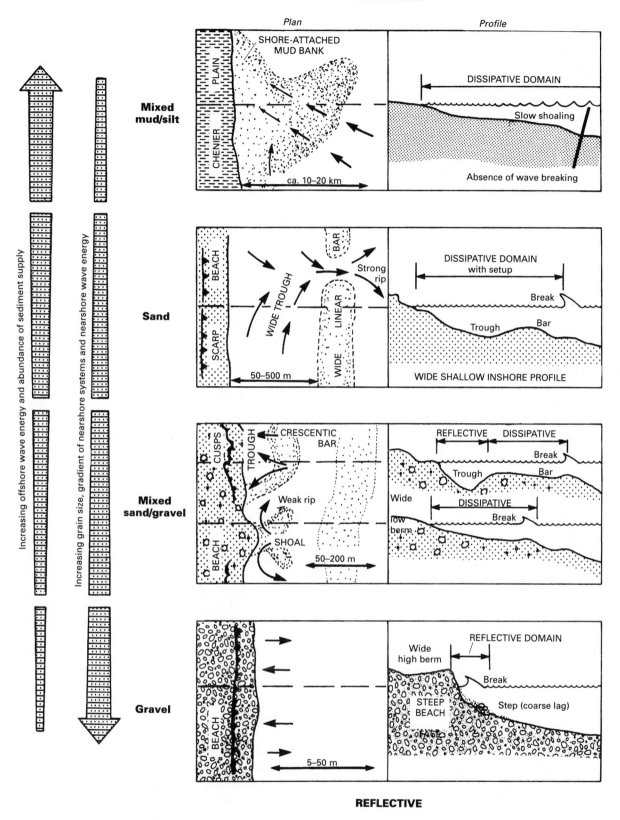

Figure 6.8 A spectrum of wave-dominated shorelines showing from reflective to dissipative, the variation depending on offshore wave energy, abundance of sediment and available grain size (after Orton & Reading, 1993).

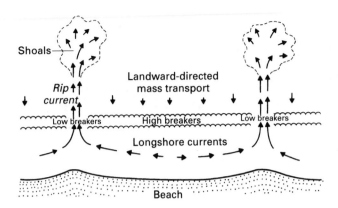

Figure 6.9 Wave-induced nearshore circulation system of longshore currents and seaward-directed rip currents (after Shepard & Inman, 1950; Komar, 1976).

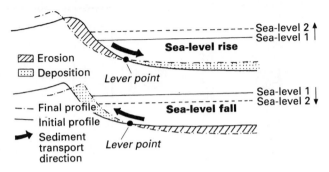

Figure 6.11 Translation of the shoreface profile by changes in relation to sea level. The lever point is the point of intersection between the initial and final profile and is normally at a shallow, subtidal level (after Bruun, 1962; from Dominguez & Wanless, 1991).

be reflective. The shoreface tends to be smooth, with no bars. The lower shoreface and offshore zones are not affected by waves and so fine-grained sediment is deposited from suspension and reworked by organisms. On the upper shoreface and foreshore, wave-induced currents, associated with the shoaling, breaker and surf zones, transport sediment landward. Little sediment is lost via seaward-directed rip currents and the beach therefore aggrades.

During *storms*, dissipative conditions prevail. High amplitude waves deepen the wave base causing much of the lower shoreface and possibly the offshore/shelf area to experience oscillatory, shoaling waves and associated processes. The upper shoreface and beach are extensively eroded, and sediment is both redeposited landward as washover fans in lagoons and swept seaward by both rip currents and wind-driven storm currents (Sect. 7.4.1).

Therefore beaches aggrade during fairweather and are eroded during storms, a process termed the *beach cycle* (Fig. 6.10) (Sonu & Van Beek, 1971). While shoreface deposition may be dominated by storms, the beach may be dominated by fairweather deposits, which are repeatedly interrupted by storm-related erosion surfaces. On a longer time scale a rise in sea level leads to erosion of the foreshore and upper shoreface and deposition

in the lower shoreface. During a sea-level fall sediment is transferred from the lower shoreface to the upper shoreface and beach (Fig. 6.11).

6.2.7 Tides

Tides result from the gravitational attraction exerted on oceanic or lake waters by the moon and the sun, with the moon having more than twice the effect of the sun. As the Earth rotates, a bulge of water, less than a metre in height in the open ocean, but with an enormous wave length, rises and falls twice daily along most coasts, to produce *semi-diurnal tides*. High tide occurs about every 12 h 26 min because the *lunar day* – the time for the moon to orbit the Earth – is longer than the solar day, and hence high tide gets progressively later by about 50 min each day. Along some coasts, however, due to the configuration of the generating basins and complexities of the shoreline, there may be *diurnal tides* with one high and one low a day, or a *mixed tide* with highs and lows of different magnitudes. When the sun and moon are aligned, that is when the moon is either new or full, the tides, known as *spring tides*, are about 20% higher than normal. When the sun and moon are at right angles relative to the Earth, the tides are about 20% lower than normal, and we get *neap tides*. This *neap–spring cycle* extends over 14 days.

Although the tidal range of open oceans is small, and even lower in enclosed seas (10–30 cm in the Mediterranean), it increases substantially as the tidal bulge encounters shallow water. In particular, where the tide is concentrated between converging shores or funnelled into estuaries, very high tides result, as in the Bay of Fundy where a range of 15.6 m has been recorded. The pattern of currents on semi-enclosed shelves is substantially influenced by the Coriolis force (Sect. 2.1.2) with a swing towards the right in the northern hemisphere. In the North Sea, for example, tides entering from the north are higher on the British coast than on the eastern side.

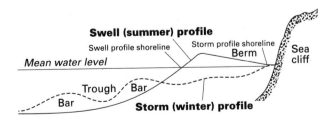

Figure 6.10 The beach cycle of alternate *swell profile* when a pronounced berm is built up and *storm profile* when sediment is shifted to the shoreface, and offshore bars are formed (after Komar, 1976).

Although tides have less effect on the transport of sediment and on coastal morphology than waves, they affect coasts in two important ways. They govern: (i) the strength and flow pattern of the regularly fluctuating tidal currents; and (ii) the amount and timing of tidal rise and fall. Tidal currents are most effective when operating in conjunction with waves.

Tidal currents, though not important along open shorelines, are very effective in estuaries and tidal channels where velocities commonly reach 1.5 m s⁻¹ with a maximum of 10 m s⁻¹ in estuarine bores (tidal waves). Such currents may generate large and complex bedforms whose nature, size and orientation reflect the spatial and temporal complexity of the current regime. Tidal currents have a unique character. Not only do they accelerate and decelerate within a few hours, but they reverse their direction of flow (Figs 6.12 & 6.13). Since we know the periodicity of tidal reversals/fluctuations we can use their sedimentary products to calculate the number of days in which certain sediment bodies were deposited (Fig. 6.14). At any one place there is nearly always an *ebb–flood tidal cycle* where one current is dominant and the other one subordinate (Fig. 6.12). The pattern of sediment transport and deposition, whether of sand or mud, depends on the relative strength and duration of the two currents. If both current phases are able to transport sand, then a fourfold *ebb–flood tidal couplet* will form (Fig. 6.13). In subtidal settings, during the dominant current stage, sand foresets form on the lee side of larger bedforms. At the succeeding slackwater stage, mud falls from suspension and drapes the foreset. The subordinate current stage may partly rework the mud-draped foreset to produce a gently sloping erosive *reactivation surface* and possibly also deposit some sand with reverse flowing ripples. During the second slackwater phase, another mud drape forms and this is then covered by the foresets of the next dominant stage. Thus each sandy foreset or *bundle*

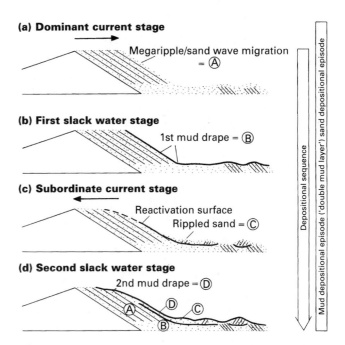

Figure 6.13 A tidal couplet that might be formed during stages (a)–(d) of the ebb–flood tidal cycle shown in Fig. 6.12 (from Visser, 1980).

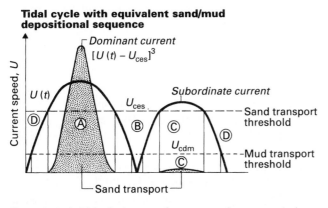

Figure 6.12 A tidal velocity curve for one strongly asymmetrical ebb–flood tidal cycle showing sand and mud transport thresholds and possible amounts of sand that might be transported during stages (A) and (C). Mud is deposited during stages (B) and (D) when velocity drops below mud transport threshold (from J.R.L. Allen, 1982b).

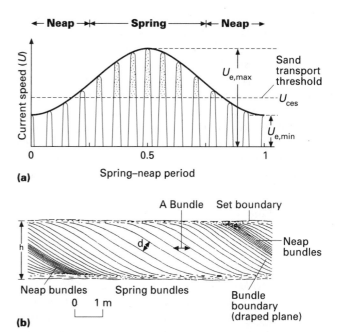

Figure 6.14 (a) Schematic distribution of current speed and sand transport (stippled area) during one neap–spring tidal cycle which has strongly asymmetrical ebb–flood tidal cycles (from J.R.L. Allen, 1982a). (b) Product of (a) showing tidal bundle sequences of cross-bed laminae composed of thick sand-dominated spring bundles and thin mud-dominated neap bundles (from Yang & Nio, 1985).

(Visser, 1980; Siegenthaler, 1982; Nio, Siegenthaler & Yang, 1983; Yang & Nio, 1985; Nio & Yang, 1991) is deposited by a single dominant tidal current and is bounded by slackwater mud drapes or, in the case of a slightly stronger subordinate current, by a mud drape and an erosional *reactivation surface* (de Mowbray & Visser, 1984). Variations on this pattern depend on the height and asymmetry of the current velocity curves and can be used to determine neap–spring cycles (Fig. 6.14) by measuring the thickness of successive bundles. Thick bundles indicate spring tides; thin bundles indicate neap tides. Ideally, 28 tidal couplets or bundles should be formed in each 2-week semi-diurnal neap–spring tidal cycle, but in many cases the number will be smaller because the threshold for sand transport is not reached during neap tides.

Tidal rise and fall produces couplets of alternate sand–mud laminae. The sand laminae may show small-scale current ripples. The mud laminae drape the troughs and crests of the sandy ripples, reflecting deposition from suspension during a stillstand, primarily of flocculated mud. Thus each sand–mud couplet represents deposition during one flood–ebb cycle. They are frequently stacked, showing a number of cyclic patterns called *tidal rhythmites* (Rahmani, 1988; Tessier, Montfort *et al.*, 1988; Williams, 1989a; Dalrymple, Makino & Zaitlin, 1991; Nio & Yang, 1991).

Vertical tidal fluctuations affect the coastline by regularly submerging and exposing the beach and intertidal flats. The different zones of the beach and shoreface are subjected to varying degrees of wave energy and populations of specialized organisms have evolved to tolerate the high levels of stress that these changes cause. Plant life is limited to algae that develop particularly complex structures in semi-arid and arid climates (Sect. 8.4) while, in the humid tropics, mangroves have adjusted to the rise and fall of the tide. In most temperate regions beaches are devoid of vegetation and are a source of wind-blown sand, especially in areas of high tidal range. Tides also move large masses of saline water landward through rivers, estuaries and delta distributaries and into lagoons. Restricted populations of specialized organisms are adapted to these salinity oscillations.

6.2.8 Wind

Wind is important at all shorelines, particularly where water levels change seasonally as in lakes or where tidal range is high. The exposed beach lacks a vegetated cover and is therefore essentially a desert that is continually fed by sand. Strong winds erode the beach and remove sand and silt either offshore or, more commonly, onshore on to the backshore, supratidal zone to form extensive complexes of aeolian dunes, which in turn supply sediment to adjacent lagoons and coastal plains. The dunes have many similarities to those that form in deserts and many of the classic studies of dunes (e.g. Oregon coast, Padre Island) have been made on coastal dunes (Sect. 5.4). In most climates soils and vegetation readily form. Wind also affects

the beach and all the structures seen in wet, interdune areas (Sect. 5.4.5) can be found on beaches.

6.2.9 Gravitational processes

Gravitational reworking is important in some coastal settings, in particular in fine-grained large-scale deltas because of sediment instability, and on steep gradient coarse-grained deltaic and other coastlines. Sediment is redistributed by the whole range of processes on unstable slopes; creep, slide and sediment flows (Sects 6.6.4 & 10.2.3).

6.3 Coastal models and classifications

There is no single way to classify coasts. As with all natural systems (Sect. 2.1) they are the response to complex controls and the variables used depend on the use for which the classification is created. We therefore need a suite of classifications that reflects the feeder system, the volume and grain size of the sediment supply, river mouth processes, reworking processes due to waves, tides and storms, gravitational reworking processes, stream gradient and basin gradient and depth, as well as changes in relative sea level.

Traditionally, geomorphologists distinguished emergent from submergent coasts on the basis of falling and rising sea levels (D.W. Johnson, 1919). Subsequently, coastal shape became an important criterion, impetus being given to the deciphering of coastal shape in ancient systems by the need to predict the shape and orientation of hydrocarbon reservoir units. Non- or low-tidal coasts were seen as a spectrum spanning linear strandplains through cuspate, arcuate, lobate to elongate deltas as the fluvial input increased relative to the straightening effects of waves (W.L. Fisher, Brown *et al.*, 1969). This process-based classification was extended by Galloway (1975) in his now classic ternary diagram for deltas in which he plotted fluvial input against tide- and wave-dominated reworking processes (Fig. 6.2). Non-deltaic, linear coasts were classified similarly, on the relative effectiveness of waves and tidal processes in controlling the development and shape of the coastline (Hayes, 1979) (Fig. 6.3).

These process-based classifications omitted three important features. They paid little attention to sediment grain size, in particular they ignored coarse-grained systems. They paid little attention to indented 'estuarine' coasts, and they were essentially static views that neglected relative sea-level changes.

Classifications of coarse-grained gravel-dominated deltaic systems have recently been proposed by Ethridge and Wescott (1984), McPherson, Shanmugam and Moiola (1987), Nemec (1990a), Postma (1990) and Wescott and Ethridge (1990). Very little attention was paid to basinal reworking processes because their effectiveness, particularly that of tides, on steep gravelly shorelines, is much less than on lower gradient fine-grained shorelines. Instead, the classifications emphasized the nature of the feeder system and the gradient and depth of the basin (Fig. 6.15).

Orton (1988), Reading and Orton (1991) and Orton and Reading (1993) combined process-based classifications and grain size by extending both the Galloway (1975) and Hayes (1979) diagrams in the third dimension to include grain size in a triangular prism. In particular, they expanded their classifications to include coarse-grained systems (Figs 6.2 & 6.19). They showed that the grain size of the system was not only related to its overall dimensions and gradient, but also that it controlled the effectiveness of many physical processes (Tables 6.1 & 6.2).

Although the importance of changing sea levels on coastline morphology has been recognized at least since D.W. Johnson's (1919) distinction between emergent and submergent coasts, it was Curray (1964) who first formalized the effects of relative sea-level conditions on the processes and facies distributions of modern shelves and coasts. He proposed a classification based both on the direction and rate of relative sea-level change and on the rate of net deposition at the shoreline. The conditions of relative sea-level change ranged from rapid fall, through slow fall to stable and then on to slow rise and rapid rise. The rate of

net deposition ranged from high rates to low rates of deposition through zero deposition to net erosion. In this way, eight groups of erosional/depositional and transgressive/regressive systems were identified using examples from modern shelves and coasts around North America. Curray's (1964) paper forms the basis for many recent models of sea-level change (e.g. Boyd, Dalrymple & Zaitlin, 1992; Posamentier, Allen *et al.*, 1992) (Fig. 6.16).

Dalrymple, Zaitlin and Boyd (1992), followed by Boyd, Dalrymple and Zaitlin (1992) extended the triangular delta diagram of Galloway (1975) downwards to include first of all a trapezoidal area of estuarine coasts where river sediment input is substantially less than in deltas, and where there is significant contribution from the sea (Fig. 6.17). They broaden the definition of estuary (Sect. 6.7.5) to include what have been called barrier/lagoonal systems, though here that term will be retained for the more wave-dominated systems. Thus estuarine coasts vary from dominantly transgressive tide-dominated embayments to wave-dominated barrier/lagoonal systems. At the base of the triangle shorelines with a marine sediment source may be composed either of strandplains, cheniers or tidal flats. These coastlines are dominantly regressive.

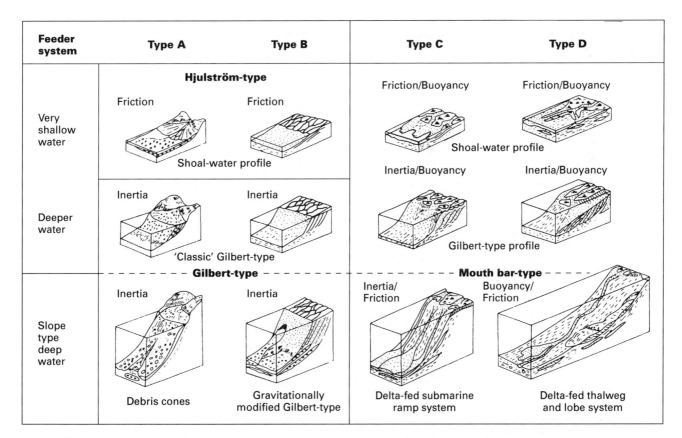

Figure 6.15 A classification for fluvial-dominated mainly coarse-grained delta systems entering very low-energy basins. The 12 categories of fluvial-dominated deltas are distinguished on the basis of: (i) four different types of feeder system; (ii) whether they enter a shallow or deepwater basin; and (iii) on the nature of the effluent entering the basin which may vary according to the rate of inflow and state of tide (from Postma, 1990).

Table 6.1 Characteristic features of alluvial and coastal depositional systems divided on the basis of grain size (from Orton and Reading, 1993). Attributes represent tendencies rather than certainties. Facies codes after Miall (1978). (Sect. 3.5.)

	1 Gravel	2 Gravel and sand	3 Fine sand	4 Mud/silt
Hinterland				
Catchment area	Small (<10^3 km^2)	Intermediate (<10^5 km^2)	Intermediate (<10^6 km^2)	Large
Relief or topography	High	Intermediate	Intermediate	Low
Climate	Arid, arctic	Temperate	Temperate	Humid, tropical
Alluvial form				
Size of stream	Small	Intermediate	Intermediate	Large
Stream gradient	Very steep (>5 m km^{-1})	Intermediate (>0.5 m km^{-1})	Intermediate (>0.05 m km^{-1})	Low
Flow velocity	High–very high	Intermediate	Intermediate	Low
Discharge	Low (<100 m^3 s^{-1})	Intermediate (<10^3 m^3 s^{-1})	Intermediate (<10^4 m^3 s^{-1})	High
Discharge variability	Very irregular	Irregular–regular	Regular–irregular	Very regular
Sediment load	Low (<10^6 tons years^{-1})	Intermediate (<10^7 tons year^{-1})	Intermediate (<10^8 tons year^{-1})	High (<10^{10} tons year^{-1})
Load/discharge ratio	High/very high	Intermediate	Intermediate	Low
Channel type	Bedload	Bedload	Mixed-load	Suspended-load
Channel pattern	Braided/absent	Braided	Meandering/braided	Straight/meandering
Bank strength	Moderate	Low–moderate	Low	High
Width/depth ratio	Intermediate	High–intermediate	High	Low
Channel mobility	Intermediate	High–intermediate	High	Low (fixed)
Delta plain				
Size	Small (<50 km^2)	Intermediate (<1000 km^2)	Intermediate (<50 000 km^2)	Large (<500 000 km^2)
Gradient	High (>5 m km^{-1})	Intermediate (0.5 to 5 m km^{-1})	Intermediate (0.1 to 1 m km^{-1})	Low (0 to 0.1 m km^{-1})
Percentage subaerial	High (>90%)	Intermediate (70–90%)	Intermediate (50–70%)	Low (<50%)
Environments	Gravel bar deposits	Braided sandflat, strandplain or beach ridges	Abandoned fluvial channels, sand flats, marsh, aeolian sand dunes, lagoon with storm washover sands	Interdistributary bays, lakes or swamps, tidal flat or chenier plain, fluvial and tidal channels, levees, crevasse splays
Lithofacies	Gm, Gms, Gp, St	Gt, St, Sp	St, Sh, Sp, Sr	Fm, Fl, St, Sr
River mouth				
Mixing behaviour	Friction–inertia	Friction	Buoyancy–friction	Buoyancy
Relative density	Homo- and hyperpycnal	Homo- and hyperpycnal	Hypopycnal–hyperpycnal	Hypopycnal flow
Salt-wedge intrusion	Never	Occasionally	Sometimes	Commonly
Effluent spreading angle	Wide	Intermediate	Intermediate	Narrow
Distributary pattern	Channel bifurcation, unchannelized	Channel bifurcation	Channel bifurcation	Single channel
Deposits	Radial bars, Gilbert-type foresets, lunate bars	Radial, middle-ground bars	Radial, middle-ground bars	Bar-finger sands
Shoreline				
Shape	Straight	Straight–cuspate	Irregular–lobate	Irregular–elongate
Gradient	Steep (>50 m km^{-1})	Intermediate (10–50 m km^{-1})	Intermediate (10–50 m km^{-1})	Gentle (0–10 m km^{-1})
Morphology	Reflective	Reflective–dissipative	Dissipative–reflective	Dissipative
Width of 'shoreface'	<50 m	50–200 m	50–500 m	>500 m
Important wave processes	Traction currents, wave-winnowing on shoreface	Traction currents, longshore, transverse, and rhythmic bars	Large–scale nearshore circulation patterns involving longshore drift and rip-currents	Easy suspension of sediment, slow shoaling of wave, nearshore trapping of sediment
Width of intertidal zone	<500 m	500 m–2 km	500 m–5 km	>5 km
Tidal influence	Low, limited ability of currents to transport sediment, intergranular seepage on foreshore prevents formation of tidal channels	Intermediate	Intermediate, tidal inlets are common	High, river channel often as mixed estuary, frequent tidal inlets and ebbtidal deltas if barrier islands form
Depositional features	Non-barred shoreface, except during storms	Barred shoreface, small barrier islands and associated lagoons, ebb-tidal deltas are rare	Barrier islands, barred shoreface, aeolian sand-dunes, tidal channels and deltas	Shore-attached mud banks, chenier plain, tidal flats
Lithofacies	St, two-dimensional wave ripples, gravel lags	St, Sp, two-dimensional wave ripples	St, HCS, tidal sandwaves	Fm, Fl, Sr
Sediment mobility	Low	Intermediate	Intermediate	High
Long-term coastline stability	Low	Intermediate	Intermediate	High
Subaqueous delta front				
Gradient	High (50–1000 m km^{-1})	Intermediate (50–500 m km^{-1})	Intermediate (5–200 m km^{-1})	Low (<20 m km^{-1})
Organization	Low	Intermediate	Intermediate	High
Frequency of resedimentation events	High	Intermediate	Intermediate	Low
Magnitude of flows	Small	Intermediate	Intermediate	Large
Depositional features	Blocky talus cones, distal avalanche blocks	'Braided' sand chutes	Branching or meandering sand chutes	Persistent mudflow gully or channel with levees, mud diapirs, growth faults
Supply mechanism	Viscous debris flows	High density turbidity currents	Turbidity currents, suspension deposition	Mudflows, slumps, suspension deposition

Table 6.2 Salient features of selected daltaic systems. Type of delta according to grain size (GR, gravel; GS, gravel and sand; FS, fine sand; MS, mud/silt) and dominant processes shaping the delta front (i, input/fluvial-dominated; t, tide-dominated; w, wave-dominated; m, mixed). Values represent mean or range unless indicated otherwise. Sediment load values do not include bedload (which may be significant) unless indicated. *Includes bedload (from Orton & Reading (1993) in which can be found references to sources of data).

| | | Source characteristics | | | | | Delta response | | | | | |
| | | | | | | | Delta plain | | Upper delta | Basin regime | | |
Delta	Type	Drainage area (10^3 km²)	Mean annual discharge (m³ s⁻¹)	Sediment load (× 10^6 tons yr⁻¹)	Load discharge ratio (g l⁻¹)	Grain size (mm)	Area (km²)	Gradient (m km⁻¹)	Slope gradient (mean or range; m km⁻¹)	Mean wave height (m)	Mean tidal range (m)	Water depth (m)
Alta	GSi					pebbly sand	10	1.5	90–700	low	1–2.5	70
Amazon	MSti	6150.0	199 634	900	0.14	0.03	467 078	0.0125	1–10	moderate	4.9	100
Bella Coola	GSiw	4.2	119	—	—	0.5–20	5	1–3.1	73–268	1.2	3.9	600
Burdekin	FS/GSm	266.7	475			0.4–1.1	2112		—	moderate	2.2	
Chachaguala	GRw	0.072	—	—	—	2.0	6	1.0	40	low	low	—
Colville	GS	59.5	491.7	—	—	0.02–10	1687	—	—	0	0.2	—
Copper	FSwt	60.0	1236	70	1.80	0.25	1920	0.6	3–11	1.5	3.4	150
Ebro	FSiw	85.8	552	6.2	0.35	0.20	325	0.38	8.7	—	0.2	100
Fraser	FSit	234.0	3549	20	0.18	0.12–0.35	480		25	—	5.0	350
Ganges/Brahmaputra	GSit	1597.2	30 769	1670	1.76	0.16	105 641	0.05–0.17	0.18	low	3.6	
Homathko	GSit	5.72	254	4*	0.51	0.14	3	1.1	20–110	1.2	4.0	550
Huanghe	MSi	865.1	1552	1080	22.62	0.02–0.06	36 272	flat	6	1.5	0.8	30–50
Irrawaddy	MSm	341.8	13 562	265	0.64	silt	20 571	low	0.6	low	4.2	<100
Jaba	GS?w	0.46	40	26	21.13	—	4	<45	—	<0.5	1.5	—
Klang	FSt	0.9	1100	—	—	0.07–0.25	1817	low	—	low	0.2	—
Klinaklini	GSi	6.5	325	18	1.84	0.10–1.0	6	1.44	20–80	1.2	3.6	350
McKenzie	FSi	1448	9100	126		silty sand	13 000	0.05	0.29	low	0.2	70
Mahakam	FSit	—	—	c.16	0.35	—	5000		6.5	<0.6	1.2	c.100
Mekong	FSm	790.0	14 168	160	0.37	vf sand	93 781	0.020	0.5	low	2.6	—
Mississippi	MSi	3344.0	15 631	349	0.73	0.014	28 568	0.020	3–15	v. low	0.4	—
Niger	FSm	1112.7	8769	40	0.08	0.15	19 135		2	moderate	1.4	100–200
Nile	FSwi	2715.6	1480	111	2.43	0.03	12 512	0.088	0.265	0.5–1.5	0.4	100
Noeick	GSi	0.562	210				0.8		80–110			250
Ord	MSt	78.0	163	22	4.30	0.176	3896	—		low	3.8	—
Orinoco	MSm	951.3	34 856	210	0.19	—	20 642	0.067	0.45	low	1.9	
Po	FSi	71.7	1484	15	0.33	0.52	13 398	0.025–0.074	3.3	—	0.7	—
Punta Gorda	GRw	0.005	—	—		100–350	0.4	14.0	40	—	—	—
Rhône	FSwi	90.0	1552	10	0.21	0.08–0.50	2540	—	3.4	—	—	50–100
Sâo Francisco	FSw	602.3	3420	6	—	—	734	—	4.3	c.1.0	2.5	—
Senegal	FSw	196.4	867.8				4254	—	7	c.2.5	1.9	
Shoalhaven	FSm	7.25	57	—	—	0.25	85		14	1.5	1.2	—
Skeidararsandur	GSw	—	400	—	1.1–3.7	0.49	600	2.0	16	2.2	2.0	—
Tiber	FSw	17.156	224			silty sand	250		4–20			150
Tunsberg Dalbre	GRi	0.136	—	0.053*	—	0.125–8.0	25	17–35	100			
Yallahs	GRwi	0.163	17.5	—	—	42.0	10.5	15	180	1.2	0.2	1100
Yangtze	MSit	1354.4	28 519	478	0.54	silt	66 669		—	1.0–1.5	2.8	50

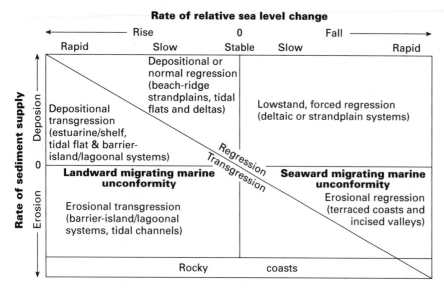

Figure 6.16 A plot of relative sea-level rise and fall against erosional and depositional rates allows discrimination between (i) transgressive and regressive coasts, with distance from the regressional/transgressional diagonal indicating increased rates of regression/transgression; and (ii) depositional and erosional coasts (modified from Boyd, Dalrymple & Zaitlin, 1992; based on Curray, 1964).

The triangle was then expanded into a prism (Fig. 6.18) to show how, during transgression, estuarine and barrier-island/lagoonal systems predominate, and, during regression, either deltas or strandplain/tidal flat systems predominate. Dalrymple, Zaitlin and Boyd (1992) suggest that coastlines transform from one type into another as sediment supply and relative sea-level change.

The triangular diagram (Fig. 6.17) and prism (Fig. 6.18) are innovative in that they combine all coastal systems into a single scheme, but they have several weaknesses.

1 The horizontal boundaries between the three divisions are not gradational, and may even be quite sharp and arbitrary. For example, wave-dominated deltas are widely separated from strandplains. Yet (see discussion in Sects 6.6, 6.6.2 & 6.7.1) they are very similar, differing only slightly in the proportion of sediment derived from rivers and by longshore drift and in the degree of wave reworking.

2 The scheme works better for small-scale systems where fluvial sediment supply is relatively limited than for large-scale deltas, such as the Ganges–Brahmaputra, Amazon and Niger, where sediment input overrides the effects of sea-level changes on the nature of the delta.

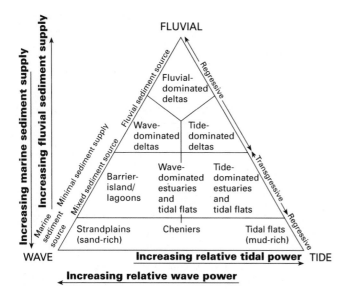

Figure 6.17 Ternary process-based classification for all coastal systems other than coarse-grained systems (modified from Dalrymple, Zaitlin & Boyd, 1992). It is extended downwards from Galloway's (1975) deltaic figure to include additional coastal systems.

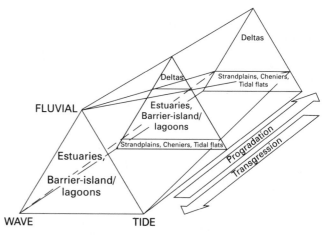

Figure 6.18 Model for the evolution of coastal environments as relative sea level and sediment supply change to yield transgressions and regressions. The central triangle is similar to Fig. 6.17. As progradation increases, estuaries give way to deltas, beach-ridge strandplains and regressive tidal flats. Transgression leads to increasing estuarine and barrier-island/lagoonal systems (modified from Dalrymple, Zaitlin & Boyd, 1992).

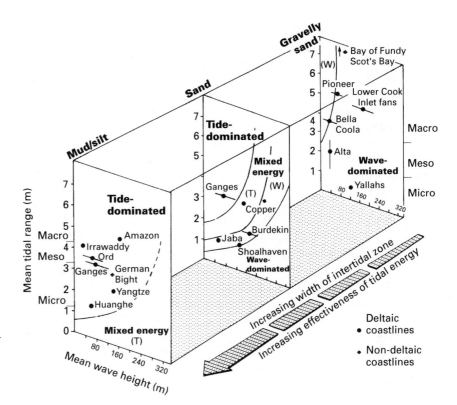

Figure 6.19 Classification of deltaic and non-deltaic coastlines based on the dominant basinal process (wave height and tidal range (after Hayes, 1979) and on grain size (from Orton & Reading, 1993)).

3 It neglects the relatively steep gradient coarse-grained systems, especially those that develop in low-energy basins. These systems only rarely have barrier-islands/lagoons or significant tidal flats.

We therefore need a suite of classification schemes, since the dominant controlling processes vary. With supply-dominated regimes, where the energy of basin processes is low, systems can be subdivided first on the nature of the feeder system and depth of water in the basin and then on the gradient of the delta profile (Postma, 1990) (Fig. 6.15). This form of classification is essentially an expansion of the fluvial portion of the Galloway diagram. It is particularly suitable for the relatively small-scale, coarse-grained steep gradient systems fed directly from the land. However, it ignores basinal processes, especially tides, and is therefore less suitable for the larger-scale 'river deltas'. For such deltas, and for examples in all higher energy basins, the Galloway process-based scheme, modified to take account of the grain size, is preferred (Orton & Reading, 1993) (Fig. 6.2). Here the gravelly sand and gravel categories embrace the first three types of the Postma classification, with the other two categories, that include all the major deltas of the world, expanding type D of Postma.

For non-deltaic coasts not directly fed by an alluvial system but where supply by longshore drift may be significant, the criteria for classification are more difficult to establish. The tidal range to wave height classification of Hayes (1979) (Fig. 6.3), extended to include grain size (Orton & Reading, 1993)

(Fig. 6.19), concentrates entirely on basinal reworking processes. It is therefore most suitable for non-deltaic siliciclastic coastlines, although the absence of input, especially from longshore drift, is a major weakness. This is redressed by Dalrymple, Zaitlin and Boyd (1992) who stress supply, both fluvial and marine, tidal and wave processes, and whether transgressive or regressive.

All classifications have strengths and weaknesses. The Postma scheme is more suitable for coarse-grained systems; the Galloway–Orton scheme for supply-dominated deltaic regimes; the Hayes–Reading–Orton scheme for non-deltaic linear and indented coastlines where accommodation space is substantial; the Dalrymple–Zaitlin–Boyd scheme for supply, marine re-working and relative sea-level change, a factor that is particularly addressed by Curray (1964) and Boyd, Dalrymple and Zaitlin (1992).

6.4 Rocky coasts

Rocky coasts have erosional shorelines where retreating cliffs, formed of bedrock, leave a series of erosional features such as headlands, sea stacks and wave-cut rock platforms. They account for more than 33% of the world's coastline today, and are in general more common on tectonically active coasts such as California (70%) and Chile (92%) than on mature passive continental margins. They are almost absent on the southern and eastern USA coasts and on the southern North Sea coast (M.E. Johnson, 1988). However, they are also common in cratonic, epicontinental settings such as Hudson Bay (30%) and on the Great Lakes (48%). They occur most commonly where loose sediment is in short supply and during transgressions.

The processes that contribute to the destruction of bedrock cliffs have been reviewed by Trenhaile (1987). They are *mechanical action, weathering* and *biological erosion. Mechanical action* by waves is the most important erosive agent, particularly where offshore water is sufficiently shallow for waves to break and dissipate their energy at the shoreline, plunging breakers being particularly effective. Wave erosion is achieved principally by hydraulic processes associated with wave impact. These involve wave shock, wave hammering and compression of air trapped between the waves and the shore, and hydrostatic pressure. In addition, abrasion takes place where wave action acts on loose particles that erode not only the cliff base but the platform as well. However, if there is too much beach debris, erosion is inhibited. The effectiveness of abrasion depends on wave energy and particle size. Large boulders can only be moved during intense storms, while sand can be moved more frequently by waves of moderate energy.

Weathering includes the same range of physical and chemical processes that occur on land, such as frost action, ice wedging, and rain wash. In addition, the presence of sea water and the cycle of wetting and drying produced by tides introduce additional processes, such as hydration, salt crystallization and dissolution, that are particularly effective around high water level and on carbonate rocky coasts.

Biological erosion takes place by leaching, rasping and boring action of organisms such as molluscs, sponges, echinoderms, barnacles and blue-green algae. On limestones, erosion may be 2–3 cm a year, but is not very important on other types of rock.

The coastal morphology of the cliffs and shore platform produced by these processes varies according to: (i) the relative importance of each destructive process, particularly wave intensity and tidal range; (ii) the nature of the bedrock, its lithology, structure, permeability, jointing and hence resistance to erosion; its susceptibility to chemical weathering, to mass movements and subaerial processes; (iii) dip and strike of the bedrock relative to shoreline orientation; (iv) gradient and width of the offshore platform; and (v) climate.

Very steep, plunging cliffs that rise abruptly from deep water suffer minimal erosion by wave action since they are affected only by swell waves which are efficiently reflected. In most cases, however, quarrying by wave erosion does take place, the form of the retreating cliffs depending partly on the relative effectiveness of subaerial and marine erosion. Where marine processes are strong and subaerial processes are weak, the cliff is undercut and the collapsed material is removed. Where both processes are moderate and in balance, collapsed material is removed slowly. Where mass flow of collapsed material is dominant over marine erosion, material is removed only intermittently.

Ancient erosional coasts occur mostly on limestone and quartzite substrates, the latter providing the best preserved coastal topography, and specialized biota have been found from the Ordovician onwards (see M.E. Johnson, 1988, table 1; 1992, appendix 1 for bibliography and examples). They are characterized by an unconformity which may be planar with a relatively extensive wave-cut platform that runs for hundreds of metres, or it may be terraced (Peters, Troelstra & van Harten, 1985; Collier, 1990) due to sea-level changes in areas of tectonic uplift. Mapping of an unconformity may indicate that the palaeo-coastline had protruding headlands, sea stacks (Whittard, 1932; Ricketts, Ballance *et al.*, 1989) and islands up to 200-m high (Radwanski, 1970; Dalziel & Dott, 1970; Dott, 1974) as well as deep-cut narrow palaeovalleys interpreted as rias, straits or fjords (Whittard, 1932; Bridges, 1975). The surface of the unconformity may be extensively bored and encrusted, as in the Miocene of Poland where rock-boring sponges, polychaetes, bivalves, cirripeds and echinoids are each found in their preferred habitat (Radwanski, 1970). Boulders also can be bored and encrusted, with a zonation from base to top (Surlyk & Christensen, 1974).

6.5 Coarse-grained gravel-rich coasts

The coarse-grained coasts described in this section are primarily fan deltas, braidplain deltas and braid deltas but many of the processes discussed apply also to gravel beaches.

An essential feature of these coasts is the direct linkage between the alluvial feeder system, the shoreline and the receiving basin, even its deepwater portion. An event within the alluvial system, such as torrential rain in the hinterland may be felt almost instantaneously in the deepwater part. Consequently the coastal system cannot be fully understood without considering both the alluvial system (Chapter 3) and the deepwater system (Chapter 10). At any one time, all these systems are closely associated in space, perhaps only 1–10 km apart, and in vertical sections their associated facies are commonly closely stacked.

Initially, following Holmes (1965), all coarse-grained deltas were known as *fan deltas*, defined as a coastal prism of sediments delivered by an alluvial fan and deposited either in the sea or in a lake. They were distinguished from alluvial fans by the presence of a basinal waterbody or by evidence for the interaction of alluvial with marine or lacustrine processes. While 'fan delta' was used for all coarse-grained systems entering a body of water and some authors (e.g. Wescott & Ethridge, 1990) still retain this usage, a wide variety of feeding patterns is recognized so that classification presents a challenge (McPherson, Shanmugam & Moiola, 1987; Orton, 1988; Nemec & Steel, 1988b, fig. 1; Nemec, 1990a; Postma, 1990). Orton (1988), for instance, distinguished between point-sourced *alluvial fan deltas, braid deltas* fed by a single braided river and *braidplain deltas* fed by a broad coastal plain of low sinuosity braided rivers, quite unrelated to alluvial fans. Postma (1990) distinguished three types of coarse-grained feeder systems (A,

B and C) on the basis of slope gradient, channel stability and nature of flow (Sect. 6.3; Fig. 6.15). Fan-delta systems can be further subdivided into three types (Ethridge & Wescott, 1984; Massari & Colella, 1988; Postma, 1990) according to the water depth of the basin and the gradient of the delta profile: (i) low-gradient *shelf-type, shallow-water* deltas; (ii) *slope-type, deepwater* deltas that have low to steep gradient, depending on the suspended load/total load ratio; and (iii) *Gilbert-type*, steep gradient deltas that form in both shallow and deep water. It must be stressed, however, that the features chosen to distinguish each type are not exactly the same and differ somewhat from author to author.

Shelf-type or *shallow-water deltas* encroach on to low-gradient shelves with very shallow water depths at and near the river mouth. They are therefore very sensitive to fluctuations of base level. They generally have only three physiographic zones, *delta plain* or *subaerial fan delta* comprising alluvial settings, *delta front* or *transition zone* affected by waves and the *prodelta*, below wave base, receiving hemipelagites. In very shallow-water basins the prodelta may not exist. There is a gradual distal diminution of grain size (Fig. 6.20) (Colella, 1988; Massari & Colella, 1988; Wescott & Ethridge, 1990) from poorly bedded, fluvial imbricated, poorly sorted coarse-grained gravels through wave imbricated fine-grained gravel and planar laminated and tabular cross-bedded sand to interbedded tabular cross-laminated sand and rippled sand giving a well-developed coarsening-upward delta front to delta plain sequence (e.g. Copper River; Galloway, 1976).

Slope-type, deep-water fan deltas have a slope separating a poorly developed delta front from the prodelta (Ethridge & Wescott, 1984). They normally include a deep-water fan system which may be mud dominated. The slope may be an inherent constructional element of the delta, a delta slope, as in deep-water fjords (Prior & Bornhold, 1990), or it may be a consequence of a faulted basin margin and therefore be separated from the delta front by a pronounced shelf/slope break (e.g. Yallahs fan delta, Wescott & Ethridge, 1980, 1990, fig. 10.4) (Fig. 6.21). In such cases subaerial fan gravels may sometimes pass seaward into a significant coastal transition zone of beach, shoreface and shelf gravels and sands before passing into deep-water mass flow deposits of the slope and base-of-slope settings.

Gilbert-type deltas have steeply inclined profiles characterized by large-scale, high-angle delta front slopes. They occur in both shallow water and relatively deep water – up to 150 m. Their development requires a high basin margin gradient, and therefore they are confined to coarse-grained systems. They consist of subaerial topset, subaqueous foreset and bottomset beds (Gilbert, 1885; Axelsson, 1967; Bogen, 1983; Colella, 1988; Massari & Colella, 1988; Nemec, 1990b; Postma, 1990; Wescott & Ethridge, 1990) (Figs 4.6 & 6.22). The topset beds are deposited by shifting channels and may be part of an alluvial fan, a braidplain or a braided river. Foreset beds form where bedload, dropped at the river mouth, continues down the delta front as grain flows or frictional debris flows. The slope gradients are commonly up to 20° and may reach 24–27° in sands and 30–35° in gravels (Nemec, 1990b) but are frequently reduced by wave action and by avalanching resulting from oversteepening of the slope by bedload deposition in the upper

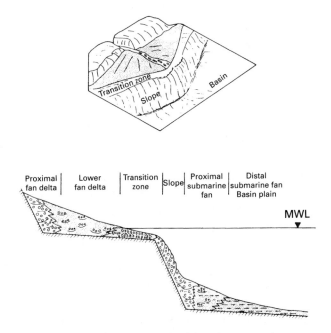

| Proximal fan delta | Lower fan delta | Transition zone | Slope | Proximal submarine fan | Distal submarine fan Basin plain |

MWL ▼

Figure 6.21 Deepwater/slope-type fan delta; plan view and cross-section. Based on Yallahs fan-delta system, Jamaica (Burke, 1967; Wescott & Ethridge, 1980) (from Wescott & Ethridge, 1990).

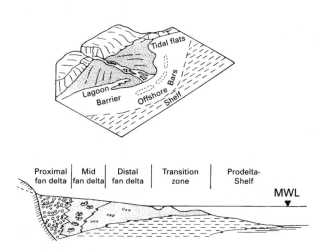

| Proximal fan delta | Mid fan delta | Distal fan delta | Transition zone | Prodelta-Shelf |

MWL ▼

Figure 6.20 Shallow-water/shelf-type fan delta; plan view and cross-section. Based on Alaskan fan deltas (Boothroyd & Ashley, 1975; Galloway, 1976; Boothroyd & Nummedal, 1978) (from Wescott & Ethridge, 1990).

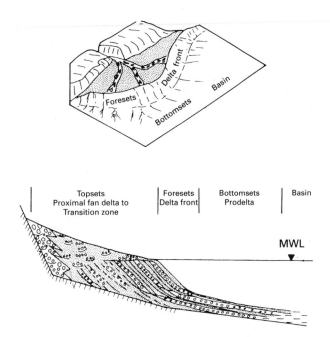

Figure 6.22 Gilbert-type fan delta; plan view and cross-section (after Gilbert, 1885, 1890; Postma & Roep, 1985; Colella, 1988; Wescott & Ethridge, 1990).

part. For avalanche foresets to form there must be: (i) a high proportion of bedload transported to the river mouth; (ii) relatively deep water immediately seaward of the mouth (i.e. a low channel/basin depth ratio); and (iii) spreading (expansion) of the effluent as an axial turbulent half-jet (Fig. 6.5) by inertia-dominated diffusion and limited wave reworking (Axelsson, 1967; Postma, 1990). The formation of Gilbert-type foresets is encouraged by homopycnal (mixed) conditions at the river mouth (Fig. 6.15), but these are not essential. Bottomset beds, deposited from a mixture of suspended load and gravity flows, form a low-gradient prodelta.

6.5.1 Feeder systems

Postma (1990) recognized three types of feeder system. The first is a true point-sourced alluvial fan, often with a very steep gradient (Fig. 6.15, type A). These fan deltas occur in areas of high relief, tectonically active regions, in fjords, and associated with volcanoes. The dominant feeder processes are ephemeral mass flows, landslides and stream floods. At the delta front very coarse gravels and sands are deposited as bedload. Hyper-pycnal flows may bypass some coarse material which combines with slope-derived mass flows and density currents to give depo-sition on the prodelta which otherwise receives hemipelagites (Sect. 10.2.3; Figs 10.53, 10.54). Unlike most coarse-grained delta systems this type can occur in semi-arid and arid climates.

The second type of feeder system (Fig. 6.15, type B) is a relatively steep-gradient braided river, well seen as fluvio–glacial

outwash, or a braidplain with a more laterally extensive feeder system. They occur in tectonically active and volcanic regions, and are characteristic of wet or glacial climates. On the delta plain, poorly confined and unconfined streams carry sediment mostly during floods. Mass flows on the delta front are rare but river flood-generated hyperpycnal flow is common on the delta front and prodelta.

The third type of feeder system (Fig. 6.15, type C) is a moderate-gradient vegetated braided river or braidplain with relatively stable channels that confine the stream flow. Bedload sediment is carried through mouth bars to the delta front where it is deposited, commonly with gravel as a subordinate com-ponent to sand. These systems are therefore transitional between 'coarse-grained deltas' and 'river deltas' described in Section 6.6.

6.5.2 Reworking at the delta front

The steep gradients and the predominance of gravel limit the width of shore zones and reduce tidal effects and the ability of basinal processes to rework sediment compared with finer-grained deltas. However, basinal processes are important in governing shoreline morphology where tidal range and wave power are high and when fluvial power is temporarily reduced.

The importance of wave reworking has been emphasized in studies around the non-tidal Mediterranean (Colella, 1988; Massari & Parea, 1990; Bardaji, Dabrio *et al.*, 1990). Not only are Gilbert-type foreset beds reworked but also topset beds (Colella, 1988) (Fig. 6.23) (cf. Fig. 4.6), and in some deltas the whole succession is dominated by beach deposits (Dabrio, 1990). All gravel beaches are reflective, although during storms dissipative conditions may prevail.

6.5.3 Resedimentation processes and slope failures

The steepness of coarse-grained delta slopes, and deposition commonly at the angle of repose, lead to abundant and frequent slope failures (Prior & Bornhold, 1989, 1990; Syvitski & Farrow, 1989; Nemec, 1990b; Postma, 1990; Figs 10.20, 10.21, 10.22). These lead to avalanche grain flows which 'freeze' lower

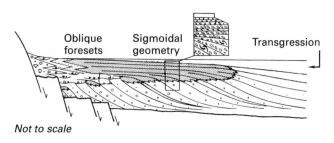

Not to scale

Figure 6.23 Wave-influenced Gilbert-type fan delta (after Colella, 1988) with wave-reworked sheet-like topset sands and gravels lying with a transitional contact on the foreset beds to give a sigmoidal geometry.

on the slope, but some mobile flows (debris flows) may extend out on to the prodelta.

6.5.4 Coarse-grained coastal facies associations

Facies associations of coarse-grained coastlines fall into three groups: (i) alluvial; (ii) deep water; and (iii) shoreline.

Alluvial facies associations are identical to the conglomeratic/sandstone facies described in Chapter 3 and are discussed in detail there. Various lithofacies schemes based on that of Miall (1978) have been constructed (e.g. Ramos, Sopeña & Perez-Arlucea, 1986; Postma, 1990 table 3). They are especially suitable for coarse-grained deltas entering low energy basins. Type A feeder systems have gravel-dominated, clast-supported, massive or horizontally stratified conglomerates with broad shallow scours and minor cross-bedded sandstones. Type B feeder systems have horizontally stratified, trough and planar cross-bedded conglomerates and sandstones with rare massive beds. Type C feeder systems have an alluvial facies of trough and tabular cross-bedded, horizontally stratified sandstones with rippled beds and only minor trough and tabular cross-bedded conglomerates.

Deep-water facies associations result from debris falls, cohesive and non-cohesive debris flows, turbidity flows, slumps and slides, and settling from suspension from buoyant plumes (Nemec, 1990b; Prior & Bornhold, 1990). Although, in many cases, especially where they are separated by evidence of a shoreline, alluvial and deep-water facies are readily distinguished, this is not always so. For example, the Pennsylvanian Haymond Formation of Texas, with boulders up to 40 m across, was originally interpreted as a turbiditic sequence (e.g. McBride, 1970). It was later reinterpreted as a prograding delta-front to delta-plain sequence formed from short-headed, steep gradient braided streams (Flores, 1975). In contrast, the Upper Jurassic Brae Formation of the North Sea (Fig. 10.63), originally interpreted as the topset beds of a series of coalescing fan deltas with conglomerates and reservoir sandstones deposited subaerially by shifting streams (Harms, Tackenberg et al., 1981) was reinterpreted as a deep-water submarine system (e.g. Stow, Bishop & Mills, 1982; Turner, Cohen et al., 1987) (Sect. 10.3.4).

The problem of discriminating alluvial from deepwater conglomerates has been addressed by Nemec and Steel (1984) and Hein (1984, table 3). They show that, at first sight, similarities are considerable and many features are identical. However, discrimination is possible by looking at the intervening strata. Coals, rootlet beds, palaeosols, non-marine faunas indicate an alluvial setting whereas hemipelagic muds and a marine fauna or bioturbation indicate a basinal environment. In alluvial environments, cross-stratification and tabular stratification are common in both conglomerates and sandstones, as is clast imbrication. In the subaqueous environment, grading, particularly normal grading, but inverse and complex grading as well, is much more common than in alluvial facies. Stratification is rare.

Conglomerates are better organized internally, with the long axis of clasts aligned flow parallel rather than flow normal as in rivers.

The *shoreline facies* associations are both unique to and diagnostic of coarse-grained coastal systems. Apart from Bluck (1967b), such systems have only been studied in recent years (Clifton, 1973, 1976; Orford & Carter, 1982; Bourgeois & Leithold, 1984; Carter & Orford, 1984; Ethridge & Wescott, 1984; Nemec & Steel, 1984; Wescott & Ethridge, 1990). Particular shoreface and beach facies have been identified (Dabrio & Polo, 1988; Massari & Parea, 1988, 1990; Marzo & Anadón, 1988; Dabrio, 1990; Bardaji, Dabrio et al., 1990). Studies of such facies have emanated from two principal areas, western North America, both the present-day west coast and the Mesozoic Western Interior Seaway, and from around the Mediterranean. The nature of these coastlines is different. The west coast of North America has everywhere some tidal range and is exposed to persistent Pacific swells. The Mediterranean coasts are non-tidal and, with the exception of parts of the Adriatic, do not face long oceanic swells. Instead they are dominated by storm-related processes (Massari & Parea, 1988).

Shoreline facies are partially recognized on the basis of their fauna, which is more likely to be preserved in shoreface than in beach facies. The high-energy environment encourages a mixed assemblage, high-density population of robust thick-shelled taxa and low faunal diversity reflecting the limited number of benthic species adapted to unstable substrates (Bourgeois & Leithold, 1984). Many shoreline conglomerates are closely associated with corals (García-Mondéjar, 1990), stromatolites and oncolites that can grow quickly at times of low terrigenous input (e.g. Sellwood & Netherwood, 1984; Dabrio & Polo, 1988). These occur both in situ and reworked as clasts. In finer-grained layers rare thin-shelled species are found. Bioturbation is rare because of grain size and the rapid rates of sedimentation and erosion. It is better developed and preserved in the interbedded sandstones.

Modern beaches are better documented than their ancient counterparts, yet the opposite is true for shoreface deposits because of ease of observation and preservation potential (Bourgeois & Leithold, 1984; Hart & Plint, 1989). Shoreline gravels are distinguished from fluvial gravels by the higher degree of sorting, the segregation of gravel from sand, the lateral continuity of beds, rounding of clasts which generally have a lower sphericity, the presence of gently dipping or horizontal lamination and seaward dipping imbrication. In contrast, fluvial facies have more high-angle cross-bedding, high-relief erosional surfaces, landward dipping imbrication, and interbedded gravelly sandstone, coals and clay (Clifton, 1973; Ethridge & Wescott, 1984; Wescott & Ethridge, 1990).

Detailed study may make it possible to distinguish beach from shoreface facies, the upper beach from the lower beach and the upper shoreface from the lower shoreface (Massari &

Parea, 1988). The offshore-transitional zone is characterized by amalgamated hummocky cross-stratified fine sandstones, rare pebbly sandstones and rare bioturbation due to its destruction by storms. The lower shoreface facies consists mainly of low-angle cross-stratified conglomerates and pebbly shelly sandstones separated by low-relief scours with pebble lags, all deposited by shore-parallel bars and rip channels developed during storms when dissipative conditions prevailed. The upper shoreface facies is characterized by interfingering tabular conglomerates, trough cross-stratified pebbly sandstones and sharp-based conglomerate/sandstone graded couplets with muddy drapes in response to alternating fairweather and storms. The lower beach facies shows thick, moderate to well-sorted, clast-supported, steeply inclined (9–19°) conglomeratic beds that wedge out up slope into the upper beach facies. Pebble shapes are blades and rods with the more rollable equant clasts and suspendable discoidal clasts rather sparse. Equant clasts tend to roll down on to the shoreface and discoidal clasts are thrown up on to the upper beachface (Orford, 1975). Well-developed clast imbrication is directed both offshore and onshore. Deposition takes place mainly during declining stages of high-energy events and under reflective conditions. The upper beach facies consists of thin, gently inclined (3–5°) planar stratification wedging out towards the lower beachface facies. Grain size, sorting and the proportion of discoidal blades and disc-shaped clasts increase upward. Cross-stratification is either landward dipping or alongshore and imbrication is dominantly seaward dipping. This facies is essentially a consequence of storm and post-storm events. Storms erode berms cutting low angle erosion surfaces, and produce the predominantly landward-directed flows with little backwash due to percolation through the gravel.

Many beaches feature a *plunge step* between the top of the shoreface and the base of the foreshore marked by accumulation of the coarsest available grain size (Fig. 6.24) (Somoza, Zazo *et al.*, 1987). Clasts are well rounded, well sorted and include any available large heavy shells (Dabrio, 1990; Bardaji, Dabrio *et al.*, 1990). This recognizably very coarse facies is a consequence of intense reworking in the breaker zone and marks the base of the lower beach facies. It is therefore a sea-level marker (Bardaji, Dabrio *et al.*, 1987).

Vertical facies sequence models have been constructed for coarse-grained fan deltas (Fig. 6.25). Shelf-type fan deltas gradually coarsen upwards from prodelta muds, either to shoreface and beach sandstones if wave activity is high, or directly into distal fan-delta sandstones if not (Fig. 6.25a). There are no conglomerates within the delta front and prodelta. Deep-water fan deltas are characterized by conglomerates in the slope facies and the reworked transitional zone (Fig. 6.25b). Gilbert-type fan deltas may have conglomerates in both the foreset and bottomset beds, as well as in the topset beds (Fig. 6.25c). Wave-dominated shorelines show a sharp break in grain size at the plunge step (Fig. 6.24).

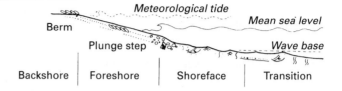

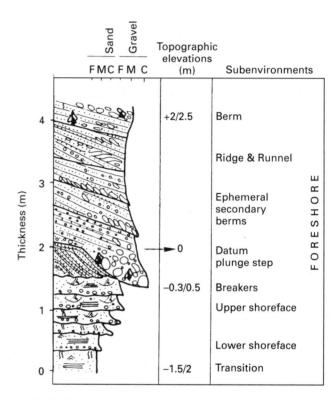

Figure 6.24 Lateral and vertical profiles through a coarse-grained beach (from Bardaji, Dabrio *et al.*, 1990).

6.5.5 Controls on coarse-grained coastal systems and sequences

Many coarse-grained coastal systems occur in fault-bounded, steep-sided basins that are filled both longitudinally and transversely (Gawthorpe & Colella, 1990; Wescott & Ethridge, 1990) (Fig. 6.26). The longitudinal, axial, fill commonly has a significantly lower gradient, is finer grained and taps a larger hinterland than the lateral feeders that are smaller, coarser, higher-gradient systems (Dunne & Hempton, 1984). These *axial deltas* may have well-developed mouth bars and the characteristics of normal river deltas. However, lateral margins also differ from each other because rift, transtensional, and strike-slip basins are asymmetric in cross-section with one steep, footwall scarp and one more gentle hanging wall dip slope. At the footwall scarp, deep-water, slope-type and Gilbert-type fan deltas tend to form, while shallow-water, shelf-type systems form

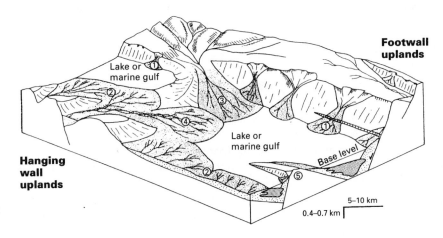

Figure 6.25 Idealized composite vertical sequences for principal fan-delta systems (from Wescott & Ethridge, 1990). The transition zone varies in the degree of development, depending on the amount of basinal reworking. In low-energy basins it may be absent. (a) Shelf-type, shallow-water fan delta. (b) Slope-type, deepwater fan delta. (c) Gilbert-type delta.

(a) Shelf-type

Interpretation
Proximal fan delta
Surge deposits (humid fan)
Debris flow & sieve deposits (arid fan)
Braided stream
Mid fan delta
Braided stream
Distal fan delta
Transition zone Tidal flats Lagoon
Beach
Upper shoreface
Lower shoreface
Prodelta

(b) Slope-type

Interpretation
Proximal fan delta
Surge deposits (humid fan)
Debris flow & sieve deposits (arid fan)
Braided stream
Lower fan delta
Braided stream
Transition zone
Beach & shoreface
Seaward imbricated gravels
Swash laminated sands (burrowed & fossiliferous)
Slope
Matrix-supported conglomerates
Fossiliferous muds
Slump folds

(c) Gilbert-type

Interpretation	
Topset beds	Proximal fan delta
	Surge flows
	Debris flows
	Sieve deposits
	Braided stream
	Mid fan delta
	Braided stream
	Transition zone
	Beach
Foreset beds	
	Delta front
Bottomset beds	
	Prodelta
	Basin

Figure 6.26 Schematic block diagram of a rift basin illustrating (1) footwall-derived coarse-grained deepwater slope-type and Gilbert-type deltas sourced from consequent drainage basins developed along the dominant border fault zone; (2) hanging-wall-derived coarse-grained shelf-type or sand-rich deltas forming a relatively continuous fringe along the hanging-wall dip slope; (3) delta sourced from a flexural transfer zone separating two *en echelon*, border-fault segments; (4) axial mouth-bar type fine-grained delta; (5) coarse-grained deltas derived from an intrabasin fault block (from Gawthorpe & Colella, 1990).

on the hanging wall dip slope (Crossley, 1984; Rosendahl, Reynolds *et al.*, 1986; Surlyk, 1989; Wescott & Ethridge, 1990). *Footwall-derived scarp systems* (Leeder, Ord & Collier, 1988; Gawthorpe & Colella, 1990) are small in area (<10 km²), prograde up to 10 km, are fan-shaped or possibly wedge-shaped, with very thick sequences that thicken towards the fault scarp. They are less affected by climatic and base-level changes than by tectonic movements. *Hanging-wall-derived systems* are larger in area (150 km⁻²) and prograde up to 15 km to develop wedge-shaped or sheet-like forms that are relatively thin and easily incised as a result of changes in base level or climate. Finally, fault relays and transfer faults transverse to the basin axis

commonly control the depocentres and the entry points of drainage systems that feed fan deltas.

Direct evidence for fault activity includes: (i) identification of a fault at the up-dip edge of the delta foreset unit; (ii) synsedimentary deformation of the foreset beds; (iii) thickening towards the fault; (iv) more frequent relative base-level changes than can be accounted for by independently derived eustatic or climatic history; and (v) decrease in syndepositional tilting upwards (Gawthorpe & Colella, 1990). In the Crati Basin, faulting and sedimentation of Gilbert-type deltas are directly related (Colella, 1988). Episodic faulting allows fluvial bedload to be transported directly over the fault scarp and build out oblique clinoforms by foreset progradation. However, the sedimentation pattern varies according to the behaviour of the fault. Where an initial large-scale slip is followed by smaller slips and periods of no slip, then oblique clinoforms alternate with aggradational, sigmoidal clinoforms and erosional surfaces are horizontal. Where there is a succession of high magnitude slips, the foreset units may be separated by erosional surfaces caused by sedimentary sliding.

Coarse-grained coastal sequences respond to several factors, which operate on a number of time scales.

First, there are the predictable annual variations in sediment supply brought about by floods due to the spring thaw, by the rainy season in semi-arid regions, and by monsoonal rains. Within the basin there may be related changes in water level, particularly where the basin is a lake. This may give ordered sequences of development as discharge and base level interact.

Second, there are catastrophic events such as river floods, flash floods, storms at sea or in lakes, landslides and earthquakes. They occur on two scales of frequency and possibly also of magnitude. River floods and storms may occur several times a year while other events are rarer and less predictable. In some semi-arid regions flash floods transport sediment into the basin once every several years to be reworked by basinal processes in the intervening intervals (Bardaji, Dabrio et al., 1990). In glacial and periglacial regions, the sudden release of water from an ice-dammed lake (jökulhlaup) may raise river discharge to many times that of the annual flood (Bornhold & Prior, 1990) (Sects 11.3.6 & 11.4.4). Earthquakes not only raise and lower relative base levels by up to 10 m instantaneously with consequent environmental changes (Colella, 1988), but set off secondary effects such as landslides and consequent sedimentation processes. Like the calving of icebergs, landslides trigger major waves in confined basins.

Third, longer-term changes reflecting tectonics, climate and base level are not easily distinguished from sedimentary evidence, but in some cases this has been attempted. For example, along the western margin of the Dead Sea, and its predecessor Lake Lisan, a classical sinistral strike-slip basin, there is no evidence of syndepositional fault movement but abundant evidence for climatic controls (Manspeizer, 1985; Frostick & Reid, 1989; Bowman, 1990). The apparently instantaneous effects of violent low-frequency, high-magnitude floods gave sheet-like, sharp-based gravel beds abruptly interbedded with fine-grained lacustrine laminites. Longer-term climatic changes affected both input of sediment and lake level. The increased rainfall appears to have encouraged vegetation and thus reduced sediment input to the small marginal fan deltas, while simultaneously increasing sediment input to larger fans and the axial River Jordan. At the same time water level in the lake rose by about 300 m in 2000 years after a fall of 500 m in about 3000 years. In very arid times lake level was low, and moderately dipping poorly sorted gravelly fan deltas were fed by high concentration catastrophic floods. As rainfall

increased, conditions for the formation of Gilbert-type deltas were initially good, but as sediment supply decreased and lake level rose, horizontally bedded topsets aggraded (Bowman, 1990).

In the marine basins of the Pleistocene of southeast Spain (Muto, 1988; Bardaji, Dabrio et al., 1990), sediment remains close to the hinterland to give coastal fan deltas during highstands. The alluvial portion is small and the fan delta progrades either as foresets, if basin processes are weak, or by beach progradation of wave reworked topsets if they are strong. During low sea-level stages, the proximal portion is widely exposed and incised and channels transfer sediment to more distal fan deltas (Bardaji, Dabrio et al., 1990).

6.5.6 Ancient coarse-grained coastal depositional systems

The examples described below illustrate depositional controls and some of the problems in interpreting ancient systems.

COARSE-GRAINED SYSTEMS IN LOW-ENERGY BASINS

Many of the best described examples of ancient shallow-water inertia-dominated coarse-grained deltas fed by alluvial fans occur at lake margins. Very thick, fault-related humid fans occur in the Carboniferous of northern Spain (Heward, 1978) where commercial coals formed after fan abandonment (Fig. 6.27). The very thin (<10 m) Devonian Domba Conglomerate of Norway shows a lateral passage from clast-supported conglomerate through matrix-supported debris flow conglomerates to well-sorted granule sandstones and fine pebbly conglomerates found at the inferred delta front (Nemec, Steel et al., 1984). Base-level fluctuations affected the architecture of a lacustrine alluvial fan – fan-delta system in the Pliocene of southeast Spain (Fernández, Bluck & Viseras, 1996). During lake-level falls, deep channels were incised and lengthened, and sediment supply to the fan delta increased. During lake-level rises, lobes formed at the channel ends and overbank processes developed, leading to the generation of radial channels, whose levees amalgamated with the levees of the main channels to give lenticular gravel bodies. Simultaneously, planar-bedded gravel sheets formed in proximal areas where there were greater amounts of sediment available.

Gilbert-type deltas that built into lakes or very restricted basins with minimal reworking include the Tertiary Coalmont Formation of Colorado (Flores, 1990). Here, two distinct types of Gilbert-type deltas have been identified: (i) vertically stacked, relatively thin (3–5 m) steeply dipping foreset beds fed from a transverse basin margin by coarse bedload braided rivers; and (ii) overlapping units with thicker (up to 19 m) foreset beds fed longitudinally by mixed load, possibly meandering streams. Homopycnal mixing and rapid deposition of sandy sediments is inferred at the river mouth.

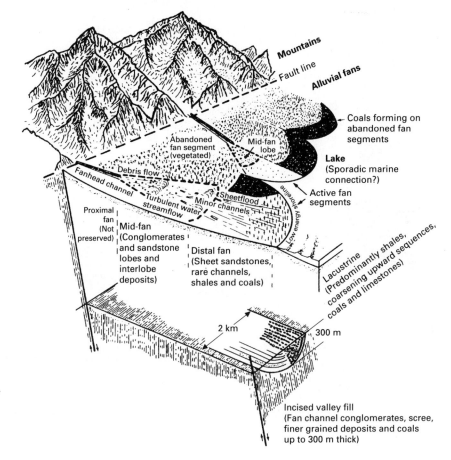

Figure 6.27 Depositional model for a lacustrine coarse-grained shallow-water fan delta debouching into a low-energy basin in the Upper Carboniferous of northern Spain (from Heward, 1978). Initially, a valley was incised and filled by up to 300 m of sediment. This was followed by repeated advances and retreats of the fan to give 200–300-m-thick progradational sequences separated by lacustrine shales and coals.

Gilbert-type deltas can also build into deep water. The Pliocene Abrioja fan delta from southeast Spain (Postma, 1984) formed in a canyon 2–4 km wide and 100–200 m deep. Numerous slump scars in the delta front and upper delta slope deposits indicate frequent slope failures (Postma & Roep, 1985) (Fig. 6.28). Progradation of the delta proceeded by: (i) primary outbuilding of a Gilbert-type delta by seasonal deposition (facies II); (ii) secondary outbuilding by liquefaction and sediment failure on the delta front and upper delta slope to produce small slumps (facies III & IV); and (iii) major slumping of the delta front to carry material (facies V) far down it on to the relatively low-gradient lower delta slope and prodelta.

In the very well-exposed Oligo-Miocene Meteora Conglomerate of Greece, the 500-m-thick conglomerate is composed of large wedge-shaped Gilbert-type delta bodies, 200 m wide and 10–20 m thick at their proximal ends and over 2000 m wide and 20–30 m thick at their distal ends. Cutting at right angles to the dip azimuths of the topset beds of the Gilbert-type deltas are channels 15–30 m deep and 80–100 m wide and mainly filled by large-scale cross-stratification dipping up to 30–40° which could easily be mistaken for Gilbert-type foresets. They, however, are inferred to be channel bars that accreted downstream. They differ from the Gilbert-type foresets by having an abrupt contact at the base, in contrast to the tangential toesets of the Gilbert-type foresets (Ori & Roveri, 1987).

WAVE-AFFECTED COARSE-GRAINED SYSTEMS

These include examples from all coarse-grained systems, regardless of how they are fed, from shallow to deep and including Gilbert-type deltas. However, the evidence for waves in most deepwater systems is limited, normally being confined to a few metres or less within relatively thick successions (e.g. Pudsey, 1984; Hwang & Chough, 1990). Shallow and Gilbert-type deltas range from those with slight wave modification to those where wave power dominated. Since they reflect the interplay of sediment input and wave activity, as well as base-level changes, the degree of wave influence in individual formations varies both spatially and temporally. In addition, the nature of the depositional system may vary widely over time, depending on position of both relative sea and lake level, precipitation, and hence sediment supply, and the tectonics of the basin margin (Fernández, Soria & Viseras, 1996).

The Oligocene Trona fan delta (Fig. 6.29), part of the 1300-m-thick Montserrat fan-delta complex, is 8 km wide and extended 6 km into a basin formed on the relatively passive southern margin of the Ebro foreland basin (Marzo & Anadón, 1988). Subaerial facies comprise proximal debris flow and braided stream conglomerates and distal, more sheet-like stream flow conglomerates. They are interbedded with finer-grained alluvial red sandstones and mudstones deposited mainly during

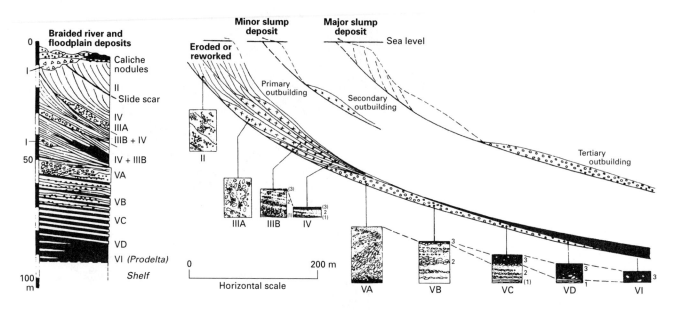

Figure 6.28 Pliocene Abrioja fan delta (from Postma, 1984). Facies types, delta sequence, and mechanisms by which a Gilbert-type delta builds into relatively deep water. Unstable, proximal delta sediments include beach, nearshore, delta front and upper delta slope sediments. Stable distal delta sediments include lower delta slope and prodelta sediments (cf. Fig. 10.54).

transgressive phases. Proximal delta front facies consist of both fluvially emplaced mouth-bar conglomerates and wave-reworked beach face conglomerates and conglomeratic sandstones. Distal delta front facies consist of mouth-bar sandstones and conglomerates and shoreface sandstones which pass seawards into prodelta and highly fossiliferous shelf mudstones and sandstones. The succession comprises a series of stacked transgressive/regressive sequences, related primarily to intermittent tectonic movement (Fig. 6.29). Conglomerates form a distinctive coarse-grained delta front facies zone in front of the relatively fine-grained distal subaerial facies which has a higher proportion of intercalated mudstones. Episodic heavily sediment-laden catastrophic floods appear to have dumped gravels at the channel mouth and left them to prolonged reworking by waves.

The Carboniferous to Permian Reinodden Formation of Spitsbergen is also a relatively shallow fan-delta system that built into a basin of moderate wave activity and minor tidal influence (Kleinspehn, Steel *et al.*, 1984). It consists of alternating sandstones and conglomerates, the latter grouped into three facies associations: fluvial, channel mouth bar and barrier/spit. The lower three sequences have relatively gradational boundaries possibly reflecting autocyclic processes and steady subsidence.

Avulsion of fluvial channels, local abandonment and erosion of the delta front, lateral migration of barriers or shoals, and aggradation of mouth bars all contributed. The upper sequences show more abrupt changes from nearshore to offshore facies at sequence boundaries, probably recording allocyclic intermittent subsidence or a rapid rise of sea level.

In the wave-influenced Gilbert-type deltas of the Albian La Miel Member (García-Mondéjar, 1990), prograding, aggrading and laterally accreting units are present (Fig. 6.30). Prograding units are formed of Gilbert-type foresets and bottomsets but with no topsets. Aggradational units are formed either of shallow marine sediments with corals in distal or interdeltaic areas or of braided fluvial sediments in proximal zones. Lateral accretion units are formed of pre-existing sediments, reworked by waves. The succession comprises a number of sequences interpreted as the result of short-term allocyclic relative sea-level fluctuations, perhaps tectonically induced. During lowstands, erosion or no sedimentation occurred in the emergent or shallow-water areas and the coarse Gilbert-type deltas built out into the basin (cf. forced regressions; Sect. 2.4). At the same time some lateral accretion deposits formed between the delta lobes. During transgressions and high sea-level stands, aggradational units formed.

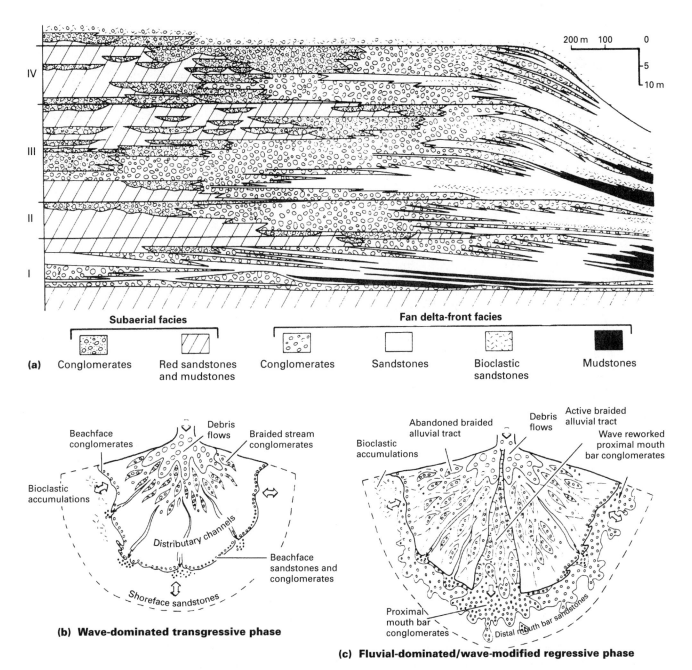

(a)

Subaerial facies — Conglomerates — Red sandstones and mudstones

Fan delta-front facies — Conglomerates — Sandstones — Bioclastic sandstones — Mudstones

(b) Wave-dominated transgressive phase

(c) Fluvial-dominated/wave-modified regressive phase

Figure 6.29 The Eocene La Trona (Montserrat) fan delta in northern Spain (from Marzo & Anadón, 1988) showing alternate phases of wave-dominated transgressions and fluvial-dominated/wave-modified regressions. (a) Cross-section of four transgressive–regressive phases shows lateral transition from subaerial distal alluvial fan through delta front to basinal mudstones. (b) Interpretation of wave-dominated transgressive phase. (c) Interpretation of fluvial-dominated/wave-modified regressive phase.

6.6 River deltas

Compared with coarse-grained gravelly systems, river deltas are characterized by an input of mud, silt and sand, and by moderate to low gradients which together allow greater effectiveness of basinal reworking processes. Although not immune from short-lived catastrophic events, they are less likely to be significantly affected by them. On the other hand, their depositional patterns strikingly reflect longer-term changes in supply and base level. Two principal types have been distinguished: (i) *shoal water* or *inner shelf deltas* which include both the very large to moderate sized deltas that protrude on to continental shelves and many smaller ones that build out into lakes, bays, lagoons and estuaries; and (ii) *shelf margin* or *shelf edge deltas*, formed during

181

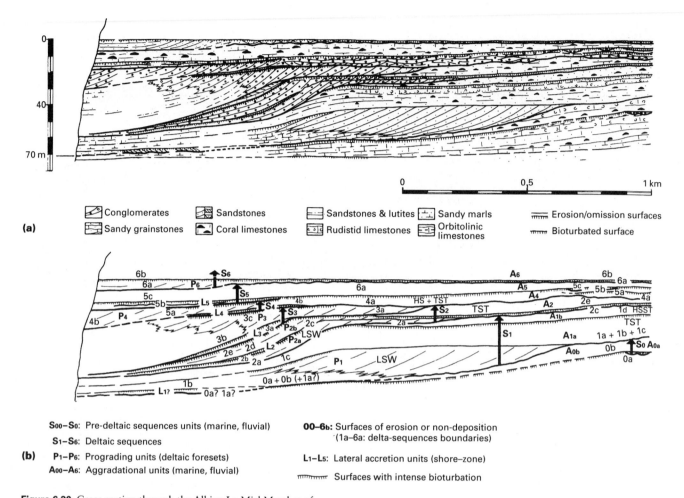

(a)

Conglomerates | Sandstones | Sandstones & lutites | Sandy marls | Erosion/omission surfaces
Sandy grainstones | Coral limestones | Rudistid limestones | Orbitolinic limestones | Bioturbated surface

(b)

S₀₀–S₀: Pre-deltaic sequences units (marine, fluvial)

S₁–S₆: Deltaic sequences

P₁–P₆: Prograding units (deltaic foresets)

A₀₀–A₆: Aggradational units (marine, fluvial)

00–6b: Surfaces of erosion or non-deposition (1a–6a: delta-sequences boundaries)

L₁–L₅: Lateral accretion units (shore–zone)

Surfaces with intense bioturbation

Figure 6.30 Cross-section through the Albian La Miel Member of northern Spain to show (a) arrangement of facies and (b) interpretation of lateral and vertical sequence in terms of deltaic sequences, aggradational units, prograding units and lateral accretion units (from García-Mondéjar, 1990).

sea-level falls at the edge of the continental shelf. These were originally thought to be closely related to growth faulting (Winker & Edwards, 1983), but high-resolution seismic profiles have shown that many are simply progradational clinoforms interpreted as deltas formed by coastal progradation and transgression when sea level was lower (e.g. Tesson, Allen & Ravenne, 1993; Sydow & Roberts, 1994).

Because the interactions of fluvial input and basin reworking processes are particularly important, river deltas are best classified by using a process-based scheme into *river-dominated deltas* where river currents reach the basin to deposit sediment beyond the shoreline, *tide–wave interaction deltas* where shoreline reworking is significant but tidal currents penetrate far inland, *wave-dominated deltas* where deposition is concentrated at the shoreline itself, and *tide-dominated deltas* where the delta plain is penetrated by large funnel-shaped estuaries (Sect. 6.7.5). This scheme has been criticized on the grounds that many coastal accumulations, commonly classified as deltas, are not deltas.

For example: (i) wave-dominated deltas (e.g. São Francisco, Senegal, Costa de Nayarit) are better considered as beach-ridge strandplains because they have a significant component of sediment supplied by longshore drift rather than a river (Sect. 6.7) (Dominguez & Wanless, 1991; Dominguez & Barbosa, 1994); and (ii) tide-dominated deltas are, in reality, tidal estuaries because they are almost exclusively transgressive systems (Walker, 1992). However, deltas, especially the larger ones such as the wave-dominated Rhône and tide-dominated Ganges–Brahmaputra, cannot easily be removed from such a classification. The problem lies partly in the scale under consideration and partly because of the gradations between coastal systems. Portions of large deltas may comprise tidal estuaries, beach-ridge strandplains or barrier-island lagoons, as well as distributary mouth bars, and thus locally can be considered as 'non-deltaic' linear coastlines. Few coastlines are exclusively supplied either from the river or from the sea. Most have a mixed supply and hence can fall into one or more category.

The process-based delta classification can be extended either by using grain size (Orton & Reading, 1993) (Fig. 6.2) where the deltas mainly fall into the mixed mud–silt and fine-sand categories or by using the evolutionary classification of Dalrymple, Zaitlin and Boyd (1992) (Fig. 6.18) where deltas increase in importance towards the prograding end of the prism. In this section we first concen-trate on prograding, regressive systems before turning to the transgressive/destructive phases and portions of deltas.

Components of prograding deltas can be identified either on the basis of the gradient of the delta profile, or according to the dominant process. At one time (e.g. Van Andel & Curray, 1960) the delta profile was divided on the basis of gradient into topset, foreset and bottomset by analogy with Gilbert-type deltas. These three zones were later referred to as: (i) *delta platform*, with a low gradient profile that is partly subaerial and partly subaqueous, and may, as in the Amazon delta, protrude 150 km from the shoreline (Nittrouer, Kuehl *et al.*, 1986) (Fig. 7.30); (ii) a relatively steep gradient *delta slope*; and (iii) a low gradient *prodelta*. This morphological terminology is now generally superseded by a process terminology: (i) *delta plain*, the largely subaerial zone dominated by rivers; (ii) *delta front*, the zone of interaction between fluvial and basinal processes; and (iii) the *prodelta*, the zone of quiet sedimentation from suspension disturbed only by gravity sliding and mass flow deposition. In addition, all deltas have a destructive or *abandonment* phase.

Although, in this chapter, the process zonation will be used for the outline, both terminologies have their uses. If sedimentary facies can be clearly defined, the process terminology is better. If the data are from seismic profiles, then the morphological, slope gradient terminology may be more suitable.

6.6.1 Delta plain

Delta plains are extensive lowland areas which comprise active and abandoned distributary channels and their associated levees, separated by shallow-water environments and by emergent or near-emergent areas. Some deltas have only one channel, but more commonly a series of distributary channels is variably active across the delta plain, often diverging from the overall basinward direction by 60° or more. Between the channels are bays, floodplains, lakes, tidal flats, marshes, swamps and/or salinas all of which are extremely sensitive to climate.

Upper delta plains are essentially unaffected by basinal process-es. They do not therefore differ substantially from the alluvial environments described in Chapter 3 except that areas of swamp, marsh and lakes are usually more widespread and channels may bifurcate downstream. Lower delta plains are also affected by fluvial processes but many are penetrated by saline water and by tidal processes. They are, however, protected from offshore waves by beach-barrier shorelines or by a wide dissipative delta front, except during extreme storms when sea water may penetrate many kilometres inland. Locally generated waves, however,

may operate in shallow-water delta-plain bays and lakes.

The delta plain can be divided into two principal components, distributary channels and interdistributary areas (Coleman, Gagliano & Webb, 1964).

Distributary channels have many of the characteristics of fluvial channels. Unidirectional flow predominates, with periodic stage fluctuations. Channel systems may have high or low sinuosity depending on gradient and grain size. Resultant deposits are erosively based sands with basal lags that fine upward through cross-bedded sands into ripple-laminated finer sands with silt and clay alternations, topped by rootlets or other features indi-cative of emergence (Oomkens, 1970, 1974; Coleman, 1981). Distributary channels differ from fluvial channels in the following ways.

1 The lower reaches are influenced by basinal processes even in low energy basins. In the Mississippi, flood tides and waves associated with strong onshore winds impound distributary discharge during low and normal river stages.

2 A saltwater wedge may penetrate the lower reaches during low river stages.

3 Switching and avulsion of channels are more frequent because shorter and steeper courses are created as the active distributary lowers its gradient through progradation.

4 The width to depth ratio of sandbodies of distributary chan-nels is lower than that of most fluvial channels mainly because of their relatively short life and limited scope for lateral migration.

Interdistributary areas of delta plains are composed of: (i) swamps, supporting a woody vegetation, which pass seaward into (ii) marshes, supporting non-woody plants such as grasses, reeds and rushes, that can be differentiated, in humid regions such as the Mississippi, into fresh, brackish and saline as the shoreline is approached (Gould, 1970, fig. 8; Tye & Kosters, 1986, fig. 3). In the Mississippi, true, potentially coal-forming peats with more than 80% organic matter form in the freshwater forested swamps and herbaceous marshes (Kosters, Chmura & Bailey, 1987). However, as in many deltas (e.g. Fraser delta; Styan & Bustin, 1984), peats in the more saline marshes of the lower delta plain have high contents of terrigenous matter and sulphur. In addition there are (iii) enclosed or partially enclosed waterbodies which are quiet or even stagnant except for locally generated wind-waves. In the upper delta plain the waterbodies are lakes. In the lower delta plain they include not only lakes, but also lagoons, estuaries and bays with some access to the sea. All these waterbodies are shallow (<5 m and often less than 1 m), yet the energy level is low and clays, silts and fine sands predominate. At the margins of lakes, extensive blanket bogs develop in humid climates while exposed surfaces with calcretes, gypsum and halite precipitates form if the climate is arid. The regime is, however, frequently interrupted by incursions of coarser clastic material brought in by floods or by diversion of distributary channels to low-lying areas. In these ways a variety of features is formed: levees, crevasse channels, crevasse splays and minor deltas.

Levees, crevasse channels and crevasse splays are discussed in Section 3.3.6. Minor deltas rapidly fill lakes (Tye & Coleman, 1989a,b), interdistributary bays open to the sea (Coleman & Gagliano, 1964; Coleman, Gagliano & Webb, 1964; Gagliano & van Beek, 1970; Coleman, 1981) (Fig. 6.31) and protected bays/lagoons (van Heerden & Roberts, 1988). Similar bay head deltas are found in the barrier/lagoonal Texas Gulf Coast (Kanes, 1970; Donaldson, Martin & Kanes, 1970; Morton & Donaldson, 1978), and at the heads of estuaries (Sect. 6.7.5).

In areas of moderate to high tidal range, tidal currents enter the distributary channels during tidal flood stage, spill over the channel banks and inundate adjacent interdistributary areas. The tidal waters are stored temporarily and released during the ebb stage. Tidal currents therefore dominate the lower channel reaches and interdistributary areas are intertidal flats (Sect. 6.7.3) (Figs 6.32 & 6.33). Tides can influence delta plains as much as 50 km inland (J.R.L. Allen, 1965c), especially where the gradient of the coastal plain is very low, even though tidal range may be small. In the Mississippi, small amplitude tides penetrate distributaries during low river stages and large areas of the lower deltaic plain are permanently covered by brackish waters, 50 km from the coast (Gould, 1970).

Tidally influenced channels (Sect. 6.7.5) typically have low sinuosity, a flared and sometimes funnel-shaped form and high

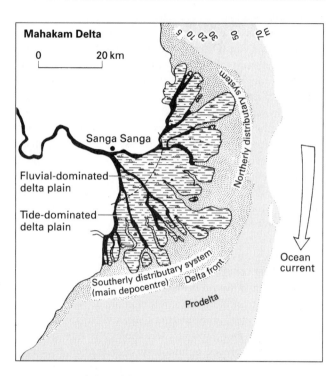

Figure 6.32 Mahakam delta, Indonesia; a fine-grained, tide-dominated delta with an extensive delta plain of tidal flats, estuarine channels and tidal creeks (after G.P. Allen, Laurier & Thouvenin, 1979).

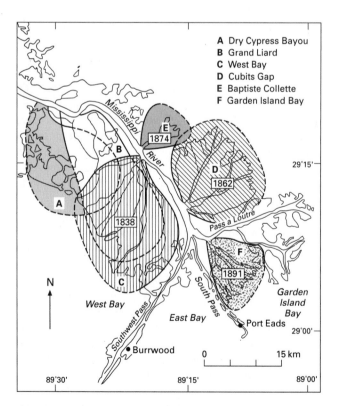

Figure 6.31 Minor subdeltas infilling interdistributary bays of the Mississippi delta (after Coleman & Gagliano, 1964). Each has built a minor mouth-bar–crevasse channel couplet (Fig. 6.43h).

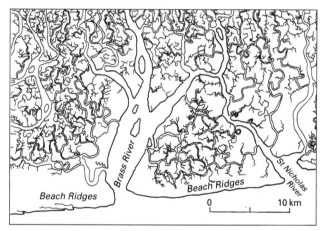

Figure 6.33 Tide-dominated lower delta plain of the wave–tide interaction Niger delta, comprising extensive mangrove swamps dissected by a maze of minor tidal creeks and tidally influenced distributary channels (after J.R.L. Allen, 1965c).

width to depth ratios. They contrast with the almost parallel-sided fluvial distributary channels. In the Niger delta more than 20 tidal inlets ranging in depth from 9 to 15 m dissect the beach-barrier shoreline (J.R.L. Allen, 1965c; Oomkens, 1974) (Fig. 6.33). Such channels have been studied in detail in the Rhine (Terwindt, 1971a; Oomkens & Terwindt, 1960; de Raaf

& Boersma, 1971; Boersma & Terwindt, 1981) and include features such as tidal sandridges and sandwaves (Sect. 6.7.5).

Tidally influenced interdistributary areas comprise tidal flats, either vegetated or evaporative (Sect. 6.7.3), and a complex of tidal channels (J.R.L. Allen, 1965c; Fig. 6.33). Sands, often muddy, are deposited by laterally migrating point bars in the tidal creeks and mangrove swamps may develop above the channel point bars (cf. Sect. 3.3.4). Thus interdistributary sediments comprise a sheet-like complex of small-scale erosive-based fining-upward facies sequences capped by tidal flat facies. Larger-scale tidally influenced distributary or estuarine channels cut through that sheet. In the Mahakam delta (Fig. 6.32) the equatorial climate enriches the fine-grained sediments with plant debris derived largely from palms and mangroves and thus yields organic rich muds (G.P. Allen, Laurier & Thouvenin, 1979). In the arid Colorado delta, at the head of the Gulf of California, the interdistributary areas are desiccated mud- and sandflats with localized salt pans (Meckel, 1975).

6.6.2 Delta front

This is the area where sediment-laden fluvial currents enter the basin and interact with basinal processes. At the distributary mouth, river flows expand both laterally and vertically and decelerate, thus decreasing flow competence and depositing sediment (Fig. 6.5). Where basinal processes are weak, fluvial processes predominate and deposit coarser sediment as distributary mouth bars, with shallow crests, and as subaqueous levees, bordering offshore extensions of the channels (Fig. 6.34). Where wave processes are important, swash bars, beach ridges, beach spits and cheniers develop, either around the distributary mouth, or alongshore in the direction of coastal current drift (Fig. 6.35). Where tidal processes are important, tidal channels with tidal sandbars and ebb and flood tidal deltas develop.

Deposition at river mouths involves a blend of inertial, frictional and buoyancy processes (Sect. 6.2.2; Fig. 6.5), the balance between them depending on the nature of the effluent and of the basin. That balance may change through time. Discharge fluctuations, the nature of the sediment load, and changes in basin water depth are particularly important. When discharge is low, river mouths may be dominated by buoyancy processes but be influenced more by frictional and inertial processes during high discharge (e.g. South Pass, Mississippi delta; L.D. Wright & Coleman, 1974) (Fig. 6.34). Thus during normal discharge, when buoyancy processes predominate and a salt wedge intrudes the channel, the distributary mouth bars of the Mississippi comprise: (i) a *bar back area* which includes minor channels, subaqueous levees and bars superimposed on a gently ascending platform; (ii) a *narrow bar crest* located a short distance offshore from the distributary mouth; and (iii) a *bar front* or *distal bar* which slopes offshore to the prodelta. During floods, discharge forces out the saltwater wedge, frictional and inertial processes predominate at the river mouth

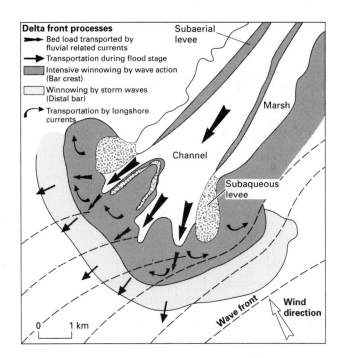

Figure 6.34 Delta front processes. The mouth bar is affected by (i) fluvial processes that are either (a) low-stage buoyancy processes which deposit sediment on to the mouth-bar crest or (b) high-stage and falling-stage inertial and frictional processes that shift sediment on to the distal bar; (ii) wave/storm processes that (a) rework the bar crest during normal periods and (b) rework the distal bar during storms (based on Coleman & Gagliano, 1965; W.L. Fisher, Brown *et al.*, 1969; L.D. Wright & Coleman, 1974).

and sediment-laden currents drive across the bar back and crest causing the bar crest to aggrade by up to 3 m. As the flood wanes this sediment is reworked by river currents and transferred to the bar front which may prograde up to 100 m. Thus the main mouth bar progradation takes place immediately after the flood peak (L.D. Wright & Coleman, 1974; Coleman, Suhayda *et al.*, 1974). Tidal fluctuations may also modify fluvial effects. For example, in the Bella Coola delta at high tide, frictional forces dominate and a large gravelly middle-ground mouth bar forms; at low tide effluent passes over the break in slope at the bar front and forms a buoyant plume that deposits a lunate, inertial bar.

The effects of waves at the delta front vary according to the grain size and amount of the river-borne sediment, the gradient of the delta front, the direction of wave approach and the relative durations and power of normal and storm waves. In the relatively fine-grained systems discussed in this section, dissipative regimes predominate (Fig. 6.8) with wide, complex surf zones in most sand-dominated deltas. Where fairweather wave energy is particularly high, as off the Niger, steep, narrow beaches form. In large muddy deltas such as the Amazon and Orinoco, shoreline wave action is strongly suppressed (Rine & Ginsburg, 1985).

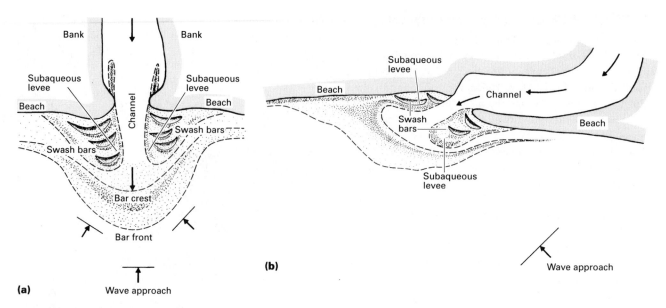

Figure 6.35 Wave-dominated river mouth settings. (a) Direct onshore wave approach. (b) Oblique-wave approach and associated longshore drift direction (after L.D. Wright, 1977).

In very fine-grained fluvial-dominated deltas, such as the Mississippi, the effect of waves is essentially to rework or winnow the sand at the point where it was deposited by the river. Only a little is transported laterally into spits. Reworking intensity is related to depth of deposition and wave energy (Fig. 6.34). Normal wave reworking affects only the mouth bar crest. During storms the bar front or distal bar is reworked.

In deltas, such as the Niger, with steep offshore profiles, high wave intensity and/or sand content, the waves may impound the river discharge and increase mixing of the water masses. Sand-grade sediment is concentrated at the shoreline while fine-grained sediment is dispersed offshore. Where waves approach the shoreline directly (Fig. 6.35a) sediment is redistributed on both sides of the river to give a cuspate wave-dominated delta. More commonly, the wave approach is oblique (Fig. 6.35b) and the effluent of the delta will act as a groyne, trapping sediment transported by the wave-generated longshore drift on the updrift side (Fig. 6.69). As fluvial discharge decreases there is an intermittent downdrift migration of the river mouth. If fluvial discharge is insufficient to give a groyne effect the river mouth migrates continuously downdrift as in the Senegal delta (Fig. 6.35b) (Dominguez & Barbosa, 1994).

Deltas affected by a combination of fluvial and wave processes (e.g. Danube, Ebro, Nile, Rhône) all of which occur in enclosed seas with moderate wave power and negligible tides, have smooth cuspate or arcuate shorelines. They prograde by beach ridge accretion (Fig. 6.36) giving a sheet of coastal sands, cut locally by distributary channel sands, beneath the delta plain.

Where wave processes redistribute most of the sediment supplied to the delta front there is only a slight protuberance of the shoreline (Fig. 6.37). Mouth bars do not form and progradation is by beach ridge accretion along the entire delta front rather than around individual input points. These deltas are always sand dominated and abandoned beach ridges on the delta plain separate shallow elongate lagoons. Aeolian dunes may also be present (Psuty, 1967; Curray, Emmel & Crampton, 1969).

Tidal currents at the distributary mouth (L.D. Wright, 1977): (i) enhance mixing of the water masses, promoting sedimentation and reducing the effects of buoyancy; (ii) can overwhelm the effects of river floods and cause bidirectional current transport for the best part of the year; (iii) extend the range of positions of the land–sea interface and of the zone of sea–river interaction both vertically and laterally; and (iv) enhance clay flocculation and thereby deposition of mud.

6.6.3 Prodelta

The prodelta is that part of the delta unaffected by wave or tidal processes though many workers include it within the delta front. In some shallow-water deltas it may not exist since the delta is building on to a storm or tide-dominated shelf. In other deltas, it comprises a relatively stable zone where mud and fine silt are deposited from suspension to form well-laminated sediments reflecting fluctuations in river sediment carried in buoyant plumes. Where bottom waters are anoxic, lamination is preserved. Where they are aerobic, laminae are disturbed by bioturbation. Occasionally these laminated and bioturbated sediments are interrupted by sharp-based graded beds emplaced by flood-generated hyperpycnal flows (cf. Fig. 6.5). The prodelta is also commonly disturbed by soft-sediment failure and, in some deltas, prodelta sediments are dominated by products of mass movement from higher on the delta front (see Sect. 6.6.4).

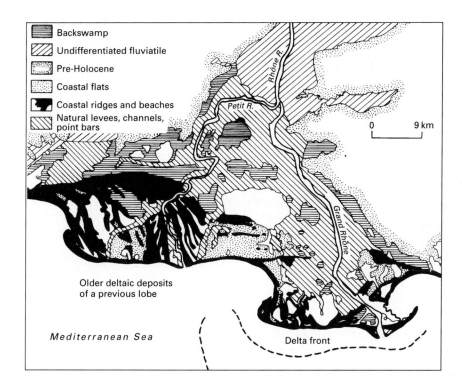

Figure 6.36 Rhône delta; a sandy wave-influenced delta with a continuous fringe of coastal barrier sands. Progradation by beach accretion is pronounced in the vicinity of the main distributary (Grand Rhône). Elsewhere the shoreline is retreating (after Kruit, 1955; van Andel & Curray, 1960).

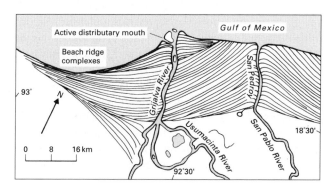

Figure 6.37 Wave-dominated Grijalva delta (after Psuty, 1967).

6.6.4 Deformational features on delta front and prodelta slope

Deformation processes affect both the prodelta and the delta front (Coleman, Prior & Lindsay, 1983; Lindsay, Prior & Coleman, 1984; Coleman, 1988) (Fig. 6.38). In the Mississippi, 40% of the sediment supplied to the delta is involved in some kind of mass movement after initial deposition (Coleman, 1981), and similar processes are known in many other deltas. For example, they have been documented from the Fraser (Mathews & Shepard, 1962; McKenna and Luternauer, 1987), Huanghe (Prior, Yang *et al.*, 1986), Itirbilung (Syvitski & Farrow, 1989), Klamath (Field, Gardner *et al.*, 1982), Magdalena (Shepard,

1973a), Niger (Weber & Daukoru, 1975), Orinoco (Nota, 1958) and within British Columbian fjords (Prior, Bornhold *et al.*, 1982; Syvitski & Farrow, 1989). However, the Amazon delta is not affected (Nittrouer, Kuehl *et al.*, 1986).

SHALLOW-WATER RESEDIMENTATION PROCESSES

The principal reasons for sediment-induced deformation are: (i) the very high sedimentation rate on the delta front, up to 2.5 m year^{-1} off the Mississippi, does not allow pore fluids to escape during burial so that the sediments are undercompacted and pore fluid pressures are high, leading to loss of sediment shear strength; (ii) biodegradation of organic debris (often 0.5–1.5%) produces free methane gas which further weakens the sediments; and (iii) shocking of accumulated sediment by storm wave action.

Rotational slides, whereby large blocks of sediment are translated downslope, are recognized on the surface of the Mississippi delta front by irregular 'stairstep' changes in gradient. The pattern changes from time to time, particularly after major river floods or wave pounding (Coleman, Suhayda *et al.*, 1974; Coleman, 1981; Fig. 6.39). Slide planes strike across the mid- to upper bar front and initially dip seawards at gentle angles of 1–4° before flattening into slope-parallel shear planes. Individual slide blocks average 90 m in width, 6 km in length and move downslope for distances in excess of 1.5 km. They are preserved intact with little flowage, and introduce seemingly anomalous shallow-water, delta front sand facies into deeper water prodelta

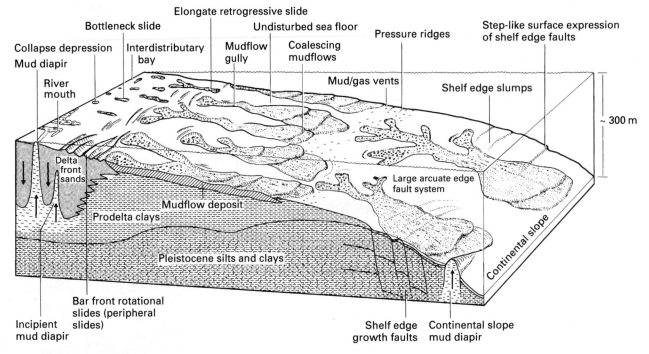

Figure 6.38 Summary of the main types of sediment-induced deformation features arising from surface instability of sediments and deep-seated flowage of overpressured clays in the Mississippi delta (after Coleman, 1981; Coleman, Prior & Lindsay, 1983).

mud–silt facies. The blocks may, however, dip landwards by up to 30°. Such rotational sliding and downslope translation is integral to progradation of the Mississippi delta and makes a substantial contribution to the final facies pattern.

Collapse depressions are bowl-shaped hollows 100 m or so in diameter and 1–3 m deep which occur in distal interdistributary bay areas and are formed by localized liquefaction/fluidization of sediment by storm waves (Coleman & Garrison,

1977; Prior & Coleman, 1978; Roberts, 1980; Coleman, 1981; Fig. 6.38). In some cases the depressions are closed, circular features rimmed by small listric fault scarps and with a chaotic mass of sediment blocks in their centres. More commonly, the depressions are open down slope and pass into '*bottleneck slides*' or '*delta-front gullies*', originally described by Shepard (1955). These gullies trend down the delta front as long, slightly sinuous, features bounded by sharp, rotationally slumped walls. They extend over several kilometres from shallow water (7–10 m) to water as deep as 100 m and are 3–20 m deep and 20–1500 m wide. They act as conduits for mudflows which emanate from

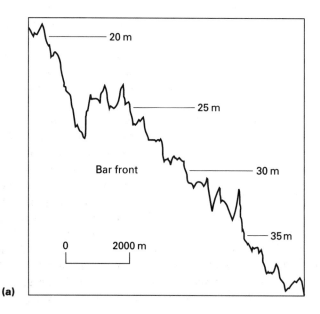

(a)

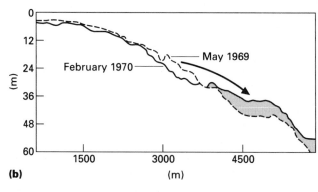

(b)

Figure 6.39 Rotational slides on the delta front of the modern Mississippi delta. (a) Irregular profile of a river mouth-bar front reflecting the presence of rotational slump planes; (b) offshore slumping associated with the slump planes, revealed by time-separated fathometer profiles (after Coleman, Suhayda *et al.*, 1974).

the collapse depressions. At the mouths of the gullies, mudflows emerge on to the prodelta surface and produce a virtually continuous fringe of coalescing *mudflow lobes* (Fig. 6.38). Individual lobes average 10–15 m in thickness and contain blocks of sediment which are commonly 150–500 m in length.

Mud diapirs occur both at river mouths and beyond the shelf-edge (Fig. 6.38). In the Mississippi delta 'mudlumps' frequently emerge near the distributary mouths to form temporary islands (Morgan, 1961; Morgan, Coleman & Gagliano, 1968). Surface exposures of the mudlumps reveal steeply dipping, stratified delta front sediments with numerous small anticlines, *en echelon* normal faults, reverse faults, radial faults and thrusts. Other features include small mud cones formed by extrusion of methane-rich muds from fault planes, and planation horizons produced by wave erosion of the exposed mudlump. Clays involved in the diapirism exhibit closely spaced fractures (Morgan, Coleman & Gagliano, 1963; Coleman, Suhayda *et al.*, 1974).

The mudlumps are considered to be thin spines superimposed on linear shale folds or ridges, with movement on large-scale, high-angle reverse faults in the mudlump crests producing most of the uplift (Fig. 6.40). Up to 200 m of uplift can be demonstrated in some cases, and rates of uplift of 100 m in 20 years have been documented. There is a close relationship between diapiric activity and distributary mouth sedimentation. The appearance of mudlumps coincides with rapid sedimentation during river floods, and the site of mudlump activity migrates seaward in concert with mouth bar progradation (Morgan, Coleman & Gagliano, 1968). Isopach maps reveal that diapirism substantially modifies the elongate bar finger geometry which would be predicted for mouth bar progradation. Instead of being linear bodies with a uniform axial thickness of approximately 70 m as originally thought (Fisk, 1961), they comprise a series of discrete sand pods up to 100 m thick separated by areas

of minimal sand thickness (Coleman, Suhayda *et al.*, 1974) (Fig. 6.40).

DEEP-WATER RESEDIMENTATION PROCESSES

Large-scale *arcuate rotational slumps* and contemporaneous *growth faults* occur on the outer continental shelf in front of the prograding delta (Figs 6.38, 6.41 & 6.42). In addition, massive shelf-edge and retrogressive failures lead to the development of submarine canyons (Fig. 6.38) (Sect. 10.3.3). These features are primarily the result of large-scale slope instability, leading to sliding, slumping and creeping (Sect. 10.2.3) combined with sand loading on shale, differential compaction, and uplift of salt or shale ridges.

Growth faults have been extensively described in the subsurface of the Gulf of Mexico, Niger delta and Mackenzie delta (Ocamb, 1961; Weber, 1971; Evamy, Haremboure *et al.*, 1978; Coleman, Prior & Lindsay, 1983; Fig. 6.42). Their two essential characteristics are: (i) the offset of individual beds increases with depth below the sea floor where offset is almost zero to a maximum at some mid-point in depth, finally decreasing as the fault plane flattens; and (ii) thicker successions, with a higher sand content in the hanging wall. Individual sediment units within the hanging wall thicken towards the fault. Rotation of the hanging wall strata towards the listric and cuspate fault surfaces produces broad anticlines (rollovers) commonly complicated by smaller, antithetic faults. As the rollovers often occur in thick, sand-dominated successions, they can form excellent hydrocarbon traps (Weber, 1971; Busch, 1975).

Massive shelf-edge failures leave sea-floor scarps up to 30 m high on the Mississippi shelf. One Pleistocene failure covers an area of more than 18 000 km² having moved downslope 8600 km³ of shelf and upper slope sediment 500 m thick

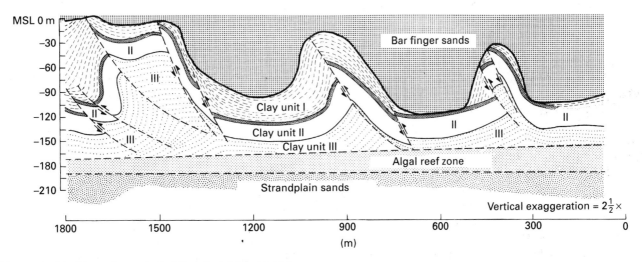

Figure 6.40 Diapiric mudlumps in the Mississippi delta with high-angle reverse faults in the diapir crest and exceptional thicknesses of distributary mouth-bar facies between the diapirs (after Morgan, Coleman & Gagliano, 1968).

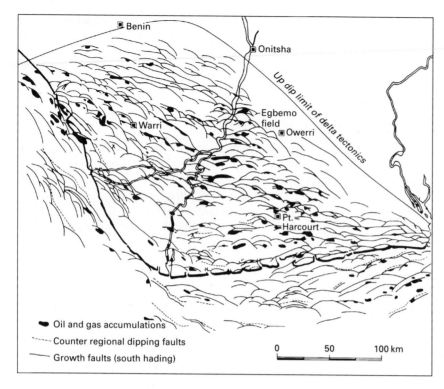

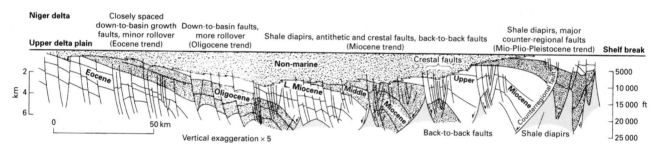

Figure 6.41 Plan view of growth faults in the Niger delta illustrating their lateral impersistence, slightly curved concave-to-basin trace, their general parallelism with the delta front and location of oil fields in the hanging wall, rollover anticlines of many faults (after Weber & Daukoru, 1975).

Figure 6.42 Cross-section through the Niger delta depocentre illustrating extensive growth faulting occurring progressively basinward and higher in the stratigraphy as progradation continues (after Evamy, Haremboure *et al.*, 1978; Winker & Edwards, 1983).

(Coleman, Prior & Lindsay, 1983). The resulting scar was then infilled by clays and silts, without any coarse detritus, showing oblique progradational seismic reflectors.

Mud diapirs or shale ridges are also found at or near the shelf edge in the Fraser, Magdalena, Niger and Orinoco deltas (Nota, 1958; Mathews & Shepard, 1962; Shepard, Dill & Heezen, 1968; Shepard, 1973a; Weber & Daukoru, 1975). In the Niger delta, for example, shale ridges that occur along the continental slope in front of the extensively growth-faulted depocentre, are probably a surface expression of overpressured clays that flowed away from the depocentre.

Submarine canyons occur on the continental shelves of many large deltas such as the Mississippi, Congo, Amazon, Magdalena and Nile and appear to have formed initially by massive retrogressive slumping of the canyon walls and to have been maintained by funnelled density flows (Sect. 10.3.3).

6.6.5 Deltaic facies sequences and their boundaries

Deltaic facies sequences vary considerably in their nature, both within and between individual deltas. The type of sequence is primarily governed by the nature of the delta, whether it is dominated by rivers, waves or tides, or builds into shallow or deep water. In addition, position of the section in relation to distributary channels influences the detail of the facies sequence. Because of their abundant supply of fluvial sediment, deltaic coastlines are characterized by depositional regressions and transgressions (Fig. 6.16). Progradational lobes alternate with somewhat thinner transgressive sediments.

Facies sequences occur on three scales. Large-scale allocyclic sequences involving the entire delta system are caused by tectonic and climatic changes in the hinterland, by major diversions of rivers upstream, and by eustatic sea-level changes. Medium-scale sequences are caused by switching of delta lobes or distributaries within a stable depocentre. Small-scale sequences result from differential subsidence on the delta plain, lacustrine delta formation, crevassing of distributary channels and migration of tidal channels. Progradational sequences are commonly terminated by abandonment/flooding surfaces whose extent and nature depend on the process controlling relative sea-level rise. The subsequent deepening gives accommodation space for further progradation. At times of low relative sea level, wide coastal plains are exposed, channel gradients increased and fluvial channels cut deeply into the coastal plain transporting coarse sediments to the continental margin where shelf-margin deltas (shelf-perched lowstand wedges) may develop to give high-angle clinoform seismic reflectors (Winker & Edwards, 1983; Suter & Berryhill, 1985; Kindinger, 1988; Tesson, Allen & Ravenne, 1993). The delta front is not only sandier than usual but suffers greater wave activity so that wave-dominated coarse-grained deltas are the norm. Little sedimentation takes place on the extensive coastal and deltaic plain during this time and mature palaeosol profiles may develop. As sea level rises, transgression at the delta front produces a series of transgressive sequences such as those described from the Rhône (Oomkens, 1970) that should be traceable over a very wide area. Increasing accommodation space leads first to infilling of incised valleys where tidal currents may be generated even on microtidal coastlines. Inundation of the coastal plain to give the initial flooding surface creates a shallow shelf across which subsequent progradations take place. Stacked progradational parasequences develop, their nature depending on whether the delta plain is fluvially dominated, wave dominated or tidally dominated (Lagaaij & Kopstein, 1964; Oomkens, 1970, 1974; Elliott, 1974b; Tye & Coleman, 1989a). Rates of relative sea-level rise will determine whether the successive parasequences prograde, aggrade or retrograde. Such stacking patterns should not be thought of in terms of purely proximal–distal two-dimensional profiles. Individual parasequences may record delta switching alongshore as accommodation space is generated.

Delta switching primarily affects fluvial-dominated deltas that build out into the sea or lake, elongate themselves, reduce their gradients and flow capacity and thus lose their gradient advantage. Such behaviour is particularly common in the Mississippi delta which has built out six major delta complexes, consisting of more than 18 smaller deltas over the last 7000 years (Frazier, 1967; Penland, Boyd & Suter, 1988; Fig. 6.46). At present, waters of the Mississippi are being diverted from a point more than 500 km upstream from the present river mouth into the Atchafalaya River that flows for only 227 km before it enters the Gulf of Mexico. The gradient advantage of the Atchafalaya River is precipitating the next major phase of delta switching leading to the abandonment of the present Mississippi. Despite man-made attempts to control this event a new bay head delta is forming rapidly in Atchafalaya Bay (van Heerden,

Wells & Roberts, 1981). The Hwanghe delta (Xue, 1993) in China has changed naturally by switching its mouth alongshore by some 500 km from a point more than 500 km upstream where every few hundred years calamitous river floods divert it either side of the Shantung Peninsula.

The smallest scale sequences are confined to the delta plain. In fluvially dominated deltas such as the Mississippi (Fig. 6.31) and fluvial-wave interaction deltas such as the Rhône (Fig. 6.36) minor mouth bar–crevasse channel couplets form through progradation of crevasse deltas into bays, lakes and lagoons (Sect. 6.6.1) (Fig. 6.43). Where there is tidal influence, as in the Niger, migrating tidal channel sequences develop.

PROGRADATIONAL SEQUENCES

The main characteristic of progradational deltaic sequences is that they shallow and coarsen upwards from offshore muds, through silts to various sand-dominated facies. However, there are considerable variations between deltas, depending on basinal water depth and on the balance of processes. Within individual deltas, the sequence varies according to the position of the advancing shoreline that it records.

Fluvial-dominated delta sequences are best documented in the present-day Mississippi (Fisk, McFarlan *et al.*, 1954; Coleman & Wright, 1975; Coleman, 1981) (Fig. 6.44). Under present-day conditions of high and stable relative sea level, progradational sequences occur on at least two scales: (i) 50–150-m-thick medium-scale sequences caused by switching of individual delta lobes due to major avulsion; and (ii) small-scale (3–10 m) sequences due to crevassing of distributary channel banks and deposition of crevasse deltas in lakes and bays (Figs 6.36 & 6.43).

In an axial position, the 50–150-m-thick sequence passes from prodelta muds, through silts, to bar front and bar crest sands. The prodelta muds, deposited below storm wave base from suspension, lack current-produced laminae, but are banded due to slight differences in grain size or colour related to discharge fluctuations or plume shifting. Bioturbation is slight because of the rapidity of deposition, but more intensely bioturbated horizons are produced when sedimentation rates decrease temporarily. The fabric of these clays comprises a framework of randomly oriented domains of clay particles with large voids and hence high initial porosity. There may be a high incidence of sediment deformation and of anomalously shallow-water facies due to slope instability. Higher in the sequence, parallel and lenticular silt laminae and eventually thin cross-laminated sands are intercalated with clays, reflecting a combination of waves, sediment-laden current incursion from the distributaries and continued deposition from suspension. Sediments deposited near the bar crest are characterized by relatively well-sorted sands with cross-lamination, climbing ripple lamination and some flat lamination, deposited principally during river floods.

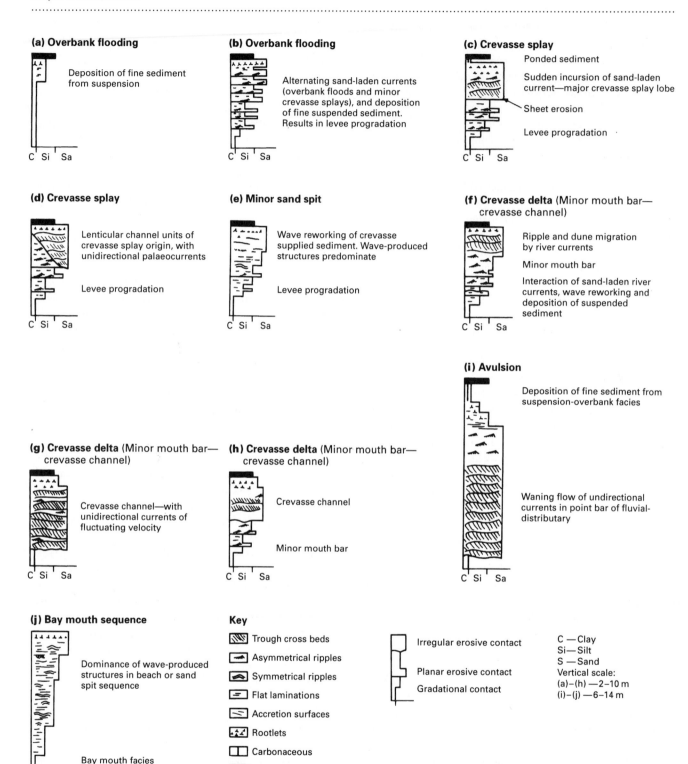

Figure 6.43 Depositional mechanisms and sequences produced in fluvial-dominated interdistributary areas (after Elliott, 1974b).

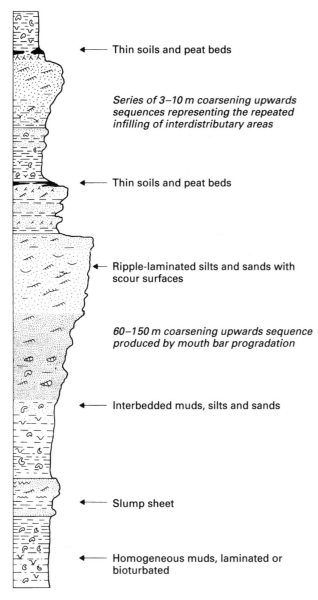

← Thin soils and peat beds

Series of 3–10 m coarsening upwards sequences representing the repeated infilling of interdistributary areas

← Thin soils and peat beds

← Ripple-laminated silts and sands with scour surfaces

60–150 m coarsening upwards sequence produced by mouth bar progradation

← Interbedded muds, silts and sands

← Slump sheet

← Homogeneous muds, laminated or bioturbated

Figure 6.44 Composite idealized sequence produced by mouth bar progradation in the Mississippi delta (after Coleman & Wright, 1975).

The upper parts of the coarsening-upward sequence may be either slightly eroded by minor channels of the bar back, or substantially eroded by the main distributary channel as a result of progradation. Upper mouth-bar sediments are often totally removed by erosion at the base of the distributary channel. The channels carry sand as large bedforms and are finally filled by a fining-upward sequence due to distributary abandonment or sea-level rise.

Away from the axis of the distributary, sand content diminishes and prodelta muds pass upwards into interdistributary bay muds and silts, with rare wave-reworked sandy lateral spits. Interdistributary bays may themselves be filled by small-scale (3–10 m) coarsening-upward sequences of minor crevasse deltas, crevasse channel couplets (Figs 6.43 & 6.44) (Sect. 6.6.1). Thus, relatively large-scale complete or partially preserved coarsening-upward sequences pass upward into shallow-water sequences of considerable lateral variability.

The sequences of *fluvial–wave interaction* deltas such as the Rhône and Ebro (Lagaaij & Kopstein, 1964; Oomkens, 1967, 1970; Maldonado, 1975; Nelson & Maldonado, 1990) are similar to those of fluvial-dominated deltas but differ in the degree of wave reworking and lateral variability towards the top. Well-sorted, horizontally bedded sands deposited by nearshore wave processes occur preferentially compared with fluvial-influenced mouth-bar facies. These laterally extensive beach-barrier sands are cut locally by distributary channel sands.

Wave-dominated deltas have facies sequences that coarsen upwards from shelf muds through silty sand to wave- and storm-influenced sands, capped by lagoonal or strandplain muds, silts and soils, their nature depending on the climate. Intense energy at the shore ensures that most fine sediment is transferred to the delta slope and prodelta. Such deltas can as well be considered to be prograding beach-ridge strandplains (Sect. 6.7.1), unless there is evidence of direct fluvial input in the vicinity.

Fluvial–wave–tide interaction deltas such as the Burdekin, Irrawaddy, Mekong, Niger and Orinoco have yet more complex sequences. These have been particularly well documented in the Niger (J.R.L. Allen, 1965c; Oomkens, 1974; Fig. 6.45). There, a coarsening-upward sequence from offshore muds into shoreface coastal barrier sands is similar in all respects to a wave-dominated delta sequence, and this sequence may be preserved. However, it is commonly truncated by tidal channels filled by tidal sands which erode the landward sides of beach ridges. Offshore and lower shoreface deposits are therefore overlain directly by fining-upward tidal channel sequences with coastal barrier sands preserved as erosional remnants between them. Eventually fluvial channel sequences may come to lie above the tidal ones, if progradation and aggradation persist.

Tide-dominated delta sequences are probably rare except in very large deltas such as the Mahakam and Ganges–Brahmaputra where sediment supply is enormous. The dominant feature is tidal current ridges that occur within the delta front and pass upward and laterally into tidal channel and tidal flat deposits similar to those of the fluvial–wave–tide interaction deltas.

TRANSGRESSIVE SEQUENCES, THE ABANDONMENT PHASE

Transgressive sequences have long been documented in the Rhône and Niger deltas (Lagaaij & Kopstein, 1964; Oomkens, 1967, 1970, 1974), though only recently (Penland, Boyd & Suter, 1988) have they been thoroughly studied in the Mississippi.

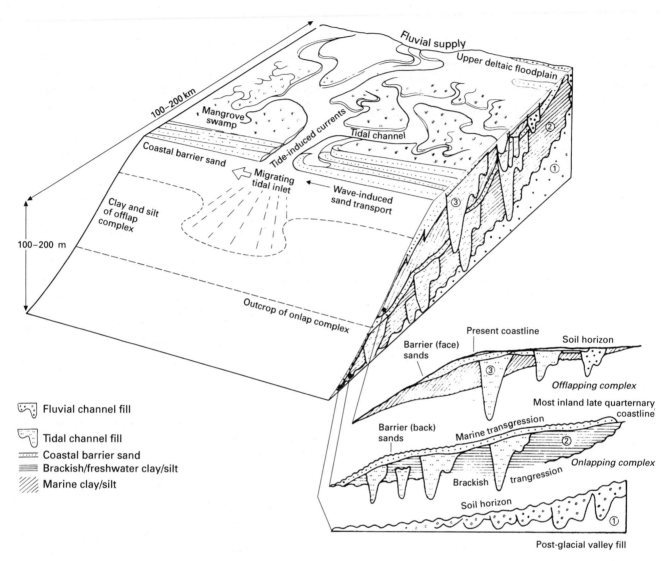

Figure 6.45 Three-dimensional view of the wave- and tide-influenced Niger delta showing wave-dominated delta front, tide-dominated lower delta plain and fluvial-dominated upper delta plain (after Oomkens, 1974) and cross-section showing type of sequences formed during three stages of sea-level change: (1) sea-level fall and early rise; (2) transgression; (3) progradation during high sea-level stand.

During the last glacioeustatic sea-level rise, two separate transgressive surfaces were formed. One occurred as the lakes, bays and lagoons transgressed over alluvial environments of the coastal plain to give rise to a *brackish transgression* (first flooding surface/transgressive surface) and a deposit of fine-grained lagoonal and swamp sediments with minimal erosion. This was followed by a *marine transgression* (ravinement surface) as a barrier shoreface advanced landward (cf. Sect. 6.7.6). A truncated coarsening-upward sequence formed largely from back-barrier washover sands may be preserved beneath the resulting ravinement surface. As with progradational sequences, the Niger differs from the Rhône in having transgressive tidal channels that locally remove part of the

regressive sequence before they themselves are cut by shoreface erosion and overlain by offshore sediments (Fig. 6.45).

Transgressive sequences formed by delta switching rather than glacioeustatic sea-level changes are best known from the Mississippi delta (Frazier, 1967; W.L. Fisher & McGowen, 1969; Penland, Boyd & Suter, 1988; Figs 6.46, 6.47 & 6.48), where several deltas have been transgressed to a greater or lesser extent depending on how long they have been abandoned.

The Mississippi delta is characterized by having: (i) fair-weather wave activity so weak that less sand is transported up the beach than is dispersed offshore by storms; (ii) a tidal range that, though low (only 30 cm), is sufficient in this very low gradient delta to play a significant role; (iii) an extremely high

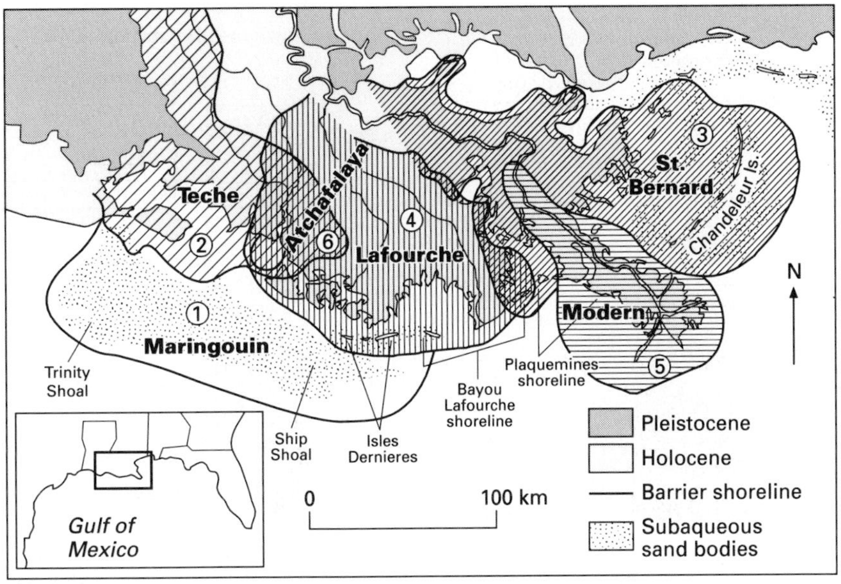

Figure 6.46 Holocene Mississippi delta plain showing sequence of major delta lobes that have prograded and been abandoned in the last 7000 years (after Frazier, 1967; Penland, Boyd & Suter, 1988). Numbers indicate abandonment sequence with the modern delta (5) about to be replaced by the Achafalaya (6) and the earlier ones showing successive stages of abandonment and subsequent transgression back to the Lafourche (4), the St Bernard (3), the Teche (2) and Maringouin (1) which has now been submerged (Fig. 6.47).

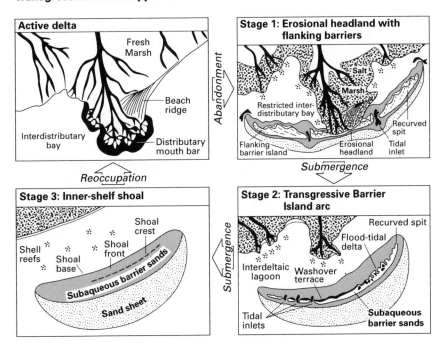

Figure 6.47 Evolution of transgressive depositional systems in the Holocene Mississippi delta as exemplified in stages of abandonment of the Lafourche (4), St Bernard (3) and Maringouin (1) abandoned delta lobes (see Fig. 6.46). Stage 1, erosional headland and flanking barriers, stage 2, transgressive barrier island arc, stage 3, inner-shelf shoals, (from Penland, Boyd & Suter, 1988).

compactional subsidence of recently deposited muddy sediment (30–60 cm 100 years^{-1}).

Following delta abandonment, a process of *transgressive submergence* takes place in which there is a horizontal component of reworking by shoreface retreat and a vertical component of submergence (cf. Sect. 6.7.6). Initially, the elongate Mississippi delta mouth bars protrude as headlands that, on abandonment, act as sources of sand to feed flanking *transgressive barriers*.

During *stage 1* (e.g. Bayou Lafourche and Plaquemines barrier systems) storms transport sand both seaward on to the inner shelf to form a sand sheet and landward as washover fans (Fig. 6.47). As the lakes and lagoons behind the barriers subside, they expand, increasing the volume of tidal exchange. Channels, initially cut by storms, transform to tidal inlets that migrate laterally, eroding and reworking the barrier sand. Thus the flanking *barrier-island sequence* (Fig. 6.48) fines upward with

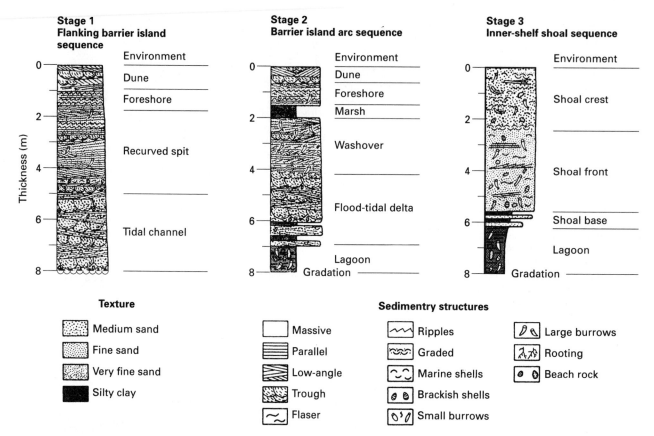

Figure 6.48 Generalized composite transgressive sequences for each stage of transgression (from Penland, Boyd & Suter, 1988). (See Fig. 6.46 for map and Fig. 6.47 for location on present Mississippi delta.) Dominant processes vary at each stage of transgression. During stage 1 they are shoreface erosion, recurved spit building and tidal channel migration. During stage 2, they are submergence and flood-tidal delta and overwash. During stage 3, they are shoreface erosion and inner-shelf reworking following barrier-island submergence.

tidal channels eroded into progradational delta deposits below and passes upward into spit sands capped by thin foreshore and dune sands.

As submergence continues, the erosional headlands disappear, the lagoons coalesce and widen, and the barrier island becomes detached (mainland detachment submergence of Hoyt, 1967) to form a *transgressive barrier-island arc – stage 2* (Fig. 6.47). Examples are the Isles Derniers of the Lafourche delta and the Chandeleur Islands of the St Bernard delta. Here there is a large lagoon, a well-developed barrier island cut by many flood-dominated tidal channels, and an offshore sand sheet. Sand is derived from reworking of stage 1 systems and is deposited mainly as flood tidal deltas and on washovers, with small amounts of inner barrier marsh, foreshore and dune sediments. The formation of barrier islands reflects a balance between rate of submergence and land retreat of the mainland shoreline versus rate of barrier shoreline retreat. Since the coastal gradients are very low (flat to 1 : 5000 or 0.2 m km^{-1}), a relative rise of sea level of 50 cm 100 years^{-1} produces greater submergence and retreat of the mainland coast (25 m year^{-1}) than of the barrier (20 m year^{-1}) and so the intervening lagoon enlarges,

transgressing the freshwater marsh. This is comparable to the brackish transgression of the Rhône and Niger, but differs from it in having a more gradational contact, because of the very low gradient. The *barrier-island arc sequence* (Fig. 6.48) thus consists of lagoonal muds that merge downwards gradationally into regressive muds and pass upwards into flood tidal delta sands and washover sands.

The final *stage 3 – inner-shelf shoals* (Fig. 6.47), exemplified by the now fully abandoned Maringouin complex, is one where only a submerged sandbar exists, several metres below sea level. This consists of a landward-facing shoal base, shoal front and shoal crest that is within the reach of both storm and fairweather processes. The shoal, which passes seawards into a sand sheet, develops from the subaerial barrier-island arc because of diminishing sand supply and subsidence. A sequence of shoal base to shoal crest sediments overlies lagoonal muds (Fig. 6.48).

While delta front sediments are reworked, other effects are felt on the delta plain itself, as the supply of clastic sediment is cut off. Upstream areas are covered by peat blankets which may extend uninterrupted for several hundred square kilometres (Coleman & Gagliano, 1964; Kosters & Suter, 1993). Seaward, a hiatus in sedimentation occurs with extensive bioturbation of previously deposited muds and silts. These stages precede eventual deepening as the creation of new accommodation space eventually outstrips organic productivity and sediment supply.

6.6.6 Ancient river deltas

Ancient river deltas are characterized by: (i) a thick predominantly clastic succession that passes upward from offshore facies into continental, fluvial facies; (ii) a sediment body of restricted lateral extent since the delta formed a depocentre fixed around a river mouth; and (iii) repetitive or cyclic successions due to repeated progradation and abandonment of the entire delta or lobes within the delta.

They can be recognized by: (i) study of the vertical succession in boreholes and at outcrops where lateral data are sparse or even absent; (ii) the establishment of lateral facies changes by areal mapping and by correlation of adjacent sections; and (iii) the identification of offlapping clinoforms on seismic sections and in very well-exposed outcrops with a low tectonic dip (Berg, 1982).

While fluvial-dominated deltas are easy to recognize because the rivers reached the coast and cyclicity is well developed, deltas influenced or dominated by waves have facies and sequences similar to those of non-deltaic siliciclastic coasts (Sect. 6.7) and tide-dominated deltaic facies resemble those of estuaries. Evidence for river processes may lie far inland and only wave, storm and tidal facies are to be seen in nearshore delta front settings. Nevertheless, within a thick vertical succession of stacked progradational units, evidence for a river is likely to be present in at least some sections.

Deltaic successions comprise five principal facies associations, delta plain, delta front, prodelta and/or shelf, delta abandonment, and incised valleys and their fills.

The *delta plain facies association* reflects primarily the source area and the climate of the delta plain. It comprises fluvial and distributary channel sequences, levee facies, swamp and marsh coals, and the facies of lakes, lacustrine and bayhead deltas, crevasse splays and crevasse deltas. In the lower delta plain of tidally influenced deltas, tidal channel sequences predominate.

The *delta front facies association* records the interplay of fluvial and basinal processes, in particular the effects of waves and tides on the fluvial input. It comprises, to a varying degree, the distributary mouth bar and its distal extension on to the prograding slope, subaqueous levee facies, beach-ridge, beach spit facies, or tidal channel and ebb or flood tidal delta facies.

The *prodelta and/or shelf facies association* records the depth, salinity, physical activity and oxygenation of the basin and offshore area, as well as the mode of sediment emplacement into the basin. It comprises quiet water laminated or bioturbated mudstones, mass flow deposits, including turbidites, and many types of deformational facies. On shallow shelves the whole range of active tidal and storm facies may be present (Chapter 8).

Abandonment facies associations reflect very slow rates of deposition when clastic input has ceased, due to relative sea-level rise resulting from major avulsion, delta lobe or distributary switching, or eustatic sea-level rise. They are extensively developed on interfluves and may include limestones, coals, omission surfaces or highly condensed horizons bioturbated by plants or animals and marine erosional unconformities (ravinement surfaces) and brackish transgressions. In addition, immediately overlying sediments may show evidence of condensation with a high abundance of fossils.

Incised valleys occur where rivers have eroded delta plain or shelf sediments as a result of a fall in relative sea level. They include type 1 sequence boundaries of Vail and Todd (1981) (Sect. 2.4). They are commonly filled by tidal and fluvial sediments. Laterally they may pass into abandonment facies.

When attempting to discriminate between delta types, it must be remembered that deltas may change character with time, even within a thin succession. The fluvial facies may change from that recording a low gradient mud/silt dominated river to a higher gradient sandy river as a result of base-level drop, or a change in climate or sediment supply. As sea level falls, wave power may change and deltas migrate from the inner shelf to the shelf margin. Tidal range also varies with changes in relative sea level.

ANCIENT FLUVIAL-DOMINATED AND FLUVIAL–WAVE-INTERACTION DELTAS

These types of ancient deltas are the most commonly recognized and documented, although they are not the most abundant today.

Fluvial-dominated and fluvial–wave-interaction deltas are particularly well exposed and studied in the Upper Carboniferous of the British Isles (D. Moore, 1959; de Raaf, Reading & Walker, 1965; Collinson, 1968, 1969, 1988; Kelling & George, 1971; Rider, 1974, 1978; Collinson & Banks, 1975; Elliott, 1975, 1976a,b,c; Jones, 1980; Pulham, 1989; Martinsen, 1990).

Progradational sequences vary in thickness from those which infill deep, pre-existing basins to thin examples, commonly in cyclic arrangements, which fill accommodation space created by contemporaneous tectonic and compactional subsidence of both block and already infilled basinal areas. The Late Carboniferous is generally regarded as a period of glacioeustatic sea-level changes (e.g. Ramsbottom, 1977, 1979) which, combined with variations in sediment supply and subsidence, gave complex patterns of relative sea-level change. In addition, basin-water salinity probably varied significantly owing to the interaction of eustatic sea-level changes and large fluvial discharges into humid tropical basins remote from the open ocean (Collinson, 1988).

In deep mudstone-dominated basins the onset of coarse clastic supply was marked by sandy turbidites (J.R.L. Allen, 1960; R.G. Walker, 1966a; Collinson, 1969, 1988; Collinson,

Martinsen *et al.*, 1991). Turbidite-bearing formations such as the Shale Grit of Derbyshire and the Ross Formation of County Clare, Ireland which are up to several hundreds of metres thick, are vertically separated from delta plain facies by siltstone-dominated units interpreted as slope deposits. In some cases (e.g. the Grindslow Shales of Derbyshire) these are cut by sandstone-filled channels which probably acted as conduits for the bypass of sand from the delta front to deeper water (Fig. 6.49a) (e.g. Collinson, 1970b, 1988; McCabe, 1978). In County Clare, the 500-m-thick slope siltstones of the Gull Island Formation are characterized by abundant slumps and slides (Fig. 6.49b) (e.g. Martinsen & Bakken, 1990; Collinson, Martinsen *et al.*, 1991).

In the Grindslow Shales, the 100-m-thick upwards-coarsening slope unit with channels passes up into the delta plain facies and there seems little doubt that the prograding slope was a delta front (e.g. Collinson, 1969, 1988). The dominant facies, particularly in the lower part, is gradational thin-bedded siltstones thought to record oscillations in the load and/or position of buoyant plumes or weak underflows off distributary mouths. The upper parts of these inferred slope units show increasing interbedding of fine-grained sandstone, bioturbation and current-generated tractional structures with palaeocurrents directed down the inferred slope. This suggests gravity-driven underflows, possibly favoured by reduced basinwater salinity.

The Gull Island Formation provides a contrast. Turbidite sandstones in the lower part of the succession, systematic upwards changes in palaeocurrent and palaeoslope indicators and the restriction of upwards coarsening to an interval of distinctive striped siltstones at the top of the succession all suggest that the slope was a true basin slope which prograded to largely infill the basin before the arrival of the first delta (Collinson, Martinsen *et al.*, 1991). On that basis the earliest deltaic interval records a shelf-edge delta (Pulham, 1989).

Where deep basins have been infilled by turbidites and slope deposits or on block areas where deep waters never existed, deltaic successions show repeated cyclothems each broadly coarsening upwards. Such cyclothems range in thickness from

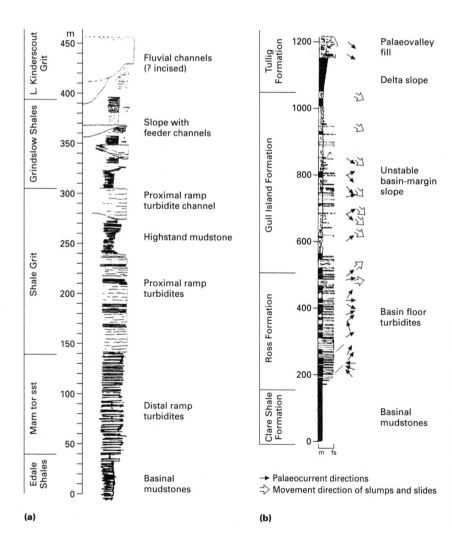

Figure 6.49 Vertical facies logs through basin-filling Namurian successions. (a) The Kinderscoutian of North Derbyshire, England where large volumes of sandy sediment apparently bypassed the slope through feeder channels. The slope interval coarsens upwards with limited evidence of syndepositional deformation. (b) The mainly Kinderscoutian of County Clare, Ireland where the main slope interval is characterized by intense syndepositional deformation. Upwards coarsening in the top probably records the advance of a shelf-edge delta. Lack of organization in the turbidites may reflect instability of the coeval slope (after Collinson, Martinsen *et al.*, 1991).

over 100 m to less than 10 m (Fig. 6.50). Thicker examples may extend over areas of thousands of square kilometres and have fossiliferous mudstone horizons ('marine bands') close to their bases. Thinner examples tend to be less extensive and less likely to have associated marine bands. Intervals between successive marine bands are, in some examples, characterized by a lower relatively thick upwards-coarsening unit and an upper section which includes one or more thin upwards-coarsening sequences.

Upwards-coarsening sequences commonly show an upwards gradation from mudstone to sandstone and an associated increase in current and/or wave-generated structures. In silt-dominated delta-front sequences, such as those in County Clare, storm wave base is identified by the incoming of thin wave-ripple form sets of slightly cleaner sand. Higher in the sequence, sandstones may be dominated by wave-generated structures such as hummocky cross-stratification and low angle flat lamination.

In other cases, discrete sandstone beds with current ripple cross-lamination or small-scale cross-bedding and wave rippled tops suggest episodic flood deposition followed by wave reworking. The lower parts of delta front sequences may also have sandstones of inferred density current origin (e.g. Martinsen, 1990). These share many of the features of turbidites but their context suggests water depths of only a few tens of metres.

Some upwards-coarsening sequences culminate in thicker sandstones which show evidence of stronger currents (Figs 6.51 & 6.52). Such mouth-bar sandstones commonly show drapes of siltstone on the floors of broad troughs which are then infilled by cross-bedded sand. Oscillation between erosion, deposition from suspension and by traction currents may record flood activity over a mouth bar and the establishment and flushing of saline wedges (Pulham, 1989). Other upwards-coarsening sequences are truncated by erosion at the bases of channel sandstones. Where these sandstones are only a few metres thick,

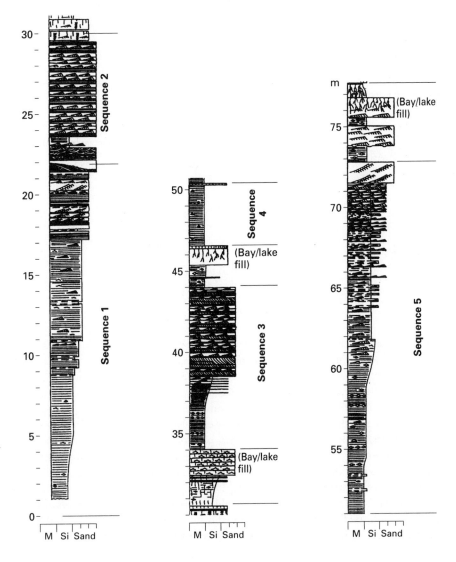

Figure 6.50 Upwards-coarsening facies sequences in the Westphalian of South Wales. The thicker sequences (1 & 5) record progradation of shoal-water deltas culminating in mouth-bar and delta-plain sediments. Thinner sequences result from crevasse deltas filling bays or lakes on the delta plain. The sand-rich sequence (2) may record a sudden avulsion (after Svela, 1988).

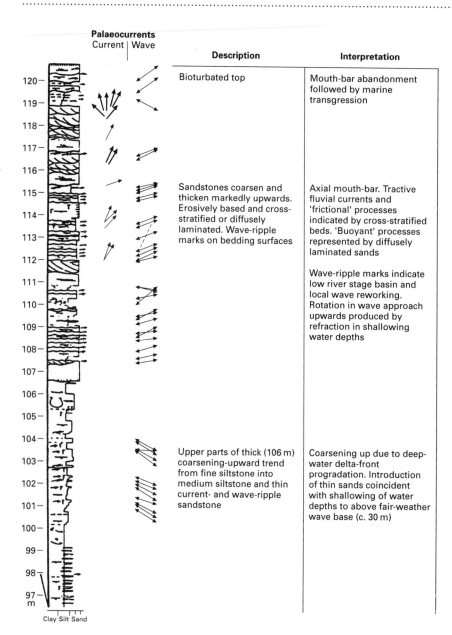

Figure 6.51 The upper part of a major upwards-coarsening succession in the Namurian of County Clare, Ireland (Tullig Cyclothem, see Fig. 6.49). The combination of wave- and current-generated structures indicates a wave-influenced, river-dominated delta front and the thick sandy upper part suggests an axial mouth-bar setting (after Pulham, 1989).

and single storey, they probably record distributary channels which form an integral part of the progradation, removing the mouth-bar crest. Where an upwards-coarsening sequence is only a few metres thick, the overlying channel may be a crevasse channel. Distributary and crevasse-channel sandstones commonly have medium-scale cross-bedding and ripple cross-lamination and their basal erosion surfaces carry a lag conglomerate of pebbles, reworked nodules and plant debris. They compare closely with sandy fluvial channel sandstones and some may show evidence of lateral accretion (Elliott, 1976b).

Thick multistorey channel sandstones, which appear out of scale with surrounding parasequences, probably fill incised palaeovalleys (Fig. 6.52). They are dealt with in more detail later.

Coarsening-upwards sequences commonly end at sharp surfaces which record abandonment and flooding. Such surfaces vary in character and, in some cases, it is possible to relate lateral facies changes to position on the surface of the delta lobe (e.g. Elliott, 1974a). Intense bioturbation may reflect abandonment and flooding in a delta front setting. Reworked quartzitic sandstones may reflect a near-delta front setting with development of transgressive barriers (Percival, 1992) or sandwave complexes (Brenner & Martinsen, 1990). Palaeosols and coal seams, on the other hand, probably record abandonment of more landward parts of a delta (Fig. 6.53) (Elliott, 1974a). Where a marine band mudstone is present in the base of the overlying cycle, it commonly occurs within about 2 m of the abandonment surface.

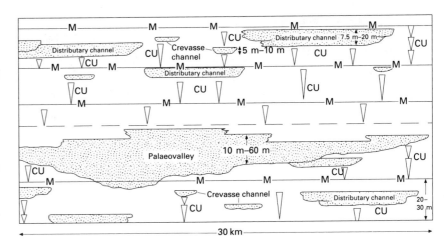

Figure 6.52 Schematic organization of Namurian deltaic sequences in the Pennines of northern England or beneath the southern North Sea. Upwards coarsening occurs at several scales between fossiliferous marine bands (-M-), often culminating in mouth-bar sandstones (not shown). Channel sandbodies range in scale from small, single-storey examples in the upper parts of cycles (crevasse channels) through larger, commonly multistorey sandbodies which erode underlying cycles and are out of scale with surrounding progradational units (palaeovalleys).

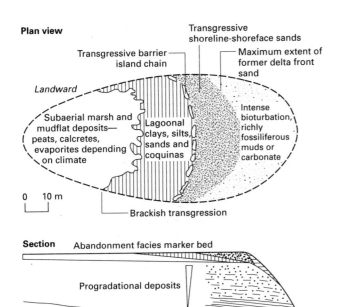

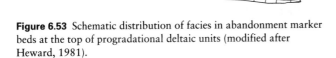

Figure 6.53 Schematic distribution of facies in abandonment marker beds at the top of progradational deltaic units (modified after Heward, 1981).

Many upwards-coarsening sequences, ascribed to deltaic progradation, can be further characterized in terms of the degree of river or wave influence on the basis of the small-scale sedimentary structures. Independent evidence of delta type can also be sought from the larger scale organization of the successions. River-dominated deltas are more likely to develop interdistributary bays and their deposits are therefore more likely to contain small-scale bay-fills. Wave-influenced deltas with more uniform coastlines should show more lateral continuity of correlative vertical sequences. In some cases lateral variation of facies and thickness may be mapped across an area to define

the type and extent of the delta. In a Namurian Yoredale cycle in northern England, Elliott (1975) mapped out discrete lobes separated by embayments of shallow-water fluvial-dominated deltas. In the Upper Namurian of the Pennine Basin, Collinson and Banks (1975) mapped highly elongate sandbodies in the upper parts of two successive upwards-coarsening cycles. The elongation axes are parallel with palaeocurrents and the sandbodies thin from around 30 m to zero over horizontal distances of about 8 km normal to palaeoflow. Upstream ends of the elongate lobes have major channels while the central and downstream parts are dominated by ripple cross-laminated sandstones thought to be delta-front and mouth-bar sediments. Inferred upper mouth bars are characterized by low-angle accretion surfaces and minor growth faults (Collinson & Banks, 1975; Bristow, 1988). In the thick Namurian deltaic cycles of County Clare, thick mouth-bar sandbodies pass laterally into less sandy, more wave influenced units (Fig. 6.54) (Pulham, 1989; Davies & Elliott, 1994).

While the characterization of delta type has been and remains an important feature of studies of ancient deltaic successions, the stacking patterns of deltaic sequences has, for many years, stimulated discussion of the causes of cyclicity (Wanless, 1950; Weller, 1956; D. Moore, 1959; Johnson, 1967; Duff, Hallam & Walton, 1967). Recent models and methodologies of sequence stratigraphy (Sect. 2.4) have extended such discussion by forcing attention on key surfaces such as flooding surfaces, palaeosols and basal erosion surfaces of channel sandbodies and also by highlighting the question of accommodation space.

Simple coarsening-upwards parasequences, bounded top and bottom by flooding surfaces and associated marine bands, are reasonably explained in terms of progradation following flooding of the preceding delta and deepening of the waterbody by a combination of relative sea-level rise and sediment starvation. Flooding, condensation and creation of most accommodation space occurred around the most rapid rise of relative sea level. The marine bands which occur above the flooding surface, often

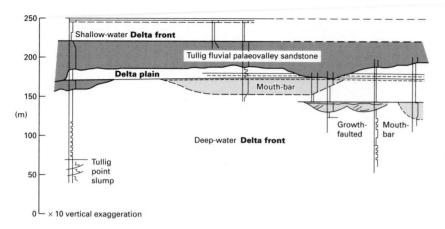

Figure 6.54 Cross-section, normal to regional palaeoflow, through a major progradational deltaic cycle in the Namurian of County Clare, Ireland showing laterally restricted mouth-bar sands and distributary channels. The large thickness and width of the multistorey Tullig channel sandstone suggests a palaeovalley fill (cf. Figs 6.49 & 6.51) (after Pulham, 1989).

providing the only clear indicator of full marine salinity, may reflect a position on the relative sea-level curve closer to the highstand. The succeeding progradational unit may then have been partly driven by a relative fall with the sequence boundary (*sensu* Exxon) not necessarily represented by a discrete surface. Alternatively, the sequence boundary may be subsumed in the upper abandonment surface, especially if this includes a palaeosol.

Most successions are considerably more complex. Not all flooding surfaces have marine bands above them and parasequences vary in thickness. Compound cyclothems, defined by marine bands, vary in the nature and stacking patterns of their constituent parasequences. Many such cycles are made up of a lower, larger-scale upwards-coarsening unit developed directly above the marine band, overlain by a series of one or more thinner parasequences. The abandonment/flooding surface of the uppermost parasequence heralds the next marine band. In such a case, the major parasequence set records highstand and post-highstand progradation with its basal surface being the initial flooding surface associated with an early stage of rising relative sea level. The minor parasequences in the upper part of the cycle could then form part of a transgressive systems tract which culminated at the maximum flooding surface beneath the next marine band. These thinner parasequences may also record shifting minor delta lobes or bay-fill deltas illustrating the coexisting autocyclic behaviour of the delta. Mapping of the lateral extent of flooding surfaces and associated parasequences provides a means of separating allo- from autocyclic controls. Parasequences which die out over a few kilometres or even tens of kilometres are more likely to record autocyclic changes.

Other thick, marine-band-defined cyclothems may be even more complex, as illustrated by the succession in County Clare, Ireland (Fig. 6.55). Here the upper parts of some cyclothems are characterized by multistorey channel sandbodies, up to 40 m thick and approximately 15 km wide interpreted as the fill of palaeovalleys. These cut into stacked, small-scale parasequences interpreted as shelf deltas. The bases of the valley fills are therefore type 1 sequence boundaries. Where palaeovalleys are

absent, mature palaeosols developed on exposed interfluves. These relationships suggest that the cyclothem defined at marine bands may reflect two cycles of relative sea-level oscillation. The palaeovalley formed during rapid sea-level fall in the later cycle and the overlying flooding surfaces and eventual marine band record the rising stages of that cycle. The earlier cycle has a more cryptic record. The rising limb of the relative sea-level curve is recorded by the initial flooding surface at the top of the main progradation and by overlying small-scale parasequences. The maximum flooding surface and any associated marine band were erosively removed by palaeovalley incision but may be present in interfluve areas beneath the lowstand palaeosol of the upper sequence. The sequence boundary of the lower cycle may be difficult to distinguish but sudden upward changes from mudstone to siltstone low in the major parasequence may be its distal expression. In that case, the greater part of the major progradational parasequence may be a lowstand wedge.

In the Namurian of the Pennines of northern England, type 1 sequence boundaries associated with thinner cyclothems are sometimes steeply incised but other major channel sandbodies such as the Rough Rock, occur as extensive sheets with no clear lateral margins (Bristow, 1988). These appear to be extensive braidplains introduced by rapid basinwards shifts of facies.

ANCIENT WAVE-DOMINATED DELTAS

Wave-dominated deltas are well described from the Cretaceous Western Interior foreland basin of North America (Rosenthal, 1988; Bhattacharya & Walker, 1991, 1992; Bhattacharya, 1993; Hadley & Elliott, 1993), the Cretaceous–Tertiary of the Gulf Coast of the USA (Weise, 1980), the Jurassic of the North Sea (Morton, Haszeldine *et al.*, 1992), and from the Tertiary of northwest Borneo (James, 1984; Ho, 1987; Johnson, Kuud & Dundang, 1989).

Outcrop studies of delta-plain facies of wave-dominated deltas show them to comprise fluvial-distributary channel sandstones and, where accommodation space was sufficient, interdistributary lagoonal/lacustrine facies with minor mouth bar and

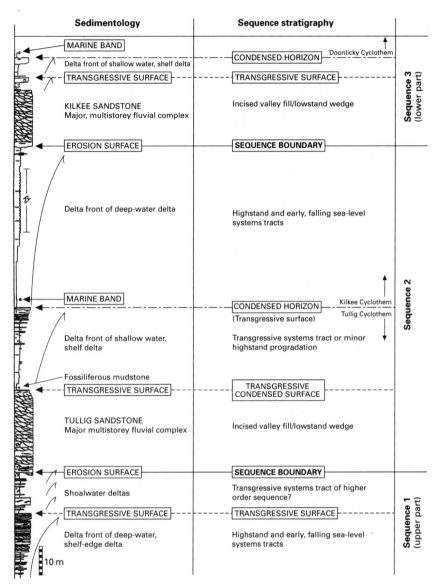

Figure 6.55 The succession of Namurian deltaic sediments in County Clare, Ireland subdivided into EPR-type sequences. Note that the maximum flooding surface of sequence 1 may have been eroded by the sequence boundary at the base of sequence 2 if the parasequences at the top of preserved sequence 1 are part of a transgressive systems tract. The upwards-coarsening unit in the top of the Tullig Cyclothem may be part of a transgressive systems tract or may have followed a minor highstand (based on Davies & Elliott, 1996).

lagoonal beach deposits. The delta front is represented by laterally extensive coarsening-upward sequences characterized by wave- and storm-dominated facies, which resemble those of prograding beach ridges (Sect. 6.7.1). The fine member at the base of the delta front sequence may include storm-generated graded beds while the mid- to upper parts are dominated by well-sorted sandstones with hummocky and swaley cross-stratification, wave ripple lamination, flat lamination and cross-bedding. Bioturbation varies in intensity but usually decreases rapidly upwards. Delta front sequences are locally eroded by fluvial-distributary channel sandstones, and upper beach face sandstones may be better developed close to these channels suggesting clustering of beach ridges around channel mouths.

In the subsurface it may be difficult to distinguish wave-dominated deltas from both fluvial-dominated deltas and prograding beach–shoreface sequences. Gamma-ray well logs all show an upward decrease in clay and, by inference, an increase in mean grain size, sand content and bed thickness. Distinction is best achieved from lithofacies analysis of cores, supported by regional isopach and sand trend maps. Compared with fluvial-dominated deltas, wave-dominated delta sequences preserve few distributary channels and record the effectiveness of both continuous wave processes and episodic storm-related sedimentation events within the delta front. Thus rhythmically bedded laminated siltstones occur in the lower delta front facies, whereas sharp-based, or amalgamated wave and current reworked sandstones occur in upper delta front facies. Bioturbation in the lower energy facies is supplanted by oscillatory wave flow in the higher energy facies. There may also be thick accumulations of woody organic material. The palaeostrandline and sandbody shape are intermediate between those of shore-normal lobate fluvial-dominated deltas and of shore-parallel elongate strandplains.

In the Cretaceous of the Western Interior, most shorelines are considered to be strandplains that have multiple fluvial feeder points. An exception is the Dunvegan Delta complex where river-dominated, wave-influenced and wave-dominated deltas have been distinguished (Bhattacharya & Walker, 1991, 1992) and key flooding and erosional surfaces identified (Bhattacharya, 1993) using sandbody geometries and vertical facies analysis from cores.

In the Gulf Coast, wave-dominated deltas have been recognized using geophysical logs and, in particular, sand isopach mapping. In some cases strike-aligned sandbodies of beach face and distributary mouth-bar origin can be traced up-dip to dip-aligned fluvial-distributary channel sands (W.L. Fisher, 1969). In the Cretaceous San Miguel Formation, Texas, a spectrum of these sandbodies, identified by isopach patterns, reflects varying degrees of wave dominance (Weise, 1980; Fig. 6.56). This formation accumulated during a relative sea-level rise as a succession of wave-dominated deltas which prograded intermittently during periods of high sediment supply. During transgression, successive deltas were offset progressively more landward. The growth and characteristics of each delta were a function of: (i) sediment supply which varied in response to hinterland tectonics; (ii) wave energy which was relatively constant in an absolute sense, but varied in effectiveness; and (iii) the rate of sea-level rise. During periods of high sediment input and a low rate of sea-level rise, waves were less able to redistribute sediment, and lobate, wave-influenced deltas resulted. During periods of low sediment input and high rates of sea-level rise, wave reworking was extensive and wave-dominated deltas were characterized by elongate, strike-aligned sandstone bodies. Wave reworking after abandonment of each delta was appreciable with much of the beach and upper shoreface facies being removed and reworked by shelf processes (Fig. 6.57). This resulted in attenuated delta front sequences, with only local preservation of fluvial distributary channel sandstones.

The Middle Jurassic Brent Group and its equivalents in the North Sea contains the largest hydrocarbon reserves in Europe (Richards, 1992; Morton, Haszeldine *et al.*, 1992). Its 300-m clastic succession is sandwiched between thick marine shales and has traditionally been interpreted as a northward prograding wave-dominated delta complex with minor tidal influence fed from a thermally domed area to the south (Eynon, 1981; Parry, Whitley & Simpson, 1981; Johnson & Stewart, 1985; Graué, Helland-Hansen *et al.*, 1987; Helland-Hansen, Steel *et al.*, 1989; Falt & Steel, 1990) (Fig. 6.58a). This interpretation has been progressively modified to suggest greater sediment supply from lateral basin margins comprising the Shetland Platform to the west and Norway to the east with redistribution by longshore drift. The provenance evidence is primarily from heavy minerals (Morton, 1985, 1992) and from the samarium–neodymium

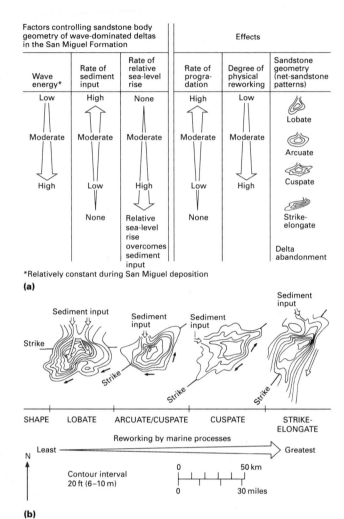

Figure 6.56 Controls and responses of the Cretaceous San Miguel delta system Texas; wave-dominated deltas deposited during a time of overall transgression (after Weise, 1980).

(Sm–Nd) ratios, which reflect compositions of the source rocks (Mearns, 1992). The originally postulated supply area to the south is very small compared with the area of the Brent delta, and gives a delta area to basin area ratio smaller than any modern delta except the Klang (Mearns, 1992). Furthermore, no major river system has been found. An alternative embayment interpretation has therefore been proposed (Richards, Brown *et al.*, 1988; Richards, 1991) that includes a major estuary to the south where the delta is shown on the original interpretation (Fig. 6.58b) with sediment derived transversely. There is evidence to support both interpretations, and the Brent Group demonstrates the complexity of any natural system and the pitfalls of applying simple models.

The succession (Fig. 6.59) can be divided into four major depositional units. The basal Broom Formation (= Oseberg Formation) comprises lowstand Gilbert-type deltas (Fig. 6.60)

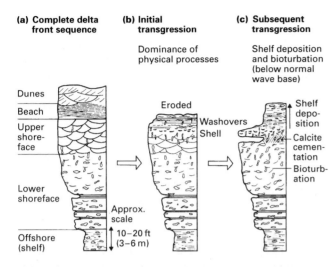

(a) Complete delta front sequence

(b) Initial transgression

Dominance of physical processes

(c) Subsequent transgression

Shelf deposition and bioturbation (below normal wave base)

Dunes
Beach
Upper shoreface
Lower shoreface
Offshore (shelf)

Eroded
Washovers
Shell

Shelf deposition
Calcite cementation
Bioturbation

Approx. scale
↕ 10–20 ft (3–6 m)

Figure 6.57 Delta-front sequences in the wave-dominated San Miguel delta system, Texas; (a) complete delta-front progradation sequence, (b) modification of the sequence during the ensuing transgression, (c) final, preserved sequence (after Weise, 1980).

derived from the tectonic uplift around the basin margin (Helland-Hansen, Ashton *et al.*, 1992; Mitchener, Lawrence *et al.*, 1992) and their reworking as transgressive sands as sea level rose. During this time estuaries may have penetrated to the south.

The second unit is the regressive, prograding beach/barrier shoreline succession that embraces the Rannoch–Etive and lower Ness Formations (Figs 6.60 & 6.61). The prograding shoreface succession shows both inferred rip channels and longshore bars indicative of a barred dissipative shoreline (Fig. 6.62) and rare microtidal inlets (Budding & Inglin, 1981; Scott, 1992). This barrier system built northward, fed by both an inferred axial river from the south (Cannon, Giles *et al.*, 1992) and longshore

drift from the margins (Morton, 1992; Mearns, 1992). Behind the barrier were wide open lagoons and a large delta plain with extensive swamps and marshes (lower Ness) yielding coals and finely laminated shales. Flood tidal deltas broke through the barrier and very small-scale distributary mouth bars prograded into the lagoon.

The third unit (Middle and Upper Ness Formation) comprises laterally extensive coals, palaeosols, and locally interfingering organic-rich mudrocks, heterolithics, and sheet-like and sharp-based ribbon-like sandstones. It has been interpreted as having been deposited in widespread swamps, lagoons and lakes, penetrated by innumerable small-scale distributary mouth bars and distributary channels (Livera, 1989; Ryseth, 1989). The Ness Formation was originally interpreted as an upper delta plain, that is partly coeval with the broadly regressive Rannock–Etive prograding beach/barrier to delta front complex (e.g. Eynon, 1981; Budding & Inglin, 1981). However, a recent sequence stratigraphic synthesis of the Brent Group argues that the widespread coals, the absence of transitional lower delta-plain facies, and basin-wide chronostratigraphic flooding surfaces suggest that this third unit is the result of a major sea-level rise which caused the southward retrogradation of the whole Brent coastal system, drove the feeder rivers landward and allowed complex aggradational sequences to form (Fig. 6.63) (Mitchener, Lawrence *et al.*, 1992). The fourth unit (Tarbert Formation), interpreted as a transgressive sand in the original model, is considered to represent the most open marine part of this retrogradational system as the shoreline continued to back-step in response to continuous relative sea-level rise.

Thus the Brent Group illustrates the complexity of a thoroughly investigated system. In part vertical facies changes result from lateral changes in the environment; in part they are due to changes in supply and base level.

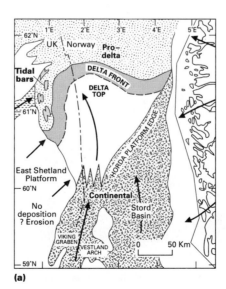

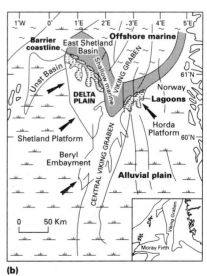

Figure 6.58 Middle Jurassic palaeogeography of the northern North Sea showing interpretation (a) as a prograding delta lobe and multiple feeder system (after Eynon, 1981), and (b) as an embayment (after Richards, 1992) as it was deposited in earliest Middle Jurassic times when supply was mainly transverse from the Shetland and Horda Platforms and estuaries penetrated southwards to the region of subsequent delta feeders.

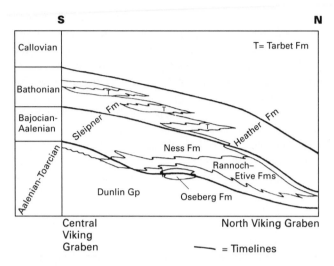

Figure 6.59 South to north schematic section of the Brent Group and adjoining formations (from Richards, 1992, based on Graue, Helland-Hanson *et al.*, 1987).

ANCIENT TIDE-DOMINATED DELTAS

Although tidal influence is now recognized in many modern deltas, ancient tide-dominated deltas have seldom been recognized. Recognition depends on the character of the mid- to upper delta-front facies and the lower delta-plain facies where tidal effects are expected to be most pronounced. In the delta front, gradational coarsening-upwards sequences of tidally influenced facies result from the progradation of ebb-tidal deltas or tidal sand ridges, which are the tidal equivalents of river-mouth bars. In delta-plain successions erosive-based tidal channel or inlet sequences are more common. The lower delta plain is likely to include tidal flat sequences, small-scale channel units produced by tidal creeks, and larger-scale estuarine–distributary channel sequences.

Tidal sediments have been recognized within deltaic systems in the subsurface Tertiary of the Niger delta (Weber, 1971), the Eocene Misoa Formation of the Maracaibo Basin of Venezuela (Maguregui & Tyler, 1991b) and at outcrop in the early Creta-

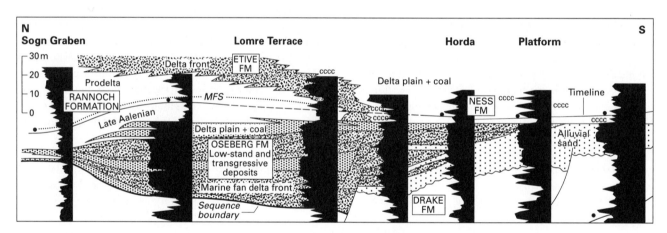

Figure 6.60 Cross-section of the eastern margin of the North Sea to show (1) early Middle Jurassic regressive phase (Oseberg Formation) progradation of lowstand Gilbert-type deltas, (2) Late Aalenian transgression and (3) progradation of Rannoch–Etive, Ness delta (from Helland-Hansen, Ashton *et al.*, 1992). MFS, maximum flooding surface; cccc, coal.

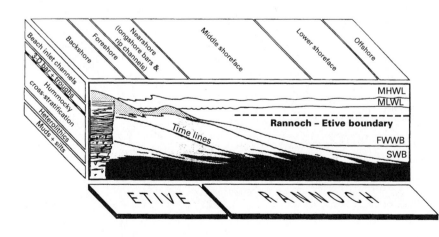

Figure 6.61 Interpretation of the prograding offshore to backshore sequence of the Rannoch–Etive Formations (from Scott, 1992). FWWB, fair weather wave base; SWB, storm wave base.

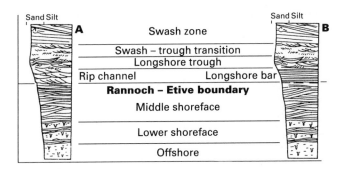

Figure 6.62 Representative logs showing details of progradational sequences through storm-dominated, barred, dissipative shorelines (from Scott, 1992).

ceous of the Saurashtra failed rift of western India (Casshyap & Aslam, 1992).

The Misoa Formation is a huge northeasterly-flowing alluvial–deltaic–shallow-marine complex that extended some 250 km from its source. The palaeogeography of a tiny part (64 km²) of this tide-dominated delta was constructed by Maguregui and Tyler (1991b) for selected time slices from the distribution of four major facies associations: estuarine distributary-channel sandstones, tidal sand ridges, tidal channel interdistributary facies and prodelta/shelf facies (Fig. 6.64a).

Estuarine distributary channels appear to have been separated by slightly higher interdistributary tide-dominated lower delta plains rather like the present Mahakam delta (Fig. 6.32). Interdistributary areas were crossed by narrow, highly sinuous, relatively muddy tidal channels (Fig. 6.33) that did not connect

with the main estuarine system but only carried ebb tide drainage (contrast the dominant flood tidal flow at the head of estuaries).

The estuarine channels were straight, passing landward into relatively stable distributaries. Estuarine distributary channel sandstones occur as elongate bodies extending seawards for 3–5 km, with a width of 300–600 m. They occur in complexes up to 1800 m wide that were stable for relatively long periods and pass seaward into tidally modified distributary mouth bars. These are composed (Fig. 6.64) of stacked sharp or erosive-based sandstone units 1–3.3 m thick showing a weak grain-size grading and a vertical trend from trough cross-stratification into wavy lamination and ripples at the top and an increasing number of shale lenses. Burrowing is common, both as isolated *Ophiomorpha*-like traces in the sandier bands and as total bioturbation of the tops of individual sandstones. Reworked gastropod, bivalve and echinoderm fragments are found in the coarser sandstones showing that marine influence extended up the distributary.

Seaward, the estuarine distributary channels diverge and mouth-bar sediments show strong tidal influence in this zone of maximum tidal current energy. Inferred tidal sandridges (Sect. 7.7.1) also occur as upward-coarsening highly bioturbated sandbodies 1–4 m thick (Fig. 6.64b,c,d) formed as a consequence of the lateral migration of higher-energy bar crests over lower-energy muds, their orientation controlled by that of the estuaries. They occur both as individual (4 m high and 600 m wide) sandbodies between prodelta/shelf shales and as large amalgamated sandbodies that pinch and swell along strike for over 2400 m (Fig. 6.65).

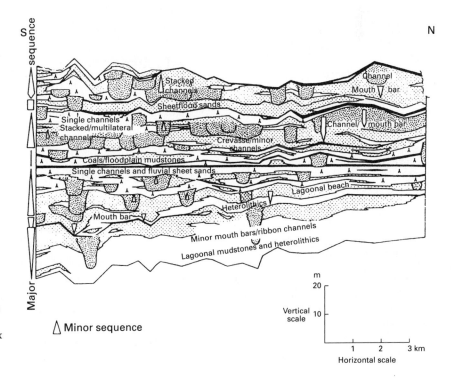

Figure 6.63 N–S cross-section through middle part of the Ness Formation to show widespread coal horizon and laterally complex pattern of lacustrine, mouth bar, and channel facies (after Livera, 1989).

(a) Estuarine distributary complex

(b) Transitional estuarine to tidally modified distributary mouth bar

(c) Proximal tidal sand ridge

(d) Distal tidal sand ridges

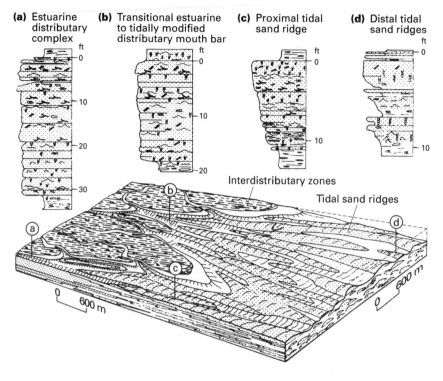

Interdistributary zones

Tidal sand ridges

Figure 6.64 Three-dimensional model of the delta front and lower delta plain of the tide-dominated delta in the Eocene Misoa Formation (from Maguregui & Tyler, 1991b) showing estuarine channels between interdistributary areas. Within the estuaries the facies sequences pass from an estuarine distributary complex (a) to a highly tidally reworked distributary–estuarine mouth bar complex (b) to a proximal tidal sand ridge that is strongly affected by tides (c) to a distal tidal sand ridge (d) where sediment supply is limited and tidal currents are weaker.

North

South

Interdistributary zone

Distributary zone

Interdistributary zone

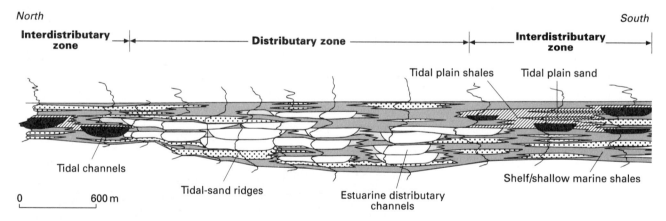

Tidal plain shales   Tidal plain sand

Tidal channels

Shelf/shallow marine shales

Tidal-sand ridges

Estuarine distributary channels

0   600 m

Figure 6.65 Strike-orientated cross-section through the Misoa Sandstone showing rapid lateral facies changes compared with dip-orientated cross-sections. In the distributary zone, shelf/shallow-marine shales are sandwiching relatively continuous amalgamated tidal-sandridges and less continuous channelized estuarine distributary channels. In the relatively fine-grained interdistributary zones are tidal-plain shales with highly sinuous shale-dominated tidal channels and crevasse splay tidal-plain sands (from Maguregui & Tyler, 1991b).

ANCIENT DELTA DEFORMATION

In addition to the deformation in the Tertiary and Holocene sediments beneath present-day deltaic depocentres, discussed in Section 6.6.4, examples are known from ancient sediments at outcrop. Most examples are from river-dominated deltas with a high content of mud and silt. In the Namurian of western Ireland, a whole suite of syndepositional deformation styles is extensively exposed. In the major basin-filling slope interval, the Gull Island Formation, most of the 500-m-thick silt-dominated interval is disturbed as a result of mass movement downslope (Martinsen, 1989) (Fig. 6.49b). Slumps and slides, some tens of metres thick and hundreds of metres wide, produce an array of compressional and extensional deformation (Martinsen & Bakken, 1990). Within displaced units, original bedding is variably preserved. In some examples, simple rotation of bedding is apparent while other units show extensional faulting, intense folding or total loss of bedding through homogenization in mobile slumps. Some slumps occur within large discordant troughs, interpreted as slump scars. Within the overlying stacked deltaic cyclothems deformation is also very common, confined mostly within 100-m-thick progradational units. Most common are listric growth faults which commonly sole out on the marine band mudstone at the base of the upwards-coarsening sequence (Rider, 1978; Pulham, 1989). Mouth-bar sandstones in the upper parts of the intervals roll over and thicken into the hanging walls of the faults. While some faults have fairly simple listric geometries, others are associated with diapirism of the mudstones in the lower part of the unit. Mud diapirs penetrate upwards into mouth-bar sands which may be broken up by small-scale extensional faults. Some diapirs broke through to the surface to generate small slumps. No diapirs appear to extend vertically beyond individual progradations, suggesting that they moved very early and under quite shallow burial.

In Triassic deltaic sediments of Svalbard a lateral succession of 10 growth faults is seen to influence four stacked upwards-coarsening units (Fig. 6.66) (Edwards, 1976). The faults are all listric and dip in the general direction of delta progradation. Thickening and rollover into the faults occur in the upper three progradational units showing that, once established, the faults were reactivated during later progradations. In some cases, large-scale cross-stratification occurs in sandy mouth-bar sediments in the hanging wall area recording small Gilbert-type deltas which built into a body of deeper water in the hanging wall, probably caused by fault movement continuing for a period following a temporary cut-off of sediment supply.

A superficially similar but radically different style of delta deformation is seen in the Lower Cretaceous of Spitsbergen (Nemec, Steel *et al.*, 1988). Here, a deltaic progradational interval, probably close to its distal limit, collapsed in a series of major cuspate slide scars, up to 1.5 km wide and more than 60 m deep. Upper delta slope and delta-top channel sandstones slid and rotated as large, internally undisturbed blocks, leaving narrow *in situ* buttresses between the scars (Fig. 6.67). The collapse was probably triggered by seismic shock from a nearby major fault. The brittle fragmentation, sliding and rotation of

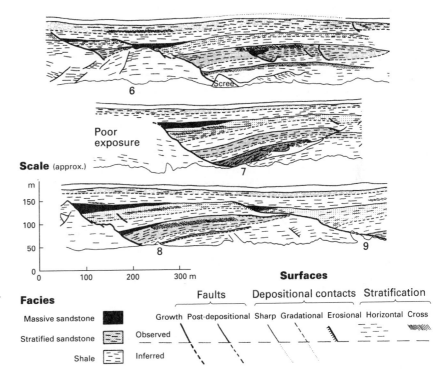

Figure 6.66 Growth faults in an Upper Triassic deltaic sequence, Svalbard, in a section parallel with palaeoflow. Note the rollover into the faults and the thickening of mouth-bar sandstones into the hanging wall area. The three panels form a continuous series from top left to bottom right (after Edwards, 1976).

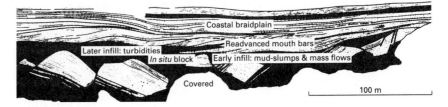

Figure 6.67 Delta-front collapse in the Lower Cretaceous of Spitsbergen. Seismic shocking probably caused the brittle fragmentation of delta-front and delta-top sediments and their sliding into major slide scars. Slumping and draping by muds preceded sandy mass-flows which helped heal the scars before mouth bars readvanced (after Nemec, Steel *et al.*, 1988).

the delta-top sediments was followed by local slumps and debris flows from the flanks of the scars and then by a period of quiescence when muds draped the irregular slump scar topography. Re-establishment of fluvial supply to the delta front led to sandy mass-flow deposits being funnelled into the slump-scar depressions, partially filling them before prograding sandy mouth bars completed the repair of the delta front.

6.7 Non-deltaic siliciclastic coasts

The siliciclastic coasts discussed in this section are those where sand, silt and mud predominate, rather than gravel, and where direct fluvial input is insignificant. Instead, the shore is fed either by longshore and coastal current drift, by upper shoreface erosion during transgression (Bruun, 1962) or by lower shoreface erosion during sea-level fall (Swift, 1976; Dominguez & Wanless, 1991). They have similarities to the wave- and tide-dominated deltas described in Section 6.6, but differ from them in that sediment supply is primarily marine, not fluvial. Since coastal gradients are lower than in gravel-dominated systems, tidal zones are wider and tides more effective.

Shorelines are generally linear, lacking the protuberances of most deltaic coasts. However, most estuarine coasts are embayed. The type of coast depends on: (i) the degree and relative power of waves and tidal activity (Figs 6.3 & 6.8); (ii) sediment grain size, whether sand, silt or mud (Fig. 6.8); (iii) marine sediment supply (Figs 6.7 & 6.8); and (iv) relative sea-level change.

Reflecting mixtures of such controls, these coasts can be divided into several overlapping types: (i) beach-ridge strandplains; (ii) chenier plains; (iii) tidal flats; (iv) barrier island/lagoons; and (v) estuaries. Beach-ridge strandplains, chenier plains and mudflat coasts are essentially regressive systems. Barrier-island–lagoonal systems and estuaries are mainly associated with transgression.

6.7.1 Beach-ridge strandplains

Beach-ridge strandplains are wave-dominated, sand-rich shorelines attached to the land and composed of both a strandplain

and a beach–shoreface. The classic beach-ridge strandplain is the Costa de Nayarit where a broadly arcuate sandy shoreline has prograded 10–15 km over a distance of 225 km since sea level stabilized after the Holocene transgression (Curray, Emmel & Crampton, 1969) (Fig. 6.68). Sand is supplied both by a number of small rivers and from the shelf and is distributed along the coast by longshore drift. However, beach-ridge strandplains that are not fed by rivers are rare today because they appear to require a fall in sea level for their formation (Dominguez & Wanless, 1991). The coast of east–northeast Brazil has experienced a relative sea-level fall during the last 5000 years of about 5 m. As sea level fell, an extensive sand-rich shelf was eroded and redeposited as a series of prograding beach ridges, backed by freshwater and mangrove swamps, that have advanced 10 km during the Holocene (Dominguez, Martin & Bittencourt, 1987), even where there is no inland river system (Fig. 6.69). Sediment supply to the protruding headland may be entirely by con-vergence of wave-induced longshore drift fed by a shallow sandy shelf. Other Brazilian beach-ridge systems, such as the São Francisco, are associated with rivers and have been interpreted as wave-dominated deltas (Galloway, 1975; Coleman & Wright, 1975), although they are fed primarily by wave-generated longshore drift (Dominguez & Barbosa, 1994). That sea-level fall is important in these Brazilian examples is shown by the way that a barrier-island lagoonal system preceded the present strandplain and was temporarily re-established each time sea level rose by as little as 2–3 m (Fig. 6.70; Dominguez & Wanless, 1991).

Wave-induced longshore drift of sediments reworked from the inner shelf during sea-level fall results in pronounced lateral asymmetry of the facies distributions and progradation rates at river mouths (Dominguez, Martin & Bittencourt, 1987) (Fig. 6.69). On the updrift side, rapid accretion of successive beach ridges gives a sand sheet composed only of relatively well-rounded marine-derived sands. On the downdrift side, progradation is slower through incorporation of sandy islands that protect mangrove swamps and form either by progradation of sandspits rooted in the downdrift margin of the river mouth or by wave reworking of mouth bars. Sands show variable

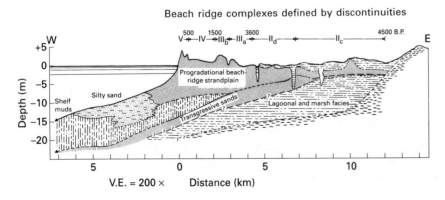

Figure 6.68 Cross-section through the wave-dominated Costa de Nayarit beach-ridge strandplain, Mexico, illustrating progradation following transgression (after Curray, Emmel & Crampton, 1969).

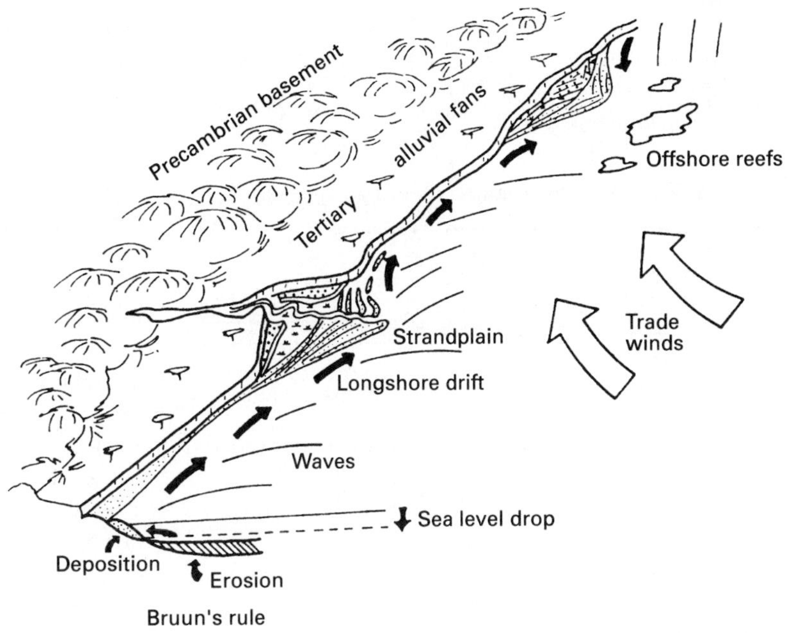

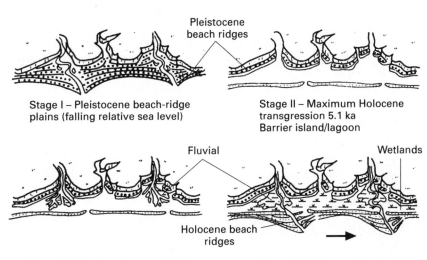

Figure 6.69 Model for sediment dispersal and accumulation along the beach-ridge strandplains of east-northeast Brazil during a small relative fall in sea level (from Dominguez, Bittencourt & Martin, 1992). Progradation occurs in areas protected by offshore reefs, in embayments and at river mouths.

Bruun's rule

Figure 6.70 Response of the east-northeast coast of Brazil to a Holocene relative sea-level rise followed by a fall of about 5 m (from Dominguez, Bittencourt & Martin, 1992).

Stage I – Pleistocene beach-ridge plains (falling relative sea level)

Stage II – Maximum Holocene transgression 5.1 ka Barrier island/lagoon

Stage III – Bayhead lagoonal deltas

Stage IV – Holocene beach-ridge plains (falling relative sea level)

roundness, with both rounded marine-derived sands and sub-angular to subrounded river-derived sands. Progradation on the downdrift side is thus affected by both longshore drift and fluvial supply. Brazilian examples show: (i) a spectrum from purely offshore/longshore fed beach-ridge strandplains to wave-dominated deltas (Sect. 6.6); (ii) that a small rise in sea level causes an emergent beach-ridge strandplain to be submerged and turned into a barrier-island lagoonal system; and (iii) that longshore drift of sediments reworked from the inner shelf provides a major supply for coastal progradation during sea-level fall.

6.7.2 Chenier plains

Chenier plains are extensive coastal mudflats with widely separated, subparallel beach ridges termed *cheniers*. These are commonly composed of sand, but may be formed of shells or vegetative matter. They rest on a silt or clay substrate and are isolated from the shore by a belt of tidal mudflats (Fig. 6.71). Otvos and Price (1979) restrict the term chenier to examples that occur in a purely littoral, that is non-deltaic, environment and lie above tidal muds. However a wider definition embraces sandy ridges that lie above prograding muddy deltaic sediments

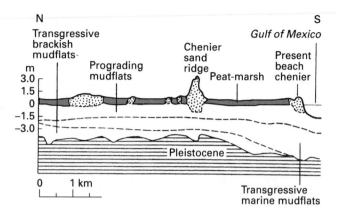

Figure 6.71 Cross-section through the prograding Louisiana chenier plain illustrating position of isolated chenier ridges above the prograding mudflats (after Gould & McFarlan, 1959).

(Augustinus, 1989). In either case, the sand ridges sit above a muddy substrate deposited by shoreline progradation; the sandy ridges were formed by minor transgression and reworking, commonly reflecting quite subtle changes in sediment supply (Fig. 6.72).

Cheniers are reported from a wide range of low-gradient coastlines (Augustinus, 1989; Shuisky, 1989) with abundant, but episodic, supply of mud. During periods of high sediment supply, mudflats are accreted. When supply diminishes, wave activity prevails, either winnowing and sorting the sediment in place, or allowing an increase in littoral transport of coarser

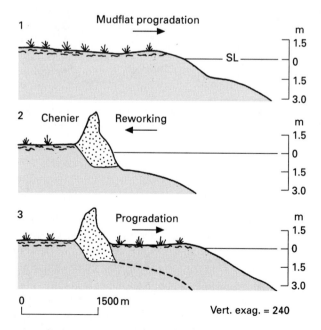

Figure 6.72 Chenier formation by means of alternating mudflat progradation and wave reworking (after Hoyt, 1969).

material. Although included here in wave-dominated coastlines, cheniers may occur in areas of high tidal range and often fringe estuarine mudflats.

Two scales of chenier system have been identified, each having a rather different model of development (Otvos & Price, 1979). *Bight-coast chenier* plains are broad coastal indentations that develop under low to moderate wave energies along open ocean shores which receive abundant fine sediment from neighbouring large-scale mud-dominated deltas but with episodic changes in sediment supply. The classical chenier system of southwest Louisiana, where the oak trees that line the crests of the ridges are termed *chene* in Louisiana French, lies downdrift of the Mississippi delta (Gould & McFarlan, 1959; Byrne, LeRoy & Riley, 1959; Hoyt, 1969; Penland & Suter, 1989). The plain is 100 km long and up to 24 km wide with several long narrow 30–450 m wide and low (max. 3 m) shelly sand ridges. Changes in supply relate to delta lobe switching within the recent Lafouche delta and to changes in local river supply (Penland & Suter, 1989). Along the Guiana coast of northeast South America the world's largest chenier plain – 700 km long and up to 50 km wide – lies downdrift from the Amazon (Fig. 7.30). Here, instead of a relationship with Amazonian sediment supply, chenier development seems to be related more to downdrift migration of large shoreface-connected mudbanks and erosive inter-mudbank areas where cheniers develop, the alternations occurring approximately every 30 years (Rine & Ginsburg, 1985). In addition, progradation has been interrupted twice in the last 6000 years by periods of erosion and formation of chenier bundles, due possibly to sea-level changes (Augustinus, Hazelhoff & Kroon, 1989).

Smaller *bay head chenier plains* develop in sheltered areas such as the Colorado delta (Thompson, 1968) where tidal ranges are high and *bay side chenier plains* occur locally on many coastlines, for example New Zealand (Woodroffe, Curtis & McLean, 1983) and the north coast of Australia (Rhodes, 1982; Chappell & Grinrod, 1984; Chappell & Thom, 1986). In addition they may occur locally along all types of coast.

Most of these chenier plains are quite small and localized and appear to be the result of many complex factors. In the Gulf of Carpentaria, Rhodes (1982) proposed that ridge construction took place during long-term dry periods when mud influx was low with mudflat progradation during wetter periods, though Chappell and Grinrod (1984) found no climatic influences. Instead they concluded that chenier formation was a consequence of the complex local morphology with mangrove forests locally trapping storm debris and shelly material and thus promoting mud progradation. In other cases, major storms build cheniers and many studies (summarized by Short, 1989) show that chenier ridges are built by storm surge reworking of intertidal shells and sands. In addition, both chenier and mudflat formation can be controlled by local basin geometry, sediment influx, trapping of storm debris and shells by mangroves, climatically induced changes in mud supply and relative sea-level oscillations (Anthony, 1989).

6.7.3 Tidal flats

Tidal flats form stretches of shoreline in all tidal settings, their extent depending on tidal range, sediment supply and shoreline gradient: they are most extensive in macrotidal settings and along muddy low-gradient coastlines, and they have an antithetic relationship with beaches. They may comprise a major portion of delta plains (Sect. 6.6.1), lagoons, estuaries, bays, cheniers, mudflats, saltmarshes and mangrove swamps. They are fully described by Weimer, Howard and Lindsay (1982). Tidal flats can be divided into two zones, supratidal flats, above normal high-tide level and intertidal flats, between high- and low-tide levels. The subtidal zone below low-tide level is usually dominated by estuarine channels.

Supratidal flats are very sensitive to climate. In temperate areas, interlaminated clays and silts provide a substrate for saltmarshes, where peats can form. The sediments are extensively disrupted by bioturbation, rootlets and nodule growth (Reineck, 1967). In arid to semi-arid regions, saltflats, called sabkhas prevail (Sect. 8.4). In humid tropical settings saline mudflats with polygonal desiccation cracks form just landward of the upper intertidal mangrove zone (Woodroffe, Chappell *et al.*, 1989).

Most *intertidal flats* are smooth, seaward-dipping surfaces dissected by tidal channels. Flood tides enter the channels, overtop the banks and inundate the adjacent flats. Following slackwater at high tide, water drains back through the channels and re-exposes the tidal flats. They can be divided into three shore-parallel zones, the high-tidal flat (exposed for at least two-thirds of the tidal cycle), the mid-tidal flat (exposed for about half of the tidal cycle) and the low-tidal flat (Klein, 1985a; Lee, Park & Koh, 1985; Frey, Howard *et al.*, 1989; Alexander, Nittrouer *et al.*, 1991) and there is always a general landward trend of diminishing grain size, reflecting fall off of wave and tidal energy towards the shore (van Straaten, 1954, 1961). However, there are departures from this general scheme. In the Wash embayment of eastern England, the trend is salt marsh → high mudflats → inner sandflats → low mudflats → low sandflats (Evans, 1965) and in Corner Island southeast Australia, seagrass, acting as a baffle to the currents, traps large patches of mud within the sand zone (Zhuang & Chappell, 1991).

The nature of tidal flats depends on tidal range, wave power and the amount and type of available sediment. Tidal flats range from those that have an abundant supply of sand, normally derived from offshore, to those that are almost completely dominated by mud.

Sand-dominated tidal flats have high-tidal flats that are characterized by muddy sediments deposited from suspension. Relatively fine-grained sediments are retained here by the processes of flocculation and settling and scour lag so that homogeneous muds are intercalated between sand layers. Mid-tidal flats are characterized by mixed mud and sand reflecting roughly equal periods of suspension and bedload deposition, with bedload increasing seaward. Thus bedding changes from

lenticular through wavy and flaser bedding (Reineck & Wunderlich, 1968) as the proportion of mud decreases. This zone may develop vertically stacked laminites or tidal rhythmites of incomplete neap–spring cycles (Sect. 6.2.7) (Tessier, Montfort *et al.*, 1988; Dalrymple, Makino & Zaitlin, 1991). In the low-tidal flat zone, current-generated sand bedforms such as asymmetrical and symmetrical ripples predominate. In macrotidal areas dunes give cross-beds as well as upper flow regime planar lamination on sandflats.

Muddy tidal flats form where there is a large supply of mud, usually from neighbouring rivers and they are the main prograding element of chenier plains. Although mud is often considered to be a feature of low-energy environments and therefore only able to form in lagoons behind a protecting barrier, it can accumulate independent of wave energy when both the amount and the concentration of mud are high (100 mg l^{-1}) (McCave, 1971a). Off the Korean coast (Frey, Howard *et al.*, 1989; Alexander, Nittrouer *et al.*, 1991) high-tidal, very mottled and homogeneous muds coarsen seaward to low-tidal highly mottled homogeneous unstratified shelly silty sand. Mid-tidal flat sediments are less mottled as a result of their greater rates of accumulation.

Although tidal currents dominate tidal flats, winds from storms can influence deposition, especially on exposed coastlines. Such random features are difficult to distinguish from those due to regular tidal rise and fall. In addition, mass movement of fluid muds has been documented on tidal flats of the chenier complex of northeast South America (Wells, Prior & Coleman, 1980). The mass movement structures comprise arcuate cracks and scarps on the landward side, disrupted and collapsed areas in the centre and elongate chutes on the seaward side. Excess pore-water pressures, built up as the tide ebbs, may initiate these movements.

Tidal channels increase in width and depth seaward. In some areas, such as the Gulf of California (Thompson, 1968) or off Korea (Alexander, Nittrouer *et al.*, 1991), the channels are stable and do not migrate. In other cases they have a highly sinuous plan form with point bars on the inner depositional bank. On some sandy mudflats these migrate both laterally and downstream at rates of several tens of metres a year (Bridges & Leeder, 1976; Barwis, 1978; Thomas, Smith *et al.*, 1987). Relatively coarse sands with shell debris and mud clasts floor the channels to form a basal lag to the channel sequence. The point bar succession is composed of thin, interlaminated clay–silt and sandbeds sometimes occurring as tidal rhythmites which form lateral accretion bedding or inclined-heterolithic stratification (Thomas, Smith *et al.*, 1987). Inclined erosion surfaces divide the point bar into discrete wedges reflecting scouring during periods of high discharge (de Mowbray, 1983).

6.7.4 Barrier-island/lagoons

In these systems (Oertel, 1985) the barrier is separated from the land by a shore-parallel lagoon and is built mainly by wave-dominated beach/shoreface processes described in Section 6.2. Although they can form as prograding systems at times of static sea level, barrier-island/lagoonal systems are generally associated with low sediment supply and relative sea-level rise and can occur within deltas during transgression (Sect. 6.6.5.).

In microtidal settings barrier islands are very extensive and have few, if any, permanent inlets for the exchange of lagoonal and open sea water (e.g. Padre Island in the Gulf of Mexico). This absence of passageways to accommodate landward-directed storm waves and wind-forced currents results in frequent inundations of the barrier, giving rise to a variety of *washover processes* (Hayes, 1967; McGowan & Scott, 1975) whose deposits have a high chance of preservation as the barrier migrates landward.

Washover channels and short-lived tidal inlets are cut by storms that erode the aeolian dune ridge which commonly caps the beach face allowing sediment-laden stormwaters to enter the lagoon (Fig. 6.73). The channels may be cut above normal sea level and hence be stranded as temporary inlets until the next storm (El-Ashry & Wanless, 1965). The deposits of washover channels and temporary inlets can form a significant portion of the depositional record of the upper beach face. At

Core Bank, South Carolina (Moslow & Heron, 1978) there are five discrete, lenticular channel units which are from 0.7 to 2.1 km wide (15–20 times greater than the width of the present channel) and fine upward. Each channel is considered to have been initiated during a storm and to have persisted for a limited period during which it migrated rapidly alongshore.

Washover fans form in the back-barrier at the mouths of washover channels and temporary inlets. Washover fans are lobe-shaped sand units formed as a result of sheet flow during storms (Hayes, 1967; Andrews, 1970). The thin, sheet-like sandbeds have planar erosive bases lined with a shell-rich lag and are usually dominated by well-defined, parallel laminae which are inversely graded in a similar way to swash-backwash laminae (Leatherman, Williams & Fisher, 1977). Slight discordances and thin, top surface, wind-winnowed lags of coarse sand often subdivide apparently uniform thicknesses of parallel laminated sand into 1–10-cm-thick depositional units (Schwartz, 1975, 1982b; Leatherman, Williams & Fisher, 1977). Subordinate structures include antidune lamination, current ripple lamination, and landward-dipping foresets which are particularly common at the distal margin of the fan where it enters the standing water of the lagoon (R.K. Schwartz, 1975, 1982; Fig. 6.74). Shortly after deposition, washover sands may be bioturbated while the sediment is still moist. Rootlets may also disrupt internal structures between storms. Many washover fans are composite bodies, either because coexisting fans overlap and merge or, more commonly, because a major breach through the barrier island is re-opened by later storms (McGowen & Scott, 1975). Washover fans associated with temporary inlets may evolve into flood-tidal deltas which spread as the inlet migrates (Moslow & Heron, 1978: Sect. 6.7.2).

Lagoons are areas of shallow water protected from the sea by barrier islands. Water salinities fluctuate, and are frequently abnormal, the salinity depending on the degree of communication with the open sea, the amount of river input and tidal range. They have many similarities with estuaries, their definition and facies overlapping in many respects. Nevertheless, they tend to be parallel rather than perpendicular to the shore (Shepard & Moore, 1960), to be wave- rather than tide-dominated, and to lack significant fluvial input (Sect. 6.7.5). In arid and semi-arid areas, hypersaline conditions occur,

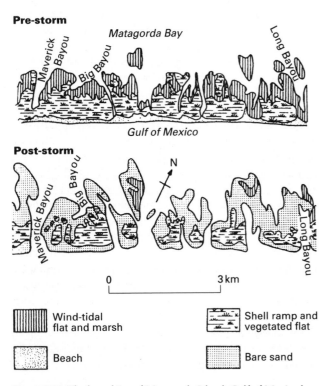

Figure 6.73 The breaching of Matagorda Island, Gulf of Mexico by Hurricane Carla, illustrating the development of numerous washover channels; many of which have re-opened previous partially plugged channels (after McGowen & Scott, 1975).

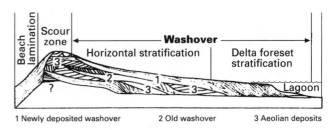

Figure 6.74 Sedimentary structures in washover fan sands (after R.K. Schwartz, 1975, 1982).

whereas in temperate and humid settings brackish water may prevail as rivers enter the lagoon from the land. In both cases, salinity may fluctuate dramatically because of increased input of fresh water during wet periods and of marine waters during storms. The diversity and abundance of faunas varies with salinity both between and within lagoons. Hypersaline conditions tend to eliminate most organisms, except some algae (stromatolites). Reduced salinities result in prolific but low-diversity assemblages whereas normal salinity, as is found in the vicinity of tidal inlets, favours high diversity.

In microtidal coasts, lagoons accumulate fine-grained terrigenous sediment deposited from suspension and/or carbonate sediment. Minor depositional systems such as washover fans, flood-tidal deltas, and bay deltas extend the range of lagoonal facies. The clays and fine silts deposited in quiet, open water areas may be finely laminated, but more commonly are structureless due to bioturbation. In *humid and temperate lagoons*, the muds are often rich in organic matter, including plant debris washed in by rivers. Water permanently covers the lagoons and wind-waves produce thin, coarse silt laminae and symmetrical ripple forms intercalated with the mud–silts. Small-scale bayhead deltas frequently occur at the landward margin of humid/temperate lagoons and produce small-scale, facies sequences which most commonly resemble those of fluvial-dominated deltas (Donaldson, Martin & Kanes, 1970; see Sect. 6.6.5). In *semi-arid lagoons*, the fine-grained sediments often have a low organic content with evidence of periodic desiccation. Laguna Madre, behind Padre Island, Gulf of Mexico, has extensive back-barrier mudflats which are exposed between major storms and subjected to wind action which produces mudcracking, deflation and dunes formed of aggregated clay pellets (Fisk, 1959; Rusnak, 1960; J.A. Miller, 1975). Extensive areas of some lagoons, especially where tidal range is high, may comprise intertidal and supratidal flats dissected by a network of tidal creeks (Phleger, 1965; Frey & Howard, 1969; Ashley, 1988).

Tidal inlets (Fig. 6.75) are developed extensively in barrier-island/lagoonal systems in high microtidal and mesotidal settings, because large volumes of water (the tidal prism) flood in and out of the lagoons and on to the tidal flats (Moslow & Tye, 1985). Even low microtidal coastlines, with tidal ranges of only 50 cm, have occasional tidal inlets (Siringan & Anderson, 1993). Inlets subdivide barrier islands into segments several kilometres or tens of kilometres in length. Each segment has a 'drumstick' shape in which the thicker portion is composed of beach spits that migrate through longshore drift (Hayes & Kana, 1976) into the tidal inlets. As tidal currents emerge from the inlets expansion and deceleration of flow results in the formation of ebb- and flood-tidal deltas.

The type of tidal inlet varies according to the intensities of tidal and wave processes. Where tidal processes are strong and wave power at the shore is low, *tide-dominated tidal inlets*, composed of deep, ebb-dominated narrow channels flanked by channel margin bars, extend seawards as ebb-tidal deltas

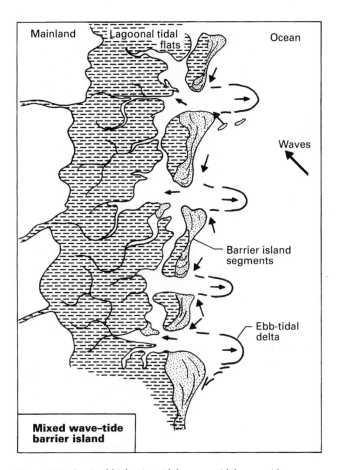

Figure 6.75 A mixed high microtidal or mesotidal wave–tide-influenced barrier-island lagoonal system with drumstick-shaped barrier islands separated by numerous tidal inlets (after Hayes, 1979).

(Figs 6.76 & 6.77). These deep inlets migrate slowly or remain fixed if they are stabilized in consolidated clay substrates or in topographic hollows. As a result, they produce lenticular, elongate sandbodies which are isolated in shoreface-sands of the barrier islands. The alongshore extent of these channel bodies is limited and their maximum elongation is normal to the shoreline. Where wave power at the shoreline is high compared with levels of tidal influence, *wave-dominated tidal inlets* occur, characterized by shallow, generally flood-dominated channels with well-developed flood-tidal deltas but weakly developed ebb-tidal deltas (Fig. 6.76). These inlets are symmetrical if wave approach is normal to the shoreline but asymmetrical if waves approach the shoreline obliquely and generate longshore currents. Wave-dominated inlets, particularly those that overlap, migrate alongshore. Migration is generally in the direction of longshore drift. Because tidal currents hinder longshore sediment transport, deposition occurs on the updrift margin of the inlet and erosion occurs on the downdrift margin (Hoyt & Henry, 1967; Kumar & Sanders, 1974). Inlets can migrate in the opposite direction

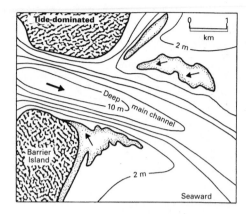 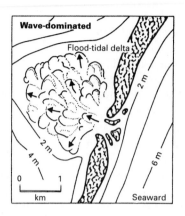

Figure 6.76 Tide-dominated and wave-dominated tidal inlets in the Georgia embayment (after Hubbard, Oertel & Nummedal, 1979).

if tidal currents scour the updrift margin of the inlet (Reddering, 1983). As inlets migrate, earlier barrier sands are eroded and sediment accretes on the depositional side of the inlet. Migration rates can be extremely high. For example, Fire Island inlet on Long Island, New York has migrated 8 km in 115 years at an average rate of nearly 70 m year^{-1} (Kumar & Sanders, 1974). Migrating inlets produce tabular sandbodies many times more extensive alongshore than the width of the active inlet (Fig. 6.78). Thus tidal inlet deposits may dominate the depositional record of a barrier island and, since the inlets scour well below sea level, their deposits have a high preservation potential.

Flood-tidal deltas are important sites of sedimentation, particularly if the tidal inlets are wave dominated (Hubbard, Oertel & Nummedal, 1979). They form a significant portion of lagoonal facies, especially if they migrate laterally with the tidal inlet. They first build into the lagoon as a series of overlapping fans or spillover lobes as, for example, in Chatham Harbour, Massachusetts, where two coalescing lobes are covered by straight and sinuous-crested dunes which are predominantly flood-orientated (Hine, 1975). With time, tidal current flow

is concentrated into channels, and mature flood-tidal deltas comprise a ramp dissected by flood-tidal channels and fringed by a series of ebb-produced spits and shields (Morton & Donaldson, 1973; Hayes & Kana, 1976). The deposits of flood-tidal deltas are dominated by sets of landward-directed planar and trough cross-bedding. Intercalated sets of ebb-orientated cross-bedding occur towards the top of the sequence.

Ebb-tidal deltas form at the seaward mouths of inlets, particularly where tide dominated (Hubbard, Oertel & Nummedal, 1979). Longshore currents are often the principal supplier of sediment (Finley, 1978). Ebb-tidal deltas conform to a general

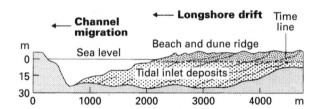

Figure 6.78 Laterally migrating tidal inlet in a mixed wave–tide influenced barrier island filled by an erosively based sandbody (after Hoyt & Henry, 1967).

pattern in which the central area is dominated by tidal channels and the flanks by wave-produced swash bars (Fig. 6.77). In the channels, ebb and flood currents dominate separate paths, with ebb currents concentrated in a deep, central channel and flood currents occupying marginal channels. Bedform type and orientation are highly variable across ebb-tidal deltas and the deposits of the systems are complex (Hine, 1975). If the inlets remain fixed in position, ebb-tidal deltas form lobe-shaped sandbodies at the inlet mouth, but if the inlet migrates laterally, they form broader sand accumulations that interfinger with the shoreface facies and may be preserved just below the erosion surface that records the base of the inlet (Fig. 6.78).

6.7.5 Estuaries

An estuary has traditionally been defined as a semi-enclosed

Figure 6.77 Sediment transport in an ebb-tidal delta in which the central area is dominated by tidal currents and the lateral margin by wave processes (after Hine, 1979).

coastal body of water which has a free connection with the open sea and within which 'sea water is measurably diluted with fresh water derived from the land' (Pritchard, 1967) (Fig. 6.79a). Within such an estuary the interaction of the tidal prism and freshwater discharge governs the circulation pattern. Where tidal range is low, less-dense fresh water extends over a saltwater wedge as a distinct layer and the water column is vertically stratified. Where tidal range is high, turbulent mixing is more pronounced and the water column becomes homogeneous both vertically and laterally.

However, most geologists consider that other features are more important than water salinity in defining an estuary. Some emphasize tides (Clifton, 1982; Nichols & Biggs, 1985; J.R.L. Allen, 1993). Dalrymple, Zaitlin and Boyd (1992) argue that it is not just the presence of tides, but a variety of marine processes, including waves, and that estuaries not only receive water from both sea and land, but are fed by contrasting sediment from both the river and the adjacent shelf. Thus they define an estuary as 'the seaward portion of a drowned valley system which receives a sediment from both fluvial and marine sources and which contains facies influenced by tide, wave and fluvial processes. The estuary is considered to extend from the landward limit of tidal facies at its head to the seaward limit of coastal facies at its mouth' (Fig. 6.79b). Estuaries are further differentiated from deltas by having a net *landward* movement of sediment in their seaward part.

All estuaries have a threefold zonal pattern: (i) a high-energy *outer zone* dominated by marine processes (waves and/or tidal currents) where sediment is derived from the sea and net bedload transport is landward; (ii) a relatively low energy *central zone* where marine energy (dominantly tidal currents) equals that of fluvial currents and there is net convergence of relatively fine-grained sediment; and (iii) a river-dominated, but marine-influenced, *inner zone* where net sediment transport is seaward (Dalrymple, Zaitlin & Boyd, 1992).

Estuaries vary according to whether tides dominate over waves, or the reverse (Fig. 6.80) and they can be grouped into one of two types, each with its own tripartite facies pattern (Dalrymple, Zaitlin & Boyd, 1992; Roy, 1994).

1 *Wave-dominated* estuaries form a spectrum reflecting tidal range and wave regime (Roy, 1984; Reinson, 1992) from microtidal 'lagoonal' estuaries identical to the barrier-island/lagoonal systems described in Section 6.7.4, through microtidal and mesotidal 'partially closed' or 'isolated barrier' estuaries to mesotidal and low macrotidal 'open-ended' or *drowned valley estuaries* (Fig. 6.80). In addition, 'river-dominated' estuaries have a high supply of fluvial sediment, yet coastal conditions where the rapid removal of sediment prevents delta progradation (Cooper, 1993).

All wave-dominated estuaries have a common facies pattern that passes landwards from a relatively coarse-grained *marine sandbody* composed of barrier, washover, tidal inlet and tidal delta deposits, the proportion of each depending on the relative strength of waves, tides and sediment supply in the same way as was described in Section 6.7.4, a central *fine-grained facies* within which sands increase in proportion to tidal energy, and an inner bay head delta *sand and gravel facies* deposited primarily by rivers, but modified by waves and tides (Bird, 1967; Roy, 1984, 1994; Zaitlin & Schultz, 1990; G.P. Allen, 1991; Nichol, 1991; Nichols, Johnson & Peebles, 1991; Dalrymple, Zaitlin & Boyd, 1992) (Figs 6.80 & 6.81). However, substantial differences in facies develop because of variations in wave and tidal power, in strength of fluvial currents, and in abundance of clastic sediment supply either from marine or fluvial sources (Dalrymple, Zaitlin & Boyd, 1992). For example, in the macrotidal Gironde estuary of France (G.P. Allen, 1991; G.P. Allen & Posamentier, 1993) tidal sandbars formed of fluvial sediment are found landward of the muddy central basin.

2 *Tide-dominated* estuaries, where tidal-current energy exceeds wave energy at the mouth, are found in the Bay of Fundy (Dalrymple, Knight *et al.*, 1990; Dalrymple, Zaitlin & Boyd, 1992), Western Australia (Wright, Coleman & Thom, 1973), Queensland (Cook & Mayo, 1977), Northern Australia (Woodroffe, Chappell *et al.*, 1989) and the Severn (Harris & Collins, 1985) (Fig. 6.82). They too can be divided into marine-

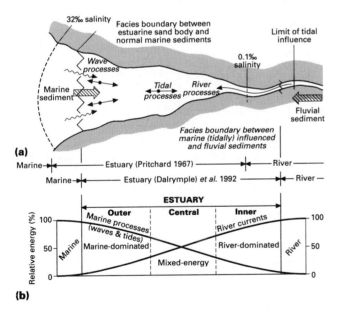

Figure 6.79 Schematic representation of (a) an estuary according to Pritchard (1967) and Dalrymple, Zaitlin and Boyd (1992) and the pattern of net, bed-material transport. Note that the facies boundary marking the landward limit of the estuary almost always lies landward of the 0.1‰ salinity value, but the facies boundary at the outer end may lie either landward or seaward of the limit of normal marine salinities (32‰). (b) Schematic distribution of the physical processes operating in estuaries, and the resulting tripartite facies zonation (from Dalrymple, Zaitlin & Boyd, 1992). The length of each zone in a particular estuary varies according to the relative magnitude of tidal and river discharges.

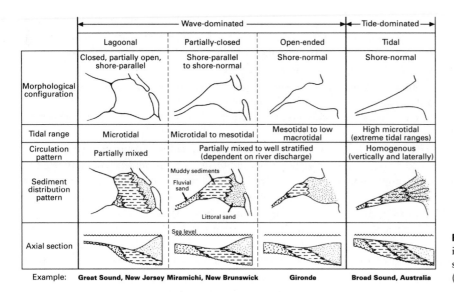

	Wave-dominated			Tide-dominated
	Lagoonal	Partially-closed	Open-ended	Tidal
Morphological configuration	Closed, partially open, shore-parallel	Shore-parallel to shore-normal	Shore-normal	Shore-normal
Tidal range	Microtidal	Microtidal to mesotidal	Mesotidal to low macrotidal	High microtidal (extreme tidal ranges)
Circulation pattern	Partially mixed	Partially mixed to well stratified (dependent on river discharge)		Homogenous (vertically and laterally)
Sediment distribution pattern		Muddy sediments / Fluvial sand / Littoral sand		
Axial section		Sea level		
Example:	Great Sound, New Jersey	Miramichi, New Brunswick	Gironde	Broad Sound, Australia

Figure 6.80 Classification of estuaries illustrating morphological, oceanographic and sedimentological characteristics of each type (from Reinson, 1992).

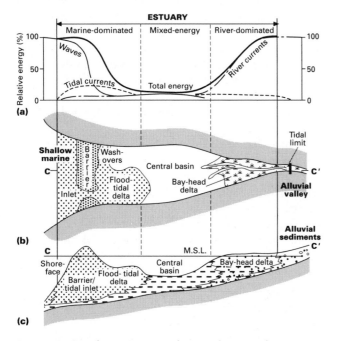

Figure 6.81 Distribution in a wave-dominated estuary of (a) energy types, (b) morphological components and (c) sedimentary facies. The barrier sand plug is shown attached to the headland, but on low-gradient coasts may not be connected and may be separated from the mainland by a lagoon. The section in (c) represents the onset of estuarine filling following a transgression (from Dalrymple, Zaitlin & Boyd, 1992). The size of the facies bodies in a particular estuary varies according to the relative input of marine or fluvial sediment.

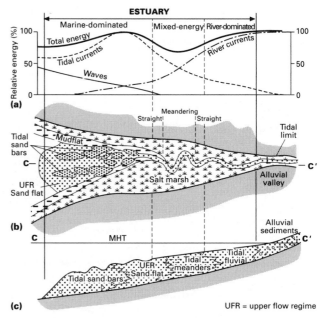

Figure 6.82 Distribution in a tide-dominated estuary of (a) energy types, (b) morphological components and (c) sedimentary facies. The section in (c) is taken along the axis of the channel and does not show the marginal mudflat and saltmarsh facies. It illustrates the onset of progradation following transgression (from Dalrymple, Zaitlin & Boyd, 1992) (cf. Fig 7.8).

dominated, mixed energy and river-dominated energy zones, with minimum energy in the centre, the whole estuary being funnel-shaped. The outer marine-dominated zone is considerably larger than in wave-dominated estuaries and tidal energy increases landward. Here a central axis of tidal sandridges is

flanked by muddy intertidal to supratidal flats and marshes that run the length of the estuary. Within the axial zone, sands are coarsest near the mouth where elongate sandridges separate ebb-dominant and flood-dominant channels (e.g. Dalrymple, Knight *et al.*, 1990). These pass landward into finer sandflats with upper flow regime planar lamination. The central, low-energy zone, shows a straight to meandering to straight progression of channel sinuosity and has the finest grain sizes. This passes into the fluvial-dominated zone, landward of which fluvial sands and gravels occur.

6.7.6 Coastal sequences

Because siliciclastic coastlines are particularly sensitive to relative sea-level changes, vertical facies changes are both frequent and complex. Facies sequences fall into three basic, essentially autocyclic styles: progradational, transgressive and aggradational (Davis & Clifton, 1987). Aggradational sequences are rare, and will not be discussed further. In addition, more dynamic, allocyclic stratigraphic models need to be considered.

PROGRADATIONAL SEQUENCES

Progradation of non-deltaic coasts occurs when sediment supply

exceeds the rate of creation of accommodation space. It is most common when relative sea level falls, but it may also occur during stable or even rising sea level if the rate of sediment supply exceeds the rate of dispersal by waves and tides (Davis & Clifton, 1987). Most progradational coasts are beach-ridge strandplains exemplified by the Costa de Nayarit (Fig. 6.68) and the east–northeast Brazilian coast (Sect. 6.7.1). They also occur as regressive tidal flats (e.g. present German Bight; Fig. 6.83), as barrier-island/lagoonal systems (Bernard, Le Blanc & Major, 1962; Fitzgerald, Baldwin *et al.*, 1992; Fig. 6.84), and within estuaries.

Vertical shoreline facies sequences have been constructed from profiles of modern prograding beaches (Fig. 6.85). The classical sequence of sedimentary structures and facies for a high-energy sandy shoreface to foreshore system has been predicted from the non-barred coast of Oregon and California (Clifton, Hunter & Phillips, 1971; Howard & Reineck, 1981) (Fig. 6.85A,B) where wave and current ripple lamination passes upwards into cross-bedding and then planar bedding, all reflecting a high

Prograding tide-dominated coast

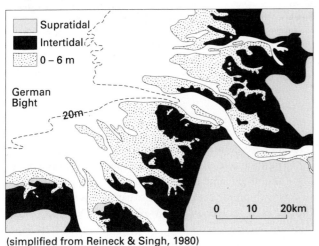

(simplified from Reineck & Singh, 1980)

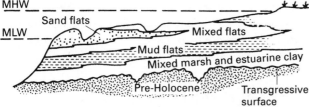

Figure 6.83 Map and cross-section through the tide-dominated German Bight showing a well-preserved transgressive sequence overlain by a currently prograding sequence (from Davis & Clifton, 1987).

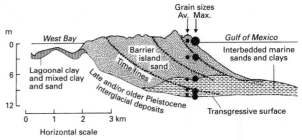

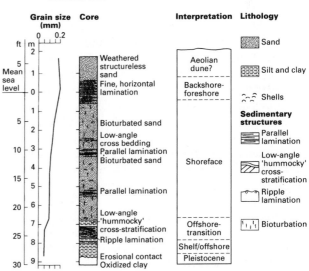

Figure 6.84 Sandbody and prograding coarsening-upward sequence of Galveston Island, Texas (after Bernard, LeBlanc & Major, 1962; McCubbin, 1982). Notice that the underlying transgressive sequence is not preserved.

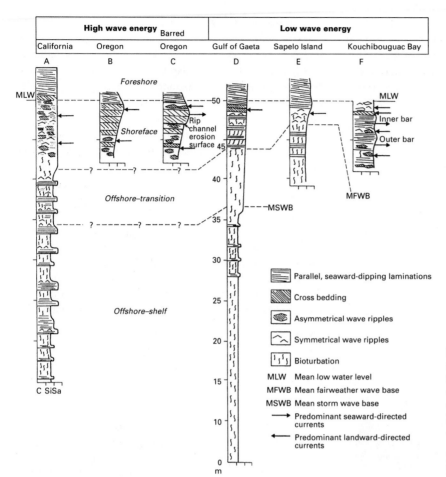

Figure 6.85 Comparative sections of coarsening-upwards sequences formed on modern wave-dominated shorelines reflecting different degrees of wave energy and existence or absence of shoreface bars (based on Reineck & Singh, 1971, 1973; Clifton, Hunter & Phillips, 1971; Howard, Frey & Reineck, 1972; Davidson-Arnott & Greenwood, 1974; Hunter, Clifton & Phillips, 1979; Howard & Reineck, 1981).

degree of physical energy. These vertical sequences contrast with that of the barred coastline of Oregon (Hunter, Clifton & Phillips, 1979) where a series of obliquely orientated nearshore bars is separated from the shore by a trough that curves seaward into a rip channel. Thus the vertical sequence predicted from progradation of such a shoreline would be interrupted by a subhorizontal erosion surface with a pebble lag (Fig. 6.85C) in contrast to a non-barred system where the grain size increases upwards with no interruption.

Low-energy shorelines occur on the margins of lakes and restricted seas, and on the leeward side of continents. Examples of non-barred shorelines include the Gulf of Gaeta in the Mediterranean (Reineck & Singh, 1971, 1973) (Fig. 6.85D) and Sapelo Island in Georgia (Howard, Frey & Reineck, 1972) (Fig. 6.85E) as well as on several beaches on the east coast of Australia (Short, 1984). A barred example is in Kouchibouguac Bay in New Brunswick (Davidson-Arnott & Greenwood, 1974, 1976; Greenwood & Mittler, 1985) (Fig. 6.85F). Such coasts tend to have lower gradients and finer grain sizes. Energy levels are lower so that cross-bedding is generally absent. However, a ridge and runnel system may develop in the more dissipative

shorelines. Bioturbation is much more intense, its nature depending on position across the shoreline profile. It is most commonly of the *Skolithos* ichnofacies, dominated by traces of suspension feeders with deeply penetrating more or less permanent domiciles, offering protection from wave energy. On Sapelo Island (Howard, Frey & Reineck, 1972) irregular burrows of *Calianassa atlantica* in the shoreface give way to vertical burrows of *Calianassa major* in the foreshore. In the backshore *Psilonichnus* ichnofacies predominated, dominated by amphibious crabs of the Ocypodidae family that excavate J-, U- or Y-shaped burrows. The depth of some of these burrows means that they penetrate depositional facies which reflect conditions quite different from those at the present-day surface. There is thus a displacement of litho- and ichnofacies.

TRANSGRESSIVE SEQUENCES

Non-deltaic transgressive coasts have a low supply of sediment and are therefore dominated by erosion (Fig. 6.16) (Demarest & Kraft, 1987; Nummedal & Swift, 1987; Reinson, 1992). They are predominantly barrier-island/lagoonal systems or estuaries.

(a) Shoreface retreat

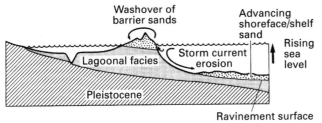

(b) In-place drowning

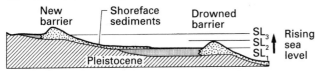

Figure 6.86 Mechanisms of landward barrier migration during a transgression by (a) shoreface retreat, (b) in-place drowning (after Fischer, 1961; Swift, 1975; Rampino & Sanders, 1980).

Barrier-island/lagoonal systems migrate landward by one of two mechanisms, by *in-place drowning* or by *shoreface retreat*.

During in-place drowning the barrier remains in place as sea level rises until the breaker zone reaches the top of the barrier. The breaker zone then jumps landward to the inner margin of the lagoon, thus drowning the barrier sandbody (Fig. 6.86b) which therefore has a high chance of preservation.

Shoreface retreat leads to erosion of sediment from the upper shoreface and its transport either to the lower shoreface–offshore area (shoreface bypassing) as storm-generated beds or to the lagoon as washover fans (Fig. 6.86a). As the upper shoreface or breaker-surf zone passes across the former barrier, it erodes barrier, lagoonal and washover facies deposited earlier. Lower shoreface facies therefore overlie more landward facies across a commonly planar erosion surface (ravinement surface). This surface is time-transgressive, since fluvial and coastal deposits found to landward and below it were deposited contemporaneously with marine strata similar to those found to seaward and above it (Fig. 6.86a). The surface does not represent a major time break but is integral to the transgressive sequence. The depth of erosion at the shoreface depends on the rate of relative sea-level rise and on wave climate. If the rise is rapid a comparatively complete transgressive sequence may be preserved, whereas with a slow relative sea-level rise erosion is more pronounced and the sequence will be truncated (Fischer, 1961; Heward, 1981) (Fig. 6.87a,b). However, greater thicknesses may be preserved if there has been some degree of incision into the earlier formed sediment. The amount of incision varies (Ashley & Sheridan, 1994). The smallest valleys are tidal inlets in barrier islands that may scour deeply into the underlying lagoonal sediment as they migrate laterally (Oertel, Henry & Foyle, 1991) (Fig. 6.87c). Larger ones include a range of incised valleys that are largely filled by estuarine and/or fluvial sediments.

(a) Shoreface erosion

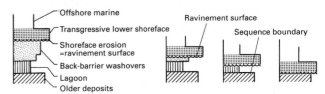

(b) Moderate tidal inlet reworking and shoreface erosion

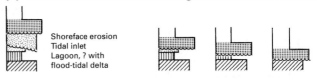

(c) Deeper tidal inlet reworking and shoreface erosion

INCREASING SHOREFACE EROSION

Figure 6.87 Sequences preserved as a result of landward migration with different degrees of shoreface erosion; (a) involves a wave-dominated barrier island, (b) and (c) involve a mixed wave–tidal barrier island with tidal inlets. Note that the sequences can only be distinguished where low levels of shoreface erosion operated during conditions of rapid relative sea-level rise (after Heward, 1981).

INCISED-VALLEY SEQUENCES

An *incised-valley system* consists of an incised valley and its fill. It has been defined (Zaitlin, Dalrymple & Boyd, 1994) as a 'fluvially-eroded, elongate topographic low that is typically larger than a single channel form, and is characterized by an abrupt seaward shift of depositional facies across a regionally mappable sequence boundary at its base. The fill typically begins to accumulate during the next base-level rise, and may contain deposits of the following highstand and subsequent sea-level cycles'.

Incised valleys pass landward into non-incised, fluvial-channel systems that feed the valleys (Fig. 6.88). They may be hundreds of kilometres long, tens of kilometres wide and hundreds of metres deep. Two varieties have been identified; piedmont incised-valley systems and coastal-plain incised-valley systems (Fig. 6.89). They provide the most complete (and at times only) evidence of lowstand to transgressive shelf-slope and/or shallow ramp, marine depositional settings. They are therefore particularly significant in determining the sequence stratigraphic history of a coast (Demarest & Kraft, 1987; Dalrymple, Zaitlin & Boyd, 1992; Dalrymple, Boyd & Zaitlin, 1994). Fluvial incision is

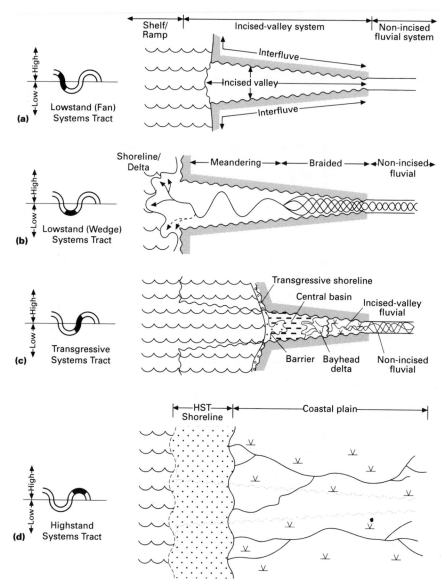

Figure 6.88 Idealized plan view of a simple, piedmont incised-valley system showing its evolution over one complete sea-level cycle (sea-level fall to subsequent highstand). (a) Lowstand (fan) time showing the incised-valley system passing headward into a non-incised fluvial-channel system. The junction between the two is the knickpoint. (b) Lowstand (wedge) showing lowstand delta at the mouth of the incised valley, and the beginning of fluvial deposition throughout the incised-valley system. (c) Transgressive systems tract time showing development of a tripartite, wave-dominated estuarine system within the incised valley. (d) Highstand time with a progradational shoreface and alluvial plain that extends beyond the margins of the buried incised system (from Zaitlin, Dalrymple & Boyd, 1994).

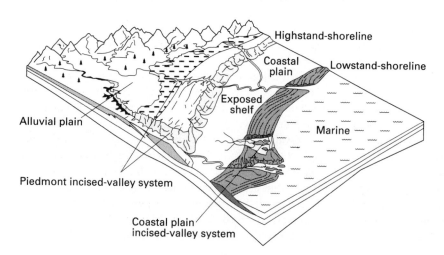

Figure 6.89 Schematic view of a coastal zone showing the distinction between piedmont incised-valley systems which have their headwaters in a mountainous hinterland and cross a 'fall line' where there is a significant reduction in gradient and coastal plain incised-valley systems that are confined to low-gradient coastal plains (from Zaitlin, Dalrymple & Boyd, 1994).

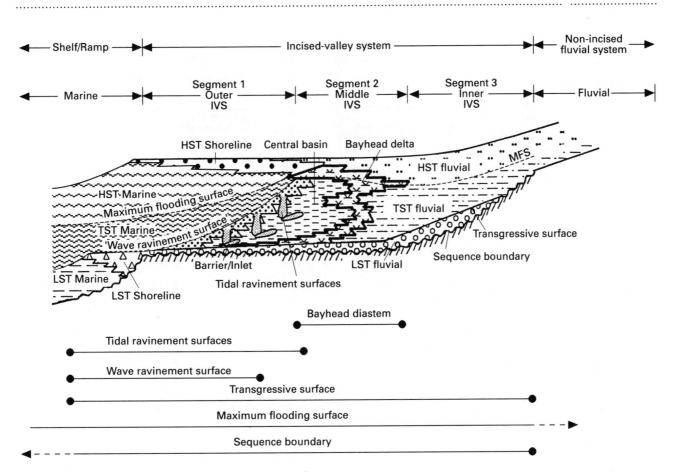

Figure 6.90 Idealized longitudinal section of a simple incised-valley system showing depositional environments, systems tracts and key stratigraphic surfaces (from Zaitlin, Dalrymple & Boyd, 1994).

promoted primarily by: (i) a relative fall in sea level caused either by a eustatic sea-level fall or tectonic uplift; and (ii) an increase in river discharge caused either by climatic change or stream capture.

Incised-valley systems have been divided into three segments (Zaitlin, Dalrymple & Boyd, 1994) (Fig. 6.90).

The *outer incised valley* (segment 1) extends from the lowstand mouth of the incised valley to the point, landward, where the shoreline stabilizes at the beginning of highstand progradation. As base level is lowered, the valley is cut by fluvial incision and the sequence boundary is formed. Sediment is bypassed to the mouth of the valley where it is deposited as either a lowstand delta and/or prograding shoreline. Fluvial deposition may also occur during this phase, and continue as sea level begins to rise so that the transgressive surface may lie within the fluvial deposits. These may be braided, meandering or anastomosing, depending on sediment supply, grain size, discharge, valley gradient and rate of transgression (cf. Fig. 3.7). As sea level rises, there is likely to be a fining upwards as the fluvial system changes from sandy braided to mixed sand/mud meandering (e.g. Fisk, 1944). Thickness of the fluvial sediments mainly depends on the ratio between rate of fluvial sediment input and rate of sea-level rise, vertical aggradation occurring when the two match each other.

As the transgression proceeds, estuarine conditions are established at the seaward end of the segment and migrate landward. In wave-dominated estuaries the full succession will be bayhead delta → central basin → estuarine barrier → shoreface → nearshore → open-marine, with possibly a tidal ravinement surface below and a wave ravinement surface above the estuarine barrier complex.

The *middle incised valley* (segment 2) extends landward to the limit of marine/estuarine conditions at the time of maximum flooding, and therefore corresponds to the area occupied by the drowned-valley estuary at the end of transgression. This segment essentially corresponds to the estuary as defined in Section 6.7.5 with its own tripartite *estuarine fill* of sand–mud–sand (Rahmani, 1988), controlled by process (Fig. 6.81). Successions vary laterally (Fig. 6.91). Near its seaward end, the succession is similar to that of segment 1 except that open marine conditions do not penetrate this segment. The barrier sediments are therefore overlain instead by highstand fluvial deposits. In the middle portion, barrier sands are absent and central basin deposits coarsen upward into progradational bayhead delta and

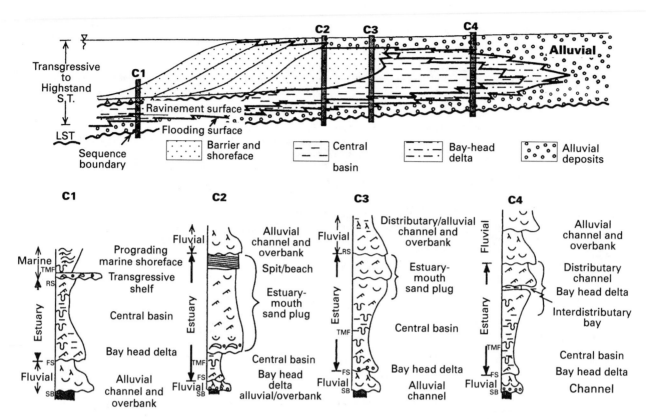

Figure 6.91 Schematic section along the axis of a wave-dominated estuary showing the distribution of lithofacies resulting from an early sea-level fall followed by transgression of the estuary, estuarine filling and shoreface progradation. The amount of transgressive sequence preserved depends on the relative rates of sea-level rise and headward translation of the shoreface (from Dalrymple, Zaitlin & Boyd, 1992).

fluvial sediments of the succeeding highstand. At the headward end, central basin deposits are absent. The landward limit of the segment is taken to be the limit of detectable marine influence, tidal features or brackish water traces (i.e. the inner end of the estuary as defined by Dalrymple, Zaitlin & Boyd, 1992).

The *innermost segment* of the incised valley (segment 3) extends from the landward limit of marine/estuarine influence to the landward limit of incision where relative sea level no longer controls fluvial style. It may extend tens to hundreds of kilometres above the limit of marine/estuarine influence (Shanley, McCabe & Hettinger, 1992; Schumm, 1993). Fluvial style may change systematically due to changes in base level and the rate of creation of accommodation space. Lowstand deposits would be relatively thin as the fluvial system would be dominantly erosional, or have acted as a bypass zone. Late lowstand to early transgressive deposits would be relatively coarse and dominated by coalescing channel sandstones that fine upwards to channel sandstone bodies isolated in and interbedded with more abundant overbank fines (Törnqvist, 1993) (cf. Sect. 3.6.9; Fig. 3.61).

The incised-valley sequences are both bounded by and divided by stratigraphically significant surfaces. At the base is the *sequence boundary* formed by a combination of fluvial incision

that erodes the valley and subaerial exposure of the interfluves. Above it is the *transgressive surface*, defined as the flooding surface separating the progradational or aggradational lowstand systems tracts from the retrogradational transgressive systems tract (Van Wagoner, Posamentier *et al.*, 1988). This surface does not necessarily separate distinctive facies and may lie within the basal fluvial deposits. A more easily recognized surface is the *initial flooding surface* commonly seen as an estuarine–fluvial contact and this may be the transgressive surface (G.P. Allen & Posamentier, 1993), but is more generally just a facies boundary of limited chronostratigraphic significance. Additional flooding surfaces may also occur. *Tidal ravinement surfaces* may be produced by a number of many different types of tidal channel. They are significant because although they are confined to the valley system, rarely being found outside it and therefore of relatively little use for regional correlation, they have a high chance of preservation. Of more regional importance laterally is the *wave ravinement surface* developed as the shoreface migrates landward and erodes not only pre-existing barrier sediments but also estuarine and fluvial deposits if wave base is sufficiently deep. However, it cannot penetrate beyond segment 1 and is absent from tide-dominated settings. The *maximum flooding surface* (MFS) corresponds to the time of maximum

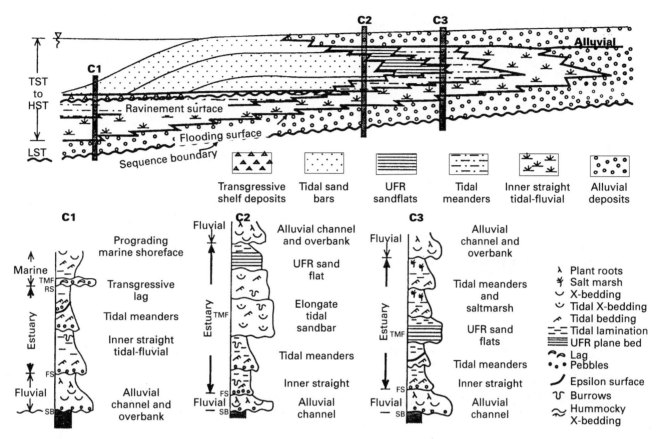

Figure 6.92 Schematic section along the axis of a tide-dominated estuary showing the distribution of lithofacies resulting from an early sea-level fall followed by transgression of the estuary, estuarine filling and progradation of sandbars and tidal flats. The amount of transgressive sequence preserved depends on the relative rates of sea-level rise and headward translation of the thalweg or the tidal channels (from Dalrymple, Zaitlin & Boyd, 1992).

transgression. Its physical expression differs markedly between the three segments. In segment 1 it occurs within marine shales and is commonly a condensed section with abundant biogenic carbonate, phosphate and above-normal levels of radioactive material (Loutit, Hardenbol *et al.*, 1988). It passes landward through the sands of the barrier and into the centre of estuarine deposits in segment 2 and into fluvial deposits in segment 3, where it may be associated with those fluvial deposits that have the most distal character and are the finest grained.

This idealized simple incised-valley model can be modified in many ways, with considerable variations in the proportions of facies deposited and removed by the various erosional processes. Changes reflect the amount, rate and continuity of sea-level change, the relative and absolute abundance of fluvial and/or marine sediment supply and the relative intensity of waves, tides and river currents, the alluvial plain gradient and sediment grain size, and whether bedrock controls the shape of the valley (Ashley & Sheridan, 1994; Nichol, Boyd & Penland, 1994).

If tidal range is particularly high, as in tide-dominated estuaries (Sect. 6.7.5), strong tidal currents promote a funnel shape, the estuary deepening, widening and migrating up-valley as trans-

gression takes place (Fig. 6.92). Erosion by tidal currents of adjacent and underlying sediments is substantial. As a result the tidal ravinement surface may bound both sides and base of the estuary. As transgression proceeds, major estuarine sandridges may be drowned to become shelf sandbodies (Stride, Belderson *et al.*, 1982; Yang & Sun, 1988) (Sect. 7.3.4) (Fig. 7.14).

6.7.7 Ancient non-deltaic siliciclastic coasts

The principal criterion for recognizing ancient linear siliciclastic shorelines is the association of coastal-plain sediments such as those of lagoons and marshes with those influenced by waves, storms and tides, together with relatively mature sandstone compositions, indicating derivation from the sea. They can be divided into: (i) progradational sequences formed either during relative sea-level highstands, when sediment flux exceeds accommodation space, *normal regression*, or when relative sea level falls and the shoreline migrates towards the basin independent of sediment flux, *forced regression* (Posamentier, Allen *et al.*, 1992); and (ii) transgressive sequences formed during relative sea-level rises.

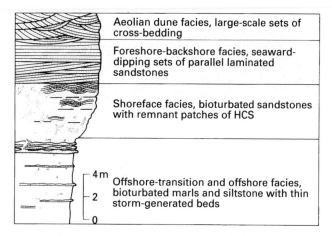

Figure 6.93 A prograding coarsening-upward storm-dominated shoreline sequence that is extensively bioturbated in the Upper Cretaceous Aren Sandstone, Spanish Pyrenees (after Ghibaudo, Mutti & Rosell, 1974).

ANCIENT PROGRADATIONAL SEQUENCES

Progradational sequences formed during normal regressions essentially coarsen upward from offshore mudstones and siltstones to nearshore sandstones. They can be divided into those dominated by storms and waves, those dominated by shoaling waves, and those dominated by longshore and/or rip currents.

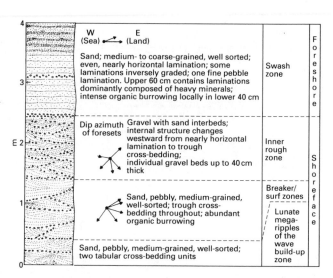

Figure 6.94 A prograding relatively coarse-grained high wave energy shoaling-wave shoreline sequence from the Quaternary of California (after Clifton, Hunter & Phillips, 1971).

Storm-dominated sequences are characterized by an offshore facies containing 1–20-cm-thick beds of storm-generated sandstones. These are sharp or erosively based and characterized by parallel lamination grading up into current ripple lamination (Sect. 7.7.2). Palaeocurrents derived from the sole marks and

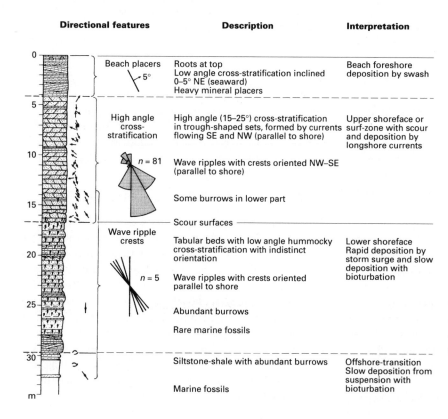

Figure 6.95 A prograding shoreline sequence in which the offshore-transition and lower shoreface zones are dominated by storm processes (34–17 m) while the upper shoreface is dominated by longshore currents (17–4 m); Cretaceous Gallup Sandstone, New Mexico (after McCubbin, 1982).

internal structures indicate seaward-directed flow (Leckie & Krystinick, 1989). The overlying offshore-transition and shore-face facies are dominated by hummocky cross-stratification (HCS) (Duke, 1990; Cheel & Leckie, 1993) (Sect. 7.7.2) and shore-normal gutter casts may be very common in some examples. HCS can grade upward within a sandbed via flat lamination into wave ripple lamination reflecting the waning storm, but the upper parts of storm-generated beds deposited below fairweather wave base are commonly bioturbated. Alternatively, successive storm-generated beds may be amalgamated. The extent of bioturbation and thus preservation of storm-generated structures records the magnitude and frequency of storms, and the overall sedimentation rate. In the Cretaceous Aren Sandstone in the Spanish Pyrenees bioturbation is intense and storm-generated structures are preserved sporadically amidst thoroughly bioturbated homogeneous sandstones (Ghibaudo, Mutti & Rosell, 1974) (Fig. 6.93). In contrast, many successions in the Western Interior Seaway are relatively weakly bioturbated, possibly reflecting a partially dysaerobic water column.

Many storm-dominated coarsening-upwards sequences pass upwards from hummocky cross-stratification to swaley cross-stratification (e.g. Leckie & Walker, 1982) which consists of superimposed shallow scours concordantly filled by gently dipping, flaggy laminae considered to have formed as a result of storm-wave deposition above fairweather wave base.

Sequences dominated by shoaling waves show only limited evidence of storm events. Instead they are largely composed of facies reflecting predominantly onshore directed shoaling waves (Sect. 6.2.4). Some sequences are from low wave energy settings lacking major storm waves but most examples record high wave energy settings (Fig. 6.94). In these cases evidence for wind-forced or geostrophic currents is sometimes present in the lower parts of the sequence below fairweather wave base but is absent higher in the sequence due to extensive reworking by high energy, fairweather shoaling waves.

Sequences dominated by longshore currents and/or rip currents superficially resemble shoaling wave sequences but are distinguished by: (i) erosion surfaces which define the base of longshore troughs or rip-channels, that occur at the base of the upper shoreface; and (ii) an abundance of current-produced cross-lamination or cross-bedding in the mid- to upper shoreface reflecting either shore-parallel flow of longshore currents, or seaward-directed flow in rip-channels (Fig. 6.85C,F). Examples are in the Cretaceous Gallup Sandstone of New Mexico (McCubbin, 1982) (Fig. 6.95), the Miocene of California (Clifton, 1981) (Fig. 6.96) and the Albian of northwest Alberta (Rahmani & Smith, 1988).

Determination of the nature of progradational storm-, wave- and tide-influenced shorelines is not easy since it must rely on careful integration of palaeocurrent data, vertical facies analysis, and compositional petrography (cf. Sect. 6.6.6). For example, Duke, Fawcett and Brusse (1991), as a result of extremely detailed examination of the well-studied Lower Silurian Medina

Group of Ontario and New York, reinterpret it as a prograding linear shoreline rather than as a delta. The shoreline, dissected by numerous shore-normal tidal channels and estuaries, was tide dominated and wave influenced. Strong longshore drift to the north-northeast resulted from oblique wave approach from the southwest. Within the overall prograding sequence were many small-scale parasequences generated by episodic aggradation and shoaling separated by transgressions. During the periods of maximum regression sinuous tidal channels were cut orientated normal to the palaeoshoreline and then filled by sand deposited by tidal currents. The channel sands were themselves incised during the following transgression and overlain by offshore mudstones and hummocky cross-stratified sandstones.

Progradational sequences formed during forced regressions are characterized by shoreface deposits that sharply overlie the underlying strata deposited during the previous highstand. Sand-bodies may either be laterally attached to the previous deposits or isolated within finer-grained offshore sediments (e.g. Ainsworth & Pattison, 1994). Proximally they have a sharp base due to wave and tidal erosion as sea level drops. Distally they may have gradational junctions with the offshore muds. In the past many of these detached sandbodies, now interpreted as 'low-stand shoreface' deposits, were interpreted as offshore bars or barrier islands. They are further discussed in Section 7.8.4.

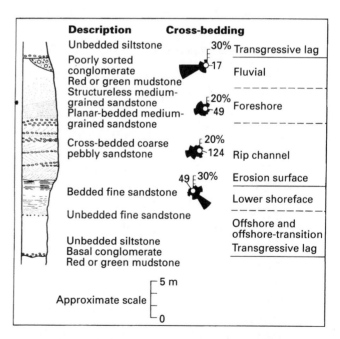

Figure 6.96 Coarsening-upward sequence produced by a prograding, barred shoreline system in the Miocene of California. Note the erosion surface in the middle of the sequence which defines the base of a rip-channel unit and the change from oblique, landward-directed southeasterly shoaling wave directions in the lower shoreface to offshore-directed westerly flow in the rip channel (after Clifton, 1981).

ANCIENT TRANSGRESSIVE SEQUENCES

Ancient transgressive sequences have been described for some time, for example from the Precambrian of Finnmark (H.D. Johnson, 1975), the Upper Silurian of Virginia, USA (Barwis & Makurath, 1978), the Lower Palaeozoic Peninsula Formation of South Africa (Hobday & Tankard, 1978) and the Cretaceous Cape Sebastian Sandstone of Oregon (Bourgeois, 1980) (Fig. 6.97). Essentially they all fine upwards. In some cases an extensive planar erosion surface, overlying coastal sediments, is directly overlain by offshore sediments, the erosion surface being covered by an extensively bioturbated surf-winnowed lag. In other cases there may be a variable thickness of sediment from a few to several hundred metres between the basal lag conglomerate and the offshore siltstones or mudstones. These sequences generally fine upward from foreshore through shoreface sandtones, often with HCS, to offshore finer siltstones and mudstones.

More recently, our understanding of transgressive sequences has been enhanced because of the increasing recognition of: (i) sequence stratigraphic concepts; (ii) tidal sediments; (iii) trace fossil assemblages reflecting brackish or stressed conditions; (iv) valley-fill sequences; and (v) that many 'offshore shelf bars' are in fact lowstand shoreface deposits.

In the Eocene of southern England (Plint, 1983, 1988a), transgressive cycles show a range of sequences that have been correlated even though they change character geographically

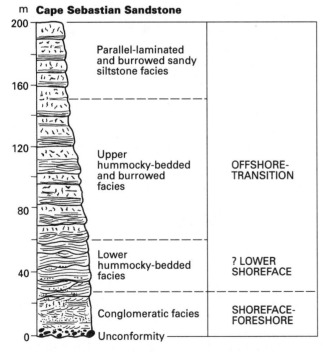

Figure 6.97 Thick, storm-wave dominated transgressive sequence from the San Sebastian Sandstone, Oregon, USA (after Bourgeois, 1980).

(Fig. 6.98). Offshore, transgressions are marked by a sharp increase in glauconite. In coastal areas estuaries are first plugged and then covered by beach sand and a pebble bed. In alluvial areas, reduction in the transport capacity of the rivers results in extensive flooding and widespread deposition of clay.

In the Pennsylvanian of the Illinois Basin the thick 'grey-shale sequences' that overlie the major coal seams had always been thought to be totally non-marine and deposited in fluvial and lacustrine environments. However, they contain substantial evidence for tidal activity such as rare herring bone cross-bedding, lenticular, wavy and flaser bedding and substantial thicknesses of rhythmites interpreted as tidal bundles (Archer & Kvale, 1993). The latter are particularly important in these mud-dominated tropical settings where tidal range is low and wave-reworking limited. They may include both parallel-laminated mudstones and parallel-laminated sandstone and mudstone couplets whose thicknesses vary in a cyclic pattern, probably related to neap–spring tidal events (Kvale & Archer, 1990). Valley-fill sequences in these Pennsylvanian rocks pass from relatively thin fluvial conglomeratic sandstones, formed in a river, through a mud-dominated quiet water 'central estuarine basin' subtidal environment, flanked by intertidal flats (some vegetated) and supratidal mires (coals) to one of higher energy subtidal to intertidal flats dissected by tidal channels (Kvale & Barnhill, 1994).

The Lower Cretaceous Woburn Sands of southern England fill a 25–30 km wide trough cut into underlying Upper Jurassic marine clays. The simplest interpretation (although see also Wonham & Elliot, 1996) is that (Johnson & Levell, 1995) (Fig. 6.99) the succession comprises a transgressive estuarine fill in a macrotidal tide-dominated estuary (cf. Figs 6.82 & 6.92) lacking any evidence for fluvial deposits. Although the sequence is marine dominated and therefore falls into the outermost division of a tide-dominated estuary (Fig. 6.82) evidence for wave or storm activity is very limited. Two rather similar sequences are superimposed on one another as a consequence either of autocyclic processes due to a shift of shoal pattern related to movement of the main tidal channels or allocyclic due to a relative sea-level change. Each sequence passes from a unit interpreted as a relatively landward portion of the marine-dominated estuary to a unit interpreted as a more seaward portion. The inner part is characterized by two distinctive laterally juxtaposed facies, a moderate to high energy, flood-dominated channel-fill sand with large-scale cross-bedding intercalated with lower-energy estuarine shoal heterolithic sands with abundant clay drapes. The outer part is composed of high energy, medium to very coarse-grained well-sorted very mature clay-free sands with carbonaceous debris and ebb-dominated palaeocurrents. Deposition was in an outer estuary open to the sea.

Trace fossil assemblages, characteristic of estuarine conditions, are extremely variable, reflecting both the lateral changes from fresh water at the landward end to normal salinities at the seaward end, and frequent fluctuations in salinity at any one place in

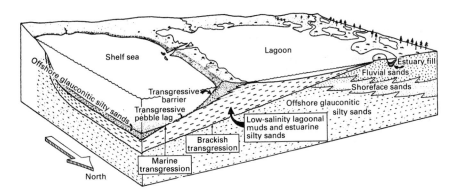

Figure 6.98 Model for 'two stage' transgression in the Bracklesham Group and upper London Clay Formation (from Plint, 1988a). A wave-built barrier sandbody developed at the beginning of the transgression, enclosing an area of sheltered lagoons, estuaries and tidal flats to landward. As sea level rose, the coastal plain was drowned by a low-energy 'brackish' transgression. The resulting accommodation caused aggradation of estuaries and permitted metres to several tens of metres of lagoonal and estuarine sediments to accumulate on the underlying lowstand erosion surface. As transgression continued, the wave-dominated barrier migrated landwards over slightly older back-barrier deposits. Wave action in the shoreface scoured the back-barrier deposits, generating a surface of 'marine' transgression, overlain by a lag gravel veneer. With continued deepening, fine-grained and often glauconite-rich sediments slowly accumulated in offshore areas.

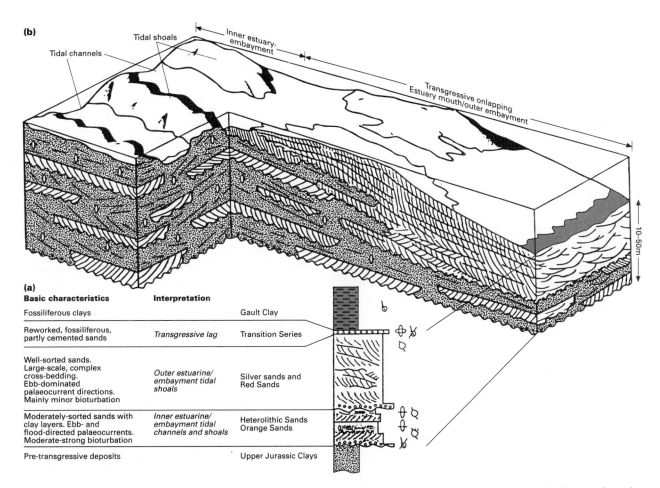

Figure 6.99 A model for a marine-dominated, tide-dominated transgressive estuarine-embayment depositional system, based on the Lower Cretaceous Woburn Sands (from H.D. Johnson & Levell, 1995). (a) Idealized vertical section showing development from the more landward to the more seaward portion. (b) Block diagram showing sandbody characteristics.

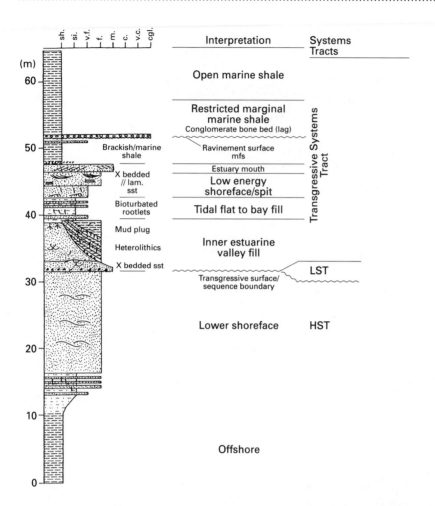

Figure 6.100 Vertical sequence, interpretation of environment, systems tracts and relative sea level for the Peace River Formation of Alberta (from Leckie & Singh, 1991).

the estuary. For example, the Viking Formation in Alberta has a trace fossil assemblage characterized by a variable and sporadic distribution of burrowing, variability in ichnogenera distribution, and dominance by simple structures of trophic generalists (Mac-Eachern & Pemberton, 1994). The suite is dominated by traces of opportunistic taxa characteristic of stressed environments, particularly those subjected to fluctuations in salinity, episodic deposition, variable aggradation rates and variability in substrate consistency. Specialized feeding and grazing structures are of secondary importance in the trace fossil suite, but record periods of fully marine, unstressed conditions. In addition, the various erosional discontinuities are marked by a substrate-controlled *Glossifungites* ichnofacies (firmground suite) dominated by vertical to subvertical dwelling structures of suspension feeding organisms such as *Diplocraterion*, *Skolithus*, *Psilonichnus*, *Arenicolites*, *Gastrochaenolites*, *Thalassinoides* and *Rhizocorallium*.

Several palaeovalley fills, some of which were once interpreted as deltaic, have now been reinterpreted as estuarine valley fills (Rahmani, 1988; Reinson, Clark & Foscolos, 1988). A typical example is the Peace River Formation of Alberta (Leckie & Singh, 1991) (Fig. 6.100) where a prograding shoreline, deposited during a highstand of relative sea level, was eroded during a

lowstand. The valley was then filled by estuarine sediment, no evidence of the fluvial episode being preserved. The lower part of the preserved fill has a brackish fauna and contains some evidence for tidal modification of meandering river channels. The bulk of the section comprises tidal sediments suggesting deposition in a relatively open ended wave-affected estuary.

Further reading

Broussard M.L. (Ed) (1975) *Deltas, Models for Exploration*, 555 pp. Houston Geol. Soc., Houston.

Carter R.W.G. & Woodroffe C.D. (Eds) (1994) *Coastal Evolution: Late Quaternary Shoreline Morphodynamics*, 517 pp. Cambridge University Press, Cambridge.

Colella A. & Prior D.B. (Eds) (1990) *Coarse-grained Deltas*, 357 pp. *Spec. Publ. int. Ass. Sediment.*, **10**. Blackwell Scientific Publications, Oxford.

Dabrio C.J., Zazo C. & Goy J.L. (Eds) (1991) *The Dynamics of Coarse-grained Deltas*, 405 pp. *Cuadernos de Geología Ibérica*, 15, 11–14, Madrid.

Dalrymple R.W., Boyd R. & Zaitlin B.A. (Eds) (1994) *Incised-valley Systems: Origin and Sedimentary Sequences*, 391 pp. *Spec. Publ. Soc. Sedim. Geol.*, **51**, Tulsa.

Davis R.A. Jr (Ed) (1985) *Coastal Sedimentary Environments*, 716 pp. Springer-Verlag, New York.

Davis R.A. Jr (Ed) (1994) *Geology of Holocene Barrier Island Systems*, 500 pp. Springer-Verlag, New York.

Davis R.A. Jr & Ethington R.L. (Eds) (1976) *Beach and Nearshore Sedimentation*, 187 pp. *Spec. Publ. Soc. econ. Paleont. Miner.*, **24**, Tulsa.

De Boer P.L., Van Gelder A. & Nio S.D. (Eds) (1988) *Tide-influenced Sedimentary Environments and Facies*, 530 pp. Reidel, Dordrecht.

Fletcher C.H. III & Wehmiller J.F. (Eds) (1992) *Quarternary Coasts of the United States: Marine and Lacustrine Systems*, 450 pp. *Spec. Publ. Soc. econ. Palaeont. Miner.*, **48**, Tulsa.

Komar P.D. (1976) *Beach Processes and Sedimentation*, 429 pp. Prentice-Hall, New Jersey.

Morgan J.P. & Shaver R.H. (Eds) (1970) *Deltaic Sedimentation Modern and Ancient*, 312 pp. *Spec. Publ. Soc. econ. Paleont. Miner.*, **15**, Tulsa.

Nemec W. & Steel R.J. (Eds) (1988) *Fan Deltas: Sedimentology and Tectonic Settings.* 444 pp. Blackie & Son, London.

Nummedal D., Pilkey O.H. & Howard J.D. (Eds) (1987) *Sea-level Fluctuation and Coastal Evolution*, 267 pp. *Spec. Publ. Soc. econ. Paleont. Miner.*, **41**, Tulsa.

Oertel E.L. & Leatherman S.P. (Eds) (1985) *Barrier Islands*, pp. 1–396 *Mar. Geol.* (Special Issue), **63**.

Shepard F.P., Phleger F.B. & Van Andel Tj.H. (Eds) (1960) *Recent Sediments, Northwest Gulf of Mexico*, 394 pp. Am. Ass. petrol. Geol., Tulsa.

Smith D.G., Reinson G.E.,. Zaitlin B.A. & Rahmani R.A. (Eds) (1991) *Clastic Tidal Sedimentology*, 387 pp. *Mem. Can. Soc. petrol. Geol.*, **16**, Calgary.

Swift D.J.P., Oertel G.F., Tillman R.W. & Thorne J.A. (Eds) (1991) *Shelf Sands and Sandstone Bodies: Geometry, Facies and Sequence Stratigraphy*, 532 pp. *Spec. Publ. int. Ass. Sediment*, **14**. Blackwell Scietific Publications, Oxford.

Whateley M.K.G. & Pickering K.T. (Eds) (1989) *Deltas: Sites and Traps for Fossil Fuels*, 360 pp. *Spec. Publ. geol. Soc. Lond.*, **41**, Bristol.

Shallow clastic seas

7

H.D. Johnson & C.T. Baldwin

....

7.1 Introduction

7.1.1 Definition

Shallow (shelf) seas occur as rims around continents and as inland seas within continental areas (Fig. 7.1). They extend from coasts dominated by shoreline processes (Sect. 6.2) to slope and bathyal environments dominated by deeper water and oceanic processes (Sect. 10.2). The main features of this environment are water depths of less than 200 m, gentle gradients (c. 1–0.1°), normal marine salinities and a wide range of physical processes (e.g. tidal currents, waves, storm-generated currents and oceanic currents). Shelf seas display a distinctive profile, which is an equilibrium surface related to wave base and characterized by a balance between deposition and erosion. This profile defines the main subenvironments of shallow seas, which are mainly related to water depth and wave energy, particularly the depth of fairweather and storm wave base (see Fig. 6.6).

Modern and ancient shallow seas occur in various shapes and sizes, mainly as a result of their geographic and tectonic settings, but two main morphological types are recognized: (i) *pericontinental seas* occur on continental margins and equate with modern continental shelves, or shelf seas, which are characterized by the classic shoreline–shelf–slope profile seen as clinoforms on aggrading continental margins (Figs 7.1 & 7.2); and (ii) *epicontinental* or *epeiric seas* form partially enclosed seas within continental areas, have shallow (or 'shelf') water depths and usually display uniformly dipping ramp profiles, although shelf–slope profiles can also occur in deeper interior basins.

The tectonic setting allows three different shelf patterns to be recognized, with each displaying characteristic thickness patterns reflecting differences in the position of the hinge zone and the location of the main sediment source areas (Fig. 7.3).

1 *Passive margins* are hinged on their landward side and develop seaward-thickening shelf prisms. Sediment is supplied from major continental drainage systems and wide, aggradational shelf areas can develop around pericontinental shelves.

2 *Convergent margins* occur in areas of subduction, are hinged on their seaward side and develop seaward-thickening shelf prisms. Shelf areas are often represented by narrow, wave-cut platforms, but in areas of high sedimentation aggrading accretionary prisms may form.

3 *Foreland basins* are also hinged on their basinward sides but maximum subsidence and sedimentation occur on the landward side adjacent to the orogenic zone. Aggrading, basinward-thinning shelf–slope profiles occur where sedimentation rates are high and extensive shelf areas may develop.

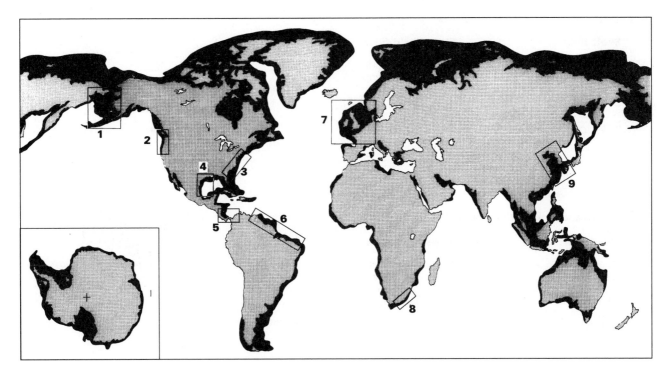

Figure 7.1 Distribution of present-day continental shelves, indicating narrow pericontinental shelves rimming continental areas (e.g. Africa, eastern North America, etc.) and broad epicontinental shelves (e.g. northern North America, Siberia, etc.). The highlighted areas (1–9) represent some of the best-studied shelves, which are referred to in the text: 1, Bering Sea; 2, Oregon–Washington; 3, Eastern USA/ Middle Atlantic Bight; 4, Gulf of Mexico; 5, northwest Colombia; 6, Amazon/Orinoco/Paria; 7, northwest Europe (North Sea, Celtic Sea, English Channel, German Bight); 8, southeast Africa; 9, Yellow Sea.

Despite these morphological differences all shallow seas are subject to the same range of physical, biological and chemical processes, which together exert the main influence on the nature of sedimentary facies in this complex environment.

7.1.2 Historical background

For many years the simple theoretical concept of a *graded shelf* with a seaward-dipping '*profile of equilibrium*' (D.W. Johnson, 1919), along which energy and grain size gradually diminish, was a widely held view of continental shelf sedimentation. This view persisted until sea-bed sampling of loose sediment demonstrated that most shelf seas are covered in a complex mosaic of texturally varied sediment, largely unrelated to present-day processes (Shepard, 1932, 1973b; Emery, 1952, 1968). These deposits were interpreted as *relict sediments* deposited in a variety of shallow-water areas during the Pleistocene lowstand of sea level, including terrestrial, coastal plain, shoreline and glacially influenced environments. The rapid rise of sea level during the Holocene caused these environments to be submerged and hence most of the sea-bed sediments appeared unrelated to their new, shallow shelf environment. Consequently most modern shelves do not display the predicted 'profile of equilibrium' but are in a state of disequilibrium with respect to present-day processes. This concept of relict sediments domi-

nated views on shelf sedimentation, especially in North America (Emery, 1968), for a long period.

The process–response concept, initiated by van Veen (1935, 1936) in the southern North Sea in the 1930s, related sea-bed characteristics to modern shelf processes. Significantly, these studies occurred in an area with modern hydrographic and meteorological data and where Pleistocene lowstand deposits had been extensively reworked both during the Holocene transgression and through to the present day. This enabled data on sediment textures and large-scale bedforms to be integrated, for the first time, with shelf hydrodynamics. The full implications of van Veen's pioneering work in the partially enclosed tidal seas of northwest Europe was delayed for several years until Stride (1963) and his co-workers added side-scan sonar techniques to the hydrodynamic data and began a systematic analysis of sediment and bedform distribution (Caston, 1976). This confirmed a close relationship with present-day processes (M.A. Johnson & Belderson, 1969; Stride, 1982) and provided an important depositional model for many ancient, tidally influenced shelf seas (e.g. Banks, 1973c; Anderton, 1976).

Process–response studies on the shelves around North America during the 1970s reached similar conclusions, although here storm-dominated conditions result in a different suite of bedforms and sediment types (Swift, Duane & Pilkey, 1972; Stanley & Swift, 1976). On other shelves (e.g. in southern Africa;

(a) Accretionary/supply-dominated shelf

Shelf floor	**Slope-basin floor**
• Graded, equilibrium surface(s)	• Gravity transport
• Basinwards-fining	• Sediment accretion
• River mouth bypassing	• Ocean currents

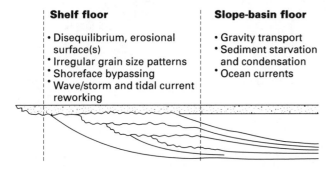

(b) Erosional/accommodation-dominated shelf

Shelf floor	**Slope-basin floor**
• Disequilibrium, erosional surface(s)	• Gravity transport
• Irregular grain size patterns	• Sediment starvation and condensation
• Shoreface bypassing	• Ocean currents
• Wave/storm and tidal current reworking	

(c) Shelf dimensions

	Average	Range	
Width	c. 75 km	(10–>100 km)	⎫ Inner shelf c. 2–50 m
Depth	c. 10–200 km	(2–500 km)	⎬
Slope	c. 0.1°	(0.001 – 1°)	⎭ Outer shelf c. 50–200 m

▨ Zone influenced by shelf hydraulic regime

Figure 7.2 Idealized profiles through accretionary shelf successions, summarizing some of the main depositional processes and dimensions. (a) An accretionary/regressive supply-dominated succession. (b) An erosional/retrogradational accommodation-dominated succession. (c) Average shelf dimensions.

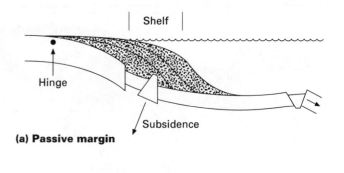

(a) Passive margin

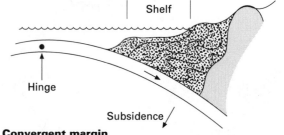

(b) Convergent margin

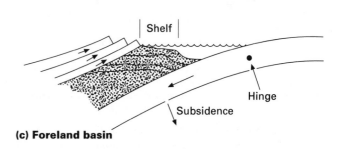

(c) Foreland basin

Figure 7.3 Variations in the structural setting of continental shelves. (a) Passive margin. (b) Convergent margin. (c) Foreland basin (after Swift & Thorne, 1991).

Flemming, 1980, 1981) reworking by oceanic currents provided an additional hydrodynamic model, while those receiving large volumes of modern suspended sediment form distinctive mud-dominated types (e.g. Amazon shelf; Rine & Ginsburg, 1985).

These studies continue to demonstrate that much of what had originally been considered to be relict sediment display varying degrees of response to present-day processes, with active reworking by both physical and biological processes, and that purely relict sediments are less common than originally thought (e.g. Creager & Sternberg, 1972; Swift, Duane & Pilkey, 1972; McManus, 1975; Stanley & Swift, 1976; Stride, 1982). They

have also provided an effective means of classifying modern shelves in terms of the dominant hydraulic regime and of providing a framework by which observations from modern shelves can be applied to the interpretation of ancient shallow marine deposits.

The importance of the type, rate and origin of sediment supply and the degree of adjustment of shelf areas to the Holocene transgression was highlighted by Swift's (1974) distinction between *autochthonous shelves*, which receive sediment almost entirely from *in situ* reworking of earlier deposits, and *allochthonous shelves*, which are partly supplied by modern sediment. These concepts have been developed further by formalizing the relationship between sediment input rate, relative sea-level changes and sediment transport in terms of an accommodation/supply ratio (Swift & Thorne, 1991). This approach reintroduces D.W. Johnson's (1919) concept of the equilibrium shelf and, among other things, provides a theoretical basis for linking aspects of modern shelf sedimentation to studies on the dynamic stratigraphic nature of ancient shelf successions (Swift & Thorne, 1991; Thorne & Swift, 1991a,b; Sect. 7.2).

Facies models of ancient shallow marine clastic deposits have partly followed the process–response phase of modern shelf studies. Reconstruction of physical processes from primary sedimentary structures and facies has been based on the identification of tidal deposits (e.g. de Raaf & Boersma, 1971), wave- and storm-dominated deposits (e.g. Goldring & Bridges, 1973; de Raaf, Boersma & van Gelder, 1977) and the relative interaction of fairweather and storm (intense high-energy or catastrophic) processes (e.g. Hobday & Reading, 1972). The apparent interbedding of offshore fairweather tidal and storm deposits highlighted the difficulty of assigning ancient facies to specific shallow marine processes, since there are very few primary sedimentary structures unique to either process (e.g. Anderton, 1976; H.D. Johnson, 1977a). One of the most diagnostic features are the mud couplets (double mud drapes) found in Recent estuarine deposits in southern Holland, which enabled Visser (1980) to demonstrate conclusively not only the presence of tidal currents but also tidal periodicity (Sect. 6.2.7). Another distinctive sedimentary structure is hummocky cross-stratification, which is interpreted as the product of a wave-related bedform (Harms, 1975). It is widely interpreted as the product of strong oscillatory processes, often with a variable but generally subordinate superimposed unidirectional flow, which forms during intense storms. The structure is commonly preserved between fairweather and storm wave base and may form a continuum with a similar but generally shallower water structure, swaley cross-stratification (Leckie & Walker, 1982) (Sect. 7.7.2). Similar, but smaller-scale, wave-formed structures have been recognized, which help distinguish between wave- and current-processes within heterolithic facies (de Raaf, Boersma & van Gelder, 1977).

During the 1980s there was further development and refinement of ancient shallow marine facies models, mostly associated with storm-dominated shelf systems, which appear to have been the most abundant over the geological record (e.g. Brenchley, 1985; Duke, 1985). This also coincided with rapid developments in sequence stratigraphy (Wilgus, Hastings *et al.*, 1988), which contributed to a major reappraisal of one of the most widely documented shallow marine facies models, namely the linear sandbodies of the Cretaceous of the Western Interior of North America. Previously, they had been almost universally interpreted as 'offshore bars' analogous to modern shelf sandridges (e.g. Campbell, 1973; Spearing, 1976; Tillman, 1985). More recently, many of these sandbodies, such as the Cardium deposits in Alberta, have been re-evaluated as shoreface deposits which became erosionally detached from the contemporaneous shoreline as a result of erosion associated with basinwide changes in relative sea level (Plint, 1988b; R.G. Walker & Plint, 1992). This emphasizes the importance of integrating ancient process–response studies, or facies analysis, with reconstructions of relative sea-level fluctuations.

This highlights the difficulty of distinguishing between shallow marine sandbodies formed in coastal environments, including inshore areas such as estuaries and tidal channels (Chapter 6), and those deposited offshore on shelves and which may have formed outside the influence of the contemporaneous shoreline. The close interaction and similarity between processes operating at the shoreline and those found on the adjacent shelf, mean that facies characteristics and vertical facies successions alone are insufficient to make this distinction. The absence of unambiguous shoreline deposits, with evidence of emergence, has been used to support shelf interpretations but this may reflect incomplete preservation of shoreline successions due to erosion during both relative sea-level lowstands and marine transgression/shoreface ravinement. Thus it is essential to understand the nature of marine erosion surfaces and the relationship of shallow marine sands to these surfaces before making an interpretation.

7.2 Clastic shelf models and classification

The dominant controls of sedimentation patterns and facies characteristics in shallow clastic seas are: (i) sediment transport (hydraulic regime); (ii) sediment supply; and (iii) relative sea-level changes. Secondary factors, which may dominate equivalent carbonate environments (Chapter 9), include: (iv) climate; (v) biological factors (animal–sediment interactions); and (vi) chemical factors (seawater chemistry and sediment composition). Ancient shallow marine successions are also influenced by: (vii) geological history/basin evolution, including tectonic controls on sedimentation; and (viii) nature of the preserved sedimentation record.

7.2.1 Process–response models and shelf hydraulic regimes

On modern shelves: (i) sediment and bedform types can be calibrated with measurements of modern shelf hydrodynamic processes; and (ii) shelf hydraulic regimes are characterized by the interaction of fairweather processes, operating on a time scale of days to months, and the frequency and intensity of storm processes, which have a periodicity of years to tens or even hundreds of years (Mooers, 1976). Storm processes are particularly important since they can enhance or completely override fairweather conditions, particularly when considered over geological time when they are likely to have higher preservation potential than fairweather events. On the basis of hydraulic regime there are four main types of shelf sea: (i) tide-dominated; (ii) wave-dominated; (iii) storm-dominated; and (iv) oceanic-current-dominated (Fig. 7.4).

Tide-dominated shelves are swept daily by powerful tidal currents, which can result in significant sea-bed erosion and bedload sediment transport in the form of large-scale linear and transverse bedforms. They are most common in partially enclosed basins and often display evidence of oppositely directed

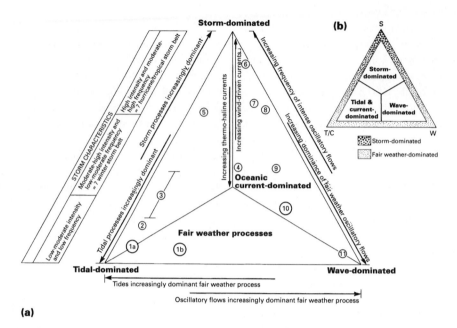

(a)

Figure 7.4 (a) Classification of the main types of modern shelf seas based on the dominant fairweather processes (i.e. tides, oceanic currents and waves) and the relative interaction with storm processes. 1a, Macrotidal embayments and estuaries (e.g. Bay of Fundy, Cook Inlet, Kuskokwim Bay, Chesapeake Bay); 1b, mesotidal embayments and estuaries (e.g. German Bight); 2, tidal straits (e.g. St George's Channel, English Channel, Malacca Straits, Taiwan Strait); 3, tidal seas (e.g. North Sea, Celtic Sea, Yellow Sea, George's Bank); 4, oceanic current-swept shelves (e.g. southeast Africa, Blake Plateau, Morocco shelf); 5, storm-dominated (wind-driven) (e.g. northwest Atlantic shelf, USA); 6, storm-dominated (wind- and wave-driven) (e.g. Oregon–Washington shelf, California Shelf); 7, storm-dominated (wind-driven) (e.g. southeast Bering shelf); 8, storm-dominated (low to moderate energy) (e.g. Gulf of Mexico, Norton Sound); 9, mud-dominated (e.g. Amazon–Orinoco–Paria shelf, Niger shelf, Yellow Sea, northwest Colombia); 10, non-tidal, low-energy embayments (e.g. Baltic Sea, Hudson Bay); 11, wave-dominated (fair weather) environments (e.g. upper shoreface, wave-built bars, etc.). (b) Simplification into the three main types of modern and ancient shelf processes and deposits.

sediment transport, both at the scale of individual bedforms and at the scale of regional sediment transport paths (Sect. 7.3). Approximately 17% of the world's shelf seas are considered to be dominated by tides, which may be augmented by storm processes to varying degrees (Swift, Han & Vincent, 1986).

Wave-dominated and *storm-dominated shelves* are combined since they represent a spectrum in hydraulic regime ranging from relatively low-energy/low-frequency wave–storm climate (e.g. in enclosed/partially enclosed basins) to high-energy/high-frequency wave–storm climate (e.g. facing major ocean basins). Both types are in effect storm-dominated, since sediment dispersal is mainly controlled by seasonal fluctuations in wave and current intensity, with maximum sediment transport typically accompanying storm activity. These shelves are dominated to varying extent by seasonal wind- and wave-generated currents, which in sand-rich areas are capable of forming large-scale bedforms, including linear sandridges with similarities to their tidal shelf equivalents. Storm processes can also enhance or dominate sediment transport associated with most of the fairweather processes. This represents the most common type of hydraulic regime on modern shelves (80%), and appears to have been the most abundant type of shelf in the geological past (Swift, Han & Vincent, 1986).

Oceanic current-dominated shelves are regularly swept by persistent unidirectional currents, which are generated in ocean basins but occasionally migrate over adjacent continental shelves. This type of shelf is typical of the narrow, pericontinental type and is, therefore, found adjacent to ocean basins (Sect. 7.5). Only some 3% of the world's continental shelves are of this type.

7.2.2 Dynamic stratigraphic models

The dynamic aspect of modern shelf sedimentation was first considered by Curray (1964, 1965), who evaluated the interaction between rate of deposition and rate of relative sea-level change in relation to transgression and regression (see Fig. 6.16). Facies patterns in several shelf and coastal areas could be related to fluctuations in eustatic sea-level changes during the Holocene transgression and to changes in sediment supply.

These concepts were extended by Swift (1969a,b, 1970), who highlighted three main facies on clastic shelves: (i) *shelf relict sand blanket* comprising a discontinuous veneer of basal transgressive sands overlying pre-Holocene deposits and in varying stages of disequilibrium; (ii) *nearshore modern sand prism* comprising a seaward thinning belt of coastal sands in equilibrium with present-day nearshore processes; and (iii) *modern shelf mud blanket* consisting of modern fine-grained suspended sediment which has spread beyond the nearshore zone to various parts of the shelf. Further distinction was then made between true *relict sediment*, representing unreworked sediment, *palimpsest sediment*, displaying characteristics of both its present and former environment, and *modern sediment*, which is supplied by present-day processes from outside the shelf area (Swift, Stanley & Curray, 1971).

The dynamic processes controlling these shelf sedimentation patterns are considered by Swift and Thorne (1991) to be determined by the interaction of five partly interdependent variables: (i) rate of relative sea-level rise (R); (ii) rate of sediment input (Q); (iii) type of sediment input (M); (iv) fluid power (P); and (v) rate of sediment transport (D). These variables interact over geological time to produce the profile of equilibrium and to determine the nature of shelf facies types and their internal organization and distribution. The ratio of accommodation space to sediment supply ($R \cdot D / Q \cdot M$) provides a theoretical framework for considering sedimentation on modern continental shelves and for emphasizing the important interactive link between shelf sedimentation patterns and coastal morphology and dynamics. Two main shelf sedimentary regimes are recognized (Table 7.1): (i) supply-dominated regimes (ratio less than 1: rate and type of sediment supply exceeds the rate of accommodation-space creation and sediment dispersal) (Fig. 7.5); and (ii) accommodation-dominated regimes (ratio greater than 1: rate of sea-level rise and sediment transport/dispersal exceeds the rate of sediment supply) (Fig. 7.6).

Supply-dominated regimes are characterized by high rates of sediment supply and by overall regression (Fig. 7.5a). The dominant mode of sediment supply to the shoreface and adjacent shelf is through the flood stage of a deltaic river mouth (*river mouth bypassing*; Figs 6.4a & 7.5a), which results in allochthonous shelf sedimentation. This is controlled by the nature and relative strength of fluvial and marine processes at the river mouth and by the resulting morphology of the delta front (Sect. 6.6.2). Offshore sediment transport is achieved mainly by fluvial flood discharge (flood stage jets; Fig. 7.5a), but other processes may be important, depending on the nature of the hydrodynamics of the shelf and littoral zone. These include rip-currents, ebb-directed tidal currents and seaward-returning (or downwelling) storm currents. Supply-dominated sedimentation results in large volumes of mainly fine-grained, river-derived sediment spreading rapidly across the shelf and building up thick, uniform accretionary units (Fig. 7.2a). The following facies characteristics are common: thick (10s–100s m), relatively homogeneous, fine-grained (including thick muddy shelf successions), variable bioturbation, high preservation of single event beds (e.g. storm layers), vertically and laterally graded shelf deposits in response to offshore increasing water depth and a corresponding decrease

Table 7.1 Coastal and shelf depositional systems, complexes and systems tracts as developed under regressive and transgressive settings (after Swift, Phillips & Thorne, 1991a).

		Regressive settings	Transgressive settings
Coastal settings		Regressive intra-coastal systems tract	Back-barrier systems tract
		Strandplain or chenier plain systems	Beach-dune–washover-fan complexes
		Deltaic-channel–mouth-bar complexes	Tidal-delta–tidal-channel complexes
Shelf settings		Regressive shelf systems tract	Transgressive shelf systems tract
		Regressive shoreface–shelf systems	Transgressive shoreface–shelf systems
		Prodelta plume systems	Sand ridge complexes
		Fine-grained deceleration sheets	Coarse-grained deceleration sheets

Supply-dominated shelf

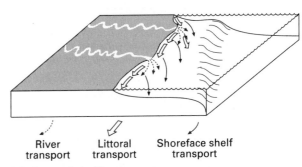

River transport   Littoral transport   Shoreface shelf transport

(a) Coastal-shelf processes

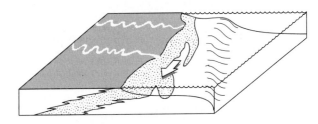

(b) Coastal-shelf morphology

Figure 7.5 Main processes and sedimentation patterns in supply-dominated shelf sedimentary regimes: (a) river mouth bypassing, through flood stage jets, transports fine sand, silt and clay on to the shelf and contributes sand to the littoral drift, and (b) resulting coastal-shelf morphology and accretionary profile (after Swift, Hudleson *et al.*, 1987; Swift & Thorne, 1991).

in fluid power, and facies trends mainly parallel to the shoreline and delta front (normal to the direction of progradation). These sediments are preserved as regressive deposits, including parasequences and parasequence sets, and are most common in highstand systems tracts (Swift, Phillips & Thorne, 1991b).

Examples of modern supply-dominated shelves include the Bering Sea (Nelson, 1982), German Bight (Aigner & Reineck, 1982) (see Fig. 6.83), Nayarit shelf (Curray, Emmel & Crampton, 1969) (see Fig. 6.68) and Washington–Oregon shelf (Nittrouer & Sternberg, 1981) as well as those receiving sediment from modern deltas such as the Louisiana Shelf (Adams, Wells & Coleman, 1982) (Sect. 6.6).

Accommodation-dominated regimes are characterized by high rates of relative sea-level rise and high rates of sediment dispersal. Transgression is therefore dominant (Fig. 7.2 & 7.6b). Large volumes of sediment are trapped in river mouths, which become estuaries, and sediment is supplied from both landward, but primarily, seaward directions (Fig. 7.6a). The shoreface and shoreline migrate landward by shoreface erosional retreat (or ravinement; see Fig. 6.86). The net landward-directed bottom stress caused by shoaling waves drives sediment in a shoreward direction, thereby creating a 'littoral energy fence' (see Fig. 6.7).

Accommodation-dominated shelf

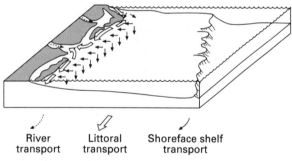

(a) Coastal-shelf processes

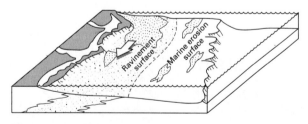

(b) Coastal-shelf morphology

Figure 7.6 Main processes and sedimentation patterns in accommodation-dominated shelf sedimentary regimes: (a) shoreface bypassing, with downwelling associated with coast-parallel storm-generated (geostrophic) currents eroding sand from the shoreface and transporting it on to the inner shelf, and (b) resulting coastal-shelf morphology, thin coarse-grained lag deposits overlying the erosional transgression (ravinement) surface and retrogradational/bevelled profile (after Swift & Thorne, 1991).

Sediment reaches the shelf by being transported through the shoreface (*shoreface bypassing*), with the littoral fence being overcome by offshore-directed storm-generated currents and by the transgressive detachment of shoreface bars and sand-ridges. Erosional shoreface retreat and shoreface bypassing result in significant reworking of underlying deposits and hence autochthonous (or *in situ*) sedimentation is dominant. The resulting shelf sediments directly overlie, or are closely associated with, transgressive surfaces, which may be either erosional or non-erosional and condensed (including flooding, maximum flooding and ravinement surfaces). The following facies characteristics are common in accommodation-dominated systems: thin (cm–m up to 10s m), variable lithology/facies, relatively coarse-grained (including winnowed lag deposits), strong bioturbation (various burrowed and bored horizons), mineralization, condensation/amalgamation and facies patterns trend normal to the shoreline (parallel to the retreat path). These deposits are most common within the lower parts of transgressive systems tracts.

Examples of modern accommodation-dominated systems include two of the world's most intensively studied shelves, namely the seas around northwest Europe (Stride, 1982) and the Middle Atlantic Bight of the eastern USA (Swift, Han & Vincent, 1986; Swift, Thorne & Oertel, 1986). They are also associated with distinctive types of transgressive shoreline.

7.2.3   Modern vs. ancient shelf seas

Present-day continental shelf seas can only constitute a limited representation of the past. This is because the Holocene glacio-eustatic rise of sea level of about 10 mky^{-1} (Sect. 2.1.4) resulted in very rapid transgression. As a result, the profile of many modern shelves is partly an erosional feature and the sediment covering this surface is wholly or partly in disequilibrium with the present-day hydraulic regime. Nevertheless, there is evidence that the theoretical profile of equilibrium, with its graded sediment cover, was common in the geological past. It is seen in the characteristic accretionary shelf–slope profile of many continental margins, including large-scale clinoforms seen on seismic sections, and in the facies and faunal distributions of many ancient shoreline–shelf transitions. Similarly, many ancient coarse-grained, sand-rich shelf sand-stones appear to have formed under accommodation-dominated regimes, when sands were emplaced on the shelf by *in situ* reworking of coastal and alluvial–coastal plain deposits. Hence modern shelf seas do provide important analogues for the interpretation of ancient successions.

In this chapter shelf seas are separated on the basis of shelf processes and facies types into: (i) tide-dominated; (ii) wave- and storm-dominated; (iii) oceanic current-dominated; and (iv) mud-dominated. The latter is treated separately because sediment supply and biochemical processes (e.g. oxygen content, fauna, mineralogy, etc.) dominate over hydrodynamic processes.

7.3 Modern tide-dominated shallow seas

Shelf seas are defined as tide-dominated when tidal range is greater than 3–4 m, with typical semi-diurnal maximum surface tidal current speeds (mean spring) ranging from 60 to >100 cm s^{-1}. These conditions are best developed in partially enclosed seas, elongate seaways and blind gulfs (e.g. North Sea, Celtic Sea, English Channel/Western Approaches, Long Island Sound, Malacca Straits, Gulf of Korea, Gulf of California (Stride 1982; McCave, 1985; Fenster, Fitzgerald *et al.*, 1990). Facies distribution is mainly related to: (i) hydraulic regime which, although tide-dominated, is also influenced by storms; (ii) nature and origin of sediment supply; and (iii) sea-level fluctuations. The relative interplay of these processes determines the particular facies of tide-dominated shelves and separates them into supply- and accommodation-dominated (Sect. 7.2.2).

Supply-dominated tidal shelves are relatively rare today but may be partially represented by tidally influenced river deltas (Meckel, 1975) (Sect. 6.6) and the proximal parts of mud-dominated shelves, such as the Amazon shelf and the Gulf of Bohai/Yellow Sea (Sect. 7.6), which are essentially subaqueous deltas.

Accommodation-dominated tidal shelves are much more extensive at present. They are closely linked with the Holocene transgression when sediment supply was provided mainly from *in situ* reworking of older deposits during relative sea-level rise and tidal current erosion. This type of transgressive/accommodation-dominated tidal shelf is described more fully below.

7.3.1 Tides and tidal currents

The mechanisms that generate tides in the open ocean and their characteristics in coastal settings are described in Section 6.2.7. This section only discusses the important modifiers of unrestricted tidal waves and the processes that generate tidal currents on shelves.

On a global scale, tides are modified by three major factors: (i) the presence of land masses and islands that interrupt the theoretical, unconstrained waterbody circulation; (ii) variations in the dimensions of the ocean basins that help to modify the simple tidal wave into a series of closed cells (amphidromic systems) that rotate around a 'central' point known as the amphidromic point (Fig. 7.7); and (iii) thermal differences in the atmosphere and hydrosphere that manifest themselves as weather (Open University Course Team, 1989a).

Vertically oscillating oceanic tidal currents propagate on to continental shelves as co-oscillations that, in conjunction with Coriolis effects, cause water particles to follow an elliptical path in the horizontal plane. In narrow seaways or in partially enclosed basins, these ellipses may be very strongly elongated to produce rectilinear reversing currents. In more open shelves the tidal ellipse has a more circular shape, which produces rotary

tidal currents. Where basins are very narrow or very shallow, frictional forces may almost completely dampen-out these co-oscillations, which results in effectively tideless seas. In contrast, other basins may have dimensions that reinforce tidal ranges and the velocities of tidal currents such as the Bay of Fundy (Knight, 1980; Dalrymple, Knight *et al.*, 1990).

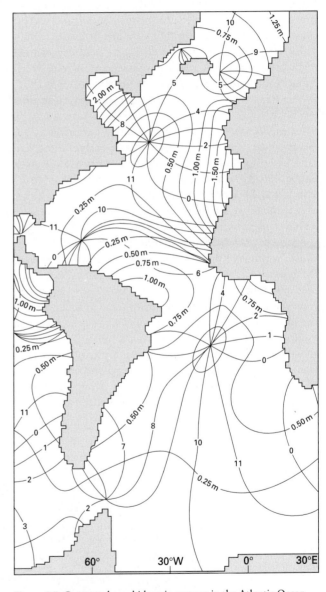

Figure 7.7 Computed amphidromic systems in the Atlantic Ocean for the dominant semi-diurnal lunar tidal component, showing how the simple, theoretical tides are modified by the distribution and geometry of the ocean basins and land areas. *Co-tidal* lines (tide is at the same phase of its cycle) radiate outwards from *amphidromic points*. The *co-range* lines (areas of equal tidal range) roughly enclose the amphidromic points (after Open University Course Team, 1989a).

7.3.2 Tide-dominated sand deposition

Modern accommodation-dominated tidal systems are characterized by three main types of sand deposition: (i) estuarine/incised-valley-fill tidal sands (Sect. 6.7.5); (ii) shoreline-associated tidal sandridges; and (iii) open shelf tidal sand sheets. These are not mutually exclusive and many of the features are shared.

1 *Estuarine/incised-valley-fill tidal sands* comprise thick (c. 40–60 m), shore-normal sandbodies of limited lateral extent (see Fig. 6.82). Elongate sandridges and tidal channels, both with superimposed large-scale subaqueous dunes (sand waves), characterize estuary mouths (Nio, van den Berg *et al.*, 1980). These sandbodies may display distinctive mud drapes (tidal rhythmites) (Sect. 6.2.7), which are characteristic of some Recent estuarine deposits (van den Berg, 1982; Nio & Yang, 1991). During transgression, estuaries and narrow embayments often evolve into blind embayments or into narrow seaways. In the latter, sandridges, originally formed within the confines of an estuary, may continue to evolve as open shelf/offshore tidal sandridges (e.g. East Bank area of the North Sea; Davis & Balson, 1992), and may be indistinguishable from the shoreline-associated tidal sandridges (see below). Hence, there is a close association between tidal estuarine and open shelf tidal deposits within transgressive depositional systems (Fig. 7.8).

2 *Shoreline-associated tidal sandridges* occur as offshore sandbanks, which owe their origin to the transgressive history of the adjacent linear shoreline. The Norfolk Banks, off eastern England, appear to have evolved by progressive detachment from the shoreline followed by growth and abandonment on the shelf (Sect. 7.3.4) in a fashion similar to storm-dominated sandridges (Sect. 7.4). The sandbodies trend parallel to the shoreline and, during shoreline retreat, coalesce to produce a laterally extensive sheet of tidal sands. Depending on their degree of preservation they exhibit planar bases (erosional flooding/ravinement surface) and an irregular, mounded top, although the latter will have a low preservation potential (Fig. 7.8). Owing

to the rotary nature of the tidal currents and high wave-effectiveness, mud drapes are more likely to reflect seasonal variations in suspended sediment distribution (e.g. post-storm mud layers; McCave, 1971a) rather than tidal periodicity/rhythmicity typical of estuaries.

3 *Open shelf tidal sandsheets* develop with distinctive bedforms along regionally extensive (>100 km) tidal current transport paths, in which tidal sandridges are also absent. They form tabular sandbodies with planar tops and bases and equate with the 'coarse grained deceleration sheets' of Swift, Phillips and Thorne (1991a). These deposits are unrelated to the adjacent shoreline but instead are in equilibrium with present-day open shelf tidal currents (Sect. 7.3.3).

7.3.3 Tidal current transport paths: processes, bedforms and facies

Regionally extensive tidal current transport paths (>100 km long) are one of the most distinctive features of modern tidal seas (Fig. 7.9). They develop a suite of bedforms and facies reflecting a downcurrent reduction in flow strength, which often accompanies an increase in water depth. They extend from zones of bedload parting, where the sea bed is eroded, to areas of bedload convergence dominated by deposition.

Individual tidal current transport paths are recognized by a combination of the following: (i) direction of maximum near-surface tidal current velocities (frequently maximum spring tides); (ii) elongation of tidal current ellipses; (iii) facing directions of sandwave and sandridge lee slopes; (iv) trend of longitudinal bedforms, including scour hollows, longitudinal furrows, obstacle marks, sand ribbons, longitudinal sand patches and tidal ridges; (v) direction of decreasing grain size; and (vi) direction of decreasing tidal elevation due to increased bottom friction (e.g. Belderson & Stride, 1966; Swift & Thorne, 1991). The long-term inequalities in sediment transport rates may be due to any one of the following: (i) variations in the strength

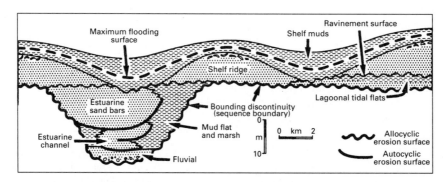

Figure 7.8 Schematic coast-parallel cross-section through a transgressive, tide-dominated estuary/shelf depositional system, highlighting the distribution of the main stratal surfaces related to either allocyclic processes (e.g. changes in relative sea-level) or autocyclic processes (e.g. tidal channel migration). Derived from a

combination of observations from modern estuaries (Cobequid Bay–Salmon River and Severn Estuary) and modern tidal shelves (southern North Sea). Note that the high degree of preservation inferred here may be significantly modified in the geological record (after Dalrymple, 1992).

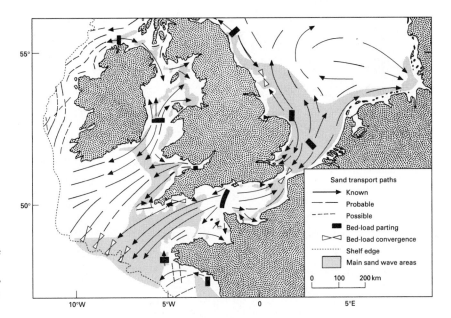

Figure 7.9 Distribution of the dominant sand transport paths and their relationship with the main sandwave areas on the northwest European continental shelf (after Stride, 1963, 1982; Kenyon & Stride, 1970; M.A. Johnson, Kenyon *et al.*, 1982).

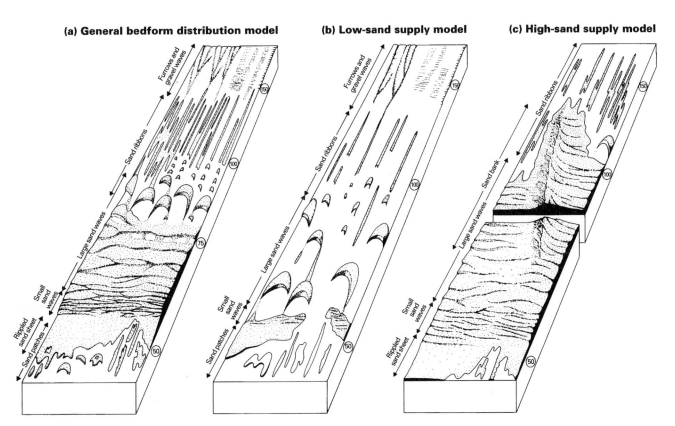

Figure 7.10 Distribution of bedform zones along tidal current transport paths: (a) general model, (b) low sand supply model, and (c) high sand supply model. The bedform zones are aligned parallel with mean spring peak near-surface tidal current velocities (shown in cm s⁻¹) (from Belderson, Johnson & Kenyon, 1982).

241

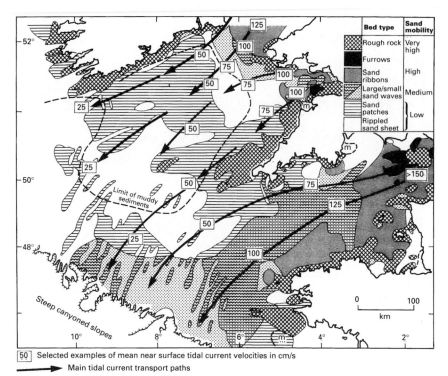

Figure 7.11 The distribution of sediment and bedform types in the Western Approaches and Celtic Sea and their relationship with major current transport paths. Note the reduction in current strength down each transport path and its close correlation with bed type and sand mobility (from M.A. Johnson, Kenyon *et al.*, 1982).

and duration of ebb and flood currents; (ii) mutually exclusive ebb and flood transport paths; (iii) lag effect of sediment entrainment associated with rotating tides; and (iv) preferential enhancement by superimposed storm-generated currents.

A complete tidal current transport path displays a predictable zonation of bedforms (Fig. 7.10a), but this varies depending on whether sand supply is low or high (cf. Figs 7.10b & c, respectively). Despite the complex nature of the hydraulic regime the bedform distribution patterns calibrate closely with near-surface mean spring peak tidal current speeds (Fig. 7.11). Five main bedform zones are distinguished in the idealized model (Fig. 7.10a).

Furrows and gravel waves occur in areas of strong tidal current velocities (>150 cm s⁻¹) and where sediment supply is sparse. Erosional features are dominant, ranging from large scour hollows in the sea floor (c. 150 km long, 5 km wide and 150 m below the surrounding sea floor; e.g. Hurd Deep, English Channel) to smaller longitudinal furrows cut into gravel floors (c. 8 km long, 30 m wide and 1 m deep) (Hamilton & Smith, 1972). The longitudinal configuration of these features is thought to relate to helical circulations in the tidal currents. Constructional features are rare, but include gravel-rich subaqueous dunes (c. 1 m high and 10 m wavelength) (Belderson, Johnson & Kenyon, 1982).

Sand ribbons consist of up to 15-km-long ribbons or strips of sand up to 200 m wide and up to 1 m thick, with intervening strips of gravel (Kenyon, 1970). Normal maximum near-surface current velocities usually exceed 100 cm s⁻¹. With higher current velocities, ribbons are made up of trains of straight-crested sand waves but in the lower ranges (c. 90 cm s⁻¹) they are made up of sinuous or barchanoid sandwaves (Belderson, Johnson & Kenyon, 1982, p. 48). Ribbons tend to broaden and, rarely, merge down the longitudinal velocity gradient.

Sandwaves (mainly *small to large subaqueous dunes* using the terminology of Ashley, 1990; sandwave terminology is retained here) are characterized by generally straight crests and well-defined lee slopes, which are most commonly less than the angle of repose (c. <15°) (Stride, 1970, 1988; McCave, 1971b; Terwindt, 1971b; Berné, Auffret & Walker, 1988). These bedforms are generally at least 1.50 m high with wavelengths between 150 and 500 m, and occur where current velocities are greater than 65 cm s⁻¹. They are characterized by complex internal structures dominated by low-angle master bedding planes and superimposed smaller-scale cross-bedding (Fig. 7.12; Sect. 7.8). The asymmetry of these bedforms, which is frequently in the direction of maximum near-surface tidal current velocities, and the nature of the internal structures indicate a response to present-day tidal currents (Stride, 1963; Houbolt, 1968). The relationship between bedform morphology and internal structure can be confidently inferred from high-resolution seismic sections

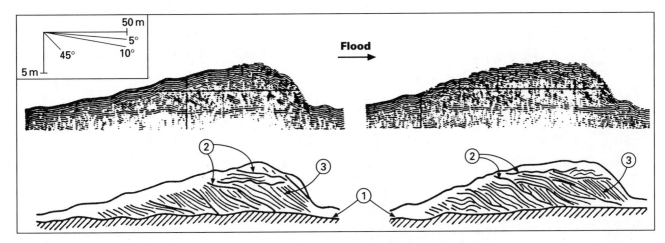

Figure 7.12 High-resolution seismic reflection profiles and line interpretations through asymmetric sandwaves from offshore Brittany, France. Interpreted sections highlight (1) major, subhorizontal bounding surface (sea bed); (2) low-angle, second-order bounding surfaces; and (3) steep foresetted intervals. (Note the exaggerated inclination of reflectors due to scale differences (from Berné, Auffret & Walker, 1988).

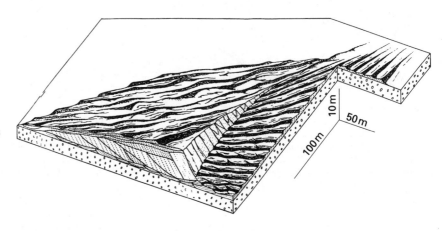

Figure 7.13 Model of the internal structure of modern large and very large subaqueous dunes (sandwaves) based on high-resolution seismic and core data (from Berné, Durand & Weber, 1991). Note the compound nature of the internal structures and their similarity with their theoretical and ancient equivalents (Sect. 7.7.1).

(Fig. 7.13; Berné, Durand & Weber, 1991), supported by the results of outcrop and modelling studies (see Fig. 7.57; J.R.L. Allen, 1980; Sect. 7.7.1).

Sand patches occur in areas of lower current velocity (c. 25–50 cm s^{-1}) which may be either longitudinal or transverse in relation to current directions (Fig. 7.11). The sand patches are commonly covered in ripples and often support a varied infauna. The relatively low tidal-current velocities probably require superimposed storm-generated currents to cause significant sand transport, resulting in waning flow sequences characteristic of storm processes (Sect. 7.11).

Mud zones are usually located at the ends of tidal current transport paths where both tidal current velocity and particularly wave activity are relatively low (Stride, 1963; McCave, 1971c, 1972a). However, mud accumulates in a wide variety of situations and transport paths for suspended fine-grained sediment may differ from tidal current bedload paths because they are influenced by both wind-drift and general circulation patterns (Sect. 7.6). These mud zones may also be considered to form the distal end of transport paths (or 'coarse-grained deceleration sheets') due to flow expansion and deceleration (Swift & Thorne, 1991).

7.3.4 Offshore tidal sandridges

Offshore tidal sandridges (Fig. 7.14), are large-scale, linear bedforms with long axes oriented up to 20° obliquely to the direction of strongest tidal currents (Off, 1963; Kenyon, Belderson *et al.*, 1981). Sandridges are typically 50 km long, 1–3 km wide, 10–50 m high and spaced up to 12 km apart and they are composed of well-sorted, medium to fine sand containing disseminated shell debris and occasional clay drapes (Houbolt, 1968, 1982; Belderson, Johnson & Kenyon, 1982).

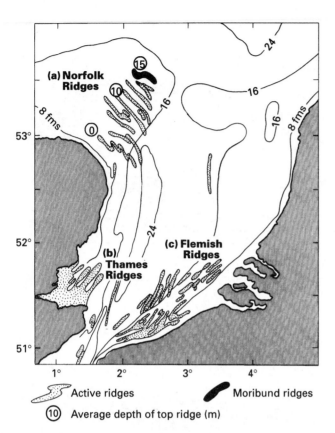

Figure 7.14 Distribution of the three main groups of sandridges in the southern North Sea: Norfolk Ridges (a) are shoreline-associated and illustrate a lateral (transgressive) sequence from nearshore active (parabolic) ridges through to offshore inactive/moribund types; Thames Ridges (b) are part of an outer estuary shoal complex; and Flemish Ridges (c) are associated with the transgressive history of the Straits of Dover and southern North Sea (from Stride, Belderson *et al.*, 1982).

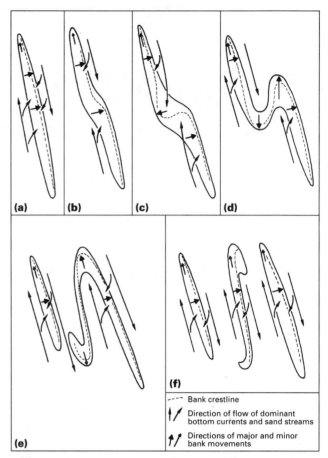

Figure 7.15 Model for the growth and development of linear tidal sandridges. A linear sandridge is built between two mutually evasive ebb and flood tidal channels (stage a). Inequality of the secondary cross shoal currents leads to the destruction of the straight crest line (stages b and c). The resulting double curve develops into an incipient pair of ebb and flood channels (stage d). The channels continue to lengthen resulting in parallelism of the centre ridge with those adjacent to it (stage e). The initial cycle is thus complete but can continue with three ridges instead of one. This sequence is based on the inner active ridges of the Norfolk Banks (see Fig. 7.14a) (from Caston, 1972).

The obliquity of most ridge orientations with respect to tidal current directions results in the transport processes on the two faces being dominated either by ebb or by flood currents (Fig. 7.15). The typical inequality of currents produces an asymmetrical cross-section on active banks, which is preserved as a series of major, low-angle (c. 3–7°) internal bedding planes separated by smaller-scale cross-stratification (Fig. 7.16b). The latter reflects ebb and flood directed sandwaves which cover active modern ridges. These sandwaves are aligned obliquely but become progressively parallel to the ridge crest, superficially indicating flow convergence towards the crestline.

Off eastern England the Norfolk Banks display a systematic change in sandridge size and geometry, becoming shorter, straighter and less parabolic when followed offshore into deeper water (Fig. 7.14), indicating a range of ridge morphologies from actively maintained to moribund (Belderson, Johnson & Kenyon, 1982; Yang, 1989). *Actively maintained ridges* are associated with tidal currents in excess of 50 cm s^{-1} and

have shallow crests, contain superimposed sandwaves and are winnowed, with relatively coarse grain sizes in crestal zones. They are asymmetrical in cross-section with steeper lee slopes inclined at up to 6° (Fig. 7.16) and they overlie lag gravel floors. Ridge crests are sharp except where they approach sea level when they display flattened tops due to wave reworking. *Moribund ridges* have rounded profiles and slopes of 1° or less (Fig. 7.16). They form part of a continuum from active, peaked forms to relatively featureless and smoothed-out sheets, occasionally with superimposed smaller-scale transverse bedforms. They are developed over sandy or muddy floors, and occur in relatively deep water where currents are less than 50 cm s^{-1}.

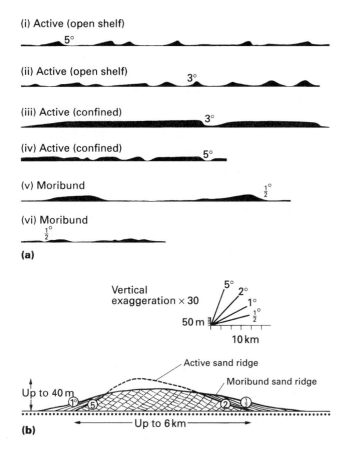

(a)

(b)

Figure 7.16 Cross-sections through modern tidal sandridges. (a) General shape of active and moribund ridges. (b) The inferred relationship of internal stratification to the external morphology of ridges (from Stride, Belderson *et al.*, 1982).

This sequence of offshore moribund to nearshore active ridges continues further onshore into a series of parabolic-shaped ridges, which are cut by mutually evasive ebb and flood tidal channels (Fig. 7.14). These ridges may be related to inshore channel-shoal features (cf. Swift, 1975b) so that a complete sequence of forms may be displayed, probably related to transgression. They do not form a predictable part of regional transport paths and, in the Norfolk Banks, occur in a zone of bedload parting where there is an additional influx of sand from the adjacent shoreline (Stride, 1974; Clayton, McCave, & Vincent, 1983). Significantly, this shows that not all sand is derived from reworked relict material.

Most other moribund sandridges, such as those in the outer Celtic Sea and in the deeper parts of the North Sea, were formed at a time of lower sea level and are now being preserved beneath a mud blanket. Active ridges are likely to merge together and be preserved as a more continuous sand sheet (Kenyon, Belderson *et al.*, 1981), which may be difficult to distinguish from the open shelf sand sheets described earlier (Sects 7.3.2 & 7.3.3).

7.4 Modern wave- and storm-dominated shallow seas

In most pericontinental shelf seas facing oceans, and in many partially enclosed seas, meteorological factors dominate the hydraulic regime, which is characterized by strongly seasonal wind- and wave-induced currents and marked alternations between fairweather and storm conditions. Wave action is most intense on those shelves that face prevailing westerly winds and are open to oceanic waves (e.g. Bering Seas, Washington–Oregon shelf, Barents Sea, northeast Atlantic shelf). Waves are less intense on shelves in the lee of prevailing winds (e.g. southeastern USA) and least intense in partially enclosed seas such as the Gulf of Mexico, Baltic Sea and western Mediterranean. Since storms and waves control the resuspension of muddy sediment and its supply to regional currents such as oceanic systems (Sects 7.5 & 7.6), they also dominate mud-rich shelves.

The main controls on long-term sediment transport, bedform development and facies distributions are: (i) the frequency and intensity of storm-induced currents; (ii) nature and origin of sediment supply; and (iii) sea-level fluctuations.

7.4.1 Wave- and storm-generated processes

Wave- and storm-generated processes are primarily the result of meteorological forces acting on the shallow parts of shelf and oceanic waters. Energy is transferred through wind shear stress and fluctuations in barometric pressure, to induce two main types of water movement (Fig. 7.17): (i) oscillatory and wave-drift currents (including longshore and rip currents in nearshore areas); and (ii) wind-driven currents.

Oscillatory and wave-drift currents are the ambient, mostly fairweather components of the storm–fairweather hydraulic regime (Sect. 6.2.4). They interact to produce a shelf surface in local dynamic equilibrium, which is characterized both spatially and temporally by instantaneous patterns of sediment erosion, transport and deposition (Swift & Thorne, 1991).

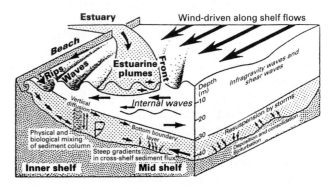

Figure 7.17 The main components of wave- and storm-dominated, inner- and mid-shelf dispersal systems (from Nittrouer & Wright, 1994).

Wave processes decrease in strength exponentially away from the surface so that at a water depth of approximately half the wavelength (wave base) there is no perceptible motion. Even for the most extreme conditions storm wave base rarely, if ever, exceeds 200 m (e.g. on the Pacific-facing Washington–Oregon shelf of North America; Fig. 7.1), while fairweather wave base is considerably shallower (c. 10 m). The seasonal and storm-influenced transit of the intersection of wave base with the gently inclined surface of most shelves provides an important determinant of one part of the dispersal system. Only at or above wave base can significant work be done by waves on the sea bed.

Water wave heights and periodicities are dependent upon: (i) wind speed; (ii) length of fetch; and (iii) the duration of the wind, and there is considerable variation in the efficiency of energy transfer from the wind to the water and eventually to the sea bed. The coupling between atmospheric circulation and the genesis of water waves can be described reasonably well with a series of semi-empirical equations. Unfortunately, the linkage between water motion, currents and the motion of sedimentary particles at the sea bed is less well known (Nummedal, 1991).

Any given wave-generating event (dominantly storms in the open ocean) will produce a wide spectrum of wave frequencies and, because the inner, feather edges of shelves are more or less permanently above fairweather wave base, practically all waves have some effect on this zone. It is referred to as the friction-dominated zone. However, it is those waves with the greatest wavelength and greatest height that affect the middle and outer parts of continental shelves. These zones, from c. –10 m to –20 m, and beyond c. –20 m to the shelf edge are referred to as the transition and geostrophic zones respectively (Swift, Figueiredo *et al.*, 1983; Swift & Thorne, 1991). The boundary between these two outer zones and the inner zone is of critical importance in determining the long-term net sediment accumulation characteristics of particular shelves. At the sea bed, water particle motions of waves on the inner shelf are strongly transformed into oscillatory, to-and-fro motions (see Fig. 6.7a), usually with a stronger onshore landward component (see Fig. 6.7b). This dominant mean landward-bottom stress results in a net shoreward transport of sand and, in conjunction with other hydraulic processes, traps sediment preferentially in the nearshore zone. This important boundary is referred to as the littoral energy fence (see Fig. 6.7). In order for sediment to 'escape' from this trapping zone it has to bypass the fence in one of two ways (Swift & Thorne, 1991): (i) river mouth by-passing (see Figs 6.4a & 7.5); or (ii) shoreface bypassing (see Figs 6.4b & 7.6; Sects 6.2.1 & 7.2).

Wind-driven currents are the result of wind shear stress on the water surface, with energy transferred through turbulent mixing (Fig. 7.17). These currents are indirectly the result of atmospheric circulation systems and operate over a wide range of temporal and spatial scales. Surface currents generated by wind stress deviate from the wind direction in response to the Coriolis effect, and at depth this deviation is intensified by the Ekman spiral effect. For example, during winter storms wind-driven currents have fluctuating near-bed velocities which are commonly greater than 25 cm s^{-1} with maxima around 80 cm s^{-1} at depths of 50–80 m (Smith & Hopkins, 1972). These unidirectional currents flow mainly offshore and fluctuations in current strength correlate with variations in wind speed.

Storm-surge is a specific storm-related condition caused by a marked reduction of barometric pressure and/or high wind stress, which produces an abnormally high water level at the coastline followed by a drastic lowering of water level. Intense wave agitation and coastal erosion accompany these surges, while the contemporaneous wind-driven seaward-returning bottom current transports nearshore-derived sand offshore on to the shelf (Morton, 1981; Sneddon & Nummedal, 1991) (Sect. 7.4.4).

7.4.2 Storm-dominated deposition on transgressive shelves

Transgressive systems are associated with accommodation-dominated regimes (Sect. 7.2) in which the shoreline migrates landward by shoreface erosional retreat and river mouths are converted to estuaries. In wave- and storm-dominated settings with micro- to macrotidal shorelines, much of the sediment stripped from the shoreface along with material derived from the littoral drift system is driven by waves into estuary mouths and tidal inlets (Sect. 6.7). Wave energy prevents the build-up of high-relief sandbodies on the inner shelf and keeps the majority of coarse-grained sediment within the shoreline–inshore environment. However, an onlapping transgressive sandsheet of subdued relief can develop seaward of the retreating shoreface (Fig. 7.18). In the Middle Atlantic Bight the nature of this sandsheet is a result of relative sea-level rise, landward retreat of coastal barrier and estuarine systems and simultaneous shelf reworking (Fig. 7.19; Swift, 1975a). Sand is supplied to this shelf as follows: (i) leakage due to the irregular nature of barrier retreat (e.g. barrier overstep); (ii) landward migration of sand-rich shoal retreat massifs aligned perpendicular or at high angles to the shoreline; (iii) transgressive retreat and reworking of incised valleys, major barrier inlets and estuary mouth shoals; and (iv) transgressive abandonment and reworking of lowstand deposits (e.g. shelf-edge and mid-shelf deltas).

Sand stored in the transgressive sand sheet is sequentially reworked during shoreface retreat, particularly in the friction-dominated inner shelf zone where storm-generated, wind-driven currents have formed extensive ridge and swale systems (Hunt, Swift & Palmer, 1977). This comprises a series of sandridges, each up to 10 m high and spaced from 2 to 4 km apart, which make an angle with the shoreline of some 25–45° and are attached to the shoreface at their downdrift end (Duane, Field *et al.*, 1972; Swift & Field, 1981). These sandbodies are initiated

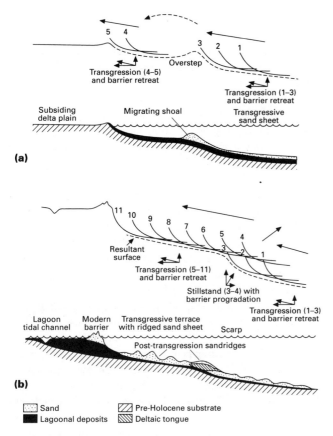

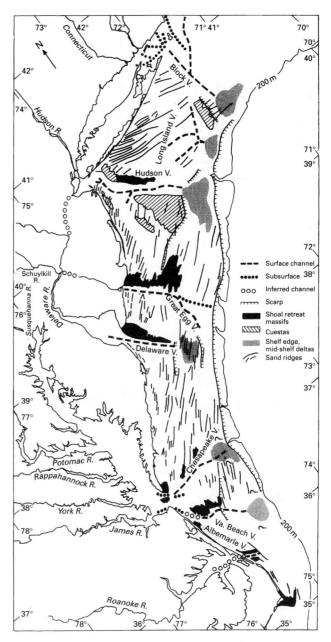

Figure 7.18 Mechanisms of landward barrier migration and the nature of the transgressive sand sheet: (a) barrier overstep with thin, smooth sand sheet, and (b) barrier step-up producing a shoreface erosion surface with an overlying stormridge sand sheet. Based on the Holocene transgression along the eastern USA (from Swift, Phillips & Thorne, 1991b).

in very shallow inner shelf to shoreface settings and continue to grow as active ridges on the inner shelf, where they are in equilibrium with the present-day storm-dominated hydraulic regime (Fig. 7.20). During transgression they become abandoned and inactive on deeper parts of the shelf (moribund ridges). This pattern resembles the development of tidal sandridges, although their internal facies characteristics will show differences such as an abundance of hummocky cross-stratification (Swift, 1975b).

The thickest sand accumulations occur in two major types of shoal retreat massif: (i) along the retreat path of estuary mouth shoals; and (ii) at the convergence of regional littoral drift systems ('cape shoal retreat massifs') (Fig. 7.19). Elsewhere on the Middle Atlantic Bight, interrupted barrier retreat has formed a series of subdued sand-prone scarps, which are the remnants of barriers that were built during short stillstand or regressive periods, and which were 'scalped' by the reestablished transgression (Figs 7.18b & 7.19).

Figure 7.19 The main morphological features of the Middle Atlantic Bight, eastern USA (from Swift, Duane & McKinney, 1973).

Conventional models of transgressive shelf–shoreline systems have been developed from tectonically inactive areas, such as the Mississippi Delta where transgressive sand shoals develop on abandoned delta lobes (Sect. 6.7.6) and the Middle Atlantic Bight, where some 10–15 m of vertical coastal erosion is attributed to shoreface retreat. In tectonically active areas, such as along the Canterbury Plains of New Zealand, erosional relief can be up to 40 m and the resulting transgressive sand sheet may be correspondingly thicker (Leckie, 1994) (Fig. 7.21).

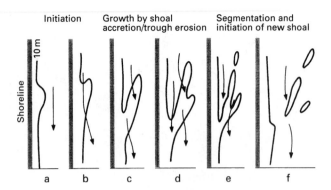

Figure 7.20 Model of progressive detachment and isolation of shoreface-attached shoals and their conversion to offshore linear sandridges. The sequence a to f of represents progressive transgression and increased shoreface detachment with time (from Field, 1980).

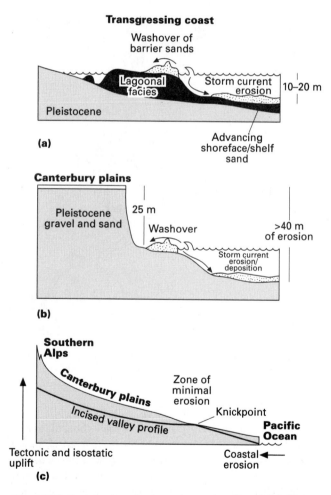

Figure 7.21 A mechanism of enhanced coastal erosion and shelf sands supply in tectonically active basins due to uplift, cliff erosion and shoreface retreat; (a) the conventional model; (b) the Canterbury Plains of eastern South Island, New Zealand; and (c) regional cross-section through the Canterbury plains (after Leckie, 1994).

7.4.3 Characteristics of modern shelf storm deposits

The nature of modern shelf storm deposits is controlled by: (i) energy level of the hydraulic regime; (ii) type of sediment available; (iii) direction of the storm-generated currents with respect to the shoreline and/or intrashelf sediment sources; (iv) amount of subsequent post-storm physical and/or biological reworking (i.e. preservation potential); (v) distance from shore-line and/or intrashelf sediment sources; and (vi) water depth.

Modern shelf storm deposits have been described from several different shelves, including the Gulf of Mexico (Hayes, 1967a; Morton, 1981; Snedden & Nummedal, 1991), German Bight (Gadow & Reineck, 1969; Aigner & Reineck, 1982), southern Brazil (Figueiredo, Sanders & Swift, 1982; Vianna, Salewicq *et al.*, 1991), the Niger Delta (e.g. J.R.L. Allen, 1965c, 1982b), Bering Shelf (Nelson, 1982), Beaufort Sea (Hequette & Hill, 1993), and on Caribbean platforms (Brackett & Bush, 1986; Wanless, Tedesco & Tyrrell, 1988). Box cores reveal that modern storm layers typically display the following characteristics (Fig. 7.22): (i) erosive base; (ii) basal lag of mud clasts, shells, plant debris and/or rock fragments; (iii) horizontal to low-angle lamination, which in three dimensions is probably hummocky-type cross-stratification (Fig. 7.22b); (iv) wave ripple cross-lamination; and (v) burrowed interval. Individual beds are around 0.50 m thick in the Transition Zone of the California shelf (Fig. 7.22d) but can reach almost 3 m in the shoreface at Fire Island (Fig. 7.22b).

Many of these inferred storm layers are graded and display waning flow sequences making them superficially analogous to turbidites (Fig. 7.22c), but shelf storm sand layers can be distinguished from turbidites by the presence of: (i) wave ripple cross-lamination; (ii) wave rippled top surface; (iii) *in situ* shallow marine shelf faunas within distal shelf muds; (iv) marked increase in the bioturbation of storm sand layers from proximal to distal settings; and (v) association with shallow-water facies (cf. Reineck & Singh, 1972; Aigner & Reineck, 1982).

In the tide- and storm-influenced German Bight proximal–distal trends show three main associations of storm deposit (Fig. 7.23): (i) coastal/shoreface sands comprise amalgamated sequences of erosively bounded storm deposits (c. 5–130 cm thick), without intervening mud layers; (ii) transition zone sands comprise well-preserved storm beds (5–100 mm thick) with single waning flow depositional sequences; and (iii) shelf mud zone contains occasional, thin (mainly 4–10 mm) and finer-grained storm sand layers.

Bioturbation commonly obliterates individual storm sand layers in shelf settings, the degree of bioturbation reflecting not only the number of burrowing organisms but the time between storm events (Sepkoski, Bambach & Droser, 1991). However, storm events may be inferred from the presence of sands within biogenically homogenized silty/sandy muds.

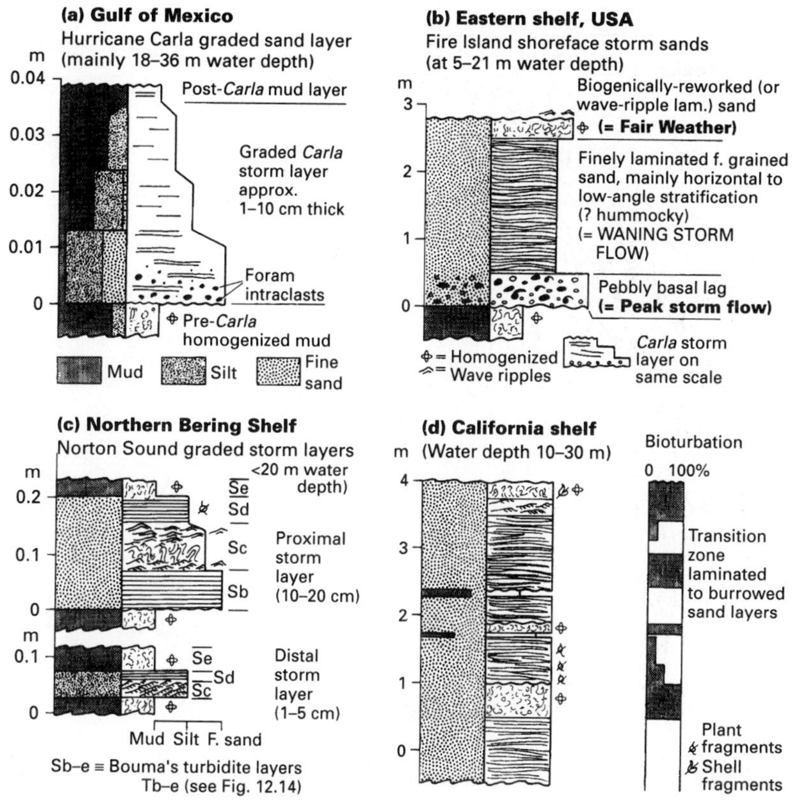

(a) Gulf of Mexico
Hurricane Carla graded sand layer
(mainly 18–36 m water depth)

Post-*Carla* mud layer

Graded *Carla* storm layer approx. 1–10 cm thick

Foram intraclasts

Pre-*Carla* homogenized mud

Mud   Silt   Fine sand

(b) Eastern shelf, USA
Fire Island shoreface storm sands
(at 5–21 m water depth)

Biogenically-reworked (or wave-ripple lam.) sand (= Fair Weather)

Finely laminated f. grained sand, mainly horizontal to low-angle stratification (? hummocky) (= WANING STORM FLOW)

Pebbly basal lag (= Peak storm flow)

= Homogenized mud
= Wave ripples

Carla storm layer on same scale

(c) Northern Bering Shelf
Norton Sound graded storm layers
<20 m water depth

Se
Sd
Sc   Proximal storm layer (10–20 cm)
Sb

Se
Sd
Sc   Distal storm layer (1–5 cm)

Mud   Silt   F. sand

Sb–e ≡ Bouma's turbidite layers
Tb–e (see Fig. 12.14)

(d) California shelf
(Water depth 10–30 m)

Bioturbation 0 100%

Transition zone laminated to burrowed sand layers

Plant fragments
Shell fragments

Figure 7.22 Four examples of modern offshore storm deposits from contrasting shelf settings (note different vertical scales). (a) Graded sand layer resulting from Hurricane Carla, offshore Texas (after Hayes, 1967). (b) Three-part subdivision of storm sands from Fire Island shoreface. Note the slightly coarser, winnowed fairweather sand layer on top (after Kumar & Sanders, 1976). (c) Proximal and distal graded storm sand/silt layers from the epicontinental Bering Shelf, adjacent to the Yukon delta. Note use of equivalent turbidite terminology (after Nelson, 1982). (d) Sequence of amalgamated storm sand layers from the Transition Zone of the California shelf (after Howard & Reineck, 1981).

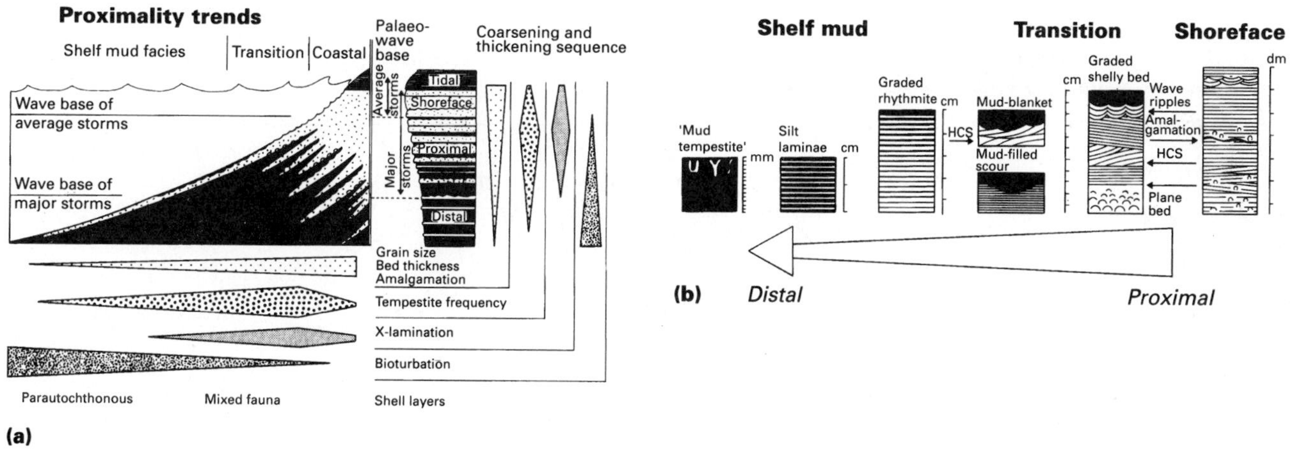

Proximality trends

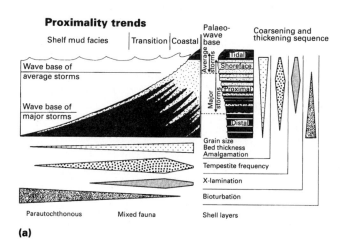

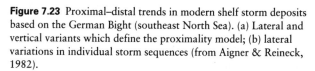

(a)

Figure 7.23 Proximal–distal trends in modern shelf storm deposits based on the German Bight (southeast North Sea). (a) Lateral and vertical variants which define the proximality model; (b) lateral variations in individual storm sequences (from Aigner & Reineck, 1982).

7.4.4 Hydrodynamic models of modern shelf storm deposits

The hydrodynamic interpretation of modern shelf storm deposits is difficult because it has never been possible to correlate precisely the physical processes accompanying major storms and processes acting on the sea bed. Processes proposed include: (i) storm waves; (ii) wind-driven currents; (iii) storm waves combined with ebbing tidal currents; (iv) storm-surge ebb currents; (v) rip currents; (vi) tsunamis; and (vii) density currents.

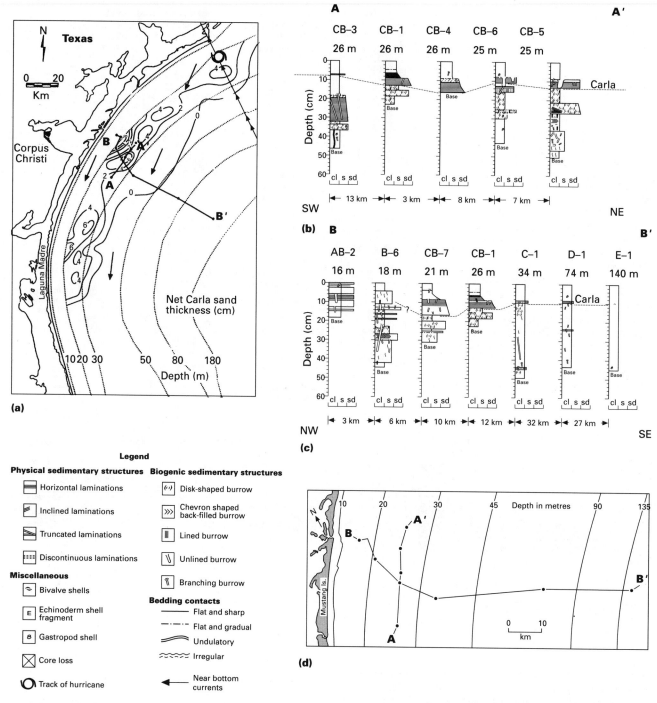

Figure 7.24 The distribution and internal characteristics of the Hurricane Carla storm sand layer on the north Texas shelf based on shallow sea-bed cores and samples (inferred along-shelf transport from northeast to southwest). (a) Net sand thickness (cm); (b) along-shelf variation in sand layer properties; (c) across-shelf variations in sand layer properties; (d) core hole locations and sea-level bathymetry (after Snedden & Nummedal, 1991).

Considerable emphasis was once placed on the analysis of Hurricane Carla, which deposited a single graded sandbed that extends continuously from the shoreface and obliquely across the shelf for some 200 km (Fig. 7.24). This was initially inter-preted by Hayes (1967) as the result of storm-surge ebb turbidity currents, but both Morton (1981) and Snedden and Nummedal (1991) argue in favour of a strong combined geostrophic (wind-driven) storm flow mechanism. Direct measurements in the

Gulf of Mexico during the passage of hurricanes and tropical storms (Murray, 1976; Forristall, Hamilton & Cardone, 1977) indicate that when the winds blow at relatively high angles or perpendicular to the coast, water in the surface layers moves onshore almost simultaneously with bottom waters flowing offshore, sometimes accompanied by strong along- shore currents. Seaward-returning bottom currents have been measured at 100–160 cm s⁻¹ with Hurricane Camille (Murray, 1976) and between 50 and 75 cm s⁻¹ for tropical storm Delia (Forristall, Hamilton & Cardone, 1977), and could have exceeded 200 cm s⁻¹ at the centre of major hurricanes such as Carla (Morton, 1981). Furthermore, maximum current velocities were found to occur shortly after maximum wind stress and storm-surge ebb currents appear to be of minimal importance. Hence, a wind-driven or geostrophic mechanism is preferred on the basis of oceanographic and modelling studies (Fig. 7.25). Both geostrophic and turbidite models have been invoked for ancient storm beds (see Fig. 7.43; Sect. 7.7.2).

The deposits of catastrophic events affecting shelf seas, such as tsunamis or exceptional storms (e.g. 100 years or greater), have never been described from modern environments and must be inferred from ancient successions which, through the detailed analysis of physical sedimentary structures, provide additional insight into all of these high-energy, episodic events (Sect. 7.7.2).

7.5 Modern oceanic current-dominated shelves

Shelves may be influenced by powerful oceanic currents that either constantly propagate on to the outer shelf edge or, due to the periodic shedding of dynamic eddies, penetrate from adjacent ocean basins (Open University, 1989b). The best known example is the southeast African shelf (Fig. 7.26).

This is a microtidal, relatively narrow (10–40 km), approximately 700 km long, oceanic current-dominated shelf characterized by an unusually steep slope (c. 12°). The powerful geostrophic Agulhas Current flows southwards just seaward of the shelf break with a surface velocity of up to 2.5 m s⁻¹.

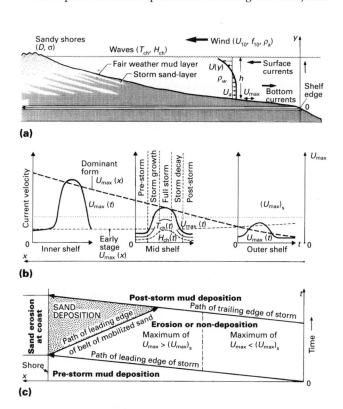

Figure 7.25 A conceptual physical model for shelf storm deposits (from J.R.L. Allen, 1982b). (a) Sandy coastal zone passing transitionally basinward into muddy shelf. Storm-generated surface winds blow onshore and have an onshore flowing surface current (U(y)) and an offshore flowing bottom current (Uₐ). Maximum orbital velocity is represented by (Umax). (b) Predicted offshore decline in wave-related current velocities (Umax(t)) at inner, mid- and outer shelf locations. Sand transport threshold is represented by (Umax)s. (c) Variation in wave conditions with time and indicating the offshore thinning sand sheet.

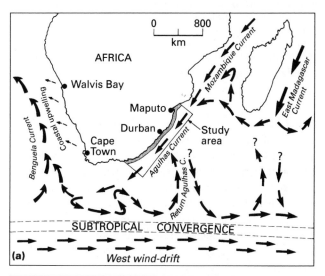

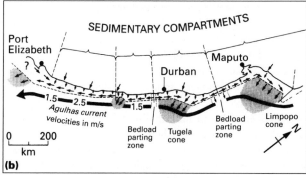

Figure 7.26 (a) General ocean current systems around southern Africa. (b) Current patterns on the Agulhas shelf and velocities of the Agulhas current in m s⁻¹. Notice current reversals associated with stepped form of shoreline which divides the shelf into a number of sedimentary compartments (from Flemming, 1980, 1981).

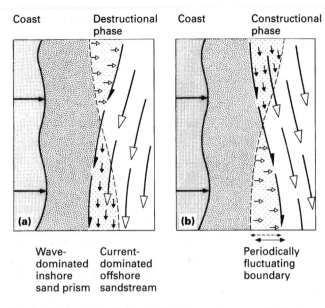

Coast | Destructional phase | Coast | Constructional phase

(a)
(b)

Wave-dominated inshore sand prism | Current-dominated offshore sandstream

Periodically fluctuating boundary

Figure 7.27 Southeast African shelf sedimentation controlled by the migration of the Agulhas oceanic current. (a) Destructional phase. Where the oceanic current sweeps inshore it erodes the seaward edge of the inshore wave-dominated zone and transports material along shelf and seaward. (b) Constructional phase. Where the oceanic current detaches from the inshore zone the sand prism advances seaward (from Flemming, 1980).

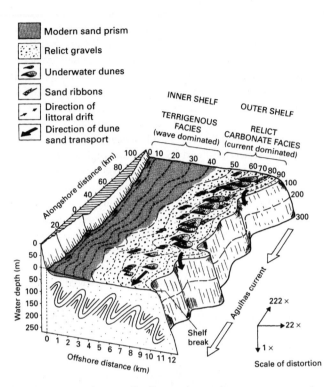

Figure 7.28 Distribution of bedforms along sediment transport paths on the southeast African shelf, and the relationship between the wave- and current-dominated zones and the Agulhas Current (from Flemming, 1980, 1981).

Periodically, this current migrates on and off the shelf (Fig. 7.27) partially in response to seasonal controls (Pages & Gili, 1992). Two possible outcomes result: (i) an outer 5–20-km-wide zone is swept clean of terrigenous sediment leaving a zone dominated by up to 80% biogenic carbonate (Flemming, 1980; Fig. 7.28); or (ii) when the current migrates to the landward margin of this outer zone (a boundary partly formed of drowned coastal dunes), it strips the outer margin of the 5-km-wide inner shelf zone of land-derived, wave-dominated clastic sediments. This sets up a mobile sand stream on which are developed, on the landward side, large sandwaves that are replaced immediately seawards by sand ribbons and sand streamers of decreasing size. Yet closer to the shelf edge are scoured lag gravels derived from *in situ* winnowing of Late Pleistocene reef carbonates and Early Flandrian estuarine terrigenous gravels. Given the 100×10^6 m^3 of sediment annually put into the inner wave-dominated shelf zone (Flemming, 1981, p. 269) the Agulhas current is effective in maintaining an equilibrium profile, particularly in those areas where the absence of the drowned coastal dunes does not promote vertical accretion of near-shore sediments. This partial equilibrium is due to the fact that the southeast African shelf has a step-like form (Fig. 7.26b) and the current either sweeps sediment over the shelf break or into the heads of submarine canyons for a net export from the shelf.

The step-like plan form of the shelf has a secondary effect on current trajectories and sediment movement on the shelf. Momentum in the current is sufficient for relatively major irregularities to promote areas of flow separation where the current ceases to hug the coastline. Thus, the shelf is divided into compartments (Fig. 7.26b) that are characterized in lee positions by clockwise flowing counter eddies that transport sediment out to the shelf edge and so supplement the normal sweep of the current. The juxtapositioning of bedload parting and convergence zones gives rise to areas with sedimentary structures that are partly comparable with those found under tidally reversing conditions (Sect. 7.3).

Most other ocean current-dominated shelves are characterized by relatively weak current systems that, while areally extensive, require significant wind enhancement in order to be able to entrain and transport even fine-grained sediment (see Sect. 7.6). Those that are influenced but not dominated by ocean currents include: (i) the outer Saharan shelf (Newton, Seibold & Werner, 1973); (ii) the Middle Atlantic Bight and the Atlantic shelf of Florida (Weibe, 1982; Churchill, Levine *et al.*, 1993) where the Gulf Stream oscillates and sheds a variety of 'warm rings' that spill on to the shelf; and (iii) the Newfoundland shelf where the relatively weak Labrador Current is able to produce transverse bedforms at water depths below normal wave base (c. –110 m) (Barrie & Collins, 1989).

7.6 Modern mud-dominated shelves

Some 70% of the sedimentary record is made up of muds, silts and lithified mudrocks. However, the processes responsible for the erosion, transportation and deposition of mud cannot be simply classified in terms of different types of shelf current, and other unique processes (e.g. oxygen level, organic content, animal activity, etc.) can be even more important (e.g. Gorsline, 1984). Hence mud-dominated shelves warrant special attention.

Modern process studies of mud deposits characteristically employ X-radiographs of box cores in order to describe macroscopic textural changes. From these it is possible to describe directly changes both in physical characteristics and relative rates of input of sedimentary particles, and to measure mineralogical and geochemical variations (e.g. Brenner & Willis, 1993). Combined with radiochemical studies (e.g. ^{210}Pb, ^{14}C, ^{238}U, and others) quantitative rate models may be developed (e.g. Kuehl, DeMaster & Nittrouer, 1986). In lithified settings, owing to the lack of obvious textural contrasts, it is often necessary to employ 'proxy-indicators', typically trace fossils and body fossils. These also include presence/absence records of physical sedimentary textures that are/are not destroyed by bioturbation, which allows reconstruction of the physicochemical conditions at the sediment–water interface. These proxy-indicators become the 'passive-markers' with which it is possible to document changes at a variety of spatial and temporal scales, ranging from varve-scale variations of a few ky, through bed-scale (Milankovitch-type cycles of 400, 100, 40 and 20 ky) and sequence-scale (1–10s of My) to megascale periods, which span several hundred My (Wetzel, 1991).

The shelf receives fined-grained suspended sediment from four sources: (i) from the land (dominant, c. 7×10^9 t year^{-1}); (ii) laterally from offshore (minor, but with seasonal variations and with ocean current and Eckman current enhancement); (iii) upward from the bottom (variable in both space and time, but may be seasonally very high on inner shelves); and (iv) downward from the near-surface organic layer (minor) (cf. Meade, Sachs *et al.*, 1975; Milliman & Meade, 1983; Milliman & Syvitski, 1992).

The bulk of this fine-grained sediment is transported on to shelves either: (i) in suspension as more-or-less diffuse sediment plumes; or (ii) as near-bed and near-surface concentrations known as nepheloid layers. Sediment is dispersed by complex combinations of currents, waves and other processes (McCave, 1984). Part of this fine sediment is deposited on shelves in areas of low energy, while the remainder is transported across shelves to accumulate in deeper water areas beyond the 'mud line' (Stanley, Addy & Beherens, 1983).

Prodigiously high fine sediment loads off the mouths of some of the world's major drainage systems including the Yellow Sea/East China Sea and the Amazon Shelf (see Fig. 7.1) build areally extensive subaqueous deltas that, in their distal setting, merge with ambient shelf sediments. In these and other examples proximal delta/inner shelf platforms are modified by waves and by wave-enhanced tidal currents and sediment is trapped inside the 'energy fence' (see Sect. 7.4.1) in inner shelf and coastal fringe environments. Other sediment slumps and slides from the fronts of these deltas and, in the case of the Hwanghe and Yangtze deltas, is incorporated in the shelf sediment in a variety of confined and unconfined hyperpycnal underflowing and overflowing plumes and plume fronts (Ren & Shi, 1986; Wiseman, Fan & Bornhold, 1986; Wright, Yang & Bornhold, 1986).

In the case of the Amazon, some local slumping and resedimentation also occurs, but fluvial discharge is so great that estuarine-like circulation is displaced out on to the shelf (Nittrouer & DeMaster, 1986). This penetrates the energy fence and bypasses sediment out to the distal delta bottomsets and beyond even the outer shelf margin. At times of peak discharge a plume of turbid water may extend some 200 km off the coast of Brazil and it may be powerful enough to generate its own backflow of bottom water on to the shelf.

Fine-grained sediments are readily transported by even relatively weak currents. In the Yellow Sea/East China Sea northward flowing, warmwater currents, derived from the Kuroshiro Current, penetrate far into this shallow (<100 m deep) and low slope (<1°) shelf (Fig. 7.29; Alexander, Demaster & Nittrouer, 1991; Congxian, Gang *et al.*, 1991; Hsueh, Wang & Chern, 1992). During the summer, these currents are intermittently enhanced by northward moving typhoons and more constant gentle winds from the south that produce only minor wave-generated turbulence; suspended-sediment concentrations remain low in the 0.5–5.0 mg l^{-1} range (Ikeva, 1988; Park, 1986; Milliman, Qin & Park, 1989). During this relatively calm period the water column on the shelf is stratified and biogenic input from the surface layer becomes important. In contrast, during the winter period frequent powerful storms moving to the south fully mix the shelf water column and significantly strengthen the four southerly permanent coldwater currents (Fig. 7.29) causing the resuspension of bottom sediments and the genesis of turbid plumes that may have sediment concentrations as high as 1–100 mg l^{-1}. This results in a sedimentation pattern where mud zones, occurring below turbid, cold southward-flowing currents, are juxtaposed against relict sand zones formed by the warm, relatively clear north-ward flows (Fig. 7.29). Significant amounts of fine sediment reach the southern end of the transport paths and bypass the shelf.

The Amazon Shelf (see Fig. 7.1) is dominated by the Guiana Current, which moves northwards over the shelf at 35–75 cm s^{-1}. This is frequently enhanced by trade-wind-driven waves which have a major effect on the inner and middle parts of the shelf shallower than −70 m. The inner zone is composed mostly of clayey-silt or silty-clay deltaic deposits with sands near river mouths, while the outer zone (to the shelf break at −100 m) is dominated by relict transgressive fine sands (Milliman, Butenko *et al.*, 1982; Alexander, DeMaster &

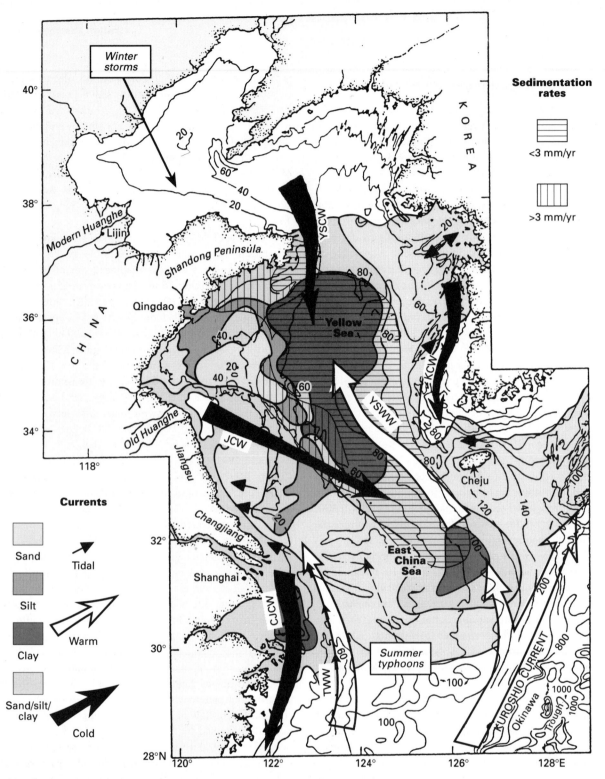

Figure 7.29 General circulation patterns and present-day surface sediment textures in the epeiric Yellow/East China Sea. The northeast-flowing Kuroshiro Current spills northward on to the shelf as the Yellow Sea Warm Water current (YSWW) and the Taiwan Warm Water current (TWW). Generally southward directed return flows are dominated by the Yellow Sea Cold Water current (YSCW), the Jiangsu Cold Water current (JCW), the Korean Coastal Water current (KCW) and the Changjiang Coastal Water current (CJCW). Note the influence of strong tidal flows on the west coast of the Korean Peninsula and near to the mouth of the Changjiang (Yangtze) River. Northward warmwater flows are enhanced by periodic summer typhoons. Southward coldwater flows are enhanced by more persistent winter storms that tract out of the northwest. Both summer and winter winds induce wave erosion of the sea bed. The extended plume in the central portion is advected under the influence of the Jiangsu Coastal Water (JCW), where the two warmwater currents (TWW and YSW) penetrate the shelf. The outer shelf is characterized by relict sands and only minor inputs of fines (after Milliman, Qin & Park, 1989).

Nittrouer, 1986). Five facies are recognized with a predictable large-scale distribution (Kuehl, Nittrouer & DeMaster, 1982; Kuehl, DeMaster & Nittrouer, 1986; Fig. 7.30): (i) physically stratified sand in the main Amazon channels, characterized by ripples, scour and fill, and rip-up clasts and apparently influenced by pulsatory, probably tidally enhanced currents; (ii) down transport path thickening, locally tidally scoured, graded sands and silts interbedded with mud layers which form a zone extending more than 100 km out on to the shelf and northwestward for over 400 km along the shelf; (iii) proximal-shelf sandy silt which occupies the topset area of the delta and is composed of homogeneous muds with thin laminae of sandy silts forming in an area of generally mixed powerful currents; (iv) faintly laminated mud which forms a suspension-fed shoal area and is characterized by rates of sedimentation so high (40–100 mm year^{-1}) that bioturbation is all but eclipsed; and (v) mottled mud which accumulates in regions of relatively low sedimentation (<40 mm year^{-1}) and is characterized by very high densities of polychaetes and extensive bioturbation. A sixth facies overprints most of these five depositional zones and is composed of black organic, varve-like laminae with spacings

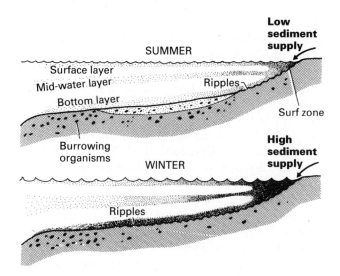

Figure 7.31 Seasonal sedimentation patterns on the Oregon continental shelf. Turbid layers are represented by stippling (from Kulm, Roush *et al.*, 1975). Turbid layer transport and sediment rippling are enhanced during winter due to high sediment input from coastal streams and long-period waves in water depths of at least 125 m.

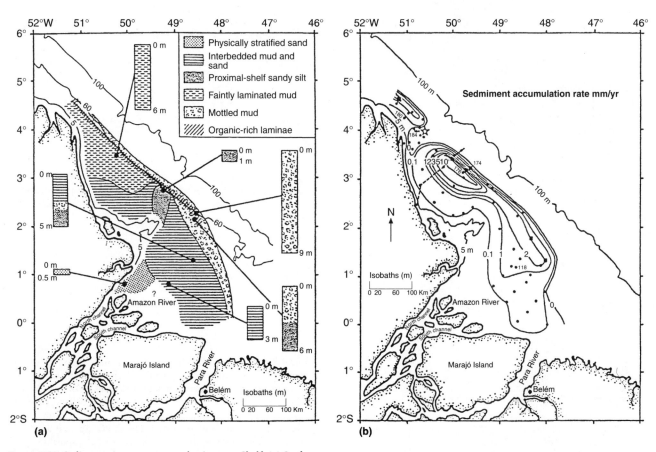

Figure 7.30 Sedimentation patterns on the Amazon Shelf. (a) Surface sediment textures and facies characteristics illustrated with representative short cores. (b) Sedimentation rates (mm year^{-1}) based on ^{210}Pb (from Kuehl, DeMaster & Nittrouer, 1986).

of 1–20 mm. This unit may either represent an annual accumulation of land-derived organic matter or derive from plankton blooms on the shelf itself, perhaps in response to seasonal upwelling or other oceanographic controls.

As with the Yellow Sea/East China Sea dispersal system the wave and seasonally enhanced current system of the c. 1600 km long, northeast South American shelf is effective in propelling some 150×10^6 m³ of sediment along the shelf and eventually over its end into the Gulf of Paria, near Trinidad. A significant portion of this load (c. 100×10^6 m³) is composed of fine-grained sediments. On the French Guyana and Surinam portions are shore-attached and estuarine mud banks (Sect. 6.7.2), spaced evenly every 30–60 km (Wells & Coleman, 1981). These are the product of deposition from shifting patches of 'slingmud' composed of phenomenally high suspended sediment loads (c. 300 mg l⁻¹) that may modify incoming water wave

trains and resolve them into solitary waves of considerable erosive and transporting power, further enhancing the regional currents.

Seasonal variations in rates of input of sediment to shelves and in the energy available in the sediment dispersal system have marked effects on mud-dominated shelves. The Oregon/North California shelf shows seasonal variability in the amount of mud it receives from river mouths, in storm reworking and in sediment transport (Kulm, Roush *et al.*, 1975; McCave, 1979; Demirpolat, 1991). A variety of seasonally moderated inflows, overflows and underflows that are jetted on to the shelf determine the seasonal position of shelf mud zones (Fig. 7.31). The Northern Colombian, Caribbean Shelf is subject to wet season/dry season variations in rates of sediment input and to wind shadowing effects by mountains (Fig. 7.32a; Pujos & Javelaud, 1991). During the wet season (May–November) the

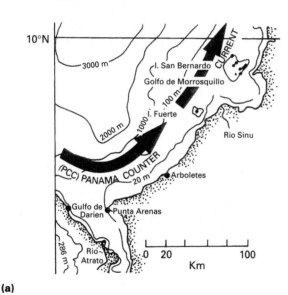

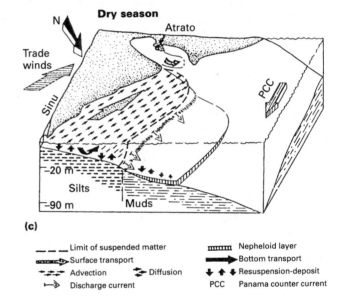

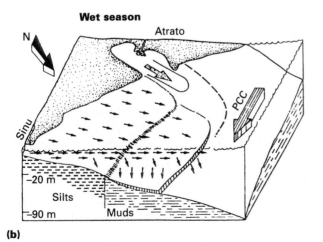

Figure 7.32 Sedimentation on the mud-dominated, seasonally ocean current-influenced northwest Colombian Shelf of the western Caribbean Sea. (a) The general track of the Panama Counter Current (PCC). (b) Wet season conditions with a powerful PCC and a strong offshore/offshelf advection of fine-grained sediment as a surface plume. (c) Dry season conditions with a weak PCC, partial wind shadowing of the trade winds on the shelf resulting in some resuspension of bottom sediments, southward alongshelf flux of fines, and only partial offshelf escape of muds as bottom-flowing nephloid layers (after Pujos & Javelaud, 1991).

shelf receives a high input of fines that, due to shadowing, are confined to an inner shelf depositional zone (Fig. 7.32b). During this season, however, the Panama Countercurrent intrudes more deeply and more powerfully on to the outer shelf and breaks up any thermohaline stratification, thus setting up a residual near-bottom current that accelerates the diffusion of fines from the shelf. During the dry season (December–April) the input of sediment is low and both vertical thermal and horizontal salinity structures become superimposed (Fig. 7.32c). This produces a relatively stable water mass shoreward of the –20 m isobath in which fined-grained sediment is advected southwards in response to a combined longshore process, ocean swells and northeast trade-wind-driven waves.

Other well-documented, highly seasonal-influenced mud-dominated shelves include those of Arctic Alaska (Sharma, 1979) and those fronting present-day or recently glaciated terrains (Molnia, 1983, 1989; see Chapter 11).

7.7 Ancient shallow clastic seas: facies recognition and interpretation

Fine-grained sediments (muds and silts) form thick and laterally extensive shallow marine mudstone successions constituting c. 70% of ancient clastic shelf sediments. In contrast, sandstones are relatively scarce, but they have received much greater attention, mainly due to their economic importance.

Ancient offshore shallow marine successions are distinguished by: (i) biological; (ii) mineralogical/geochemical; and (iii) sedimentological/stratigraphical data (H.D. Johnson & Baldwin, 1986). No single criterion can be used, although biological data (palaeontology, palynology, palaeoecology and ichnology) generally provide the most reliable means of identifying holomarine conditions and estimating shelf palaeobathymetries. The biological and mineralogical/geochemical aspects of ancient shelf deposits are beyond the scope of this chapter and are only referred to in those cases where they contribute to reconstructing depositional processes and environments.

Since there is no single sedimentological/stratigraphical criterion diagnostic of shelf environments, emphasis must be placed on assemblages of data, including texture, sedimentary structures, sandbody geometry, palaeocurrent patterns and stratigraphic relationships. The sequence stratigraphic framework of inferred shelf sandstones may be especially important, notably by establishing both their relationship to contiguous shoreline deposits and the significance of bounding erosional surfaces in relation to relative sea-level changes. This information may be important in distinguishing between offshore, shoreline and inshore sandbodies, which may not be possible by facies analysis alone. For example, the sedimentary structures of inshore/estuarine and offshore tidal sandbodies may be virtually identical. Yet they may be distinguished by their sandbody geometries and orientation, which can be quite different in each environment.

At the facies analysis scale the physical process-based classification of modern shelf seas (Sect. 7.2), combined with the type of available sediment, provides the best means of distinguishing between different ancient clastic shelves. This scheme identifies three main types of ancient clastic shelf deposit (Fig. 7.33): (i) *tide-dominated*; (ii) *storm-dominated*; and (iii) *mud-dominated*.

Tide- and storm-dominated facies are both potentially sand-dominated and, although they are initially considered here as separate deposits, they may be preserved in close association with each other. The most notable absentee in this scheme compared with modern shelves is oceanic current-dominated deposits. On theoretical grounds they are likely to be difficult to distinguish from tidal shelf deposits, notably in cases where they are associated with sand-rich successions dominated by bedload transport. In such cases abundant cross-bedding would be predicted as the main product. However, these deposits are rare on modern shelves (c. 3–5%), they have only been sporadically inferred in the ancient and, as a result, they are not considered any further.

7.7.1 Ancient tide-dominated offshore facies

Tidal sandstones are documented throughout the geological record and many have been assigned to a subtidal offshore setting. Offshore tidal facies differ from tide-dominated sediments deposited within deltas, inlets and inshore/estuarine environments because they are influenced by: (i) rotary tides; (ii) less distinct ebb/flood tidal cyclicity; and (iii) storms and other related offshore processes. However, most of the classic criteria used for recognizing tidal deposits (e.g. herringbone cross-bedding, lenticular/flaser bedding, double clay drapes, tidal bundles, etc.) have come from inshore settings, notably from tidal flats and estuaries where tidal action may be the only significant physical process (Nio & Yang, 1991). Many of these criteria may not be directly applicable to the offshore shelf environment where hydrodynamic conditions are generally more complex.

The importance of this can be illustrated by the reinterpretation of several shelf sandstones as estuarine deposits (e.g. Roda Sandstone: Nio, 1976; Yang & Nio, 1989; Woburn Sands: Bridges, 1982; H.D. Johnson & Levell, 1980, 1995; Wonham & Elliott, 1996). Hence, although this section is primarily concerned with offshore tidal deposits, it will be necessary to consider diagnostic criteria for other tidally influenced deposits and to evaluate what, if any, assemblages of facies and sedimentary structures can be considered indicative of the tidal shelf environ-ment. The most abundant and best documented are found in the late Precambrian, Cambrian and early Palaeozoic, where several thick (c. 100s–1000s m) and laterally extensive quartzitic sandstone units, usually of extremely high mineralogical and textural maturity, occur. Similar deposits are found in younger successions but these are invariably much thinner, less extensive and volumetrically less important.

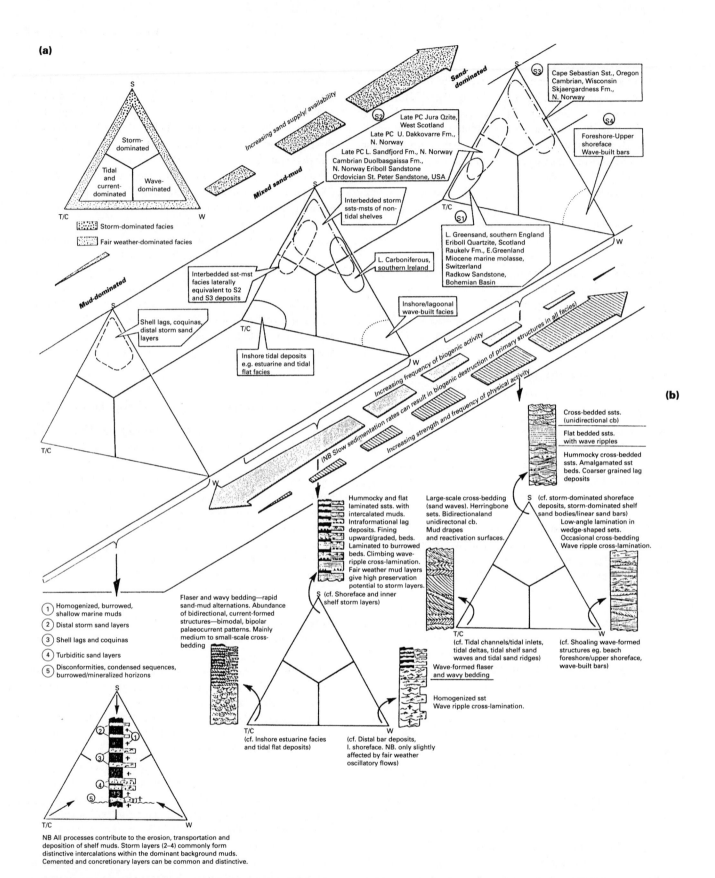

Figure 7.33 (a) Schematic summary of shallow marine siliciclastic facies based mainly on *sand/mud content* (reflecting sediment supply/availability) and inferred *depositional processes* (relative interaction of fairweather and storm processes). (b) Simplified characteristics of the main end-member facies types. See Fig. 7.4 for comparison with modern shelf equivalents.

SEDIMENTARY STRUCTURES AND FACIES IN OFFSHORE TIDAL DEPOSITS

The main sedimentary features common to many of these ancient deposits are: (i) tidal sandbars; (ii) tidal cross-bedding (sandwaves/dunes); (iii) mud layers/drapes; (iv) erosion surfaces/ lag deposits; and (v) mineralogical and textural maturity.

Tidal sandbars (or *banks*) and *sandridges* are recognized by a combination of the following criteria: (i) elongate sandbody orientation/thickness trends; (ii) large-scale, low-angle (<6°) accretion surfaces dipping perpendicular to the sandbody axes; (iii) medium- to large-scale cross-bedding (dm–m thick sets) reflecting smaller, transverse bedforms (sand waves and dunes); (iv) palaeocurrent patterns within the cross-bedded sandstones indicating bedload transport mainly parallel, or slightly oblique to, the sandbody axis (cf. transverse accretion in sand wave facies); (v) evidence of fluctuating current strength and direction (e.g. reactivation surfaces, bipolar palaeocurrents, mud drapes, etc.); and (vi) coarsening/thickening-upward vertical facies profiles (bar migration/progradation).

Examples of inferred ancient sandridges include the Eocene Vlierzele and Diest Sands of Belgium (Houbolt, 1982; Houthuys & Gullentops, 1988) and the Eocene Misoa Formation, Maracaibo Basin, Venezuela (Maregui & Tyler, 1991). The sandridges in the Vlierzele Sand are identified on the basis of sandbody thickness trends (10–20 m thick in axial zones), coarsening-upward facies profile with mud-draped sands, low-angle surfaces interpreted as ridge flanks and palaeocurrents trending mainly parallel to the ridge axis and with bipolar patterns. These sands were deposited in a narrow (<20 km wide), elongate embayment. The Diest sandridges overlie an eroded, pebble-strewn surface, which is incised into older deposits and parallel the trend of the underlying scours, suggesting that the ridges formed within a relatively narrow, elongate seaway or estuary mouth, following a major flooding event (Houbolt, 1982). The Misoa Formation sandbodies form a series of elongate, coarsening-upward sequences, which were deposited within the mouth of a tide-dominated delta (see Sects 6.6.1 & 6.6.6; Figs 6.64 & 6.65).

Tidal sandridges have been inferred from coarsening-upward facies successions (Fig. 7.34) which pass from heterolithic beds with storm sand layers into trough and tabular cross-bedded sandstones (H.D. Johnson, 1977a; Bridges, 1982; Mutti, Rosell *et al.*, 1985; Simpson & Eriksson, 1991; McKie, 1990a). The flanks of the sandbars are preserved as low-angle (c. 5–10°), seaward-dipping accretion surfaces, which are separated by smaller-scale cross-bedding (sandwaves and dunes) indicating bedload transport parallel, or slightly oblique, to the strike of the bar flank (Fig. 7.35). An absence of wave- and storm-related structures, such as hummocky cross-stratification, and lack of extensive bioturbation distinguishes them from storm-generated sandridges (Pozzobon & Walker, 1990) and prograding, storm-dominated shoreface deposits (Sect. 7.7.3).

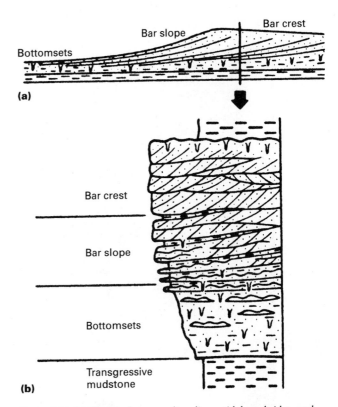

Figure 7.34 (a) Model of a prograding, linear tidal sand ridge, and (b) schematic coarsening/thickening-upward facies succession. Based on the Eocene of the Ager Basin, northern Spain (after Mutti, Rosell *et al.*, 1985).

Palaeocurrent patterns from individual bar flank facies are invariably unidirectional and, as in modern tidal sandridges (Houbolt, 1968, 1982), appear to preserve just one part of the bidirectional sand transport system. Occasionally these bars are dissected by tidal channels, in which bidirectional palaeocurrent patterns from large-scale cross-bedding indicate the presence of mutually evasive ebb and flood tidal currents (H.D. Johnson, 1977a).

Extremely large-scale cross-bedding (set size c. 10–50 m) with avalanche foresets, interpreted to be too large to be transverse bedforms, represents another type of shoreline-attached, tidal sandbank (see Fig. 7.51) (Surlyk & Noe-Nygaard, 1991; Sect. 7.8.2).

Tidal cross-bedding is the product of large flow-transverse bedforms (*sandwaves* or *medium to very large subaqueous dunes*; Ashley, 1990) and forms one of the best known ancient subtidal sandstone facies (e.g. Narayan, 1971; Anderton, 1976; Nio, 1976; Levell, 1980b; P.A. Allen & Homewood, 1984; Teyssen, 1984; Richards, 1986; Kreisa, Moiola & Nøttvedt, 1986; Surlyk & Noe-Nygaard, 1991). This comprises a variety of cross-bedding, which is commonly of medium- to large-scale (0.5–10 m set thickness) and ranges from simple avalanche sets

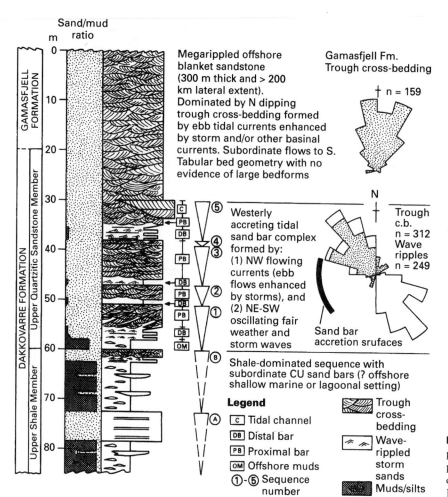

Figure 7.35 Vertical facies succession and palaeocurrent patterns from westerly prograding late Precambrian linear sandbar deposits in north Norway (after H.D. Johnson, 1977a).

through to compound cosets with low-angle surfaces separated by smaller-scale downslope- and upslope-dipping sets. In shelf deposits, cosets are commonly bounded by laterally extensive, sheet-like erosion surfaces, which emphasizes continuity of the cross-beds (e.g. Levell, 1980b). The physical origin of these structures is synthesized in J.R.L. Allen's (1980) analysis of tidal sandwaves, which relates the wide range of cross-bedding types in subtidal sands mainly to the time–velocity asymmetry of rectilinear tidal currents (Figs 6.12 & 7.36).

Mud drapes are often considered one of the most distinctive features of tidal deposits. At times of very high preservation, mud drapes may line foresets in a systematic way and define tidal bundles, which may allow detailed reconstruction of the tidal regime (Visser, 1980; J.R.L. Allen, 1982a; Terwindt & Brouwer, 1986; Sect. 6.2.7). Tidal bundles and sand/mud couplets have only been positively identified in the tidal deposits of inshore areas and coastal embayments and this may enable distinction between these and offshore environments. Mud drapes may be common in offshore tidal shelf deposits (e.g. Houbolt, 1968; Stride, 1982) but their precise distribution has

not been established. In addition to thin mud drapes, many ancient sandstone-dominated tidal shelf deposits contain mud layers up to 5–10 mm thick. Most of the mud layers preserved in these offshore deposits probably do not reflect short-term (diurnal/semi-diurnal) tidal periodicities as in inshore deposits but are more likely to reflect seasonal variations in suspended-sediment concentration. The latter increases dramatically after winter storms due to the reworking of muddy shelf areas, erosion of muddy coastal areas and higher suspended-sediment loads passing through river mouths (McCave, 1970, 1985). The occurrence of prominent mud layers above winnowed lags in several ancient tidal shelf deposits (see below) also supports a post-storm origin (Anderton, 1976; Levell, 1980b). Mudflake layers highlight the probability that most mud layers in sand-rich zones have a low preservation potential and will be eroded and/or resuspended.

Erosion surfaces occur on a wide range of scales in tidal deposits, but the larger features (i.e. 1000s m lateral extent) may be unique to the offshore shelf environment, and allow discrimination between high-energy tidal sands in shelf and

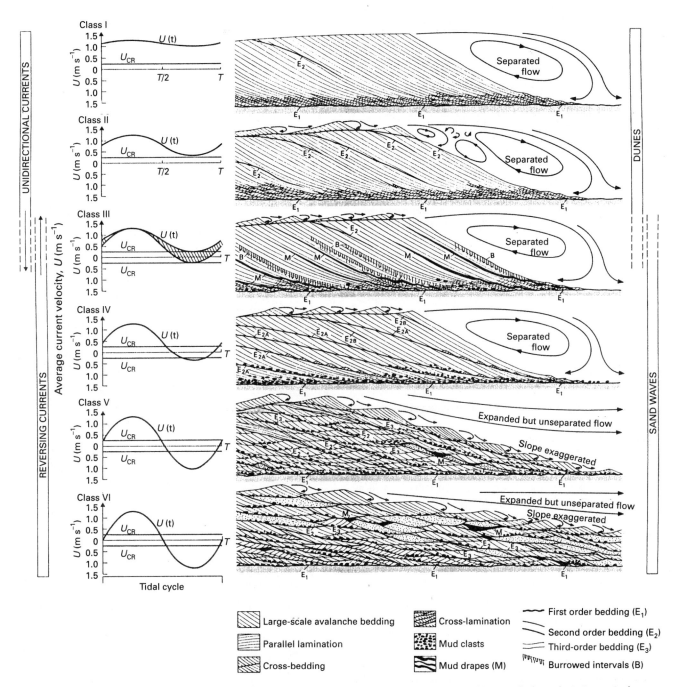

Figure 7.36 A theoretical model to explain the relationship between sandwave (large to very large subaqueous dunes (Ashley, 1990)) morphology and the inferred types of internal sedimentary structure (after J.R.L. Allen, 1980). Classes I–III form in response to effectively unidirectional currents, since the reversing current remains below the critical threshold to allow sediment transport, resulting in bedforms with unidirectional, angle-of-repose foresets. Classes IV–VI form under increasingly bidirectional, rectilinear tidal currents. The large bedforms are less asymmetric; they have gentler lee slopes and are covered by smaller-scale bedforms, which results in increasingly compound sedimentary structures, often with more evidence of flow reversals. The profiles were constructed assuming equilibrium bedforms with the following characteristics: grain size = 0.25 mm, height = 4.25 m, length = 210 m and average water depth = 24.5 m. Abbreviations: U_{CR} = critical velocity to initiate grain movement, $U_{(t)}$ = total current velocity, $T/2$ = half tidal cycle, T = single tidal cycle.

Tidal sand sheets

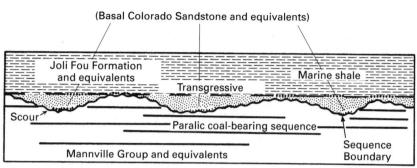

Figure 7.37 Offshore shallow marine, tidally influenced sandstones, overlying an inferred tidal current scoured sequence boundary and separating coal-bearing coastal plain deposits from offshore muddy shelf sediments. Based on the Upper Cretaceous Basal Colorado Sandstone and its equivalents in the Western Interior Seaway of North America (Cessford Field, Alberta) (after Banerjee, 1991).

inshore/estuarine settings. In tidal shelf deposits these surfaces are, in addition to their regional extent, characterized by low relief (mainly <1 m), thin, sheet-like layers of pebbles or granules (e.g. often only 1–5 grain diameters thick) and may be overlain by thin mud or silt layers (Anderton, 1976; Levell, 1980b). In tidal shelf sands they often form bounding surfaces above and below cross-bedded sandstone units (1–15 m thick), which contributes to the tabular, laterally extensive geometry of the sandbodies.

These surfaces are attributed to winnowing by tidal currents, possibly enhanced by storms, when large shelfal areas, originally lined with subaqueous dunes and sandwaves, were planed-off and degraded. Winnowing by waves and currents formed a gravel pavement, analogous to those associated with aeolian dune fields in deserts, while the overlying fine-grained layer preserves the post-storm suspended-sediment load, which is unique to the shelf environment.

Earlier interpretations of these surfaces considered their origin to be essentially autocyclic, formed during extremely high-energy conditions and possibly reflecting bedload parting zones (erosional furrows and sand ribbons) (Sect. 7.3). However, some are so extensive that they may be due to allocyclic processes, such as a fall in relative sea level. For example, some of the regional erosional surfaces within the Lower Cretaceous Viking Formation and Basal Colorado Sandstone (Alberta, Canada), interpreted as resulting from subaqueous tidal scour in a shelfal setting (Banerjee, 1991; Reynolds, 1994), display a rolling topography and are overlain by shallow marine mudstones and prograding tidal shelf sandsheets (Fig. 7.37). The basinward shift in facies and overlying tidal sediments suggests that the erosion surfaces were cut by tidal scour initiated by small-scale falls in relative sea level.

The recognition of tidally influenced sandbodies *above* a subaqueous shelfal erosion surface provides one of the best means of distinguishing between a tidal shelf and an inshore/estuarine setting (Fig. 7.8).

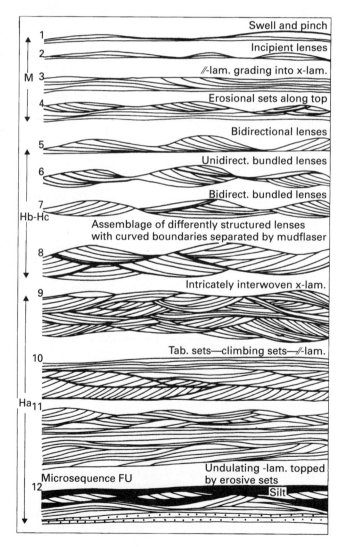

Figure 7.38 Some characteristic wave-dominated structures from Lower Carboniferous shallow marine deposits, Co. Cork, southern Ireland (after de Raaf, Boersma & van Gelder, 1977).

7.7.2 Ancient wave- and storm-dominated offshore facies

Ancient storm deposits show many of the characteristics of those described in Section 7.4. However, the typical, geologically significant storm event may be much rarer than any event observed today. The frequency may have been of the order of one in several thousand years (c. range 1 : 400–15 000 years and average 1 : 5000 years; Brenchley, Newall & Stanistreet, 1979; Kreisa, 1981; Aigner, 1982). Hence, while storms observed and measured by oceanographers may be representative of the type of physical process, they may not always reflect the magnitude of ancient events. In addition, seismically induced tsunamis may create an event that is orders of magnitude greater than the most powerful storm.

SEDIMENTARY STRUCTURES AND FACIES IN
WAVE-AND STORM-DOMINATED DEPOSITS

The dominant primary sedimentary structures associated with storm-dominated processes are: (i) wave ripples/wave ripple cross-lamination; (ii) hummocky cross-stratification; (iii) swaley cross-stratification; (iv) tabular/trough cross-bedding; and (v) graded beds (Aigner, 1985).

Wave ripples and *wave ripple cross-lamination* are common in many shallow-water environments affected by oscillatory waves, including large lakes and lagoons, but they are particularly well developed in shelf seas where wave height and wave fetch are at their largest.

De Raaf, Boersma and van Gelder (1977) recognized a suite of sedimentary structures attributed to oscillatory flows (Fig. 7.38). Features that distinguished wave-generated cross-lamination from that formed by unidirectional flows, are: (i) a less trough-like shape; (ii) an irregular and undulating lower bounding surface; (iii) bundle-wise upbuilding of foreset laminae; (iv) swollen lenticular sets; and (v) offshooting and draping foresets. Since in nature most oscillatory flows are highly variable and display some component of net flow, most of these structures display some degree of asymmetry, often directed onshore.

Hummocky cross-stratification has been defined as a form of medium- to large-scale cross-stratification, in which the undulating and gently dipping laminae preserve a three-dimensional bedform comprising large amplitude (1–5 m), low relief (0.1–0.5 m) mounds and troughs. A hummocky cross-stratified set typically displays the following features (Fig. 7.39): (i) erosional lower set boundary with dips mainly below 10° (maximum 15°); (ii) overlying laminae concordantly draping the basal erosion surface and passing laterally into convex-upwards, accretionary hummocks; (iii) systematic lateral variations of laminae thickness (i.e. pinching and swelling of individual laminae); (iv) highly variable dip directions of both set boundaries and internal laminations; (v) sole marks (mainly prod and drag marks) preserved on basal erosion surfaces where these overlie clay layers; and (vi) a flat, gently undulose or wave rippled top surface. This structure is mainly developed in coarse silt to fine sand, with secondary components typically comprising abundant mica, carbonaceous debris and scattered intraclasts of mudstone and/or marine shells. Textural changes from fine sand through

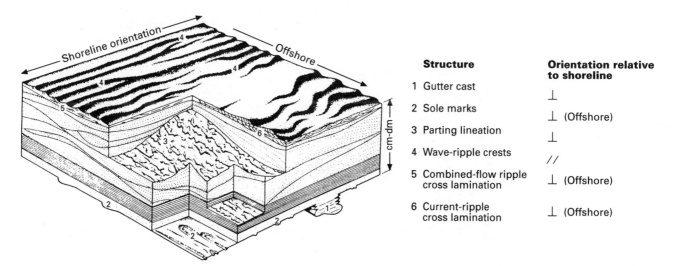

Structure	Orientation relative to shoreline
1 Gutter cast	⊥
2 Sole marks	⊥ (Offshore)
3 Parting lineation	⊥
4 Wave-ripple crests	//
5 Combined-flow ripple cross lamination	⊥ (Offshore)
6 Current-ripple cross lamination	⊥ (Offshore)

Figure 7.39 Characteristic features of an idealized storm sandstone bed. The bed is a hummocky cross-stratified sandstone bed and a frequently observed palaeocurrent relationship in coastal/nearshore associations is shown (after Leckie & Krystinick, 1989). Note that different palaeocurrent patterns can occur in more open shelf situations or where coast-parallel currents dominate.

to coarse-grained, conglomeratic sandstone are often accompanied by a change from hummocky cross-stratification to large-scale wave ripples, with more steeply dipping cross-bedding (Fig. 7.40; Bourgeois & Leithold, 1984; Leckie, 1988; Cheel & Leckie, 1993).

Hummocky cross-stratification is often associated with a distinctive suite of trace fossils, mainly representing infaunal suspension feeders, which can provide additional information on substrate conditions, palaeobathymetry and relative rate of storm sedimentation (Fig. 7.41; Ekdale, Bromley & Pemberton, 1984; Frey & Pemberton 1984; Vossler & Pemberton, 1988).

Hummocky cross-stratification was originally considered to form primarily in response to high-energy currents with a strong oscillatory component and orbital velocities greater than 0.5 m s^{-1} and to represent a large-scale, wave-formed bedform (Harms, Southard & Walker, 1982). Evidence for oscillatory flows also includes: (i) random distribution of dip directions of mainly low-angle laminations; (ii) dominance of wave ripples and wave ripple cross-lamination which commonly overlie hummocky cross-stratification in single, waning flow sequences; (iii) rarity of current-formed sedimentary structures; and (iv) flume experiments (Arnott & Southard, 1990; Southard, Lambie *et al.*, 1990). Evidence for simultaneous unidirectional flows includes: (i) irregular basal erosion surfaces with dm relief; (ii) basal sole marks; (iii) gutter casts (Fig. 7.39) (Whitaker, 1965; Bridges, 1972); and (iv) primary current lineation (e.g. Banks, 1973c; Dott & Bourgeois, 1982). In many ancient successions, these latter data often point to shore-normal, offshore-directed

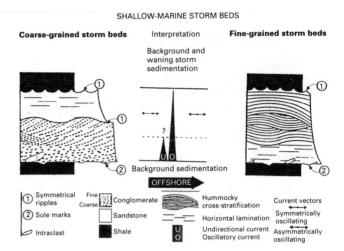

Figure 7.40 Comparison of sedimentary structures and vertical profiles in coarse- and fine-grained storm beds (after Cheel & Leckie, 1993).

currents and support the interpretation of combined (oscillatory/unidirectional) flows (e.g. Cheel, 1991).

No equivalent bedforms have ever been observed in the process of formation on modern shelves but similar structures have been recorded from sporadic core data, including storm-dominated sandridges (e.g. Oertel, 1983; Rine, Tillman *et al.*, 1991; Sects 7.4.3 & 7.4.4). Experimental results indicate that hummocky cross-stratification without any preferred dip direction

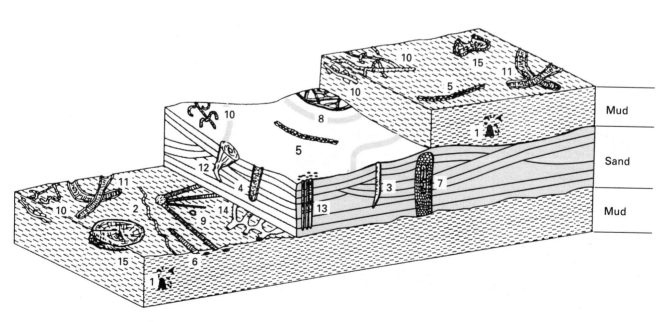

Figure 7.41 Typical trace fossil assemblages within offshore storm (hummocky cross-stratified) sand layers and their bounding mud units. 1, *Chondrites*; 2, *Cochlichnus*; 3, *Cylindrichnus*; 4, *Diplocraterion*; 5, *Gyrochorte*; 6, *Muensteria*; 7, *Ophiomorpha*; 8, *Palaeophycus*; 9, *Phoebichnus*; 10, *Planolites*; 11, *Rhizocorallium*; 12, *Rosselia*; 13, *Skolithus*; 14, *Thalassinoides*; 15, *Zoophycus* (from Ekdale, Bromley & Pemberton, 1984).

can be produced by strong, purely oscillatory flows (Southard, Lambie *et al.*, 1990), while the superimposition of even weak unidirectional flows leads to lateral bedform migration and the formation of cross-stratification inclined in a downcurrent direction (Arnott & Southard, 1990).

Swaley cross-stratification displays similarities with hummocky cross-stratification and probably has some genetic similarities. The structure, as originally defined by Leckie and Walker (1982), is restricted to relatively thick sandbodies (>2 m) and comprises shallow scours ('swales' c. 5–50 cm deep), with circular to elliptical shapes in plan view. In cross-section the scours are concordantly filled by low-angle to undulatory laminae, which thin and flatten both upwards and outwards. The main difference with hummocky cross-stratification is that in swaley cross-stratification the laminae rarely pass laterally into accretionary 'hummocks'. Swaley cross-stratification commonly overlies hummocky cross-stratification within shallowing-upward sequences and may be characteristic of storm-dominated shoreface deposits (see Sect. 6.7.7). However, the structure has not been described from modern deposits and could be present in other high-energy storm deposits, such as sandridges. Where this structure is amalgamated within thick sandbodies there is no preservation of fairweather deposits and it may represent the highest energy deposit in a continuum from wave ripple cross-lamination, through hummocky and into swaley cross-stratification.

ANCIENT OFFSHORE STORM SAND MODELS

Storm sandstone beds have been described from inferred shelf deposits from throughout the geological record (e.g. Dott & Bourgeois, 1982; Marsaglia & Klein, 1983; Brenchley, 1985; Cheel & Leckie, 1993). Physical process models conclude that deposition occurred under intense storm conditions in which oscillatory flows were dominant but with a superimposed unidirectional component of debatable strength (Hunter & Clifton, 1982; Swift, Figueiredo *et al.*, 1983; P.A. Allen, 1985; Nøttvedt & Kreisa, 1987; Swift & Nummedal, 1987; Duke, 1990; Duke, Arnott & Cheel, 1991; Scott, 1992).

Despite local variations it is possible to consider storm deposits in terms of an idealized vertical succession of textures and sedimentary structures representing single storm–fairweather events. The main process–response features of an idealized storm sequence are (Fig. 7.42): (i) *storm erosion*: a basal erosion surface, cut by combined oscillatory and unidirectional flows, which may be flat to undulatory (relief up to 0.40 m), with gutter casts, sole marks and intraclasts of pebbles, shells or mudstone; (ii) *main storm deposition*: main hummocky cross-stratified interval, possibly with a parallel (horizontal to subhorizontal) laminated layer with parting lineation directly overlying the basal erosion surface, and deposited under continuing combined flow conditions; (iii) *waning storm deposition*: wave ripples (straight-crested to bifurcating patterns)

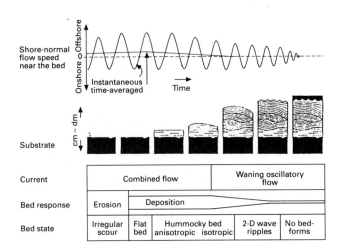

Figure 7.42 Physical process model for the deposition of discrete storm sand layers formed by geostrophic flows during waning storms (after Cheel, 1991; Duke, Arnott & Cheel, 1991).

indicating a return to lower flow regime oscillatory currents, although occasionally unidirectional current ripple cross-lamination is present; and (iv) *post-storm/fairweather mud deposition*: reflects either the final suspension fall-out of storm-derived sediment (i.e. post-storm mud) or the return to normal, background sedimentation (i.e. fairweather mud). This idealized sequence may be disrupted to varying degrees by bioturbation, which in extreme cases can obliterate physical evidence of storm deposition (Fig. 7.41).

The idealized storm sequence frequently shows the following proximal–distal relationships (e.g. Fig. 7.23): (i) amalgamated and thicker sandstone beds occur in proximal settings, which grade laterally into more discrete beds with more frequent and thicker mudstone intercalations in more distal areas; (ii) reduction in grain size and sand percentage; and (iii) increase in gutter casts, thickness of the basal parallel laminated layer, wave ripples, normally graded layers and intensity of bioturbation.

Interpretation of palaeocurrent data has resulted in two main models to explain the emplacement of storm sands in shelf successions: (i) turbidite model (Fig. 7.43a); and (ii) geostrophic model (Fig. 7.43b). One idealized storm sequence highlights structures indicating unidirectional transport (e.g. sole marks, gutter casts and parting lineation), which are often oriented perpendicular both to the shoreline and to wave ripple trends in the upper, waning storm deposition layer (Fig. 7.39; Leckie & Krystinik, 1989). This is most commonly recorded in ancient inner shelf to shoreface sequences and has been used to infer seaward-flowing turbidity currents (e.g. Hamblin & Walker, 1979; Walker, 1984; Brenchley, 1985; Rosenthal & Walker, 1987; Higgs, 1990; Plint, 1991; Chiocci & Clifton, 1991). The applicability of this model has been extensively debated (e.g. Duke, 1990; Cheel, 1991). For example, in elongate

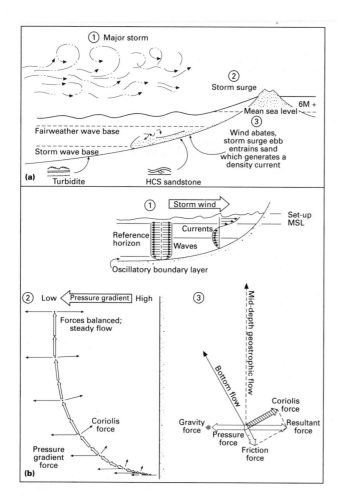

Figure 7.43 Two postulated models to explain the transport of sands from the coast to shelf: (a) turbidity currents, initiated by storm surge erosion in coastal areas, flowing perpendicular to the shoreline (after Hamblin & Walker, 1979), and (b) seaward-returning geostrophic (wind-driven) flows entraining coastal sands during storm set-up and flowing mainly parallel, or oblique, to bathymetric contours (after Duke, 1990).

storm-dominated shelf seas the dominant unidirectional flow is frequently parallel to the shoreline and late-stage wave reworking may produce wave ripple trends orthogonal to these currents (e.g. Levell, 1980a). Furthermore, Duke (1990) argues that sole marks associated with storm beds reflect instantaneous flow conditions and cannot be used to predict longer-term, time-averaged flow and sediment transport conditions. The latter are better indicated from larger-scale, high-angle cross-bedding, which more commonly reflects shore-parallel flows and is more consistent with modern shelf observations (Sect. 7.4). Geological evidence supports a range of shore-normal to shore-parallel storm transport paths with the geostrophic flow model (Fig. 7.43b) more consistently combining both ancient and modern observations.

7.7.3 Ancient mud-dominated offshore facies

Mud-dominated facies are the most abundant of all ancient clastic offshore deposits. However, they have mostly been studied from their palaeoecological and ichnological aspects (see Dodd & Stanton, 1981), and strictly sedimentological information – excluding the clay mineralogy – is often sparse. Thick, mud-dominated shelf deposits probably accumulated most rapidly in areas of high sediment supply (Swift & Thorne, 1991) and under the influence of the various dispersal systems reviewed earlier (Sect. 7.6). Significant offshore or distal mud facies are associated with both allochthonous, supply-dominated shelf dispersal systems and their autochthonous, accommodation-dominated equivalents and they characterize the trailing portions of transgressive systems. Thus, mudstone facies tend to bury and seal proximal shelf and shoreline sand facies. Because of their economic importance, considerable emphasis has been placed on mud facies in terms of their hydrocarbon source potential, particularly those of the Mesozoic and Cenozoic (e.g. Schlanger & Jenkyns, 1976; Schlanger & Cita, 1992).

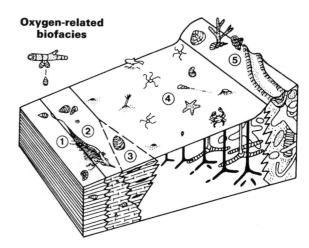

① Anaerobic	Well-laminated strata lacking *in situ* macro- and microbenthic body fossils and microbioturbation; may contain well-preserved remains of nektonic vertebrates, planktonic, epiplanktonic, or otherwise transported invertebrates, and faecal material of planktonic and/or nektonic origin.
② Quasi-Anaerobic	Laminated strata containing microbenthic body fossils (e.g. foraminifera), but lacking *in situ* macrobenthic body fossils; transported vertebrates and invertebrates and faecal material may be common; laminae disrupted slightly by microbioturbation.
③ Exaerobic	Laminated strata similar to that of the anaerobic or quasi-anaerobic biofacies but containing *in situ* epibenthic macroinvertebrate body fossils (e.g., bivalves, molluscs and brachiopods).
④ Dysaerobic	Bioturbated strata characterized by a low diversity assemblages of relatively small, poorly calcified macrobenthic body fossils or absence of body fossils altogether; diversity, size, and depth of penetration of burrows decrease systematically with declining oxygenation within this realm.
⑤ Aerobic	Bioturbated strata (where physical processes do not dominate) containing diverse assemblages of relatively large, heavily calcified macrobenthic body fossils; trace fossils and ichnofabric are variable as a function of environmental energy, substrate consistency, salinity, etc.

Figure 7.44 Depositional model of oxygen-related biofacies. Note the retention of physical structures (laminae) in the low oxygen facies (1, 2 & 3) and the increasing depth of bioturbation in the dysaerobic facies (4). The vertical partitioning of trace fossils represents oxygen-controlled tiering (after Savrda & Bottjer, 1991).

Shallow marine mudstones are frequently preserved as laterally extensive blanket-type deposits which covered large epicontinental areas in response to eustatic rises in sea level; examples include the Upper Jurassic Kimmeridge Clay and Lower Cretaceous Gault Clay of northwest Europe (e.g. Hallam & Sellwood, 1976; Vail & Todd, 1981; Ziegler, 1982), Mesozoic of the Western Interior of North America (e.g. Kauffman, 1974; Simpson, 1975; Plint, Walker & Bergman, 1986) and the Palaeozoic sequences of Africa, Europe and North America (Berry & Wilde, 1978).

The most recent studies of mud facies have been directed towards developing models of anoxia, including the stratigraphic evolution of atmospheric and ocean water column oxygenation ('ocean ventilation') (Wilde & Berry, 1992), and the nature and origin of anoxia events (e.g. Tyson & Pearson, 1991; Einsele, Ricken & Seilacher, 1991; Sageman, Wignall & Kauffman, 1991).

Physical and biogenic structures are often subtle and can only be adequately analysed on polished and unweathered surfaces or in cores. Lamination (mm scale) and primary sedimentary structures may be preserved where the sea bed was anoxic and benthonic organisms scarce, or where sedimentation rates were particularly high. Thin (cm scale) graded layers probably reflect deposition from distal storm-generated flows, which alternate with muds deposited from suspension during fairweather. However, most palaeoenvironmental information on shallow marine mud rocks comes from the study of ichnofabrics and the palaeoecology of *in situ* benthonic organisms. In Cambrian

and younger shelf mudstones physically formed structures display varying degrees of disruption due to the burrowing and grazing activities of organisms.

SHELF MUDSTONE FACIES OXYGENATION MODELS

Black shale facies are characterized by variations in oxygen conditions under which they accumulated. Much of this work builds upon the tripartite model of Rhoads and Morse (1971) who identified three main biofacies: (i) aerobic (>1 ml l^{-1} O$_2$); (ii) dysaerobic (1.0–0.1 ml l^{-1} O$_2$); and (iii) anaerobic (0.1–0.0 ml l^{-1} O$_2$). The specific oxygen boundary classes have subsequently been modified (Fig. 7.44) but the general model has been supported through analyses of a wide range of Phanerozoic deposits (Wilde & Berry, 1992).

The dominant controlling parameters relate to (i) the degree of oxygenation of the shelf water column (Tyson & Pearson, 1991; Einsele, Ricken & Seilacher, 1991); (ii) the stratification of oxygenated and oxygen-deficient water layers; and (iii) oxygenation changes through time and space in the near-surface layers of accumulating sediment (Rhoads & Morse, 1971; Rhoads, Mulsow et al., 1991). Both spatial (Fig. 7.45) and temporal, including seasonal (Fig. 7.46), differentiation of facies is possible by backtracking from the faunal and ichnological indicators.

The evolution of the palaeopatterns of oxygen distribution in the near-surface layers of sediments can be assessed through the concept of organic tiering. At any given time there is a gradient of decreasing oxygen away from the normally oxygenated water column down into the sediment. While the surface mixed or active layer, characterized by a very high water content, will preserve little in the way of physical structure (Fig. 7.45) the underlying

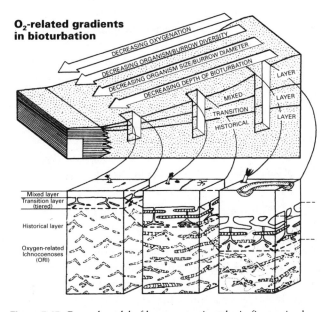

Figure 7.45 General model of burrow stratigraphy in fine-grained shelf sediments. Oxygen-related ichnocoenocese (ORI) are generated in the tiered transition layer and stored or fossilized in the historical layer. Note the gradual thickening and deepening of the lower interface of the transition layer as it is traced up the oxygen gradient (after Savrda & Bottjer, 1991).

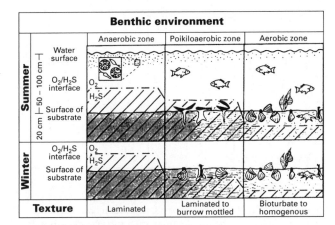

Figure 7.46 Schematic facies model of muddy shelf sediments based on the seasonal variation in the position of the O$_2$/H$_2$S interface. Note that this interface is above the sediment surface and within the water column in the anaerobic zone. In the poikiloaerobic (roughly equivalent to dysaerobic) zone the interface migrates in and out of the sediment surface (from Oschmann, 1991).

transition layer displays a vertical partitioning of fauna and, hence, biogenic structures that correspond to tiered oxygen-related ichnocoenoses (ORI) (Savrda & Bottjer, 1991, Savrda, Bottjer & Seilacher, 1991). The progressive burial of this layer and its subsequent incorporation in the underlying historical layer (Ekdale, Bromley & Pemberton 1984) provides a record (Fig. 7.47) of oxygenation and deoxygenation gradients and events and provides a remarkably sensitive tool to describe the dynamics of the transit of the redox surface into and out of the sediment as well as data on the periodic nature of sedimentation and erosion 'events'. An important product of all such analyses is a measure of degrees of bioturbation of mud rocks, from which precise descriptive indexes of bioturbation have been employed in facies differentiation (Bottjer & Droser, 1991).

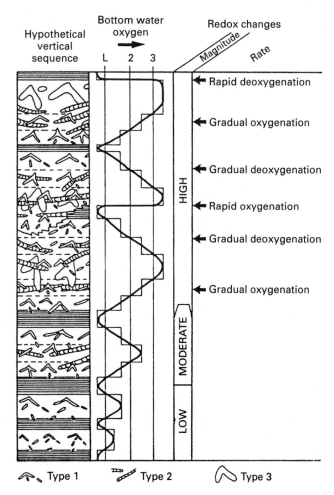

Figure 7.47 Hypothetical stratigraphic sequence model describing possible relationships between bottom-water oxygenation and styles and degrees of bioturbation. In the oxygenation column, the 'L' line represents low-oxygenation or anaerobic conditions that lead to laminated sediments. Lines '2' and '3' correspond to the threshold conditions for the genesis of progressively larger forms of burrows that are organized in oxygen-controlled tiers (from Savrda, Bottjer & Seilacher, 1991).

7.8 Ancient offshore shallow marine clastic sedimentation patterns: interaction of physical processes and relative sea-level fluctuations

The major controls on ancient offshore clastic facies were: (i) sediment supply (type and rate of supply); (ii) hydraulic regime (type, intensity and frequency of reworking); (iii) biogenic activity; (iv) water depth; and (v) relative sea-level fluctuations.

The nature of ancient shelf sediments is closely related to: (i) the sediment supply mechanism (external supply versus *in situ* reworking); and (ii) rate of accommodation space creation. Together these determine whether the shelf was supply- or accommodation-dominated (Sect. 7.2.2).

The hydraulic regime of ancient shelves commonly defines two main end-members: (i) tide-dominated; and (ii) storm-dominated. Some ancient shelves were significantly tidal, depending on the magnitude of the tidal range, and were reworked on a daily basis by fairweather tidal currents. Others were effectively tideless and were dominated by other physical (e.g. storms) and biogenic processes. On these shelves, biogenic activity predominated during fairweather periods and physical reworking was restricted to intense, but short-lived, storm conditions.

In addition to the autocyclic processes of the shelf hydraulic regime, allocyclic processes may be of equal or greater significance, particularly in terms of the preservation potential of shelf successions. The most important allocyclic processes are tectonic activity and relative sea-level fluctuations, which exert a major influence on: (i) the size, depth, geometry and orientation of shallow marine basins and seaways (including communication with oceanic basins); (ii) sediment supply patterns; and (iii) the nature and lateral extent of regional, intrabasinal erosion surfaces. The latter may be due to both relative rises and falls of sea level and can be a major factor determining the distribution, geometry and preservation of shallow marine sandstones.

Because of this variability a range of representative depositional models is considered, which evaluates the relative importance of both autocyclic and allocyclic processes, including local basin conditions and changes in relative sea level.

7.8.1 Tide–storm interaction on Late Precambrian and Lower Palaeozoic shelves

The shallow shelf seas which bordered the Iapetus Ocean in the late Precambrian to early Cambrian were characterized by numerous, laterally extensive sand-rich shelves, which were affected by tides and storms (Banks, 1973c; Anderton, 1976; H.D. Johnson, 1977a,b; Levell, 1980a,b; Hiscott, James & Pemberton, 1984; Eyles, 1988a; McKie, 1990a, b; Bryant & Smith, 1990). Their deposits form part of the distinctive 'orthoquartzite–carbonate suite' (Pettijohn, 1957) or 'Sauk Sequence' (Sloss, 1963) of the North American craton, which includes many Lower Palaeozoic transgressive sequences worldwide (e.g. Hereford, 1977; Tankard & Hobday, 1977; Hobday & Tankard, 1978; Dott & Byers, 1981; Driese, Byers & Dott, 1981; Cotter, 1983; Dott, Byers et al., 1986b; Cudzil & Driese, 1987; Fedo & Cooper, 1990; Simpson & Eriksson, 1991). These cratonic depositional systems in general, and their shelf environments in particular, differ from later (post-Silurian) environments for several reasons: (i) continents were devoid of land vegetation and, therefore, experienced different weathering, erosional and coastal processes (e.g. less cohesion and more wind erosion); (ii) continental deposits are, as a result, dominated by braided fluvial sandstones with significant, but generally subordinate, aeolian deposits; (iii) continental–shoreline–shelf transitions are often less distinctive (e.g. thinner and/or fewer shoreline and deltaic successions); (iv) epeiric seas were much larger and of different configuration to modern shelves; and (v) aquatic faunas are either absent or, at best, imprecise for palaeoenvironmental reconstructions (e.g. no

accurate bathymetric indicators (e.g. Dott & Byers, 1981). Hence depositional models of early Palaeozoic and Proterozoic epeiric seas display some unique features, which provide particular insights into shelf sedimentation patterns. For example, the lack or paucity of bioturbation results in exceptional preservation of primary sedimentary structures.

Facies analysis of these blanket sandstones may enable distinction between a relatively wide range of sedimentary environments, including fluvial, aeolian, coastal and shelf deposits (e.g. H.D. Johnson, 1975; Levell, 1980b; Dott, Byers et al., 1986; Cudzil & Driese, 1987; McKie, 1990b) (Fig. 7.48).

Shelf sandstones are distinguished by combinations of: (i) high textural and mineralogical maturity, particularly relative to contemporaneous fluvial deposits; (ii) paucity of mud rocks; (iii) considerable lateral extent (several hundred kilometres); (iv) abundant medium- to large-scale cross-bedding in medium- to coarse-grained sandstone (often with compound structures and numerous reactivation surfaces); (v) evidence of reversing

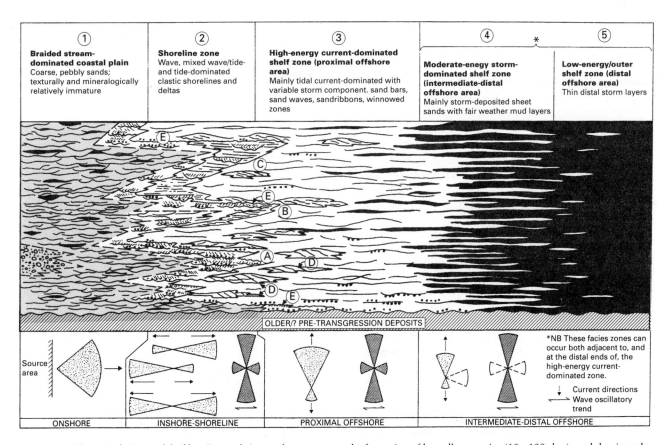

Figure 7.48 Schematic facies model of late Precambrian–early Cambrian current- and storm-dominated shallow marine sandstones and their inferred relationship to adjacent deposits. The model envisages sand-dominated coastal plains with braided rivers passing basinwards into sand-rich coastal and shelf deposits (see also Fig. 7.50). Repeated coastal and delta plain progradation and erosional transgression of tide- and storm-influenced shallow seas, resulted in

the formation of laterally extensive (10s–100s km) sand-dominated shelf successions. The schematic palaeocurrent patterns, dominated by coast-parallel currents in the shelf deposits, reflect those commonly observed through these successions, but is based here on the c. 5-km-thick late Precambrian to early Cambrian succession of north Norway (Banks, 1970, 1973c,d; Banks, Edwards et al., 1971; H.D. Johnson, 1975, 1977a,b; Johnson, Levell & Siedlecki, 1978).

currents (with one mode often dominant); (vi) laterally extensive, low-relief erosion surfaces overlain by thin gravel or granule lags, or silt/mud drapes (Levell, 1980b); (vii) fine- to very-fine-grained feldspathic sandstone and siltstone, characterized further by hummocky cross-stratification and storm/waning flow layers; (viii) glauconite; and (ix) variable bioturbation (e.g. *Planolites*, *Skolithos*, *Cruziana*, *Rusophycus* and *Monomorphicnus*) in Cambrian and younger deposits; and (x) an absence of structures unique to aeolian processes (e.g. adhesion structures, large ripple index, fine climbing translent lamination, grainfall and grain-flow stratification; Sect. 5.4; Dott, Byers *et al.*, 1986).

OFFSHORE TIDE–STORM DEPOSITIONAL MODELS

Tidal shelf depositional models have been proposed for the Lower Cambrian Duolbasgaissa Formation (Banks, 1973c), the late Precambrian Jura Quartzite (Anderton, 1976), the late Precambrian Lower Sandfjord Formation (Levell, 1980a) and part of the Cambrian Eriboll Sandstone Formation (McKie, 1990a,b). In general, they comprise extensively cross-bedded sandstones which are interbedded with finer-grained, heterolithic deposits (intercalated sandstone–mudstone facies) displaying a wide range of wave- and storm-generated structures (cf. Sect. 7.7.2). Vertical profiles often comprise thick (10s–100s m), amalgamated successions of similar facies, usually without any predictable facies sequence. Occasionally, smaller-scale (e.g. 5–10 m) coarsening-upward trends are also developed, which are interpreted as representing the lateral migration of tidal sand banks or ridges (e.g. H.D. Johnson, 1977a; McKie, 1990a). The larger-scale, lateral facies relationships show that most of the tabular, erosively based, cross-bedded sandstone bodies pass laterally, along the dominant palaeocurrent direction, into finer-grained deposits with increasing evidence of storms (Fig. 7.48).

The cross-bedded sandstones are interpreted as subaqueous sandwave and dune deposits formed within the sand-rich parts of palaeotidal current transport paths, while the finer-grained/heterolithic deposits represent lower-energy, deeper-water facies in which storm-generated processes were dominant. The lateral facies relationships are predicted both from outcrop observations and from analogy with the distribution of bedforms and facies found along present-day tidal current transport paths (e.g. northwest European shelf; Sect. 7.3.3). For example, the facies in the Jura Quartzite are interpreted by Anderton (1976) in terms of four hydraulic regimes: fairweather, moderate storm, intense storm and post-storm conditions (Fig. 7.49).

Fairweather conditions comprise a downcurrent decrease in grain size within four main depositional zones: (i) winnowed gravel; (ii) dunes and sandwaves; (iii) current-rippled sands; and (iv) mud (Fig. 7.49a). Bedform migration was relatively slow, reaching a maximum during spring tides and a minimum during neap tides. Cross-bedded and cross-laminated sands are the main preserved features.

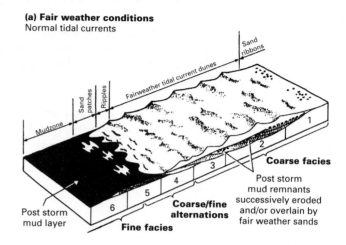

(a) Fair weather conditions
Normal tidal currents

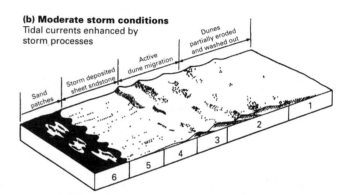

(b) Moderate storm conditions
Tidal currents enhanced by storm processes

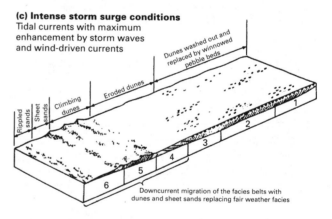

(c) Intense storm surge conditions
Tidal currents with maximum enhancement by storm waves and wind-driven currents

Figure 7.49 Process–response model of a late Precambrian tide-dominated and storm-influenced shelf sea (Jura Quartzite of northwest Scotland) highlighting three main phases of deposition: (a) general model depicting the inferred fairweather (tide-dominated) distribution of bedforms along a c. 10s–100s km long sediment transport path (zones 1–6); (b) inferred response during moderate storm conditions, when sediment transport rates increase; (c) inferred response during intense/exceptional storm conditions, when erosion may have dominated large areas of the sea bed (after Anderton, 1976).

Moderate storms were characterized by increased sediment transport rates and downcurrent migration of the fairweather zones (Fig. 7.49b). Dunes were partially eroded in the proximal zone while distally a thin downcurrent-thinning and -fining sand layer may be deposited.

Intense storms represent the highest energy conditions and sediment transport rates reach a maximum when this coincides with spring tides (Fig. 7.49c). In the proximal zones erosion produces shallow channels, planar erosion surfaces and winnowed pebble lags. Downcurrent zones experience rapidly migrating bedforms, including climbing dunes, with thinner storm sand layers deposited distally. The ensuing sediment sheet has a high chance of preservation.

Post-storm conditions represent a transitional period between the height of a storm and the return of fairweather conditions (Fig. 7.49a). Prior to the re-establishment of fairweather bed-

forms, laterally extensive mud layers may be deposited, if suspended-sediment concentrations were sufficiently high. The mud and silt layers, which frequently drape planar erosion surfaces, are probably of this origin.

In other late Precambrian and Lower Palaeozoic shelf successions only storm-dominated facies are present, suggesting that fairweather tidal currents were absent or too weak to regularly rework the sea bed (e.g. Banks, 1973d; Brenchley, Newall & Stanistreet, 1979; Levell, 1980a; Brenchley, Romano & Guiterrez Marco, 1986; Brenchley, Pickerill & Stromberg, 1993). Cross-bedded sandstone facies in some of these deposits (up to c. 1–2 m height) remain difficult to interpret since they require relatively persistent currents. If tidal processes are discounted, they could represent subaqueous dunes which migrated in response to episodic storm/wind-driven currents. Dott, Byers *et al.* (1986) conclude that the parallelism between aeolian and

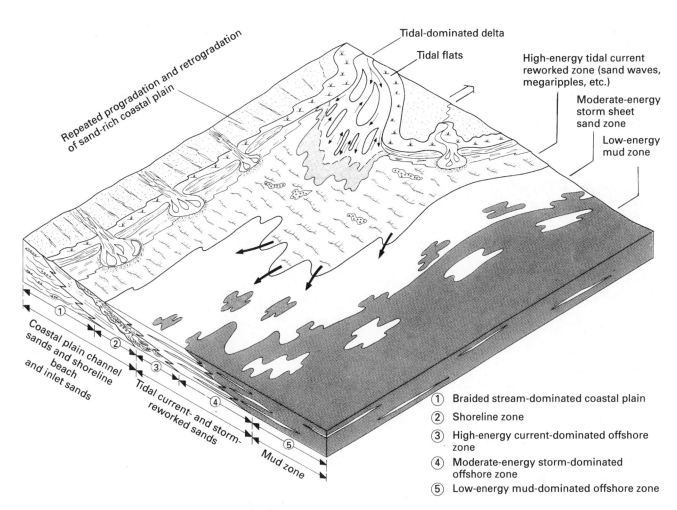

Figure 7.50 Depositional model for the development of thick, laterally extensive, late Precambrian sandstones in which braided-stream-dominated coastal plain sands pass basinwards via shoreline sands into transgressive, tidal- and storm-reworked shelf sands (modified from Levell, 1980b). Syndepositional faulting may result in

the formation of thick, amalgamated successions comprising similar facies types (e.g. Lower Sandfjord Formation of north Norway). Sand supply to the shelf is mainly through erosional transgression and reworking by a high-energy shelf hydraulic regime.

shelf cross-bedding in the Lower Palaeozoic Wonewoc and St Peter sandstones supports high-intensity wind-driven shallow marine currents, generated by tropical storms in a low-latitude location (southern hemisphere trade winds belt).

SAND SUPPLY TO LATE PRECAMBRIAN–LOWER PALAEOZOIC SHELVES

The reasons for these large volumes of shallow marine sandstone, which are unique in the geological record, include: (i) extensive sand-rich continental areas (non-vegetated braidplains with variable aeolian influence); (ii) high-energy shelf hydraulic regime, which is frequently tide-dominated with superimposed storms; and (iii) repeated relative sea-level changes.

Dott, Byers *et al.* (1986) suggest that the sheet-like geometry of some of these deposits may be largely due to non-marine, rather than marine processes. Their depositional model of the Lower Palaeozoic Wonewoc and St Peter deposits envisages a sand-dominated coastal plain, comprising both braided rivers and subordinate aeolian dunes, passing rapidly basinwards through a narrow shoreline into a sand-dominated inner shelf covered in small subaqueous dune bedforms. Repeated eustatically driven, marine transgression and reworking contributed to the accumulation of extensive shelf sandstones, with additional contributions inferred from both net seaward

dispersal of sand during each phase of shoreline regression, and more episodic (e.g. catastrophic storms) supply of sand during periods of erosional shoreface retreat.

Levell's (1980b) analysis of the 1.5 km Lower Sandfjord Formation, in north Norway also concluded that the most likely mechanism for emplacing enormous volumes of sand on to the shelf was by repeated transgression of sand-rich coastal plains dominated by braided streams, with adjacent sand-dominated coastal environments including wave- and tide-dominated deltas (Fig. 7.50). This model also envisages a high-energy offshore tidal current regime, which provided the most effective mechanism for the reworking and offshore transportation of the transgressively abandoned nearshore sands. In addition, the stacking of such thick sequences, with substantial net vertical accumulation of shelf sands, implies long periods of relative sea-level rise. The apparent stable position of the shoreline suggests that this was tectonically controlled, probably by syndepositional faulting.

7.8.2 Shallow marine sand deposition in an elongate, intracratonic rift basin: Upper Jurassic–Lower Cretaceous of East Greenland

Basin geometry, subsidence, sediment supply rates and eustatic sea-level fluctuations all had a major influence on the deposition of Upper Jurassic–Lower Cretaceous shallow marine sandstones

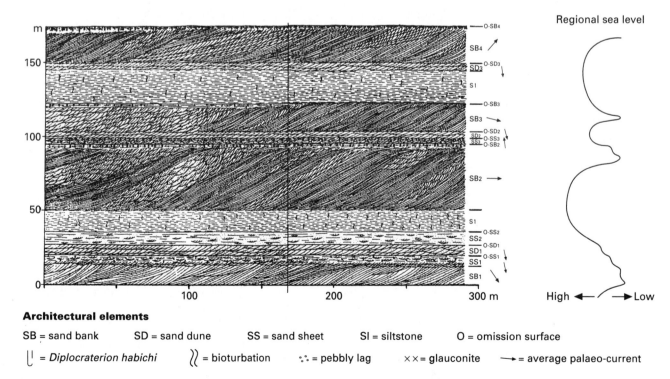

Architectural elements

SB = sand bank   SD = sand dune   SS = sand sheet   SI = siltstone   O = omission surface

$\lfloor^!$ = *Diplocraterion habichi* $\langle\langle$ = bioturbation   ∴ = pebbly lag   ×× = glauconite   → = average palaeo-current

Figure 7.51 Schematic illustration of sedimentary structures formed by sandbanks and sand dunes, the vertical facies succession and their inferred relationship to regional sea-level fluctuations (in the Upper Jurassic–Lower Cretaceous Raukelv Formation of East Greenland (after Surlyk & Noe-Nygaard, 1991). Notable features include the thickness of the cross-bedding, the extreme sheet-like geometry of all the units and the continuity and correlatability of omission surfaces.

in the 200–300 m thick and exceptionally well-exposed Raukelv Formation of East Greenland (Surlyk & Noe-Nygaard, 1991). Deposition occurred in a narrow (125 km wide), tectonically active, north–south trending, intracratonic seaway and resulted in a laterally extensive (>900 km²), tabular-shaped unit of mainly high-energy shallow marine sandstones, which are most notable for the very large-scale of cross-bedding (10–50 m thick tabular sets) (Fig. 7.51). Four facies, referred to as 'architectural elements' have been distinguished (Surlyk & Noe-Nygaard, 1991).

1 *Sandbank (SB) element* is the dominant sandstone facies (80%) and is characterized by single, tabular cross-beds (15–50 m thick) which form extensive, sheet-like units containing huge volumes of sandstone (c. 30 km³). It is composed of coarse-grained, poorly sorted, texturally immature quartz sand, with transported assemblages of bivalves, belemnites, ammonites and crinoids. The structures range from simple avalanche sets, with foresets dipping at up to 25°, through to more complex compound structures consisting of smaller intrasets (Fig. 7.51). The top and bottom surfaces of each unit are sharp and planar, occasionally erosional (maximum c. 1 m at the top), and are frequently burrowed and glauconitized. Gravel veneers are common along the top surface. Based partly on size, these structures are interpreted as linear, north–south trending sandbanks which were attached to a shoreline in the west and prograded eastwards across the full width of this narrow seaway. The smaller intrasets indicate secondary currents flowing obliquely to the south across the avalanche sets. Processes responsible for sand emplacement are uncertain, but reworking by tidal currents is evident in the other facies and presumably had a major influence on the sandbank element.

2 *Sand dune (SD) element* forms thinner (2–8 m thick) but equally extensive, sheet-like units. It consists mainly of compound cross-bedding (set thickness c. 0.5–4 m) with unimodal palaeocurrent directions to the south, but with rare sets dipping towards the north. This element is bounded by planar surfaces, which are glauconitized and burrowed, but without gravel veneers. There is abundant evidence of tidal activity (Sect. 7.7.1), including double mud drapes and occasional oppositely dipping cross-bedding. This element is interpreted as extensive fields of linear to slightly sinuous subaqueous dunes, which were transported southwards by coast-parallel tidal currents.

3 *Sand sheet (SS) element* is 2–10 m thick with a sheet-like geometry and sharp boundaries. It consists of medium- to coarse-grained sandstone with a wide range of structures, including low-amplitude trough cross-bedding, cross-lamination and horizontal lamination. These sandstones are glauconitic, strongly burrowed (*Diplocraterion habichi*) and richly fossiliferous (bivalves, crinoids, ammonites and belemnites). Palaeocurrents are directed mainly to the south, but in places also to the north. This element was deposited more slowly than the previous deposits, but by similar mainly southward flowing, coast-parallel tidal currents, possibly augmented by storms.

4 *Marine siltstone (SI) element* forms 2–20 m thick, laterally extensive (>100 km²), wedge-shaped sheets with planar or sharp boundaries (Fig. 7.51). It consists of bioturbated (*Curvolithus* burrows) silty or sandy mudstones and is richly fossiliferous. It was deposited slowly in a low-energy, well-oxygenated, offshore marine environment.

The vertical succession comprises an alternation of four thick sandstone units (10s m thick comprising SB, SD and SS elements), usually separated by thinner fine-grained (SI) intervals, but the individual architectural elements are not organized into a predictable vertical profile (Fig. 7.51). Each sandstone unit is dominated by the sandbank element, which represents major eastward progradation and downlap of a straight-faced, north–south trending sandbar or bank over fine-grained offshore siltstone deposits. Assuming that the large cross-bedding represents progradational units then basinal water depth in front of the sandbank was at least c. 15–50 m. The extreme sheet-like geometry of all the sandstone units and the lack of both wave scour and wave-formed structures within the sandbank suggest that the top of the bank was several tens of metres below sea level (up to c. 50 m). The subordinate and smaller-scale cross-bedded sandstones were deposited by tidally influenced currents that flowed mainly to the south, parallel both to the trend of the coastal zone and the strike of the sandbank. This north–south trend is also the orientation of the underlying extensional faults, which define the geometry of the basin.

This Mesozoic sandbank system is without a direct modern analogue, although there are some similarities with the lateral migration of linear tidal sand ridges (Sect. 7.3) and a tide-dominated seaway model provides the most plausible physical mechanism of sand transport and deposition. However, a depositional model has been proposed which relates bank initiation, growth and abandonment to the shelf hydraulic regime and relative changes in sea level (Fig. 7.52).

Bank initiation was accompanied by relative sea-level fall, causing a sudden and rapid influx of coarse-grained sediment. The mechanism of sand emplacement is uncertain but presumably included significant direct fluvial supply and reworking by powerful coastal currents within a narrow, tide-dominated seaway.

Bank growth was marked by a relatively rapid (c. 10 cm year⁻¹), basinwards shift in the coarse-grained sandbank element, which downlapped over the underlying condensed beds. In some cases the sea floor in front of the eastward prograding sandbank was covered by sheets of southward migrating dunes (Fig. 7.52). Each sandbank element took some 1.25 My to migrate from the western to the eastern basin margin and occurred during a period of sea-level rise leading to a highstand. Irregular rates of bank progradation are indicated by the lateral variations in cross-bed types, intraset glauconitized horizons and fluctuations in faunal content.

Bank abandonment is inferred to have coincided with a decrease in the rate of sea-level rise and hence reduction in

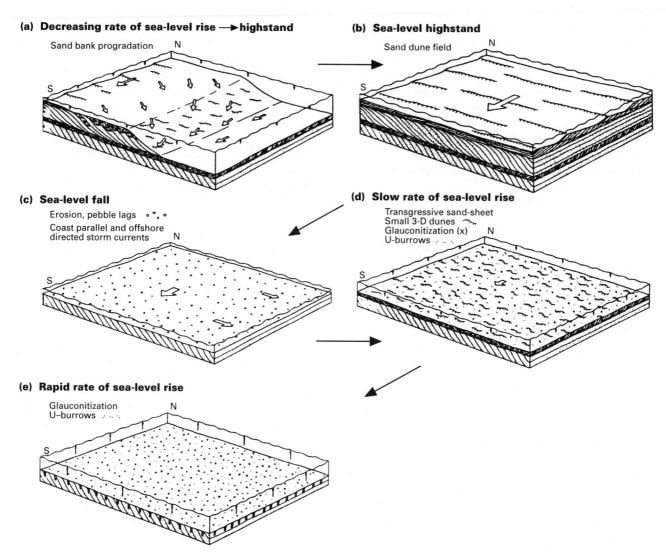

(a) Decreasing rate of sea-level rise ➔ highstand

Sand bank progradation

(b) Sea-level highstand

Sand dune field

(c) Sea-level fall

Erosion, pebble lags
Coast parallel and offshore
directed storm currents

(d) Slow rate of sea-level rise

Transgressive sand-sheet
Small 3-D dunes
Glauconitization (x)
U-burrows

(e) Rapid rate of sea-level rise

Glauconitization
U–burrows

Figure 7.52 Depositional model of the development and abandonment of the shallow marine sandbanks in the Raukelv Formation, interpreted as being controlled mainly by fluctuations in the rate of relative sea-level change (after Surlyk & Noe-Nygaard, 1991).

the rate of accommodation space creation. Simultaneously, bank tops underwent non-deposition (glauconitization and bioturbation), sheet erosion (truncation of sandbank tops) and formation of gravel veneers.

From this analysis, it is concluded that relative sea-level fluctuations were the main process in controlling sandbody evolution. The extreme tabular, sheet-like geometry of all the units implies that other factors, such as sediment supply and subsidence, were relatively uniform. The primary sedimentary structures indicate that the basin's hydraulic regime was current dominated with vast quantities of sand moving as coarse-grained bedload over the shelf floor. The dominant southward-flowing basinal currents were probably tide-dominated, given their apparent persistence over long periods, but may have been enhanced by storms and wind-driven currents. The north–south tectonic grain of this post-rift, thermally subsiding basin determined the geometry and orientation of the seaway, which was a blind gulf to the north and probably allowed amplification of the tidal wave as in modern narrow seaways (Sect. 7.3).

7.8.3 Shallow marine sand deposition along an active fault-controlled shelf margin: Bohemian Cretaceous Basin, Central Europe

Large-scale, cross-bedded shallow marine sandstones occur in the Bohemian Cretaceous Basin of Central Europe, where syndepositional tectonic activity strongly influenced sandbody geometry and distribution (Jerzykiewicz & Wojewoda, 1986).

The Radków and Szczeliniec Sandstones form tabular bodies of shallow marine sandstone (each up to c. 150 m thick), which extended for at least 60 km along a fault-controlled, depositional strike, but wedge-out basinwards into offshore calacareous mudstones and limestones within 3 km (Fig. 7.53). The sandstones are separated by c. 100 m organic-rich, bioturbated mudstones deposited during the mid-Turonian rise in sea level.

Both sandstones mainly comprise vertically stacked large-scale, cross-bedding in sets 3–17 m thick, with foresets dipping at 15–25° and showing small-scale grain-size alternations (up to coarse sand and granules). Individual sets display good lateral continuity and are bounded by major, laterally extensive erosion surfaces, which are lined with shells and granules. In places, simple avalanche cross-bedding passes downdip into smaller-scale, compound structures. The dominant palaeocurrent pattern is unidirectional offshore (Fig. 7.53). These large structures are interbedded with, and pass laterally into, both smaller-scale cross-bedded sandstones displaying bidirectional structures (offshore mode dominant) and poorly to non-stratified sandstones, which are texturally varied and contain granule conglomerates and coquinas (e.g. comprising thick-walled bivalve shells of *Exogyra*, *Lima* and *Pecten* up to 120 mm diameter). Individual trace fossils occur throughout these sandstones, including *Ophiomorpha*, *Thalassinoides* and *Chondrites*, but close to major stratal surfaces bioturbation is intense and primary structures obliterated.

The principal evidence for deposition in an offshore shallow marine environment includes body and trace fossils, intercalation with marine mudstones, and the prominent sandbody continuity along depositional strike. Evidence of tidal current activity is supported by the bidirectional palaeocurrents, with mainly seaward- but occasionally landward-directed cross-bedding. Tidal currents provide the most effective and persistent fairweather system for transporting these very large volumes of sand offshore, but they may have been enhanced by other seaward-flowing currents (e.g. geostrophic currents, rip currents, etc.). Support for offshore, rather than inshore/estuarine, conditions include: (i) absence of deep erosional or channel-fill features; (ii) sheet-like geometry of intrasandstone erosion surfaces; (iii) lack of mud drapes; and (iv) frequency of storm deposits (e.g. coquinas).

The two major influxes of sand on to the shelf were associated with source area uplift and rapid deposition of sand on a narrow, tectonically defined shelf/basin margin. There are no obvious modern analogues for these large-scale bedforms. They appear to represent some form of offshore-prograding sandbank, which coalesced to form major sandbodies with large lateral continuity. As in the Raukelv Formation, the absence of wave-formed structures implies deposition below fairweather wave base (several 10s m water depth), but in this case sand accumulation was more directly controlled by extensional faulting. It is inferred that seaward-flowing tidal current transport paths, enhanced by other seaward-flowing currents, deposited sand on the lee side of fault scarps (Fig. 7.53).

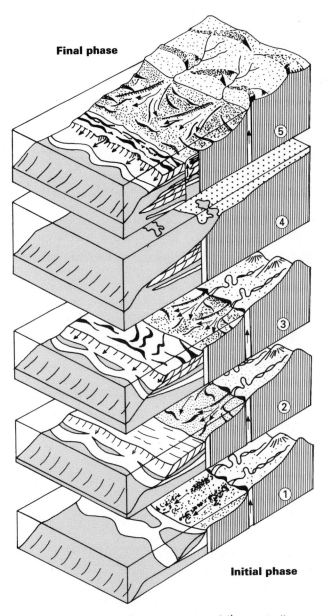

Figure 7.53 Palaeogeographic reconstruction of the tectonically controlled shelf in the northeast margin of the Bohemian Cretaceous Basin and the distribution of the two main shallow marine sandstone units: (Radków Sandstone (Phases 1–4) and Szczeliniec Sandstone (Phase 5)). Note the rapid shore-normal, basinwards interfingering with shelf mudstones across syndepositional normal faults (after Jerzykiewicz & Wojewoda, 1986).

The sandbanks, with their angle-of-repose slopes, developed in response to continued fault scarp movement and to a rapid and regular bedload supply of sand across the shelf. Frequent reworking by storms modified the front of the sandbanks and deposited coquinas over the extensive planar erosion surfaces. The maximum relief at the front of these sandbanks was probably close to the thickest cross-bed sets (c. 17–20 m), which is substantially smaller than those in the Raukelv Formation. The adjacent upthrown fault blocks are dominated by smaller subaqueous dunes, which would have been the feeders to the larger, seaward sandbanks.

7.8.4 'Offshore bars' vs. lowstand shoreface deposits in the Cretaceous Western Interior Seaway of North America

One of the most widely quoted depositional models of ancient shelf sandstones is that of the 'offshore bars', which has been developed almost exclusively from the Cretaceous rocks of the Western Interior Seaway of North America. The model is an interpretation of a distinctive type of elongate shallow marine clastic body encased in shelf mudstones, which occurs at various stratigraphic horizons (but mainly Upper Cretaceous) throughout this long (1000s km) and narrow (c. 500 km) seaway. The configuration of these sandstones has provided ideal stratigraphic traps for hydrocarbons resulting in a major oil exploration concept along the whole length of this seaway, which largely explains the abundance of published data on this type of shallow marine sandbody. The combination of outcrop, but particularly the dense subsurface well control, has resulted in a unique data base for these sandstones. However, recent re-evaluation of part of this vast data base, most notably the Cardium Formation in Alberta (Plint, 1988b; Plint and Walker, 1992), provides compell-ing evidence for an alternative interpretation to the original 'offshore bar' model, which may be applicable to many, if not all, similar elongate sandbodies in this seaway.

SANDBODY CHARACTERISTICS

The main features of these sandbodies are: (i) well-sorted, glauconitic and quartzose sandstones, occasionally conglomeratic; (ii) physical and biogenic structures indicative of shallow marine conditions; (iii) mainly coarsening-upward facies successions; (iv) primary sedimentary structures ranging from sandstone beds displaying wave- and storm-generated structures, including hummocky cross-stratification (Sect. 7.7.2), through to tabular and trough cross-bedding in the shallower/sandier facies; (v) distinctive size and geometry (10–30 m thick, mainly 5–10 km wide and several tens of kilometres, up to c. 160 km, long); (vi) palaeocurrents, derived from cross-bedding, are unidirectional and trend parallel, or slightly obliquely, to sandbody orientation; (vii) location tens of kilometres (up to 100 km) seaward from, and oriented parallel to, the palaeoshoreline; and (viii) encasement in shallow marine mudstones, which physically separate these 'offshore' sandbodies from contemporaneous shoreline sand bodies (e.g. Fig. 7.54).

The coarsening-upward vertical facies sequence within these 'offshore bars' was seen to be remarkably similar to that of regressive shoreface deposits (e.g. Fig. 6.85; Sect. 6.7.7; Tillman,

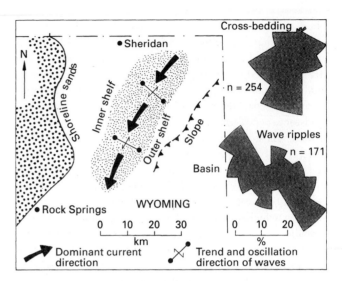

Figure 7.54 Palaeogeographic interpretation of the Upper Cretaceous Shannon Sandstone in Wyoming and the relationship between sandbody trend, palaeocurrents and wave trends (after Spearing, 1975, 1976). The apparent shelf location of these sandbodies led to the development of the 'offshore bar' depositional model for these deposits (Tillman & Martinsen, 1984; 1987; Gaynor & Swift, 1988). A more recent reinterpretation suggests that these are more likely to represent detached, lowstand shoreface deposits (Walker & Bergman, 1993; Bergman, 1994).

1985), but the apparently unique offshore palaeogeographic location of these sandbodies suggested a different depositional origin to those deposited at the shoreline. Facies interpretations based on primary sedimentary structures emphasized the intermittent nature of sand deposition and concluded that storm-related processes were dominant throughout this seaway (e.g. La Fon, 1981; Swift, Hudelson et al., 1987; Gaynor & Swift, 1988; Krause & Nelson, 1991).

'OFFSHORE BAR' FACIES MODEL

The basis of the 'offshore bar' model assumed, not unreasonably, that the present-day distribution of the sandbodies, notably their apparent isolated occurrence seaward of the contemporaneous shoreline, reflected their position at the time of deposition (Campbell, 1973; Berg, 1975; Spearing, 1976; Tillman & Martinsen, 1984; Siemers & Ristow, 1986; Tye, Ragnanathan & Ebanks, 1986). Hence the model proposed that the bars grew *in situ* on a muddy shelf floor, either as flat or very gently inclined, migrating sheets (e.g. Seeling, 1978) or as topographically elevated, linear bars or ridges (Tillman & Martinsen, 1987). The apparent dominance of storm-generated processes, combined with the elongate sandbody geometry, resulted in frequent analogies with the modern storm-dominated sandridges off the eastern USA shore (e.g. Tillman & Martinsen, 1987; Gaynor & Swift, 1988; Sect. 7.4.2). One of the main difficulties with this comparison was that most modern shelf sandridges

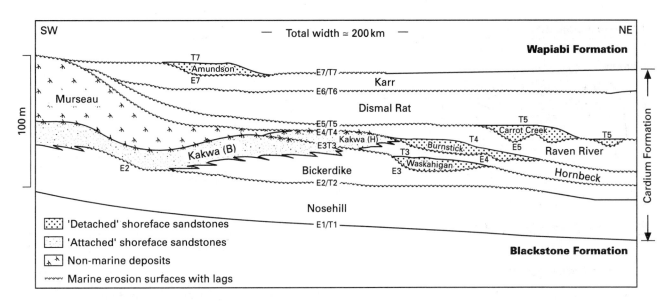

Figure 7.55 Allostratigraphic framework of the Cardium Formation of Alberta across a c. 200 km long, SW–NE (left to right) shoreline-shelf transition (from Plint, Walker & Bergman, 1986). The main stratigraphic units are bounded by lowstand erosion surfaces (E1–E7) and by transgressive surfaces (T1–T7). Two main types of shallow marine sandbody are present: (i) one laterally extensive, sandier/coarsening upward unit (Kakwa); and (ii) four discontinuous, erosionally-based conglomerate bodies (Waskahigan, Burnstick, Carrot Creek and Amundson).

(both storm and tidal varieties) form by the *in situ* reworking of older, often immediately underlying, sand-rich deposits during transgression. In contrast, the Western Interior sandbodies invariably occur within mud-dominated successions and far removed from sand sources (e.g. Fig. 7.55). Hence it seemed that sand and gravel had to be transported considerable distances (10s km to > 100 km) to their present offshore position.

Walker (1984) was among the first to discuss major difficulties with this model, particularly the question of sand supply and the mechanics of bar development, and three questions were posed: (i) How were sand and gravel transported large distances offshore from the shoreline? (ii) What shelf process could concentrate this coarse-grained sediment into such long and narrow bodies within an otherwise mud-dominated shelf? (iii) Why should the vertical facies succession be upward coarsening? Although explanations for these questions have been offered, often appealing to subtle tectonically-induced topographic variations on the shelf floor to explain the location of specific sandbodies (Campbell, 1973; Slatt, 1984), they had never been adequately explained, nor fully supported by modern analogues. For example, the elegant 'shelf sand plume model', which envisaged storm-current erosion of deltaic headlands followed by offshore-directed sand transport is no longer supported by detailed analysis of the inferred modern analogue from the Nile delta (Scheihing & Gaynor, 1991).

SHOREFACE SEQUENCES IN RELATION TO
RELATIVE CHANGES IN SEA LEVEL

Recent studies have concluded that the original 'offshore bar'

model is no longer tenable. Plint (1988b) was the first to question the initial assumptions and to demonstrate that isolated, linear, shore-parallel sandbodies can result from the erosional detachment of shoreface sands due to fluctuations in relative sea level. This model was initially based on detailed subsurface studies on the Cardium Formation in Alberta (Plint & Walker, 1987; Bergman & Walker, 1987; Plint, 1988b), but similar interpretations were later proposed for sandbodies within the Mowry Shale (Davis & Byers, 1989) and for the Shannon Sandstone (Walker & Bergman, 1993; Bergman, 1994). However, a shelf sandridge interpretation for the Cardium Formation at the Crossfield–Cardium oil field is still preferred by Krause & Nelson (1991).

The Cardium Formation is a prograding, storm-dominated shoreface and shallow shelf deposit in which there are two main types of shelf to shoreline facies successions: (i) 'normal', gradational-based, shoreface, in which bioturbated muddy sandstones pass through hummocky cross-stratified (HCS) sandstones and into swaley cross-stratified (SCS) sandstones capped by a root horizon; and (ii) sharp-based, shoreface, in which mud-free SCS sandstones are abruptly separated from muddy offshore sediments by regionally extensive erosion surfaces and HCS intervals are thin or absent (Fig. 7.55). These erosion surfaces are invariably overlain by thin veneers of conglomerate which, tens of kilometres seaward of the progradational limit of the Cardium shoreface, thicken into a series of narrow, elongate conglomeratic sandstones traditionally interpreted as 'offshore bars' (Swagor, Oliver & Johnson, 1976; Bergman & Walker, 1987, 1988).

Several regionally extensive erosion surfaces, resulting from

(a) Open marine deposition

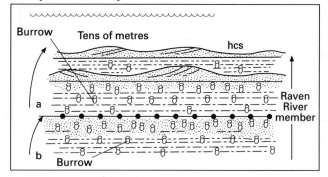

(b) Maximum lowstand of sea level

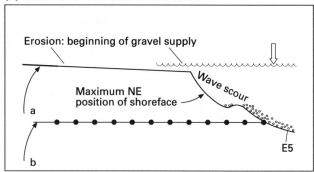

(c) Stillstand

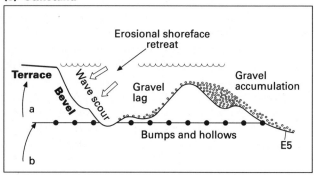

(d) Stillstand

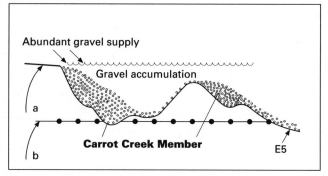

(e) Early stages of transgression

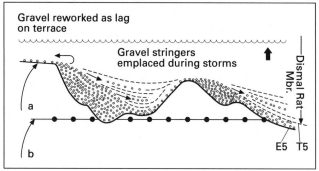

(f) Open marine deposition

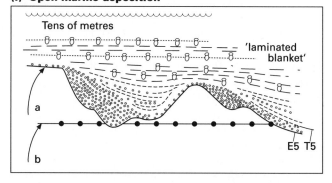

Figure 7.56 Depositional and erosional history of the Cardium Formation at Carrot Creek and the inferred fluctuations in relative sea level (after Bergman & Walker, 1987). (a) Two sequences of basin aggradation in an open marine/shelf setting with periods of non-deposition; (b)–(f) effects of a relative sea-level fall followed by a relative sea-level rise. (See text for further discussion.)

erosion during combinations of relative sea-level lowstands, stillstands and transgression, divide the Cardium Formation into a series of allomembers (Fig. 7.55). These surfaces are asymmetric in cross-section, with up to 20 m erosional relief, and in plan view they are oriented parallel to depositional strike and palaeoshoreline trend (Bergman & Walker, 1987). The geometry of these surfaces suggests that they were cut by shoreface erosion, probably following a period of rapid lowering of relative sea level, which also resulted in conglomeratic sediment reaching the shoreface (Fig. 7.56). As sea level falls the more steeply-inclined bevelled (seaward-facing) surface was

eroded by wave scour at the shoreface and an initial supply of gravel reached the shoreline across an (inferred) emergent platform (Fig. 7.56b). During a limited period of shoreface retreat following stillstand (Fig. 7.56c,d), both the erosion and supply of gravel by longshore drift increased when gravel was laterally reworked along the shoreface and led to the formation of elongate conglomerate bodies. Subsequent transgression halted new gravel supply, but continued reworking resulted in the formation of gravel stringers during storms and lag veneers across transgressive horizons, which frequently merge with the inferred lowstand erosion surfaces (Fig. 7.56e). Finally, mud blanketed the whole unit (Fig. 7.56e).

Hence, the narrow, elongate geometry of these shallow marine deposits appears to be mainly controlled by the shape and orientation of the bounding marine erosion surfaces, formed initially during a period of rapid fall of relative sea level followed by further modification during erosional transgression, rather than by any primary depositional (accretionary) sandbody geometry (Fig. 7.57). Sand emplacement on to the shelf is, therefore, explained by fluctuations in the rate of relative fall in sea level, resulting in movement of the shoreline for 10s–150 km rather than the bedload transport of sand. When the rate of relative sea-level fall is only slightly greater than the rate of subsidence, then: (i) shoreface progradation is accelerated; and (ii) shoreface sandbodies are thin. However, if the rate of sea-level fall greatly exceeds the rate of subsidence, then (i) there is no remaining accommodation space; and (ii) the shelf becomes an emergent, non-depositional surface. The shoreline continues to move rapidly basinwards until the rate

of sea-level fall is reduced sufficiently to enable wave erosion to cut a new shoreface profile through erosional shoreface retreat. This 'forced regression' (Posamentier, Allen *et al.*, 1992) provides additional basinwards accommodation space which, following renewed alongshore supply of sand or gravel, will be filled by a prograding, sharp-based shoreface deposit.

The previously unanswered questions of the 'offshore bar' model become obsolete and many, if not all, of these Western Interior sandbodies may also be reinterpreted as similar lowstand shoreface deposits. Many of the differences noted by previous workers in these deposits may reflect local variations in: (i) the rate of relative sea-level change; (ii) the rate of sediment supply; and (iii) the degree of marine erosion. Hence a wide range of vertical facies successions may be developed, which preserve different parts of the idealized profiles predicted in the two end-member shoreface models (Plint & Walker, 1992): (i) prograding shoreface (gradationally based); and (ii) incised shoreface (sharp-based). This emphasizes the value of being able

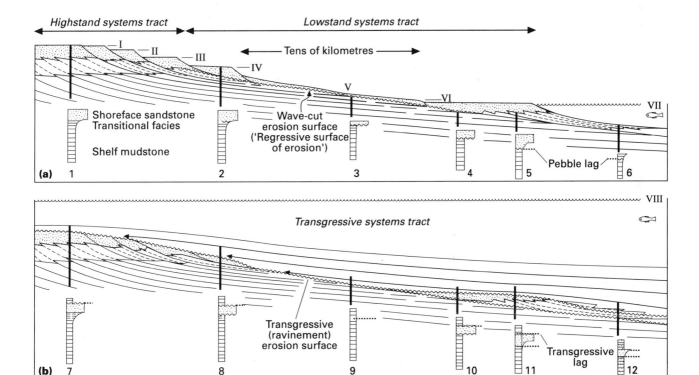

Figure 7.57 Generalized depositional model relating the origin of isolated/detached shallow marine sandbodies ('offshore bars') to fluctuations of relative sea level (based on Plint, 1988b; from Plint & Walker, 1992). (a) Section across a prograding wave-dominated shoreline, initially forming part of a highstand systems tract (sea level I) but subsequently followed by successive falls of relative sea level forming a lowstand systems tract (sea levels II–VII). Rapid relative sea-level falls (IV–VI) result in erosively-based shoreface sandbodies (vertical profiles 2–4). As the rate of relative sea-level fall decreases and accommodation space increases a gradationally based, prograding shoreface is re-established (sea level VII/vertical profiles

5–6); (b) relative rise in sea-level terminates the lowstand systems tract and results in the development of a transgressive systems tract. The shoreline migrates landward and the accompanying process of shoreface erosion forms a gently undulating erosion surface (ravinement surface), which truncates and may partially remove underlying lowstand deposits (vertical profiles 7–12). The only recognition of this important event may be a thin (cm scale) transgressive lag (e.g. thin pebble layer; vertical profile 9). Continued transgression results in widespread deposition of shelf muds, which onlap and partly encase the earlier sand deposits.

to describe and interpret not only the vertical facies successions but also the nature, geometry and extent of any bounding discontinuities in ancient shallow marine clastic deposits. Together they provide perhaps the best means of establishing the dynamic relationship between time-equivalent shelf and shoreline depositional systems.

Further reading

Dalrymple, R.W. (1992) Tidal depositional systems. In: *Facies Models: Response to Sea Level Change* (Ed. R.G. Walker and N.P. James), pp. 195–218. Geol. Ass. Can., Waterloo, Ontario.

Flemming, B.W. & Bartholomä, A. (Eds) (1995) *Tidal Signatures in Modern and Ancient Sediments. Spec Publ. int. Ass. Sediment.*, **24**, 358 pp. Blackwell Science, Oxford.

Knight, R.J. and McLean, J.R. (Eds) (1986) *Shelf Sands and Sandstones. Mem. Can. Soc. petrol. Geol.*, **11**, 347 pp. Calgary.

Nummedal, D., Pilkey, O.H. & Howard, J.D. (Eds) (1987) *Sea-level Fluctuations and Coastal Evolution. Spec. Publ. Soc. econ. Paleont. Miner.*, **41**, 267 pp. Tulsa.

Open University Course Team (1989a) *Waves, Tides and Shallow-Water Processes*, 187 pp. Pergamon Press, Oxford.

Open University Course Team (1989b) *Ocean Circulation*, 238 pp. Pergamon Press, Oxford.

Smith, D.G., Reinson, G.E., Zaitlin, B.A. & Rahmani, R.A. (Eds) (1991) *Clastic Tidal Sedimentology. Mem. Can. Soc. petrol. Geol.*, **16**, 387 pp. Calgary.

Stanley, D.J. & Swift, D.J.P. (Eds) (1976) *Marine Sediment Transport and Environmental Management*, 602 pp. Wiley, New York.

Stride, A.H. (Ed.) (1982) *Offshore Tidal Deposits: Processes and Deposits*, 222 pp. Chapman & Hall, London.

Swift, D.J.P., Oertel, G.F., Tillman, R.W. & Thorne, J.A. (Eds) (1991) *Shelf Sand and Sandstone Bodies: Geometry, Facies and Sequence Stratigraphy. Spec. Publ. int. Ass. Sediment.*, **14**, 532 pp. Blackwell Scientific Publications, Oxford.

Walker, R.G. & Plint, A.G. (1992) Wave- and storm-dominated shallow marine systems. In: *Facies Models: Response to Sea Level Change* (Ed. R.G. Walker & N.P. James), pp. 219–238. Geol. Ass. Can., Waterloo, Ontario.

Marine evaporites: arid shorelines and basins

8

A.C. Kendall & G.M. Harwood

....

8.1 Introduction

Most ancient evaporites are interpreted as marine precipitates; precipitates formed by the evaporation of sea water, but perhaps contaminated by meteoric, ground- or hydrothermal waters. It is commonly believed that: (i) only sea water can supply the necessary amounts of salts; and (ii) non-marine evaporites are invariably small and differ compositionally from marine evaporites. Hardie (1984) challenged these views, noting that non-marine evaporites can be as large or as thick as, or may be compositionally similar to, marine evaporites. In 1990, Hardie further argued that most potash-evaporites were products of the evaporation of hydrothermal waters or of sea water substantially modified by hydrothermal brines. Non-marine evaporites and the origin of potash salts are topics that are not covered here, the former being dealt with in Sects 4.7.4 & 4.14 and additional reference should be made to Hardie (1984, 1990), Lowenstein and Spencer (1990) and Smoot and Lowenstein (1991).

Evaporites only *form* in climates where evaporation losses exceed precipitation (rain and snow). Evaporites only *accumulate* in settings where concentrated brines generated by evaporation are not diluted by excessive influx of less saline waters (fresh or sea water). This requirement confines marine evaporite accumulation to sites that are isolated from, or have greatly restricted access to, the ocean. Today, marine evaporites are confined to coastal supratidal settings, and to sites where marine waters seep into low-lying pools and small basins.

Some ancient marine evaporites did form in settings like those of today, but these are but a small proportion of the evaporite geological record. Throughout the Phanerozoic, thick and extensive evaporites ('saline giants') formed but have no modern equivalent. One of the larger modern analogues is Lake Macleod, Western Australia (100×50 km) with up to 9 m of Holocene evaporites (Sect. 8.5). This pales in comparison with the Messinian Mediterranean basins which extended across some 2400×600 km, and have evaporite fills up to 2 km thick (Fig. 8.1). Pleistocene–Holocene evaporites of the Dead Sea Basin may be 11 km thick, approaching the maximum thickness of halite that could be expected in basins subject to thermotectonic subsidence. Some saline giants contain so much salt that their formation must have caused significant reductions of oceanic salinity. Stevens (1977) calculated that Permian salts are equivalent to 10% of the salt in present-day oceans, suggesting pre-Permian ocean salinities may have been 10% higher than at present.

The absence of modern depositional analogues of comparable scale, coupled with a lack or sparsity of core in many evaporites,

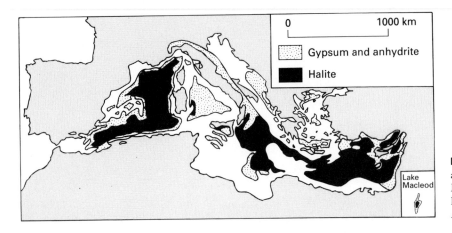

Figure 8.1 Size comparison between ancient and modern evaporites. Messinian of the Mediterranean (after Rouchy, 1980) and Holocene of Lake Macleod, Western Australia.

has meant that determining the depositional conditions of saline giants is highly contentious. To make matters worse, evaporites are also prone to wholesale alteration after deposition. The early growth of displacive evaporites frequently distorts and, in places, obscures depositional fabrics. On burial, hydrous salts (such as gypsum) lose water, commonly with textural loss. Evaporites are also sensitive to tectonic stresses and may totally recrystallize; they may also be deformed, becoming contorted, brecciated and, where thick sequences occur, can form salt pillows and diapirs. Here, even the depositional thicknesses are difficult to determine. Furthermore, evaporites dissolve near the surface, with the more soluble salts being preferentially removed. Where dissolution is pervasive, dissolution breccias and collapsed sediments from overlying strata may be all that remains as witness to hundreds of metres of former evaporites. The ambiguities in evaporite fabrics, resulting from pervasive alteration, combine with the lack of modern extensive evaporite-precipitating environments to heighten debate on the origin of ancient evaporites.

The interpretation of ancient evaporite sequences has changed tremendously in the last three decades. The earliest interpretation was that most evaporites were products of silled, deep, hypersaline basins (Ochsenius, 1877; King, 1947; Scruton, 1953, Sloss, 1953, 1969; Briggs, 1957; Schmalz, 1969). This view was undermined by the discovery of supratidal evaporites in the Trucial Coast (Curtis, Evans *et al.*, 1963; Kinsman, 1966, 1969; Shearman, 1966; Butler, 1969; Bush, 1970). But unfortunately another dogma was created with, it would seem at times, even the slightest hint of a nodular anhydrite being used to promote a sabkha origin for the evaporite in question. In recent years more and more of these nodular anhydrites have been reinterpreted as former subaqueous gypsums based upon models developed from studies of modern salt lakes (Arakel, 1980; Warren, 1982; Logan, 1987) and artificial salt ponds (Geisler, 1982; Orti Cabo, Pueyo Mur *et al.*, 1984; Handford, 1990). Possibly this pendulum has shifted too far again and

some shallow-water gypsums may have formed in desiccated environments subject to infrequent floods.

A major reconsideration of evaporite depositional settings came with the identification of over 2 km of Miocene evaporites beneath the floor of the Mediterranean by deep-sea drilling (Hsü, 1972; Ryan, Hsü *et al.*, 1973). Miocene evaporites currently exposed in Spain, Italy, Crete, Cyprus and North Africa were seen as marginal expressions of an enormous saline giant. The majority of the evaporites, now beneath the floor of the Mediterranean, were apparently deposited in shallow-water or sabkha-like environments. A revolutionary hypothesis was proposed: the Mediterranean had almost completely desiccated, with substantial drawdown of the brine level within the Mediterranean basin (Ryan, Hsü *et al.*, 1973). Recognition of this Mediterranean salinity crisis stimulated acceptance of other evaporites as having formed in desiccated basins, a proposal made previously by Maiklem (1971).

There are several modern compilations on evaporites that emphasize sedimentological aspects (Schreiber, 1986, 1988; Warren, 1989; Melvin, 1991; Kendall, 1992); whereas an approach more closely associated with brine evolution has been compiled by Sonnenfeld (1984). Today, the evaporite researcher has a larger range of facies models and recent analogues to work with than ever before, but still must rely upon non-actualistic models to interpret 'saline giants'. Perhaps more than for any other group of sediments, the iconoclastic view of the past has been transformed into the conventional wisdom of the present (and may be rejected tomorrow ?). In this chapter, we have tried to illustrate the considerable debate that surrounds the interpretation of many evaporite sequences (including some in which we have been participants). We have had to leave out, for lack of space, many well-known evaporites, among which those from the Permian, Michigan, Elk Point Basins and Mediterranean are notable. Studies of these evaporites are no less important than those we have chosen to use in this chapter.

8.2 Conditions of marine evaporite formation

8.2.1 Marine brines and their precipitates

The composition of sea water varies little and appears not to have changed significantly over the last 2000 My (Schopf, 1980). Twelve constituents are present in concentrations greater than 1 ppm (Table 8.1), but, of these, Cl⁻ and Na⁺ account for 85% of the total, thus explaining the volumetric importance of halite in ancient marine-derived evaporites.

The most common marine evaporite minerals are gypsum, anhydrite and halite, although many others occur (Table 8.2). Sea-water evaporation causes the precipitation of an ordered sequence of minerals of increasing solubilities (Usiglio, 1849; reviews by Braitsch, 1971; Holser, 1979; Harvie, Weare *et al.*, 1980). Gypsum precipitation starts when sea water has been concentrated 3.8 times; halite when concentrations exceed 10 times sea-water values; magnesium sulphate salts appear at c. 70 times concentrations, and potassium salts only when concentrations exceed 90 times that of sea water (McCaffrey, Lazar & Holland, 1987) (Fig. 8.2). The mineral succession is complex after halite precipitation and this is partly due to the varying extent to which brines can react with earlier-formed precipitates. Two main types of evaporite sequence are recognized. '*Normal*' marine sequences contain magnesium sulphates and complex salts such as polyhalite, kainite and langbeinite, but lack primary sylvite. In contrast, the so-called '*modified marine*' sequences contain primary sylvite and lack magnesium sulphates and complex salts. This 'abnormal' type (which actually is the more common) may also contain calcium chloride minerals, as products of the last stages of evaporation. These cannot be obtained from simple sea-water evaporation (Hardie, 1990).

Table 8.2 Common marine evaporite minerals.

Anhydrite	$CaSO_4$
Aragonite	$CaCO_3$
Bassanite	$CaSO_4 \cdot \frac{1}{2}H_2O$
Bischofite	$MgCl_2 \cdot 6H_2O$
Bloedite (astrakhanite)	$Na_2SO_4 \cdot MgSO_4 \cdot 4H_2O$
Calcite	$CaCO_3$
Carnallite	$MgCl_2 \cdot KCl \cdot 6H_2O$
Dolomite	$CaCO_3 \cdot MgCO_3$
Epsomite	$MgSO_4 \cdot 7H_2O$
Glauberite	$CaSO_4 \cdot Na_2SO_4$
Gypsum	$CaSO_4 \cdot 2H_2O$
Halite	$NaCl$
Hexahydrite	$MgSO_4 \cdot 6H_2O$
Kainite	$MgSO_4 \cdot KCl \cdot \frac{11}{4}H_2O$
Kieserite	$MgSO_4 \cdot H_2O$
Langbeinite	$2MgSO_4 \cdot K_2SO_4$
Leonhardtite	$MgSO_4 \cdot 4H_2O$
Leonite	$MgSO_4 \cdot K_2SO_4 \cdot 4H_2O$
Loewite	$2MgSO_4 \cdot 2Na_2SO_4 \cdot 5H_2O$
Magnesian calcite	$(Mg_xCa_{1-x})CO_3$
Pentahydrite	$MgSO_4 \cdot 5H_2O$
Polyhalite	$2CaSO_4 \cdot MgSO_4 \cdot K_2SO_4 \cdot 2H_2O$
Sylvite	KCl
Syngenite	$CaSO_4 \cdot K_2SO_4 \cdot H_2O$
Tachyhydrite	$CaCl_2 \cdot 2MgCl_2 \cdot 12H_2O$

The thickness of the various evaporite minerals precipitated from sea water depends on the rate and mode of replenishment of the evaporated sea water. If a column of sea water 1000 m deep were allowed to evaporate completely, only about 14 m of evaporites (mostly halite) would form. Thus, any substantial thicknesses of evaporites require the basin brines to be repeatedly replenished with additional batches of sea water. Sites of thick marine evaporite accumulation are thus sites where large

Table 8.1 Composition of modern sea water referred to a standard chlorinity of 19‰ (from Braitsch, 1971).

Ion	‰	mol 1000 mol H_2O	Dead Sea ‰
Na⁺	10.56	8.567	34.74
Mg²⁺	1.27	0.976	41.96
Ca²⁺	0.40	0.186	15.8
K⁺	0.38	0.181	7.56
Sr²⁺	0.008	0.002	–
Cl⁻	18.98	9.988	208.02
SO₄²⁻	2.65	0.514	0.54
HCO₃⁻	0.14	0.043	0.24
Br⁻	0.065	0.015	5.92
F⁻	0.0013	0.001	–
B	0.0045	0.008	–
$H_4SiO_4^-$	0.001	–	–

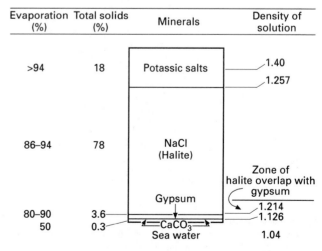

Figure 8.2 Relative proportions of evaporite components in sea water (after Borchert & Muir, 1964).

volumes of sea water can be introduced and evaporated, but sea-water access cannot be so free as to substantially dilute the basinal brines.

The ratio between halite and anhydrite of an evaporite precipitated by complete sea-water evaporation should be 22 : 1 (by volume). In most ancient evaporites, however, the ratio is nearer 3 : 1; it commonly is lower than 1 : 1 (Fig. 8.3), and many evaporites lack any halite. If sea water was the influx, the low ratios in ancient evaporites imply either: (i) incomplete sea-water evaporation and export of much of the halite (and more soluble salts) as a result of some kind of reflux of brines out of the basin; or (ii) later preferential dissolution of the more soluble salts.

Evaporation, and thus brine concentration, occurs only at the brine–air interface. This is maximized in artificial salt ponds by having large surface area to brine volume ratios. In basins with deep brines, the ratio is smaller and consequently the time necessary to evaporate sea water to obtain high salinity brines is relatively long. Deep-water evaporites thus need stable conditions in which to form. Once the high salinities have been

Table 8.3 Rates of evaporite deposition from marine waters and marine-fed marginal deposits (from Schreiber & Hsü, 1980).

Sediment type	Area of formation	Observed rates of deposition
Sulphates and carbonates	Sabkha	Thickness of 1 m per 1000 years with 1–2 km progradation per 1000 years
Sulphates (usually gypsum)	Subaqueous (observed in solar ponds)	1–40 m 1000 years over entire basin
Halite	Subaqueous (solar ponds)	10–100 m per 1000 years over entire basin

achieved, however, evaporites can accumulate with great rapidity. Evaporite deposition is potentially more rapid than that of any other sediment, with rates the equivalent of up to 100 m ky^{-1} being observed in salt ponds (Table 8.3). Gypsum crystal growth, in restricted areas of continued water flow, may surpass 5–6 cm year^{-1} (Schreiber & Hsü, 1980); whereas halite crystal growth averages 10–50 cm year^{-1}. During the Miocene of the Mediterranean, up to 2 km of evaporites formed in half a million years – a rate (3–4 m ky^{-1}) comparable with those in salt ponds.

8.2.2 Climatic controls on sea-water evaporation

Marine evaporites today occur within the subtropics, usually between latitudes 15 and 30° (Borchert & Muir, 1964). They may also occur leeward of orographic barriers, theoretically at any latitude. Areas of high evaporation rates coincide with deserts (see Fig. 5.1) and are determined by the belts of subtropical high pressure that result from descending air masses (see Fig. 2.1). Evaporation rates of up to 3.5 m year^{-1} have been measured from the Red Sea and Gulf of Suez, and maximum possible rates are probably near 4 m year^{-1}.

Only dry, undersaturated air can evaporate sea water. As the salinity of a brine rises, the extent to which the air must be undersaturated prior to evaporation occurring also increases (Kinsman, 1976). Relative humidities <76% are necessary to evaporate brines saturated with respect to halite, and relative humidities <67% are needed before brines capable of precipitating potassium salts can be generated (Fig. 8.4). Today, low-latitude coastal regions have average relative humidities between 70 and 80%, with lower values occurring over continents. Although daytime relative humidities may be as low as 30% in very arid coastal zones, during the night they may increase to near 90% (Bush, 1973). At these higher humidities, brines become diluted by absorbing atmospheric moisture. This dew dissolves precipitated salt crusts, especially those composed of deliquescent

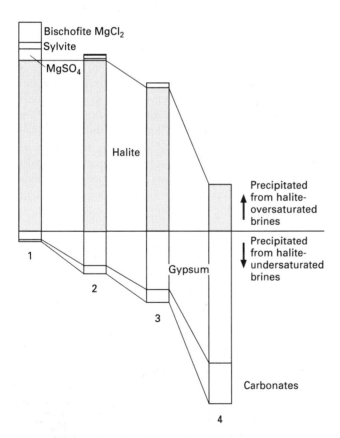

Figure 8.3 Quantities of precipitates formed from sea-water evaporation (1), compared with the composition of ancient evaporite deposits; (2) Zechstein evaporites; (3) the average for evaporite basins; (4) anhydrite-dominated sequences (after Borchert & Muir, 1964).

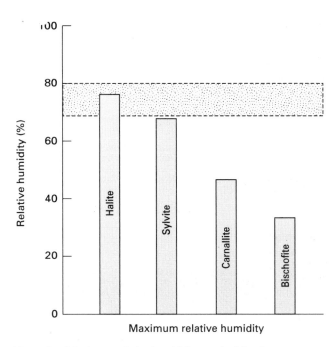

Figure 8.4 Maximum relative humidities required for the precipitation of certain evaporite minerals. Brines, saturated for a given precipitate, will cease to evaporate at higher relative humidities. The range of average low latitude coastal humidities is **stippled** (modified from Sonnenfeld, 1984).

evaporite minerals. Thus many evaporites are dominated by gypsum (or anhydrite) simply because the maximum humidities at the time of deposition are, or were, too high for more saline minerals to survive. Halite and potassium salts only accumulate in areas of very low relative humidities typical of intracontinental environments. When found in ancient marine evaporites, they imply that the marine basin must have been almost entirely surrounded by land in a region of abnormally low humidity. The Gulf of Suez is a modern example. This is a rifted basin with a hyperarid climate. Its long, narrow shape, with relatively small surface area, does not substantially increase the relative humidity of the area. It has mean annual relative humidities of only 38–53% (Evenari, Gutterman & Gavish, 1985). Even so, occasional summer humidities may exceed 76%. Because very saline evaporites rapidly dissolve, the mineralogy of evaporites that accumulate may be controlled more by the frequency and magnitude of short-lived high humidity events than by the mean low relative humidities of an area.

8.2.3 Controls on evaporite deposition within basins

The main controlling factors affecting the types and distributions of evaporites within marine-fed basins are: (i) the magnitude of the water influx; (ii) the amount of brine reflux out of the basin; and (iii) the ratio between input and output of water within the basin (Logan, 1987; Sanford & Wood, 1991). The last factor determines the length of time water can spend in a basin and

thus the length of time evaporation affects it. This *residence time* is a major determinant of brine salinity and it is salinity that controls evaporite mineralogy.

Thick sequences of marine evaporites require a large inflow of sea water into isolated or highly restricted basins. Sea water can be introduced through channels crossing the barrier; or by seepage through a continuous but permeable barrier. The size of any channel is critical. In order for brines of a certain salinity to form, there must be a high ratio between the surface area of the evaporite basin A_o and the cross-sectional area of any inlet A_t (Fig. 8.5). Lucia (1972) concluded that basin brines can only reach gypsum saturation when any channel present has a geographically insignificant cross-sectional area; at least six orders of magnitude smaller than the basin surface area (Fig. 8.5). If sea water also entered the basin by seepage, gypsum saturation can only be maintained if the inlet size is further reduced. For brines to reach halite saturation, the inlet area must be eight orders of magnitude smaller than the basin surface area, and complete surface disconnection from the ocean is almost certainly required (Lucia, 1972; Kendall, 1988). This is because such small inlets are unlikely to be stable over the period of time required to precipitate the basin evaporites.

Seepage through a permeable barrier is a more realistic means of sea-water inflow, especially for basins containing halite or even more saline salts. Seepage inflow occurs today at Lake Macleod, Western Australia (Logan, 1987), Lake Assal, Djibouti (Perthuisot, 1989) and smaller evaporitic pools (Aharon, Kolodny & Sass, 1977; Gavish, 1980; Perthuisot, 1980; Friedman, Sneh & Owen, 1985). The rates at which seepage inflow can occur, even where highly permeable barriers are present, are sometimes

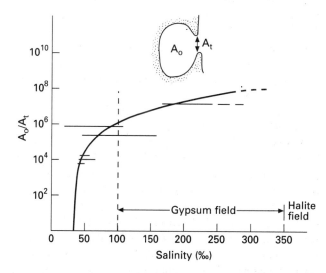

Figure 8.5 Relationship between the ratios of the cross-sectional area of an inlet A_t and the surface area of the evaporite basin A_o, and the maximum salinity developed within the basin, with data plotted from modern restricted marine settings (after Lucia, 1972).

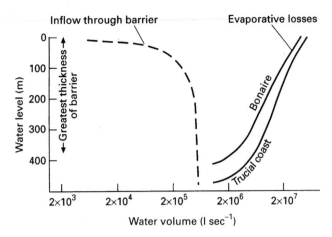

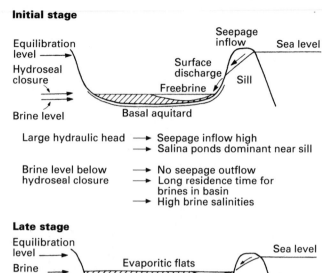

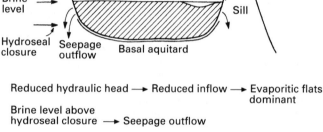

Figure 8.6 Comparison between rates of seepage inflow through the permeable barrier and probable rates of evaporative water loss from the Middle Devonian Elk Point Basin (western Canada). Inflow rate assumes a very high value for the average permeability of the barrier (73 darcys) and water losses are calculated for climatic conditions similar to those of Bonaire and the Trucial Coast. Evaporative water losses are higher than water inflow and isolated basins exhibit tendencies toward desiccation (after Maiklem, 1971).

Figure 8.7 Models of isolated evaporite basins subject to evaporative drawdown, showing evolution of basins with time. Initial stage with large hydraulic head (between sea and brine levels) and brine levels located below elevation of basal seal closure. Late stage of basin evolution with reduced hydraulic head and brine levels above basal seal. Consequences of these changes are shown with respect to rates of seepage reflux, brine residence times, and possible brine salinities. Based on work of Logan (1987) (after Kendall, 1992).

orders of magnitude lower than the rates of potential evaporative water losses from the basin surface (Fig. 8.6). Thus, unless something else intervenes, the volume of water remaining in an isolated basin decreases and its brine level falls: this is the *evaporative drawdown* of Maiklem (1971). The increasing hydraulic head, caused by the increasing elevation difference between the ocean and the falling brine level in the basin, promotes greater rates of sea-water seepage through the barrier. Changes in relative sea level will also change the magnitude of the hydraulic head and hence the rate of sea-water influx (Fig. 8.7).

Isolated basins with greatly depressed brine levels, even if fed by sea-water seepage, are also likely to be sites of formation-water inflow from laterally adjacent or underlying sediments (Kendall, 1988). Evaporites formed in such basins will have been precipitated from sea water modified to some extent by mixing with these other waters.

We have seen that, in order to precipitate halite within them, basins must be isolated. Yet the absence or reduced amounts of the more soluble salts within ancient evaporites requires that the basin could export these more soluble constituents. In isolated basins this loss can only occur through the basin floor or its margins (Logan, 1987; Kendall, 1992). On the other hand, were the permeability of the evaporite basin floor sufficiently high to allow easy seepage reflux of these unwanted brines, no brines would have been generated! As soon as they began to form, brines would have seeped down through the permeable basin floor, or would have been diluted by upflowing formation waters impelled by the local hydraulic head. Evaporite basins

must have basal aquitards (Logan, 1987) to prevent such brine losses or dilutions. They must also allow some brine export (or storage within basin sediment porosity) to account for the overall composition of their evaporitic fills, but these losses cannot be too large or else brines would not be generated in the first place.

Calculations by Brantley and Donovan (1990) indicate that basins are capable of generating and exporting, by seepage outflow, large volumes of brines – volumes able to fill rock porosity to depths of many kilometres beneath the evaporite basin. The geochemistry of subsurface waters in many sedimentary basins suggests that they are, in part, brines generated and exported from evaporitic basins (e.g. Land & MacPherson, 1992).

The salinity of basin brines, and thus the mineralogy of evaporites precipitated, is largely controlled by the length of time sea water remains in the basin before being exported (Logan, 1987; Sandford & Wood, 1991). If fluid inflow and outflow rates are both low relative to the rate of evaporative water loss, then, in climates having sufficiently low relative humidities, all brines will be evaporated or lost to the basin by seepage outflow, and the basin will desiccate leaving only a crust of evaporite minerals. Residence times of waters in the basin will be short and no surface bodies of brine will be present within which to form subaqueous evaporites. Continued sea-water seepage outflow will introduce additional salts but these will most likely be precipitated within existing sediments. Similarly, when seepage reflux rates are high, relative to evaporative losses, brines also have low residence times and are exported from the basin before evaporation has time to concentrate them sufficiently to precipitate the more saline components. In these basins only evaporites of low solubility are precipitated, even in areas of great aridity. Where reflux of brines is low, residence times of basin waters are long and, in very arid climates, evaporation can generate brines of high salinity, allowing precipitation of halite or even more soluble salts.

Anywhere evaporites are precipitated, the rates of inflow, reflux and evaporation will change with time. These changes alter the residence times of brines, their salinities and, therefore, the mineralogy of the precipitated evaporites. Variations in evaporite mineralogy are to be expected anywhere that evaporite deposition continues for an extended period, and are not always climate or sea-level related.

In subaqueous evaporitic environments the different evaporite minerals are precipitated from brines of significantly different density. To precipitate different minerals in separate parts of a brine-filled basin would require the brines to exhibit a marked horizontal density variation. Where brine depths exceed a few metres, such a marked horizontal variation is hydrodynamically unstable (Fig. 8.8) (Shaw, 1977; Kendall, 1988). The brine body quickly comes to equilibrium by becoming vertically stratified. It will convert to one with horizontally uniform brine densities. The same evaporite minerals will be precipitated over the entire depositional area and any salinity change will affect the entire basin at the same time and cause simultaneous changes in evaporite mineralogy over the entire basin. Vertical changes in subaqueous evaporite sequences reflect basin-wide changes in the precipitating brine bodies, and do not represent lateral migration of lateral facies (James & Kendall, 1992). Apparent lateral facies changes indicate either that deposition occurred in extremely shallow environments where bottom friction prevented equilibrium conditions being reached, or that the different facies are not, in fact, time equivalents of each other. Thus, Walther's Law (see Sect. 2.2) does not apply to subaqueous evaporites (Kendall, 1988). The use of evidence from overlying and underlying facies to constrain the interpretation of subaqueous evaporites is thus always suspect.

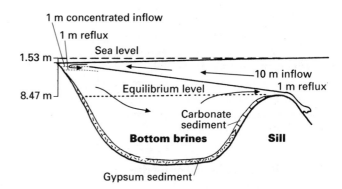

Figure 8.8 Hydrostatic conditions in a silled evaporite basin at the point when halite is just about to be precipitated (based on Shaw, 1977). At distal parts of the basin the inflow has been reduced to a tenth of its original depth. In order to achieve hydrostatic balance across the basin, additional bottom brines must be present in distal areas. Difference in water depths across a basin in hydrostatic balance would require (for a 10-m influx) unreasonable elevation differences of the water surface (1.53 m) and of the surface between the inflow and the bottom brines (8.47 m) (after Kendall, 1988). Note that the equilibrium level here differs from that in Fig. 8.7.

8.3 General features of evaporites

8.3.1 Basin-marginal and basin-central evaporites

The distribution of evaporites in basins is of major importance to the correct identification of evaporitic environments (Kendall, 1992) because the same or very similar facies may occur in very different environmental settings. Basin-marginal and basin-central evaporite distributions (Fig. 8.9) may be distinguished (Krumbein & Sloss, 1963). Basin-central evaporites occupy the entire depositional basin, whereas basin-marginal and shelf evaporites pass laterally into normal marine strata in the basin centre. All present-day marine-fed evaporite environments are basin marginal. No modern representatives of *marine* basin-central evaporites exist, although similar shallow-water facies and facies patterns do occur in the larger salinas and in non-marine settings. Non-actualistic models are required to interpret deepwater basin-central evaporites, although the Dead Sea (Sect. 8.6.2) is a poor, non-marine analogue.

Basin-marginal evaporites, usually dominated by sulphates, may form in saline lagoons (*salinas*; Sect. 8.5.2), the margins of which are surrounded by supratidal deflation surfaces. Similar salt-encrusted surfaces, or *sabkhas*, also occur along protected arid coastal flats. They are commonly confined to the landward parts of shelves and platforms. If, however, a continuous barrier is formed at the shelf/platform edge, the entire inner platform may become evaporitic or evaporite-accumulating. There are no modern equivalents of such wide evaporitic shelves.

Basin-central evaporites form during sea-level falls, when a basin becomes isolated and its rim is exposed; they are represented on the basin rims by non-sequences. The magnitude of brine drawdown, however, need bear little resemblance to the amount of external sea-level fall (cf. Tucker, 1991) because the brine levels are determined by basin size and the permeability of its floor and barrier (Fig. 8.7). It is in these basins that the more soluble evaporites may accumulate, including potash salts.

These two major modes of evaporite distributions have different sequence stratigraphic implications (Sarg, 1988; James & Kendall, 1992) (Sect. 8.10).

8.3.2   Brine depths and evaporite facies

Evaporites can accumulate in brine bodies of any depth and also form within existing supratidal sediments. The greater volume, however, appear to form in shallow settings where evaporation is most effective and where brine concentration is enhanced. Different sedimentary facies are discriminated on the basis of their mineralogy and sedimentary structures. Evaporites can be divided into: (i) accumulations of surface-nucleated crystals (crystal cumulates); (ii) bottom precipitates (crusts); (iii) clastic accumulations of evaporite particles; and (iv) diagenetically emplaced precipitates which can replace, displace or incorporate the host sediments (Fig. 8.10).

Variations in brine concentration control the nature of the precipitating minerals, whereas changes in the rate of evaporation control evaporite crystal size. The presence (or absence) and strength of currents and wave activity affecting the bottom brines, perhaps modified by microbial-mat fixation, determine the types of sedimentary structures formed. Once precipitated, evaporite mineral grains may be subject to the same physical processes of erosion, transport and deposition that operate in siliciclastic and carbonate depositional environments. Many evaporites contain sedimentary bedforms common in other environments; however, the brines that transport and deposit evaporite grains are denser than sea water, and both gypsum and halite are less dense than siliciclastic grains (Karcz & Zak, 1987). Care must therefore be taken in extrapolating current velocities from non-evaporitic to evaporitic environments.

Perhaps the most important (and the most disputed) environmental parameter to be established about ancient evaporites is the brine depth during deposition. Three main environmental settings for subaqueous evaporites have been identified on the basis of sedimentary structures present (Schreiber, Friedman *et al.*, 1976; Fig. 8.11). The high-energy intertidal and shallow subtidal environments are identified by the presence of waves and current structures. The absence of these structures, together with the presence of algal structures, indicate a deeper environment, but one still within the photic zone; whereas laminated sediments lacking current/wave and algal structures are believed to characterize the deeper, subphotic environments.

These criteria are not always easily used. Stromatolites can grow in protected water settings; thus algal-modified evaporites lacking current structures may occur in the shallowest settings. It may also be impossible to distinguish planar algal mats from

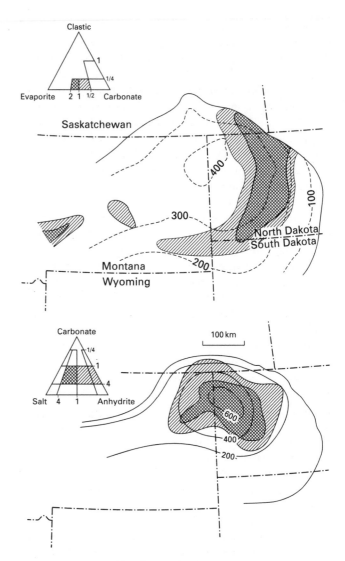

Figure 8.9 Evaporite distributions for the Mississippian (Lower Carboniferous) of Williston Basin. (a) Basin-marginal lower part of the Charles Formation; (b) basin-central upper part of the Charles Formation (after Krumbein & Sloss, 1963).

organic films formed by a passive accumulation of organic matter; the latter having no depth significance. Widespread even laminations also do not always identify deepwater environments. Similar lamination is produced in modern subaerial, ephemerally flooded environments (see Fig. 8.32), or in shallow brine pools.

Other criteria used to assess depositional depths are the presence of dissolution surfaces, bottom-growth crystal crusts or sediment-displacing crystals. Dissolution surfaces form when influxes of less saline water dissolve the uppermost part of the previously deposited evaporite. For this to occur, the inflow must dilute the whole brine column (Adams, 1969; Warren, 1982), implying original brine volumes were small and, consequently, brine depths were shallow. Growth of crystals as

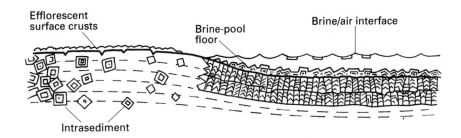

Figure 8.10 Main types of evaporite (halite) (modified from Handford, 1991).

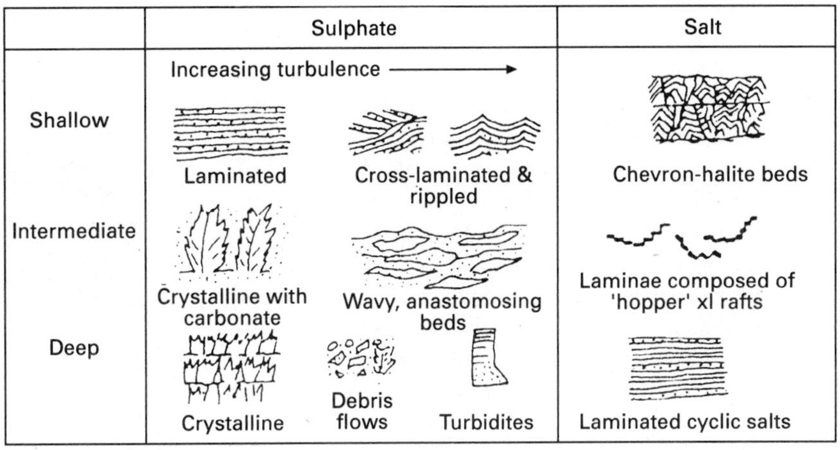

	Sulphate			Salt
	Increasing turbulence ⟶			
Shallow	Laminated	Cross-laminated & rippled		Chevron-halite beds
Intermediate	Crystalline with carbonate	Wavy, anastomosing beds		Laminae composed of 'hopper' xl rafts
Deep	Crystalline	Debris flows	Turbidites	Laminated cyclic salts

Figure 8.11 Environmental interpretation of subaqueous evaporite facies (after Schreiber, Friedman *et al.*, 1976).

bottom crusts or within previously deposited sediments, on the other hand, implies that these precipitates were in contact with supersaturated brines. Evaporative concentration leading to such supersaturation occurs only at the brine–air interface. Deep bodies of brine almost always have temperature and/or salinity stratification which prevent descent of the concentrated surface brines. Thus bottom crusts should not occur in deep stratified bodies of brine. Unless supersaturation can be reached by means other than surface evaporation, or brine stratification is destroyed (thus allowing surface brines to descend), evaporite crusts and intrasediment-grown crystals should record precipitation from brines, at most only a few tens of metres deep. Gypsum crusts in the Castile Formation, however, appear to have precipitated from brines several hundred metres deep (Sect. 8.9).

Distinction of shallow from deep evaporite-precipitating environments cannot be made using single criteria alone. All possible evidence must be used, including the geologic setting, the range of sedimentary structures present and, where possible, the lateral facies relationships.

The next three sections discuss the evaporite facies of three main environmental settings: sabkhas, shallow-water and deep-water environments. In each, a short description of the general features of the environment is followed by descriptions of characteristic facies (both recent and ancient), and by examples of present-day evaporitic settings.

8.4 Sabkhas

Sabkhas are low-lying salt-encrusted surfaces sometimes adjacent to perennial, or ephemeral, bodies of brine. Only those in coastal settings are considered here. They occur along the shores of the Arabian (Persian) Gulf, Gulf of Suez and the Red Sea, west and southern Australia, North Africa and Mexico. Many are prograding supratidal flats adjacent to protected lagoons and are composed of sediment transported from both the sea and the continental hinterland. Groundwaters may have been introduced by marine flooding, by marine seepage, by inflowing continental groundwaters or by rare and ephemeral sheet floods. This means that some sabkhas, like those along the coastal plain of the western Gulf of Sirte in Libya (Rouse & Sherif, 1980), are marginal marine in terms of geography but are supplied by continental brines. They have non-marine evaporites and geochemistries. Even the most classic of marine sabkhas, those in the Abu Dhabi area (Arabian or Persian Gulf), are associated with marine-derived brines on their seaward sides but continental brines away from the coast. A coastal sabkha may pass without noticeable change in surface morphology into a continental sabkha. Coastal sabkhas may also evolve into continental sabkhas as the coastal plain progrades and marine-derived brines are replaced by continental waters.

Sabkhas are subaerially exposed for most of the time, and formed largely by wind deflation. They contain similar evaporite facies to those in mudflats around continental playa lakes (see Sect. 4.7.4), although the evaporite mineralogy of these non-marine settings may differ if the continental brines differ compositionally from sea water. Coastal sabkhas may be underlain by siliciclastic and/or carbonate sediments and these become partially cemented, replaced or displaced by evaporite minerals that form in the capillary zone above a saline water table. Only dry uncemented sediment can be eroded by wind, so that the level down to which deflation can erode is usually determined by the top of the capillary zone, itself controlled by the level of the underlying water table. Since this brine level is essentially flat, sabkhas are equally planar and constitute the flattest of all landforms.

As a sabkha progrades, the height of the water table on its landward side may fall, allowing the uppermost sabkha sediments to desiccate and be removed by the wind (Patterson & Kinsman, 1981). This occurs because water is continuously being lost by evaporation and these losses cannot be entirely offset by sea water or continental groundwater seepage through the relatively low permeability sabkha sediments. In the interior of wide sabkhas, flooding from the sea occurs too infrequently to compensate for these evaporative losses. Deflation stops when erosion has progressed down to cemented intertidal sediments – that is, when all of the supratidal sediments have been removed. The preservation potential of an extensive sabkha containing only ephemeral halite is therefore low.

Coastal sabkhas are found in four settings: (i) as prograding supratidal flats above sea level, along protected coastlines; (ii) as depression sabkhas, where salt pans occur between dunes or beach ridges; (iii) as salina-marginal sabkhas, associated with salt lakes fed by marine seepage through barriers; and (iv) on top of evaporite-filled salinas. Only the first two are discussed in this section.

8.4.1 Sabkha evaporite facies

Evaporite minerals in sabkhas are precipitated either within sediments by the evaporation of capillary brines, or as ephemeral surface crusts. Only intrasediment evaporites can be preserved and it is they that characterize ancient sabkha sequences. Many features of coastal sabkhas are identical to those in mud- or sandflats around inland playa systems, and the sabkhas can only be distinguished by their palaeogeographic position, their relations with adjacent facies, and perhaps by their mineralogy and geochemistry. Two main facies occur, dry and saline mudflats. They correspond to the facies distinguished both in non-marine playas (Hardie, Smoot & Eugster, 1978; Smoot & Lowenstein, 1991) and in salinas (Handford, 1991).

In *dry mudflat facies*, evaporites are not preserved. Here, sedimentary layering or lamination forms during and after floods, but subsequently is partially disrupted by desiccation cracks and by repeated cycles of growth and dissolution of evaporite minerals. The only evaporites present are efflorescent saline crusts. These grow on the mudflat surfaces during intervals between floods, but are ephemeral. The next flood begins to dissolve the crusts, and coarser sediment, introduced by the same flood, is trapped in dissolution depressions. Repeated formation of these irregular coarser-grained areas produces a sandpatch fabric (Smoot & Castens-Seidell, 1994), characteristic of dry mudflats (Fig. 8.12). Any salts that survive floods, dissolve upon burial because groundwaters are undersaturated.

In *saline mudflats*, original lamination and desiccation structures are destroyed by the growth of abundant intrasediment evaporite crystals within the capilliary zone (Figs 8.13 & 8.14)

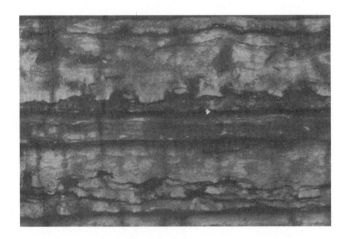

Figure 8.12 Sandpatch fabric from basal Red River Formation, Saskatchewan. Coarser-grained areas of dolomite (lighter in colour) are inferred to have originally filled dissolution depressions on former saline crust.

Figure 8.13 Gypsum crystal mush disrupting high intertidal algal mat, Abu Dhabi sabkha. Photograph taken from core discussed in Butler, Harris and Kendall (1982).

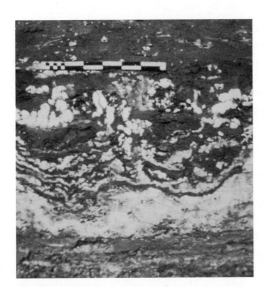

Figure 8.14 Nodular (chicken-wire) anhydrite, generated by a replacement of an earlier gypsum crystal mush and addition of sulphate from pore waters causing further expansion, jacking up of the sabkha surface and the formation of folds and diapirs, Abu Dhabi sabkha. Additional nodules and nodule layers have grown within supratidal sediment above main replacement layer (photograph courtesy of R. Park).

and, locally, within the groundwater zone. Evaporite crystals close to the surface may dissolve during floods, resulting in crystal moulds. Repeated cycles of evaporite precipitation, dissolution and sediment collapse into the crystal moulds produces a chaotic mix of evaporite crystals within an unlaminated mud matrix. Growth of evaporite crystals, may be so extensive that the host sediment becomes little more than a sparse matrix between adjacent crystals, and all original sedimentary structures are destroyed by plastic flowage. The prolific growth of evaporites within the sediment also causes the mudflat surface to rise. The uplifted sediments dry out (since they are now above the capillary zone) and, unless cemented, are blown away by the wind, until the original level is reached.

In ancient sequences, evaporites are only likely to be preserved within saline mudflats. These, however, may be intimately interbedded with (and pass laterally into) dry mudflat deposits. Such interstratification records variation in the extent to which the intrasediment evaporite crystals survive dissolution, variations that may reflect only slight environmental changes.

In coastal sabkhas the evaporites that form are gypsum, anhydrite and halite; more rarely, some polyhalite replaces earlier calcium sulphates. Gypsum crystals commonly have a discoidal habit and may be complexly twinned and intergrown to form crystal rosettes and clusters. In algal mats and fine-grained sediment, displacive growth is common (with crystal mushes forming where gypsum crystals are abundant, Fig. 8.13), whereas in sands the growing gypsum usually encloses sediment to form large cement crystals.

Anhydrite is less common than gypsum and, where present, is confined to the capilliary zone. Here it may be a primary precipitate, but more commonly is a replacement of earlier-formed gypsum (Butler, 1970; Butler, Harris & Kendall, 1982). It forms as discrete nodules and as bands of coalesced nodules that displace and replace the host sediment (Fig. 8.14). Commonly, layers are complexly contorted (enterolithic layering) either because of continued anhydrite growth within the sediment (Shearman, 1970) or because the anhydrite replaces gypsum crusts that grew upon contorted algal mats (Hardie, 1986a). With sufficiently high temperatures (>34°C), gypsum may slowly dehydrate to anhydrite over some months; but lower temperatures, a single rainstorm or a marine flood can rehydrate the anhydrite in a matter of hours.

Halite commonly forms either as an ephemeral efflorescent crust on saline mudflats or as displacive halite within the sediments. Displacive crystals commonly grow preferentially along cube corners and edges to generate hopper-like pyramidal hollows on each cube face (Fig. 8.15). In extreme cases, skeletal and dendritic crystals form. Sediments with abundant displacive halite cubes are termed haselgebirge (Arthurton, 1973).

8.4.2 Modern supratidal sabkhas

Modern supratidal sabkhas develop where shorelines have prograded seawards over previously deposited subtidal, commonly restricted lagoonal, sediments. This generates shoaling-upward successions of subtidal, intertidal and supratidal sediments, of which the supratidal portion is seldom thicker than 1 m.

Supratidal sabkhas form in protected environments: along carbonate coastlines, where protected by a series of barrier islands (Abu Dhabi; Warren & Kendall, 1985); in delta top settings in hyperarid areas (Shatt el Arab delta; Purser, Azzawi *et al.*, 1982); and on alluvial fan margins where protected by offshore islands (Mehran, Iran: Baltzer, Conchon *et al.*, 1982). They are restricted to microtidal and low mesotidal regimes because a larger tidal range tends to eliminate the protecting barriers (see Sect. 6.7.4) with little opportunity for supratidal flats to form. Recent summaries by Handford (1991) and Warren (1991) describe modern and ancient sabkhas and their settings, and Purser (1985) emphasizes the range of modern sabkha types, especially those in the Middle East. The carbonate sedimentology of sabkhas and related lagoons is discussed in Sect. 9.4.1. Of all modern sabkhas that of Abu Dhabi is the most fully studied (Curtis, Evans *et al.*, 1963; Shearman, 1966; Kinsman, 1966, 1969; McKenzie, Hsü & Schneider 1980; Hsü & Siegenthaler, 1980; Patterson & Kinsman, 1981) and it has been used (sometimes inappropriately) for the interpretation of most ancient sabkha sequences.

The Abu Dhabi sabkha is part of a belt of sabkhas that extends along the shores of the United Arab Emirates for more than 300 km and is up to 15 km wide. Those southeast of Abu Dhabi City (Figs 8.16, 8.17 & 8.18) are the best known, contain the

Figure 8.15 Cubic, hopper and skeletal crystals of displacive halite within structureless siliciclastic mudstone matrix; Lower Elk Point clastics, Alberta. Greater concentrations of displacive crystals (bottom of core) form 'Haselgebirge'.

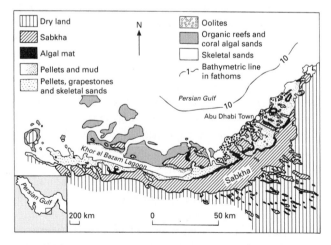

Figure 8.16 Map of coastal carbonate facies in Abu Dhabi (Trucial Coast) illustrating protected environment of sabkha progradation (from Butler, Harris & Kendall, 1982).

greatest variety of minerals and sedimentary structures, but now are increasingly being covered by urban and industrial development. In this arid setting a complex carbonate depositional coastline containing beaches, barrier islands, tidal deltas and channels, and backed by intertidal and supratidal flats and by coastal dunes, has developed over the last 7000 years (see Sect. 9.3.4). The sabkha has prograded at approximately 1.5 km per thousand years, following a maximum transgression 4000 years ago.

Relative air humidities in the region are variable; varying from 30 to 95% within 24 h during the summer (Bush, 1973). On the sabkha, high evaporation rates (up to 150 cm year^{-1}) and low, sporadic rainfall (averaging 4–5 cm year^{-1}) raise groundwater salinities and cause the precipitation of gypsum, anhydrite and halite. Constant jacking up of the sabkha surface by this evaporite growth, and subsequent wind deflation, creates a planar erosion surface with a slight seaward slope (dipping at rates of only 1:2000 and 1:3000; Butler, 1970). Locally, the surface is interrupted by small hillocks of older Tertiary and Pleistocene carbonates, by low amplitude tidal channels and by planed-off, earlier Holocene beach ridges (Kendall & Skipwith, 1969; Butler, Harris & Kendall, 1982; Warren & Kendall, 1985); gypsum and halite form in any slight depression, especially along abandoned tidal channels.

Evaporite minerals occur in the upper intertidal and supratidal zones of the sabkha. Microbial mats characterize the upper intertidal sediments, forming in areas too hostile for grazing cerithid gastropods. Below the sabkha surface older microbial mats are compacted into peats (Park, 1976). Discoidal gypsum crystals occur in these sediments (Fig. 8.13), with local cementation by aragonite, magnesite and protodolomite (Butler, Harris & Kendall, 1982). Supratidal sediments overlie the algal peats and are predominantly formed of reworked shallow marine carbonates (skeletal grains, hardened pellets, carbonate mud) but siliciclastic sands may be locally important or even dominant. These sediments are introduced by winds or by onshore storm surges. Rare inland-derived floods also bring siliciclastic components on to the sabkha surface. It is these sediments that host gypsum and anhydrite nodules, beds and enterolithic structures (Fig. 8.14). Away from the shore, microbial mats/peats become increasingly disrupted by growth of lensoid crystals of gypsum, which may, in places, form a distinct layer of crystal mush (Fig. 8.13). Decimetre-scale poikilotopic gypsum crystals also grow and incorporate algal mat, host carbonate and siliciclastic sediment, and earlier-formed lensoid gypsum crystals.

The amount of evaporite in sabkha sediments is controlled by the sabkha hydrology. Marine flooding, driven by onshore winds, recharges sabkha groundwaters. Where marine flooding occurs at intervals of less than once a month, anhydrite nodules appear within the supratidal sediments, sediments which by now have been largely replaced by magnesite and dolomite. The anhydrite nodules increase in both size and number landwards, forming interlocking sulphate polygons 0.3–1 m in

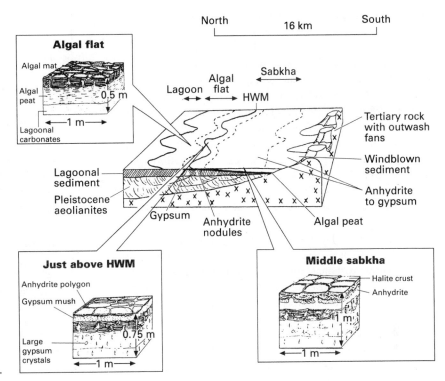

Figure 8.17 Schematic block diagrams illustrating carbonate sediment and diagenetic evaporite distribution in the Abu Dhabi sabkha (from Butler, Harris & Kendall, 1982).

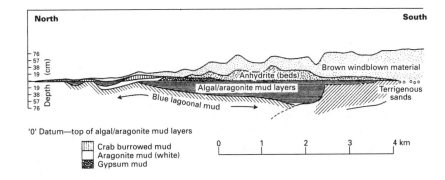

Figure 8.18 Cross-section of supratidal sediments along a north–south traverse of Abu Dhabi sabkha (Fig. 8.16). Note the great vertical exaggeration (vertical scale in centimetres; horizontal in kilometres). Anhydrite is confined to sediments above the brine table (from Butler, Harris & Kendall, 1982).

diameter, and up to 70 cm thick, with displacement and replacement causing the original sediment to be confined to partings between anhydrite nodules. Most of this mosaic anhydrite, however, has probably replaced earlier-formed gypsum (Butler, 1970; Bush, 1973) but, because of sediment flowage and the continued growth of primary anhydrite within and around the replacive anhydrite, pseudomorphs after the original gypsum crystals are uncommon.

On the sabkha surface, ephemeral crusts of halite with minor gypsum also occur. Exposure to the sun may alter surficial gypsum to bassanite, with diurnal fluctuations occurring between the two minerals. During hot dry periods, bassanite may partially alter to anhydrite. The halite is not an accumulative phase in this sabkha.

The Abu Dhabi sabkha sequence (Figs 8.17 & 8.18) is thus characterized by microbial peats containing discoidal gypsum crystals, overlain by structureless dolomite- and magnesite-rich sediments containing displacive and replacive gypsum/anhydrite. An erosion surface at its top truncates sedimentary and diagenetic structures in the underlying sediments.

The large quantities of anhydrite present in parts of the Abu Dhabi sabkha are somewhat atypical. Only 40 km away from the classic area described by Kinsman and others, sabkhas contain no evaporites, presumably because brines escape by seepage reflux (Baltzer, Kenig et al., 1994). Warren (1992) describes the large variation of supratidal sabkha types present in the Trucial Coast. Those associated with barrier islands and lagoons, like that at Abu Dhabi, are widest, but narrower sabkhas are associated with beach–dune barriers, tidal estuaries and distal portions of fan deltas. None is likely to prograde significantly seaward.

Coastal supratidal sabkhas associated with siliciclastic sediments are known from the Persian (Arabian) Gulf, Western Australia (Davies, 1970a; Caldwell, 1976), the Sinai Peninsula (Gavish, 1980), and within hypersaline lagoons such as Laguna Madre, Texas (Miller, 1975). Handford (1988a) reviewed these modern examples of evaporite–siliciclastic associations. In all cases evaporites are present as either ephemeral efflorescent crusts or displacive gypsum and halite crystals, not all of which are accumulative phases. These evaporite occurrences are probably better considered as simply diagenetic additions to other facies.

Sabkhas developed within carbonates differ from those formed in siliciclastic sediments. In both types, precipitation of displacive and cementing carbonate and gypsum removes calcium from interstitial brines, increasing their Mg/Ca ratios. In carbonate sabkhas subsequent dolomitization (or replacement by magnesite) of carbonate host sediments releases calcium back to the brine and this reacts with dissolved sulphate to precipitate additional gypsum and anhydrite within the sabkha sediments. In siliciclastic-hosted sabkhas, however, dolomitization is absent or much reduced (because of an absence or sparsity of original carbonate in the sediment), less calcium is added to the brine and, consequently, less calcium sulphate is precipitated.

8.4.3 Modern depression sabkhas

Depression siliciclastic sabkhas form immediately behind sedimentary barriers, although some may initially have prograded across lagoons. They have been described from Kuwait (Picha, 1978; Gunatilaka, Saleh *et al.*, 1987) and Qatar (Shinn, 1973); the Sinai coast of the Gulf of Suez (Gavish, 1974, 1980) (although these are salina-related), the Mediterranean coasts of Sinai (Levy, 1977, 1980) and Libya (Rouse & Sherif, 1980), and the Nile Delta coast of Egypt (West, Ali & Hilmy, 1979). Some, such as at Umm al Qaywyan, northwest of Dubai (Warren, 1991), are characterized by wide tidal channels, lined with sabkhas.

On the north coast of Sinai, sabkhas occur shorewards of protected lagoons formed by barrier elongation (Levy, 1977, 1980). Coastal sabkhas are adjacent to the present lagoon, but inland sabkhas, many now below sea level, are present behind older barrier complexes. In the coastal sabkhas the amounts of displacive and incorporative gypsum are small. Halite is only found in the landward depressions, which are dominated by continental brines. There is a gradation from the more marine, sulphate-dominated margins of the modern coastal sabkhas, to the more inland sabkhas, with decreasing gypsum and increasing amounts of halite, both as ephemeral crusts and, to a lesser extent, as displacive crystals and cements.

Similar sabkhas occur further to the west, between the coastal ridges of the Nile Delta (West, Ali & Hilmy, 1979) (Fig. 8.19).

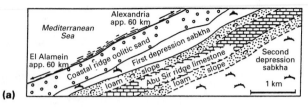

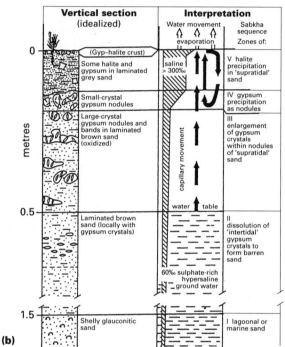

Figure 8.19 Siliciclastic depression sabkhas marginal to the Nile Delta, near El Hammam. (a) Map showing location of sabkhas in interdune depressions; (b) vertical section through the more coastal depression sabkha with inferred groundwater flow and the resultant evaporite mineralogies (from West, Ali & Hilmy, 1979).

Coastal plain sediments are dominantly carbonates, with ridges of ooid and skeletal beach and dune grainstones. Sabkha sediments, however, are mixed carbonate and siliciclastic muds and sands, deposited in earlier lagoons when the depressions were receiving sediment from a Nile distributary. The youngest depression sabkha, nearest the coast, contains a few displacive gypsum nodules. An older, more extensive depression to the south contains more nodules, some coalesced or forming enterolithic structures in the upper 0.5 m of sediment. It is noteworthy that these are primary gypsum nodules, indicating that not all nodular calcium sulphate in ancient evaporites originally need have been composed of anhydrite. Halite occurs as displacive hopper crystals and as an ephemeral crust.

A general model of depression sabkhas shows them to be isolated from open marine water, but fed by marine and continental seepage. Their extent is limited, both by the size of the depression and by the level of the water table. Although marine flooding may occur, perhaps by storm overwash, this is much less frequent than on supratidal sabkhas. The sabkha sediments are commonly fine grained, and may be wind-blown, lagoonal or, where associated with deltaic environments, riverine silts and clays. Rare examples of coarse-grained sabkhas occur (de Groot, 1973). Sediments may be siliciclastic, carbonate, or mixtures of the two. Capillary evaporation leads to calcium sulphate growth below the water table, initially as displacive lensoid crystals. Above the water table, these coalesce to form nodules, which rarely are converted to anhydrite. Sulphate growth is rarely sufficient to form enterolithic or ptygmatic structures. The upper sabkha surface is erosional, deflation removing sediments beyond the range of capilliary evaporation; thus the sabkha proper is limited to a thickness less than 1 m, even less in coarse-grained sediments. The scarcity of flooding leads to enhanced development of halite on the depressed sabkha surface, compared with that on supratidal sabkhas. Thus, as the sabkha matures, halite crusts with polygonal pressure ridges develop, together with displacive halite crystals in the uppermost sediments. This halite has, however, a low preservation potential.

8.5 Shallow-water evaporites

Modern shallow-water evaporites form subaqueously within both marine and non-marine fed salt lakes (salinas). Salinas usually form part of much larger complexes in which sabkhas may cover a much larger area than the salina itself. Within ephemeral salinas the primary subaqueous evaporites may be substantially modified and added to during desiccation phases.

Three types of salina can be distinguished; each having a different style of evaporite deposition. *Desiccated salinas* are rarely flooded and are kept moist only by inflowing groundwater. *Ephemeral salinas* dry out seasonally or for longer periods. Both desiccated and ephemeral salinas produce brine-pan sequences, the first dominated by diagenetic alteration so that sediments resemble those of sabkhas, whereas depositional features, including those produced during floods, are more important within ephemeral lake deposits. *Perennial salinas*, on the other hand, generate continuously subaqueous sequences. Filling of salina depressions with evaporites generates successions that may successively progress through the perennial lake – ephemeral lake – brine pan stages. The mineralogy of evaporites is commonly determined by the permeability of the salina floor. The amount and rate of brine reflux determines the salinity of brines that can be held within the salina.

Most modern marine-fed salinas are small and, because evaporite accumulation is rapid (see Table 8.3), they have been filled. This has converted them into brine pans or sabkhas where continuously subaqueous evaporites are no longer forming. Thus most of our understanding of subaqueous evaporites has come, not from examples of natural Recent marine environments, but from studies of artificial salt ponds, non-marine saline lakes (see Sects 4.7.4 & 4.14), older parts of modern salina sequences, and inferences about ancient evaporites.

8.5.1 Shallow-water evaporite facies

The abundance of shallow-water clastic textures and sedimentary structures, together with the presence of desiccation features, crystal crusts and microbial mats, makes identification of most well-preserved shallow-water evaporites relatively easy. Identification is less certain when original sedimentary features are obscured by later diagenesis or by overprinting, as when the depositional environment dries out and is subjected to sabkha processes. Some evaporites possess subaqueous features but were deposited on evaporitic flats which only flooded infrequently during storms or particularly high tides. Because in these settings all deposition occurs during floods, the evaporites appear to be entirely subaqueous, despite having formed in environments that were subaerially exposed for most of the time.

EVAPORATIVE CARBONATES

Many carbonates in evaporitic basins are highly altered and dolomitized normal marine sediment, within which original grains and fossils may still be discerned. Other carbonates that are closely associated with anhydrite, are fine-grained, unfossiliferous, sometimes organic-rich and laminated. These may be true evaporative carbonates, precipitated from brines concentrated to just below gypsum saturation, and possibly modified by biological activities. Ponds in solar salt works, with brines just below and above gypsum saturation, support luxurient microbial mats, some of which are associated with biologically induced micrites and pelletoidal muds.

Shallow carbonate-precipitating areas are prone to short-lived episodes of desiccation. When brines become highly concentrated, ephemeral precipitation of displacive halite can occur and evaporative carbonates not uncommonly contain halite moulds or pseudomorphs.

GYPSUM CRUSTS

Gypsum crystals precipitated on the floors of shallow bodies of brine may form thin crusts or, with time, superimposed beds of near-vertically standing, elongate and commonly swallow-tail twinned crystals, arranged in a pallisade or as upwardly radiating conical clusters. Crystals vary in size from prismatic needles only a few millimetres long to giant crystals over 5 m in height. Crystals may be separated by micritic carbonate or fine-grained clastic gypsum, or they may form an interlocking crystal mosaic.

Some crystals exhibit bizarre growth and twinning behaviours and suffer crystal-splitting that generates palmate clusters of subparallel crystals (Figs 8.20 & 8.21) that may themselves split or exhibit curvature. Twinning occurs at a variety of angles, influenced by incorporation of organic material on the crystal faces. Sometimes only one twin grows. Crystals commonly contain carbonate and anhydrite inclusions, which pass through the crystalline beds more or less parallel to bedding (Sect. 8.5.2).

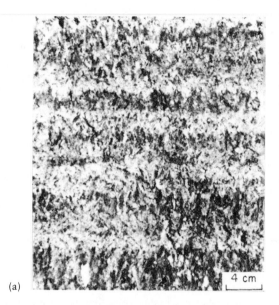

(a)

4 cm

(b)

Figure 8.20 Shallow-water, bottom-grown gypsum crusts. (a) Grass-like gypsum crystal crusts interbedded with algal carbonate layers and laminae (Passo Funnuto, Sicily); (b) acute 'twin' angle, each segment of which is really a separate subcrystal – see Fig. 8.21 (Messinian, Cyprus) (photographs from B.C. Schreiber).

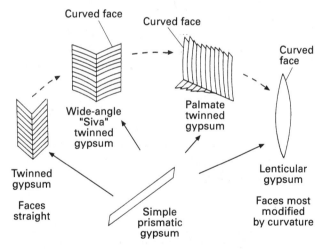

Figure 8.21 Diagram showing variations in primary gypsum morphology. Formation of the different twin forms is the product of growth modification of certain crystal faces, which become progressively more curved as a result of changes in temperature, pH, and the presence of organic impurities (from Schreiber, 1988).

Other crystals have invaded and incorporated cyanobacterial mats.

The largest gypsum crystal crusts form in environments of constant and active brine flow. Artificial brine ponds that are drained during wet seasons only develop clastic gypsum layers or crusts composed of centimetre-sized (or smaller) crystals. Crusts grown in shallower settings are also characterized by internal dissolution surfaces and by having incorporated microbial mats.

When altered to anhydrite during burial gypsum crusts may be difficult to identify. They are replaced by mosaic anhydrite (Rouchy, 1976; Loucks & Longman, 1982; Kendall & Harwood, 1989) which can closely resemble displacive nodular anhydrite, perhaps formed in sabkha environments. Where gypsum crystals incorporate carbonate laminae, the resultant nodular anhydrite may retain this lamination and can be mistaken for a replacement of laminated clastic gypsum (Warren, 1985).

Gypsum crusts may be replaced by halite, sylvite or polyhalite; the pseudomorphs being outlined by anhydrite. This type of replacement has not been identified in recent sediments but appears to have occurred in only the shallowest of brine ponds (Hovorka, 1992; Schreiber & Walker, 1992).

CLASTIC GYPSUM

Every type of shallow-water siliciclastic sedimentary feature has been recognized in gypsum deposits, and they may exhibit a range of grain sizes from silt through sand to pebbly sands. The particles originally grew as bottom crusts, later broken and reworked; or were precipitated at the brine surface and sank. Particles are eventually overgrown by cement and the laminae converted into interlocking crystal mosaics. Sedimentary features include cross-bedding, current and wave ripples, basal scoured surfaces and rip-up clasts. Their presence indicates that gypsum can be eroded, transported and deposited in the same fashion and environments as other clastic sediments, providing the depositional environment had gypsum-saturated brines, or the transported gypsum was quickly buried, preventing its dissolution. Reverse graded laminae are common and form during storm floods when an upward segregation of coarser particles occurs within highly concentrated, flowing sand sheets.

HALITE

Two crystal growth habits are present in halite crusts: those composed of cornet-shaped crystals (with one cube face, with its hopper-like depression, uppermost), and chevron halite in which crystals grow with corners uppermost (Figs 8.22 & 8.23)

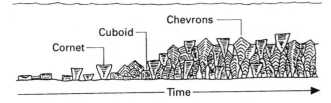

Figure 8.22 Bottom-precipitated halite may consist of crystals with cuboid, chevron (with corners facing up) or cornet (cube face with well-developed hopper depression facing up) growth habits. Crystal crowding and competition for space (see also Fig. 8.23) eliminate cuboids and generate vertically oriented fabrics that reflect preferred survival of cornets or chevrons (from Handford, 1991).

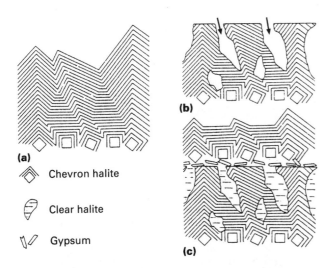

Figure 8.23 Diagrams illustrating formation of layered halite rock. (a) Overgrowth of crystal seeds with different orientations and the progressive elimination of less favourably oriented crystals produces a fabric of vertically oriented crystals with corners (coigns) uppermost (chevron halite) within which growth layering is shown by variations in inclusion density; (b) formation of planar dissolution surface and dissolution voids (preferentially located along crystal boundaries; arrowed) by a flood; (c) evaporation concentrates floodwaters, precipitating a thin lamina of gypsum crystals, overgrown by another halite crystal crust. Dissolution voids are filled by inclusion-free halite cement (from Shearman, 1970).

(Shearman, 1970; Arthurton, 1973; Gornitz & Schreiber, 1981; Logan, 1987). Brine-filled inclusions are concentrated into layers parallel to cube faces so that, in vertical section, this zoning appears as chevrons with upwardly pointing apices.

Upper surfaces of halite crusts may either be formed of crystal faces or are planar to irregular dissolution surfaces associated with dissolution cavities. Halite crusts are composed of two types of halite: a primary, inclusion-zoned type and a clear halite that fills pores between the primary crystals or dissolution cavities (Fig. 8.23) (Shearman, 1970; Brodylo & Spencer, 1987; Casas & Lowenstein, 1989). The presence of dissolution features and of halite cements indicates environments in which brines varied considerably between supersaturation and undersaturation. This only occurs when brine volumes are small, implying the halite crusts are very shallow-water precipitates.

Detrital halite is probably more abundant than previously recognized. It appears to be particularly susceptible to recrystallization. Where it preserves its original characters it is composed of fragmentary surface-grown hopper crystals and small cubes that are: (i) overgrown hoppers; (ii) crystals that precipitated by the mixing of different brines; or (iii) reworked clastic material from older halite crusts. Detrital halites exhibit cross-stratification, are ripple-marked, and include other detrital material. Around the Dead Sea, halite ooids have formed in high energy areas (Weiler, Sass & Zak, 1974). Commonly, crystal growth continues after deposition, partially obliterating the detrital textures. Only sedimentary structures remain to identify the detrital origin.

POTASH SALTS

Not much is known about the origin of *marine* shallow-water potash and magnesia salts or even if they really exist (Sect. 8.1). With the exception of polyhalite replacements of earlier calcium sulphates in some sabkhas, they are not known from modern *marine* settings. Most ancient examples are markedly recrystallized and have changed their mineralogies, sometimes so much so that some are better considered metamorphic rocks than sediments. Preserved primary or early diagenetic features have, however, been identified in some potash evaporites. A primary origin of sylvite crystals in layered halite–sylvite rocks is suggested by their intimate association with halites that preserve subaqueous textures (Fig. 8.24a) (Lowenstein & Spencer, 1990). The sylvite layers were accumulations of surface-nucleated crystals that crystallized as a result of surface-brine cooling. Wardlaw (1972) described crusts of bottom-grown carnallite crystals interbedded with detrital halites.

Most potash–magnesia salts appear to be replacements of, or additions to, earlier halite or sulphate sediments (Fig. 8.24b), commonly producing rocks in which textures and fabrics are diagenetic. Lowenstein (1982) identifies primary and early diagenetic textures in halite and anhydrite (originally gypsum) from potash ore beds of the Permian Salado Formation of west

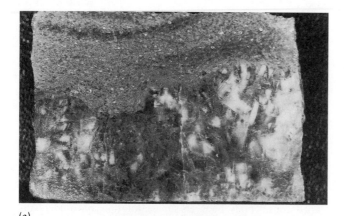

(a)

(b)

Figure 8.24 Fabrics of potash–halite rocks. (a) Chevron halite crust with inclusion-rich crystals (below) overlain by sylvite cumulate crystal layers (Oligocene, Rhine Graben); (b) muddy halite rock with euhedral halite cubes set within sylvite cement (McNutt Potash Zone, Salado Formation, New Mexico) (photographs courtesy Tim Lowenstein).

Texas–New Mexico. In the Salado and the Devonian Prairie Evaporite of Saskatchewan (western Canada) sylvite and some carnallite are early diagenetic cements that filled intercrystalline porosity and cavities within brine-pan halite crusts. Others cement loose accumulations of cubic halite crystals. They are

comparable with Recent occurrences from continental brine pans from China (Casas, Lowenstein *et al.*, 1992) where carnallite cements are precipitated. The complexity of potash deposits is illustrated by the Prairie Evaporite where multiple generations of halite, sylvite and carnallite occur. Recognition of primary and early diagenetic generations from those that are later diagenetic modifications is difficult.

8.5.2 Modern salinas and Holocene salina sequences

Salinas within the coastal plain of the Gulf of Elat, Sinai have been studied in detail (Gavish, 1980; Kushnir, 1981; Friedman, Sneh & Owen, 1985; Gavish, Krumbein & Halevy, 1985). They have sedimentary facies similar to those of ancient shallow-water evaporites but the small size of the salinas, only a few hundred metres across, makes them inappropriate analogues.

Coastal salinas of South Australia are much larger, up to 20 km × 12 km across, and are sites of gypsum precipitation with sequences up to 10 m thick of laminated gypsarenites and bottom-grown crusts (von der Borch, Bolton & Warren, 1977; Warren, 1982, 1985; Warren & Kendall, 1985). In plan view, the salinas exhibit a 'bullseye' sediment pattern with a concentric zonation of gypsum at the centre and carbonate as a broad marginal rim (Fig. 8.25). Salinas occur in isolated depressions below sea level, between Quaternary beach-dune ridges. During

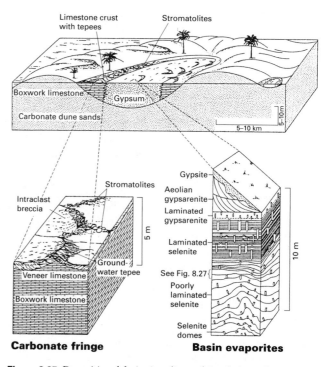

Figure 8.25 Depositional facies in salinas of South Australia with bedded subaqueous gypsum surrounded by carbonates (from Warren & Kendall, 1985).

the hot dry summers, some salinas dry up completely, but Deep Lake has a shallow (<0.5 m) perennial water cover. Summer evaporation lowers dune and salina water tables below sea level allowing sea water to seep into the salinas. Increasing water salinities permit aragonite, gypsum, and then halite, to be precipitated at the sediment surface. Gypsum is also precipitated within sediments. During the winter, rainfall on the salinas causes water levels to rise, a decrease in salinity, and the dissolution of halite and some of the gypsum.

Although modern salinas desiccate and only have water depths of a metre or less in winters, when they first formed 5000–6000 years BP they were perennial brine lakes up to 10 m deep. Gypsum accumulation decreased brine depths and converted most lakes from perennial into ephemeral features, within which evaporites now partially dissolve during wet seasons. The evaporite fill of salinas is a shallowing-upward sequence in which the influence of winter lower salinity waters increasingly affected the depositional surface (Figs 8.25 & 8.26).

Surficial salina sediments contain sand-sized, etched and corroded crystals of gypsum. In the subsurface, beneath the zone affected by fluctuating water levels, pristine crystals occur and the gypsarenites are laminated. Millimetre-thick laminae of gypsum crystals, some in growth position, others reworked, alternate with laminae of aragonite peloids (ostracod and brine-shrimp fecal pellets and micritized remnants of algal tubules) which probably represent the former presence of microbial mats. Some gypsarenites are ripple cross-stratified. These laminated gypsarenites formed in a perennial lake or when the brine surface did not descend beneath the surface as it does today during dry seasons.

Beneath the laminated gypsarenite (Fig. 8.25) occur beds of large, 20-150 mm, coarse, twinned gypsum crystals. They grew as bottom crusts in the deeper parts of the brine pools. Long axes of crystals are oriented normal to bedding and concentrations of aragonite peloids define a lamination within crystals that cross-cuts the vertical crystal alignment (Fig. 8.27) (Warren, 1985). Laminae in the uppermost crystals are flat. They record carbonate deposition on to planar dissolution surfaces that truncated the underlying gypsum crystals. Renewed growth of the gypsum, in crystallographic continuity with that below, poikilitically enclosed the laminae. Laminae in lower parts of the salina sequence define former crystal faces. They indicate that gypsum crystals were never affected by undersaturated brines. Carbonate was deposited on to the crystal faces. Weakly laminated domes of radiating crystals occur in the lowermost parts of the sequence (Fig. 8.25).

In contrast to the well-preserved fabrics of the bedded sulphates, carbonates are composed of 'boxwork' limestones – earlier carbonates and gypsum altered by inflowing marine and non-marine waters. This highly porous carbonate contains microbial structures, and pseudomorphs after gypsum crystals. Above the boxwork carbonate occur subaerial microbial tufas, subaqueous stromatolites and microbial mats or fenestral carbonates with tepee structures. The tepees form above fractures used by resurging seepage waters (Warren, 1982) and contain pisoids, laminar cements and internal sediments. Gypsum also occurs on the surface and the periphery of the salina, forming dunes and lunettes.

Salina Ometepec, Baja California occurs within a modern, active siliciclastic sabkha near the mouth of the Colorado River adjacent to the northwest Gulf of California (Fig. 8.28) (Thompson, 1968; Shearman, 1970; Lowenstein & Hardie, 1985; Hardie, 1986a). This is a large (100 km²), shallow (<2 m) sabkha depression, bordered on its landward side by alluvial fans and seawards by high, dry supratidal flats. The tidal flats are periodically inundated by storm-driven marine waters and, because their sediments are relatively impermeable, floodwaters do not interact with or become part of the groundwater regime. Instead they collect in the depression, evaporate and produce an ephemeral saline lake from which gypsum initially precipitates. Each flood generates a mud-layer–gypsum-crust couplet. Gypsum grows vertically as a 'grass-like' crust on microbial mats and as crystal muds and sands. When the gypsum lake desiccates it leaves a small halite pan in its lowest part. During these periods, gypsum continues to grow within the sediments from groundwater brines. This additional diagenetic gypsum causes gypsum crusts to be polygonally cracked and thrust. Brines also occur within muddy laminated supratidal sediments that surround the gypsum pan. Growth of gypsum crystals disrupts original sedimentary structures and forms a saline mudflat surrounding the gypsum pan.

Dehydration of the gypsum in the pan and saline mudflat sediments generates nodular anhydrite; isolated nodules in the

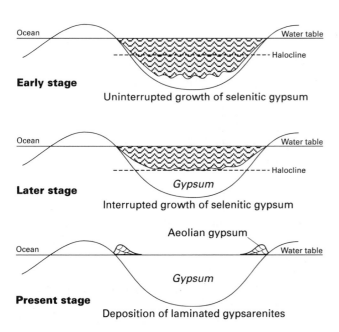

Figure 8.26 Evolution of South Australian salinas (from Warren, 1982).

299

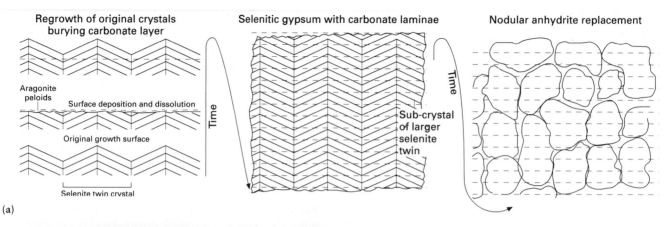

Regrowth of original crystals burying carbonate layer

Selenitic gypsum with carbonate laminae

Nodular anhydrite replacement

Aragonite peloids

Surface deposition and dissolution

Original growth surface

Selenite twin crystal

Sub-crystal of larger selenite twin

Time

(a)

(b)

Figure 8.27 Carbonate laminae within large, twinned gypsum crystals. (a) Diagrams showing origin of lamination. Flat upper surface of crystals is the product of seasonal dissolution when the former growing surface is covered by brackish water. Aragonite, deposited on the dissolution surface, is buried within the gypsum crystals as they continue to grow in crystallographic continuity with the earlier precipitated gypsum (after Warren, 1982). Replacement by nodular anhydrite (far right diagram) gives false appearance that anhydrite has replaced laminated gypsum; (b) photograph of aragonite-laminated, coarsely crystalline gypsum.

mudflat, and contorted nodular layers in the gypsum pan. The latter resemble nodular and enterolithic anhydrites of the Abu Dhabi sabkha. Nodular fabrics originate by distortion of pseudomorphs as the original gypsum crystals are replaced by loose aggregates of easily disturbed anhydrite laths.

In the halite pan most of the halite observed by Shearman (1970) was initiated as floating pyramidal hopper crystals at the brine–air interface. The floating hoppers aggregate together into 'rafts'. These sink or are blown to the side of the pan and serve as the nucleii for additional crystal growth, forming chevron halite crusts. Repeated flooding and subsequent re-precipitation of halite following renewed evaporative brine concentration produces successive layers of halite. Each minor flood causes some dissolution of the underlying halite, truncating the upper surfaces of crusts and forming small solution pipes along crystal boundaries (see Fig. 8.23). These pipes are later filled by growth of clear halite cement.

The halite pan is ephemeral. Each major flooding dissolves all the halite and it is flushed off the sabkha, leaving the less soluble gypsum to accumulate preferentially. The Salina Ometepec clearly illustrates the necessity for a closed system if halite is to be a persistent phase. Holocene halite has only accumulated in settings, such as Lake Macleod and Sabkha el Melah (see below), where there has been an effective barrier between the depositional site and the sea.

West Caicos Island (Turks and Caicos) is located on the western margin of an isolated carbonate platform, Caicos Bank (Fig. 8.29). In the east of the island is a broad, flat desiccated salina (East Salina) bounded by Pleistocene carbonate beach-dune ridges to the west and by Holocene oolitic dune ridges up to 16 m high in the east (Perkins, Dwyer *et al.*, 1994). The salina surface lies less than half a metre below sea level and is covered by a rubbery microbial mat encrusted by gypsum and ephemeral halite. Gypsum also precipitates within the mat and in the older mats beneath.

Up to 2.4 m of unconsolidated sediments are found beneath the salina and were deposited as two restrictive-upward cycles (Fig. 8.29). Each cycle begins with a bioturbated marine lagoonal

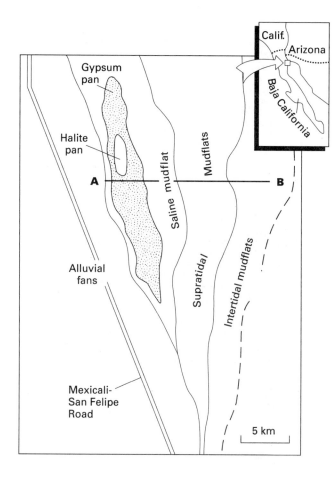

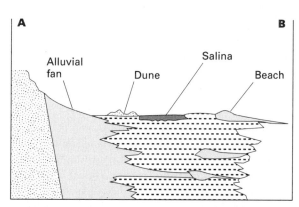

Figure 8.28 Siliciclastic sabkha of northwest Gulf of California, Baja California, Mexico. Map of the sabkha and associated environments (from Hardie, 1986a) and schematic cross-section (from Handford, 1991).

carbonate unit followed by highly restricted, microbally laminated carbonates. These restricted carbonates were deposited from waters with variable salinities, ranging from brackish to hypersaline. The upper cycle is terminated by a gypsum mush layer, up to 0.7 m thick and composed of displacive discoidal to lenticular gypsum crystals within a sparse (<10%) carbonate

matrix. It is currently forming. The upper cycle is very similar to both the sequence generated by the prograding tidal flats of the Trucial Coast (Perkins, Dwyer et al., 1994) and to ancient shallow-water evaporites interpreted as sabkhas.

The salina is fed by marine waters that pass through underlying karsted Pleistocene limestones and up through the salina sediments (Fig. 8.30). These sediments, however, act as an aquitard. The supply by upward seepage is less than evaporite losses so that water never reaches the surface. Microbial mats only flourish when the salina is inundated with rainwater. Evaporation of surface and shallow groundwaters generates a denser brine that sits within the uppermost sediments above a zone containing the ascending less-saline waters. The rising waters dissolve gypsum in the lower part of the gypsum mush layer, and earlier evaporites have been removed. The only evidence for their former presence are crystal moulds in cemented carbonate crusts. Dissolution of gypsum from the lower cycle probably also occurred during the deposition of the marine part of the succeeding cycle, when the salina was flooded with marine waters.

Lake MacLeod, Western Australia lies in a graben at the northern end of Shark Bay (Fig. 8.31). During the Pleistocene and early Holocene this depression was a marine embayment of the Indian Ocean but around 5000 BP it became isolated and began to fill with up to 12 m of halite, gypsum and carbonates. The present lake surface is several metres below sea level and sea water enters the salina by seepage through a karstified limestone barrier, the Quobba Ridge. Some of the seepage areas support perennial brine pools – Cygnet and Ibis Ponds – but the greater part of the salina is a dry, flat and salt encrusted plain (termed a *majanna* by Logan, 1987), locally with solution pits and depressions. Episodically, the majanna is inundated by river floods or by thin brine sheets transported by winds from the perennial ponds. Each flood deposits a thin bed or laminae of detrital and bottom-grown gypsum (Fig. 8.32), some of which are traceable over much of the evaporite basin. This deposit thus has many of the characteristics of so-called 'deepwater' evaporites, yet formed in an environment which is, for the most part, dry.

Carbonates, up to a metre thick and containing tepee structures and pisolites (Handford, Kendall et al., 1984), are precipitated from thin sheets of water in seepage inflow areas. Restricted marine carbonates (oolites and cemented pelleted grainstones) also form in the most proximal pond. Over the greater part of the basin gypsum is being deposited, as subaqueous crystal crusts in more distal ponds or as clastic gypsum layers and laminae on the majanna. In the west and south, the gypsum interfingers with clastics introduced by ephemeral rivers. Precipitation and reworking of the gypsum sediment occur during floods. Most of the time, however, the surface is dry and covered with an ephemeral halite crust, although halite is not accumulating today. Brines are being exported from the basin by seepage outflow especially along the western margins of the basin.

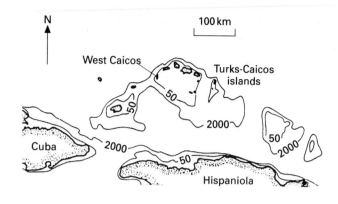

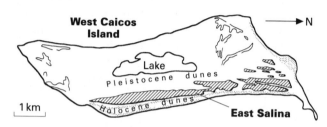

Figure 8.29 Location map of West Caicos Island with position of East Salina between Pleistocene and Holocene dune ridges, and idealized sedimentary sequence of East Salina sedimentary sequence (from Perkins, Dwyer *et al.*, 1994).

The northern and western parts of the Holocene basin fill largely consist of layered gypsarenites (Figs 8.31 & 8.32) up to 5 m thick, the Ibis Gypsite, which formed in the same environment as the present-day majanna. In the central parts of the basin, beneath a thinner Ibis Gypsite, is a bedded halite, up to 6 m thick, composed of pyramidoidal and columnar (chevron) halite crystals with gypsum laminae, the Texada Halite unit.

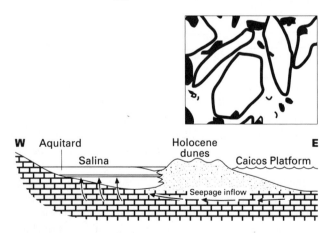

Figure 8.30 Inferred hydrology of East Salina and adjacent parts of West Caicos Island. Inset is a drawing of gypsum crystal moulds from carbonates immediately beneath the gypsum unit. Gypsum dissolution is caused by undersaturated groundwaters rising toward the salina surface (after Perkins, Dwyer *et al.*, 1994).

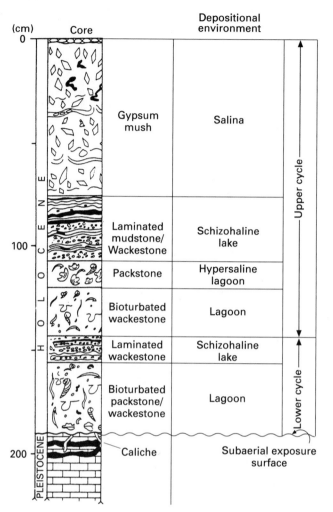

Figure 8.29 (*continued*)

Beneath both the gypsite and halite is the Cygnet Carbonate unit (up to 1 m thick), composed of aragonitic pellet sand, aragonitic mud, lithoclastic and skeletal sands and clastic gypsite layers. The uppermost carbonate sand is gypsum cemented, making it an effective aquitard. This retains brines in the basin and has allowed evaporites to form and accumulate.

Lake MacLeod presents us with a range of modern analogues of different evaporitic facies, beautifully illustrated by Logan (1987). More significantly, it also provides a model for how ancient evaporite basins might operate, in particular explaining how different evaporite facies are deposited at different times (see Fig. 8.7). Immediately after basin isolation, waters entering the MacLeod Basin became concentrated by evaporation and the Cygnet Carbonate became cemented by gypsum and formed an efficient aquitard. Evaporative losses and the confinement of the inflow to seepage through the Quobba Ridge caused evaporative drawdown and allowed concentrated brines to form. The basal aquitard prevented both seepage loss of brines

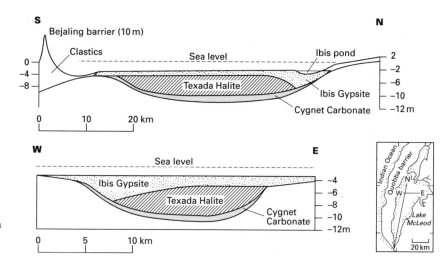

Figure 8.31 Lake MacLeod, Western Australia and a schematic cross-section through its Holocene fill (after Logan, 1987).

Figure 8.32 Layered gypsarenites, Lake MacLeod.

and dilution of these same brines by sea water that otherwise would have seeped upwards into the partially desiccated basin. At the time the Texada Halite formed, deposition was confined to the deeper, central parts of the basin which were entirely floored by the Cygnet Carbonate aquitard. Brines were retained in the basin by this hydroseal long enough to be evaporated and concentrated beyond halite saturation. At a later time, because of depositional onlap on to basin flanks (where the aquitard was absent), brines began to be lost to the basin by seepage reflux, as they are today. Sea-water inflow had also been reduced by the reduction in hydraulic head caused by sediment aggradation in the basin. Water was retained in the basin for shorter periods. Eventually it spent so short a time in the basin that evaporation could no longer concentrate it to halite saturation. The Texada Halite stopped forming and only gypsum, precipitated from the more dilute brines, covered most

of the basin floor, as it does today. The more dilute brines also dissolve some of the earlier halite to form solution depressions on the majanna surface.

Many of the features of the Lake MacLeod closely resemble those of *Sebkha el Melah* – a 150-km² flat plain below sea level, on the eastern coast of southern Tunisia. The evaporite basin beneath it is well known because of numerous boreholes drilled for potash (Perthuisot, Floridia & Jauzein, 1972; Perthuisot, 1974, 1975, 1989). Unlike the largely desiccated basin-fill sequence of Lake MacLeod, the Sebkha el Melah basin sequence has been interpreted as the product of a perennial lake basin in which different primary mineral facies (carbonate, gypsum and halite) were deposited contemporaneously. However, a closer comparison of the two basin fills may demonstrate they are more similar than published interpretations would suggest. Important differences between the two localities apparently do exist and include the fabrics of the halite (cumulates(?) in Sabkha el Melah; bottom-grown crusts at Lake MacLeod) and the presence of concentrated brines at Sabkha el Melah which altered earlier-formed gypsum to polyhalite.

8.6 Deep-water evaporites

Identification of deep-water evaporite facies is based entirely upon ancient examples. Whereas Dead Sea bottom sediments do include evaporites, they are atypical. Nevertheless, this basin is important in that it provides a model for the hydrodynamics of ancient evaporite basins.

8.6.1 Deep-water evaporite facies

Deep-water evaporites are the product of the large size and volume of the brine body. Hence they are usually identified by their vertical and lateral continuity. Large bodies of brine are not expected to fluctuate in their composition or concentration as a result of short-term changes, as shallow brine bodies do. On the other hand, the common interlamination of minerals of different solubility (carbonate with calcium sulphates (Fig. 8.33); calcium sulphates with halite, etc.) records short-term changes in the rates of influx or evaporation and their effect on the concentration of *surface* brine layers. Stratigraphic units composed of laminated carbonates, sulphates and halite may be 10s–100s m thick and extend laterally across an entire depositional basin (10s–100s km). This lateral and vertical continuity of facies is the most common argument used to infer the existence of a deep body of brine. Turbidites and other mass-flow deposits, composed entirely or in part of resedimented shallow-water evaporites, may also occur associated with the laminated deep-water facies.

Most nucleation and growth of evaporite crystals is believed to occur at the brine surface. These crystals sink and settle through the brine column as a pelagic rain to form cumulate deposits. Some halite and gypsum within inferred deep-water evaporites, however, grew on the bottom, although the mechanism that allows this type of precipitation is uncertain. Evaporite crystals may also precipitate during diffusive mixing of different layers of brine (Raup, 1982).

Laminar sulphates (originally gypsum), alone or in couplets or triplets with carbonate and/or organic material, are by far the most common deepwater evaporite facies (Fig. 8.33). It may pass up or down into *laminar carbonates* which have the same characters and are probably also of evaporitic origin. Laminae are thin, usually 1–2 mm thick, and have sharp, even boundaries. They are commonly interpreted as seasonal or annual increments (varves). However, it has never been demonstrated conclusively that the laminae are truly annual, and evaporitic carbonate laminae in Recent bottom sediments of the Dead Sea are deposited only once every 3–4 years.

Laminar and bedded halite are usually difficult to recognize as deepwater deposits because most examples have suffered extensive recrystallization, obliterating any proof of their deepwater origin. Inferred deepwater halite is commonly thin-bedded, containing anhydrite laminae similar to those of deepwater laminated sulphates. Even finer lamination occurs in some halite layers, defined by variations in inclusion content. The fineness of this lamination indicates that the halite, prior to recrystallization, must have been an accumulation of very fine grained cumulate crystals. Other presumed deepwater halite is composed of larger, bottom-nucleated crystals (Fig. 8.34) (Nurmi & Friedman, 1977; Beyth, 1980; Raup & Hite, 1992). A deepwater setting here is suggested by the clear, inclusion-free nature of the halite crystals and by the absence of any dissolution surfaces associated with anhydrite laminae. Anhydrite laminae drape over and define crystal terminations of the former halite crusts. In the Silurian Salina A salts of the Michigan Basin, this facies is only found in the basin centre, again suggesting its deeper-water origin (Nurmi & Friedman, 1977).

Gravity-displaced evaporites occur with, or within, laminated evaporites as slump, mass flow, density current or turbidity current deposits. Gypsum and anhydrite turbidites are similar to their non-evaporite equivalents: sometimes an entire Bouma sequence is present (Schreiber, Friedman *et al.*, 1976) but more commonly the beds are only composed of graded units or have parallel lamination in their uppermost parts (Schlager & Bolz, 1977). Mass flow deposits are breccias composed of clasts of reworked shallow-water sulphates, either alone or with carbonate fragments. They are commonly intimately associated with slump-folded beds. Resedimented halites have only rarely been identified but can form spectacular slump deposits with clasts exceeding 10 000 m³. Their origin is clearly established only when they contain non-evaporite clasts (Czapowski, 1987).

8.6.2 The Dead Sea – a modern deep-water evaporite basin

The Dead Sea is a large, closed, perennial saline lake, in a rift

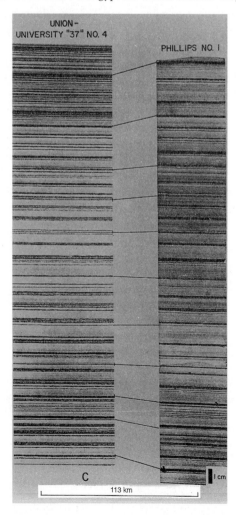

Figure 8.33 Correlation of Castile Formation laminated anhydrite between wells 113 km apart; dark layers are calcite, light layers anhydrite (from Dean & Anderson, 1982).

Figure 8.34 Crusts of elongate, inclusion-poor halite crystals, with flat terminations, outlined by layers of fine-grained anhydrite, Paradox Formation, Utah.

became a homogeneous body (Steinhorn, 1985) although, as will be seen, this is a seasonal phenomenon and for most of a year the waterbody is still stratified.

The predominant sediment in the Dead Sea basin is a clayey silt (Fig. 8.35; Garber, Levy & Friedman, 1987). Evaporitic minerals include aragonite, gypsum and halite. Calcite, previously considered a diagenetic alteration of gypsum (Friedman, 1965), is now interpreted as a clastic component (Garber, Levy & Friedman, 1987). Aragonite needle precipitates form white sediment laminae alternating with detrital laminae. Gypsum occurs as hard layers of lenticular crystals, as crystal clusters, but predominantly as dispersed isolated crystals. Its distribution is highly variable but generally in concentrations less than 20%. The presence of gypsum in bottom sediments discredits an earlier view that this mineral cannot accumulate in deepwater anoxic sediments because of the activities of sulphate-reducing bacteria (Friedman, 1965; Neev & Emery, 1967). Halite, other than a shallow-water halite 'slush' in the shallow southern basin, never constitutes more than 10% of the bottom sediment and is present as disseminated cubic crystals and as crystalline blocks.

Prior to 1979, evaporation during summer months concentrated the surficial part of the upper, more dilute, water mass.

valley more than 400 m below sea level, and fed predominantly by waters from the River Jordan. It is bounded on both sides by either steep cliffs or by coarse-grained alluvial fans (see Sect. 6.5.5) and to the north the River Jordan has built out a delta. It is the only Recent analogue of ancient deep-water evaporites, but evaporites are currently not the predominant sediment present and the composition of its brine differs considerably from that of sea water. Sulphate and bicarbonate account for only 0.4% of total ions; chloride is the dominant anion, followed by bromide. The dominant cations are magnesium and sodium, followed by calcium. The abundance of calcium (nearly 16% of all cations) is noteworthy.

The lake is divided into two distinctly different basins: a shallow southern basin (never deeper than 4.5 m, now largely dry) and a deeper northern one which has large areas with brine depths greater than 200 m. Before 1979, the Dead Sea was a chemically and thermally stratified water body. Differences of several per cent in density occurred across a pycnocline at 40–50 m. The lower water mass was near, or at, halite saturation; whereas the upper layer was undersaturated with respect to both halite and gypsum, except at the air–brine interface. This situation has now changed. Reduced inflow from the Jordan River has allowed the brine level in the Dead Sea to fall, the southern basin to dry up, and the density of the upper water layer to increase until it became the same as the bottom layer. At this point overturn occurred and the entire water mass

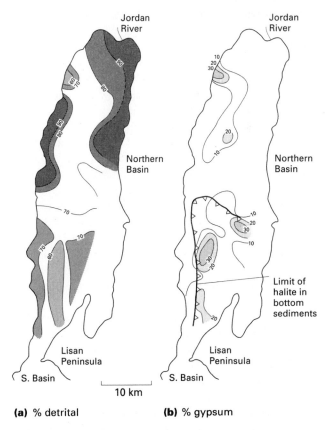

(a) % detrital   **(b)** % gypsum

Figure 8.35 Distribution of detrital (a) and evaporite (b) components in the bottom sediments of the Dead Sea (after Garber, Levy & Friedman, 1987).

Aragonite and, to a lesser extent, gypsum and halite crystals precipitated at the surface and settled through the undersaturated upper water mass. Halite dissolved, whereas aragonite and some of the gypsum reached the bottom and became incorporated into the terrigenous bottom sediments. Most of the aragonite, however, occurs in millimetre-thick laminae which form during short-lived but extensive precipitation events that turn surface waters turbid. These 'whitings' are formed by intense evaporation that accompanies aperiodic episodes of very high temperatures. They are not annual events, so that the aragonite laminae are not varves.

Homogenization of the Dead Sea brine column and its further concentration by evaporation should cause a massive precipitation of evaporites (mainly halite) but this has yet to occur. Brines are now precipitating salts, mostly at the lake periphery but also in the lake interior (Anati & Stiller, 1991).

Each winter there is a freshwater inflow, the amount varying from year to year. The fresh water mixes with the uppermost few metres of lake water to form a buoyant layer floating on the Dead Sea brine. A stable halocline thus forms at a depth between 5 and 15 m and prevents the downward penetration of turbulence. Spring heating reinforces the stability of this stratification, because all heat is trapped in the upper water layer making it even less dense.

When the winter freshwater influx exceeds the amount that will be evaporated during the succeeding year, the lake remains strongly stratified (*meromictic* conditions) with both temperature and salinity stabilizing the density stratification. In these years the salinity of the bottom brine layer is not increased.

In contrast, when winter inflows are low, the entire upper water layer is concentrated by evaporation during the summer and, in mid-December, the density contrast disappears and the entire brine column is uniform. This *holomictic* condition lasts for about 2 months, until a new seasonal pycnocline begins to form in late February. During these holomictic years, a less stable stratification occurs in the summer and autumn. Here, slightly more saline surface brines lie above less saline. This unexpected situation exists because the density increase resulting from higher salinity is more than offset by a density decrease caused by higher temperature of the surface-heated brines. When the salinity difference between the two brine layers reaches a critical value, however, this stratification breaks down. In this process (double diffusive mixing) parts of the upper warm brine lose heat to the lower cooler brine and become denser. They become unstable and descend, gradually mixing with the lower brine until the descending brine finger reaches the same temperature as the surrounding deep brine body. In this way salt (and heat) is conveyed into the lower water mass and its overall salinity increases (Anati & Stiller, 1991).

The change between holomictic and meromictic conditions in the Dead Sea may provide an explanation for the presence of both laminated evaporite intervals and bottom-grown evaporite crusts in some ancient deepwater evaporites. Laminated intervals

would record times when the basin was meromictic and all evaporites formed in upper brine layers. Bottom-grown crusts, in contrast, form during holomictic conditions, either during short intervals of overturn, or from supersaturated fingers of descending brine generated during double diffusive mixing in hot dry seasons. Alternations between bottom crusts and intervening laminated intervals, as in the Castile Formation (Sect. 8.9), could thus reflect variations in the amount of water influx between different years; large influxes inducing meromictic conditions and deposition of laminae, small influxes causing holomictic conditions and formation of bottom crusts.

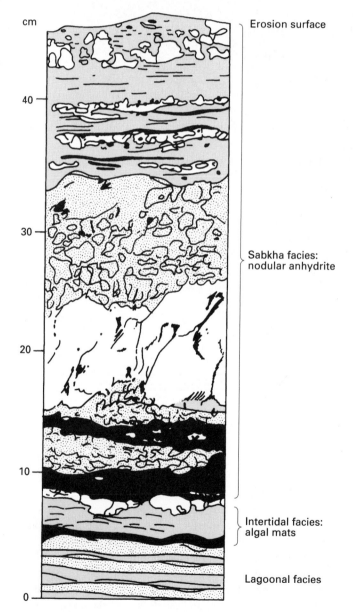

Figure 8.36 A complete sabkha cycle in the Lower Purbeck Beds of the Warlingham borehole (after Shearman, 1966).

8.7   Ancient sabkha/salina evaporites

The significance of modern supratidal sabkhas as analogues for ancient evaporites has been overemphasized. Modern marginal-marine supratidal sabkhas form in protected microtidal/low mesotidal environments and evaporites that form in them are less than a metre thick. Ancient examples therefore should be thin and overlie restricted marine or continental facies. Many examples cited in the literature document sabkha deposition above open marine sediments. These are either not sabkha sequences or some sabkhas developed differently from those of today. Other interpreted ancient sabkha evaporites are many metres thick and are laterally extensive. These thickness differences are commonly explained by a continuous relative sea-level rise but this explanation cannot explain both the excess thickness and the wide distribution. In fact, many ancient evaporites have been interpreted as sabkhas simply because there was no alternative explanation for the origin of nodular anhydrite. For example, Jurassic anhydrites in the Middle East were some of the first to be identified as being products of prograding sabkhas (Wood & Wolfe, 1969), some even of enormous size (Leeder & Zeidan, 1977). Many of these evaporites have now been reinterpreted as subaqueous deposits (Warren & Kendall, 1985; Alsharhan & Kendall, 1994; Saner & Abdulghani, 1995).

Ancient prograding supratidal sabkhas should not be identified solely by the presence of nodular anhydrite. Their identification requires that their evaporites are thin (less than 1 m), have evidence of erosion and exposure at their tops, and occur in appropriate settings (basin margins). Perhaps best known are the first to have been identified – in the Jurassic *Purbeck* of southern England (Shearman, 1966) (Fig. 8.36). It is interesting that, although Shearman originally compared these Jurassic sabkhas with the anhydrite-rich modern sabkhas of the Trucial Coast he had worked upon, earlier work on the diagenesis of the Purbeck sulphates by West (1964) suggests the Purbeck sulphate nodules were originally gypsum, not anhydrite, and thus were more similar to those of northern Egypt (Sect. 8.4.3).

The lower part of the Upper Permian *Bellerophon Formation* of the Dolomites of northern Italy is a spectacularly exposed cyclic gypsum–dolomite sequence, up to 600 m thick, developed over a 16 000 km² area (Bosellini & Hardie, 1973). It overlies continental redbeds and is overlain by shallow-marine shelf-lagoonal carbonates. Cycles were deposited in a marginal marine setting, and each is composed of three elements (Fig. 8.37): a lowermost thin-bedded dolomite with fenestral structures and a restricted marine fauna; a middle unit of sandy dolomite with scattered gypsum nodules, locally with flat and cross-lamination; and an upper unit of gypsum consisting of thin beds of coalescent nodular gypsum, alternating with shaly-dolomites with abundant isolated gypsum nodules. Bedding is undulatory and locally contorted into enterolithic folds. Some cycles have an uppermost layer of laminar gypsum, commonly with highly deformed bedding, some even isoclinally folded.

The cycles were first interpreted as products of the progradation of sabkhas (upper units) over evaporitic sandy intertidal–supratidal zones (middle units), which in turn prograded into restricted penesaline lagoons (lower units) (Bosellini & Hardie, 1973). Evidence for this included the presence of marine fossils, the burrowed basal carbonate mud, and the close similarities of the upper unit to layered nodular anhydrite in the Abu Dhabi sabkha (Sect. 8.4.2). A revised interpretation (Hardie, 1986a), concluded that the upper units show greater resemblance to bedded gypsum from a gypsum pan in Baja California (Sect. 8.5.2). A spectrum of gypsum pan (and their anhydrite replacements) deposits may be identified, with most of the

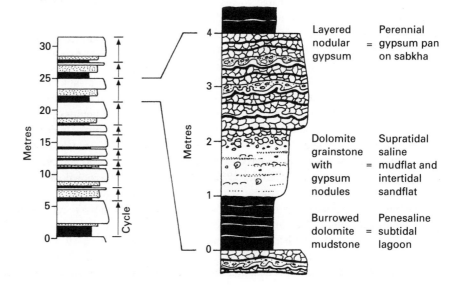

Figure 8.37 Shallowing-upward cycles, with anhydrite component interpreted as deposits of perennial gypsum pans, Bellerophon Formation of the Dolomites area, Italy (from Hardie, 1986a).

Bellerophon Formation representing one extreme where layered sequences are formed in frequently flooded pans. In contrast, diagnetic modifications (enterolithic folding, tepee structures, microdiapirs and the erosional truncation of these features) dominate in pans that are flooded infrequently. Some of the deformation of Bellerophon gypsum beds, however, may also be of tectonic origin (Schreiber, Roth & Helman, 1982).

Another evaporite succession (*Baumann Fjord Formation*) from the Ordovician of Ellesmere Island, Arctic Canada, has been interpreted in a similar fashion (Mossop, 1974). Here cycles are terminated by laminated anhydrite that Mossop interpreted as tectonically deformed nodular sulphate, originally formed within prograding sabkhas. Photographs of remnant nodular anhydrite, however, show that the 'nodules' contain deformed laminae. It is possible that it is the lamination that is original and that the deformed 'nodules' are a (localized) tectonic overprint. This would imply Baumann Fjord laminated sulphates were deposited in environments more similar to the majanna of Lake MacLeod than the prograding sabkhas of Abu Dhabi.

Mississippian (Lower Carboniferous) evaporites of the Williston Basin (see Fig. 8.9) form part of a carbonate ramp composed of a wide range of carbonates (from deep water-basinal to inner shelf and lagoonal). The innermost shallow shelf and peritidal facies prograded westwards, eventually to fill the basin. Evaporites deposited after basin-filling have basin-central distributions (see Fig. 8.9). The basin-margin sequence contains thin, regionally developed marker beds (usually 0.3–1 m thick) of argillaceous and sandy sediment which cross-cut facies boundaries. Abrupt lateral shifts in facies belts occur across these marker beds, suggesting they are essentially isochronous in nature and represent short-lived and abrupt regressive episodes.

The range of facies is similar to that of the Permian of the Guadalupe Mountains (see Sect. 9.3.3), but there are no basin-marginal reefs. Inner shelf carbonate facies include a pisolitic-tepee belt, behind which laminated to massive fine-grained dolomites with restricted faunas were deposited. These carbonates were developed as 1–3-m-thick cycles (Elliott, 1982). Patterned dolomite occurs atop the cycles and probably records intervals of former gypsum (Kendall, 1977) precipitated during times of lagoonal restriction and desiccation, but which dissolved on renewed lagoonal flooding (cf. West Caicos Island salina, Sect. 8.5.2). The restricted dolomites interfinger with and are overlain by nodular, nodular bedded and laminated anhydrite (Fig. 8.38). Nodular anhydrites are interpreted as both sabkha and salina sediments developed on the inner side of a lagoon (Fig. 8.39; Lindsay & Roth, 1982). Much of the anhydrite is developed as poorly preserved pseudomorphs of vertically oriented gypsum crystals in a dolomite matrix (Kendall, 1979; Lindsay & Roth, 1982; Schreiber, Roth & Helman, 1982), but these are commonly later deformed by displacive anhydrite nodules (Kendall, 1979). These nodules grew when the lagoon or salinas filled or became desiccated, and the area converted

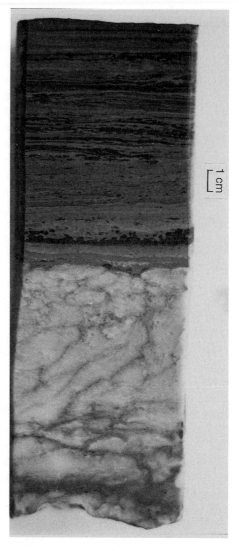

Figure 8.38 Upper part of sabkha–salina cycle from Frobisher Evaporite (Mississippian) from Saskatchewan, with elongate anhydrite nodules, probably after former subaqueous gypsum crystals, truncated by erosive sabkha surface and overlain by laminated (schizohaline lagoonal ?) deposits of the succeeding cycle (from Kendall, 1979).

to a sabkha. Other evaporite units are entirely composed of laminated to thinly bedded anhydrites, anhydritic dolomites and dolomites (commonly patterned). They were deposited on the floor of large salinas, similar to the majannas of Lake MacLeod.

Anhydrite units are markedly cyclic with erosive tops (Fig. 8.38) and cycle thicknesses are similar to those in the adjacent lagoonal carbonates. There is a hierarchy of different cycles. Small, 1 to 3 m thick, cycles shallow upward but commonly contain representatives of only one or two facies (restricted dolomites and/or anhydrites). Small cycles combine into larger ones, 10 to >30 m thick, that also shallow upwards and are terminated by marker beds. These mark abrupt basinward shifts in depositional conditions.

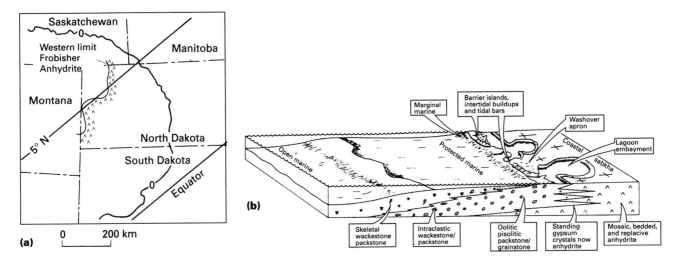

Figure 8.39 Depositional model of Mississippian facies in Williston Basin. (a) Palaeogeographic setting; (b) model, with anhydrites deposited within hypersaline lagoons and prograding subtidal sabkhas (after Lindsay, 1985).

One of the few examples of an ancient depression sabkha is recognized in the Permian *Upper Minnelusa Formation* of the Powder River Basin in Wyoming (Achauer, 1982). The region was a broad shallow marine shelf, at times emergent, with aeolian dunes. Cycles begin with shallow subtidal carbonates overlain by desiccation-cracked laminated dolomicrites, quartz

sandstones (interpreted as aeolian) and thin (<0.5 m) sabkha facies composed of nodular anhydrite (Fig. 8.40). The depression sabkha interpretation is based on their areal distributions: highly elongated areas several hundred metres wide and several kilometres long, extending between former aeolian sandridges.

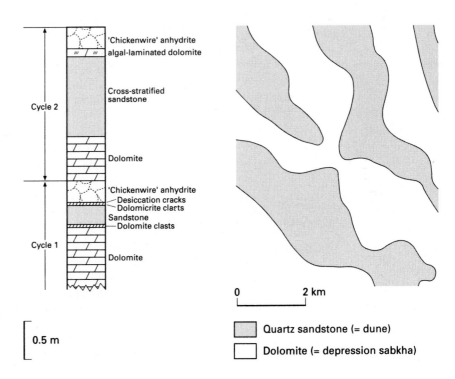

Figure 8.40 Log of two depositional cycles in the upper Minnelusa Formation, South Rozet Field, Wyoming, and map of dolomite distribution between elongated aeolian sandstones (after Achauer, 1982).

8.8   Ancient basin-marginal platform evaporites

Stratigraphic study of shallow-water and sabkha evaporites of wide shelves or platforms demonstrates that whereas some formed by progradation of marginal marine environments essentially similar in character (and size) to those of the present day, most formed simultaneously in evaporitic settings much larger than anything found today. These evaporites must have been generated in vast expanses of evaporitic lagoons and mudflats, reaching 10 000s–100 000s km^2 in extent, over which the brine depths were at most only a few metres. In the absence of a modern analogue it is difficult to explain how evaporites can be deposited in these nearly uniform, but vast, areas of shallow brine. Warren (1991) speculates that wide bodies of brine precipitated subaqueous gypsum under near-equilibrium conditions. Evaporation kept brine surfaces below sea level, and brine depths restricted to only a few tens of centimetres. An underlying aquitard, relatively high air humidities (for an area of evaporite deposition), and frequent resupply of sea water from storms and seepage is supposed to have kept these environments from desiccating. However, it is still difficult to see, in such shallow bodies of brine, how uniform salinities could have been maintained over such large areas. We lack viable depositional models for these evaporites.

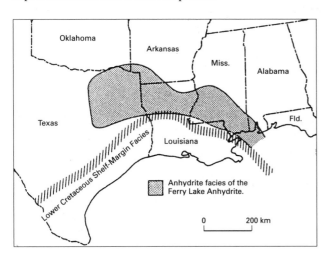

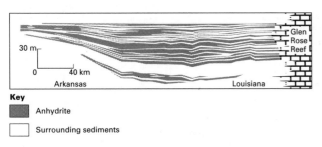

Key

▨ Anhydrite

☐ Surrounding sediments

Figure 8.41 Map and north–south cross-section of the Lower Cretaceous Ferry Lake Anhydrite (from Loucks & Longman, 1982; Pittman, 1985).

The *Lower Cretaceous Ferry Lake Anhydrite* is a 90-m-thick sequence of stacked argillaceous carbonate–anhydrite cycles (Loucks & Longman, 1982) that extends from east Texas to Alabama (Fig. 8.41). Each cycle passes upward from marine carbonates into saline evaporites and is terminated by low-intertidal carbonate or, more rarely, by a sabkha carbonate/evaporite unit. Deposition occurred within a shallow subtidal lagoon, up to 260 km wide in east Texas, interpreted to have been barred from the sea by shelf-margin rudist reefs and banks. Immediately behind the rudist bank barrier the lagoon is believed to have had normal marine salinities but landward it became more restricted and hypersaline (Loucks & Longman, 1982). Further landward still, the lagoonal sediments interfinger with alluvial plain conglomerates and redbeds. Lagoonal carbonate units are mainly mudstones and wackestones. Overlying anhydrites are up to 7 m thick and can be traced, using geophysical well logs, for distances greater than 150 km (Pittman, 1985). At Fairway Field in east Texas, Loucks and Longman (1982) interpreted vertically or near vertically oriented anhydrite nodules within the evaporite beds as pseudomorphs after subaqueous gypsum crystals. They proposed that the greater part of the Ferry Lake anhydrites were of subaqueous origin. Shallow restricted subtidal to low intertidal carbonates with microbial mats, anhydrite pseudomorphs after lenticular gypsum and intraclasts were deposited in shoal areas that prograded over the subaqueous evaporites.

The depositional model proposed by Loucks and Longman for the Ferry Lake Anhydrite has some anomalies. If the rudist banks caused the restriction of the lagoon, then fully marine carbonates would not be expected in its seaward parts. If there had been a pronounced lateral salinity gradient in the lagoon (from normal marine waters to brines capable of precipitating subaqueous gypsum) it must have been exceedingly shallow to prevent brine mixing, yet the gypsum developed is characteristic of stable, deeper-water environments. Perhaps the Ferry Lake anhydrites represent episodes of isolation, when the lagoon became affected by evaporative drawdown and the normal marine carbonates are different in age from the anhydrites.

Fabrics and stratigraphic geometries of bedded halites in the *Permian San Andres Formation* of the Palo Duro Basin (Texas Panhandle) indicate they formed in vast platform brine pools (Hovorka, 1987) on a broad, low-relief shelf separated from the marine limestones and sandstones of the neighbouring Midland Basin (Fig. 8.42) by a rim of oolite and other shallow-water carbonates. Here, intertidal and supratidal facies are subordinate and discontinuous, suggesting the barrier between the normal marine shelf to the south and the evaporites in the north was a discontinuous mosaic of islands. This barrier was in part located and influenced by the Matador Arch, a structural feature between the two basins. To the far north clastic sediments were deposited in aeolian and mudflat settings.

The San Andres Formation contains more than 20 cycles composed of carbonates, evaporites and clastics. These range from 1 m to more than 100 m thick and are traceable over more than 10 000 km² with only minor changes in thickness and facies. Each cycle consists of a transgressive thin, anhydritic mudstone overlain by a regressive sequence of carbonate, anhydrite and bedded halite (Fig. 8.43). The continuity and similarity of facies over wide areas indicate deposition in a setting with little topographical relief.

Carbonate depositional environments were predominantly subaqueous (Bein & Land, 1982; Fracasso & Hovorka, 1986) with thicker beds being formed in normal marine waters, and thinner beds being deposited within more restrictive, hypersaline environments. Anhydrite beds contain abundant halite or anhydrite pseudomorphs after vertically oriented gypsum crystals (Hovorka, 1992) and record deposition within very shallow brine pools. Individual anhydrite beds extend over areas up to 26 000 km² (Fracasso & Hovorka, 1986), again with only minor changes in thickness or facies.

Different subfacies in the bedded halites are recognized (Hovorka, 1987). Cumulate halites nucleated on surfaces of brine bodies and accumulated on their floors. These deposits commonly recrystallize to colour-banded halite rocks. Bottom-nucleated halite grew as crystal crusts, generating layers of chevron halite. The shallowness of the depositional site is demonstrated by the common presence in these crusts of dissolution surfaces and microkarst pits (later filled with coarse halite and geopetal fills of muddy anhydrite or mudstone). Chaotic mixtures of structureless mudstone and displacive halite cubes are also common. They formed when the brine pool desiccated frequently: surficial halite crusts were repeatedly dissolved by floods, and the halite was reprecipitated within the introduced clastic sediment. Individual units of brine-pool halite, only a few metres thick, have been traced over areas greater than 1000 km².

The depositional environment was an areally extensive, low-relief shelf covered by a shallow perennial brine pool. Halite deposition was, however, interrupted by episodes when brines evaporated away and brine-level fell beneath the sediment surface. Floods during these desiccation events generated microkarst and

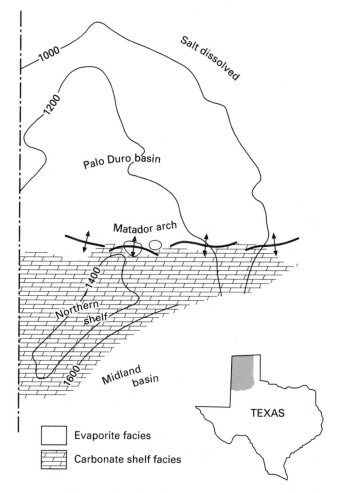

Figure 8.42 Isopach map of Permian San Andres Formation and distribution of evaporites in Palo Duro Basin (after Hovorka, 1987).

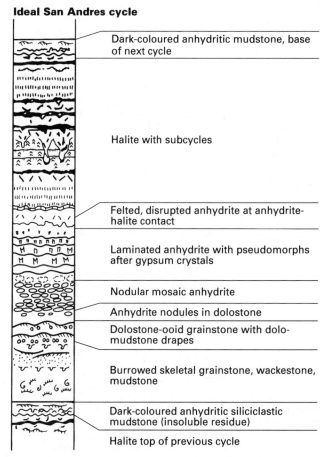

Figure 8.43 Vertical facies sequence of an idealized cycle, San Andres Formation, Palo Duro Basin (after Fracasso & Hovorka, 1986).

dissolution surfaces on the exposed salt pans. Longer desiccation episodes were associated with progradation of distal alluvial mudstone facies across the desiccated halite flat, converting the area into a dry mudflat.

Marine waters could flood such a flat shelf when sea level rose only slightly. The predominance of evaporites within Palo Duro Basin, however, indicates that evaporite precipitation (and consequent sediment aggradation) must usually have been fast enough to offset the effects of basin subsidence and sea-level rise. This necessitates an abundant supply of solutes into and across the wide evaporitic shelf and a restriction at the Matador Arch that was both sufficient to allow brines to concentrate and halite to precipitate, yet open enough to permit frequent influx of marine water into the brine pool. The question remains, however, as to whether a barrier formed of a mosaic of islands over a Matador 'shoal' (Hovorka, 1987), would be sufficiently closed to allow brines to reach halite saturation.

8.9 Ancient basin-central evaporites

Three types of basin-central marine evaporites may be identified (Fig. 8.44): (i) *shallow basin–shallow-water evaporites*, deposited in basins that had little depositional relief and from shallow

bodies of brine or within saline mudflats; (ii) *deep basin–deep water evaporites* formed in basins with substantial depositional relief (10s–100s m) which were largely filled with brine; and (iii) *deep basin–shallow-water evaporites* precipitated in deep basins that were subject to substantial evaporative drawdown; evaporites were formed in shallow-water or mudflat environments and resemble those of the first type.

In many large basin-central evaporites the basin evolves from one type to another. Brine levels in a deep isolated basin change rapidly or fluctuate in response to even slight changes in the rates of inflow, outflow and evaporative brine loss. Deep evaporite

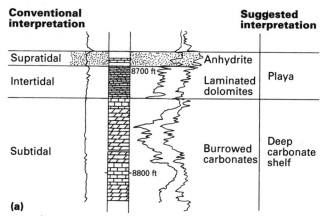

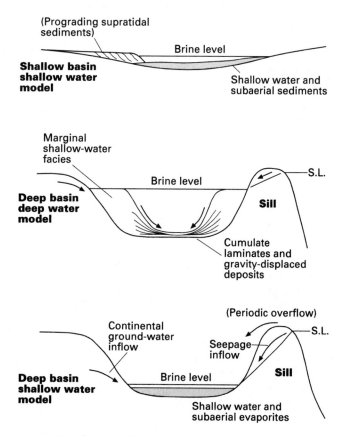

Figure 8.44 Alternative depositional models for basin-central evaporites (modified from Kendall, 1979).

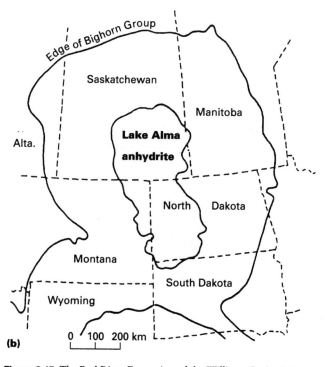

Figure 8.45 The Red River Formation of the Williston Basin. (a) Lowermost sedimentary cycle; (b) basin-central distribution of lower cycle (Lake Alma) anhydrites (after Kendall, 1988).

basins may thus alternate between deepwater evaporite deposition and shallow water/mudflat evaporites when the basin is desiccated. Similarly, as a deep basin is filled with evaporites, the evaporites will pass up into the shallow-basin–shallow-water type.

In the following, evaporites of the Red River Formation illustrate shallow-basin–shallow-water evaporites, those of the Otto Fjord Formation are an example of deep-basin–shallow-water evaporites, and the Castile Formation is presented as the classic deep-basin–deepwater evaporite. Zechstein evaporites exemplify the complexities of enormous evaporite bodies.

The *Upper Red River Formation* of the Williston Basin contains thin but widely distributed, basin-central anhydrites (Fig. 8.45) which are interpreted differently by different authors. The anhydrites form the upper parts of basin-wide sedimentary cycles that begin with bioturbated, fossiliferous carbonates and progress through laminated, unfossiliferous dolomites into laminated dolomitic, and nodular anhydrites. Many have been impressed by a similarity of the Red River laminated dolomites and nodular anhydrites to facies in intertidal and supratidal environments (respectively) of the Trucial Coast (G.B. Asquith, 1979; Clement, 1985; Ruzyla & Friedman, 1985). They interpret Red River

cycles as prograding intertidal and arid supratidal (sabkha) environments over marine subtidal carbonates.

However, this interpretation ignores: (i) the basin-central location of the Red River anhydrites; prograding sabkhas would be best developed at basin margins; and (ii) the excessive thickness of the inferred intertidal and supratidal deposits relative to those in the Trucial Coast; the laminated dolomites in the lower cycle are almost 20 m thick in the basin centre (Kendall, 1976, 1988).

The Red River anhydrites form 'brining-upward' cycles and were deposited in a shallow basin from shallow brines (Kendall, 1976, 1988; Kohm & Louden, 1978; Longman, Fertel & Glennie, 1983). Kendall (1992) reinterprets the laminated dolomite units as carbonate dry mudflat deposits (some with sandpatch fabric; see Fig. 8.12). Many strikingly resemble the schizohaline pond deposits of the West Caicos Island salina (Sect. 8.5.2). Some anhydrites resemble gypsum saline mudflat deposits but others are nodular replacements of large bottom-grown selenite crystals. The last mentioned must have been deposited in an enormous shallow salina, fully 550 km across (Fig. 8.45).

The *Otto Fjord Formation* is magnificently exposed on high

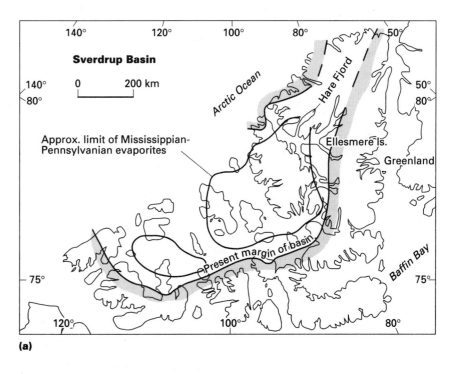

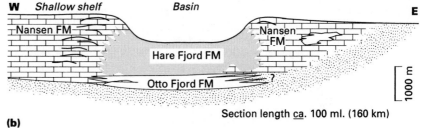

Figure 8.46 Otto Fjord Formation, Canadian Arctic Archipelago. (a) Subsurface distribution of evaporites in Sverdrup Basin; (b) schematic restored cross-section of northern Sverdrup Basin showing axial distribution of Otto Fjord evaporites and major shelf-to-deep basin facies transitions (after Davies, 1977).

relief walls of fjords in northwestern Ellesmere Island, Arctic Canada. Its evaporites formed on the floor of a developing deep basin which is estimated to have been at least 400 m deep at the end of Otto Fjord deposition (Fig. 8.46) (Wardlaw & Christie, 1975; Davies, 1977; Nassichuk & Davies, 1980). The lateral relationships between evaporites and shelf-marginal and reef carbonates provide criteria whereby deepwater and desiccated basinal evaporites may be distinguished.

The Otto Fjord Formation occupies an axial position in the Sverdrup Basin and is overlain by deepwater, basin-central, argillaceous marine carbonates and shales of the Hare Fjord Formation (Fig. 8.46b). Both formations pass laterally into coeval shelf-marginal and shallow-shelf carbonates of the Nansen Formation. The Otto Fjord Formation is up to 500 m thick and consists of about 20 cycles of limestones and anhydrite, with the limestones locally containing reefs. Anhydrite units, 3 m to over 60 m thick, are mainly composed of nodular–mosaic anhydrite. Some are bedded and contain pseudomorphs of vertically oriented gypsum crystals.

Cycles begin with a sharp contact between limestone and the anhydrite of the preceding cycle. Lowermost carbonates are dark, unfossiliferous and laminated but pass upward into skeletal wackestones. Uppermost limestones are again laminated and have gradational contact with overlying laminated anhydrites. This interlaminated anhydrite–carbonate grades into calcareous nodular–mosaic anhydrite, and this into pure nodular–mosaic anhydrite, some containing the gypsum pseudomorphs.

Reefs pass laterally into thinner carbonate beds that are sandwiched between thick anhydrites, the upper anhydrite onlapping and pinching out at reef flanks. This relationship suggests the original gypsum only grew or accumulated in inter-reef areas and thus probably only formed in shallow (<30 m) brines.

A basinal evaporite to shelf carbonate facies change, modified by local reefs, is beautifully exposed near Girty Creek (Hare Fjord; Fig. 8.47). A predominantly anhydrite section of the Otto Fjord Formation is interrupted by a 40–50-m-thick discontinuous limestone bed that locally contains reefs. This limestone unit drastically thins eastwards into the basin. Westwards it thickens and is traceable to the base of a steeply dipping limestone tongue that climbs up for several hundred metres into the Nansen Formation. Anhydrite units pinch out westwards between tongues of steeply dipping limestones. The depositional relief demonstrated by these carbonate foreslopes increased during the accumulation of the Otto Fjord: relief was initially only a few metres but increased to more than 400 m for the uppermost anhydrite–limestone cycles. From published photographs and interpretive diagrams, it appears that the anhydrite units of the Otto Fjord Formation only climb a few tens of metres up the flanks of the Nansen foreslope limestones (Fig. 8.47). This suggests evaporite deposition occurred from brines only a few tens of metres deep. Thus Otto Fjord anhydrites were deposited from relatively shallow bodies of brine located within a basin, which interbedded carbonates indicate must have been at least 400 m deep toward the end of the Otto Fjord depositional episode. During times of evaporite deposition the basin suffered extensive evaporative drawdown, whereas it became reflooded at the beginning of each carbonate phase.

The Otto Fjord Formation demonstrates clearly that gypsum may be deposited in a deep basin setting, yet be of relatively shallow-water origin. The critical evidence for this resides in exposures showing the stratigraphic relationships of evaporites to the topographic relief on underlying carbonates. Such evidence is usually unavailable in other basins where the data come entirely from the subsurface.

The *Castile Formation* (Permian of Texas and New Mexico) is the quintessence of a deepwater evaporite. It was deposited entirely within the Delaware Basin: a 260 × 140 km depression, surrounded by a carbonate–evaporite shelf rimmed by the earlier

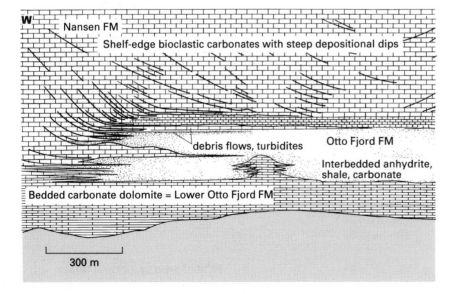

Figure 8.47 Carbonate–anhydrite facies relationships along the north wall of Hare Fjord, Ellesmere Island, showing anhydrites of the Otto Fjord Formation climbing only partially up depositional slopes of the Nansen Formation (after Davies, 1977).

Capitan reef complex (see Fig. 9.13). Prior to Castile deposition the basin had a depositional relief of 500–600 m and within it, deepwater 'starved-basin' sediments were deposited, interrupted by turbidites and mass-flow deposits (the Bell Canyon Formation).

Castile evaporites are predominantly laminated carbonate–anhydrites (see Fig. 8.33) but also contain minor limestones, non-laminated anhydrite and, in the eastern and more deeply buried part of the basin, thick intercalations of halite. The laminated sediments form an essentially uninterrupted sequence beginning with organic-rich laminated siltstones and mudstones of the uppermost Bell Canyon Formation and passing up via a thin basal laminated limestone into the interlaminated calcite–anhydrites of the main part of the Castile Formation (Fig. 8.48). Anderson, Dean *et al.* (1972) developed a time-series for the Castile (and overlying Lower Salado) based on measurements of these couplets, assuming they were annual events. Over 260 000 couplets were recorded. The average couplet thickness remains remarkably constant (1.9 mm) throughout a 500-m-thick Castile section (Dean & Anderson, 1978). Individual laminae are traceable between cored wells for at least 113 km across the basin (Figs 8.33 & 8.48), and thus are probably basin-wide. Calcite–anhydrite and anhydrite laminae constitute 97% of the Castile couplets, the remainder being formed of thicker anhydrite–halite alternations.

Beds of brecciated anhydrite occur in the western part of the basin and correlate in stratigraphic position with halite beds in the east. They formed by the subsurface dissolution of halite and thus record the former presence of halite in areas where it has been removed. Originally the halites had nearly basin-wide distributions. Correlations of anhydrite laminae beneath halites (or their equivalent dissolution breccias) indicate halite began to accumulate throughout the Delaware Basin simultaneously.

Nodular anhydrite intervals occur between laminated anhydrites and overlying halite. Some nodular intervals contain pseudomorphs after bottom-grown gypsum crystals (Kendall & Harwood, 1989) suggesting many other nodular anhydrites had a similar precursor. Some also replaced beds where gypsum crystals grew displacively within laminated anhydrite. Nodular intervals are traceable between cored wells and the onset of nodule development, relative to the laminite time series, is constant throughout the basin.

Stratigraphic successions of laminar anhydrite, nodular anhydrite and halite (or equivalent brecciated anhydrite) are cyclically arranged. Cycles reflect basin-wide salinity increases within a deeply filled basin (Dean & Anderson, 1978). The ideal cycle begins with thin calcite–anhydrite couplets in which calcite is dominant. Couplets thicken upward by the appearance of additional sulphate. Near the top of the anhydrite, nodular beds appear and are overlain by halite with anhydrite laminae. The succeeding cycle appears abruptly above the halite. Few cycles are complete; in most the basal calcite-rich part is poorly developed and many cycles lack the uppermost halite member. Cycles are composed of between 1000 and 3000 couplets. There

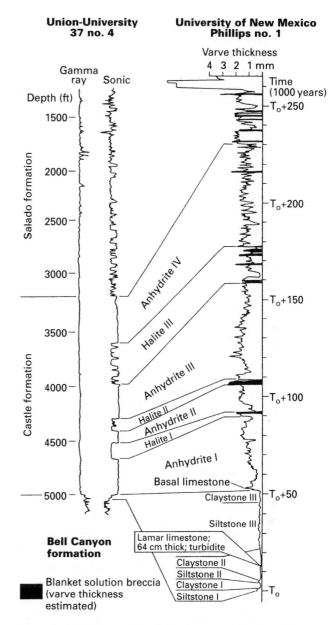

Figure 8.48 Thickness and cyclicity of measured Castile and Lower Salado calcite–anhydrite couplets in the University of New Mexico Phillips No. 1 borehole, and correlation with the sonic log of a second well (Union-University '37' No. 4) (from Anderson, Dean *et al.*, 1972).

is no systematic change in the number of couplets per cycle throughout the Castile, suggesting cycles had similar durations. Younger cycles, however, exhibit greater variability in couplet thickness, and are more likely to contain halite. The salinity of brines during a cycle was evidently more variable later during Castile deposition.

Brine depths during Castile deposition must have been sufficient to preserve the remarkably fine and even laminae. Dean and Anderson (1978) suggested that the rapid, synchronous changes in salinity (recorded in basin-wide mineralogical and geochemical changes) could only have occurred by basin-wide changes in water chemistry within a deep body of brine. Trace element geochemistry of beds also varies only slightly across the basin and vertically (Dean, 1978). To maintain this constancy, the body of brine must have changed relatively little, implying the brine body was large and deep.

As the basin shallowed (as a result of sediment accumulation) there was a progressive decrease of brine volume. This is reflected in the change in the amplitude of salinity cycles upwards through the Castile. The smaller the brine volume, the less the evaporite system is buffered and the greater is the possibility of large fluctuations in salinity. The greater changes in anhydrite lamina thickness and the increasing preservation of halite in younger cycles suggests salinity fluctuations increased with time as the basin shallowed.

Whereas deposition of Castile laminites in a deepwater setting accounts for most features of the formation, it is unlikely that deposition always occurred from a completely brine-filled basin. During halite deposition the basin was almost certainly isolated and subject to evaporative drawdown.

The presence of former gypsum crusts (now nodular anhydrites) at intervals throughout the Castile (even within 3 m of the base) is seemingly inconsistent with deposition from a deep body of brine. Their presence signifies that bottom brines were supersaturated with respect to gypsum, yet supersaturation is thought only to be achieved by evaporation at the brine surface. Comparison with the hydrology of the Dead Sea suggests that stratification within a deep brine would prevent the surface brines from reaching the basin floor. Kendall and Harwood (1989) therefore used the presence of former gypsum crusts to argue that they, and the remainder of the Castile evaporites, must have been deposited in brines only a few tens of metres deep. It is now believed, however, that an unknown mechanism (perhaps double diffusive mixing associated with changes from meromictic to holomictic conditions; Sect. 8.6.2), allowed gypsum to grow on the floor of a basin containing a deep brine column.

Zechstein evaporites accumulated in a huge, post-Hercynian intracratonic basin that extended from the British Isles to eastern Poland and Lithuania (Fig. 8.49). Structural highs separated smaller, intercommunicating basins within which the sedimentary fill was slightly different. The entire basin originated as an area of inland drainage which lay below sea level. Following an initial transgression, deposition in the central parts of the basin began in a deepwater setting, perhaps 200–300 m deep. Synsedimentary subsidence led to the accumulation of more than 2 km of Zechstein carbonates, evaporites and siliciclastics, over a period of 5 million years (Menning, Katzung & Lutzner, 1988). Locally, 3000 m accumulated in troughs where differential subsidence may have originated from strike-slip movements (Ziegler, 1989).

The Zechstein is divisible into five main depositional cycles, the lower two of which are believed to contain deepwater

Figure 8.49 Generalized sketch map of the Zechstein Basin showing the main structural units (from Füchtbauer & Peryt, 1980; D.B. Smith, 1980).

evaporites, the upper cycles probably being, at least in part, non-marine. The ideal cycle begins with a thin clastic unit that passes up through marine carbonates into anhydrite, halite, and potash and magnesium salts, reflecting a progressive salinity increase in the basin.

The lowest cycle (Z1) began with deposition of a thin, highly radioactive and sapropelic shale (Marl Slate, Kupferschiefer). It passes up into carbonates (Zechsteinkalk or Lower Magnesian Limestone) with a normal marine fauna. Sediment accumulation was greatest in marginal parts of the basin where shallow, warm aerated conditions allowed the formation of a prograding carbonate wedge, commonly with a reefal front (Fig. 8.50). In the basin itself the Zechsteinkalk is a 'starved basin' deposit – a few metres of dark, pyritic and argillaceous carbonate. Stratigraphic relationships between platform and basin sections suggest that water depths at the foot of the marginal slope were 200 m, with depths in the basin centre presumably somewhat deeper.

The same depositional pattern, with thicker basin-marginal deposits thinning abruptly basinwards into a thin stratigraphic equivalent, was repeated in the succeeding Z1 evaporites, and in the carbonates and evaporates of the overlying Z2 cycle (Fig. 8.51). Each sedimentary wedge developed basinward of its predecessor and marked an episode in the progressive lateral filling of the deepwater Zechstein basin. By the end of the Z2 cycle, however, a thick halite sequence had been deposited in the basin centre, so eliminating most of the original basin topography. Succeeding cycles appear to be largely composed of shallow-water and brine-flat deposits.

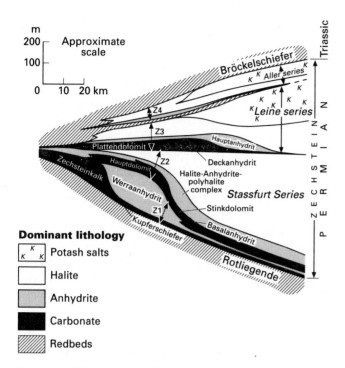

Figure 8.50 Diagrammatic shelf–basin profile of the Zechstein sequence in UK southern North Sea, giving German nomenclature and showing successive prograding carbonate and evaporite wedges of the Zechsteinkalk to Stassfurt Halite–Anhydrite–Polyhalite complex and the Plattendolomit and Hauptanhydrit, and the basin-filling salts of the Stassfurt and Leine Series (from Taylor, 1990).

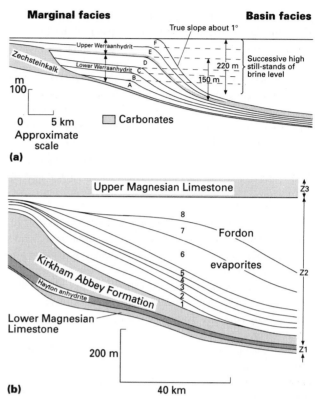

Figure 8.51 Sketch cross-sections across marginal evaporite wedges in Zechstein of North Sea and adjacent parts of England. (a) Werraanhydrit, note restriction of older units to basin floor (implying evaporative drawdown), gradual onlap of older carbonate depositional slope and estimated brine depths (after Taylor, 1980); (b) Basal Stassfurt (Forden Evaporites) halite–anhydrite–polyhalite complex, with similar basinward and shelfward thinning of units 1–7, and basin-filling halite of unit 8 (after Colter & Reed, 1980).

The Z1 evaporite (Hartlepool, Hayton or Werraanhydrit) sharply overlies the basinal Zechsteinkalk, probably marking a sudden salinity increase. It occurs in two main facies (Fig. 8.51a). A marginal platform facies, 25 km wide and up to 200 m thick, girdles the basin and is composed of nodular and enterolithic anhydrite set within a sparse, brown and microcrystalline dolomite, some of which has been interpreted as stromatolitic. This facies is interpreted to have formed in a marginal complex of shoal, sabkha and salina environments with halite beds locally deposited in salt pans (Taylor, 1980, 1990; D.B. Smith, 1989). Where fabrics are well preserved, pseudomorphs after bottom-grown gypsum crusts are common. It terminates abruptly landward against the former depositional slope of the Zechsteinkalk (Fig. 8.51a). It also thins rapidly into the basinal facies (only 20 m thick) forming a depositional slope similar to that of the earlier Z1 carbonates (Taylor & Colter, 1975; D.B. Smith, 1980; Taylor, 1980, 1990). Within the slope, turbidites and mass-flow deposits may be common. The basinal facies is a fetid, evenly laminated carbonate and anhydrite sequence containing several intercalated nodular anhydrite horizons (Taylor, 1980; Richter-Bernburg, 1985). Each of the depositional cycles (Taylor, 1980) starts abruptly with nodular/enterolithic anhydrite and grades up, via laminar anhydrite, into laminar carbonate. Each cycle is traceable over the basin floor and into

the thicker marginal facies. Younger cycles overlap older to rest upon the older Zechsteinkalk (Fig. 8.51a). The stratigraphic relations of Z2 evaporite cycles to each other and to the older Zechsteinkalk indicate that water levels fell at least 100 m prior to evaporite deposition. Deposition was initially confined to the depression lying basinward of the Zechsteinkalk depositional slope. Brine depths must have been at least 150 m at the foot of this slope and increased to over 200 m at the end of the Werraanhydrit (Fig. 8.51a). Brine levels rose in stages, each producing a depositional cycle, possibly interrupted by temporary falls.

The nodular anhydrites at the base of cycles are interpreted to have formed during initial stages of brine recharge when conditions were shallowest and most saline (Taylor, 1980). Nodular anhydrite formed at the edge of an expanding brine body, whereas overlying laminated anhydrite accumulated on the floor of the basin. This would imply that facies relationships are diachronous. In the basin centre, however, nodular intervals occur at precisely the same intervals relative to a laminite time sequence (Richter-Bernburg, 1985) – they are not diachronous.

The second Zechstein depositional cycle began in the basin with a thin laminated carbonate similar to that of the Z1 cycle. It passes shoreward into a marginal wedge of slope carbonates surmounted by shallow-water and intertidal dolomites (Haupt-dolomit, Kirkham Abbey Formation) that interfingers with continental mudstones with evaporites around the basin edge. The evaporite part of the Z2 cycle, perhaps 1400 m thick, is composed largely of a basin-central halite overlying a complex of laminated anhydrite, halite, polyhalite and kieserite (the Stassfurt Salz or Forden Evaporites). The lower complex, 50-90 m thick in the basin, thickens rapidly at the basin margin and has foresetting internal stratigraphic relationships (Fig. 8.51b; Taylor & Colter, 1975; Colter & Reed, 1980). This again implies that basin and lower slope equivalents were deep-water deposits. This conclusion poses a considerable interpretive problem because the beds in question contain polyhalite and kieserite. A fine preservation of laminae in them indicates that they are primary or syndepositional reaction products, and thus must have been formed in contact with brines saturated with respect to potash salts. However, brines of this type can only be generated from seawater by a 60-fold increase in concentration. During this concentration they would have precipitated an equal or greater volume of calcium sulphate and halite (Schmalz, 1969; Valyashko, 1972) so eliminating the basin. This problem remains unsolved.

At the top of the Z2 evaporites is a widespread thin layer of potash minerals, suggesting an extensive level surface and considerable lateral conformity in sedimentation. This more uniform depositional pattern is repeated in succeeding cycles. These are thinner, contain only thin to absent, restricted marine carbonates (commonly algal laminated) at their base, and are separated by redbeds. Evaporites are variable, ranging from potash salts in the basin centre to microbial mats and nodular anhydrites on the margin; all deposited in shallow to emergent conditions (D.B. Smith, 1973, 1989).

Zechstein anhydrites and halites have been most recently studied in northern Poland (Czapowski, 1987; Peryt, Orti & Rosell, 1993; Peryt, 1994) and much of this work could be usefully applied to interpretation of Zechstein evaporites elsewhere. A more iconoclastic interpretation of Zechstein sulphates was given by Langbein (1987) including the view that the main facies differences between basin and margin originate diagenetically during burial compaction.

One of the more important messages given by the Zechstein evaporites is that 'saline giants' of this type are complex and no single depositional model can be applied to them. They vary enormously in space and time, with the depositional environment varying from marginal sabkhas and salines on basin edges to deepwater evaporites in the basin centre. They are also cyclic, forming in a number of stages with varied depositional conditions.

Similar statements may be made about the evaporites of the *Messinian (Miocene) of the Mediterranean*. Seismic studies of the Mediterranean floor identified widely traceable reflectors that indicate the presence of laterally continuous evaporites, which range from a few metres at basin margins and on topographic highs to more than 2 km in the basin centres (Ryan, Hsü et al., 1973). These evaporites are underlain and overlain by deepwater marine sediments and so mark the desiccation of the entire Mediterranean Basin. Some of the evaporites are interpreted as sabkhas (Friedman, 1973), but most formed in shallow subaqueous environments (Garrison, Schreiber et al., 1978) whereas others contain evenly laminar beds that Dean, Davies and Anderson (1975) suggest are deeper-water facies. Studies of Messinian evaporites in Italian basins suggest that some formed in very shallow water (Hardie & Eugster, 1971), whereas others were deposited in deep water (Parea & Ricci-Lucchi, 1972). In Sicily, the full range of depositional environments is present (Hardie & Eugster, 1971; Decima & Wezel, 1973; Schreiber, Friedman et al., 1976; Schreiber, Catalano & Schreiber, 1977; Schreiber, McKenzie & Decima; 1981).

8.10 Evaporites and sequence stratigraphy

The application of any type of stratigraphy to evaporites and evaporite basins is in its infancy (Kendall, 1988). Evaporites are usually treated as orphans, to be appended to sequence stratigraphic considerations of carbonates (Sarg, 1988; Tucker 1991; James & Kendall, 1992; Handford & Loucks, 1993). This is probably reasonable when applied to basin-marginal evaporites because there they are only parts of a larger facies mosaic and are affected by sea-level changes in somewhat similar ways to laterally contiguous sediments. But basin-central evaporites are different and their relationship to sea-level changes is far from straightforward. They are highly dependent upon the hydrology of the basin and this is also controlled by factors other than sea-level change.

8.10.1 Basin-marginal evaporites

Sabkha, salina and platform evaporites on basin margins are characteristic highstand and (less importantly) transgressive systems tract deposits in arid and some semi-arid settings. Prograding strandline terrigenous clastics can isolate salinas and sabkhas behind beach or dune ridges, but it is the reduced accommodation space and a propensity to develop poor circulation on carbonate platform tops that is the ideal setting for thin shelf and peritidal evaporite accumulation. Diachronous units of sabkha and salina evaporites may be widespread. Well-developed, continuous rims to platforms act as barriers to water circulation and, during short-term falls in relative sea level, may produce large hypersaline lagoons or shelves, within which laterally extensive subtidal evaporites form. During lowstands these basin-marginal evaporites are exposed and affected, to varying degrees, by erosion.

Both supratidal and depression sabkhas are significantly affected by small changes in relative sea level. As sea level and the sabkha brine table rise, evaporites are dissolved and reprecipitated but, if the coastal barrier is high, continuous and permeable enough, the sabkhas may be replaced by salinas. Conversely, as sea level falls, the sabkha brine table is depressed, sabkha sediments dry out, become susceptible to wind deflation and are eroded. Warren and Kendall (1985) suggest subtidal–intertidal cycles in the Permian backreef sediments of the Guadelupe Mountains (West Texas) may have originally contained supratidal sabkha sequences that were truncated.

Sea-level changes of greater magnitude destroy sabkhas. Sea-level falls cause sabkha abandonment, and establishment of a new offshore barrier in equilibrium with the new sea level. The abandoned sabkha has a low preservation potential. A large rise in sea level, causing coastal retreat, will flood sabkha surfaces, destroying any surficial evaporite crusts. The displacive intrasediment evaporites are only retained if they were deposited in and are overlain by sediments of low permeability.

The preservation potential of coastal salina evaporites is somewhat similar to that of sabkhas. Large sea-level rises drown the barriers that separate the salina depressions from the ocean, converting them into marine embayments. If the volume of salina evaporites is substantial, however, only the uppermost, or the most soluble, will be dissolved to leave much of the evaporite sequence intact. In a similar fashion, relative sea-level falls, particularly of short duration, will not necessarily remove all the evaporites from an emergent salina. The large volume of evaporites commonly present, the aridity of the hinterland (preventing substantial erosion), and the fact that many salina evaporites were deposited as crystal crusts, which are more resistant to erosion than many sabkha sediments, are all factors enhancing the preservation of at least part of the salina sedimentary record. Some salina sequences do exhibit evidence of repeated erosion and dissolution events (Powers & Hassinger, 1985; Hovorka, 1987) and Vai and Ricci Lucchi (1977) interpret clastic gypsum deposits from the Messinian of the Apennines as fluvially reworked evaporites

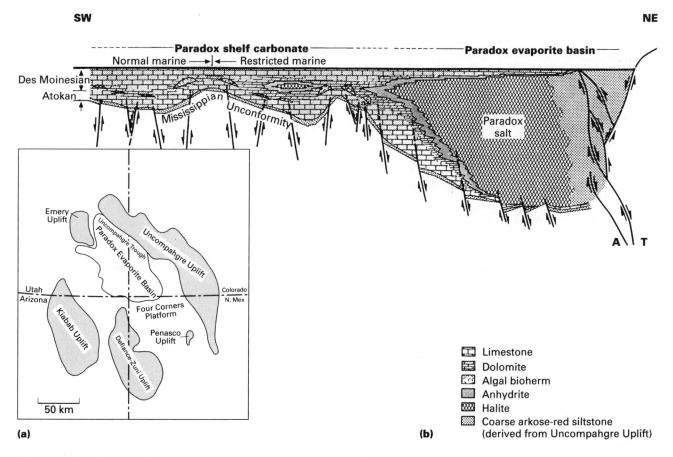

Figure 8.52 Paradox Basin, Four Corners Region. (a) Location map of present-day extent of Paradox halites and surrounding uplifts (after Peterson & Hite, 1969); (b) generalized SW–NE cross-section across basin showing overall facies relationships between Middle Pennsylvanian shelf carbonates, basinal evaporites and coarse clastics proximal to the rising Uncompahgre Uplift (from Goldhammer, Oswald & Dunn, 1994).

319

derived from the destruction of earlier formed shallow-water evaporites that became exposed by falling brine levels within the basin.

Gypsum precipitated subaqueously in salinas that were subjected to repeated and episodic periods of emergence may be altered to nodular anhydrites, perhaps losing much of its original nature. This process has been termed, rather horribly, 'sabkharization' (Hussain & Warren, 1989) and has been identified within modern non-marine playa sediments as well as in Mississippian salina anhydrites (Sect. 8.7).

The link between the types of evaporites formed in salinas, the salinity of the salina brines and the relative rates of water influx and outflow has already been discussed (Sect. 8.2.3), especially in connection with Lake MacLeod (Sect. 8.5.2). A rise in sea level increases the hydraulic head between the ocean and the brine level in the salina. This promotes greater rates of inflow and, in the absence of increased seepage outflow, brine

residence times of basin brines increase and more saline evaporites should be precipitated. In reality, however, rates of vertical accretion of evaporites (see Table 8.3) usually exceed any possible rate of sea-level rise. Thus the hydraulic head will diminish, even during episodes of sea-level rise. Depositional onlap of salina margins by evaporites may thus occur during a rising sea level, a stillstand or during falling sea level. It continues so long as sea level lies above the brine level in the basin, inducing seawater seepage influx. Depositional onlap thus cannot be used to infer sea-level rise in evaporite basins. If sea level falls to elevations below the brine level in the basin, brine may be exported from the basin, evaporite deposition will stop, and earlier-formed evaporites begin to be eroded or dissolved.

8.10.2 Basin-central evaporites

Basin-central evaporites, often of great thickness, are lowstand

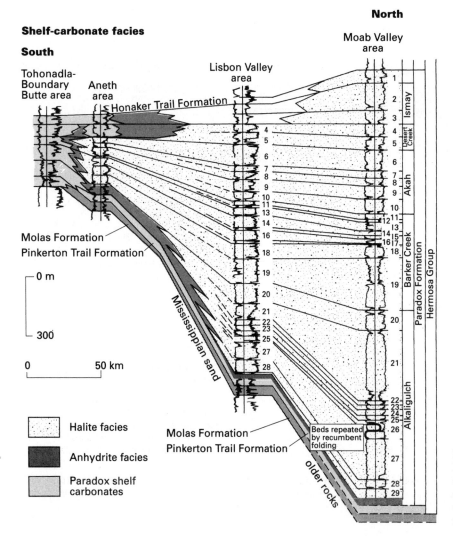

Figure 8.53 Correlation of well logs across Paradox Basin illustrating 29 evaporite cycles in basin and relations between basinal halite, anhydrite and shelfal carbonate facies (from Goldhammer, Oswald & Dunn, 1994; modified from Peterson & Hite, 1969).

deposits within *marine* basins. They form when the basins become isolated or greatly restricted by lowered sea level. Marine evaporites within otherwise *continental* basins, on the other hand, could represent highstand deposits that formed when the sea-level rise initiated marine seepage into the basin.

The extent of water-level drop in basins is greatly magnified by evaporative drawdown. Because large volumes of water have to be evaporated to form enough high-salinity brines to fill an entire deep basin, deep-water basins most commonly do not generate brines capable of precipitating halite. These partially desiccated basins develop marginal wedges of shallow-water gypsum and deep-water laminated gypsum/evaporitic carbonates on basin floors (Tucker, 1991). But, as some Zechstein halites demonstrate (Colter & Reed, 1980; Czapowski, 1987), deep-water basins filled with halite-saturated brines, or even those of greater salinity, are possible.

Basins subjected to nearly complete desiccation have shallow-water or sabkha-like evaporites on their floors. These are commonly dominated by brine-pan halite and may contain potash salts. In deep basins where brine levels were depressed more than a few tens of metres below sea level, it will have been impossible for dense brines to reflux out of the basin. The absence, or paucity of the more soluble salts in their fills is therefore puzzling. It may be that the concentrated brines (which constitute only a tiny fraction of the original seawater volume) can be stored within the porosity of halites, to be flushed out of the basin later when the basins are refilled with marine waters.

Because evaporite deposition occurs in nearly isolated basins, the elevation, relative to sea level, of the lowest points on the basin rim is usually more important than the elevation of the depositional site. Only where sea level falls below the depositional surface and there is no influx (unlikely in very deep desiccated evaporite basins) will evaporite deposition cease. Basin-wide beds of fine-grained (aeolian ?) clastics in some desiccated deep-basin evaporites may record such events.

For basins with relatively high rims, even slight falls of sea level, may allow basins to become disconnected from the sea, desiccate, and accumulate evaporites. On the other hand, if sea-level rises surmount basin rims, drowning basins with seawater, evaporite deposition is abruptly terminated. An evaporite basin that was especially prone to the effects of sea-level change is the Late Carboniferous (Pennsylvanian) Paradox Basin of the Four Corners Region, western USA. Repeated sea-level changes produced a markedly cyclic sequence of basinal evaporites, although their interpretation is contested.

Evaporites of the *Paradox Formation* accumulated in a rapidly subsiding asymmetric evaporite basin (Fig. 8.52). Its deepest part was an elongate trough partially filled by a thick wedge of arkose derived from the rising Uncompaghre Uplift to the northeast that passed basinwards into a 2-km-thick sequence, mostly of evaporites (Fig. 8.52b). The latter consists of more than 40 cycles (Fig. 8.53), 33 containing halite and 19 also

with sylvite or carnallite. Halites are separated by thinner interbeds of anhydrite, silty dolomite and black, organic-rich 'shale'. The number of evaporite cycles decrease southwards as lower ones onlap shelfal carbonates or pass into shelfal carbonate–shale cycles (Fig. 8.53).

An idealized Paradox basinal evaporite cycle consists of: (i) a disconformity on top of halite; (ii) laminated and nodular anhydrite; (iii) silty dolomite; (iv) black, laminated, sapropelic, silty calcareous shale, with a high gamma-ray log response; (v) dolomite; (vi) anhydrite; and (vii) halite, with potash salts in some cycles, terminated by another dissolution disconformity (Peterson & Hite, 1969; Hite, 1970; Raup & Hite, 1992). The black shales of the basin are traceable shelfward into equivalent carbonate cycles (Fig. 8.53). An idealized Paradox shelf cycle consists of a black 'shale' (an argillaceous sapropelic dolomite, equivalent to the shale unit of basinal cycles) overlain by a shallowing-upward sequence of shelfal carbonates. Evidence for subaerial exposure occurs atop the carbonates.

Evaporite cycles are responses to decreasing and increasing salinities in the basin. These changes were initiated by alternating periods of relative rise and fall of sea level which affected, in one interpretation (Peterson & Hite, 1969; Hite, 1970; Raup & Hite, 1992) a restricted, barred evaporite basin, across which a strong salinity gradient occurred. This gradient allowed contemporaneous deposition of carbonates,

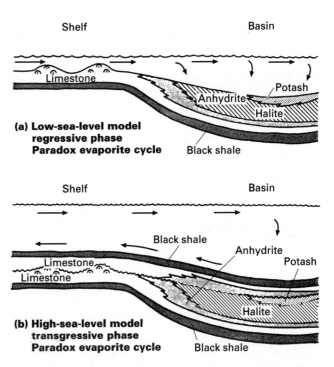

Figure 8.54 Interpretation of Paradox depositional cycles (from Hite, 1970).

sulphates, halite and potash salts (Fig. 8.54). The upward change from anhydrite to dolomite and then to black 'shale' was attributed to increasing water depths as well as to decreasing salinities.

More recent work (Kendall, 1988, unpubl.; Williams-Stroud, 1994) has questioned the deepwater, barred basin setting of the Paradox evaporites, suggesting the strong salinity gradient within the basin was implausible (Sect. 8.2.3). In addition, the evaporites display features characteristic of shallow-water or subaerial environments, rather than deeper-water settings. Anhydrites beneath and within halites contain anhydrite–halite pseudomorphs after vertically oriented gypsum crystals (Raup & Hite, 1992; Williams-Stroud, 1994) that suggest deposition occurred from very shallow perennial brine ponds. Euhedral halite crystals draped by fine-grained anhydrite laminae (see Fig. 8.34) characterize lower parts of Paradox halites. Precipita-

tion occurred from brines sufficiently deep to prevent dissolution affecting the halite crusts. These features are absent in upper parts of Paradox halites where anhydrite laminae have flat or irregular contacts with underlying halite. The anhydrite laminae originally covered dissolution surfaces that truncated the now-recrystallized halite crusts. Sylvite also fills former dissolution cavities. These dissolution features indicate deposition occurred from much shallower brines, perhaps within brine pans. Anhydrites above halites are laminar and nodular, some displacive and sabkha-like, others replacing large gypsum crystals. The overlying dolomites and anhydritic dolomites contain sedimentary structures indicating deposition on desiccated but occasionally flooded mudflats: desiccation cracking (sometimes so intense as to be mistaken for bioturbation; see Fig. 8.56; Williams-Stroud, 1994), sand-patch fabric, centimetre-bedding with erosive bases (flood

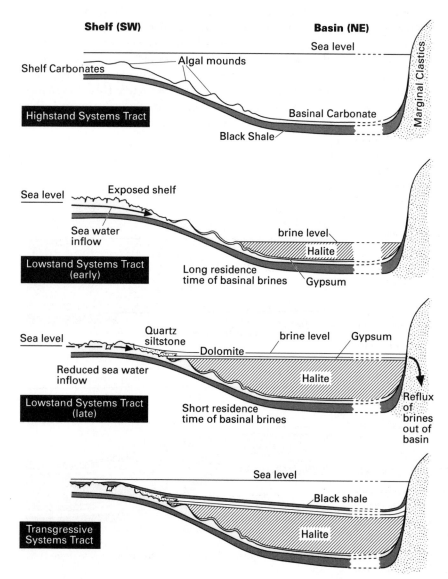

Figure 8.55 Simplified depositional model for the development of depositional cycles in Paradox Basin illustrating the cyclic and reciprocal style of shelf vs. basin accumulation and the relationship between degree of brine-reflux through marginal clastics and the solubility of the evaporites precipitated (based on Kendall, 1988 and Goldhammer, Oswald & Dunn, 1994).

Figure 8.56 Knife-sharp contact between desiccation-cracked, silty dolomites and overlying laminated and *Zoophycus*-burrowed black 'shale', Akah–Desert Creek boundary (see Fig. 8.52), Paradox Formation. The contact records the abrupt drowning of the uppermost Akah salina and conversion of the area into a deep euxinic basin.

deposits), anhydrite pseudomorphs of small, displacive lenticular gypsum crystals *in situ* or reworked into clastic laminae, and halite crystal moulds casted in overlying sediment (Kendall, 1988, 1992). Except for the salinity of their brines, carbonate deposition occurred in mudflat and shallow subaqueous environments that were essentially the same as those of the underlying anhydrites and halites.

The characteristics of the Paradox evaporites suggest deposition on the floor of a deep but desiccated basin (Kendall, 1988). Evaporite precipitation occurred entirely after carbonate deposition had ceased. In this interpretation the black 'shales' are the deepest-water sediments of cycles and their upward passage into carbonates is attributed to a sea-level fall. However, during the latter part of the sea-level fall the surrounding carbonate shelf was exposed and acted as a topographic barrier that greatly restricted influx of seawater into the basin. Evaporation of basin waters caused: (i) drawdown and partial to complete basin desiccation; (ii) brine concentration; and eventually (iii) evaporite deposition; initially in

the basin centre but later, onlapping the basin flanks. The black 'shales' of the basin centre acted as aquitards, allowing waters to be retained in the basin for sufficiently long periods for evaporation to concentrate them to halite saturated brines (Fig. 8.55). As deposition extended on to the area of arkosic sediments flanking the Uncompahgre Uplift (Fig. 8.52), however, loss of dense brines through the permeable sands became possible. Increasing brine losses progressively reduced the time that waters spent being evaporated and their salinities became progressively lower (Fig. 8.55). Eventually, they became undersaturated with respect to halite; earlier precipitated halite began to dissolve (producing a dissolution contact) and only gypsum (now anhydrite) could accumulate. With further brine dilution, only evaporitic dolomites were deposited. The different Paradox evaporites and evaporitic dolomites accumulated in essentially the same desiccated environment but from brines of differing salinity. The vertical facies changes within the evaporites are not the product of increasing water depth. Evaporite deposition terminated when, during a relative sea-level rise, the Paradox shelf was drowned and the desiccated Paradox basin flooded. These drowning events produced abrupt contacts between desiccation-cracked silty dolomites and overlying euxinic laminated black 'shales' (Fig. 8.56). Cycle boundaries are best placed at these dolomite-black 'shale' boundaries rather than at the dissolution surfaces atop halites. The latter merely record partial dissolution of the uppermost halites as basin brines became less saline.

The position of the most important sequence stratigraphic boundaries in basins with basin-central evaporites and flanking carbonate shelves, such as the Zechstein and Paradox basins, is discussed by Tucker (1991). He proposes that in the basins they occur between evaporites and *underlying* carbonates. Previously cycle boundaries have been positioned between evaporites and *overlying* carbonates. Tucker's argument is that carbonate–evaporite boundaries are traceable onto basin flanks where they correspond to major sedimentary breaks. The shelf disconformities correspond with episodes of basin desiccation. This interpretation is the view from the carbonate shelf. In evaporite basins, exemplified by the Paradox Basin, the most significant contacts may occur above the evaporites or evaporitic carbonates. They record the reflooding of basins. They are abrupt and commonly juxtapose sabkha-like sediments with deepwater (and in the case of the Mediterranean, even oceanic) sediments. They also are traceable shelfward to the same major sedimentary breaks that Tucker (1991) identifies as the most important.

Further reading

James N.P. & Kendall A.C. (1992) Introduction to carbonate and evaporite facies models. In: *Facies Models, Response to Sea Level Change* (Ed. by R.G. Walker and N.P. James), pp. 265–275. Geol. Assoc. Can., Waterloo, Ontario.

Kendall A.C. (1992) Evaporites. In: *Facies Models, Response to Sea Level Change* (Ed. by R.G. Walker and N.P. James), pp. 375–409. Geol. Assoc. Can., Waterloo, Ontario.

Logan B.W. (1987) *The Lake MacLeod Evaporite Basin, Western Australia*, 140 pp. *Mem. Am. Ass. petrol. Geol.*, 44, Tulsa.

Melvin J.L. (Ed.) (1991) *Evaporites, Petroleum and Mineral Resources*, 556 pp. Elsevier, Amsterdam.

Schreiber B.C. (Ed.) (1988) *Evaporites and Hydrocarbons*, 475 pp. Columbia University Press, New York.

Warren J.K. (1989) *Evaporite Sedimentology*, 285 pp. Prentice Hall, Englewood Cliffs.

Shallow-water carbonate environments

9

V.P. Wright & T.P. Burchette

9.1 Introduction

9.1.1 History of research

The study of carbonate rocks has both academic and economic importance. Not only do limestones and dolostones comprise large parts of the stratigraphic record, so that their study helps us understand the history of the Earth, but they are also the hosts to extensive ore deposits, are useful as building materials, and they act as reservoirs for at least 40% of the world's oil and gas reserves.

Carbonate rocks have generated serious scientific interest since the time of Henry Clifton Sorby in the mid-nineteenth century (Folk, 1973). The major stimulus, however, for the study of carbonate depositional systems and facies, and the one that has facilitated the most significant progress, has come more recently from the petroleum industry. Classic studies by oil company staff on the modern carbonate areas such as the Bahamas and Florida (e.g. Ginsburg, 1953, 1956; Illing, 1954; Newell, Imbrie et al., 1959) provided major stimuli to thought and enabled geologists investigating ancient carbonate rocks to comprehend and interpret facies relationships in the rock record. Further studies of the Bahama Banks (e.g. Purdy, 1961, 1963a,b) were followed by publications on other carbonate provinces such as the Persian (Arabian) Gulf, northwest Yucatan, Belize and Shark Bay.

After this period when the large-scale facies patterns for individual carbonate provinces were documented, emphasis switched to refining individual facies models. The drive for much of this detailed work, like that of the regional studies, was for the better recognition and understanding of reservoir-prone facies assemblages. The peritidal model, with or without sabkha evaporites (see Sect. 8.4) was developed, based especially on Andros Island (Shinn, Lloyd & Ginsburg, 1969; Hardie, 1977). Such models were readily applied to the geological record (see examples in Ginsburg, 1975). Platform interiors were also extensively studied, particularly those with distinctive mud mounds which might be analogues for late Palaeozoic reef mounds. Important studies of platform interiors were carried out on the Florida shelf (Enos & Perkins, 1977), Shark Bay (Logan, Davies et al., 1970; Logan, Read et al., 1974) and Belize (Wantland & Pusey, 1975). Models for ancient oolites were developed from the study of the extensive platform margin oolite shoals of the Bahamas, although they are probably not directly analogous to many ancient oolites (Ball, 1967; Hine, 1977; P.M. Harris, 1979). Interest continued to move towards the platform margin culminating in the seminal studies of Bahamian margins by Mullins and coworkers (e.g. Mullins, Heath et al., 1984),

325

and finally to the carbonate apron facies models (Mullins & Cook, 1986).

Concurrent with these developments was the refinement of reef models and the growing appreciation of the high degree of variability due to biological changes through time (Heckel, 1974; James, 1983).

Studies of ancient carbonates paralleled the development of better defined modern systems. However, the great diversity of ancient carbonate deposystems, especially those differing from the simple Florida–Bahamian model, required new concepts. Ahr (1973) stressed the importance of ramps as a viable model for many (most?) ancient carbonate sequences, but it took some years before this model became widely used. J.L. Wilson (1975) provided a source of ideas and information on the range of ancient carbonate deposystems which has proved invaluable to a generation of sedimentologists.

The development of improved, genetic classifications of limestones by Folk (1959), Leighton and Pendexter (1962) and Dunham (1962) further facilitated describing and interpreting limestones. The awareness of the crucial importance of understanding diagenetic features and the overprinting of primary depositional fabrics by secondary ones was a crucial development. The seminal work of Bathurst, culminating in his treatise of carbonate diagenesis (Bathurst, 1975) marked a major stage in the study of limestones. Since then our understanding of diagenetic processes has improved with the advent of cathodo-luminescence and electron microscopy, and with sophisticated geochemical techniques.

The interest in large-scale patterns continued and with the increasing demands for applied sedimentology to relate to the seismic and sequence stratigraphic scale, many studies in the 1980s looked at the very large scale growth of platforms, both at outcrop (e.g. Bosellini, 1984) and from active platforms (Eberli & Ginsburg, 1989). However, the development of small-scale facies models, the essential tools for facies sedimentology, all but ceased and few, if any, refinements have taken place in many of the simple models developed through the 1960s and 1970s.

The next step forward has been the development of simple computer simulations of both actual sequences (e.g. Osleger & Read, 1991; Goldhammer, Oswald & Dunn, 1991) and of whole platforms (e.g. Aurell, Bosence & Waltham, 1995). These developments offer opportunities to conduct 'experiments' on the controls on sequence and platform growth.

The most recent innovations have related to the increased awareness of temperate or even near-polar seas, first stressed by Lees (1975) as sites of extensive carbonate deposition (e.g. Nelson, 1988; Boreen & James, 1993). Such new models are being applied to the geological record (e.g. Lavoie, 1995).

This chapter reviews carbonate deposystems. Modern systems are integrated with ancient analogues to provide the widest possible range of 'models', together with a review of the factors controlling them, to aid in interpreting ancient sequences. Every deposystem is a unique process domain, and carbonate systems are extensively influenced by biological processes. The profound changes which have taken place in the biotas of carbonate-forming environments through time mean that a strict uniformitarian approach to ancient carbonate sequences has severe limitations.

9.1.2 The role of organisms in carbonate systems

Under optimal conditions, organisms that produce rigid calcareous skeletons may generate extensive framework-reefs. These form the substrate for a host of calcareous plants (e.g. red and green algae) and animals with calcareous shells or skeletons that contribute directly to the particulate component of the sediment and may be reworked as carbonate sands. Other organisms may induce carbonate precipitation due to metabolic processes, although this appears to be more prevalent in non-marine carbonate environments (see Sect. 4.7.1).

Shallow-water carbonate sediments are modified, postdepositionally, by a wide range of non-calcareous organisms, the effects of which are to break down, redistribute, or retexture the sediment. Carbonate substrates can be weakened or destroyed by a range of organisms including microsponges, algae, fungi, and grazing macro-organisms that bore, rasp or ingest and break down carbonate rock. Such bio-eroders are a critical factor in the destruction of reef frameworks. Deposit feeding and grazing organisms commonly aggregate carbonate mud into faecal pellets, another prominent component of carbonate deposits. Burrowing organisms are important in mixing sediments and changing the original depositional fabric.

Algal or cyanobacterial mats and sea grasses are important as both sediment stabilizers and trappers. Cyanobacterial mats can stabilize extensive areas of the subtidal sea floor and may form stromatolites in some modern shallow marine environments. In the Precambrian and earliest Palaeozoic, when grazing and burrowing organisms were absent, subtidal stromatolites dominated carbonate depositional systems. Throughout much of the Cenozoic, sea grasses too have played a major role in the trapping, stabilization and production of carbonate sediments. Mangroves have similarly influenced sediment accumulation since the Miocene through the baffling action of their roots.

9.1.3 Major components of carbonate sediments

Carbonate sediments and limestones consist of six main components: (i) biogenic carbonate (as detrital *bioclastic* material or as non-detrital, *in situ* calcareous organisms); (ii) peloids; (iii) coated grains (e.g. ooids); (iv) aggregates; (v) lithified and reworked sediment clasts (litho- and intraclasts); and (vi) matrix (mud-grade carbonate). Such primary sedimentary materials can be modified by synsedimentary diagenetic processes.

Biogenic carbonates are the main components in most limestones and consist of the remains of calcareous protozoans, metazoans and plants. This calcareous material is broken down by physical, chemical and biological processes with each kind of skeletal or calcareous plant material behaving differently. Most of this biogenic material ends up as disarticulated, abraded and fragmented detrital bioclasts but some, especially the larger skeletons of colonial organisms or calcareous algae, can remain *in situ*. Such material, commonly encrusted by other organisms or cemented by carbonate cement, forms the framework of some types of reef (Sect. 9.4.5). As a consequence of biotic diversity, evolution and extinction, the variety of biogenic material is enormous. Each stage of the geological column has a different biota. Many of these organisms are palaeoenvironmental indicators, and biogenic grains are among the most useful indicators of depth and energy conditions.

Peloids (synonymous with pelloids) are sand-sized grains of mud-grade carbonate, and are polygenetic. Some represent lithified faecal pellets of mud ingestors – that is, are pellets. Others, perhaps the most common group, represent *micritized* bioclasts or ooids that have been infested by boring micro-organisms (endoliths), such as cyanobacteria and fungi. The microbores, the diameters and lengths of which are measured in microns, became filled by very fine-grained carbonate (micrite) resulting in the progressive replacement of the bioclast by micrite. Peloids also form by the fragmentation of finely crystalline calcareous algae, and the breakdown of lithified carbonate mud. Peloids are not diagnostic of any particular environment but are the main components of very shallow-water subtidal sediments, especially lagoonal sands and muds.

Coated grains are also polygenetic and include both abiogenic and biogenic varieties (Tucker & Wright, 1990). Encrusting algae and micro-organisms play a role in the formation of some forms, such as oncoids and rhodoliths, but by far the most important type are ooids (synonymous with ooliths). These are sand-sized grains with distinctive concentric coats of carbonate typically coating a shell fragment or quartz grain. Their microstructures have been the focus of considerable research (Tucker & Wright, 1990). Ooids typically form in very shallow, warm, agitated waters, saturated or supersaturated with respect to calcium carbonate. Their concentric coats reflect episodic accretionary growth and they form in areas where both wave and tidal action move grains (Sect. 9.4.2.). The tidal regime facilitates the retention of the grain in the ooid-precipitating environment. Ooids, while mainly forming in waters a few metres or less in depth, can be transported considerable distances. Some of the thickest oolitic limestones in the geological record are resedimented turbidites (Sect. 9.4.4). As with peloids, coated grains, as a group, are not diagnostic of a particular environment.

Aggregates are sand-sized particles that have been agglutinated to form compound grains. The process involves microbial binding and is commonly found associated with mucilaginous biofilms with cyanobacteria, algae and fungi. Such films or mats are found in lower energy, shallow-water settings, such as protected bays and lagoons. These micro-organisms bind, encrust and typically bore into the particles, resulting in micritization. One of the most common types of aggregate are micritized ooids called *grapestones* (Fabricius, 1977). Grapestones occur in sheltered areas behind oolite shoals into which the ooids have been transported by storms and stabilized by the microbial mats (Sect. 9.3.2).

Lithoclasts are recognizable clasts of a lithified, pre-existing carbonate sediment dissimilar to its host sediment or to sediments associated with its host. The simplest case is where clasts of a limestone of say Carboniferous age, occur in a Jurassic limestone. *Intraclasts* are clasts composed of sediments which are represented either in the host sediment or in associated sediments. A simple example would be a clast of oolitic limestone in an oolitic host. Such intraclasts are relatively common and are a consequence of early lithification (cementation). Surfaces formed by such early, subaqueous cementation are termed *hardgrounds*. Another common example consists of flakes of carbonate mud produced on desiccated mudflats. Such flake-breccias are common in peritidal deposits (Sect. 9.4.1).

Carbonate mud is a major component of carbonate sediments and limestones and is also polygenetic. Calcareous algae in shallow waters, particularly green algae such as *Penicillus*, are capable of producing vast quantities of aragonite mud as their calcified thalli disintegrate on death. In present-day areas the potential for such production of mud is enormous but there is growing evidence that much is derived by direct precipitation (Shinn, Steinen *et al.*, 1989; Milliman, Freile *et al.*, 1993). Mud derived from green algae and direct precipitation is made of aragonite but calcitic mud is also produced as a result of the bioerosion of calcitic material (Adjas, Masse & Montaggioni, 1990; Reid, Macintyre & Post, 1992). In deeper-water settings, calcitic mud is produced largely by the fragmentation of calcitic planktic Foraminifera and coccolithophores.

9.2 Controls on carbonate production and sedimentation

The most important controls on carbonate sediment production are *temperature*, *salinity* and *light intensity*: these determine the type and abundance of carbonate-producing organisms (Lees, 1975) and whether or not carbonate is likely to be precipitated inorganically. Carbonate production is also influenced, on both regional and local scales, by factors such as tectonic setting, behaviour of relative sea level, climatic stability, degree of turbidity, nature of the substrate, nutrient flux, and wave/current regimes. The absence of significant terrigenous sediment input is also a factor. The major sites of tropical-type carbonate deposition at present , the Bahamas–Caribbean–Yucatan area, the Arabian Gulf, Western Australia, Indonesia and the Pacific atolls, are largely isolated from siliciclastic input or occur in arid areas.

However, the main controls on carbonate depositional systems and carbonate platform geometries are climate and water depth.

Climate controls water temperature and salinity and the wave energy of the environment, factors that influence the rate of carbonate production and its distribution. It therefore has a major influence on the character of a carbonate deposystem.

Modern carbonate sediments occur in both high and low latitudes, but production is greatest in warm waters between latitudes 30°N and 30°S. This was also broadly true for most ancient carbonate deposits, although climatic belts have varied in extent during time. Large carbonate bodies are thus restricted to those areas of the world that were tropical or subtropical at the time of deposition. However, while climate is broadly latitude related, its expression is also complicated by the presence of landmasses (e.g. microclimatic effects) and whether the environment is isolated or attached to a land area. Because most carbonate sediments form in shallow subtropical waters they are also particularly strongly influenced by storms and hurricanes.

In low latitudes, where shallow-water temperatures are consistently over 20°C, and salinities 32–40‰, carbonate sediment producers are diverse, comprising calcareous green algae (e.g. *Halimeda*), hermatypic, symbiont-bearing corals, and molluscs. This group of sediment producers constitutes the *chlorozoan* association of Lees (1975). However, many of the most prolific carbonate-producing organisms cannot tolerate high salinites. Where this occurs, biotas may become impoverished and production of carbonate sediment is reduced. Corals do not survive at elevated salinities whereas green algae continue to produce sediment forming the *chloralgal* association (Lees, 1975). Enhanced temperatures and salinities, however, may result in supersaturation of sea water with respect to calcium carbonate so that ooids and certain mud-grade carbonates may form by precipitation.

Where ambient temperatures commonly fall below 20°C, and particularly below 15°C, organisms such as the hermatypic corals and green algae are absent, leaving the skeletal biota dominated by bryozoans, Foraminifera, coralline red algae and molluscs, constituting the *foramol* association (Lees, 1975). Inorganic carbonate precipitation is also significantly reduced in such settings. Long-term changes in climate, caused by continental drift or other global effects, may thus have a profound effect on the character of a carbonate depositional system. The phenomenon is also well illustrated by Cenozoic sediments of the eastern and southern Australia shelves (Davies, Symonds *et al.*, 1989; James & Bone, 1991).

Water depth influences light intensity, which depends also on the turbidity of the water column. Many key carbonate-producing organisms, particularly Cenozoic forms, are or were light dependent. Some rely solely on light as the main energy source (phototrophs, such as calcareous algae) or rely partly on photosynthetic symbiots (i.e. mixotrophs such as corals and some Foraminifera). Both these types typically require high light intensities, as also do sea grasses, which indirectly produce

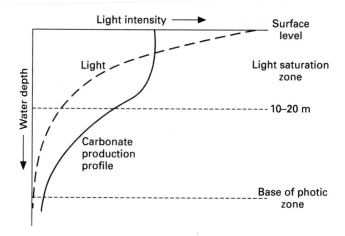

Figure 9.1 Organic production, light and water depth (after Bosscher & Schlager, 1992).

carbonate. Light intensity decreases exponentially with depth (Fig. 9.1) and carbonate production is a function of light intensity. In the top 10–20 m, the 'light saturation zone' of Bosscher and Schlager (1992), light is not the limiting factor and little change in organic production with depth takes place. Beyond this zone there is a rapid decrease in organic growth, with most carbonate production taking place in the top 120 m of the water column.

Oceanographic controls include both the ambient and exceptional tide and wave regimes which influence the depositional environment. These are related to factors such as shelf width, area of adjacent ocean (fetch), and the carbonate platform orientation with respect to dominant winds. Particularly important is the extent of upwelling in the vicinity of the carbonate environment since this can introduce cold, nutrient-rich oceanic bottom waters into the shallow marine setting. Most large modern coral reefs grow in relatively nutrient-deficient waters and elevated nutrient levels can result in corals being displaced, through competition, by fast-growing non-calcareous algae or sponges (Hallock & Schlager, 1986), or poisoned by periodic 'blooms' of micro-organisms. High nutrient areas of the southwestern Pacific, are characterized by the widespread development of *Halimeda* banks (Roberts & Phipps, 1988). Identifying the effects of oceanographically controlled nutrient supply in ancient successions is difficult but its likely importance in controlling sediment production in ancient systems has been emphasized (R. Wood, 1993; V.P. Wright, 1994a). Flooding of exposed carbonate platforms may also lead to local reworking of nutrients from soils and may be responsible for the low productivity of some carbonate systems during transgressions.

9.2.1 Variations in carbonate production and accretion

Carbonate production rates in shallow, clear tropical waters may be prodigious (Fig. 9.2), but vary due to a wide range of physical factors. Production rates may also have varied through geological time due to changes as carbonate-producing organisms have evolved or been replaced (Fig. 9.3) (Heckel, 1974; James, 1983). Carbonate production rates in modern environments are representative only of the late Cenozoic (?Miocene–Holocene). In the early Cenozoic, in contrast, coral reefs were scarce and larger benthic forams were the dominant shallower-water carbonate producers, especially in the deeper photic zone. In the late Mesozoic, rudist bivalves were major sediment contributors at platform margins (Ross & Skelton, 1993). Carbonate production profiles during the Palaeozoic were probably different to those of today, with significantly more carbonate production in intermediate water depths (Fig. 9.4) (Sect. 9.5). Late Cenozoic systems were dominated by phototrophs and mixotrophs such as corals, calcareous algae and sea-grass epibionts including symbiotic foraminifers (Fig. 9.4a). In contrast, during the Mississippian (Lower Carboniferous) (Fig. 9.4b) sediment production was mainly by heterotrophic suspension feeders such as crinoids, brachiopods and bryozoans. Cyanobacteria may have played a role in forming deeper-water mud mounds (Sect. 9.4.5) but such mounds were not major sediment exporters. The differences

in the productivity profiles which resulted from shifts in the nature and depth location of these sediment producers influenced the dynamics of carbonate platform growth, with progradation rather than aggradation being favoured by Mississippian platforms.

Short-term carbonate production rates in Recent environments average 1 m ky^{-1} (Schlager, 1981) (Fig. 9.2) and individual coral growth rates can be much higher. In temperate systems, carbonate production rates are generally much slower, ranging from 0.1 to 0.01 m ky^{-1}, although some warm-temperate sea-grass banks have production rates comparable to those of some tropical systems (James & Bone, 1991; Bosscher & Schlager, 1992) (Fig. 9.2).

In discussions of carbonate sedimentation, it is important to distinguish between the rates of *production* and *accumulation*. Long-term rates of carbonate sediment accumulation – that is, carbonate platform growth – are much lower than suggested by measurements of carbonate production made in modern seas (Fig. 9.2). Accumulation rates for ancient carbonates appear, superficially, to be closer to those of Holocene to Cenozoic cool-water systems, rather than those of warm-water regimes. However, the latter are expressed as short-term production rates, and are typically two or more orders of magnitude greater

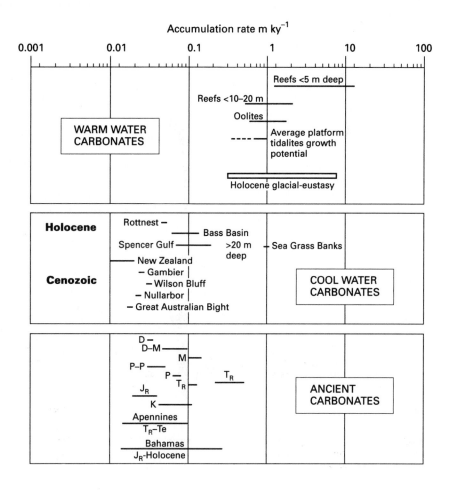

Figure 9.2 Calculated accumulation rates of modern and ancient carbonate systems (modified from James & Bone, 1991). The warmwater carbonate field refers to short-term rates, calculated by Schlager (1981) for shallow-water systems. The coolwater values include deeper-water settings. The values for ancient carbonates reflect longer-term rates which for all geological phenomena are much lower than for short-term rates. Superficially, the values from ancient systems resemble those of coolwater systems but the data sets are not directly comparable.

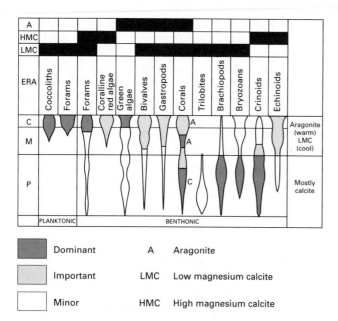

	Coccoliths	Forams	Forams	Coralline red algae	Green algae	Bivalves	Gastropods	Corals	Trilobites	Brachiopods	Bryozoans	Crinoids	Echinoids	
A														Aragonite (warm)
HMC														LMC (cool)
LMC														
ERA														
C								A						
M								A						
P								C						Mostly calcite
	PLANKTONIC				BENTHONIC									

▨	Dominant	A	Aragonite
▨	Important	LMC	Low magnesium calcite
☐	Minor	HMC	High magnesium calcite

Figure 9.3 Main plant and animal groups through the Phanerozoic (P, Palaeozoic; M, Mesozoic; C, Cenozoic) important for carbonate production. The composition of the contributors also controls the diagenetic potential of the carbonates (modified from Wilkinson, 1982, and Jones & Desrochers, 1992).

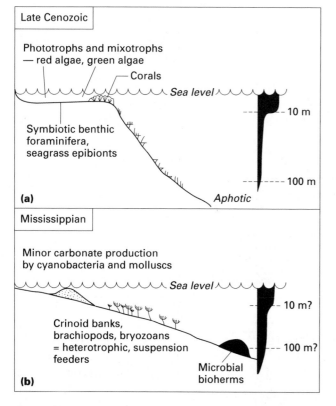

Figure 9.4 Contrasting locations and contributions made by different organisms in platform development for (a) late Cenozoic systems and (b) Mississippian systems.

than those estimated from pre-Quaternary sequences. This phenomenon reflects the incompleteness of the geological record due to multiple pauses in sedimentation, sediment removal by currents, compaction, or even periodic demise of the system due to subaerial exposure or drowning (Wilkinson, Opdyke & Algeo, 1991).

Carbonate production rates are not only depth dependent, but may vary areally within a platform. Coral growth rates, for example, are highest along windward margins of modern carbonate platforms, while the associated lagoons have slower sediment production rates. The differential production rate contributes to the raised rims and bucket-like morphology of many platforms (Kendall & Schlager, 1981).

9.2.2 Sea-level changes and carbonate production/accumulation

Carbonate platforms commonly follow a non-linear vertical growth 'profile', controlled largely by relative sea-level changes. Platforms terminated by subaerial exposure and then reflooded, commonly show a delay between submergence and the onset of strong carbonate production. It appears that a minimum water depth, probably a few metres (which may require several thousand years to be achieved as sea level rises), is necessary before the primary carbonate producers colonize the new environment and begin production. Following this lag period is the *start-up* stage (Kendall & Schlager, 1981). The lag phase lasted a few thousand years after sea levels rose in the Holocene but, following major extinction events, a longer 'recovery' effect may have occurred (Hottinger, 1989; Schlager, 1992). Under optimal conditions, as water depth and circulation increase, production increases so that sediment accumulation commonly catches up with sea level (*catch-up* stage). The system may reach its full growth potential as the environments stabilize and diversity increases as new niches develop. Once strongly established, sediment accumulation may maintain pace with rising relative sea level, although by this stage the *rate* of relative sea-level rise is commonly waning. Eventually carbonate production rate outstrips the rate of relative sea-level rise, and the carbonate platform expands in area through progradation (*keep-up* stage). Space (controlled in part by water depth) becomes limited and productivity may stabilize (Fig. 9.5) (Schlager, 1992). These 3 stages are recognizable on various scales, from metre-scale shallowing-upward cyclothems and decametre-scale reefal sequences, to whole carbonate platforms (Kendall & Schlager, 1981; Neumann & Macintyre, 1985). Holocene platforms, for example, often show an initial low sediment production rate which increases rapidly until the platform achieves apparent equilibrium with relative sea level, the productivity profile following a sigmoidal curve (Fig. 9.5). These growth patterns represent one of the basic criteria used in sequence stratigraphic interpretations of carbonate platforms (Sect. 9.5).

The high rate of *in situ* sediment production in tropical

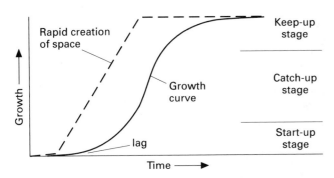

Figure 9.5 Growth curve for carbonate systems (modified from Schlager, 1992).

carbonate systems is the most important factor that distinguishes carbonate from siliciclastic depositional systems. Using rates of carbonate production in modern environments as a rough guide to rates of carbonate sediment accumulation in both modern and ancient systems, Schlager (1981) suggested that carbonate platform growth should be able to compensate, through vertical growth, for relative sea-level rises of any conceivable origin other than glacio-eustatic events. Rates of sea-level change caused by subsidence, for example, are generally within the range of 0.01–0.045 m ky^{-1} (Schlager, 1981; Algeo & Wilkinson, 1988), around an order of magnitude less than average short-term carbonate production rates. Glacio-eustatic sea-level rises, however, at 1–10 m ky^{-1}, may outpace carbonate production, particularly if combined with other factors (see below). Sea-level rises caused by intraplate stress may also cause rapid sea-level rises (Cloetingh, 1988). Rates of sediment supply in siliciclastic systems are, in contrast, generally sufficiently low for third-order rises to cause drowning.

However, in carbonate systems third-order rises are mostly too slow to cause drowning, and most high-frequency, rapid eustatic sea-level changes are generally of too low an amplitude (and may be followed by sea-level *falls* of similar magnitude) to submerge the carbonate system to significant depths. Most small-scale changes in relative sea level are represented in carbonate platforms not by significant drowning, but by variations in the nature of the shallow-water facies, creating complex sequences as discussed below. Drowning of carbonate systems does occur where large, rapid sea-level rises result from glacial ice-melt as at the end of the Pleistocene. Lower rates of relative sea-level rise can lead to drowning if carbonate production is significantly reduced by environmental stress (Schlager, 1991) (Sect. 9.3.5).

9.3 Carbonate platforms

The term carbonate platform is used both *morphologically* for a three-dimensional structure and *stratigraphically* for thick sequences of shallow-water carbonates. In distinguishing the different types of platforms problems arise because of the absence of some types at the present day, models having

to be developed intuitively from the rock record, and also because more than one morphological system can occur in an ancient succession. Yet four end-member carbonate platform morphologies can be recognized: *epeiric platforms, isolated platforms, shelves,* and *ramps* (Fig. 9.6). The term *drowned platform* is commonly used to denote a shallow-water carbonate platform which became so deeply submerged that growth ceased and it became capped by distinctive condensed facies or buried by deeper-water sediments.

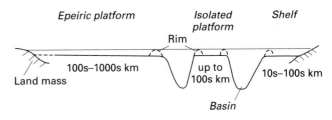

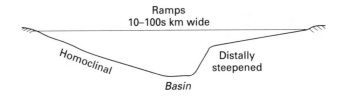

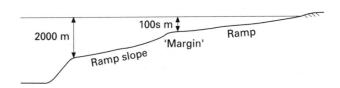

Figure 9.6 Platform types (see text).

9.3.1 Epeiric platforms

Although there are no modern examples, epeiric carbonate platforms were widespread during the Infra-Cambrian (e.g. in the Middle East), the Lower Palaeozoic (e.g. in North America), and the late Palaeozoic (e.g. the Triassic of Europe, the Permian of the Middle East). Though many so-called epeiric platforms probably represent the interiors of major shelves or ramps (see below), some were hundreds or thousands of kilometres across and covered millions of square kilometres. They developed in relatively stable cratonic interiors or on wide, flooded continental shelves. Subsidence over such large areas is seldom uniform and many ancient systems were characterized by distinctive sags or *intraplatform basins*. Many large platforms have probably developed, if small Cenozoic examples can be scaled up, by rapid progradation and lateral growth including the incorporation of smaller platforms by coalescence (Eberli & Ginsburg, 1989; Hine, Harris *et al.*, 1994).

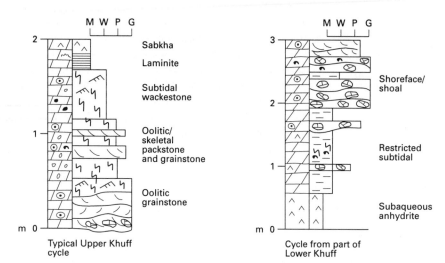

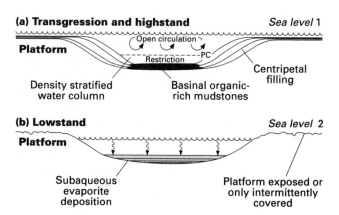

Figure 9.7 Representative intervals from parts of the late Permian Khuff Formation in the Arabian Gulf, an epeiric platform interior.

Without good modern analogues, the nature of these vast, broad shallow areas is difficult to understand. Sea-water circulation was not necessarily highly restricted and there commonly appears to have been connection to the open sea. Storms and winds strongly influenced circulation, and the huge fetches over these platforms would have caused water to pile up in some areas, leading to intermittent exposure in others. Tidal activity should have been a major process on such platforms, however, the case for significant tidal currents on ancient epeiric platforms has not yet been proved, although numerical models for tidal activity in such settings can be devised (e.g. Slingerland, 1986).

Many epeiric successions are dominated by shallowing-upwards cyclothems, representing low-energy, shallow subtidal to intertidal deposits (Sect. 9.4.1). Evidence suggests that these cyclothems formed as shorelines prograded for hundreds of kilometres or more across the platform. However, the Lower Ordovician succession of Newfoundland appears to represent deposition on and around scattered, ephemeral, low-relief islands whose shorelines prograded into the surrounding shallow subtidal areas to generate areally restricted shallowing-upwards cyclothems (Pratt & James, 1986) (Sect. 9.4.1).

A large proportion of the carbonate succession of the Arabian Gulf and surrounding areas can be characterized as epeiric complexes. For example, much of the thick Permian and Tertiary carbonate successions of this area were deposited in the interiors of extensive carbonate platforms which bordered the passive-margin of the Arabian Craton. The Permian Khuff Formation, in particular, is cyclic on several scales. This 1500-m-thick carbonate platform succession was initiated as a ramp in the late Khazanian or early Tatarian, but evolved into an epeiric platform, probably the interior of a major restricted shelf, during the later Permian and earliest Triassic. The succession consists of numerous, stacked shallowing-upwards cyclothems, each 2–10 m thick. Although there is stratigraphic variation in the character of the cycles, in the Upper Khuff most are oolite-

dominated at the base and grade upwards into laminites and sabkha anhydrites at the top (Fig. 9.7). The succession is punctuated by intervals of subaqueous evaporites and in the middle of the Khuff there is a thick, probably subaqueous, evaporitic interval, the Middle Anhydrite. Cyclothems are arranged into larger cyclothems, on several orders of scale in stacking patterns that indicate transgressive and regressive events.

Intraplatform basins in epeiric platform settings are typically rimmed by low gradient, ramp-like margins and water depths rarely seem to have exceeded 100–200 m. They are surrounded by extensive, shallow-water areas. During sea-level falls they may become sufficiently restricted to lead to evaporite pre-

Figure 9.8 Intraplatformal basins. (a) During sea-level highstand the basin becomes density stratified with the development of cyclic suboxic or anoxic basin centre sediments. These commonly correspond with small cyclic sequences in shallower waters. The margins of the shallow basin typically centripetally prograde during this stage. (b) During sea-level lowstands drawdown may lead to isolation of the basin and evaporite deposition; the surrounding platform will be exposed. PC = pycnocline. (After Burchette & Wright, 1992.)

cipitation. In some cases anoxic bottom waters developed during highstand or transgressive phases (Fig. 9.8). Organic-rich deposits may form in this setting and examples of such deposits are known from a number of hydrocarbon provinces, particularly in the Jurassic and Cretaceous of the Middle East (Droste, 1990; Burchette, 1993).

9.3.2 Isolated platforms

Isolated, or 'unattached' carbonate platforms are morphologically variable. They range in size from a few to, more rarely, hundreds of kilometres across and the larger platforms may build sediment piles hundreds of metres thick. The margins of isolated platforms may be ramp-like, but, more typically, are steeper than 15°. Larger isolated platforms, such as the Great Bahama Bank, have flat tops where sediment is deposited in less than 10 m of water. The platform margin may be rimmed either by reefs or by skeletal or oolitic sand shoals while the interior is generally mud-dominated and may have restricted circulation with the open ocean. Because isolated platforms are surrounded by deeper water on all sides, their margins are strongly influenced by the orientation of dominant winds, waves, and tidal currents. Distinct windward and leeward margins can commonly be recognized.

Reefs develop preferentially along windward margins where wave and current activity are highest and burial by mobile sediment is least likely. This is clear from modern settings such as the Great Bahama Bank (Hine, Wilber & Neumann, 1981), and oceanic atolls (Guilcher, 1988), but is more difficult to establish for ancient examples. The high growth potential of reefal margins, and their widespread early cementation, means that they commonly grow slightly higher than the platform interior to create a wave-baffling rim. Marginal shoals, beaches and aeolian dunes on windward margins may become exposed and cemented to form 'cays', islands that persist for long periods and modify sedimentation within the platform. They may also act as nucleii (antecedent topography) for subsequent reef growth or shoal development.

MODERN EXAMPLE – GREAT BAHAMA BANK

The Great Bahama Bank (Fig. 9.9a) is one of a chain of 16 carbonate platforms that stretches for 1500 km from the Little Bahama Bank off Florida to the Navidad Bank near the Dominican Republic in the Caribbean Sea. Average water depth over their interiors is 3–10 m, but they are flanked by steep margins c. 200–300 m high which eventually fall away to the basin floor at depths of up to 4500 m. On most of the platforms, shallow-water carbonate deposition began in the Jurassic, and aggradation has produced limestone and dolomite packages up to 5 km thick. General reviews of the region can be found in Bathurst (1975) and Tucker and Wright (1990).

The Great Bahama Bank is the largest of the platforms, its northwestern bank portion having formed by the coalescence of three smaller platforms due to strong leeward progradation (Eberli & Ginsburg, 1987, 1989). Since the late Tertiary the platform has undergone vertical aggradation of approximately 1500 m and the leeward margin has prograded more than 25 km (Hine & Neumann, 1977; Eberli & Ginsburg, 1987), giving a ratio of lateral to vertical growth of between 10 : 1 and 17 : 1, although during shorter periods of maximum progradation this ratio can exceed 50 : 1 and even 80 : 1 (Fig. 9.10).

Over most of the flat top of the Bank, water depth averages about 7 m. Tidal activity is weak, with a range of less than 1 m at the platform margins, decreasing towards the interior. Near the ends of deeper-water embayments and between the islands, shoals and reefs, however, tidal currents are strong enough to form extensive tidal oolite and bioclastic sand shoals (Sect. 9.4.2).

The major control on background wave energy is the northeasterly trade winds, which dominate wind patterns from May to October (Fig. 9.9b). In the northern Bahamas, wind speeds are generally less than 10 knots but these become stronger to the south. Tropical cyclones, which can reach hurricane strength, regularly affect the region and strongly influence sediment distribution and dispersal, particularly along leeward margins.

The protective effect of the margin reefs, islands and shoals results in restricted water exchange between the platform and the ocean so that during the warmer summer months elevated salinities develop in the platform interior (Fig. 9.9b). Many ancient platform interiors exhibit restricted biotas, probably reflecting comparable effects. Sediment movement across the platform is predominantly towards the west. Leeward sediments therefore contain high proportions of platform-derived mud (hemipelagic mud) and grains such as peloids, grapestones and micritized skeletal grains. This transport pattern has been a key factor in reducing the amount of reefal growth on the leeward side of the platform since reefs that grew in this location were smothered by sediment following bank flooding (Hine & Neumann, 1977; Hine, Wilber & Neumann, 1981). Windward margin-derived sediments, in contrast, are grainier, consisting of bioclasts and ooids which have had only a short residence period in the lower energy platform interior.

The Great Bahama Bank (Fig. 9.9c,d) exhibits a range of environments and biofacies related to local variations in wave energy, temperature, salinity, nutrient supply, turbidity, and substrate across the platform. Subtidal sediments on the platform can be divided into two associations: (i) the platform margin; and (ii) the platform interior.

Platform margin sediments, dominated by *coralgal* and *oolite* lithofacies, form in zones affected by wave turbulence and tidal currents. The coralgal lithofacies occurs down to water depths of about 50 m, largely on the windward platform margins, and comprises reefs and grainstones of coral and coralline algal

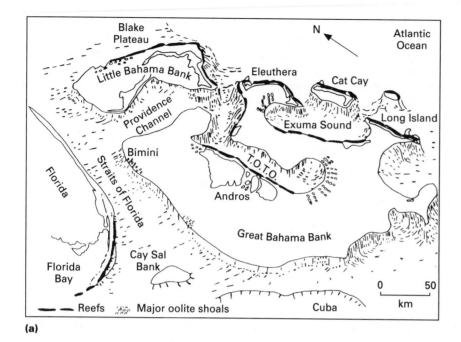

(a)

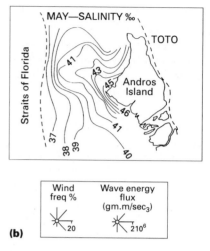

(b)

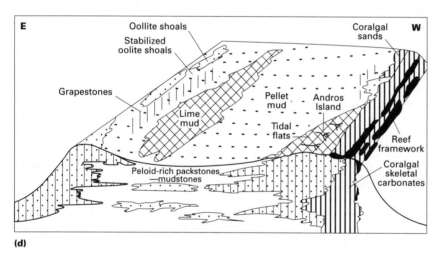

(d)

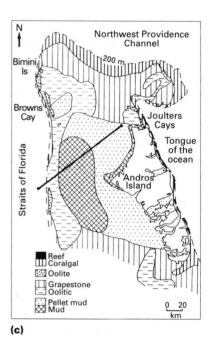

(c)

Figure 9.9 Bahama Banks. (a) Position of major banks and channels: Tongue of the Ocean (TOTO). (b) Salinity, wind frequency and wave energy for the northeastern part of the Great Bahama Bank. (c) Major lithofacies on the Great Bahama Bank. Line is position of section shown in Fig. 9.10. (d) Schematic diagram of morphology and architecture of the Great Bahama Bank ((a)–(c) from Gebelein, 1974; Tucker & Wright, 1990).

fragments, molluscs and other skeletal components. The reefs are not continuous and are dissected by channels that allow tidal exchange between the platform and the ocean and facilitate sediment distribution into the platform interior. Large areas of the platform margin, in a belt averaging 5 km wide but up to 40 km in exposed settings (e.g. northeast Great Bahama Bank), are dominated by mobile bioclastic sediment (Sect. 9.4.2). The oolite lithofacies is best developed in wave and tidally influenced areas where water depths are less than 3 m. Oolitic sandbodies occur as sandwaves and subaqueous dune fields up to 50 km in

length (Sect. 9.4.2). The export of such sediment off the platform forms an important contribution to the platform margin aprons (Sect. 9.4.4).

Platform interior sediments of the Great Bahama Bank consist of three principle sediment facies: the *grapestone, pellet mud* and *mud* facies (Fig. 9.9c,d).

The grapestone facies covers a much larger area than the active oolite sands of the platform margin, and occurs down to depths of 9 m. Because wave and current energy are slightly higher in these areas than in the rest of the platform

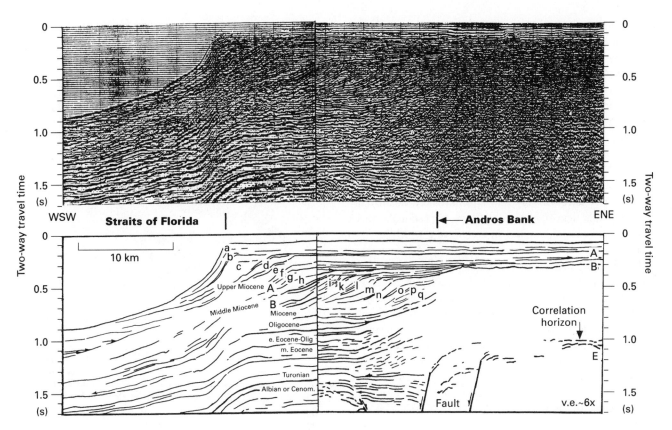

Figure 9.10 Interpreted seismic section running from just north of Andros Island (see line on Fig. 9.9c) showing progradation across the platform from mid-Oligocene to Recent times (from Eberli & Ginsburg, 1989).

interior, mud does not accumulate and grains are stabilized by mats of cyanobacteria, soft algae, or sea grasses and calcareous algae (Bathurst, 1967; Scoffin, 1970; Neumann, Gebelein & Scoffin, 1970). The calcareous algae are major sediment contributors of both sand-grade material (e.g. *Halimeda*) and carbonate mud (e.g. *Penicillus*, *Udotea* and *Rhizocephalus*). Sea grasses not only stabilize the sediment in this environment, but are also hosts for coralline algae, foraminifers and serpulids.

The pellet mud and mud facies occur in the lowest-energy, protected parts of the platform interior, in water generally less than 4 m deep. The sediments consist of highly bioturbated, aragonitic mud rich in faecal pellets (Sect. 9.1.3). Biotic diversity is lower than in the other facies, with a predominance of molluscs. Much of this mud is transported on to tidal flats around Andros Island or is flushed off the leeward side of the platform (Neumann & Land, 1975) (Sect. 9.4.1).

The Great Bahama Bank illustrates some important principles pertinent to the interpretation of ancient isolated carbonate platforms. It suggests that, although large platforms of this kind commonly develop reef or shoal rims, they are not necessarily continuously rimmed. At least for the late Pliocene to the present

day the platform architecture of the Great Bahama Bank has resembled an atoll, with an interior of peloid-rich packstones, wackestones and calcimudstones and probable rims of reefs or oolitic and skeletal sands (Beach & Ginsburg, 1980), although the view that the margins were reefal has been challenged by Mullins, Hine and Wilber (1982).

Today, subtidal deposition dominates the interior of the Great Bahama Bank, while intertidal and supratidal environments constitute only a small component of platform sediments. For platforms in the past, the relationship between subtidal and peritidal sediments in the platform interior has changed with time. During periods of high-frequency, low-amplitude relative sea-level fluctuations, many ancient platform interiors were dominated by peritidal deposits, while during periods of high-frequency, high-amplitude (i.e. glacio-eustatic) sea-level fluctuations, platform interiors were characterized by predominantly subtidal sequences.

ANCIENT EXAMPLE – THE TRIASSIC OF THE DOLOMITES, ITALY

(1) *Background*. Early- to mid-Triassic carbonate platforms in

the Dolomites of the southern Alps developed as a complex group of isolated carbonate platforms and attached shelves in a broadly extensional basin situated at the margin of the early Tethyan Ocean (Bosellini, 1984, 1991; Fig. 9.11). All the platforms were steep-sided, flat-topped structures which possessed vertical relief above the sea floor of up to 800 m and were separated in places during growth by deep, narrow seaways. Superficially, therefore, the Triassic carbonate platforms of the Dolomites possess many features in common with modern platforms of the Bahamas, although none had the same lateral extent. Those platforms, such as the Sella and the Latemar, which can be unequivocally shown to have been isolated, are seldom greater than 10 km across. Closer inspection, however, shows some basic differences in architecture which serve to illustrate the variety in carbonate platforms of different ages.

The Triassic platforms developed in two phases, in the Ladinian and the Carnian. The earlier Ladinian phase was founded upon a carbonate ramp which had become disrupted by extensional faulting during the late Anisian. Where seen in the Dolomites, the Anisian ramp facies consist of relatively deepwater highly bioturbated lime mudstones, wackestones and packstones, although the environment appears to have shallowed gradually towards the south. The tectonic event appears to have coincided with a significant relative sea-level rise, possibly driven by a strong component of subsidence, so that wide areas of the Anisian ramp were drowned. Locally, however, possibly in tectonically elevated areas, shallow-water carbonate deposition persisted or was re-established so that strong vertical growth continued through much of the Ladinian, a period of 3–4 million years. Late in the Ladinian, growth on some of the margins became strongly progradational and the fact that, in some platforms, this appears to be assymetrical suggests that palaeoslope, possibly related to the underlying extensional faults, may have played a role in controlling platform growth. The hiatus between the first and second phases of platform growth was marked by widespread platform exposure and volcanism, which partially or completely buried many platforms with coarse terrigenous volcaniclastic sediments. Carnian platforms seeded upon relief created by the Ladinian platforms, and, in the shallower basin, prograded more strongly. The characteristics of the two platform phases are, however, broadly similar.

(2) Platform margin and slope. The margins of the Dolomite Triassic carbonate platforms were abrupt and bioconstructed or grainstone facies are now not prominent at outcrop. The *in situ* high-energy facies form a narrow belt, seldom wider

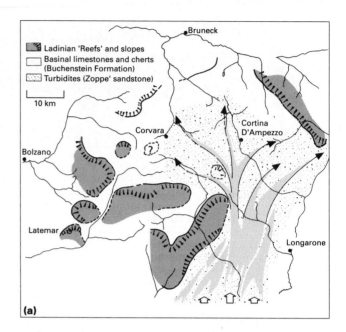

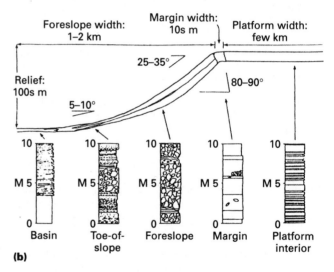

Figure 9.11 (a) Map of Dolomites region of north Italy during the Middle Triassic (Ladinian) showing the distribution of carbonate platforms (after Bosellini, 1991). (b) Facies characteristics of the Latemar platform and slope apron (from Goldhammer & Harris, 1989).

Platform interior	Bedded, open to restricted shallow subtidal and cyclic peritidal carbonates with tepèes
Platform margin	Very narrow, nearly continuous rim. Peloidal skeletal grainstones/packstones and algal (*Tubiphytes*) boundstones with sponges. Diverse biota, extensive marine cements
Foreslope	Slope of 30–35°; breccia and megabreccia sheets (clinothems) of margin-derived clasts
Toe of slope	Slope of 5–10°; breccias replaced by graded beds (<1 m thick) of peloidal, skeletal grainstones of material from platform margin and interior
Slope	Mudstones (<10 cm) graded, carbonate turbidites

than 100 m but usually much narrower, at the break in slope and die out rapidly both towards the platform and the basin (Bosellini, 1984). Debate has commonly centred around the volume of sediment which such narrow margins were able to contribute to platform growth. Although most Triassic carbonates in the Dolomites show wholesale dolomitization, examples of platform marginal facies have been preserved as calcite either in incompletely dolomitized platforms, such as the Latemar, or in 'cipit' blocks which are isolated within the impermeable basinal facies. These show the margin to have consisted largely of bioconstructed carbonates comprising a framework-building assemblage of corals, sponges, calcareous algae and Problematica such as *Tubiphytes* (Fürsich & Wendt, 1977; Brandner, Flügel & Senowbari-Daryan, 1991). A wide range of ancillary organisms, including molluscs and brachiopods, is also present. Sediments in the margin appear to have been bound by large quantities of fibrous carbonate cement which filled most inter- and intraskeletal cavities causing almost complete induration prior to burial (Harris, 1993). In this manner, most fine sediment generated at the platform margin may have been captured, which may be the reason that slope sediments comprise largely blocky material.

The slopes of the Dolomite Triassic carbonate platforms were uniformly steep, ranging from 30 to 40°. They comprise for the most part coarse breccias of platform-margin and platform-interior debris which, at the toe of slope, degrade into a turbiditic fringe containing stringers and isolated debris flows of shallow-water material. The turbiditic facies penetrates only a relatively short distance into the basin which itself was relatively starved. This suggests that the quantity of fine-grained material produced by these platforms was relatively small or that mechanisms for its redistribution were poorly developed. M.T. Harris (1994) has shown that graded grainstones at the base of slope correspond only with periods of platform submergence while the volumetrically more important breccias were deposited during episodes of both exposure and submergence. Locally within the adjacent basinal facies lie large (locally house-sized) blocks of cemented carbonate which rolled in from the platform margin (Biddle, 1980; Brandner, Flügel & Senowbari-Daryan, 1991). Despite their height and steepness, nowhere do the slopes appear to have undergone wholesale autoerosion as in the Bahamas (Sect. 9.4.4) and have collapsed only where faults were active, locally generating megabreccias at the base of slope.

The slope deposits of both Ladinian and Carnian platforms show prominent accretion surfaces (clinoforms), which can be traced almost from the flat platform surface to the base of slope. Internal downlap and onlap patterns record significant stratigraphic information. Major clinoforms appear to separate packages of prograding sediment which intercalate with basinal facies at the toe of slope. Some clinoforms are erosional surfaces and appear to represent shear planes related to slope failures (M.T. Harris, 1994). A variety of clinoform shapes have been recognized (Bosellini, 1984) which allow conclusions to be drawn about the relative rates of vertical and lateral platform growth and the rates of sedimentation in the adjacent basins.

(3) *Platform interior*. The Triassic platform interiors consist of flat-lying successions up to 800 m thick, which consist almost entirely of metre-scale peritidal cycles. Each cycle comprises a lower unit of bioturbated peloidal and skeletal wackestone or packstone and an upper, thinner unit of coarse sand- to gravel-grade oncolitic–lithoclastic grainstone in which tepee structures are locally prominent. Many cycles are capped by a thin dolomitic subaerial cap indicating periodic exposure of the whole platform surface. This platform facies has been most intensively studied and set in a sequence stratigraphic framework in the Latemar build-up (Goldhammer & Harris, 1989) where over 500 stacked cycles have been logged. Few studies have been carried out on this facies elsewhere in the Dolomites and it is unknown whether this exact pattern of cyclicity is repeated in the other build-ups.

The Latemar cyclicity is complex and, in addition to the small-scale cyclicity, several lower orders of cyclicity have been identified, interpreted broadly as the result of interaction between subsidence and eustatic sea-level changes. Within the broad 'third-order' alternation between transgressive and highstand systems tracts, there appears to be a 'fourth-order' cyclicity which has bundled the metre-scale 'fifth-order' cycles into upward-shallowing groups of five (Goldhammer, Dunn & Hardie, 1990; Hardie, Wilson & Goldhammer, 1991). In addition, the expression of individual cycles varies systematically depending on stratigraphic position, ultimately a response to the rate at which relative sea level varied during platform accretion. Thus there are zones of cycles dominated by submergent features and others dominated by emergent features. The main tepee-dominated (emergent) interval has been interpreted as occurring around a sequence boundary in the middle of the succession (Hardie, Wilson & Goldhammer, 1991), while the transgressive systems tracts are characterized by acyclic intervals or subtidally dominated cycles.

9.3.3 Shelves

A carbonate shelf is a shallow-water, land-attached platform with a pronounced break in slope into deeper water (see Fig. 9.6). The break of slope is typically a wave-agitated margin with a more or less continuous rim of reefs and/or skeletal or oolitic sand shoals (a rimmed shelf). In some cases no rim exists and the break in slope may be at depths below the main zone of wave agitation, forming *open shelves*. The rim, if present, restricts circulation in the landward area, creating a shelf lagoon. The width of such lagoons rarely exceeds 100 km. Such shelves differ from their siliciclastic counterparts in being typically flat-topped (high sediment production allows shelves to keep up with any sea-level rises) and have a distinct rim which acts as a restraining margin. The possession of this rim is a critical factor and determines whether a ramp or a flat-topped, steep-margined shelf develops. Modern rimmed shelves commonly possess a reefal margin of corals and coralline algae, but different organisms have occupied this niche in the past, for example stromatolites in the Precambrian, stromatoporoids in the early and mid-Palaeozoic, and, to a lesser extent, rudists in the Cretaceous.

The Florida Shelf is an example of a rimmed shelf for which detailed reviews are given in Bathurst (1975), Enos (1977), and Tucker and Wright (1990). The inner shelf margin is marked by the Florida Keys, an emergent Pleistocene reef–oolite-shoal complex, behind which is sheltered Florida Bay bounded on its northern margin by the coastal swamps of the Everglades. The Florida shelf margin is 5–10 km wide and rimmed by a reefal belt up to 1 km wide and 200 km long (Fig. 9.12). The shelf break occurs in water depths of 8–18 m, from which the slope descends at an angle of up to 10° until it flattens out at depths of up to 1000 m in the Florida Straits. The shelf margin complex also contains coral–algal patch reefs, sand shoals, and mud-rich zones. Mobile skeletal sands occur in a belt several kilometres wide behind the reefal rim. Along the inner shelf margin, in less turbulent areas, the sea floor is extensively colonized by sea grasses and calcareous algae. The presence of a restricted Florida Bay to the west of the Keys exerts some control on reef development since reefs are well developed only in those areas unaffected by water leaving the Bay (Fig. 9.12). A major feature of the shelf is numerous mud mounds composed of skeletal-rich muds (Bosence, Rowlands & Quine, 1985; Wanless & Tagett, 1989) (Sect. 9.4.5).

The Permian (Guadalupian) of the Delaware Basin in west Texas and southern New Mexico provides an example of an ancient reef-rimmed shelf margin whose rocks form hydrocarbon reservoirs in the subsurface and are also exposed. The Permian Basin covers an area of more than 100 000 km^2 and was almost completely encircled by a reefal margin, the Capitan and Goat Seep Reefs, which rimmed an extensive evaporitic shelf (Fig. 9.13; see Sect. 8.8; Figs 8.42 & 8.43). The massive carbonate build-up was initiated during the Wolfcampian as a low-angle ramp which later underwent lateral and vertical accretion (Mazzullo & Reid, 1989). These relationships have been linked to alternations between relative sea-level stillstands or slow rises, during which the margins rapidly prograded, and rapid rises during which margins stepped back or accreted vertically. These cycles can be correlated around the whole basin and provide a powerful predictive exploration tool (Silver &

Todd, 1969; Ward, Kendall & Harris, 1986). The deep-water basin contains both clastic and carbonate resedimented material.

On the shelf, outer lagoonal deposits form a belt up to 13 km across, dominated by widely-correlatable shallowing-upwards cycles of lagoonal dolomite mudstone and wackestone, each capped by anhydritic laminites (Ward, Kendall & Harris, 1986). There is an abrupt transition (over 200 m) into an inner lagoon and marginal coastal playa facies belt over 40 km wide, comprising intercalated dolomite mudstones and siltstones, with evaporite units up to 7 m thick; a largely subaqueous origin has been suggested for the evaporites (Sarg, 1977).

The rim to the Permian shelf comprises a complex of reefs and shoals arranged in linear belts paralleling the shelf margin. Reefal deposits consist of massive, fractured carbonates 1–200 m thick, overlying a bedded foreslope sequence up to 500 m thick (Ward, Kendall & Harris, 1986). Lack of bedding in the massive facies masks depositional slopes but associated slope deposits exhibit dips of 35°, and depositional dips in the underlying Goat Seep Reef are 25–30° (Ward, Kendall & Harris, 1986). There is an upward transition within the Capitan reefs from massive marine-cemented carbonate mudstone to massive fossiliferous carbonate containing sponges, bryozoa, brachiopods and *Tubiphytes* bound by calcareous algae and marine cements; a change in framework organisms and the character of the marine cements occurred over time and corresponds with the evolution from a low to a high angle of slope (Ward, Kendall & Harris, 1986). Most of the organisms involved in Permian reef construction were filter feeders and it seems unlikely that there was more than moderate wave turbulence. Major subsidence fractures developed parallel to the oversteepened shelf margin and were filled by internal sediments or marine cements and algae (Schmidt, 1977; Yurewicz, 1977).

Between the reefs and the shelf-lagoonal sequences, 400 m–8 km from the shelf-slope break, is a belt of carbonate sandbodies 800 m–3 km wide comprising coarsening-upwards sequences of cross-laminated to structureless fine-grained carbonate grainstones which grade upwards into subaerial pisolites. These sediments have been interpreted as barrier shoreface deposits (Assereto & Kendall, 1977). The implication, therefore, is that this Permian reefal tract grew in moderately deep water compared with modern coral reefs, possibly at around wave base (?30 m) and that the high-energy shelf-margin facies belt consisted of a linear trend of barrier islands and shoals.

The carbonate apron, from toe of slope to the shelf-break, is c. 2 km wide (Ward, Kendall & Harris, 1986). Slope and basin-margin carbonates are massive to thinly-bedded and

---

Figure 9.12 (Opposite) The Florida rimmed shelf. (a) Map of southern Florida showing the Florida Keys, a line of Pleistocene reefs and oolite shoals, the modern reef belt along the shelf break, mud banks at Tavernier and Rodriguez and Florida Bay with its network of banks and 'lakes'. (b) Bathymetry of the south Florida Shelf platform (after Ginsburg, 1956; Sellwood, 1986). (c) Sediment texture on the south Florida Shelf platform (after Ginsburg & James, 1974; Sellwood, 1986). (d) Major carbonate grain types on the south Florida Shelf platform (after Ginsburg, 1956). (e) Distribution of sediment grain size and type across the Florida Shelf and Florida Bay (after Ginsburg, 1956). (f) Simplified facies belts of the south Florida Shelf platform (after Enos, 1977; Sellwood, 1986).

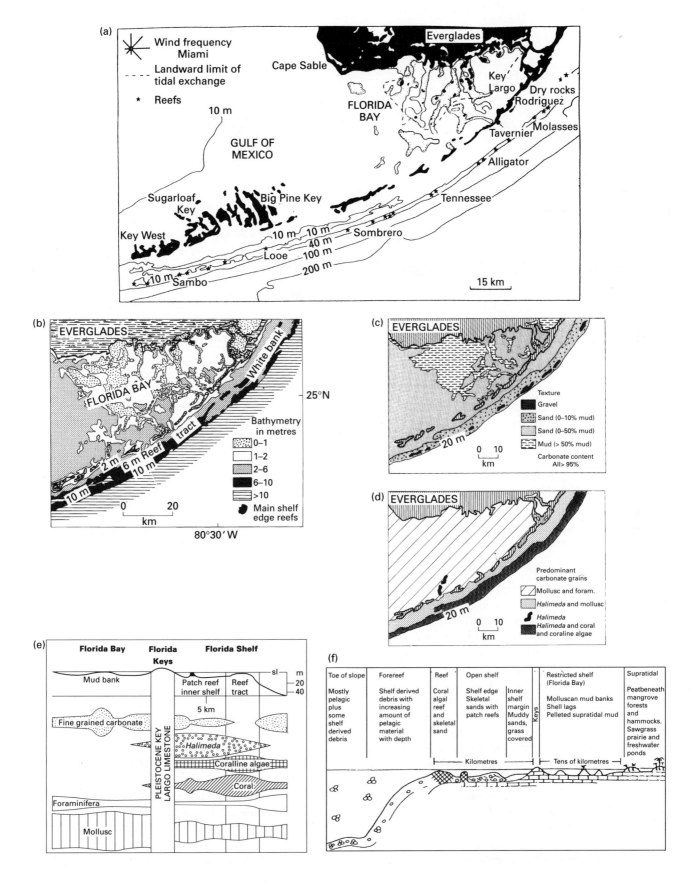

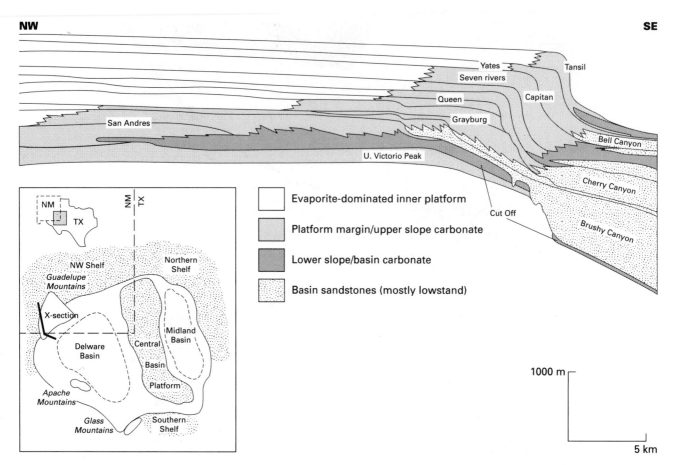

Figure 9.13 Schematic cross-section of the Permian carbonate platform in the Guadelupe Mountains of West Texas and southern New Mexico (inset maps) showing general stratigraphy and depositional facies belts. Note the initial, strongly progradational ramp-like foundation stage of the platform in the San Andres Formation and its evolution to a rimmed shelf with a steep margin several hundred metres high as accommodation space increased. Basinal sandstones were derived from material that bypassed the platform during periods of relative sea-level lowstand, mostly at sequence boundaries. The Permian platform interior was dominated by cyclic low-energy packstones that become increasingly evaporitic internally (see Figs 8.42 & 8.43) (after Kerans, Fitchen et al., 1993).

comprise slumps of shelf material, breccias and debris flows (Melim & Scholle, 1995) with local channelling. On shallower slopes, small build-ups developed downslope. Basinal clastics consist of channelized, well-sorted sandstones and siltstones that form widespread sheets which thin towards the basin margin and lap on to the carbonate slope; sandstone channels in the basin may be up to 6 km wide and the fill 35 m thick. These trend normal to the platform margin and received siliciclastic sediments which had bypassed the carbonate shelf via canyons. Channels in the platform margin were probably cut during relative sea-level lowstands or have been initiated by platform-margin collapse. Some may have been cut by saline density currents flowing from the shelf margin (Williamson, 1977). Some Permian Basin margins were erosional in origin (see Fig. 9.37). Deep-water deposits of the Permian Basin consist of both carbonate and siliciclastic sediments. The carbonates represent slope and toe-of-slope deposits, which thin away in several tongues from the reefal margin. One example, the Lamar

Limestone, thins from 90 m at the basin margin to less than 2 m several kilometres from the slope.

9.3.4 Ramps

A carbonate ramp is a gently sloping surface with a slope of less than 1°. The wave agitated zone is close to the shore, not at a shelf break some distance seaward of the shoreline, as in the case of rimmed shelves. Early studies of modern carbonate settings, such as the Bahama Banks and Florida shelf, led to the development of a simple facies model for carbonate platforms which placed a shoal or reefal belt at the platform margin and a lagoonal area behind. It soon became apparent that many ancient carbonate successions did not possess steep slopes and platform margins near the break of slope, and Ahr (1973) proposed a new model, which he termed the *carbonate ramp*. At about the same time, Evans, Murray *et al.*, (1973) provided a detailed modern analogue for such a system from the Trucial Coast of the Persian (Arabian) Gulf.

The original ramp model (Ahr, 1973) envisaged a uniform slope from shoreline to basin, later termed the *homoclinal ramp* by Read (1982, 1985). However, some ramps appear to have a slope break in deeper water and these are referred to as *distally steepened ramps* (Read, 1982).

Although ramps are known from a wide variety of tectonic settings (Burchette & Wright, 1992), the largest develop along passive margins and in foreland basins where flexural subsidence dominates. Many ramp depositional systems thus appear to have inherited their morphology from an underlying substrate. Biological and sedimentary controls, particularly the absence of well-developed reef communities, are also important in maintaining the ramp morphology. Ramps may be of long duration and some evolve into other platform types (Read, 1982). This may be due to differential sedimentation and the growth of an organic or shoal rim, or to tectonism (e.g. extensional faulting) (Gawthorpe, 1986; Burchette & Wright, 1992).

Several ramp classifications have been offered. Read (1985) recognized six types based on the character of the highest-energy facies and the distribution of shallow-water facies. Burchette and Wright (1992) have proposed a classification based on the degree of wave, tide and storm activity. However, ramps show a great deal of spatial variability in relation to energy level and features such as tidal influence. Furthermore, many ramp *successions* contain the deposits of several individual ramps, each of different character (Burchette, Wright & Faulkner, 1990).

Most ramp profiles can be divided into three parts with reference to dominant depositional processes (Fig. 9.14). The *inner ramp* is that zone above fairweather wave base where wave and current activity are almost continuous. As with siliciclastic shorelines, shoreface, beach, lagoonal and tidal flat environments may be identifiable. On high-energy ramps, this zone is characterized by sand shoals or possibly minor reefal-growth. The *mid-ramp* zone lies between fairweather wave base and storm wave base so that storm processes dominate (Sect. 9.4.3). Graded beds (tempestites) and hummocky cross-stratification are the main depositional styles. Distinct proximal to distal trends in the abundance and degree of amalgamation of storm beds can be recognized in many ancient mid-ramp sequences (Aigner, 1984; Burchette, 1987; Faulkner, 1988).

The *outer ramp* zone extends from below normal storm wave base to the basin floor. Storm-generated currents can lead to the deposition of graded units (Aigner, 1984; Calvet & Tucker, 1988), or to local erosion or reworking of sediment (Miller, 1991). In some small ramp systems, water depth may be so shallow that abundant storm beds might occur even in the outer ramp zone. Some outer-ramp environments, particularly in intraplatform basins, may have restricted circulation so that bottom waters may become suboxic or anoxic. In such settings organic-rich laminites may develop, particularly during relative sea-level rises (e.g. Droste, 1990).

The deeper parts of ramps appear, from the examples documented in the geological record, to be predominantly low-energy settings. However, in some exceptional cases, for example the Florida ramp slope, contour currents generate zones of winnowed sand at depths of 400–600 m (Mullins, Gardulski & Hinchey, 1988).

Basin-floor sediments associated with homoclinal ramps contain little allochthonous shallow-water sediment because ramp slopes are mostly too gentle (typically from 400 to 1000 : 1) to generate significant gravity flows. Although slumps and slides may occur on such slopes, they are relatively scarce, generally affecting deeper-water facies and seldom reworking inner-ramp deposits. Resedimentation by gravity flow does occur adjacent to distally steepened ramps, but in such cases the resedimented materials are typically of outer ramp lithologies unless related to sea-level lowstands.

While shallow-water reef-builders are scarce on ramps, small, isolated biological build-ups are common (Burchette & Wright, 1992). Many of these structures consist of mud-dominated mounds that initially grew in mid- to outer-ramp areas. Some were constructed largely by deeper-water biotas (e.g. some Waulsortian and other mud-mounds), while others formed in shallow water (e.g. many Cretaceous and Tertiary build-ups).

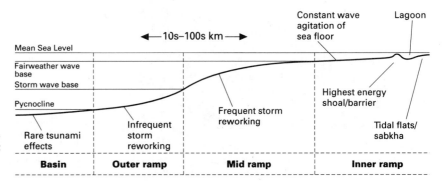

Figure 9.14 The main environmental subdivisions on a homoclinal, carbonate ramp. The pycnocline is not always originally present or identifiable in the rock record (from Burchette & Wright, 1992).

Isolated build-ups on ramps appear to develop most commonly during the transgressive stages of ramp sequences. As in other settings, their development is also favoured by areas of topographic relief. Isolated build-ups on ramps may stack one upon the other, reflecting either a persistent tectonic or halokinetic control, or compaction over previous build-ups. Examples of the latter phenomenon occur in the Silurian of the Michigan basin (Mesolella, Robinson *et al.*, 1974), and the Upper Devonian of Europe (Burchette, 1981) and the West Canada Basin (Chevron Standard, 1979).

The Trucial Coast of the Arabian (Persian) Gulf is a widely discussed modern example of a carbonate ramp (Fig. 9.15)

(Purser, 1973). This depositional system grades from the shoreline northwestwards into water depths of over 80 m. The ramp is dotted with a number of offshore shoals and islands, their locations being influenced by halokinesis in the deeply buried Proterozoic Hormuz salt. The Gulf itself has a fairly restricted circulation, which in the hot, arid climate leads to elevated salinities (>40‰ for most of the Gulf, >70‰ in coastal lagoons).

The offshore zone (mid- to outer-ramp) accumulates skeletal sandy mud, mixed with siliciclastic material introduced from the Tigris–Euphrates rivers to the northwest. The skeletal component is mainly molluscan (bivalves) and Foraminifera

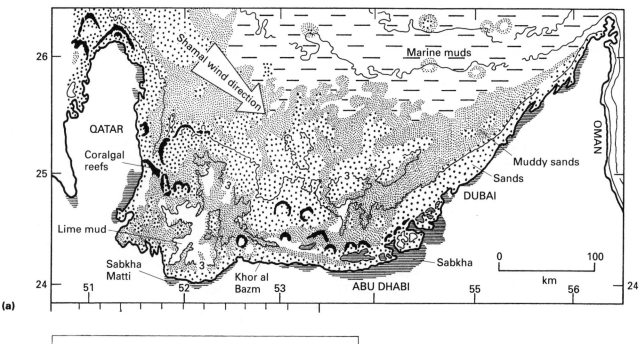

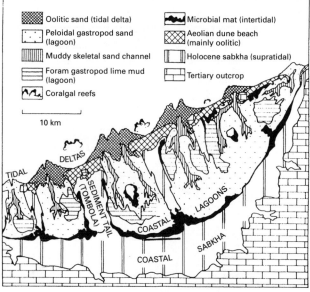

Figure 9.15 (a) Major sediment types on the southern Arabian Gulf ramp coast (Trucial Coast). (b) Schematic map of the Abu Dhabi region of the central part of the Trucial Coast showing coastal depositional environments and sediments (after Purser, 1973).

(Wagner & Van der Togt, 1973). Storms are the major factor influencing deposition offshore although, in such shallow waters, most of the sediments are thoroughly bioturbated and little evidence for tempestites remains.

The inner-ramp zone is characterized by extensive subtidal sand shoals in the west (the Great Pearl Bank), which are seeded on a Pleistocene erosional feature. This widens towards the east, forming a group of islands around Abu Dhabi (Fig. 9.15) where Pleistocene outcrops acted as foci for oolite deposition. Small ebb tidal deltas (Fig. 9.28), the sites of much of the oolite production, have developed between the islands where tidal water is exchanged between the Gulf and the back-barrier lagoons (Fig. 9.15). The growth of spits due to strong longshore currents, influenced by seasonal northwesterly Shamal winds, is a prominent feature of deposition on this coastline. Small coralgal patch reefs have developed in places along the seaward

margin of the barrier system, but rich coral growth is limited by water salinity and periods of reduced temperatures.

The barriers and shoals shelter extensive saline coastal lagoons up to 30 km wide (Fig. 9.15). These are only a few metres deep and characterized by muddy peloidal and bioclastic sediments. High salinities limit the diversity of the biota, although there is an active infauna which homogenizes much of the lagoonal sediment by burrowing. In the less-restricted Khor al Bazm lagoon (Fig. 9.15a), mud does not accumulate and the bottom sediments are sand dominated (Kendall & Skipwith, 1969). The source of the lagoonal muds, whether of biogenic or abiogenic origin, has not been resolved (Ellis & Milliman, 1986).

The most studied of the Trucial Coast environments is the intertidal and low supratidal zone bordering the lagoons, which over much of the area around Abu Dhabi is one of tidal flats, dominated by microbial mats, and sabkhas (see Sect. 8.4). The

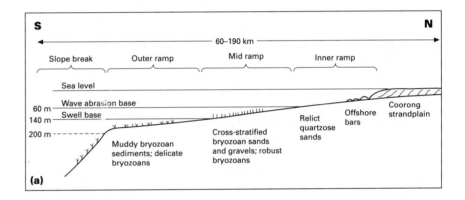

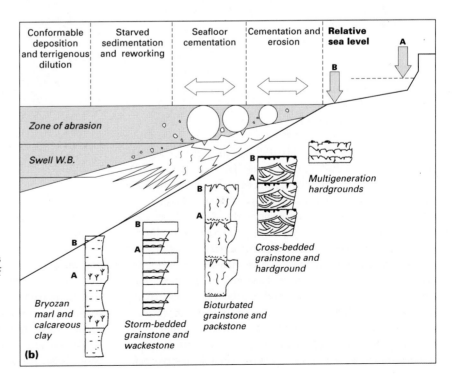

Figure 9.16 (a) Schematic cross-section across the high-energy temperate carbonate system of southern Australia (based on James, Bone *et al.*, 1992). (b) Facies model for temperate, wave and swell-dominated carbonate cyclothems based on Tertiary sequences from southeastern Australia. (A) and (B) are different sea-level positions (from Boreen & James, 1995).

more open shorelines in the northwest of the area (e.g. Sabkha Matti, Fig. 9.15) are formed by wave-influenced sandy beaches and spits.

The Southern Australian Shelf is a platform appearing to be undergoing incipient drowning. Although large quantities of sediment are being produced, these are prevented from accumulating by strong current activity. Here temperate-water sediments are being deposited (James, Bone *et al.*, 1992) to form a distally steepened ramp, extending to a break in slope in water depths greater than 250 m. The sea floor, swept by oceanic swells to depths of 140 m, is veneered by Holocene and Late Pleistocene mixed carbonate–siliciclastic sediments. The cool average water temperature (<18°C), excludes tropical carbonate sediment producers, but coralline algae are abundant below 70 m and bryozoans to depths of 350 m. Factors that have influenced the morphology of this ramp have been the complex Pleistocene sea-level fluctuations, its high-energy character, and the high rate of offshore carbonate production.

The ramp can be divided into three zones (Fig. 9.16a)
1 An inner ramp zone, above 60 m water depth where the rate of carbonate accumulation is low and sediments are predominantly relict quartzose sands.
2 A mid-ramp zone between 60 and 140 m, below the photic zone and beyond the depth of wave abrasion, but still affected by swells. This is a zone of coarse-grained, cross-bedded bryozoan bioclastic sands. The bryozoans comprise mostly robust forms and the sands are organized into symmetrical, linear or bifurcating, sharp-crested dunes with wavelengths up to 0.6 m and amplitudes up to 0.3 m. Depositional slopes in this zone are approximately 1 : 500. This example suggests that it is unwise to regard bioclastic sands as being restricted to the shallowest water environments.
3 An outer ramp zone between 140 and 200 m. Slopes are steeper and the sediments comprise bryozoan-rich muds.

Significant carbonate production by non-phototrophs is occurring in deep water. Carbonate production rates from this southern Australian Cenozoic shelf average 25 mm ky^{-1}, around two orders of magnitude less than for tropical shallow-water carbonate systems (see Fig. 9.2).

Ancient analogues for this deposystem occur in Tertiary sequences in southwestern Australia (James & Bone, 1991; Boreen & James, 1995) (Fig. 9.16b). Metre-scale cyclothems in these limestones, which are entirely subtidal, are interpreted as allocyclic in origin. Changes in the depths of critical interfaces (abrasion depth and swell and storm wave base) control facies type. Climate-related changes in deeper water productivity and terrigenous dilution also affected cyclothem development. The similarity of this model to many Palaeozoic ramp settings, with the exception of the role of swells and abrasion, is striking. Whether this reflects the importance of cool bottom waters in all such ancient systems or that they share a dominance of heterotrophic benthos, because of a sufficient nutrient supply, is unclear but interpreting all Palaeozoic echinoderm–bryozoan–

brachiopod limestones as cool-water deposits is not justifiable. Their formation in the past, as today, may reflect a combination of temperature and nutrient conditions caused by upwelling, and appreciating the influence of oceanographic factors on carbonate platform style is critical.

The Yucatan 'Shelf', northern offshore Yucatan, is a drowned flat-topped shelf of Cretaceous to Tertiary age (Fig. 9.17). Modern and Quaternary relict and fresh sediments which veneer the shelf (Logan, Harding *et al.*, 1969) form a distally steepened ramp. In the zone between 60 m water depth and the slope break at 200 m, the sea floor is covered in relict oolitic, peloidal and intraclast sands, but at present the main contributors to the sediments in this zone are pelagic and benthonic Foraminifera. Mixing of modern and relict sediment is taking place due to periodic winnowing and burrowing down to depths of 110 m. In water depths <60 m, molluscan sands are currently forming over wide areas.

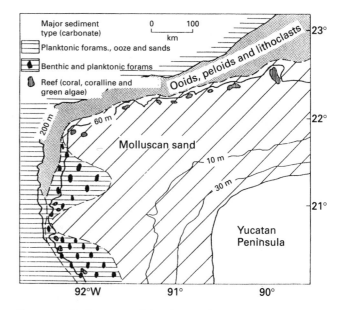

Figure 9.17 Bathymetry and distribution of major carbonate sediment types, Yucatan Shelf, Mexico (after Logan, Harding *et al.*, 1969).

The west central Florida margin is an example of an outer-ramp transition into deepwater, basinal facies (Mullins, Gardulski & Hinchey, 1988) (Fig. 9.18), as a *ramp slope*, and where oceanic currents may strongly influence sedimentation. This distinctive outer ramp zone, which extends from 200 to 2000 m depth with a slope of 1–2°, is bordered by a steep basinward margin. Some of this topography was inherited from the underlying Tertiary or Late Cretaceous depositional systems.

Mollusc-rich siliciclastic sands dominate the inner ramp (slope 0.4–1.6 m km^{-1}) down to 60 m, passing downslope into coralline algal carbonate sand (Fig. 9.18b). At 80–100 m, relict oolitic sands with pelagic and benthonic Foraminifera dominate (Ginsburg & James, 1974) as on the Yucatan (Campeche) Shelf. The ramp slope starts at about 200 m water depth and down to 400 m the sea floor is characterized by hardgrounds, coralline algal ridges, intraclastic grainstones, and rhodolith mudstones. Between 400 and 600 m the surface sediments comprise planktonic forams, intraclasts, glauconite and reworked phosphorite grains with local ahermatypic (deepwater) coral mounds. Between 600 and 2000 m is a zone of bioturbated pelagic oozes. The hardground covered areas and the winnowed grainstones reflect the activities of the Loop Current, a contour current with velocities of 20–30 cm s⁻¹ at 500 m, and 4 cm s⁻¹ at 1000 m. Although turbidites are rare on this slope, there is evidence of common large-scale slope failures. The significance

of this example lies in that it shows how deepwater currents can modify outer ramp sediments.

Carbonate ramps were particularly widespread during the early Carboniferous, particularly the early Mississippian (Ahr, 1989; V.P. Wright & Faulkner, 1990). The extinction of many reef-building organisms during the late Devonian was an important factor in this respect, but other oceanographic factors may also have been significant (Burchette & Wright, 1992; V.P. Wright, 1994a).

The *Carboniferous Limestone of southwest Britain* (Fig. 9.19) provides insights into the complexity of styles of sedimentation during the accumulation of a thick ramp sequence (a ramp stack). The succession forms a wedge, comprising largely outer-ramp muds and bioclastic limestones with abundant crinoid and brachiopod remains. Storm deposits are wide-

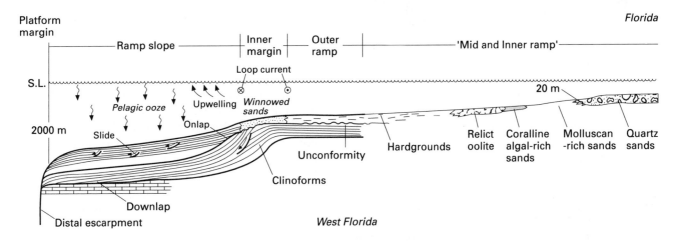

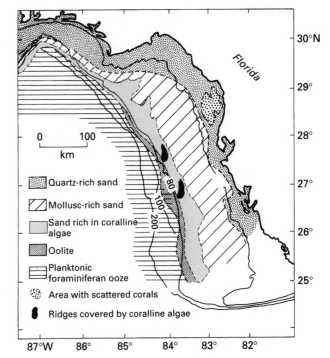

Figure 9.18 Schematic summary of the carbonate ramp and ramp slope of central west Florida. The inner ramp area zone has a mixture of relict and actively forming sediment (see Ginsburg & James, 1974). I, modern sequence with a seaward-thickening wedge; II, prograding clinoforms of an earlier phase (modified from Mullins, Gardulski & Hinchey, 1988). Inset map shows the bathymetry and facies distribution of shallower ramp region (after Ginsburg & James, 1974; Sellwood, 1986).

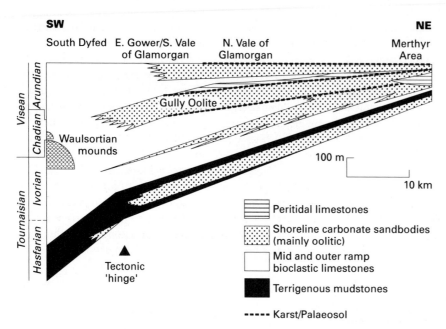

Figure 9.19 Early Carboniferous ramp stack, south Wales (after Burchette, Wright & Faulkner, 1990).

spread in the mid-ramp, with a range of tempestites and hummocky and swaley cross-stratification (Faulkner, 1988). Crinoidal wackestones and packstones in this zone probably represent the sites of offshore crinoid banks, while much larger Waulsortian reef mounds, some with depositional relief of 200 m, grew in outer-ramp settings (e.g. Lees & Miller, 1985). As judged from studies of various sequences, the roles of waves, tides and storms appear to have varied through this time interval of perhaps 10 million years, and is expressed in variation in the geometries and architectures of the main oolitic or bioclastic sandbodies (Burchette, Wright & Faulkner, 1990). During highstands, tongues of oolitic grainstones prograded basinwards as shoals or beach-fronted strandplains capped by exposure surfaces. During the transgressive phases, landward migration of bioclastic shoals or beaches allowed the deposition of extensive sheet sands overlying shoreface erosion (ravinement) surfaces. During some transgressive phases, oolitic and barrier islands prograded short distances, only to be overtaken by rising sea level to produce shoe-string sands buried by offshore muddy sediments. Behind the barriers, lagoons and tidal flats developed but these comprise only a small part of the overall thickness of the ramp succession. Corals are uncommon, and were mainly solitary, while calcareous algae appear to have played only a minor role in inner ramp settings.

9.3.5 Drowned platforms

When growing under optimal environmental conditions, carbonate platforms are persistent and resilient structures which can accrete vertically at rates that match or exceed those of most processes known to cause relative sea-level rises. They are able to recover from numerous setbacks, any one of which might

potentially have resulted in their demise. To 'drown' a carbonate platform thus requires either an extraordinary rapid sea-level rise or a significant change in the productivity of the carbonate system which is also coincident with, or precedes, a relative sea-level rise (Schlager, 1989).

However, *drowned carbonate platforms*, which are overlain by deeper water sediments and thus appear to owe their demise to rapid submergence to significant depths, are widespread in the geological record. In fact, many factors only indirectly related to 'drowning' may contribute to the demise of carbonate platforms.

1 A carbonate platform may be prepared for drowning when carbonate production is switched off or inhibited by exposing part or all of the sediment-producing area, following a sea-level fall. When sea level rises again carbonate production cannot resume until sea water reaches a depth of a few metres, but conditions conducive to high rates of production may require slightly greater depths. Thus full productivity is delayed typically by several thousand years (the 'lag time'). If relative sea level rises rapidly during this critical lag period, a platform may be flooded to depths of many metres, possibly taking it below the zone of highest carbonate productivity (e.g. below 10–20 m; see Fig. 9.1). Even if production resumes, the platform may lose its race with rising relative sea level.

2 Off-bank sediment transport may reduce a carbonate platform's ability to keep pace with rising relative sea level. If the platform rims do not form effective barriers to waves and currents, sediment may be transported off the platform, hindering its ability to 'keep up' with relative sea-level rise. This process appears to have terminated the growth of the Serranilla Bank on the Nicaragua Rise (Triffleman, Hallock & Hine, 1992). Other platforms in this area, while veneered with relict

sediment at depths of 15–30 m, still produce copious quantities of sediment (1.3–2.0 m ky⁻¹), largely *Halimeda* grains, although most of it is exported to the adjacent basin (Glaser & Droxler, 1991).

3 In many build-ups, particularly those with gently sloping margins such as ramps and reef-mounds, the locus of carbonate production may be displaced landwards during a relative sea-level rise to a position of slightly higher elevation and shallower water where it can re-establish and prograde as soon as the rate of rise diminishes. Many ancient sequences show clear evidence for 'back-stepping' of this kind which involves a local shift to deeper water, while elsewhere growth apparently continues or resumes after only a brief pause. This leads to 'tongues' of deeper-water sediments which break up the vertical continuity of the shallow-water carbonate succession. Such layers are generally less porous and permeable than the shallow-water carbonates.

In some cases this process does not completely terminate platform growth, so that benthic carbonate production continues at a reduced rate. This phenomenon, seen in some modern carbonate platforms such as Cat Island in the Bahamas (Dominguez, Mullins & Hine, 1988), has been termed *partial* or *incipient drowning*. Cat Island platform has an area of some 790 km², largely at depths greater than 30 m, and is veneered by thin (<4 m), coarse-grained carbonate sediment which is mostly older than 2000 years. The flat platform surface developed at a lower relative sea-level stand and was flooded earlier in the last major relative sea-level rise than other Bahamian platforms, at a time when the rate of rise was at its highest.

In contrast, the term *drowning* has been restricted to those cases where the growth of a carbonate platform has *apparently* been terminated by a relative sea-level rise. However, carbonate platforms may succumb to a relative sea-level rise for several reasons, not all of which are directly related to the deepening

event. It is impossible, from the morphology of a platform alone, to determine the exact cause of its demise (Schlager, 1989). For this reason the morphological relationship between a shallow-water carbonate platform and any overlying, onlapping deeper-water deposits, which may be significantly younger than the platform carbonates, has been termed the *drowning unconformity* (Schlager, 1991). To determine the sequence of events which contributed to the termination of a carbonate platform and the reasons for subsequent burial by deeper-water deposits requires a detailed study, such as that of Ehrlich, Barrett and Guo (1990), of the final phases of platform growth and the unconformity surface itself, and commonly of more regional tectonic or depositional events within a basin.

There are numerous examples of ancient drowned carbonate platforms in the geological record. In southern Europe many Jurassic platforms were dissected by rifting, and subsided during development of the Tethyan passive margin (Bernoulli & Jenkyns, 1974; Winterer & Bosellini, 1981; Bosellini, 1989). The drowning events within these platform segments are considered to have been largely due to tectonic subsidence (Schlager, 1981). They are typically marked by condensed intervals, in places ranging over several ammonite zones, which consist of nodular hardgrounds veneered by iron and manganese crusts (the *Ammonitico rosso* facies). In some of these Jurassic examples the crusts developed over karsted platform carbonates (Fig. 9.20). Few of the platforms seem to show incipient drowning, although some were capped prior to drowning by coarse crinoid grainstones. It is also possible that in some cases off-bank transport stripped sediment formed during this stage from the platform tops (Bice & Stewart, 1990), as on the Nicaragua Rise. Zempolich (1993) has suggested that increased nutrient supply and low oxygen conditions may have played a key role in reducing sediment production on some Tethyan Jurassic platforms.

Figure 9.20 Drowning histories of Lower–Middle Jurassic platforms of southern Spain. (a) Deposition of large tidally influenced oolitic sandwaves. (b) Rifting resulted in uplift, exposure and karstification, especially along footwall highs. Vadose cements and speleothems formed in fractures and cave systems. (c) No direct evidence of the incipient drowning phase has been noted, perhaps reflecting the loss of sediment by offbank transport. (d) The karsted surface is veneered by an iron–manganese hardground, overlain by red, nodular pelagic limestones (*Ammonitico Rosso* facies, based on data in Vera, Ruiz-Ortiz *et al.*, 1988 and authors' observations). Apparently not all drowning surfaces were subaerially exposed (Winterer & Sarti, 1994).

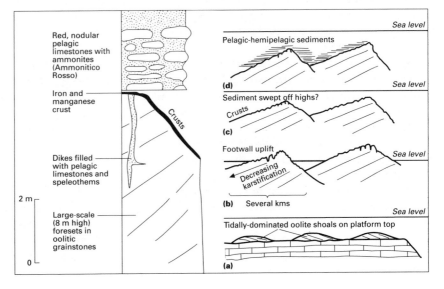

A number of apparently global carbonate platform drowning events occurred during the early and mid-Cretaceous (Valanginian to Turonian) (Schlager, 1991). Since major ice sheets were absent during this time, an alternative explanation to the glacio-eustatic (e.g. Plio-Pleistocene) model for sea-level changes is required. The Cretaceous drowning events correspond with periods of ocean anoxia and it is possible that periodic inversion of the oceanic water column, in which the bottom waters were anoxic and therefore rich in nutrients, could have contributed to the termination of reefal growth on some carbonate platforms. This might have occurred as certain non-calcifying organisms, their growth favoured by high ambient nutrient levels, flourished at the expense of the reefal biotas (Hallock & Schlager, 1986; Schlager, 1991).

9.4 Carbonate depositional environments

Large-scale carbonate platforms of the types described above can be subdivided into depositional subenvironments, based on the dominant depositional processes ('process domains') and the sediment types deposited: (i) platform interiors (low energy); (ii) platform interiors (high energy) and platform margin sand shoals; (iii) offshore systems; (iv) slope and base of slope; and (v) reefs.

Owing to evolution and ecological substitution, the organisms which are characteristic of each platform subenvironment have changed and it is not possible here to detail the types found for each geological period.

9.4.1 Platform interiors (low energy)

Platform interiors include embayments, subtidal lagoons, beaches, and tidal flats. It is one of the most varied carbonate domains, encompassing environments ranging from shelf lagoons, tens or hundreds of kilometres across, to back-barrier or back-reef lagoons just a few hundred metres or a few kilometres across. Water salinities in such settings may be predominantly normal marine, hypersaline, periodically brackish, or they may vary seasonally between these states.

Lagoons are typically shallow-water depositional environments, <10 m deep, which are protected from strong wave action. They occur in the broad interiors of rimmed shelves or behind inner ramp shoal belts. The extent to which they are affected by winds, storms, or water exchange with the sea, varies considerably. In situations where circulation is highly restricted, water temperatures and salinities may become highly elevated. In contrast, where significant fresh water is introduced (e.g. in parts of Florida Bay adjacent to the Everglades) brackish conditions may develop. Such effects may be seasonal. Deeper shelf-lagoons, where water depths of 60 m or more may be achieved, also occur rarely (e.g. in the Belize shelf; Purdy, Pusey & Wantland, 1975).

Lagoons on ramps range from relatively narrow (a few

kilometres) coastal lagoons, to broad expanses (tens of kilometres) of shallow water which may become highly restricted. On flat-topped, rimmed shelves lagoons may be even more extensive, covering thousands or even tens of thousands of square kilometres on some ancient platforms. In such cases it is commonly possible to identify an 'open-' or outer-lagoonal zone, in which the sediments and biotas indicate more normal marine salinities, and an inner-lagoonal zone of more restricted subtidal conditions (J.L. Wilson, 1975). The outer lagoonal areas receive sediment from platform-margin reefs or shoal belts (Ginsburg, 1957; Pusey, 1975; Enos, 1977; Wanless & Dravis, 1989). Carbonate production in lagoons is typically dominated by phototrophs (organisms dependent on high light intensity) and since the early Cenozoic such environments have been extensively colonized by sea grasses. However, prior to the Cenozoic, production was by a very varied range of organisms.

Patch reefs (Sect. 9.4.5) are common in deeper, open lagoons. These show a large range in size, some of those on the Great Barrier Reef shelf interior reaching up to 10 km in diameter. Such reefs influence sedimentation in their immediate vicinity so that sediment close to larger examples (the 'talus haloe') consists of poorly sorted, sand to boulder size coral, molluscan, and *Halimeda* material (Scoffin & Tudhope, 1985). The coarsest sediments commonly lie on the windward side of such patch reefs, as in the Belize shelf lagoon (Rutzler & Macintyre, 1982). More distal sediments comprise reef-derived sands which are heavily bioturbated by shrimps.

Of perhaps more importance in platform interiors are reef mounds. It is possible that many ancient interiors were covered in low relief mounds or banks comparable to those found in the Florida shelf lagoon (Sect. 9.4.5).

Lagoonal sediments are typically peloidal, comprising faecal pellets generated by mud ingestors and grains micritized by endolithic micro-organisms (Reid, Macintyre & Post, 1992). On platforms with grainstone shoal-dominated margins, carbonate sand is commonly redistributed into the lagoons by storms over wide areas of the platform interior. Ooids transported into the lagoons may be cemented into small clusters in the quiet platform interior to form grapestones (Sect. 9.1.3). Although mostly sheltered, platform interiors are subject to episodic storm disturbance. The preservation potential of sediments deposited by such events is low due to the high degree of bioturbation which occurs in most lagoons. The importance of biogenic reworking, mostly burrowing, in lagoonal settings since the mid-Palaeozoic cannot be over-emphasized since it obliterates many depositional fabrics, modifies grain populations, and retextures sediment by mixing (Tudhope & Scoffin, 1984; Wanless, 1991). Open burrows in firm muddy substrates commonly collect coarser, sorted, storm-mobilized sediments, as described by Wanless, Tedesco and Tyrrell (1988) from the Caicos Platform, West Indies, to leave a cryptic record of this process.

The innermost platform areas are commonly restricted or

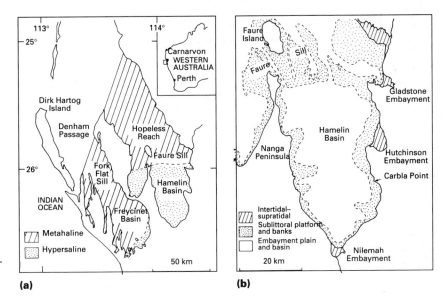

Figure 9.21 Major depositional environments in Hamelin Basin, Shark Bay, Western Australia. (a) Location, salinity zones and sills. (b) Detail of Hamelin Basin (after Logan, 1974).

hypersaline, and support only a low diversity biota. However, individuals of single species tolerant of such conditions may occur in vast numbers. In the later Mesozoic and Cenozoic, for example, the tests of miliolid foraminifers are a major component of such sediments. Ostracods, dasycladean algae, oncoids, calcispheres and gastropods, and molluscs, are also common grain types.

Where highly restricted, platform interiors in arid areas may become hypersaline due to high evaporation rates and poor water circulation. This is particularly well illustrated by parts of the Western Australia coast (e.g. Shark Bay), where extensive sea-grass banks restrict circulation so that salinities of up to 70‰ develop in some coastal lagoons (Fig. 9.21). The baffling and stabilizing effects of sea grass, and the high productivities of its associated carbonate-producing biota, has created banks up to 7 m thick of grainstone, packstone and wackestone, which have built out as shoreline-attached platforms. In the eastern part of Shark Bay (e.g. Hamelin Basin) the banks run parallel to the coast for about 129 km, and cover an area over 1000 km². Sediments in the centre of the lagoon ('embayment plain', which is 9–10 m deep) comprise mainly lime mud with molluscan bioclasts, *Penicillus* (a green alga), and Foraminifera. Gastropods dominate the biota in hypersaline subtidal areas and extensive microbial mats, locally forming large stromatolites, occur in shallow subtidal and intertidal zones (Davies, 1970a; Logan, 1974; Read, 1974; Playford & Cockbain, 1976).

SHORELINE DEPOSYSTEMS

Shoreline carbonates are referred to as peritidal (term meaning 'around the tides'; Folk, 1973) deposits. Peritidal carbonates include all of those deposited in marginal areas of (tropical or subtropical) marine water bodies, including nearshore, very

shallow subtidal zones (i.e. the shallower parts of lagoons), tidal flat and supratidal zones (including sabkhas and mangals or mangrove forests), and coastal marshes. Owing to their very shallow-water origins, peritidal carbonates are liable to freshwater invasion, leading to early diagenetic alteration by meteoric processes, and dolomitization.

In most successions of peritidal rocks, a consistent triad of depositional subenvironments can be recognized.

1 The *subtidal zone* is the permanently submerged, very shallow-water area seaward of the intertidal zone. It may be strongly influenced by wave action and tidal currents and have a biota whose abundance and diversity depends mainly on salinity and temperature, largely controlled by the degree of tidal exchange.

2 The *intertidal zone* lies between the normal low- and high-tide levels and is alternately flooded by sea water and exposed. Climate and environmental 'energy' strongly influence the character of intertidal sediments and a wide variety of sedimentary and diagenetic structures may develop (Figs 9.22 & 9.23). Low-energy intertidal zones commonly comprise mud or microbial flats, or low-energy beaches; higher-energy intertidal zones commonly have beaches or spits with associated supratidal beach ridges and beach rock development. *Criteria for recognizing intertidal deposits* are those related to exposure and include desiccation cracks, fenestrae (small irregular to laminar pores), evaporite mineral growth, soils and related features.

3 The *supratidal zone* lies above high tide level and is only flooded during high spring tides and storms. On low-relief prograding coastlines it can be many kilometres wide. Its character is also controlled largely by climate and in semi-arid and arid settings this zone may be the site of extensive evaporite precipitation and wind deflation (see Sect. 8.4). Marshes, as in the Everglades or on the western side of Andros Island (Monty

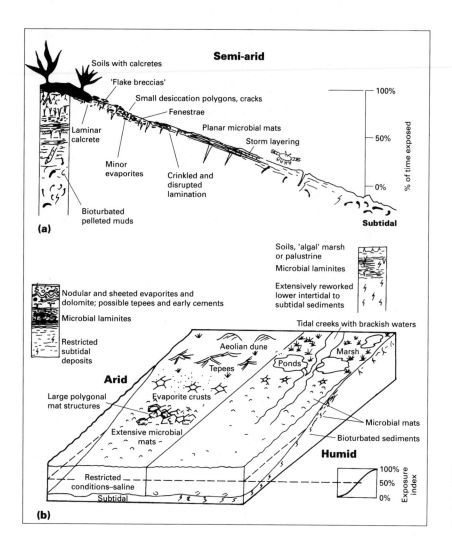

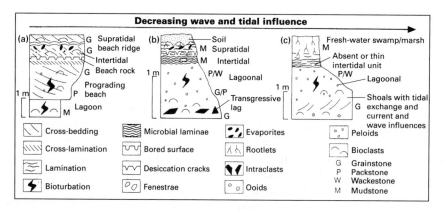

Figure 9.22 (a) Variations in sedimentary structures across a carbonate tidal flat, related to exposure index (right). (b) Contrast in environments and resulting sediments across an arid and humid tidal flat zone.

Figure 9.23 Spectrum of peritidal shoaling sequences reflecting differing degrees of wave and tidal influence. (a) High-energy type based on Khor Duwahine, extreme western part of Abu Dhabi (Fig. 9.15); although this area is protected from the Shamal winds it still lacks protection by a barrier system and is wave influenced. Microbial mats only occur on the upper part of the intertidal zone and are associated with evaporites (gypsum). These sandflats are prograding seawards with an accretion slope showing seawards-inclined bedding (Purser & Evans, 1973). (b) Protected, lower-energy sequence. The initial coarse transgressive lag horizon is followed by a shoaling phase showing well-defined intertidal facies. Based on Lower Carboniferous peritidal deposits from south Wales

(V.P. Wright, 1986); similar sequences are shown by James (1984). (c) Sequence from highly restricted setting; shoaling from a grainstone facies, with wave and tidal influences, into lagoonal and finally freshwater facies (supratidal–terrestrial). Intertidal deposits are thin or absent (based on data in Palmer, 1979 for the Middle Jurassic White Limestone Formation). This sequence is broadly comparable to the transition across the present-day Florida Shelf lagoon. Wave action and tidal exchange are greatly reduced over the shelf which results in nearshore areas being effectively tideless. In contrast to (a) and (b) this sequence has formed under a more humid climate and lacks evaporites. Diagram modified from V.P. Wright (1984).

& Hardie, 1976), may form in humid supratidal settings. These are mostly shallow, seasonal environments which may generate freshwater lime or terrigenous muds ('palustrine' sediments – see Sect. 4.7). Peritidal deposits intercalated with such palustrine sequences are known from the Cretaceous (Martin-Chivelet & Gimenez, 1992).

One technique used to classify modern carbonate peritidal environments and associated sediments consists of making observations at various locations of the amount of time exposed during the tidal cycle – that is, the 'exposure index' (Ginsburg, Hardie *et al.*, 1977) (Figs 9.22 & 9.24c). Values assigned to common sedimentary structures, or assemblages of structures, can then be used to zone and categorize the peritidal environments. For example, sediments with crinkly lamination, a common feature in microbially-covered intertidal sediments, have an exposure index of 40–90% (lower intertidal zone) in this scheme, while small desiccation cracks would have an exposure index of 90–100% (upper intertidal to supratidal

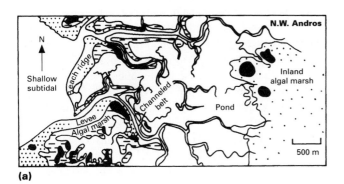

(a)

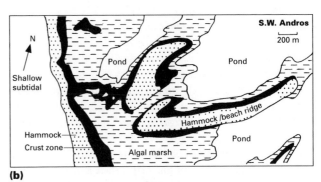

(b)

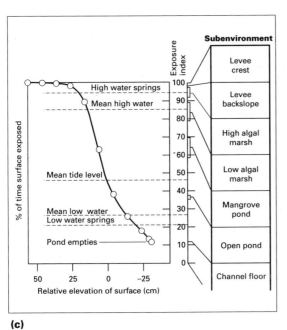

(c)

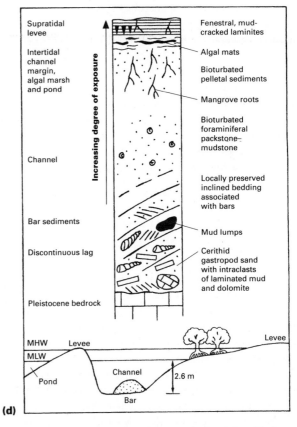

(d)

Figure 9.24 Tidal flat environments and sequences of western Andros, Great Bahama Bank. (a) Northwest Andros, with well-developed channelled belt (after Hardie, 1977). (b) Southwest Andros, with beach ridges (after Gebelein, Steinen *et al.*, 1980). (c) Exposure index (percentage of time the tidal flat surface is exposed) and subenvironments of western Andros (after Ginsburg, Hardie *et al.*, 1977). (d) Andros tidal channels; section shows major environments; log shows schematic section through a channel migration sequence (after V.P. Wright, 1984).

zone). Since peritidal sediments are relatively consistent in their expression throughout the geological record, the above scheme is applicable to ancient peritidal deposits. Lithofacies successions in which pertitidal deposits occur commonly exhibit an upwards increase in the degree of exposure, a consequence of shoreline progradation and aggradation.

Lamination is a distinctive feature in intertidal deposits and reflects alternations of sediment input and microbial activity. Sediment transported on to intertidal flats from adjacent subtidal areas, principally during storms, is commonly deposited as millimetre alternations of mud and silt to sand-grade carbonate material. Coarser layers are commonly rippled or graded and mostly peloid-rich (Davies, 1970b; Park, 1977; Hardie & Ginsburg, 1977; Wanless, Tyrrell et al., 1988). Where tidal flats are overgrown by mats of cyanobacteria and algae (most common in elevated salinities), these trap and bind fine material to generate a laminated sediment (*microbial laminite*; also called *cryptalgal laminite*, *fenestral laminite* or *loferite*).

Microbial communities on tidal flats produce flat mats, or ones with tufted, crinkled or pustular surface morphology (e.g. Logan, Hoffman & Gebelein, 1974; Kinsman & Park, 1976). Under suitable conditions microbial mats can form domal structures or *stromatolites*; those with relief of more than a few centimetres generally appear to have formed in lower intertidal to subtidal areas (Tucker & Wright, 1990). Irregularities in mat morphology can also be caused by desiccation, mat expansion during growth, evaporite mineral precipitation, or blisters and 'fenestrae' caused by gas generated during decomposition. Mat morphology is less regular in the upper intertidal and supratidal zones where desiccation is severe and fine lamination is rarely preserved. On the lower intertidal flats, in less restricted areas, grazing and burrowing organisms may disrupt or prevent mat growth (Fig. 9.22).

PERITIDAL CYCLOTHEMS AND STACKING

The peritidal deposits of many carbonate platform interiors occur in thick successions comprising hundreds of discrete metre-scale shallowing-upwards sedimentary sequences (reviewed by Hardie, 1986a). Such *cycles* or *cyclothems* are usually asymmetric and have been referred to as *PACS* (punctuated aggradational cycles; Anderson & Goodwin, 1978). A wide variety of platform-interior sedimentary cycles has been documented in the geological record (Ginsburg, 1975; Pratt, James & Cowan, 1992), the lithofacies variations reflecting different energy levels, tidal ranges, biotas, and climates (Fig. 9.23). The causes for the regular stacking of these shallowing-upwards cyclothems continues to be the subject of considerable discussion. Any model needs to explain two key features. First, most cyclothems begin with a subtidal unit overlying, commonly erosively, the top of the underlying unit; this represents a relative rise in sea level. Second, the upper part of the cyclothem

represents progressive shallowing, which may reflect a fall in sea level or purely sedimentary processes. The debate over the causes of multiple cyclothems has revolved around whether both rises and falls are required and what (tectonism or eustasy) has caused these sea-level oscillations.

We can divide the possible controlling factors into autocyclic (i.e. sedimentary) and allocyclic (tectonic or eustatic) controls.

Autocyclic models. In carbonate platform-interior settings two processes may generate shallowing-upwards sedimentary cyclothems: (1) channel migration and (2) shoreline progradation. 1 *Tidal channels.* The tidal flats of northwestern Andros Island (Fig. 9.24a), for example, are dissected by laterally migrating tidal channels, 3 m or less deep and up to 100 m wide. The channels become shallower and narrower landward and terminate in the upper intertidal zone. The channel floors contain gastropod-bearing muds and lags of intraclasts (Fig. 9.24d). The channel-fill deposits may preserve inclined bedding which represents the point bar surfaces. Since the channel is an extension of the subtidal zone into the tidal flats it lacks exposure features, but the upper part of the channel-fill sequence comprises fine-grained levee deposits (Fig. 9.24c,d) which are almost continuously exposed and exhibit fine lamination (including storm layering) and fenestral fabrics. As the channel migrates, therefore, a sheet-like shallowing-upwards unit develops. Despite the fact that this process is widespread in modern carbonate peritidal depositional environments, few examples of channelled tidal flats have been identified in the geological recod (V.P. Wright, 1984; Mitchell, 1985; Cloyd, Demicco & Spencer, 1990). The recognition of the point bar structures is critical for their identification but these are prone to destruction by bioturbation. Most examples of ancient tidal channel carbonates (see above) are from Precambrian and early Palaeozoic sequences, when bioturbation was less. 2 *Shoreline progradation.* Most small-scale inner platform carbonate depositional cycles appear to represent rapid infill of shallow subtidal areas, followed by progradation of the intertidal and supratidal zones across the filled lagoons. Modern examples are the tidal flats of the Trucial Coast which have prograded about 5 km in 4000 years (G. Evans, Schmidt et al., 1969) (see Sect. 8.4.2), and the shoreline of southwest Andros Island (Fig. 9.24b) which has prograded 5–20 km in 1000 years (Gebelein, Steinen et al., 1980; Hardie, 1986a,b). The sediment is supplied by the adjacent subtidal area (Gebelein, 1977). Cyclothems are largely asymmetric because, following the flooding of a subaerial or intertidal surface, there is a 'lag time', possibly as long as several thousand years, until sufficient water depth exists to support carbonate sediment producers. During this time, little or no sediment is deposited and pre-existing sediments may be extensively reworked. By the time sediment production commences, subtidal conditions are well established. Most cyclothems therefore commence with subtidal sediments and shallow upwards.

The lag effect is well displayed in the Holocene peritidal sequence of Abu Dhabi (Fig. 9.25a). Here the record of the initial transgression, c. 5000 years ago, is an intertidal unit just 0.1 m thick, although the tidal range at the time was probably 1–2 m (Kinsman & Park, 1976). Most peritidal cyclothems do not preserve even this much of the transgressive intertidal phase, but one ancient example occurs in the Lower Jurassic Calcare Massiccio Formation of the central Apennines (Fig. 9.25b) (Colacicchi, Passeri & Pialli, 1975). The tops of peritidal cyclothems represent periods of subaerial exposure and commonly exhibit features indicative of soil development (V.P. Wright, 1994b).

The tidal channel model is no longer widely used to explain peritidal cyclothems but progradation is a key element in most autocyclic explanations for cyclicity. However, the process only explains the shallowing-upwards component of each cyclothem and some other explanation is needed to explain the creation of new space for the subsequent units.

In one of the simplest explanations (Fig. 9.26), the stacking of peritidal cyclothems is regarded as the result of changes in sediment supply combined with linear *subsidence* (Ginsburg, 1971; Hardie, 1986a,b; Pratt, James & Cowan, 1992). Since the sediment supply is from the adjacent subtidal factory, the infilling of the lagoon, as the tidal flat progrades, progressively decreases the sediment source, starving the tidal flats, which cease to prograde. With continued subsidence, the sea inundates the area, but the lag period allows net water depth to increase without any sediment accretion or progradation. Once subtidal sedimentation resumes, deposition of a new shallowing-up cyclothem commences.

Although this model is attractive by itself, it fails to explain some critical characteristics of stacked cycles (Goldhammer, Oswald & Dunn, 1991). Many show clear evidence of subaerial

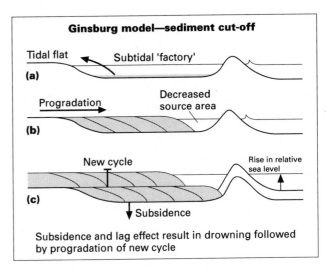

Figure 9.26 Model for the generation of stacked peritidal cycles by 'autocyclic' processes.

exposure surfaces or vadose alteration. Because sea level would never actually fall in the autocyclic model, it would only allow the development of very wide supratidal zones (Strasser, 1991). Also, since subsidence would be the major control on the creation of accommodation space, only cyclothems of uniform thickness would form. However, most peritidal successions show wide variations in cyclothem thicknesses and in some cases the changes appear to be ordered (e.g. Goldhammer, Dunn & Hardie, 1987). More significantly, subsidence rates, such as those on passive margins (0.01–0.04 m ky^{-1}) would generate cycles that require many tens of thousands of years to form; many ancient peritidal cycles appear to have formed in shorter intervals.

Figure 9.25 (a) Stratigraphic sequence from the sabkhas of Abu Dhabi (modified from V.P. Wright, 1984). The transgressive phase deposits are thin, a result of low sediment supply. Owing to a greater sediment supply the regressive intertidal unit is thicker (based on data in Kinsman & Park, 1976; cf. Figs 8.16, 8.17 & 8.18). (b) Peritidal cyclothem from the Calcare Massiccio Formation (Lower Jurassic) of the central Apennines. Note the thin transgressive intertidal unit compared with the much thicker regressive (progradational) upper intertidal unit. These thickness differences possibly reflect a lower rate of sediment supply during the early transgressive (start-up) stage (based on Colacicchi, Passeri & Pialli, 1975 and Sellwood, 1986).

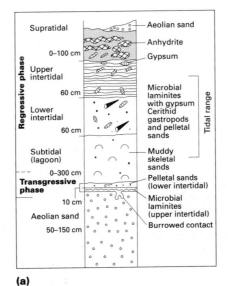

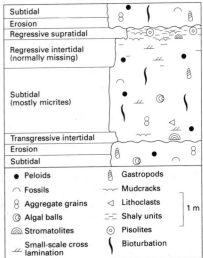

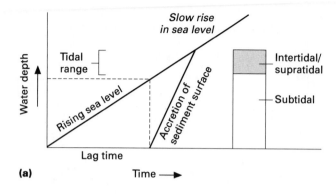

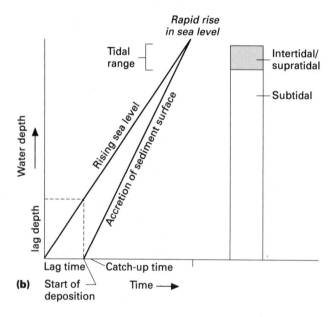

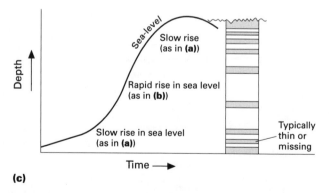

Figure 9.27 Contrast between cyclothem types (a) when the rate of relative sea-level rise is lower (with thinner subtidal units) and (b) when the rate is higher (with thicker subtidal units). (c) Pattern of small-scale autocyclic peritidal cyclothems during a rise and minor fall in sea level. There are no significant subaerial exposure surfaces capping the cyclothems; and thicknesses have changed as the rate of relative sea-level rise varied.

If, however, there was a significant rise of relative sea level, with variable rates, autocyclic processes could generate cyclothems of variable thickness (e.g. Drummond & Wilkinson, 1993; Goldhammer, Lehmann & Dunn, 1993) (Fig. 9.27). Such cyclothems would lack evidence of significant subaerial exposure.

Another autocyclic model has been suggested to account for low lateral continuity in peritidal cycles of the Lower Ordovician Cow Head Group of Newfoundland (Pratt & James, 1986). In this model, shallowing-upwards sequences of relatively small areal extent might be generated by the growth of small islands scattered across a platform during a period of constant sea-level rise. Island shoreline progradation continues until sediment supply from the nearby subtidal area ceases. However, where lateral cycle continuity is large, it would be necessary to invoke island progradation over larger distances than seems resonable.

Allocyclic models. Arguments that invoke tectonic controls on small-scale peritidal cyclicity have found little support, but have been reviewed by Goldhammer, Lehmann and Dunn (1993). One 'yo-yo' model proposes rise and subsidence of the platform, but lacks a credible example. Normal faulting might be a cause, but would neither be effective over large areas, nor allow cyclothems with prominent exposure surfaces to develop. Variation in intraplate stress is a conceivable cause of base-level fluctuations at rates of 0.01–0.1 m ky^{-1} (Cloetingh, 1988), but seems unlikely to directly control the development of small-scale, high-frequency carbonate peritidal cyclothems (Goldhammer, Lehmann & Dunn, 1993).

The hypothesis that eustasy is a major control on cyclothem development has been widely discussed due to the introduction of Milankovitch theory (see Sect. 2.1.5). Evidence for Milankovitch-style oscillation takes several forms. Cyclothem duration has been calculated by dividing the total duration of the succession by the number of cycles it contains, an operation that consistently provides durations on several scales, namely 20–40 ky, 100–200 ky, or in some cases 200–400 ky (e.g. Tucker & Wright, 1990, pp. 62–64). These conveniently fit the theoretical Milankovitch periodicities. A counter argument, however, is that the thickness ranges of proposed eustatic cyclothems correspond closely to the thickness of sediment that might be accommodated anyway during these time intervals by subsidence alone (Algeo & Wilkinson, 1988).

Key evidence quoted in support for Milankovitch-band eustatic changes is that in many successions cyclothems are grouped into 'bundles'. In the platform interior of the Triassic Latemar build-up in the Dolomites, northern Italy, for example, five small, thinning-upwards cyclothems are grouped into one larger cycle (Goldhammer, Dunn & Hardie, 1987, 1990). Similarly, in the Carboniferous of the Paradox Basin, USA (Goldhammer, Oswald & Dunn, 1991) the ratio is 9 : 1 and in the Cambrian 4 : 1 (Osleger & Read, 1991). The fivefold bundling is interpreted as representing the interaction of the precession and eccentricity cycles. A 4 : 1 pattern might represent the short (100 ky) versus the long (400 ky) eccentricity cycles. The 9 : 1 case is more problematic, but possibly represents the obliquity versus the long eccentricity cycle. Whether or not 'Milankovitch' processes control peritidal cyclicity is a complex problem and a field of intense debate (e.g. see Hardie, Wilson & Goldhammer, 1991; Goldhammer, Lehmann & Dunn, 1993).

9.4.2 Platform interior (high-energy) and platform margin carbonate sandbodies

Carbonate sandbodies are a prominent feature of high-energy subtidal to intertidal environments in many platform settings, but have not been intensively studied. Some develop along platform margins; others in high-energy settings in platform interiors, for example inner shelf and ramp shorelines. Modern carbonate sandbodies are small and depositional models are few though this has not prevented their application to interpreting the geological record with insufficient thought. The section below uses modern examples where possible, but includes some ancient examples to illustrate the diversity of carbonate sandbody types. Wanless and Tedesco (1993) have provided a review of modern oolitic sandbodies in the Bahamas and Caribbean, stressing the influence of climate.

PLATFORM INTERIOR SANDBODIES

Carbonate sandbodies may develop along shorelines, in platform interiors or, less commonly, form shallow, offshore banks. Inden and Moore (1983) and Tucker and Wright (1990) have provided reviews of carbonate shoreline sand deposystems. Shoreline carbonate sandbodies may develop with various morphologies, but possess similar characteristics to siliciclastic sand accumulations in comparable settings, including barrier complexes (shoreface to backshore, tidal inlets, deltas and washovers), strandplains (beach-fronted prograding shorelines), and nearshore tide or storm-generated sand sheets or shoals (e.g. Burchette, Wright & Faulkner, 1990). However, no macrotidal carbonate shorelines, ancient or modern, have ever been recorded. Where platforms are land-attached, mobile platform-interior sediment may be largely siliciclastic.

Barrier and beach sandbodies (Sects 6.2 & 6.7) typically develop in inner-ramp settings and along high-energy inner shelves where longshore drift is important. Modern examples of carbonate barrier systems include the coast of northeastern Yucatan (Ward & Brady, 1979; Ward, Weidie & Back, 1985) and the Trucial Coast of the Persian Gulf (Purser, 1973) (see Fig. 9.15). As in siliciclastic systems (Reinson, 1992; see Sect. 6.7.4), carbonate mesotidal barrier complexes (tidal range >2 m) are likely to contain inlet fills, tidal and tidal delta deposits (Fig. 9.28). Microtidal barriers (tidal range <2 m) generally

Figure 9.28 Aerial photo-mosaic of the area around Sadiyat island showing details of the oolitic tidal delta and lagoonal complex (from Schreiber, 1986).

comprise simple shoreface, beach and washover sands. However, in carbonate environments, as in the Trucial Coast, barrier beach faces and ebb tidal deltas are important sites of sediment, mostly oolite, generation and the systems are to a large extent therefore *self sourcing*. Morphologically, the Trucial Coast ebb tidal deltas are extensive sand shoals, several kilometres across and dissected by channels. The delta fronts and tops are covered with bars, sandwaves and dunes while the channels are floored with sand waves and dunes of mixed peloidal, oolitic and skeletal sand.

Well-developed shoreline sandbodies form only where sediment supply exceeds removal by erosion and where adequate accommodation space exists. When relative sea level is static or falling, back-barrier areas fill, and beach-fronted strandplains may prograde large distances forming sheet-like sandbodies (cf. Sect. 6.7.1). The Pleistocene of northeastern Yucatan provides a good example in which episodic progradation has constructed a series of ridges which mark successive shoreline positions (Ward & Brady, 1979; Ward, Weidie & Back, 1985). The result is a coarsening-upwards succession c. 10 m thick (Fig. 9.29), capped by a prominent subaerial exposure surface. Examples have been documented from the early Carboniferous (Mississippian) of southwest Britain (Burchette, Wright & Faulkner, 1990) and from the early Cretaceous of Texas (Inden & Moore, 1983). A sandbody, the Gully Oolite, from the first example consists of several stacked 10–20 m thick,

shallowing-upwards oolite shoreface sandbodies which extend for over 30 km downdip and can be traced laterally for 150 km (Figs 9.19 & 9.30a). Comparable oolitic strandplains occur on the Caicos Platform (Lloyd, Perkins & Kerr, 1987; Wanless & Tedesco, 1993).

Under conditions of slowly rising relative sea level, and where the rate of sediment supply exceeds the rate of sea-level rise, barrier–lagoonal systems may prograde. In such cases, back-barrier deposits onlap the barrier sediments as the system evolves (Fig. 9.30b).

Barriers migrate landward during a transgression in one of two ways (see Sect. 6.7.6). The barrier may drown, leaving a string-like sandbody overlain by offshore deposits (Fig. 9.30c). Examples of oolitic–bioclastic sandbodies of this type have been recorded from Mississippian ramp successions of southwest Britain (Burchette, Wright & Faulkner, 1990). More commonly, barriers or beaches migrate landwards by shoreface retreat, whereby erosion in the shoreface zone removes the barrier, and commonly even back-barrier deposits, leaving a prominent erosion surface, or ravinement (Fig. 9.30d), overlain by a shoreface sand sheet. Controls on these processes are the rate of sediment supply, erosion, and the rate of relative sea-level rise. Examples have been documented from the Mississippian of southwest Britain (Burchette, Wright & Faulkner, 1990), illustrating the potential variability of sandbody types in a stacked ramp succession.

Where tidal inlets are prominent, such as under higher tidal ranges, the transgressive sequence may contain 2 erosion surfaces. The lower surface represents a scoured tidal inlet channel base, while the upper is the shoreface ravinement, the two being separated by channel-fill deposits (Sect. 6.7.6 and Fig. 6.87).

(1) *Tidally influenced subtidal sands.* Although many carbonate sediments were undoubtedly deposited in tidally influenced regimes, tidal carbonate sandbodies are not as well documented as those from siliciclastic depositional settings (see Sects 7.3 and 7.7.1). The best-known modern tidal carbonate sands are found at the margins of major platforms, such as the Great Bahama Bank (see below). Examples of Mississippian linear, tidal oolitic sandbodies have been described by Smosna and Koehler (1993), and Kelleher and Smosna (1993) from West Virginia, USA (see Fig. 9.33). The elongate form of these sandbodies is a characteristic feature.

(2) *Storm-influenced subtidal sands.* Most modern shelves are storm-dominated. Evidence suggests that this was also true of many ancient carbonate depositional environments, particularly on the gentle slopes of ramps. Carbonate sands deposited in such settings are therefore strongly storm influenced and exhibit a range of sedimentary structures and textures better known from siliciclastic settings, including graded and laminated beds, hummocky cross-stratification, shell lags,

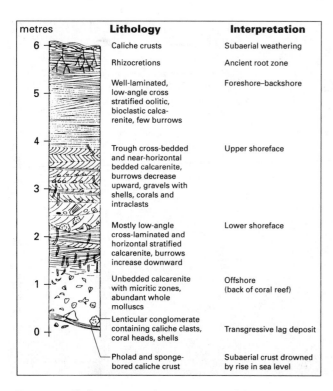

metres	Lithology	Interpretation
6	Caliche crusts	Subaerial weathering
	Rhizocretions	Ancient root zone
5	Well-laminated, low-angle cross stratified oolitic, bioclastic calcarenite, few burrows	Foreshore–backshore
4	Trough cross-bedded and near-horizontal bedded calcarenite, burrows decrease upward, gravels with shells, corals and intraclasts	Upper shoreface
3		
2	Mostly low-angle cross-laminated and horizontal stratified calcarenite, burrows increase downward	Lower shoreface
1	Unbedded calcarenite with micritic zones, abundant whole molluscs	Offshore (back of coral reef)
0	Lenticular conglomerate containing caliche clasts, coral heads, shells	Transgressive lag deposit
	Pholad and sponge-bored caliche crust	Subaerial crust drowned by rise in sea level

Figure 9.29 Shallowing-upwards regressive strandplain sequence from the Upper Pleistocene, northeast Yucatan, Mexico (after Ward & Brady, 1979).

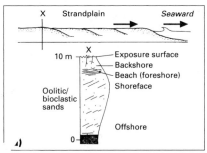

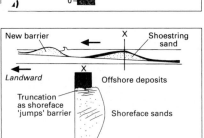

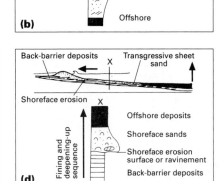

Figure 9.30 Barrier-related carbonate sands. (a) Regressive, progradational strandplain creating a largely coarsening-upward sequence capped by a prominent exposure surface. Such sequences are generated during late highstand and lowstand system tracts. Examples include the Gully Oolite, Mississippian of southern Britain (Fig. 9.19). (b) Sequence generated when accommodation space was being created, allowing the maintenance of a back-barrier zone, the deposits of which onlap the barrier. No prominent exposure surface caps the carbonate sandbody. (c) A transgressive barrier bypassed ('drowned') and a new barrier developed in a more landward position. The drowned barrier may have been truncated and is overlain by lower shoreface to offshore deposits. The resulting sandbody is likely to have shoestring or ribbon-like geometry. (d) A relatively thin transgressive carbonate sand resulting from the landward migration of the barrier by shoreface retreat. Unlike (a)–(c) no actual barrier is preserved but a fining- and deepening-upwards sheet sand is produced. Examples of (a), (c) and (d) have been documented in southwest Britain by Burchette, Wright & Faulkner (1990) (cf. Figs 6.86 & 6.87).

and scours and channels (Sects 7.4 & 7.7.2). Successions of these sediments commonly also show cyclicity, expressed as thickening- and coarsening-upwards (or the reverse) progressions, related to climate or relative sea-level changes. Broad banks, and extensive sheets of oolitic sands affected by intermittent high-energy wind-driven currents (such as hurricanes) are a feature of the platform interior of the Caicos Platform in the Bahamas (Wanless & Tedesco, 1993). The intermittent movement results in ooids having irregular, non-spherical morphologies.

PLATFORM MARGIN SANDBODIES

Carbonate sandbodies made up of bioclastic and oolitic sands occur extensively at the margins of carbonate platforms of all ages, reflecting the dissipation of most wave and tidal energy at such margins (Tucker, 1985; Wanless & Tedesco, 1993). The key factors that control the nature of such accumulations are topography (including the presence of reefs or islands), orientation with respect to dominant winds and waves (windwardness–leewardness), and tidal range (Fig. 9.31).

Where islands are present at the platform margin, tidal channels and deltas may develop between islands, as between

some of the Florida Keys (Ebanks & Bubb, 1975) or along the western side of Exuma Sound, Bahama Banks. Such deltas are predominately flood-tidal and may consist of oolitic or bioclastic sediment. Tidal bar belts of bioclastic sediment, dissected by tidal channels, also form behind and between the discontinous rocky Pleistocene rim (Ball, 1967). Much of the sediment in these banks appears to be mud-supported with textures heavily altered by bioturbation (Wanless & Tedesco, 1993), although the seaward portions have grainy, storm or hurricane-reworked caps (Aigner, 1985).

Carbonate sandbodies on the windward and leeward margins of the Bahama Banks, termed *marine sand belts*, exhibit different geometries and sediment compositions (Hine, Wilber & Neumann, 1981; Tucker, 1985). Oolitic sandbodies, in particular, dominate the windward, open margins where sediment transport is mostly on to the shelf. On leeward margins (Fig. 9.31d), net sediment movement is off-shelf so that much of the material is derived from the platform interior, largely micritized peloidal, skeletal and oolitic grains. Such differences in ancient platform margin deposits can be used to differentiate windward from leeward orientations (e.g. Blendinger & Blendinger, 1989).

Cat Cay on the leeward margin of the Great Bahama Bank (Ball, 1967), the Berry Islands (Hine, Wilber & Neumann,

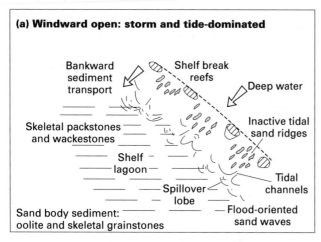

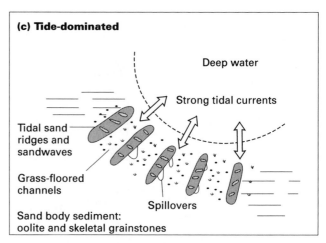

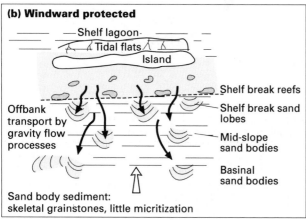

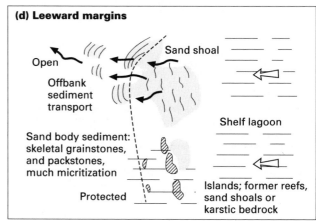

Figure 9.31 Simple facies plans for platform margin sandbodies. The controlling factor is the platform margin's orientation relative to waves, storms and tidal currents (modified from Tucker & Wright, 1990).

1981), and Lily Bank on Little Bahama Bank (Hine, 1977) are examples of marine sand belts of peloidal and bioclastic sand shoals, with widths of 1–4 km, and lengths of up to 75 km (Fig. 9.32), cut by deep channels covered with sea grasses. The shoals have asymmetric forms, with a 'flood ramp' covered in asymmetrical flood-directed sand waves (with heights of 1–2 m) on the platform side which become more symmetrical in the 'shield' zones on the shoal crests (Fig. 9.32). Wave and tidal currents control the movement of sediment over the shoals and the shield protects the ramps from ebbing currents, which are confined to the channels. During major storms carbonate sand migrates into the stable sea-grass covered platform interior. The sediment is deposited as spillover lobes up to 500 m wide, with avalanche faces up to 1.5 m high. The Joulters Cay oolite shoal is one of the best-documented examples of an active shoal and occurs near the northern tip of Andros Island, Great Bahama Bank and was initiated 3–4000 years ago as a marine sand belt. This belt developed spit-like extensions and channels, eventually building up to form an extensive sandflat capped by

beach and aeolian dune deposits (Harris, 1979). Ramp–shield sediments would be detected in the rock record by being dominated by medium- to small-scale cross-stratification, with a change in current vectors from the ramp to shield zones (Fig. 9.32). However, the lobes display more uniform and potentially larger scale cross-stratification (Fig. 9.32). The resulting sandbodies are typically lensoid in shape and parallel to the platform margin. A possible ancient analogue has been described from the Mississippian St Louis and Sainte Genevieve Limestones of Kansas (Handford, 1988b).

Tidal bar belts develop in areas where tidal energy re-entrants are dominant (Ball, 1967; Dravis, 1977; Palmer, 1979), and are particularly well developed at the ends of deep oceanic re-entrants into shallow-water platforms, such as Tongue of the Ocean and Exuma Sound in the Great Bahama Bank (see Fig. 9.9). In areas of strong tidal flow (peak currents >1 m s^{-1}), sands are fashioned into elongate shoals or bars orientated perpendicular to the platform margin. The tidal shoal trend along the southern rim of Tongue of the Ocean stretches for

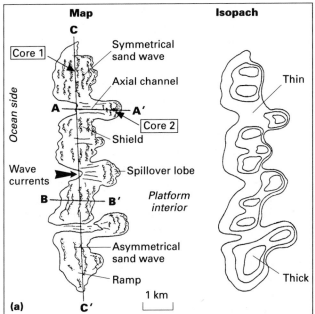

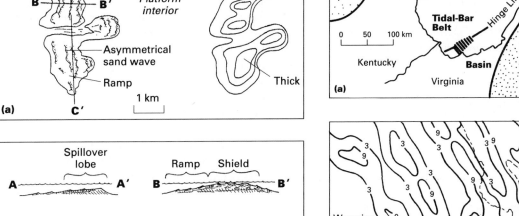

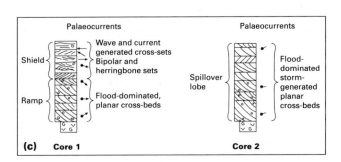

Figure 9.32 Major features of marine sand belt. (a) Plan view with hypothetical isopachs. (b) Cross-sections. (c) Inferred vertical sequences for cores (modified from Handford, 1988, based on data in Ball, 1967 and Hine, 1977).

over 100 km as a complex of linear tidal bars organized in parallel sets 0.5–1 km across and 12–20 km long and with amplitudes of 3–9 m, comparable to siliciclastic tidal sandridges (Sect. 7.3.4). The interbar areas are stabilized zones 1–3 km wide and up to 7 m deep with sea grasses, microbial mats and stromatolites, and sparse sandwaves, while the ridges are covered in sandwaves with a variety of orientations.

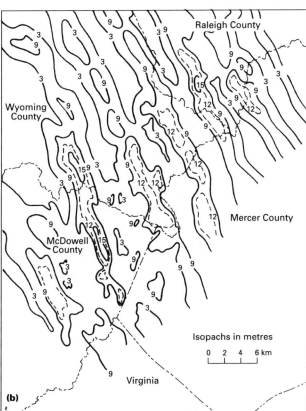

Figure 9.33 Mississippian oolitic tidal bar sandbodies from the Viséan of the central Appalachians. (a) Palaeogeographic reconstruction of the Greenbrier Gulf. The oolitic tidal bar belt (the Union Oolite member of the Greenbrier Group) is situated along a hinge-line separating the basin from a shelf-like shallow platform. (b) Isopach map of the oolite, interpreted as northwest–southwest trending tidal bars. The oolite thins to 0 m between some bars (after Kelleher & Smosna, 1993).

Oolitic tidal bar sandbody hydrocarbon reservoirs have been documented from several Mississippian sequences in the USA (Keith & Zuppann, 1993; Kelleher & Smosna, 1993) (Fig. 9.33). Some developed at changes of slope (Fig. 9.33). Not all ancient tidal bar sandbodies are oolitic and Dabrio and Polo (1985) and Molina (1987), have described large mixed oolitic and crinoidal tidal sandwaves, with amplitudes of over 8 m, from the Lower Jurassic of southern Spain.

Aeolian carbonate sandbodies are a striking feature of modern carbonate shorelines and marine deposition for a carbonate sand should not be assumed uncritically. Details of Quaternary aeolian carbonate sands have been given by McKee and Ward (1983) and Caputo (1993). Ancient analogues have been documented from the Mississippian (Hunter, 1993) and Pennsylvanian and Permian (Rice & Loope, 1991) of North America. Criteria for differentiating ancient marine and aeolian carbonate sands have been discussed by Dodd, Zuppann *et al.* (1993).

9.4.3 Offshore carbonate deposystems

The common high mud content of rocks deposited in mid- and outer-shelf settings resulted initially in the assumption that they were mostly deposited under low-energy conditions, and that the biotas which they contain are more useful for environmental interpretation than their rock fabrics. However, storms produce a wide range of stratification types in offshore siliciclastic regimes, and have a major effect in carbonate and mixed siliciclastic–carbonate sediments, particularly in ramp settings (e.g. Aigner, 1984, 1985; Handford, 1986; Faulkner, 1988; Calvet & Tucker, 1988). In some successions, cyclicity based on the 'packaging' of storm beds can be recognized. The source of carbonate muds which, volumetrically, dominate such successions remains problematic. Organic carbonate production in these deeper-water settings has also varied through time and the relationship between *in situ* sediment production and allochthonous shallow-water carbonate is poorly understood.

Many outer- to mid-ramp sediments appear to have been deposited within the photic zone, yet biotas are typically dominated by suspension feeders such as crinoids, brachiopods, bryozoans, sponges and bivalves rather than framebuilders. Crinoids were prominent in such environments throughout the Ordovician to Triassic and their decline had a major impact on the nature of mid- to outer-ramp deposits. Jurassic ramps are largely mud-dominated. The appearance of the rudist bivalves and other organisms such as orbitolinid foraminifers in the Cretaceous led to another change in the style of mid- and outer-ramp deposition which has continued, with other benthic foraminifers, into the Tertiary (Buxton & Pedley, 1989). It is, therefore, noteworthy that, while much attention has been given to the significant changes that occurred in shallow-water framebuilding communities during the Phanerozoic (e.g. James & Bourque, 1992), equally significant changes have occurred in the deeper water 'shelfal' or 'offshore' skeletal biotas.

As in siliciclastic settings (Sect. 7.7), criteria used to characterize offshore carbonate deposits are bed thickness, grain size, sedimentary structures, and ichnofaunas. In shoreface environments down to fairweather wave base, the sea floor is regularly disturbed by waves, tidal, and other currents. Bedform migration generates mostly tabular and trough cross-stratification which have high preservation potential because burrowing is inhibited. Below fairweather wave base, storms are the dominant control on sediment movement. A distinctive feature of proximal offshore to shoreface settings in some carbonate ramps is hummocky cross-stratification, or HCS (Sect. 7.7.2). Several types exist, representing flow regimes ranging from oscillatory, to combined oscillatory, and unidirectional. Limestones displaying HCS commonly give way proximally to planar laminated or swaley cross-stratified grainstones (SCS) (e.g. Handford, 1986; Faulkner, 1988; Sami & Desrochers, 1992), commonly organized in amalgamated units (Dott & Bourgeois, 1982; Faulkner, 1988). Distally, sediment becomes finer grained and beds thin, forming discrete packages separated by mudstones (Aigner, 1985; Faulkner, 1988) (Fig. 9.34). Storm events may also be represented by graded or laminated beds. A common feature is winnowed shell lags alternating with bioturbated suspension-load deposits. The coarse fraction may fill burrows to produce tubular tempestites (Wanless, Tedesco & Tyrrell, 1988). In many Precambrian and Cambrian ramp deposits, storm layers comprise lags of rounded intraclasts (Mount & Kidder, 1993), probably reworked cemented layers, something which is less common in younger rocks where deeper bioturbation hinders widespread cementation (Sepkoski, 1982). A distinctive feature of some deeper ramp sequences are finely laminated muddy carbonates (e.g. Faulkner, 1988; Calvet & Tucker, 1988). Their origin remains problematic but they may indicate the past activities of contour currents.

The dynamics of sediment movement during storms can be complex and are poorly understood (Gagan, Chivas & Herczeg, 1990) and so the extent to which currents generated by storms transfer sediment across carbonate ramps is unclear. The presence of shallow-water clasts (e.g. ooids) in offshore deposits, or the ratios of allochthonous to autochthonous biotas, can be used to assess this (Aigner, 1985). Handford (1986), for example, has suggested that tropical storms or hurricanes over Mississippian ramps in Arkansas transported sediment for as much as 30 km offshore. Recently, Aurell, Bosence and Waltham (1995) have suggested storm-related offshore transport of 25–40 km for Upper Jurassic ramps in Spain.

In sediments infrequently disturbed by storms, biological reworking ('bio-retexturing') is a prominent process. In Oligo-Miocene outer ramp deposits from Sicily and Malta, burrowed zones remained uncemented while adjacent unburrowed areas became lithified. This resulted in apparently random variations in sediment fabric, with highly irregular intraformational contacts, some of which resemble subaerial exposure surfaces (Pedley, 1992). Bioturbation and local cementation

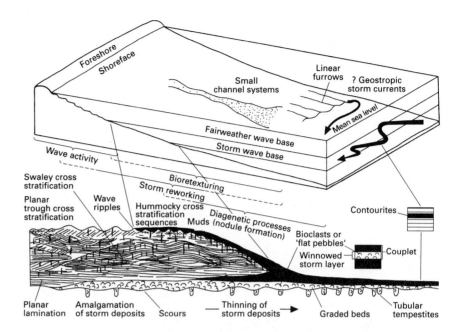

Figure 9.34 Processes and products in offshore ramp systems (see text).

also cause nodular bedding in mud-rich intervals which may be further enhanced by compaction. Deep burrowers, such as some shrimps, not only homogenize beds of different lithology, but can generate muddy layers of waste derived from subsurface excavations.

Proximal to distal trends in deeper-water carbonate depositional facies signify changes in environmental energy that are related to wave-base depth and therefore commonly to sea-level stand (Jennette & Pryor, 1993). This concept has been applied to the analysis of relative sea-level changes in Triassic ramp carbonates in western Europe (Aigner, 1984, 1985; Calvet & Tucker, 1988; Calvet, Tucker & Henton, 1990). Many distal ramp sequences comprise limestone–marl or limestone–shale alternations (Einsele & Ricken, 1991). Although this phenomenon has a variety of origins, including diagenesis (e.g. Raiswell, 1988), in many cases it may provide evidence of orbital forcing (e.g. Weedon & Jenkyns, 1990; Burchell, Stephani & Masetti, 1990; Elrick & Read, 1991). Elrick, Read and Coruh (1991) have also recognized possible 1–3 ky climatic cycles in rhythmically interbedded limestones and shales in outerramp and ramp-slope deposits of Mississippian age from Montana.

Whereas most detailed studies have demonstrated that these mid- to outer-ramp or deeper shelf settings were predominantly low-energy depositional environments, punctuated by occasional storm events, some, such as the Middle Devonian ramps of New York State, exhibit evidence of significant erosion by marine currents below normal storm-wave base. These generated parallel linear ridges 1.5 m high and 30–50 m wide separated by troughs containing bioclastic lags (Miller, 1991). Quine and Bosence (1991) have documented similar erosional features from Cretaceous deeper water chalks.

9.4.4 Carbonate slope deposystems

On high-relief, steep-sided carbonate platforms, there is an abrupt transition at the shallow-water platform margin to a transitional slope facies in which the bulk of the sediment has been resedimented. Slope deposits, although highly variable in character, provide much information on the nature of the adjacent productive platform and can preserve a sensitive record of the relative sea-level changes which affected the platform. In some ancient carbonate successions, shallow-water platform carbonates may be largely absent due to erosion or tectonism so that the nature of the platform and its history must be deciphered from the remaining slope deposits.

Carbonate slope heights range from 10s to 1000s m. The slope angles of active and fossil carbonate platforms range from about 1° to almost vertical. Slope profiles are highly variable, mostly being concave upwards, although they may also be planar or convex, with the angle of slope typically increasing with slope height (Schlager & Camber, 1986; Kenter, 1990). The steep profiles of many carbonate slopes, compared with siliciclastic slopes, are due to a combination of factors, namely the high shear strengths of carbonate sediments (Kenter, 1990), the presence of framebuilding organisms, early cementation of granular sediment, and the early lithification of carbonate muds (Kenter & Schlager, 1989). A critical control on slope angle is sediment composition. Grainy, non-cohesive mud-free sediments such as carbonate sands and conglomerates are able to construct steeper slopes than muddy sediments (Kenter, 1990) (Fig. 9.35). On the forereef slope of the Capitan Formation (Sect. 9.3.3), the upper slope deposits had primary dips of 30°, and are composed of intraclastic packstones and rudstones. The middle slope had lower primary dips (15–30°) with finer-grained rudstones. The

○ Well documented examples
□ Examples lacking precise control on geometry
△ Flanks stabilized by organic framebuilding or cementation

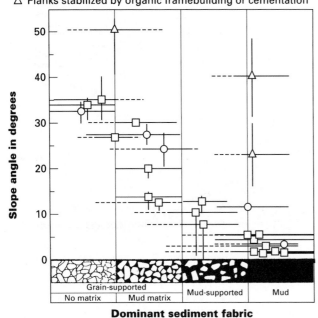

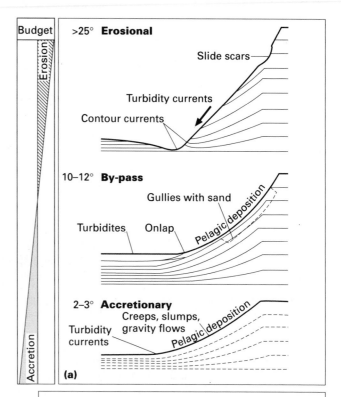

Figure 9.35 Plot of carbonate slope angle against the dominant sediment fabric. The error bars represent the range of fabrics and slope angles for each margin. The data set comes from a range of active and fossil slopes (after Kenter, 1990).

lower slope, with dips of 5–10°, are richer in mud-supported carbonates (Melim & Scholle, 1995). However, differential compaction may also cause an increase in slope angles (Saller, 1996).

CARBONATE SLOPE TYPES

Three main types of carbonate slope have been identified from the modern platform margins of the Bahamas (Fig. 9.36), an area where such settings have been intensively studied: erosional, bypass, and accretionary (or depositional) (Schlager & Ginsburg, 1981).

Erosional slopes are generally steep, >25° (Fig. 9.36), and represent exposed submarine rock walls or slopes truncated by collapse of large sections of the platform margin or scoured by sediment gravity flows or contour currents (e.g. Freeman-Lynde & Ryan, 1985; Mullins, Gardulski & Hine, 1986). Erosional platform margin retreat of many kilometres has been document-ed (Mullins, Gardulski & Hine, 1986). An ancient example of an erosional margin may be present in the Permian Grayburg Formation of the Guadalupe Mountains, Texas (Fig. 9.37), where the erosion marked a shift in depositional style from a ramp-like to a steep, vertically accreting shelf margin (Franseen, Fekete & Pray, 1989). In this case slumping was probably the major cause of erosion, coupled with erosion by debris flows and bottom currents.

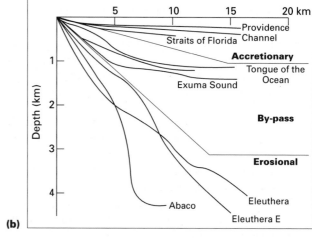

Figure 9.36 Bahamian slope types. (a) Slope profiles (schematic) with major processes. (b) Actual profiles (vertical exaggeration is four times over horizontal) (originally from Schlager & Ginsburg, 1981). See also Fig. 9.42.

Bypass slopes are relatively steep (>10–12°) (Fig. 9.36) and, although they may accumulate drapes of pelagic sediment, material derived from the shallow-water platform edge bypasses this zone as sediment gravity flows. Low sedimentation rates on bypass and erosional margins promote extensive cementation and hardground generation (e.g. Land & Moore, 1977).

Accretionary slopes have low angles (Fig. 9.36a,b), and are built of sediment gravity flow deposits. On the Bahamian margins the main slopes are separated from the shallow

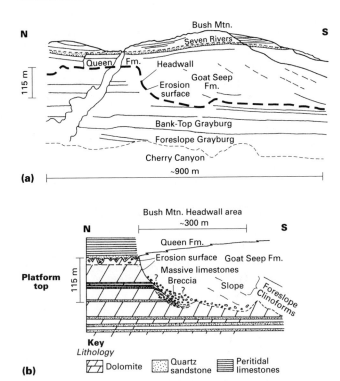

Figure 9.37 Submarine erosion surface, Permian Grayburg Formation (Bush Mountain) Guadalupe Mountains, west Texas (see Fig. 9.13). (a) Overall view of Bush Mountain section showing scale of erosion surface. (b) Detail of surface (after Franseen, Fekete & Pray, 1989).

platform edge by a marginal escarpment, at depths of 20–40 to 150–180 m. The accretionary slopes are developed below this level and that of the northern Little Bahama Bank has been extensively documented (Mullins, Heath *et al.*, 1984) (Fig. 9.38). As in other Bahamian slopes two zones can be identified: an upper slope which is gullied, and a lower slope or basin margin rise, which constitutes an apron of resedimented material.

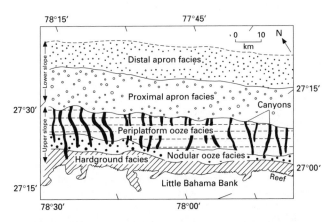

Figure 9.38 Map of near-surface facies of the northern slope of the Little Bahama Bank (modified from Mullins, Heath *et al.*, 1984).

The gullied zone occurs at depths of 200–900 m. The gullies or canyons are V or U shaped features 1–3 km across and 50–150 m deep, which terminate at the base of the upper slope or, more rarely, continue as broad channels across the proximal part of the apron. Coarse bouldery sediments lie in the gully floors, while the walls are coated with Fe and Mn oxide crusts and are extensively bored. The gullies are probably initiated as slides and slumps, possibly during platform margin steepening, the resultant scars subsequently being further scoured by sediment gravity flows. The intergully areas are covered with periplatform ooze, planktonic foraminiferal and pteropod muds, with much platform-derived aragonitic and high-Mg calcite mud. The composition of the coarser fraction in these sediments varies markedly with the orientation of the margin to the prevailing wind (see below). On the upper portion of the slope, sediments have been cemented to form nodular oozes and hardgrounds (Mullins, Neumann *et al.*, 1980; Droxler, Schlager & Whallon, 1983).

The major site of deposition is the lower slope apron (Fig. 9.38). Proximally, the succession consists of interbedded mud-supported debris flows up to 5.5 m thick and coarse turbidites up to 2.5 m thick. The former thin over 20–30 km so that distal successions comprise grain-supported debris flows <1 m thick and variable coarse to fine turbidites. The coarse fraction in these deposits consists largely of resedimented, cemented periplatform oozes. Sediment gravity flows in other areas of the Bahamas, as in Tongue of the Ocean and Exuma Sound, contain abundant shallow-water grains (e.g. Schlager & Chermak, 1979; Crevello & Schlager, 1980). The slopes are prone to large-scale rotational slippage and detachment (Harwood & Towers, 1988) and some major debris flows, which cover large areas of the lower slope and basin floor, represent platform margin collapse.

Sediment supply to carbonate platform slopes may come from a number of sources. Fine-grained carbonate may be supplied from the shallow platform and from planktonic fall out to produce 'hemipelagic' periplatform oozes. Coarser, shallow-water grains are also supplied from the platform top and margin and mix with material eroded from the slope and redeposited in deeper water. Slope sediments accumulate in extensive, basinwards-thinning wedges, commonly with prominent internal accretion surfaces (clinoforms) which may be identifiable on seismic profiles (e.g. Brooks & Holmes, 1989). Such wedges appear to be generated mostly during relative sea-level highstands (Wilber, Milliman & Halley, 1990; Schlager, 1991; Glaser & Droxler, 1991; Wilson & Roberts, 1995), when rates of deposition as high as 2–2.5 m ky^{-1} may be achieved on the shallower portions of slopes (Brooks & Holmes, 1989, 1990; Glaser & Droxler, 1991). This phenomenon, known as *highstand shedding* (Mullins, 1983; Schlager, Reijmer & Droxler, 1994), is the opposite of that seen in most siliciclastic slope-basin systems where sediment is normally retained on the shelf or coastal plain during relative sea-level highstands and fans of eroded material develop during lowstands (see Sect. 2.4).

During highstands the carbonate platforms are largely submerged and carbonate production is turned on, so that during these intervals significantly larger volumes of sediment are produced (Fig. 9.39) and thicker and more frequent sediment gravity flows occur (e.g. Droxler & Schlager, 1985).

On flat-topped platforms, shallow-water carbonate production may be switched on or off during even relatively small sea-level oscillations. Changes in the sources of carbonate sediment with different sea-level stands means that many carbonate slope deposits are commonly cyclic, with variations in both mud mineralogy (Droxler, Schlager & Whallon, 1983) and the coarser sediment fraction (Haak & Schlager, 1989). More aragonite mud is introduced on to the slope from the shallow platform top during highstands than during lowstands, when any mud appears to be mostly calcitic (Kier & Pilkey, 1971). During lowstands the coarser fraction is dominated by bioclastic grains derived from organisms living on the bank margin and calcitic, largely low magnesian calcite (LMC), muds form (with coccoliths). When platforms are flooded, abundant non-skeletal grains (ooids, peloids) may be produced (Fig. 9.40). In the geological record such differences may be expressed in different diagenetic histories of interbedded carbonate mudrocks in slope and near-platform settings. Triassic calciturbidites in the eastern Alps have revealed such compositional variation, reflecting periods of shedding, flooding and exposure on the adjacent platforms (Reijmer, Ten Kate *et al.*, 1991). An additional complication to this model is that grain types in these redeposited sediments may also strongly reflect the windward or leeward orientation of platform margins (Fig. 9.40). On protected windward margins, mainly skeletal material is redeposited from the platform margin (with less muds than leeward margins). On more open windward margins most shallow-water sediment is transported on to the platform top. On open leeward margins extensive offbank transport results in the redeposition of ooids, peloids and micritized grains derived from the platform interior. On tidally dominated margins extensive oolite resedimentation may take place.

On the gently sloping south Florida continental margin, the slope wedge, deposited in 100–600 m water depth, is characterized by seaward-dipping clinoforms. The sediments are strongly cyclic, but reveal a complex history related to relative sea-level changes (Brooks & Holmes, 1989) which suggests analogies with many ancient distally steepened ramps. During the later regressive (highstand) and early transgressive phases, when the shelf was shallowly submerged, offshelf transport was vigorous and the wedge prograded. During the later transgressive to early highstand phases, when the shelf was covered by over 100 m of water, less offshelf transport occurred. During lowstands, enhanced erosive activity by the Florida Current, a platform-margin-parallel contour current (Sect. 9.3.4) resulted in erosion of the previously deposited sediments, even at depths of greater than 330 m, enough to generate erosional unconformities. The sediments consist of

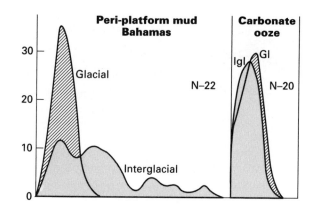

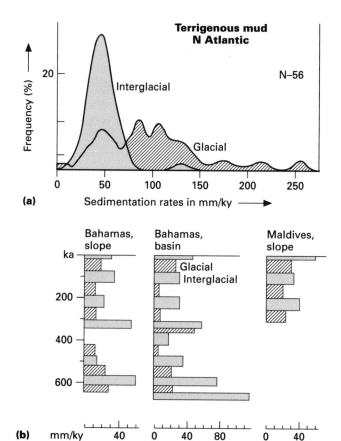

Figure 9.39 Carbonate deposition rates. (a) Sedimentation rates during sea-level highstands and lowstands in various deposystems. Periplatform muds are produced at highest rates during interglacial (highstands) periods (highstand shedding). The reverse is true for terrigenous muds, with lowstand shedding. Pelagic carbonate ooze production is low and unaffected by glacioeustatic sea-level changes (modified from Schlager, 1991). (b) Highstand shedding from the Bahamas and Maldives from ODP sites, shown by increased sedimentation rates. Key as above. The interglacial intervals are defined by oxygen isotope stratigraphy and probably represent longer intervals than the duration of platform flooding. This has the effect of reducing the apparent differences between high- and lowstand rates (modified from Schlager, 1991, and other sources).

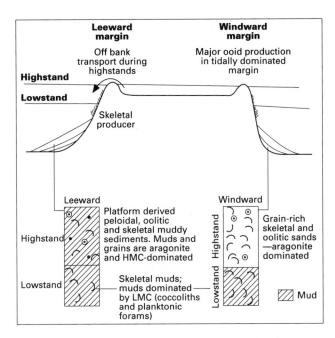

Figure 9.40 Contrasting compositions of resedimented carbonates during highstand and lowstand and with leeward or windward positions (see text).

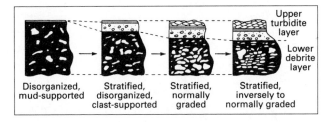

Figure 9.41 Two-component debrite–turbidite layer, showing change from proximal to distal settings. The base may be erosional or conformable and sole structures are rare (after Krause & Oldershaw, 1979; Tucker & Wright, 1990).

breccia (debrite), probably produced by a debris flow, while the top may grade into a turbidite.

FACIES MODELS

Two basic facies models are available for interpreting ancient slope-associated carbonates: the apron and fan models. The latter has no modern analogue and ancient examples are few in number (see reviews by Tucker & Wright, 1990; Coniglio & Dix, 1992). As a consequence of the linear nature of carbonate platform margins, they tend to supply sediment to the slope from a *line source* or from multiple, closely spaced sources (gullies), generating wedge-shaped aprons of sediment.

Although these two simple models, the fan and apron, are readily applicable to the geological record, modern slope systems can be variable. Mullins and Neumann (1979) and Mullins (1983) have recognized several major types of Bahamian slope facies associations, of which some are likely to be applicable to the geological record (Fig. 9.42). The significance of these Bahamian margins is that they illustrate that a single platform can have margins of more than one style.

CARBONATE FANS

Most ancient carbonate fans appear to have developed in tectonically active areas, and there commonly seems to be a strong structural control on their location. In the Jurassic of Portugal (Wright & Wilson, 1984), carbonate fans appear to have inherited their morphology from antecedent siliciclastic fans and the platform margin may have been dissected by major channels which localized transfer of both siliciclastic and carbonate material into the basin.

Ancient carbonate fans show similarities in lithofacies and architecture to siliciclastic fans (see Chapter 10), although documented examples suggest that their internal organization is somewhat simpler, which is possibly the effect of grain size. The similarities of carbonate fan sequences to those of siliciclastic fans allows them to be differentiated from apron deposits.

fine sand- to mud-grade carbonates, with the coarser fraction containing grains derived from both shallow-water and open marine environments, the former represented especially by molluscs and benthic foraminifers.

LITHOFACIES

As in siliciclastic systems, slides, slumps, debris and grain flows, and turbidity flows, all occur on carbonate slopes, and modification of slope deposits by contour currents is also well documented (Cook & Mullins, 1983; Tucker & Wright, 1990; Coniglio & Dix, 1992). Differences between carbonate and siliciclastic lithofacies do occur where coarse sediments are involved since many resedimented carbonates contain prominent boulders and clasts, from metres to hundreds of metres in size (olistoliths), and are widely termed megabreccias (e.g. Hiscott & James, 1985; Garcia-Mondéjar & Fernández-Mendiola, 1993). Such clasts may comprise reworked, indurated slope material or cemented platform-top sediments. Most large boulders lie in a coarse talus zone close to the platform margin. Although huge parts of platform margins may collapse as a result of inherent slope instability, tectonic activity is probably a trigger in many cases (e.g. James, Stephens *et al.*, 1989). Sea-level falls, by removing hydrostatic support, may also contribute to the collapse of carbonate escarpments.

Many carbonate slope deposits exhibit metre-scale, two-component debris sheets (Krause & Oldershaw, 1979) (Fig. 9.41), the lower part of each unit consisting of mud-supported

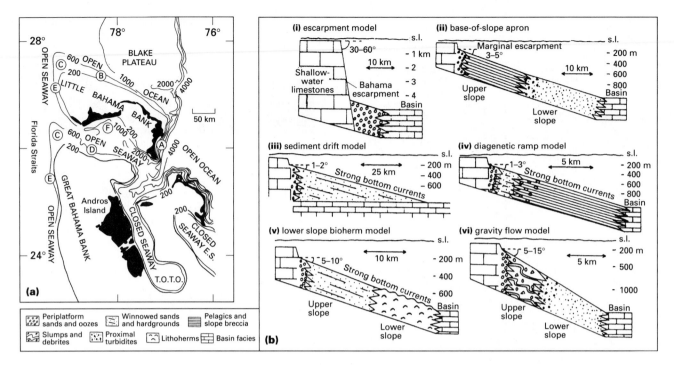

Figure 9.42 Bahamian slope models. (a) Positions of various slope facies types. (b) Facies distributions (originally after Mullins & Neumann, 1979 and Mullins, 1983). Sediment drifts are discussed by Mullins, Neumann *et al.*, (1980).

SLOPE APRONS AND BASE-OF-SLOPE APRONS

Mullins and Cook (1986) have recognized a range of margin slope types which has two simple end-members, namely *slope aprons* and *base-of-slope aprons* (Fig. 9.43). These are comparable to ramp and, confusingly, slope apron models respectively, in siliciclastic systems (Reading & Richards, 1994) (Sect. 10.3). Slope aprons develop on relatively gentle slopes (<4°) and are continuous with the shallow-water platform margin from which they are supplied. Sediment supply is from a line source and little sediment bypasses the slope. There appear to be no large slope aprons related to modern carbonate platforms, a consequence of relatively recent high rates of relative sea-level rise, but numerous ancient examples existed. Base-of-slope aprons (Fig. 9.43) accumulate adjacent to steeper slopes (4–15°) so that sediment bypasses the upper slope, and is transferred to its foot via numerous gullies (Fig. 9.43). Small channels may cross the upper proximal apron.

The proximal, inner apron facies of both slope and base-of-slope aprons consist of debris flow material, matrix-poor megabreccias, and thick, proximal turbidites and periplatform oozes. In base-of-slope aprons, however, large-scale gullies may be present, and there may be intervals of periplatform ooze with slumps, slides and truncation surfaces. Sediments in the distal apron zone comprise finer-grained turbidities. Where the gullies that supply base-of-slope aprons are deeply incised, reworked lithified platform margin sediment may be abundant in the apron deposits, while slope apron deposits may contain

few platform limestone clasts. Base-of-slope apron deposits may show thickening-up packages of sediment gravity flows, reflecting changes in gully evolution or the progradation of discrete depositional lobes, but slope apron successions are typically poorly organized.

The Triassic platforms of the Dolomites, northern Italy (Sect. 9.3.2) provide spectacular examples of platform margin geometries (see Fig. 9.11), with slope aprons. The Latemar build-up, which is some 700 m thick and had a depositional relief of 400 m at times, has 40–50% of its volume composed of foreslope deposits. Adjacent to the platform margin reefal boundstones are steeply dipping, more proximal foreslope talus breccias, which consist of lobate breccia beds, 2–5 m thick, a few tens of metres across and extending hundreds of metres downslope. Planar clinoforms extend the entire height of the slope, delimiting packages of sediment termed *clinothems* (Harris, 1994). The clinoforms possibly relate to shear surfaces formed during large slope failures. In the toe of the slope (Fig. 9.11b) graded grainstones, <1 m thick, occur together with some extensions of the foreslope breccia units. Debris flow deposits are absent from these slope deposits, in contrast to modern foreslope systems. This lack of mud may reflect the small size of the source platform (4 km wide) which was probably swept by high-energy currents (Harris, 1994).

Base-of-slope aprons have been extensively documented from the geological record, and are particularly well developed along the faulted escarpment margins of European Triassic and Jurassic platforms (e.g. Eberli, 1987; Abbots, 1989). The Jurassic

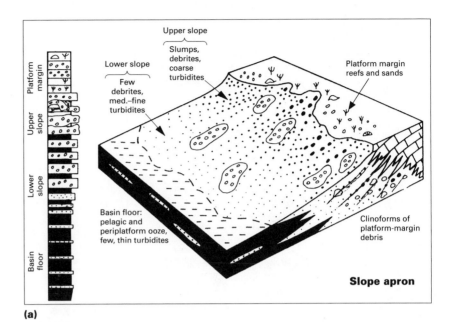

(a)

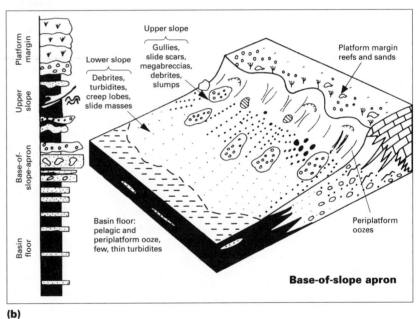

Figure 9.43 Apron facies models (after Mullins & Cook, 1986).

(b)

Vajont Oolite of the Venetian Pre-Alps (Southern Alps) in northern Italy (Abbots, 1989), forms a wedge 800–1000 m thick, 100 km long and up to 50 km wide, and may represent deposition from the early Bajocian to the early Callovian (over 12 million years). The resedimented oolite has an estimated volume of 1200–2500 km³ yet was deposited in a linear rift-basin several hundred metres deep (the Belluno Trough), possibly a Jurassic equivalent to Tongue of the Ocean (Bosellini, Masetti & Sarti, 1981), between the Fruili and Trento platforms (Fig. 9.44a). The limestones comprise oolitic and peloidal turbidites in amalgamated beds ranging from a few centimetres to several metres in thickness. Palaeocurrents suggest that the major

source for the apron was the Fruili Platform to the east and its complex internal architecture, with intervals of thickening and coarsening-upwards beds, suggests deposition as lobes (Abbots, 1989) (Fig. 9.44). However, fan-related characteristics, such as channels, are absent and sedimentary units are otherwise disordered suggesting aggradation rather than systematic progradation. The base-of-slope apron may have been confined by tectonic activity, with an eastward basin floor gradient coupled with active subsidence serving to restrict its basinward progradation (Abbots, 1989) (Fig. 9.44b). Internal thickness variations probably represent the influence of subsidiary faults within the basin.

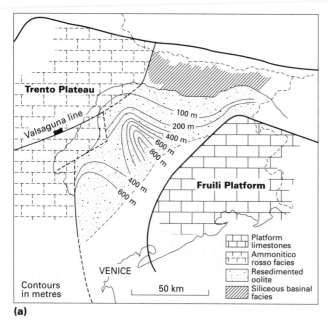

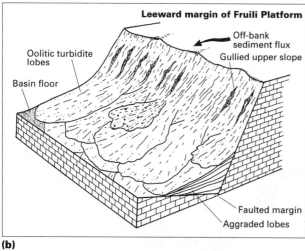

Figure 9.44 Middle Jurassic Vajont Limestone, north Italy.
(a) Isopach pattern (modified from Bosellini, Masetti & Sarti, 1981).
(b) Proposed base-of-slope apron model (after Abbots, 1989).

9.4.5 Reefs

In reefs, biological carbonate sediment production and environmental modification are realized to maximum extent. The term *reef* is used here simply to denote any biologically influenced carbonate accumulation which was large enough to have developed topographic relief above the sea floor. Discussions on reef terminology have been given by Heckel (1974), Longman (1981) and Geldsetzer, James and Tebbutt (1988).

REEF CLASSIFICATION

Reefs are complex structures that generate a whole spectrum

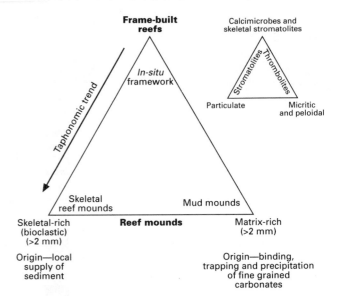

Figure 9.45 Classification of reef types. Frame-built, or skeletal reefs are produced by large, *in situ* calcareous metazoans, calcareous algae or microbialites. Reef mounds lack this framework but exhibit a wide range of structures reflecting varied origins. Some frameworks are destroyed by bioerosion and other processes (taphonomic loss) forming framework-depleted reef mounds. The small triangle represents a similar classification for microbialites.

of structures of geological significance and the classification outlined in the following section addresses only the end-members in this series. A fundamental distinction is between reefs that possess a calcareous framework (frame-built reefs) and those that lack a rigid structure (reef mounds) (Figs 9.45 & 9.46).

Frame-building organisms have not always been present and neither have they always been effective as reef constructors. In addition, local conditions inhibited frame-building organisms, and reef mounds were the only build-up type that could develop. Such build-ups characterize areas today where salinities, temperature, significant depth and, apparently, levels of nutrient supply prevent coral growth. In the past, but particularly in the early Carboniferous and early Jurassic, such mounds were important features of deeper ramp settings (Burchette & Wright, 1992).

Reefs can be characterized on the nature of the dominant organisms which constructed them (e.g. algae, stromatoporoids, rudists, corals), or on their morphologies (e.g. atoll, faro, barrier and fringing reefs) fashioned by environmental factors (see discussion in Tucker & Wright, 1990). Reef growth itself may produce large, wave-resistant reef complexes with sufficient relief to create a range of depth- and wave-energy-dependent subenvironments (e.g. forereef, reef front, crest and back reef). Smaller *patch reefs* may develop in the shallow-water interiors of major platforms and atolls, or on the shallower parts of ramps. Both reef complexes and patch reefs are typically classifiable as frame-built reefs.

REEF PROCESSES

The character of reef types has generally been determined by four basic, but interrelated, processes: construction, destruction, sedimentation and cementation – the emphasis among them changing during the evolution of a single reef.

Constructive processes represent the rate and style of carbonate production by a wide range of reefal organisms (Fig. 9.46). Primary frame builders (Scoffin & Garrett, 1974), mostly forms with heavily calcified skeletons such as colonial corals, stromatoporoids and calcareous algae (particularly red algae), constitute the rigid elements of frame-built reefs. The frameworks are consolidated by secondary frame builders which encrust and bind. In reefs from the mid-Tertiary to the present day, for example, crustose coralline algae have played the latter role, along with serpulids, vermetid gastropods, bryozoans and encrusting foraminifers in subsidiary roles. Such organisms are also commonly abundant in cavities in reefs.

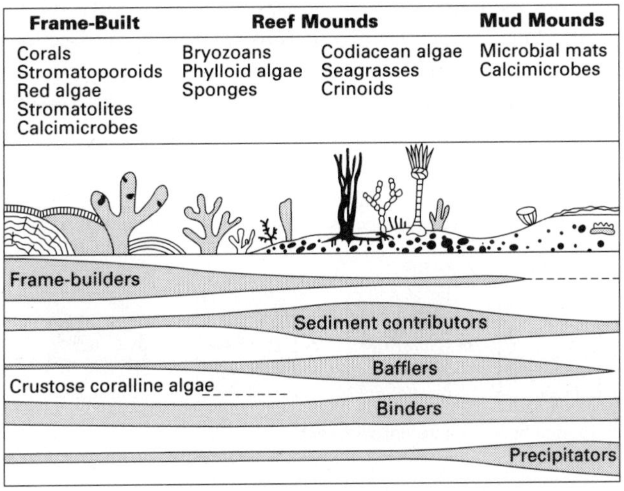

Frame-Built	**Reef Mounds**		**Mud Mounds**
Corals Stromatoporoids Red algae Stromatolites Calcimicrobes	Bryozoans Phylloid algae Sponges	Codiacean algae Seagrasses Crinoids	Microbial mats Calcimicrobes

Figure 9.46 Spectrum of reef types and the roles of the various organisms involved (after Tucker & Wright, 1990).

Frameworks of this type are an important component of many reefs but need not be volumetrically dominant, even in apparently frame-built reefs. Cores through modern coralgal reefs show that large volumes of the structure are occupied by cavities and loose sediment. The cavities are generated by the irregular growth of frame builders and by their subsequent skeletal degradation due to bioerosion. In shelf-edge reefs of St Croix in the Caribbean, for example, only 41% of cores through the reef are composed of framework material, with some 40% of the volume consisting of loose sediment and 19% of cavities (Hubbard, Miller & Scaturo, 1990). At any given time, only about 20–50% of the reef had active coral growth, the rest was undergoing bioerosion or was being covered by sediment. In core examination, identifying reefal facies can be extremely difficult, as demonstrated by disagreement over the reefal or non-reefal nature of some Bahamian platform margins (Beach & Ginsburg, 1982; Mullins, Hine & Wilber, 1982). At times during the Palaeozoic and in the early Triassic, calcimicrobes, organisms of problematic affinity such as *Renalcis, Epiphyton* and *Tubiphytes,* constructed significant reefs which consist essentially of 'micro-frameworks' (e.g. James & Gravestock, 1990) (Fig. 9.45). These represent a hybrid reef type with affinities to microbial reef mounds.

Frame building is not the only constructive process which occurs in reefs. Many epibenthic organisms that live on reef surfaces or within growth cavities also generate large volumes of sediment. *Halimeda*, a calcareous alga, is an important sediment contributor on modern reefs and, in slightly deeper water where corals are absent, may create extensive skeletal banks (Roberts & Macintyre, 1988). Other organisms simply mediate carbonate precipitation on the reef or mound surface and within cavities, as in mud mounds. Modern platform banks such as those of the Florida shelf (Sect. 9.3.3) are the products of localized skeletal accumulations, although strongly modified by bioerosion and bioturbation (Bosence, 1989).

Organisms play other constructive roles, but indirectly. Sea grasses, for example, build a range of skeletal banks at the present time, where the calcareous epibionts living on their blades contribute to the coarser sediment fraction but the grasses also trap the sediments by their blades baffling finer grain sizes and their rhizomes binding the loose sediment together.

Destructive processes, particularly the rate at which these operate relative to constructive processes, are extremely important in reef development. They include not only physical destruction (storms, wave and tidal action), but also biological destruction, or bioerosion. Major storms, although relatively infrequent, are a significant influence on reef ecology and sedimentology, because over short periods they can remove large quantities of reefal material and disrupt organic growth. However, the recognition of storm events in reefal sedimentary successions is particularly difficult because the effects that they produce may be masked, or 'healed', by subsequent organic growth or bioturbation (Bonen, 1988).

There are several sorts of biological erosion. Some organisms (endoliths) bore into hard calcareous substrates and many corals have rinds, up to several centimetres thick, that consist of multiple borings (Jones & Pemberton, 1988). Such bioeroders include bivalves, polycheates, cyanobacteria, fungi, sponges, barnacles and echinoids. Clionid sponges are particularly active and appear to be responsible for 90% of bioerosion on some modern reefs; over half the potential framework carbonate produced on the St Croix reefs may be reduced to sediment in this fashion (Hubbard, Miller & Scaturo, 1990), mostly carbonate silt or very fine sand which are major components of many reef sediments (Acker & Risk, 1985). Some organisms, for example gastropods and echinoids, graze and rasp the living surfaces of the reef. Others, such as parrot fish, break off and ingest fragments of living corals. Burrowing organisms may also disrupt less robust low-energy reef structures to produce skeletal packstones and wackestones (Bosence, 1985).

The main effect of bioerosion, besides supplying sediment, is to weaken the reef framework, making it more susceptible to physical destruction, thus contributing to the gradual loss of reef fabric (Longman, 1981; Friedman, 1985). At times, the rate of bioerosion may exceed the rate of framework growth. So widespread are these destructive processes that when well-preserved frameworks *are* encountered in cores or outcrop, this usually appears to be the result of rapid burial or encrustion of the reef before bioeroders were able to dismantle it (Macintyre, 1984; Hubbard, Miller & Scaturo, 1990).

Sedimentation has several consequences for reef growth. A minimum amount of sediment is required to fill the extensive cavity systems that develop within reefs, but excessive sedimentation can result in their complete burial. Sources of sediment include the products of physical and biological destruction of the framework organisms as well as the skeletons of other benthos associated with the reef. On modern reefs the latter include *Halimeda*, coralline algae, molluscs, echinoderms and foraminifers. Active transport of sediment through reefs due to wave and tidal action results in the distribution of fine sediment to even the smallest pores and cavities. Internal sediments are thus a characteristic feature of reefal limestones (Fig. 9.47).

Cementation is a pervasive process in reefal carbonates where there is active water circulation (Schroeder & Purser, 1986; Tucker & Wright, 1990). Some types of reef cement are poorly understood. Peloidal 'cements', for example, are abundant in many reefs, but, while generally confined to simple intraparticle cavities or as thin crusts on modern reefs (Camoin & Montaggioni, 1994), they formed a major proportion of many ancient framework reefs and reef mounds (Bridges & Chapman, 1988; Sun & Wright, 1989). Such cements are commonly associated with microbial activity and stromatolitic structures. A more unusual reefal cement comprises botryoids of fibrous carbonate. Such cements constitute up to 80% of the volume of some Late Palaeozoic and Triassic 'cementstone reefs' (Stemmerik, 1991). Algae have been implicated in the formation of some forms (see discussion in Tucker & Wright, 1990).

A combination of the above processes, commonly with multiple phases of development, results in complex reefal fabrics at all scales, from that of outcrops to that of thin sections and generates highly distinctive organism–sediment mosaics (Figs 9.47 & 9.48).

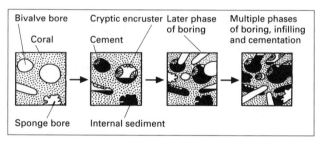

Figure 9.48 Fabric destruction of a reef framework due to multiple phases of bioerosion, sedimentation and cementation.

CONTROLS ON REEF GROWTH

The great diversity of organisms that have contributed to reefs over geological time, and the varied settings in which reefs occur, make it difficult to determine a simple set of requirements that are necessary for reefal growth. Nevertheless, the absence of reef complexes associated with hypersaline, brackish, turbid, or cool conditions suggests that ancient shallow-water reef-building organisms preferred similar conditions to those in which modern coral reefs grow. However, the area of the Earth's surface suitable for reef growth has not remained constant, so that at times (e.g. the Mesozoic), reefs and other carbonate depositional environments have apparently been much more widely distributed than at present.

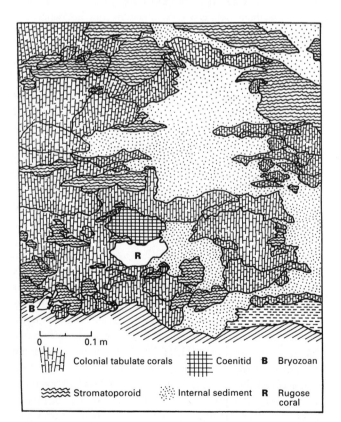

Figure 9.47 Organism–sediment mosaic from a vertical exposure of a Silurian (Wenlock) Hogklint reef, Gotland, Sweden (from Watts, 1981).

Modern corals grow best at depths of less than 100 m in waters of near normal salinities which vary little in temperature outside the range 25–29°C. However, some modern (and many ancient) reef-mound builders live for various reasons under cooler, deeper-water conditions. Water depth is a major control on the distribution of phototrophs, such as calcareous algae, or organisms, such as hermatypic corals, which are partly dependent on photosynthetic symbionts. It seems likely that Palaeozoic rugose and tabulate corals were also light dependent, as perhaps were some stromatoporids or even a few of the rudist bivalves. Such symbiotic relationships appear to enable the rapid growth and thus very high rates of carbonate sediment production. Under optimal shallow-water and climatic conditions, short-term growth rate for some corals is 100 m ky^{-1} (Buddemeier & Kinzie, 1976). Averaged growth rates of 9–15 m ky^{-1} have been recorded from the Caribbean (Adey, 1978) and 7–8 m ky^{-1} from the Great Barrier Reef (Marshall & Davies, 1984). Bosscher and Schlager (1992) have reviewed growth rates for corals and have used computer simulations to demonstrate the dependency of reef growth on water depth. The role of nutrients in inhibiting reef growth is another consideration (Hallock & Schlager, 1986). It is possible that the absence of large reef systems at intervals in the past might reflect more nutrient-rich seas (Wood, 1993).

Reefal communities capable of constructing significant platform margins have not existed continuously throughout the Phanerozoic (Longman, 1981; James, 1983; Fagerstrom, 1987). During periods when such organisms were absent or suppressed, most notably the early Carboniferous and early Jurassic, carbonate platforms were commonly ramp-like. Small reef mounds or patch reefs were the main reef types and, where distinct margins or build-ups developed, organisms played a less direct role. It has also been suggested that many Cretaceous platform margins may have been of this latter sort (Ross & Skelton, 1993).

Many reefs exhibit prominent biotic and sedimentological zonation which is controlled by changes in wave energy, light intensity, degree of exposure and sedimentation rate (Graus & Macintyre, 1989). Thus, significant lateral biotic zonation exists between the seaward and landward sides of reefs and vertically in the composition and morphology of the biota on reef fronts and slopes. The growth forms of a single species may vary with respect to these factors, or one species or genus may substitute for another. Modern corals with robust forms, commonly with hemispherical or thick dendroid morphologies, occur in high-energy settings. Delicate, branching forms occur preferentially in lower-energy settings, where erect growth hinders sediment smothering. Platy or tabular forms occur in deeper water, where increased surface area is an advantage in low light intensities. With the exception of some modern branching coral species (e.g. *Acropora palmata*), only encrusting forms such as coralline algae can withstand the high wave stress of the surf zone (Fig. 9.49a,b). Similar morphological zonation has been observed in

many Tertiary reefs (Fig. 9.49c). Although some Siluro-Devonian stromatoporoid reefs and Precambrian stromatolite 'reefs' do appear to show comparable features (James & Bourque, 1992), care is required when attempting to apply the principle to older reefs (Stearn, 1982; Fagerstrom, 1987, pp. 219–223). Many ancient reefs, particularly smaller reef-mounds or patch reefs, exhibit a consistent vertical succession of organic assemblages, commencing with a basal bioclastic accumulation, followed by colonization by frame builders which become increasingly diverse upwards, and culminating in a low-diversity unit of encrusting organisms (K.R. Walker & Alberstadt, 1975; Cooper, 1988).

Growth form		Environment	
		Wave energy	Sedimentation
	Delicate, branching	Low	High
	Thin, delicate, plate-like	Low	Low
	Globular, bulbous, columnar	Moderate	High
	Robust, dendroid branching	Mod- high	Moderate
	Hemispherical, domal, irregular, massive	Mod-high	Low
	Encrusting	Intense	Low
	Tabular	Moderate	Low

(a)

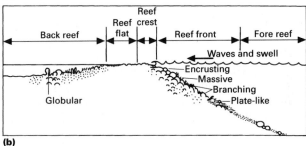

(b)

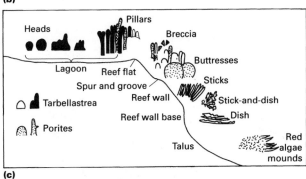

(c)

Figure 9.49 (a) Growth forms of reef-building metazoans and metaphytes and their relation to wave energy and sedimentation. (b) Cross-section of a hypothetical reef illustrating the different morphologies of the reef builders ((a) and (b) based on James, 1984). (c) Morphologies of corals from the Miocene Reefal Unit of Mallorca, Spain (based on Pomar, Fornos & Rodriguez-Perea, 1985; from Tucker & Wright, 1990).

REEF FACIES AND ENVIRONMENTS

Two basic framework reef types are considered here: *reef complexes* and *patch reefs*. In addition there are reef mounds that display a much broader spectrum of fabrics and morphologies and it is more difficult to provide simple, representative facies models for these structures. Although studies of modern reefs reveal numerous reefal subenvironments (e.g. Graus, Macintyre & Herghendroder, 1984; Guilcher, 1988), in ancient reefs 'filtering' by taphonomic and diagenetic processes means that generally only a much smaller range of facies is evident. Using criteria such as rock fabric, organism types and morphologies, and the distribution of early diagenesis, broad patterns can be recognized in outcrop or core. The most prominent among these are: reef crest and flat, backreef lagoon, reef front and forereef (Fig. 9.49).

Reef complexes are typically large reef tracts such as those that develop at rimmed shelf margins. Thus they may extend continuously or discontinuously for hundreds or even thousands of kilometres along strike. The Australian Great Barrier Reef and the Florida Shelf margin are modern examples.

The reef crest and front are the main productive zones of an active reef and extend from the highest point on the reef (the crest) to a point where frame construction ceases. Because of the light dependency of many reefal organisms, this zone extends, with productivity decreasing downwards, to 70–100 m below sea level (the photic zone), but variations occur related to the turbidity in the water column. Some organisms produce carbonate below this depth, but none is a significant framework constructor.

The reef crest is exposed to continuous wave action and even periodic exposure (Adey, 1978), so that the degree of breakage, abrasion, bioerosion and cementation is highest here (Macintyre, 1984) (Fig. 9.50). Organisms in the crestal zone of many reefs,

ancient and modern, adopt encrusting, wave-resistant growth forms, resulting in reef rocks with low species diversity, dominated by bindstones and some framestones. In slightly lower-energy settings on modern reefs, the elk horn coral *Acropora palmata* is prevalent.

Below the reef crest, many modern reefs exhibit a zone of buttresses (spurs) and chutes or channels (grooves) (Fig. 9.50). The spurs are coral and hydrozoan- (*Millepora*) covered extensions of the reef front (Shinn, Hudson *et al.*, 1981; Shinn, Hudson *et al.*, 1982) and the grooves are the carbonate sand-filled depressions between them. The former have reliefs of up to 7 m, widths of 5–10 m and lengths of 150 m and are orientated towards the direction of dominant wave approach. Spurs have been recognized in some ancient reefs (e.g. Pomar, Fornos & Rodriguez-Perea, 1985).

Below this zone is the main reef front, typically steep on modern reefs, which is veneered with thickets of corals, algae (including *Halimeda*) and sponges. The steep 'wall' zones, or forereef escarpments, of many modern reefs may, in part, owe their morphology to the rapid sea-level changes that occurred in the late Quaternary (James & Ginsburg, 1979; James, 1983). The reef front is capable of rapid lateral growth; for example Hubbard, Burke and Gill (1986) have measured lateral growth rates of 0.84–2.55 m ky^{-1} on St Croix. On steep-fronted reefs this may induce spalling of the upper wall (the 'brow' zone). Some of the debris may then be fixed on the upper forereef slope by cementation, the combined processes leading to progressive steepening (Ginsburg, Harris *et al.*, 1991). The sediments generated in this zone comprise largely 'reef rock' with a framestone or bafflestone fabric, with local bindstones (Fig. 9.50).

The *forereef* slope extends from the reef front to the basin floor and is fed by sediment derived from the reef through collapse, gravity flows, storms etc. The steep escarpments of

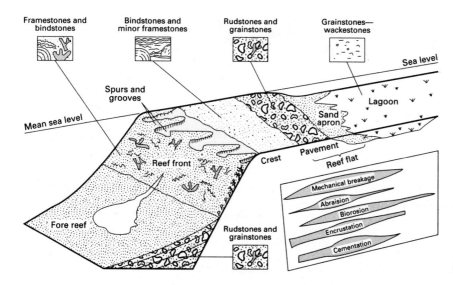

Figure 9.50 Major features of a reef complex.

many modern reefs are largely bypassed by loose sediment and the forereef talus effectively represents a base of slope apron (Sect. 9.4.4). In many ancient reefs the forereef is continuous with the reef crest and thus analogous to a slope apron.

Behind the crest is the *reef flat* which can be broadly divided into a *pavement* and a *sand apron* (Fig. 9.50). The former lies immediately behind the reef crest and is a narrow zone, ranging from a few to a hundred metres wide, over which water depth

is at most a few metres. The area is thus commonly exposed during very low tides. On modern reefs this zone is veneered by calcareous algae, small coral growths and 'micro-atolls' with storm debris from the reef front. Bioerosion can be marked. The resulting reef-rock exhibits a low-diversity biota, with fabrics such as bindstones, rudstones and sparse framestones.

The sand apron extends behind the pavement into the platform interior to depths of 10 m or so and can be narrow or,

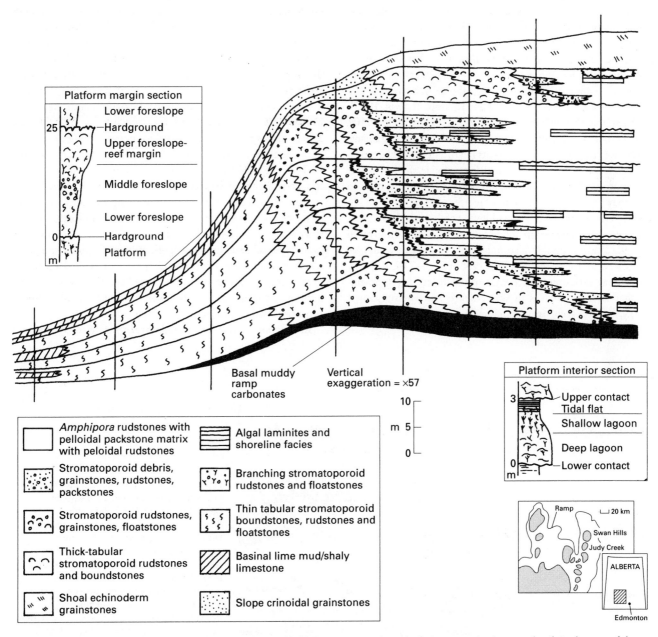

Figure 9.51 Cross-section through the reef-dominated margin of an isolated Devonian carbonate platform, the Swan Hills build-up, in the West Canada Basin, Alberta. The platform grew in several stages in which the margins initially stepped forward and then backstepped. In this platform, stromatoporoids formed the principal constructors. Insets show the facies succession in more detail. In the area of the foreslope the depositional cycles are separated by hardgrounds, while in the platform interior, subaerial surfaces mark this stratigraphic position. Vertical lines represent wells which were used to generate the cross-section. Compiled from Viau (1983) and Wendte (1992a,b,c).

as in the case of the Florida reef tract, very broad. Sediments deposited in this zone comprise largely skeletal grainstones and rudstones, grading to more matrix-rich lithologies towards the platform interior. In some modern reefs the area between the reef flat and the platform interior is marked by a steep, angle-of-repose or avalanche margin (Vernon & Hudson, 1978), in others the transition possesses little relief (Shinn, 1980).

The sediments of *backreef lagoons* vary with water depth and the degree of shelter afforded by the reefal rim. They typically include bioclastic packstones, wackestones and subordinate grainstones with a shallow-water biota. In less-restricted lagoons, minor patch reefs may occur. In modern lagoons, calcareous algae, such as *Halimeda* and *Penicillus,* as well as sea grasses, strongly influence the nature of the sediments generated. Lagoonal sediments of all types are typically burrowed and micritized (Tudhope & Scoffin, 1984; Tudhope & Risk, 1985).

The simple facies characteristics outlined above (Fig. 9.50), such as sediment texture, and the role of the major organisms can be used to identify environments in ancient reefs (Fig. 9.51).

Patch reefs are isolated reefs which develop in shallow-water environments such as platform interiors or on inner ramps. In scale they range from a few tens of metres to as much as 10 km across. Owing to their accessibility, patch reefs have been intensively studied and the models developed for their growth (Fig. 9.52) are probably also applicable to ancient examples (see review in Tucker & Wright, 1990, pp. 214–216.

Regional variations in the style of patch reef development, which reflect differences in water depth and environmental energy, can be seen on both modern and ancient platforms (Maxwell, 1968; Riding, 1981).

A principle control on the distribution of patch reefs in modern reef complexes appears to be the availability of suitable substrates, namely bare Pleistocene limestone or skeletal-rich sediments. Initially coral growths are scattered and generate a loose, cavity-rich framework, but eventually individual coral growths coalesce to form a tighter framework and the patch reefs aggrade into shallower, wave-agitated water. Depending on environmental conditions, windward–leeward margins may develop, with the higher-energy windward zone dominated by encrusters. Once the reef has aggraded to sea level, it expands with most rapid growth to leeward (Scoffin, Stoddart *et al.*, 1978; Marshall & Davies, 1982). As with larger reefs, a reef flat with micro-atolls and gravelly washover sediment may develop behind a windward algal crest. On the leeward margin, sediment dumped at the confluence of wave sets refracted around the reef may produce a sand cay or island (Fig. 9.52). The windward rampart and the leeward cay may become subaerially exposed and colonized by mangroves. The evolution of one such system on the Great Barrier Reef has been discussed by Tudhope (1989) and a comparative study with Silurian reefs has been made by Scoffin and Tudhope (1988). Windward–leeward effects can be recognized in many ancient patch reefs, as in the Silurian of the Great Lakes region of North America (Ingels, 1963; Shaver, 1977) (Fig. 9.53). In any depositional cycle, the backreef lagoon progressively infills, but small patch reefs may continue to develop until the lagoon is filled. However, much may be destroyed by physical and biological erosion, leaving the bulk of the lithosome composed of 'lagoon-processed' sediment (Tudhope, 1989).

Reef mounds (and mud mounds) are by far the most abundant

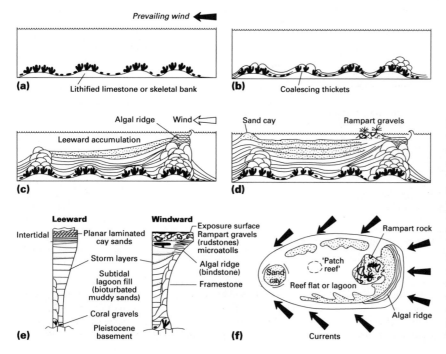

Figure 9.52 Growth of modern patch reefs. Growth begins on lithified Pleistocene or skeletal banks (a). Vertical growth is dominant with isolated thickets coalescing with continued growth (b). As growth continues into the zone of wave action (c), differentiation between the windward and leeward margins begins to develop with sediment accumulating on the leeward side. Wave refraction results in deposition around the leeward zone and sand cays may develop ((d) and (f)). Storm-generated gravels may form on the windward margin. (e) Shows different stratigraphic sequences which develop across the reef (based on data in Scoffin, 1987; Scoffin & Tudhope, 1988; Tudhope, 1989).

reefal structures in the geological record. They are typically matrix-rich, frame-deficient, lensoid (biohermal) or tabular (biostromal) structures (James, 1983). There is a wide diversity of types of reef mound, grading from ones largely composed of bioclastic accumulations to ones predominantly composed of carbonate mud or peloidal matrices (Figs 9.45 & 9.46). However, they are also subject to the same environmental controls that govern all reef growth. Reef mounds also occur in some modern environments. They range from very shallow-water skeletal mud banks on the Florida shelf, to somewhat deeper water (30 m) *Halimeda* banks of the Great Barrier Reef and the Kalukalukung platform (Roberts & Macintyre, 1988), or the deepwater *Lophelia* banks of the Atlantic (Scoffin, Alexandersson *et al.*, 1980; Mullins, Newton *et al.*, 1981) and the lithoherms of the Straits of Florida constructed by alcyonarian, crinoid, sponge and hydrocoral communities (Messing, Neumann & Long, 1990).

Reef mounds are constructed by organisms that lack significant frame-building potential. Most are therefore essentially banks of skeletal sediment, but where a frame-building component existed, the presence or absence of a framework seems to have been controlled by the balance between constructional and destructional processes (Bosence, 1985; Kiene, 1988; Scoffin, 1992).

Reef mounds on the Florida shelf (see Fig. 9.12), are of two types (Bosence, Rowlands & Quine, 1985; Wanless & Tagett, 1989; Bosence, 1995; Wanless, Cottrell *et al.*, 1995). The first comprises discrete mounds up to 3 km long, 1 km wide and 4 m high along the inner shelf margin, typified by Rodriguez Bank (Turmel & Swanson, 1976) and Tavernier Key (Bosence, Rowlands & Quine, 1985). These developed initially in sheltered depressions where muddy substrates supported *Thalassia* sea-grass communities. The mounds are composed of bioturbated mud derived from the breakdown of codiacean (green) algae and *Thalassia* epibionts, and are capped by skeletal

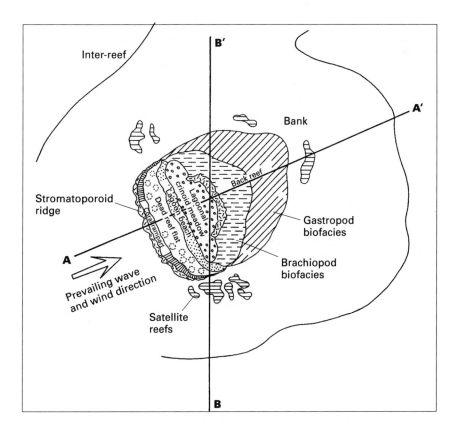

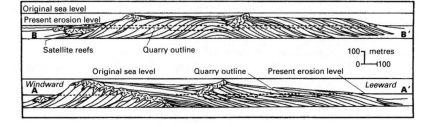

Figure 9.53 Palaeogeographic reconstruction of the Silurian Thornton reef system, northeastern Illinois, USA and cross-sections showing a typical windward–leeward facies distribution (modified from Ingels, 1963).

sand and gravel. In their present form, the mounds do not export sediment into the adjacent subtidal areas.

The second mound type is restricted to Florida Bay. In the west of the Bay, the mounds are broad and irregular in shape and account for 75% of the sediment cover, but in the east they are narrow and discontinuous, forming only 13% (Enos & Perkins, 1979; Wanless & Tagett, 1989; Wanless, Cottrell et al., 1995). In the central area of the Bay they have developed as a network of ridges 3–4 m in height separated by 'lakes' 3–4 m deep. Some of the mounds are subaerial and capped by mangrove islands. These shelf-lagoonal mounds nucleated on antecedent topography such as coastal levees and mangrove peats as sea-level inundated the area between 4500 and 3000 BP (Bosence, 1989). There is some evidence that the mounds are migrating 'downwind', towards the west. The mounds show marked variations in morphology and internal character across the Bay (Wanless & Tagett, 1989; Wanless, Cottrell et al., 1995). Their internal stratigraphies of layered muds reflect episodes of storm erosion (producing packstone–grainstone lags) and mud deposition, alternating with sea-grass colonization and bioturbation (Wanless & Tagett, 1989; Wanless, 1991).

From examples, such as the Florida Shelf, much can be inferred about the role of mounds in ancient carbonate shelf interiors, particularly the possibility that sediment mounds may have played a much more significant role than hitherto suspected (Wanless, 1991). Such features may have been widespread in shelfal environments and strongly influenced surrounding sedimentation. Although mounded, they have amplitudes of no more than a few metres and cover large areas so that in cross-section they appear essentially sheet-like. At the scale of most limestone outcrops, such features would appear as sheets of muddy bioclastic packstones and wackestones. Palaeozoic shelf-interior limestones commonly comprise great thicknesses of such sediments, possibly representing similar accumulations (Wanless & Tagett, 1989). Accumulations of larger benthic Foraminifera (e.g. *Nummulites*, alveolinids) are prominent in Cenozoic carbonates, but little attempt has been made to determine whether these might also have been deposited by such processes.

Some of the most enigmatic ancient reef mounds are those dominated by muds (Monty, Bosence et al., 1995), such as the late Devonian 'récifs rouges' of Belgium, Waulsortian mounds of the early Carboniferous (of north Africa, Belgium, Britain, Ireland, USA and Canada), and middle Triassic reefs of Spain. *Mud mounds* of this sort consist largely of micritic limestones, typically peloidal, with variable amounts of skeletal material and rare framestones (Figs 9.45 & 9.46). Some such structures possessed depositional relief of 200 m or more and developed as swarms that covered thousands of square kilometres (Lees & Miller, 1985). Smaller mounds commonly amalgamated to form larger structures. They tended to develop in deeper water, mainly in outer ramp settings, in some cases

probably below the photic zone. Many Waulsortian mounds show regular facies zonation related to water depth and level of turbulence, and in many cases appear to have accreted vertically into the photic zone (e.g. Lees & Miller, 1985, 1995). The clotted and peloidal micrites in their cores probably represent microbially mediated precipitates; sponges may also have played a role (Bourque & Boulvain, 1993; Bourque, Madi & Mamet, 1995). Similar microbial precipitates are widespread in many other reef types, but particularly those of the early Mesozoic (Reid, 1987; Sun & Wright, 1989; Leinfelder, Nose et al., 1993).

Most ancient deeper-water mud mounds appear to have developed as *in situ* carbonate accumulations and neither imported nor exported sediment (Lees & Miller, 1985, 1995; Miller, 1986; Bridges & Chapman, 1988). Studies of modern shallow-water mud mounds of Florida Bay and the Florida Shelf (Bosence, Rowlands & Quine, 1985), however, have shown that variations in their composition and fabric reflect changes in the sediment budget and the rate of sedimentation. The same mound can act variously as a sediment 'sink', as a closed system, or as a sediment 'exporter'. Shifts from one phase to another may occur in response to subtle changes in environmental energy, currents, or possibly even the degree of bioturbation. Mud banks of this sort represent local accumulations of sediment strongly modified by bioerosion and bioturbation (Bosence, 1989). Sea grasses contribute to their growth by acting as sediment baffles, while their rhizomes bind loose sediment together and calcareous epibionts on their blades augment the coarser sediment fraction.

REEFS VS. SEA LEVEL

Because the growth potential of reefal organisms is strongly related to water depth (Bosscher & Schlager, 1992), relative sea-level changes exert a major control on the geometry and architecture of reefal sequences. Using observations from outcrops, seismic sections and computer simulations, models can be generated that predict how carbonate platforms are likely to react during different relative sea-level stands and to different rates of relative sea-level change. Observations on the rates of carbonate sediment production and accretion in modern coralgal reefs imply potential rates of growth that exceed by almost an order of magnitude most likely rates of sea-level rise. This has been termed the 'paradox of carbonate platform growth' (Schlager, 1981). To drown a *healthy* modern reef would require either extremely high rates of relative sea-level rise or inhibition of carbonate production due to some environmental factor (e.g. Hallock & Schlager, 1986).

Many modern reefs were initiated as platform surfaces reflooded following ice sheet melting at the end of the Pleistocene period. However, there appears to have been a 'lag period' of 500–2500 years prior to the establishment of maximum carbonate production (Bosscher & Schlager, 1992) (see Fig. 9.5).

The causes for this are still unclear, but may be related to a threshold depth for optimal coral growth. Where, for some reason, the onset of carbonate production lags significantly behind flooding, water depths of 20 m or more may be achieved prior to significant coral growth. Once higher production begins (*start-up* stage), however, reefs may become established and grow rapidly to sea level in only a few thousand years (*catch-up* stage) (Neumann & Macintyre, 1985). The reef is then either able to maintain pace with rising relative sea level (*keep-up* stage) or is not able to keep up with the rate of relative sea-level rise (*give-up* stage) (Fig. 9.54A).

The responses of reefs to the balance between accretion rate and sea level is complex (Fig. 9.54). Give-up reefs (Fig. 9.54A) are only one response where the rate of sea-level rise exceeds the accretion rate. Some reefs 'struggle' to keep up, creating a

cone-like or even wedding-cake like structure (Fig. 9.54B). Large-scale structures like this are known from the northeast Australian shelf (Davies, Symonds *et al.*, 1989). If the accretion rate is slow enough for all the reef to become re-established in shallower water the reef retreats (Fig. 9.54C). In other situations the reef retreats in a series of discrete 'jumps' (back-stepping). Ancient reefs displaying this are well documented from the Devonian of Canada and Western Australia (Bassett & Stout, 1967; Playford, Hurley *et al.*, 1989). Figure 9.51 illustrates in its late stages, a back-stepping Devonian reef. If the rate of sea-level rise and accretion rate are balanced, the reef grows vertically to produce a pinnacle or reef-front–forereef escarpment (Fig. 9.54E). If the rate of rise is slow enough the reef will prograde over its own talus (Fig. 9.54F,G). Examples of these types of prograding, or advancing reefs have been documented by Bloxsom (1972) and Bay (1977) from the Lower Cretaceous of Texas (see also Fig. 9.51). If sea level is static the reef may prograde (spill-out) horizontally (Fig. 9.54H).

During falling sea level the reef may migrate downslope as downlapping units (Fig. 9.54I), and examples have been documented from the Messinian (Miocene) of Spain (Dabrio, Esteban & Martin, 1981). However, if the rate of fall is rapid or pulsed, the reefs may down-step (Fig. 9.54J) (e.g. Santisteban & Taberner, 1983).

In reefal successions of long duration, it may be possible to detect a range of responses to relative sea-level changes (e.g. Mesolella, Sealy & Matthews, 1970; Viau, 1983; Playford, Hurley *et al.*, 1989) (Fig. 9.74). A well-documented example of a small reefal system which shows such variations occurs in the Upper Miocene of Mallorca, Spain (Pomar, 1991, 1993) (Fig. 9.55). This barrier reef complex prograded for some 20 km across a shallow shelf. Detailed studies of a 6-km portion have revealed a number of erosion surfaces that correspond with vertical shifts in the reefal facies belt. These are interpreted as responses to relative sea-level oscillations of several orders. Each accretionary reefal package shows:

1 horizontally-bedded lagoonal facies consisting of bioturbated bioclastic grainstones, coral patch reefs, oolites and brackish inner-lagoonal sediments with mangroves;

2 reef-core and front deposits 30–50 m thick comprising framestones in which the corals display a range of growth morphologies (see Fig. 9.49c);

3 inclined-bedded reef-slope facies consisting of grainstones and coral–*Halimeda* rudstones, passing into more distal, lenticular grainstones;

4 a muddy 'open shelf' or shallow basinal facies containing planktonic forams and echinoids.

During relative sea-level lowstands and highstands the reefs prograded, downlapping on to the open platform deposits. During relative sea-level rises the reefs and lagoons aggraded vertically (Fig. 9.55).

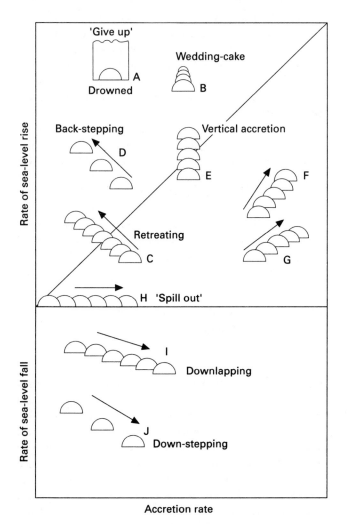

Figure 9.54 Schematic diagram showing the responses of reefs to sea-level rises and falls. Accretion rate refers to the actual accretionary rate of the whole reef and not the coral growth rate.

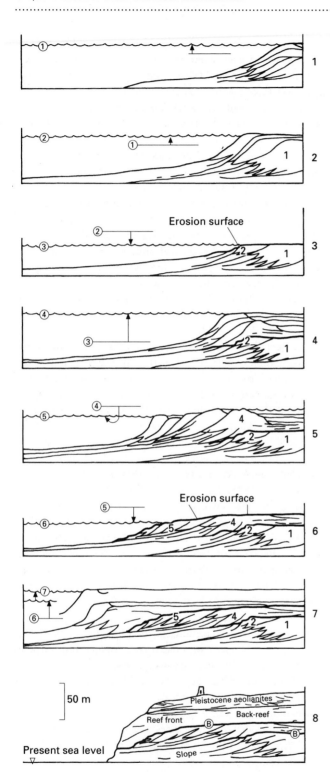

Figure 9.55 3-My sea-level cycles interpreted from late Miocene reefal sigmoid sets at Cap Blanc, Mallorca. Aggrading sigmoid sets occur during sea-level rises (1, 4, 7), while during highstands (2 & 5) and lowstands (1 & 4), sets of reefal sigmoids prograded over the open platform. Erosion surfaces developed during relative sea-level falls (3 & 6) (from Pomar, 1991).

9.5 Carbonate platforms and relative sea-level changes

The late acknowledgement that siliciclastic and carbonate depositional systems exhibit markedly different responses to relative changes in sea level (see Sect. 2.4) resulted in a long delay between the publication of the intial sequence stratigraphic literature (Vail, Mitchum & Thompson, 1977a,b) and the development of integrated sequence stratigraphic models for carbonate rocks. Consequently, in recent years, a major task has been to reconcile sequence stratigraphic models for these two sediment families and the literature on this aspect of carbonate sedimentation has now burgeoned (see e.g. Sarg, 1988; Crevello, Wilson *et al.*, 1989; Tucker, Wilson *et al.*, 1990; Loucks & Sarg, 1993; Simo, Scott & Masse, 1993).

The intial stages of this revision entailed the abandonment of a number of preconceptions about the manner in which carbonate sediments were deposited (e.g. Schlager, 1981; Kendall & Schlager, 1981) and the acknowledgement of the fundamental principles of carbonate sedimentation outlined in Sect. 9.2. Particularly contentious problems have been how carbonate sediment deposition is 'phased' with respect to relative sea-level changes and the nature of the 'drowning' events that commonly seem to mark the end of platform growth (Sect. 9.3.5).

Sequence stratigraphic studies are contributing to our understanding of the way in which carbonate platforms as a whole grow and are proving to be a most effective tool for clarifying the distribution of porosity and permeability heterogeneities in carbonate petroleum reservoirs (Grant, Goggin & Harris, 1994; Kerans, Lucia & Senger, 1994). However, it is first useful to recapitulate the main controls on carbonate productivity and accumulation (Sect. 9.2) as these relate to relative sea-level changes, since it is these factors that govern the character of carbonate sequences.

9.5.1 General controls on carbonate sequence geometry

CARBONATES VS. SILICICLASTIC SYSTEMS

In both clastic and carbonate marine systems sequence geometries are directly related to the morphologies of the depositional systems. Depositional morphology is controlled by interaction between the underlying topography, sediment type, and the rate of sediment accumulation as these relate to relative sea-level changes. Differences in the morphologies of sequences formed by each type of system arise from the different *manners* in which sediment is supplied, and the different *locations* of maximum sediment accumulation within the depositional systems. Thus, it is the *in situ* manufacture of carbonate sediment at or just below sea level (Sect. 9.2) that makes carbonate depositional systems highly sensitive to relative sea-level changes and can be identified as the single most important factor that accounts for these morphological differences.

Most morphologies encountered in carbonate depositional systems (Sect. 9.3) are foreign to siliciclastic systems and this is reflected in their wide range of responses to relative sea-level fluctuations. Although carbonate ramps respond, in many respects, similarly to siliciclastic shelves (Burchette & Wright, 1992), isolated build-ups or rimmed shelves that have flat tops and steep, high basinward slopes behave very differently. The internal stratal architectures of such carbonate platforms reflect the frequency and magnitude of relative sea-level changes much more effectively than most shallow-marine siliciclastic systems.

PRODUCTIVITY VARIATIONS WITH SEA-LEVEL CHANGES

Under optimal conditions, carbonate platforms tend to expand through sediment overproduction and most platforms experience this growth phase not only once, but commonly many times, in their 'life cycle'. It can only occur, however, when the rate of carbonate production for the whole platform exceeds that required to maintain the 'carbonate factory' in the shallow-water zone of optimal growth by infilling the deeper-water areas around it and corresponds with the keep-up condition discussed in Sect. 9.2.2. With respect to the sequence stratigraphic interpretation of carbonates, the significant fact is that the bulk of carbonate sediment production occurs in the top 100 m of the water column, with by far the highest production rates at depths of less than 20 m. If submerged below this depth, rates of carbonate production, and so the ability of the carbonate system to cope with increases in relative sea level, are significantly impaired (Sect. 9.2.1). It follows, that the ability of a carbonate platform to expand will be more effective in shallow-water settings, or at times of reduced rates of relative sea-level rise (where there is less sediment 'wastage'),

than in deeper-water or at times of rapidly increasing relative sea level. The rate of basinal sedimentation due to other sediment sources (e.g. pelagic, siliciclastic or volcaniclastic material) is clearly also of importance in this context since it can influence basinal water depth, and therefore the available 'accommodation space', independently of the sea-level curve.

Because high rates of carbonate production are limited to such shallow water, they vary enormously with relative sea-level stand (Sect. 9.2.2) and are most often discussed with reference to an idealized relative sea-level curve (Fig. 9.56). On most carbonate platforms, sediment production is greatest during relative sea-level highstands, when the whole platform surface is shallowly submerged ('keep-up' phase), than during the intervening lowstands or transitional phases (Fig. 9.57) (Haak & Schlager, 1989; Schlager, 1991). This phenomenon is commonly termed 'highstand shedding' (Schlager, Reijmer & Droxler, 1994). More rapid submergence to deeper water depths (incipient drowning, Sect. 9.3.5) commonly initiates collapse of the shallow-water environment (Ehrlich, Longo & Hyare, 1993) from which recovery may require a significant time, the 'lag time' (Sect. 9.3.5). A relative sea-level rise of just a few metres may be sufficient to precipitate such an event.

In contrast, a relative sea-level fall of just a few metres, on the scale of many 4th- and 5th-order cycles (Sects 9.2 & 9.4.1), may expose the whole interior of flat-topped platforms, potentially over hundreds of square kilometres. Although the carbonate factory can persist under such lowstand conditions, it is restricted to the platform slope (Fig. 9.57) and operates at a substantially reduced rate unless broad areas of the lower slope and basin floor are forced into the photic zone by the changes. The reduced supply of platform-derived sediment during relative sea-level lowstands means that, where siliciclastic sediments are absent, pelagic carbonates may become relatively more prominent in slope areas due to lack of dilution by shallow-

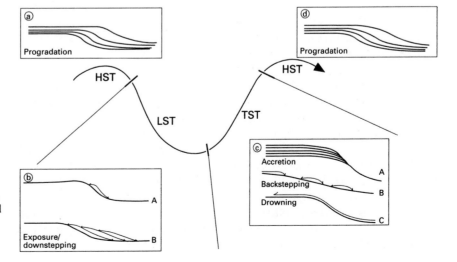

Figure 9.56 Ideal symmetrical relative sea-level curve (cf. Fig. 2.17) showing typical carbonate sequence behaviour during the various systems tracts. a & d, Highstand systems tract (HST): progradation dominated. b, Lowstand systems tract (LST): (A) lowstand wedge; (B) lowstand progradational system. c, Transgressive systems tract: (A) accretion dominated; (B) backstepping; (C) drowning.

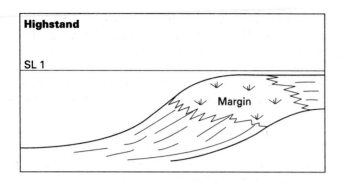

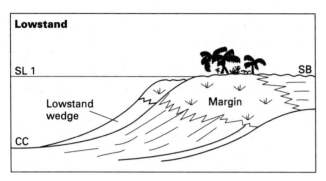

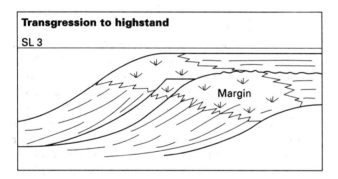

Figure 9.57 Organic carbonate platform margin showing behaviour during a simplified relative sea-level cycle (after Eberli & Ginsburg, 1989).

water carbonate sediments, a phenomenon that has been effectively demonstrated in a number of Tertiary build-ups including the western Great Bahama Bank (Eberli & Ginsburg, 1989) and the Miocene Natuna build-up in Indonesia (Rudolph & Lehmann, 1989). Minor falls in relative sea level, which do not expose the platform surface, have relatively little effect on the depositional system, but may stimulate progradation or the development of a shelf-margin wedge. Rimmed shelves show particular asymmetry in the rates of carbonate sediment production with respect to relative sea-level high- and lowstands, which is commonly reflected in an alternation in the composition of adjacent basinal deposits (e.g. Droxler & Schlager, 1985; Boardman, Neumann *et al.*, 1986).

The behaviour of carbonate platforms during relative sea-level lowstands has formed one of the main foci of debate

with respect to their sequence stratigraphic interpretation, since it is here that carbonate platforms exhibit some significant differences when compared with siliciclastic systems. In reality, only a limited number of examples have been documented in detail and the precise behaviour of carbonate platforms during relative lowstands of sea level is still a subject of some debate. In carbonate-only systems, deposition should be relatively predictable, according to the simple rules outlined in previous sections. In mixed carbonate–siliciclastic systems more complex relationships exist and these are discussed further in Sect. 9.5.6. The contrasts in responses to base-level changes, particularly during lowstands, between the different end-member platform morphologies, ramp and rimmed shelf, can be illustrated using modern examples as shown in Fig. 9.58 (Burchette & Wright, 1992). Note, however, that modern carbonate platforms offer no good analogues for highstand sedimentation on ramps (Sect. 9.1).

A 1–2 m fall in relative sea level will expose a significant area of a mature, flat-topped platform such as the Great Bahama Bank, but would have little effect on a 'homoclinal' ramp such as the southern Arabian Gulf or Shark Bay (Fig. 9.58). A 10-m fall would expose the whole of the platform surface of a major rimmed shelf and relegate shallow-water carbonate production to a narrow, unstable belt along the old platform margin (Cook & Taylor, 1991) (Fig. 9.58). On a carbonate ramp, this would expose a 20–50 km tract of the inner and mid-ramp, while much of the mid- and outer ramp would remain in or enter a favourable environment for lowstand shallow-water carbonate production. Basinal restriction might increase. Since many ancient ramps also appear to have been flat-topped during highstand progradational phases (Burchette & Wright, 1992), it seems that even quite small relative sea-level falls would have exposed extensive tracts of their platform interiors.

Large, rapid relative sea-level falls, of the order of 1–200 m, typical of glacial events, but probably atypical of much of the Earth's Phanerozoic history (Sect. 9.2), would entirely expose modern carbonate ramps (Fig. 9.58). Shallow basins, such as the Arabian Gulf and Shark Bay, would empty completely to become lacustrine or fluvial systems (e.g. Purser & Evans, 1973). Major rimmed shelves adjacent to deep basins, as in Belize (James & Ginsburg, 1979) and the Caribbean (Fig. 9.59) (Eberli & Ginsburg, 1989), would continue growth during lowstands as narrow shelves rimming the old platform slope or escarpment (Fig. 9.58).

Modern, distally-steepened ramps lie adjacent to deep basins and are likely to behave as ramps during minor base-level falls, but might become rimmed shelves during larger falls if shoreline facies belts migrate out to the break in slope (Fig. 9.58).

Drowned highs in deeper basins may become the sites of shallow-water carbonate production during relative sea-level lowstands of all scales.

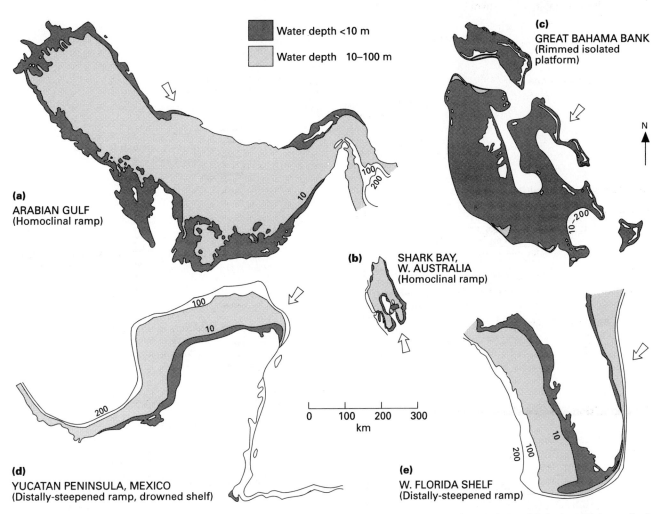

Water depth <10 m

Water depth 10–100 m

(a)
ARABIAN GULF
(Homoclinal ramp)

(b)
SHARK BAY,
W. AUSTRALIA
(Homoclinal ramp)

(c)
GREAT BAHAMA BANK
(Rimmed isolated
platform)

(d)
YUCATAN PENINSULA, MEXICO
(Distally-steepened ramp, drowned shelf)

(e)
W. FLORIDA SHELF
(Distally-steepened ramp)

0 100 200 300
km

Figure 9.58 Modern carbonate platforms drawn to the same scale, showing the areas exposed by relative sea-level falls of 10 m (black), 100 m (stippled), and 200 m (blank). Arrows represent dominant wind directions. Note that for a relative sea-level fall of 10 m, only the marginal areas of ramps, but the whole of a flat platform top, will be subaerially exposed. For a 100 m fall, basins with homoclinal ramps would be drained, while the margins of rimmed shelves and distally steepened ramps would retain some substrate for lowstand sediment production (after Burchette & Wright, 1992).

In all the above cases, there would be a basinward shift in coastal onlap, with shallowing of offshore depositional environments. However, it is important to be aware that a whole range of responses is possible, depending on platform morphology (i.e. ramp, rimmed shelf, or isolated build-up), magnitude of the relative sea-level changes and the rate at which these occur, and the responsiveness/productivity of the carbonate factories which depend largely on their geological age.

SEQUENCE AND PARASEQUENCE STACKING PATTERNS

The basic building blocks of carbonate platforms of all styles are small-scale shallowing-upwards cycles or parasequences (Sect. 9.4.1). Within any 3rd-order sequence these commonly vary systematically in character and are organized into distinctive 'stacking patterns' (Fig. 9.60) which reflect their

position in the relative sea-level cycle as well as the predominant growth style of the platform (Handford & Loucks, 1993; Goldhammer, Harris *et al.*, 1993). In the same manner, sequences may be arranged into 'sequence sets' within higher-order cycles. The latter generally reflect longer-term variations in relative sea level and for the most part are driven by subsidence. Typical configurations are backstepping or 'retrogradational', aggradational, and progradational or 'out-building' sequence sets (Fig. 9.61). Variable subsidence may result in variations in sequence geometry laterally within a basin and contribute to problems in correlating sequences over wide areas, particularly where sample points are widely spaced.

Where relative sea-level behaviour is dominated by high rates of subsidence, relative sea level may continue to rise strongly during the early highstand systems tract so that the platform aggrades rapidly or backsteps and shows little over-

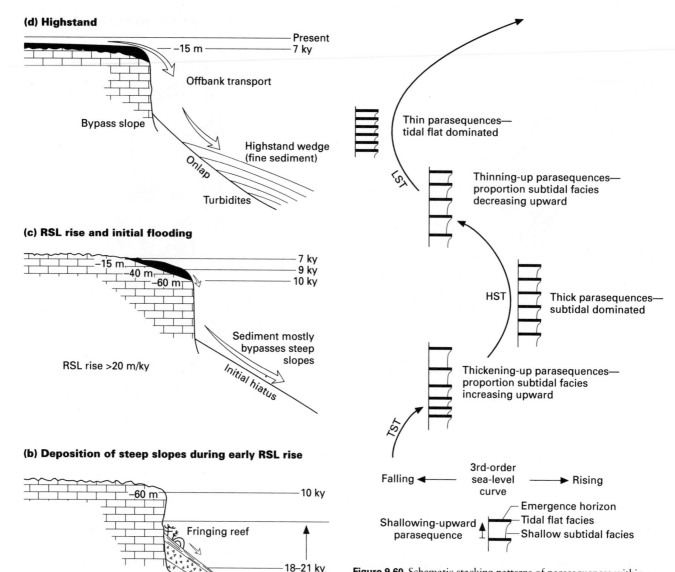

(d) Highstand

Present
−15 m — 7 ky
Offbank transport
Bypass slope
Highstand wedge (fine sediment)
Onlap
Turbidites

(c) RSL rise and initial flooding

−15 m — 7 ky
−40 m — 9 ky
−60 m — 10 ky
Sediment mostly bypasses steep slopes
RSL rise >20 m/ky
Initial hiatus

(b) Deposition of steep slopes during early RSL rise

−60 m — 10 ky
Fringing reef
18–21 ky
Slope sediment derived from wall (coarse, angular clasts)

Thin parasequences— tidal flat dominated
LST
Thinning-up parasequences— proportion subtidal facies decreasing upward
HST
Thick parasequences— subtidal dominated
Thickening-up parasequences— proportion subtidal facies increasing upward
TST

Falling ◄ 3rd-order sea-level curve ► Rising

Shallowing-upward parasequence
Emergence horizon
Tidal flat facies
Shallow subtidal facies

Figure 9.60 Schematic stacking patterns of parasequences within a carbonate platform developing under conditions of high-frequency (4th/5th-order, 10–100 thousand years) relative sea-level changes superimposed upon a longer-term (3rd-order, 1–10 million years) higher amplitude cycle (after Tucker, 1993).

(a) Lowstand

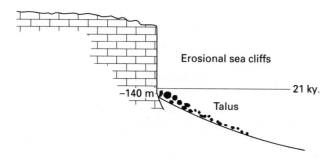

Erosional sea cliffs
−140 m — 21 ky.
Talus

Figure 9.59 Development of the margin of the Great Bahama Bank in the area of Tongue of the Ocean, from lowstand c. 20 ky ago to present highstand (modified after Grammar & Ginsburg, 1992). RSL, relative sea level.

all progradation during the subsequent highstand (Fig. 9.62a). In areas of moderate subsidence, growth may be strongly aggradational during the transgressive systems tract and moderately to strongly progradational under highstand conditions (Fig. 9.62b). In areas of low subsidence, relative sea-level behaviour is dominated to a greater extent by eustasy, so that little additional accommodation space is created during the transgressive phase and pronounced progradation commences early in the highstand systems tract (Fig. 9.62c).

In the following sections the changes that might be expected in platforms of different morphologies during each phase of a 'standard' symmetrical relative sea-level cycle are examined (Sect. 2.4) and the typical sequence architectures that might arise are discussed. These summaries follow several recent reviews by Sarg (1988), Tucker (1991), Burchette and Wright

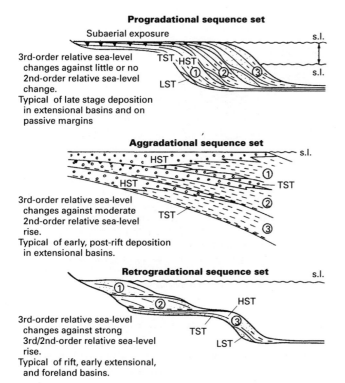

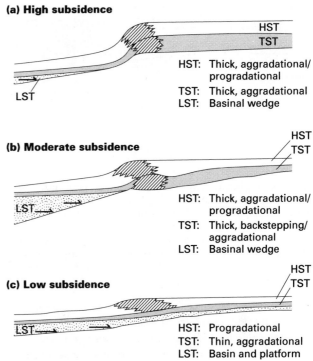

Figure 9.61 Sequence stacking patterns in carbonate platforms under different conditions of long-term increases in accommodation space, largely driven by subsidence (after Tucker, 1993). HST, highstand systems tract; LST, low stand systems tract; TST, transgressive systems tract.

Figure 9.62 The effect on carbonate platform sequence geometry of varying subsidence rates along a basin margin. In this example lowstand deposits comprise siliciclastic sediment (after García-Mondéjar & Fernández-Mendiola, 1993). HST, highstand systems tract; LST, lowstand systems tract; TST, transgressive systems tract.

(1992), Handford and Loucks (1993), and Hunt and Tucker (1993).

9.5.2 Ramps

As in all carbonate platforms, the basic building blocks of ramps comprise small-scale shoaling-upwards depositional cycles or parasequences (cf. Elrick & Read, 1991; Read, Osleger & Elrick, 1991; see Sect. 2.4). The overall thickness of ramp sequences is rarely more than 2–300 m ,reflecting the limited accommodation space available. Low rates of sediment production (Elrick & Read, 1991), mean that individual ramp sequences exhibit relatively little 'keep-up' growth, except locally as small, isolated organic build-ups (e.g. Read, 1985). Ramp successions, however, do show cumulative lateral and vertical accretion through a process of repeated incipient drowning, recovery and progradation in response to long-term (2nd- or 3rd-order) relative sea-level rises (e.g. Burchette & Wright, 1992). This generates layered successions ('ramp stacks') with a range of subtle geometries, comprising stacked or shingled progradational ramp sequences separated by incipient drowning or backstepping events characterized by deeper-water facies.

Because carbonate accumulation rates on ramps are low, flooding at cycle boundaries is commonly marked, as in siliciclastic systems, by abrupt vertical shifts from inner-ramp shoreline carbonates to outer-ramp terrigenous mudstones, argillaceous carbonate muds, or mixed sediments which then shoal upwards to the next inner-ramp unit. Identification of flooding surfaces at all scales in such successions is less problematic than in successions where one shallow-water carbonate unit rests directly upon another. Examples occur in the Frasnian Nisku Formation of the West Canada Basin (Watts, 1988), the early Mississippian ramp of southwest Britain (Burchette, 1987), and the late Jurassic and early Cretaceous of the Vercor, southern France (Jacquin, Arnaud-Vanneau *et al.*, 1991).

SYSTEMS TRACTS

(1) *Transgressive systems tract.* On ramps with high wave or tidal energy, cumulative minor fluctuations within a longer-term relative sea-level rise may generate a set of stacked or backstepping and onlapping (retrograde) parasequences (Fig. 9.63) consisting of beach, barrier-island, or shoal grainstones and associated shoreface and transitional sediments (e.g. Elrick & Read, 1991; Burchette & Wright, 1992). Such units are generally a few metres or tens of metres thick, commonly preserved during drowning, and dominated by bioclastic sediments (e.g. Aigner, 1984).

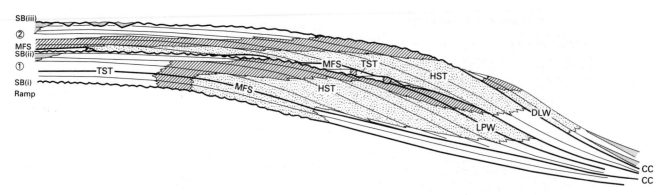

Figure 9.63 Carbonate ramp depositional sequences showing different sequence architectures controlled by the balance between sediment supply and the rates of sea-level rise/fall. 1, High sediment supply or slow rise/fall. Minor vertical accretion in the TST (transgressive systems tract) followed by strong progradation in the HST (highstand systems tract). Shallow-water facies (cross-hatched) track falling sea level during the subsequent LST (lowstand systems tract) of sequence 2 to form a lowstand prograding wedge (LPW). The sequence boundary (SB) can be traced into a correlative conformity (CC) in the basin. 2, Low sediment supply or rapid rise/fall. Backstepping parasequences during the TST followed by strong progradation during the HST. The subsequent lowstand is represented by a lowstand wedge detached from the highstand inner ramp (DLW). Stipple represents siliciclastic sediment.

Shoreline progradation for several kilometres may occur even within transgressive systems tracts. The style and continuity of backstepping depends on the *rate* and form of relative sea-level rise and the amount of sediment removal, and situations occur in which backstepping parasequences are widely spaced in dip section, closely spaced, or vertically stacked, or even replaced by simple ravinements in a similar style to Holocene siliciclastic sediments on the eastern shelf of the USA (see Sect. 6.7.6).

On ramps with low wave and tidal influence, parasequences in the transgressive systems tract consist largely of packstone and wackestone sediments. High-energy grainstones occur only as local shoals. Such sequences may be dominated by production of offshore bioclastic sediment (e.g. echinoderm–bryozoan, or large foraminiferan packstones), reflecting slower rates of sediment production at the shoreline. This depositional style occurs in the foundation stages ('banks') of many late Devonian isolated build-ups in the West Canada Basin (e.g. the Cooking Lake and Swan Hills platforms; Wendte, 1992a) and the Devonian 'Schwelm' facies of Europe (Burchette, 1981), the Mississippian of Wyoming and Montana (Elrick & Read, 1991), and the Triassic (Muschelkalk) of Spain (Calvet, Tucker & Henton, 1990).

In transgressive settings, where drowning at parasequence boundaries entails strong landward shifts of the shorelines, depositional cycles may be capped by thin condensed sections (Fig. 9.63) comprising distinctive facies such as black shales, phosphatic mudstones, or sedimentary ironstones, possibly associated with concentrated biotic assemblages (Burchette & Wright, 1992). Exposure at parasequence boundaries, characterized by minor palaeokarst, palaeosols, or dissolution, is commonly seen in the upper portions of barrier successions or in inner ramp settings.

(2) *Highstand systems tract.* During relative sea-level highstands, ramp sediments prograde strongly seaward (Fig. 9.63), are commonly grainier than those in the transgressive systems tracts (e.g. Aigner, 1984; Burchette, Wright & Faulkner, 1990), and are more commonly oolitic. Shoal deposits tend to dominate over beach or barrier-island depositional systems and restricted or lagoonal facies make up a larger proportion of the inner ramp than at other times. On progradation, ramps also tend towards the form of a flat-topped platform, but commonly maintain a low-gradient seaward margin. The potential for slope steepening and the development of clinoforms is greater at this stage than at any other, however, because sediment production must infill accommodation space created during the transgressive and early highstand systems tracts (Fig. 9.63). A vertical section through a highstand succession may show stacked, upward-shallowing, coarsening and thickening-upwards parasequences which each begin with a flooding surface and culminate in shoreface grainstones or thin lagoonal sediments capped by karstic surfaces that become more pronounced upwards within the stack as the upper sequence boundary is approached (Sect. 9.4.1).

(3) *Lowstand systems tract.* If a relative sea-level fall is small and does not widely expose the inner ramp (i.e. beneath a 'type 2' sequence boundary), facies belts on a 'homoclinal' ramp may simply prograde strongly. During a larger relative sea-level fall, ramp facies may undergo a wholesale basinwards shift. Both are examples of forced regressions in a carbonate system. Although, in the latter case, a tract of the inner ramp may become exposed, a broad substrate remains over the previous mid- and outer-ramp to form the shallow-water portion of a strongly progradational lowstand wedge (Fig. 9.63). Because of the low slope angle, there may be little change in the character

of shoreface ramp sediments between highstand and lowstand systems tracts and these may be amalgamated. It can thus be difficult to distinguish shoreface or shoal sediments of the highstand systems tract from those of any subsequent lowstand succession, unless revealed by the distribution or maturity of karst, calcretization, or zones of meteoric diagenesis. Moreover, under such lowstand conditions, lagoonal facies are unlikely to track very far seaward with falling sea level and the lowstand shoreface will be attached.

A rapid relative sea-level fall, exceeding the depth of fair-weather wave base, is necessary for lowstand inner-ramp sediments to be completely separated from those of the previous highstand systems tract (Fig. 9.63), but the expression of such a 'forced regression' will be strongly dependent on the *rate* of the fall, the gradient of the slope, and the depth of the basin. Where relative sea-level fall is rapid, there may be abrupt shallowing of the depositional environment over the mid- and outer ramp, indicating the development of a detached lowstand prograding wedge. Probable examples of this have been documented in oolitic grainstones in the early Carboniferous of Belgium (Van Steenwinkel, 1990) and outer-ramp Waulsortian build-ups in southwest Britain (Faulkner, 1988; Burchette, Wright & Faulkner, 1990). A possible example from a distally steepened ramp has been documented from the early Cretaceous of the Vercours, southern France (Jacquin, Arnaud-Vanneau *et al.*, 1991).

During a 3rd-order lowstand, the whole of the inner ramp becomes exposed and karstified and fluvial siliciclastic sediments may overlie or incise into the previous highstand inner-ramp sediments. A documented example of this phenomenon occurs in the early Carboniferous of South Wales (V.P. Wright, 1986). Hinterland relief is generally low in the settings in which ramps develop. However, if the rate of siliciclastic sediment supply is high, terrigenous muds and silts may bypass the inner ramp to augment sedimentation in the outer ramp and may interfere with carbonate productivity across the ramp. Because ramp slope angles are low, lowstand fans or aprons of transported and resedimented material seldom develop, although mass failure may be a feature of some steeper or distally steepened ramps (cf. Jacquin, Arnaud-Vanneau *et al.*, 1991; Pedley, Gugno & Grasso, 1992). In shallow, restricted basins, lowstands may lead to the deposition of sabkha and subaqueous evaporites, as in the Jurassic Hanifa Formation, Arabian Gulf (Droste, 1990).

The small isolated build-ups which occur in some carbonate ramp systems appear to start mostly during the lowstand systems tract, when outer ramp areas may enter the photic zone, or experience greater turbulence (Burchette & Wright, 1992). They may become swamped by progradation of adjacent ramps during the highstand systems tract. Where this does not occur, isolated build-ups may continue to develop through several ramp sequences (e.g. Watts, 1988).

9.5.3 Rimmed shelves

Rimmed shelves are the most productive of carbonate depositional systems. Under favourable conditions they aggrade to sea level and then expand to form flat-topped, tabular edifices. Rapid sedimentation, coupled with the ability of carbonate sediments to support steep depositional slopes, allows the construction of a distinct, commonly abrupt, shelf-break between the flat platform top and the slope down to the basin floor (Sect. 9.3). This allows potential for an enormous contrast in the style of sedimentation between various stages of the relative sea-level cycle (Schlager, Reimer & Droxler, 1994) since the slope height and angle are critical in determining the lowstand behaviour of the platform and can strongly influence sequence architecture. In some cases, extrinsic factors, such as the local tectonic environment, may also influence the height and steepness of carbonate shelf margins. In extensional settings, for example, the behaviour (e.g. progradation, accretion) of carbonate platforms seeded over fault block crests commonly varies with the throw along the strike of a major fault, since the height of the fault scarp determines the slope angle and amount of accommodation space in front of the platform and thus controls its ability to accrete or prograde (Fig. 9.64). This phenomenon has been clearly demonstrated in outcrop of the early Miocene of the Gulf of Suez (Burchette, 1988) and the Silurian of northern Greenland (Hurst & Surlyk, 1984) and in the subsurface of the Devonian of the Canning Basin (Kemp & Wilson, 1990). Similar responses can be caused by less abrupt topographic features, relief over a previous carbonate platform for example, or simply by local variations in basinal water depth.

A major problem in the sequence stratigraphic interpretation of rimmed shelves occurs when the area of investigation contains

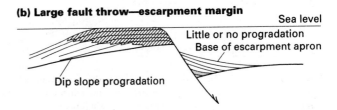

(a) Small fault throw—progradational margin

(b) Large fault throw—escarpment margin

Figure 9.64 The effect of variable throw on a normal fault on carbonate platform progradation.

no platform margin, as in the interiors of extensive rimmed shelves (including many 'epeiric' platforms – Sect. 9.3.1). Here, the absence of dipping substrate means that stratigraphy is essentially 'layer-cake'. Techniques that are used to determine the nature of stratigraphic events in such settings include examining platform-interior cycle stacking patterns and the maturity of weathering surfaces (e.g. soils and palaeokarsts) and their relationship to unconformities (Fig. 9.65) (Goldhammer, Dunn & Hardie, 1990; Osleger & Read, 1991; Goldhammer,

Oswald & Dunn, 1991; Saller, Dickson & Boyd, 1994; Sect. 9.4.1). Because the interior of a rimmed shelf is only depositionally active when flooded, it clearly represents in most cases just the record of late transgressive and highstand deposition even where the relative sea-level changes are of relatively small amplitude. During relative sea-level lowstands the exposed platform surface may be karstified or incised by streams, particularly at its margins (Handford & Loucks, 1993).

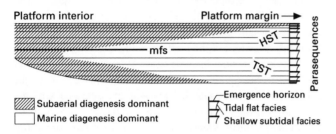

Figure 9.65 Schematic section across a carbonate platform showing the organization of parasequences within systems tracts and the distribution of dominant diagenetic regimes. Note how the parasequences thicken and become more marine upwards in the transgressive systems tract (TST) up to the maximum flooding surface (mfs) and thin and become more restricted upwards during the highstand systems tract (HST) (modified after Tucker, 1993).

SYSTEMS TRACTS

(1) *Transgressive systems tract.* Rimmed shelves show variable responses to transgression that are entirely dependent on the relationship between the rates of relative sea-level rise and carbonate accumulation. In situations where the rate of rise is significantly greater than carbonate accumulation over the platform surface, the carbonate factory may be forced to back-step along with the shoreline (Fig. 9.66), possibly several times, to shallower-water positions which allow temporary recovery prior to the next drowning event or longer-term recovery and progradation during a succeeding highstand (Kendall & Schlager, 1981; Saller, Armin *et al.*, 1993). Condensed deposits or basinal mudstones may be deposited over the outer, drowned platform or over the defunct platform surface (Fig. 9.66)

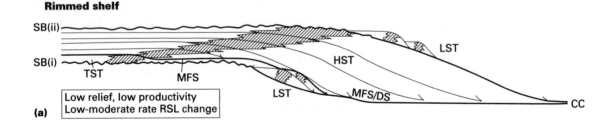

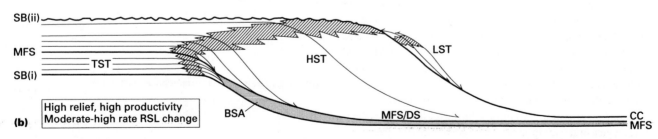

Figure 9.66 Rimmed shelf depositional sequences showing a range of responses to relative sea-level (RSL) changes. (a) Low-relief, low productivity platform (e.g. many Palaeozoic and late Jurassic platforms) developed under low to moderate rates of relative sea-level change. Note pronounced backstepping during the transgressive systems tract (TST) which on a flat-topped platform may leave little record. In this example the maximum flooding surface (MFS) lies

close to the sequence boundary. (b) A high-relief, high-productivity platform developed under moderate to high rates of relative sea-level change. Note strong vertical accretion during the transgressive systems tract with the development of a pronounced base of slope apron due to sediment bypass. In this example, the maximum flooding surface in the proximal portion of the platform lies within the platform interior succession. CC, correlative conformity; DS, drowning surface.

(Kendall & Schlager, 1981; Schlager, 1989; Ehrlich, Longo & Hyare, 1993).

Where the rate of relative sea-level rise exceeds carbonate production in all depositional environments except the carbonate factory, usually located at the shelf margin (Sect. 9.1), this may survive and grow vertically, while other environments experience incipient drowning. Isolated build-ups and patch reefs associated with the shelf may grow strongly vertically at this stage.

Where the rate of relative sea-level rise approximately equals that of carbonate accumulation, the platform may aggrade strongly (Fig. 9.66b), as in the initial stages of Triassic platforms in the Dolomites (Sect. 9.3.2), possibly steepening its basinward margin in the process (e.g. Schlager & Camber, 1986). This can lead to bypass of whatever material is shed from the platform top, with resedimentation as a base of slope apron that may mimic the geometry of lowstand deposits (Fig. 9.66). Determination of features critical to sequence stratigraphic interpretation in the platform interior, such as the maximum flooding surface, may be difficult in this context since the depositional record is an entirely shallow-water one.

The sedimentary record of transgression in the basin and the shelf-top may be condensed, particularly in instances where backstepping has occurred.

(2) *Highstand systems tract.* As the rate of relative sea-level rise decreases, sediment production eventually becomes more than sufficient to maintain vertical growth of the platform. Under these conditions the platform cannot only aggrade, but also prograde rapidly, commonly for several tens of kilometres, due to sediment overproduction (Fig. 9.66), except in situations where the shelf faces a deepwater environment. Carbonate margins that have backstepped may prograde out to the previous shelf margin and beyond. As noted, this is also the time of maximum export of shallow-water carbonate material to adjacent basinal environments (Schlager, Reijmer & Droxler, 1994).

As the rate of relative sea-level rise declines towards zero and the platform progrades, small-scale cyclic flooding events within the platform interior become less pronounced, and sub-aerial exposure surfaces more pronounced. Where evidence for platform margin and slope facies is absent, such stacking patterns of platform-interior cycles can be used to identify events of sequence stratigraphic significance (Goldhammer, Harris *et al.*, 1993; Sect. 9.4.1). In slope areas, pelagic influence becomes less prominent as deeper-water material becomes strongly diluted by sediment derived from the shallow-water platform.

(3) *Lowstand systems tract.* The behaviour of a rimmed shelf during a relative sea-level lowstand depends on the geometry of the system (i.e. principally slope angle) and the nature of the carbonate factory (whether grainstone or reefal). The scale of any relative sea-level fall is important, since a fall below the platform rim will drastically reduce the amount of substrate available for carbonate production and also limit the supply of shallow-water material to the adjacent basin (Fig. 9.66). Significant lowstand deposition is more likely on platforms that have low-angle basinward slopes than on those with steep slopes or escarpments (Fig. 9.66). The duration of the lowstand phase of the relative sea-level cycle is also important since a long period of lowstand allows a significant lowstand platform to develop. Long periods of lowstand also permit modification of the platform surface, with a greater likelihood that there will be incision at the margins or karstification and missing intervals ('beats') in the cyclic deposits of the platform surface. A sea-level fall may lead to increased restriction or isolation of the basin, which may limit carbonate productivity or result in the deposition of evaporites.

Lowstand sediments on rimmed shelves with steep, cemented margins may be bouldery, developing as wedges of talus at the base of slope, or as minor, possibly unstable fringing reefs (Grammar, Ginsburg & Harris, 1993). Fine-grained, unindurated material is likely to be significantly reduced in quantity compared with slope sediments deposited during highstands. On rimmed shelves attached to a coast, siliciclastic material may also be transported at this stage across the carbonate shelf, via streams or rivers that incise into the platform surface, or by winds (Sect. 9.5.5). These may form fans or sheets in front of the carbonate slope.

One school of thought has suggested that large-scale collapse of steep platform margins occurs preferentially during periods of lowstand, due to loss of buoyant support to the platform margin afforded by the adjacent water column (H.U. Schwarz, 1982; Cook & Taylor, 1991; Aby, 1994), submarine erosion by margin-parallel currents (Pinet & Popenoe, 1985) or chemical corrosion by exiting formation water (Back, Hanshaw *et al.*, 1986). However, this relationship has not been demonstrated to be universal and tectonism or loading are just as likely to be the cause of collapse. These factors may occur during other phases of the relative sea-level cycle (see wider discussion by Handford & Loucks, 1993).

9.5.4 Isolated build-ups

The variable sizes, depositional morphologies, and locations of isolated build-ups means that, as a group, they show a wide range of responses to relative sea-level changes. Among the most critical factors unique to isolated build-ups appear to be the area of the build-up and whether or not it possesses a flat top.

Problems are commonly encountered when attempting to track or date sequence boundaries within isolated build-ups on seismic data, particularly where the build-up margins do not interfinger effectively with basinal sediments. Because platform interior and reefal biotas possess low stratigraphic resolution, absolute age determination is commonly impossible, and it becomes difficult to relate internal stratigraphic boundaries to regional sequence stratigraphic interpretations. This is particularly true where the build-ups have been buried and isolated from attached shelf complexes by deeper-water siliciclastic deposits or where only platform interior sediments have been sampled, as in a petroleum exploration well.

Isolated build-ups rarely appear to be as strongly progradational as attached platforms. The main reason for this is that they tend to develop initially in situations of limited substrate availability, for example on extensional fault blocks or other sea-floor highs, surrounded by deeper water, and are commonly stressed by moderate to high rates of subsidence. During basin development, small- and intermediate-sized isolated build-ups characteristically exhibit an overall aggradational or backstepping history (e.g. Waite, 1993).

SYSTEMS TRACTS

(1) *Transgressive systems tract.* Flat-topped isolated build-ups tend to drown rapidly during a major relative sea-level rise. Following such events, the isolation of such platforms commonly leads to their capping by open marine deeper-water facies or grainstone shoals (e.g. Judy Creek build-up; see Fig. 9.50), depending on water depth and energy, or by onlapping or draping hemipelagic or condensed pelagic facies (Fig. 9.67). Numerous examples of such drowning facies associations have been identified in the geological record, but good illustrations are provided by Tertiary isolated build-ups in the South China Sea (Ehrlich, Barrett & Guo, 1990; Ehrlich, Longo & Hyare, 1993). A backstepping carbonate factory on a flat-topped isolated build-up is less likely to encounter a shallower-water shoreline which might form the locus for subsequent growth than in attached rimmed shelves where a distinct shoreline exists. Shallow-water carbonate sedimentation may resume during a

subsequent relative sea-level fall if the platform has not been irrevocably submerged below the photic zone.

During smaller relative sea-level rises, the margins of isolated carbonate build-ups aggrade (Fig. 9.67), step back to a point of higher elevation on the platform surface (Fig. 9.67), or do both (e.g. Cucci & Clarke, 1993). Aggradation tends to increase water depth in the interior of build-ups with lagoons, as demonstrated by the Quaternary history of many modern pacific atolls. Partial drowning may be more pronounced on one side than on the other (Fig. 9.67), such that the build-up aggrades or backsteps on one side and progrades on the other, and depends on the nature and productivity of the marginal facies with respect to oceanographic or tectonic processes. The late Devonian Swan Hills build-up (see Fig. 9.51) of the West Canada Basin, for example, shows multiple asymmetrical back steps, with greatest retreat on the northeastern, probably windward, side (Wendte, 1992b). In build-ups with resistant bioconstructed margins (e.g. some South East Asian Tertiary build-ups), it is the lower-productivity, lower-energy margins that appear to drown more readily.

In areas of significant subsidence, or during a long-continued, cyclic eustatic sea-level rise, margins may step back a number of times during the transgressive systems tract, with periods of partial recovery in the intervening highstands, the build-up each time becoming smaller in area until finally it terminates with a drowning succession of increasingly deeper-water deposits (Fig. 9.67), which itself may be cyclic. This phenomenon is more common in mound-like build-ups than in isolated platforms with flat tops and is a factor that is presumably linked to the overall increase in accommodation space and the decrease in area of the available substrate for carbonate production, as well as the potential for high rates of sediment wastage in open basinal environments.

(2) *Highstand systems tract.* As in other platforms, aggradation mostly continues during the initial part of the highstand systems

Isolated platform

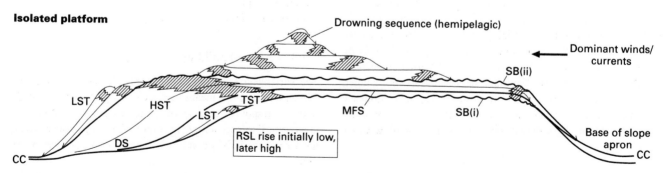

Figure 9.67 Depositional sequences within an isolated platform under conditions of relative sea-level (RSL) change which were initially low and later high. Asymmetry can be due to the effects of dominant winds or to variations in substrate slope. After sequence boundary

(SB) (II), the dominant motif is of a retrogradational sequence set in which the area of the platform progressively decreases in a 'wedding cake' fashion. This configuration is commonly the product of high-amplitude relative sea-level rises reinforced by subsidence.

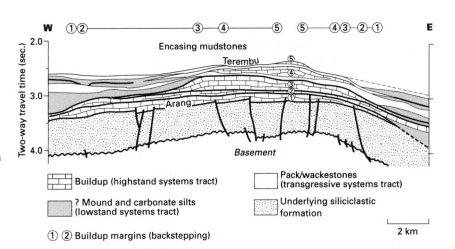

Figure 9.68 Schematic seismic interpretation of the Miocene Natuna isolated platform, South China Sea, showing the distribution of lowstand-, transgressive-, and highstand-systems tracts within five depositional sequences. The platform nucleated on a faulted basement high and retreated further on the western, shelfwards side than on the eastern side. Position of platform margins indicated by numbers linked at the top of the diagram. Each sequence boundary is characterized by exposure and erosion (after Rudolph & Lehmanns, 1989).

tract, but as sea-level rise decelerates, platform sequences may begin to expand in area through progradation (Fig. 9.67). This may be limited in a deep basin by a rapid increase in water depth away from the build-up (e.g. due to location upon some elevated feature). As with drowning in the transgressive systems tract, highstand progradation within a sequence may be asymmetrical due to windward and leeward effects (Sect. 9.3.2). A steep windward margin may continue to build vertically with practically no progradation, while the leeward margin progrades strongly (Fig. 9.67). This phenomenon may be cumulative over many sequences and is particularly well demonstrated by the eastern and western margins of the Great Bahama Bank (Eberli & Ginsburg, 1989; Fig. 9.9). Highstand platform morphology may also be strongly influenced by topography on the substrate upon which the build-up grows. An isolated platform growing on a tilted fault block, for example, may prograde more readily in the gentle dip direction than over the steep scarp of a block bounding fault.

(3) *Lowstand systems tract.* The development of a well-defined lowstand systems tract in isolated build-ups depends, as in other carbonate systems, on relative sea-level behaviour (rate of fall, duration of lowstand, etc.) and the gradient and area of the substrate below the slope break (Fig. 9.67). Clastic sediments are introduced preferentially into many basins during lowstand periods and this may also influence the effectiveness with which the carbonate factory can continue to produce, particularly if it is impoverished or partially exposed by the relative sea-level fall. During lowstands, restriction may increase in enclosed basins so that carbonate production on smaller isolated build-ups in internal basins (cf. the Devonian Keg River bioherms behind the Pres'quille barrier in the West Canada Basin) may be completely shut down due to hypersalinity or even buried by evaporites during these periods. Lowstands, of course, may be represented by karstification or soil development over large tracts of the platform top.

The Miocene Natuna carbonate platform in the east Natuna-Sarawak basin, South China Sea, shows seven depositional sequences, each consisting of a lowstand systems tract, a transgressive systems tract, and a highstand systems tract (Rudolph & Lehmann, 1989). Interpretation is based largely on seismic data tied to a few wells, but lowstand deposits appear to consist largely of mounded reefal facies that encircle the platform at the base of slope (Fig. 9.68). The transgressive systems tracts comprise deepening-upwards successions and are characterized by progressive backstep of the platform margins so that the build-up decreased in area by increments (Fig. 9.68). In the highstand systems tract, reefal shallow-water facies accreted vertically and are thickest over the highest parts of the platform (Fig. 9.68). The retreat of the Natuna build-up occurred in an asymmetrical fashion, by c. 8 km on the low-productivity western side and only 1.5 km on the higher-productivity eastern, open margin which also shows the best-developed progradation (Rudolph & Lehmann, 1989).

9.5.5 Relative sea-level lowstands – carbonate, evaporite and siliciclastic sediment partitioning

Where siliciclastic sediment is supplied coevally to a carbonate environment, deposition of the two sediment types is generally mutually exclusive, the 'carbonate factory' favouring or retreating to areas that are free of polluting siliciclastic material. A wide range of examples of the interrelationship between carbonate and siliciclastic deposition is given in Doyle and Roberts (1988) and Budd and Harris (1990). To some extent the relationship between evaporites and shallow marine 'carbonate factory' sediments is similar, the two sediment types rarely being closely juxtaposed at any one time (Sect. 8.10).

In siliciclastic-only depositional systems, a relative sea-level fall commonly rejuvenates siliciclastic source areas and enhances siliciclastic sediment input to the basin. Siliciclastic sediments of the previous sequence are commonly incised and cannibalized during this process (Sect. 2.4). However, many large carbonate platforms developed adjacent to land areas when the supply of siliciclastic material was either slow, or only episodic. In such cases, there is often marked 'partitioning' in the dominant sediment type and deposition style with respect to relative sea-level stand. Studies of such mixed systems suggest that carbonate sedimentation dominates during relative sea-level highstands and rises, while siliciclastic sedimentation dominates during lowstands when carbonate sedimentation is largely shut down.

Where a siliciclastic hinterland exists adjacent to an attached carbonate platform of any type, therefore, a relative sea-level lowstand represents a time when the amount of siliciclastic sediment transported to the basin may increase. Transport may be fluvial, via incisions in the platform margin, by winds over the exposed platform surface, or laterally by longshore or contour currents from another part of the basin margin. Siliciclastic material commonly bypasses the exposed carbonate system to be deposited on the adjacent sea floor as fans, wedges, or sheets that onlap the defunct carbonate slope. This relationship is clearly shown by the Cenozoic history of parts of the Australian Great Barrier Reef (Fig. 9.69) (Davies, Symonds *et al.*, 1989).

The height and steepness of the underlying carbonate slope is an important factor in determining the character of any siliciclastic sandbodies formed during periods of lowstand (e.g. Melim & Scholle, 1995). In the absence of significant amounts of siliciclastic sediment, basinal environments may be starved of sediment while carbonate production is shut down due to exposure of the carbonate factory.

Lowstand siliciclastic-dominated sediments thus commonly alternate with highstand carbonate-dominated sediments on

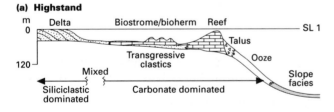

(a) Highstand

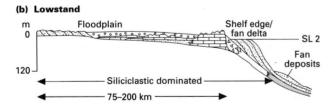

(b) Lowstand

Figure 9.69 Cross-sections of the northeast Australian shelf in the area of the central Great Barrier Reef showing the control exerted by (a) highstand, and (b) lowstand conditions on the distribution of carbonate and siliciclastic facies. Siliciclastic sediments, restricted to the platform interior during the highstand, advance to the shoreline during the lowstand (after Davies, Symonds *et al.*, 1989).

attached carbonate platforms. This phenomenon, also termed 'reciprocal sedimentation', was recognized some years before modern sequence stratigraphic models for mixed siliciclastic carbonate systems were in vogue (Stoakes 1980). However, one must not assume that all siliciclastic sediments deposited at the base of carbonate platform slopes are related to lowstand episodes since other processes, including tectonic events or changes in climate or oceanographic conditions with no clear relationship to relative sea-level stand, may introduce large quantities of siliciclastic sediment into a basin otherwise dominated by carbonate deposition.

The character of lowstand siliciclastic sediments depends largely on proximity to the siliciclastic source area and on depth of deposition. In the late Devonian of the Canning Basin, outcrop information provides little evidence for the presence of lowstand deposits and suggests that in proximal areas local extensional tectonics may have complicated interpretation. However, subsurface data from the outer margin of the Lennard shelf and the Fitzroy trough (Fig. 9.70) demonstrate the existence of several hundred metres of siliciclastic basinal and slope sediments deposited below and onlapping the shelf margins, apparently during multiple periods of relative sea-level lowstand (Southgate, Kennard *et al.*, 1993). Three styles of basinal siliciclastic deposition have been documented: (i) basin-floor fans; (ii) slope fans; and (iii) prograding complexes. Although each lowstand unit is complex and locally cyclic, the earliest sediments are generally the coarsest and fine upwards to mudstones. This reflects initial stream rejuvenation and subsequent waning supply of sediment from sources in the Kimberly Block as equilibrium was reached or renewed flooding occurred.

A similar reciprocal relationship between highstand carbonate and lowstand siliciclastic sedimentation existed in the Delaware Basin of west Texas (Saller, Barton & Barton, 1989; Brown & Loucks, 1993; Sonnenfeld & Cross, 1993) albeit in a situation where a strongly aggradational carbonate platform created high, steep slopes (Fig. 9.71). Lowstand deposition is dominated by sand- and siltstone which were transported across the emergent shelf by aeolian and fluvial processes and deposited as wedges that onlap the toe of the carbonate slope (Fig. 9.71). During the early transgressive systems tract, mixed carbonate and siliciclastic sediments were deposited while siliciclastic sediment continued to be transported across the exposed platform and renewed carbonate production along the platform margin contributed carbonate boulders and other debris to the slope (Brown & Loucks, 1993). During the late transgressive systems tract and the highstand systems tract, siliciclastic sediments were ponded on the shelf or diluted by high carbonate production rates.

The late Oligocene to Miocene of the Luconia province offshore Sarawak (Malaysia) provides a good example of the interaction between carbonate and siliciclastic sediments on a regional scale where there have been both eustatic and structural

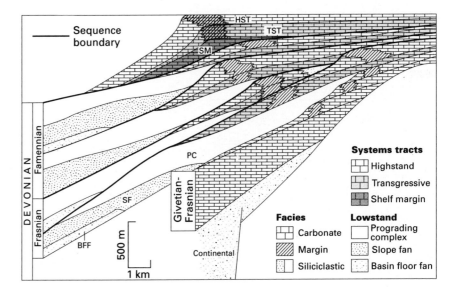

Figure 9.70 Sequence stratigraphic model for the Frasnian–Famennian carbonate platforms in the subsurface of the Canning Basin. Note the apparent partitioning of carbonate and siliciclastic sediments within each sequence between the highstands and lowstands respectively (after Southgate, Kennard *et al.*, 1993).

Delaware Basin

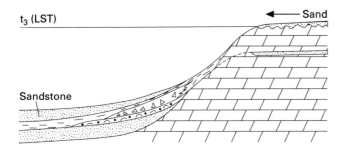

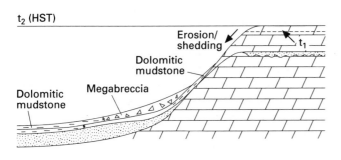

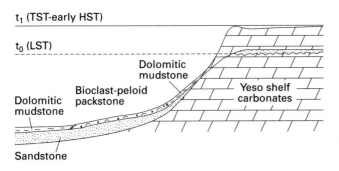

Figure 9.71 Relative sea-level control on the distribution of carbonate and siliciclastic sediments in the early Permian of the Delaware Basin, Texas (modified after Saller, Barton & Barton, 1989).

controls (Fig. 9.72). The sequence stratigraphy for the area has been outlined by Ho (1978) and, although initially based on recognition and correlation of maximum flooding surfaces (cf. Galloway, 1989a) which define eight sequences, this has been reconciled with the sequence stratigraphic approach of Van Wagoner, Posamentier *et al.* (1988) in that the main reference points are now unconformities.

Tertiary sediments in the offshore region of central Sarawak consist of a wedge of siliciclastic coastal plain and mixed siliciclastic/carbonate shelf sediments which overlies a faulted substrate of Palaeogene and older deeper-water sediments. The succession represents a transgressive–regressive supersequence upon which smaller-scale cyclicity is superimposed (Mohamed & Meng, 1992).

The early part of the supersequence, Sarawak sequences 1–3 (Fig. 9.72), comprises a transgressive sequence set, the stacking of which was dominated by regional subsidence. In basin-margin areas, the sequences are delimited by tectonically-enhanced, angular unconformities. The regressive portion of the supersequence, Sarawak sequences 4–8 (Fig. 9.72), comprises a progradational sequence set which is punctuated by distinct lowstand subaerial exposure surfaces. Although subsidence continued, late Miocene to Recent glacio-eustatic sea-level changes appear to have played a much more prominent role in sequence development (Mohamed & Meng, 1992). Carbonate growth is symmetrically arranged around a zone of maximum regional flooding and strongly influenced by proximity to siliciclastic sediment supply. This was controlled by uplift and erosion in the Borneo hinterland and increased progressively through the supersequence. Build-ups seeded on numerous fault blocks (Fig. 9.72), in many cases initially as ramps, and evolved into a range of tabular and pinnacle forms (Epting, 1980). Landward carbonate deposition started later and ended earlier than in distal areas, while basinward carbonate deposition started earlier and finished later. Build-ups were progressively buried from southeast to northwest as the clastic wedge prograded (Fig. 9.72). The most distal build-ups remained uncovered and

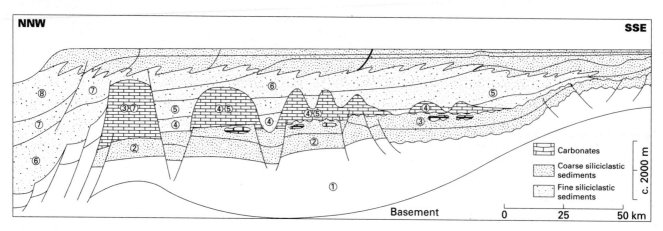

Figure 9.72 Schematic section across the Luconia Province, offshore Sarawak, Malaysia showing isolated carbonate platforms which developed in a complex, but probably reciprocal relationship with large-scale siliciclastic sedimentation (modified after Epting, 1980).

continued to grow. Many of the build-ups, particularly those with mound-like morphology, exhibit numerous thin lateral stringers which intercalate strongly with the surrounding largely siliciclastic basinal sediments. These indicate that carbonate growth and the deposition of siliciclastic background sediment alternated on a fine, possibly 4th-order (parasequence) scale.

9.5.6 Evolutionary trends between platform types

Most large carbonate platforms are the cumulative end products of numerous relative sea-level changes on several orders of scale. The Great Bahama Bank, for example, has a history that spans some 200 My, from the Triassic to the present day, while other large Mesozoic shelves along the margins of the Tethyan Ocean existed for over 150 My. Some large Palaeozoic isolated platforms in the North Caspian basin developed intermittently for over 100 My, from the late Devonian to the early Permian, constructing carbonate successions some 2000 m thick. Such

platforms will clearly encompass many 3rd-order sedimentary sequences, and a number of megasequences, while the dominant control on sequence character may change fundamentally several times during the platform history. Over such long histories, platform morphologies or individual sequence geometries can also change significantly in response to both tectonism and eustatic sea-level changes. Ramps commonly evolve into rimmed shelves, as in the Permian basin of west Texas (see Fig. 9.13) or the Jurassic of the Neuquén Basin, Argentina (Fig. 9.73), or into isolated build-ups as in the Devonian of the West Canada Basin (Wendte, 1992a) or many Tertiary build-ups in the South China Sea (e.g. Epting, 1980). A progradational rimmed shelf may evolve into an escarpment margin in response to rapid subsidence or erosion, or it may revert, following drowning and backstepping, to a ramp or distally steepened ramp depositional style as in the Mid-Cretaceous Austin Chalk of the USA Gulf Coast which inherited a steep slope in parts of the outer ramp from the underlying Stuart City reef trend. Where initially

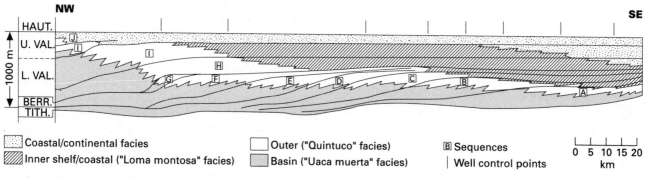

Figure 9.73 Section across the subsurface late Jurassic to early Cretaceous carbonate platform of the south-central Neuquén Basin, Argentina showing the evolution of the platform from an initial ramp (A) to a strongly progradational and aggradational rimmed shelf (after Mitcham & Uliana, 1985).

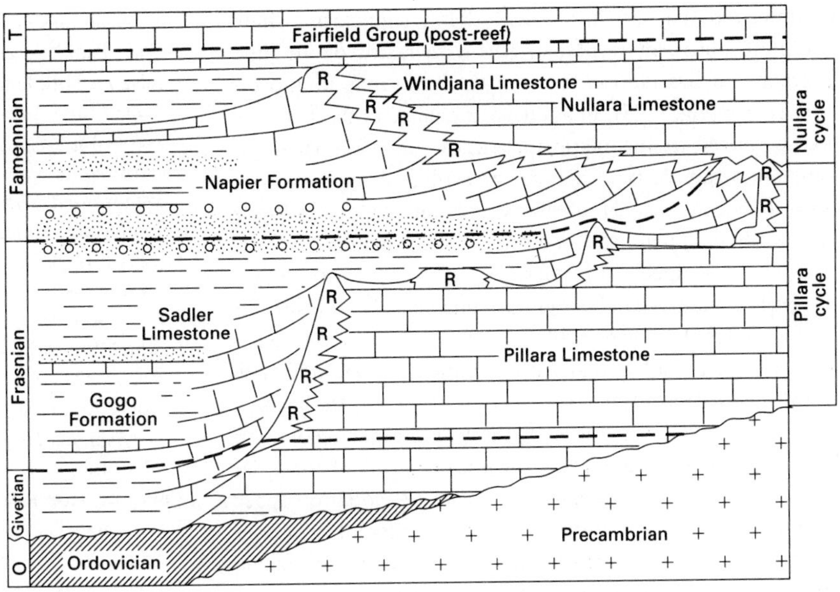

Figure 9.74 Schematic stratigraphic section through late Devonian carbonate platforms along the northeast margin of the Canning Basin. Note how the platform morphology changes from predominantly aggradational escarpment-margins in the Frasnian to a strongly progradational and aggradational rimmed shelf in the Famennian. Long-term changes of this character are usually driven by changes in the rate of subsidence, but may also be due to changes in the nature of the dominant carbonate producing organisms (after Playford, Hurley *et al.*, 1989).

high rates of subsidence decrease, as for example in the Devonian of the Canning Basin (Fig. 9.74) (Playford, Hurley *et al.*, 1989), aggradational or backstepping sequences may give way to a strongly progradational style.

Indeed there seems to be discernible 'evolutionary trends' in platform morphology (Read, 1982, 1985) which are commonly related to the subsidence style within the basin of growth (Fig. 9.75). Steepening of the basinward slope of a carbonate platform may have several causes; it may be: (i) tectonically driven (e.g. through differential subsidence, extensional faulting);

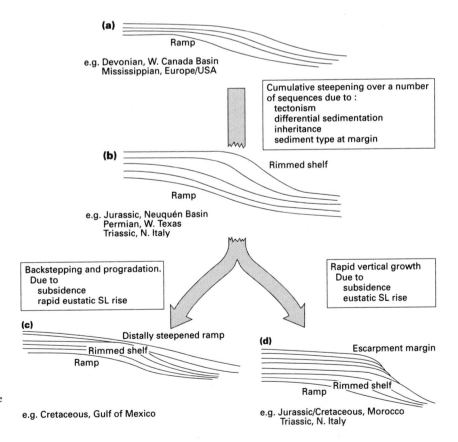

Figure 9.75 A commonly observed, 'evolutionary tree' for carbonate platforms. Ramps (a) commonly evolve into rimmed shelves (b) which may then flood, in which case a subsequent platform (if any) may be very different (c), or evolve into an escarpment-bound shelf (d). The transition (a)→(b) may have a range of causes, while the transitions (b)→(c) and (b)→(d) are mostly related to relative sea-level rises.

(ii) inherited (e.g. an antecedent delta, siliciclastic shelf, carbonate platform slope); or (iii) as a result of intrinsic differential sedimentation between the basin margin and the basin centre (Fig. 9.75). In addition, a change in the sediment type deposited at the platform margin, for example if the dominant sediment type changes or evolves from grainstone shoals to bioconstructed framework, as occurred globally during the late Mississippian, can also bring about a similar response since coarse-grained or framework sediments will support steeper slope angles than finer-grained unconsolidated sediments (see Fig. 9.35) (Kenter & Schlager, 1989; Kenter, 1990). Such processes are cumulative and are commonly associated with overall deepening of a basin, so that slope angles become accentuated from one sequence to the next. Steepening of ramp slopes in particular appears to occur within individual 3rd-order sequences as highstand deposits infill accommodation space generated during the transgressive stage. Different responses might also be expected between windward and leeward facing ramps or where an outer ramp has a higher sedimentation rate due to pelagic input, as in some late Jurassic and Cenozoic examples (cf. Mullins, Gardulski & Hinchey, 1988).

Further reading

Crevello P.D., Wilson J.L., Sarg J.F. & Read J.F. (Eds) (1989) *Controls on Carbonate Platform and Basin Development*, 405 pp. *Spec. Publ. Soc. econ. Paleont. Miner.*, **44**, Tulsa.

Geldsetzer H.J., James N.P. & Tebbut G.E. (Eds) (1988) *Reefs, Canada and Adjacent Areas*, 775 pp. *Mem. Can. Soc. petrol. Geol.*, **13**, Calgary.

Harris P.M. & Kowalik W.S. (1994) *Satellite Images of Carbonate Depositional Settings*, 147 pp. *Am. Ass. petrol. Geol.*, Methods in Exploration Series, **11**, Tulsa.

Loucks R.G. & Sarg J.F. (Eds) (1993) *Carbonate Sequence Stratigraphy: Recent Developments and Applications*, 545 pp. *Mem. Am. Ass. petrol. Geol.*, **57**, Tulsa.

Scoffin T.P. (1987) *An Introduction to Carbonate Sediments and Rocks*, 274 pp. Blackie and Son Ltd., Glasgow.

Scholle P.A., Bebout D.G. & Moore C.H. (1983) *Carbonate Depositional Environments*, 708 pp. *Mem. Am. Ass. petrol. Geol.*, **33**, Tulsa.

Simo J.A., Scott R.W. & Masse J.-P. (Eds) (1993) *Cretaceous Carbonate Platforms*, 479 pp. *Mem. Am. Ass. petrol. Geol.*, **56**, Tulsa.

Tucker M.E. & Wright V.P. (1990) *Carbonate Sedimentology*, 496 pp. Blackwell Scientific Publications, Oxford.

Tucker M.E., Wilson J.L., Crevello P.D., Sarg J.R. & Read J.F. (Eds) (1990) *Carbonate Platforms: Facies, Sequences and Evolution*, 328 pp. *Spec. Publ. int. Ass. Sediment.*, **9**, Blackwell Scientific Publications, Oxford.

Wilson J.L. (1975) *Carbonate Facies in Geologic History*, 471 pp. Springer-Verlag, Berlin.

Deep seas

D.A.V. Stow, H.G. Reading & J.D. Collinson

10

10.1 Introduction

Investigation of deep-sea sediments began with the voyage of HMS *Challenger* (1872–76) which established the general morphology of the ocean basins and the types of sediment they contained. The volume emanating from this voyage (Murray & Renard, 1891) became the cornerstone of deep-sea sedimentology for a long time. The authors believed that only pelagic clays and biogenic oozes were found in the deep sea and thus, during the 19th century, and for most of the first half of the 20th century, oceans were considered to have quiet, undisturbed floors where only pelagic and hemipelagic sediments could be deposited. All sandstones were thought to have been deposited in shallow water.

Soon after the *Challenger* expedition, geologists working in Scotland, the Caribbean, the Alps and elsewhere, in the late 19th and early 20th centuries, claimed recognition of deep-sea deposits on land (see Jenkyns & Hsü, 1974 for review). Contenders for such recognition included Tertiary *Globigerina* limestones from Malta, radiolarian deposits from Barbados, and red radiolarian clays from Timor, Borneo and Rotti, some with sharks' teeth and manganese nodules. Mesozoic limestones and radiolarites from certain parts of the Alps were also proposed as analogues for Recent biogenic

oozes. Steinmann (1905, 1925) further noted the common association of these cherts with ophiolitic mafic and ultramafic igneous rocks. However, there was much opposition to the interpretation of these ancient sediments as truly oceanic, and the controversy lasted until the 1960s, when the advent of the theory of Plate Tectonics allowed an interpretation of ophiolite assemblages as oceanic. The occurrence of oceanic sediments within continental settings was established.

A similar debate centred around the occurrence of sands and sandstones which were considered initially almost universally to be shallow water. Although density underflows had long been known in lakes (Lake Geneva and Lake Meade) and Daly had suggested in 1936 that density currents (first called turbidity currents by D. Johnson in 1938) could cause erosion of submarine canyons during periods of lowered sea level, nothing was said about the role of turbidity currents as a transporter and depositer of sands in deep water.

This concept had to await the publication of one of the most important advances in sedimentology. In 1950, Kuenen and Migliorini brought together the flume experiments on turbidity currents (Kuenen, 1937, 1950) and Migliorini's observations on graded sand beds in the Italian Apennines to suggest that many graded sandstones in ancient successions had been deposited by high-density turbidity currents. Their ideas initiated a revolution in sedimentology (R.G. Walker, 1973) and immediately explained many apparent anomalies that had been arising in modern oceans where sands were increasingly being discovered in very deep water (Ericson, Ewing & Heezen, 1951). They also stimulated a spate of papers interpreting flysch deposits as the result of turbidity currents (e.g. Natland & Kuenen, 1951).

For more than 10 years the turbidity current hypothesis held sway as a cause of graded beds. Generally, the term used for the rock was 'greywacke' and for the formation 'flysch' as it was not until about 1960 that the word 'turbidite' was introduced for the product of a turbidity current. In addition, all ancient turbidites were considered to have been deposited on basin plains similar to those of the Atlantic, for example the Aberystwyth Grits in Wales (Wood & Smith, 1959), the Martinsburg Formation in the Appalachians (McBride, 1962) and all Alpine and Carpathian flysch whether in Poland, Romania, Switzerland or Italy (e.g. Dzulynski, Ksiazkiewicz & Kuenen, 1959). Because *all* graded beds were considered to have been deposited by turbidity currents, many formations, which were subsequently interpreted as shallow-water storm and flood deposits, were misinterpreted as deep-water formations.

In 1962, Bouma reported on his analysis of hundreds of graded beds in the flysch of the Peira Cava of southeast France. He showed that there was a preferred sequence of sedimentary structures in each graded bed although many had the top or bottom of the sequence missing. The Bouma sequence was then interpreted hydrodynamically (R.G. Walker, 1965; Harms & Fahnestock, 1965) and, from then on, all deep-water sands and gravels were interpreted in terms of their transporting, depositional and postdepositional processes (e.g. Stauffer, 1967).

However, research through the 1970s showed that the Bouma sequence was only strictly applicable to medium-grained sand–mud turbidites. Consequently, parallel sequence models were developed for both coarse-grained (conglomeratic) turbidites (Lowe, 1982), and fine-grained (mud-rich) turbidites (Stow & Shanmugam, 1980).

While this work was being carried out on the description and interpretation of turbidite facies, a third important line of progress centred on the distribution of deep-water sediments and the formulation of environmental depositional models.

In 1959, Gorsline and Emery delineated the three principal environments of deposition that we still recognize: basin floor, submarine fan and slope apron (Fig. 10.1). This was followed, during the 1960s, by work in California (Menard, 1960; Normark, 1970; Haner, 1971) which gradually increased our knowledge of modern submarine fans, showing that turbidites could be deposited on them as well as on the basin floors that Atlantic workers had stressed. In the mid-1960s the submarine fan model was first applied to ancient turbidites (R.G. Walker, 1966a; Jacka, Beck *et al.*, 1968). Mutti and Ricci Lucchi (1972) then formulated a more general deep-water depositional model that could be applied to a wider range of ancient sequences. This suggested the types of sequence to be expected not only in

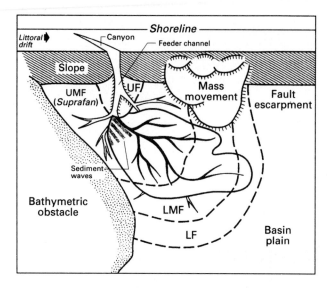

Figure 10.1 Composite model for late Quaternary mud–sand-rich submarine fan development in the California Continental Borderland showing feeder channel, canyon and shelf and adjoining slope apron and basin plain. Note that direction of fan growth is controlled by a bathymetric obstacle (after Nardin, 1983).

an ancient submarine fan, where inner fan, middle fan and outer fan facies associations were recognized, but also in ancient slopes and in deep-water basin plains (see Fig. 10.35).

Over the ensuing years, numerous attempts have been made to describe the variability of deep-water systems in order to provide predictive models from outcrop and subsurface analysis (see detailed review in Reading & Richards, 1994). Normark (1974, 1978) divided fans into coarse-grained, canyon-fed and fine-grained, delta-fed systems, whereas Mutti developed the idea of efficient (mud-rich) versus inefficient (sand-rich) fans (Mutti & Johns, 1978; Mutti, 1979). In spite of these and other models, the single 'all-purpose' fan model of R.G. Walker (1978) was still widely used.

In the 1980s, the universality of the single point source model was questioned with the development of multiple-sourced submarine ramp models (Chan & Dott, 1983; Heller & Dickinson, 1985). Stow (1985, 1986) proposed three fan types (the elongate or mud-rich, radial or sand-rich, and fan delta or gravel-rich fans), but also emphasized the importance of different types of slope-apron and basin-plain systems.

Over the past 25 years, starting in the 1960s, an alternative to the turbidite hypothesis was put forward as an explanation for deep sea sands. This hypothesis proposed that sands could be transported and deposited by deep-ocean bottom currents that travelled transversely along the continental rise, following the contours (Heezen, Hollister & Ruddiman, 1966; Hollister & Heezen, 1967). The sediments deposited by bottom currents became known as contourites and specific sedimentological criteria were proposed for their identification (Hollister & Heezen, 1972).

Many attempts were made to discover contourites in the ancient record, but some reports were almost certainly misinterpretations of either shelf sediments or distal turbidites. As giant contourite drifts were drilled and cored it was appreciated that they were generally composed of fine-grained sediments. Facies models for sandy and muddy contourites were put forward by Stow and Lovell (1979) and Stow (1982), and a composite facies model for contourite sequences was developed by Stow, Faugères and Gonthier (1986). Today, the recognition of outcrop examples of contourites and ancient drift systems is still a matter for debate (e.g. Faugères & Stow, 1993; Shanmugam, Spalding & Rofheart, 1993).

10.2 Processes and products

Within the deep sea, three main groups of processes are capable of eroding, transporting and depositing terrigenous, biogenic, volcanigenic and other particulate materials (Fig. 10.2): *pelagic settling, semi-permanent bottom currents* and

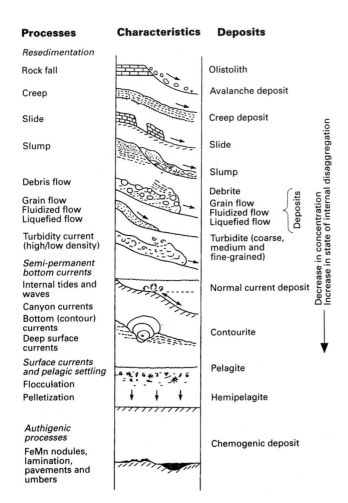

Figure 10.2 The range of processes that operate in the deep sea and their products (from Stow, 1994).

episodic resedimentation processes. In addition, some deep-sea deposits are formed by *in-situ* authigenic processes. Several attempts have been made to classify resedimentation processes based on the mechanical behaviour of the flow, the transport mechanisms and clast-support system (Sect. 10.2.3), so that a plethora of terminology and confusing (partial) synonyms exists. There is always a temptation to simplify the nature of processes operating in the deep sea and to relate these directly to resulting deposits. However, the more we learn from measurement and observation, the more the true complexity of processes, from initiation to deposition, and their relationship to depositional products becomes apparent (Fig. 10.3).

Depositional products span a wide range of types and several facies schemes have been erected. In this section we follow broadly the Mutti and Ricci-Lucchi (1972) scheme as modified by Stow (1985). This is discussed in Sect. 10.2.4 and illustrated in Fig. 10.29.

10.2.1 Pelagic and hemipelagic sedimentation

Pelagic is used here for sediment which is generated in the open sea. Pelagic sediment is chiefly composed of biogenic material diluted by a proportion (<25%) of non-biogenic components. In areas close to continental margins and in enclosed basins where clastic supply is more abundant, the rain of biogenic debris is further diluted by a silt- and clay-sized terrigenous component. This sediment is regarded as hemipelagic.

PELAGIC FALLOUT

In the open ocean, the everyday process of sedimentation is one of slow and relatively steady fallout of biogenic debris generated largely in the upper part of the water column. This material includes the microscopic skeletal remains of calcareous and siliceous planktonic flora and fauna as well as their soft organic tissue. This may be supplemented in some areas by fine volcanic ash, wind-blown dust and dilute plumes of terrigenous clay.

In the absence of bottom currents or turbidity currents these particles settle vertically under the influence of gravity. The rate of settling for single clay-sized particles and open-framework microskeletons is inordinately slow, and most pelagic material reaches the sea floor as larger composite particles known as 'marine snow'. This is formed by the electrophysical process of flocculation, whereby minute electrical charges on clays cause mutual attraction between flakes, and by the biological process of pelletization, whereby zooplankton ingest smaller particles and excrete loose organically bound fecal pellets. Elongate algal filaments and mucous strands with attached organic and inorganic particles also contribute to the marine snow.

There are three principal factors that control this pelagic flux through the water column and determine the nature and

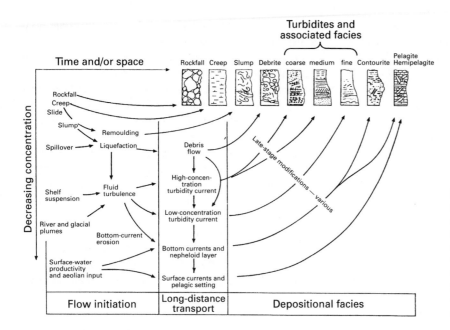

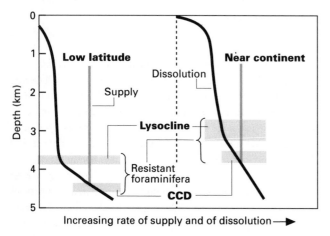

Figure 10.3 Deep-sea process interaction and continuum in relation to main sedimentary deposits. Framework is one of time and/or space, and concentration of flows. Idealized facies models that result from deposition by the different processes are also shown. Post-depositional modification can involve current reworking, liquefaction and bioturbation (from Pickering, Stow *et al.*, 1986; modified after Walker, 1978).

distribution of the resulting sediments. These are: (i) dissolution; (ii) productivity; and (iii) masking.

(1) *Dissolution.* Below a few hundred metres water depth, sea water is undersaturated with respect to all forms of calcium carbonate, yet major dissolution of calcareous tests typically takes place at depths of several kilometres. This lack of reaction in the upper levels is most probably due to the presence of thin organic coatings that impede dissolution. At some critical level of undersaturation, solution rates of calcium carbonate increase markedly. This point is known as the lysocline (Fig. 10.4), which may correspond to the upper boundaries of cold corrosive bottom-water masses (Sliter, Bé & Berger, 1975). The oceanic

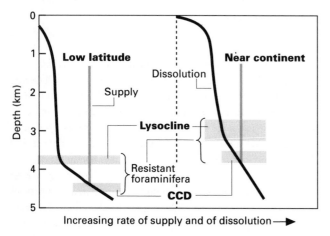

Figure 10.4 Model showing the relationship between pelagic supply of skeletal carbonate and dissolution in the southeast Pacific (from Jenkyns, 1986).

lysocline is thus the level separating well from poorly preserved carbonate fossil assemblages in deep-ocean areas away from continental margins (Berger, Bonneau & Parker, 1982; Diester-Haass, 1991). Generally a few hundred metres deeper than the lysocline is the calcite compensation depth (CCD), below which calcite does not accumulate on the sea floor. This is the level at which the rate of supply of pelagic carbonate is balanced by its rate of dissolution.

Aragonite skeletal matter is relatively more soluble than calcite so that the aragonite compensation depth (ARD) typically occurs about 1 km above the CCD. Sea water is also undersaturated with respect to biogenic silica. However, thin organic coatings coupled with slow rates of dissolution allow the accumulation of siliceous oozes below the CCD in the deepest parts of the ocean basins.

(2) *Productivity.* In normal open ocean conditions, most planktonic organic matter is destroyed during its transit through the water column by bacterial oxidation. Typically, at depths between 300 and 1500 m, where this process dominates, an oxygen minimum layer is developed. This layer is also characterized by a maximum in carbon dioxide and nutrients (phosphates and nitrates). Upwelling from this part of the ocean leads to high primary productivity in the surface waters, subsequent intense oxygen minima and an abundant supply of biogenic matter to the sea floor. If oxygen levels are sufficiently low and rates of supply particularly high, then organic matter can be preserved in the sediment. The formation of sedimentary phosphates may also be favoured under these conditions.

Upwelling is most common at oceanic fronts, where different water masses meet, in a peri-equatorial belt and at

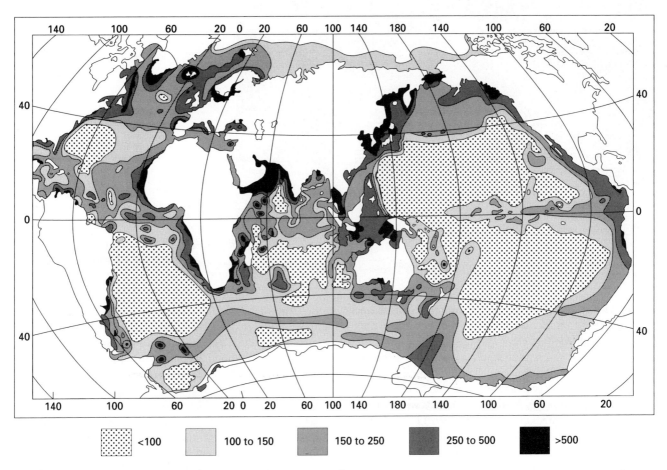

Figure 10.5 Global phytoplankton productivity (mgCm⁻² day⁻¹) (from Pelet, 1987).

northern and southern latitudes between about 50 and 60°. It also occurs adjacent to some continental margins, particularly on the western sides of the continents. These zones are clearly evident on a world map of primary productivity (Fig. 10.5) and are generally underlain by silica-rich sediments and oozes (Fig. 10.6).

(3) *Masking*. Where any one of the several components of pelagic sediments occurs in particular abundance then it may obscure or mask the presence of the others. For example, siliceous material dominates under high productivity belts, and calcareous oozes dominate on elevated regions of mid-ocean ridges and plateaus above the CCD. In barren regions of the oceans below the CCD, red clays, derived mainly from aeolian, volcanic and cosmic sources accumulate by default.

The oceanwide distribution of modern pelagic sediments is determined by these interacting controls (Fig. 10.6) as is their overall rate of accumulation (Table 10.1).

PELAGITES

Pelagic sediments (pelagites) and hemipelagic sediments (Facies

G.1; see Fig. 10.29) have been variously defined and classified (Jenkyns, 1978). The most comprehensive scheme is that of Berger (1974). A simpler scheme was introduced by the Shipboard Scientific Party during DSDP Leg 75 (Hay, Sibuet *et al.*, 1984), based on the three-component system, non-biogenic, siliceous biogenic and calcareous biogenic (Fig. 10.7). This scheme includes not only the relatively pure calcareous and siliceous oozes and hemipelagic clays but also mixtures between these components (Dean, Leinen & Stow, 1985). They are known as marls, sarls and smarls. These arls are typically interbedded with the more pure oozes in beds from about 20 to 150 cm in thickness that show gradational bed boundaries and are thoroughly bioturbated throughout. Burrow types in deep water commonly include *Zoophycos, Planolites* and *Chondrites* (ichnogenera).

Long periods of extremely low accumulation rates, sea-floor dissolution and/or distinct erosive events can result in hiatuses in the sedimentary record and in the formation of hardgrounds through early cementation. In some cases hardgrounds become coated with phosphate crusts, ferromanganese pavements and act as sites for the slow growth of nodules. They are then subject to boring by molluscs and other active benthos.

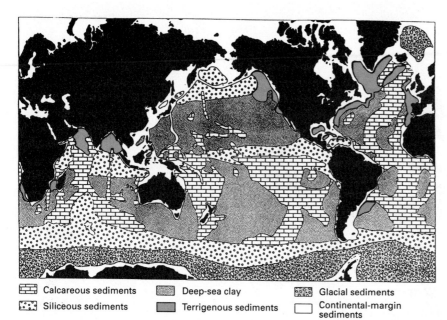

▦ Calcareous sediments	▨ Deep-sea clay	▨ Glacial sediments	
⠿ Siliceous sediments	▩ Terrigenous sediments	☐ Continental-margin sediments	

Figure 10.6 Global distribution of principal types of pelagic and other sediments on the ocean floors (after Davies & Gorsline, 1976).

Non-biogenic pelagites may also include rare ferromanganese nodules or horizons, but otherwise mostly comprise remarkably uniform, bioturbated red clays. In reality, these are brownish in colour with more or less green mottling where the Fe^{3+} has been locally reduced to Fe^{2+}. Pelagic sediments lying directly over newly formed oceanic crust at mid-ocean ridges are locally rich in Fe–Mn oxyhydroxides, Fe-rich smectites, as well as a range of other metal sulphides and sulphates (Sect. 12.11).

In areas of high organic input and/or reduced bottom-water oxygenation, any of the above pelagic facies may become more organic-rich (see below).

In addition to any early diagenesis on the sea floor, the transition from initial pelagic deposits to ancient pelagites (limestones, cherts, etc.) involves greater changes than for most clastic sediments. Calcareous ooze gradually converts to chalk at a burial depth of around 200 m and then to true limestone from around 600 m. This process involves progressive microfossil disintegration and breakage and replacement by micrite and microspar cements. Aragonite and high-Mg calcite cements are particularly unstable, but low-Mg calcite is also readily soluble in subsurface conditions. Irregular dissolution seams (stylolites) lined with insoluble clays, and nodular beds formed by differential dissolution and cementation, are characteristic of pelagic limestone facies.

Siliceous microfossils are made of opal, a highly unstable, reactive amorphous form of silica. On burial, this transforms to quartz via cryptocrystalline cristobalite. Dissolution and remobilization of biogenic silica may take place at all stages of burial diagenesis resulting in the formation of chert and flint nodules, regular radiolarite, diatomite and chert beds, and in the partial or complete silicification of adjacent limestones. Volcaniclastic components of pelagites are also unstable on burial. Alteration of basaltic material gives rise to common smectites in pelagic red clay facies.

Facies models for ancient pelagites are shown in Fig. 10.8. Structureless pelagic limestones and chalks occur in four main facies types, all of which are bioturbated, although bioturbation is often cryptic: (i) regularly bedded, with thin to medium-thick beds and more or less distinct breaks between beds; (ii) irregularly bedded, in thin wavy to lenticular beds separated by clay-lined dissolution seams; (iii) regularly/irregularly bedded with hardgrounds, showing thin to medium bed thickness, wavy dissolution seams, common Fe–Mn and phosphate crusts and nodules and borings; and (iv) nodular limestone, occurring in highly irregular thin bands, typically with extensive clay-lined dissolution seams between nodules and strong evidence of boring.

Table 10.1 Rates of accumulation of Recent and sub-Recent pelagic facies (after Berger, 1974).

Facies	Area	mm ky^{-1}
Calcareous ooze	North Atlantic (40–50° N)	35–50
	North Atlantic (5–20° N)	40–14
	Equatorial Atlantic	20–40
	Caribbean	c. 28
	Equatorial Pacific	5–18
	Eastern Equatorial Pacific	c. 30
	East Pacific Rise (0–20° S)	20–40
	East Pacific Rise (c. 30° S)	3–10
	East Pacific Rise (40–50° S)	10–60
Siliceous ooze	Equatorial Pacific	2–5
	Antarctic (Indian Ocean)	2–10
Red clay	North and Equatorial Atlantic	2–7
	South Atlantic	2–3
	Northern North Pacific (muddy)	10–15
	Central North Pacific	1–2
	Tropical North Pacific	0–1

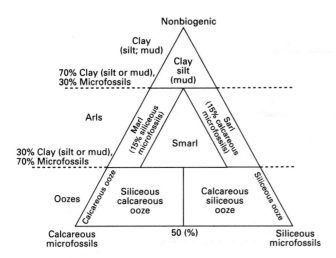

Figure 10.7 Classification of the three-component system (non-biogenic, siliceous biogenic and calcareous biogenic) that makes up pelagic and hemipelagic sediments (after Hay, Sibuet *et al.*, 1984).

Laminated limestones and chalks are invariably organic-carbon rich, and occur in two main types: (i) well-laminated (fissile lamination), generally with regular thin to medium bedding that is more or less distinct, and an absence of bioturbation; and (ii) moderately well-laminated (fissile lamination), thin to medium bedded, with some bioturbation especially by forms such as *Chondrites* indicative of low oxygenation. Laminated limestones and chalks with low organic carbon contents have most likely been reworked or resedimented by bottom currents or turbidity currents.

Cherts show a less varied spectrum of facies types than their calcareous equivalents. Bedded and nodular facies are both recognized, as well as partially or completely silicified limestones. In many cases evidence for sedimentary structures or bioturbation has been lost. Black organic-rich cherts are not always laminated. Laminated but non-organic-rich cherts, like

their calcareous counterparts, are most likely to be the result of bottom current or turbidity current activity rather than true pelagic processes. Pelagic red claystones are monotonous, poorly bedded, colour mottled and highly bioturbated.

Many of these pelagic facies occur interbedded with each other and with hemipelagic facies. Even where such interbedding is not evident, some vertical fluctuation in compositional or other properties is characteristic of most pelagic sediments (see Sect. 10.4).

HEMIPELAGIC ADVECTION

Up to 75% of hemipelagic sediment is of biogenic derivation and settles vertically through the water column. Its formation and occurrence are governed by the same controls – dissolution, productivity and masking – as for pelagic sedimentation.

The other part is fine-grained terrigenous material derived from a variety of sources (Fig. 10.9): (i) fluvial sediment carried out into the basin as freshwater plumes off river mouths where it falls from suspension as the plumes mixes and decelerates; (ii) fine-grained sediment stirred up in shelf areas by storms, or eroded from muddy shorelines; (iii) fine glacial sediment carried out to sea by floating ice; (iv) wind-blown dust from arid and semi-arid areas; and (v) fine volcanic dust.

In addition, the finest portions of low-concentration turbidity currents are in some cases stripped off at density discontinuities within the water column. Large eddies that become detached from bottom currents can further contribute their very dilute suspensions to the ambient hemipelagic fallout.

Once thrown into suspension, some of the material is dispersed by tidal and other near-surface currents and eventually settles through the water column. Other sediment suspensions may have sufficient excess density to sink as very dilute, slow-moving density flows (or turbid layer flows), which move down a slope sometimes penetrating as far as the basin floor. In other cases they detach from the bed and flow out into the basin as

Structureless limestone

 Regular-bedded ± distinct thin-thick ± dissolution

 Irregular-bedded wavy-lenticular clay dissolution seams ± Fe Mn nodules

 Regular-irregular bedded + hardgrounds ± dissolution seams ± Fe Mn nodules ± Phosphate crusts

 Nodular highly irregular extensive clay-dissolution seams ± boring

Laminated limestone

 Well-laminated ± well-bedded no bioturbation Organic-C-rich

 Moderately laminated ± well-bedded some bioturbation Organic-C-present

Pelagic cherts

 Bedded chert no structures ± silicified limestone

 Nodular chert no structures

 Laminated chert Organic-C-rich

Pelagic red clays

 Indistinct bedding Colour mottling Bioturbated Fe Mn nodules and lamination

Figure 10.8 Facies models for pelagic sediments (based on Stow, 1986).

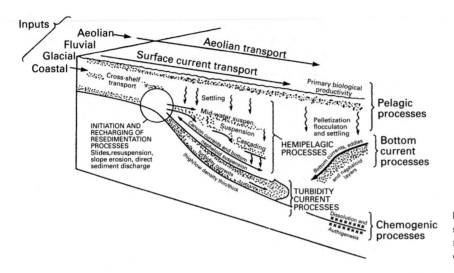

Figure 10.9 Processes of fine-grained sedimentation in the deep sea indicating the range of processes contributing to hemipelagic deposition (after Stow, 1985).

an interflow at a density interface within the water column. Such flows decelerate and material settles vertically, perhaps to regroup at a deeper density interface and flow further basinwards before slowing and settling once more. This process is known as suspension cascading.

All these processes may contribute to a nepheloid layer, which itself may be maintained and dispersed along slope by deep bottom currents (Sect. 10.2.2). This combination of processes (Fig. 10.9) has been documented by various authors (McCave, 1972b; Drake, Hatcher & Keller, 1978; Stow, 1985) and is referred to here as hemipelagic advection. It represents part of the process continuum that includes very dilute turbidity currents and deep bottom currents.

The distribution of hemipelagic sediments closely follows the principal supply pathways of the terrigenous component. Hemipelagites dominate on many continental slopes, in marginal or semi-enclosed basins, and at high latitudes. They are closely interbedded with turbidites in the distal and marginal parts of submarine fans and on abyssal plains. The global distribution of sediment types shown in Fig. 10.6 does not adequately represent the very large hemipelagic contribution. Rates of accumulation of hemipelagic sediments vary considerably, depending on the rates and proportions of contributions of the different components. Typical rates are between 5 and 15 cm ky^{-1}.

HEMIPELAGITES

Hemipelagic deposits (Facies G.2; see Fig. 10.29) are fine-grained sediments with both biogenic and terrigenous components. In extreme cases of high input from floating ice or from sediment plumes of major rivers in flood, the biogenic component may be <10%. They are commonly strongly bioturbated, with *Zoophycos, Planolites* and *Chondrites* being the most common ichnogenera in deep water. Regular and irregular bedding types are recognized as for pelagic sediments. In more restricted basins with anaerobic bottom conditions, lamination

is preserved together with a relatively high proportion of organic carbon. Phosphates are commonly associated with hemipelagites deposited in oxygen-deficient settings (Burnett, 1977; Bentor, 1980), while glauconites occur in more shallow-water hemipelagites (Odin & Matter, 1981). Only a very generalized facies model for hemipelagites can be formulated (Hesse, 1975; Stow, 1986; Coniglio & James, 1990).

The diagenetic changes that transform soft hemipelagic sediments into compacted hemipelagites are similar to those that affect pelagites. However, when compacted, the mixed clay-rich composition typically results in the development of a fissile lamination or fissility (Stow & Atkin, 1987).

CYCLICITY AND THE MILANKOVITCH MECHANISM

A pervasive feature of both pelagites and hemipelagites is cyclicity (Schwarzacher & Fischer, 1982; Einsele & Ricken, 1991). Pleistocene sedimentary cycles, reflecting fluctuations in clay and carbonate contents, are well developed in the major ocean basins where they correlate with and were probably caused by glacial–interglacial variations. However, Atlantic Mesozoic deep-sea sediments, deposited when the Earth is generally considered to have been ice-free, contain a variety of sedimentary cycles showing regular changes in clay, carbonate and organic carbon content (Dean, Gardner *et al.*, 1978; Fischer, de Boer & Premoli Silva, 1990). Thus glaciation is not a prerequisite for cyclicity.

On land, cyclicity is present in most pelagic facies in rocks of all ages from the early Palaeozoic (e.g. the Ordovician limestone–marl cycles of the Baltic Shield) to the Neogene (e.g. the diatomaceous Monterey Formation of California and the Trubi Marls of Sicily) (Einsele, Ricken & Seilacher, 1991; de Boer & Smith, 1994a; House & Gale, 1995). Similarly well-known examples are the mid-Cretaceous limestones, marls and black shales (black cherts) that occur all around the former Tethys and early Atlantic Oceans (de Boer, 1982a,b; Schwarzacher & Fischer, 1982).

These cycles are frequently interpreted to be a consequence of climatic changes caused by astronomical forcing (see Sect. 2.1.5) known as Milankovitch processes (Berger, 1988). In the oceans, late Pleistocene glacial–interglacial periodicity correlates most obviously with the 100 000-year cycle (Imbrie & Imbrie, 1980). Polar icecaps generate cold bottom waters that may raise the level of the lysocline and the CCD, increase the area of sea floor affected by carbonate dissolution and decrease the carbonate/clay ratios in the resultant deposits. Thus deep-sea sediments deposited during glacial periods should be carbonate-poor. This is so for Pleistocene Atlantic cycles, but the reverse correlation holds in the Indian and Pacific Pleistocene thus revealing the complexity of the systems (Gardner, 1975). Changing wind and marine current patterns, equally climate-controlled, govern influx of fine-grained terrigenous material and may produce cycles through varying the dilution of carbonate by clay. Another potent variable is upwelling with its dramatic effects on plankton productivity. For example, regular movements of the equatorial high-productivity zone could account for some deep-sea organic-carbon-rich sedimentary cycles, which are diachronous (de Boer, 1982a; Denis-Clocchiatti, 1982.)

BLACK SHALES AND THEIR ORIGIN

Black shale is the general term for any dark-coloured, fine-grained, organic carbon-rich sediment. Total organic carbon (TOC) contents typically range from 1 to 15%. Many black shales are hemipelagites; others, such as black cherts and organic-rich limestones, are pelagites; whereas still others are fine-grained turbidites (Arthur, Dean & Stow, 1984; Stow, 1987).

Deposition of black organic-rich shales is favoured if surface-water productivity is high and/or terrestrial higher-plant material is introduced in abundance. Whether much organic carbon is preserved depends on its rate and mode of transit to the sea floor, the oxygen content of the mid- and bottom waters, sediment particle size, and sedimentation rate (Demaison & Moore, 1980). These factors suggest that models of black-shale deposition can be divided into two end-member types: one of enhanced supply, the other of enhanced preservation. Examples of the former are the Gulf of California and the Santa Barbara Basin, which are areas of high plankton

productivity, intense oxygen-minima, and accumulation of organic-rich anoxic sediments. An example of the latter is the Black Sea, a salinity stratified and largely anoxic water body, whose surface water productivity is relatively low, but which nevertheless accumulates organic-rich sediments since degradation of the organic carbon is incomplete. At the present time, organic-carbon-rich sediments are not accumulating in the central parts of major ocean basins, a situation demonstrably different in the past.

Anoxic black shales have been cored from the Mesozoic, particularly the Cretaceous of the Atlantic, Indian and Pacific Oceans; not only from basins but also from seamounts, aseismic ridges and plateaus – that is, a wide range of water depths in oceans of different shapes and sizes. There is also a substantial record from Cretaceous outcrops on land (Jenkyns, 1980). Palaeozoic deep-sea graptolitic shales are a globally developed organic-rich facies but their original sites of deposition are difficult to ascertain. What is clearly established, both for Palaeozoic and Mesozoic examples, is that deposition of organic-rich black shales or similar facies took place over wide areas at specific times (Fischer & Arthur, 1977; Jenkyns, 1980; Leggett, McKerrow et al., 1981; Thickpenny & Leggett, 1987). These favoured intervals have been taken to record so-called Oceanic Anoxic Events (Schlanger & Jenkyns, 1976) during which the amount of dissolved oxygen in certain levels of the world ocean, most particularly the oxygen-minimum zone, was unusually low (Fig. 10.10). Although controversy exists over whether these events are coeval and truly global or due to a combination of local environmental factors, detailed stratigraphic studies reveal the synchroneity of at least one such Cretaceous event across much of the globe (Schlanger, Arthur et al., 1987). Major periods within which deposition and preservation of organic carbon in the deep-marine environment were commonly favoured are the mid- Cambrian, the early mid-Ordovician, the early Silurian, the late Devonian, early Carboniferous, early and late Jurassic and the mid- and late Cretaceous (Brooks & Fleet, 1987). Isotopic evidence confirms that global perturbations of the carbon budget, related to burial of enhanced quantities of organic matter, took place during some of these intervals (Fig. 10.11) (Scholle & Arthur, 1980; Arthur & Jenkyns, 1981).

Causal factors behind such events are thought to be sluggish bottom-water oxygen renewal in the absence of polar icecaps,

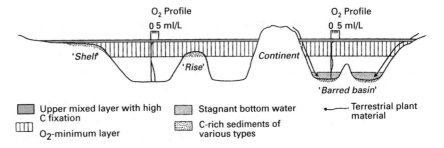

Figure 10.10 Model of the world's oceans with (relative to today), an intensified and expanded oxygen-minimum layer favouring deposition of organic-rich sediments (after Schlanger & Jenkyns, 1976).

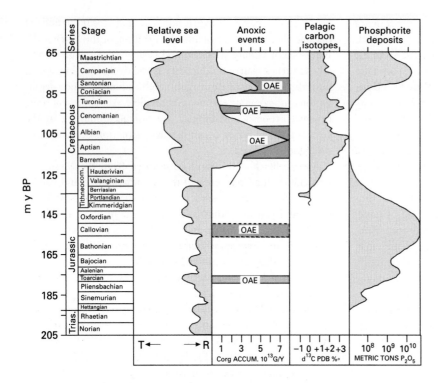

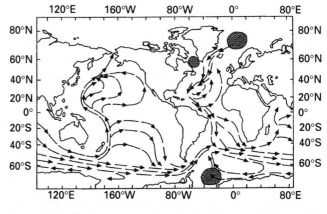

Figure 10.11 The relationships between relative sea level, Oceanic Anoxic Events, carbon isotopes and major deposits of phosphate (after Arthur & Jenkyns, 1981).

warmer ocean waters containing less dissolved oxygen, and increased organic productivity and/or preservation. There is good correlation between 'Anoxic Events' and widespread marine transgressions throughout the geological column. Flooding of continental shelves may have produced fertile moderately deep epeiric seas which, through bacterial oxidation of phyto- and zooplankton, could have produced intense oxygen minima (Jenkyns, 1980). We may therefore distinguish two end-member palaeo-oceanographic modes: the 'Icehouse State', a glacial Earth, with low sea levels and a well-mixed, oxygenated ocean and the 'Greenhouse State', a non-glacial Earth, with a less stirred ocean, higher sea levels, recycling of nutrients on the shelves and a tendency towards anoxia (Fischer, 1981).

10.2.2 Semi-permanent ocean bottom currents (contour currents)

Fine-grained sediment delivered to the ocean floor may be subjected to reworking by semi-permanent ocean bottom currents which are the deep-water expression of oceanic thermohaline circulation.

Deep-ocean bottom water is formed by the cooling and sinking of surface water at high latitudes (Gill, 1973; Killworth, 1973) and the deep, slow thermohaline circulation of these polar water masses throughout the world's ocean (Neuman, 1968) (Fig. 10.12). Highly saline but warm water also flows out of the Mediterranean Sea as an intermediate-level water mass.

Antarctic Bottom Water (AABW), the densest and deepest water in the oceans, forms in the region of the coast-hugging

Figure 10.12 Global pattern of abyssal circulation. Shaded areas are regions of production of bottom waters (after Pickering, Hiscott & Hein, 1989).

westward-directed surface polar current, with localized areas of major generation, such as the Weddell Sea. This water flows down the continental slope, circulates eastwards around the Antarctic continent, perhaps several times, and then flows northwards into the Atlantic, Indian and Pacific Oceans (Fig. 10.13).

A major source of Arctic Bottom Water appears to be the subpolar gyre in the Norwegian and Greenland Seas, although it remains partly trapped by an irregular topographic barrier known as the Scotland–Iceland–Greenland Ridge. Once the Norwegian–Greenland Seas basin is filled with this cold, dense water mass then intermittent overflow to the south occurs through narrow channels across the ridge.

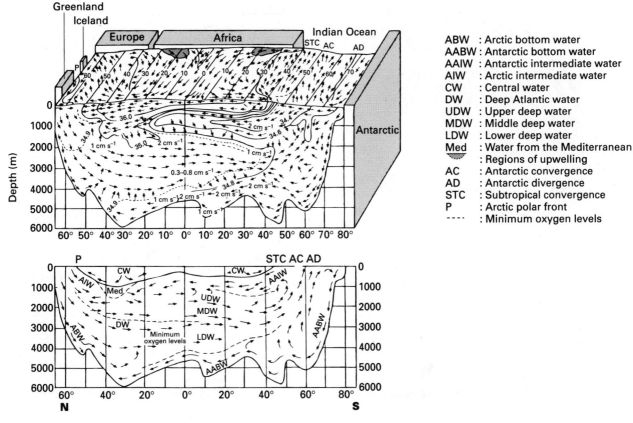

Figure 10.13 Bottom water masses in the North Atlantic Ocean (from Stow, 1994).

These slowly moving water masses are affected by the Coriolis Force (see Sect. 2.1.2), due to the Earth's spin. The result is that deep-water flows are banked up against the continental slopes on the western margins of northern ocean basins. These are unable to move upslope against gravity and so become restricted and intensified, forming Western Boundary Undercurrents (WBUC). This movement is compensated by oceanic upwelling and by oceanic surface currents. Bottom currents are also locally intensified where restricted by narrow passages on the deep-sea floor, for example Fracture Zone gaps in the Mid-Ocean Ridge system.

Whereas much of the deep-sea floor is swept by very slow currents (<2 cm s^{-1}), the western boundary currents commonly attain velocities of 10–20 cm s^{-1} and exceed 100 cm s^{-1} where the flow is particularly restricted (Stow & Lovell, 1979; McCave & Tucholke, 1986). Although these bottom currents are more or less continuous and sufficiently competent in parts of the ocean to erode, transport and deposit sediment, they are also highly variable in both velocity and direction (Luyten, 1977; Richardson, Wimbush & Mayer, 1981; Nowell & Hollister, 1985). Large-scale eddies peel off and move at high angles to the main flow, and average velocity decreases from the core to the margins of the current. Both seasonal (Shor, Lonsdale *et al.*, 1980) and tidal (McCave, Lonsdale *et al.*,

1980) periodicities have been recorded, and current reversals are common. Variation in ocean floor kinetic energy (Richardson, 1983; Cheney, Marsh & Beckley, 1983) results in alternations of short (days to weeks) episodes of erosion, associated with high velocity currents, and longer periods (weeks to months) of deposition associated with lower velocity. Episodes of high current velocity, called 'benthic', 'abyssal' or 'deep sea' storms (Gardner & Sullivan, 1981; Hollister & McCave, 1984), correspond to high surface kinetic energy, due to local storms. During these episodes, large volumes of material are resuspended and contribute to a high-density nepheloid layer (Fig. 10.14). This material may then be transported over long distances in the WBUC before eventual deposition in contourite drifts.

CONTOURITES

Two clastic contourite facies end-members result from deposition by bottom currents: *muddy contourites and sandy contourites* (Fig. 10.15). Facies models for these types have been developed from Tertiary to Recent contourite drifts in the deep sea (Stow & Lovell, 1979; Stow, 1982), but it has proved particularly difficult to confidently recognize ancient contourites at outcrop (Lovell & Stow, 1981; Faugères & Stow, 1993).

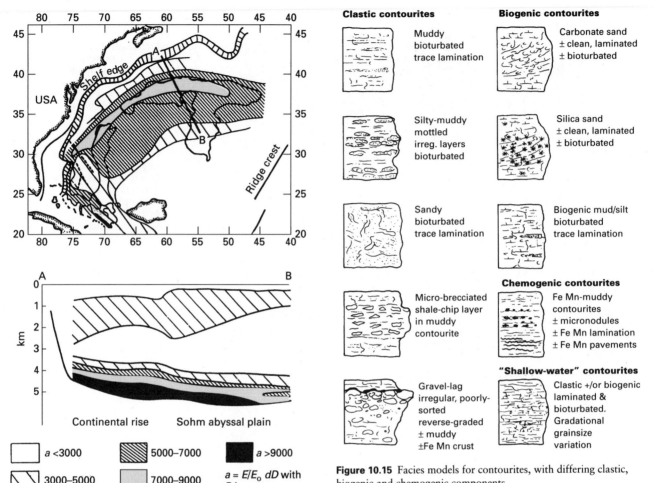

Figure 10.14 Horizontal and vertical distribution of concentration of suspended matter in the nepheloid layer off the eastern USA. The relative concentrations range from a maximum of about 0.1 ppm (0.2 mg l^{-1}) near the sea floor to a minimum of about 0.01 ppm (0.02 mg l^{-1}) in the mid-column clearwater zone (after Eittreim & Ewing, 1972).

Figure 10.15 Facies models for contourites, with differing clastic, biogenic and chemogenic components.

Muddy contourites (Facies E.1.2 and D.1.3; see Fig. 10.29; Fig. 10.15) are composed of fine-grained, poorly-sorted clay- and silt-sized biogenic and terrigenous sediment with up to 15% sand fraction. They are mainly homogeneous or structureless and thoroughly bioturbated, more rarely having irregular layering, lamination and lensing. They range from finer-grained homogeneous muds to siltier mottled silts and muds. With increasing terrigenous content they pass into hemipelagites which they closely resemble.

Sandy contourites (Facies C.1.2; see Fig. 10.29; Fig. 10.15) occur either as thin irregular layers (<1–5 cm) or thicker beds (5–25 cm), that are either structureless and thoroughly bioturbated or have some primary horizontal and cross-lamination preserved. They can show either negative or positive grading, or both, and have sharp or gradational bed contacts. Grain size

is commonly fine sand, more rarely medium sand, with poor to moderate sorting. In many cases the mean grain size is in the coarse silt grade and the facies may be more accurately termed 'silty to fine sandy' contourites. The composition is variable, commonly mixed terrigenous and biogenic. The facies may sometimes be confused with fine-grained turbidites.

Muddy and sandy contourites commonly occur together in characteristic vertical 'sequences', in some ways analogous to Bouma turbidite sequences (Stow, Faugères & Gonthier, 1986). A complete sequence (Fig. 10.16) shows a negative grading from a fine homogeneous mud, through a mottled silt and mud, to a fine sand and then positive grading back to a mud. The grain-size changes and concomitant changes in sedimentary structures and composition probably relate to long-term fluctuations in the mean current velocity over 2000–10 000 years for a 50-cm sequence.

The effects of *winnowing* and *reworking* by bottom currents result in contourite facies with rather different characteristics (Stanley, 1993). Thin, irregular, poorly-sorted, structureless, mixed-composition, iron–manganese coated, coarse-sand and *gravel-lag contourites* (Fig. 10.15) are formed by the winnowing

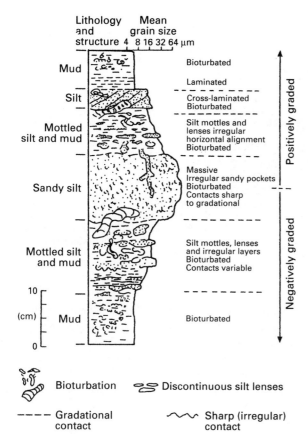

Figure 10.16 Combined facies model for muddy, silty and sandy contourites arranged in typical negatively to positively graded sequence (from Gonthier, Faugères & Stow, 1984; Stow, 1994).

of all fines from a coarse-grained sediment by powerful bottom currents. The reworking more or less *in situ* of sandy turbidites can result in bottom-current modified turbidite sands which are thought common on continental slopes and rises. In the central parts of ocean basins, bottom currents construct large sediment drifts out of almost pure biogenic material (Stow & Holbrook, 1984: Kidd & Hill, 1986). These *biogenic contourites* are often very similar to pelagites (Fig. 10.15).

10.2.3 Resedimentation processes and deposits

These are the main processes whereby large volumes of sediment are transported into deep water from an original shallow-water setting (Einsele, 1991). They comprise a complex suite, much of it a continuum, between the end-members of dilute turbidity currents and subaqueous slides (e.g. Nardin, Hein *et al.*, 1979) (see Figs 10.2 & 10.3). All resedimentation depends upon the down-slope component of gravity acting upon material stored on or moving down a slope. All the processes may be classified as subaqueous mass movement and most of them as sediment gravity flows. Only rock falls and slides fall outside such a classification.

Sediment gravity flows may be subdivided either on their rheological behaviour or on the nature of their clast-support mechanism (Figs 10.17 & 10.18). The rheological classification distinguishes between plastic and fluidal flows (Lowe, 1979). Plastic flows have a discrete shear strength which must be overcome before movement may occur (Fig. 10.19). A consequence of this is that on deceleration, plastic flows 'freeze' as a critical shear threshold is passed. Fluidal flows, on the other hand, have no inherent strength and continue to move as long as the applied shear exceeds zero. They therefore decelerate continuously and gradually drop their sediment load until it is exhausted.

Sediment gravity flows are all mixtures, in varying proportions, of water and sedimentary particles. In order for these mixtures to move and deform internally, particles must be dilated to move relative to one another and the various clast-support mechanisms provide a further basis for subdivision (Fig. 10.18) (Middleton & Hampton, 1976). There are, theoretically, five modes of clast support, not all of which are mutually exclusive: (i) matrix cohesive strength; (ii) intergranular collision (dispersive pressure); (iii) excess pore fluid pressure (liquefaction); (iv) upwards escape of pore water (fluidization); and (v) turbulent suspension. In addition, buoyancy operates in association with all the processes as the sediment–water mixture, being denser than water on its own, causes particles to displace a greater mass of matrix (Hampton, 1979). The transport of the large particles, in particular, is favoured by buoyant forces.

Most downslope movement results from general or local loss of shear strength so that the sediment–water mass or some particular layer within it is no longer able to resist the downslope gravitational shear. Such instability is favoured by: (i) high slope gradients; (ii) high rates of sedimentation, particularly of fine-grained sediment; (iii) repeated cyclical stress, commonly due

Flow behaviour	Flow type		Sediment support mechanism
Fluid	Fluidal flow	Turbidity current	Fluid turbulence
		Fluidized flow	Escaping pore fluid (full support)
		Liquefied flow	Escaping pore fluid (partial support)
Plastic (Bingham)	Debris flow	Grain flow	Dispersive pressure
		Mudflow or cohesive debris flow	Matrix strength / Matrix density

Figure 10.17 Nomenclature for sediment gravity flows indicating both flow behaviour (rheology) and dominant clast-support mechanism (after Lowe, 1979).

407

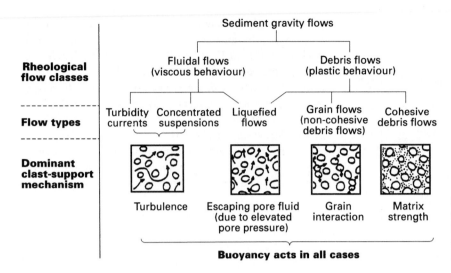

Figure 10.18 Sediment gravity flows, showing the dominant grain-support mechanisms. Buoyancy is especially important in higher concentration flows (after Middleton & Hampton, 1976).

to seismic shocking but also due to storm and wave action; (iv) high biological productivity and/or bottomwater anoxia leading to high organic content in the sediment; which in turn may lead to (v) gas generation due to organic decay and clathrate decomposition.

Here gravity-driven resedimentation is discussed largely in terms of clast-support mechanisms.

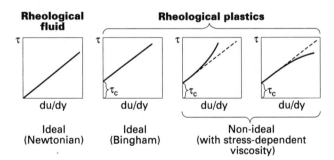

Figure 10.19 Schematic plots of applied shear stress (τ) against strain rate (du/dy) for flows with different rheological properties (after Nemec & Steel, 1984).

ROCK FALLS

Rock falls are short-lived events in which blocks of lithified material fall down a steep subaqueous slope with minimal interference between falling clasts. Large clasts are usually involved and, on arriving at the foot of the slope, they collide with and bounce over similar clasts that have fallen earlier, making up a subaqueous talus apron. Subaqueous slopes steep enough to allow this type of movement are confined to active fault scarps, the margins of carbonate platforms and the heads of some steeply incised canyons. Some rock falls are triggered by external events such as earthquakes or major storms.

Rock-fall deposits are wedges of chaotic breccia with clasts that may be very large. The breccias commonly have very high initial porosities which tend to be filled by pelagic and hemipelagic sediment. Large single clasts may fall into finer-grained sediment at the foot of slope as *olistoliths* and disturb the original structure and fabric of that material. (Facies F.1; see Fig. 10.29).

SEDIMENT CREEP

Sediment creep is a process of slow strain due to the downslope weight of sediment. On land the analogous process is soil creep on hillslopes. The process may occur over short or long intervals from hours to thousands of years and may be continuous or intermittent. Continuous or rheological creep results from the breaking and re-establishment of electrochemical bonds between particles. Intermittent creep results from physical or biological displacement normal to the slope with downslope readjustment. While freeze–thaw is a major agent in subaerial settings, on subaqueous slopes, burrowing organisms may provide the main agent. When shear stress is high, creep may be a precursor to creep rupture and the initiation of a slump or slide.

Creep deposits are poorly documented but might be inferred from subtle surface morphology on modern slopes. In ancient sediments, mudstones in a slope association on the Antarctic Peninsula which show near-surface, bed-parallel extension, syndepositional faulting and wavy bedding have been interpreted as creep deposits (Whitham, 1993).

SLIDES AND SLUMPS

Slides involve large blocks of lithified or partially lithified sediment that move in isolation or as clusters in continuous contact with underlying sediments. The blocks move without internal deformation as all the shear is concentrated on basal slip surfaces. Slides occur across a wide range of scales and displacement, the largest involving blocks measured in many tens of kilometres along slope and moving over several kilometres. Some local slides may be largely rotational in movement with strongly curved slip surfaces. Such slides are transitional to listric faults, for example in delta-front settings (see Sect. 6.6.4) (Crans, Mandl & Haremboure, 1980; Martinsen, 1989). Others are translational where rather flat, slope-parallel slide surfaces allow material to move over longer distances. In the upslope part of a slide, extensional deformation leads to families of listric faults which sole out at the level of basal décollement (Fig. 10.20). In downslope areas, compression is more common with thrusting and bulldozing of basin-floor sediments (Lewis, 1971; Dingle, 1977). With downslope movement, slide masses may show progressive fragmentation.

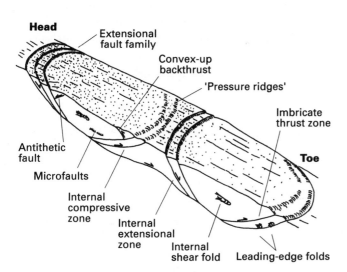

Figure 10.20 Idealized model for the distribution of deformational structures within a slide in semi-consolidated material (after Martinsen, 1989).

Slide deposits (Facies F.2.1; see Fig. 10.29) may make major contributions to the deposits of continental slopes and rises and are characterized by large, often isolated blocks of material, often limestones, in otherwise fine-grained background sediments. They are commonly imaged on seismic (Barnes & Lewis, 1991; Schwab, Lee & Twichell, 1993). Entire slide masses may extend over several hundreds of square kilometres with individual olistoliths up to many kilometres along strike. Detailed structural style may vary quite markedly along strike within major slide complexes.

Slumps share many of the features of the slides and are gradational with them. They may occur over a wide range of scales with the largest slump layers covering many hundreds of square kilometres. They may occur on very low-gradient slopes, especially where sedimentation rates of fine-grained sediment are high (see Sect. 6.6.4).

As with the slides, the bulk of displacement is concentrated on a basal slip surface, which may be listric at its upslope end. However, the mass of moving material is commonly unconsolidated and undergoes complex internal deformation as it moves downslope. The degree and style of internal deformation varies with position in the moving layer and with the strength and heterogeneity of the slumping material. In a simple two-dimensional model (e.g. Lewis, 1971), the head areas of slumps are dominated by extensional structures in the form of normal, commonly listric faults, while the downslope part of the slump, where movement ceases, tends to be dominated by compressional structures such as folds and thrusts which ramp up from the basal complex as a result of internal components moving relative to each other. Some components may be more akin to slide units with little deformation compared with surrounding material. This leads to internal zones of extension, compression and shearing with a far more complex distribution (Fig. 10.21). In addition to deformation related to lateral translation of the slumping layer, it is likely that there will also be internal loading owing to density inversions between layers of weak sediment.

Because slumping involves plastic deformation, a slump will freeze once applied shear falls below some critical value. This freezing may not occur at the same time throughout a slumping sheet and certain parts will move after other parts have stopped, adding to the complexity of internal deformation. Within slump sheets, therefore, it is common to find a very wide variety of styles of deformation, ranging from brittle to highly ductile, sometimes in quite close proximity (Fig. 10.21) (e.g. Martinsen & Bakken, 1990).

The areas from which the largest slumps are removed may become sites for the development of slope channels and conduits

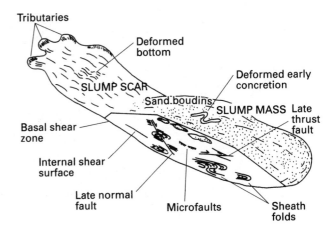

Figure 10.21 Idealized model for a slump in poorly consolidated material showing the variety and possible distribution of different styles of internal deformation (after Martinsen, 1989).

for the transfer of later sediments to deep water. Following the initial slump movement, retrogressive failures in the head area may lead to a succession of smaller, later flows (cf. Pickering, 1979). It is probable that one of the main consequences of slumping is to act as a starting process for a whole range of other sediment gravity flows which develop through the downslope acceleration, dilution and transformation of slumps (Figs 10.2 & 10.22).

Slump deposits (Facies F.2.1, F.2.2 & F.2.3; see Fig. 10.29) vary in thickness from over 100 m to a few centimetres. They are recognized by the presence of pervasive deformation structures within which it is still possible to recognize remnants of original bedding and lamination. Slump folded units are distinguished from the products of tectonic deformation by: (i) the presence of undeformed beds above and below; (ii) the erosional truncation of folds on the top surface; and (iii) immediately overlying beds which rapidly eliminate any relief on the top surface of the slump and restore a horizontal surface. Folding varies in its complexity with sheath folds being common.

Only where folding is very simple may the orientation of the fold axes be used to infer the overall movement direction of the slump. In many cases structures reflect local internal deformation and careful and widespread analysis may then be needed to arrive at general conclusions (Woodcock, 1979).

Highly mobile slumps, where fragmentation has led to a distinction between 'clasts' and 'matrix', are transitional to a cohesive debris flow and their deposits may then be approximate Facies A2 of Mutti and Ricci-Lucchi (1972) (Fig. 10.23, 1–3).

DEBRIS FLOWS

These are plastic flows in which sediment and water are fully mixed and where any original bedding and lamination is largely destroyed. They may be subdivided into cohesive and non-cohesive types based on the clast-support mechanism.

Cohesive debris flows (mudflows) occur where the cohesive strength of the matrix is the dominant clast-support mechanism,

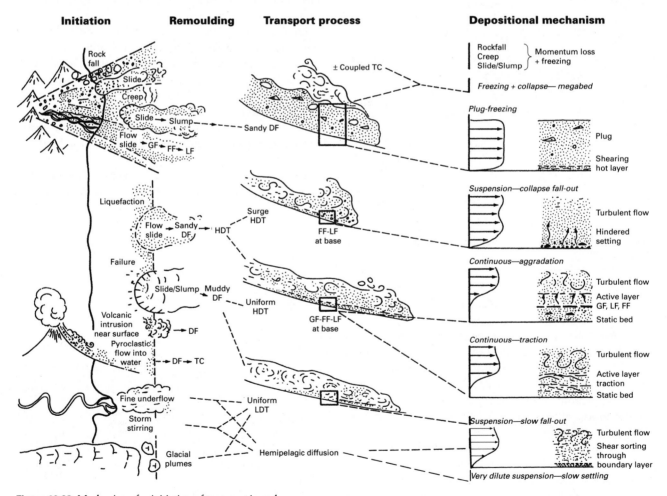

Figure 10.22 Mechanisms for initiation of transport in and deposition from debris flows and turbidity currents (modified from Stow, Brackenburg & Johansson, 1994).

considerably supplemented by buoyancy (A.M. Johnson, 1970; Nemec & Steel, 1984). Much of our knowledge of their likely behaviour in subaqueous settings comes from the analogy with the subaerial mudflows common on some alluvial fans (see Sect. 3.3.8). They usually have a fine-grained matrix with a significant content of clay-size material, although movement can take place where muddy matrix content is quite low. The deformation within a moving debris flow may be laminar or turbulent depending upon the viscosity of the matrix and the speed of the flow. Turbulent flows tend to be more dilute and may be transitional to turbidity currents.

Many debris flows show internal shearing throughout the flow but, in more viscous flows, internal shear stresses may only be large enough to overcome matrix strength close to the bed. The core of the flow then moves as a rigid plug of 'frozen' sediment (Fig. 10.22). At the lateral edges of a debris flow, where the layer is thinner, shear stresses may again be too small to overcome viscous matrix strength. Levees may then develop to confine a more mobile central zone. Such flows tend to be rather narrow tongues, elongate downslope while more mobile flows spread laterally to give extensive sheets. Cohesive debris flows are able to move on small gradients and, because of their fine-grained and therefore low-permeability matrix, they tend to dewater slowly, sustaining their mobility. They are able to spread over wide areas and transport sediment for quite long distances. In addition, the viscous strength and the density

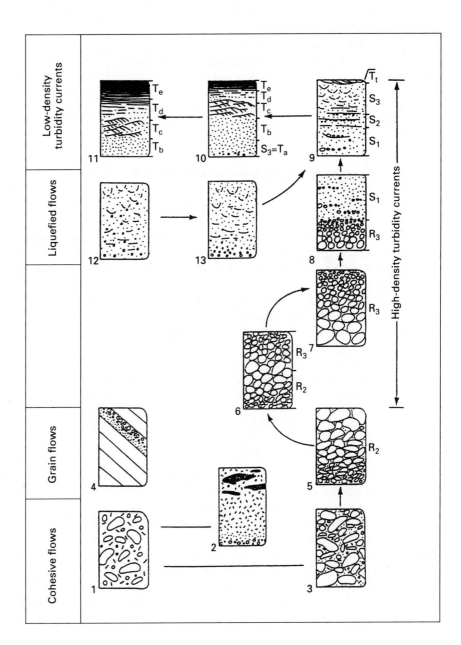

Figure 10.23 Spectrum of deposits produced by the full range of sediment gravity flows. Lines without arrows indicate a continuum while lines with arrows indicate possible evolutionary stages within individual flows (after Lowe, 1982).

of the matrix mean that they are able to carry large clasts, some of which tend to float close to the upper surface of the flow.

Debris flow deposits of cohesive flows are commonly termed *debrites* (Facies A.1.1; A.1.2; F.3.1 & F.3.2; see Fig. 10.29). Because these flows are rheological plastics they 'freeze' once shear stresses can no longer overcome internal shear strength with the result that their poorly sorted texture is preserved (Fig. 10.23, 1–3). Large clasts may 'float' in a finer-grained matrix and overall the deposit shows little or no internal bedding or lamination. In some sequences, maximum clast size appears to correlate positively with bed thickness (Nemec & Steel, 1984). Some beds show poorly defined grading; inverse grading is less common. In addition, late-stage widespread matrix strength means the leading edge of a debris flow may freeze as a steep front from which secondary sliding and slumping may occur.

Non-cohesive debris flows (grain flows) derive their mobility from intergranular collisions which create 'dispersive pressure' as a result of shearing (Bagnold, 1954). Grain flows are best developed in well-sorted sand and gravel and only occur in a pure form on steep slopes (Fig. 10.23, 4) typically as a grain avalanching when the angle of rest is exceeded. The relevance of such processes to major resedimentation is minimal though grain flows of sand occur in the steep heads of submarine canyons.

However, intergranular collision is important in layers of well-sorted sediment subjected to powerful shearing on a bed by an overriding current. Grain flow-like processes may then operate in conjunction with powerful turbidity currents flowing on low gradients where the current sustains a traction carpet of colliding grains close to the bed (Fig. 10.24). One property of grain flows is that, through the operation of dispersive pressure and so-called 'kinetic sieving', larger particles work their way into higher parts of the shearing layer. As grain flows have the rheological properties of a plastic, this distribution of clast sizes is commonly preserved as inverse grading, which is a common feature of the lower parts of conglomeratic turbidite beds (Facies A.2.3; see Fig. 10.29; Fig. 10.23, 5).

Grain flows therefore are of little significance on their own and equivalent processes may occur in association with other types of current.

LIQUIFIED AND FLUIDIZED FLOWS

Liquefied and fluidized flows both depend upon the behaviour of pore fluid in granular systems. Though neither is important in sustaining long-distance sediment transport, both may be active during initiation of movement and during deposition from decelerating turbidity currents.

Sediment liquefaction occurs when a metastable grain packing texture is suddenly disturbed, commonly from shock due to storm waves or an earthquake. The sudden shift towards a closer grain packing creates excess pore fluid and, until the fluid escapes, intergranular friction is broken down as a result of the

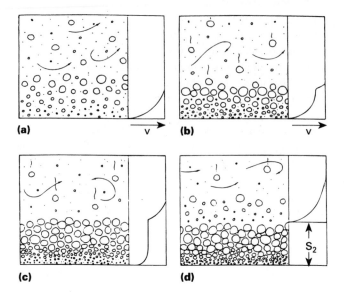

Figure 10.24 Deposition from a traction carpet. (a) High-density flow with inverse grading in base due to dispersive pressure. (b) Fallout from suspension increases clast concentration with all clasts supported by intergranular dispersive pressure. (c) Freezing of traction carpet as a plug in the upper part of the carpet. (d) Final freezing and establishment of a new traction carpet above the frozen one. Schematic average velocity profile on the right (after Lowe, 1982).

excess fluid pressure. In reasonably well-sorted sediments, where such reorganization is most likely, loss of excess pore fluid will be rapid and strength will be quickly restored. However, during the period when strength is lost, the sediment–water mixture behaves as a rheological fluid. If it is on a slope it will begin to flow. Depending on the gradient, the thickness of the liquefied layer and the rate of fluid loss, the moving layer will either refreeze or it will accelerate and transform into a more dilute flow such as a turbidity current. Very short-lived liquefaction will lead to deformation of any original lamination while more sustained loss of strength may lead to total homogenization. During deceleration of a turbidity current, when grains are arriving rapidly on the bed, sediment liquefaction may again occur, leading particularly to convolute lamination in the upper part of the Bouma sequence. Liquefaction may also occur during the immediate post-depositional history of thick, massive sands deposited by high-density turbidity currents giving rise to features of rapid dewatering such as sheets, pipes and dish structures (Facies B.2.3; see Fig. 10.29; Fig. 10.23, 12–13). However, such structures do not necessarily involve liquefied flows but rather record post-depositional *in-situ* liquefaction.

Fluidization is often closely associated with liquefaction in that the resulting upward movement of fluid then supports the grains. It is thus a process that might occur during dewatering of a liquefied layer. Fluidization only lasts as long as pore fluid is available and once that is exhausted it ceases. Fluidization in deep-sea sands acts therefore as an adjunct to liquefaction and evidence for it lies in the same range of structures that liquefaction produces (see Fig. 10.23, 12–13) (e.g. Lowe & LoPiccolo, 1974; Lowe, 1975).

TURBIDITY CURRENTS

Turbidity currents are suspensions of sediment that are sustained by fluid turbulence. They are the most important transporters of coarse-grained sediment into deep water and, while they have not been observed directly in deep-marine settings, they have been inferred through observations of sequential breaks in submarine telegraph cables (Heezen & Hollister, 1971). The classic example is the flow triggered by the Grand Banks earthquake of 1929. This flow developed from an enormous slump and travelled downslope for hundreds of kilometres attaining a maximum velocity of 70 km h^{-1} (25 m s^{-1}) (Heezen & Ewing, 1952; Piper, Shor & Hughes Clarke, 1988; Hughes Clarke, Shor *et al.*, 1990). However, turbidity currents are known directly from lakes, from laboratory experiments and they have a well-founded theoretical basis. Much of our knowledge of turbidity current deposition is inferred from extensive studies of their deposits, turbidites.

Within an active turbidity current the upwards components of turbulent fluid motion provide the main grain support mechanism and this behaviour can be sustained over long distances through a feedback loop known as *autosuspension* (Bagnold, 1962; Southard & Mackintosh, 1981). In this dynamic equilibrium: (i) turbulence is generated by the flow; (ii) flow results from the excess density of the suspension; (iii) excess density results from the suspended load; and (iv) the suspended load is maintained by turbulence. To maintain flow and keep the loop intact, the energy loss through friction and settling must balance the gain in energy as the flow travels downslope. Turbidity currents develop autosuspension if density exceeds some critical value (Pantin, 1979). Below that value a current will settle and subside, while above it will 'ignite', increasing in density and velocity to achieve autosuspension.

Gradient is also a major control on flow behaviour, with currents accelerating and eroding or decelerating and depositing sediment as gradients change. They are able to operate on very low gradients and may transport sediment of a wide range of grain sizes from gravel to the finest clays and at a wide range of grain concentrations. This last aspect has led to the recognition of high- and low-density currents though in reality there is a complete continuum.

High-density turbidity currents result in sand and gravel being resedimented into deep-water settings. They are initiated by a variety of processes (Normark & Piper, 1991). Most result from the dilution and transformation of episodic slumps and debris flows high on a slope or from the transformation of grain flows in the upper parts of canyons. Some may be derived directly from rivers during floods when the effluent is so turbid that its density exceeds that of the basinal waters. Such hyperpycnal flow is most common in freshwater lakes (see Sect. 4.6) but it may also occur in some marine settings fed by relatively small- or medium-sized rivers either annually or as a rare catastrophic event (Mulder & Syvitski, 1995). While some river-generated underflows may be sustained for long periods, depending on river discharge, turbidity currents triggered by mass movements may have a more limited duration, though this may be tens of hours (Normark & Piper, 1991).

Pyroclastic and atmospheric density flows (see Sect. 12.4.4) and experimental turbidity currents, all of which are analogues for full-scale turbidity currents, develop a series of longitudinal zones (Figs 10.22 & 10.25). The head zone has a steep leading edge behind which fluid velocities at the bed exceed the propagation velocity of the flow front (Middleton, 1966; Middleton & Hampton, 1976). In plan view, the head has a pattern of regularly spaced lobes and clefts that facilitate mixing (J.R.L. Allen, 1971). In the head, the suspension is swept forward and upward to mix with the overlying basin water. Coarsest grains are swept into the head and high near-bed velocities lead to erosion of sole marks. In the main body of the flow, sediment is carried in suspension through friction with the bed and the overlying clear water; flow thickness is uniform. The tail of the flow thins rapidly and becomes more diluted. Fine sediment falls from suspension as the flow fades away.

Deposition results from deceleration of the flow and occurs most commonly from the body and tail. Deceleration may occur in both time and space. A flow decelerates in time through the waning of the catastrophic surge-type event while deceleration in space occurs through the flow expanding laterally, as at a channel mouth, or by encountering a change of gradient, as at the foot of a slope. Clearly there are several combinations of temporal and spatial velocity change under which flows may deposit and these may influence the pattern of deposition (Kneller, Edwards *et al.*, 1991; Edwards, Leeder *et al.*, 1994). There are four principal depositional mechanisms from high-density turbidity currents (Fig. 10.22).

1 *Collapse fallout* from a turbulent flow as it loses momentum, becomes unstable and rapidly deposits. Sediment may pass briefly through a near-bed layer of hindered settling where modified grain-flow, fluidized-flow and liquefied-flow support mechanisms operate. Short-lived surge-type flows or more steady flows encountering an abrupt change of gradient typically deposit in this way.

2 Evolution of a turbidity current into a sandy debris flow followed by *freezing* of the shearing layer beneath the main plug as it advances downslope.

3 *Continuous aggradation* beneath a sustained steady or near-steady current with the sediment passing through an active basal layer of hindered settling (traction carpet) (Kneller & Branney, 1995). Grain-flow, fluidized-flow and liquefied-flow support mechanisms may all operate while a freezing front moves upward as deposition proceeds (cf. Fig. 10.24).

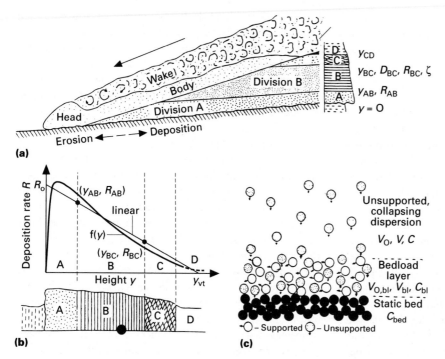

(a)

(b)

(c)

Figure 10.25 Features of a turbidity current and its deposits. (a) Streamwise profile of a flow showing its major zones and the likely sites of deposition of the Bouma intervals. (b) Depositional rate and its relationship to the Bouma sequence. (c) Possible layer structure of the flow during deposition (after J.R.L. Allen, 1991).

4 *Continuous traction* beneath a sustained steady or near-steady current. Bedload traction occurs where the rate of sediment supplied to the basal zone is not so high as to swamp the tractional processes. These processes may be especially important during deposition of clast-supported framework gravels.

Transformations between these mechanisms appear possible with single depositional events involving both turbidity current and debris flow processes (Stow, Brackenburg & Johansson, 1994).

Deposition from low-density turbidity currents or from the dilute, late stages of high-density turbidity currents can occur via one of two very different mechanisms.

1 *Slow fallout* from dilute suspension that may be accompanied by traction, forming silt ripples, and shear sorting of silt grains from clay flocs in a boundary layer, forming graded silt-laminated mud turbidites (Stow & Bowen, 1980).

2 *Concentration* of the dilute suspension as flow decelerates with damping of turbulence, flow stratification and formation of a slow-moving, hyperconcentrated flow (50–100 kg m⁻³) or muddy debris flow which may freeze as an ungraded mud layer (McCave & Jones, 1988).

Theory suggests that, for many average turbidity currents, most coarse sediment is deposited in a matter of hours though complete settling of the fine-grained tail may take a week (Stow & Bowen, 1980; J.R.L. Allen, 1991). Eventually as particles are deposited, the current becomes more buoyant, ceases to move forward and ascends as a plume in a process known as *lift-off* or *flow lofting* (Sparks, Bonnecaze *et al.*, 1993). Fine sediment from such lofted flows may become much more widely dispersed than the deposits of the main current. On

the Bengal Fan, it appears that such a process is common in distal turbidite settings (Stow & Wetzel, 1990). The 'dying' turbidity current discharges its suspension into the water column up to more than 1000 m above the ocean floor and further material is added to the suspension cloud as the tail of the current arrives in the area over a period of, perhaps, several days. This cloud settles slowly to the sea floor both above and beyond the distal feather edge of the muddy turbidite deposited by the original current. The dis-tribution of bioturbation throughout the layer suggests a time scale of a few weeks to many months for deposition of a thick layer termed a *hemiturbidite* (Fig. 10.26) (Stow & Wetzel, 1990).

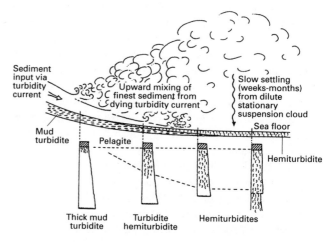

Figure 10.26 Model for flow lofting and hemiturbidite deposition (after Stow & Wetzel, 1990).

TURBIDITES

The deposits of turbidity currents, *turbidites* (Fig. 10.23, 6–11; Fig. 10.25) are widespread in many deep-water settings and form successions up to kilometres in thickness. Individual turbidite beds range in thickness from a few millimetres to several metres and grain size ranges from mud to gravel. To accommodate the range of grain sizes, three turbidite facies models have been proposed, coarse-grained, medium-grained and fine-grained, each having a distinct sequence of internal structures within individual beds. Complete sequences are rarely encountered and partial sequences are the norm with top-absent or bottom-absent beds being typical. Grading is a common feature of all three models, though reverse grading and multiple grading also occur, the former particularly at the base of the thick coarse-grained beds, the latter in silty fine-grained beds.

The basal surfaces of many turbidite beds are decorated by erosional sole marks, probably generated during the passage of the head of the flow. Tool marks, resulting from the impact or scraping of objects carried in the flow, are more common on thinner beds while flute marks and other products of turbulent scour tend to be associated more often with thicker beds. Load casts, related to gravitational foundering of the coarser layer into the muddier layer beneath, are also common features of these surfaces.

The classic model for medium-grained sand–mud turbidites, presented by Bouma (1962), involves sharp-based, parallel-sided sandstone beds and with an ordered sequence of internal lamination (Fig. 10.23, 10; Fig. 10.25). Five intervals are recognized in the complete sequence. The uppermost intervals, D and E, are products of settling from suspension; in the case of the D interval with very delicate grain sorting on the bed. The C interval, ripple cross-lamination, reflects fallout of sand or silt from suspension while lower flow regime current ripples were moving on the bed. Where fallout was rapid, climbing ripple cross-lamination occurred. This interval may also include convolute lamination attesting to short-lived liquefaction. The B interval, parallel lamination, often with primary current lineation, results from plane bed transport in the upper flow regime. The sequence B to E therefore records flow deceleration through well-established flow regimes (Fig. 10.25) (R.G. Walker, 1965; Komar, 1985).

The lowest interval ('A') lacks depositional lamination and may show grading. Extrapolation of the hydrodynamic interpretation of intervals B to E, might suggest that A intervals result from flows in the upper part of the upper flow regime (e.g. R.G. Walker, 1965). However, one critical factor in preventing lamination is thought to be sedimentation rate (Walton, 1967). Sediment arrived at the bed too quickly for it to be reworked into any bedform or lamination (Fig. 10.25). Rapid deposition also leads to unstable initial grain packing and liquefaction may then follow as suggested by the common occurrence of water escape structures such as dish or pillar

structures and dewatering pipes and sheets. A second possibility is the occurrence of a sustained flow through which sand accumulated continuously through a basal layer of hindered settling (Kneller & Branney, 1995). Turbidite sandstones showing typical Bouma intervals and interbedded with hemipelagic fine-grained sediments would normally fall into Facies C of Pickering, Stow *et al.* (1986) (see Fig. 10.29).

The model for coarse-grained turbidites is an extension of the Bouma sequence (Fig. 10.23, 6–9; Fig. 10.27) (Lowe, 1982). The Bouma A division is equivalent to the S_3 interval in which water-escape structures are extensively developed. The S_2 and S_1 intervals record phases of gravelly bedload transport beneath a powerful current. R_2 and R_3 intervals record gravel transport and deposition from suspension. The inverse grading of R_2 records the existence of grain-flow-like processes in a

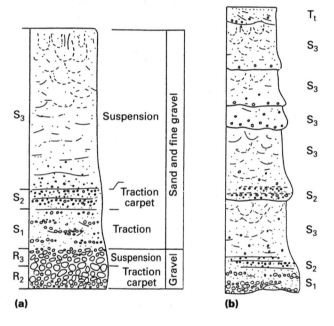

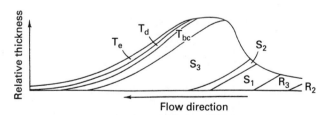

Figure 10.27 Traction and suspension deposits of high-density coarse-grained turbidity currents and (a) their relationship with higher sandy parts of the bed. S_3 corresponds with Bouma interval A. (b) Deposits of a surging high-density flow where successive pulses give discrete graded units. (c) Suggested extension of the Bouma sequence, showing the likely sequence of internal intervals along the length of a flow, valid provided a sufficiently wide range of grain sizes is available in the source area (after Lowe, 1982).

gravel traction carpet beneath a very powerful flow (Fig. 10.24). The comparative rarity of intervals S_2 and below reflects sediment availability rather than the strength of turbidity currents generally. Sequences dominated by beds made up primarily of S_3 intervals would normally be classified as Facies B1 or C1 while coarser, gravelly units might be Facies A1 (see Fig. 10.29).

The fine-grained turbidite model (Fig. 10.28) is dominated by Bouma intervals D and E, the deposits resulting from low-concentration flows transporting mainly silt- and clay-sized material (e.g. Piper, 1978; Stow, 1979; Stow & Shanmugam, 1980; Kelts & Arthur, 1981). The complete sequence is deposited by suspension fallout and traction (T_0–T_2), shear-sorting of silt grains and clay flocs in the bottom boundary layer (T_3–T_5), and suspension fallout (T_6–T_8). Many fine-grained turbidites develop only partial sequences (Piper & Stow, 1991), and excessively thick mud turbidites appear to show a more complex sequence including multiple repetitions of certain divisions (Porebski, Meischner & Gorlich, 1991). The units might normally fall into Facies D or F (Fig. 10.29). The deposits of the finest low-density flows may be virtually indistinguishable from hemipelagic background sedimentation, especially in outcrop examples and in calcilutites (Stow, Wezel et al., 1984; Bromley & Ekdale, 1987; Coniglio & James, 1990).

Figure 10.27(c) presents a simple model of along-flow change in the deposit of a flow for which a wide range of grain sizes was initially available. Such a model is the basis of the idea of 'proximality' in turbidite sequences (R.G. Walker, 1967). The idea is that in upcurrent (proximal) areas a higher proportion of beds will begin with lower intervals of the sequence compared with the situation in downcurrent (distal) areas. As the flows that construct a succession are of different magnitude, then such an approach can only be valid on a statistical basis, and then only for internal comparison within the same system. The idea is also complicated by the fact that similar changes may take place transverse to flow from 'axial' to 'lateral' settings and by the fact that local changes in basin-floor gradient can cause anomalies.

Thick (>4 m), structureless beds of both sand and mud are quite common in deep-water successions, often interbedded with normal turbidites, and their interpretation can be problematic. Stow, Brackenburg and Johansson (1994), on the basis of an extensive worldwide study, concluded that massive sands were most commonly deposited by: (i) freezing of sandy debris flows; (ii) collapse fallout from; or (iii) continuous aggradation beneath a high-density turbidity current with sediment passing through an active basal layer of hindered settling involving grain-flow, fluidized-flow and liquefied-flow processes. The restricted range of grain sizes requires input from a sand-dominated source and the apparent thickness of some beds (>25 m in some examples) commonly results from the amalgamation of several thinner (2–5 m) beds.

Very thick ungraded mud beds which include *unifites* (Feldhausen, Stanley et al., 1981) and *homogenites* (Cita, Camerlenghi et al., 1984) are probably best explained by the ponding of low-concentration turbidity currents in confined basin plains. Tails of incoming currents provide more and more sediment so that the concentration builds up and the flow transforms into a muddy debris flow and eventually freezes (Rothwell, Weaver et al., 1994). Alternatively, extreme flow dilution as a result of reversing buoyancy and flow lofting (Sparks, Bonnecaze et al., 1993) may lead to very slow deposition of thick, continuously bioturbated hemiturbidites (Stow & Wetzel, 1990).

In addition to the sedimentary structures and lamination types that make up the Bouma sequence and its variants, other characteristics also aid the identification and interpretation of turbidites. Normal (positive) grading is common in turbidites across a wide spectrum of grain sizes while inverse (negative) grading is a feature of the base of some coarse-grained beds. More precise measures of grain-size distribution and derivative statistical parameters have been thought to further characterize turbidites (e.g. Kranck, 1984). Orientation of elongate particles (e.g. plant fragments, graptolites, etc.) is often preferentially

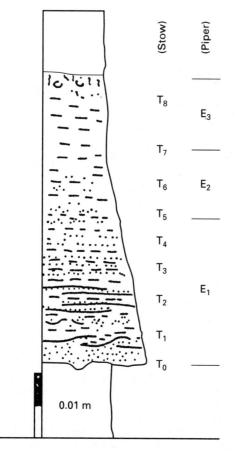

Figure 10.28 Terminology for the various intervals recognized in fine-grained turbidites.

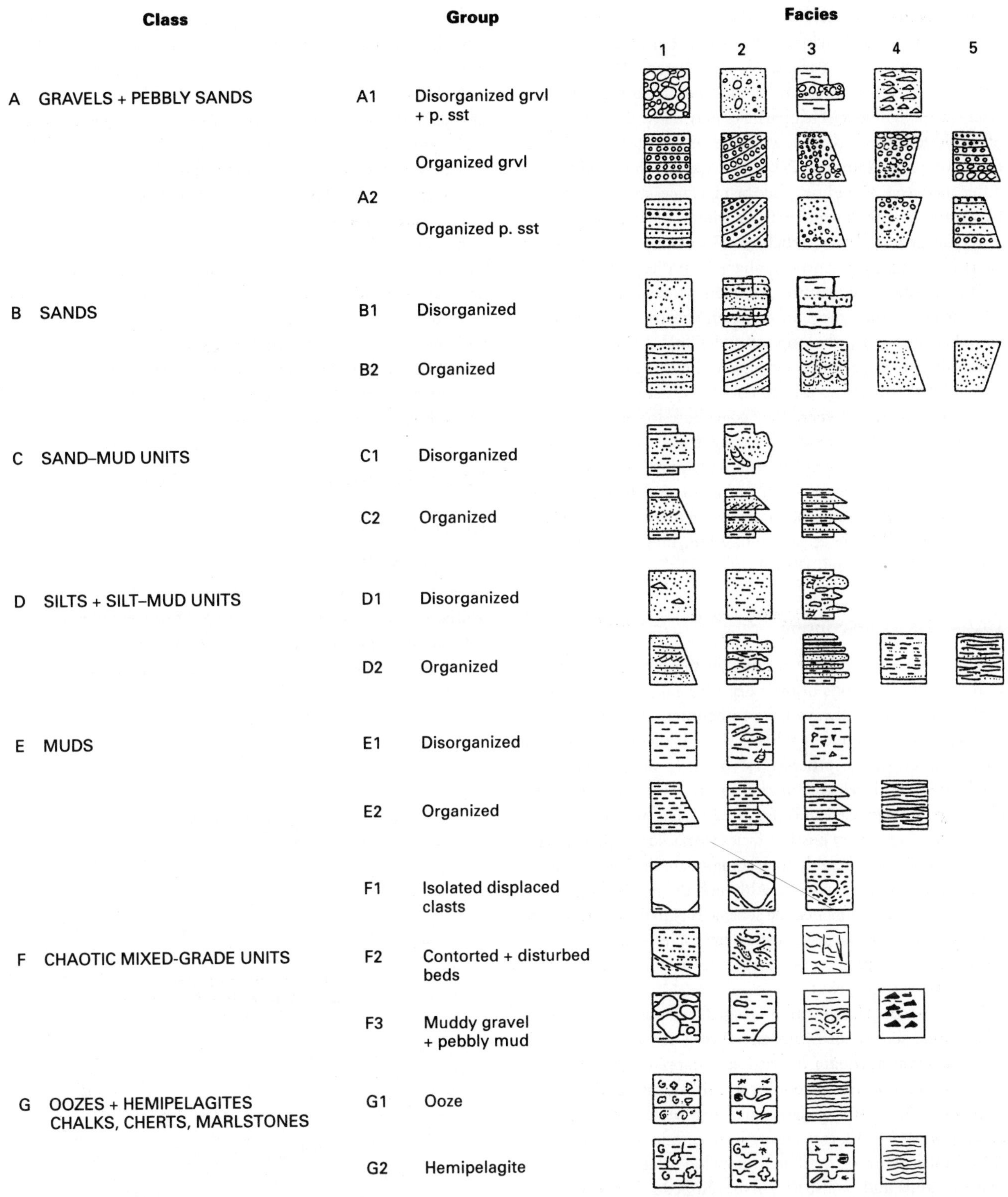

Class	Group		Facies				
			1	2	3	4	5
A GRAVELS + PEBBLY SANDS	A1	Disorganized grvl + p. sst					
		Organized grvl					
	A2	Organized p. sst					
B SANDS	B1	Disorganized					
	B2	Organized					
C SAND–MUD UNITS	C1	Disorganized					
	C2	Organized					
D SILTS + SILT–MUD UNITS	D1	Disorganized					
	D2	Organized					
E MUDS	E1	Disorganized					
	E2	Organized					
	F1	Isolated displaced clasts					
F CHAOTIC MIXED-GRADE UNITS	F2	Contorted + disturbed beds					
	F3	Muddy gravel + pebbly mud					
G OOZES + HEMIPELAGITES CHALKS, CHERTS, MARLSTONES	G1	Ooze					
	G2	Hemipelagite					

Figure 10.29 The main classes and groups of sediment facies recognized in the deep sea (from Stow, 1985; Pickering, Stow *et al.*, 1986; modified after Mutti & Ricci Lucchi, 1972). The facies classes are distinguished on the basis of grain size (Facies Classes A–E), internal organization (Facies Class F) and composition (Facies Class G). Facies groups are distinguished mainly on the basis of internal organization of structures and textures. Individual facies (subgroups 1–5) are based on internal structures, bed thickness and composition. Subdivision into individual facies given here is not definitive: the scheme is flexible so that more or fewer facies can be defined depending on the nature of the study.

aligned parallel to palaeoflow and such alignment may diverge increasingly upwards within a bed from that of its basal sole marks (Scott, 1967; Parkash & Middleton, 1970). Grain imbrication, where present, is inclined upcurrent. In silty sediments, alignment of silt particles may allow discrimination of turbidites, due to flow down slope, and contourites, due to flow along slope (Stow, 1979). It also appears that mud fabrics differ between turbiditic and hemipelagic muds (O'Brien, Nakazawa & Tokuhashi, 1980), turbidites having larger, more randomly arranged clay particle clusters whereas hemipelagites have more single clay particles aligned parallel to bedding.

Trace fossils are features of many turbidite sequences. Dwelling or resting traces are commonly restricted to, or more abundant towards, the top of individual beds or within interbedded hemipelagites. Grazing and feeding surface traces and interfacial burrows are commonly best seen on the bases of the turbidite beds where they contribute to the assemblage of sole marks. The control on trace fossil assemblage in a turbidite sequence is a function of bathymetry, salinity, grain size and composition of both turbidites and hemipelagites, turbidity frequency and geological age (Werner & Wetzel, 1982; D'Alessandro, Ekdale & Sonnino, 1986; Frey, Pemberton & Saunders, 1990; Wetzel, 1991a).

10.2.4 Facies description

The sedimentary products outlined in the foregoing sections were discussed in terms of their depositional processes which, in the rock record, always have to be inferred. Any comprehensive, descriptive facies scheme should, therefore, also aid interpretation of the processes involved in deposition. Such a scheme was developed in the 1980s (Stow, 1985; 1986; Pickering, Stow et al., 1986) (Fig. 10.29). This scheme was not intended to show every possible facies that occurs in deep water, but should be used flexibly to accommodate the level of detail required in a particular study. Although the terminology used is for siliciclastic sediments, the scheme is equally applicable to bioclastic and volcaniclastic deep-water facies, simply by using the appropriate prefix (e.g. carbonate muds or calcilutites, volcaniclastic sands, etc.).

The first-order classification is based, essentially, on the still widely used scheme of Mutti and Ricci Lucchi (1972, 1975) which is often adequate for regional mapping or reconnaissance work. For the second-order classification the facies classes A to E are subdivided into *disorganized* and *organized* facies *groups* (A1, A2, etc). Disorganized groups essentially lack clear stratification or grading and include thick structureless gravels, sands and muds; irregular, thin-bedded gravel lag or coarse sandlayers; and bioturbated, massive or irregularly-layered, silty muds. Organized facies groups show some degree of stratification or marked grading and include regularly laminated, cross-laminated, rippled and graded layers of variable bed thickness and grain size.

Facies class F is mainly disorganized and can be subdivided into three groups: exotic clasts, ranging from giant rock-fall boulders to small glacial dropstones (F1); contorted and disturbed slumps and slide masses (F2); and pebbly muds or muddy gravels (F3). Facies class G comprises both the pelagic biogenic sediments, the calcareous, siliceous and muddy oozes, and the silty biogenic sediments or hemipelagites.

The third-order classification into individual facies shown in Fig. 10.29 is an *example* of the range of facies recognized in modern and ancient deepwater sediments, and simply represents points on a facies continuum. Detailed studies of a particular class of sediments typically identify many more subtle facies distinctions. For example, Stow, Bishop and Mills (1982) described eight different mud and silt turbidite facies within Group D2 from the North Sea Kimmeridgian and over 20 facies were recognized within the massive sands of Group B1 (Stow, Brackenburg & Johansson 1994).

10.3 Deep-water clastic systems

10.3.1 Controls on deep-water sedimentation
(Fig. 10.30)

Deep-sea sedimentation is ultimately controlled by plate tectonic configurations, climate and sea level. However, more directly, sedimentary supply is probably the most important single factor in governing depositional style. This includes the grain size and

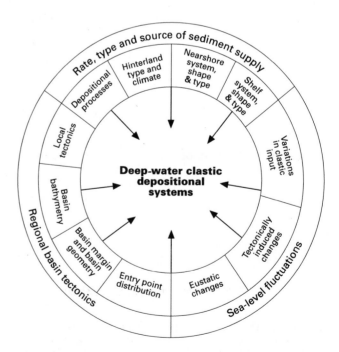

Figure 10.30 Schematic diagram illustrating the range of controls that influence deep-water sedimentation (by courtesy of M. Richards).

composition of the sediment, the volume, rate and frequency at which it is made available for deposition, and the number and position of input points. It is governed by the tectonics and morphology of both the distant and immediate source areas, by climate, in particular its influence on the supply of sediment to the shelf and shoreline, and by the regularity, magnitude and frequency of flows from the source region. Gravel-rich systems may be directly affected by rainstorms and ice-melt effects.

Tectonic setting controls sedimentation by affecting: (i) the regional stress regime, which can affect the relative rates of uplift and subsidence in the source region as well as in the basin; and (ii) rates of uplift and denudation and hence sediment supply from the source region, drainage pattern, width and gradient of coastal plain and shelf, the slope gradients, gross sediment budgets, the morphology of the receiving basin, and position and switching of depocentres.

The magnitude and frequency of seismicity may be important. However, there is no direct relationship between volume of turbidite bed and magnitude of earthquake (except perhaps for megabeds). All things being equal, the more frequent the earthquakes, the less time there is for sediment to accumulate in the source area and the smaller the ensuing turbidity currents. Infrequent earthquakes, on the other hand, as on passive margins such as surround the Atlantic Ocean, yield the largest turbidites (Pilkey, Locker & Cleary, 1980).

Sea-level fluctuations influence sedimentation in deep-sea systems by affecting the immediate nearshore source areas, and thus sediment supply. During periods of low sea level, sediment sources such as rivers and deltas may have direct access to the shelf edge. During periods of high sea level, access is more likely to be from littoral cells on the shelf carrying sediment to canyon heads (Burke, 1972). As sea level falls, delta-fed deep-sea systems in particular have very large volumes of recently deposited loose sand and mud available for remobilization as large-scale slumps and turbidity currents (Mutti, 1985). In many circumstances, it is during periods of sea-level fall that the largest and most far-travelled turbidity currents are formed. Rising sea level is the time of starvation of deep-sea systems and blanketing by seal-forming hemipelagic mud. However, particular situations may arise. For example, the present Mississippi Fan is considered to have received the highest rate of sediment input during the maximum rise of sea level in the Holocene, not during the lowstand, because North American glaciers were melting at this time and discharging large volumes of glacial meltwater into the Gulf of Mexico (Kolla & Perlmutter, 1993) (Sect. 10.3.3).

Accommodation space in deep water is not generally an important depositional factor. However, accommodation space in the source area, particularly at the coast and shelf, is critical because it controls the extent of storage, bypassing and erosion, and therefore supply to deep water.

Although the above factors are the prime control on deep-sea sedimentation, and govern the broad patterns of fan, ramp or slope-apron growth, the actual facies distribution that is encountered is a consequence of many other more local factors which operate primarily within the sedimentary basin itself. These factors include the shape and size of the basin (e.g. Nardin, 1983), depth of water, basin salinity, organic supply, gas production within the sediment, erosional and depositional features of the sediments, synsedimentary tectonics, diapiric movements, especially salt movements and differential compaction. Examples cited later will show that these local factors need to be taken into account before the actual sedimentary facies patterns can be fully understood.

10.3.2 Classifications, terminology, elements and sequences

Several reviews of deep-sea systems have appeared in the last decade. In the previous editions of this book, the three principal environments of clastic deposition identified by Gorsline (1978) (Fig. 10.1), *slope aprons, submarine fans* and *basin plains* were differentiated (Sect. 10.1). In this edition we also recognize *submarine ramps*, intermediate between submarine fans and slope aprons, and *contourite drifts* as important constructional mounds that occur both independently and as part of the other environments. *Channel systems* considered as distinct by Pickering, Hiscott and Hein (1989) and Clark & Pickering (1996), are here treated as important elements of ramp and fan systems.

A concise, but very thorough review, that emphasizes reservoir properties of deep-sea systems, is that of Shanmugam and Moiola (1988). They separate deep-sea systems into four types based on tectonic setting: (i) immature passive-margin fans (North Sea type); (ii) mature passive-margin fans (Atlantic type); (iii) active-margin fans (Pacific type); and (iv) mixed-type settings. However, they admit that both the immature passive-margin type and the active-margin type are similar in that they are typically sand-rich. Their 'mixed' category contains fans that cannot be classified into one of the other three types and includes the Bengal and Indus fans, and the Orinoco. They also admit that the Magdalena fan, though forming along an active margin, exhibits characteristics of a passive-margin fan. In all, the attempt demonstrates that tectonic setting cannot be used satisfactorily to categorize fans.

Based on experience of the highly elongate, longitudinally supplied basins of the south Pyrenees and steering clear of the term 'submarine fan' as used in numerous earlier papers (e.g. Mutti, 1979), Mutti (1985) identified three types of turbidite depositional system (Fig. 10.31), which he related closely to the volumes of gravity flows. These, he believed, were largely a function of relative sea-level position.

Type 1 turbidite depositional systems consist of thick, extensive lobe deposits that are correlated with large-scale erosional features cut into adjacent shelf edges that may form from large-scale slope failures of unconsolidated shelfal mud and sand

(Mutti, 1992). They compare with 'efficient' mud-rich fan systems (Mutti, 1979).

Type 2 turbidite depositional systems consist of channel-lobe transition deposits that grade basinwards into subordinate lobe deposits. They are correlated with large-scale erosional features cut into adjacent basin-margin strata and compare with 'poorly efficient' sand-rich systems (Mutti, 1979).

Type 3 turbidite depositional systems are very small. They are formed entirely of thin-bedded and graded 'mud-laden' units, deposited in small sandstone-filled channels that are restricted to the inner parts of the system. Major erosional features are missing.

Disillusionment with existing classifications of deep-sea systems in the late 1980s led to two separate approaches. One was to treat sedimentary successions in a hierarchical sense in the same way as was being done for fluvial and aeolian sequences (Mutti & Normark, 1987, 1991; Mutti, 1992; Ghosh & Lowe, 1993, 1996). The other way was to develop further an environmental classification of deep-sea systems (Reading, 1991; Reading & Orton, 1991; Reading & Richards, 1994). Both approaches are complementary.

Type I: Channels with detached lobes

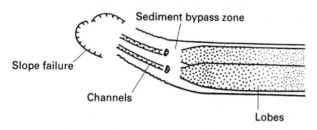

Type II: Channels with attached lobes

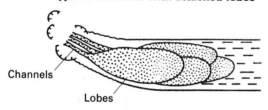

Type III: Channel–levee complex without lobes

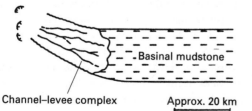

Figure 10.31 Plan views of the three main types of turbidite depositional system of Mutti (1985).

HIERARCHICAL SCHEMES AND ELEMENTS

The hierarchical scheme of Mutti and Normark (1987, 1991) established a number of elements, each bounded by a distinctive surface and containing a particular facies. They followed sequence stratigraphers rather than facies modellers in making the first order the largest scale. Although they suggested durations for each hierarchical order, they warned against a too rigid adherence to a time frame because turbidite systems, particularly oversupplied systems, can accumulate within a period that is one or two orders of magnitude faster than can be resolved by biostratigraphy.

Ghosh and Lowe (1993, 1996), on the other hand, followed the facies modellers such as J.R.L. Allen (1983) and Miall (1985a) for fluvial sequences and Brookfield (1977) for aeolian sequences in starting with the smallest unit, though the numbers do not equate (Fig. 10.32; Table 10.2). In this case a first-order architectural element is the Bouma division. Second-order architectural elements comprise discrete sedimentation units deposited by individual flows. Third-order architectural units are packets of similar flow units, interpreted in terms of the major features recognized in deep-sea environments: channels, levees or lobes. They correspond to the 'primary elements' of Mutti and Normark (1987, 1991). Fourth-order architectural elements consist of associations of contrasting third-order elements. They represent channel–levee–overbank complexes or interbedded channel–levee sequences. Fifth-order architectural

Table 10.2 Comparison of hierarchies of architectural elements of Ghosh and Lowe (1995) and elements of Mutti and Normark (1987, 1991). In the architectural elements approach, vertical trends (+ve/−ve cycles) are not important for the recognition of facies and facies associations. This approach is based solely on process.

Mutti & Normark (1987, 1991)	Ghosh & Lowe (1995)
	1st Order S1, S2, Ta, Tb, etc.
	2nd Order Single, sedimentation unit
5th Order Turbidite Substage macroform	3rd Order Channel infill
4th Order Turbidite System minor lobe, channel-levee	4th Order Channel complex
3rd Order Major Lobe	
2nd Order Turbidite Complex Depositional System	5th Order Multi-storey channel stack
1st Order Fan Complex	6th Order Fan complex

elements are at the level of formation and may comprise several stacked channel–levee and overbank fourth-order elements. The highest order architectural element is the sixth-order element, comprising several fan complexes or large slope aprons.

Each and every scale of architectural element is bounded by a surface that represents a change in conditions (Fig. 10.33). First-order bounding surfaces represent changes in conditions of unsteady flows. Second-order surfaces form between flows,

separating sedimentation units. They may show scouring. Third-order and fourth-order surfaces separate distinctive groups or packets of sedimentation units and, in some cases, are major erosional features.

Erosional features are important because they may be genetically linked with depositional events downcurrent (Mutti, 1992). Thus, in the deep sea, there are a number of major depositional and erosional features that can be recognized

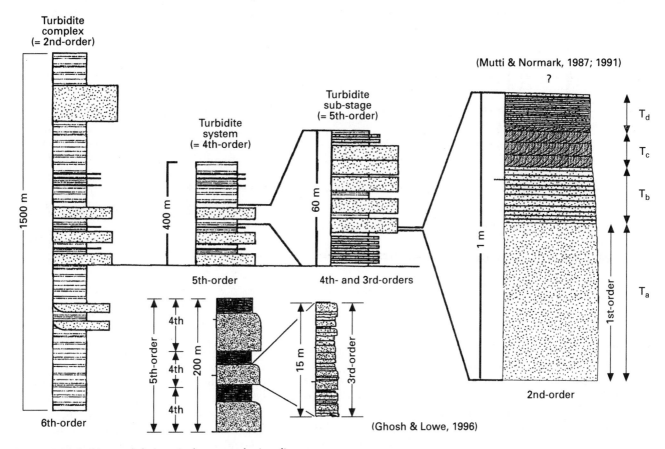

Figure 10.32 Architectural elements in deepwater clastic sediments (from Ghosh & Lowe, 1993, 1996). Figure is based on that of Mutti and Normark (1987, 1991) whose elements are shown in brackets.

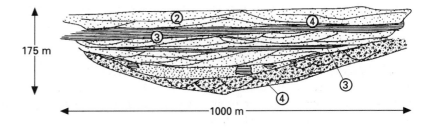

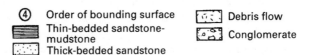

Figure 10.33 Cross-section through the Turonian Venado Sandstone, California showing scales of erosional bounding surfaces (from Ghosh & Lowe, 1993).

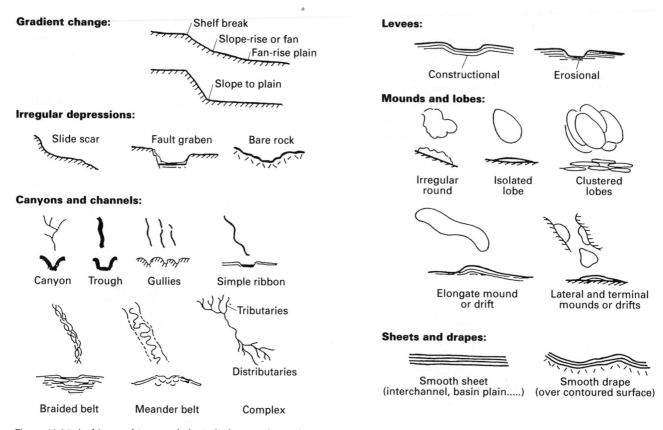

Figure 10.34 Architectural (or morphological) elements that make up deep-water depositional systems.

that include the third-order architectural elements of Ghosh and Lowe (1993, 1996), the fourth-order elements of Mutti (1992), as well as some additional ones (Fig. 10.34): (i) major erosional features; (ii) overbank deposits; (iii) depositional lobes; (iv) irregular mounds; (vi) elongate drifts; (vii) sheets and drapes in open slopes and basin plains; (viii) megaturbidites and megadebrites; and (ix) tectonic features such as growth faults, diapirs and compressional fault mounds.

These elements are commonly a few hundred metres to several kilometres in width, a few metres to a few hundred metres in relief or thickness and may be approximately equidimensional or markedly elongate.

Major erosional features include shelf-edge failures, slope failures, intrabasinal failures and canyons. In modern systems they show large-scale negative relief and, because of their size, they commonly have clear seismic expressions in buried basins. In outcrop studies their recognition always requires extensive and careful mapping and detailed stratigraphic and facies analyses. Most commonly they are filled by chaotic deposits and thin-bedded turbidites, with locally developed coarse-grained, usually lenticular bodies, particularly in the basal part of the fill (e.g. Morris & Busby-Spera, 1990).

Many slope failures act as major sources for turbidite accu-

mulations, mainly through retrogressive sliding. Some failures may involve the removal of up to 5000 km^3 of sediment in extremely short periods of time (10^3–10^4 years). Correlation within turbidite systems that form through these rapid near-instantaneous resedimentation processes can only be based on physical surfaces, since the time span involved in their formation is less than conventional biostratigraphic resolution.

Channels include both the large, leveed valleys and smaller, commonly unleveed channels. Scales range from kilometres to metres. In modern systems they show negative relief, and usually contain a coarse-grained basal lag. In ancient systems the channel-fill bodies are lens-shaped and rest on well-expressed erosional surfaces. Most channels are filled by the coarsest sediment in the system, which includes basal lag deposits, thick-bedded sandstones and pebbly sandstones, thin-bedded turbidites, particularly at the channel edges and chaotic deposits. Differential compaction of muds around coarse channel fills can lead to distortion of channel cross-sections and the development of convex-upwards top surfaces. Mud-filled channels, however, do occur. Over a period of time, channels may be dominantly erosional, depositional or a mixture of the two. Although meandering channels are common in modern fans, they are seldom recognized in ancient systems except on 3-D seismic.

Overbank deposits are essentially marginal wedges formed of thin-bedded turbidites. In modern systems they are components of channel–levee complexes that show levee relief decreasing downstream. In ancient systems they are represented by channel-related overbank deposits, overbank wedges lacking sizeable channel-fill units but containing discrete, metre-thick packages of thin-bedded and fine-grained sandstone or monotonous successions of graded mudstones. Slumps are common. Marginal wedges of thin-bedded, predominantly muddy turbidites form a major element in most large mud-rich fans. They appear to be less common in ancient fans, possibly because they are not recognized and in some cases mistaken for 'distal' turbidites.

Lobes present a major problem of definition, and their terminology is particularly confusing. The term 'lobe' has been used in many conflicting senses and on many different scales (see review by Shanmugam & Moiola, 1991) often for features not at all like lobes (e.g. the fanlobes of Bouma, Stetling & Coleman 1985) that embrace the entire fan system of the Mississippi. Here we restrict the term to lobe-shaped, low-relief depositional mounds generally located at the distal ends of channels.

Lobes on modern fans are defined by morphology (i.e. lobate in plan view) and vary considerably in size and shape from kilometres for the Crati Fan through tens of kilometres for the Navy to 100 km for the Monterey and several hundred kilometres for the Mississippi Fan. Little is known about the thickness and facies composition of modern fan lobes. On the other hand, lobes is ancient sequences have been defined as roughly tabular non-channelized bodies, 3–25 m thick, composed of well-graded medium- to thick-bedded classic turbidites, commonly developed as very small-scale thinning and thickening-upward cycles which are thought to be compensation cycles resulting from the progressive smoothing of subtle depositional relief produced by the stacking of individual sandbeds during upbuilding of each lobe (Mutti & Normark, 1987, 1991). Although scours occur, they are not deep and are less common than in some other elements. Scour junctions often cause amalgamation, showing just as a line of rip-up clasts. Interbedding of lobe facies and interlobe facies leads to vertical successions several hundreds of metres thick. Lobes constitute the most important element in ancient small basin-fills, both in area and volume of sediment.

Irregular mounds are formed essentially of slide, slump and debris flow masses that have come to rest in a lower slope or proximal basin position. In modern systems they show an irregular positive relief and a transparent or chaotic internal seismic facies. In ancient systems they are recognized by a typically fine-grained, disturbed or chaotic facies, in some cases with larger irregular clasts.

Elongate drifts have a more regular mounded relief, often parallel to the contours in modern systems, and are constructed by long-term bottom current flow. A number of different types of drifts are now recognized (Sect. 10.3.7).

Sheets and drapes are parallel-sided, laterally extensive depositional elements that make up large areas of interchannel, open slope and basin plain systems. Sheet systems that gradually fill up depositional lows are composed of fine-grained turbidites and interbedded pelagites. Purely pelagic and hemipelagic sedimentation forms regular drape deposits over pre-existing bottom topography.

Megaturbidites and megadebrites are very thick instantaneous deposits within the sheet-like fill of basin plains. They also occur in channels and across slope-apron systems. They represent smoothed and organized downslope deposits resulting from major erosive or foundering events in an upslope location.

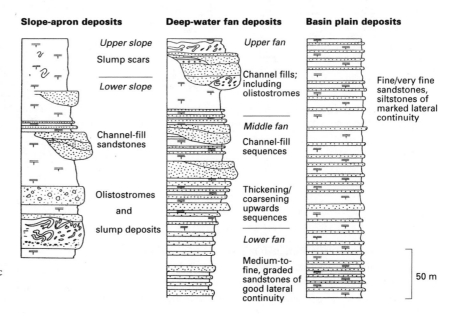

Figure 10.35 Facies associations characteristic of different deep-water depositional systems (modified after Mutti & Ricci Lucchi, 1972).

Within ancient systems, apart from the last mentioned element, these architectural elements are most easily identified on the basis of *facies associations* and *vertical facies sequences*.

Facies associations (Mutti & Ricci Lucchi, 1972) are genetically related groups of facies that co-occur in particular environmental settings (Fig. 10.35). Furthermore, they are considered to be stacked in characteristic *vertical sequences*, of which several different types are recognized (Fig. 10.36) (Ricci Lucchi, 1975; Shanmugam, 1980; Stow, 1985). Valuable though this approach is, there is a danger in jumping too directly from a vertical sequence (e.g. one that thickens up) to a morphological element (e.g. a lobe).

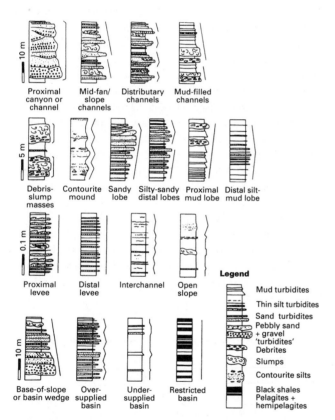

Figure 10.36 Vertical sequences of turbidites and associated sediments for various morphological elements in the deep sea. Fining (thinning)-upward, coarsening (thickening)-upward, blocky, symmetrical and irregular sequence types are indicated by the lines to the right of lithological columns (from Stow, 1985).

AN ENVIRONMENTAL CLASSIFICATION

The environmental classification for basin margin turbidite systems (Reading & Richards, 1994) is based on: (i) the volume and grain size of the available sediment (mud-rich, mud/sand-rich, sand-rich and gravel-rich). These control the size, slope gradient, flow frequency and patten of morphological elements and facies of the systems; and (ii) the nature of the supplying system (a point-source *submarine fan*, a multiple source *ramp*, and linear-source *slope apron*). This controls the feeder channel stability, organization of the depositional sequence and downcurrent length/width ratio (Table 10.3).

It is important to stress that there is a gradational continuum between the systems differentiated (Reading & Orton, 1991) and there will, inevitably, be examples that fall between two classes. Some examples may even fall into one class on one criterion and another class on another criterion. In addition, systems may change quite rapidly from one class to another and back again, as controlling factors such as sediment supply, sea level and subsidence rates alter.

10.3.3 Submarine fans

Submarine fans are distinctive constructional features on the sea floor that develop seaward of a major sediment point source, such as a river, delta, alluvial fan or glacial tongue, or beyond a main cross-slope supply route, such as a canyon, gully or trough at the base-of-slope (see Bouma, Normark & Barnes (1985) for the most extensive coverage of submarine fans). Four types of submarine fan are recognized on the basis of their dominant grain size (Table 10.3).

Modern *mud-rich submarine fan systems* (Fig. 10.37) include all the large, elongate, high-efficiency, major deep-sea fans of the present oceans. All, except the Zaire Fan, and the Laurentian Fan which is mostly fed by glacial till and glaciomarine sediments, are fed by a major river. As with most fans, their size is related, in the long term, to the amount of river sediment available from a terrestrial source (Wetzel, 1993). Their radii range from 100 km to nearly 3000 km. Their areas range from 3 000 000 km² for the Bengal Fan, through the Indus (1 000 000 km²), the Amazon and Mississippi (330 000 and

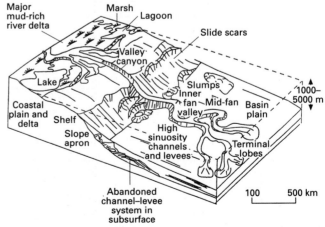

Figure 10.37 Depositional model for mud-rich submarine fan (from Reading & Richards, 1994).

Table 10.3 Classification of deep-water, basin margin turbidite systems based on: (i) volume and grain size of the available sediment; and (ii) the nature of the supplying system (after Reading & Richards, 1994 in which references to systems can be found). Many systems fall on or close to the boundary between compartments and, over quite short periods, may move from one compartment to another.

	Mud-rich systems	Mud/sand-rich systems	Sand-rich systems	Gravel-rich systems	
Submarine fan point source	Mississippi, Indus, Amazon, Bengal Nile, Magdalena Laurentian Monterey Mozambique Astoria Valencia	La Jolla Navy, Rhône Peira–Cava Gannet Stevens Kongsfjord	Avon Calabar Redondo Frigg	Bear Bay Yallahs Gulf of Corinth North Brae Cap Enragé Marambio	Increasing dominance of a single feeder system, feeder channel stability, organization of depositional sequence, downcurrent length/width ratio, 'life time' of source area
Multiple source ramps	Cap Ferret Wilmington Catskill Hareelv (Long)	Ebro C. America Trench San Lucas Crati Butano Forties	Matilija Campos Claymore–Galley Tyee Shale Grit	Mweelrea Curavacas Huriwai Brae Wagwater	
Slope apron linear source	Nova Scotia Highstand NW Africa SW Africa Gull Island	Nova Scotia Lowstand Hareelv (Trans) Hunghae Alba	Sardinia–Tyrrhenian Tonkawa	Helmsdale Marambio Wollaston Forland	

Increasing size of source area, depositional system, size of flows, tendency for major slumps, persistence and size of fan-channels, channel–levee systems, tendency to meander, thin sheet-like sands in lower fan and basin plain

Decreasing grain size, slope gradient, frequency of flows, tendency for channels to migrate laterally

300 000 km² respectively) to smaller ones such as the Magdalena (53 000 km²) and Astoria (32 000 km²) which are only a tenth the size of the Mississipi, itself a tenth the size of the Bengal Fan. Estimated volumes of sediment are proportional to area, ranging from some 4 000 000 km³ in the Bengal Fan, through the Mississippi with 290 000 km³, to the Astoria Fan with 27 000 km³.

The slope gradients are all relatively low and range from 2.4–0.74 m km⁻¹ in the Bengal Fan through the Indus (5.0 m km⁻¹) and the Amazon (8.5–2.1 m km⁻¹) to the Mississippi (18–1 m km⁻¹) and Astoria (18–3 m km⁻¹). Thus the largest fans have the lowest gradients.

Since these systems are commonly enormous, they have to be fed by 'efficient' large-volume, far-travelling, relatively low-density turbidity currents that were generated by major events, probably slumps initiated in the fronts of large deltas. The upper fan is characterized by a large fan valley that contains many smaller channels and may be filled mainly by slumps from the valley walls. The middle fan is made up of large persistent meandering channel–levee systems confined by high levees, due to the muddy nature of the turbidity flows. The meander pattern and the superposition of channel–levee systems are best

documented from the Amazon Fan (Damuth, Flood *et al.*, 1988; Flood, Manley *et al.*, 1991) (Fig 10.38). Towards the toe of the fan the sand/shale ratio generally increases, as very large-volume, efficient turbidity currents spread out from the broad, confining channels. Since the dominant sediment load comprises mud and only small quantities of sand, the resulting terminal lobes comprise varying proportions of thin-bedded sandstones, siltstones and mudstones, depending on the volume, inertia and transport capacity of individual turbidity currents.

In the Indus Fan (Kolla & Coumes, 1985, 1987; Droz & Bellaiche, 1991; McHargue & Webb, 1986; McHargue, 1991) there is a clear downfan decrease in the influence of channelized and overbank turbidity currents with an increase in unchannelized sands (Fig. 10.39). At any one time there was one feeder canyon on the shelf–slope and only one channel on the upper fan but over a period of time in the past, there were several feeder canyons.

Each of the fan channels evolved from an initial erosional phase by lateral migration and aggradation as flow volume and velocity were progressively reduced (Fig. 10.40). Thus size of channels decreased, preservation of overbank fines increased and amalgamation of channel sands and their continuity

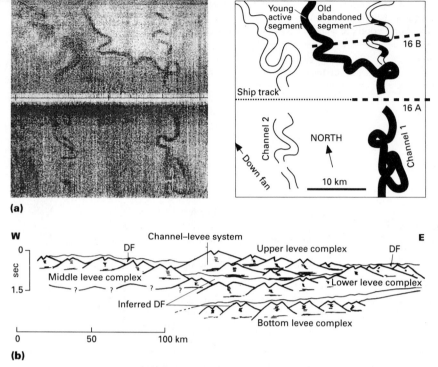

(a)

(b)

Figure 10.38 The middle Amazon Fan.
(a) GLORIA sonograph and interpretation of two meandering channels. Channel 2 is older; channel 1 has bifurcated by avulsion to the west (from Damuth, Flood *et al.*, 1988).
(b) Cross-fan strike section of individual channel–levee systems that overlap and coalesce to form levee complexes which are separated by large debris flows (DF) generated by updip diapiric activity, sediment failure or other processes that can bury large areas of channel–levee systems (from Flood, Manley *et al.*, 1991).

decreased up the sequence. This development is typical of other mud-rich fans such as the Mississippi and Amazon. The cause of the sequences is uncertain. At one time an autocyclic explanation such as lateral channel migration (cf. fining-upward fluvial sequences, see Sect. 3.6.5) would have been offered. While that cannot be ruled out, it seems unlikely on the scale envisaged. Allocyclic causes such as climatic change, tectonics and sea-

level changes have also been suggested, with a preference for relative sea-level changes, lowstands initiating each channel–levee complex (see McHargue, 1991 for discussion). However, there is no stratigraphical evidence to support any particular hypothesis.

The Mississippi Fan (Fig. 10.41) is the most fully described and discussed modern mud-rich submarine fan (Bouma, Stetling

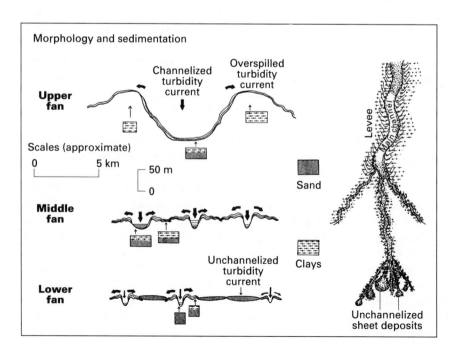

Figure 10.39 Sedimentation model for the Indus Fan. Notice decreasing influence of channelized and overbank turbidity currents, increasing influence of unchannelized currents and increase in sand down the fan (from Kolla & Coumes, 1987).

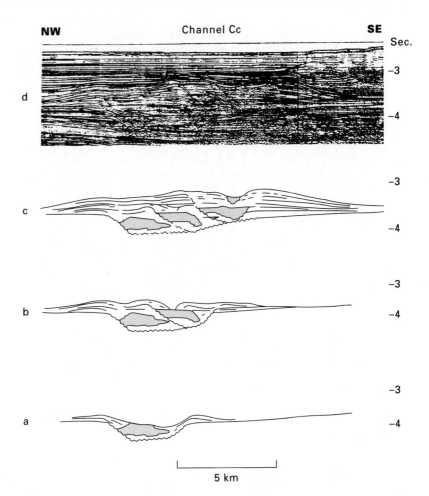

Figure 10.40 Evolution of one fan channel in the mud-rich Indus Fan. An initial channel with an erosional base and small levees 5 km wide, filled by 250 m of sediment (a) was followed by lateral migration of the channel (b). The channel was 7 km wide, with levees 175 m thick and filled by 440 m of sediment. Lateral migration continued (c) with considerable aggradation. At abandonment the channel was 10 km wide, 550 m thick, with levees 275 m thick (from McHargue, 1991).

& Coleman, 1985; Goodwin & Prior, 1988; Weimer, 1990, 1991; O'Connell, Ryan & Normark, 1991; Twichell, Kenyon *et al.*, 1991) with extensive seismic coverage and a significant number of DSDP and ODP cores. The general longitudinal pattern is the same as the Indus Fan with a diminution of channel–levee relief down fan. The modern fan, deposited during the late Pliocene and Pleistocene, consists of 17 discrete channel–levee systems each of which comprises one or more minor individual channels (Fig. 10.42). The channel–levee systems migrated a distance of about 250 km from west to east during this depositional phase (Fig. 10.43) and their migration demonstrates that, in this case, distinction between a single point fan and a multiple sourced ramp depends largely on the time scale

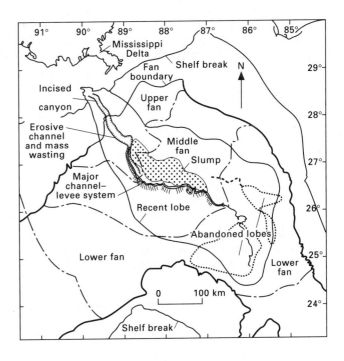

Figure 10.41 Map of present Mississippi Fan. The large incised canyon was formed by large-scale slumping. The upper fan has a gentler channel, with a channel–levee system on the middle fan. A large slump is present in the middle fan (based on Bouma, Stetling & Coleman, 1985).

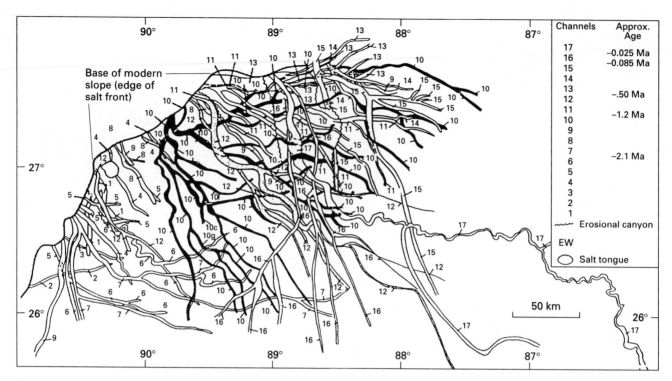

Figure 10.42 Map of 17 channel–levee systems that have developed over the last 4 million years. Numbers from 1 to 17 are oldest to youngest. Channel 10 was about 1.2 million years old, channel 16 about 0.025 million years old (from Weimer, 1990).

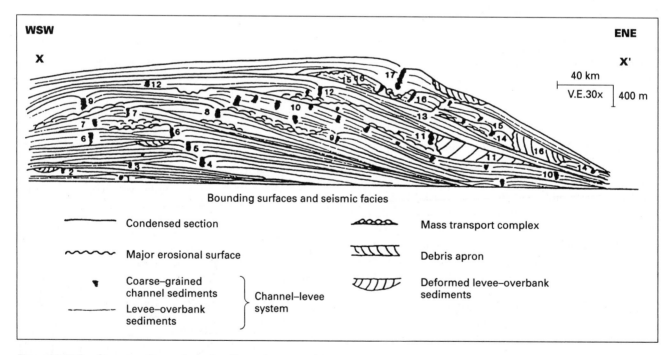

Figure 10.43 Southwest–northeast schematic strike section across the Mississippi Fan showing migration of channel–levee systems from southwest to northeast. Note vertical exaggeration is ×30 (from Weimer, 1990).

on which a system is examined. At any point in time the Mississippi Fan is a single point source fan. Observed over a period of only a few hundred thousand years – that is, the time period during which many formations were deposited in ancient sequences – it would appear multiple sourced and therefore would be classified here as a mud-rich ramp.

Pliocene–Pleistocene eustatic cycles are interpreted to have been the major factor controlling the timing and style of sedimentation (Bouma, Coleman *et al.*, 1989; Weimer, 1990). During periods of highstand, a thin layer of hemipelagic sediment was deposited on the fan surface. As sea level fell, mass transport complexes were deposited, derived from retrogressive slumping during excavation of submarine canyons on the upper slope and outer shelf (Fig. 10.44a). At the lowest point of sea level (Fig. 10.44b), sediment derived from deltas was transported into the deep basin via submarine canyons and deposited as channel–levee systems. Initially these channel–levee sediments were slightly coarser grained and more poorly organized than later when they became better organized and finer grained. Turbidite fan deposition is thought by most authors to have ceased as the rise of sea level began (Bouma,

Coleman *et al.*, 1989) or just afterwards (Weimer, 1990). However, Kolla and Perlmutter (1993), making use of detailed calculations of the timing, volume and rate of sediment supply from the Mississippi River over the last 30 000 years argue that, in the canyon area of the Mississippi margin, substantial (but decreasing) sandy turbidite deposition continued well into the Holocene (12 000–11 000 years BP) (Figs 10.44c,d & 10.45). While this was the period of sea-level rise, it was, more significantly, a period of most rapid deglaciation (Twichell, Kenyon *et al.*, 1991) when meltwater discharge and sediment load were perhaps 10 times as much as that during the maximum low-stand when the source area was heavily glaciated. This large sediment supply to the Gulf could have resulted in instability of the outer shelf and upper slope sediments, causing the failures that supplied large volumes of sediment to the Mississippi Fan.

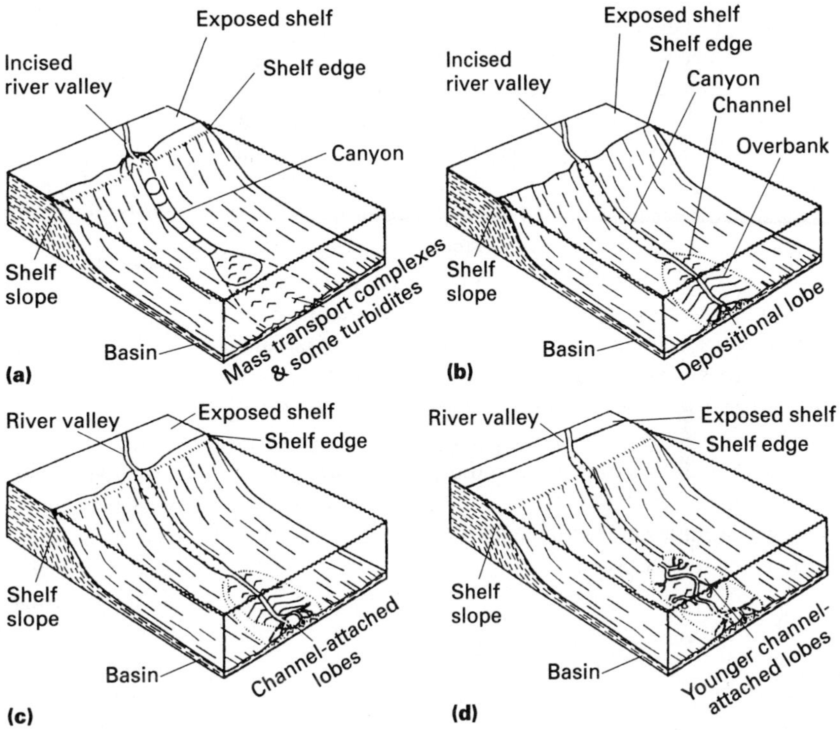

Figure 10.44 Schematic block diagrams showing evolution of the canyon-fed Mississippi Fan (from Kolla & Perlmutter, 1993). (a) Falling sea level. The canyon formed on the upper continental slope; the mass transport complex was deposited in the basin. (b) Maximum lowstand. The river valley and the canyon were connected; mainly turbidity current transport and deposition. (c) Early sea-level rise. (d) Middle–late sea-level rise, retreat of shoreline and locus of lobe deposition. According to Kolla and Perlmutter (1993) significant turbidite deposition continued because of increase in river supply following deglaciation in the source area.

Figure 10.45 Different ideas for the timing of Mississippi Fan sedimentation with respect to sea level: (a) according to Posamentier and Vail (1988), Weimer (1990); (b) according to Kolla and Perlmutter (1993) showing continuation of turbidite sedimentation during a significant rise of sea level (from Kolla & Perlmutter, 1993).

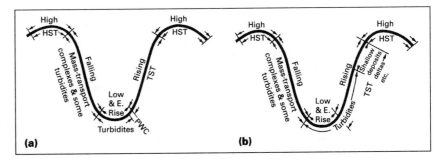

Thus, in this case, deep-sea fan sedimentation is thought to have been significant during transgression. On the other hand, where canyons are lacking, or do not penetrate far landward from the upper slope, sandy turbidite sedimentation probably ceases prior to transgression and the more conventional sequence stratigraphic models (e.g. Posamentier & Vail, 1988) apply (Kolla, 1993; Kolla & Perlmutter, 1993).

Mud/sand-rich submarine fans (Fig. 10.46) are represented by many of the classic submarine fans of the Californian deep-sea basins (Normark, 1970; Piper & Normark, 1983) and closely resemble the fan model of R.G. Walker (1978). They include a spectrum from more muddy to more sandy types. The size of these fans is moderate, but varies according to the nature of the trapping area and available sediment volumes. Typical areas range from 70 000 km² (Rhône) through 44 000 km² (Delgada), 1200 km² (La Jolla) to 560 km² for the Navy Fan, and volumes of sediment range from 40 000 to 75 km³. Their radii are between 450 and 10 km and they are lobate rather than elongate in shape when not constrained by local bathymetric complexity. They are fed either from a moderately sized mixed-load delta (e.g. Rhône), a coastline, or an active sediment-rich shelf cut by a canyon. Slope gradients range from 18 to 6 m km⁻¹, similar to those of smaller mud-rich fans.

One of the most fully studied modern fans in this group is the Navy Fan (Piper & Normark, 1983; Normark & Piper, 1985) (Fig. 10.47). It lies at the sandy end of the spectrum and is fed from a narrow, seismically active shelf characterized by little longshore transport. The size and composition of turbidity currents varies as the source characteristics change. As sea level rose during the Holocene, conditions changed from river mouth discharge at the shelf edge and upper slope (with large basin-penetrating sandy turbidity currents probably generated by slope failures triggered by high discharge) to a canyon at the shelf edge receiving little sand (with smaller muddy turbidity currents generated by seismic events).

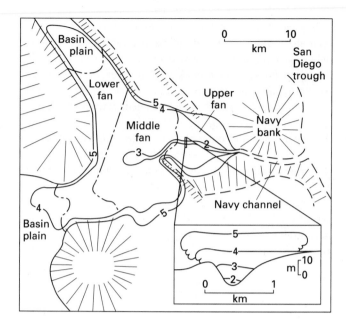

Dies at birth	1
Dies in channel	2
Flows through channel to lobe	3
Spills out, deposits levees and deposits on lower fan	4
Spills out, erodes levees, and deposits on lower fan	5

Figure 10.47 Schematic map of the Navy Fan and cross-section through the upper fan valley to show thickness and extent of turbidity currents of classes 2–5 (after Piper & Normark, 1983; Normark & Piper, 1985).

Piper and Normark (1983) showed that behaviour of a turbidity current, in terms of erosion and deposition, depends on both volume and grain size, with turbidity currents ranging from those that die at birth to those that spill out over the levees and deposit on the lower fan (Fig. 10.47). Large muddy turbidity currents overflow on to and cause aggradation of the upper fan levees, providing a large amount of sediment to the lower fan and basin plain. Large sand-carrying turbidity currents can erode levees, form mesotopography of the type observed on the mid-fan, and deposit much of their sediment in distal environments away from the active lobes. Much of the sediment deposited proximally on fans comes from smaller currents. Fast, relatively thin, sand-carrying turbidity currents may remain within the channel system. Slow, thick, muddy currents have quite independent pathways as a result of flow stripping – that is, the process by which the upper part of a current is stripped off and overflows levees – particularly at sharp bends.

Flow frequency and bed thickness are inversely related. In the Navy Fan, Class 1, 2 and 3 turbidity currents are the most common but deposit beds of very irregular thickness. Class 4 deposits most of the mid-fan sediment. Class 5 is largely

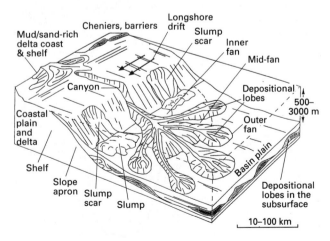

Figure 10.46 Depositional model for mud/sand-rich submarine fan (from Reading & Richards, 1994).

erosional in mid-fan areas and deposits mostly in the lower fan and basin plain where deposition is least frequent.

The architecture of these systems is governed by the presence of both sand and mud in the same region, in varying quantities, but not the enormous volumes of mud-rich systems. In the upper fan, a single feeder channel is sharply demarcated from the confining levee and overbank sediments that pass laterally into the slope apron. The feeder channel is filled by sands that link shoreward with shoreline, shelf or deltaic sands. In the upper middle fan, the main channel divides into distributaries and aggradation takes place to form a suprafan bulge. In the lower middle and lower fan the channels may advance over prograding lobes. Well-developed and persistent channel–levee systems lead to stable banks and discrete, unconnected linear channels and depositional lobes. Successive lobe elements are generated during periodic avulsion of updip middle and lower fan channels. An increase in sand content within the fan leads to a decrease in levee stability, increased channel switching and greater potential for lobe connectivity. In the lower fan and basin plain, hemipelagic sediments predominate over sheet sands, in contrast to the situation on mud-rich submarine fans, since sandy turbidity currents seldom reach that far. However, very small channel-lobes may sometimes be found.

The largest and most fully described outcrop example of a fan of this type is the Precambrian Kongsfjord Formation (Pickering, 1981) which lies at the muddy end of the spectrum. Its lateral extent cannot be determined, but it is 3200 m thick, with channel–levee systems estimated to be some 30 km wide, and channels 30 m deep can be seen. It passes upward into a 2500–3500-m-thick basin slope and deltaic system and is thus thought to have been fed by a large delta.

Ideally, the shape of these systems should be radial. However, since they mainly develop in small, confined, tectonically active basins, they tend to be constricted by basin topography and syn-depositional tectonic movement as is the case for the Navy Fan. In the ancient record, the subsurface Miocene Stevens Fan of California (MacPherson, 1978), with a radius of 25 km has been shown (Scott & Tillman, 1981) to either onlap on to structural highs when the structural grain is normal to sediment transport direction or be deposited in confined synclines to give channel-like geometries when the structural grain runs parallel

to palaeoflow direction (Fig. 10.48). In the North Sea Eocene Gannet Fields (Armstrong, Ten Have & Johnson, 1987) the pattern of the fan deposition was governed by a combination of switching point sources (Figs 10.49 & 10.50) and salt diapirs that deflected turbidity currents around them.

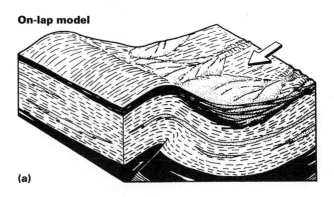

On-lap model

(a)

Confinement model

(b)

Figure 10.48 Fan models developed for Miocene Stevens Sandstone Formation, San Joaquin Basin, California showing effects of synsedimentary tectonic control on deposition. (a) Onlap model showing a series of sandbodies that lap on to and stack vertically against a rising anticline. (b) Confinement model showing a series of sandbodies that accumulate in a synclinal low between adjacent anticlines. Vertical stacking and along-axis progradation can occur (from Scott & Tillman, 1981).

Figure 10.49 Cross-section through the Eocene Tay Sands of the Gannet area of the North Sea. Cross-section is approximately west–east. In the Lower and Middle Sands it is a strike section through the fan. In the Upper Sands it is a proximal to distal section through the fan. Note influence of salt diapirs. For location of section see Fig. 10.50 (after Armstrong, Ten Have & Johnson, 1987).

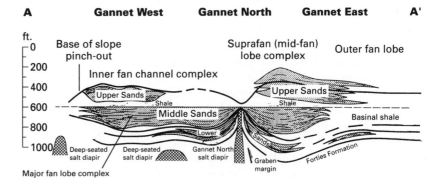

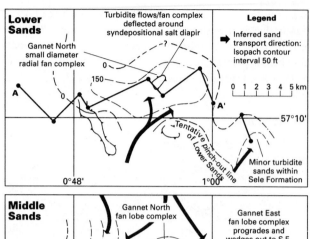

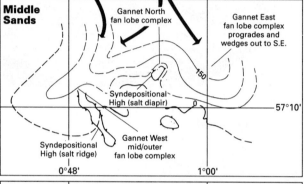

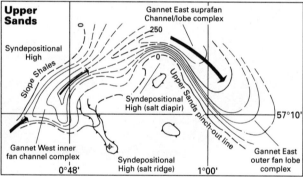

Figure 10.50 Palaeogeographies of the Lower, Middle and Upper Tay Sands within the Gannet area of the North Sea. Notice changing point sources and influence of salt diapirs on the fan patterns (from Armstrong, Ten Have & Johnson, 1987).

Sand-rich submarine fans (Fig. 10.51) are completely dominated by sand, even compared with gravel-rich systems. They are moderate in size, with radii of 5–10 km and tend to be radial rather than lobate in shape. Sand is supplied from the neighbouring sand-rich shelf and coastal system and carried into the basin via canyons by one of two methods: (i) major incisions during lowstands or failures of relict sands formed from earlier deltaic systems, particularly during sea-level rises or highstands, may liberate very large volumes of sediment; (ii) direct access to coeval littoral drift sediment may provide sand to the fan, the volumes of sediment depending on the size and tectonics of the basin, together with the width of the shelf.

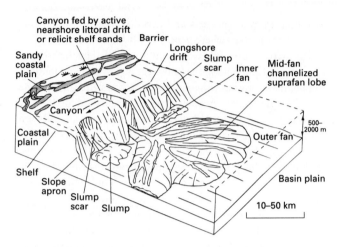

Figure 10.51 Depositional model for sand-rich submarine fan (from Reading & Richards, 1994).

In typical strike-slip Californian Borderland Basins, the volume of sediment is small and only small fan systems such as the Redondo (Haner, 1971) can develop. In contrast, on passive Atlantic-type margins, the length and width of the shelf together with the size of the ultimate feeder system may be considerable, leading to a very large volume of sediment accumulation. This is amply demonstrated in the Gulf of Benin where a million cubic metres of sediment accumulate every year in the heads of both the Avon and the Calabar submarine canyons (Burke, 1972), enough to produce a 1-m-thick layer over an area of 1000 km² every 1000 years (Fig. 10.52).

The low efficiency turbidity currents do not travel far into the basin. Instead they deposit thick massive sands that aggrade upwards to form a suprafan. The absence of well-developed

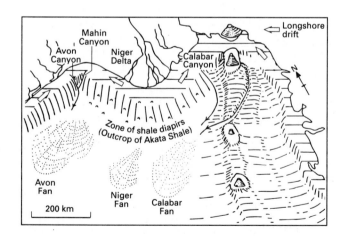

Figure 10.52 Sketch map showing relationship of Niger Delta to submarine fans in the Atlantic Ocean. The currently active Avon and Calabar fans are fed by submarine canyons at corners of the delta that receive sediment by converging longshore drifts. The Niger Fan was fed by the Niger Delta from the Niger Trench at times of low sea level, but is probably inactive at present (from Burke, 1972).

levees leads to unconfined channels that switch constantly across the fan surface and to the formation of a low sinuosity to braided system. This lack of channel confinement and impersistence of channel systems inhibits the formation of depositional lobes in the outer fan.

Outcrop examples of these systems include the Eocene Rocks Sandstone (Link & Nilsen, 1980), the classic sand-rich Grès d'Annot of southeast France (Stanley, Palmer & Dill, 1978) that covers an area of at least 5000 km², the much smaller Cengio system of northwest Italy (Cazzola, Mutti & Vigna, 1985) that has a radius of 6.4 km, a width of 4.8 km and an estimated gradient of 43–90 m km⁻¹, and a beautifully exposed fan in the Tabernas Basin, Spain, estimated to be 15–25 km² with channels less than 1 km wide (Kleverlaan, 1988). Similar systems are well represented in the North Sea where there are sand-rich oil reservoirs with a high degree of lateral and vertical connectivity, well illustrated by the Jurassic Magnus and Miller fans (De'Ath & Schuyleman, 1981; Garland, 1993), the Eocene Frigg and Gryphon fans (Héritier, Lossel & Wathne, 1979; McGovney & Radovich, 1985). A characteristic of the latter is spectacular deformed massive sands injected after deposition (Newman, Reeder *et al.*, 1993).

Gravel-rich submarine fans (Fig. 10.53), commonly comprising the subaqueous portion of coarse-grained fan deltas (see Sect. 6.5), have been extensively studied in the past 10 years following their previous relative neglect (see papers in compilations by Nemec & Steel, 1988a; Colella & Prior, 1990). Most coarse-grained submarine fans have a radius of 1–5 km or at most 10 km, but larger ones such as the Var system near Nice in the Mediterranean (Savoye, Piper & Droz, 1993) may

have developed in large tectonically active basins. Areal extents range from 1 to 50 km² and slope gradients from 250 (14°) to 20 m km⁻¹ (1–2°), with some deep-sea cones having local gradients of 450 m km⁻¹ (27°).

Deep-water fans have been studied in the Aegean, particularly in the Gulf of Corinth graben (Ferentinos, Papatheodorou & Collins, 1988; Piper, Kontopoulos *et al.*, 1990) where gravelly fan deltas have built out into 300–800 m water depths with slope gradients of 270–840 m km⁻¹. However, the most thorough description of deep-sea morphology, sedimentary processes and facies of modern fans comes from fjords in North America, especially British Columbia (Prior & Bornhold, 1988, 1989, 1990; Syvitski & Farrow, 1989). These submarine fans are fed directly from alluvial cones and fan deltas entering either from the heads of the fjords or from sidewalls. The processes include debris avalanching, inertia turbidity flows and various slope instability processes (Fig. 10.54). The events that carry sediment to the subaqueous fan range from relatively frequent rainstorms that last a few hours, to annual floods lasting days to weeks, to rarer, very high magnitude events related to glacier or other natural dam-breaks when significant volumes of water and sediment are discharged at rates equivalent to 10 times the annual peak flood. Thus for gravel-rich systems, variations in sediment discharge and water volumes on land directly and immediately affect the deep-water fan itself, a feature in contrast to the finer-grained submarine fans described above.

Ancient gravel-rich submarine fans do not appear to be common since most coarse-grained examples that have been described have a multiple source. Dabrio (1990) summarizes the sedimentological and tectonic patterns of a range of fan

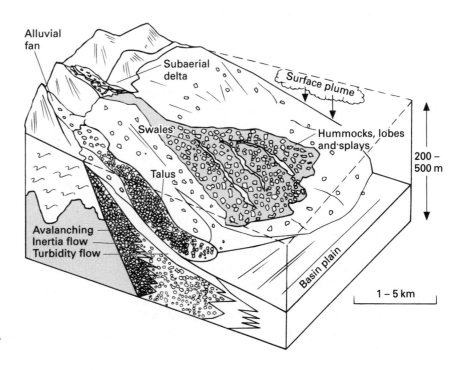

Figure 10.53 Depositional model for gravel-rich submarine fan (from Reading & Richards, 1994; based on Prior & Bornhold, 1988, 1990).

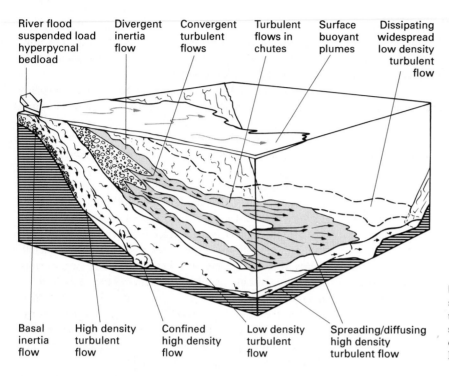

River flood suspended load hyperpycnal bedload

Divergent inertia flow

Convergent turbulent flows

Turbulent flows in chutes

Surface buoyant plumes

Dissipating widespread low density turbulent flow

Basal inertia flow

High density turbulent flow

Confined high density flow

Low density turbulent flow

Spreading/diffusing high density turbulent flow

Figure 10.54 Schematic representation of sediment transport processes operating over the subaqueous portion of a gravel-rich submarine fan as exemplified by Bear Bay fan delta at maximum river flood stage (from Prior & Bornhold, 1989).

deltas of Neogene and Quaternary age in the Betic Cordillera of Spain, some of which have a substantial deepwater component. He separates 'isolated' from 'coalescent' (i.e. multiple sourced) fan deltas. Sequences range up to 1000 m and the fans have radii of about 10 km with coarse-grained facies reaching about 5 km into the basin and calculated slope gradients being about 100 m km^{-1}. Talus breccias, including a megabreccia, turbidite lobes and slumps dominate the fans with the highest rates of subsidence. Fans with somewhat lower, but still high subsidence rates compared with those that prograde into shallow water, have slumps forming debris flows and submarine channels (chutes) feeding coarse-grained lobes. The lobes pass distally into finer-grained turbidite lobes fed by turbidite channels resembling the chutes of British Columbian fjords.

10.3.4 Submarine ramps

The term 'submarine ramp' was introduced by Heller and Dickinson (1985) to distinguish sandy deep-sea systems fed by deltas that have prograded to the shelf-slope break from those fed by a submarine canyon. Ramps therefore show relatively short-lived progradational and aggradational sequences derived from a multiplicity of switching channels. Here, however, we follow Reading and Richards (1994) in using the term 'ramp' in a broader sense for all broad constructional areas of the slope and base-of-slope that have multiple sources in the upslope region whether the source is a delta or well-supplied shelf. Siliciclastic ramps differ somewhat from the ramps of carbonate platforms (Sects 9.3.4 & 9.5.5).

Mud-rich ramps (Fig. 10.55) are not well known at the present day, but may occur in a number of settings. They may be fed by: (i) a fluvial-dominated delta such as the mud-rich Mississippi (Figs 10.42 & 10.43) or silt-rich Huanghe (Yellow River) that progrades to the shelf edge, and then shifts frequently due to either overextension or upstream avulsion; (ii) a large muddy shelf and chenier plain, itself receiving sediment by longshore drift from a major mud-dominated delta such as the Mississippi or Orinoco; and (iii) a multiple channel system from the

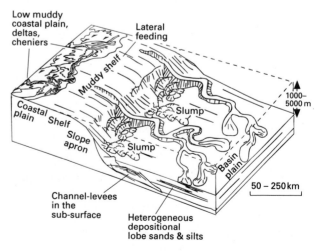

Low muddy coastal plain, deltas, cheniers

Lateral feeding

Coastal plain

Shelf

Muddy shelf

Slope apron

Slump

Slump

Channel-levees in the sub-surface

Heterogeneous depositional lobe sands & silts

Basin plain

1000– 5000 m

50 – 250 km

Figure 10.55 Depositional model for mud-rich ramp (from Reading & Richards, 1994).

continental shelf edge as the Cap-Ferret Fan in the southeast Bay of Biscay (Cremer, Orsolini & Ravenne, 1985).

One well-exposed ancient example is at the top of the Precambrian Kongsfjord Formation of north Norway. Here, 600 m of 'fan lateral margin deposits' are characterized by a very high proportion of fine-grained sandstone and siltstone turbidites, relatively small channels orientated at various angles to the regional slope, lobes associated with the small channels, clastic dykes and other evidence of soft-sediment deformation (Pickering, 1983). Another example is the deep-water portion of the Devonian Catskill delta (Woodrow & Isley, 1983; Lundegard, Samuels & Pryor, 1985) where multiple-fed turbidity currents flowed downslope with a gradient estimated to be 2.5 m km^{-1}.

Mud/sand-rich ramps (Fig. 10.56) are generally fed by a mixed sand–mud delta that itself may prograde directly across a gently sloping shelf or may feed the basin via multiple slope valleys. Sedimentation may be active during periods of either rising or falling sea level. These systems are distinguished from fan systems by the presence of several feeders that are active more or less simultaneously and from slope aprons by the presence of discrete channel lobe systems between portions of slope apron.

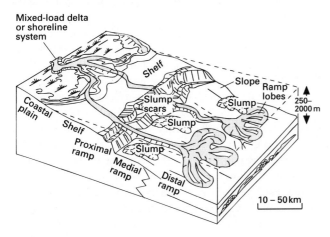

Figure 10.56 Depositional model for mud/sand-rich ramp (from Reading & Richards, 1994).

The tectonically controlled deep-sea Ebro depositional system in the northwest Mediterranean (Nelson & Maldonado, 1988; Alonso & Maldonado, 1990; Alonso, Canals *et al.*, 1991) is fed by a medium-sized delta and covers an area of about 5000 km^2 (Fig. 10.57). With a radius of 50 km and a width of 100 km, it is twice as wide as it is long and has a complex pattern of alternating channel–levee complexes and base-of-slope aprons alongside each other, each with very distinctive features. Gradients are similar to those of mud/sand-rich submarine fans (35–7 m km^{-1}, with an average of 18 m km^{-1}). It

passes downcurrent into the Valencia trough where sediment is transported parallel to the margin along the 200 km long Valencia Valley to debouch at the mud-rich elongate Valencia Fan (Maldonado, Got *et al.*, 1985). In contrast to the Ebro delta-fed system, off the east coast of Corsica, small partially coalescing fans form giving up to 1.8 km of Plio-Quaternary sediments, derived from the insular drainage system of Corsica (Bellaiche, Droz *et al.*, 1994).

In the North Sea, some of the most prolific petroleum reservoirs occur in mud/sand-rich ramps. At the muddy end of the spectrum lies the Agat Field (Gulbrandsen, 1987), where the reservoir section comprises a series of stacked, elongate or lobate sandbodies deposited from a multiple source. The

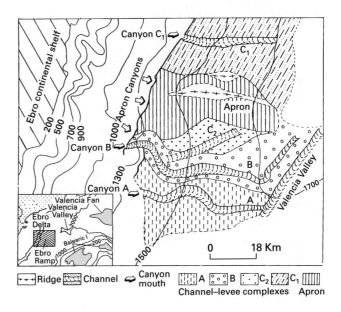

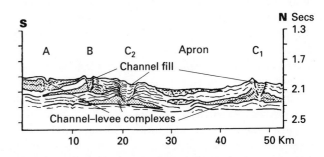

Figure 10.57 Ebro mud/sand-rich ramp. (a) Map showing channel–levee complexes A to C (A is youngest) and base-of-slope apron that has a line source of canyons, but no channels upon it. Channels have a low sinuosity, do not bifurcate and are diverted at their distal ends by the Valencia Valley. (b) Interpretation of seismic profile across the Ebro system in the middle continental rise. Channel A is youngest and erosional. Other channels are partly filled. Apron consists mainly of chaotic mass flows (from Nelson & Maldonado, 1988).

Palaeocene Forties Fan (Parker, 1975; Carman & Young, 1981; Whyatt, Bowen & Rhodes, 1991) built 130 km into the Central Graben, with a width of 80 km. Its major reservoirs occur in 50–100-m-thick units of amalgamated sandstones that formed in channels 2.5–3 km wide separated from each other by 500-m-wide shale-dominated interchannel areas (Fig. 10.58).

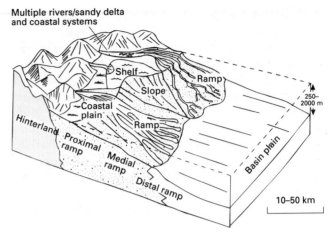

Figure 10.59 Depositional model for sand-rich ramp (from Reading & Richards, 1994).

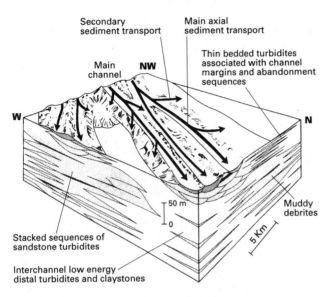

Figure 10.58 Depositional model for the Palaeocene Forties Formation in the North Sea showing multiple narrow elongate channels, typical of a mud/sand-rich ramp. Scales are approximate (from Whyatt, Bowen & Rhodes, 1991).

Sand-rich ramps (Fig. 10.59) are not always easy to distinguish from the sandier members of mud/sand-rich ramps. Yet the distinction is important because the lack of silt and mud inhibits the formation of levees and discrete channels and leads to sheet-like sandbodies with significant lateral continuity of the sands. They are fed along a broad front by sandy deltas, by coastal plains with a relatively narrow shelf, or from rapidly uplifted fault blocks. Multiple channels linked to river and distributary channels cut the slope and divide the sediment inflow so that penetration into the basin is limited. The distal ramp and basin plain receive little sediment.

Systems that fit this category include Atlantic margins either at the early immature rifted stage of development or during rising sea level when relict sands provide the bulk of the sediment. Alternatively, where the source is a relatively narrow shelf between an ocean and a granitoid basement close to the coast, substantial volumes of sand can be produced.

The Claymore–Galley systems of the North Sea formed during a period of active faulting and show the complexities in detail of many sand-rich ramps (Boote & Gustav, 1987; O'Driscoll,

Hindle & Long, 1990) (Fig. 10.60). They extend basinwards about 10–20 km, have widths of up to 40 km and are composed of sand packets of well-sorted ungraded, structureless and frequently amalgamated sandstones up to 400 m thick. They were derived by the erosion of earlier Jurassic deltaic sands, some of which had been reworked into mature quartzose sands. They are followed by gravel-rich ramp and fan systems as uplift continued and basement rocks were exposed.

The Campos Basin off Brazil contains a large number of productive turbidite petroleum reservoirs up to 250 m thick of Cretaceous and Tertiary age (Guardado, Gamboa & Lucchesi, 1989). Although they formed in a number of deep-water environments, on flat basin plains, in submarine fans and as multiple sourced systems, they all have in common a sand-rich source region, very similar to the present-day coast of Brazil (see Sect. 6.7.1, Fig. 6.69), that led to poorly efficient depositional systems of limited downslope extent. The source region was a narrow fault-controlled shelf where sands derived from a neighbouring coastal range were reworked and matured. In the late Oligocene a moderately large and complex turbidite system, with multiple feeders, extended 100 km along the slope and 50 km into the basin (Fig. 10.61).

Ancient outcrop examples include the very sand-rich Eocene Tyee Formation of Oregon (Chan & Dott, 1983; Heller & Dickinson, 1985), the Matilija Formation of California (Link & Welton, 1982), and the Carboniferous Shale Grit Formation of England (see Fig. 6.49), all of which were dominated by high-density turbidity currents supplied by sand-rich deltas. This latter example was the first fully described ancient deep-water system interpreted as a submarine fan (R.G. Walker, 1966a; Collinson, 1988). The system contains up to 60 small-scale straight 'channels' (R.G. Walker, 1966b) similar in size to the linear 'chutes' of present-day coarse-grained fan-delta systems (Prior & Bornhold, 1988) (see Fig. 10.54). Similar systems are also probably abundant in volcanic terrains. For

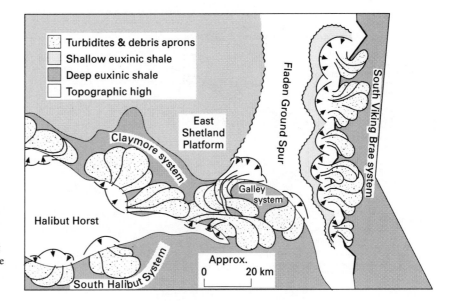

Figure 10.60 Simplified facies distribution of the Claymore–Galley Sands in the North Sea showing complex pattern of sand-rich ramps derived from a variety of uplifted blocks and highs (from Boote & Gustav, 1987). Contrast the gravel-rich ramps of the Brae system in the South Viking Graben, a model for which is shown in Fig. 10.63.

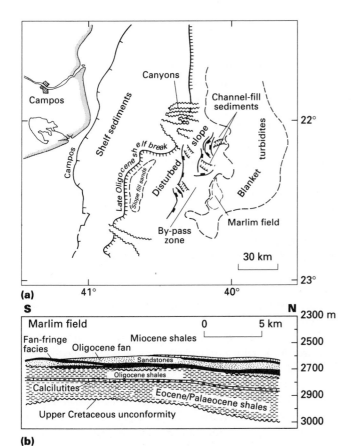

Figure 10.61 The late Oligocene Campos Basin of northeast Brazil. (a) Palaeogeographic map during a lowstand. Sands are transported from a relatively narrow shelf either directly on to the slope or via canyons, through channels to multiple relatively small sand-rich fans. (b) Downfan cross-section through the Marlim Field that covers an area of only 160 km² and yet contains 8 × 10⁹ barrels of oil in place (from Guardado, Gamboa & Lucchesi, 1989).

example, the Archaean Beardmore–Geraldton basin (Fig. 10.62) contained a series of fans fed from a shelf by multiple canyons that received sediment from braidplains across which sediment was transported from an active volcanic terrain (Barrett & Fralick, 1989).

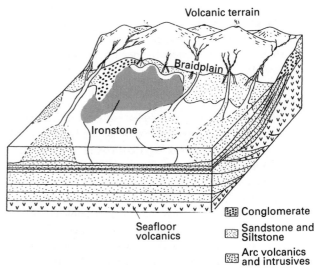

Figure 10.62 Interpretation of the Beardmore–Geraldton Formation showing a sand-rich ramp fed by a braidplain/braid delta sourced from a volcanic terrain (from Barrett & Fralick, 1989).

Gravel-rich ramps form in response to two rather different types of feeder system, either through large-scale switching of a gravelly braided delta or from a series of alluvial fans forming along a faulted basin margin. The sedimentary facies and processes of these two types are similar, resulting in proximal to distal facies relationships comparable to those described for

437

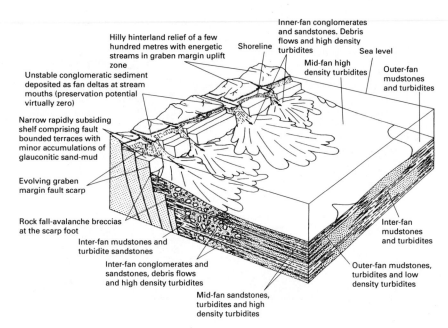

Inner-fan conglomerates
and sandstones. Debris
flows and high density
turbidites

Hilly hinterland relief of a few
hundred metres with energetic
streams in graben margin uplift
zone

Shoreline

Sea level

Mid-fan high
density turbidites

Outer-fan
mudstones
and turbidites

Unstable conglomeratic
sediment
deposited as fan deltas at stream
mouths (preservation potential
virtually zero)

Narrow rapidly subsiding
shelf comprising fault
bounded terraces with
minor accumulations of
glauconitic sand-mud

Evolving graben
margin fault scarp

Rock fall-avalanche breccias
at the scarp foot

Inter-fan mudstones and
turbidite sandstones

Inter-fan conglomerates and
sandstones, debris flows
and high density turbidites

Mid-fan sandstones,
turbidites and high
density turbidites

Inter-fan
mudstones
and turbidites

Outer-fan mudstones,
turbidites and low
density turbidites

Figure 10.63 Depositional model to illustrate facies relationships and palaeogeography of the Brae Formation, a gravel-rich ramp. The facies patterns are a consequence of a restricted source area, high hinterland relief, narrow shelf and active faulting both parallel and perpendicular to the graben margin (see Fig. 10.60 for location) (from Harris & Fowler, 1987).

gravel-rich submarine fans. However, talus debris flows are probably absent from the delta-fed systems, which tend to be lobate in shape, in contrast to those that form along a faulted basin margin to produce a linear belt of partially coalescing fans. No good modern example of either type has been described, though the 'coalescent' fan deltas of Dabrio (1990) (Sect. 10.3.3) may fit this category. They are probably rare today as a result of the Holocene rise of sea level which has led to the accumulation of the coarser material within delta plains and subaerial alluvial fans.

However, many ancient examples are known. The fault-bounded Brae Formation of the North Sea has been very well studied and interpreted both as a coalescing multiple ramp and a slope-apron system (Stow, 1985; Harris & Fowler, 1987; Turner, Cohen et al., 1987; Stephenson, 1991; Cherry, 1993) (see Fig. 10.63). In detail, it comprises several fields and reservoirs and more than one interpretation is possible depending on the scale and location of the study. Delta-fed examples include the Carboniferous Curavacas Formation of northern Spain (Colmenero, Agueda et al., 1988), the Eocene Wagwater Formation of Jamaica (Wescott & Ethridge, 1983), the Jurassic Huriwai Group of New Zealand (Ballance, 1988) though this is somewhat less conglomeratic. In Antarctica, a Cretaceous backarc basin was filled from a faulted linear source by laterally alternating mud-rich slope-apron deposits and those of coarse-grained submarine fans in a similar fashion to the Brae Formation (Fig. 10.63) (Ineson, 1989). In the Californian Continental Borderlands the Miocene Blanca turbidite system is a 1200-m-thick conglomeratic sequence (McLean & Howell, 1985) that extends parallel to its source for at least 40–50 km, yet the inner fan mass-flow conglomerates form a belt that

extends only 10 km into the basin and the mid-fan sandstones and conglomerates a further 15 km.

10.3.5 Slope aprons

Slope aprons lie between the shelf or land area and the basin floor, surrounding both small shelf basins and large ocean basins. They include the continental slope–rise couplet, the flanks of oceanic ridges, plateaus and isolated seamounts, the margins of carbonate platforms and of small basins in every tectonic setting. Various authors have used 'debris-apron' and 'ramp' as synonymous with slope apron. Slope aprons are distinguished from ramps by being fed from an essentially continuous linear source. However, the distinction is not easy, particularly at the coarser end of the spectrum. Their extent into the basin ranges from 2 to 200 km with relatively high gradients (10–150 m km^{-1}).

Mud-rich slope aprons (Fig. 10.64) range from mainly constructional, with a smooth convex–concave profile that has been built upwards and outwards by slope progradation, to mainly destructional with slumping and sliding leading to a less regular profile. In either case, the slope may be cut by canyons or gullies that feed isolated lobes in the base-of-slope area, or it may be smoothed and moulded by contour-following bottom currents. An irregular distribution of mainly finer-grained sediments is the norm, with coarser-grained sands and gravels confined to sand spillover from the shelf break, the axes of channels and base-of-slope lobes. Overall, coarsening-upward sequences may result from distinct phases of slope progradation.

Modern examples of mud-rich aprons are well known from many parts of the world (Gorsline & Emery, 1959; Nardin,

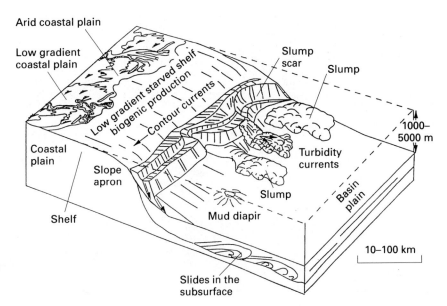

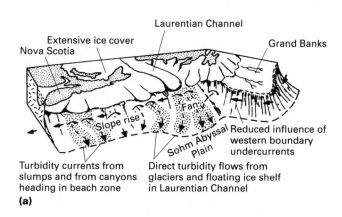

Figure 10.64 Depositional model for mud-rich slope apron (from Reading & Richards, 1994).

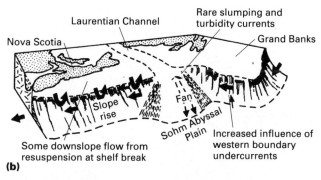

Figure 10.65 The Nova Scotia slope apron off eastern Canada showing late Quaternary sedimentation history. (a) Glacial: lowered sea level, sediment supply direct to upper slope, turbidity current resedimentation dominant, slumping common, reduced bottom current activity. (b) Postglacial: rapid rise in sea level, retreat of ice, broad submerged shelves, sediment resuspension at shelf break, hemipelagic sedimentation dominant, bottom currents active (after Stow, 1981).

Hein *et al.*, 1979). High-latitude slopes may be dominated by debris flow lenses such as those off northeast Newfoundland (Aksu & Hiscott, 1992), or by an extensive network of tributary channels as on the Labrador margin (Hesse, 1992), as well as by a significant input of glaciomarine sediment (Stoker, Leslie *et al.*, 1994). The mid-latitude slope apron off Nova Scotia was characterized by mass movements and turbidity currents (Hill, 1984) at times of glacially lowered sea level and is now being remoulded by alongslope bottomcurrents (Stow, 1981; Fig. 10.65). The northwest African slope apron is also much affected by slide and debris flow processes (Masson, Kidd *et al.*, 1992) (Fig. 10.66) and is periodically influenced by marginal upwelling and accumulation of organic-rich sediments.

A number of ancient deep-water successions have been interpreted as slope-apron deposits mainly on the basis of a mud-rich facies associated with a lack of an obvious feeder system. The Carboniferous Gull Island Formation has spectacularly exposed slides and slumps (Martinsen, 1989) (see Figs 10.20 & 10.21). The Plio-Pleistocene Rio Dell Formation of northern California (Piper, Normark & Ingle, 1976) is a thick prograded sequence of basinal turbidites overlain by slope-apron and then shelf deposits. The slope facies include reworked shelf-depth foraminifera and shallow slump scars. In the Precambrian Gowganda Formation of Canada (Miall, 1985b) and Yakataga Formation of Alaska (C.H. Eyles, 1987), glaciomarine material contributes to otherwise mud-rich and hemipelagic slope deposits.

Mud/sand-rich slope aprons (Fig. 10.67) are associated with low-gradient basin margins that form either at passive margins or at times of relatively inactive fault movements in pre-rift or post-rift stages. They receive a moderate amount of sand from a broad coastal plain that feeds the system via a wide mud–sand shelf. In addition to extensive slumping, multiple

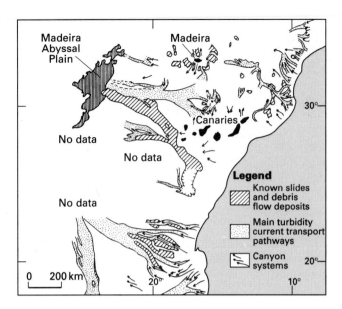

Figure 10.66 Map of northwest African continental margin and Madeira Abyssal Plain showing the numerous slides and debris flow deposits and turbidity current transport pathways (after Rothwell, Pearce & Weaver, 1992; Masson, Kidd *et al.*, 1992) (see also Figs 10.74 & 10.75).

channel systems may deliver thin- and medium-bedded turbidites to a series of lobes at the foot of the slope. The Hareelv Formation of Greenland (Fig. 10.68) has been interpreted by Surlyk (1987, 1989) as a 'non-organized, line-sourced ramp-like, slope apron to basinal turbidite system'. It consists of numerous sand-filled gullies, up to hundreds of metres wide, more than 5 km long and up to 50 m thick, that are cut into a thick slope mudstone and pass basinward into laterally extensive sandstones. It was fed from a shelf rich in well-sorted sand during a period of high sea-level rise, and low tectonic activity.

Sand-rich slope aprons (Fig. 10.69) are associated with

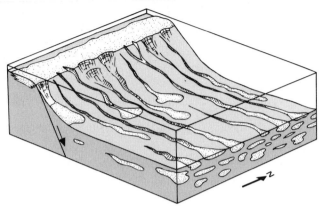

Figure 10.68 Model for an early syn-rift linear source slope apron at sea-level highstand and without tilting of the hangingwall block. Based on the Oxfordian–Kimmeridgian Hareelv Formation of east Greenland. Notice numerous gully and lobe sandstones up to 50 m thick, hundreds of metres wide and some more than 5 km long, separated by mudstones. Total system extends for several tens of kilometres from the basin margin (from Surlyk, 1989).

fault-bounded margins that receive so much sand that the deep-water system consists of small-scale fans that coalesce to form a complex linear sheet of sandstones that do not extend far into the basin. They are not well known at the present day but may have been much more common in the past before the Devonian or at times of falling or low sea level (see Fig. 10.65).

The sands may be derived: (i) from a current-swept shelf, where sediment is today being trapped and flushed out of a series of closely spaced channels as on the tectonically active Sardinia–Tyrrhenian margin (Wezel, Savelli *et al.*, 1981); (ii) from a sandy coastal braidplain such as occur along many actively uplifted continents or islands, typical of Southeast Asia or New Zealand. A fall in relative sea level or change in the position of faults and increase of gradient, would lead to these braidplain accumulations yielding large volumes of easily

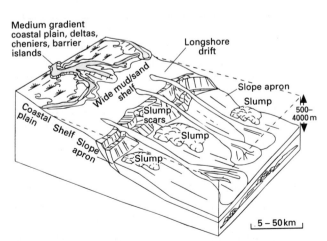

Figure 10.67 Depositional model for mud/sand-rich slope apron (from Reading & Richards, 1994).

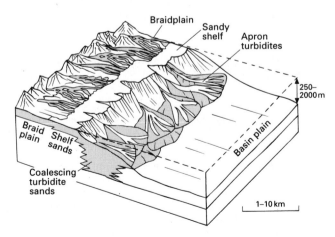

Figure 10.69 Depositional model for sand-rich slope apron (from Reading & Richards, 1994).

eroded sands for transfer to deeper water; or (iii) from a passive continental margin such as off Nova Scotia on the east coast of North America during periods of lower sea level when the coastal plain reached close to the present shelf edge.

An ancient example is the Pennsylvanian middle Tonkawa Sandstone of Oklahoma which Kumar and Slatt (1984) interpret as having been deposited on a slope apron that extended 200 km parallel to the slope with a gradient of less than 17 m km^{-1} (1°). Unlike ramp systems there is no evidence for cut and fill channels.

Gravel-rich slope aprons (Fig. 10.70). As with sand-rich slope aprons, modern gravel-rich aprons are not well known partly because they occur off tectonically active islands that have not been studied so that our models are mainly derived from ancient examples. They typically form across an active synsedimentary fault margin, and commonly have a step-like profile with relatively steep parts alternating with gentler gradients and perched basins. There is an abrupt change of gradient at the base-of-slope to a flat basin floor, with little development of a lower slope or rise. A very thick but narrow wedge of sediment may accumulate at the foot of the slope as a result of continued downfaulting. Slump scars, slump and debris-flow masses and short-lived shallow channels commonly form part of a series of small fans and lobes that overlap along the length of the slope. Although these aprons are called 'gravel-rich', they may contain much mixed-grade sediment.

The Wollaston Forland Group of east Greenland (Surlyk, 1978, 1984, 1989) (Fig. 10.71) is up to 3 km thick and exposed for 10–20 km across strike and 10s km along strike. Here a linear fault-controlled belt of alluvial fans and fan deltas extended some 15 km into the basin with water depths estimated to have been about 1000 m. Many ideas and questions are raised by Surlyk (1989) where he separates 4 stages in the development of the system: an early submarine talus breccia stage, a prograding gravel-rich slope apron, an aggrading fan delta to submarine fan stage that has elements of a multiple-source ramp (Fig. 10.71) and finally, a mud-rich fan that fits the sand-rich category defined here. He discusses whether small-scale fining-upwards cycles are the result of alternate progradation and retreat of small-scale fans or of surging flows. He then considers controls: (i) climate, which was warm and wet and apparently constant throughout and thus was not a control on facies changes; (ii) tectonics, which was critical; and (iii) sea level and the presence or absence of a shallow marine shelf. The latter two controls affect sediment supply and in particular the sand:mud ratio (Fig. 10.72). Strong tilting of the hangingwall block promotes development of steep footwall slopes, of a well-defined fault-parallel basin axis, and of a seawards emergent or submergent barrier. Flows tend to be deflected parallel to strike in the basin axis. Gentle tilting allows flow to continue downslope for a considerable distance. Low sea level favours development of very coarse-grained clay-poor aprons. High sea level favours line-sourced disorganized systems with thick prominent draping clays which form significant permeability barriers and good correlative horizons.

Volcaniclastic sandstone and sandstone/conglomerate dominated slope-apron successions are well known from active arc-related settings such as the Miocene backarc margin now exposed in southeast Korea (Chough, Hwang & Choe, 1990; Chough, Choe & Hwang, 1994). However, mostly the coarser-grained facies are closely associated with fine-grained resedimented and fall-out tuffs and tuffaceous hemipelagites (e.g. White & Busby-Spera, 1987).

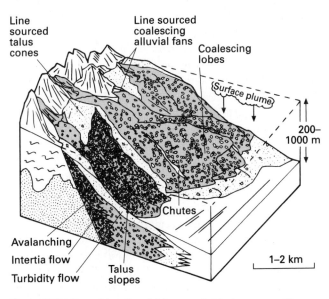

Figure 10.70 Depositional model for gravel-rich slope apron (from Reading & Richards, 1994).

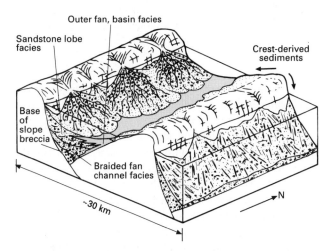

Figure 10.71 Model for synrift slope apron deposition showing one stage in the development of the Wollaston Forland Group, east Greenland. Coalescent gravel-rich fan deltas to submarine fans at sea-level lowstand and with strong tilting of the hangingwall block. They aggrade rather than prograde and the subaerial fan delta has a low preservation potential. Sediment may also be supplied from the crest (from Surlyk, 1989).

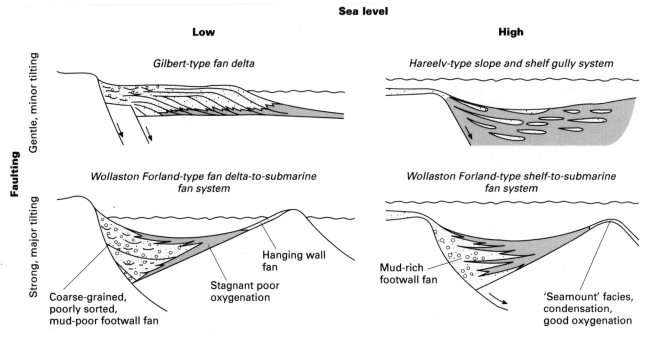

Sea level

Low　　　　　　　　　　　　　　　**High**

Gentle, minor tilting

Gilbert-type fan delta

Hareelv-type slope and shelf gully system

Faulting

Strong, major tilting

Wollaston Forland-type fan delta-to-submarine fan system

Wollaston Forland-type shelf-to-submarine fan system

Hanging wall fan

Stagnant poor oxygenation

Coarse-grained, poorly sorted, mud-poor footwall fan

Mud-rich footwall fan

'Seamount' facies, condensation, good oxygenation

Figure 10.72 Suggested influence of sea-level stand and tectonic style on the development and configuration of marine syn-rift depositional systems. These systems may follow each other in any one place, or form simultaneously in different parts of a basin (from Surlyk, 1989).

10.3.6 Basin plains

Basin plains are flat and relatively deep. Relief is low because topographic irregularities are smoothed and buried by turbidite fill and they are therefore within the reach of a significant sediment supply, normally from a continent. Sediment thicknesses are mostly a few hundred metres, though in some fault-bounded basins synsedimentary subsidence can lead to thicknesses of several kilometres.

Abyssal plains are a specific type of basin plain located on true ocean floor where the original abyssal hills have now been largely submerged by the infilling of the deeper areas (Weaver, Thomson & Hunter, 1987) (Fig. 10.73). Thus they exclude the flat floors of trenches and of small basins such as the strike-slip generated basins off California (Gorsline, 1978), and the many small areas of ponded turbidites such as occur on the mid-Atlantic ridge (Van Andel & Komar, 1969). Some 75 abyssal plains have been identified. They range in size from the Alboran in the Mediterranean (2600 km²) through the Madeira (54 000 km²), the Hatteras in the North Atlantic (460 000 km²), the Angola in the South Atlantic (1 001 000 km²) to the Enderby in the Antarctic (3 703 000 km²). Because a large sediment input is an essential condition of an abyssal plain, and most of the Pacific Ocean is surrounded by deep-sea trenches, abyssal plains are rare in the Pacific Ocean being confined to the northeast where sediment supply from the North American continent has overridden the adjacent trenches. In contrast, the largest abyssal

plains surround the Antarctic where trenches are absent and ice erosion is enormous.

Basin plains can be distinguished using the criteria of: (i) composition (e.g. terrigenous versus carbonate); (ii) basin restriction (open vs. enclosed; that is, whether they have an outlet or not as with lakes); (iii) fill geometry (progradational, mounded, onlap and drape fills); (iv) depth (above and below CCD); and (v) sediment supply (mud-rich versus sand-rich or undersupplied vs. oversupplied) (Stow, 1985).

As with basin-margin systems, basin-plain sediments vary widely, depending primarily on the volume and grain size of the source sediment and the distribution of the feeder points. Relatively small basin plains, surrounded by sand-rich margins, may be dominated by pelagic and hemipelagic muds because turbidity currents do not reach far into the basin (Gorsline, 1978). Large basin-plain, especially abyssal-plain, sediments are dominated by turbidites, hemipelagic muds usually comprising less than 10–20% of the sediment column; debris flows are less important (Pilkey, 1987). Accumulation of sediment and its distribution is a function of: (i) the nature of the basin margin and proximity of the shoreline, in particular the position of sea level; (ii) tectonic framework, whether the basin and surrounding areas are seismically active or passive; (iii) ratio of drainage-basin area to area of basin-floor (DA/BA) (Pilkey, Locker & Cleary, 1980); and (iv) arrangement of entry points around the edge of the basin plain (Pilkey & Hokanson, 1991), whether these are radial (from all around the basin), semi-radial (from

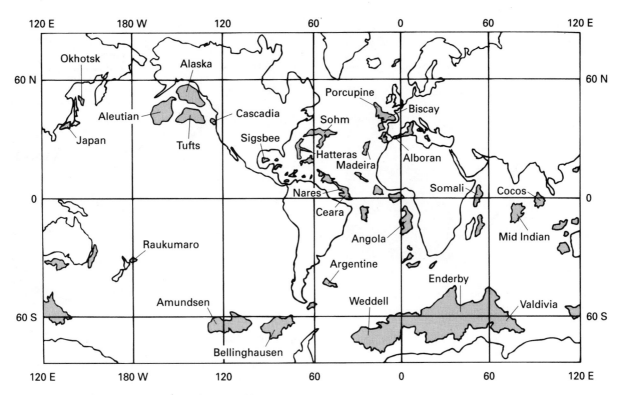

Figure 10.73 Distribution of present abyssal plains. In addition, there are about 10 in the Arctic Ocean (from Weaver, Thomson & Hunter, 1987).

two sides of the basin), longitudinal (from one end) or lateral (from one side).

The major characteristics of *undersupplied* basin plains were summarized by Pilkey (1987), whose studies are based primarily on the basin plains of the western North Atlantic. Turbidites dominate the sediment column except in backarc basins such as the Grenada Basin where there are significant volcaniclastic debris flows (Sigurdsson, Sparks *et al.*, 1980). The nature of fans or slope aprons bounding the plain is important, in particular the amount and grain size of the sediment supply. Relatively sand-rich feeder systems such as the Rhône deposit sand locally at the margins and little sand reaches the basin floor. In contrast, mud-rich margins can provide very large volume flows which are more widely distributed. For example, on the Hatteras Abyssal Plain the volume of the Black Shell Turbidite was over 100 km³ (Elmore, Pilkey *et al.*, 1979). Individual turbidity currents produce tongue-shaped turbidites elongated in the direction of flow but this pattern may be modified if material arrives simultaneously from multiple sources or flow is restricted by basin morphology. Most sediment arrives via point sources such as channels or submarine canyon mouths, though line sources may be important in more seismically active basins. Pre-existing bathymetry within the plain is an important control on flow paths and turbidite thickness. Even where regional relief is very gentle (10–20 m) individual turbidites thicken

slightly in depressions, thus tending to flatten the sea floor. Channelization of flows is unimportant, and so is erosion since basin plains are, by definition, depositional features. However, if a very large turbidity current, perhaps containing coarse sediment, follows deposition of finer material, scouring can occur.

In contrast to the undersupplied systems of the western North Atlantic, the Madeira Abyssal Plain is *oversupplied*. It lies at a depth of 5400 m and covers an area of 68 000 km², accumulating sediments derived from a drainage basin, including its subaerial continuation, of 3 360 000 km² (Fig. 10.74), giving a DA/BA of about 50, an order of magnitude greater than the abyssal plains of the western North Atlantic such as Hatteras (4.3) (Pilkey, 1987). It is the most fully studied abyssal plain and was initially investigated by an international group to examine the feasibility of disposal of high-level radioactive waste (see papers in Weaver & Thomson, 1987; Rothwell, Pearce & Weaver, 1992). Some 100 30-m cores have been taken. Not only has the sedimentology (Weaver & Rothwell, 1987), especially the geochemical character (e.g. Pearce & Jarvis, 1992), and micropalaeontology of the coccolith pattern (Weaver, 1994) been studied, but the surrounding source areas of the continental rise have been mapped by GLORIA side-scan sonar, and sediment cores taken (Masson, 1994).

The turbidites are mud dominated and fall into three groups: (i) volcanic-rich, derived from the oceanic islands of the Canaries

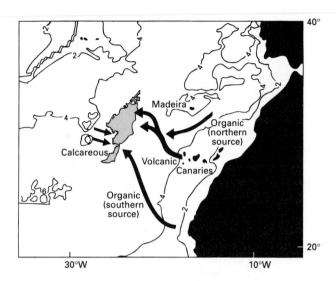

Figure 10.74 Madeira Abyssal Plain and sources of turbidites. For detailed map of pathways see Fig. 10.66 (from Weaver, Rothwell *et al.*, 1992).

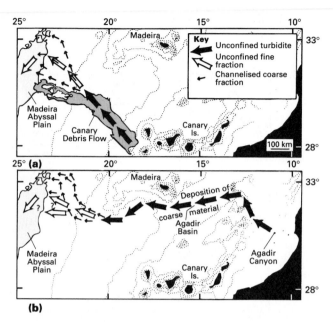

Figure 10.75 Suggested pathways followed by two turbidites on their way to the Madeira Abyssal Plain (MAP). (a) A large (125 km³) volume volcanic turbidite derived from the Canary Debris Flow or Slide. Coarse basal sediment was diverted into pre-existing channels. The fine fraction continued to travel northwest until interaction with the topography of the MAP and loss of momentum caused it to turn towards the southwest across the MAP. (b) Large-volume organic turbidite derived from the African margin, showing flow stripping. Coarse sediment was deposited where the ocean floor gradient was essentially zero (from Masson, 1994).

and Madeira; (ii) organic-rich, derived from the African continental margin, both from a northern and a southern source; and (iii) calcium-carbonate rich, derived from seamounts to the west, which are of very minor importance.

Over the last 300 000 years, 600 km³ of turbidites have been deposited over the plain, compared with only 60 km³ of pelagic sediments. Individual turbidites, whose thicknesses vary from a few centimetres to over 5 m and volumes from 3.5 to 190 km³, appear to have been emplaced about every 20–30 000 years, the timing controlled by sea-level and/or climate changes. However, the exact nature of the control is uncertain.

Within the basin, the smaller turbidites appear to thin down-current, covering only a part of the basin, their distribution being controlled by their source direction and the microtopography of the plain (Rothwell, Pearce & Weaver, 1992). In contrast, the geometry of the large-volume turbidites does not relate to source direction and they may even thicken down current because of ponding and filling of deeper topography. The sedimentary sequences of individual flows are very complex, with composite bases and repeated graded intervals. This is a consequence both of ponding effects and reflection from topographic highs in the basin and of varied rates of arrival of flows into the basin due to division of flows on the continental rise where pathways to the basin floor are complex (Masson, 1994). Smaller turbidites were delivered from a point source at the northeast edge of the plain. These flows were wholly confined to channels, never being more than 20 m thick (the channel depth) and 2 km wide with no significant overbank deposition. Larger flows transported their coarse fractions within channels, either by being captured by pre-existing channels (Fig. 10.75a) or by remaining within the channel system

(Fig. 10.75b) while the finer fraction separated from it by the process of flow stripping (Piper & Normark, 1983; see Fig. 10.47). This separation of flow into different paths explains why the coarse fraction forms a wedge along the eastern margin of the abyssal plain while the fine fraction enters the plain in the northeast before flowing to the southwest.

Ancient basin-plain sediments are not easy to identify as they may have many similarities to those of outer fan/ramp systems, especially those from mud-rich systems where the proportion of sandstone turbidites increases downfan. In addition, many workers, use the term 'fan' for all ancient turbidite systems and appear to think of submarine fans and turbidite systems as synonymous (e.g. the book entitled *Submarine Fans and Turbidite Systems*; Weimer, Bouma & Perkins, 1994). This is in spite of the formulation of 'non-fan' models (e.g. Macdonald, 1986) and the replacement of the term 'fan' by 'turbidite system' (Mutti, 1985), as well as the evidence from present-day basin plains. Consequently, many ancient systems have been interpreted as fans when a basin-plain model would probably have been more appropriate.

The principal features that characterize ancient basin-plain sediments are (i) the sheet-like nature of the bedding; (ii) relative lack of channels, apart from small-scale localized erosion surfaces, which are common, (iii) abundant compensation cycles, since any topographic irregularities are quickly smoothed out; (iv) pelagic sediments; (v) proximal to distal changes in bed thickness, sand/shale ratio and internal bed structures commonly can be traced axially along the basin (Hsü, Kelts & Valentine, 1980; Macdonald, 1986) (see Fig. 10.27c); however, complications arise because most basin plains are fed from several sources and show, at their simplest, in elongate examples, both axial and lateral flows, with lateral ones turning axially as they enter the deeper parts of the basin; (vi) sequences tend to aggrade rather than prograde and therefore remain relatively constant in any one place; and (vii) some single beds may be very large indeed, an order of magnitude larger than the 'normal' turbidite beds and therefore represent an 'exceptional' event (see Sect. 2.1.11) – these megabeds may form widespread 'key' stratigraphic markers within the basin (Ricci Lucchi & Valmori, 1980).

The Ross Sandstone was deposited in the axial zone of the Clare Basin in Ireland (see Fig. 6.49b). It has all the characteristics of a basin plain (Collinson, Martinsen *et al.*, 1991) although it has been interpreted as a submarine fan (Chapin, Davies *et al.*, 1994). It is composed of up to 380 m of fine-grained sandstones with bed thicknesses from less than 1 cm to around 8 m for some amalgamated units. The beds are dominantly sheet-like, with small sand-filled channels and mud-filled scours occurring towards the top. Sequences of systematic changes in grain size or bed thickness are lacking, but packets of thicker and thinner beds occur. Slump sheets occur towards the top of the formation.

In the Lower Cretaceous backarc basin of South Georgia, dominantly volcaniclastic turbidites have been interpreted as the deposits of a linear trough that show proximal to distal changes in character of the sandstones and reduction in sandstone bed thickness (MacDonald, 1986). The classic Oligo–Miocene foreland basins of the periadriatic region of Italy (Fig. 10.76) include well-documented basin-plain successions (Ricci Lucchi & Valmori, 1980; Ricci Lucchi, 1985). The overall dimensions of these basins were large (400 × 50 km) and they were filled rapidly by relatively coarse-grained and thick-bedded turbidites as well as the associated finer-grained facies. Correlation of individual turbidites, and especially of megabeds, is possible over wide areas. Palaeocurrents indicate both longitudinal and lateral supply. Lobe and channel sequences have also been identified and are interpreted as parts of submarine fans that fed laterally into the basin.

Pilkey (1987) has emphasized that basin plains are distinctive features independent of submarine fans and should not be considered to be the distal portions of fans. This is true both for abyssal plains and for basin plains that are fed from numerous entry points. Nevertheless, there may be an overlap between fans and basin plains and their respective facies associations, particularly where a submarine fan is either the main or a significant supplier of sediment to the basin plain.

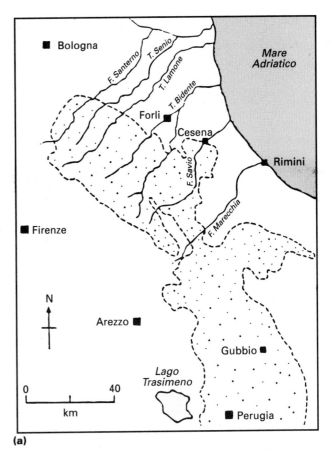

(a)

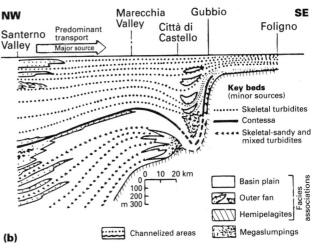

(b)

Figure 10.76 Outline map and schematic longitudinal cross-section through the Miocene Marnosa–Arenacea basin plain of northern Italy. Some key beds can be traced over 100 km. The Contessa Bed is up to 16 km thick (after Ricci Lucchi & Valmori, 1980).

10.3.7 Contourite drifts

Bottom (contour) currents create bedforms that occur at a range of scales from large sediment waves and erosional furrows to smaller-scale dunes, ripples, lineations, scour and tail marks,

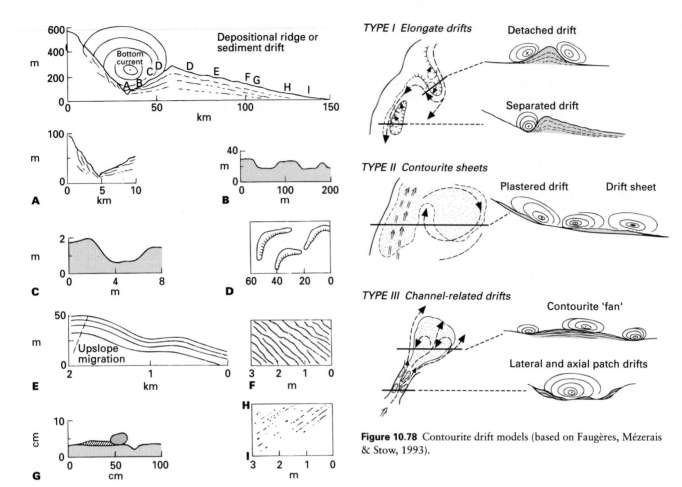

Figure 10.77 Bottom (contour) current bedforms (from Stow, 1994). Composite diagram showing position of A, moat; B, large furrow; C, small furrow; D, dunes and waves; E, giant sediment waves; F, ripples; G, scour and tail; H, lineation and coarse lag; I, smoothed sediment (from Stow, 1994).

Figure 10.78 Contourite drift models (based on Faugères, Mézerais & Stow, 1993).

etc. (Fig. 10.77) (Heezen & Hollister, 1971; Stow, 1982; Nowell & Hollister, 1985; McCave & Tucholke, 1986). Bedform assemblages have been related to bottom-current intensity (Hollister & McCave, 1984), although the exact processes of formation still remain questionable for some of them, especially for giant sediment waves.

Sediment waves in different areas may be parallel, perpendicular or at an angle to current flow; they may propagate downstream or upstream, and if they are formed on a gentle slope, they can migrate downslope or upslope (Hollister, Flood *et al.*, 1974; Asquith, 1979; Lonsdale & Hollister, 1979; Embley, Hooje *et al.*, 1980; Kolla, Eittreim *et al.*, 1980; Flood & Shor, 1988). Furthermore, these sediment waves are similar in dimensions and morphology to the sediment waves formed by turbidity currents (e.g. Marani, Argnani *et al.*, 1993).

Contourite accumulations can be divided into three main classes (Faugères, Mézerais & Stow, 1993): (i) *giant elongate*

drifts, (ii) *contourite sheets*, and (iii) *channel-related drifts* (Fig. 10.78).

Giant elongate drifts (Fig. 10.78) are very large mounded sediment bodies that extend parallel to current flow and parallel the bottom topography (i.e. continental margin, mid-ocean ridge, etc.). They can occur anywhere from the upper slope to abyssal depths, depending on the depth at which the bottom current travels. They are particularly common in the North Atlantic Ocean (Stow & Holbrook, 1984; McCave & Tucholke, 1986), but also occur in the South Atlantic and elsewhere (Flood & Shor, 1988; Faugères, Mézerais & Stow, 1993). They are 10s–100s km long, tens of kilometres wide, and range from 0.1 to more than 1 km in relief above the surrounding sea floor. Their accumulated thickness may locally exceed 2 km. Some, but not all, are partly covered by large fields of giant sediment waves (Fig. 10.78).

Typically, sedimentation rates are moderately high, averaging between 2 and 10 cm ky^{-1} and the facies are relatively fine grained, including silty, muddy and biogenic contourites, with only very rare coarse-grained sandy contourites. In many cases, they are lithologically very similar to hemipelagites (calcareous silty clays) or pelagites (clayey foraminifera/nannofossil ooze).

Contourite sheets (Fig. 10.78) are less easily recognized and hence less known than elongate mounded drifts. They comprise extensive low-relief accumulations that either fill abyssal basins or are plastered against the continental margin. In the Argentine Basin, flat fields of sediment waves extend over an area of about 1 000 000 km²; the waves have heights of 10–80 m, wavelengths of 1–10 km, and show a steady migration (normally upcurrent) with time.

On the South Brazilian Basin across the continental rise and abyssal plain, a large contourite sheet system is swept by the Antarctic bottom water (Damuth & Hayes, 1977). The sedimentation rate on the rise averages 2–3 cm ky⁻¹ where the deposits are silty clays with low carbonate contents and cyclic variations in silt content. The rates of deposition decrease at greater depths, with a progressive change towards carbonate-free manganiferous brownish-coloured clays (Massé, Faugères *et al.*, 1991).

In the northeast Atlantic, a complex system of bottomcurrent and downslope accumulation includes elongate drifts and contourite sheets covered in parts by fields of sediment waves (Howe, Stoker & Stow, 1994).

Channel-related drifts (Fig. 10.78) are related to deep channels or passages through which the bottom circulation is so constricted that flow velocities are markedly increased. Such deposits have been described from deep passages in the Atlantic, Indian and Pacific oceans (e.g. Kane Gap, Samoan Passage, Almirante Passage), and include irregular, discontinuous sediment bodies on the floor of the channel, on the flanks, and at the downcurrent exit of the channel.

The Vema Channel through the Rio Grande Rise (Le Pichon, Ewing & Truchan, 1971; Gamboa, Buffler & Barker, 1983), connects the Argentine Basin and the Brazilian Basin and allows the Antarctic bottom water to flow northward beneath the southward flowing North Atlantic deep water (Richardson, Biscayne *et al.*, 1987). The channel floor deposits consist of irregular patches of unconsolidated sediment and manganiferous muds and nodules that accumulated very slowly. Channel flank silts and clays occur on terraces along both margins ornamented with sediment waves. Accumulation rates are about 2–3 cm ky⁻¹. At the channel exit a cone-shaped drift some 250 m high and 100 km wide, has accumulated at rates of about 3–4 cm ky⁻¹.

The identification of ancient contourites both in the subsurface and on land remains controversial. Based primarily on studies of present-day Atlantic contourites, contourite facies models have been produced by Gonthier, Faugères and Stow (1984) (see Figs 10.15 & 10.16). These models suggest (e.g. Stow, 1994) that an ancient drift deposit should be mostly fine grained and comprise mud, silt and fine sand arranged in cyclic microsequences. Bioturbation would be intense throughout. Ferromanganese encrustations, layers and micronodules may occur locally, together with intraformational shale-clast conglomerates. Widespread hiatuses and hardgrounds would indicate periods of current intensification. The most common

associated facies would be hemipelagites and pelagites. Examples of such contourites include the carbonate-rich contourites from the Cretaceous Talme Yafe Formation in Israel (Bein & Wéiler, 1976) and the Ordovician continental margin of the Yangtze Terrane in southern China (Duan, Gao *et al.*, 1993). Stanley (1988a,b, 1993) demonstrated the potential of bottom currents to have reworked fine-grained turbidites in the Upper Cretaceous rocks in St Croix in the Caribbean and applied this model to other successions, including the Niesenflysch, Switzerland and some Annot sediments, southern France.

Mutti (1992) argues that many sandbodies interpreted as turbidite fan systems are formed of original turbidite sands that have been strongly winnowed, reshaped and redeposited by bottom currents. Shanmugam, Spalding and Rofheart (1993) suggest that many traction bedforms, such as cross-lamination, horizontal lamination and inverse grading, as well as internal erosion surfaces, that are not normally associated with contourites, are the result of bottom-current reworking. They also consider bioturbation to be rare in contourites. They therefore interpret some deep-sea sands as contourites that most workers would interpret simply as the consequence of the natural fluctuations in turbidity flows at the depositional site. Since both turbidity currents and bottom flows are very complex in nature, and both types of current can show very similar features, differentiation between turbidites and contourites will always be an uncertain task.

10.4 Deep-water pelagic and hemipelagic systems

Pelagic and hemipelagic sediments occur in every part of the world's oceans (see Fig. 10.6) and are also well represented in ancient deposits exposed on land. They occur interbedded with turbidites and other resedimented clastic deposits in each of the fan, ramp, slope apron and basin plain settings outlined in Section 10.3. Pelagites and hemipelagites are also found in association with contourite drifts, though distinction between muddy contourites and associated hemipelagites is very difficult (e.g. Kidd, Ruddiman *et al.*, 1986).

However, in addition to their intercalation with terrigenous turbidites and contourites where pelagic sediments form only a small proportion of the sediments (Sect. 10.3), there are areas removed from turbidity current or bottom-current supply in which pelagites and hemipelagites form all, or nearly all, of the sediment column. Although no specific 'models' have been constructed for these systems in the same way as for clastic systems, it is possible to recognize four different settings in which they dominate: (i) spreading ridges (see Sect. 12.11.1); (ii) aseismic volcanic structures (see Sect. 12.11.3); (iii) ocean basins and abyssal hills; and (iv) continental margins, banks and basins.

In addition, pelagites, such as the Upper Cretaceous chalks of northern Europe, are known from shallow-water shelf settings that have no modern equivalent. Spreading ridges and aseismic volcanic structures include much volcaniclastic sediment and so are covered in Sect. 12.11. The other two are discussed below.

10.4.1 Ocean basins and abyssal hills

Basin plains, including abyssal plains, that receive a significant proportion of terrigenous sediment, have already been discussed in Section 10.3.6. Here, we are concerned with those ocean basins that are either too far from a source of terrigenous sediment, as they are in most of the Pacific Ocean, or have a relief that puts them beyond the reach of turbidity currents (abyssal hills). In addition, some small, marginal basins may have local pockets and highs that are starved of terrigenous sediment.

Sediments that accumulate below the CCD comprise red clay, radiolarian and diatomaceous ooze. Ferromanganese nodules are a common accessory, as are whales' earbones and sharks' teeth.

The red clay of the barren ocean comprises a variety of clay minerals, chiefly illite and montmorillonite, plus lesser and local amounts of kaolinite and chlorite. In addition, X-ray-amorphous iron-manganese oxyhydroxide, authigenic zeolites such as phillipsite and clinoptilolite, local amounts of palygorskite, and cosmic spherules may be present (e.g. Berger, 1974; Chamley, 1989). The latter include not only black magnetic spherules of nickel-iron but also chondrules of olivine and pyroxene. Associated detrital material includes feldspars, pyroxenes and quartz. Feldspar, pyroxene and montmorillonite derive from volcanic intra-oceanic sources, the latter by sub-marine degradation of basalt. Chlorite is detrital, being derived from low-grade metamorphic terrains. Quartz, illite and to a lesser extent kaolinite are thought to be aeolian winnowing products delivered as fallout from high-altitude jet streams. The aeolian contribution to deep-sea clay is probably between 10 and 30%, though, exceptionally, aeolian dust may comprise up to 80% of the total sediment, the Sahara Desert supplying clay not only to the eastern Atlantic but to the Caribbean. The Indian and North Pacific Oceans probably derive their aeolian clays from the Asian mainland; in the South Pacific, Australia is a likely source (Windom, 1975; Pye, 1987).

Siliceous oozes, derived from Radiolaria, diatoms, silico-flagellates and sponge spicules, occur today in three major areas: (i) a global southern belt; (ii) a North Pacific zone including the backarc basins; and (iii) a near-equatorial belt which is better developed in the Pacific and Indian Oceans than in the Atlantic (see Fig. 10.6). In the northern and southern areas diatoms are particularly important; in the equatorial zones Radiolaria locally dominate (Lisitzin, 1971, 1985).

Ferromanganese nodules commonly occur in association with red clays and the two occur together over much of the deep North Pacific floor. Populations of nodules, distinct in terms of morphology, chemistry and mineralogy, occur in different parts of the oceans (Cronan, 1980). Particularly

striking is their intimate relationship with evidence of Recent erosion such as ripples, scours, lineations and disconformities (Kennett & Watkins, 1975; Pautot & Melguen, 1976). Because of their slow growth, nodules develop preferentially under conditions of nil or even negative sedimentation (Lonsdale, 1981; Faugères, Mézerais & Stow, 1993). In the southeast Indian Ocean, where bottom water movement is vigorous, a manganese-nodule pavement covers an area of some 10^6 km^2, whereas where sluggish currents flow over a broad area of abyssal basin floor in the southwest Atlantic Ocean, pelagic red clays are deposited in association with fine manganiferous lamination and micronodules (Mézerais, Faugères et al., 1993).

In parts of the deep-sea floor that are distant from terrigenous influx, but where the water depth does not exceed the CCD, the dominant pelagic component is calcareous skeletal matter. This is dominated by foraminifers and nanofossils, but also includes a wide range of other biogenic material, some siliceous debris and fine terrigenous clays and silts.

Many boreholes have now been drilled through thick oceanic sections of such pelagic carbonates during the DSDP and ODP. In many cases, there is a marked cyclicity of more and less carbonate-rich facies, in Recent through to Jurassic sequences. This cyclicity may be due to variations either in carbonate productivity, terrigenous dilution or carbonate dissolution (Einsele & Ricken, 1991). The increased clay content of carbonate pelagic cycles in some Atlantic sequences away from the continent is considered to result either from dilution by ice-rafted terrigenous influx during glacial periods or variations in equatorial upwelling (Diester-Haass, 1991). Some sequences out of reach of terrigenous supply show no rhythmic carbonate changes. Similar cycles in Pacific sediments that show a similar Milankovitch time scale, are thought to be more the result of dissolution, suggesting fluctuations in the degree of saturation of deep water with respect to calcite (Grötsch, Wu & Berger, 1991).

Areas such as the Red Sea, Gulf of California, Mediterranean, or the Pacific backarc basins differ from major oceans in their oceanography and sedimentary environments. They are variously referred to as 'small ocean basins' or 'marginal seas' and provide examples of ocean basins that are generally relatively oversupplied with clastic sediments, as well as being generally above the CCD. True pelagites only occur on highs and in parts of the basins shielded from major terrigenous input, whereas hemipelagites are well represented throughout.

The pattern of water circulation in a basin is particularly important in determining depositional patterns. If deep-water inflow and shallow-water outflow prevail ('estuarine circulation' of Berger, 1970), then nutrient-rich water is introduced which, on upwelling, promotes high productivity and abundant skeletal material. If the reverse flow pattern applies, as is the case for mildly evaporative basins, then productivity is very low. Should the mouth of the basin be constricted, as with the Mediterranean and Red Seas at the present time, then a eustatic fall in sea level or tectonic readjustment could result in severe restriction with the possible onset of stagnation, favouring deposition of organic-rich sediments and, in the most extreme case, precipitation of evaporites. Formation of freshwater basins, such as the Pleistocene Black Sea, is another possibility. Pelagic sediments laid down in 'small ocean basins' may therefore be stratigraphically associated with facies of highly contrasting character.

The Mediterranean, like the Red Sea, is today characterized by anti-estuarine circulation (shallow-water inflow, deep-water outflow). It has salinities slightly above those of the open ocean and relatively low plankton productivity. Calcareous sediments are, however, widely distributed in areas beyond the influence of terrigenous input.

In the Gulf of California circulation is estuarine: deep-water inflow and removal of surface water by offshore winds causes vigorous seasonal upwelling and high plankton productivity. Diatoms and, to a lesser extent, Radiolaria dominate the central part of the Gulf; clay-rich sediments are important on its eastern side where rivers draining the Mexican mainland debouch into the sea (Calvert, 1966). Glauconite occurs in many non-depositional areas. Laminated diatomaceous organic-rich facies, with alternating diatom-rich and clay-rich laminae, are confined to those parts of the basin slopes affected by the oxygen-minimum zone (300–1200 m) (Curray, Moore *et al.*, 1982).

The North Pacific backarc basins are also sites of intense silica production, particularly the Bering Sea, Sea of Okhotsk and, to a lesser extent, the Sea of Japan (Lisitzin, 1985). Most of these marginal basins were apparently initiated in the Late Oligocene to Early Miocene and they have records of diatomaceous sedimentation that extend back to this period. Massive and laminated diatomaceous sediments, associated with small and local amounts of nannofossil and *Globigerina* carbonate, interbedded with terrigenous and volcanogenic turbidites, have been recorded from the Miocene and Pliocene of the Sea of Japan and Bering Sea (Scholl & Creager, 1973; Karig, Ingle *et al.*, 1975; Pisciotto, Ingle *et al.*, 1992). The Pliocene and Pleistocene sequence recovered from the Oki Ridge in the Japan Sea shows a series of light–dark cyclic pelagic deposits that imply rapid and extreme variations in the dissolved oxygen content of the bottom water, perhaps related to periodic isolation of the Japan Sea from the Pacific (Tamaki, Suyehiro *et al.*, 1992).

The Lau Basin in the central western Pacific Ocean, between the Tofua Arc and the Lau Ridge, contains several small, elongate, fault-bounded, partially sediment-filled sub-basins (Hawkins, Parson *et al.*, 1994). The Neogene succession in these basins, directly overlying late Miocene to early Pliocene oceanic crust of a backarc-spreading system, consists of clayey nanofossil oozes (pelagites/hemipelagites) interbedded with resedimented oozes and volcaniclastics. The hemipelagic sections are generally monotonous and bioturbated rather than cyclic,

and punctuated by the tectonically induced resedimented facies (Rothwell, Weaver *et al.*, 1994).

A number of pelagic successions on land can be interpreted as having been deposited in ocean basins. However, the sediments of deep open oceans are less readily incorporated into continental crust than those of small marginal basins.

Deposits preserved in accretionary prism complexes, such as the oceanic sediments on Barbados (Pudsey & Reading 1982; Speed & Larue 1982), those of the Shimanto Belt on Honshu, Japan (Taira, Okada *et al.*, 1982), and the lower Palaeozoic forearc system of the Southern Uplands, Scotland (Leggett, McKerrow & Casey, 1982), all have oceanic pelagic/hemipelagic components.

Around the margins of the Pacific Ocean, stretching from Korea, Japan and the Soviet Far East to western USA, there are abundant Miocene and Pliocene diatomaceous deposits. Typically, the siliceous sediments are found between continental and shallow-marine clastics below, and deep-marine clastics above. A lateral passage to terrigenous clastics is also seen. The depositional sites of these marginal Pacific diatomites probably included backarc regions comparable with the present-day Japan Sea, simple rifts, and, as suggested for the Monterey Basin, pull-apart basins related to transform motions along the proto-San Andreas Fault (Ingle, 1981). The best-documented example is the Monterey Formation of California (Garrison, Douglas *et al.*, 1981; Isaacs, Pisciotto & Garrison, 1983).

The total thickness of the Monterey Formation varies considerably but locally extends to 1.5 km. The Monterey pelagic rocks may be divided into a basal calcareous facies, a middle transitional phosphatic and glauconitic member and a thick upper unit of siliceous rocks. The siliceous rocks are light-coloured and vary from quartzose chert, through opal-CT or quartz porcellanite to soft opaline diatomite. Generally, the more lithified diagenetically advanced varieties occur towards the base of the section where they are typically developed as centimetre-scale beds separated by paper-thin partings and thus superficially resemble ribbon-bedded radiolarian cherts in all but colour (Pisciotto & Garrison, 1981; Jenkyns & Winterer, 1982).

Palaeogeographic reconstructions suggest that the Monterey Formation was deposited in irregular basins, locally deep, and separated by shallow marine sills or by land areas, the present-day analogue being the Gulf of California. Only around the basin margins were clastics and land-derived faunas deposited in important amounts. In deeper central regions sedimentation was essentially a function of phyto- and zooplankton production, initially dominantly calcareous and then, as fertility and productivity increased, siliceous.

Many well-known carbonate successions, particularly those of Jurassic and Cretaceous age, display a distinctive limestone–marlstone cyclic alternation (Einsele & Ricken, 1991). These were deposited in a number of settings including: pelagic–hemipelagic marine environments from outer shelf to deep water (above CCD), shallow marine carbonate platforms, and lacustrine basins. Although the cycles vary in thickness and nature from shallow- to deep-water settings, the Milankovitch control on their origin is commonly evident (e.g. Fischer, 1991; Eicher & Diner, 1991). Pelagic black-shale–limestone cycles are commonly interpreted in the same way (e.g. de Boer, 1991), although Arthur, Dean and Stow (1984) have demonstrated that the cyclic preservation of black shales, in a variety of Mesozoic examples, may depend on a complex interaction of controls, namely supply of organic matter, sedimentation rate and deep-water oxygenation which, to some extent, vary independently.

10.4.2 Continental margins, banks and basins

Many continental margins possess a submarine topography of shallower seamounts, banks and plateaus interspersed with deeper-water basins. These structural features are generally non-volcanic and are considered to rest on attenuated continental crust with perhaps some overlap on to oceanic crust. The presence of pelagic sediments in such settings depends on a virtual absence of terrigenous material due to primary lack of supply and/or the presence of an intervening clastic-sediment trap. Prior to the formation of the adjacent ocean basin, these continental margins initially lay close to sea level, and many of them display a prepelagic record of shallow-water deposition, typically platform carbonates. These include the Oman margin in the northwest Indian Ocean (Prell, Niitsuma *et al.*, 1989, 1990) and the Bahama Banks in the western North Atlantic.

To the north of the Bahamas lies the Blake Plateau which has an area of some 228 000 km² and an average depth of about 850 m (Pratt & Heezen, 1964). A DSDP core on the Blake Nose demonstrated that this part of the Blake Plateau lay near mean sea level (i.e. it was a carbonate platform like the Bahamas) until early Cretaceous time when it abruptly drowned. Birdseye limestones of tidal-flat origin are overlain by red pelagic carbonates containing goethitic pisolites and crusts. Drowning of the main body of the Plateau probably took place later.

The sediment facies and their distribution over the Bahama Banks and Blake Plateau area represent a complex interplay of pelagic, hemipelagic, bottom current and resedimentation processes (Sheridan & Enos, 1979; Mullins, Neumann *et al.*, 1980; Ogg, Haggerty *et al.*, 1987).

A feature of deep-water outer margins of continents, and all basins not too far from land is that they are affected by fluctuations in terrigenous input and cyclicity is closely related to climatic changes that affect run off and erosion from continents (Dean & Gardner, 1986; Einsele & Ricken, 1991). For example, on the continental slope south of New Zealand, greenish carbonate-poor interbeds represent glacial periods, whereas greenish-white carbonate-rich interbeds represent inter-

glacial periods (Diester-Haass, 1991). On the Exmouth Plateau off northwest Australia, the late Cretaceous succession comprises pelagic cyclic sediments: light-coloured nanofossil chalks alternating with dark-coloured clayey nanofossil chalks on a decimetre scale. Cycle periods are around 21 and 41 ky (Huang, Boyd & O'Connell, 1994).

The carbonate-bank–basin style of pelagic and related sedimentation is commonly encountered in ancient successions preserved on land. Several of these were carefully reviewed in the previous edition of this book (Jenkyns, 1986) including Palaeozoic pelagic series from Europe, North America and North Africa and the equivalent Mesozoic series from the Tethyan region.

The latter consists of red, grey and white limestones, manganese nodules, pelagic 'oolites' and cherts of Triassic and Jurassic age, which overlie and laterally interdigitate with

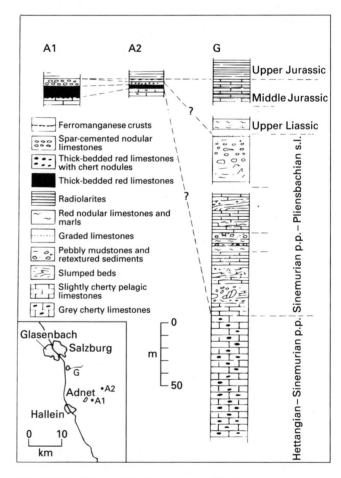

Figure 10.79 Columnar section of an expanded Jurassic sequence at Glasenbach (G) and coeval condensed sequences at Adnet (A1, A2) in the eastern Alps of Austria. The condensed sequences are rich in fauna and impregnated with Fe–Mn oxyhydroxides; the expanded sequences contain graded beds and show evidence of hard- and soft-sediment deformation. The condensed sequences are interpreted as having formed on a topographic high, the expanded sequence in a deeper basin (after Bernoulli & Jenkyns, 1970).

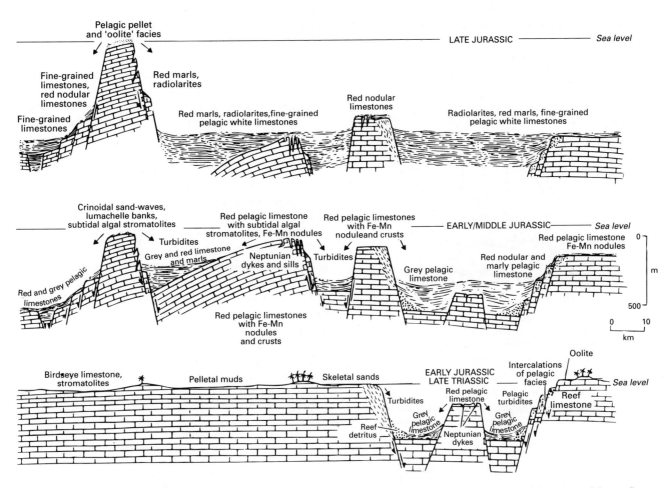

Figure 10.80 Sketch of the palaeogeographic evolution of the Tethyan continental margin during the Triassic and Jurassic. Block-faulting and differential subsidence affected most (but not all) carbonate platforms and gave rise to a seamount-and-basin topography on which coeval condensed and expanded sequences

were deposited. A general smoothing and deepening of the sea floor is indicated for the Late Jurassic except for those seamounts that were accumulating pelagic 'oolites'. Some seamounts and plateaus persisted through much of the Cretaceous (after Bernoulli & Jenkyns, 1974).

shallow-water platform carbonates (Bernoulli & Jenkyns, 1970, 1974). These Mesozoic pelagic rocks can be divided into stratigraphically condensed and expanded facies (Fig. 10.79).

The condensed facies is typically developed as a metre or so of red nodular biomicrite containing an abundant fauna. Mineralized hardgrounds and ferromanganese nodules are also present in some cases. The red biomicrites pass upwards into a still more nodular and marly lithology, and then into grey cherty micrites and radiolarian cherts.

The expanded facies are considerably thicker and comprise burrow-mottled limestone–marl interbeds, some more red nodular marls (equivalent to those of the condensed series), and common evidence of resedimented carbonate facies, including slumps, debrites and calciturbidites.

The interpretation of these Triassic and Jurassic pelagic sequences is illustrated in Fig. 10.80. During the Cretaceous in the Tethyan region, this bank and basin system persisted for a

considerable time in some areas, for example the Scaglia Rossa Formation in the Umbrian–Marchean Apennines of central Italy (Stow, Wezel et al., 1984), but became gradually subdued topographically. Tertiary tectonic activity led to the rapid filling of the formerly pelagic/calciturbidite basins with coarse clastic turbidites.

10.4.3 Chalk

Extensive deposits of chalk formed over cratonic shelf areas of northwest Europe and the Western Interior Seaway of North America during the Upper Cretaceous and Palaeocene. Deposition was in water depths of between 50 and a few hundred metres and they formed thicknesses of a few hundred metres to 1200 m of mainly resedimented chalk in the Central Graben of the North Sea.

Chalk is a very fine-grained (micritic) limestone that is largely composed of coccoliths and rhabdoliths with some planktonic foraminifera and calcispheres. Sedimentation probably took place as faecal pellets, ejected by small crustaceans (copepods), and/or pelagic tunicates. Sedimentary rates may have been at times as high as 15 cm ky^{-1}. Macrofossils include a pelagic fauna

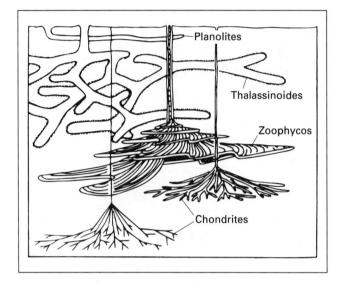

Figure 10.81 The cross-cutting effects of an endobenthic fauna due to the different depths to which they burrow in the chalk (from Bromley & Ekdale, 1986). These effects are commonly seen as flint nodules that have diagenetically replaced the burrow systems.

of ammonites and belemnites and a benthic fauna of crinoids, echinoids, bivalves, sponges, brachiopods and bryozoa. These large fossils are frequently encrusted with epizoans. Trace fossils are abundant, with *Thalassinoides* common in shallow-water chalks and absent in deep-water chalks (Ekdale & Bromley, 1984) and *Chondrites*, *Zoophycos* and *Planolites* occurring in both shallow-water and deep-water chalks. Since the organisms that produced these traces lived at different depths below the sea floor, the burrows tend to cross cut each other, those that burrowed deeper cutting the shallower burrowing ones (Fig. 10.81). Chert nodules (flints) are ubiquitous, the more slender varieties mimicking the form of burrows in many cases (Bromley & Ekdale, 1984).

Locally interrupting the succession are nodular chalks, omission or discontinuity surfaces, and lithified hardgrounds (Kennedy & Garrison, 1975). Omission surfaces may only be apparent from varying trace-fossil assemblages and a difference in sediment above and below. Lithified hardgrounds may be mineralized, bored and encrusted and may correspond to a distinct palaeontological gap. The variety of surfaces reflects the varying effects of one or more cycles of early diagenesis, exhumation, erosion and burial (Fig. 10.82).

The chalk may be all CaCO$_3$ or it may be contaminated with a proportion of clay. It is generally very evenly bedded, the stratification being due to vertical variations in clay content,

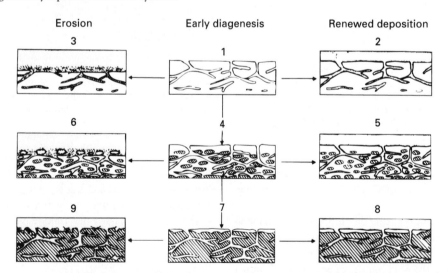

Figure 10.82 Flow diagram showing proposed relationship between diagenesis, erosion, burial and the formation of nodular chalks and hardgrounds. (1) A pause in sedimentation leads to the development of an omission suite of trace fossils (*Thalassinoides*). (2) If buried this is preserved as an omission surface. (3) With scour, an erosion surface is formed; burrows are filled with calcarenitic chalk and the surface is overlain by a shelly lag. (4) Early diagenesis associated with a longer pause in deposition leads to the growth of calcareous nodules in soft sediment. Burrow systems are extended, the animals avoiding the nodules. (5) Burial at this stage leads to a nodular chalk. (6) If eroded, nodules are reworked and burrows truncated; the nodular chalk is overlain by a terminal intraformational

conglomerate, the pebbles of which may be mineralized and bored. Burrows are filled by, and pebbles embedded in a winnowed calcarenitic chalk. (7) Prolonged diagenesis leads to a coalescence of nodules to form a continuous lithified subsurface layer; later burrows are entirely restricted to sites of pre-lithification burrows. (8) If buried, this rock band, with no signs of superficial mineralization, becomes an incipient hardground. (9) With erosion, the rock band is exposed on the sea floor, and a true hardground develops. It may become bored (borings shown in black), encrusted by epizoans and mineralized at or below the sediment–water interface. All these processes also affect the walls of burrows (after Kennedy & Garrison, 1975).

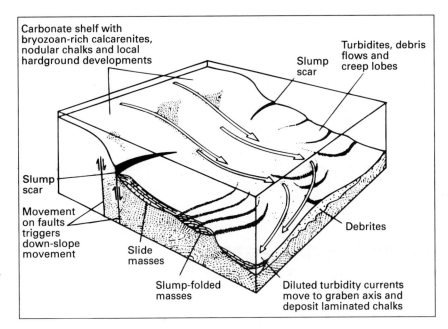

Figure 10.83 Depositional model for resedimented chalk that formed along a linear slope apron, based on the Cretaceous Tor Formation of the North Sea (from Kennedy, 1987).

degree of bioturbation and diagenesis that are manifested as alternations in colour and hardness. These small-scale rhythmic alternations on a decimetre to metre scale have been termed a periodite facies (Kennedy, 1987) and attributed to Milankovitch-driven climatic fluctuations that govern organic productivity, water temperature and clastic and freshwater input (Hart, 1987; Gale, 1989), although the degree to which each of these factors was responsible remains controversial (Ditchfield & Marshall, 1989).

Although chalk occurs largely as a bioturbated authochthonous sequence formed by pelagic fallout of the nanofossil material, in the Central Graben of the North Sea (and locally onshore) it is also found in a wide range of redeposited facies, including slides, slumps, debris-flows and turbidites, as well as both laminated and microlaminated chalk and homogeneous chalks (Nygaard, Lieberkind & Frykman, 1983; Bromley & Ekdale, 1987) (Fig. 10.83). These redeposited facies were emplaced by mass-flow processes emanating from a nearby linear shelf, rather than through canyons and other localized pathways (Hatton, 1986; Kennedy, 1987). They therefore have all the features of a mud-rich slope apron (Sect. 10.3.5).

Further reading

Bouma A.H., Normark W.R. & Barnes N.E. (Eds) (1985) *Submarine Fans and Related Turbidite Systems*, 351 pp. Springer-Verlag, New York.

Einsele G., Ricken W. & Seilacher A. (Eds) (1991) *Cycles and Events in Stratigraphy*, 955 pp. Springer-Verlag, Berlin.

Hsü K.J. & Jenkyns H.C. (Eds) (1974) *Pelagic Sediments: On Land and Under the Sea*, 477 pp. *Spec. Publ. int. Ass. Sediment.*, **1**. Blackwell Scientific Publications, Oxford.

Mutti E. (1992) *Turbidite Sandstones*, 275 pp. Agip Instituto di Geologia Università di Parma, Milano.

Pickering K.T., Hiscott R.N. & Hein F.J. (1989) *Deep Marine Environments: Clastic Sedimentation and Tectonics*, 416 pp. Unwin Hyman, London.

Stanley D.J. & Kelling G. (Eds) (1978) *Sedimentation in Submarine Canyons, Fans, and Trenches*, 395 pp. Dowden, Hutchinson & Ross, Stroudsburg, PA.

Stow D.A.V. & Piper D.J.W. (Eds) (1984) *Fine-grained Sediments: Deep-water Processes and Facies*, 659 pp. *Spec. Publ. geol. Soc. Lond.*, **15**. Bristol.

Weaver P.P.E. & Thomson J. (Eds) (1987) *Geology and Geochemistry of Abyssal Plains*, 246 pp. *Spec. Publ. geol. Soc. Lond.*, **31**. Bristol.

Weimer P., Bouma A.H. & Perkins B.F. (Eds) (1994) *Submarine Fans and Turbidite Systems*, 440 pp. Soc. econ. Paleont. Miner., Gulf Coast Section, Austin, TX.

Weimer P. & Link M.H. (Eds) (1991) *Seismic Facies and Sedimentary Processes of Submarine Fans and Turbidite Systems*, 447 pp. Springer-Verlag, New York.

Glacial sediments

11

J.M.G. Miller*

....

11.1 Introduction

The concept of extensive ice sheets and glaciers existing during periods of colder climate, and transporting and depositing sediment, originated in continental Europe, primarily during the 1830s. This 'glacial theory' explained the origin of superficial deposits that contain large erratic boulders and which cover broad areas of Europe and North America. By the latter part of the 19th century, the glacial theory had replaced the earlier notion of a universal flood, the Biblically inspired 'diluvial' theory, and its offshoot, the 'drift' theory which invoked extensive iceberg rafting to distribute foreign boulders and other rock debris derived from the far north (Flint, 1971, pp. 11–15). Recognition of pre-Quaternary ice ages also dates back to the mid- to late 19th century with the interpretation of Permo-Carboniferous rocks in Gondwana as ancient glacial deposits (John, 1979). Most poorly sorted, till-like sediments (boulder clays) were interpreted as the deposits of glaciers or icebergs, and therefore as indicators of ancient ice ages. Consequently,

ice ages were identified in almost all geological periods (Coleman, 1926).

In the 1950s, recognition of the role of turbidity currents and of slumping and mass movement in deep oceans provided an alternative mechanism for the deposition of till-like sediments, such as pebbly mudstones (Crowell, 1957). The realization that sediments produced by non-glacial processes can be so similar to those deposited by glaciers led to controversy over the accurate interpretation of many of these unsorted deposits (Dott, 1961). Criteria characteristic of glacial conditions, in addition to diamictite, were highlighted by Harland, Herod & Krinsley (1966) and Flint (1975) (see Sect. 11.5.1). However, heated discussions continued as to whether the widely recognized diamictite-bearing units truly record cold climates or whether many, particularly those of Late Proterozoic age, reflect deposition in regions of active tectonics (Schermerhorn, 1974).

Rigorous sedimentological study and re-examination of many ancient inferred glacial deposits, and more thorough investigation of the sedimentary processes and products of many modern glacial and glaciomarine environments, have taken place in the last few decades in response to this phase of scepticism. Through improved communication between those working

* Parts of this chapter are based on the second edition, written by Marc B. Edwards. Sections and figures that have been changed very little are designated with a footnote.

on modern and Pleistocene sediments and those studying ancient glaciogenic rocks, our knowledge and understanding of glaciogenic sedimentary processes and products is now a great deal more sophisticated.

Non-genetic terms *diamicton* (unlithified) and *diamictite* (lithified) were introduced by Flint, Sanders and Rodgers (1960a,b) to describe sediment 'consisting of sand and/or larger particles dispersed through a muddy matrix' (p. 508, Flint, Sanders & Rodgers, 1960a). In contrast, the genetic terms *till* (unlithified) and *tillite* (lithified) are used for sediment that has been transported and deposited primarily by glacier ice with subordinate modification by other processes (Dreimanis & Lundqvist, 1984; Dreimanis, 1989). Note that the original definition of diamicton (diamictite) states that the sediment (rock) is matrix-supported, in contrast to some recent uses of the term that include clast-supported sediment and rock (Frakes, 1978; N. Eyles, Eyles & Miall, 1983). Diamicton(ite) should be restricted to matrix-supported sediment (rock) and the terms are used in that sense here. *Diamict* is a general term for both consolidated and unconsolidated deposits (Harland, Herod & Krinsley, 1966).

Accurate recognition and interpretation of glacial sediments is extremely important because ice sheets, which are continental-sized glaciers, influence sedimentation over the whole globe by causing changes in climate, sea level, and oceanic circulation patterns. Glacial sediments are very varied, however, and the extent of glacial influence is typically difficult to define. This is because glacial sedimentary processes are commonly superimposed upon processes characteristic of other environments and are themselves modified by other processes, for example fluvial or marine reworking, downslope sediment failure. In these respects, this chapter interconnects with much information provided elsewhere in this book.

11.2   Characteristics of glaciers

A glacier is a mass of ice which deforms and moves due to the force of its own weight. Mass – that is, ice and/or water – and heat are constantly exchanged between the glacier and both the atmosphere above and the bed, or waterbody, below. This continual transfer of mass and heat, which ultimately responds to variations in climate, controls the balance between erosion and deposition and the nature of many sedimentary processes in glacial environments.

Glacial ice may extend from the land into the sea, over areas of low and high relief, at low and high altitudes, and may be subject to small or enormous seasonal fluctuations in temperature. Thus glacial deposits can be preserved in both terrestrial and marine settings, and under a variety of tectonic and climatic conditions. About 10% of the Earth's surface is covered by glacial ice today. During the Quaternary glaciation, maximum coverage was about 30% (Flint,

1971, p. 80), and the resulting sediments were deposited over a large portion of the Earth's surface. In addition to till, glaciofluvial, glaciolacustrine, glacioaeolian and glaciomarine sediments cover substantial areas marginal to glaciated regions.

11.2.1   Mass balance

In general, the formation of glaciers requires low temperatures and high precipitation. Both high latitudes and high altitudes are conducive to glacial growth. Glaciers are nourished in the *accumulation zone* (Fig. 11.1), where snow is buried and compacted by subsequent snowfalls and locally by the refreezing of percolating water after a summer thaw. The density increases as air is gradually squeezed out. The resulting glacial ice consists of interlocking ice crystals with isolated air bubbles and is effectively impermeable to both air and water (Paterson, 1969). Material is removed from the glacier in the *ablation zone* (Fig. 11.1) by surface melting, basal ice melting, meltwater runoff, sublimation, and iceberg calving.

The algebraic difference between the amounts of accumulation and ablation over a given time is the *net mass balance*. A glacier in which accumulation is equal to ablation (net mass balance = 0) will have a constant mass, and a corresponding constant thickness and area. A change in mass balance changes the dimensions of the glacier, determining whether the margin is advancing, retreating, or stationary.

11.2.2   Thermal regime

Thermal regime refers to whether most of the ice is above or below the *pressure melting point* (the temperature at which ice

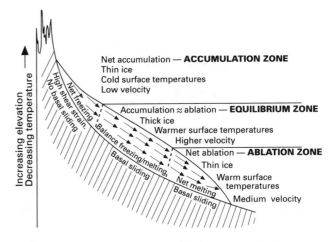

Figure 11.1 Composite model of the thermal regimes of a valley glacier showing velocity vectors, the variety of possible conditions, and controlling parameters (modified after Paterson, 1969; Drewry, 1986).

in a glacier melts, dependent upon the overburden pressure) and is largely determined by climate, mass balance, and ice thickness and velocity (Boulton, 1972b). Thermal regime defines boundary conditions at the glacier sole which have an important effect upon glacial and sedimentological processes. Basal ice temperature is the most important parameter; it is controlled by surface temperature, geothermal heat flux, and heat generated by friction within and at the base of the ice. Three basal thermal regimes can be defined. If the basal ice temperature is above the pressure melting point, there is net melting at the base and the ice is *warm-based* or *wet-based*. The ice slides over the substrate and may be separated from it by a thin layer of water. If the basal temperature is below the pressure melting point, the ice will be frozen to its bed and is *cold-based* or *dry-based*. The adhesive strength of the frozen glacier–bed contact is greater than the shear strength of the ice. An intermediate regime exists in which there is a balance between melting and freezing; the glacier slides but no excess meltwater is produced (Boulton, 1972b).

A cold climate, such as in Greenland and Antarctica, surrounds a glacier with cold air and consequently the glacier is likely to be cold-based. In contrast, a warm climate, such as in southern Alaska and western Norway, leads to summer thaws during which water percolates into the glacier or through it via tunnels, and the glacier is likely to be warm-based. As a result, warm-based glaciers are also called *temperate* glaciers, cold-based ones *polar*, and intermediate ones *subpolar*, but note that these terms are not necessarily indicative of latitude and so can be misleading. Note also that different thermal regimes may exist in different parts of the same glacier. A glacier may be cold-based at high elevations near its source and warm-based near its snout (Fig. 11.1), or it may be warm-based up-glacier where the ice is thicker and cold-based near its terminus.

11.2.3 Types of glaciers

Modern glaciers fall into two main types.

1 *Ice sheets* and ice caps, or continental glaciers, cover large areas and are unaffected by topography (Fig. 11.2). The ice sheets of Antarctica and Greenland today extend over 12.5×10^6 km^2 and 1.7×10^6 km^2, respectively (Fig. 11.3). Ice caps cover less than 50 000 km^2, for example Vatnajökull in Iceland extends over 12 000 km^2 (Flint, 1971). *Outlet glaciers* form where the ice moves through mountains. A *marine ice sheet* is an ice sheet that is *grounded* below sea level, that is its base is in contact with the sea bed and submerged. Large floating *ice shelves* exist locally in polar latitudes where an ice sheet extends into the sea. Ice shelves 200–1300 m thick border parts of the Antarctic coastline today (Fig. 11.3). The effects of erosion and deposition by an ice sheet are geograpically widespread; deposits may be laterally continuous for hundreds of miles.

2 *Valley*, mountain or alpine, *glaciers* are constrained by topography. They originate in highland icefields or cirques and typically form a dendritic pattern similar to that of river systems. Because more surface area of ice is in contact with bedrock and because debris falls on to the glacier surface from adjacent slopes, valley glaciers carry proportionally more debris than ice sheets. Deposits of valley glaciers are geographically restricted and tend to show more rapid facies changes than those of ice sheets.

All types of glaciers comprise three zones. The *basal* or *subglacial zone* is the lower part of the glacier which is influenced by contact with the bed (Fig. 11.2). Both erosion and deposition can take place within this zone. The *englacial zone*, or interior of the glacier, is primarily a region of passive transport of entrained sediment. The *supraglacial zone* includes the seasonally influenced upper surface of the glacier as well as detached masses of stagnant glacial ice.

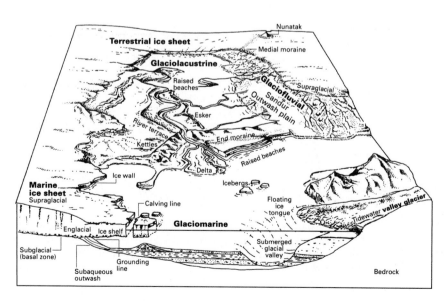

Figure 11.2* Types of glaciers, glacial environments and glacial landforms. *See footnote on p. 454.

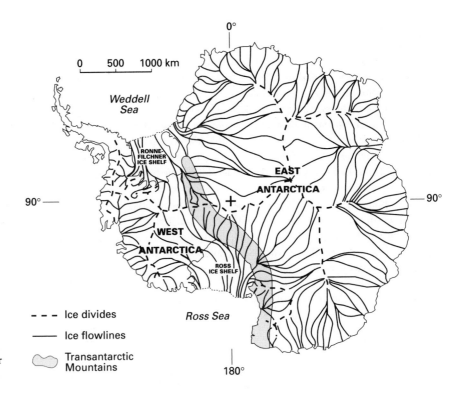

Figure 11.3 Geographic map of Antarctica showing East and West Antarctic ice sheets separated by Transantarctic Mountains, and ice drainage for the continent (modified after Drewry, 1983).

0°

0   500   1000 km

Weddell Sea

RONNE-FILCHNER ICE SHELF

90°

EAST ANTARCTICA

90°

WEST ANTARCTICA

ROSS ICE SHELF

Ross Sea

180°

– – – Ice divides

——— Ice flowlines

Transantarctic Mountains

11.3 Processes

Processes associated with *moving ice* are unique to glacial environments. *Moving water*, *wind* and *resedimentation* also play important roles. Moreover, *low temperatures* have an overriding, if sometimes subtle, effect upon many sedimentary processes. For example, cold water is more viscous, sediment may be frozen, biological activity is changed, and frost action is important (Sect. 11.4.7).

11.3.1 Mechanics of ice flow

Ice moves by two main processes: *internal deformation* or *creep*, and *basal sliding*. These processes affect the ways that debris is eroded, transported and deposited by glaciers. Creep occurs at all levels within a glacier and consists of slippage within and between ice crystals, due to stress caused by the weight of the ice. Creep velocity depends primarily upon ice thickness and slope angle. Basal sliding is concentrated in the basal layers where three main processes occur: (i) *enhanced basal creep* is caused by local stress concentrations within the ice due to bed irregularities; (ii) *regelation* (pressure melting and refreezing) occurs as ice melts on the upstream side and refreezes on the downstream side of an obstacle, enhancing movement of the ice mass; and (iii) *slippage over a water layer* increases ice velocity because of the reduction in friction caused by a layer of water between rock and ice. Another important mechanism of glacier flow is movement within soft, deformable substrate sediment; this can contribute 90% of the total basal movement

of the glacier (Boulton, 1979).

Total surface ice velocity is the sum of velocity due to creep and basal sliding. In warm-based glaciers, basal slip is generally dominant. In cold-based glaciers, creep and substrate deformation usually dominate. Longitudinal flow within an ideal glacier follows vectors which are inclined to the glacier surface. In general, there is a downward velocity component within the accumulation zone, and the flow is described as extending (Fig. 11.1). In the ablation zone, there is an upward component of movement and the flow is compressive. Local variations caused by topography are superimposed upon this pattern. The direction of ice flow in an ice sheet depends upon the size and shape of the ice mass (Fig. 11.3). *Ice streams* are narrow zones within an ice sheet along which flow is much faster than in adjacent broader areas. The distribution of flow lines (slip lines) within ice affects dispersal and transport of englacial debris.

Normal velocities for valley glaciers vary between 3 metres year[-1] and 7 kilometres year[-1] (Boulton, 1974; Drewry, 1986). Temporary, catastrophic increases in basal sliding velocity result in glacial *surges* when velocities may increase 10-, 100-, or even 1000-fold.

11.3.2 Glacial erosion

Ice can be an extremely powerful agent of erosion. Even the lowest measured rates of abrasion alone are twice the world average for erosion of lowland river basins (Boulton, 1974). Thus both large erosional landforms and huge volumes of sediment are produced during glaciations.

Ice erodes primarily by plucking and abrasion. Cold-based ice can only pluck. Large slabs can be quarried, plucked and carried forward by the glacier if weaknesses exist within the substrate and/or the cohesion of the subglacial material is overcome (Boulton, 1972a).

In the presence of meltwater, plucking or debris entrainment may occur either by plastic flow of ice around a block until the block is encased and carried away within the ice, or in association with regelation: debris is frozen on to the glacier base in the lee of bedrock protuberances. Ice within the basal debris-rich zone below warm-based glaciers tends to be melted at the next bump of comparable size. Consequently, basal debris layers of warm glaciers are thinner than those of cold glaciers.

Below warm-based ice, debris carried in the basal layers abrades the bedrock producing *rock flour*, composed of fresh mineral fragments generally smaller than 100 µm (Sugden & John, 1988). Abrasion produces polished surfaces, striations and gouges (Fig. 11.4). Crescentic cracks and gouges form by crushing or fracturing of bedrock below debris-laden ice and can indicate ice flow direction (Fig. 11.4).

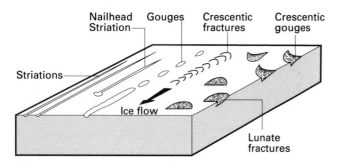

Figure 11.4 Forms of glacial abrasion (modified after Embleton & King, 1975; Shaw, 1985).

11.3.3 Glacial sediment transport

Debris may be carried within any part of a glacier, but the largest volume is generally within the basal zone. The basal zone is usually less than 1 m thick (Boulton, 1972a; Sugden & John, 1988), although it may be as thick as 15 m (Matanuska Glacier, Alaska; Lawson, 1979). Within it, sediment concentrations are variable, averaging 25% but ranging up to 90% by volume for warm glaciers (Drewry, 1986). Since much shearing and abrasion takes place here, the sediment may be layered but it is not sorted. Rock flour is abundant. Clasts are commonly striated and faceted but, due to abrasion, have higher roundness values than supraglacial debris (Fig. 11.5; Boulton, 1978). The development of striations depends primarily upon clast lithology (Kuhn, Melles *et al.*, 1993). Estimates of striated stone abundance range from 0.1 to 28% (Drewry, 1986). Clasts within the basal zone generally show a preferred orientation with their long axes parallel to flow and a small up-glacier dip (Fig. 11.6a;

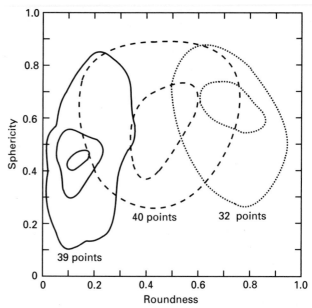

- ⬭ Supraglacially derived boulders in high level transport
- ⬭ Boulders from the zone of traction
- ⬭ Boulders embedded in lodgement till

Contours at intervals of 2, 4, 6 points per 1% of area

Figure 11.5 Positions of clasts in high level transport, from the zone of traction, and deeply embedded in lodgement till, in Krumbein's sphericity/roundness matrix. Clasts are taken from Breidamerkurjökull in southeastern Iceland (Boulton, 1978).

Lawson, 1979). Under highly compressive flow, long axes may be transverse (Boulton, 1971).

Englacial debris is, for the most part, highly dispersed. It is derived from both the supraglacial and subglacial zones and, since it is not usually modifed during transport, shows textures characteristic of those zones. Supraglacial debris is derived either directly from nearby slopes, in which case the clasts are angular (Fig. 11.5), or from en- or subglacial positions and brought to the ice surface along upward-tilted flowlines. As the debris moves passively on the ice, clast shape may be modified slightly by weathering or abrasion by supraglacial meltwater.

11.3.4 Glacial deposition

Sediment is deposited directly from either moving (active) or stagnant (passive) ice by very different processes. Below moving ice lodgement processes dominate – that is, plastering of basal debris on to the substrate. Above or below stagnant ice deposition occurs during melting.

Lodgement till forms: (i) when the frictional drag on clasts in the basal zone is equal to or less than the tractional force exerted upon them by the moving ice; and (ii) when pressure melting below moving ice allows small particles to be freed and lodged

(a) Debris-rich basal glacier ice

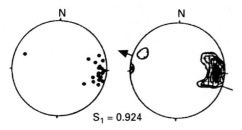

$S_1 = 0.924$

(b) Basal melt-out till

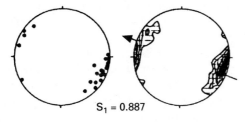

$S_1 = 0.887$

(c) Lodgement till

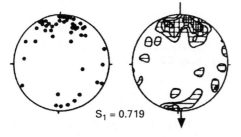

$S_1 = 0.719$

(d) Sediment gravity flow (flow till)

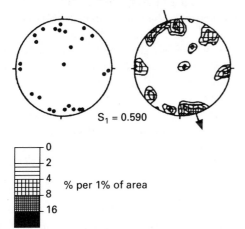

$S_1 = 0.590$

0
2
4
8
16

% per 1% of area

Figure 11.6 Schmidt equal-area projections of clast orientations in glacier ice, tills, and a sediment gravity flow deposit. Arrows indicate direction of ice flow (a,b,c) or of sediment flow prior to deposition (d). Data sources: (a), (b), (d) from Matanuska Glacier, Alaska (Lawson, 1979); (c) from Skalafellsjökull, Iceland (Sharp, 1982). S_1 eigen value for each sample summarizes fabric strength (from Dowdeswell, Hambrey & Wu, 1985).

or plastered on to the glacier bed (Boulton, 1975; Sugden & John, 1988). The latter process is most common beneath warm-based ice. Both processes are affected by irregularities in the subglacial bed, leading to variations in the distribution and thickness of lodgement till.

Melt-out till forms subglacially or supraglacially as stagnant ice melts. Since it is largely a passive process, fabrics from the debris-laden ice are preserved, with slight modification because of volume loss and meltwater movement. In very cold, dry regions, sublimation till rarely forms from debris-laden ice when the ice slowly changes directly into water vapour (Shaw, 1985).

11.3.5 Glaciotectonism

Ice movement and static ice loading can deform consolidated or unconsolidated substrate (Aber, Croot & Fenton, 1989). Deformation occurs where stresses transferred from the glacier are greater than the strength of the stressed material. Ice exerts both drag or shear stress, due to its movement over the bed, and vertical stress, due to its weight. Both ductile and brittle deformation occur and resultant structures are very varied. Much deformation is caused by a lateral pressure gradient. For example, folded and faulted substrate with minor displacement is caused by ice-push at the glacier terminus or by shearing along the glacier sole. Large blocks (up to kilometres across) can be transported long distances from their source and deformed, if, after subglacial plucking, they are carried up into the ice by compressive flow and then moved with it (Moran, 1971).

11.3.6 Related processes: water, resedimentation, wind

Meltwater is a powerful agent of erosion, transport and deposition of glacial sediment, particularly in the region of the glacier terminus and around warm-based ice. Water is present in sub-, en-, supra- and proglacial environments. Because of high sediment load, seasonal fluctuations, and sudden drops in velocity (e.g. where confined flow from a subglacial tunnel enters standing water), sedimentation and aggradation rates can be very high.

Subglacial water flows either in channels or in sheets. Channels are incised either downward into the substrate ('Nye channels') or upward into the ice ('Röthlisberger channels'; Shaw, 1985; Drewry, 1986). Nye channels are stable under moving ice, whereas Röthlisberger channels are only stable under stagnant ice. When subglacial channels cannot accommodate the discharge (Shaw, 1985), or when the basal water pressure is equal to or greater than the ice pressure (Drewry, 1986), subglacial sheet flow occurs. The importance of this sheet flow is controversial. Walder (1982) maintained that sheet flow is probably only quasi-stable at thicknesses <4 mm, whereas Shaw (1985) proposed that large erosional forms can be cut into both bedrock and overlying ice by high magnitude subglacial sheet floods. Ponding of water in subglacial depressions can produce subglacial lakes (Shreve, 1972; Drewry, 1986).

In proglacial environments water exists in rivers, lakes and oceans. One unique feature is catastrophic floods, called *jökulhlaups* (Icelandic). These are caused by sudden discharge of waterbodies that were ponded by ice and can move enormous volumes of sediment (see Sect. 11.4.4).

Sediment gravity flows of all types (see Sect. 10.2.3) are important in most glacial environments. They vary primarily in their water content and exhibit a continuum from slope failure through plastic to fluid behaviour (Lawson, 1979; Fig. 10.3). They are particularly common at and near the glacier terminus in both subaerial and subaqueous settings. In the terminal region of the Matanuska Glacier, Alaska, Lawson (1979) estimated that 95% of the sediments are the result of resedimentation.

Strong *winds* are common in glaciated regions because large ice caps influence atmospheric circulation. Aeolian activity is enhanced by sparse vegetation and the availability of easily eroded sediment. Winds erode and deposit sediment, particularly within the zone fringing the glacier terminus, although *loess* (wind-blown silt) may be distributed over a belt hundreds of kilometres wide and approximately parallel to the ice margin (Brodzikowski & van Loon, 1991).

11.4 Modern glacial environments and facies

Glacial and related environments include a variety of subenvironments each characterized by a distinctive set of processes and sedimentary facies. The *glacial environment* proper embraces all areas which are in direct contact with glacial ice, and includes the *basal* or *subglacial*, *englacial* and *supraglacial* zones (Sect. 11.2.3; Fig. 11.2). The *proglacial environment* occurs around the margin of the glacier and includes: (i) the *ice-contact zone* immediately adjacent to the glacier, where buried stagnant ice (ice no longer flowing with the glacier) is commonly preserved; (ii) *glaciofluvial*; (iii) *glaciolacustrine environments*, which receive glacial meltwater; and (iv) the *glaciomarine environment*, in which glacier ice floats over, or is adjacent to, the sea and/or glacial meltwater influences the marine water structure. Overlapping with the proglacial environment is the *periglacial environment*, which is not directly affected by ice but is influenced by distinctive climatic zones adjacent to an ice sheet.

In the following discussion, descriptions of each glacial environment and facies are based as far as possible upon studies of modern examples, which provide information about processes and spatial configuration of deposits. Sedimentary facies details also come from descriptions of selected Pleistocene deposits. Obviously, some inferences are made here: once the ice has melted we cannot be certain that we know the conditions under which the deposit formed. However, inferences based upon facies analysis of Pleistocene sediments and pre-Pleistocene glacial rocks give important information about stratigraphy and facies preservation and thus provide a valuable contribution to the reconstruction of modern environments and processes, particularly those that are difficult to reach today.

11.4.1 Subglacial zone

The relative importance of ice and water as agents of erosion, transport and deposition within this zone depends upon the thermal regime of the ice and specific position below the glacier. Some direct observations have been made below glaciers, despite obvious logistic difficulties (Kamb & La Chapelle, 1964; Boulton & Vivian, 1973).

Erosional features formed below glaciers depend upon the nature of the substrate. On bedrock surfaces, striae, grooves, crescentic marks (Fig. 11.4) and *roches moutonnées* form beneath warm-based glaciers. Where the substrate is unlithified, deformation of unconsolidated subglacial sediment can contribute significantly to glacier flow, and flutes may form at the ice–substrate interface (Boulton, 1979). This subglacial deformation is characterized by longitudinal extension superimposed on simple shear, resulting in folding, flattening, attenuation and boudinage of original structures. As deformation continues, the till or substrate sediment becomes horizontally laminated and eventually homogenized to produce massive till. Deformation intensity decreases with increasing depth (distance below the glacier base), but may be concentrated along décollement surfaces (Boulton, 1987a). Thus *deformation till* describes weak rock or unconsolidated sediment detached from its source, its primary structures deformed or destroyed, and to which foreign material has been added (Elson, 1989).

Lodgement and melt-out till are important subglacial deposits. Both occur as moraine – that is, ridges or mounds of debris deposited or pushed up by a glacier. Moraine formed subglacially may be fluted, shaped into drumlins (streamlined hillocks generally elongated parallel to ice flow), or form transverse ridges. Lodgement and melt-out till are both composed predominantly of diamicton, although lodgement till may be finer grained due to comminution of debris in the basal zone of the ice. *Lodgement till* is commonly structureless, thrusting and shearing within and below the basal ice causing it to be very compact and sometimes fractured along shear planes and joints that dip up-glacier (Fig. 11.7). Clasts within the till, and deformed sedimentary inclusions if present, are generally orientated so that their long axes are parallel to ice movement, with a gentle up-glacier dip (Fig. 11.6c; Dowdeswell, Hambrey & Wu, 1985; Dowdeswell & Sharp, 1986). Striated and faceted clasts are common as well as bullet-shaped ('flatiron' or 'elongate pentagonal') clasts positioned with their blunt end down-glacier. Lodgement till usually has a sharp erosional basal contact and forms widespread units up to only a few metres thick (Dreimanis, 1989). However, individual units,

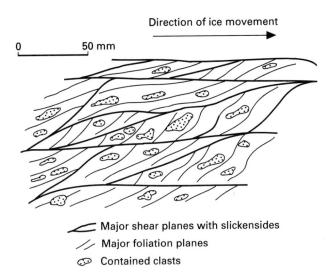

Direction of ice movement

0 ——— 50 mm

⌐ Major shear planes with slickensides

/⁄ Major foliation planes

⬭ Contained clasts

Figure 11.7 Schematic diagram showing the shear structures in lodgement till which may result from stress beneath moving ice (Sugden & John, 1988, modified after Boulton, 1970).

sometimes petrographically distinct, may be superimposed upon one another during a single glacial cycle (N. Eyles, Sladen & Gilroy, 1982). Internal erosion surfaces, sometimes delineated by stone (boulder) pavements, within lodgement till probably form by lodgement, deformation or erosional processes (Hicock, 1991).

Melt-out till is also commonly structureless, but it can contain subhorizontal laminae with gradational contacts derived from debris stratification in the ice, and/or lenses, layers or pods of sorted sediment deposited by escaping meltwater (Lawson, 1981; Shaw, 1982, 1985). The sorted layers may drape over larger clasts due to volume loss as the ice melts. Clasts are orientated parallel to ice movement but usually show a lower dip than in lodgement till due to settling during melt-out (Fig. 11.6b). Because melt-out till is deposited from stagnant ice, it may contain undeformed clasts of unlithified sediment (Shaw, 1982, 1985). Melt-out till units are usually up to a few metres thick, but may be stacked (Dreimanis, 1989). However, Paul and Eyles (1990) emphasize that melt-out till is only of local significance, and is commonly deformed by shear or water escape after deposition.

Subglacial meltwater deposits fill channels aligned subparallel to ice movement which may be tens of kilometres long and are either cut into substrate as *tunnel valleys* or cut upward into the ice to form *eskers*. The former are associated with active, moving ice and the latter with active or passive ice. Tunnel valleys commonly have a trough-like form, an undulating longitudinal profile, and can be a few hundred metres deep and up to 3 km wide. They are filled with sand, gravel, silt and clay and may terminate in a sand–gravel fan at the ice margin (Woodland, 1970; Ehlers, 1981; Patterson, 1994). Eskers are continuous or segmented ridges which can be tens of metres

high and hundreds of metres wide (Warren & Ashley, 1994). They are typically composed of massive or stratified boulder gravel to sand with sedimentary structures similar to fluvial deposits (see Sect. 3.2.2) except that, because flow was constrained within a tunnel, antidunes are absent. Scour and fill structures are common. Esker sediments may show fining-upward cycles up to a few metres thick; they may also fine outwards, and show faulting and rotation of beds in the marginal zones (Banerjee & McDonald, 1975; McDonald & Shilts, 1975; Shaw, 1985; Brennand, 1994).

Catastrophic subglacial meltwater sheet floods have recently been recognized as important agents of erosion and deposition. Their effects include areas of giant flutings, tunnel valleys, scoured bedrock tracts, and possibly the deposition and shaping of drumlins (Shaw, Kvill & Rains, 1989; Rains, Shaw *et al.*, 1993).

11.4.2 Supraglacial zone

Many varied subenvironments can exist on top of glaciers, including thick, living vegetation. Once ice is covered by a layer of debris it is protected from solar radiation and so the rate of ablation is reduced. Deposition can occur from melting ice, running or standing water, mass movement or aeolian activity. However, all these sediments have a low preservation potential.

Supraglacial facies formed by passive melting and by mass movement are the most important, amidst a variety of facies types (Fig. 11.8; Brodzikowski & van Loon, 1991). *Supraglacial melt-out* (ablation) *till* is most abundant in areas of mountain glaciation because of the proximity of debris derived from valley walls. It is commonly intercalated with proximal outwash deposited by meltwater (N. Eyles, 1979). Rarely, *sublimation till* forms and delicate englacial structures are preserved distorted only by loss of ice volume (Shaw, 1985, 1989). Various types of *sediment gravity flow deposits*, similar to those described in the proglacial zone, are abundant. This resedimentation is enhanced by the presence of meltwater and an irregular supraglacial topography. Deformation structures, for example collapse features, normal faults and slumps, are ubiquitous in supraglacial facies because these sediments overlie ice that subsequently melts.

11.4.3 Ice-contact proglacial zone

This zone is characterized by very irregular and hummocky topography. End or push moraines parallel the glacier front, whereas differential melting of buried ice blocks and slumping of wet till produce irregular highs and lows, referred to as ice-disintegration topography (Fig. 11.8). Important processes include glaciotectonism adjacent to the glacier snout, resedimentation by sediment gravity flow and mass movement, melting of stagnant ice, and erosion and deposition by running water.

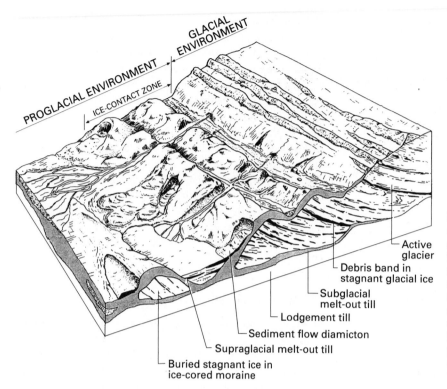

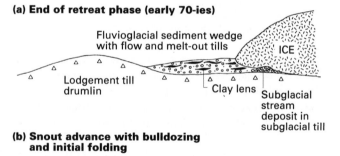

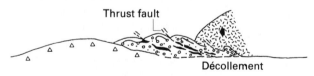

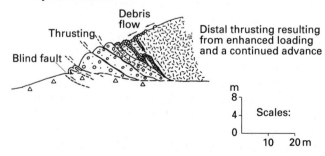

Figure 11.8* Sedimentation in the supraglacial and ice-contact proglacial zones of a slowly retreating, subpolar glacier. Subglacial material is brought into the glacier by basal freezing and thrusting. This debris is released at the surface as the enclosing glacial ice gradually melts. The till is rapidly reworked by flowing meltwater, and may slump and flow downslope to form flow diamictons. Diamicton beds can be intercalated with proglacial stream or lake deposits, and may be extensively reworked (modified from Boulton, 1972a). *See footnote on p. 454.

(a) End of retreat phase (early 70-ies)

Fluvioglacial sediment wedge with flow and melt-out tills

ICE

Lodgement till drumlin

Clay lens

Subglacial stream deposit in subglacial till

(b) Snout advance with bulldozing and initial folding

Thrust fault

Décollement

(c) 1986-situation in western part of push moraine

Debris flow

Thrusting

Blind fault

Distal thrusting resulting from enhanced loading and a continued advance

Scales:

m
8
4
0
10   20 m

Figure 11.9 Selected stages of a schematic model of push moraine formation at the snout of the Turtmannglacier, Switzerland (modifed from Eybergen, 1987).

Proglacial deformation is characterized by compressive structures, for example folding and thrust faulting at various scales. Till or proglacial outwash sediments may be deformed. Glaciotectonic structures are probably most common in glaciers frozen to their bed in the terminal zone (Sugden & John, 1988), but a glacier in the Swiss Alps formed a 5–10 m high push moraine complex composed of deformed older glacial sediments when it readvanced about 100 m between 1971 and 1986 (Fig. 11.9; Eybergen, 1987).

Resedimentation processes are facilitated by local slopes, copious meltwater, and abundant rapidly deposited often clay-rich sediment. Most flows are initiated by backwasting of slopes composed of sediment and stagnant ice. Four types of sediment gravity flows have been recognized at the terminus of the Matanuska Glacier, and the deposits range from massive diamicton to massive to graded sand to sandy silt (Lawson, 1979). Diamictons formed in this way have been called *flow till* (Hartshorn, 1958; Boulton, 1972a), although this term is somewhat misleading since deposition is primarily by resedimentation, which strictly is a non-glacial process (Lawson, 1982). Clast fabrics from sediment gravity flow deposits differ from lodgement and melt-out diamictons in showing either no preferred orientation, a multimodal pattern, or near horizontal long axes aligned parallel to flow (Fig. 11.6d). Other forms of mass movement include talus cones, which accumulate at the base of ablating ice-cored slopes, and slope failure by slumping or spalling (Lawson, 1979; Shaw, 1985).

The labels in Figure 11.8:

Active glacier

Debris band in stagnant glacial ice

Subglacial melt-out till

Lodgement till

Sediment flow diamicton

Supraglacial melt-out till

Buried stagnant ice in ice-cored moraine

GLACIAL ENVIRONMENT

PROGLACIAL ENVIRONMENT

ICE-CONTACT ZONE

Melt-out till (Sect. 11.4.1) is deposited in the ice-contact zone, and meltwater flows generally deposit well-sorted silt to silty sand which may be massive, normally graded, parallel stratified, or cross-stratified if deposited over a change of slope. Sorted meltwater deposits are often interbedded with sediment gravity flow deposits (Lawson, 1979).

11.4.4 Glaciofluvial environments

An outwash plain, or *sandur* (Icelandic, pl. *sandar*), generally forms in front of the glacier terminus, beyond the ice-contact zone (Fig. 11.2). Glaciofluvial environments and facies are almost identical to those of non-glacial braided fluvial systems (see Sect. 3.3.2). Glaciofluvial systems differ in that they are affected by fluctuating positions of the ice margin and by buried and transported blocks of ice; they show strong seasonal and weather-dependent discharge variations; and they carry a high sediment load, typically lack vegetation, and contain rare till clasts (N.D. Smith, 1985). Polar sandar differ from temperate sandar in that they grow more slowly due to lower energy and lower magnitude meltwater discharges (Rains, Selby & Smith, 1980).

Sandar may form a wide plain or outwash fan (Fig. 11.2) or may be restricted by topography to form a valley fill (valley train). Outwash fans are typically subdivided into a proximal zone, where flow is confined to a few main channels that are relatively deep and narrow, an intermediate zone, with a complex network of wide, shallow, distinctly braided, shifting channels, and a distal zone, in which flow in shallower, ill-defined channels may merge to form a single sheet during periods of high discharge (see Fig. 3.31). In the proximal zone the sandur may bury patches of stagnant ice, sometimes forming obstacle marks in surrounding sediment (Russell, 1993). When the ice melts, kettle-holes or pits form that may be up to 40 m wide and 6.6 m deep (Price, 1971), and faults may form in the sedimentary strata (McDonald & Shilts, 1975). During periods of low discharge, large inactive areas on the sandur may become vegetated or subject to strong aeolian influence. Although braided streams predominate on outwash fans, meandering and anastomosed channel patterns are also present, particularly in distal areas where silt and clay are abundant (Boothroyd & Ashley, 1975; Boothroyd & Nummedal, 1978; N.D. Smith, 1985).

Facies in the proximal zone are dominated by massive to crudely horizontally bedded gravels, which are locally cross-bedded, with thin and sparse fine-grained units. Downstream, these give way to sandier deposits with abundant tabular cross-beds as well as ripple-drift cross-lamination, trough cross-beds and horizontal stratification (N.D. Smith, 1985). Clast size and gradient decrease downstream (Boothroyd & Ashley, 1975). However, downstream-fining is most pronounced when aggradation rate is high (N.D. Smith, 1985).

Evidence for highly fluctuating discharges shows through frequent vertical changes in grain size and sedimentary structures,

abundant scour surfaces, reactivation surfaces, and common fine suspension deposits (clay, silt, fine sand) that also occur as intraclasts (N.D. Smith, 1985). Jökulhlaups transport huge volumes of sediment, and produce distinctive deposits including sand and gravel cross-beds up to 10 m high deposited in channels, graded silt and sand deposited by turbidity currents (Baker, 1973; N.D. Smith, 1985), and sequences of massive homogeneous gravel deposited by hyperconcentrated flows overlain by erosional surfaces, boulder lags and cross-bedded and horizontally bedded sand and gravel (Maizels, 1989).

Vertical facies profiles of glaciofluvial deposits have been characterized by three different types (Miall, 1977, 1983): Scott (proximal), Donjek (medial) and Platte (distal). Miall (1983) emphasized the cyclic nature of Donjek type deposits, whereas N.D. Smith (1985) questioned whether cyclicity is typical of vertical profiles of glaciofluvial outwash. Cyclicity on a scale of several metres in sandar may be rare because of continuously fluctuating discharge, despite Miall's claims. However, some cyclicity would be expected on a larger scale as the outwash facies respond to fluctuations in the position of the ice margin. A change from dominantly braided, steep, low-sinuosity, high width/depth ratio bedload channels to deeper, more sinuous, low-gradient, single thread channels occurs in response to long-term deglaciation (Maizels, 1983). Coarsening-upward sequences 10 m or more thick may form as a result of glacial advance, while fining-upward sequences would result from glacial retreat (Miall, 1983; Ashley, 1988).

11.4.5 Glaciolacustrine environments

Lakes are very common in glaciated landscapes. They form as a result of damming of river courses by ice, formation of irregular topography by glacially deposited or eroded landforms, or on a larger scale by the reversal of regional slope due to isostatic depression caused by glacial build-up. Unique to glacial lakes is the fact that they receive a substantial proportion of their annual water and sediment budgets from melt-water (Ashley, 1989), and it is these 'glacier-fed' lakes that are discussed here. Glacier-fed lakes are subdivided into *ice-contact* lakes, in which some portion of the lake is in direct contact with glacial ice, and *non-contact* or *distal* lakes, which are located some distance from the ice and are fed by outwash streams.

The main factors that affect physical processes and sedimentation in glacier-fed lakes are proximity to the ice, thermal stratification of the lake water (see Sect. 4.3; Fig. 4.2), and seasonality of inflowing meltwater and ice cover (Ashley, 1989). Thermal stratification is best developed in non-contact lakes where it is most pronounced and stable in mid-summer (see Fig. 4.3). Mixing of the water layers, or 'overturning', can occur during the autumn (see Sect. 4.3). In ice-contact lakes, thermal stratification is inhibited by the continuous supply of meltwater at 0°C. Glacier-fed lakes may also show sediment, or density stratification due to a gradual increase in suspended sediment content with depth (Gustavson, 1975).

In ice-contact lakes, meltwater streams may enter near the surface of the lake or at or near the lake bottom from englacial or subglacial channels. In non-contact lakes, all meltwater enters at the lake surface. As they move into the lake, these cold influent streams form overflow, interflow, or underflow plumes depending upon the relative density of the influent versus the standing lake water (see Fig. 4.5; Sect. 4.4.2). Meltwater input into glacial lakes shows marked fluctuations over hours and days or weeks and months. This causes pulses or surges of inflow which overprint the quasi-continuous flows into the lake (N.D. Smith & Ashley, 1985).

Glaciolacustrine environments are subdivided into marginal (proximal) and basinal (distal) ones. Deposition in marginal subenvironments is dominantly by mass movement and underflows. Deltas form at the margins of both ice-contact and non-contact lakes (Fig. 11.10). Gilbert-type deltas (see Fig. 6.22) form where coarse-grained debris is supplied under relatively high-energy conditions to a deep lake margin (see Sects 4.6.2 & 6.5). Where a lower energy, finer-grained system enters a shallow lake basin, a delta with gently dipping foresets (<20°) will form. Density underflows typically build lobes on the delta front, and the deposits show evidence of rapid sedimentation and dewatering (Ashley, 1988). Sequences of ripple drift and draped lamination characterize regions between the lobes (Gustavson, Ashley & Boothroyd, 1975).

Subaqueous fans are built where meltwater enters a lake at or near the base of the ice. Coarse gravel typically forms the

proximal core of the fan which is surrounded by sand. Some sands are massive or show chaotic bedding and/or dewatering structures, and may contain scattered large clasts. They were deposited by sediment gravity flows, filling channels near the apex or forming sheets farther out on the fan. Other sands are cross-bedded, ripple cross-laminated or parallel laminated, deposited by traction currents (Rust & Romanelli, 1975; Rust, 1977; Ashley, 1988).

Ice cliffs or ice ramps (where lake water covers stagnating ice) exist along ice-contact lake margins (Fig. 11.10; Holdsworth, 1973; Barnett & Holdsworth, 1974). In both cases debris accumulates and locally forms a sublacustrine moraine as the ice calves or melts; massive to bedded diamicton is interbedded with massive or cross-bedded sand and mud. In large Late Pleistocene glacial lakes which occupied moderate- to high-relief basins, debris flows deposited thick (<10 m) and extensive (<3 km) diamicton units (N. Eyles, 1987). Debris was contributed by slope failure along lake margins as well as by glaciers. Near the margins of large non-contact lakes, wave and storm reworking may deposit hummocky cross-stratified sand (N. Eyles & Clark, 1988). Shoreline glaciolacustrine sediments can also be deformed by pressure from lake ice (Brodzikowski & van Loon, 1991).

Basinal subenvironments are similar in all glacier-fed lakes and sedimentation processes are more regular because meltwater discharge fluctuations are dampened by the lake body (Ashley, 1988). The effects of seasonal fluctuations are most pronounced here and rhythmites and varves form (see discussion in Sect. 4.8; Figs 4.14 & 4.15). Sediment is deposited from suspension and by density currents, turbidity currents and mass flow (see Sect. 4.6.3). A typical resultant succession contains rhythmically laminated or varved fine sand to clay interrupted by turbidites with complete or incomplete Bouma sequences (see Sect. 10.2.3), current rippled sand deposited by bottom-hugging flows, and massive or graded diamicton beds deposited by mass flows. Soft-sediment deformation, for example loading and slumping, is quite common (Ashley, 1975; Sturm & Matter, 1978; Smith & Ashley, 1985).

A distinguishing feature of ice-contact lake deposits is the presence of ice-rafted debris which interrupts regularly bedded or laminated sediment (Fig. 11.11). Dropstones bend, penetrate, ruck and rupture laminae (see Table 11.2). Dump structures, conical mounds of sediment, form when large quantities of debris are released by the break-up or overturning of icebergs (Thomas & Connell, 1985). Clots of frozen, poorly sorted sediment dropped from floating ice and embedded in laminated sediment are called till pellets, clots or clasts (see Table 11.2; Ovenshine, 1970).

Lastly, biogenic activity may modify glaciolacustrine sediment, for example through burrowing (Ashley, 1975; Gibbard & Dreimanis, 1978). It may also contribute to it: pelletization enhances the rate of suspension settling in some glacial lakes (N.D. Smith & Syvitski, 1982).

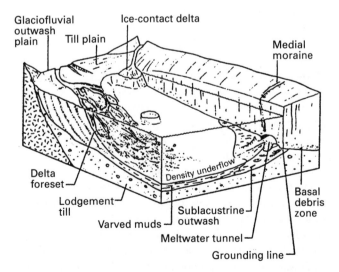

Figure 11.10* Sedimentation in a glacial lake. The three sources of sediment shown here are a sublacustrine outwash fan, an ice-contact delta, and a delta supplied by a glacial meltwater stream. Most of the fine-grained sediment is carried in suspension in a density underflow, and is deposited on the lake bottom. *See footnote on p. 454.

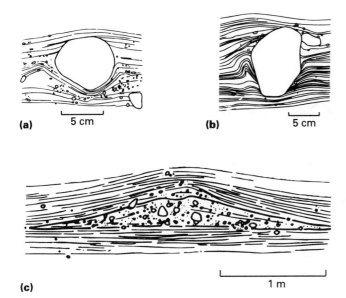

(a) 5 cm

(b) 5 cm

(c) 1 m

Figure 11.11 Dropstone structures (a, b) and dump structure (c) from Pleistocene glacial lake deposits (Thomas & Connell, 1985).

11.4.6 Glaciomarine environments

In marine settings glacial processes are superimposed upon normal marine processes, their effectiveness diminishing away from the ice terminus. Dropstones are the most obvious indicator that ice was present, as well as iceberg keel scour marks (see below; Andrews & Matsch, 1983; Drewry, 1986). Glacial thermal regime affects sedimentation, as discussed below. Ice sheets enter the sea as ice cliffs (walls, faces), floating ice shelves or ice tongues. Valley glaciers enter with tidewater fronts or floating ice tongues (see Fig. 11.2; Drewry, 1986). Ice cliffs and tidewater fronts form where ice terminates at the *grounding line* or *zone* at which ice entering a waterbody comes afloat.

Glaciomarine environments can be subdivided into: (i) subglacial zone; (ii) grounding line zone; (iii) ice shelf and ice tongues; (iv) fjords; and (v) open ocean, distal, or iceberg zone. The first four zones can only exist as far seaward as the continental shelf break, but the iceberg zone can extend over the continental slope and deep ocean floor.

The *subglacial zone* lies below sea level and landward of the grounding line. Processes of sediment transport and deposition are identical to those in the terrestrial subglacial zone described above (Kellogg & Kellogg, 1988).

Sedimentation in the *grounding line zone* varies with glacial thermal regime. Under polar conditions sediment supply is low because glacial erosion is less effective and meltwater is scarce or absent; grounding line sediment is derived only from the base of the ice. Where an ice sheet terminates in ice cliffs, waves erode the cliffs and bottom currents redistribute the finer sediment leaving only coarse ice-rafted debris and bioclastic material (J.B. Anderson, Brake *et al.*, 1983; Domack, 1988;

J.B. Anderson, 1989). Where ice shelves and tongues exist, most sediment is deposited near the grounding line. At the termini of ice streams, below which deformation till has been recognized, *diamict aprons* form by deposition of subglacial debris and downslope slumping of this sediment. These aprons (also called 'till deltas') are tens of metres thick and tens of kilometres long and include topset, foreset and bottomset beds (Alley, Blankenship *et al.*, 1989; J.B. Anderson & Bartek, 1992; Hambrey, Barrett *et al.*, 1992).

The grounding line zone of temperate glaciers is the one where the largest volume of sediment of all glaciomarine environments is deposited. As the ice terminus advances and retreats, grounding line sediments can be spread over a large area. At a tidewater front or ice cliff, sub-, en- and supraglacial debris melts out or falls during calving or continued melting. A subaqueous moraine forms, composed of a complex mixture of diamicton, gravel, sand and mud (Fig. 11.12a) that is commonly redistributed by mass movement or deformed glaciotectonically during ice advances. Submarine end moraines can be hundreds of metres high, kilometres wide and hundreds of kilometres long (King, Rokoengen *et al.*, 1991). Wedge-shaped deposits, termed *till tongues*, often occur on the distal side of continental shelf moraines and are interbedded with stratified glaciomarine sediment (King, Rokoengen *et al.*, 1991; Anderson & Bartek, 1992; King, 1993). Till tongues are commonly 25–50 m thick in their root area, may extend laterally for tens to hundreds of kilometres, and are characterized by acoustically incoherent seismic reflections (King, Rokoengen *et al.*, 1991). Those formed near the terminus of tidewater glaciers may be cut by tunnel valley deposits (King, 1993).

At the exit point of sub- or englacial streams, grounding line (subaqueous outwash) fans form when the ice terminus is stable (Fig. 11.12b; Cheel & Rust, 1982; Powell, 1988, 1990). As the incoming meltwater jet suddenly decelerates, a downcurrent fining of traction current deposits may form, namely imbricate gravels, fine granule gravels, and then sands which may be massive, horizontally or cross-stratified, inversely or normally graded, or ripple-drift cross-laminated. At a certain distance from the influx point, the jet will detach from the sea bottom, dump sediment which builds a gravel–sand bar, and form a buoyant plume. Sediment gravity flows may be generated from oversteepened slopes of the bar. Facies on grounding line fans resemble those of coarse-grained fan deltas (see Sect. 6.5) in that traction current deposits are abundant. However, dropstones may be present, as well as diamicton layers in interchannel areas and deformation caused by melting of buried ice (Cheel & Rust, 1982). Should the ice terminus remain stable for a long period of time, a grounding line fan may aggrade to sea level and form an ice-contact delta with wave- or tide-dominated topset beds and slides and slumps on the delta front. Alternatively, if the grounding line migrates rapidly, sheet-like deposits form. In areas adjacent to the ice terminus

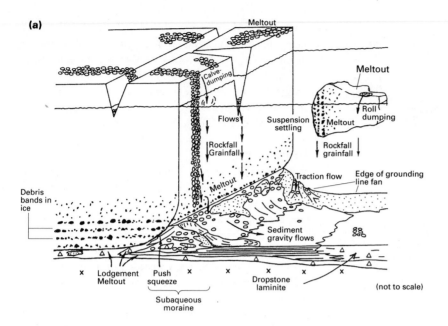

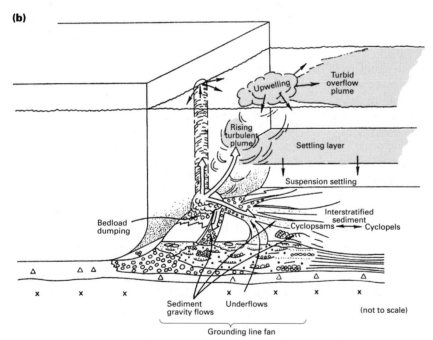

Figure 11.12 Processes acting, and sediment deposited, at and in close proximity to the grounding line zone of a wet-based tidewater glacier (a) where a subaqueous moraine is forming (the margin of a neighbouring grounding line fan interfingers with the moraine), and (b) with submarine discharge of a subglacial meltwater stream forming a grounding line fan (after Powell, 1988).

but distant from active calving or meltwater effluxes, low-energy conditions dominate and mud and ice-rafted debris accumulate. Laminae may result from seasonal fluctuations in meltwater and sediment supply (Powell, 1988, 1990).

Ice shelves and ice tongues are both composed of floating ice attached to and partly fed by land-based ice, but ice tongues are smaller and are long and narrow. The Ross Ice Shelf (see Fig. 11.3) is 420 m thick where drilled (Webb, Ronan *et al.*, 1979) and extends more than 400 km from the grounding line. Ice shelves dampen water movement and currents al-

though thermohaline circulation occurs below them, particularly near the grounding line (Robin, 1979; Jacobs, Gordon & Ardai, 1979). Most sediment is deposited at or near the grounding line because significant basal melting occurs there (see above). Seaward of the grounding line, bottom currents may sort and winnow the sediment (Fig. 11.13).

Beneath floating ice shelves very little sediment is deposited because little debris exists within this ice and sedimentation may be inhibited by basal freeze-on. Massive diamicton with subtle stratification and/or a benthic foraminiferal assemblage

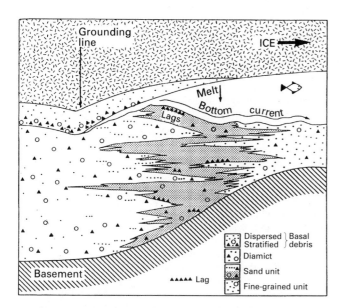

Figure 11.13 Sedimentary facies generated at the inner grounding zone of a floating ice shelf or glacier tongue undergoing grounding-line oscillations (due to sea-level or upstream mass balance changes) but without experiencing major erosion of sea-floor deposits (after Drewry, 1986) Note that below large polar ice shelves, current reworking is minimal and the proportion of massive diamict would be higher than shown here.

is thought to accumulate (Fig. 11.14; J.B. Anderson, Brake *et al.*, 1983; J.B. Anderson, Kennedy *et al.*, 1991). Sub-ice shelf sediments can be thicker if the ice shelf is fed by nearby mountains. Ice shelves only exist in high latitudes, and if ice shelf retreat is rapid, sub-ice shelf deposits may be virtually absent. Consequently, the vertical facies association of massive diamicton (subglacial or sub-ice shelf deposit) directly overlain by open ocean (iceberg) zone glaciomarine sediment with a significant biogenic component (see below) and without intermediate meltwater deposits, provides strong evidence for a polar or subpolar climate (J.B. Anderson, Kennedy *et al.*, 1991).

Fjord sedimentation occurs primarily during glacial retreat and differs from other glaciomarine settings only because of the shape of the basin. In temperate regions, submarine moraines and outwash form in the grounding line zone (Fig. 11.12); homogeneous muds containing ice-rafted debris commonly dominate elsewhere, with sandy gravelly mud on slopes and the outer shallow sill (Elverhøi, Lønne & Seland, 1983). Peak meltwater discharges and sediment gravity flows, commonly generated by slope failures, deposit coarse laminae. If the glacier terminates on land, sand and gravel outwash delta deposits prograde into the fjord and sand intertongues distally with mud, which may be laminated and rarely contains ice-rafted debris (Powell, 1981). Biogenic sedimentation in temperate fjords is minor because of high terrigenous sedimentation rates (Powell & Molnia, 1989). In polar and subpolar fjords, biogenic facies are more common and may dominate over terrigenous sediment with increasing distance from the fjord head (Syvitski, LeBlanc & Cranston, 1990; Domack & Ishman, 1993).

Laminated sediment, including *cyclopsams* (sand–mud laminae) and *cyclopels* (silt–mud laminae), is deposited from over- and interflow plumes 0.5 to several kilometres from the grounding line of tidewater glaciers in Alaska (Mackiewicz, Powell *et al.*, 1984). The couplets have sharp basal contacts and are normally graded (Powell, 1988). Two couplets form per day, due to the interaction of the suspended sediment plume with tidal currents (Cowan & Powell, 1990). Sedimentation rates can be extremely high, for example up to 15.4 cm dry sediment (cyclopsams) deposited in 19 h (Powell & Molnia, 1989). These marine couplets look similar to glaciolacustrine varves, but the gradational upper contact of the marine coarse laminae should distinguish them from freshwater varves in which the coarse–fine contact is commonly sharp because the two varve layers had different modes of deposition (see Sect. 4.8).

The *open ocean* or *iceberg zone* is affected by glaciers in several ways. Normal marine sedimentation is modified by deposition from icebergs and a varied supply of terrigenous sediment. Moreover, the continental shelf is typically depressed by the weight of the ice sheet and has a proglacial depression (Fig. 11.14) and an irregular bathymetry (Sect. 11.5.4).

Polar glaciomarine shelf environments are characterized by accumulation of biogenic material, favoured by low terrigenous

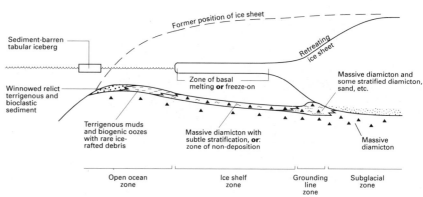

Figure 11.14 Schematic facies model for sedimentation at and around ice shelves associated with polar marine ice sheets during ice-sheet retreat. Note that continental shelf deepens towards continent due to weight of ice sheet. Not to scale, but with vertical exaggeration (adapted from J.B. Anderson, Kennedy *et al.*, 1991).

influx (Orheim & Elverhøi, 1981; Domack, 1988). The Antarctic shelf averages about 500 m deep (Dunbar, J.B. Anderson *et al.*, 1985). Below about 300 m are terrigenous muds and siliceous muds and oozes (biogenic silica concentrations of up to 30–40%), with minor ice-rafted debris (J.B. Anderson, Brake *et al.*, 1983). Carbonate bioclastic material or winnowed relict terrigenous sediment exist at shallower depths (Domack, 1988; J.B. Anderson, 1989).

In temperate regions, such as the Gulf of Alaska and the Beaufort and southern Barents seas, terrigenous sediment dominates (Barnes & Reimnitz, 1974; Vorren, Hald & Thomsen, 1984; Powell & Molnia, 1989). Bioclastic carbonate material is restricted to lag deposits on banks. Although these regions reflect sedimentation during an interglacial, the higher silici-clastic to biogenic ratio is probably indicative of temperate versus polar glaciated shelves.

Iceberg-rafting is common in shelf environments bordered by glaciated coasts. Isolated clasts and dump structures are similar to those described from glacial lakes (Fig. 11.11) (Gilbert, 1990). Icebergs have recently been recognized as important agents in scouring and reworking shelf sediment to depths as great as 500 m around Antarctica (Barnes & Lien, 1988). Surface features include furrows, typically a few metres deep and tens of metres wide, cut by iceberg keels (Fig. 11.15), and subcircular depressions (30–150 m across) formed by grounding of iceberg keels. Ice keels plough and rework shelf sediment to produce structureless diamicton which is termed 'ice-keel turbate' (Vorren, Hald *et al.*, 1983; Barnes & Lien, 1988).

During glacial maxima, large ice sheets transport voluminous unsorted sediment to the shelf break. This sediment moves by sediment gravity flows down the continental slope. On Antarctic continental slopes, slide and slump deposits near the top are replaced by non-stratified, non-sorted debris flows downslope and then by turbidites near the base (J.B. Anderson, Kurtz & Weaver, 1979; R. Wright & Anderson, 1982; R. Wright, Anderson & Frisco, 1983). On the northeast Newfoundland slope, thin but extensive shingled Quaternary debris flow lenses, derived from a line source at the shelf edge, are locally interbedded with stratified hemipelagic sediments (Aksu & Hiscott, 1992).

On the deep ocean floor, biogenic sedimentation with an ice-rafted component is typical during glaciation. Cores from the North Atlantic contain specific, 'Heinrich' layers, which have unusually high ratios of ice-rafted debris to Foraminifera shells and record massive influxes of icebergs and/or surges of the Laurentide ice sheet (Bond, Heinrich *et al.*, 1992). In the central Arctic Ocean, uniformly alternating layers of silty and sandy lutites accumulated over the last 5 My (Clark, Whitman *et al.*, 1980). A large submarine fan exists on the abyssal ocean floor of the eastern Weddell Sea (see Fig. 11.3). This fan probably formed during a prior, more temperate glacial setting of Antarctica, when terrigenous sediment supply was higher and before the shelf was lowered by glacial erosion and isostasy (J.B. Anderson, Wright & Andrews, 1986).

11.4.7 Periglacial zone

The periglacial zone is not directly affected by ice but is characterized by intense frost action and at least seasonally snow-free ground. It is frequently underlain by permanently frozen ground, *permafrost*. Aeolian, fluvial and lacustrine processes are also important in this zone.

Frost action leads to formation of many periglacial features, some of which may be preserved in the rock record (Washburn, 1980). Ice wedges and sand wedges, associated with polygons on the ground surface, form by frost cracking in a permafrost environment (Black, 1976). Ice wedges grow laterally and expand upwards, usually causing bedding in adjacent sediments to be bent upwards (Fig. 11.16a). After the ice melts, the wedge is filled with collapsed sediment. Although ice wedges grow preferentially in fine-grained material, they are best preserved in gravels. Sand wedges only form under arid conditions, and the frost cracks are filled repeatedly and incrementally by loose sand. Sand wedges can be distinguished from ice-wedge casts by the different orientation of the infilling laminae (Fig. 11.16b).

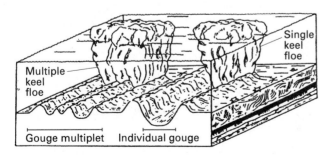

Figure 11.15 Scouring by single and multiple keel icebergs (floes) to produce an idealized ice gouge and gouge multiplet (from Barnes, Rearic & Reimnitz, 1984).

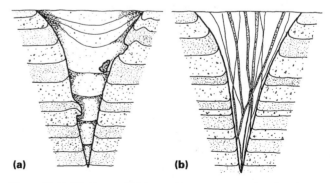

Figure 11.16 Diagrams of (a) ice-wedge-cast cross-section showing slump structures in cast and upturned beds in enclosing material; (b) sand-wedge cross-section showing vertical fabric in wedge and upturned beds in enclosing material (Black, 1976).

Irregular contortions, including deformation, folding and interpenetration of pre-existing strata and termed periglacial *involutions* (cryoturbations), are produced by frost action. Other rarely preserved periglacial features include fossilized pingos (Beuf, Biju-Duval *et al.*, 1971).

Soils of cold climates can show a variety of features caused by ice in the soil including brecciated zones where frozen soil has fallen into position after tabular bodies of ice melted (Retallack, 1990). Buried weathering profiles on Pleistocene tills show vertical grain size variations due to solution of carbonates, movement and alteration of clay minerals, and disaggregation and decomposition of less stable silicate minerals (Willman, Glass & Frye, 1966). Original depositional features may be fundamentally changed through drainage and consolidation, and fines are transported down through the profile by percolating water or lost by wind action (Boulton & Dent, 1974). Palaeosols on tills can be extremely useful for distinguishing different till sheets and interglacial periods (White, Totten & Gross, 1969).

Sand dunes, sand sheets and loess are the main periglacial aeolian deposits. Most dune-forming mechanisms are similar to those in non-glaciated regions (see Chapter 5). However, periglacial sand is usually more heterogeneous in texture and composition than in typical aeolianites, and the dunes migrate more slowly because of moisture freezing in the sand and burial of the dunes by snow (Ahlbrandt & Andrews, 1978). Sand sheets are composed of subhorizontal to low-angle cross-stratified sand and form in areas where limited availability of loose, dry sand inhibits dune formation (Lea, 1990). The inclusion of snow or ice layers and ice-cementation leads to distinctive deformation structures when the ice melts (Calkin & Rutford, 1974; Ahlbrandt & Andrews, 1978). Where fluvial and aeolian processes interact, sediments deposited as channel fills and aeolian sandsheets are truncated by laterally persistent, planar deflation surfaces where permafrost restricts the depth of aeolian reworking (Good & Bryant, 1985).

Loess is composed dominantly of silt, most of which is produced by glacial grinding (glacial flour). Both grain size and deposit thickness decrease with increasing distance from the source (Brodzikowski & van Loon, 1991), whereas the roundness of quartz silt grains increases downwind (Mazzullo, Alexander *et al.*, 1992). Loess deposits may be massive or show faint parallel or undulating laminae. Palaeowind directions may be deduced from the preferred orientation and imbrication of silt grains in loess (Matalucci, Shelton & Abdel-Hady, 1969).

Some periglacial lakes, for example in the 'dry valleys' region of Antarctica, exhibit a characteristic association of sulphate and carbonate evaporites, stomatolitic sediments and sands (Walter & Bauld, 1983). In the ice-covered lakes, small quantities of sediment percolate slowly through the porous ice cover and water-filled, vertical gas channels, and algae living on the lake bottom are not overwhelmed by sedimentation (Nedell, Andersen *et al.*, 1987). The resulting facies association may be an important modern analogue for certain problematic ancient associations of carbonates with glacial facies (Walter & Bauld, 1983) (See Sect. 11.5.5).

11.5 Ancient glacial facies

11.5.1 Characteristics and recognition

Lithofacies deposited in glacial environments are very diverse; all terrigenous sedimentary rock types can be present, and carbonate rocks exist locally (Table 11.1). The diversity results from many glacial facies being hybrids of glacial and other sedimentary processes, for example glaciofluvial, glaciolacustrine deposits. *High sediment supply* is typical in all proximal, and some distal, settings. Whether the sediment is dumped by the glacier or redistributed by meltwater, the rate of sedimentation can be extremely high and the volume of sedimentary deposits large. In addition, *marked seasonal fluctuations* in meltwater and sediment supply characterize glaciation in temperate regions and give a distinctive signature to some glacial facies (Sect. 11.4).

To infer accurately a glacial depositional environment, all sedimentary rocks, structures and contacts as well as the overall facies context must be carefully examined. No individual feature is uniquely diagnostic of glacial processes. None the less, certain specific criteria are useful and are discussed briefly in Table 11.2. Comprehensive discussions of sedimentary features characteristic of glacial deposits are given by Harland, Herod and Krinsley (1966), Schermerhorn (1974), Hambrey and Harland (1981), and Hambrey (1994).

Many authors have used a facies code, particularly one adapted by N. Eyles, Eyles and Miall (1983) from one originally proposed for fluvial sediments (Miall, 1977; see Sect. 3.5), as a shorthand for describing glacial facies. Lithofacies codes should be purely descriptive, and one shortcoming of N. Eyles, Eyles and Miall's (1983) scheme is that it incorporates genetic interpretation of some diamictites facies. Moreover, their distinction between matrix-supported and clast-supported diamictite is inappropriate because diamictites was originally defined as matrix-supported (see Sect. 11.1). The variation in the clast concentration of diamictites can be included through use of a lithofacies code that incorporates 'apparent clast packing density' (Visser, 1986) or a revised classification for poorly-sorted sedimentary rocks that quantifies clast-rich versus clast-poor diamictites (Moncrieff, 1989). Pitfalls of using any lithofacies codes are that they encourage simplistic pigeon-holing such that important variations and details may be overlooked, as may be the nature of contacts between facies. Any scheme should be individually tailored to suite the specific needs and goals of each project (see Sect. 2.2.1).

11.5.2 Glacio-eustasy and glacio-isostasy

The waxing and waning of ice sheets causes important eustatic

Table 11.1 Lithology, bedding characteristics, and sedimentary structures of common glacial lithofacies.

Lithology	Bedding and sedimentary structures
Diamict	Massive Stratified – Bedded (>~1 cm) – Massive 　　　　　　　　　　　　　　　　Inversely or normally graded 　　　　　　　　　Laminated (<~1 cm) Foliated (splits preferentially along subparallel, closely spaced surfaces 　　which are conformable with regional bedding)
Conglomerate and/or sandstone	Massive Stratified – Horizontal beds 　　　　　　Planar cross-beds 　　　　　　Trough cross-beds 　　　　　　Graded beds 　　　　　　Ripplemarks 　　　　　　Ripple drift cross-laminae 　　　　　　Hummocky or swaley cross-stratification Deformed – soft-sediment folds, slumps, etc.
Siltstone, claystone, and/or fine sandstone	Massive Laminated – Rhythmically laminated 　　　　　　*or* randomly laminated 　　　　　Graded beds　　　　　　　　　　} with or without 　　　　　Ripplemarks　　　　　　　　　　　isolated clasts Deformed – Soft-sediment folds, etc.
Carbonate rocks	Massive Laminated, unfossiliferous – Parallel laminae 　　　　　　　　　　　　　Wavy laminae Algal/stromatolitic laminae Bioclastic

and isostatic responses that affect facies types and distribution both close to and distant from the ice centre. *Glacio-eustasy* affects global sea level, because large volumes of water are removed from the oceans and locked up in ice caps. Global sea level has risen about 130 m since the last Pleistocene glacial phase (Lambeck, 1990). Glacio-eustatic effects are therefore widespread. Late Palaeozoic cyclothems of North America and Europe are interpreted as due to fluctuations in the size of ice caps in Gondwana (Veevers & Powell, 1987), and glacio-eustatic sea-level changes can be linked with 200–500 ky, or fourth-order, cycles in sequence stratigraphic terminology (Miall, 1984).

The effects of *glacio-isostasy* vary depending upon ice volume and proximity to the ice mass. Beneath the ice, the Earth's crust is depressed by the weight of the ice and rebounds after it melts. Around an ice sheet, a rising forebulge forms during glaciation due to flexural rigidity of the crust. Both magnitude and rate of postglacial isostatic rebound are high. In Fennoscandia, up to 830 m of uplift has occurred in the last 13 000 years, with peak rates during deglaciation of 50–500 mm year^{-1}. Meanwhile, the forebulge area, including the North Sea, has subsided as much as 170 m (Mörner, 1980).

Within a glaciated region, glacio-eustasy and glacio-isostasy combine to alter elevation compared with sea level. The net result at any specific location depends upon the magnitude of each factor and proximity to the ice centre. Inside the *glacial maximum*, the position of the ice margin at the time of greatest glacial extent, there is generally a lowering of sea level during glaciation, followed by postglacial eustatic sea-level rise and widespread transgression, and then rapid isostatic rebound and a period of regression. Areas outside the glacial maximum but close to the ice margin are commonly dominated by isostatic depression and rebound (i.e. submergence followed by emergence), while more distal areas are dominated by eustatic changes (i.e. emergence followed by submergence) during the same time period (Andrews, 1978).

Climatic cyclicity, which largely controls the waxing and waning of ice sheets, is an earmark of glacial episodes. Pleistocene deposits are cyclic on several orders, ranging from 10 000s to 100 000s years. Changes in global sea level during the Pleistocene, because they primarily record global ice-volume fluctuations, are documented by the deep-sea planktic foraminiferal $\delta^{18}O$ record (Matthews, 1984). Much cyclicity of the $\delta^{18}O$ record corresponds to systematic variations in the Earth's orbit around the sun, as predicted by the Milankovitch theory (Imbrie, Hays *et al.*, 1984; see Sect. 2.1.5). Orbital variations therefore are a likely cause of some cycles in glacial sedimentary successions. The $\delta^{18}O$ record, however, also shows

Table 11.2 Selected criteria useful for recognizing glacial depositional processes, and possible ambiguities.

Individual criteria	Comments	
	Glacial origin	Possible ambiguities
Diamictite	One of the commonest glacial facies, produced by a variety of glacial processes (Sects 11.3 & 11.4)	Can be produced by non-glacial depositional processes, e.g. debris flows
Striated surface	Can be good evidence of glacial erosion, particularly if it shows two or more intersecting sets of subparallel striae, and/or associated glacial erosional features (Sect. 11.3.2), and if overlain by glaciogenic diamictite	Can be produced by tectonic or mass flow processes, but those surfaces show different characteristics
Striated and faceted stones (clasts)	Preferably show two or more intersecting sets of subparallel striae on one or more surfaces. Glacial striae and facets must be unrelated to internal features of rock, e.g. stratification, fractures	Rarely produced by tectonic or mass flow processes, but with different characteristics
Dropstones	Diameter of stone should greatly exceed thickness of surrounding laminae, in order to discount emplacement of stone by traction currents. Laminae may be disrupted by stone as it fell into place (Fig. 11.11).	Large stones can be carried by fairly thin sediment gravity flows, and can be rafted by sea ice, algae or plants
Till pellets (clasts)	Composed of unsorted debris identical to matrix of nearby till beds. Pellets are typically 2–3 mm in diameter, commonly with diffuse boundaries, and flattened and elongated parallel to bedding, owing to deposition while unconsolidated. Pellets result from freeing of sediment which filled interstices between ice crystals, a process unique to glaciers (Ovenshine, 1970)	
Varves	(See Sects 4.8 & 11.4.5)	
Ice-/sand-wedges	(See Sect. 11.4.7)	
Periglacial involutions	(See Sect. 11.4.7)	
Surface textures of quartz/garnet grains	(Krinsley & Doornkamp, 1973; Gravenor, 1982)	

cycle asymmetry which reflects progressive cooling, implying gradual sea-level fall, culminating in a cold stadial, followed by a rapid transition to the next warm interstadial, implying rapid sea-level rise (Bond, Broeker *et al.*, 1993).

Ice sheet dynamics can also cause sea-level fluctuations during glacial periods, which may be independent of external climatic forcing. Decoupling of marine ice sheets from the sea floor leads to rapid collapse of those ice sheets and rapid sea-level rise (J.B. Anderson & Thomas, 1991; Sect. 11.5.3). In addition, sporadic ice-sheet collapse and rapid sea-level rise have been linked to: (i) oscillations in the flow of an ice sheet caused by changes from frozen to thawed, and deformable, subglacial till (MacAyeal, 1993); and (ii) catastrophic release of meltwater reservoirs (Blanchon & Shaw, 1995).

11.5.3 Glacial facies zones*

Patterns of glacial facies distribution can be divided into two: those that form under terrestrial conditions (Fig. 11.17) and those that form under marine conditions (Fig. 11.18), although thermal regime and relief are also important influencing factors. Since erosion generally dominates in subglacial areas during glacial advance, most deposition occurs during glacial retreat.

TERRESTRIAL GLACIAL FACIES ZONES

During terrestrial glaciation, the glaciated area and adjacent proglacial zone are above sea level. Pleistocene deposits of North America and northern Europe provide well-documented examples of these facies in a continental, low-relief setting and the gross, regional facies distribution can be delineated from many publications and large-scale glacial geological maps (Flint, 1945, 1959; Prest, Grant & Rampton, 1968; Woldstedt, 1970, 1971).

Three main facies zones are differentiated (Fig. 11.17; Sugden & John, 1988). Surrounding an inner erosional zone with thin, sporadic till deposits is the *subglacial facies zone*, which appears on Pleistocene geological maps as a fluted or drumlinized till plain, and is composed predominantly of lodgement till. Till may be the only deposit across much of this zone, though subglacial meltwater facies can be present. Locally, the subglacial facies may be overlain by stratified glacial retreat facies, including widespread varved mud. The upper surface of these deposits is subject to frost action and/or soil development (Sect. 11.4.7). However, with repeated glacial advance and retreat,

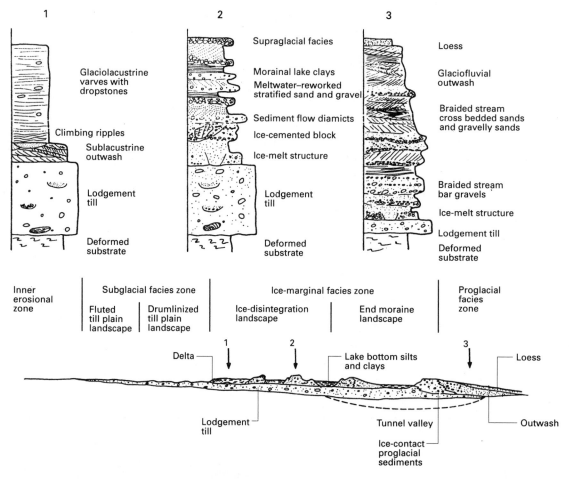

Figure 11.17* Schematic representation of the zonation of terrestrial glacial landforms (from Sugden & John, 1988) and facies deposited by a wet-based ice sheet in a low-relief setting, and characteristic vertical profiles that would be deposited in each zone. The radius of the ice sheet could have been up to 2000 km. The resulting deposits are typically 5–50 m thick. *See footnote on p. 454.

* See footnote on p. 454.

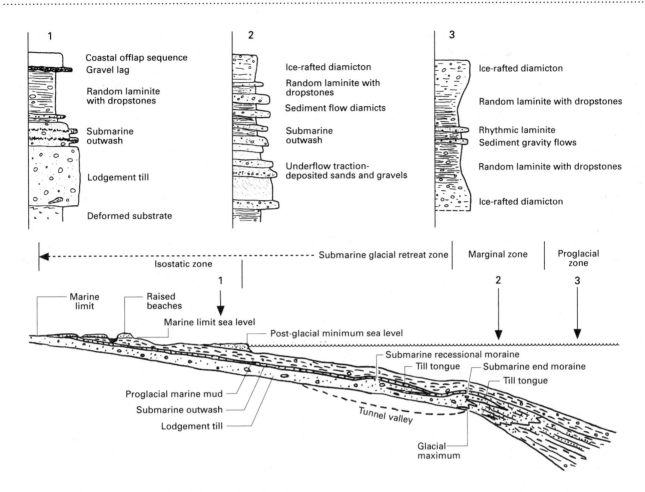

Figure 11.18* Schematic representation of the zonation of marine glacial facies deposited by a wet-based marine ice sheet in a low-relief setting, and characteristic facies sequences that would form in each zone. * See footnote on p. 454.

deposits formed during one retreat may be stripped away during an ensuing advance, so that ultimately a sequence of lodgement tills with thin or no intervening facies is built up.

The *ice-marginal facies zone* occurs in the outer parts of glaciated regions, with two distinct landscapes: end moraines and ice-disintegration topography. Three facies associations occur in this zone: (i) lodgement till and other subglacial facies are deposited throughout, resting upon a regional erosion surface. The underlying deposits may show glaciotectonic deformation; (ii) above are supraglacial and ice-contact proglacial deposits, occurring as end moraines in the outer part of the zone, and as ice-disintegration topography in the inner part; and (iii) widespread varved lacustrine muds may rest on any of these facies. A comparable facies association can develop where the glacier terminus is grounded in a large lake (Landmesser, Johnson & Wold, 1982). In the course of several glacial phases, a complex succession of lodgement tills with thin interstratified outwash sediments can accumulate (White, Totten & Gross, 1969). The region towards the edge of the glaciated area is particularly sensitive to ice-margin fluctuations which may generate a complex stratigraphic record, in some cases including glacial advance as well as retreat successions, and tunnel valley fills. In contrast, more proximal areas remain ice-covered and a simple succession results.

The *proglacial facies zone* includes some ice-contact deposits and all proglacial deposits such as glaciofluvial sands and gravels, lacustrine muds, and windblown sand and silt. The area is largely beyond the zone of till deposition. Apart from the rare occurrence of striated clasts, it may be impossible to deduce glaciation from these deposits alone. Outside of major drainage channels, where valley trains accumulate, these deposits thin rapidly away from the end moraine complex.

MARINE GLACIAL FACIES ZONES

Glaciomarine facies are deposited below sea level and at elevations up to the *marine limit*, the maximum extent of the sea reached during postglacial eustatic sea-level rise. The generalized model presented in Fig. 11.18 is a composite compiled from several Pleistocene examples. Because the bulk of Pleistocene glaciomarine sediments lie below modern sea level and are studied principally by seismic profiling and shallow coring, they are less well known than terrestrial deposits. Moreover, glaciomarine facies associations may be more complex than their terrestrial counterparts, especially where deposition was influenced by isostatic and eustatic effects and the resulting deposit shows the effects of terrestrial, marine and glacial agents.

The critical boundaries that control facies distribution and contacts at any particular time are: (i) glacial maximum; (ii) marine limit; and (iii) postglacial minimum sea level (Fig. 11.18). Because of the number of parameters involved, many scenarios can be reconstructed for glacial retreat sedimentation near sea level. Some of these involve a combination of marine and terrestrial processes because the glacial maximum may rise above sea level during continued glacial recession. To simplify the discussion here three assumptions are made: (i) the glacial maximum was below contemporaneous sea level; (ii) isostatic depression exceeded eustatic fall; and (iii) eustatic rise preceded isostatic rebound and relative fall in sea level. Four glaciomarine facies zones can then be defined, moving from distal to proximal locations: proglacial, marginal, submarine retreat, and isostatic (Fig. 11.18).

The *proglacial marine facies zone* occurs seaward of the glacial maximum. Facies of the open ocean and iceberg zone (Sect. 11.4.6), for example dropstone laminites, sediment gravity flow deposits, ice-rafted and ice-keel reworked diamictons, are gradationally to sharply interbedded with non-glacial facies. Typically, sediment gravity flow deposits are more abundant closer to the ice terminus, and ice-rafted diamictons are more important distally (Boulton & Deynoux, 1981). Therefore, during an advance–retreat cycle a vertical profile such as (3) in Fig. 11.18 may result.

The *marginal glaciomarine facies zone* is formed close to the glacial maximum, generally during major glacial stillstands. It consists of grounding-line-zone sediment, including submarine end moraines, grounding-line fans, and accumulations of diamicton that may form diamicton aprons and/or till tongues (Sect. 11.4.6; Figs 11.12 & 11.13). These facies may rest on thin deposits of basal till and be overlain and interfinger distally with facies of the proglacial zone.

Facies associations in both the marginal and proglacial zones vary depending upon whether the ice is wet- or dry-based. Under wet-based conditions, supply of terrigenous sediment is likely to be high; if the ice is dry-based, terrigenous sediment is probably sparse and biogenic facies may dominate (Sect. 11.4.6).

The marginal facies zone may be areally widespread if the grounding line migrates over time, for example in response to changes in mass balance. Also, submarine outwash may occur locally as recessional moraines, above lodgement till, inside the glacial maximum up to where the ice terminus becomes subaerial

during glacial retreat. However, marginal glaciomarine facies can only accumulate as far oceanward as the continental shelf break, the limit for ice sheet grounding, whereas proglacial marine facies can accumulate on the shelf, slope or nearby deep ocean floor.

The *submarine glacial retreat facies zone* occurs between the marine limit and the glacial maximum. It includes the isostatic facies zone, discussed below, and so may be eroded away above the postglacial minimum sea level. The sequence begins with a lodgement till which rests on a subglacial erosion surface. Pleistocene till, deposited in this zone during lowered sea level, is widely recognized on modern high-latitude continental shelves from drilling and geotechnical and seismic properties (Cooper, Barrett *et al.*, 1991; Hambrey, Barrett *et al.*, 1992). Till is overlain by glaciomarine sediments, in many cases dropstone laminite. Under wet-based conditions, submarine outwash will likely be abundant between the till and laminite facies. However, near a retreating dry-based ice sheet basal till may be abruptly overlain by marine proglacial or non-glacial deposits. Following deglaciation of adjacent land areas, normal marine sediments should cap both successions.

Rate of glacial retreat affects facies types and distribution in the submarine retreat zone and the relative abundance of retreat zone versus marginal zone facies. Rate of retreat, or changing position of the ice sheet grounding line, depends upon mass balance, sea level and slope of the substrate near the ice margin (Thomas, 1979; J.B. Anderson & Thomas, 1991). If either mass balance becomes negative or sea level rises and the bed slopes away from the ice centre, then the grounding line will slowly rise and retreat and a gradual submarine retreat sequence will form (Fig. 11.19). Alternatively, if the bed slopes toward the ice centre, the grounded outer part of the ice sheet will become unstable. The ice sheet will rapidly rise to an ice shelf configuration and extensive calving will occur; the grounding line will quickly retreat to a point at which the substrate is shallow enough for the ice margin to stabilize (Fig. 11.19). A large, though short-lived, area of sub-ice shelf sedimentation (Sect. 11.4.6) may be created which is included within the submarine glacial retreat facies zone. Such rapid retreat will result in a catastrophic submarine retreat sequence. Homogeneous diamicts, deposited by rain-out during rapid ice shelf disintegration, may overlie basal tills, with no record of the usually intervening marginal facies zone, and be abruptly overlain by proglacial or non-glacial facies. This ice-sheet decoupling and rapid retreat should be common during large-scale marine deglaciation because most glaciated continental shelves have an irregular bathymetry and proglacial isostatic depression (Fig. 11.14). On a smaller scale, rapid retreat can occur in a fjord from a grounding line situated at a sill (a submarine rise), as demonstrated recently by the Columbia Glacier in Alaska (Hambrey & Alean, 1992).

The *isostatic glaciomarine facies zone* lies between the marine limit and the postglacial minimum sea level, below which

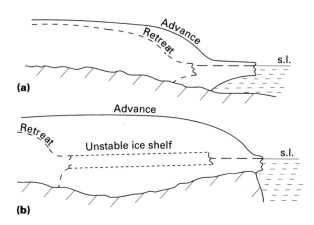

Figure 11.19 Simplified models to show effect of bed slope on rate of marine ice-sheet retreat, holding sea level constant. (a) Bed slopes seaward, away from ice centre: negative mass balance causes gradual retreat and a limited shift in grounding-line position. (b) Bed slopes landward, towards ice centre: negative mass balance causes catastrophic retreat and a large shift in grounding line position; dotted line shows intermediate, unstable ice shelf configuration (modified from Thomas, 1979).

marine sediments are not subject to subaerial erosion. Within this zone submarine glacial retreat zone facies are deposited first and may be capped by muds formed during eustatic sea-level rise, which generally occurs before major isostatic rebound (Boulton, Baldwin *et al.*, 1982). Later, when relative sea level falls due to isostatic uplift overtaking eustasy, progradational shoreface sequences form, either as a blanket or locally, as well as thin lags or deep scours formed by subaerial to coastal erosion of the exposed sea-floor deposits. At any one place the resultant succession will depend upon the balance between sediment supply, eustasy and isostasy: it may range from predominantly glacial products (Domack, 1983; McCabe, Bowen & Penney, 1993), through a balance (Miller, 1982), to predominantly coastal marine products (Nelson, 1981). Raised beaches and glaciomarine deltas may record old sea levels, and sediments may be modified by soil-forming and periglacial processes.

11.5.4 Stratigraphic architecture

The stratigraphic architecture, or large-scale three-dimensional stratal geometry, of continental margins bordering large ice sheets differs from that typical of other continental margins for several reasons. The continental shelf around an ice sheet is deeper than most shelves and has a reversed bathymetric profile (deepens from the shelf edge towards the coastline) due to isostatic loading (Fig. 11.14). Glacio-eustatic sea-level fluctuations, which strongly affect the architecture of low-latitude continental margins, will not therefore subaerially expose this overdeepened shelf, but will affect the distribution of grounded ice and thus of erosion and sedimentation on the shelf. Shelves

bordering ice sheets also commonly show rugged relief, including troughs and mounds formed by glacial erosion and deposition when grounded ice covered the shelf (e.g. Boulton, Thors & Jarvis, 1988). Moreover, progradation of the shelf is typically along a broad front, owing to the ice sheet terminus lying close to the shelf edge, as opposed to the point sources of most margins (Larter & Barker, 1989; Cooper, Barrett *et al.*, 1991; Hambrey, Barrett *et al.*, 1992; Larter & Cunningham, 1993).

Glaciomarine facies architecture, as affected by isostatic and eustatic relative sea-level changes, has been modelled by Boulton (1990) and compared with actual patterns of shelf and slope sedimentation formed during the last glaciation. Boulton proposes that glacier expansion to the shelf edge produces upper slope progradation, because of the large sediment supply to and over the shelf edge (Fig. 11.20). This creates a type of complex sigmoid–oblique clinoform (see Fig. 2.4). During glacial retreat the shelf edge is relatively starved of sediment and reworked by coastal currents. Sediment will collapse and slump from the upper slope to accumulate lower on the slope. Reworked and slumped units onlap the underlying strata (Fig. 11.20). In a comparable model, Larter and Barker (1989) place sequence boundaries at the floating of a previously grounded ice sheet as sea level is rising during a glacial recession.

Seismic stratigraphic studies coupled with drilling on the Antarctic and northern Norwegian continental margins support these generalizations. Prograding sequences that build the continental shelf outward are common; they have complex sigmoidal geometries and progradation is typically from a line source. Troughs and mounds lie parallel and perpendicular to the shelf edge. Aggradational sequences underlie and overlie

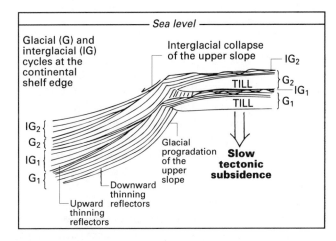

Figure 11.20 Idealized structure of a shelf edge and upper slope in which glacier expansion to the shelf edge produces upper slope progradation, and phases of glacier retreat permit collapse of the upper slope and build-up of sediments lower on the slope. Tectonic subsidence permits a vertical sequence to build up which would otherwise be removed by energetic shelf processes. Units labelled 'till' could represent till tongues (Boulton, 1990).

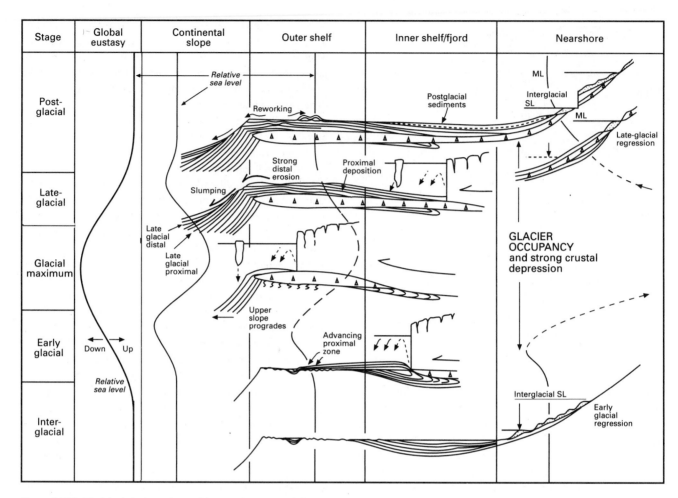

Figure 11.21 Model of glaciomarine architecture in space and time showing facies in different continental margin locations which accumulate through a whole glacial cycle. Relative sea-level changes appropriate to each zone are shown (Boulton, 1990).

the progradational ones. These may have a simple geometry and gently dipping reflections, or may be flat-lying and composed of overcompacted diamict deposited from grounded ice (Vorren, Lebesbye *et al.*, 1989; Cooper, Barrett *et al.*, 1991; Alonso, Anderson *et al.*, 1992; Hambrey, Barrett *et al.*, 1992). Note, however, that stratigraphic architecture is also affected by tectonic setting, particularly rate of subsidence, and differs between passive and active glaciated continental margins (Gipp, 1994).

Few attempts have been made to model, or reconstruct, the effects of glacio-eustasy and glacio-isostasy upon continental margin architecture (from continent across the shelf to the slope) throughout a glaciation, that is from preglacial through glacial to the postglacial stage. Figure 11.21 shows a simple model of this type. Reading and Walker (1966) attempted a large-scale reconstruction for the Late Proterozoic glacials of north Norway, but much has been learnt about glacial sedimentology and stratigraphic architecture since then. Research and modelling

of this type are needed, perhaps for the Pleistocene Laurentide ice sheet and/or for older strata where a broad view of the basin through space and time is available.

11.5.5 Ancient glacial facies associations

Glacial facies associations in the rock record are affected by tectonic setting, local relief, and glacial thermal regime as well as their environment of deposition. Cyclicity may also be visible due to repetition of advance or retreat successions. The influences of these parameters, which are commonly interrelated, are illustrated in selected examples discussed briefly below.

Distinguishing between deposition from ice sheets and valley glaciers is extremely difficult because most glacial processes are common to both. The best evidence for large ice sheets are distinctive far-travelled clasts and geographically widespread deposits.

TERRESTRIAL ASSOCIATIONS

Spectacular exposures of terrestrial glacial rocks deposited in a *low relief setting* on a *stable craton* exist within the Upper Ordovician of northwest Africa. Their documented areal extent is approximately 6–8 million km², or about half the area covered by the Quaternary Laurentide ice sheet (Biju-Duval, Deynoux & Rognon, 1981; Deynoux & Trompette, 1981b). Topographical expression of facies in plan view is excellent, allowing three-dimensional reconstructions.

Important regional facies variations are evident which broadly agree with the zonation of Pleistocene continental glacial deposits (Biju-Duval, Deynoux & Rognon, 1981).

1 In the south, massive diamictites, interpreted as tillites, are abundant, sequences are comparatively thin, and internal erosional unconformities are numerous, have low relief, and attest to successive glacial phases related to ice sheet fluctuations.

2 Farther north, there is greater relief and deposits show rapid lateral thickness changes, reaching about 200 m where they fill large palaeovalleys. Outwash sandstones and other marginal facies are abundant, and massive diamictite is relatively scarce (Fig. 11.22).

3 North of the outcrop area, boreholes indicate a gradual increase in marine strata, with thicknesses up to several hundred metres and fewer signs of glacial erosion.

Facies cyclicity demonstrates repeated episodes of glacial advance and retreat in places within the central zone. A typical succession, comprising an unconformity overlain by massive diamictite (interpreted as till), channelled sandstone (glaciofluvial), and argillaceous sandstone (glaciolacustrine or glaciomarine), is repeated three or four times to fill palaeovalleys (Fig. 11.22; Deynoux & Trompette, 1981b; Deynoux, 1983). Above these valley-fills is massive diamictite with sparse clasts (waterlain diamictite) conformably overlain by shales containing dropstones together recording the transition from glacial to postglacial marine conditions (Fig. 11.22; Deynoux, 1985). Well-sorted aeolian sandstones are present elsewhere, interbedded locally with glaciofluvial deposits, as well as excellently preserved periglacial features, including sandstone wedges and fossilized pingos (Beuf, Biju-Duval *et al.*, 1971; Biju-Duval, Deynoux & Rognon, 1981).

Another well-preserved *low relief*, terrestrial glacial succession exists within the Upper Proterozoic of western Mauritania. Lateral facies relationships are varied and quite complex (Fig. 11.23). Again the sequence is thin, reaching 50 m where it fills a wide, shallow depression (Fig. 11.23; Deynoux, 1983, 1985). Polygonal structures and sand wedges cap the younger glacial deposits (Deynoux, 1982). Overlying them is 3–5 m of baryte-bearing calcareous dolomite with sandy or shaly intercalations, interpreted to mark the beginning of the postglacial

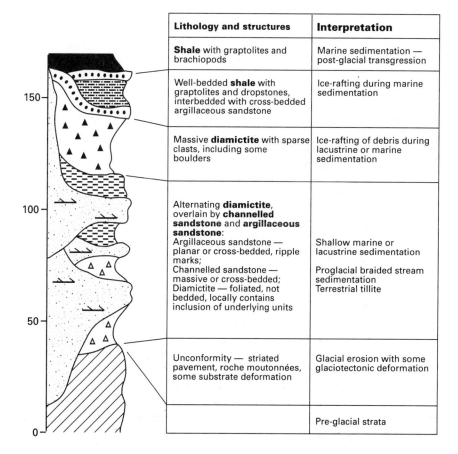

Lithology and structures	Interpretation
Shale with graptolites and brachiopods	Marine sedimentation — post-glacial transgression
Well-bedded **shale** with graptolites and dropstones, interbedded with cross-bedded argillaceous sandstone	Ice-rafting during marine sedimentation
Massive **diamictite** with sparse clasts, including some boulders	Ice-rafting of debris during lacustrine or marine sedimentation
Alternating **diamictite**, overlain by **channelled sandstone** and **argillaceous sandstone**: Argillaceous sandstone — planar or cross-bedded, ripple marks; Channelled sandstone — massive or cross-bedded; Diamictite — foliated, not bedded, locally contains inclusion of underlying units	Shallow marine or lacustrine sedimentation Proglacial braided stream sedimentation Terrestrial tillite
Unconformity — striated pavement, roche moutonnées, some substrate deformation	Glacial erosion with some glaciotectonic deformation
	Pre-glacial strata

Figure 11.22 Section showing facies cyclicity and abundance of sandstones in terrestrial glacial deposits in a low-relief setting, Upper Ordovician, southeast Mauritania (modified from Deynoux, 1985).

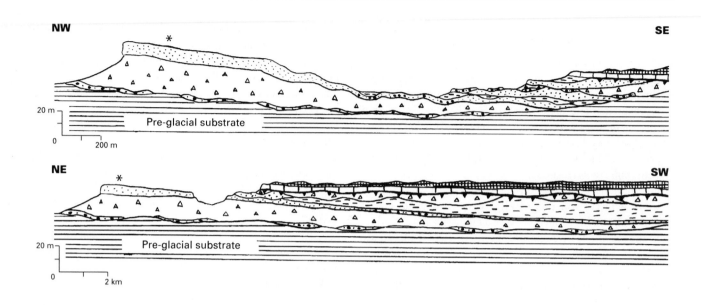

NW ⋆ **SE**

20 m

Pre-glacial substrate

0   200 m

NE ⋆ **SW**

20 m

Pre-glacial substrate

0   2 km

Explanation and interpretation of units

	Bedded chert and shale	Transgressive marine sedimentation
	Baryte-bearing calcareous dolomite and associated facies	Lacustrine, lagoonal or marine
	Polygonal structures and sandstone wedges in pebbly sandstone	Periglacial
	Sandy, calcareous dolomite	Deltaic
	Shale, locally containing fine sandstone and dropstones	Lacustrine or marine
	Sandstone, commonly cross-bedded, locally conglomeratic or glauconitic	Fluvial or deltaic; locally subglacial fluvial
	Diamictite, locally includes pebbly shale	Subglacial (terrestrial tillite)

Figure 11.23 Schematic perpendicular cross-sections showing complex lateral facies relationships in terrestrial glacial deposits in a low-relief setting, Late Proterozoic, Adrar of Mauritania. Ice movement in area was from approximately NNW to SSE. Vertical facies succession shows two glacial advances (tillites) each overlain by interglacial deposits. Note different horizontal scales. *Marks approximate intersection point of sections (modified after Deynoux, 1983, 1985).

transgression (Deynoux & Trompette, 1981a). Such carbonate caps are a common and somewhat problematic feature of many Late Proterozoic glacial successions (see below). Spectacular periglacial aeolian deposits representing ergs peripheral to outwash plains exist farther south in western Mali (Deynoux, Kocurek & Proust, 1989).

Ancient examples of terrestrial glacial deposition in *high-relief* settings are rare. Where preserved, the glacial record may be patchy, incomplete and largely resedimented (Collinson, Bevins & Clemmensen, 1989).

Wet-based glacial retreat sedimentation in a terrestrial setting is found within the Upper Proterozoic Smalfjord Formation in Finnmark, north Norway (Fig. 11.24; Edwards, 1975, 1984). At the base of a low-relief palaeovalley, massive diamictite with intraformational clasts (lodgement or melt-out till) is overlain by interbedded diamictite, sandstone and conglomerate (supraglacial and/or proglacial sediment) followed by interbedded conglomerate and sandstone (glaciofluvial deposits) indicating progressive glacial retreat (Fig. 11.24). The upper 75 m of the palaeovalley fill comprises largely massive sandstones with some conglomerate, diamictite and shale, deposited by

sediment gravity flows into a lake or fjord after ice left the area (Edwards, 1984).

Repeated glacial advance and retreat near the non-marine terminus of a *wet-based glacier* has caused marked *facies cyclicity* in Permo-Carboniferous strata in the Transantarctic Mountains (Miller, 1989). Sequences differ, depending upon whether advance was subaerial or into standing water (Fig. 11.25). Subaerial ice advance is evidenced by massive diamictite overlying a sharp basal contact with erosional relief and/or glacial grooves and striations. Subaerial ice retreat is recorded by: (i) massive or sheared diamictite which may contain tilted sandstone lenses (lodgement till); overlain by (ii) diamictite containing conformable, discontinuous stringers of sorted sandstone and conglomerate (melt-out till); followed by (iii) massive diamictite interbedded with planar cross-bedded or massive sandstone, with usually planar but occasionally erosive bed contacts (proglacial resedimented till interbedded with glaciofluvial outwash). Glacial advance under subaqueous conditions places diamictite (commonly clast-poor) gradationally above massive or laminated shale which may contain dropstones. Glacial retreat leads to a gradational contact between usually massive

Observations

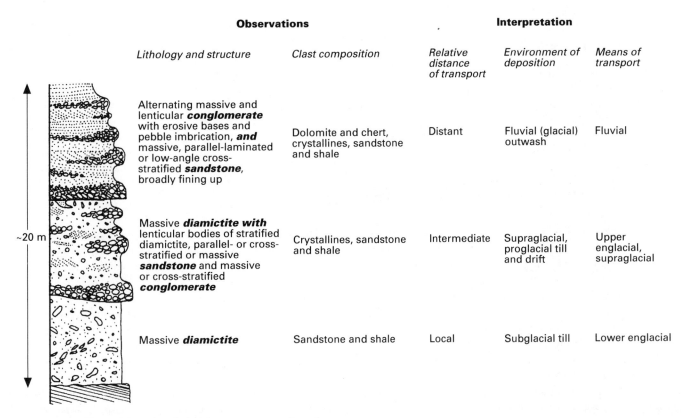

Lithology and structure	Clast composition	Relative distance of transport	Environment of deposition	Means of transport
Alternating massive and lenticular **conglomerate** with erosive bases and pebble imbrication, **and** massive, parallel-laminated or low-angle cross-stratified **sandstone**, broadly fining up	Dolomite and chert, crystallines, sandstone and shale	Distant	Fluvial (glacial) outwash	Fluvial
Massive **diamictite with** lenticular bodies of stratified diamictite, parallel- or cross-stratified or massive **sandstone** and massive or cross-stratified **conglomerate**	Crystallines, sandstone and shale	Intermediate	Supraglacial, proglacial till and drift	Upper englacial, supraglacial
Massive **diamictite**	Sandstone and shale	Local	Subglacial till	Lower englacial

Interpretation (spanning the Relative distance of transport, Environment of deposition, Means of transport columns)

~20 m

Figure 11.24 Terrestrial wet-based glacial retreat sequence from the late Proterozoic Smalfjord Formation, north Norway (after Edwards, 1975).

diamictite, which may contain sandstone lenses, and overlying shale with sparse clasts and some thin rippled sandstone lenses and beds (Fig. 11.25). In much of this area the succession is remarkably complete and three to six advance–retreat cycles exist in sections 100–200 m thick (Fig. 11.25; Miller, 1989).

In the subsurface in Oman, Upper Palaeozoic *glacio-lacustrine* shales with varve-like laminations and dropstones are associated with diamictites, sandstones and conglomerates, in an oil-bearing succession which has been thoroughly examined using both cores and wireline logs (Levell, Braakman & Rutten, 1988). Varvites are also documented within the Upper Palaeozoic Itararé Subgroup of the Paraná Basin, Brazil (Rocha-Campos, Ernesto & Sundaram, 1981). Rhythmite sequences 30 m or

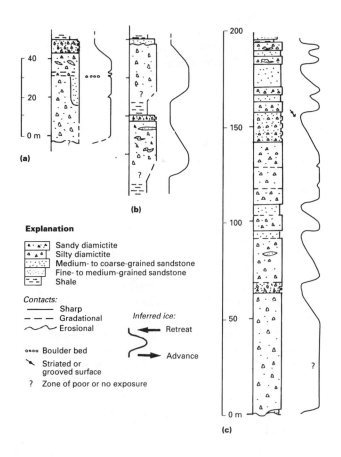

Explanation

- Sandy diamictite
- Silty diamictite
- Medium- to coarse-grained sandstone
- Fine- to medium-grained sandstone
- Shale

Contacts:
— Sharp
– – – Gradational
⌇ Erosional

Inferred ice:
← Retreat
→ Advance

∘∘∘∘ Boulder bed

⬦ Striated or grooved surface

? Zone of poor or no exposure

Figure 11.25 (a,b) Parts of measured sections through the Permo-Carboniferous Pagoda Formation, central Transantarctic Mountains, showing glacial advance followed by retreat under (a) subaerial (grounded ice) and (b) subaqueous conditions. (c) One complete section showing approximately six advance–retreat cycles, deposited in this case under dominantly subaerial conditions. Small kinks on advance–retreat interpretative curve signify minor fluctuations of ice margin and/or changes in basal ice dynamics or pauses in sedimentation (Miller, 1989).

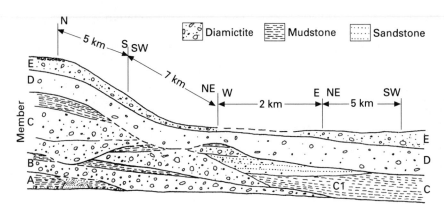

Figure 11.26 Schematic cross-section showing facies relationships during several ice sheet advances and retreats in a marginal low-relief, stable continental shelf setting, late Proterozoic Smalfjord Formation, north Norway (after Edwards, 1984).

more thick are associated with diamictite and sandstone. Annual (seasonal) control upon rhythmite deposition is indicated by concentration of pollen and spores in the coarse layers with the fine layers practically barren, and spectral analysis of thickness variation and palaeomagnetic data from three rhythmite sequences.

Deposition in a *marginal, low-relief, stable continental shelf* area where terrestrial and marine glacial conditions alternated during several advances and retreats of an ice sheet is demonstrated by the upper Smalfjord Formation, Upper Proterozoic, north Norway. Five lithologically distinct diamictite units can be mapped over areas of 10s–100s km², and occur within a repetitive vertical sequence of: (i) erosion surface; (ii) generally massive diamictite (lodgement till) which may show banding formed by glacial shearing near its lower contact (Edwards, 1984, 1986); (iii) rhythmic to randomly laminated mudstone, with rare to abundant dropstones and rare intervening sandstone (shallow glaciomarine deposits; Fig. 11.26; Edwards, 1984). Deposition took place near the margin of grounded ice sheets in an area where sea water could encroach (Edwards & Føyn, 1981).

However, at one place in the upper Smalfjord Formation faintly stratified siltstone, with a high silt and low clay content, overlies diamictite (basal till), and is interpreted as indurated wind-deposited silt or '*loessite*' (Edwards, 1979).

MARINE ASSOCIATIONS

Glaciomarine deposits are typically thicker than their terrestrial counterparts and have a higher potential for preservation (N. Eyles, 1993). They are widespread in the rock record, both in space and time (J.B. Anderson, 1983; Andrews & Matsch, 1983). A marine depositional environment is established by the presence of marine fossils; in the absence of fossils it has to be inferred from the facies and facies associations. As with terrestrial deposits, tectonic setting, local relief, glacial thermal regime and climatic cyclicity influence facies associations, as well as the specific depositional environment, but in marine environments glacio-eustatic and glacio-isostatic effects may also be evident.

The effects of *relief* and *thermal regime* upon glaciomarine deposition are both clearly demonstrated in the Late Palaeozoic Dwyka Formation of the Karoo Basin, southern Africa (Visser, 1991). The tectonic setting was a foreland basin near the margin of Gondwana.

Contrasts between southern and northern facies associations are due to differences in *local relief* (Fig. 11.27). The 'shelf facies association', exposed in the south, is widespread, lithologically fairly homogeneous, and up to about 800 m thick. Massive clast-rich diamictite is commoner at the base and massive clast-poor diamictite near the top of the sections (Fig. 11.28). Stratified diamictites increase in abundance northward – that is, nearer the unstable ice front (Fig. 11.28, Elandsvlei section). Mudrock facies can be traced for distances of 400 km and have sharp contacts with adjacent diamictite units. The vertical and lateral homogeneity of the massive diamictites suggests deposition below the inner parts of disintegrating ice shelves. Stratified diamictites were probably deposited near the grounding line. Widespread mudrock units indicate interglacial periods. Most sedimentation occurred during short periods of time when rapidly rising sea level caused the ice sheet and shelf to decouple and become self-destructive (Visser, 1991).

In contrast, the 'valley facies association', exposed in the north and restricted to palaeovalleys (fjords), is up to 200 m thick and characterized by rapid facies and thickness changes. It consists predominantly of massive and stratified diamictite, rhythmite, argillite with isolated clasts ('lonestones') and mudrock, with minor sandstone and conglomerate (Fig. 11.28). Deposition was by tidewater glaciers while mudrocks were being deposited on the shelf (Fig. 11.27; Visser & Kingsley, 1982; Visser, 1991).

The influence of *thermal regime* is shown in the Dwyka Formation through an upward transition from polar to temperate climatic conditions (Visser, 1991). Eroded and glaciotectonized bedrock underlies the formation in most locations, suggesting that initially dry-based grounded ice existed and left little or no sedimentary record. Alternating diamictite and mudrock facies in the shelf association suggest fluctuations between polar and subpolar conditions. A polar regime is required by the inferred existence of ice shelves. Abundant meltwater deposits of the valley facies association and mudrocks at the top of the shelf association indicate deposition under temperate climatic conditions.

Tectonic setting played an important part in glaciomarine sedimentation of the late Miocene to Recent Yakataga Formation in the Gulf of Alaska. The formation is 5 km thick and was deposited in the collision zone between oceanic and continental crust (C.H. Eyles, Eyles & Lagoe, 1991). *Thermal regime* and *local relief* were also important: very high sedimentation rates (as much as 10 m ky^{-1}) were due to abundant meltwater from coastal temperate tidewater glaciers carrying sediment derived from rapidly uplifting coastal mountains.

The lower and middle parts of the formation comprise laminated mudstone, with clast-rich bands and sandstone

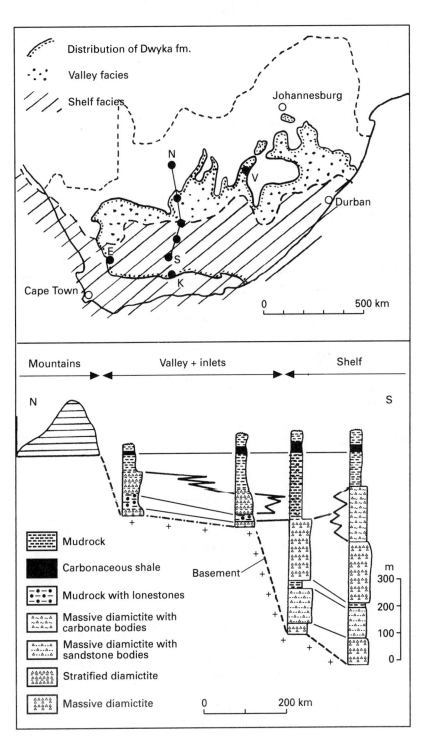

Figure 11.27 Map and section illustrating distribution of and relations between the valley and shelf facies in the Late Palaeozoic Dwyka Formation, Karoo Basin, Republic of South Africa. Note pinch-out of the lowermost shelf units against the palaeoescarpment. N–S, location of section; E, Elandsvlei; K, Klaarstroom; V, Virginia sections shown in Fig. 11.28 (modified after Visser, 1991).

Valley facies   **Shelf facies**

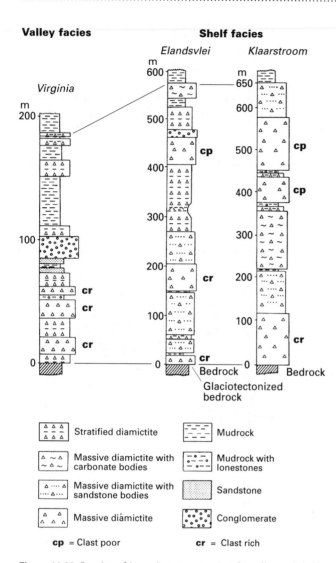

Figure 11.28 Stratigraphic sections representing the valley and shelf diamictite facies of the Dwyka Formation. See Fig. 11.27 for locations of sections. Note different vertical scale for each section (Visser, 1991).

and diamictite interbeds, overlain by sandstone turbidites, massive mudstones, swaley cross-stratified sandstones, and a thick section of broadly lenticular diamictites. Deposition was by turbidity currents and debris flows in upper slope environments followed by downslope collapse of inner neritic glaciomarine and marine facies into deep water. This succession shows how difficult it can be to recognize a glacial imprint upon sedimentation in open ocean environments, particularly in a region of active tectonism. The glacial influence here is inferred from an increase in arctic benthic Foraminifera and the arrival of large volumes of heterogeneous sediment due to initiation of glaciation around the Gulf (C.H. Eyles, Eyles & Lagoe, 1991).

The upper part of the Yakataga Formation is a 1.25-km-thick section exposed on Middleton Island in the Gulf of Alaska

(Fig. 11.29). Gravels near the base were deposited by channelized sediment gravity flows which cut into the outer continental shelf and upper slope. Some striated, faceted, and occasionally outsize clasts indicate a glacial sediment source and background ice-rafting (C.H. Eyles, 1987; C.H. Eyles & Lagoe, 1990). In the remainder of the section (Fig. 11.29), massive locally fossiliferous diamictons in extensive, sheet-like beds up to 100 m thick are consistent with deposition by suspension settling and ice-rafting on a low-relief outer shelf. Some stratified diamictons were deposited by sediment gravity flows, possibly triggered by earthquake shocks due to the active tectonic setting. Interbedded striated boulder pavements formed when a grounded marine ice sheet abraded a submarine boulder lag surface during episodic ice advances to the shelf edge (C.H. Eyles, 1988b; C.H. Eyles & Lagoe, 1990). Coquinas (shell beds) record sediment-starved, ice-free conditions during low relative sea level. Bioturbated muds record quiet-water deposition during interglacial conditions with higher relative sea level (C.H. Eyles, C.H. Eyles & Lagoe, 1991). The section therefore records temperate, glacially influenced sedimentation on a low-relief outer shelf with fluctuations of sea level and the ice margin.

Many Upper Proterozoic glaciomarine successions were deposited in active rifts and thus demonstrate the effect of a different *tectonic setting*. Sections commonly show abrupt lateral thickness changes due to contemporaneous faulting, sometimes coupled with glacial erosion. For example, glaciogenic strata in parts of the Adelaide 'geosyncline', South Australia show a change in thickness from less than 200 to 5000 m over about 16 km (Young & Gostin, 1989). Many of the diamictites are crudely stratified and show evidence for resedimentation, another feature typical of glaciomarine deposits in active rifts which complicates recognition of a glacial influence. This succession contains two diamictite-dominated to laminated mudstone-dominated cycles interpreted as deposited during two glacial advance–retreat cycles. A temperate climatic regime is inferred because meltwater deposits are abundant (Young & Gostin, 1991).

Cyclicity on different scales is recognized using a sequence stratigraphic approach in Upper Proterozoic marine to non-marine intracratonic glacially related strata of West Africa (Proust & Deynoux, 1994). Facies deposited in a range of continental to marine environments form progradational, continental and transgressive wedges and are grouped into depositional genetic units bounded by regionally correlative, maximum flooding or erosional bounding surfaces. Cyclicity of depositional genetic units on the smallest scale is related to glacial advance and retreat rhythms with a duration of about 0.1 My. Stacked depositional genetic units comprise cycles which may represent glacial to interglacial stages, and larger-scale cycles reflect either glacial epochs or tectonic processes.

Glacio-eustatic and glacio-isostatic influences have been emphasized in an Upper Proterozoic glaciogenic succession in Central Australia, where the facies associations and sequence

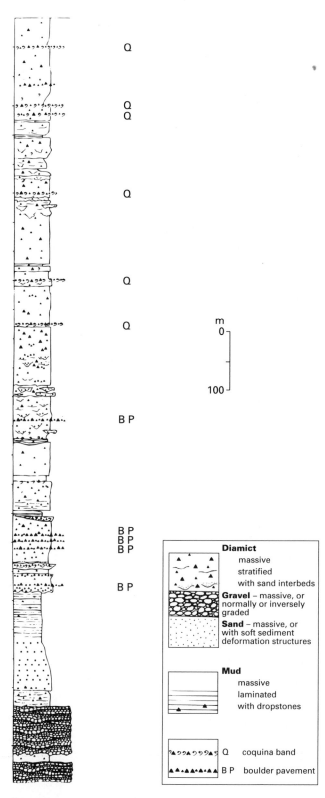

Figure 11.29 Sedimentological log through facies of temperate, glacially influenced sedimentation on a low-relief outer shelf, Yakataga Formation, Middleton Island, Gulf of Alaska (after C.H. Eyles, 1987).

boundaries follow a pattern determined by basin growth and initial fall and subsequent rise of sea level as an ice sheet grew and declined (Lindsay, 1989). Erosion followed by subglacial deposition occurred during lowered sea level. Ice proximal then ice distal and marine sedimentation occurred during glacio-eustatic sea-level rise. Erosion followed due to postglacial isostatic rebound. Diamictites in a younger formation are probably reworked sediment from the underlying formation and belong to a lowstand wedge or fan in the overlying depositional sequence, as opposed to recording a second glaciation.

In many Upper Proterozoic, mostly glaciomarine, successions a *carbonate cap* abruptly overlies glacial diamictite. The limestone or dolomite may be massive or laminated, and may show algal or stromatolitic laminae (Williams, 1979; Fairchild & Hambrey, 1984). If the carbonates record warm conditions, then abrupt climatic changes are implied (Williams, 1979; Fairchild, 1993). However, cold-water carbonates are known to be associated with some glaciomarine facies (Rao, 1981; Sect. 11.4.6). Alternative explanations for this carbonate–diamictite juxtaposition include: (i) carbonate deposition occurred during periods of sea-level rise and deglaciation when terrigenous input was reduced (Tucker, 1986); or (ii) associated carbonate facies may form where detrital carbonate debris exists within the associated glacial deposits (Fairchild, 1993).

11.6 Ice ages in Earth history

Glacial facies are concentrated in certain geological periods, reflecting episodes of cold climates or ice ages during Earth history (Frakes, 1979; Frakes, Francis & Syktus, 1992). The oldest probable glacial rocks occur within the approximately 2.9 Ga Witwatersrand Supergroup of South Africa (Von Brunn & Gold, 1993). Early Proterozoic glacial deposits occur in North America (including the Huronian Supergroup, about 2.3 Ga), southern Africa, Australia, India, and Finland. Late Proterozoic glacial facies are geographically widespread and have been found on all continents, possibly including Antarctica. Glaciation may have peaked during four glacial periods between 900 and 600 Ma (Harland, 1983). Late Ordovician to Early Silurian glacial facies exist in North Africa and South America, and Late Devonian glacials in South America (Caputo & Crowell, 1985), whereas Permo-Carboniferous glacial deposits are well represented on all Gondwanan continents. Cenozoic glacial facies are widespread on all northern continents, on and around Antarctica, and in the Andes (Hambrey & Harland, 1981).

Causes of ice ages include both terrestrial and extraterrestrial factors (Young, 1991). The position of continents and oceans with respect to air–ocean circulation is commonly important (Crowell, 1978). Variations in solar insolation, caused by Earth's orbital variations (the Milankovitch effect; see Sect. 2.1.5), are the most likely driving force for short-term climatic change (i.e. <100 000 years), and for glacial–interglacial cycles (Boulton,

1987b). Particular problems surround the Late Proterozoic glaciation because the deposits are so widespread, some appear to have been deposited at low palaeolatitudes, and a number are closely associated with carbonate facies (Sect. 11.5.5).

Glacial sedimentology has become more sophisticated and accurate in recent years. Frontier areas include further application of sequence stratigraphic concepts and improved recognition and understanding of the significance of discontinuities within glacial successions. Accurate recognition, understanding and interpretation of glacial facies are paramount to elucidating Earth's climatic history.

Further reading

Anderson J.B. & Ashley G.M. (Eds) (1991) *Glacial Marine Sedimentation: Paleoclimatic Significance. Spec. Pap. geol. Soc. Am.* **261**, 232 pp. Boulder, CO.

Ashley G.M., Shaw J. & Smith N.D. (1985) *Glacial Sedimentary Environments. 246 pp. Soc. econ. Paleont. Mineral., Tulsa, Short Course* **16**,

Dowdeswell J.A. & Scourse J.D. (1990) *Glacimarine Environments: Processes and Sediments*, 420 pp. *Spec. Publ. geol. Soc. Lond.* **53**, Bath.

Drewry D. (1986) *Glacial Geological Processes*, 276 pp. Edward Arnold, London.

Hambrey M.J. (1994) *Glacial Environments*, 296 pp. UCL Press, London.

Hambrey M.J. & Harland W.B. (Eds) (1981) *Earth's Pre-Pleistocene Glacial Record*, 1004 pp. Cambridge University Press, Cambridge.

Menzies J. (Ed.) (1995) Modern Glacial Environments: Processes, Dynamics and Sediments. *Glacial Environments*: Vol. 1, 621 pp. Butterworth-Heineman, Oxford.

Menzies J. (Ed.) (1996) Past Glacial Environments: Sediments, Forms and Techniques. *Glacial Environments*: Vol. 2, 598 pp. Butterworth-Heineman, Oxford.

Sugden D.E. & John B.S. (1988) *Glaciers and Landscape: A Geomorphological Approach*, 376 pp. Edward Arnold, London.

Volcanic environments

G.J. Orton

12

12.1 Introduction

Volcanism is the manifestation at the surface of a planet of internal thermal processes that emit solid, liquid or gaseous products. A *volcano* is the site at which this material reaches the surface of the planet, its shape and form depending largely on the mechanism and rate of extrusion. Products of volcanism vary from coherent lavas and intrusions to fragmental volcanic rocks, termed tephra or volcaniclastic (Fisher, 1961).

At the present day, volcanoes only comprise a tiny portion of the Earth's surface, with most volcanism restricted to within a few hundred kilometres of plate margins. At individual volcanic centres eruptions represent less than 1% of the history of the volcano, with intervening repose periods usually spanning 1000s–10 000s years. The geological significance of volcanism, however, is far greater than suggested by the localized distribution of volcanoes and/or their long periods of quiesence. Whereas in non-volcanic siliciclastic terrains sediment supply is limited by rates of weathering and erosion in the source area, volcanism can provide not only a unique and locally abundant source of sediment to adjacent basins but virtually instantaneous volumes of sediment several orders of magnitude greater. It has been estimated that volcaniclastic rocks comprise 27% of post-Archaean sedimentary rocks (Fisher & Schmincke, 1984, table 1.6).

Effects of volcanic eruptions can be far reaching. Mass flows can affect regions far beyond the immediate vicinity of the volcano, extending for hundreds of kilometres. Particles produced by explosive expansion of a gas phase can be transported somewhat independently of slope or base level; for instance, airborne clasts can overtop topographic barriers and deposit a layer simultaneously in different basins and environments. Fine-grained volcanic silt layers transported by wind represent the most precise geological material for stratigraphic correlation on a regional, even global scale, allowing correlation between marine, terrestrial and glacial environments. Finally, because of high rates of sedimentation, general absence of bioturbation and the large range of clast sizes and types available, physical processes are often well displayed in volcanic successions. Thus the study of volcaniclastics can illuminate clastic sedimentation in general (e.g. Sorby, 1859, 1908).

Volcanic terrains have the most complex surface environments on Earth and ancient volcaniclastics are notoriously difficult to interpret. Some of the reasons are: (i) the wide range of grain sizes (from 1 km blocks to fine dust), vesicularity, and compositions (from basalt to rhyolite) that can be produced during a single eruption; (ii) a spectrum of highly contrasting eruptive and transport systems, many of which overlap and/ or grade into each other; (iii) the great variability in the hydrodynamic and aerodynamic properties of particles produced; (iv) the susceptibility of volcanic glasses to alteration and diagenesis; (v) the episodic and catastrophic nature of sediment supply, with the facies preserved in the sedimentary record also being determined by ambient conditions; (vi) the build-up of volcanic landforms transforms topography, determines patterns of sediment dispersal, isolates depositional basins and influences local climates; and (vii) the rapid volcano-tectonic movements that cause complex patterns of uplift and subsidence and rapidly changing lake or relative sea levels.

12.1.1   Development of concepts

Scientific research in volcaniclastic sedimentation has advanced in surges, partly as new techniques developed, but mostly as interest was stimulated by catastrophic volcanic eruptions (Fig. 12.1).*

Volcanic processes occurring during eruptions have attracted more attention than the sedimentary processes that occur within repose periods. The eruption of Krakatau in 1883 brought into focus the relation of calderas to explosive volcanic activity (Fouqué, 1879; Veerbeek, 1885) and it was not long before an analogous ancient caldera had been described (Glen Coe caldera of Scotland: Clough, Maufe & Bailey, 1909). Although Wolf (1878) had earlier described pyroclastic flows of Cotapaxi volcano (Ecuador), the significance of the process was not fully appreciated until the devastating eruptions of Mt Pelée (Martinique) and La Soufrière (St Vincent), described by Anderson and Flett (1903) and Lacroix (1904). Lacroix introduced the term *nuée ardentes*, literally meaning glowing clouds, to describe the transport process observed on Mt Pelée but also recognized that a denser underflow was hidden beneath the voluminous but dilute, turbulent cloud. Although deposits from ancient *nuée ardentes* were soon recognized (Dakyns & Greenly, 1905), it was not until the 1960s, that detailed study of the transport and deposition of pyroclastic flows took place. Up until then, and even today, study of volcanic products had been dominated by igneous geochemists and it was a tremendous advance for volcanologists to see volcanic rocks as sediments, and describe them in terms of physical processes. Influential early papers, including those by R.L. Smith (1960a,b), Fisher (1966a), Fisher and Waters (1970), G.P.L. Walker (1973a), and Sparks, Self and Walker (1973), were followed in the 1970s and 1980s by a plethora of publications, each describing a new sedimentary structure, assigning it a new name, and attributing it to a new volcanic phenomena. This created much confusion, even among those familiar with the deposits. It is only recently that some consensus has been reached regarding the significance of various facies and facies relations. This is partly due to systematic three-dimensional analysis of well-exposed deposits, partly because

careful study of volcanic particles has allowed unambiguous origins to be assigned to many clast types, and partly because 'volcanologists' have increased their understanding of the fluid dynamics of particulate dispersions and sedimentary phenomena such as high-density turbidites.

Sedimentary processes in volcanic terrains have generally received less attention. The volumetric importance of sedimentary deposits around volcanoes has been recognized at least since the development of plate tectonics, when ancient volcanic arcs were studied to determine their tectonic setting (e.g. Dickinson, 1968, 1974b; Mitchell, 1970; Mitchell & Reading, 1971; references in W.R. Hamilton, 1988). Yet it is only in the last decade that volcaniclastics have received significant attention from sedimentologists (see Further reading). This is due both to the maturing of other fields of sedimentology, and to events such as the 1985 eruption at Nevada del Ruiz (Columbia) where 20 000 people died in a lahar (volcanic debris flow) resulting from an otherwise insignificant eruption.

The eruption of Mount St Helens is significant in terms of volcaniclastic research for at least two reasons: (i) it renewed interest in laterally directed volcanic explosions and their deposits. Although such explosions had been suspected at least since the eruption of Mt Pelée, and had been documented in other volcanic eruptions (Diller, 1916; Gorshkov, 1959), the view that all pyroclastic flows were produced by collapse of vertically directed eruption columns dominated thinking through the 1960s and 1970s; (ii) it refocused attention on the significance of volcanic landslides and resultant debris avalanches with subsequent discovery of numerous analogous deposits elsewhere (cf. Francis & Wells, 1988). Although debris avalanches were first described following the 1888 eruption of Bandai-san (Sekiya & Kikuji, 1889) and were later found to be associated with volcanoes in Indonesia (Neuman van Padang, 1929), up until 1980, at least in the English literature, large volcanoes were generally regarded as stable constructional features.

Historic eruptions, however, do not yield as much information on physical processes as one might hope. Most of the world's volcanoes are in developing countries which have only rudimentary geological and geophysical survey programmes. More often than not, the region of the volcanic eruption, transport and sedimentation is hidden from view by an enveloping ash cloud. Eruptions are hazardous, often lethal, to observe close up. Initial eruptive products are usually buried by later ones and may, or may not be exposed by later erosion. Products from the Vesuvius, Tambora, Krakatau, Mt Pelée and Katmai eruptions (Fig. 12.1) were not studied in detail until the early 1980s by which time regrowth of vegetation and/or erosion had obscured many facies relations. Moreover, it must be stressed that all observed eruptions are quite small (Fig. 12.2), with the volume of ejecta erupted at Mount St Helens being four orders of magnitude less than the 3000 km³ produced by the largest known explosive eruption 27.8 Ma (Fish Canyon Tuff, USA). Observed eruptions also do not represent many of

---

* These eruptions are sometimes cited elsewhere in the text without supporting references given in this figure.

Year	Volcano	Volume of magma erupted (km^3)	Type of magma	Syn-eruptive processes* and the volume of their deposits (km^3)	Ref.
BC 1450	Santorini (Greece)	c. 29	rhyolite	fall (25); flow & surge (29); caldera	1
AD 79	Vesuvius (Italy)	3–4	phonolite	fall (6), surge (4)	2
AD 186	Taupo (New Zealand)	c. 35	rhyolite	fall (35); flow & surge (30); caldera	3
1815	Tambora (Indonesia)	50	trachy-andesite	fall (5), flow and surge (65); co-ignimbrite fallout (90); caldera	4
1883	Krakatau (Indonesia)	9–10	dacite	fall (1); flow & co-ignimbrite fallout (18–21); caldera	5
1902	Pelée (Martinique)	<0.2	andesite (basalt-dacite)	block-and-ash flows	6
1912	Katmai (Novarupta) (USA)	13	rhyolite to andesite	fall (17), flow (11)	7
1963–	Surtsey (Iceland)	c. 0.5	basalt	fall, surge, lahars	8
1965	Taal (Philippines)	0.09	basalt	surge, fall	9
1980–	Mt. St Helens (USA)	0.18	dacite	lateral blast (0.2), debris avalanche (2.5), lahars (>0.1), ash flows (0.2), lava dome (0.07)	10
1985	Nevado del Ruiz (Columbia)	0.002	andesite	flows and surges (0.004), lahars (0.09)	11
1991	Unzen (Japan)	0.06	dacite	lava flows (0.03), block-and-ash flows (<0.05)	12
1991	Pinatubo (Philippines)	c. 10	dacite	fall (3); flows (6); lahars (>4)	13
Ongoing	Hawaii (USA)	<1	basalt	lava flows	14
Ongoing	Stromboli (Italy)	<1	basalt	scoria fall & lava flows	15

Figure 12.1 Selective volcanic eruptions in the last 4000 years. A few recent sources of additional data are: (1) Druitt, Mellors *et al.*, 1989. Druitt and Francaviglia, 1992; (2) Sigurdsson, Carey *et al.*, 1985. Barberi, Cioni *et al.*, 1989; (3) C.J.N. Wilson, 1985; (4) Sigurdsson and Carey, 1989; (5) Simkin and Fiske, 1983; Francis and Self, 1983; Mandeville, Carey *et al.*, 1994; (6) Fisher and Heiken, 1982; Lajoie, Boudon and Bourdier, 1989; (7) Fierstein and Hildreth, 1992; (8) Kokelaar, 1983; (9) Moore, Nakamura and Alcarez, 1966; Waters and Fisher, 1971; (10) Lipman and Mullineaux, 1981 (and many others); (11) Pierson, Janda *et al.*, 1990; (12) Yamamoto, Takarada and Suto, 1993; Nakada, Miyake *et al.*, 1995; (13) Pierson, Janda *et al.*, 1992; Torres, Self and Martinez, 1995; (14) Decker, 1987; Rowland and Walker, 1990; and (15) Giberti, Jaupart and Sartoris, 1992.

*Processes listed in order of occurrence. Fall, surge and flow indicate pyroclastic fall, pyroclastic surge and pyroclastic flow(s) respectively.

the eruption types described from the stratigraphic record. And no deep marine eruptions have been observed directly.

Analysis of ancient volcanic facies and facies relations has therefore been equally important in advancing our understanding of volcanic and sedimentary processes. Unfortunately, the study of ancient volcaniclastics has long been cursed by the problem of alteration, with the texture and composition of volcaniclastic particles being increasingly difficult to discern as deposits increase in age. The availability of petrographic microscopes has been essential in distinguishing ancient deposits. Marshall (1935) was the first to realize that widespread lava-like rocks (welded tuffs) were actually composed of glass shards deposited at high temperatures, and the discovery of such rocks radically altered interpretation of many ancient volcanic fields (e.g. Rast, 1958). Conversely, it is now recognized that many supposed pyroclastic rocks within metamorphosed volcanic successions are in fact altered silicic lavas (R.L. Allen, 1988). An additional problem in studying ancient terrains is that the depositional environment may be ambiguous, and may even change in conjunction with large eruptions. Study of both volcanic facies *and* the bounding sedimentary rocks is usually needed before credible interpretations can be made of either.

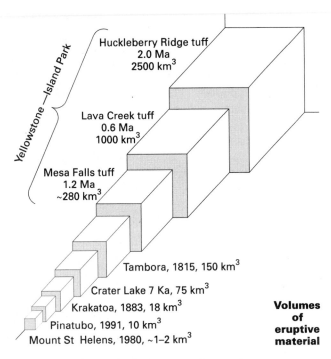

Figure 12.2 Volumes of Yellowstone's explosive volcanic eruptions compared with other well-known volcanic eruptions of the world (from R.B. Smith & Braile, 1994).

12.2 Distribution and products of volcanism

12.2.1 Relation to global tectonic processes

About 90% of volcanism takes place along the boundaries of lithospheric plates (Fig. 12.3), with four distinct tectonic settings recognized:

1 *Divergent plate boundaries*, including mid-ocean ridges where oceanic crust is formed, are volumetrically the most important sites of magma generation (Fig. 12.4). Volcanism is almost entirely submarine, except at ridge-centred hotspots (e.g. Iceland) and during the earliest stages of continental rifting (cf. R. White & McKenzie, 1989). In most cases, however, water depths generally exceed the partial pressures of magmatic gases (Sect. 12.4), such that most of the extruded products are non-explosive and volcanism has little direct impact on the sedimentary record or atmosphere. Oceanic crust also has a low preservation potential because of subduction.

2 *Convergent margins* are more important from the point of view of the effects of volcanism on sedimentology. They include about 80% of the Earth's subaerial volcanoes and the resultant volcanic arcs are typically 200–300 km wide and stretch for thousands of kilometres. The position of the volcanic arc with respect to the plate boundary, and its behaviour, is controlled by at least four independent variables (Cross & Pilger, 1982; Uyeda, 1982; Jarrard, 1986; Kanamori, 1986): (i) the relative

plate convergence rate; (ii) the direction and rate of absolute upper plate motion; (iii) the age of the subducting plate; and (iv) the nature of the subducting plate, whether it is composed of aseismic ridges, oceanic plateaux or intraplate islands/seamount chains. Lower subduction angles result from combinations of rapid absolute upper plate motion towards the trench, rapid plate convergence, and subduction of low-density or young oceanic lithosphere. Consequences are a wide arc–trench gap, perhaps 600 km or more, a compressional tectonic regime across and behind the arc, and 'strongly coupled' subduction, with many great earthquakes. In contrast, steeper subduction results from slow or retrograde absolute upper plate motion, slow relative rates of plate convergence and subduction of old dense lithosphere. The volcanic arc develops close to the trench (as little as 100 km or less), extensional tectonics is induced within and behind the arc, and the subduction zone is 'weakly coupled' with less seismicity.

3 *Oceanic intraplate* volcanism is best illustrated by the Hawaiian islands. They lie at the end of a segmented 2000-km-long chain of extinct volcanic islands and seamounts that have been explained in terms of motion of the Pacific plate over a 'fixed' *hotspot* within the upper mantle (J.T. Wilson, 1963).

Additional topographic highs within ocean basins include oceanic plateaux and aseismic volcanic ridges (Fig. 12.3). Oceanic plateaux reach several hundred square kilometres in size, are elevated more than 1000 m above the surrounding sea floor, have thick sedimentary covers, and today cover about 10% of the ocean floor. The majority are regions of thickened oceanic crust formed at triple junctions, along fracture zones, or at hotspots. Aseismic ridges rise up to 4000 m above the surrounding ocean floor, and reach up to 400 km in width and 5000 km in length. They comprise 25% of the ocean floor and appear to represent subsided chains of volcanic islands. Both of the above features, by virtue of their increased thickness, are comparatively buoyant and either tend to resist subduction, becoming accreted on to the overriding plate margin, or if they are subducted, may lower the subduction angle and cause volcanism to cease.

4 *Continental intraplate* volcanism is the least common but most varied type of volcanism. Most is clearly associated with extensional tectonics and rifting (e.g. East African Rift, Rhine Graben, Basin and Range province). However, it is usually very difficult to decide whether asthenospheric upwelling causes uplift of the lithosphere and rift formation or the rift forms first, followed by magmatic activity and uplift of its flanks (cf. Keen, 1985). There is probably a complete spectrum between these two limiting cases. Although volumetrically insignificant, volcanism within continental rifts is understood better than oceanic volcanism as its products have an extremely high chance of being preserved in the geological record. Continental volcanism can also involve hotspots as discussed below.

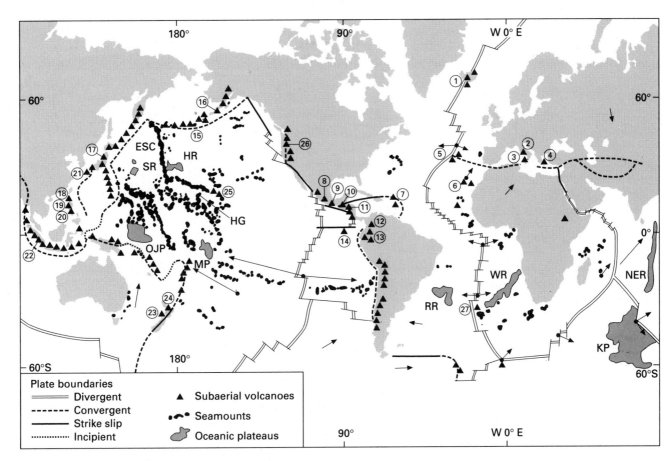

Figure 12.3 Distribution of volcanism with respect to the boundaries of major lithospheric plates. The length of the spreading and closing rate vectors is proportional to the rate. Names of active subaerial volcanoes are: 1, Surtsey; 2, Vesuvius; 3, Stromboli; 4, Santorini; 5, Azores; 6, Canary Islands; 7, Mt. Pelée; 8, Paricutin; 9, Santa María; 10, El Chichóna; 11, Fuego; 12, Nevado del Ruiz; 13, Cotopaxi; 14, Galapagos Islands; 15, Aleutian Islands; 16, Katmai; 17, Fuji; 18, Taal; 19, Mayon; 20, Pinatubo; 21, Unzen; 22, Krakatau; 23, Ruapehue; 24, Taupo; 25, Hawaii; 26, Mount St Helens; 27, Tristan da Cunha. Names of seamounts, guyots, oceanic plateaus, ridges and rises are: ESC, Emperor Seamount Chain; HG, Horizon Guyot; HR, Hess Rise: KP, Kerguelen Plateau; MP, Manihiki Plateau; NER, Ninety-East Ridge; OJP, Ontong Java Plateau; RR, Rio Grande Rise; SR, Shatsky Rise; WR, Walvis Rise.

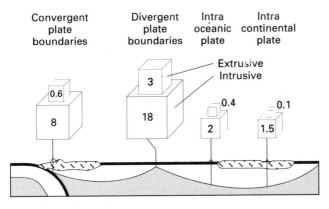

Figure 12.4 Cenozoic rates of magma production (in km³ year⁻¹) at various tectonic settings (from Fisher & Schmincke, 1984).

12.2.2 Types of magma and their origin

Volcanic rocks may be divided into two major magma series, alkalic and subalkalic, each containing rocks ranging from ultrabasic (mafic) to acid (silicic or felsic) in composition (Fig. 12.5). The subalkalic field can be further subdivided, on the basis of K_2O versus SiO_2 diagram into a *tholeiitic* and a *calc-alkaline* magmatic series (Fig. 12.6).

Nearly all magmatism is fundamentally basaltic in the sense that partial melting (between 5 and 30%) of mantle material produces primary mafic magmas, and that subsequent processes, including fractional crystallization, diffusive processes within the liquid phase, magma mixing, and varying degress of crustal assimilation generate a range of more differentiated (more SiO_2-rich) compositions as the primary magmas progress upwards from their depth (typically 50–100 km) of origin (Fig. 12.7). The extent to which primary magma is modified during its upwards ascent depends on

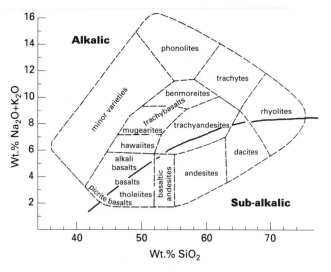

Figure 12.5 Nomenclature of normal (non-potassic) volcanic rocks (devised by Cox, Bell & Pankhurst, 1979).

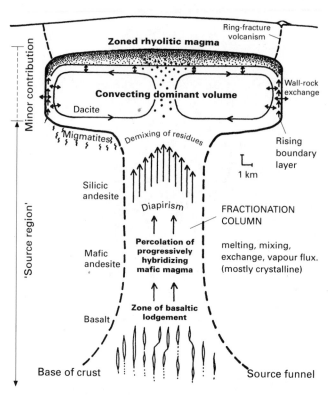

Figure 12.7 Model of a large mature magmatic system beneath thick continental crust (from Hildreth, 1981). Note that the 'source' of the magma that forms any high-level magma chamber consists of contributions from the mantle and from many levels of the crust, in unknown but variable proportions.

the rate magma is supplied from the asthenosphere and the thickness, age, composition and state of stress of the crust through which the primary magma must rise. These control the rate at which magma ascends to the Earth's surface, its residence time in the crust, and the ratio of intruded to extruded products. These parameters vary significantly between the different plate tectonic settings (see M. Wilson, 1989).

Magma generated at mid-oceanic ridges is the least evolved and simplest, consisting almost entirely of tholeiitic basalts with variation depending on the rate of sea-floor spreading. Fast-spreading ridges (e.g. East Pacific Rise) have large magma chambers in which open system fractionation combines with efficient magma mixing to generate comparatively homogeneous magma batches whereas at slow spreading ridges (e.g. Mid-

Atlantic Ridge) small, discontinuous high-level magma reservoirs undergo more extensive differentiation and erupt a wider range of basalt types.

Island-arc volcanic suites vary according to the thickness and composition of the overriding lithosphere (Fig. 12.6) (Gill,

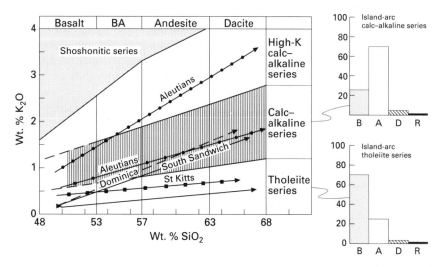

Figure 12.6 Plot of wt% K_2O versus wt% SiO_2 showing major subdivisions of island arc volcanic suites, with trend lines for individual volcanoes (or groups of volcanoes). Histograms on right depict relative amounts of basalt (B), andesite (A), dacite (D), and rhyolite (R) within tholeiite and calc-alkaline volcanic rock suites (from M. Wilson, 1989).

1981). At oceanic island arcs (e.g. the Izu-Marianas) mantle-derived magmas are relatively unimpeded in their ascent, resulting in eruption of very fluid, often aphyric, tholeiitic basalts and basaltic andesites. As the arc develops, the thickening arc massif depresses the oceanic crustal layer. Once the crust has thickened to some 20 km it may start to act as a density filter, and primary magma may become ponded in a series of interconnected high-level magma chambers. The upward ascent of magma, particularly under the centre of the arc, is a slow and fitful process and differentiation processes generate calc-alkaline and more intermediate magmas such as andesite (e.g. Aleutians). Early studies of convergent-margin magmatism, largely based on studies of the Japan arc, suggested that erupted magmas should become more alkaline away from the trench, apparently reflecting the increasing depth to the source of the magmas along the Benioff zone (Fig. 12.8) (Kuno, 1959; Dickinson, 1975). However, many arcs do not follow this simple pattern, probably due to source heterogeneity, variable depths and degrees of partial melting, and anomalous tectonic features such as fracture zones.

Along convergent continental margins, the situation is the most complex, largely due to passage of magmas through thick continental crust, with one of the most conspicuous differences being the greater abundance of more silica-rich magmas (dacites and rhyolites) relative to oceanic island arcs. Arc-transform strike-slip faults may ease the ascent of magma and control the type of magmatism, causing large variations over short distances along the volcanic arc (e.g. Skulski, Francis & Ludden, 1992).

Oceanic intraplate volcanoes erupt both tholeiitic and alkalic magmas. Large oceanic islands display a general evolutionary sequence from an early tholeiitic shield-building stage to later more alkalic phases, which often postdate prolonged periods of dormancy (e.g. Hawaii). Presumably early eruptive products represent relatively uncontaminated partial melts of the mantle plume source whereas later products reflect small degrees of partial melting of oceanic lithosphere. The more alkalic ocean island magmas have comparatively high volatile concentrations, yield hydrous minerals (e.g. amphiboles) and thus eruptions are more often explosive.

There are two end-members of intracontinental rift volcanism (Barberi, Santacroe & Varet, 1982; M. Wilson, 1989); volcanically active rifts are characterized by more voluminous magmatic activity, higher rates of crustal extension, mildly alkalic basalts, and bimodal distribution of basic and acid magma types. Those that initially form as rifts have localized centres of alkaline volcanism (e.g. the Rhine Graben), are characterized by relatively small volumes of erupted products, low rates of crustal extension, discontinuous volcanic activity, and a wide spectrum of basaltic and more differentiated magma compositions.

Large areas of continents have also been covered by vast thicknesses of laterally extensive tholeiitic basalt lava flows. These have been termed continental flood basalt provinces. The youngest example is the Columbia River Plateau, northwestern USA, which was active from 17 to 6 My ago. This province is linked geographically with more recent volcanic activity in the Snake River Plain–Yellowstone Park region, which appears to be related to the passage of the North American plate over a mantle hotspot, currently located beneath Yellowstone (R.B. Smith & Braile, 1994). Volcanism here is bimodal, with basalt magmas derived from the mantle plume either erupting, or interacting with the silicic continental crust to produce large volumes of rhyolitic partial melts that were erupted in the caldera-forming volcanism of Yellowstone. When mantle plumes rise beneath continental crust, their effects also include kimberlite eruptions (Crough, Morgan & Hargraves, 1980) and, on a larger scale, continental rifting. Hence most ancient continental flood basalt provinces can be traced towards a modern active hotspot by a linear chain of volcanic islands or seamounts (Cox, 1978).

12.3 Magmatic processes and their effect

Magma has three primary components, or phases; a silicate melt, a variable proportion of crystals, and a volatile or gas phase, of which H_2O, CO_2 and SO_2 are most abundant. Variable proportions of foreign rocks (xenoliths) and crystals (xenocrysts), incorporated from the country rock, may also be present. Each component may influence whether magma reaches the Earth's surface or forms intrusions, and the type, size and frequency of volcanic eruptions.

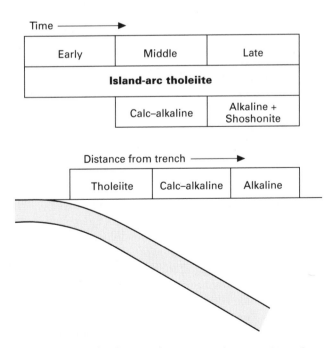

Figure 12.8 Spatial and temporal variation in the composition of island-arc magmas (from M. Wilson, 1989).

12.3.1 The physical behaviour of magma

The viscosity and volatile content of a magma are the most important controls on its eruptive behaviour.

Viscosity is affected by the temperature, chemical composition, crystal content, concentration of dissolved volatiles, and presence of bubbles. All fluids become more viscous as they get colder, although at the same temperature silicic melts are always more viscous than basaltic (or peralkaline) ones. Crystals increase viscosity, and, because magmas containing more crystals are too viscous to erupt, the crystal content of effusive volcanic rocks rarely exceeds 50% (Ewart, 1979). Volatiles have a more complex effect on viscosity, with dissolved water lowering viscosity but dissolved carbon dioxide having the opposite effect.

Volatiles comprise only a small weight per cent of most magmas, with water content ranging from <1% in basalts to about 6% in rhyolite magmas. Although volatiles are dissolved at great depths, they can exsolve to form *bubbles* or *vesicles* if the confining pressure on the magma is reduced, as during its rise to shallow crustal levels. Exsolution commences when vapour pressure of the dissolved volatiles equals the confining pressure and is initiated at greater depths in a volatile-rich magma. Bubbles may also occur when crystallization in a cooling magma concentrates volatiles in the remaining liquid (Burnham, 1979). Once formed, bubbles may grow larger by coalescence, diffusion of gas through the magma into the bubble and/or a further decrease in confining pressure. Bubbles cease growing when the liquid magma is too viscous to be forced through the intricate channels and along the thin liquid films between adjacent bubbles that are trying to expand. At this stage, gas bubbles typically comprise about 80%, by volume, of the magma (Sparks, 1978).

Growth of vesicles affects the magmatic system in at least three ways: (i) bubble growth can generate enormous pressures within a magma chamber, thereby increasing the potential for an explosive eruption; for instance, vesiculation of an andesite magma containing only 2.8 wt% water increases the volume of the system by as much as 50%; (ii) the presence of bubbles in the melt causes its viscosity to further increase; and (iii) volatiles diffuse relatively easily through magmas to accumulate towards the top of the magma chamber in a volatile-rich layer of foam whose behaviour controls eruptive style (Sect. 12.4). This foam layer becomes more stable with increasing melt viscosity (i.e. more siliceous magmas) because: (i) the liquid melt is unable to drain along the thin films between bubbles; and (ii) the rate that bubbles rise relative to the rate of magma rise, slows (Jaupart & Vergniolle, 1988; Jaupart & Allegre, 1991).

12.3.2 Ascent and storage of magma

Magma moves up through the lithosphere or a volcanic edifice due to its buoyancy. The factors controlling whether a magma erupts or intrudes, and the depth at which it intrudes to form a magma chamber are: (i) the density of the magma relative to that of the rocks composing the Earth's lithosphere; (ii) the viscosity of the magma; (iii) the strength of the lithosphere and crust; and (iv) the tectonic stress regime beneath and within the volcano (cf. Nakamura, 1977; Ryan, 1987; Takada, 1994a). Most magma does not reach the surface but remains resident in the crust to either solidify in the form of batholiths or sills, or further fractionate to generate more evolved magma that is later erupted.

Basaltic or peralkaline magmas, by virtue of their low viscosity, have considerable freedom to move through a volcanic edifice to seek a position of neutral buoyancy. Because of their fluidity, a series of interconnected reservoirs typically extends from the zone of magma generation to storage zones within a few hundred metres of the surface, such as Kilauea (Hawaii) whose primary conduit extends upwards from about 35 km depth, with the shallowest chamber occupying a 2–4 km depth interval (Ryan, 1988). Fissures and dykes are also important avenues of conveyance of low-viscosity magmas, particularly with large parental magma reservoirs (Parfitt, Wilson & Head, 1993). Although silicic magmas are less dense than mafic magmas and most crustal rocks, thereby favouring their buoyant, often diapiric ascent, their high viscosity tends to inhibit magma from reaching the Earth's surface. The tops of large silicic magma chambers typically reside at depths of 3–10 km and are believed to be less than 10 km in thickness (R.L. Smith, 1979).

Magma in most high-level chambers is stratified with respect to its density, composition, temperature, and volatile content (Fig. 12.7) (Hildreth, 1981). Gradients in these properties can be set up directly during filling or replenishment, or by various fractionation or diffusion mechanisms driven by the interaction of the cooling and crystallizing magma with the boundaries of the chamber (Sparks, Hubbert & Turner, 1984). Lighter, more volatile-rich, and more silicic magmas typically lie on top of hotter, compositionally more mafic, and more phenocryst-rich magmas. The time scale for development of compositional zonation is proportional to magma eruption volume, with large-scale and more silicic systems (magma volumes $>10^3$ km^3) evolving on time scales of 10^5–10^6 years, and small systems (magma volumes <10 km^3) on time scales of 10^2–10^4 years. Quite commonly, the magma system breaks up into a series of independently convecting layers, bounded by sharp diffusive interfaces. Wide compositional gaps, up to about 15% SiO_2, may exist between layers. This phenomena may form when a layer of hot dense magma underlies a layer of cooler less dense magma, as would occur when a new pulse of primitive magma is injected into the base of a chamber. Heat is transferred between the layers faster than chemical components, thereby driving convection in both layers and maintaining a sharp interface between them (Turner & Campbell, 1986).

As the volatile-rich uppermost layers of a magma chamber are the first to erupt, the early stages of an eruption are often more violently explosive than later stages. During an eruption, it is also common for bulk SiO_2 to decrease, magma temperature and total phenocryst content of the magma to increase (by as much as 30%), plagioclase to increase at the expense of quartz and sanidine, the percentages of ferromagnesian minerals to increase strongly and, due to volatile gradients, the mineralogy to progress from hydrous phenocrysts (e.g. hornblende, biotite) towards anhydrous equivalents. However, magma drawdown during an eruption, particularly from small magma chambers, may not be orderly (Spera, 1984) with less viscous, hotter, and more mafic magma sometimes surging from deeper levels to penetrate and fountain through the silicic capping magma that is venting concurrently. The extreme case is provided by eruptions in which rhyolitic and basaltic magmas are ejected together from the same vent (e.g. Askja, Iceland) forming compositionally banded pumice.

During the lifetime of even a modest volcano (c. 10^5 years), magma must be regarded as an open system, with periodic input of fresh batches of relatively primitive magma from the underlying mantle source into the root of the system, periodic compositional and thermal losses of more evolved magmas from the top of the magma chamber during eruptions, and continuous fractionation. The composition of erupted magma will only approach a steady state in unusual situations where the composition and amount of parental magma supplied in each cycle is balanced by rate of magma crystallization and magma escape (e.g. Stromboli; Giberti, Jaupart and Sartoris, 1992). More commonly, numerous cycles of replenishment, fractionation, and eruption can lead to such extreme zonation that the products of different eruptions can appear to be derived from different magmatic sources. Ascending basaltic magmas are normally unable to pass up through silicic magmas (Hildreth, 1981), and large volcanoes commonly possess a central 'shadow zone' within which basaltic vents are rare or absent. Eruptions on the flanks of volcanoes (e.g. Etna, Ruapehu), can be particularly distinctive, being often fed by magmas which rise rapidly from the mantle with no intervening high level storage.

12.4 Eruption processes and facies

There is a spectrum between explosive or pyroclastic eruptions and effusive eruptions. Both the nature of an eruption and the type of explosivity depend on the properties of the magma and how it is stored (Sect. 12.3), the rate magma rises, and ambient conditions such as the confining pressure on the magma body and whether magma has access to an external supply of water.

Pyroclastic eruptions are produced by the generation and rapid expansion of a gas phase that disrupts magma and/or surrounding wall rock or sediment. For explosive fragmentation to occur, the gas phase must be able to expand instantaneously against the confining pressure, whether it be a subterranean lithostatic load, the hydrostatic load of a water column, subaerial atmospheric pressure, or a combination thereof.

The gas phase driving explosions may originate in one of two ways: (i) through rapid exsolution and decompression of magmatic volatiles in 'dry' *magmatic* eruptions; and (ii) through interaction between the magma and external water, with or without accompanying vesiculation of magma, in *hydrovolcanic* eruptions. As erupting volcanoes are almost impossible to study, the relative importance of the two gas sources has been assessed by detailed study of tephra deposits.

Explosive volcanic eruption commonly involves three stages: fragmentation of the magma and/or country rock (Sect. 12.4.1), eruption of the fragmented mass through a vent into the surrounding atmosphere or hydrosphere (Sect. 12.4.2), and transport and deposition of the volcanic particles (Sects 12.4.3–12.4.5), either to sites away from the vent or within depositional basins such as calderas that form as a result of the eruption (Sect. 12.4.6).

Effusive eruptions producing lavas occur if the volatile content of the erupting magma is low relative to ambient pressure. The volatile content of the magma may have been low originally, or it may be low because the magma chamber has slowly degassed or volatiles have been used in earlier explosive phases. Low viscosity basaltic magmas can easily degas and more commonly reach the surface as lavas. Silicic magmas degas through the steady loss of volatiles through fractured, permeable conduit walls (Fink, Anderson & Manley, 1992) or when explosive activity removes the vesiculated froth from the top of the magma chamber. This is because of the higher stability of silicic foams (Sect. 12.3.1). Hence, silicic lavas typically only occur late in an eruptive cycle following a large explosive eruption or in deep water where explosive eruptions are impossible.

12.4.1 Volcanic fragmentation processes and their products

Volcanic fragments are called *essential* or *juvenile* when generated from the erupting magma, *accessory* or *cognate* when broken from deposits of earlier eruptive episodes at the volcano, and *accidental* (or *foreign*) when derived in the vent from underlying basement rock of any composition, or entrained from surficial deposits during transport. The terms volcanic *dust* (<63 µm), *ash* (63 µm–2 mm), *lapilli* (2–64 mm) and *blocks* and *bombs* (>64 mm) are a set of terms erected by volcanologists to describe the size of pyroclastic deposits and can be equated with grain-size terms used for non-volcanic siliciclastics (Fig. 12.9). *Scoria* (if basaltic) and *pumice* (if more evolved) are ill-defined terms for glassy vesicular fragments in the lapilli and bomb size range.

Volcaniclastic fragments can also be named according to the mechanism of fragmentation. The term *pyroclastic fragments*, or *pyroclasts*, is used both in the general sense for any fragment

Grain size	Pyroclastic deposits	
	Unconsolidated tephra	Consolidated pyroclastic rock
<1/16 mm	Fine ash	Fine tuff
<1/16–2 mm	Coarse ash	Coarse tuff
2–64 mm	Lapilli tephra	Lapillistone (or lapilli tuff or tuff-breccia)
>64 mm	Bomb (fluidal shape) tephra Block (angular) tephra	Agglomerate (bombs present) Pyroclastic breccia

Figure 12.9 Grain-size based nomenclature for common types of volcaniclastic rocks (based on Fisher, 1961).

formed in a volcanic explosion, and in a specific sense to refer to clasts produced by explosive magmatic. *Hydroclasts* form by magma–water interactions, either explosive eruptions or non-explosive quench fragmentation. Less abundant particles include *autoclasts* formed by the mechanical friction of moving lavas or autobrecciation, and *alloclasts* formed by disruption of pre-existing volcanic rocks by subterranean igneous processes. In addition to these volcanic processes, volcaniclastic fragments may be produced by the less dynamic surface processes (i.e. mechanical and chemical weathering) that typify other sedimentary environments. These are termed *epiclasts*.

Explosive magmatic eruptions involve the exsolution and expansion of magmatic volatiles. Eruptions vary according to magma composition, especially its volatile content and viscosity, the depth of the magma chamber, and vent geometry. In general, magma that is relatively rich in volatiles or of high viscosity has the highest chances of being erupted explosively. This can be seen in subaqueous settings. Magmatic fragmentation only occurs above water depths of about 100–200 m in tholeiitic magmas (J.G. Moore & Schilling, 1973), about 1 km for more alkalic mafic compositions (Staudigel & Schmincke, 1984), and perhaps 2 km for silicic magmas (Klug & Cashman, 1992). However, pyroclasts may apparently be generated from deeper water volcanoes when hot vesiculating bombs are transported upwards in a submarine eruption cloud (Sect. 12.5.2) into shallower waters to undergo secondary fragmentation (Lackschewitz, Dehn & Wallrabe-Adams, 1994).

Basaltic to silicic magmas tend to exhibit characteristic styles (Figs 12.10 & 12.11) (G.P.L. Walker, 1973a; L. Wilson, 1980). At least four poorly defined types of eruption are

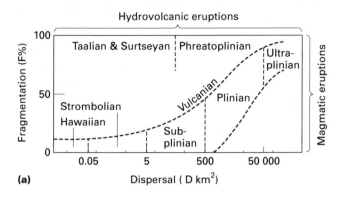

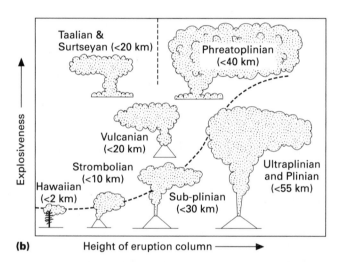

Figure 12.10 (a) Classification of explosive volcanic eruptions based on the degree of fragmentation (F) and area of dispersal (D) of pyroclastic fall deposits (largely based on G.P.L. Walker, 1973a). F% is the weight of deposit finer than 1 mm along the dispersal axis where it is crossed by the isopach line which is 10% of the maximum thickness (0.1 T_{max}). The dispersal index is the area enclosed by the 0.01 T_{max} isopach line. T_{max} is usually obtained by extrapolation. (b) Cartoon explaining F–D plot in terms of column height and explosiveness (Cas & Wright, 1987).

recognized, named after subaerial volcanoes where this type of volcanic activity was first recognized or particularly common, and widely used for subaerial eruptions and their products. *Hawaiian* activity is the least explosive and usually associated with basaltic magmas. Long periods (weeks) of quite outflow of gas-poor lava alternate with short periods (a few hours to days) when gas-rich liquid sprays high into the air, in extreme cases >1 km, as fire fountains (L. Wilson & Head, 1981). *Strombolian* activity involves more viscous (often calc-alkaline) magmas. Eruptions are largely a result of intermittent bursting of large bubbles at the top of the magma column in an open conduit, and extreme discharges can produce eruption columns more than 1 km high (Blackburn, Wilson & Sparks, 1976).

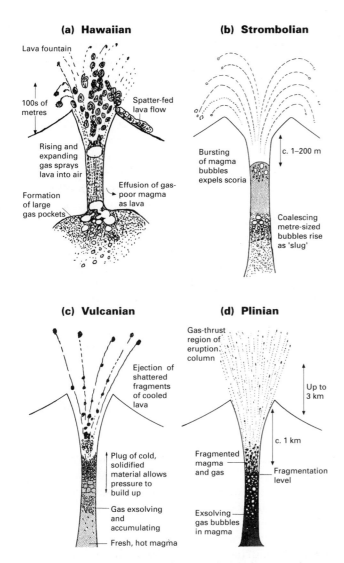

(a) Hawaiian

Lava fountain

100s of metres

Spatter-fed lava flow

Rising and expanding gas sprays lava into air

Effusion of gas-poor magma as lava

Formation of large gas pockets

(b) Strombolian

Bursting of magma bubbles expels scoria

c. 1–200 m

Coalescing metre-sized bubbles rise as 'slug'

(c) Vulcanian

Ejection of shattered fragments of cooled lava

Plug of cold, solidified material allows pressure to build up

Gas exsolving and accumulating

Fresh, hot magma

(d) Plinian

Gas-thrust region of eruption column

Up to 3 km

c. 1 km

Fragmented magma and gas

Fragmentation level

Exsolving gas bubbles in magma

Figure 12.11 Types of magmatic eruptions (largely from L. Wilson, 1980). The most crucial aspect of eruptions is the behaviour of gas and liquid phases of magma relative to each other. In Hawaiian eruptions, bubbles in the low-viscosity magmas easily coalesce and may rise upwards faster than the rate magma is rising; eruption of gas-rich magma as fire fountains alternates with effusion of gas-poor magma as lava flows. Plinian eruptions are the other end of the spectrum; bubbles and melt remain in close proximity and are expelled together when depressurization fragments the magma. Diagrams are not all at the same scale. In general the depth to the fragmentation level and height of the gas-thrust portion of the eruption column increases from (a) to (d).

Vulcanian eruptions are small volume (less than 1 km³) but explosive eruptions involving higher viscosity, commonly crystal-rich andesites. Explosions result when gas, confined beneath a plug of solidified lava, is suddenly released. The gas pressures result from exsolution of volatiles from rising fresh magma and/or partial vaporization of groundwater (Self, Wilson & Nairn, 1979). *Plinian* eruptions are the most explosive mag-

matic eruptions, producing high convecting eruption columns and involve high viscosity, mostly silicic, vesiculated magmas (G.P.L. Walker, 1981).

Products of magmatic explosions, termed pyroclasts, are pumice or scoria, glass shards, crystals, magmatic bombs and accidental lithic fragments. If a magma is fragmented at the peak of its vesiculation and erupted quickly, a uniform assemblage of juvenile clasts containing approximately 75–83% vesicles should result (Sparks, 1978; Houghton & Wilson, 1989). Vesicles are initially spherical but may be deformed during the eruption into long tubes and oblate shapes (Heiken & Wohletz, 1991). *Glass shards* are pieces of vesicle walls, and range from cuspate glass plates to hollow needles depending on the shape of the vesicles within the erupted magma. As magma viscosity and eruption rates become lower, erupted fragments increasingly include poorly vesicular material which has had time to degas partially during residence in the vent. With extremely low viscosity magmas, the shape of particles is additionally influenced by the effects of surface tension, with small fragments falling to the ground as elongate tear-drop shapes known as Pele's tears or 'achneliths', and large fragments forming spindle-shaped bombs.

With magmatic eruptions, clasts sometimes retain sufficient heat to sinter or *weld* together (R.L. Smith, 1960b). In welded zones, glass shards define a planar foliation called *eutaxitic* texture, and pumice is usually flattened to form flame-shaped structures called *fiamme*. The rate and degree of welding depends on the viscosity and yield strength of the hot glass particles (Branney & Kokelaar, 1992). Low viscosity clasts will agglutinate upon immediate impact, forming spatter cones, welded pumice falls, or again coalescing to form lavas. Particles of intermediate viscosity or temperature may agglutinate as they are transported away from the vent to form lava-like and rheomorphic ignimbrites (Sect. 12.4.4). With these intensely welded tuffs, pyroclasts sometimes retain enough magmatic volatiles to exsolve vesicles after being deposited. More viscous or cooler pyroclasts weld after deposition, with the degree of welding reflecting the length of time a particle remains above the threshold temperature *and* the lithostatic load, dependent on the deposit thickness, that was placed on the particles (Riehle, 1973). 'Warm' non-welded pyroclastic deposits, especially the upper zones of thick flow deposits, may also be cemented by *vapour-phase crystallization*. This involves growth of new low-temperature minerals such as tridymite in pore spaces, with mineral species derived from continued exsolution of magmatic gases and/or heated groundwater rising through the deposit (R.L. Smith, 1960b; Sheridan, 1970).

Explosive hydrovolcanic eruptions result from the interaction of hot volcanic materials with the sea, a lake, a glacier, when rising magma intersects a groundwater aquifer, or when lava or a hot pyroclastic flow moves over or into water or water-saturated sediment. The driving force for fragmentation is the large increase in the volume of heated water during the liquid to gas (i.e. steam) transition. At the critical pressure of water, which corresponds to a depth of 3.1 km for sea water or about 1 km for crustal rocks, there is essentially no difference between the liquid and vapour phases of water and hydrovolcanic explosions are not possible.

In addition to the formation of steam, hydrovolcanic eruptions depend on:

1 *The viscosity of the magma. Rhyolites* are less prone than basalts to interact explosively with water because viscous magma mixes less readily than more fluid magma (Wohletz, 1986).

2 *The rate of extrusion or mixing.* Even in very shallow water, explosive interaction between magma and water requires vigorous interaction between magma and water (P. Kokelaar, 1986). At Hawaii, for instance, when lavas entered the sea at very high rates, violent steam explosions produced scoria and lithic fragments (R.V. Fisher, 1968) whereas slower moving flows descended into the sea to form pillow lavas (Moore, Phillips *et al.*, 1973). It is therefore possible to produce a transition from lava-dominated to clastic-dominated successions that has nothing to do with water depth, or magma type, but merely the rate of extrusion (Carlisle, 1963).

3 *Containment of steam produced.* When rising magma comes into a contact with an aquifer, heat can transfer to surrounding water without explosive fragmentation if aquifer units are permeable enough to allow the heated water to escape rapidly (Delaney, 1982). In subaerial settings the escaping groundwater forms hot springs.

Small amounts of superheated water coming into contact with hot volcanic rock can cause *hydrothermal* or *phreatic* explosions (e.g. Old Faithful, Yellowstone; Muffler, White & Truesdell, 1971). Although common, they are not large, and little or no solid ejecta is produced. However, phreatic explosions may be important for precious metal, particularly gold, mineralization (Hedenquist & Henley, 1985), and they often precede larger magmatic eruptions. Phreatic explosions may even trigger a large eruption by weakening the overburden and/or opening the volcanic vent, thereby depressurizing the magma chamber (Sheridan, Barberi *et al.*, 1981).

When larger amounts of water interact with magma, steam can fragment and eject the magma and/or surrounding country rocks. These are termed *phreatomagmatic* explosions. When magma first comes into contact with water, steam bubbles coalesce and form a thin insulating film along the contact surface (Mills, 1984). However, magma may be explosively torn apart if vigorous oscillations of this film fragment the magma, mix it with water, and the resultant mixture boils and/or larger amounts of water or wet sediment become engulfed within magma and converted to steam (Sheridan & Wohletz, 1983; P. Kokelaar, 1986; Wohletz, 1986). For optimal release of explosive energy the interacting mixture should have a water/magma mass ratio close to 0.35 (Fig. 12.12). Such activity is best referred to as *Taalian* and usually results from steam explosions at an aquifer or along a fissure; that is, the hydrovolcanic activity starts on land. This contrasts with *Surtseyan*

hydrovolcanic activity at shallow-water volcanoes where ascending magma must pass up through earlier-formed water-saturated volcaniclastic deposits, possibly partly fluidized by steam; the water/magma mass ratio of erupted mixtures is higher, explosions are less intense, and coarser fragments are produced.

The term *phreatoplinian* was introduced by Self and Sparks (1978) for larger volume eruptions, usually silicic, driven by both vesiculation of magma and magma–water interactions. These are the Earth's most violent explosions, and most commonly occur when rhyolite magma rises beneath caldera lakes, such as the 22 500-year-old Oruanui eruption of Taupo Volcano (C.J.N. Wilson, 1994).

Hydrovolcanic deposits are characteristically finer grained, but more poorly sorted, than those from magmatic explosions. The median diameter produced in the most explosive eruptions is often less than 1 mm (G.P.L. Walker, 1973a), with rhyolitic fragments smaller than basaltic ones due to the low surface tension of more silicic magmas. Glass shards are typically more blocky and less vesicular than magmatically formed shards and often possess a baked mud coating, although a great diversity of morphologies is present (Heiken & Wohletz, 1991). Juvenile clasts are commonly recycled through the vent and refragmented by successive explosions, often many times (Houghton & Smith, 1993). A large proportion of non-juvenile lithic clasts also may be present varying from between 5 and 50% for 'dry' phreatomagmatic eruptions and Vulcanian explosions (Self, Wilson & Nairn, 1979; Houghton & Schmincke, 1989), 60–90% for 'wetter' phreatomagmatic eruptions (Houghton & Nairn, 1991), and almost 100% in phreatic explosions (Mastin, 1991). These include admixed sedimentary rock fragments from the aquifer where explosions were initiated, lava that has solidified in the volcanic vent, and ultramafic rock fragments (and diamonds!) thought to be derived from the mantle.

Peperites are 'frozen' admixtures of magma and sediment that occur along the margins of intrusions and at the basal contact of lava flows. Textures vary from blocky to globular according to the properties of intruding magma and the host-sediment, the confining pressure (rock or water overburden) and the availability of water (Kokelaar, 1982; Busby-Spera & White, 1987). The formation of peperites can involve auto-brecciation, quenching, or phreatomagmatic explosions, or combinations of these processes. In cases where a specific interpretation is impossible, the general term *hydroclastic* is useful (Hanson, 1991).

In hydrovolcanic eruptions, much of the thermal energy of the magma is used to heat water and turn it to steam producing 'wet' transport and depositional systems. Thus hydrovolcanic products are never hot enough to weld during or after deposition. However, wet ash can clump together during transport to form various aggregates. *Accretionary lapilli* are spherical concentrically layered clasts, typically 5–15 mm in size, that form when ash particles collide with each other

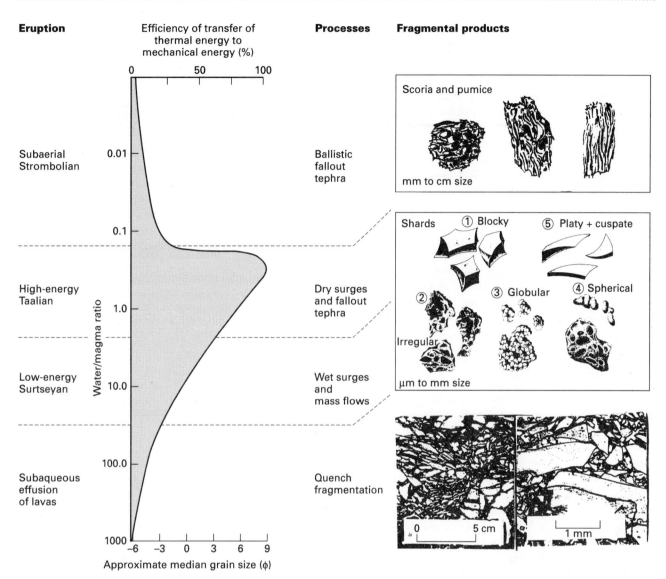

Figure 12.12 Interrelations of explosive energy, water–magma ratio, style of volcanic activity and volcaniclastic fragment in basaltic hydrovolcanic eruptions (largely after Wohletz & Sheridan, 1983). The smallest fragments are produced in Taalian eruptions when most thermal energy is transferred to mechanical energy. The shape of shards produced (1–5) depends on the viscosity of the magma and its degree of vesiculation: blocky shards (1) of poorly vesicular magma are most common; irregular, globular and spherical shards (2–4) indicate fluidal melts; platy and cuspate shards are part of vesicle walls and develop if vesiculated, generally more viscous magma interacts with water. Non-explosive quench fragmentation can occur in any environment. For instance, views of hyaloclastites are from the Mid-Atlantic ridge (Schmincke, Robinson *et al.*, 1978) and a Permian example where lava flowed over unconsolidated nearshore marine sediments (Cas & Wright, 1987).

in turbulent ash clouds (Schumacher & Schmincke, 1995). Although accretionary lapilli usually only form in subaerial eruptions, the aggregates can be indurated enough to be preserved in subaqueous environments. Subaerial hydrovolcanic deposits also contain millimetre to centimetre size cavities (Lorenz, 1974) formed by trapping of air during coalescence of mud aggregates (Rosi, 1992). These vesiculated tuffs are commonly well indurated compared with neighbouring beds, probably because the muddy ash is comparatively enriched in

acidic volcanic aerosols capable of causing a higher degree of post-depositional cementation.

Autobrecciation is caused by non-explosive fragmentation of flowing lava. Autoclasts vary from irregular (e.g. basaltic aa lava) to blocky (e.g. rhyolite lavas) in shape as the viscosity of the magma increases (see Sect. 12.4.6). Clasts are generally cobble- to boulder-sized, but fine debris (and pyroclastic flows) can be generated if a hot lava flow disintegrates in a rockfall. The temperature of the parental lava lobe is important in producing fine debris; hot blocks shatter easily (Mellors, Waitt & Swanson, 1988) whereas rockfalls of cold or degassed lava roll down without disintegrating to form steep talus slopes.

Quench fragmentation is a non-explosive hydroclastic process caused by thermal stresses in magma undergoing rapid cooling, combined with stress imposed on the chilled outer part of lava flows and intrusions by continued movement of the ductile interior (Carlisle, 1963; Honnorez & Kirst, 1975). The process is closely connected with autobrecciation. The products of quench fragmentation, known as *hyaloclasts*, are typically highly angular, splintery to blocky fragments, ranging from fine sand to large brecciated blocks (Fig. 12.12). The shape of large clasts reflects the magma viscosity, with low-viscosity magmas forming pillow-shaped breccias and more silicic magmas forming angular polyhedral blocks (Sect. 12.4.5). Quench clasts often have glassy rims and may contain tiny fractures along their outer surfaces. If not removed from their site of fragmentation by resedimentation processes, large clasts commonly fit together as in a jigsaw.

12.4.2 The eruption and dispersal of pyroclastic sediment

Explosive eruptions are directed toward the position where the surrounding rocks are weakest. In most cases this is directly above the magma chamber, resulting in vertically directed explosions and primary eruption columns. However, where a volcano rises above the surrounding landscape the weakest point may be adjacent to the magma chamber rather than above it and a small but powerful laterally directed explosion may result. Subaqueous eruption columns may also occur.

Vertically directed subaerial eruption columns have three regions based on the relative importance of momentum and buoyancy (Sparks, 1986; Fig. 12.13). Immediately above the vent is a *gas thrust* region in which particles are propelled upwards at high, often supersonic velocities by momentum derived from decompression and concomitant expansion of the exsolving magmatic gases and/or vaporized water. The height of the gas thrust region largely depends on the initial exit velocity, and ranges from a few hundred metres to 4–9 km for sustained Plinian columns. Acceleration of the magma reservoir into the surrounding atmosphere can produce a *shock wave* that moves upwards and outwards ahead of the ejected fragments (Wohletz, McGetchin *et al.*, 1984). In addition to producing its own characteristic surge-like facies (Sect. 12.4.4), the shock wave can reduce the drag forces on ballistic blocks ejected at the onset of an eruption, thus creating problems when trying to use maximum clast size to constrain eruption dynamics.

Upon exit from the conduit, the density of the eruption column is progressively reduced by fallout of large clasts and admixture and heating of entrained air. If the resultant mixture becomes less dense than the atmosphere, convective uprise occurs (Fig. 12.13). The height of the *convective* region is

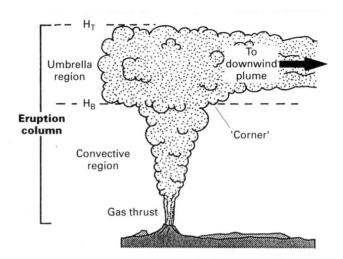

Figure 12.13 A schematic volcanic eruption column, showing the variation of velocity with height and relative importance of buoyancy and momentum. Buoyancy carries column to height H_B; lateral spreading takes place above H_B. Momentum drives some material upwards to a maximum height H_T (from Self & Walker, 1994).

determined by the rate at which heat is transferred from particles to the entrained air (L. Wilson, Sparks & Walker, 1980). The smaller the fragments, the more rapid the heat exchange, and the higher the convective column; thus the convective region comprises 90% of the height of Plinian columns but may be largely absent with Strombolian or Hawaiian eruptions. In phreatomagmatic eruptions, a great deal of thermal energy is utilized in converting water to steam, so eruption columns are generally lower. An eruption column rises convectively until it reaches a level where its density is the same as the surrounding atmosphere (Hb, Fig. 12.13). Where not influenced by wind, it then spreads radially sideways, with vertical momentum driving some material in the centre of the column upwards to a maximum height (Ht) (Sparks, Carey & Sigurdsson, 1991). The region between Hb and Ht is known as the *umbrella* region and gives eruption columns their distinctive mushroom shape. Upwards-directed velocities decrease rapidly once the flow is diverted laterally and the column has a well defined 'corner' (Fig. 12.13) where support for particles abruptly diminishes.

Pyroclastic fragments can be transported away from the eruption column in two ways depending on whether pyroclasts descend individually or *en masse* as flows or currents. *Pyroclastic fall* results when pyroclasts descend from the eruption column independent of each other *and* the surrounding medium (magmatic gas, water vapour, hot air, or water). This usually means fallout from overhanging parts of the eruption umbrella, but includes fragments explosively ejected on ballistic trajectories. Fall deposits typically maintain a relatively uniform thickness while draping all but the steepest topography (Fig. 12.14). *Pyroclastic gravity currents* are sediment gravity flows

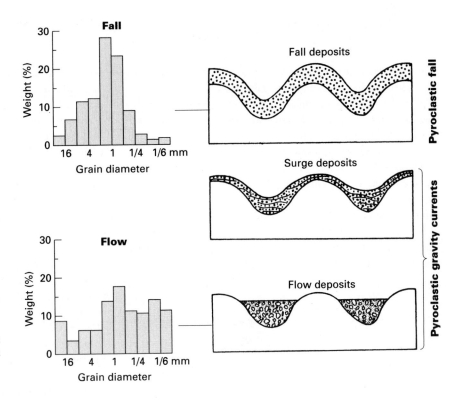

Figure 12.14 Characteristic features of pyroclastic deposits. Grain-size data are from the Upper Bandelier Tuff, New Mexico (from J.V. Wright, Smith & Self, 1980; Cas & Wright, 1987, fig. I.1).

of volcanic fragments, often hot, that spread outwards around the volcano, partially filling valleys and mantling intervening ridges. Pyroclastic gravity currents generated by vertically directed explosions occur when the density of the eruption column (or parts of it) becomes greater than that of the surrounding atmosphere, such that the hot fragments, combined with the surrounding medium, collapse downward under gravity. Thus pyroclastic gravity flow deposits are generally less well sorted than those from pyroclastic falls (Fig. 12.14).

Most eruptions involve both fallout and flow dispersal mechanisms. Both pyroclastic flow and fallout deposits may form penecontemporaneously from different levels or sections of a sustained eruption column (e.g. Fierstein & Hildreth, 1992). More commonly, an initial Plinian eruption column that produces pumice fall deposits is followed by a collapsing column producing pyroclastic gravity currents (Sparks, Wilson & Hulme, 1978). This change can arise because: (i) the rate of magma discharge increases; (ii) exit velocity decreases due to eruption of hotter or volatile-depleted magma from deeper levels of a magma chamber; (iii) widening of the eruption conduit; or (iv) density of phreatomagmatic eruption columns increases due to condensation of water vapour (Fig. 12.15). However, column collapse is not a simple transformation from clasts falling as individuals to *en masse* transport. There is commonly an intervening period in which low concentration gravity currents are produced (Fisher, 1979; Wohletz, McGetchin *et al.*, 1984). However, once full-scale collapse begins, the core of the column becomes protected from mixing with the surrounding air, which in turn further reduces column height and the amount pyroclasts cool. Consequently, the redevelopment of Plinian activity is difficult without a break in eruptive

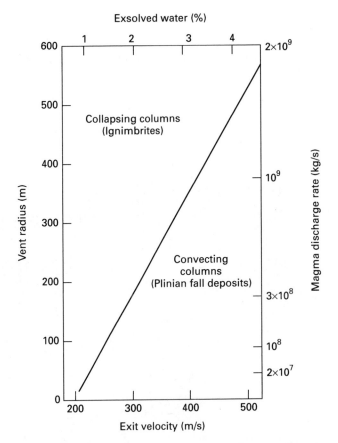

Figure 12.15 Stability fields for convecting and collapsing columns in terms of vent radius and magmatic volatile content. Magma discharge rate (right side) is largely a function of vent radius (left side) whereas exit velocity (bottom) is largely a function of volatile content (top) (from Wilson, Sparks & Walker, 1980).

activity, a decrease in vent radius (which is unusual) or a major increase in explosive activity caused by tapping of magmas with higher volatile contents and/or introduction of water to the system. Pyroclastic gravity currents thus usually transport hotter clasts as the eruption continues.

Laterally directed explosions produce destructive high-velocity pyroclastic flows, from even small volume eruptions, which can persist for large (>10 km) distances away from the vent. One of the best recent examples is the 18 May 1980 *lateral blast* of Mount St Helens (Fig. 12.16) (Kieffer, 1981). This occurred when a volcanic landslide catastrophically depressurized a high-level magma chamber. After leaving the vent, the blast quickly expanded over a 150° sector north of the volcano,

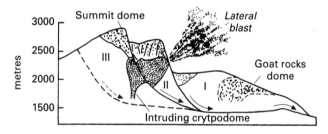

Figure 12.16 Development of lateral blast at Mount St Helens. An earthquake triggered retrogressive detachment of at least three slide-blocks. The landslide depressurized the volcano's magmatic and hydrothermal system, resulting in a laterally directed explosion and the blast deposit (Fig. 12.17), as well as producing a debris-avalanche deposit with a volume of 2.5 km³ (Fig. 12.32).

devastating the conifer forests covering 500 km² (Fig. 12.17). Unsteady supersonic flow conditions were maintained for about 10 km. For nearly half of its extent the flow was accelerating with velocity increasing to about 1100 km h⁻¹ and erosional furrows carved 3.5–9 km from the crater. However, as the flow diverged, its internal pressure dropped, and once flow velocities decreased to subsonic, the blast flow increasingly followed or was deflected by the topography ('channelized blast zone' of Fig. 12.17). As it travelled, decompression, deposition of solid fragments and heating of air decreased the density of the blast and it became buoyant, rising into the atmosphere (see below) and burning the 'singed zone' vegetation as it rose.

All pyroclastic gravity currents travelling downslope are concealed by turbulent curtains of fine ash that roll upwards above them, mimicking the giant convecting clouds of the main eruption column. This ash cloud can become buoyant relative to the base of the basal gravity current, ascending vertically as a *secondary eruption column*. Although these secondary plumes have no gas-thrust region, they can still disperse fine ash great distances away from the volcano. For example, at Mount St Helens the centre of a rising umbrella cloud more than 25 km high developed 12 km north of the volcano when the laterally moving blast transformed into a vertically ascending column (Sparks, Moore & Rice, 1986; Figs 12.18 & 12.19). Large secondary eruption columns can also be produced by violent explosions during entry of pyroclastic gravity currents into the sea, as during the 1815 eruption of Tambora (Sigurdsson & Carey, 1989).

Subaqueous eruptions differ from subaerial eruptions in

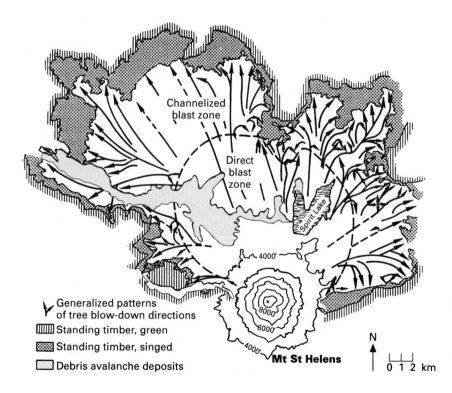

Figure 12.17 The area devastated by the lateral blast surrounding Mount St Helens (from Kieffer, 1981). The lateral blast lofted to form the giant eruption cloud of Figs 12.18 & 12.19. Debris avalanche deposits in the North Toutle valley are further detailed in Fig. 12.32.

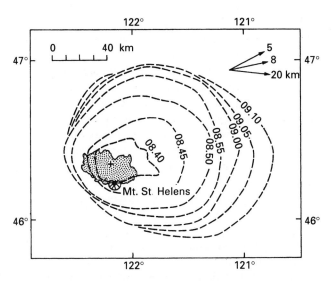

Figure 12.18 Outline of the giant umbrella cloud at 5-min intervals as observed from geostationary satellites. Cross shows initial centre of cloud ascent; stippled area shows extent of blast deposit; arrows at top right are wind directions at 5, 8 and 20 km altitude, measured at Spokane, Washington (from Sparks, Moore & Rice, 1986).

that expansion of magmatic volatiles and/or ingested external water is suppressed because of higher confining pressures (cf. Cashman & Fiske, 1991; Kokelaar & Busby, 1992) (Fig. 12.20). Consequently, the vertical extent of the gas-thrust

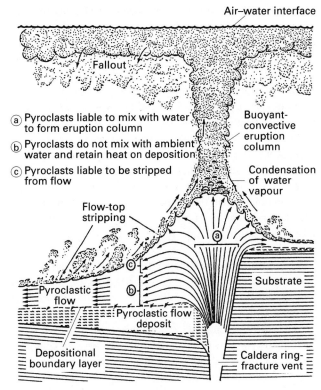

Figure 12.20 Schematic diagram illustrating processes of a high mass-discharge subaqueous explosive eruption. No relative scales are implied (from Kokelaar & Busby, 1992).

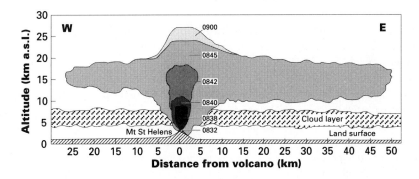

Figure 12.19 E–W profile showing early vertical growth and lateral expansion of plume from the 18 May 1980 eruption of Mount St Helens, compiled from visual and satellite observations (*upper*). Wind directions at selected altitudes (2.2–26.5 km) and average wind speed profile at Spokane, Washington during the eruption (*lower right*). Isochron map showing maximum downwind extent of ash from airborne ash plume, carried by fastest moving wind layer as observed on satellite photographs (*lower left*). Four digit numbers are Pacific Day Time (compiled from Sarna-Wojcicki, Shipley *et al.*, 1981).

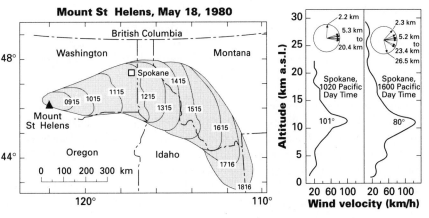

region of a subaqueous eruption column, the facility for early mixing with the surrounding water, and consequent heat exchange and expansion, are reduced. In deep water and/or at high mass discharge rates, only the outermost parts of a gaseous particulate flow may be able to mix with ambient water. A protective cupola of steam may largely enclose the subaqueous column, and ejecta erupted within the cupola may not come into contact with liquid water until well after its deposition and will thus show signs of a high emplacement temperature such as welding fabrics or plastic deformation of ballistic fragments. In contrast, shallow-water, unsteady or intermittent eruptions (i.e. the norm) lead to great turbulence and increase the probability that the erupted material mixes with water and becomes cold. Transport of material through the sea surface into the air is possible during eruptions in shallow water, typically less than 50 m.

A buoyant–convective eruption column, consisting of warm water and entrained solids may also arise from sea water mixed into the gas-thrust region and/or buoyant fluids stripped from the top margins of laterally moving pyroclastic flows (Kokelaar & Busby, 1992). The umbrella region at the sea surface would be thin compared with subaerial eruptions because continued upwards motion is inhibited by the extreme density contrast between sea water and air.

12.4.3 Pyroclastic fall

Pyroclastic fallout can result from primary or secondary eruption columns. Although particles can fall out of any level of an eruption column, it is useful to divide sedimentation into two regimes: an inner proximal region involving ballistics and fallout from plume margins and an outer region beyond the corner controlled by fallout from the radially spreading cloud (Fig. 12.13).

Very large particles (>10 cm) follow trajectories that are more or less ballistic, little affected by wind and only slightly by the expanding eruption cloud. The distribution of these clasts is generally restricted to a few kilometres from the vent. Ballistic fragments often produce impact structures, termed *bedding sags* in subjacent sediment and prove useful as their maximum size allows explosion energy to be estimated (L. Wilson, 1972). If crustal stratigraphy is known, explosion depth can be estimated from the rock type.

Intermediate-sized particles (c. 1–10 cm) travel up into the convective region but fall out from its margins to accumulate as 'proximal' fall deposits.

Finer particles, and most tephra from large-scale explosive eruptions, fall out from the expanding umbrella region at the top of the eruption column. Particles fall out from the base of the laterally spreading plume, and are advected by local winds until they reach the ground. Very fine-grained tephra, generally less than 250 µm, have a long atmospheric residence time, and with eruption columns that penetrate the tropopause the smallest

particles may circle the Earth many times before settling. Differences in settling velocity between crystals and glass shards can cause downwind changes in the bulk composition of these distal fallout ashes, a process termed *aeolian fractionation* (Lirer, Pescatore et al., 1973). Very small particles can also become agglutinated by moisture and electrostatic forces and fall out prematurely (Carey & Sigurdsson, 1982). Winds too can distort and divert the eruption plume. In a layered atmosphere the wind commonly blows in different directions and at different velocities at different altitudes. As large eruption columns penetrate 40 km into the atmosphere, distribution patterns can be quite complex as different atmospheric layers are traversed. Furthermore, at a volcano with more than one style of explosive activity, the various ashes produced can have distinct and separate dispersal patterns because of the different heights of their respective eruptive columns.

Subaerial fallout deposits often exhibit an exponential decrease in thickness and grain size, and improvement in sorting, with increasing distance from source. This can be explained by a sedimentation law governing the loss of particles from a turbulent radially spreading gravity current (Sparks, Carey & Sigurdsson, 1991; Sparks, Bursik et al., 1992). Because of the exponential radial thinning or fining behaviour, plots of isopach or isopleth data using log thickness versus area$^{1/2}$ coordinates are straight lines (Pyle, 1989). Data on such plots are often segmented, with a steep proximal slope and a more gradual distal one (see Fig. 12.21). The break in slope has been attributed to the transition from sedimentation from the convective region to sedimentation from the umbrella region (Bursik, Sparks, et al., 1992), with the inflection point representing the 'corner' between these two regions (Figs 12.13 cf 12.21).

The geometry of lithic and pumice isopleths can be used quantitatively to infer both ancient wind patterns and wind speeds and eruption column height (Roobol, Smith & Wright, 1985; Carey & Sparks, 1986). As first recognized by G.P.L. Walker (1973a), as eruptions become more powerful, fall deposits become more widely dispersed and/or grain size becomes finer grained. There is a continuous spectrum from small volume (c. 0.01 km³), weakly explosive strombolian eruptions building steep-sided scoria cones to larger (volumes 0.1–50 km³) plinian or phreatoplinian eruptions producing laterally extensive sheets (Fig. 12.21).

Proximal subaerial fallout deposits commonly display weak bedding due to fluctuations in the energy and discharge rate of the eruption and direction and strength of winds. Reverse, coarse-tail and normal grading within deposits can be attributed to a change in the dynamics of the eruptive system (e.g. column height, vent radius, vent flaring) (Carey & Sigurdsson, 1989). Compositionally zoned plinian pumice-fall deposits also occur and invariably these show an upwards increase in the proportion of more mafic juvenile components. *Distal subaerial fallout* deposits may be thin, or thick and homogeneous (caused

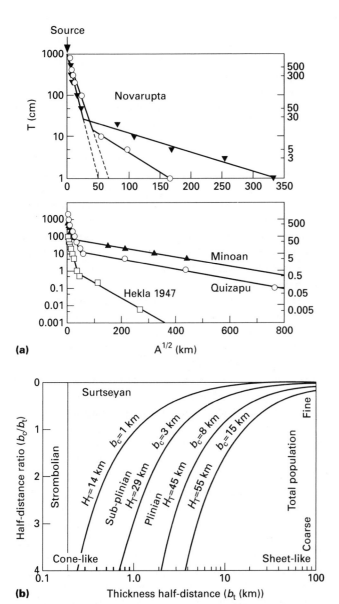

(a)

(b)

Figure 12.21 Sedimentation from pyroclastic fallout. (a) Variation in tephra fall thickness (log-scale) with distance from source (isopach area$^{1/2}$) for selected eruptions (from Fierstein & Hildreth, 1992). Straightline segments indicate exponential thinning. Note abrupt change in slope in proximal regions of the eruption column. This inflection point represents the 'corner' between the convective and umbrella regions of the eruption column (see text and Fig. 12.13). (b) Classification of pyroclastic fallout based on maximum clast size and deposit thickness (from Pyle, 1989). This is a refinement of the approach by George Walker (Fig. 12.10a). Parameters b_c and b_t are the average distances over which maximum clast sizes and the deposit thickness decrease by half respectively. The figure is contoured for various eruption column heights; irrespective of original size distribution of particles b_c should be the same for a given column height.

by bioturbation and other soil-forming processes) resembling wind-blown loess (Fisher, 1966b).

Subaqueous fallout deposits may be derived either from distant subaerial eruptions or from submarine eruption columns. Subaerially derived ash layers are initially dispersed by wind and are typically less than 10 cm thick. Beds typically show excellent normal grading (Pederson & Surlyk, 1977) due to a combination of the slow rates of particle settling and great difference in settling velocities of the different grain populations. Size grading data have been used to calculate the duration of eruptions (Ledbetter & Sparks, 1979) or place constraints on water depth in ancient successions (Lowe, 1988).

Submarine fallout from submarine eruption columns is less well understood, although experimental and theoretical studies suggest their deposits are distinctive (Cashman & Fiske, 1991). The small density difference between water-saturated pumice and the surrounding water results in significantly lower terminal velocities, with hydraulically equivalent assemblages of pumice and lithics having diameter ratios of 5:1 to 10:1 (cf. 2:1 to 3:1 for fallout from subaerial eruptions). Provided that post-depositional alteration and deformation have not destroyed original textures, the ratio of pumice to lithics may be used to indicate a subaqueous origin where depositional environment is otherwise uncertain.

12.4.4 Pyroclastic gravity currents: flow and surge deposits

Pyroclastic gravity currents are hot gas–particle dispersions whose density generally exceeds that of the atmosphere or hydrosphere into which they are introduced. The gas comprises magmatic volatiles exsolved prior to and during the eruption, volatiles released from pyroclasts during flowage, steam from vaporized ground and surface waters, ingested air, and gas from combustion of vegetation. They can have velocities near the vent in excess of 1000 km h^{-1}, travel more than 100 km from their source, and surmount mountainous barriers over 500 m high. The volume of pyroclastics transported in one event can be as little as 100 m^3 or over 1000 km^3, with thickness of deposits ranging from a few centimetres to a few kilometres. Temperatures of subaerial gravity currents vary from <100°C for those from phreatomagmatic activity to more than 700°C for currents from magmatic eruptions (McClelland & Druitt, 1989).

Pyroclastic gravity currents can be formed in several ways. Large (>1 km^3) gravity currents may originate by collapse of an overloaded vertical eruption column whereas smaller gravity currents can arise from laterally directed explosions (Sect. 12.4.2), 'boiling over' of gas-charged magma from an open vent, decoupling of single currents into separate sediment gravity flows (Fisher, 1995), and even collapse of secondary eruption columns rising from the gravity currents themselves (Yamamoto, Takarada & Suto, 1993). A volcanic explosion may not be necessary. Small gravity currents can also be generated by gravitational collapse of the front of a gas-charged lava flow or dome (Sect. 12.4.1). Secondary pyroclastic flows, termed drain-down deposits (G.P.L. Walker, Hayashi & Self, 1995), can be generated while the primary pyroclastic flow is still moving forward due to slumping or gravitational draining of pyroclastic material deposited on steep slopes (cf. Fisher, Glicken & Hoblitt, 1987). Pyroclastic material can remain loose and hot (i.e. >100°C) for months to even years after being deposited (Ragan & Sheridan, 1972). A significant proportion of rainwater or groundwater that seeps into the porous interior of these primary pyroclastics is vaporized; fluvial incision or earthquakes can destabilize this hot pressurized debris and set in motion additional secondary, and even tertiary, hot, gas-supported pyroclastic gravity currents (e.g. Mount Pinatubo; Torres, Self & Martinez, 1995).

PYROCLASTIC TRANSPORT AND DEPOSITION: GENERAL CONSIDERATIONS

Pyroclastic gravity currents have traditionally been divided into two categories: *pyroclastic surges* and *pyroclastic flows*, interpreted as low- and high-concentration end-members of a flow spectrum based on particle concentration (Sparks, 1976; Carey, 1991). As studies of subaerial pyroclastic flows and surges have progressed, it has become clear that there is considerable diversity in the nature of flows at both ends of the particle concentration spectrum.

It is important to recognize the difference between the active pyroclastic current and the deposit it generates. In general, the bedding and sorting characteristics of the deposit only reflect depositional processes in the *basal* part of the transport system during the final moments of sedimentation, and most pyroclastic gravity currents involve separate transport and depositional systems that are neither spatially nor temporally uniform. Owing to the enormous variation in the dimensions, composition and textures of pyroclastic gravity currents it is unlikely that a single model will adequately account for all the facies variation observed in deposits.

None the less, there is growing appreciation that many facies associations can be explained if pyroclastic gravity currents are treated as density stratified turbidity currents (Valentine, 1987; Druitt, 1992), following the ideas first proposed in the 1960s (Fisher, 1966a; Aramaki & Ui, 1966). Conditions favouring turbulent transport include a high velocity, a thick flow, a low viscosity, a small particle size and a low rate of suspended sediment fallout. These parameters are initially established by the type of eruption (Sect. 12.4.1) and the mechanism of formation of the pyroclastic gravity current (Fig. 12.22). For instance, a fast-moving gravity current from a laterally directed explosion (e.g. Mount St Helens lateral blast) may involve turbulence throughout the entire gravity current for most of its travel distance whereas slow-moving (<50 km h^{-1}) gravity currents with large poorly vesicular particles from gravitational collapse of a rhyolite dome will only involve turbulence on their uppermost surfaces.

During transport, pyroclastic currents become segregated into a high concentration lower part (underflow or basal avalanche) and an overriding more dilute turbulent flow (ash cloud surge) (Fig. 12.23a). This occurs because: (i) heavier particles settle (Fisher & Heiken, 1982); (ii) air becomes mixed into the upper part of the flow (Denlinger, 1987); (iii) fines are elutriated and gases lost from the base of the flow (C.J.N. Wilson, 1980); and (iv) particles aggregate during transport. Extremely hot particles may agglutinate, thereby enhancing deposition or creation of a basal high-concentration flow (Branney & Kokelaar, 1992; Freundt & Schmincke, 1995). Wetter particles (e.g. a phreatomagmatic eruption) may also coalesce subaerially, creating a wet slurry – effectively a lahar. Particle support mechanisms in the higher concentration basal portions include fluidization from upward flow of gas (either exsolved magmatic gas or incorporated air or steam), buoyancy, mechanical collisions between particles, acoustic mobilization where particles move as part of elastic waves (G.P.L. Walker, Hayashi & Self, 1995), and self-lubrication along a basal layer of agitated particles (see also Sect. 12.5.1). In contrast, smaller fragments within the upper portions of

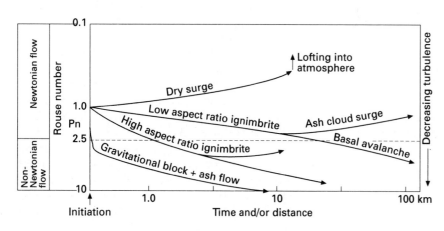

Figure 12.22 Transport mechanisms within pyroclastic gravity currents in terms of transport rate (specifically shear velocity) and particle settling velocity (based on ideas in Fisher, 1983; Valentine, 1987). Lines depict flow paths for a few types of gravity currents.

(a) Transverse

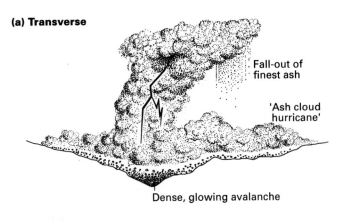

Fall-out of finest ash

'Ash cloud hurricane'

Dense, glowing avalanche

(b) Distal   **Proximal**

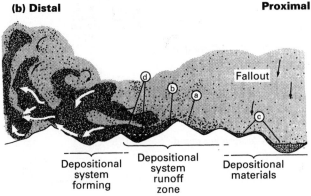

d   b   a

Fallout

c

Depositional system forming | Depositional system runoff zone | Depositional materials

Figure 12.23 Effects of topography on pyroclastic gravity currents. Diagrams depict transverse (a) and proximal–distal (b) facies relations (based on Fisher & Heiken, 1982 as in Francis, 1993; Fisher, 1995). In ancient successions one rarely has a complete cross-section of valley fill deposits. Consequently, establishing whether surge deposits that are interstratified with flow deposits represent overbank deposits of high-concentration gravity currents or deposits from separate and unrelated low concentration gravity currents is very difficult. In (b), letters denote: a, blocking of high concentration basal part of transport system; b, downslope drainage of pyroclastics to form depositional system; c, secondary pyroclastic flows from gravitational drain-back; d, streamlines denoting density stratification in transport system.

the transport system or dilute currents may be carried by traction or saltation or suspended solely by turbulence. Because the lower and upper parts of a stratified flow have different densities and particle support mechanisms they have different rheologies, and usually decouple with one part stopping before the other (Fisher, 1995). If the high concentration basal part is thick enough, it may be able to travel large distances under its own momentum as a dense non-turbulent flow, comparable to a grain flow or cohesive debris flow. However, the basal part is affected by topography, becoming ponded within valleys and blocked by small topographical obstacles. In contrast, the upper part may travel over higher ground.

Pyroclastic currents can deposit sediment, *grain-by-grain*

or *en masse*, the particular facies at any site depending on the concentration of sediment at the base of the flow *during deposition*. With low concentrations, traction sedimentation and/or fallout from small suspension clouds will result in thinly stratified or cross-stratified facies. High-concentration basal dispersions yield poorly sorted structureless beds, deposition occurring: (i) owing to rapid gradual or incremental aggradation directly from suspension (Fisher, 1966a; Druitt, 1992; Branney & Kokelaar, 1992; Kneller & Branney, 1995); or (ii) because the depositional system of the flow, which may be all or only part of the gravity current, increases its yield strength due to friction and/or adhesion between particles and 'freezes' in a manner similar to cohesive debris flows (Sparks, 1976; J.V. Wright & Walker, 1981; Freundt & Schmincke, 1986). Irrespective of its origin, this structureless facies is more prevalent proximally and within valleys and is the dominant facies of large volume pyroclastic flows.

SUBAERIAL PYROCLASTIC SURGES

Pyroclastic surges are low-concentration, high-velocity currents that move in a dominantly turbulent fashion. They are small volume (<1 km³), rarely reach more than 10 km from their source, and are best known from hydrovolcanic eruptions involving basaltic or, more rarely, rhyolitic magma (Moore, 1967; Fisher & Waters, 1970). As hot, small-volume turbulent suspensions are almost impossible to sustain under water, surges usually deposit subaerially. In arid climates, surge deposits may be interstratified with aeolian volcaniclastics; distribution, palaeoflow directions, grain size, and depositional structures permit discrimination of these two similar facies (G.A. Smith & Katzman, 1991).

Because of their low density and momentum, pyroclastic surges form thin but extensive deposits that mantle topography (Fig. 12.14). In proximal settings, where the shear velocity of the surge is the highest, erosion may occur with U-shaped channels or longitudinal furrows being carved into earlier deposits. Deposition occurs when velocities are less than that required to support particles. Lateral and vertical facies sequences of individual depositional units generally indicate a proximal to distal and/or temporal decrease in flow velocity and/or the rate of suspended sediment fallout (Fig. 12.24). With distance from source, or upwards, grain size decreases, the wavelength and amplitude of bedforms decreases, and antidune structures are sometimes replaced by lower flow regime dunes. Proximally, or within channels, where sediment concentrations and/or rate of suspended-sediment fallout at the base of the surge cloud are extremely high, structureless or faintly laminated deposits typically form (Crowe & Fisher, 1973; Druitt, 1992). At lower sediment concentrations and/or sedimentation rates, tractional processes result in well-developed stratification, with bed thicknesses of a few millimetres up to tens of centimetres,

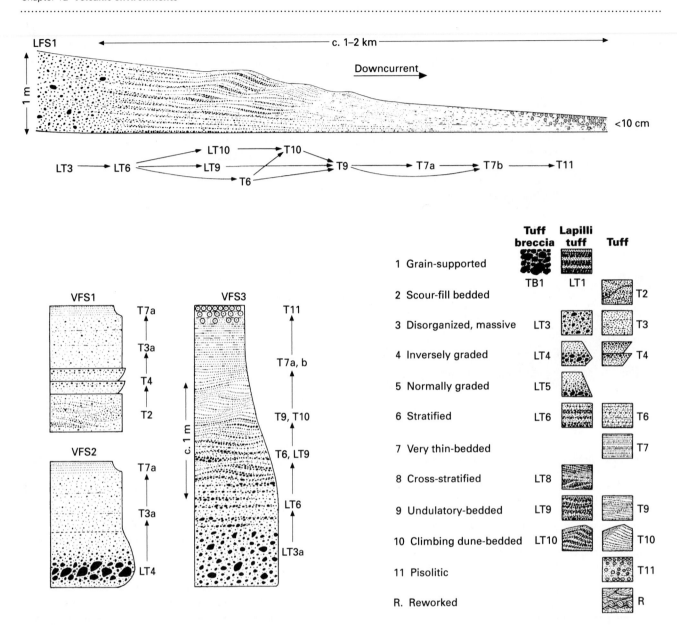

Figure 12.24 Proximal to distal (LFS) and vertical (VFS) facies variation in pyroclastic surge beds (i.e. single depositional units) on Songaksan tuff ring (from Chough & Sohn, 1990). The lateral facies sequence (LFS1) was distilled from common downcurrent facies transitions in flank deposits whereas vertical facies sequences (VFS) are distilled using Harper's (1984) method of facies sequence analysis. VFS1 and VFS2 are from proximal near-vent deposits (S1 of Fig. 12.50), probably where short-lived pyroclastic surges were overladen by suspended sediment fallout (compare with Lowe, 1988). LFS1, and its vertical expression (VFS3) indicates downcurrent *decrease* in particle concentration, grain size, and suspended-load fallout rate with a resultant increase in traction and sorting processes.

and sedimentary structures such as dune bedforms or plane-parallel lamination (e.g. Schmincke, Fisher & Waters, 1973).

Transport and depositional processes within surges also vary according to whether or not condensed water occurs in the erupted mixture. *Dry surges* become more dilute as they travel as entrained air is heated and expands. There comes a point where the surge cloud is less dense than ambient air and rises into the atmosphere, a process referred to as lofting, an example being the 1980 lateral blast at Mount St Helens (Sect. 12.4.2). Proximal to distal variation in depositional facies records the decrease of suspended-load fallout rate, a decreasing velocity, and an increase in traction sedimentation (Sigurdsson, Carey & Fisher, 1987; Chough & Sohn, 1990; Druitt, 1992). *Wet surges* are complicated by the effect of moisture which facilitates particle adhesion and particle settling through the current, reduces the amount that entrained air is heated, and may result

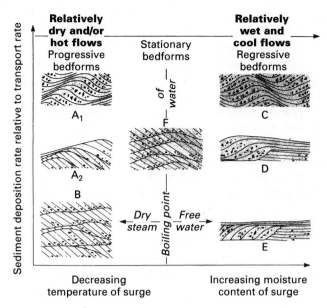

Figure 12.25 Theoretical effect of temperature and moisture content on surge bedforms (from J.R.L. Allen, 1984).

1 *Pumice flows* or *ash-flows*, and their deposits called *ig-nimbrites* or *ash-flow tuffs*, consist predominantly of glass shards and crystals, with additional pumiceous lapilli and small blocks. Most result from Plinian eruption of dacitic to rhyolitic magma, large ones travelling over 100 km from their source. However, andesitic or basaltic ash flows, involving smaller volumes of magma, are also reported (e.g. Robin, Eissen & Monzier, 1994; Freundt & Schmincke, 1995). Three distinct depositional facies can be recognized in subaerial ash flows (Fig. 12.26a,b).

Layer 1 deposits are thin (<100 cm) and extremely variable layers occur locally at the base of some ignimbrites. They may include fine-grained layers, zones of crystal or pumice concentration, coarse-lithic breccias, and even surge-like cross-bedding. Explanations include a turbulent flow that preceded the main pyroclastic flow (Sparks, Self & Walker, 1973), blast phenomena (Wohletz, McGetchin *et al.*, 1984), boundary layer turbulence at the base of the flow (Valentine & Fisher, 1986) and sedimentation at the turbulent front of the flow (C.J.N. Wilson & Walker, 1982).

Layer 2 is thickest, ranging from a few centimetres to tens of metres. The base of Layer 2 (Layer 2a) is often depleted of fine clasts, somewhat better sorted than the rest of the deposit, and reversely graded; it is attributed to shearing at the base of the main flow against the substrate during late stages of emplacement. Most of the deposit (Layer 2b) is structureless, with the two possible mechanisms of deposition discussed earlier. Normal grading in dense lithic fragments and reverse grading in pumice or scoria lapilli sometimes occur. Grading indicates relative movement of particles during transport and/or deposition according to their settling velocities, with light and/ or small particles being swept upwards. With some types of pyroclastic flows, such sorting is believed to reflect partial fluidization of the depositional unit due to vertical flux of an externally derived source of gas up through the granular material (C.J.N. Wilson, 1980, 1985). Commonly invoked sources of gas have included exsolution of magmatic volatiles, air ingestion, combustion of vegetation, and entrapment of groundwater. Hindered settling, in which the sedimentation of large particles causes a sufficiently strong vertical fluid flux, may reinforce external gas sources and act as an effective self-fluidization mechanism (Druitt, 1995).

in the regressive upstream migration of bedforms due to plastering of particles on to the stoss side of the dunes (Fig. 12.25). Proximal to distal facies variations may show evidence for deflation of the surge cloud and an increase in particle concentration at surge base (Wohletz & Sheridan, 1979). Wet surges are associated with mudflow and sheetwash deposits. Accretionary lapilli, bedding slumps, induration of the deposits, and vesiculated horizons are also characteristic.

SUBAERIAL PYROCLASTIC FLOWS

Pyroclastic flows are hot, comparatively high-concentration ground-hugging, highly mobile gas-particle dispersions. They are more voluminous and travel further than pyroclastic surges. A spectrum of flow types can be recognized, based on juvenile clast type, emplacement temperature, and the probable method of flow formation.

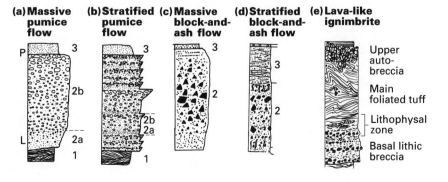

Figure 12.26 Idealized products of subaerial pyroclastic flows. P, pumice clasts; L, lithic clasts. Note concentration of pumice at the top of the flow unit and lithics at the base in all cases (from Sparks, Self & Walker, 1973; Sheridan, 1979; Fisher & Heiken, 1982; Branney, Kokelaar & McConnell, 1992). Note absence of Layer 1 deposits from block and ash flows and lava-like ignimbrites.

Layer 3 deposits, termed *co-ignimbrite ashes* (Sparks & Walker, 1977), are thin, fine-grained ashes that cap the ideal depositional unit. In most cases they represent fine ash derived from the pyroclastic flow that was deposited as ash cloud surge and fallout deposits.

Ignimbrites also show proximal to distal and flow-transverse variation in grain size and emplacement temperature. In proximal locations (i.e. <5–15 km) a coarse breccia, termed *lag or co-ignimbrite breccia*, often grades laterally into the base of the 2b layer of the ignimbrites and consists mainly of pyroclasts that were too large and heavy for the eruption column or derivative pyroclastic flow to support (J.V. Wright & Walker, 1977; Druitt & Sparks, 1982). Lithic breccias can also be deposited at changes in slope and preferentially occur in the base of valley fills. Post-emplacement welding is most complete for thicker and more proximal deposits (Fig. 12.27). Welding can obscure recognition of the depositional units that record emplacement processes and if several flows are emplaced in rapid succession, cooling structures can be continuous across compositional and depositional boundaries, producing what is referred to as a compound cooling unit.

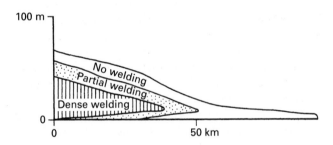

Figure 12.27 Ideal lateral variation of welding zones in a simple cooling unit of ignimbrite (from R.L. Smith, 1960b).

2 *Fountain-fed lava flows, lava-like ignimbrites* and *rheomorphic ignimbrites* result when the viscosity of pyroclasts is so low that they coalesce during transport and/or deposition. Deposits take on the characteristics of lava flows, with thick upper autobreccias, less common basal autobreccias and central columnar-jointed or flow-banded zones (Fig. 12.26e; Branney, Kokelaar & McConnell, 1992; Henry & Wolff, 1992). Evidence for a pyroclastic origin in such deposits is often obscure, with vitroclastic textures preserved only at the base of the deposit. Broken crystals, which do not occur in lava flows, are additional evidence for explosive volcanism.

3 *Block and ash flows* consist of dense, microvesicular lapilli and ash, usually porphyritic andesite or dacite. They result from fragmentation of silicic lava flows and domes by Vulcanian explosions or gravitational collapse. Aprons form from many small (<0.1 km³), separate, but closely spaced (i.e. over weeks to years) eruptions whose total volume is usually much less than 1 km³. Individual deposits vary according to whether an explosion occurred and its intensity. Block and ash flows from gravitational collapse or weak explosions produce thick, valley confined, high-concentration high-yield strength flows with well-defined deposit fronts and margins. The deposits are generally massive, ungraded or reversely graded, matrix to clast-supported; fine-grained deposits corresponding to the ash cloud are present but poorly developed (e.g. Fisher & Heiken, 1982; Boudon, Camus *et al.*, 1993) (Fig. 12.26c). Facies suggest accumulation from a concentrated suspension of cohesionless solids exhibiting non-Newtonian behaviour (i.e. debris flow or density-modified grain flow). By contrast, block and ash flows from violent explosions are more similar to pumiceous pyroclastic flows, consisting of a dense valley-confined flow and an overriding low-density ash cloud surge (e.g. Lajoie, Boudon & Bourdier, 1989). Deposits are often normally graded over the entire bed and stratified, suggesting transport and deposition from a turbulent suspension (Fig. 12.26d).

SUBAQUEOUS PYROCLASTIC FLOWS

Subaqueous pyroclastic flows may also form, either generated by subaqueous eruptions or from subaerial pyroclastic flows that have entered the sea. Unlike their subaerial counterparts, subaqueous pyroclastic flows quickly transform to (or start as) water-supported flows, and usually become more expanded and dilute (rather than deflating) with distance from source. Owing to the slower settling rate of particles within water, distal subaqueous deposits are better sorted and more crystal-rich than distal subaerial ones (Cas, 1983; Stix, 1991).

Proximal *deposits from subaqueous eruptions* are non-sorted and massive and may resemble subaerial pyroclastic flow deposits (Fig. 12.28). In several cases there is evidence, such as thermoremanant magnetization or welding textures, for a high emplacement temperature (Tamura, Koyama & Fiske, 1991; Kokelaar & Busby, 1992; Kano, Orton & Kano, 1994). However, most subaqueous pyroclastic flows seem to transform into water-supported density currents within a distance up to about 5 km. During the transition, the flow may consist of a mixture of clasts, steam and water. As it may take days for the cores of large blocks (i.e. >25 cm) of lava to cool below 100°C, fracturing of clasts may generate a continual supply of steam during transport. Medial–distal deposits from subaqueous eruptions commonly have a relatively coarse-grained basal massive portion and an upper bedded, finer-grained portion (Fig. 12.28). The upper portion may be doubly graded, with density graded individual beds within a total succession of size-graded beds, with each overlying bed progressively finer grained than the underlying one. These doubly graded units can be interpreted either as a succession of events from one waning eruption column (Fiske & Matsuda, 1964) or as sorting within high-density turbidity currents unrelated to eruptive events (Yamada, 1984).

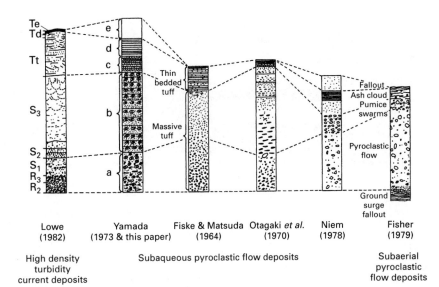

Figure 12.28 Comparison of vertical sequences from subaqueous pyroclastic flow deposits. The deposits of high density turbidites and of subaerial pumice flows are shown for comparison (from Yamada (1984) and references listed therein).

Lowe (1982) — High density turbidity current deposits

Yamada (1973 & this paper) — Fiske & Matsuda (1964) — Otagaki *et al.* (1970) — Niem (1978) — Subaqueous pyroclastic flow deposits

Fisher (1979) — Subaerial pyroclastic flow deposits

A more complex succession of events appears to occur when *land-generated pyroclastic flows enter the sea*, partly depending on flow rheology. Dense, non-turbulent pyroclastic flows, or non-turbulent bases of more dilute flows may maintain their integrity as hot pyroclastic flows for several kilometres past the coastline (e.g. Reedman, Howells *et al.*, 1987). When more turbulent, land-generated, hot pyroclastic flows enter the sea, they quickly mix with water and probably either continue directly into the sea as aqueous density currents, become quenched, pile up, and later slump, or explode at the shoreline to generate additional massflow deposits. Owing to the multitude of possible nearshore events, offshore deposits often have complex facies relations and consist of a succession of sediment-gravity flow units which rarely correlate from section to section (Tassé, Lajoie & Dimroth, 1978; Whitham, 1989).

12.4.5 Lava flows

Low viscosity lavas (e.g. basalts) generally flow large distances (10s–100s km), cover large surface areas and form thin, low profile sheets. In contrast, high viscosity lava flows (dacites–rhyolites) generally form short (<10 km), thick, and steep angle lava flows or domes (Fig. 12.29). However, some anomalous far flowing (10s km) subaerial and subaqueous silicic lavas of greater volumes (up to 200 km³) have also been described (e.g. Cas, 1978; De Rosen-Spence, Provost *et al.*, 1980; Henry & Wolff, 1992); this requires a low viscosity magma, either due to a low initial water content or, in deep-water settings, due to the influence of high hydrostatic pressures on volatile exsolution and hence flow rheology. There is little difference in the gross morphology and facies of subaerial and subaqueous effusive products, except that hydroclastic debris is more common in subaqueous settings.

The structure and facies of low-viscosity lavas are determined by the volumetric flow rate and viscosity (Bonatti & Harrison, 1988; Rowland & Walker, 1990; Griffiths & Fink, 1992). At slow effusion rates and/or extremely low viscosities, restriction of flow by rapid growth of chilled crusts causes the lava to advance as small digital lobes, producing subaerial tube-fed *pahoehoe* or submarine *pillow* lavas (Fig. 12.29d). In contrast, at higher volumetric effusion rates or with slightly more viscous magmas (e.g. basaltic andesites), lava continues to flow after it has cooled significantly, and repeated tearing of the cooled lava crust produces a rough, spinose or clinkery surface. These lavas, termed *aa*, (Fig. 12.29c) are generally thicker (several metres up to 20 m) than pahoehoe, and move by plug flow, bulldozing obstacles in their path. *Sheet lavas* form at even higher discharge rates, as in subaerial flood basalt fields (Sect. 12.9.4) or submarine outflow lavas (Lipman, Clague *et al.*, 1989). However, sheet lavas can also arise when thin layers of pahoehoe lava coalesce or inflate due to injection of lava under a chilled crust (Hon, Kauahikaua *et al.*, 1994).

Gas vesicles are ubiquitous in pahoehoe lava flows. Bubbles tend to rise and grow by coalescence, and in general the thicker a lava flow, and hence the longer it takes to solidify, the more complete is the loss of bubbles from the lower and middle parts of the flow (G.P.L. Walker, 1993). However, aa flows are almost totally non-vesicular due to deformation and progressive elimination of vesicles by shearing. Even more viscous andesite and dacite lavas typically have scoriaceous to blocky upper and front surfaces, (Fig. 12.29b) and in subaqueous examples the blocks are sometimes referred to as pseudo-pillows (Yamagishi, 1991; Kano, Takeuchi *et al.*, 1991).

With silicic lavas, ramping leads to blocky, rough and ridged upper surfaces. Flows are never glassy all the way through, however, and commonly display an internal textural stratigraphy (Fink & Manley, 1987). Crystallization within the centre of

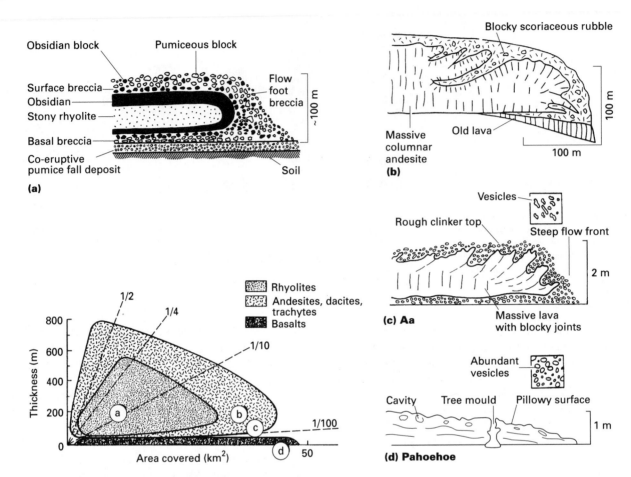

Figure 12.29 Aspect ratio (thickness divided by area), depositional slopes, and facies for lavas of different compositions: (a), rhyolite; (b), andesite; (c) and (d), basalt (collated from G.P.L. Walker, 1967, 1973b; Cas & Wright, 1987).

the flow releases dissolved magmatic volatiles which rise through microcracks to form a distorted coarsely vesicular pumice which thickens downflow. The accumulation of volatiles pressurizes the lava flow, and distal flows can sometimes explode to form block and ash flows (Fink & Manley, 1989).

With all types of magma, energetic subaqueous lava fountains and fissure eruptions facilitate mixing between the extruded lava and surrounding water, and may result in thick (30–200 m on average), voluminous (up to 20 km³), composite mass–flow deposits (Bergh & Sigvaldason, 1991; Mueller & White, 1992) (Fig. 12.30).

12.4.6 Explosion craters, calderas and volcanotectonic basins

Volcanoes are unique in that processes that create and transport sediment can also create several types of repositories in which sediment can be deposited.

Craters result when explosions during a volcanic eruption excavate a depression within the surrounding country rock. Cratering is most commonly associated with maar-volcanoes (Sect. 12.9.2), where craters range up to about 3 km in diameter and over 500 m in depth. They may also occur as volcanic elements of other types of volcanoes, such as those pitting the floor of Crater Lake caldera (Nelson, Bacon *et al.*, 1994).

Calderas are volcanotectonic basins, more or less circular in shape, that form by rapid subsidence resulting from partial evacuation of the underlying magma chamber. They range up to 100 km wide and about 4 km deep. The volume of magma erupted in caldera-forming eruptions is typically >10 km³ at intermediate volcanoes (Sect. 12.12) and may exceed 50–1000 km³ at rhyolitic volcanoes (Sect. 12.13), with caldera volumes approximating the volume of magma erupted (Spera & Crisp, 1981). Calderas are usually associated with large explosive eruptions of silicic systems, with Crater Lake (Oregon), Santorini (Greece), and Long Valley (Fig. 12.31) being well

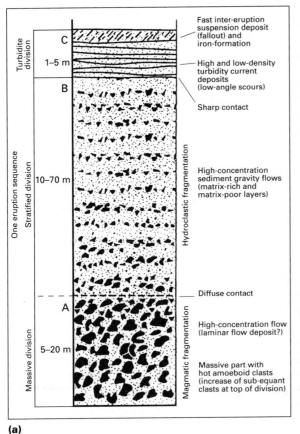

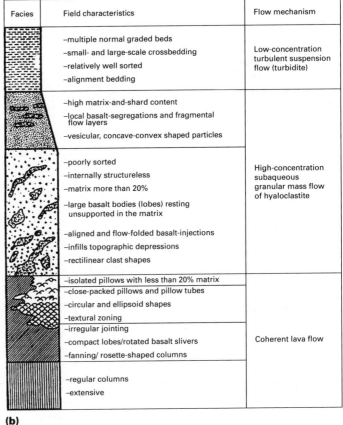

(a)   **(b)**

Figure 12.30 Comparison of subaqueous composite mass-flows arising from (a) deep-water rhyolitic lava fountains (from Mueller & White, 1992), and (b) shallow-marine basaltic fissure eruptions (from Bergh & Sigvaldason, 1991). The basaltic deposition units vary from 35 to 140 m in total thickness and have estimated volumes of c. 2–30 km³. Note similar facies sequence despite widely different magma compositions.

known 'recent' examples. Caldera collapse generally occurs above the source magma chamber. Rarely, this may differ from the vent location, as with the 1912 eruption of Novarupta where a small caldera formed at the summit of Mount Katmai 10 km east of the eruption site. Calderas are known from both subaerial and subaqueous settings.

The rate of magma discharge, the total erupted volume, and the strength of the magma chamber roof, control whether and where the collapse occurs and the timing of collapse. Large eruption rates favour collapse of eruption columns to form pyroclastic flows (Sect. 12.5.2) and also favour caldera collapse. Caldera collapse can occur slightly before, during, or after the main phase of an eruption and it is sometimes difficult to distinguish cause and effect. An eruption may not be necessary to cause caldera collapse and vice versa. Subsidence at the summit of a basaltic shield volcano may be purely isostatic and caused by sinking of the great prism of high-density intrusive rocks and cumulates underlying the caldera into the thermally weakened lithosphere at the root of the volcano (G.P.L. Walker, 1988).

Caldera floors may subside as: (i) a relatively coherent piston-like block bounded by discrete steep ring faults; (ii) a piecemeal breccia; (iii) a funnel shape into a central area much smaller than the surficial caldera diameter; and (iv) a broad downsag of the landscape (Lipman, 1984; G.P.L. Walker, 1984). The type of collapse that occurs depends on several interdependent controls, including the strength of the subjacent lithosphere, the size and shape of the near-surface magma chamber, and the rate of magma discharge. The mechanism of caldera floor collapse influences the nature of the concurrent and later volcanic eruptions. Few calderas conform to a single model and most have a complicated history involving several mechanisms of collapse and more than one event. For example, nested volcanotectonic depressions of the Vulsini district (Italy) formed along an array of pre-existing extensional faults due to incremental subsidence during a series (at least seven) of moderate-sized eruptions that spanned several hundred thousand years and involved piecemeal, downsag and piston-like collapse in different sectors of the caldera floor (Nappi, Renzulli & Santi, 1991). Large silicic calderas (e.g. Santorini:

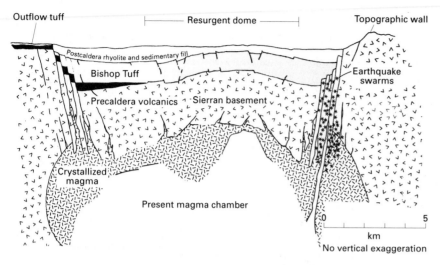

Figure 12.31 Cross-section of Long Valley Caldera based on surface exposures, drilling and geophysical data (from Hill, Bailey & Ryall, 1985). No vertical exaggeration. Note relative thickness of intracaldera (Bishop Tuff) and outflow tuffs. The topographic wall of the caldera is larger than its structural boundary due to gravitational collapse of unstable caldera margins. Earthquakes are attributed to the intrusion of magma along caldera margin faults.

Druitt & Francaviglia, 1992; Valles: Hulen, Nielson & Little, 1991) can also develop in several stages. Because they involve more widely spaced eruptions, it may be easier to separate individual caldera-forming events in the stratigraphic record.

Volcanotectonic processes are easy to overlook unless some sort of circular collapse depression can be found. But many calderas are not circular, their shape being controlled by local and/or regional fault systems and stress fields. Volcanism can also affect pre-existing regional structures. For instance, although Permo-Carboniferous basins in the Catalan Pyrenees were formed by regional strike-slip faults, their subsidence pattern was regulated by volcanism (Martí, 1991). Others note that accelerated and/or differential basin subsidence occurs during or shortly after volcanic eruptions (cf. Orton, 1995). Magmatism in highly extended terranes can also rotate principal stress orientations, thereby facilitating and perhaps initiating the low-angle detachment faults that often characterize these regions (Parsons & Thompson, 1993).

CALDERA FACIES

Pyroclastics produced during explosive caldera-forming eruptions can be divided according to their site of deposition and are described here since comparable facies are associated with both polygenetic intermediate and silicic volcanoes (Sects 12.12 & 12.13). *Intracaldera* deposits occur within the topographic wall (Fig. 12.31) of the caldera. Outflow sheets represent pyroclastic gravity currents formed when turbulent intracaldera pyroclastics 'escape' or 'boil-over' from the caldera container, or when portions of the eruption column collapse outside the caldera boundary.

Intracaldera deposits consist of thick (1–2 km) ash flow tuffs interleaved with volcanic breccias (Lipman, 1984). Many intracaldera tuffs show intense welding and rheomorphic flowage. Pyroclastic facies of subaqueous calderas differ little from their subaerial counterparts, particularly intracaldera

deposits which are lithologically rather uniform throughout and include welding and columnar jointing. However, if there are pauses in the eruption, subsidence can intermittently allow vents to become flooded so that high-temperature magmatic pyroclastic eruptions are punctuated by violent phreatomagmatic activity and deposition of fine-grained fallout ashes (Branney & Kokelaar, 1994).

Volcanic breccias within calderas are of three types.
1 Thick (up to 2 km) wedge-shaped breccias containing blocks hundreds of metres long that thicken toward the caldera wall or intracaldera faults. They pass distally into thinner (c. 50 m) laterally extensive (up to 10 km) tabular debris avalanche (Sect. 12.5.1) and debris flow deposits of decimetre to metre-sized clasts (Lipman, 1976). These breccias form by the collapse of unstable scarps created by large vertical displacements on caldera-margin or intracaldera faults.
2 Pyroclastic lag breccias that arise from fallout of large lithic fragments near the vent (Druitt & Sparks, 1982) (Sect. 12.4.4).
3 Intrusive breccias and tuffs emplaced along ring fractures during caldera subsidence.

Outflow sheets typically extend 50–200 km from the caldera as a comparatively thin (up to 200 m) but widespread deposit. They generally thin away from the caldera, but because of pre-existing topography can display considerable thickness variation due to ponding and secondary flow. Thick structureless deposits (Layer 2b of ignimbrite of Fig. 12.26) partially infill valleys while a thinner veneer of thinly stratified and cross-stratified facies, representing deposition from more expanded parts of the gravity current (Layers 1 or 3), mantles intervening ridges (cf. Fig. 12.23). Whereas the intracaldera ashflow sheets are often welded, the outflow facies are often not, an exception being thick, proximal valley-confined facies.

If the caldera occurs near or within the sea, dense portions of the pyroclastic gravity current may enter or be contained within water, whereas the upper part may remain less dense than sea water and travel >50 km across open water, as evidenced by the 1883 eruption of Krakatau. An ancient example of a subaqueous outflow is the Ordovician Lower Rhyolitic Tuff Formation caldera of North Wales, UK (Howells, Reedman & Campbell, 1986). The tuff, up to 55 m thick, can be traced >25 km from its source caldera and indicates deposition from a high-density turbidity current.

CALDERA TECTONICS

Only a small portion, perhaps as little as 10%, of the magma-chamber is erupted during a caldera-forming eruption, with the remaining magma taking hundreds of thousands to millions of years to cool fully. The caldera floor is extremely sensitive to movement of this magma, and numerous episodes of intracaldera uplift and subsidence are to be expected (Newhall & Dzurisin, 1988).

Vertical ground movements are best recorded at calderas located by the sea where elevation changes, forming land, are very conspicuous. A unique example is provided by the 2000-year historic record from Campi Flegrei caldera (Italy), first described by Lyell (see Dvorak & Gasparini, 1991). Volcanotectonic changes were most pronounced during the sixteenth century. For instance, a tiny eruption (0.03 km³) of ash in 1538 was preceded by 40 years of episodic uplift and land formation culminating in 7 m of uplift in the 2 days preceding the eruption at a site (Pozzuoli) several kilometres away.

Over longer 10^3–10^5 year time scales, renewed rise and vesiculation of magma can lift up the central core of the caldera as a *resurgent dome* (Fig. 12.31) (R.L. Smith & Bailey, 1968). The total amount of uplift can be dramatic, with over 500 m of uplift documented. However, with calderas smaller than c. 10 km in diameter, the magma may solidify before the caldera experiences resurgence (Marsh, 1984). Also, if the caldera floor was disrupted during subsidence (i.e. piecemeal collapse; common at stratovolcanoes) or if remaining magma is of lower viscosity, magma may leak through fractures rather than raise the cauldron block. The type and structure of the crust is also important. Evidence for resurgent doming is absent in the Taupo volcanic zone centres where young faulted crust is thought to prevent formation of high-level magma chambers (C.J.N. Wilson, Rogan *et al.*, 1984).

12.5 Sedimentary processes in volcanic terranes

Most volcanoes are active only for short periods. During intervening repose periods a succession of temporally and spatially overlapping processes carry volcaniclastic fragments away from the volcano. These include gravity-driven processes (debris avalanches) that are often associated with eruptions and processes in which water is the dominant agent of sediment transport (lahars, hyperconcentrated floods). If the volcano

is close to or within water, sediment can be transferred into deeper water, either directly or after a period of storage on the continental shelf, by submarine slides and sediment gravity flows.

12.5.1 Volcanic landslides and debris avalanches

Volcanic landslides range from small or moderate rock slides to massive slope failures, termed *sector collapse*, where a large sector of the volcanic edifice has collapsed to form *debris avalanches* (Fig. 12.32). Several historic subaerial debris avalanche deposits exceed 1 km³ and at least four avalanches larger than 10 km³ have occurred during the Holocene, one of the largest being the Socompa debris avalanche (Wadge, Francis & Ramirez, 1995). Subaerial avalanches typically move 200–300 km h⁻¹ (50–80 m s⁻¹) on steep proximal slopes and can travel more than 50 km from the volcano. Most debris avalanches are emplaced in a cold state, and at temperate volcanoes can include snow and ice blocks. Subaqueous volcanic land-slides may also form. Those on the flanks of the Hawaiian islands attain volumes of several thousand cubic kilometres and extend more than 200 km seaward from the present coast-lines to depths of nearly 5 km (J.G. Moore, Normark & Holcomb, 1994). In addition, numerous medium-sized landslides, having volumes of tens of cubic kilometres, are common in shallower water but have not been adequately mapped.

Most volcanic landslides are probably triggered by earth-quakes. Sector collapse also depends on the height and composition of the volcano and on climate. In subaerial settings it is most likely to occur (Francis & Wells, 1988): (i) above a threshold height of 2000–3000 m; (ii) on steep-profile dacitic stratovolcanoes rather than gentler more mafic volcanoes; and (iii) in wetter climates where volcanic rocks will more rapidly weather or be altered hydrothermally to form low-yield-strength clays. Oversteepening and slope failure is further promoted by intrusion of magma within the volcano and/or by tectonic movements (Duffield, Steiltjes & Varet, 1982; Siebert, 1984). Thus debris avalanches preferentially develop on volcanic slopes which are perpendicular to regional tectonic elements (i.e. trend of the arc).

Sector collapse produces a horseshoe-shaped amphitheatre that typically occupies a 40–70° sector of the volcanic edifice. The detached mass initially slides as large coherent blocks which can pile in the mouth of the amphitheatre as large backwards rotated blocks. In the Socompa debris avalanche, the largest block is more than 3 km long and 1 km broad and is bound by scarps more than 400 m high. The material is largely volcanic, but extremely large slides can include non-volcanic material from the underlying basement.

During sliding, debris blocks may dilate along jigsaw fractures that were formed by deformation during intrusion of the magma and/or stresses set up during associated explo-sions and initial sliding. Dilation results in a loss of strength,

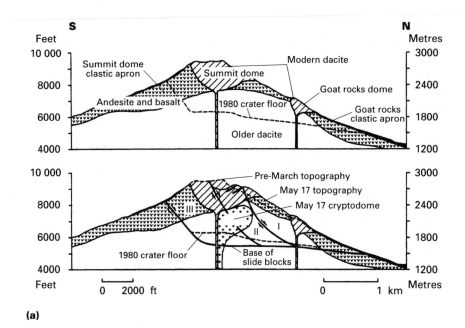

(a)

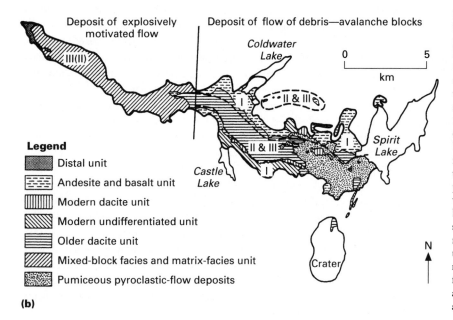

Legend

▨	Distal unit
☰	Andesite and basalt unit
▥	Modern dacite unit
◩	Modern undifferentiated unit
☰	Older dacite unit
▨	Mixed-block facies and matrix-facies unit
⠿	Pumiceous pyroclastic-flow deposits

(b)

Figure 12.32 (a) N–S cross-section of Mount St Helens just before the eruption showing pre-1980 geology and position of slide blocks I, II and III within the mountain. (b) Lithological map of debris avalanche deposit in the North Fork Toutle River valley with resting places of slide block fragments shown by Roman numerals (from Glicken, 1991): The lateral blast (Figs 12.16 & 12.17) occurred when the exploding cryptodome burst through slide block II. Slide block III failed as the cryptodome was still exploding; the continued explosions disrupted large portions of the slide block and intermixed it with fragments from the cryptodome and slide block II. This assemblage overtook the earlier slide blocks and flowed westward down the river valley to form the mixed-block facies unit (also labelled as deposit of explosively motivated flow). The pumiceous pyroclastic flow deposits were formed by later 'eruptions' and have no genetic relation to the debris avalanche deposit.

facilitating continued sliding, disaggregation of some large blocks into smaller fragments, and eventual transition to a debris avalanche flow. The maximum travel distance (L) of an avalanche is related to its vertical drop (H) and the volume of debris involved (Ui, 1983). Large avalanches travel further, not because they fall from a greater height, but because the greater avalanche volume decreases the rate vibrational (acoustic) energy is released from its basal layer during emplacement (Hsü, 1975; Melosh, 1987). They may even move uphill. For example, the distal reaches of some subaqueous debris avalanches off

Hawaii have travelled tens of kilometres up the slope of the Hawaiian Arch (J.G. Moore, Normark & Holcomb, 1994).

Fragments from the original slide block, termed *primary components*, can be mixed with *secondary components* consisting of anything ripped up during transport. The fine-grained component of a debris-avalanche is comparatively mobile and tends to flow further. The larger and more coherent blocks left behind give rise to the irregular hummocky topography characteristic of proximal debris avalanche deposits, with small conical hills and/or ridges up to 200 m high (hummocks)

(a) Lithofacies parallel to major flow direction

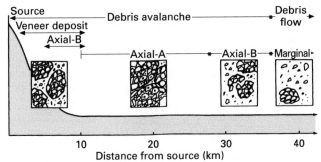

(b) Lithofacies transverse to major flow direction

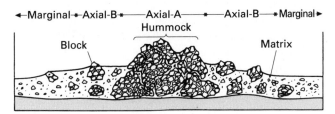

Figure 12.33 Characteristic lithofacies change in New Zealand debris avalanche/lahar deposits on the Egmont ring plain (from Palmer, Alloway & Neall, 1991).

separated by basins and shallow depressions (Fig. 12.33). Hummocks are cored by one or more coherent blocks (termed *debris blocks* or *megaclasts*) up to several hundred metres across and usually consisting of primary components and surrounded by a mixture (termed *mixed-facies* or *matrix*) of smaller fragments (Glicken, 1991; cf. Palmer, Alloway & Neall, 1991). The two facies often differ in composition. In the Socompa avalanche, for instance, large debris blocks are composed of strong materials such as lavas but most of the 'mixed-facies' is formed of stream gravels and ignimbrites.

The rate at which a volcanic landslide transforms to a debris avalanche flow depends on the strength of the slide block, which in turn depends on its composition, the amount of tectonic and volcanic fracturing, and the amount of weathering. As impressive debris avalanches occur in very dry environments (Moon, Mars, Socompa), water is not an essential ingredient for transport. It is, however, an important control on the type of facies formed. Two end-member rheologies are: dry, coherent and wet, incoherent debris avalanches.

Coherent debris avalanches develop from landslides of dense competent (hence dry) rocks, such as unfractured impermeable dacite lava (Siebe, Komorowski & Sheridan, 1992; Wadge, Francis & Ramirez, 1995). Flow motion is propelled mainly by stress transmitted due to collision between individual grains (i.e. Bagnold's grain flow). Although elsewhere such granular flows only occur on steep inclines such as aeolian sand dunes (Sect. 5.3.3), the high velocities developed in landslides allow

the mechanism to be effective on almost any slope, provided dense clasts are involved. Theoretical considerations show that almost all collisional activity will take place in a thin but dilute basal layer of highly-agitated and dispersed particles (high speed granular flow) while the main body of the avalanche rides above in a comparatively passive manner (C.S. Campbell, 1989). Consequently, little deformation appears to take place within the body of the debris avalanche, and subtle stratigraphic relations that were originally present in the volcanic edifice and successive retrogressive slide events can be mapped within avalanche deposits (Fig. 12.32). With coherent debris avalanches, sliding can prevail for a long distance from the source, hummocks and ridges are very steep (with slopes of 25–30°) and of high relief (forming internal ramps 40–50 m high), and even distal deposits have features indicating a high yield strength, including levees, sharply defined flow margins, terminal scarps (up to 50 m) and huge (tens of metres) surficial boulders. The avalanche stops suddenly when the basal shear strength falls below the yield strength of the flow. When grinding to a halt it can transmit much of its final momentum into underlying deposits. If these are unlithified, folding and thrust faulting can extend for tens of metres beneath the debris avalanche deposits (e.g. Siebe, Komorowski & Sheridan, 1992).

Incoherent debris avalanches usually contain more water, probably greater than 20%, and more matrix. Their development is promoted by fine-grained source materials, the presence of hydrothermal and/or magmatic fluids, pre-failure fracturing within the edifice, and incorporation of water-saturated sediment or ice/snow during transport. This leads to more tongue shaped and flat-surfaced (less hummocky) deposits, lower and shallower flow fronts, and marginal or distal, transformation to matrix-rich lahar deposits (e.g. Endo, Sumita *et al.*, 1989; Palmer, Alloway & Neall, 1991). Wet subaerial debris avalanches stop more slowly and travel further because of their lower yield strength.

The distribution of either type of debris avalanche is also influenced by the volcano's topography. Except in proximal areas where avalanches may overtop ridges and spill into adjacent drainages, debris avalanches are likely to travel down major valleys. With channelized avalanches, hummocks are often absent because matrix material is prevented from dispersing and megablocks may undergo additional shearing and fragment into smaller blocks. In contrast, when not confined by valley walls and in areas of low relief, the avalanche can spread laterally and thin such that the large megaclasts become grounded and matrix is allowed to 'drain' away.

Unlike their subaerial counterparts, the margin of modern subaqueous debris avalanches, at least off the Hawaii islands, consists of a broad zone of isolated and detached blocks. Within the toe of the South Kona landslide, giant blocks up to 10 km long and 500 m high are separated by a 10–15 km belt of numerous smaller 1–3 km long slide blocks (J.G. Moore, Bryan et al., 1995). These blocks apparently moved 60–80 km from their site of origin, gliding about 40 km across relatively flat sea floor in advance of the debris avalanche. Isolated blocks also characterize the distal lobes of the Alika landslide of Hawaii (Lipman, Normark et al., 1988). These isolated blocks may be giant analogues of the large 'out-runner' clasts in the toe zone of gravelly steep-face deltas (cf. Nemec, 1990b). Alternatively, the intervening and missing matrix material may have been dispersed further downslope as turbidity flows. For instance, turbidite deposits over 10 m thick and more than 300 km west of Hawaii are thought to be a distal facies of one of its landslides (Garcia & Hull, 1994).

The hummocky surface topography used to identify modern debris avalanches is preserved for a significant period of time (i.e. hundreds of thousands of years) only in arid climates where erosion rates are low. Thus, ancient debris avalanches are best distinguished by the presence of abnormally large, atypical or brecciated clasts, and an unusually large (>100 m) thickness or lateral extent (e.g. Ballance & Gregory, 1991; Fortuin, Rope et al., 1992; Macdonald, Moncrieff & Butterworth, 1993).

Large landslides cannot only reduce the load on a volcano, triggering a volcanic eruption and altering the topography of the pre-existing landscape, but they can also generate tsunamis if the resultant debris avalanche crashes into an ocean or lake. These tsunamis may have severe consequences. For example, giant tsunami waves generated by the Hawaiian landslides are thought to have swept blankets of coral–basalt–breccia–conglomerate 60–100 m above present-day sea level on islands of Hawaii about 100 ky ago and 240–200 ky ago (J.G. Moore, Bryan & Ludwig, 1994), and crossed the Pacific, causing catastrophic erosion on the eastern coast of Australia (Young & Bryant, 1992).

12.5.2 Lahars

Lahar is an Indonesian term originally used to describe mud-flows containing volcanic debris, but the term has recently been extended to describe all rapidly flowing mixtures of rock debris and water (other than normal streamflow) from a volcano (G.A. Smith & Lowe, 1991). They differ from debris avalanches (Sect. 12.5.1) in that water is an essential ingredient in sediment transport. Lahars are *flood events* that involve a number of different flow processes covering a wide spectrum of sediment/water ratios. The definition also recognizes that some aspects of these processes are unique to volcanic areas, but are not always directly related to eruptions, a link that is difficult or impossible to establish in ancient successions. Although largely used in subaerial settings, the term lahar has also been used for subaqueous volcanogenic mass flows (e.g. Mitchell, 1970).

Lahars require an abundance of both loose sediments and water. Steep slopes and sparseness of vegetation are commonly, but not always, important contributing factors. Some lahars are a direct result of eruptive activity. Others are not temporally related to eruptions.

During eruptions large amounts of water and/or sediment can be liberated by:
1 Ejection of crater-lake water (cf. Major & Newhall, 1989).
2 Melting of snow and ice by hot pyroclastic density currents, including subglacial pyroclastic eruptions (Maizels, 1989). For example, the 1985 eruption of Nevado del Ruiz, with less than 5 million m³ of magma, generated 20 million m³ of water which transformed into 90 million m³ of lahar deposits (Pierson, Janda et al., 1990). Such eruptions can also generate a hybrid type of mass-flow, termed *mixed avalanches*, when snow avalanches incorporate additional glacier ice and rock debris (Pierson & Janda, 1994).
3 Liquefaction of just-deposited debris-avalanche deposits. For example, the largest recent lahar at Mount St Helens (100 million m³) was initiated approximately 5 h after the 18 May 1980 eruption and avalanche emplacement occurred at a site about 18 km from the volcano when comminuted interstitial ice melted rapidly and saturated the host avalanche debris (Scott, 1988a).

Between eruptions lahars can be triggered exclusively by rainfall (Rodolfo & Arguden, 1991). Seepage of rainfall or groundwater into hot pyroclastic debris can also cause violent steam (phreatic) explosions that further mantle the landscape with loose pyroclastic debris, which is later mobilized as lahars (Pierson, Janda et al., 1992). Extremely large post-eruptive lahars can also be triggered by breakouts from transient lakes dammed by valley-fill deposits. The largest historic flood event at Mount St Helens, for instance, formed 2500 years ago when ancestral Spirit lake breached a debris avalanche dam. The resultant flood is calculated to have had a peak discharge more than three times the largest recorded flood peak of the Mississippi (Scott, 1988b).

Flowing sediment–water mixtures can be classified on sediment concentration and/or thresholds in rheological behaviour (Beverage & Culbertson, 1964; Pierson & Costa, 1987), and range from dilute streamflows through hyperconcentrated to debris flows (Figs 12.34 & 12.35).

With *normal streamflow*, turbulence is important and fluid and granular phases act independently, sand or gravel particles being free to settle grain by grain to develop traction structures or bedforms. As the concentration of suspended particles increases, a point is reached where these particles begin to interact and the fluid acquires a yield strength (Fig. 12.34). The concentration at which this threshold is crossed is highly dependent on the amount of clay and the clay mineralogy. Flocculated suspensions of smectite can develop measurable yield strengths at concentrations of 3% by volume whereas coarse pumiceous suspensions can remain a Newtonian fluid until frictional interaction begins – about 50% by volume.

Hyperconcentrated flows are intermediate in character, having sediment concentrations of 20–60% by volume in mud-rich

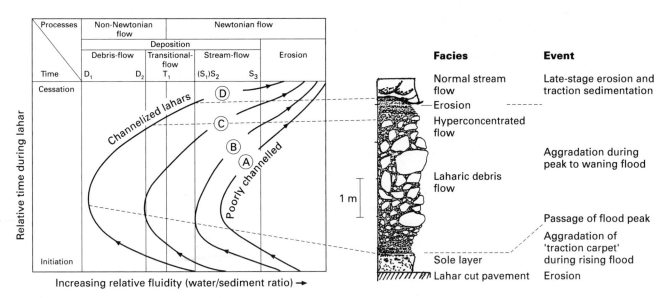

Figure 12.34 Conceptual model showing the changes in water:sediment ratio, and the sequence of depositional and erosional processes observed in modern lahars. Bold lines and associated arrows illustrate the relative timing of facies deposition, resulting stratigraphic sequence, and relative timing and amount of erosion for various lahar events. Lines A and B represent flow regime paths for streams. poorly channelized lahars (e.g. jökulhlaups) that remain fluid throughout the lahar event. Whereas lines C and D represent channelized lahars which bulk up to form cohesive debris flows. Column on right depicts an ideal lahar (line D) and how it might be interpreted in this context. See text for further discussion. Idea and diagram on left modified from Wells and Harvey (1987).

streams. Turbulence is dampened by the higher sediment concentrations, with dispersive pressures generated by intergrain collisions and/or buoyancy playing a significant role in particle support. Sand and low-density gravel are typically carried in suspension whereas larger outsized boulders (up to 1 m) are transported as bedload (Pierson & Scott, 1985). Owing to the high sediment concentrations, sand aggrades on channel beds as well as being deposited on floodplains, although channel deposits are usually reworked by later streamflow. Hyperconcentrated flow deposits are generally massive or very crudely stratified, reflecting the rapid and generally uninterrupted deposition from suspension (Fig. 12.35) (G.A. Smith, 1986; Maizels, 1989). Cross-stratification is absent, and scour structures, reflecting localized erosion, are rare, even where outsized clasts are deposited with sand. Depositional units are typically tens of centimetres to metres thick, and often normally graded with stratification better developed in upper fine-grained parts. These subaerial deposits resemble those of subaqueous high-density graded–stratified turbidity currents (e.g. Postma, Nemec & Kleinspehn, 1988; see Sect. 10.2.3).

Debris flows, with water contents ranging from about 10 to 25%, are non-Newtonian fluids in which clasts are largely supported by buoyancy and matrix strength. Flows move as fairly coherent masses in what is thought to be a predominantly laminar fashion. In general terms, volcanic debris flows and their deposits are not radically different from those generated in non-volcanic settings. The two principal differences are clay content and scale (G.A. Smith & Lowe, 1991; Pierson, 1995).

Volcanic debris flows are: (i) generally clay-poor, few being mudflows, because most form by mobilization of fragmental pyroclastic and autoclastic debris soon after eruption; and (ii) very large, travelling faster and farther than other terrestrial debris flows. For example, a lahar from the 1877 eruption of Cotapaxi travelled 270 km from its source to the Pacific. This is in contrast to non-volcanic debris flows on alluvial fans that rarely extend more than 10 km from their source (G.A. Smith, 1988). However, resultant deposits are invariably much thinner than the flows. For instance, the largest lahar generated by the 1985 eruption at Nevado del Ruiz was more than 45 m deep; deposits are at most 1–2 m thick. Owing to the high velocity of their emplacement, lahars may also leave deposits that mantle rather than infill preflow topography.

Facies of volcanic debris flows vary between channels and unconfined sites such as floodplains or alluvial aprons (Fig. 12.35). *Channel debris flow facies* contain large clasts that: (i) accumulate as linear clusters of outsized boulders aligned parallel to the direction of flow (Pierson, Janda *et al.*, 1990; cf. Brayshaw, 1984), sometimes forming longitudinal streamlined bars of poorly sorted inverse to normally graded clast-supported gravels ('whale-back bars'; Scott, 1988a); or (ii) are dumped more randomly as depositional terraces of coarse, matrix-supported debris flow deposits (Rodolfo, 1989). Channel facies of the largest lahar events sometimes include a basal, comparatively well-sorted 'sole-layer' or 'ball-bearing' bed, comprising 10–15% of the deposits. It consists of inversely graded crudely stratified pebbly sands or clast-supported pebbles,

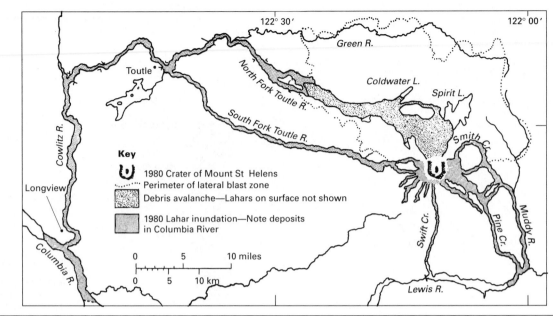

(a)

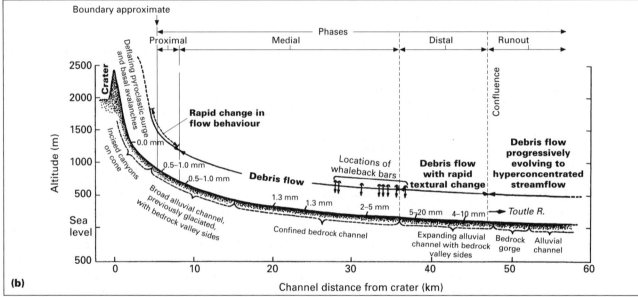

(b)

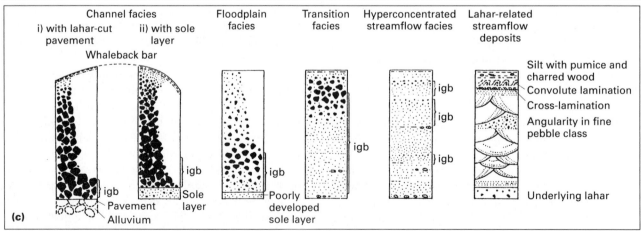

(c)

Figure 12.35 (a) Distribution of lahars about Mount St Helens. (b) Downstream transformation of the lahar of the South Fork of the Toutle river. (c) Typical lahar facies (largely from Scott, 1988a); 'igb' denotes inversely-graded bedding. Lateral blast-zone is further detailed in Fig. 12.17 whereas debris-avalanche deposit is detailed in Fig. 12.32.

attributed to cataclasis and dispersive pressure within a basal shear zone, and is thought to be deposited near peak discharge (Scott, 1988a). Debris flow deposits at individual channel cross-sections are typically coarser at lower levels and fine progressively upwards toward and on to the floodplain, indicating that lahars were not homogeneous but density-stratified during transport.

Unconfined debris flow facies are usually matrix supported, and are either ungraded or display normal grading of the larger clasts (Scott, 1988a; Arguden & Rodolfo, 1990; Best, 1992). They range from 1 to 6 m in thickness, with thickness proportional to the size of the flows. Sole-layers are finer grained, thin or absent, and outsized clasts are common on tops of beds.

Once initiated, a single lahar or a succession of lahars can change in character, either with time and/or distance downvalley, from debris flow to hyperconcentrated flow and even normal stream flow and vice versa (Figs 12.34 & 12.35). These rheological changes largely reflect variation in the sediment/ water ratio due to addition of either sediment (bulking) or fluid (dilution).

Bulking is caused by erosion of channels, which can be several tens of metres deep, into unconsolidated sediments (Rodolfo, 1989). If lahars erode to bedrock, they can produce an abraded pavement similar to glacial pavements. Eroded material is incorporated in the flow, thus increasing flow volume, sediment concentration and peak flow rate. It is estimated that 50–90% of the particles in lahars are added to the flows by erosion (Scott, 1988a; Pierson, Janda *et al.*, 1990).

Dilution occurs when streamwater is added to the flow head, the saturated streambed is eroded and incorporated, or coarser particles are deposited upstream (Pierson & Scott, 1985). The depositional record of this dilution is a bipartite unit with basal hyperconcentrated flow deposits grading upwards and upstream into debris flow deposits (Fig. 12.35). Dilute flows can also occur late in a lahar event due to passage of the dilute tail of the main flood wave and/or run-off from dewatering of debris flow deposits (Scott, 1988a; Maizels, 1989). These flows often erode the underlying peak flood deposits, and produce hyperconcentrated or cross-stratified streamflow facies (Fig. 12.34).

Most lahars change their flow state several times. Debris flows may 'freeze' to block the channel, only to be soon remobilized as hyperconcentrated flows if streamflow ponds behind and breaches this temporary dam. Streamflows from tributaries dilute a debris flow; conversely, slumps from channel walls locally bulk a hyperconcentrated flow. If the lahar is split by obstacles *en route*, the individual parcels may arrive at a distal location at different times. Erosion by early lahar pulses may remove channel roughness elements (e.g. trees) and create hydraulically smooth channels that allow later lahars or lobes to travel more rapidly and overtake earlier ones. The result of all this complexity is that at any one site a single lahar may leave a stack of very diverse deposits, often separated by unconformities.

Most of the facies observed in modern lahar deposits have been recognized in their ancient counterparts (Fig. 12.36) (e.g. Schmincke, 1967; G.A. Smith, 1987a, 1988; Walton & Palmer, 1988). Ancient lahar deposits, however, even those a few thousand years old, typically have high clay contents because of the susceptibility of volcaniclastic particles to diagenesis. They may resemble pyroclastic flow deposits, with distinction complicated by the fact that some lahars carry hot rock fragments. In fact, the recent events at Mount St Helens and Pinatubo reveal that pyroclastic flows and lahars are end-members

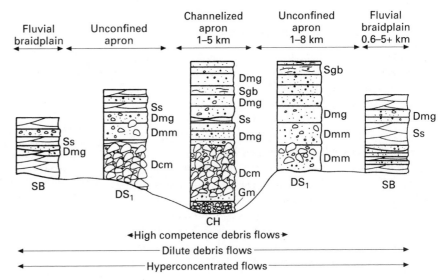

Lahar-dominated ring-plain system

Figure 12.36 Lateral relations among architectural elements of a lahar dominated ring-plain, Egmont Volcano (from Palmer & Neall, 1991). These facies relations can occur either transverse to flow according to distance away from channel axis, as depicted, or with distance away from the volcano. Architectural elements are: CH; channel; DS, diamicton sheet; SB, sandy bedform sheet. Facies codes are: Dcm, clast-supported ungraded diamicton; Dmm, matrix-supported ungraded diamicton; Dmg, matrix-supported graded diamicton; Ss, low-angle, trough cross-bedded sand; Sgb, poorly bedded sand.

of a complete spectrum of pyroclast-rich sediment gravity flows, ranging from hot flows in which any water exists as steam, to cold lahars in which water is a liquid, with many intervening scenarios. The best criteria to distinguish between the two end-member types is often provided by associated facies. Intercalated beds of primary fallout ash and/or surge deposits, carbonized plant fragments, or exceptionally thick or laterally extensive marker beds suggest pyroclastic flow deposits.

12.5.3 Coastal processes

Pyroclastic material introduced into shallow subaqueous settings is sorted by wave and tidal processes. The type of coastline can change rapidly as the size, density and amount of the sediment available is modified. For instance, the formation of Holocene beach ridges along the Caribbean coast of Costa Rica can be related to discontinuous sand supply by andesitic volcanic eruptions that compensated for, or even reversed, eustatic sea-level rise (Nieuwenhuyse & Kroonenberg, 1994).

An ancient example is provided by the Ordovician Lower Racks Member, North Wales (Fig. 12.37), which records sedimentation following emplacement of a largely welded vitric-rich (95% glass shards) ash flow tuff into a high wave energy epicontinental sea. Following emplacement, its unwelded top was removed by subaerial erosion with the reworked pyroclastics 'dumped' into the marine realm. Lowermost reworked deposits are structureless, and probably represent deposits from subaerially generated debris flows. The abundance of fine-grained low density sediment in the sea favoured development of a low-gradient dissipative coastline dominated by suspension processes. However, the grain size and density of nearshore sediment increased with time as lighter glass shards were removed by winnowing. This increased the reflectivity of the shoreface, and eventually allowed development of an intermediate beach state involving rip-currents and traction sedimentation.

12.6 The stratigraphic record of volcanism

In volcanic terrains, volcanic and sedimentary processes combine to generate diverse volcaniclastic deposits. Clasts that are generated and initially deposited by volcanic processes may be rapidly or subsequently reworked by non-volcanic surface processes. Volcaniclastic sediment may be additionally supplied to adjoining basins by weathering and erosion during periods between eruptions, which may be quite long.

A stratigraphic framework for this sedimentation is best provided if it can be determined when volcanism was active, and if one tries to group deposits accordingly (Fig. 12.38). An *eruption* is the volcanic event that forms the most fundamental focus for studying deposits. A single eruption usually lasts a few days to months, large eruptions for a few years to decades. Eruptions are often subdivided into: (i) *eruptive pulses* that may

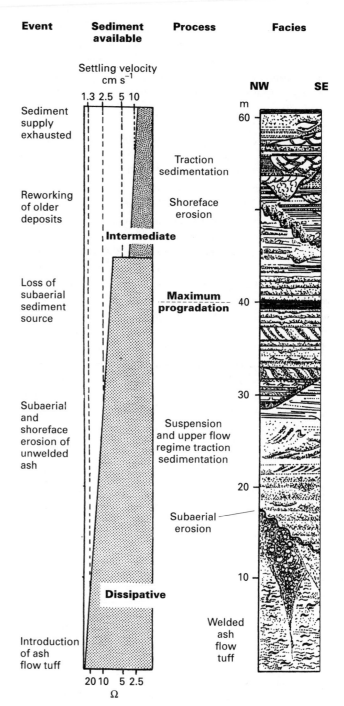

Figure 12.37 Dynamics of sedimentation following emplacement of an Ordovician ignimbrite into shallow water. Settling velocity (top) of available sediment in cm s^{-1}. Ω is a dimensional index incorporating breaker wave height, wave period, and sediment characteristics. Observations of L.D. Wright and Short (1984) from modern coastlines suggest Ω must be greater than 6 for a beach to remain fully dissipative.

last a few seconds to minutes; and (ii) *eruptive phases* that last a few hours to days and consist of numerous eruptive pulses. For convenience modern eruptions are typically defined in terms of the length of the quiet periods between eruptions (Simkin, Siebert *et al.*, 1981); thus an eruption which follows its predecessor after a few months is considered a phase of the earlier eruption. Fisher and Schmincke (1984) introduced the term *eruption unit* to describe the volcanic material (i.e. sediment and lava) deposited during a single volcanic eruption whereas the term *tephra event* (Fig. 12.38) has been used to embody both the period of the eruption and the immediate posteruptive period when considerable resedimentation occurs.

It is sometimes useful to conceptualize longer time periods of volcanic activity when considering large volcanic landforms or volcanic fields. For example, large volcanoes are constructed by many eruptions; these are not spaced equally but cluster within *eruptive periods* or *stages*, as at Mount St Helens (Hopson & Melson, 1990) or Santorini (Druitt, Mellors *et al.*, 1989), separated by significant time intervals. There may be cyclic compositional or eruption patterns within or between eruptive periods due to evolution of the underlying magma body. Experience shows that interpretation and recognition of volcanic activity is more easily done for active or very young volcanoes, but becomes increasingly difficult for older sequences where mappable stratigraphic units typically contain products from many phases or eruptions.

12.6.1 Syn-eruptive versus inter-eruptive periods

Within the above context, sedimentary sequences associated with active volcanoes are often considered to consist of two elements (G.A. Smith, 1991): (i) syn-eruptive sequences, which result from primary volcanic processes and immediate post-eruptive reworking; and (ii) inter-eruptive sequences, which record the deposition without significant influence of volcanic activity where normal sediment delivery processes are dominant (Fig. 12.38).

Syn-eruptive deposits are a direct consequence of a volcanic eruption. They include not only volcanic and contemporaneous sedimentary processes during the eruption but also penecontemporaneous sedimentary processes following volcanism while hydrological and hill-slope conditions remain disturbed (Figs 12.39 & 12.40). In humid climates, areas without vegetation and under-water syn-eruptive conditions may continue for a long time, at least one or two decades, until most loose debris from the eruption is completely removed.

Large eruptions typically involve most of the volcanic and sedimentary processes that were described earlier (Sects 12.4 & 12.5). Volcanic processes during an eruption largely depend on physical and chemical processes within the magma reservoir feeding the volcano (Sect. 12.3). These determine the style of volcanic activity or mechanism of clast formation (Sect. 12.4.1) and can change extremely rapidly either because volatile and

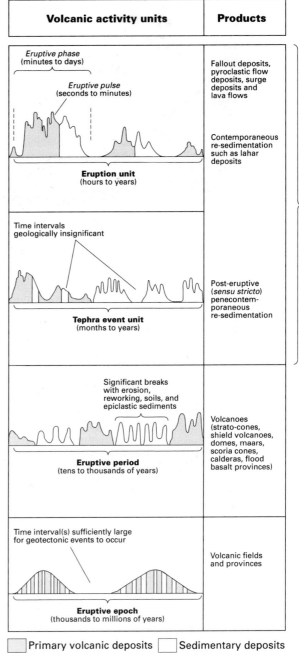

Figure 12.38 Hierarchy of volcanic activity units and associated products (after Fisher & Schmincke, 1984; modified from Schmincke & Bogaard, 1991). Sedimentary deposits (white areas) include syn-eruptive deposits from resedimentation during and immediately after the volcanic eruption as well as inter-eruptive pyroclastic (*sensu* R.V. Fisher) and epiclastic deposits from later reworking.

chemical gradients within the magma chamber allow magma properties to change or because boundary conditions (e.g. the amount of water in a surrounding aquifer) may change. In

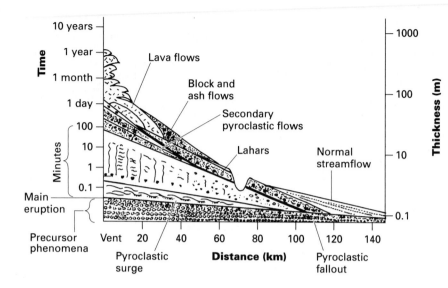

Figure 12.39 Schematic illustration of syn-eruptive volcanic and sedimentary processes associated with a volcanic event. Vertical axis is time and/or thickness. Lateral axis is distance from vent. Note that the time axis is logarithmic so that the rapid succession of events, and their deposits, during the main phase of the eruption and close to the volcano can be displayed.

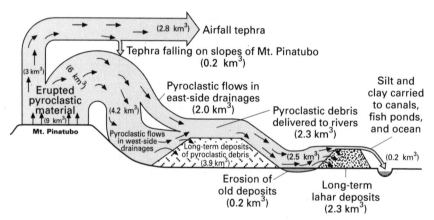

Figure 12.40 Schematic portrayal of transport and distribution of pyroclastic material produced by the June 1991 eruptions of Mount Pinatubo. Thickness of flux arrows is proportional to the estimated volume of material. Lateral scale is about 30 km (from Pierson, Janda *et al.*, 1992).

general, as an eruption proceeds, it becomes less explosive, more mafic and the temperature of erupted clasts increases. The decrease in explosivity is generally matched by a decrease in the distance primary volcanic deposits extend away from the vent (Fig. 12.39). The final phases of many eruptions involve extrusion of lavas, and intrusion of magma, directly above, within, and immediately adjacent to the vent. At any one time slice of the eruption, transport processes also change away from the vent with hot pyroclastic and volcanic processes giving way distally to cold sedimentary processes.

During and following subaerial eruptions, sediment erosion and delivery rates are high, not only because of the large volumes of unconsolidated material, but because sediment-stabilizing vegetation is often destroyed. During the year following the eruption of Mounts Pinatubo and St Helens, for instance, the amount of sediment eroded was comparable with the annual suspended sediment load of the Mississippi and several orders of magnitude greater than pre-eruption conditions. However, rates of sediment supply decrease exponentially (Pierson, Janda *et al.*, 1992); at Mount St Helens, more than half of the sediment that would eventually wash down to lowland rivers did so within the first 5 years. The eruption of one volcano can also temporarily increase erosion rates of neighbouring volcanoes by at least one order of magnitude (C.A.L. Wood, 1980b). The eruption of Volcan Paricutin in Mexico blanketed surrounding older cones with fine-grained tephra, killing the slope-stabilizing plants, and causing gullies to erode into pre-eruption slopes (Segerstrom, 1950).

There are two types of volcanic deposit that can be resedimented. *Landscape-mantling deposits*, including tephra fall-out and some pyroclastic surge deposits, are deposited in comparatively thin sheets of unwelded pyroclastics that mantle topography and/or cap eruption sequences. These sediments are quickly eroded by sheetwash or rill and gully erosion and a distinctive topography emerges with relics of the original post-eruptive surface separated by a dendritic maze of narrow gullies. Sediment yield from these sources declines rapidly within a few years as the gullies reach permeable and less erodible substrates and resistent surface crusts develop (Collins & Dunne, 1986). *Valley-fill deposits*, such as debris avalanche, pyroclastic flow and lahar deposits erode not from the surface, but from the side (e.g. channel widening, bank erosion and collapse) with sediment yields less affected by post-eruptive re-establishment of vegetation.

A historic example of syn-eruptive resedimentation is provided by the 1902 eruption of Santa María (Kuenzi, Horst & McGhee, 1979) (Fig. 12.41). This generated at least 5.5 km³ of loose andesitic–dacitic pyroclastic debris that was initially deposited as a 20–30-m-thick mantle on the steep upper slopes of the volcano. Erosion of this sediment caused the lower reaches of the Samalá River to aggrade 10–15 m, damming several pre-1902 tributaries to produce elongate lakes which were filled by organic-rich muds or prograding deltaic deposits. In addition, over the next 20 years, a delta at the mouth of the Samalá, 60 km from the summit, prograded nearly 7 km seaward in response to deposition of about 4 km³ of volcaniclastic sediment. Subsequently, deltaic sands have been redistributed laterally to prograding shoreface and beach environments.

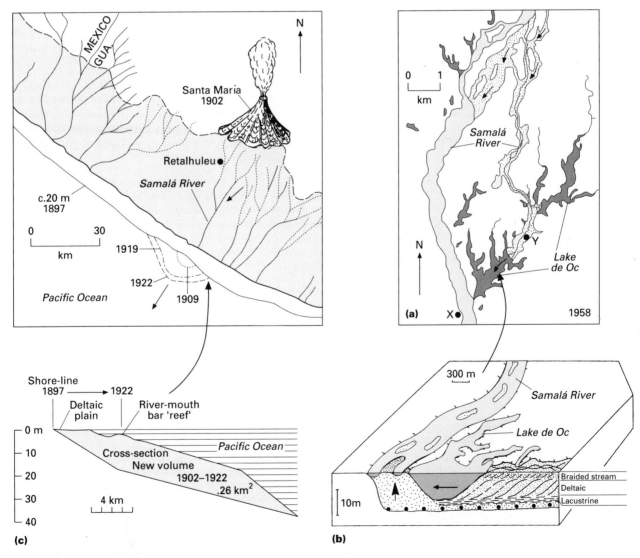

Figure 12.41 Effects of the 1902 eruption of Santa María. (a) Aggradation of braided stream deposits within the Samalá River blocks tributaries, forming lakes, which are then (b) infilled by small lacustrine deltas. Marine deltas (c) also developed at the coastline (from Kuenzi, Horst & McGhee, 1979).

Syn-eruptive sedimentation is strongly aggradational, with deposition occurring within depressions (e.g. calderas) created by the volcanism, on aprons of sediment surrounding the volcanic centre, or in distal sites receiving fallout tephra. Syn-eruptive stratigraphy is an event stratigraphy with time planes essentially parallel to facies boundaries. Owing to rapid change in the type of fragment produced, stratigraphic units represent events spanning time frames of only seconds to months. Syn-eruptive facies have a high preservation potential because they are relatively thick and laterally extensive, and because the volcanic eruption often creates accommodation space by growing higher, forming calderas, or driving basin subsidence.

During *inter-eruptive periods*, sediment delivery is greatly diminished and non-volcanic sediment transport and deposition processes are dominant. Almost any depositional environment and facies could conceivably occur, depending on the topography, relief, type of bedrock, climate, subsidence history, etc. of the volcanic field. Subaerial high-relief volcanoes are commonly drained by gravel-bedload streams (Vessel & Davies, 1981). Low-relief volcanic fields are characterized by alluvial plains, lakes, and playa environments where fine-grained mud is deposited (Buesch, 1991; Besly & Collinson, 1991). Inter-eruptive deposition, particularly in subaerial settings, depends on accommodation space being created by base-level adjustments. Sedimentation is thus characterized by erosion and channel incision, particularly in proximal settings, combined with localized distal aggradation if storm-induced floods or waves are able to erode sediment. New volcaniclastic particles created solely by surface weathering and erosion, termed *epiclasts*, and biogenic particles may also be deposited at this time. However, weathering and erosion of pre-existing poorly or non-welded syn-eruptive deposits can simply release the original pyroclasts or autoclasts and rapidly provide large volumes of recycled material. As a result only a small proportion of particles present in most volcano-sedimentary systems are true epiclasts (in the grain origin sense).

12.6.2 Geometry of volcanic successions and their controls

The overall geometry of a volcanic succession is controlled by the relative importance of syn-eruptive versus inter-eruptive conditions, which in turn is controlled by the *frequency and magnitude of volcanic eruptions, rates of tectonic or volcanic subsidence, rates of erosion*, and *position with respect to the centre of the volcano* (Vessel & Davies, 1981; G.A. Smith, 1991; Palmer, Purves & Donoghue, 1993).

The magnitude and frequency of volcanic eruptions are the most important controls. High eruption frequencies and large eruptions allow little or no depositional record of inter-eruptive periods (Fig. 12.42). The magnitude of historic volcanic eruptions is commonly categorized using a volcanic explosivity index (VEI) which closely parallels the volume of erupted tephra and composition of magma (Newhall & Self, 1982). It is a logarithmic scale and increases by an order of magnitude with each VEI integer. Like earthquakes, the frequency of eruptions decreases with increasing size (Fig. 12.43). In general, the frequency of eruptions at an individual volcano decreases as the silica content of the volcanic system increases. Basaltic systems have small eruptions every 1–10 years, andesite–dacite systems have intermediate-sized eruptions every 100–10 000 years, and silicic systems have large eruptions every 10 000–1 000 000 years.

Low rates of basin subsidence also inhibit the accumulation of thick inter-eruptive deposits. This is especially true in continental settings (G.A. Smith, 1991). However, at subaqueous volcanoes, or volcanoes near oceans, the surrounding basin is usually so deep and wide that it never becomes filled during even the largest volcanic events; thus there is usually accommodation space available to accumulate inter-eruptive facies, although not necessarily in proximal settings.

Rates of erosion determine how quickly volcaniclastic deposits are resedimented between volcanic eruptions. In subaerial settings this is chiefly determined by climate. Deep valleys can

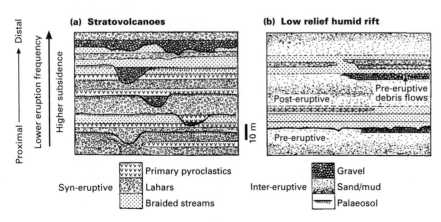

(a) Stratovolcanoes   **(b) Low relief humid rift**

Distal ← → Proximal

Lower eruption frequency

Higher subsidence

Post-eruptive

Pre-eruptive debris flows

Pre-eruptive

10 m

Syn-eruptive
- Primary pyroclastics
- Lahars
- Braided streams

Inter-eruptive
- Gravel
- Sand/mud
- Palaeosol

Figure 12.42 Effect of eruption frequency and subsidence rate on facies geometries. At high relief stratovolcanoes (a) laterally extensive syn-eruption sheets are incised during inter-eruptive periods, with the lateral extent and thickness of inter-eruptive deposits greatest where eruption frequency is low relative to the rate of basin subsidence. In rapidly subsiding continental rifts (b) inter-eruptive deposits are comparatively thick; debris flows and incision may sometimes occur *beneath* syn-eruptive sheets due to volcano-tectonic uplift before the eruption (from G.A. Smith, 1991; Orton, 1995).

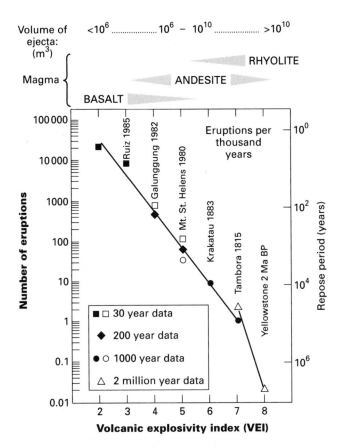

Figure 12.43 Magnitude and frequency of Holocene eruptions. Data show cumulative number of eruptions for each VEI class, which are labelled using well-known eruptions. The VEI scale is logarithmic since each integer represents a 10-fold increase in the volume of tephra erupted and the effect of the eruption. The 1980 eruption of Mount St Helens (VEI 5) was a *local* catastrophe, the 1883 eruption of Krakatau was *regional*, whereas the 1815 eruption of Tambora, and presumably Yellowstone, caused cooling on a hemispheric to *global* scale. Time intervals between eruptions at *individual* volcanoes are suggested on right (modified from Simkin, 1993).

be carved solely by fluvial processes, particularly in tropical climates. Two months after the Krakatau eruption of 1883, gullies 40 m deep had been cut into the pyroclastic deposits (Verbeek, 1886). In humid climates a high proportion of pyroclastic deposits is redistributed by fluvial processes. For instance, much of the 'normal' sedimentary load of streams on Ruapehu volcano comprises products reworked from small (<0.1 km³) strombolian and phreatomagmatic events (Hackett & Houghton, 1989); on Fuego volcano only eruptions greater than 0.06 km³ were capable of triggering sedimentary events large enough to modify the landscape (Vessel & Davies, 1981). Thus the stratigraphic record in humid (and subaqueous) settings will only clearly preserve the effects of volcanic events larger than a certain geomorphic threshold.

Glaciers affect sedimentation on high-standing volcanoes in humid climates by providing a summit reservoir of water that can be utilized in volcanic eruptions and through glacial erosion. Glaciers develop mostly above latitudes of 35° (e.g. Mounts Rainier and St Helens) but also on high (>4000 m altitude) equatorial volcanoes constructed on an elevated basement (e.g. Cotapaxi, Ecuador; Kilamanjaro, Kenya; Nevado del Ruiz, Columbia). Some of the most voluminous and catastrophic lahars and floods result when pyroclastic flows scour and melt summit snow and ice (Major & Newhall, 1989; Pierson, 1995). The deep fluvioglacial valleys dissected around such volcanoes contain permanent streams and act as conduits through which lahars can transport sediment to depositional sites 50–500 km beyond the flanks of the volcano. Lahar deposits are confined to these valleys and occur as radial belts ranging from 5 to 40 km in width (G.A. Smith, 1988). These lahars are extremely erosive and obtain most of their sediment through bulking (Sect. 12.5.2), including boulders previously fragmented by and/or transported by glaciers. Thus widespread diamictites in volcanically active regions containing striated and glacially shaped boulders may record lahars rather than being till from episodes of extensive glaciation (C.H. Eyles & Eyles, 1989).

In contrast, with more arid climates, the volcanic edifice is flanked by a relatively featureless alluvial plain with less pre-existing topography to guide accumulation. Lahars consist largely of unchannelized debris flows, and construct broad, arcuate aprons extending only 20–40 km from source (e.g. Palmer & Walton, 1990; Waresback & Turbeville, 1990; Turbeville, 1991). Aeolian processes may also mantle the volcano with fine ash, further smoothing topographic irregularities. Although most volcanic events make an impact on the sedimentary environment, so do high-discharge events from storms. Little channel incision occurs between volcanic events and the absence of erosion surfaces makes distinction between inter-eruptive and syn-eruptive sedimentary deposits difficult (G.A. Smith, 1991).

High-standing volcanic landforms may show a range of climate zones from subtropical to polar (see Sect. 2.1.2) and the opposing sides of a volcano can have marked differences in climate because of orographic effects (e.g. Tenerife, Canary Islands). Moist maritime summit climates often generate perennial streams, some of which flow landward across semi-arid to arid basins that lie in the rainshadow of the arc (e.g. Cascades, USA).

The rate of degradation of subaqueous volcanoes depends on wave and tidal processes. The most important control on the strength of these processes, from the point of view of a volcano, is the water depth of the volcaniclastic successions. Wave and tidal forces weaken with depth and, in more than about 200 m of water, resedimentation depends primarily on gravity-driven processes. Thus syn-eruptive deposits of deep-water volcanoes have a higher preservation potential.

Finally, at both subaerial and subaqueous volcanoes the relative abundance of syn-eruptive and inter-eruptive deposits varies according to position (Fig. 12.44). Rates of erosion are

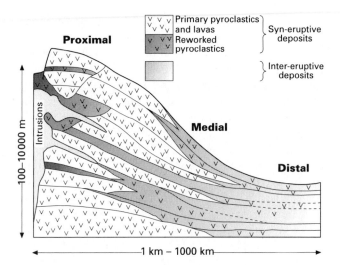

Figure 12.44 Simplified cross-section through a subaerial or subaqueous volcano showing geometry and relative abundance of syn-eruptive versus inter-eruptive deposits at proximal, medial, and distal settings. Note various vertical and horizontal scales over which facies transitions could occur. Boundaries of syn-eruptive deposits are time lines.

generally highest in proximal and near-vent settings, and decrease distally. Much of the original proximal sedimentary and volcanic record is lost, with resistant units, usually primary volcanic deposits such as welded tuffs or lavas, disproportionately represented in the rock record. Unconformities may be common, but of local significance. Lateral facies and age relations are complex, and change abruptly. In contrast, in distal parts of basins, away from vents, slopes are usually lower, facies are more laterally continuous, and more consistent stratigraphic relations should occur. However, the total succession in distal sites is relatively thin. Deposits largely consist of reworked syn-eruptive and inter-eruptive volcaniclastics, and it is usually difficult either to recognize separate volcanic events, or to correlate their products. Although products reworked from extremely large eruptions may be recognized, they can be spread over several metres of vertical sequence. Because of the incompleteness of the proximal record, and the difficulty of interpreting the distal record, sites at intermediate distances are the best place to reconstruct the eruptive and sedimentary history of a volcano.

12.7 Classification of volcaniclastic deposits

Given the complexity of volcaniclastic succession, deposits should first be described on the basis of their lithofacies (e.g. a crystal-rich sandstone, a laminated rhyolitic shard-rich mudstone, a poorly sorted volcanic breccia). The adjective volcanic (e.g. volcanic sandstone) is used to indicate a deposit consisting largely of volcanic fragments, irrespective of their origin. The term *tuff* was originally used, at least since the mid-nineteenth century, in a similar manner to refer to a lithified volcanic sandstone.

To provide information on eruption and/or emplacement processes and establish the penecontemporaneity of volcanism, genetic classifications must be attempted. At least three general approaches have evolved (Fig. 12.45). Unfortunately, the same terms have very different meanings within the three classification schemes, creating much confusion. The terms pyroclastic, epiclastic, tuff and tuffaceous are particularly troublesome, as outlined below.

The most established and widely used nomenclature is based on grain origin (Fisher, 1961; Fisher & Schmincke, 1984). The terms pyroclastic, hydroclastic, and autoclastic are applied both to individual grains (Sect. 12.4.1) and to deposits composed of these grains. Reworked pyroclastics are sometimes referred to as *secondary* pyroclastic deposits whereas undisturbed deposits emplaced by volcanic processes are *primary*. The term *epiclastic* is used for volcaniclastic fragments formed by 'weathering and erosion of pre-existing rocks'. *Tuff* is used for volcanic sandstones composed largely (>75%) of volcanically formed particles, with further classification according to the mechanism of transport (fallout tuff, ash-flow tuff, aeolian tuff) or dominant component (e.g. crystal tuff). *Tuffaceous* (since Hay, 1952), or less commonly *tuffite* (since 1893) is used to imply the presence of both pyroclasts (or autoclasts) and epiclasts, whereas *tuffisite* (since 1941) is occasionally used to refer to intrusive tuffs.

Although this first classification is potentially the most practical, ambiguity arises from the term pre-existing which may have a different sense for each researcher. Second, the definition of epiclastic, where it concerns weathering and erosion, does not take into account subaqueous settings. Finally, in many cases, particularly reworked deposits and coarse volcanic breccias, it is extremely difficult to assign a specific origin to individual grains. A single bed commonly includes clasts generated by several fragmentation processes, and may even include clasts deposited by several mechanisms, such as simultaneous fallout and surge processes. A.E.H. Wright and Bowes (1963) introduced the term *alloclastic* for deposits formed by the fragmentation of pre-existing rocks by volcanic processes, and this term may also be useful for volcanically produced fragments and deposits of multiple or uncertain origin (e.g. a debris avalanche deposit).

A second perspective, partly introduced by Hay (1952) and refined by Cas and Wright (1987), shifts emphasis to the processes of transport and deposition. 'Pyroclastic' (and related terms such as tuff) is restricted to deposits 'formed by explosive volcanic activity *and* deposited by transport processes resulting *directly* [my emphasis] from this activity' (Cas & Wright, 1987, p. 8). Although the second part of the definition is potentially vague, the authors later make it very clear that they only consider

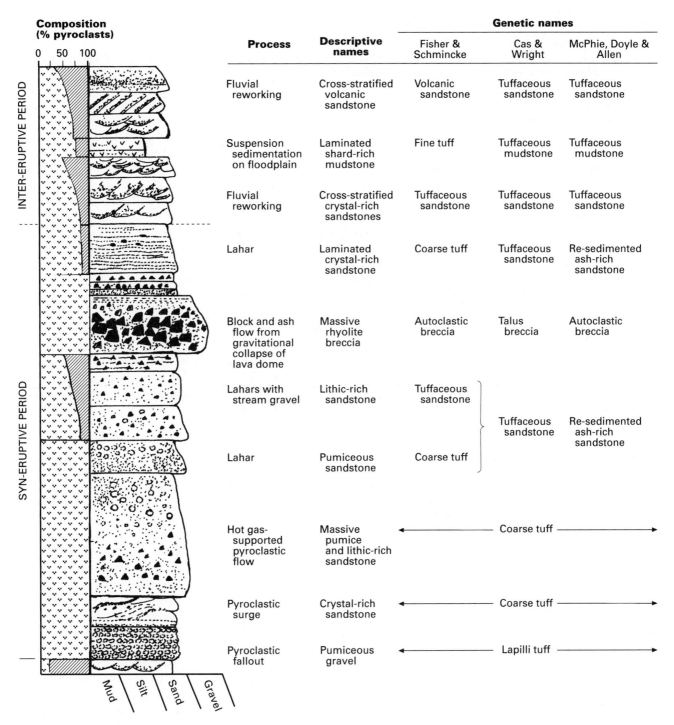

Figure 12.45 Hypothetical sequence of processes during and after a volcanic eruption. Descriptive and genetic names that might be applied to these deposits are tabulated on right. It is assumed that process(es) responsible for formation, transportation, and deposition of clasts are known. Although all three schemes agree for primary pyroclastic deposits at bottom, they differ in names applied or meaning attached to names (e.g. tuffaceous) when pyroclasts become transported by surface processes.

a deposit to be pyroclastic if the transport processes are volcanic (e.g. a hot gas-supported pyroclastic flow) (also Cas & Wright, 1991). Epiclastic is also redefined to include all facies deposited by normal sedimentary surface processes irrespective of fragment origin. There are some inconsistencies, however. The deposits of a pyroclastic flow that entered the sea (and transformed into a water-supported mass flow) would be categorized as epiclastic mass flow deposits whereas coeval pyroclastic fallout that rained down into the sea, becoming water-settled fallout, are called ash deposits or tuff rather than an epiclastic suspension deposit.

McPhie, Doyle and Allen (1993) steer a middle ground between the above two schemes, classifying deposits on the basis of clast-forming processes *and* transport and depositional processes. One advantage of their approach is that it recognizes the uniqueness of the syn-eruption period (see Sect. 12.6) and attempts to classify deposits accordingly (see McPhie, 1995, for an application). Pyroclastic appears to be used in the sense of Cas and Wright, whereas epiclastic (or *volcanogenic sediments*) is largely used in the grain-origin sense of Fisher, although it is extended to include sediments derived from unlithified volcaniclastic deposits but 'redeposited long after eruptions'. New terms are devised for syn-eruptive volcaniclastic deposits in the grey area between volcanic and sedimentary processes (e.g. lahars, debris avalanches), although whether these should all be called 'resedimented' has to be questioned. Conceptual confusion might also arise because they use clast-forming and transport processes within the same scheme. For instance, tuff (*sensu* Cas and Wright) refers to deposits resulting from pyroclastic fragmentation and transport processes whereas tuffaceous is used in the grain-origin sense for any deposit containing pyroclasts.

12.8 Volcanic landforms

Although each volcano is unique, two principal groups can be discerned: *monogenetic* and *polygenetic* (Fig. 12.46). These can be further divided according to the principal *type of magma*

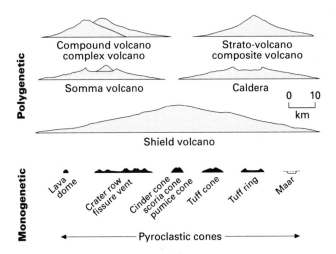

Figure 12.46 Types of volcanic landforms. Vertical exaggeration 2 to 1 (polygenetic) and 4 to 1 (monogenetic). Relative sizes are only approximate (after Simkin, Siebert *et al.*, 1981).

that is erupted, which both determines eruption processes and influences sedimentary processes.

A *monogenetic* volcano is the product of a single eruptive phase or eruption. This may last a few hours, a few years (Surtsey and Heimay), or several decades (Paricutin, Santiaguito) but once the eruption has ceased, the plumbing connecting the vent to its magmatic source freezes over and the volcano never erupts again. Owing to their short lifespan, monogenetic volcanoes are typically only a few hundred metres high. They can occur on their own in a monogenetic volcanic field, but commonly they are the building blocks of larger polygenetic volcanoes. *Monogenetic volcanic fields* contain tens or hundreds of small to mid-sized widely distributed monogenetic cones that are active more or less contemporaneously. Some monogenetic fields are exclusively basaltic (e.g. Auckland, New Zealand); others are bimodal (Higashi-Izu, Japan) or dominated by silicic volcanoes (Chichinautzin, Mexico) (G.P.L. Walker, 1993 and references therein).

Polygenetic volcanoes erupt more than once from the same vent(s). They generally possess a high-level magma chamber, situated at a level of neutral buoyancy, which stores magma and modulates its delivery to the volcano and to subvolcanic intrusions. Deep storage reservoirs may also exist. These volcanoes may show lithological discontinuities due to major changes in magma chemistry, volcanotectonic events, or long erosional intervals, and may last >10 My. Over long time spans, polygenetic volcanoes are also elements of larger volcanic edifices; Vesuvius, for example, lies within the Monte Somma caldera complex. Polygenetic volcanoes can be subdivided into at least three types based on the number and location of vents (Francis, 1993). *Simple cones* (e.g. Fuji-san, Japan; Mayon, Philippines) are the result of repeated, closely spaced eruptions, usually of basaltic magmas, from a single summit vent. They are characterized by a radial symmetry, and embody the image most people have of a volcanic landform. *Composite cones* (e.g. Vesuvius, Etna, Mount St Helens) still retain an overall radial symmetry but have experienced several episodes of construction and destruction (e.g. sector collapse, caldera formation) and possess several vents, with most eruptions from a central and summit conduit but also from innumerable flank or satellite vents. *Compound* or *multivent volcanoes* (e.g. Hawaii, Santorini) result from spatial and temporal overlap of products from several different, but closely spaced, vents and can include products of numerous smaller monogenetic or polygenetic volcanoes.

Polygenetic volcanoes reach impressive sizes. Mauna Loa in Hawaii is the largest oceanic volcano rising nearly 9 km from the Pacific floor and having an estimated volume of 40 000 km³. Land-based volcanoes are comparatively small, with Fuji, the highest volcano in Japan only rising 3700 m above its base with an estimated volume of 1400 km³. The height of polygenetic volcanoes is limited because: (i) volcanoes are constructed on moving lithospheric plates, and erupt for geologically short periods before being moved away from the source of their magma; (ii) once a certain height is reached, every additional increment of growth requires a huge additional increase in volume, with each additional metre requiring the eruption of tens of millions of cubic metres of magma; (iii) destructive processes (caldera formation, landslides) 'behead' the volcano; (iv) as height increases, erosion becomes more effective and an increasing proportion of magma is intruded rather than erupted; and (v) large volcanoes (Olympus Mons, Hawaii, Etna) creep, spread or slide laterally due to their enormous weight, causing deformation (e.g. thrusts) in surrounding strata (e.g. Borgia, Ferrari & Pasquare, 1992).

The dominant control on the type, size and chemical composition of a volcano is the *plate tectonic setting*. This controls magma supply and output rates, the composition and thickness of the crust, and regional stress regime, in particular the amount of extension (Nakamura, 1977; Fedotov, 1981; G.P.L. Walker, 1993; Takada, 1994b). Polygenetic volcanoes occur if magma concentrates during its ascent, such that a stable and sustained magma path is formed, and are favoured by a large magma supply rate and a small tensile stress (Fig. 12.47). Monogenetic volcanoes and/or more widely distributed volcanism occur when separate magma-filled cracks within the lithosphere are unable to coalesce, and are favoured by a small magma supply rate and/or an extensional stress field to facilitate magma ascent.

In many volcanic fields, polygenetic and monogenetic volcanoes co-exist within small (<100 km²) areas and during short (<0.1 My) time spans, indicating variation in stress regime or magma supply. Due to their short lifespan, monogenetic volcanoes are only affected by local conditions; magma output may be low because magma supply throughout the entire volcanic field is low (Auckland or Honolulu volcanic fields), because the volcano developed peripheral to the main conduit of magma ascent (e.g. flank volcanoes), or because the volcano developed after an episode of relatively voluminous and/or explosive volcanism (e.g. silicic domes atop stratovolcanoes).

Polygenetic volcanoes, in contrast, are more dependent on the regional plate-tectonic framework. Basaltic systems largely occur at divergent (e.g. spreading ridge) and derivative intraplate (e.g. hotspot) settings, usually on oceanic crust, intermediate systems largely on convergent margins, usually on continental crust, and silicic polygenetic volcanoes usually lie within continental rifts.

12.9 Monogenetic basaltic volcanoes

Most monogenetic basaltic volcanoes are short lived and small, the total volume of tephra in each volcano generally being less than 0.15 km³. Both type of eruptions, and volcanic landforms, are controlled by whether or not water is present at or near the surface at the time of the eruption (White, 1991). *Scoria cones*

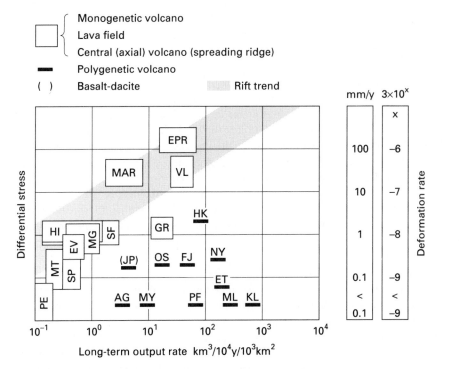

AG, Aogashima, Japan
EPR, East Pacific Rise
ET, Etna, Italy
EV, Eifel VF, Rhine graben, Germany
FJ, Fuji, Japan
GR, Great Rift, Idaho
HI, Higashi-Izu VF, Japan
HK, Hekla, Iceland
(JP, Polygenetic volcano, Japan)
KL, Kilauea, Hawaii
MAR, Mid-Atlantic Ridge
MG, Michoacan Guanajuato VF, Mexico
ML, Mauna Loa, Hawaii
MT, Mt. Taylor VF, New Mexico
MY, Miyakejima, Japan
NY, Nyamuragira, Zaire
OS, Izu Oshima, Japan
PE, Posterosional volcanism, Hawaii
PF, Piton de la Fournaise, Réunion
SF, San Francisco VF, Arizona
SP, Springerville VF, Arizona
VL, Veidivötn-Laki, Iceland

Figure 12.47 Control of regional stress regime and long-term magma output rate on occurrence of monogenetic and polygenetic basaltic volcanoes (from Takada, 1994b).

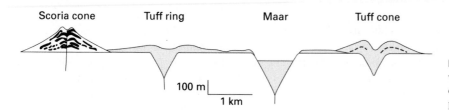

Figure 12.48 Types of basaltic monogenetic volcanic landforms. Stipple is hydrovolcanic deposits whereas white areas are magmatic pyroclastic deposits. Note scale.

and lava flows develop in dry subaerial upland settings, such as the flanks of large shield (e.g. Mauna Kea) or stratovolcanoes (e.g. Etna): *maars and tuff-rings* form from hydrovolcanic eruptions in wet poorly drained sedimentary basins or coastal regions; *surtseyan tuff cones* develop in shallow (<200 m) subaqueous settings (Fig. 12.48).

Flood-basalt volcanic fields are an exception. Although they are also monogenetic in the sense that individual vents erupt only once, they can involve enormous volumes (>10^5 km^3) of quickly erupted basaltic magma. Owing to the high rate of magma output the ambient environment has little effect on eruption style.

12.9.1 Scoria cones and lava fields

Scoria or *cinder cones* are steeply sloping piles of poorly bedded, very coarse-grained, poorly vesicular pyroclastic debris that form during subaerial strombolian eruption of slightly gas-charged magma (e.g. Wood, 1980a) (Fig. 12.49). Fragments tend to follow ballistic paths and the bulk of the material falls back near the vent. Thus fresh scoria cones have a steep angle of repose and outward-dipping slopes of about 33° around a shallow bowl-shaped crater. Finer-grained ejecta, however, may be dispersed tens of kilometres by wind and the volume of these distal fallout deposits can far exceed that of the cone itself. As the cone builds up, a large proportion of the scoria is resedimented downslope by avalanching, rolling and grain flows to form inversely graded deposits with lenticular bedforms. In humid regions, floods can strip sediment from the cone during the eruption.

Central deposits of scoria cones include bombs of lava spatter, welded spatter, clastogenic lava flows from coalescence of spatter, and plugs of dense coherent lava (e.g. Houghton & Schmincke, 1989). These facies can become more common toward the end of eruptions when hotter or gas-poor magma is ejected. Coherent lava can either pond within the crater, or form lava flows if radial dykes break through to the surface low on the flanks of the tuff cone. Lava eruptions can also occur from vents not associated with scoria cones. Although the upstanding scoria cones are the most visible result of this type of basaltic volcanism, the volume of lava flows is typically 10 times that of associated scoria cone and fallout products (White, 1991).

When a scoria cone eruption has ceased, climate, particularly the amount of precipitation, determines the rate of cone degradation (Wood, 1980b). Initially, scoria deposits are porous and permeable and able to absorb water like sponges, giving erosion little chance, and the original morphology may remain recognizable for millennia. For instance, in the Mojave Desert it requires 15 000 years for gullying to occur (Dohrenwend, Wells & Turrin, 1986). However, plant growth and alteration of scoria to clayey soils, and to a smaller extent accumulation of aeolian silts, can decrease surface permeability, thereby promoting gullying and mass wasting. At Vulcan in New Guinea, gullies were eroded only 30 years after the cinder cone stopped erupting (Ollier & Brown, 1971). Once erosion gets under way, the end result is similar in all climates. All loose scoria is eroded and transported into neighbouring depressions (cf. Segerstrom, 1950) or beyond the volcanic field. Thus the rock record of scoria cone fields consists almost entirely of overlapping lava flows, or residual necks of lava and lithified tuff breccias (Kieffer, 1971), such as Ship Rock in New Mexico (USA) or Arthur's Seat in Edinburgh (Scotland).

12.9.2 Tuff rings and maars

Tuff rings are small volcanoes, up to 3000 m wide, with a very wide central crater up to 500 m deep, surrounded by a low (<100 m) broad cone (or 'rim') of outward-dipping ejecta. *Maars* are similar, except that the base of the central crater lies below the level of the original ground surface. These volcanoes are produced by explosive interaction of rising magma with surface or groundwater (Heiken, 1971; Wohletz & Sheridan, 1983), and the deposits may contain very little, or no, juvenile ejecta. Because of its comparatively fine grain size, a large proportion of the ejecta may be carried completely away from the eruption site, such as Nilahue maar where fallout occurs 1500 km downwind (White, 1991).

The wide central crater gradually widens and deepens to a depth of 3–4 km due to a progressive downward migration of the eruption focus (fragmentation level) as groundwater readjusts to the vent topography. This may be indicated by a gradual change in the age of the accidental fragments in the near-vent stratigraphy, such as the Quaternary Laacher See (Bogaard, Schmincke *et al.*, 1990). This downward coring is most effective if the active vent remains isolated from direct

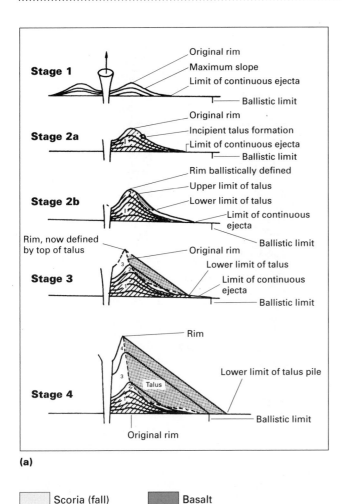

(a)

Scoria (fall)

Scoria (redistributed)

Basalt

Basement

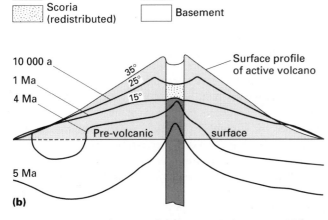

(b)

Figure 12.49 (a) Development of cinder cones in four stages: (1) low rimmed scoria rim; (2) build-up of rim and development of exterior talus; (3) slumping and volcanic explosions destroy original rim; (4) talus extends beyond ballistic limit of ejecta (after McGetchin, Settle & Chouet, 1974). (b) Erosion of a cinder cone in central France; time lines show changes in slope from original (35°) unweathered cinder cone (after Kieffer, 1971).

influx of lake water for a long time. Within the vent, and underlying the crater floor, a pipe-like conduit or *diatreme* is filled with volcaniclastic debris, usually a breccia composed of country rocks and volcanic plugs fragmented and deposited *in situ* and slumped from crater walls.

The surrounding rim extends outward up to 10 km from the central crater. It is largely constructed by 'dry' pyroclastic surges originating from collapsing phreatomagmatic eruption columns (Sect. 12.4.4) (Fig. 12.50). In general, proximal disorganized and crudely stratified lapilli tuffs pass outwards into thin, graded and dune-bedded tuffs. Rim deposits also include accidental or cognate lithics ejected by phreatic crater-forming explosions; these are typically much larger than juvenile fragments and form thin layers of clast-supported and openwork breccias emplaced largely as near-vent fallout deposits (Sohn & Chough, 1989).

The eruptive histories of these volcanoes generally record an overall *decrease* in water/magma ratio (Fig. 12.51). Typically, the lowermost deposits are coarse-grained, poorly sorted, massive to weakly bedded breccias with a low juvenile content from vent opening phreatic explosions. Overlying phreatomagmatic deposits progressively become coarser grained and thicker bedded upwards, reflecting an overall change from 'wetter' to relatively 'dry' base surges. Superimposed on this overall trend, surge deposits may display numerous smaller fining upward 'dry' to 'wet' cycles, or alternate with scoria or pumice (fall and grain flow deposits) from strombolian or plinian eruptions due to fluctuations in the influx rates of magma, vent geometry and amount of external water (e.g. Houghton & Hackett, 1984; Houghton & Schmincke, 1989; Sohn & Chough, 1989). A scoria cone may be eventually built inside the maar crater or tuff-ring. Volcanism typically ends abruptly with retreat of magma from the vent. However, if water can reach the retreating lava column, late-stage phreatomagmatic eruptions may also occur.

During, and mostly after, the pyroclastic eruptions, crater rim deposits may be eroded and redeposited in adjacent lowland areas as lahar, fluvial and lacustrine facies. The enclosed depression typically fills with water to form a crater lake. This is usually hydrologically closed (Sect. 4.1), and commonly density stratified with anoxic, often CO_2-rich, bottom waters. Immediately after the main eruption, coarse volcanic breccias are typically deposited in the crater, due to additional volcanic eruptions, rapid erosion and slumping of the crater walls prior to stabilization by vegetation (R.M.H. Smith, 1986; White, 1992). This is followed by slower and more fine-grained lake sedimentation, with occasional influx of coarse-grained deltaic sediments stretching inward of larger drainage systems. Mass-flow events, as well as small earthquakes, can 'overturn' lake stratification causing asphyxiation and mass mortality of lacustrine or subaerial fauna, such as the unfortunate Lake Nyos disaster in Cameroon (Kling, Clark *et al.*, 1987).

Maar volcanoes are commonly preserved in the geological

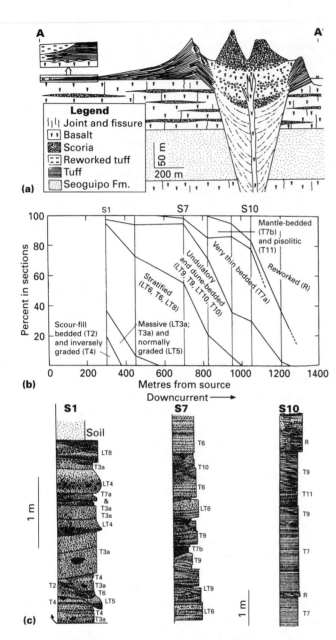

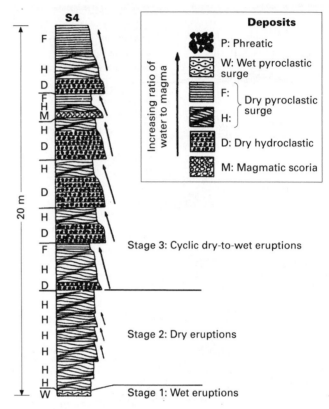

Figure 12.50 (a) Facies relations and schematic cross-section of the Songaksan tuff ring, Korea (collated from Chough & Sohn, 1990). (b) The relative abundance of sedimentary facies with distance away from the centre of the volcano. (c) Representative sections from several positions. Deposits are interpreted to result from fallout (Facies 1), pyroclastic surges (Facies 2–10), co-surge ash falls (Facies 11), debris flows (variants of Facies 3), and nearshore reworking of the distal flanks of the volcano (Facies R). Explanation of lithofacies codes and deposits of individual surges are detailed in Fig. 12.24.

Figure 12.51 Simplified section from Suwolbong tuff ring, Korea, showing eruptive conditions based on interpretation of individual facies and facies sequences (modified from Sohn & Chough, 1989). It comprises an overall wet-to-dry trend with several smaller dry-to-wet cycles in it, suggesting an overall decrease in abundance of external water and fluctuation in the rate of magma rise.

12.9.3 Tuff cones

When rising magma encounters more abundant water, hydro-volcanic eruptions may be only moderately explosive (Fig. 12.12). 'Wetter' and denser eruption columns give rise to poorly expanded and less mobile pyroclastic surges, phreatomagmatic fallout deposits and ballistic breccias, and produce cones of steeply dipping, comparatively coarse-grained, weakly bedded tuff (Fig. 12.52). Abundant water is present within tephra upon deposition, as attested by vesiculation of tuff after deposition, vertical accretion or 'plastering together' of beds, and syn-eruption 'slurries' or 'mudflows' (Ross, 1986; Sohn & Chough, 1992). Tuff cones can form in water-rich subaerial environments, including where basaltic lavas enter the sea (i.e. littoral cones: Fisher, 1968), but many tuff cones comprise the emergent portions of volcanoes (i.e. surtseyan tuff cones) initiated in subaqueous settings, both marine and lacustrine.

Surtsey (Iceland) is a good example of a subaqueous to emergent tuff cone (Kokelaar, 1983). In general, explosive processes became more vigorous as the volcano shallowed

record because hydrovolcanism is linked to subsiding lowland settings, and craters that may be excavated hundreds of metres deep act as repositories for sediment.

and hydrostatic pressure decreased, resulting in an increase in the vesicularity and decrease in the grain size of juvenile fragments. Because of the enclosing water, ballistic dispersal of clastic materials was initially very restricted. The submarine part of a surtseyan succession thus consists of diffusely bedded, water-settled tuffs and lapilli tuffs, ballistic breccias, and resedimented tuffaceous turbidites, debris flow, and slide deposits (Cas, Landis & Fordyce, 1989). As Surtsey approached the water surface, tephra-laden jets erupted into the air, ballistic dispersal became much wider, and a convectively uprising plume carrying fine-fragments formed. Partly because of the wider dispersal of pyroclasts, but mainly because of wave erosion, a surtseyan volcano tends to broaden at the expense of an increase in height, and a vent may persist close to the water surface for a considerable time. Subaerial syn-eruptive facies include proximal poorly sorted, thin-bedded massive lapilli tuffs and breccia layers from near-vent fallout, medial thick-bedded massive lapilli tuffs from debris flows or expulsion of vent filling slurries, and distal thinly stratified lapilli tuffs formed by traction and suspension sedimentation from wet base surges. If the vent is isolated from surrounding sea water by a resistant tephra rim, it can dry out, in which case eruptions become dominantly or purely magmatic. In the case of Surtsey, this involved Hawaiian-type lava fountaining to form lava flows.

After the eruption ceases, volcanic cones are rapidly destroyed by subaerial and/or wave erosion. Rapid degradation continues until erosion is inhibited by consolidation of the tuff cone deposits, formation of a vegetation cover, or armouring of the tuff cone by boulder beaches. The subaerial and shallow-water deposits of small tuff cones may be completely removed, such as Surtla which was lowered from near sea level to a depth of 45 m in only 15 years (Kokelaar, 1983). Removal of the tops of these volcanic centres produces a fairly flat wave cut plateau and widens the volcanic edifice through deposition of a surrounding apron of volcaniclastic and bioclastic talus and mass-flow deposits (e.g. Cas, Landis & Fordyce, 1989) (Fig. 12.52). Wave-cut benches, steep cliffs, and reworked littoral deposits have also been described from sublacustrine tuff cones (Oviatt & Nash, 1989), but the degree of reworking is much lower than for marine examples because of the lower energy of waves within lakes and the possibility for variations in lake level over time. Thus both subaqueous and emergent parts of the volcanoes can be preserved, such as

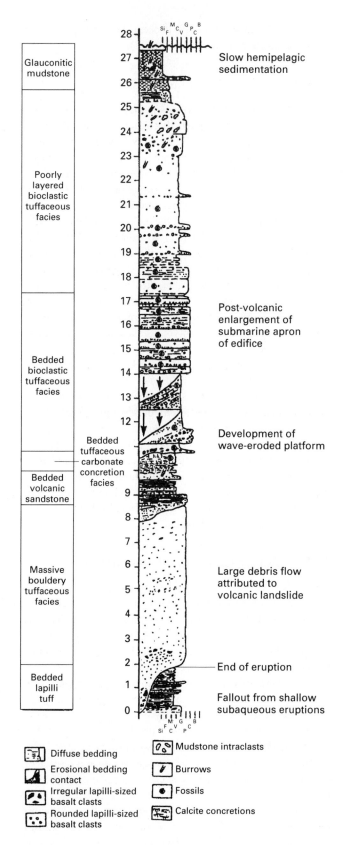

Figure 12.52 Representative section from a basaltic Eocene surtseyan volcano, New Zealand (from Cas, Landis & Fordyce, 1989). Figure largely shows uppermost reworked volcaniclastics; 20–120 m of bedded Lapilli tuffs, similar to the bottom interval, lie beneath the log shown.

surtseyan volcanoes from the Pliocene Lake Idaho of Utah (Godchaux, Bonnichsen & Jenks, 1992).

12.9.4 Flood-basalt volcanic fields

Giant flood-basalt fields have volumes in the range 10^5–10^7 km^3. They are distributed throughout geological time at average intervals of 32 My (Rampino & Stothers, 1988), each presumably forming at the inception of a hotspot (Richards, Duncan & Courtillot, 1989). Examples are the 16 My ago Columbia River Basalts in the northwestern USA (Reidel & Hooper, 1989) and the 65 My ago Deccan Traps of peninsular India that may be implicated in the biological mass-extinction events at the Cretaceous–Tertiary boundary. Lava flows cover wider areas than in monogenetic volcano fields, overlap, are superposed to form parallel-stratified successions, and have much greater volumes. Vents, marked by spatter cones and welded scoria, are apparently widely scattered (e.g. Swanson, Wright & Helz, 1975). However, vents are emphemeral features, easily eroded, often covered by younger lava flows, and difficult to recognize.

Sill swarms, which can rival flood lavas in total volume, underlie many flood-basalt fields. For instance, a swarm of Jurassic dolerite sills up to 300-m thick outcrops over 25% of Tasmania, and can be found on disrupted portions of Gondwanaland in Antarctica and South Africa (Carey, 1958). The injection of sills into poorly consolidated sediments commonly generates soft-sediment deformation structures in the country rocks or results in hydrothermal explosions and peperites. Basaltic sills can also become pillowed. Because of their density, lava can also reinvade, and locally flow as shallow sills under wet surface sediments.

12.10 Monogenetic silicic volcanoes

Small silicic volcanoes occur when volatile-poor or partially degassed, rhyolite to andesitic magma is able to rise either to the Earth's surface or into near-surface aquifer systems. Though there are rare exceptions (Sect. 12.4.5), silicic lavas do not generally flow far, piling up to form thick (10s –100s m), short (<5 km) and small volume (<1 km^3) *domes*.

Lava domes grow through alternating periods of extrusion (and eruption) and intrusion (and inflation) spread over years to decades, with intrusion becoming more important as the dome gets bigger and effusion rate declines (Rose, 1987; Swanson, 1990). Over long time spans (10^4 years), periodic eruption from closely spaced vents can result in a cluster of overlapping domes and associated clastics (e.g. Lipari: Sheridan, Frazetta & La Volpe, 1987). During emplacement of a lava dome its hot interior typically deforms and flows between rigid cooler layers. Movement of the flow interior causes fragmentation of the top and base resulting in a carapace of lava blocks in a granular matrix (i.e. autobreccia) on the flow

margins. Volcaniclastic deposits may also form due to explosive eruptions or quench fragmentation.

At *subaerial silicic domes* (Fig. 12.53), plinian and phreatomagmatic eruptions commonly precede emplacement due to the original volatile gradients in the magma chamber and/or interaction of magma with a surrounding aquifer. On the Lipari and Vulcano volcanoes eruptive cycles begin with hydrovolcanic

Tuff ring and tuff cone
with overflowing lava plug

Coalescing domes with phreatic
and phreatomagmatic carapaces

Explosion crater

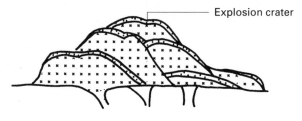

Plinian, far-field, pumice falls and
flows, peléean avalanche deposits

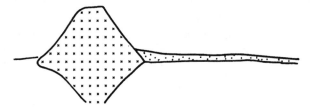

Polygenetic dome growth
with intercalated flank deposits
(composite cone)

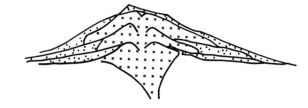

Figure 12.53 Common occurrences of dome-related tephra deposits (from Heiken & Wohletz, 1987).

breccias (partly exploding a previous lava plug), wet surges and fallout deposits. Dry surge deposits become more common upwards and each cycle ends with pumice falls and lava flows (Sheridan, Frazetta & La Volpe, 1987). During emplacement, the dome may be disrupted or destroyed by explosive vulcanian eruptions (Sect. 12.4.1) (e.g. Mont St Pelée, 1902), gravitational collapse of unstable dome margins (e.g. Unzen), explosions from vesicular zones at the front of lava flows (Sect. 12.4.5), or phreatic explosions due to local interaction of the dome with groundwater and/or snow pack (Heiken & Wohletz, 1987). Pyroclastic fragments can form fallout deposits, be carried away in block and ash flows, accumulate as aprons of talus breccia and can trigger small lahars (see also Brooker, Houghton *et al.*, 1993).

If degassed silicic magma erupts in *shallow water*, phreato-magmatic, explosive magmatic and effusive activity combine with syn-eruptive mass flow resedimentation and reworking. In the 1953–1957 eruptions of Tuluman volcano in the Bismarck Sea, Papua New Guinea, effusion of rhyolitic lava, from a vent in about 130 m of water, was followed by explosive eruptions which built several small partly emergent pumice cones (Reynolds, Best & Johnson, 1980). The Tuluman eruption has been used as an analogue for part of a volcaniclastic sequence of the Devonian Bunga beds, Australia (Cas, Allen *et al.*, 1990) (Fig. 12.54). Coherent rhyolite and rhyolite breccia at the base of the Bunga bed sequence record emplacement of a rhyolite dome into wet unconsolidated sea floor accompanied by quench fragmentation, autobrecciation and resedimentation. Shoaling

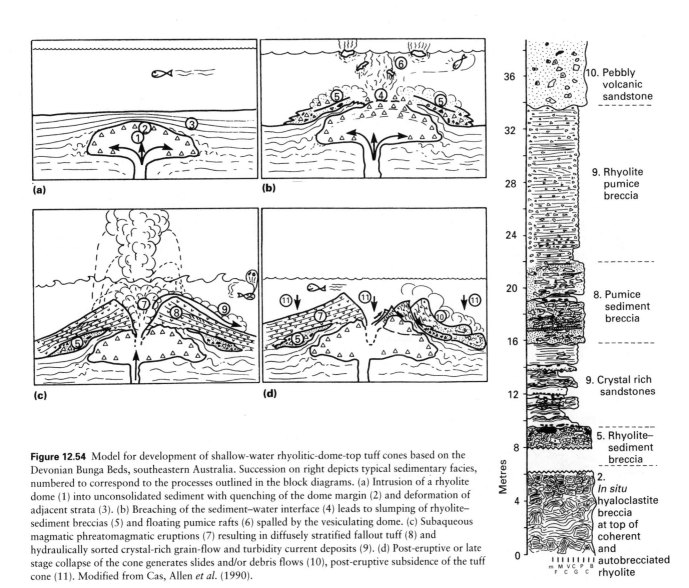

Figure 12.54 Model for development of shallow-water rhyolitic-dome-top tuff cones based on the Devonian Bunga Beds, southeastern Australia. Succession on right depicts typical sedimentary facies, numbered to correspond to the processes outlined in the block diagrams. (a) Intrusion of a rhyolite dome (1) into unconsolidated sediment with quenching of the dome margin (2) and deformation of adjacent strata (3). (b) Breaching of the sediment–water interface (4) leads to slumping of rhyolite–sediment breccias (5) and floating pumice rafts (6) spalled by the vesiculating dome. (c) Subaqueous magmatic phreatomagmatic eruptions (7) resulting in diffusely stratified fallout tuff (8) and hydraulically sorted crystal-rich grain-flow and turbidity current deposits (9). (d) Post-eruptive or late stage collapse of the cone generates slides and/or debris flows (10), post-eruptive subsidence of the tuff cone (11). Modified from Cas, Allen *et al.* (1990).

of the dome permitted several pyroclastic eruptions, as indicated by rhyolite pumice and crystal-rich breccias (submarine mass flows) and diffusely stratified crystal-rich to crystal-poor sandstones (water-settled fallout). Debris flows and turbidites were produced by resedimentation during and after the pyroclastic eruptions.

Deep-water silicic lavas form not only massive lava flows and domes, but lobes, pods and pillow-like bodies. This is probably because the confining pressure of the water column inhibits exsolution of dissolved volatiles, thereby ensuring a lower magma viscosity. The confining pressure also inhibits explosive magmatic or phreatomagmatic eruptions. None the less, a volcaniclastic carapace can still surround coherent lava facies (Fig. 12.55). In many cases, the lava flows and domes mix with subjacent wet sediment to produce a texturally complex lava–sediment breccia (peperite: Sect. 12.4.1). Above are masses of angular breccia (*in situ* hyaloclastite) and stratified volcanic breccia from mass-flow redeposition of *in situ* hyaloclastite. Resedimented hyaloclastite may be carried away from the dome. Polymetallic Pb-rich volcanogenic massive sulphide (VMS) deposits may result from hydrothermal circulation within and beneath such rhyolite dome complexes (e.g. Urabe & Marumo, 1991; R.L. Allen, 1992).

The high-confining pressure in deep water also allows magma to rise to very shallow stratigraphic crustal levels. If the magma encounters an interval of weak, poorly consolidated sediments, it may intrude. These high-level intrusions, termed *crypto-domes*, cause up-doming of the overlying sediments or rocks. They appear to be especially favoured in subaqueous settings characterized by interstratified volcanic and sedimentary rocks. Partially extrusive cryptodomes locally break through the cover and emerge at the surface. Contact relationships, in particular the distribution of resedimented hyaloclastite and autoclastite facies, is the basis for determining whether the silicic lava was intruded, extruded, or both.

12.11 Polygenetic basaltic volcanoes

Polygenetic basaltic volcanoes mostly develop on oceanic or thinned continental crust, and are usually *tholeiitic*. Such volcanoes form oceanic spreading ridges and oceanic crust, and include oceanic intraplate seamounts, plateaux and lava-shields. Their chances of being preserved on land are small because the majority are destined to be consumed at convergent plate margins. Large basaltic volcanoes also occur at oceanic island arcs where they are associated with more silicic (e.g. andesite)

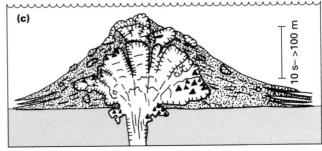

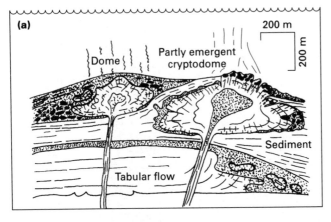

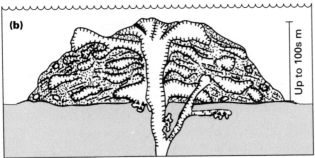

Figure 12.55 Various forms of subaqueous silicic lavas and domes. (a) Intrusive and partially extrusive domes. (b) Vent-top submarine dome. (c) Lava lobe–hyaloclastite complex (collated by Cas, 1992).

rocks. These are described with other convergent margin volcanoes in Section 12.12.

12.11.1 Oceanic spreading ridges and abyssal plains

Spreading ridges are loci of active seismicity, voluminous basaltic volcanism and high heat flow. As the crests of active spreading ridges typically lie at depths of several kilometres, explosive eruptions are generally not possible, an exception being where spreading ridges coincide with hotspots, as in Iceland. The size and persistence of magma chambers beneath the ridge axes as well as the morphology and facies of spreading ridges vary considerably, depending on spreading rates (K.C. MacDonald, 1982) (Fig. 12.56).

Slow rates of spreading, such as at the Mid-Atlantic ridge, yield a ridge with pronounced topographic relief and a 2-km-deep axial rift along its centre. Magma chambers beneath the ridge are small and discontinuous, thereby allowing for some magma differentiation and eruption of a comparatively wide range of basalt types, with the ridge axis consisting of numerous very small, but coalesced seamounts (D.K. Smith & Cann, 1992). Slow-spreading ridges are typically regularly segmented by transform fault zones with large offsets and deep nodal basins; accommodation zones between adjacent segments have opposing half-graben polarities (Karson, 1991). The stratigraphy is dominated by sheet and pillowed lavas, although talus breccia units 50–85 m thick can develop by gravitational collapse within the axial rift. Pillow breccias and hyaloclastite sandstones can develop on the margins of lava flows and at

least locally form layers tens of metres thick in Layer 2 of oceanic crust (Schmincke, Robinson et al., 1978).

Fast-spreading ridges, such as the East Pacific Rise, have a much more subdued topography and lack the prominent axial rift. They have large continuous magma chambers that generate comparatively homogeneous magma batches. Owing to the higher rates of magma discharge, sheet lavas are more common. Because of the comparatively smooth surfaces of sheet lavas (Sect. 12.4.5), and the subdued topography of the rift, volcaniclastic sediments are rarely formed.

Spreading ridges are typically covered by patches of dark brown sediment enriched in Fe, Mn and a host of other metals. Accumulation rates of these metalliferous sediments are apparently related to rates of sea-floor spreading so that on the East Pacific rise this facies is particularly thick (Boström, 1973). Considerable insight into the origin of these sediments has come from observations made by submersibles. At various points on the East Pacific Rise, hydrothermal mineral-rich solutions, with temperatures as high as 350°C and with a pH as low as 4, jet from vents along the ridge axes and precipitate a range of columnar structures called chimneys built of various sulphide and sulphate minerals (Fig. 12.57). Two basic types exist: the high-temperature fast-growing (8 cm per day) 'black smokers' belching clouds of finely disseminated pyrrhotite plus sphalerite and pyrite, and cooler (up to 300°C) 'white smokers' emitting particulate amorphous silica, barite and pyrite (Haymon, 1983; Parson, Walker & Dixon, 1995). In addition, they are populated by prolific indigenous faunas, including crabs, giant clams and tube worms,

Figure 12.56 Sediment distribution on (a) a fast-spreading ridge of East-Pacific-Rise type. (b) A slow-spreading rifted ridge of Atlantic type. Note (i) the enhanced development of Fe–Mn-enriched basal sediments, in part derived from oxidation of sulphides, on the fast-spreading ridge; mineralized crusts on the rifted ridge change from hydrothermal to authigenic as they move away from the vents in the median valley; (ii) the difference in igneous–sedimentary rock relationships and sediment geometry on the two types of ridge: on the Atlantic type, sediments are highly lenticular and rest on a variety of mafic and ultramafic rocks. The calcite compensation depth (CCD) (Sect. 10.2.1) controls the change in facies down the ridge. Not to scale (modified from Garrison, 1974; Davies & Gorsline, 1976).

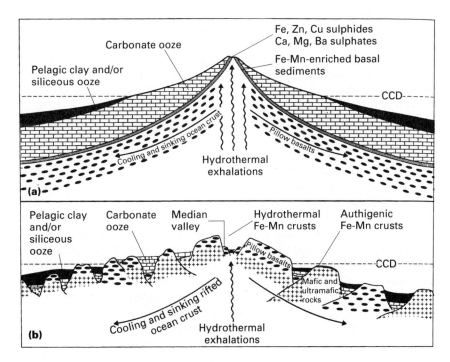

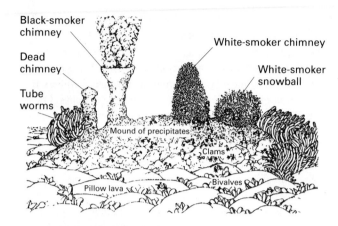

Figure 12.57 Idealized sketch illustrating the hydrothermal field on the crest of the East Pacific Rise at 21°N (after MacDonald & Luyendyk, 1981). Diagram copyright 1981 Scientific American Inc.: all rights reserved.

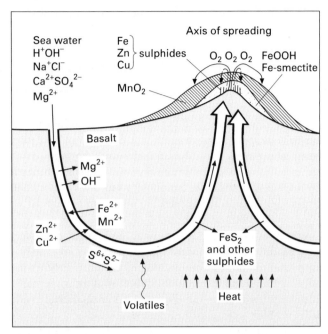

Figure 12.58 Model illustrating basalt–sea-water interaction, hydrothermal circulation and metallogenesis at oceanic spreading centres. Removal of Mg^{2+} and OH^- into smectite clay minerals keeps circulating solutions acid and capable of significant leaching (Motti & Seyfried, 1980). Major sulphide precipitation is assumed to take place within the igneous basement. Exhaled fluids are converted directly, or via a sulphide phase, to oxyhydroxides and clay minerals. Fe-rich phases, being readily oxidized, may occur near the hydrothermal vents, Mn-rich species tend to be precipitated further afield (after Bonatti, 1975).

that have sulphur-oxidizing bacteria at the base of their food chain.

Mature black smoker chimneys have a characteristic concentric zonation, primarily due to temperature decrease across the wall, from chalcopyrite on the inside, via a zone dominated by pyrite, sphalerite and wurtzite, to an exterior zone of anhydrite with minor sulphides (Fig. 12.57).

White smokers are chimneys in which hydrothermal precipitation has sealed orifices and decreased permeability of the chimney walls, thus allowing more intrachimney cooling by sea-water entrainment and heat conduction. These less active chimneys sustain the most prolific growth of tube worms, which may become so densely packed as to form a spherical mass of white tubes ('snowballs'). Ultimately chimneys become mechanically unstable and collapse to form a *mound* of chimney talus, on which chimney growth begins again. Sea-floor sulphides are chemically very unstable in modern sea water, and rapidly oxidize upon cessation of hydrothermal activity to form ochreous deposits dominated by hydrated iron oxides, some reacting with silica to form iron-rich smectites (Alt, Lonsdale *et al.*, 1987).

The current consensus is that the hydrothermal fluid that emanates from modern submarine spreading ridges is sea water that has circulated through the oceanic crust within a convection cell and leached metals from the rocks along its flow path (Fig. 12.58). Thus the metal content of the hydrothermal fluid reflects the trace element composition of the subjacent source rocks. Most black smokers, as well as sulphide deposits where the dominant lithology at the time of ore deposition was of mafic composition (i.e. oceanic crust), are Cu-rich (Cu–Zn type VMS deposits). In contrast, at spreading sites in modern or ancient backarc basins, particularly rifted continental crust, volcanism can be bimodal with extrusion of dacite or rhyolite domes in addition to the basalt volcanism;

hence volcanogenic massive sulphide deposits in these settings are comparatively Pb rich (e.g. Kuroko or Zn–Pb–Cu VMS deposits; see Urabe & Marumo, 1991).

Away from the crestal portions of spreading ridges, biogenic sediments assume importance, particularly in depressions (Jenkyns, 1986). On the Mid-Atlantic Ridge they are dominated by carbonate derived from planktonic Foraminifera, pteropods and calcareous nanofossils, locally lithified by high-Mg calcite cements, with subordinate low-Mg calcite oozes and aragonite cements and freshwater diatoms, testifying to an aeolian source for some fine-grained material. Further away from the ridge, sediments thicken, particularly in valleys. Fine-grained turbidites derived from calcareous pelagic deposits of the neighbouring hills are about 500 m thick (van Andel & Komar, 1969). Redeposited material may also include ferromagnesian minerals and breccias of serpentinized peridotite, derived from upthrust ultramafics exposed in fracture zones, and set in low-Mg clacite and aragonite cements. Finally, as the ridge flank and its overlying sedimentary cover dive below the calcite compensation depth (CCD), carbonates change to siliceous oozes and red clays (Fig. 12.56).

Away from turbidity currents, the deep ocean floor receives little sediment, rates of sediment accumulation being typically only a few millimetres per thousand years. *Siliceous ooze*, derived from Radiolaria, diatoms, silicoflagellates and sponge spicules, and phosphates form in areas of upwelling and high productivity (Sect. 10.2.1). In barren regions outside these high-productivity zones, *red clays*, chiefly derived from volcanic, aeolian and cosmic sources, accumulate by default. Feldspar, pyroxene, montmorillonite and smectite derive from intraoceanic volcanoes. Chlorite is detrital, being derived from low-grade metamorphic terrains. Quartz, illite and, to a lesser extent, kaolinite are assumed to derive as fallout from high-altitude jet streams; the aeolian contribution to deep-sea clay is probably between 10 and 30%. The Saharan desert, for instance, supplies clay to the Atlantic, with material from African dust storms extending to the Caribbean. Cosmic constituents include black magnetite spherules of nickel–iron and chondrules of olivine and pyroxene. Ferromanganese nodules commonly occur on red clays, and because of their slow growth, nodules develop better under conditions of nil sedimentation or erosion.

Ancient spreading ridges are usually preserved as stratigraphic elements of oceanic crustal slices, which are usually tectonically emplaced as *ophiolites* in subduction-related accretionary prisms or obducted on to continental margins. Most ophiolites probably represent oceanic crust from back-arc, or even strike-slip (Sarewitz & Lewis, 1991) basins. The Troodos ophiolite, for example, was once believed to represent a mid-ocean ridge spreading centre, but is now considered to have formed by spreading during the early stages of intraoceanic subduction prior to development of a well-defined magmatic arc, on geochemical evidence of arc-related volcanism and comparison with modern intraoceanic arcs, particularly the Marianas (Stern, Bloomer *et al.*, 1989). These have been termed supra-subduction zone ophiolites (Pearce, Lippard & Roberts, 1984). Although there is no *a priori* reason why the volcanic morphology of these ridges should be significantly different from their mid-ocean counterparts, marginal basins may also contain volcaniclastic debris and montmorillonitic clays derived from the volcanic chain, terrigenous clastics deposited by turbidity currents and continental margin slumps, slides and debris flows (Sect. 12.12).

Numerous ancient examples of oceanic spreading ridges have been described from the Late Palaeozoic–Recent Tethyan ocean, now exposed in the Mediterranean region (cf. Robertson, 1994). A well-preserved ophiolite in the Ligurian Apennines of Italy has been identified as a slow-spreading rifted Atlantic-type ridge (Barrett, 1982), on the basis of rapid changes in thickness and evidence for synsedimentary faulting; pillow lavas change laterally to talus breccias and volcaniclastic turbidites. In contrast, other ophiolites are more similar to the fast-spreading and hydrothermally vigorous East Pacific Rise (Fig. 12.56), examples including the Troodos Massif

(Fig. 12.59) and the ophiolite in Oman (MacLeod & Rothery, 1992). There is a lack of synsedimentary breccias and pillow basalts are overlain and occasionally interbedded with metalliferous sediments, including Fe- and Mn-rich mudstones termed *umbers*, and Fe-rich sediments termed *ochres*. Sulphide deposits of these ophiolites contain chimney fragments and fossil tube worms (Oudin & Constantinou, 1984; Haymon, Koski & Sinclair, 1984) which resemble the modern chimney structures of the East Pacific Rise. In the Troodos Massif of Cyprus (Robertson & Hudson, 1974) the basal sediments pass up into regularly bedded and finely laminated radiolarian cherts which are locally overlain by bentonitic (illite–montmorillonite) clays, interpreted as a hemipelagic product from nearby island-arc volcanoes.

In both the Troodos Massif and the Ligurian Apennines, siliceous radiolarites pass upwards into pelagic nanofossil chalks or limestones, then bedded marls or calcilutites. This change from siliceous to carbonate facies, rather than the reverse as on modern ridges, may be the result either of tectonic uplift above

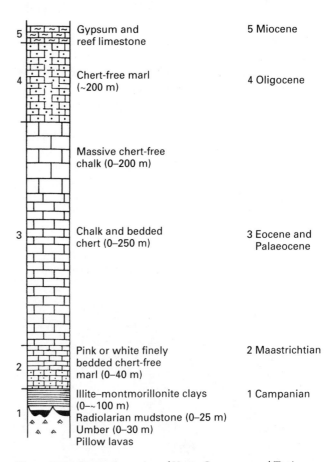

Figure 12.59 Composite section of Upper Cretaceous and Tertiary oceanic sediments of Cyprus with capping of Miocene gypsum. Umber is a chocolate-brown metalliferous mudstone. The igneous basement is everywhere pillow basalt (after Robertson & Hudson, 1974).

the CCD or palaeoceanographic factors unique to small ocean basins such as increased surface-water productivity or oxygenation levels (Sect. 10.4.3).

12.11.2 Oceanic seamounts and lava-shields

Volcanic seamounts are conical features that rise from the ocean deeps and are particularly widespread in the Pacific (Fig. 12.3) and to a lesser extent the Atlantic. There are at least three types of volcanic seamount: (i) small seamounts within oceanic plates; (ii) larger seamounts above present-day or former positions of oceanic hotspots; and (iii) seamounts associated with convergent margins and island arcs, discussed in Section 12.12.

Most of the first type of seamount form at or near ocean ridges, such as those located on oceanic crust 1.5–7.5 My in age on the flanks of the East Pacific Rise (T.L. Smith & Batiza, 1989). These range to tens of kilometres in diameter and several kilometres in height. Because of their depth of formation, volcanoes are dominated by sheet and pillow lavas (Fig. 12.60). Circular to elliptical craters and calderas, up to 5 km in diameter and several hundred metres deep, commonly occur on the summits of these seamounts. As on subaerial lava-shields, summit collapse is probably caused by incremental withdrawal of roof support due to migration of magma outward to feed fissure eruptions. Large volumes of non-vesiculated and blocky hyaloclastite debris can be formed at the summits of these seamounts when lava fountains up through the sea floor and is rapidly chilled and fractured. This sediment can be dispersed around the vent by density currents.

Oceanic hotspot volcanoes are well represented by the Hawaiian–Emperor volcanic chain stretching nearly 6000 km across the North Pacific (Fig. 12.3) and consisting of more than 100 volcanoes. Volcanism and sedimentation change systematically due to variation in magma-supply rate as plate motion conveys the volcanic system over and then away from the hotspot focus (Fig. 12.61). Hawaii volcanoes, as presently understood, evolve through seven stages (Fig. 12.62): (i) deep marine stage; (ii) shield-building stage; (iii) capping stage; (iv) erosional stage; (v) renewed volcanism stage; (vi) atoll stage; and (vii) late seamount stage.

The *deep marine* stage is well represented by Loihi seamount, the most recent addition to the Hawaiian chain of volcanoes (Malahoff, 1987; Fournari, Garcia *et al.*, 1988). Although its base is at a depth or more than 4 km, submarine eruption of tholeiitic–alkalic basalts has already built its summit to within 1000 m of sea level. Fresh pahoehoe, pillow, and aa lava flows are erupted on the upper flanks of the volcano but break up to form talus on the steep lower slopes. Mass wasting and landsliding on the lower slopes maintain the steep profile.

The *shield-building* stage is characterized by frequent and voluminous eruption of tholeiitic basalts from summit vents, or from radial fissures and rift zones which tend to radiate from a common point (Fiske & Jackson, 1973). It probably

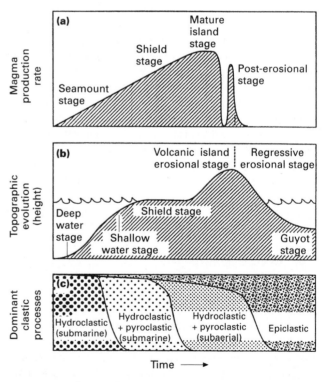

Figure 12.61 Changes of (a) magma production rate; (b) topographic height; and (c) dominant volcanic and clastic processes during the evolution of volcanic oceanic islands. The difference in gradient in (a) and (b) mainly results from the abundant production of volcaniclastic material during shallow submarine and initial subaerial shield stages (combination of hydroclastic and pyroclastic processes) (from Schmincke & Bogaard, 1991).

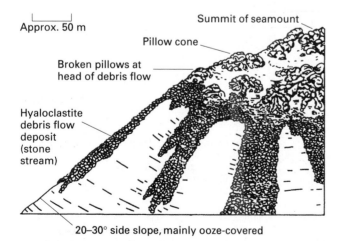

Figure 12.60 Sketch of summit area of a seamount near the East Pacific Rise (from Londsale & Batiza, 1980).

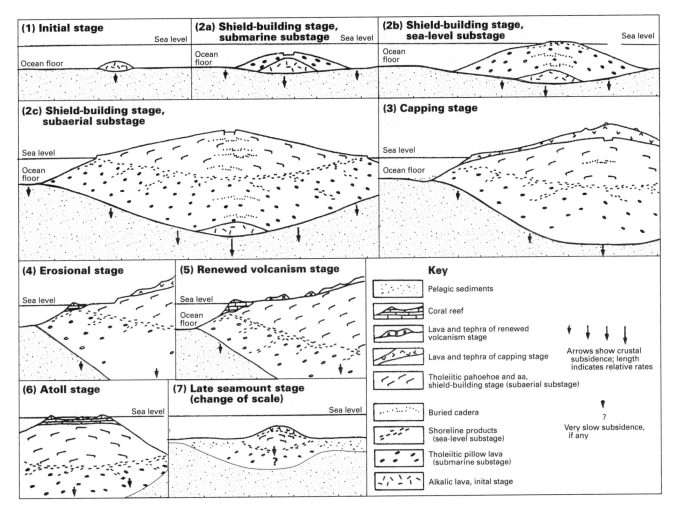

Figure 12.62 Successive stages in the evolution of a Hawaiian volcano (from Peterson & Moore, 1987). Changes between stages are normally transitional and gradational. For simplicity feeder conduits and rift zones are omitted. Note substantial crustal subsidence due to volcanic loading, particularly through growth stages 1–3. Vertical exaggeration varies from c. 2× to c. 4×.

accounts for >95% of the volume of each volcano. Three substages can be recognized (Figs 12.62, 12.63). *Deep-marine* pillow lavas initially build a moderately sloping submarine edifice. As basaltic seamount grows into *shallow water*, explosive phreatomagmatic eruptions may produce surtseyan tuff cones (Sect. 12.11.3). Even larger amounts of volcaniclastic debris can be liberated by subaqueous fissure eruptions (Fig. 12.30). Once a basaltic seamount becomes *subaerial* and *emerges*, additional clastic debris can be generated in coastal regions in at least three ways: (i) a land-derived lava flow crossing the shoreline undergoes non-explosive quenching and phreatomagmatic explosions – the resultant pyroclasts, hyaloclasts, and pillow lava tongues tend to accumulate as a Gilbert-type delta that progrades offshore, and are capped by a platform of subaerially 'frozen' lava (Fig. 12.64) (Jones & Nelson, 1970; J.G. Moore, Phillips *et al.*, 1973; Porebski & Gradzinski, 1990); (ii) tuff-rings, such as Diamond Head (Hawaii), develop on

land where basaltic magma interacts with coastal aquifers; and (iii) wave processes erode cliffs in newly formed land to form boulder and sand beaches.

Despite the numerous possible scenarios to generate volcaniclastic fragments at/or near the coastline, volcaniclastic sediments typically comprise a small proportion of the total succession of large basaltic seamounts. Once the volcano is fully emergent above sea level, volcanism becomes dominated by effusion of pahoehoe and aa lavas. During a typical eruption, a large fraction of lava (even 80%) flows through lava tubes; this extends the distances the flows can reach (perhaps by a factor of 10) by limiting the extent flows chill by radiant cooling, and enables flows to reach the sea and extend coastlines. A myriad of thin lava flows only a few metres thick build a gently sloping (<10°) convex-up shield volcano whose height is typically about 1/20 of its basal diameter. Mauna Loa in Hawaii is the largest having an estimated volume of 40 000 km³,

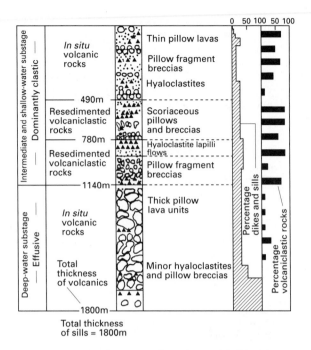

Figure 12.63 Stratigraphic section through an uplifted basaltic seamount on La Palma, Canary Islands (after Staudigel & Schmincke, 1984). Substages correspond to 2a and 2b of Fig. 12.62.

rising nearly 9 km from the Pacific floor, and has a keel that projects well below it into the isostatically depressed oceanic crust. Weight of this edifice causes substantial subsidence of

the subjacent oceanic lithosphere forming a flexural moat around the volcano.

Intrusions account for a significant fraction of the total volume of shield volcanoes. For instance, at Kilauea volcano, from 1963 to 1983 there were 53 rift zone intrusive events, 60% of which took place without an accompanying eruption (Decker, 1987). Intrusion forms basaltic dyke complexes 3–5 km wide, which are exposed by erosion at rifts on older shield volcanoes (G.P.L. Walker, 1992), and are accommodated by lateral slip (gravity sliding) on deep faults which may coincide with the contact of the base of the volcano with weak sea-floor sediments (Duffield, Stieltjes & Varet, 1982). Landslides are another important aspect of large basaltic volcanoes, particularly during the shield-building stage when the volcanoes are close to their maximum size. A seismic reflection survey across the Hawaiian flexural moat indicates that fill is dominated by four major debris avalanche units, each up to 700 m thick and is related to a landslide event (Rees, Detrick & Coakley, 1993).

The *capping stage*, exemplified by Mauna Kea (Hawaii), marks a reduction in magma supply rate by several orders of magnitude. Mauna Kea has not erupted in historic times. Magmas during this stage are transitional or alkalic basalts, and eruptions tend to be more explosive, building large cinder cones. Lavas are mostly aa.

When active, construction of shield volcanoes keeps ahead of erosion but between eruptions red oxidized soil horizons may develop with minor downslope resedimentation of epiclastic material. Once eruptions cease, however, stream and

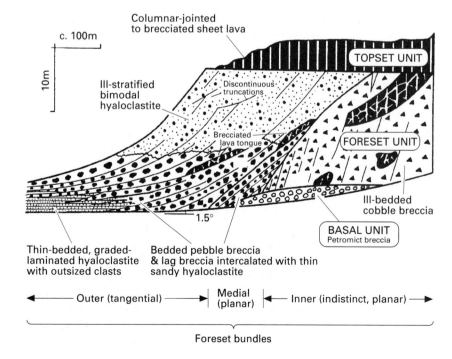

Figure 12.64 Composite cross-section through a lava-fed Gilbert-type delta, Oligocene, West Antarctica (from Porebski & Gradzinski, 1990).

wave erosion cuts valleys and cliffs. This is the *erosional stage*. Because basaltic lavas and the scoriaceous horizons between them are so porous, surface drainage does not easily establish on lava plateaux. Instead, water seeps downward, moving laterally when it reaches clay-rich horizons, eventually to soak out as springs. Erosion of basalt plains initially takes place by *sapping*, rather than stream incision, producing deep steep-sided, dead-end canyons on the older Hawaiian Islands. Headwater erosion of these canyons takes place as springs undermine or sap the cliff line, so that it eventually collapses. Talus is removed by chemical disintegration and solution of the basalt, or mechanical transport when flow rate is great enough. During a long eruptive hiatus, wave erosion can cut inward from the coastline and steepen slopes of the shield volcano, as on Volcán Ecuador, Galápagos islands (Rowland, Munro & Perez-Oviedo, 1994).

Some shield volcanoes subsequently enter a *renewed volcanism* stage with eruptions producing small monogenetic cinder and spatter cones, surtseyan tuff-rings near the coast, and short flows of silica-poor lava (alkalic basalts, basanite, nephelinite). In the Hawaiian Islands this volcanism is exemplified by the Honolulu Volcanic Group on Oahu, which includes some notable scenic landmarks such as Diamond Head and Hanauma Bay, about 1 My following cessation of shield-building activity. Resumption of volcanism seems to occur at a nearly constant distance of about 200 km from the contemporaneous active shield volcanism, as flexural compressive stresses along the axis of their chain becomes released (ten Brink & Brocher, 1987).

During and after the erosional stage, Hawaiian shield volcanoes (and other volcanic islands) often become surrounded by a perimeter reef. With the cessation of active volcanism and continuing erosion, the original volcanic island shrinks in size, subsides and eventually submerges beneath the waves. At the same time the perimeter reef grows upwards, keeping pace with island sinking (Darwin, 1842). The combination of submergence and reef growth produces a central lagoon that expands and is surrounded by a ring-shaped reef, termed *atoll*. If reefs become drowned, the submerged edifice is termed a *guyot*. They generally have basal subaerial lavas and volcaniclastic sediments (e.g. ODP Leg 143) capped by intertidal and shallow-water limestones overlain by pelagic chalks and oozes, indicating drowning of the guyots (e.g. Lonsdale, Normark & Newman, 1972).

Oceanic seamounts are rarely preserved in a recognizable state in the geological record, with most ancient examples showing some association with convergent margins and island arcs. For instance, the *Crescent Formation* (Washington, USA) contains a thick section of Eocene Hawaiian-type tholeiites, characterized by pillow lavas intercalated with red pelagic limestones, suggesting deep-water hotspot volcanism (Garrison, 1973) although perhaps sited within a rifted continental margin (Babcock, Burmester *et al.*, 1992). The volcanics occur as two separate centres, a 15-km and a 5-km thick pile, which pass

laterally and upwards into graded terrigenous clastics. Andesitic flows, pyroclastics and breccias are interbedded with the overlying clastics suggesting a change to convergent-margin volcanism.

12.11.3 Oceanic plateaux and aseismic ridges

Oceanic plateaux are vast (hundred of thousands of square kilometres) elevated areas of relatively thick oceanic crust that commonly rise to within 2–3 km of the ocean surface and rest on an abyssal ocean floor that lies several kilometres deeper. They include the Ontong–Java and Manihiki Plateaux in the Pacific and the Kerguelen Plateau in the Indian Ocean. Aseismic ridges are long linear features that rise to similar heights above the abyssal plain.

Modern oceanic plateaux show a general history of subsidence of a subaerial volcanic edifice, partly offset by the counteracting accumulation of the thick sedimentary sections, and punctuated by phases of uplift represented by unconformities. The Kerguelen Plateau was apparently elevated above sea level for 30–40 My (Sykes & Kidd, 1994). On the Manihiki Plateau, more than 250 m of basaltic volcaniclastics pass upwards into shallow-water limestones, overlain by pelagic chalks and claystones (Jenkyns, 1976; Beiersdorf, Bach *et al.*, 1995), the nature and thickness of the pelagic sediments depending on oceanic productivity and water depth.

Aseismic ridges show a similar subsidence history. Some were subaerially exposed, with brown coal and possibly lagoonal dolomites lying atop the Ninety-East Ridge. Many show faunal or sedimentary evidence for initial shallow-water, sometimes volcaniclastic deposition followed by deep-water pelagic carbonates (e.g. Rio-Grande, Walvis and Ninety-East Ridges). Other aseismic ridges (e.g. Carnegie Ridge, Cocos Ridge, eastern equatorial Pacific) have an entirely pelagic sedimentary history.

Ancient oceanic plateaux or aseismic ridges are impossible to recognize due to their large size. For instance, at Ballantrae, west Scotland, upward-coarsening clastic units interpreted as lava 'deltas' and indicating deposition close to sea level are locally capped by black shales and radiolarian cherts (Bluck, 1982b). However, it is still disputed whether the sequence represents parts of a subsided island-arc system, a seamount, or a slice of an aseismic ridge or oceanic plateau.

12.12 Polygenetic intermediate volcanoes

These volcanoes are the characteristic volcanic landform of convergent margins. Although andesite is the most common magma type, eruptive products range in composition from basalt to rhyolite. The range of eruptive activity is thus extremely varied making this the most complex volcanic environment. Individual volcanoes (e.g. Fuji, Mount St Helens) seem to have been constructed in 100 000 years. However, multivent volcanic centres (e.g. Santorini) show evidence for volcanic eruptions for almost a million years. Furthermore, most volcanic islands or volcanic arcs include several closely spaced composite volcanoes with eruptions spread over at least 10–30 My, albeit with repose periods between construction of individual edifices.

Textbook diagrams for intermediate volcanoes usually show a symmetrical cone surrounded by an apron of steeply dipping pyroclastic deposits interbedded with short lava flows, with average slopes ranging from 15 to 33°. However, the shapes, eruptive behaviours, and facies of intermediate volcanoes are extremely variable. The two primary controls on this complexity, and the basis for the different types of intermediate volcano, are: (i) regional stress regime; and (ii) composition of magma erupted, which itself depends on stress regime, crustal foundation and longevity of the arc.

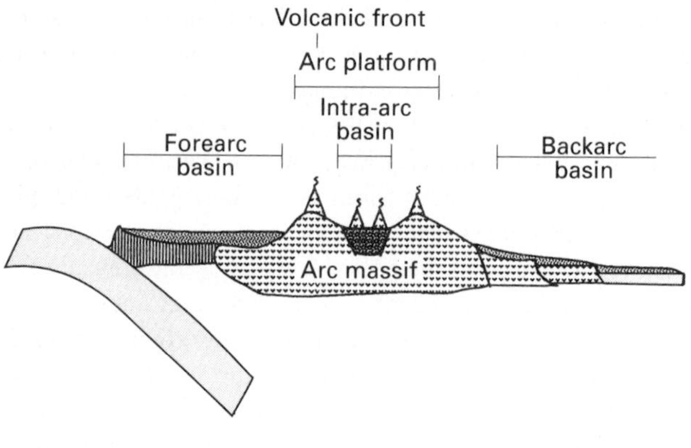

12.12.1 Sites of deposition

Several broad sites of volcaniclastic deposition can be recognized on convergent margins (Fig. 12.65). The *volcanic arc* forms an important frame of reference. The trenchward side of the arc is characterized by a sharply defined *volcanic front*, which typically corresponds to the axis of the largest and most active volcanoes, typically spaced at distances of 30–50 km. A diffuse zone of volcanism extends behind this axis away from the trench, with most volcanism concentrated within 100 km of the volcanic front. The *arc platform* is the topographic feature, typically of positive relief, formed by superimposed or overlapping volcanic edifices (G.A. Smith & Landis, 1995). The *arc massif* refers to that region underlain by crust generated by arc magmatic processes, and in long-lived

arcs the arc crust may underlie a broader area than can be associated with volcanism for short time intervals.

The volcanic arc is separated from the deep sea *trench*, by the *arc–trench gap* which is up to 400 km wide (Fig. 12.66) (Dickinson, 1974b). The arc–trench gap in some arcs includes an uplifted ridge or *outerarc*, sometimes rising above sea level. The depositional basin between the volcanic arc and the outer arc is termed the *forearc basin*. Behind the volcanic arc, there is sometimes a *backarc basin*. This could be a small area of oceanic lithosphere between two island arcs, or a basin between the arc and a continent. A remnant-arc ridge may occur within the backarc basin, its presence and position depending on the position of arc rifting, forearc or backarc (B.

Figure 12.65 Structure and depositional basins of convergent margins (from Smith & Landis, 1995). Detail or forearc side of arc platform is in Fig. 12.66.

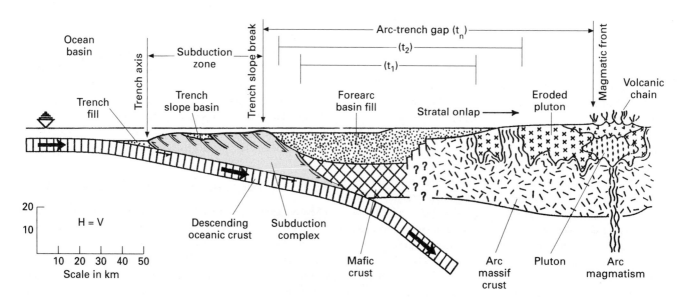

Figure 12.66 Structure of forearc basins (from Dickinson, 1995).

Taylor & Karner, 1983). In extreme instances, where backarc rifting initiates on the rear flank of the volcanic arc, no remnant arc is produced (Karig, 1972). The term *retroarc* is used to describe foreland basins behind compressional arcs. Although 'retroarc' and 'backarc' are literally synonomous, the latter is used for extensional and neutral arc–trench systems.

Additional sedimentary basins, termed *intra-arc basins*, may occur between volcanoes along the arc platform. Two end-member types are recognized (G.A. Smith & Landis, 1995). *Volcano-bounded* basins are low-areas on the arc platform, either between volcanoes or as linear troughs between older and newer arc axes such as on the Marianas Ridge (Karig, 1971). These basins have a high preservation potential only below sea level, generally in oceanic arcs. In contrast, *fault-bounded* intra-arc basins represent rapidly subsiding segments of the arc platform. Mechanisms for generating subsidence include tectonic processes which cause extension within or across the arc (Sect. 12.12.2), as well as isostatic flexural adjustment of the lithosphere to the weight of the volcano (Hildebrand & Bowring, 1984).

Along many convergent margins, particularly extensional margins, the spatial separation of arc settings may be poorly established. The Taupo Volcanic Zone, a Quaternary volcano-tectonic structure within the New Zealand subcontinent is a good example (Fig. 12.67). Onshore, two linear and parallel volcanic units are generally recognized, a southeastern andesitic arc and northwestern backarc graben of rhyolitic volcanic centres (Cole, 1990). However, the presently active rhyolitic centres only lie some 5–20 km behind the 'arc', and considerable extension can be recognized within the 'arc'. Offshore, continuation of these volcanotectonic trends and the distinction between arc and backarc settings appears even less applicable; numerous stratovolcanoes occur within the 'backarc' region; tectonism, although pervasive, increases from northwest to southeast, so that some of the greatest extension and subsidence occurs on the 'forearc' side of the volcanic arc nearest to the subduction zone (I.C. Wright, 1992) (e.g. near Whakatane of Fig. 12.67).

Distinguishing the various tectonic elements of convergent margins is further complicated by the spatial migration, over

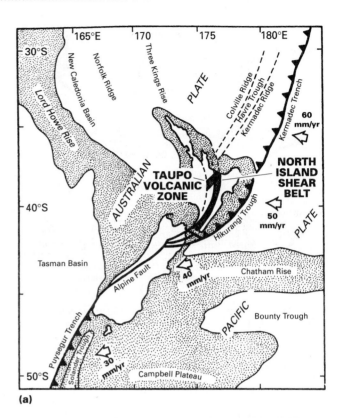

(a)

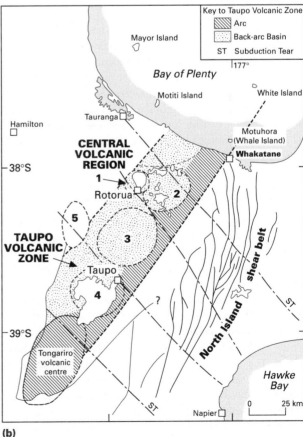

(b)

Figure 12.67 Taupo Volcanic Zone 'backarc' basin, New Zealand (simplified from Cole, 1990). (a) Location of Pacific–Australian plate boundary in New Zealand region. *Stippled area* represents continental crust; arrows show motion of Pacific plate relative to the Australian plate. The Havre Trough is a zone of intraoceanic backarc spreading that is offset sinistrally 45–50 km from the northern Taupo Volcanic Zone. (b) Details of the volcanic zone. Rhyolitic volcanic centres are: 1, Rotorua; 2, Okataina; 3, Maroa; 4, Taupo; 5, Mangakino. Onshore note close juxtaposition of arc and backarc elements. Immediately offshore, the greatest subsidence occurs on the eastern side of the Taupo Volcanic Zone near Whakatane (in b), along strike from the 'arc'.

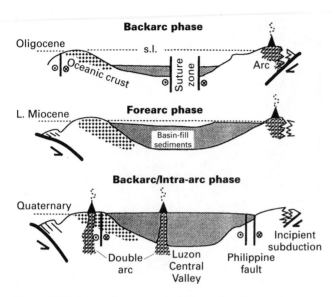

Figure 12.68 Schematic cross-sections illustrating the tectonic evolution of the Luzon Central Valley (Philippines). Reversals of arc polarity and shifting arc positions produced sequential backarc, forearc and intra-arc stages in basin evolution, which is also influenced by strike-slip motion along the Philippines fault (from Smith & Landis, 1995).

time, of the arc axis. A single depositional basin may lie between two volcanic arcs, or may change between intra-arc, backarc, and forearc positions, such as the Luzon Central Valley and Marinduque basins in the Philippines (Fig. 12.68) (Bachman, Lewis & Schweller, 1983; Defant, DeBoer & Oles, 1988). Where arc polarity cannot be established, as in many ancient successions, it may be difficult to assign a unique tectonic location to arc-related volcaniclastic successions (e.g. Houghton & Landis, 1989).

12.12.2 Effect of regional stress regime

Volcanic arcs or arc–trench systems can be classified as *extensional*, *neutral* or *compressional* (Dewey, 1980). Causes of arc extension are poorly understood but include transtension and concomitant block-rotation along oblique convergent margins (Burkhart & Self, 1985; Geist, Childs & Scholl, 1988), impingement of allogenic extensional terranes (Guffanti, Clynne *et al.*, 1990), expansion of backarc spreading into the arc, and gravitational collapse of overthickened magmatic arcs, which has been hypothesized for the Andes (Sebrier & Soler, 1992). The development of extension within a convergent margin can also be characterized in terms of the relative motion of the overriding and subducting plates and the dip of the subducting slab (Sect. 12.2.1), with extension favoured where trench rollback is faster than the trenchward migration of the overriding plate.

Extension influences the number, size, shape and longevity of volcanic centres, as well as the type of volcanism. Compressional arcs include significant silicic magmatism, have high relief, abundant sediments and shallow trenches. In contrast, extensional arc systems are characterized by more basaltic magmatism, have low relief, thin sediments and deep trenches. Extension also affects the depositional environment of the proximal volcanic arc: compressional arcs tend to be mountainous, neutral ones just emergent and shelfal, and extensional arcs intraoceanic and largely submerged.

In *compressional arcs*, magma finds it more difficult to ascend and tends to reuse the same pathway to the surface, resulting in high-standing simple or composite cones. Calderas rarely form. The entire volcanic arc may be uplifted by crustal underplating, the formation of high-level plutons and/or compressional tectonics (e.g. thrusting). In the Central Volcanic Zones of the Andes, for example, 90% of its volcanoes reach about 6000 m (de Silva & Francis, 1991). In compressional systems, emplacement of plutons may be aided by localized extension, such as along strike-slip faults.

Neutral arcs include the Aleutians, Indonesian and Central American arcs. In each arc oblique convergence results in through-going strike-slip faults, variable amounts of extension, and the development of numerous arc-parallel and arc-transverse intra-arc basins along the volcanic arc. Extension facilitates rise of magma and provides new pathways. Consequently, individual volcanic centres generally tend to have a shorter lifespan, resulting in small, low relief, more widely distributed calc-alkaline stratocones. Localized extension allows basalt magmas to rise to shallow crustal levels producing large volumes of silicic magmas that are erupted to form large calderas. An arc-parallel graben may develop with calc-alkaline basaltic and basaltic andesitic magmas erupted within the depression while larger and often silicic volcanoes concentrate along faults or on adjacent unfaulted crust (Dengo Bohnenberger & Bonis; 1970; Guffanti, Clynne *et al.*, 1990). In El Salvador and Nicaragua nearly all the arc volcanoes lie within such a fault-bound depression (Burkhart & Self, 1985) (Fig. 12.69), with the forearc region also characterized by weak extension, including block-faulting (Bourgeois, Pautot *et al.*, 1988). More pervasive extension in these volcanic arcs, such as the peninsular Aleutians (Geist, Childs & Scholl, 1988; Singer & Myers, 1990) or southeast Guatemala can even result in bimodal volcanic suites.

Extensional arcs are typified by the western Pacific intra-oceanic arcs, especially the Izu–Bonin–Marianas arc. Continued extension has resulted in splitting of the largely submarine volcanic arc, with dispersal of arc fragments over an area as wide as 450 km from the trench, and backarc spreading. Volcanism is largely basaltic and effusive. Strike-slip faulting appears to be less common than in continental arcs, perhaps due to the greater strength of oceanic lithosphere (Ingersoll & Busby, 1995).

environments, and felsic–intermediate caldera volcanoes; and (iii) strong compression since 4 My ago has resulted in crustal shortening and further uplift, with high-standing mafic strato-volcanoes becoming the dominant volcanic landform. A similar evolutionary trend, from highly extensional to compressional stress regimes, is seen in the Mesozoic of California (Fig. 12.70) and of South America (Fig. 12.71), which Ingersoll and Busby (1995) suggest may be typical of arc systems facing large ocean basins.

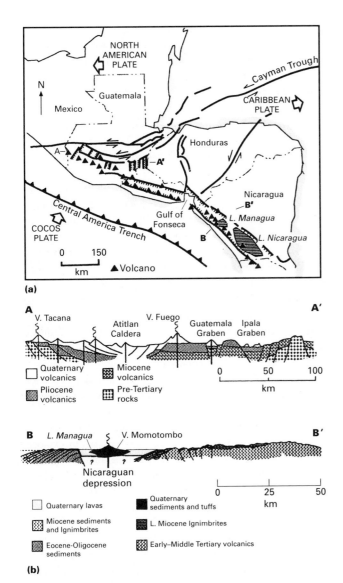

(a)

(b)

Figure 12.69 (a) Map shows relationship of arc-transverse pull-apart basins, in Guatemala, and the arc-parallel Medial Trough–Nicaraguan depression to relative plate motions in Central America. (b) Sections A–A′ and B–B′ show schematic cross-sections through these depressions (largely after Burkhart & Self, 1985).

It is important to remember that the stress regime of convergent margins may change rapidly and often. Hence a volcanic and sedimentary succession may superimpose products from quite different sources. For instance, three tectono-stratigraphic phases are recognized in Cenozoic successions of northeast Japan (Sato & Amano, 1991): (i) opening of the Sea of Japan between 22 and 14 My ago resulted in extension, development of arc-parallel grabens, subsidence to bathyal depths, and bimodal volcanism; (ii) a neutral or transitional stress regime between 14 and 4 My ago was characterized by dome-like uplift of the arc, shallow-marine to subaerial

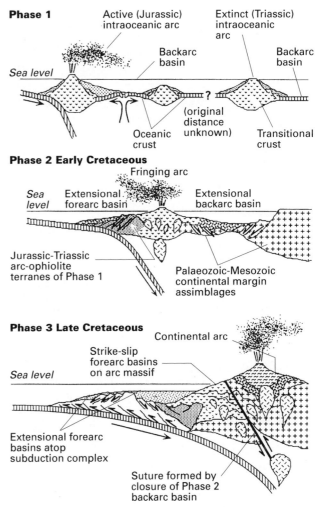

Figure 12.70 Evolutionary model for arc systems facing large oceans, based on Baja California, Mexico (from Ingersoll & Busby, 1995): Phase 1 – highly extensional intraoceanic-arc system with small rifted basins. Phase 2 – mildly extensional fringing arc system with narrowing backarc basin and extensional forearc. Phase 3 – compressional continental-arc system with closure of the backarc basin, forearc strike-slip basins, and a high-standing continental arc. Note long time span (150 My) of the three phases. Compare with Fig. 12.71. Northeast Japan displays a similar sequence of events since 15 My related to development of the Japan Sea (Sato & Amano, 1991) and may now be at the beginning of Phase 3.

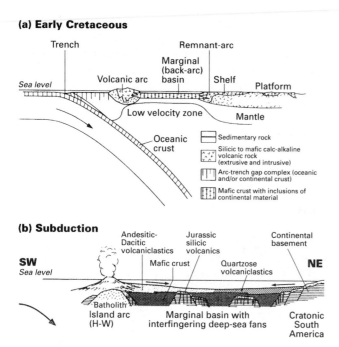

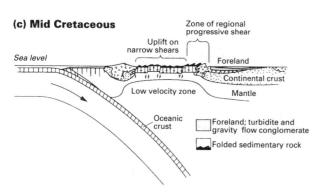

Figure 12.71 (a) Cross-section of Chilean subduction complex in Early Cretaceous times (from Bruhn & Dalziel, 1977). (b) Cross-section of Chilean and early Cretaceous marginal backarc basin (from Winn & Dott, 1978). (c) Cross-section of Chilean arc system in middle Cretaceous times (from Bruhn & Dalziel, 1977).

12.12.3 Types of intermediate volcano

For simplicity, only three end-member types of volcano are described here, although given the above complexity there are undoubtedly many variations and hybrids.

Basalt–basaltic andesite stratovolcanoes are the simplest. They mostly develop on oceanic or thinned continental crust, building either up from the sea floor to form oceanic islands, or subaerially at major fractures. They usually involve tholeiitic or low-K calc-alkaline magmas. Examples include the South Sandwich islands, Stromboli and Etna (Italy), Fuego (Guatemala), Fuji and Izu–Oshima (Japan) and Mayon (Philippines). They are essentially very symmetrical overgrown shield volcanoes or

scoria cones and are almost continuously active. Numerous small (<0.1 km³) eruptions produce scoria-fallout deposits, small pyroclastic flows, and lava (Vessel & Davies, 1981). During and between eruptions, innumerable small lahars redistribute sediment to the lower flanks of the volcano, such as on Mayon where channels tens of metres deep can be carved during a single typhoon (Rodolfo, 1989). Small landslides, such as those on Fuji, or the Sciara del Fuoco collapse scar on Stromboli (Kokelaar & Romagnoli, 1995), can also redistribute volcani-clastic sediment basinward.

Andesite–dacite stratovolcanoes occur on 'mature' island arcs and convergent continental margins, such as much of the circum-Pacific, the Lesser Antilles and Indonesia. They usually develop on comparatively thick continental crust that is already elevated above sea level. Recently active examples include Mount Egmont and Ruapehu (New Zealand), Mount St Helens (USA), Nevado del Ruiz (Columbia) and Pinatubo (Philippines). These volcanoes are less symmetrical because: (i) volcanic eruptions are more varied, in terms of size, composition and explosivity; (ii) steep-sided spines, domes or cryptodomes, formed from comparatively viscous magma, resist erosion and/or displace strata during their emplacement; and (iii) volcanic eruptions are less frequent, thereby allowing deep valleys to be carved into syn-eruptive deposits. Volcanic processes include: (i) small strombolian or phreatomagmatic eruptions from summit, flank and satellite vents; (ii) effusion of aa- and block-lavas; (iii) gravitational or explosive (vulcanian eruptions) disruption of summit domes to produce block and ash flows; (iv) small to large sub-plinian to plinian eruptions to produce scoriaceous to pumiceous pyroclastic gravity currents and fallout tephra; and (v) laterally directed blast surges. Sector collapse and formation of debris avalanches is extremely common, probably because dome-forming magmas steepen the summit of the volcano. The volcanic edifice builds up through numerous episodes of aggradation (small effusive or explosive eruptions) and degradation (sector collapse) (cf. Palmer & Neall, 1991). Repose periods between large (>0.1–1.0 km³) volcanic events are 100s–1000s years, allowing fluvial and glacial processes to dissect the landform.

Low-lying intermediate caldera volcanoes lack a central cone and usually possess one or several calderas. They preferentially develop at neutral to weakly extensional continental margin volcanic arcs, particularly where deep-seated basement fracture systems cut across the arc (e.g. Santorini; Valles: Self, Goff *et al.*, 1986; Toba in Sumatra), but also occur atop mature oceanic island stratovolcanoes (e.g. St Lucia; Wohletz, Heiken *et al.*, 1986). Santorini is the best understood because of the well-studied stratigraphy displayed on its caldera walls (Druitt, Mellors *et al.*, 1989). Two major mafic to silicic cycles can be recognized. Each cycle commenced with explosive eruptions of andesite or dacite, accompanied by interspersed effusion of basalt to andesite lavas to form shield complexes and steeper stratocones, and culminated in a pair of major dacitic to

Figure 12.72 Hydrothermal processes at stratovolcanoes (from Hedenquist & Lowensterm, 1994). Active volcano-hydrothermal systems extend from the degassing magma to fumaroles; they range from porphyry copper deposits formed adjacent to or hosted by intrusions to epithermal deposits formed at relatively low temperatures and shallow depths. Epithermal deposits incorporate high-sulphidation types formed by acidic springs near volcanoes and low sulphidation types formed by near-neutral pH reduced fluids as found in geothermal systems.

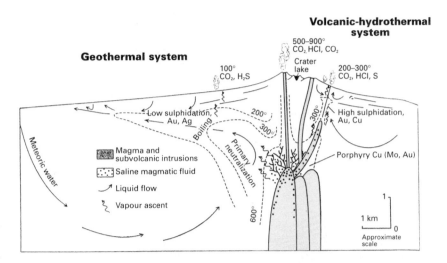

rhyodacitic eruptions forming calderas. Other modern examples of this type of volcano that are less well understood include Krakatau and Toba (Indonesia), Rabaul (Papua New Guinea) and Monte Somma (Italy).

Hot magma drives hydrothermal circulation cells within and beneath each type of intermediate volcano (Fig. 12.72), analogous to the processes operating at mid-ocean ridges. Hot water circulates through fissures and cracks, carrying various metals in solution, and may ultimately re-emerge as hot springs. Economic benefits of these hydrothermal processes include porphyry copper and epithermal Au–Ag deposits in ancient successions, and geothermal energy at modern volcanoes.

12.12.4 Subaerial convergent-margin sedimentation

Subaerial convergent-margin volcaniclastic facies largely occur on continental margins around andesite–dacite stratovolcanoes or intermediate caldera volcanoes. Subaerial settings also develop on the emergent portions of more mafic intraoceanic stratovolcanoes, but these environments are rarely preserved in the ancient record.

The upward growth of a volcano, combined with tectonic uplift or widening of the arc system (Dickinson, 1974b), results in a prograding coarsening upward succession, typically hundreds of metres to >1 km thick, in basins within or adjacent to the arc. Volcaniclastic sediment is thickest near the volcanic cone, and around isolated volcanoes forms circular alluvial plains, termed *ring plains*. More commonly, linear *aprons* of volcaniclastic sediment develop adjacent to the volcanic arc. The term apron has also been used in ancient successions where it cannot be demonstrated that comparable volcani-clastics accumulated on all sides of the volcanic centre. Proximal to distal relationships may be complicated by more mafic monogenetic satellite volcanoes on the flanks of the stratovolcano.

HIGH-STANDING STRATOVOLCANOES WITHOUT CALDERAS

Terrestrial volcaniclastic successions at stratovolcanoes have three general facies associations which preserve different portions of the volcano's history (Fig. 12.73): (i) a *proximal facies* of lava flows, autoclastic and pyroclastic breccias, and high-level intrusions; (ii) a *medial apron* association of pyroclastic flow, debris-avalanche, debris-flow and hyperconcentrated flow deposits (lahars) and fluvial conglomerates; and (iii) a *distal* facies association dominated by fluvial sands, either braided or meandering, overbank alluvium and pyroclastic fall deposits. The proximal facies association reflects extrusive or poorly explosive phases of volcanic activity, and builds up a volcanic cone. Intrusions, often a radial framework of dykes and sills, comprise about 60% of mafic stratovolcanoes and perhaps 90% of more silicic stratovolcanoes. Medial and distal associations include fragments from more explosive volcanism and reworked proximal and medial material.

On the backarc side of a volcanic arc, the volcaniclastic apron usually occurs within a retroarc foreland basin. Consequently, the distribution of sediments is additionally controlled by compressional tectonics, particularly the effect of thrust faulting on sediment load and isostatic flexure of the lithosphere (Jordan, 1995). Sediments include andesitic–dacitic volcaniclastics from contemporaneous eruptions, and polymict detritus from uplift and erosion of the arc massif (e.g. Van der Weil, Van der Bergh & Hebeda, 1992). Streams draining the landward side of continental volcanic arcs commonly act as tributaries to a trunk river that flows parallel to the volcanic arc due to confinement by inactive older arcs, basement uplifts, or thrust sheets (cf. Mathisen & Vondra, 1983; G.A. Smith, 1988). The position of this trunk river will be forced landward following episodes of prolonged volcanism and ring-plain aggradation. Thicker arc-adjacent successions can also occur

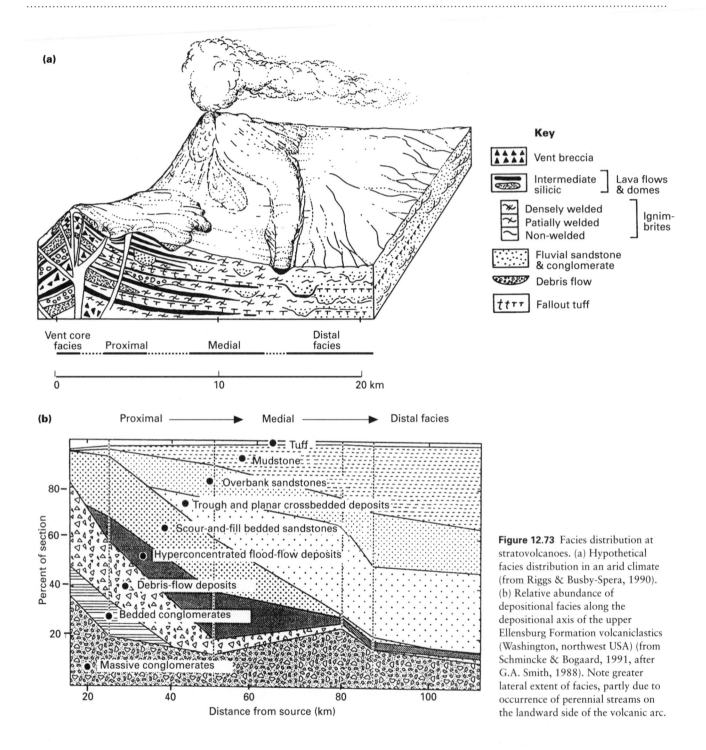

(a)

Key

▲▲▲▲	Vent breccia
⬮⬯⬮	Intermediate silicic — Lava flows & domes
✕	Densely welded
∿	Patially welded — Ignimbrites
⁓	Non-welded
∴	Fluvial sandstone & conglomerate
⬮⬯⬮	Debris flow
⸓⸓⸓	Fallout tuff

Vent core facies   Proximal   Medial   Distal facies

0   10   20 km

(b)   Proximal ⟶ Medial ⟶ Distal facies

- Tuff
- Mudstone
- Overbank sandstones
- Trough and planar crossbedded deposits
- Scour-and-fill bedded sandstones
- Hyperconcentrated flood-flow deposits
- Debris-flow deposits
- Bedded conglomerates
- Massive conglomerates

Percent of section: 80, 60, 40, 20

Distance from source (km): 20, 40, 60, 80, 100

Figure 12.73 Facies distribution at stratovolcanoes. (a) Hypothetical facies distribution in an arid climate (from Riggs & Busby-Spera, 1990). (b) Relative abundance of depositional facies along the depositional axis of the upper Ellensburg Formation volcaniclastics (Washington, northwest USA) (from Schmincke & Bogaard, 1991, after G.A. Smith, 1988). Note greater lateral extent of facies, partly due to occurrence of perennial streams on the landward side of the volcanic arc.

if retroarc basins are partially extensional. For instance, the Upper Cretaceous–Eocene Purilactis Group of northern Chile contains over 4 km of continental strata deposited in an arid climate (Hartley, Flint *et al.*, 1992). Although there was synchronous arc activity, it apparently did not contribute significantly to basin fill. Instead, most sediments were derived from uplift and unroofing of the volcanic arc, as indicated

by a systematic change from andesitic to granodiorite detritus.

Although the supracrustal parts of stratovolcanoes have a very low preservation potential, intra-arc extension, due to a change in arc stress regime, sometimes allows preservation of thick proximal successions. For instance, a Late Carboniferous terrestrial intra-arc graben in Chile, formed on top of pre-existing composite volcanoes, contains 200–600 m of lacustrine and fluvial volcaniclastics (Breitkreuz, 1991). Presumably much of the near-vent stratigraphy of the older stratovolcanoes is buried beneath these graben deposits.

More typically, facies associations are preserved 30–100 km from the inferred volcanic arc, in adjacent basins. At many, perhaps even most, large volcanoes only one flank (the subsiding one) of the volcanic edifice ever becomes preserved, usually as a linear volcaniclastic apron hundreds of kilometres in length. Ancient examples of these aprons include Carboniferous conglomerates in Australia (McPhie, 1987), 'molasse' facies of the Andean foreland basin (Houten, 1976; Van der Wiel, Van der Bergh & Hebeda, 1992), the Eocene strata of Yellowstone National Park (Smedes & Prostka, 1972), Miocene successions of Washington and Oregon (USA) (G.A. Smith, 1987a,b, 1988)

and Pliocene deposits of the Philippines (Mathisen & Vondra, 1983). Most of these aprons are less than 1 km in thickness and represent the continental side of the volcanic arc (Fig. 12.74). There is often evidence, such as folding of strata, erosional unconformities, or raising of marine facies, for regional uplift during deposition of volcaniclastics. Long-term rates of sediment accumulation are therefore comparatively low for volcanic terranes, most less than 20 cm 1000 years^{-1}.

Slightly different to the above, volcaniclastic aprons of the Oligocene–Miocene Mount Dutton Formation of Utah occur *between* and around several volcanoes (Palmer & Walton, 1990), in a manner similar to modern ring plains. Presumably weak extension resulted in more widely distributed andesite volcanism, rather than a linear volcanic front. Most lahars only reach 20–30 km from the volcanoes. Although the small runout of the volcaniclastics might be related to the semi-arid climate, it seems probable that the volcanoes were only of moderate height.

More disorganized, sometimes chaotic, accumulations of andesitic–dacitic volcaniclastic rocks, lavas and subvolcanic hypabyssal intrusions occur within low-relief intra-arc grabens (e.g. Riggs & Busby-Spera, 1990; White & Robinson, 1992). Although successions may be >4 km thick and extend over 100 km laterally, there is commonly no systematic facies variation and both 'proximal' and 'distal' facies are closely juxtaposed (Fig. 12.75). Numerous volcanoes, represented by eroded plugs or lava flows, may be identified, although these lack the zones of hydrothermal alteration associated with major stratocones. Furthermore, near-vent facies assemblages are onlapped by pyroclastic flow deposits and/or craton-derived aeolian deposits. Debris flow deposits are rare. Together these

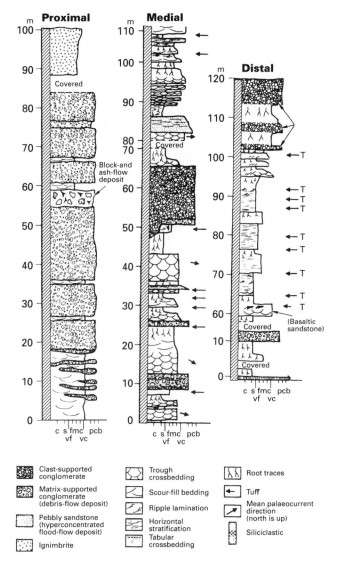

Figure 12.74 Representative logs from the volcaniclastic apron of the upper Ellensburg Formation (after G.A. Smith, 1988). See Fig. 12.73.

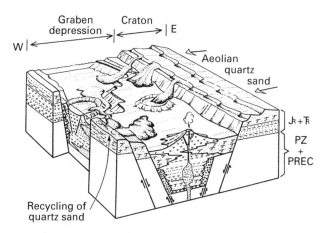

Figure 12.75 Diagram illustrating the Mesozoic arc–graben that formed at the western margin of the North American craton (from Busby-Spera, 1988c). This intra-arc basin was filled with as much as 11 km of continental and marine volcanic and volcaniclastic strata, including small composite volcanoes and caldera complexes. Aeolian quartz sands onlap volcanoes.

features are interpreted to record an abundance of small- to medium-sized volcanoes erupting within a weakly extensional sedimentary lowland (Riggs & Busby-Spera, 1990; White & Robinson, 1992). Explosive silicic volcanism and calderas have also been described from these basins (see below).

INTERMEDIATE CALDERA VOLCANOES

The basic evolutionary stages of large silicic calderas are (R.L. Smith & Bailey, 1968) (Fig. 12.76): (1) Precaldera volcanism and sedimentation over at least 10^4–10^6 years; (2a) caldera-forming eruptions and caldera formation over <10 years, often days to weeks; (2b) syn-eruptive resedimentation over 10–10^3 years; and (3) postcaldera deposition, including additional volcanism and possibly resurgent uplift, over 10^4–10^6 years. As there are only minor differences between caldera-forming eruptions (Stage 2a) at convergent-margin volcanic arcs (this section) and within silicic volcanic fields (Sect. 12.13), the facies of this stage have been largely described above (Sect. 12.4.6).

Precaldera (Stage 1) volcanism reflects the tectonic setting of the volcanic field. In many convergent-margin volcanic arcs basaltic to andesitic lavas extrude to form low-angle shield volcanoes, clusters of small andesitic stratocones which may include dacitic domes, and smaller monogenetic basaltic volcanoes. However, precaldera volcanics may not always be present, such as the Aira caldera in southwest Kyushu which rests on an older accretionary prism.

After the eruption (Stage 2), the unfilled caldera depression forms a depositional basin. The caldera basin may eventually be filled with water, either gradually over a few tens of years due to precipitation or immediately if the caldera develops near sea level. However, at subaerial calderas in arid climates (e.g. Valles) a permanent lake may never form. At calderas that develop atop stratovolcanoes, the unfilled posteruptive topographic depression is usually deep, signifying that the caldera collapsed late in the eruption. For instance, Crater Lake, the second deepest lake in North America, occupies a 1200-m-deep basin of about 10 km diameter. Extremely thick postcaldera successions can be preserved within this unfilled depression, such as 1000 m of volcanic and lacustrine strata within the Pleistocene Onikobe caldera of Japan (Yamada, 1988). Caldera walls are characteristically steep, rising hundreds of metres to >1 km above the top of the intracaldera tuff at angles of 45°. Early infill of these calderas consists of subaerial talus breccias and debris aprons (cf. silicic volcanic centres, Sect. 12.13), and at Crater Lake are thought to include phreatic debris produced when water interacted with hot intracaldera tuff. Early lake beds at Crater Lake are apparently coarse-grained and thick-bedded turbidites, whereas thin-bedded, finer-grained, basin-plain turbidites have been deposited during the volcanically quiescent period of the past 4000 years (Nelson, Bacon et al., 1994).

Postcaldera activity (Stage 3) may include continued volcanism within or near the caldera, resurgent uplift of the caldera related to renewed rise of the subjacent magma body to shallow levels, fine-grained sedimentation within the caldera basin and hydrothermal activity and mineralization associated with interactions between meteoric water and thermally disturbed

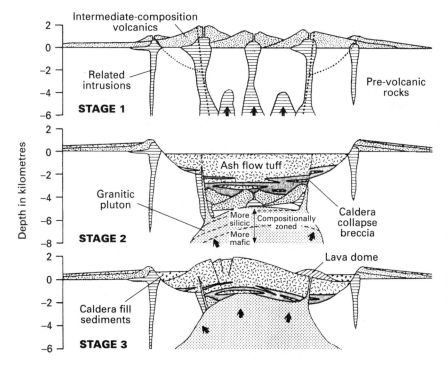

Figure 12.76 General caldera cycle (after Lipman, 1984). Stage 1 – precaldera volcanism develops clusters of small intermediate stratovolcanoes, Stage 2 – eruption of zoned magma chamber deposits develops caldera. Ash flow tuffs interfinger with caldera collapse breccias within the caldera whereas a thin outflow sheet extends outward from the caldera, Stage 3 – postcaldera deposition of volcanics and sediment and resurgent doming. See text and Section 12.4.6 for further discussion.

country rocks adjacent to the shallow magma. Owing to the multitude of volcanic and sedimentary events, with sources all around the caldera margin, postcaldera facies relations within calderas can be quite complex (e.g. Hulen, Nilsen & Little, 1991). Although postcaldera volcanics (Stage 3) are typically less silicic than the caldera-forming ash flow tuffs, they are commonly more evolved than precaldera volcanics. At calderas atop stratovolcanoes, intracaldera volcanics include andesite lava, andesitic tuff breccias and lapilli tuffs, rhyodacite lavas, and dacite pumice tuffs (Lipman, 1984; Yamada, 1988). An andesitic to rhyolite stratocone, such as Wizard Island in Crater Lake, or Vesuvius at Monte Somma (Italy) may build up within the centre of the caldera. If a central stratocone develops or the

caldera floor is resurgently domed (Sect. 12.4.6), the caldera basin becomes a narrow moat or horseshoe-shaped depression confined on its outer edge by the topographic caldera wall. Further updoming can tilt alluvial or lake beds, invert the topography and cause complete disappearance of any caldera basin.

The development of a caldera provides a local mechanism through which proximal deposits of large intermediate volcanic centres can be preserved in the stratigraphic record. Rapid synvolcanic subsidence associated with caldera formation lowers proximal precaldera strata below the general landscape surface and prevents their removal by erosion. In the Borrowdale Volcanic Group of the English Lake District (Fig. 12.77),

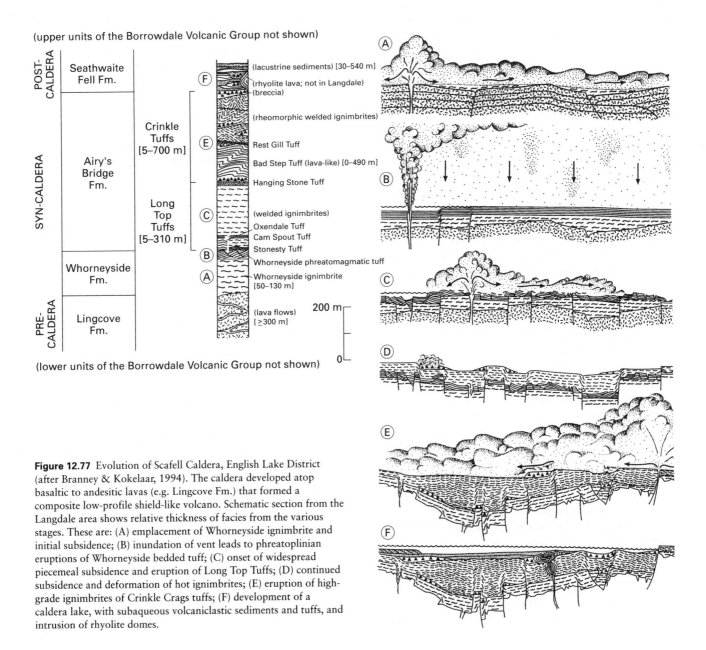

Figure 12.77 Evolution of Scafell Caldera, English Lake District (after Branney & Kokelaar, 1994). The caldera developed atop basaltic to andesitic lavas (e.g. Lingcove Fm.) that formed a composite low-profile shield-like volcano. Schematic section from the Langdale area shows relative thickness of facies from the various stages. These are: (A) emplacement of Whorneyside ignimbrite and initial subsidence; (B) inundation of vent leads to phreatoplinian eruptions of Whorneyside bedded tuff; (C) onset of widespread piecemeal subsidence and eruption of Long Top Tuffs; (D) continued subsidence and deformation of hot ignimbrites; (E) eruption of high-grade ignimbrites of Crinkle Crags tuffs; (F) development of a caldera lake, with subaqueous volcaniclastic sediments and tuffs, and intrusion of rhyolite domes.

553

precaldera volcanic activity is recorded by laterally impersistent basaltic to andesitic lava and peperitic sills interleaved with thin lenticular pyroclastic and volcaniclastic sedimentary rocks (Petterson, Beddoe-Stephens *et al.*, 1992). The effusive rocks were ponded in topographic depressions and produced a low-angle shield or plateau-like volcano. These strata are overstepped by laterally extensive tuffs which record voluminous (>400 km³) andesitic to rhyolitic explosive eruptions and associated caldera collapse (Branney & Kokelaar, 1994). Water periodically gained access to venting magma, and subaerial phreatomagmatic tuffs separate welded caldera-fill ignimbrites (Branney, 1991). Postcaldera deposits include silicic lavas and their breccias, subaqueous volcanogenic sediments, and small phreatomagmatic eruptives (Sorby, 1859, 1908).

Medial–distal (i.e. extracaldera) deposits may also be preserved because caldera volcanoes preferentially occur within neutral to weakly extensional volcanic arcs, often within subsiding intra-arc depressions. For instance, in Kyushu (southwest Japan) more than 5000 km³ of Pliocene–Pleistocene largely andesitic magmatic material, but including rhyolitic products from large calderas (Aso, Ata, Aira), erupted within several volcano-tectonic grabens (Aramaki, 1984; Kamata, 1989). As recognized by Busby-Spera (1988c), early Mesozoic andesitic–dacitic volcanic rocks of the southwest Cordilleran United States, including multiple collapse calderas, occurred within a 1000-km-long graben depression (see also Riggs & Busby-Spera, 1990). The arc-graben depression acted as a long-lived (more than 40 My) trap for craton-derived aeolian quartz sands which funnelled across the width of the depression and onlap the low-lying volcanic centres (Fig. 12.75). In more eroded volcanic terranes, intragraben calderas may contain the only supracrustal remnants of the magmatic arcs.

12.12.5 Marine sedimentation within and adjacent to volcanic arcs

Modern oceanic volcanic arcs are surrounded by large *volcaniclastic aprons*, kilometres thick, whose volume may far exceed that of the volcanoes. Although the foundation of the arc may be dominated by lavas, most of the apron consists of pyroclastic fragments generated by explosive volcanic activity in shallow water or on land and/or reworked volcaniclastics. The submarine slopes of stratovolcanoes are especially vulnerable to resedimentation because of steep regional gradients, high seismicity, and rapid sediment accumulation. In addition, shallow-level intrusions into arc aprons on steep slopes may cause sediment failure and remobilization. The result is frequent, small, shallow slides.

The Ocean Drilling Program/Deep Sea Drilling Project has penetrated the marine record of modern arc-adjacent basins at more sites than any other sedimentary–tectonic domain. Yet drill hole data are largely available only from the distal portions of the basins (Klein, 1985b), sites dominated by hemipelagic

mud, distal fallout tephra, and thin-bedded turbidites. Furthermore, resolution of studies in most modern marine arcs remains at dimensions comparable only to regional-scale field studies. Our understanding of arc-adjacent sedimentation thus relies equally on uplifted ancient successions (e.g. Mitchell & Reading, 1971).

Arc-adjacent deltaic and shoreface-to-shelf facies are generally characterized by an apron of amalgamated sand-rich pebbly debris flow and turbidite deposits with minor fallout tephra (e.g. Chan & Dott, 1983; Ballance, 1988; Orton, 1988; Scasso, Olivero & Buatois, 1991), with the overall successions resembling those of coarse-grained fan deltas. Turbidite facies may be directly overlain by alluvial facies, and shorelines may be difficult to pick (Sect. 6.5.6).

Deeper-water volcaniclastic deposits include: (i) volcanics themselves, such as subaqueous pyroclastic flows or fallout; (ii) syn-eruptive mass flows from subaerial, shallow marine or deep marine eruptions; and (iii) mass flows, unrelated to specific volcanic events, from deposition of volcaniclastic material by subaerial erosion and/or submarine slumping. Unusually large-volume mass-flow deposits of juvenile debris are presumably eruption-related although one can rarely be certain. It is also usually not clear from clast petrography alone whether these volcaniclastics were derived from subaerial or submarine eruptions. Coarse-grained volcaniclastic aprons share similarities with many deep-sea systems, particularly gravel-dominated ramps and slope aprons with numerous 'immature' and rapidly shifting source locations and repeated variation in supply due to the active volcanism (Cas, Powell *et al.*, 1981; Macdonald, 1986; Houghton & Landis, 1989). Volcaniclastic facies, particularly distal ones, interdigitate with normal basinal facies including resedimented carbonate reef talus, biogenic siliceous oozes, biogenic pelagic carbonates, hemipelagic and pelagic clays, and wind-blown continent-derived dust (Karig & Moore, 1975; Klein, 1985b).

The overall succession within a volcaniclastic apron is a product of at least six competing effects: (i) the distribution of transport processes around an arc is usually asymmetric, due to prevailing wind patterns, different submarine arc slopes and ocean currents (Fig. 12.78); (ii) as the island arc develops, older and more mafic volcaniclastics, including scoria, hyaloclastics and phreatomagmatic shards, are succeeded by products of more differentiated magmas, including abundant pyroclastic material (Fig. 12.79); dormancy of the island arc can result in blanketing of the volcaniclastic apron with silty epiclastic turbidites; (iii) the position of the island arc or individual volcanoes can migrate, often oceanward towards the trench or along the arc; (iv) as the volcano grows into shallow water and emerges, eruptions become more explosive, with ash dispersed greater distances from the volcano; (v) sedimentary processes continually sort volcaniclastic fragments by grain size/density into a proximal coarse-grained facies (pillow breccias, debris-avalanche and lahar deposits), a medial debris flow

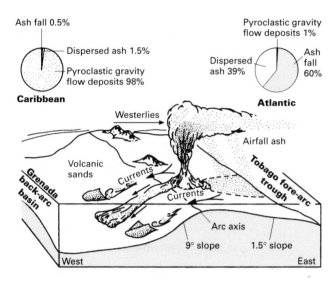

Figure 12.78 Schematic illustration of the Lesser Antilles volcanic arc to show distribution of volcaniclastic deposits within forearc and backarc basins (from Sigurdsson, Sparks *et al.*, 1980). Explosive eruptions reach into the troposphere (8–17 km above sea level) where powerful westerly winds blow the ash over the Tobago forearc basin. The absence of sediment gravity flow deposits in this basin has been related to the gentle flank slopes on this side causing pyroclastic flows to disintegrate and deposit before reaching deep water.

facies, and a distal facies consisting of thin distal turbidites and fallout ashes; and (vi) the rates of arc volcanism and backarc spreading can vary, with periods of continuous arc volcanism and reduced spreading favouring progradation of the volcaniclastic apron. Although obviously related to subduction, there is no *a priori* reason why backarc spreading should be synchronous with arc volcanism (Carey & Sigurdsson, 1984).

DEEP-SEA TRENCHES

Deep-sea trenches, 6000–11 000 m deep, receive sediment from: (i) oceanic-plate detritus passively conveyed by plate motion into a trench during subduction; (ii) lateral input from the forearc; and (iii) axial transport along the trench, possibly from distant areas (Underwood & Moore, 1995).

Trenches receive only minor amounts of volcaniclastic sediment compared with other tectonic settings along convergent margins. Oceanic-plate sediments are normally a thin veneer of pelagic or hemipelagic muds associated with Fe- and Mn-rich precipitates (Sect. 12.11.3). However, in the Indian Ocean the oceanic plate is buried beneath several thousand metres of terrigenous turbidites of the Bengal deep-sea fan (G.F. Moore, Curray & Emmel, 1982), and a comparable clastic succession is being funnelled northwards from the Amazon on to the Atlantic crust before being subducted westward at the Lesser Antilles Trench (Damuth, 1977). Lateral input from the

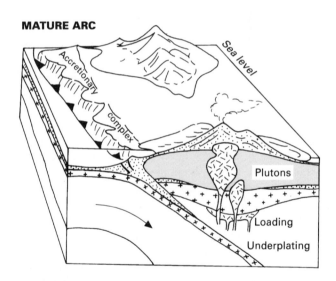

Spacing of volcanic centres about 50–70 km

Figure 12.79 Construction of basaltic island arc volcanoes (after Larue, Smith & Schellekens, 1991). Early arc growth features lava flows, breccias and hyaloclastites with no explosive volcanism until volcanoes shoal sufficiently. In mature island arcs crustal thickening, through underplating, intrusions, extrusion of volcanics, and deposition of arc flanking sediments, changes the composition of the magma erupted. Explosive eruption of andesite–dacite magmas produces pumice, lapilli and abundant sediment.

volcanic arc may be funnelled via small submarine canyons, such as those documented from the Middle America Trench (Fig. 12.80) (Underwood & Bachman, 1982). Ancient forearc submarine canyons are filled mainly with pebble to boulder conglomerates, such as the Cretaceous San Carlos canyon of Mexico (Morris & Busby-Spera, 1988). In the Central Aleutian Trench, at sites not associated with well-developed canyons, the occurrence of Holocene volcaniclastic sand layers suggests

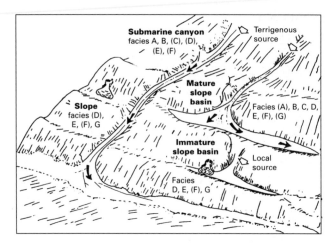

Figure 12.80 Conceptual diagram of turbidite facies and palaeocurrents in a trench-slope system (after Underwood & Bachman, 1982). For deep-sea facies terminology see Fig. 10.29; subordinate facies are shown in parentheses.

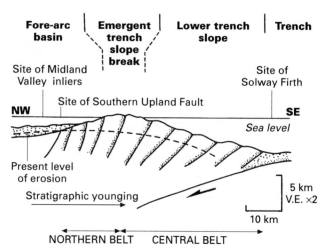

Figure 12.81 Accretionary prism from Southern Uplands of Scotland in mid-Silurian times (after Leggett, McKerrow & Eales, 1979). The youngest sequence is to the southeast but sediments themselves young to the northwest. If the detailed stratigraphical palaeontology were unknown, a succession several tens of kilometres thick might be estimated as the depositional thickness.

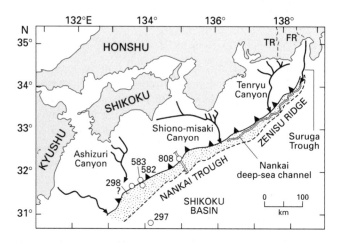

Figure 12.82 Simplified map of the Nankai Trough and vicinity (from Underwood & Moore, 1995). Submarine canyons are highlighted by arrows. The Tenryu (TR) and Fuji (FR) rivers are important sources for trench sediments. Numbers refer to DSDP/ODP drill sites.

unconfined non-channelized turbidity currents may provide another means for coarse sediments to bypass the forearc and reach the trench (Underwood, 1986). Terrigeneous and volcaniclastic trench sediment can be redistributed by geostrophic currents along the trench axis, as in the Chile Trench (Thornburg & Kulm, 1987).

The nature of the subduction process does not favour preservation of relatively intact trench stratigraphies in the oceanic record. Instead, ocean floor and trench sediments are scraped off the subducting plate above a low-angle thrust and tectonically accreted to the overriding plate as an *accretionary prism* (Fig. 12.81). With continued arrival of younger sediments in the trench, and progressive accretion and rotation, an imbricate wedge of flysch and minor pelagic sediment 1–2 accumulates (Karig & Sharman, 1975). Such accretionary prisms best develop where there is an abundant supply of sediment to the downgoing plate or trench. A modern example is the Nankai accretionary prism of Japan (Fig. 12.82). Most of the trench deposits were volcaniclastic sands first derived from the Tenryu and Fuji Rivers and associated fan deltas (Soh, Tanaka & Taira, 1994), then funnelled westward through the fault-controlled Suruga Trough, before streaming hundreds of kilometres as unconfined sheet flows down the trench axis.

Ancient accretionary prisms can be identified by a tilted succession of tectonic slices with palaeontological evidence for younging 'down succession', towards the trench, but sedimentological evidence for upwards younging, towards the arc (Fig. 12.81). Well-described examples include the Makran (Platt, Leggett *et al.*, 1985) or Shimanto accretionary prisms (Taira & Ogawa, 1988). During accretion wet sediments often become liquefied and fluidized to produce mud diapirs, ridges and 'volcanoes' (e.g. Brown & Westbrook, 1988). The largest

injections, comprising a fine-grained matrix with incorporated blocks of rock, raise the sea floor hundreds of metres, even to the point of forming islands. In ancient successions, with limited exposures, such chaotic layers could easily be misinterpreted as debris flows or even 'tectonic' mélanges.

FOREARC BASINS

Forearc basins range from relatively small ponds of sediment, also known as *trench slope* basins, above accretionary prisms (Fig. 12.80), to large basins such as the Tobago Basin in the Lesser Antilles forearc or the Sunda forearc with more than 5 km of sediment fill (Fig. 12.83). Most forearc basins are 50–100 km wide and can be more than 500 km long, with strings of linked forearc depocentres extending for 2000–4000 km along modern arc–trench systems. Basins tend to become wider and shallower with time (Dickinson, 1973), owing partly to accretion at trenches. Forearc sediments are derived from three sources: the outer arc, the volcanic arc, and in some cases longitudinally from an adjacent continent.

The tectonic style of the forearc region varies, with sedimentary basins developing in response to both compressional and extensional tectonics (Dickinson & Seely, 1979; Dickinson,

1995). For instance, extension, including normal faulting, characterizes the forearc off Peru with the compressional part of the accretionary prism only 15–85 km wide (Bourgeois, Pautot *et al.*, 1988). The Lesser Antilles arc, on the other hand, has an accretionary prism, the Barbados ridge, which is about 260 km wide and seismic evidence suggests back-thrusting within the Tobago forearc basin (Westbrook, 1982; Speed, Torrini & Smith, 1989). In the modern Andaman forearc basin, strike-slip tectonics may also be important.

Sea-level changes can also affect sedimentation. In Japan, vertical movements on the forearc side are small when compared with the active tectonics of the backarc region (Sato & Amano, 1991). Hence Pliocene–Pleistocene shallow-marine forearc successions accumulated under the influence of eustatic sea-

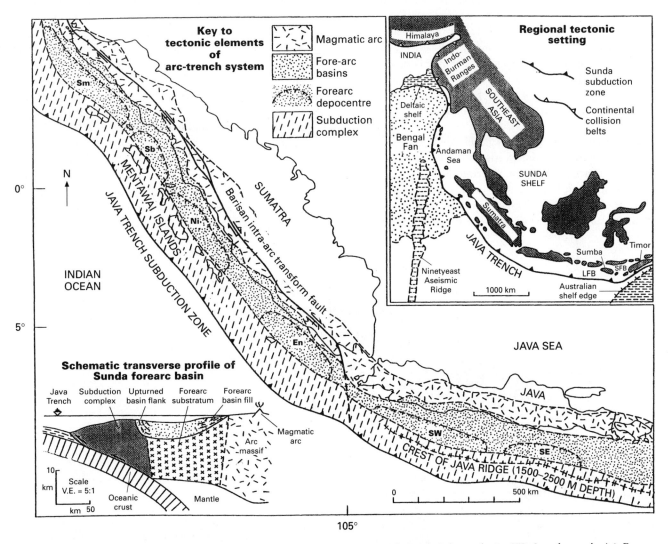

Figure 12.83 Configuration of elongate forearc basin, with multiple depocentres, in Sunda arc–trench system of Sumatra and Java (from Dickinson, 1995). Inset shows relation of Sunda subduction zone to Himalayan suture zone and incipient arc–continent collision at Timor (LFB, Lombok forearc basin; SFB, Savu forearc basin). Forearc depocentres off Sumatra–Java: Sm, Simeulue; Ni, Nias; Sb, Siberut; En, Enggano; SW, southwest Java; SE, southeast Java.

level fluctuations (Ito, 1995), with volcaniclastics more abundant in transgressive systems tract sediments. This was related to greater destruction and decreased preservation potential of unstable volcanic lithics during episodes of high-stand coastal progradation, although the possibility of episodic fluctuation in the rate of volcanism was not discounted. Relative sea-level changes were also invoked to explain cyclic alternations of pelagic carbonates and epiclastic volcaniclastic turbidites or tuffs within a distal deep-water forearc apron in the Himalayas (Robertson & Degnan, 1994). During periods of high sea level the arc interior would have been flooded, thereby reducing terrigenous run-off and favouring pelagic accumulation.

Ancient forearc basins were first recognized by Dickinson (1971b). In California, the late Mesozoic to Paleogene Great Valley Sequence was deposited between the magmatic arc of the Sierra Nevada and the outer arc mélange of the Franciscan complex (Dickinson & Seely, 1979; Dickinson, 1995). It overlies oceanic crust and contains up to 15 km of sediment, mostly volcaniclastic turbidites, but including deltaics, and shallows towards the top (Fig. 12.84). The Eocene Tyee Formation, a 3-km-thick succession deposited in a small tectonically active forearc basin, similarly shallows from a deep-marine basin, through submarine fan and delta successions (Chan & Dott, 1983; Heller & Dickinson, 1985).

More extensive forearc basins include parts of the Cretaceous Chugach Terrane of southern Alaska (Nilsen & Zuffa, 1982) and the Cretaceous–Neogene Shimanto Belt extending across southwest Japan (Fig. 12.85) (Taira, Okado *et al.*, 1982; Tairo & Ogawa, 1988). Both of these have a linear continuity of about 2000 km. The Shimanto belt records the episodic

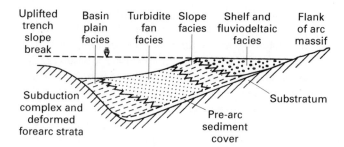

Figure 12.84 Facies framework of an ideal forearc basin shown in transverse profile (from Dickinson, 1995). There is no scale and the influence of intrabasinal structures is not depicted.

formation of accretionary prisms and crustal accretion associated with oblique subduction of the Philippine Sea plate first beneath Eurasia, then the Japanese Islands. One of the last episodes of accretion involved the mid-Miocene to Pleistocene collision between the Izu–Bonin arc and mainland Japan (Honshu). The forearc part of the arc, such as the Miura–Boso terrane, comprises Miocene–Pliocene volcaniclastic deep- to shallow-marine sediments and dismembered ophiolitic rocks (Soh, Pickering *et al.*, 1991). More complex but relatively short-lived remnants of foreland basins, which are filled by overall coarsening up successions 1000–6000 m deep, developed between the collision of volcanic-island terranes and the Honshu arc (Ito & Masuda, 1986). Large volumes of conglomeratic detritus were shed from uplifted orogenic highlands and accumulated in fan-delta and submarine-fan environments. These deposits overlie and interfinger with andesitic–dacitic intrusions and

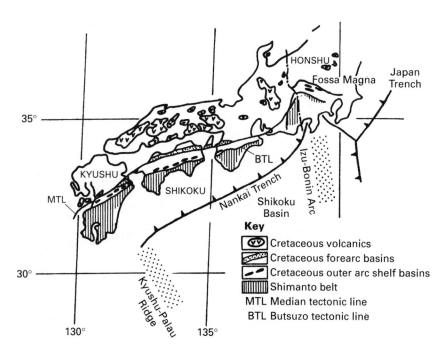

Figure 12.85 Sketch map of southwest Japan showing main tectono-stratigraphic units (after Taira, 1985). The Shimanto Belt represents a Cretaceous–Neogene accretionary prism, largely composed of deep-marine clastics and basaltic igneous rocks.

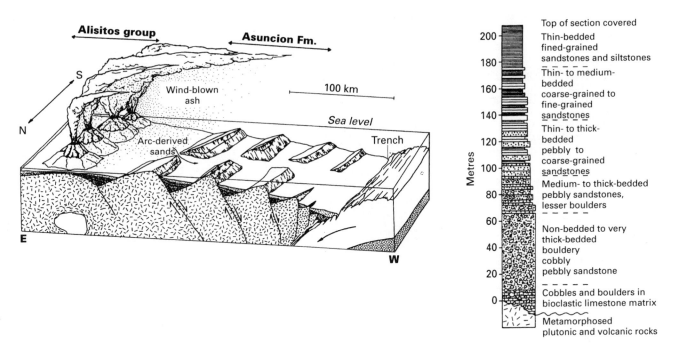

Figure 12.86 Palaeogeographic reconstruction of the early Cretaceous arc–forearc region of Baja California. Fan-deltoid aprons formed on the flanks of island volcanoes and along fault scarps of the extensional forearc (from Busby-Spera, 1988b). Arc-derived sands were only locally important. Section shows typical upward-fining and upward-thinning sequence that was deposited in each graben, recording erosional retreat and gradual denudation of the adjacent fault scarp and horst block.

volcaniclastics, including subaqueous pyroclastic flow, debris flow and turbidite deposits, derived from the colliding arc or intrabasinal volcanoes.

All of the above ancient forearc successions coarsen upwards, and contrast with a Cretaceous forearc apron (Asuncion Formation) of Baja California (Busby-Spera, 1988b). Here forearc sedimentation was initiated as numerous small grabens foundered below sea level to bathyal water depths (Fig. 12.86). Clastic slope aprons fringed the horst blocks which formed islands. Most of the detritus was derived locally, although turbidity currents and pyroclastic fallout supplied volcaniclastics from the island arc to the east. The upwards fining is interpreted to reflect a single phase of rapid down-dropping of the graben floors, followed by gradual erosion of the horst blocks, as inferred for graben infills associated with Jurassic–Cretaceous rifting in Greenland (Surlyk, 1984).

ISLAND ARCS AND INTRA-ARC BASINS

Island arcs are extremely varied, depending on the type of oceanic crust and whether the volcanic arc is extensional, neutral or compressive. Only a few modern oceanic volcanic arcs have been described, and most of our understanding is based on uplifted and eroded ancient arcs.

A volcaniclastic succession can be considered intra-arc if it includes arc volcanoes or their eroded remnants and/or lies within a structural basin shared by arc volcanoes. Abundant synsedimentary shallow-level igneous intrusions provide the strongest evidence that volcaniclastic strata were deposited in intra-arc, rather than forearc or backarc, basins. For instance, volcaniclastic facies of an Early Miocene volcanic arc on Malekula island in the New Hebrides are intruded by basaltic and andesitic sills of calc-alkaline composition, and by plutons (Mitchell, 1970). Basaltic sills are also common within intra-arc strata of the Alistos Group of Baja (California), mixing with sediment to form peperites, and sometimes remobilizing unconsolidated sediments (White & Busby-Spera, 1987).

Extremely thick intra-arc successions may develop if the arc position remains stable. In the Klamath Mountains of California, volcanism and sedimentation persisted from the Early Devonian to Middle Jurassic time, about 220 My, and yielded thick successions of water-laid volcaniclastic debris and lavas, interstratified with carbonate and siliciclastic sequences (M.M. Miller, 1989). Late Palaeozoic sequences are over 20 km thick. Volcanic episodes, represented by basaltic, andesitic and dacite volcaniclastic aprons and mafic intrusions, were punctuated by epiclastic sedimention and/or establishment of shallow-water platforms (Fig. 12.87).

Successions of individual volcanoes differ according to proximity to the eruptive vent(s) and changes in water depth as the volcano shoals (Fisher, 1984; Staudigel & Schmincke, 1984;

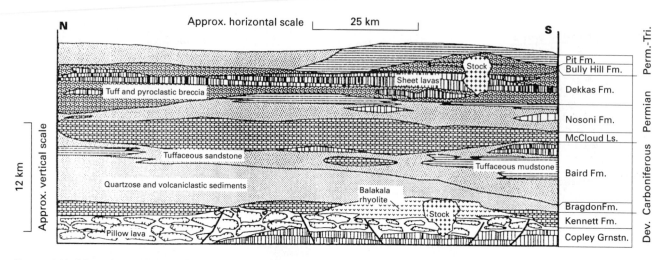

Figure 12.87 Schematic cross-section through mid–late Palaeozoic volcanic, volcano-sedimentary, and carbonate rock assemblages of the Klamath Mountains (after Miller, 1989 as in Smith & Landis, 1995). Note vertical thickness and lateral continuity of defineable rock units, albeit with considerable local variation, in an oceanic arc terrane. Volcanic units would represent products from an entire eruptive epoch (*sensu* Fig. 12.35) and several volcanoes.

Busby-Spera, 1988b; McPhie, 1995; Sohn, 1995). The lower deep-water parts of island arcs are similar to other submarine basaltic volcanoes (see Figs 12.61 & 12.63) with volcaniclastics scarce (<25%) relative to sheet and pillow lavas. A large proportion of the clastics represents material brought from shallower levels by sediment gravity flows. Hydrothermal manganese deposits may also occur, as described from modern submarine seamounts of Izu–Ogasawara Arc (Japan) (Usui & Nishimura, 1992).

Shallow-water parts of island arc seamounts often include more silicic and more volatile-rich magmas, that may erupt explosively to produce subaqueous pyroclastic flows. These explosive subaqueous eruptions and mass wasting processes (erosion, sector collapse) between eruptions may destroy the volcanic edifices, with new volcanoes building up on dissected remnant edifices. For instance, a Miocene seamount complex in Japan comprises three or more superimposed volcanic piles, each several kilometres wide and several hundred metres in thickness (Kano, Yamamoto & Takeuchi, 1994).

The uppermost volcaniclastic deposits may include evidence for emergence of the seamount and formation of a volcanic island. This includes the appearance of terrestrial plant debris, unequivocal epiclastic fragments, and well-rounded clasts with a history of fluvial or littoral abrasion (Mitchell, 1970; Fisher, 1984; McPhie, 1995). The submarine to subaerial transition, if it occurs, is likely to be marked by the presence of *in situ* primary pyroclastic deposits, lavas with oxidized or scoriaceous margins, and the absence of subaqueous facies. The subaerial portions of the volcano are small relative to the total volcanic edifice and are usually destroyed by erosion (but see Sohn, 1995).

BACKARC BASINS

Most backarc basins form by extension and thinning of the crust behind oceanic or continental volcanic arcs, although the exact reason for extension remains enigmatic (cf. Taylor & Karner, 1983; Marsaglia, 1995). Many backarc basins go through an evolutionary cycle similar to the J.T. Wilson (1966) cycle of: (i) oceanic opening by sea-floor spreading; (ii) oceanic closure by subduction; and (iii) continental collision. However, the average lifespan of backarc basin spreading systems is less than 25 My (Tamaki & Honza, 1991) compared with perhaps 200 My for the Wilson cycle.

In the evolution of oceanic backarc basins isolated from a continental sediment influx, four general phases have been recognized (Carey & Sigurdsson, 1984) (Fig. 12.88): Phase 1, rifting of an intra-arc basin, with steep unstable basin margins mantled by various mass-flow deposits; Phase 2, backarc spreading and island arc volcanism; unstable faulted margins tend to be smoothed by sediment gravity flows; Phase 3, basin maturity and a decrease in arc volcanism and/or backarc spreading rates – this favours transgression and preservation of finer-grained hemipelagic and pelagic sediments over the coarse volcaniclastic apron; and Phase 4, basin inactivity dominated by pelagic sediments although epiclastic debris may be eroded from the volcanic arc.

The early rift phase of *intraoceanic backarcs* is well illustrated by the Sumisu Rift within the Izu–Bonin arc system (Taylor, Brown *et al.*, 1990). This rift is characterized by high-angle normal faults at all scales, the major ones exhibiting up to 1000 m of bathymetric relief and up to 2500 m throw. Opposing master faults and/or rift-flank uplifts are linked by oblique transfer zones. Rift volcanism concentrates along these transfer zones and is largely basaltic and effusive, although silicic lava flows and explosive eruptions forming submarine calderas also occur. Sedimentation has approximately matched subsidence, with extremely high rates, up to 6000 m My^{-1}. Sediments are primarily derived from the silicic arc calderas, and deposited as mass-flow units of pumiceous gravels up to 100 m thick (Nishimura, Marsaglia *et al.*, 1991). Basalt volcanic centres, flanked by coarse talus slopes and debris aprons, are locally important.

Once sea-floor spreading has commenced, the isolated remnant arc, which may have been originally emergent, isostatically subsides, forming a submerged chain of seamounts or a ridge. Backarc and major ocean basins have similar crustal structures and magnetic-lineation patterns, indicating similar modes of crustal origin. With both, mean heat flow values and basement elevations slowly decrease with age due to thermal subsidence. Mature basins are generally asymmetrical with depths increasing towards the remnant arc (Fig. 12.88).

Most oceanic backarc basins are eventually subducted and their sedimentary fill partly preserved in remnant basins, as imbricate slices in outer arcs, or as nappes in collision belts. The Gran Cañon Formation of Baja California represents a fragment of a Jurassic backarc basin (Busby-Spera, 1988a).

Ophiolites, generated by rifting of arc basement, were immediately followed by progradation of a deep-marine volcaniclastic apron contemporaneous with the growth of island-arc volcanoes (Fig. 12.89). Lithofacies include fine-grained basaltic tuffs from deep marine eruptions, lapilli tuffs and tuff breccia turbidites representing resedimented hyaloclastic and scoriaceous debris, dacite pyroclastic flow deposits, and laterally extensive basalt flows. The volcaniclastic apron was blanketed with volcanic lithic silts and sands eroded from the extinct arc segment within 10 My of formation of the backarc basin.

Continental backarc basins are similar, although backarc spreading may be limited or absent and oceanic crust is not always formed. They stretch around southeast Asia from the Andaman Sea behind the Sumatra–Java arc to the Sea of Japan. The Okinawa Trough is an example of an incipient continental backarc basin, and displays graben structures similar to the Sumisu Rift. The Japan Sea is the best example of a mature continental backarc basin. It is presently inactive, with some evidence for incipient subduction of the Japan Sea along its eastern margin (Kikuchi, Tono & Funayama, 1991).

Facies relations are more complex in backarc basins along continents, partly because of varied terrigeneous input. Some, such as the Andaman Sea, East China Sea and South China Sea basins are fed by large rivers, and basins contain up to 8 km of

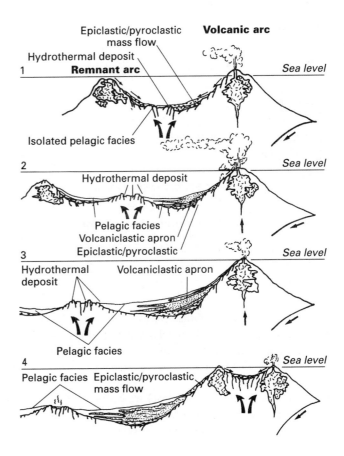

Figure 12.88 Distribution of volcanogenic sediments in an evolving back-arc basin (after Carey & Sigurdsson, 1984). See text for discussion of stages.

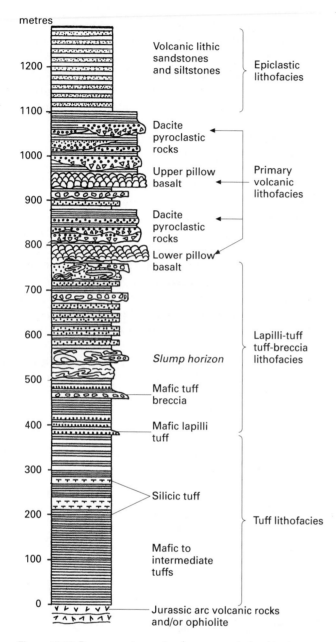

Figure 12.89 Representative section from progradational backarc apron, Gran Cañon Formation, Cedros Island (from Busby-Spera, 1988a). Tuff used *sensu* Fisher and Schmincke (see Sect. 12.7 and Fig. 12.45).

are known from the Ordovician Lachlan foldbelt of southeast Australia. Facies patterns of the Late Ordovician have been compared with those of the present day Andaman Sea (Cas, Powell & Crook, 1980); deep-marine upper Silurian to Devonian volcaniclastic aprons that interdigitate with epiclastic terrigenous sediments have been compared with the interarc Havre trough where it impinges on the North Island of New Zealand (Cas, Powell *et al.*, 1981). Similarly, coarse-grained sediment gravity flows, forming submarine fans with a total thickness of 7–8 km, dominate the fill of a Mesozoic marginal basin at the southern tip of South America (see Fig. 12.71). Turbidites entered this basin from both sides, with distinctive compositions. Andesitic–dacitic volcaniclastics derived from an active island arc to the west interfinger with quartzose volcaniclastics eroded from older silicic ignimbrites to the east (Winn & Dott, 1978). In South Georgia, turbidites within the backarc basin flowed west-northwest parallel to the basin margin (Storey & MacDonald, 1984), collecting in a long linear fault-controlled trough comparable with the Grenada basin.

12.13 Polygenetic silicic volcanoes

The largest silicic volcanoes are found on subaerial or submerged, but shallow-water, continental crust associated with major crustal extension. Pleistocene examples include La Primavera (Mexico) (Fig. 12.90), Long Valley, Valles and Yellowstone (USA), and Taupo (New Zealand). Although they are among the largest volcanic landforms on Earth, they generally have extremely low relief, with flank slopes varying from <1° to 5°. In some cases the land surface slopes towards the centre of the volcano (e.g. Taupo volcano, New Zealand). This subdued topography arises because: (i) any pre-existing upstanding volcanic edifice collapses into the space created through evacuation of the top part of the underlying magma chamber; (ii) eruptions are extremely powerful, and disperse pyroclastic fragments a long way from the vent; and (iii) volcanoes often develop in extensional settings which are already subsidence prone. Owing to their low relief, silicic volcanoes have little effect on local climates, and pre- and post-eruptive facies are influenced more heavily by the ambient climate than at high-standing stratovolcanoes.

Although the repose period between eruptions is much longer than at polygenetic intermediate volcanoes, there is considerable variation. Silicic volcanism at Taupo is similar to the Yellowstone system in terms of size, longevity, thermal flux and magma output rate. However, Taupo contrasts with Yellowstone in the exceptionally high frequency, every 10^4–10^5 years, but small size, mostly 30–100 km^3, of caldera-forming eruptions (Wilson, Houghton *et al.*, 1995). This contrast reflects the thin rifted nature of the Taupo crust which precludes the development of long-term magmatic cycles.

As with the intermediate caldera volcanoes discussed in Section 12.12.4, volcanism and sedimentation can again be

continental-margin sediment. Other basins, such as the Sea of Japan, are relatively starved and have a large biogenic component. In addition, rhyolitic volcaniclastics, derived either from the volcanic arc or from within the marginal basin, are more important.

Unlike their oceanic counterparts, backarc basins sited on continental crust have a high preservation potential and may be only moderately folded. For example, several backarc basins

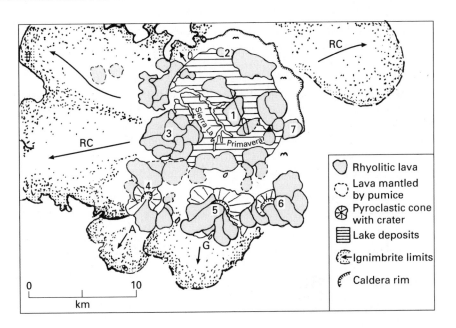

Figure 12.90 Geology of La Primavera Volcano (simplified from G.P.L. Walker, Wright *et al.*, 1981). Note poor definition of caldera subsidence structure, particularly the limited extent of the caldera rim. Some rhyolite domes (2, 3, 7) lie along this caldera rim whereas others (4, 5, 6), which are large enough to be volcanoes in their own right, may lie on an outer younger ring structure. Lake deposits are only 100 m thick (cf. c. 1 km at many calderas at stratovolcanoes, see Sect. 12.12.4). RC (Rio Caliente) indicates ignimbrite deposits from the caldera-forming eruption. A and G are ignimbrites from smaller pre- and postcaldera eruptions.

(Legend): Rhyolitic lava; Lava mantled by pumice; Pyroclastic cone with crater; Lake deposits; Ignimbrite limits; Caldera rim

viewed in the context of the caldera-forming eruption. Precaldera activity in rhyolitic volcanic terranes records the rise and fractionation of a large shallow batholith. Tectonic, volcanic and sedimentary effects may all be recognized. The regional signature of the Yellowstone hotspot is a 600-m-high topographic bulge centred on the caldera, that extends across a 600-km-wide region (R.B. Smith & Braile, 1994). Thick wedges of sedimentary and volcanic breccias can develop due to movement, uplift, and instability on precaldera fault zones (e.g. Howells, Reedman & Campbell, 1991). Volcanically driven uplift can also cause progradation of deposition systems (e.g. Orton, 1991). Volcanic leaks from the evolving magma body commonly produce clusters of volcanoes over the magma chamber, sometimes in linear or arcuate patterns over radial and ring faults, at other times controlled by regional structures. Precaldera volcanism is typically bimodal. For instance, the 0.6 My ago Yellowstone caldera was preceded for 0.5 My ago by eruption of a bimodal assemblage of basaltic or rhyolitic lavas, and an arcuate trend of rhyolite domes at Mono Craters (California) is thought to represent volcanism above an incipient precaldera magma body. Additional premonitory activity, during a shorter time span, includes earthquakes, small volume pyroclastic eruptions, and changes in the amount of fumarolic activity (Newhall & Dzurisin, 1988).

Syn-eruptive deposits of rhyolitic calderas lack many of the facies found at intermediate calderas (cf. Sect. 12.12.4) due to the subdued topography of silicic volcanic fields and the comparatively fine size of the pyroclastic debris produced by the most powerful of volcanic eruptions. Rhyolitic calderas, particularly those formed in extensional basins, are often broad but shallow sags with only the central part experiencing significant subsidence. For instance, Lake Taupo is now only a few hundred metres deep, with extremely gentle margins, and is being filled by deltaic and shoreface deposits (R.C.M. Smith, 1990). Owing to its low relief the landscape surrounding the caldera may be more dramatically transformed by the eruption. Thick ash flow tuffs flatten topographies (cf. Buesch, 1991; R.C.M. Smith, 1991), and local stream gradients after eruptions may be so low that only the finest grained material, if any at all, is delivered to areas previously receiving coarse alluvial clastics. Differential compaction of the almost flat-lying tuffs can create isolated depressions which may be filled by lacustrine and playa deposits. Unwelded portions of the outflow sheet, particularly its thin veneer that mantles ridges, combined with additional pyroclastic material from smaller postcaldera eruptions will be rapidly reworked into upwards-coarsening and thickening aprons, hundred of metres thick, and composed largely of sand-sized pyroclastic debris. These include sand-rich alluvial fans (Walton, 1986) or submarine fans (Busby-Spera, 1985; Fritz, Howells *et al.*, 1990). Owing to the dominance of sand-grade material and the low gradient of the volcano's flanks, sediment will be largely redistributed by sheet-like, unconfined flows, such as hyperconcentrated stream floods and turbidity currents.

Postcollapse volcanism in silicic volcanic terranes includes clusters of rhyolite domes within the caldera, with more mafic magmas such as basalt dykes reaching the surface outside the caldera rim. Large blocks of rhyolite pumice sometimes occur within lacustrine muds from extrusion of rhyolite domes underwater. Resurgent doming, as discussed earlier (Sect. 12.4.6), may also occur.

Several ancient silicic volcanic systems occur within the Ordovician marginal basin of Wales. Although this may have been a backarc basin, it contains no apparent arc-derived sediment and few unequivocal deep marine deposits (Howells, Reedman & Campbell, 1991). Presumably the island arc at this time, possibly represented by the Lake District of England, was of low relief or arc detritus was trapped in more proximal basins. In North Wales, a 3-km-thick succession of shallow marine and subaerial volcaniclastic deposits was derived from several intrabasinal volcanoes. Volcanism was essentially bimodal, and

included several large rhyolitic calderas (Sect. 12.14, Fig. 12.91), subaqueous–emergent basaltic volcanoes (Kokelaar, 1993), and numerous smaller volcanic centres, represented only by sheets of reworked pyroclastics within adjoining basins (e.g. Orton, 1995). The distribution of volcanic centres was controlled by an orthogonal array of deep-seated basement fractures (Kokelaar, 1988), and the 'polarity' of basin fill switched many times due to the influence of silicic volcanism on basin subsidence.

12.14 Analysis of ancient volcanic successions

Methods used to study volcaniclastic successions are similar to those used to study sedimentary rocks and have similar purposes: establishing correlations, vertical sequences, determining facies changes and the like in order to establish features such as basin history, climate and tectonics. In addition, we hope to learn something about the size and shape of the original volcanoes, the eruptive history of the volcano, the amount of volcanic material removed and transported to adjacent sedimentary basins, and the relation between volcanism and climate change. Unlike other sedimentary environments, large volcanoes indicate a specific plate-tectonic setting and directly depend on the movement of lithospheric plates.

Modern and ancient volcanic systems provide different understandings of how a volcano works and its controls. Research on modern volcanoes typically focuses on the physical processes responsible for the deposition and distribution of an individual facies, usually within the context of the most recent volcanic event at that volcano. Whether this event is typical and a common occurrence at the volcano, representing the norm for that type of volcano, can only be established through study of ancient volcaniclastic successions.

12.14.1 Reconstructing volcanic landscapes

To reconstruct an ancient volcanic landscape, and separate volcanic from background controls, we need to establish: (i) the types of volcanic fragments present and their abundance; (ii) when volcanism was active; (iii) a chronostratigraphy; (iv) depositional environments; and (v) the position and size of volcanic centres.

After coming to rest, *volcanic fragments* and their host deposit are subject to alteration that results in a change in their mineralogy, chemistry and texture. Processes include devitrification, hydration, diagenetic and hydrothermal alteration, diagenetic compaction, metamorphism, and tectonic deformation. Although any sedimentary rock can be altered by these processes, they are more common in volcaniclastic successions because: (i) many deposits consist of chemically unstable and reactive volcanic glasses; (ii) syneruptive products are often deposited in a hot state; (iii) the parental magma chamber

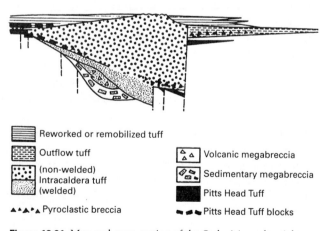

Figure 12.91 Map and cross-section of the Ordovician subaerial to subaqueous Lower Rhyolitic Tuff caldera, North Wales (modified from Howells, Reedman & Campbell, 1986).

provides heat to drive regional hydrothermal circulation; and (iv) most successions preserved in the ancient record were deposited along convergent plate margins where tectonic and metamorphic processes are already active.

Each alteration process overprints and modifies original textures, thus hampering the recognition of volcanic products. Lavas preferentially alter along permeable fracture networks. Patchy alteration of coherent lavas can produce apparent or pseudoclastic textures typical of volcaniclastic mass-flow deposits (R.L. Allen, 1988), including particles that closely resemble pyroclastic glass shards (Fig. 12.92). In contrast, the porous and permeable nature of unwelded pyroclastic deposits allows early, rapid and pervasive diagenetic alteration, with chalcedony, opal, zeolite and clay minerals as alteration by-products (Surdam & Boles, 1979; Mathisen & McPherson, 1991). Mafic frag-ments are more prone to alteration than silicic ones, and thus more difficult to recognize in ancient successions. During diagenesis and lithification, pumice and scoria-clasts can be mechanically flattened to produce a bedding parallel compaction fabric that closely resembles the eutaxitic fabric of welded ignimbrites (Branney & Sparks, 1990). Thus attention needs to be paid to the shape and texture of glass shards rather than pumice to establish a hot state of emplacement.

Establishing when volcanism was active is essential to the understanding of volcanism and volcaniclastic sedimentation.

(a) Classical perlite

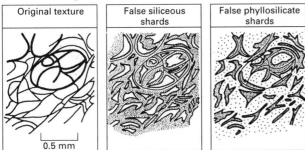

(b) Banded perlite

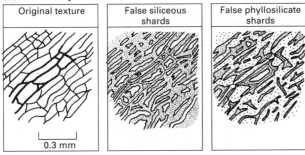

Figure 12.92 (a) Original fracture patterns for classical perlite (a) and (b) banded perlite. False unwelded (a) and welded (b) vitroclastic textures result from selective alteration to phyllosilicates along fractures (modified from R.L. Allen, 1988).

This task becomes easier as eruptions become larger and less frequent. Typically only the largest volcanic events can be distinguished. Many features, including the geometry of the deposit, its sedimentary facies, the variability of its composition, shape and sorting parameters, and the extent that sediment composition differs from the 'norm' for sediments within the basin, are needed to establish penecontemporaneous explosive volcanism. Syn-eruptive deposits are often a large volume, widespread sheet of monolithologic sand-grade sediments. Products from smaller eruptions are usually completely re-worked. In subaerial settings the end result of sorting and alteration processes is usually crystal-lithic sandstones, with framework grains set in a fine-grained, variably altered and/or recrystallized, nondescript matrix (e.g. Cather & Folk, 1991; G.A. Smith & Lotosky, 1994). These sandstones are often interbedded with siliceous mudstones, which could represent a separate source of pyroclastic detritus, or fines winnowed from interbedded coarser-grained lithologies. In subaqueous settings, distinguishing primary pyroclastic deposits generated by single eruptions from secondary deposits that have been reworked is even more difficult because of a lack of adequate unambiguous criteria (Stix, 1991).

Deposits from pyroclastic eruptions, particularly fallout layers, represent excellent *chronostratigraphic* marker horizons because most tephra layers differ in colour, thickness, grain-size characteristics, sediment structures, mineralogical, chemical and/or isotopic composition from enclosing non-volcanic deposits. Ash layers can be used for correlation over long distances or between different basins, and are one of the few geological phenomena that allow stratigraphic correlation between the subaerial and marine sedimentary record (e.g. Sarna-Wojcicki, Morrison *et al.*, 1987). However, separate ash layers may have very similar microscopic and geochemical features, being derived from the same or similar magma chambers (see Hildreth & Mahood, 1985). Also, fallout tuffs are rarely preserved on the volcano itself, where stratigraphic correlation is most challenging. In general, establishing a stratigraphy is easiest with more silicic volcanic systems. This is because eruptions are larger and less frequent, and their products less commonly altered beyond recognition. In con-trast, the stratigraphy of lahar-dominated successions that surround andesitic stratovolcanoes is particularly problematic, although clast petrography can sometimes be used to group lahar deposits and identify different sources (e.g. Palmer & Walton, 1990). Large basaltic landforms are also problematic, as the resolution of radiometric dating techniques is only suf-ficient to erect an extremely general stratigraphy for prehistoric successions. These mafic volcanic landforms also include a multitude of small intrusions. Cross-cutting relationships and the occurrence of peperite on the uppermost surface allow tabular sheet-like intrusions (i.e. sills) to be distinguished from surficial lava flows, although these features are often not present (Branney & Suthren, 1988).

Individual facies seldom identify the host depositional environment unequivocally. Pyroclastic surge deposits should be subaerial. Pillow lavas can occur in a few metres or a few kilometres of water. Welding is more common, but not unique to subaerial settings. Accretionary lapilli usually indicate subaerial expression of the eruption column but final deposition could be in subaqueous settings. As a consequence, volcanic environments can only be defined by facies associations. However, determining depositional environments requires more detailed observation of process and facies relationships than is necessary in most non-volcanic successions. Some of the reasons are:

1 Fauna and their fossils are often absent, even in marine successions, due to high rates of sedimentation. Or, if faunas are present, they are often displaced downslope within mass flows so the onshore–offshore faunal zonation that characterizes non-volcanic shelves is absent (Lockley, 1990).

2 Facies models developed elsewhere may not be applicable. The settling velocity of pyroclastic debris often varies by several orders of magnitude, creating difficulty in using existing bedform stability diagrams. During the syn-eruptive period, rates of suspended sediment fallout are so high that near-bed turbulence and tractive sediment transport are suppressed, resulting in massive beds. Consequently, the sedimentary structures that are normally used to establish flow energies of the ambient currents and water depths are absent from the depositional record of syn-eruptive deposits.

3 Sedimentation is dominated by mass flows, with only minor differences between subaerial facies and their subaqueous counterparts. Close attention needs to be paid to the background facies.

4 Mass flows may be tens or even hundred of metres thick. One depositional event may overwhelm and radically change pre-existing environments.

5 Rapid uplift and subsidence, sometimes on the order of hundreds of metres, is common in volcanic terrains, and can juxtapose disparate shallow marine and subaerial facies. Few transitional facies may be preserved.

Direct evidence of a *volcanic centre* includes a high concentration of intrusions or lava flows, coarse volcanic breccias, more intense hydrothermal alteration, and recognition of a caldera. Occurrences of thick homogeneous ignimbrite sheets are the best evidence for calderas because extracaldera outflow sheets are rarely more than 100–200 m thick. Indirect evidence is provided by thickness or grain-size relations. Individual syn-eruptive facies will each generally fine away from the source, although complications may arise due to the pre-existing topography. As in other siliciclastic systems, reworked facies fine downslope. This is often, but not always (e.g. Lake Taupo), away from the centre of the volcano. Adjacent beds, particularly in intra-arc basins, are often from different volcanoes and need to be considered individually. As a general rule, the volcanic landscape will be more complex than even the most imaginative reconstructions.

12.14.2 Facies models in volcanic settings

The conceptual basis for such an analysis is similar to that used in other sedimentary environments (cf. R.G. Walker & James, 1992). A useful eruptive or facies model for a volcanic landform must: (i) integrate the various magmatic, volcanic, sedimentary and tectonic processes (see also G.P.L. Walker, 1993); (ii) act as a framework and guide for future events at active examples of this type of volcano; and (iii) be able to predict process or facies relations in new geological situations.

Developing such models requires studying numerous volcanic events at several similar volcanoes, followed by a distillation process in which local details are boiled away and essential features are highlighted. Facies models for most other sedimentary environments, except for those environments dominated by mass-flow processes, can be developed largely from study of modern environments. In contrast, volcanoes rarely erupt and processes during and between eruptions are markedly different. In a few cases, such as at frequently erupting basaltic volcanoes or short-lived monogenetic volcanoes (Paricutin), the historic record may provide enough volcanic events from which to distill some sort of model. However, a longer deposition record of 10 000s–1 000 000s years is usually needed.

One common element in the approach to the development of facies models in both non-volcanic and volcanic terranes is the need to develop a focal element about which sedimentation is discussed (Cas & Wright, 1987). This is usually the source of sediment: in fluvial successions the focal element is the river channel, in deltas it is the river mouth, and in volcanic terranes it is the eruptive centre.

In volcanic terranes there is an additional focal element: the volcanic eruption (Orton, 1995). During syn-eruptive periods each bed represents a unique *historical* event with no direct genetic relation to the bed before; Walther's law is less applicable and sedimentology essentially loses some of its predictive capability and to an extent its ability to define environments. In contrast, during inter-eruptive periods, facies changes reflect changing energy levels within the depositional setting, as in non-volcanic terranes.

This additional focal element is the curse and blessing of volcaniclastic successions. Because of the large number of controls on the timing, intensity, and volumes of material supplied during volcanic events, it is doubtful whether generalized facies models can be developed to the same extent as in other sedimentary environments. On the other side of the coin, the high resolution chronostratigraphy provided by volcanic eruptions provides an unparalleled framework in which other controls on deposition, such as eustasy and basin tectonics, can be assessed.

Further reading

Busby C.J. & Ingersoll C.V. (1995) *Tectonics of Sedimentary Basins*, 579 pp. Blackwell Science, Oxford.

Cas R.A.F. & Wright J.V. (1987) *Volcanic Successions: Modern and Ancient*, 528 pp. Allen & Unwin, London.

Fisher R.V. & Schmincke H.U. (1984) *Pyroclastic Rocks*, 472 pp. Springer-Verlag, Berlin.

Fisher R.V. & Smith G.A. (1991) *Sedimentation in Volcanic Settings*, 257 pp. *Spec. Publ. Soc. econ. Paleont. Miner.*, **45**. Tulsa.

Francis P.W. (1993) *Volcanoes: A Planetary Perspective*, 443 pp. Oxford University Press, Oxford.

Heiken G. & Wohletz K. (1985) *Volcanic Ash*, 246 pp. University of California Press, Berkeley.

McPhie J., Doyle M. & Allen R. (1993) *Volcanic Textures: A Guide to the Interpretation of Textures in Volcanic Rocks*, 198 pp. Centre for Ore Deposits and Exploration Studies, University of Tasmania.

Additional information

Every year volcanic eruptions shed new light on volcanic and sedimentary processes at volcanoes. Events at volcanoes are also often quite spectacular and photogenic. Hence, volcanologists have found the combination of text and images possible with the 'World Wide Web' of Internet to be an ideal, quick, and free medium to disperse new information and ideas. Below are the addresses of several useful 'home pages' followed by a few words on their content.

http://magic.ucsb.edu/~fisher/ recent thoughts of R.A. Fisher plus tabulation of resource material on volcanic environments.

http://www.geo.mtu.edu/eos/ focuses on NASA research on surface and atmospheric effects of recent eruptions such as Mount Pinatubo bases on satellite data.

http://www.geo.mtu.edu/volcanoes timely information about on-going eruptions, particularly Central American volcanoes.

http://vulcan.wr.usgs.gov/home.html information about on-going studies of the Cascades Volcanic observatory including the 1980 eruption of Mount St Helens.

Problems and perspectives

13

H.G. Reading

Sedimentary geology underpins three crucial areas of the earth sciences: (i) Earth history; (ii) economic deposits (oil, gas, coal, aggregates, building materials and certain minerals); and (iii) environments, both modern and ancient. The interaction of these three areas is continually increasing and, as a consequence, the multidisciplinary application of sedimentary geology is likely to expand dramatically in the near future. Indeed, the need for sedimentary geologists to contribute to the solution of multidisciplinary problems within the energy, mineral resources and environmental areas has probably never been greater.

13.1 Historical review

The foundations of sedimentology were laid in the 19th century. By 1894, Walther had established his 'Law of Succession of Sedimentary Facies' showing how the vertical sequence may be used in environmental analysis. Sorby, between 1851 and 1908, had initiated studies of sedimentary petrography, quantitative petrography, sedimentary structures, palaeocurrent analysis, diagenesis and shell structures and his work on petrology of chertified limestones pre-dated the first paper on igneous petrology. By the turn of the century Van't Hoff, a leading physical chemist, was using experimental chemistry and an

understanding of the phase rule to explain evaporites. He experimented on marine evaporites before the first experimental laboratories for the study of igneous and metamorphic rocks were established.

Nevertheless, although the principal directions for sedimentological research had been mapped out, very little progress was made in the first half of the 20th century.

This was partly because, in the early part of this century, the study of sedimentary rocks was largely undertaken by stratigraphers who were mainly concerned with correlation and used fossils as zonal indices rather than as environmental indicators. Because the stratigraphically 'useful' fossil was largely independent of environment, facies-bound fossils tended to be ignored, despite their environmental significance. Sedimentary petrography, apart from a few diagenetic studies, became a tool for correlation, with concentration on the analysis of heavy minerals and microfacies.

In the 1930s and 1940s the more outward-looking stratigraphers and sedimentary petrographers were mainly concerned with tectonic aspects of sedimentation. In particular they used sedimentary petrography and sedimentary facies such as 'flysch' and 'molasse' as indicators of syn-sedimentary tectonics and of stages in the development of geosynclinal basins and orogenic belts; others emphasized cyclicity and sought to explain sedimentary cycles as due to intermittent uplift or downwarp. The sediments that made up the cycles were underplayed and there was little appreciation of the role of sedimentary factors in generating cycles.

Modern sedimentology was initiated in the 1950s with the turbidity current hypothesis. This not only solved the problem of deep-sea sands and graded beds but also demonstrated the importance of integrating discoveries in modern oceans, in ancient rocks and through experiments. It focused attention on process, and forced geologists to examine every facet of a rock, each grain and sedimentary structure, and to wonder what produced them. During the next three decades process sedimentology dominated sedimentary geology. Models that were largely autocyclic or intrinsic were developed with extrinsic causes only involved as a last resort. The broader stratigraphic context of these depositional models was often ignored. This was then changed in the late 1970s with the sequence stratigraphy revolution which reminded many older sedimentologists of their geological roots and gave the younger generation, who had rarely learnt the basic concepts of stratigraphy, new insights into external controls of sedimentary patterns. Within that revolution, there has been a marked shift from an early emphasis on eustasy and global controls to a more mature and sophisticated appreciation of the interplay of sea-level change, basinal tectonics and sediment supply and we are now at a position where the main thrust of sedimentary geology has moved towards trying to decipher the relative significance of changing sea level, climate, tectonics and sediment supply on observed sedimentary patterns.

It is somewhat ironic that, in the past 15 years, probably the greatest impact on sedimentary geology was triggered by seismic interpreters. This occurred in two quite different ways. Firstly, by reinventing stratigraphy through the concepts of seismic stratigraphy and its offspring sequence stratigraphy, it has led to a renewed appreciation of the stratigrapic significance of both marine flooding surfaces and of unconformities. The importance of identifying these critical surfaces, obscure though they may be, in outcrop and in the subsurface is now realized. The establishment of sequence stratigraphic methodologies has led to significant progress, but it is important to distinguish the methods from the models, with which they are closely linked. The danger, as with all new fashionable models, is that they are applied too literally, without regard to their true applicability. Recourse to models in a simplistic way, without fully appreciating the complexity of sedimentary processes, may prove almost as dangerous as the way in which many process sedimentologists of the previous era ignored stratigraphy.

Secondly, modern seismic acquisition and processing has given us three-dimensional patterns that, for the first time outside a few very closely drilled petroleum reservoirs and coal mines, have shown us the true nature of facies patterns. We can now see the complex pattern of spaghetti-like channels, levees and lobes in deep-sea clastic systems, not only in petroleum

provinces such as the Gulf of Mexico and the North Sea but in the Quaternary where the controls of sea level and climate on sediment supply are well constrained.

13.2 Economic aspects

A continuing stimulus to research in sedimentary geology has been economic, in particular the necessity to discover and exploit resources of coal and oil. In the 1930s and 1940s much fruitful discussion centred on the cyclic arrangement of coal-bearing strata in the Carboniferous. Later decades saw the growth of sedimentological research groups within the major oil companies as they recognized the need to improve methods of interpretation that would lead to earlier prediction of subsurface porosity trends and reservoir distribution.

In recent years, improvements in the tools, interpretation techniques and types of data employed, most notably seismic and well data, have provided an unrivalled insight into the three-dimensional organization of sedimentary rocks at scales that range from whole basins to individual reservoirs. Furthermore, much of the data collected during hydrocarbon exploration and development have provided research opportunities that have added considerably to the understanding of depositional processes, basin evolution, diagenesis and fluid migration.

This trend continues as both the quality and quantity of subsurface data increases. Foremost amongst this trend is the move to exploration-scale three dimensional seismic surveys and the stitching together of earlier field-specific surveys with infill data. The resulting 'wall-to-wall carpeting' of three-dimensional seismic data in some of the World's most well-established petroliferous basins, in particular the Gulf of Mexico and North Sea, is providing dramatic improvements to the understanding of both regional- and field-scale stratigraphic architecture, particularly the distribution of reservoir rocks. Not only do three-dimensional seismic data add much greater understanding to those intervals of immediate economic interest but also to the full stratigraphy of the basin under investigation. The application of modern concepts in seismic and sequence stratigraphy further adds to the quality of modern basin and reservoir studies.

An additional source of high resolution stratigraphic data is found in mature oil and gas fields, some of which contain the most detailed information on the three-dimensional geometry of sedimentary rocks. For example, in the West Canadian sedimentary basin of Alberta, a combination of high resolution core, well log and dynamic reservoir data (e.g. pressures, production rates, etc.) from densely drilled wells has provided a three-dimensional description only rivalled by that from coal mines. Not only has it provided a wide range of models for stratigraphic traps in sedimentary rocks but new stratigraphic concepts have been developed and tested.

Quantitative modelling studies of both sedimentary basins and reservoirs developed rapidly during the 1980s, driven by:

(i) the need to improve exploration for subtle traps and to optimize oil recovery from existing accumulations; and (ii) the availability of powerful computers capable of processing large volumes of data quickly and cheaply.

Basin modelling systems vary in detail but share the common requirement to simulate stratigraphic architecture. The models are essentially two dimensional cross-sections derived from seismic and well correlation sections in which the sedimentary architecture is investigated by varying the change in accommodation space creation. This is achieved mainly through variation in the rates of subsidence, sediment supply and eustasy. These quantitative forward models can be calibrated against known basins and provide valuable learning sets by testing the impact of changing input parameters on the resulting stratigraphic models. Ideally, they should also give insight into predicted reservoir–source–seal configurations in underexplored areas.

Quantitative field- (or reservoir-) scale models are primarily aimed at providing the best geological representation of the interwell rock volume. Traditional, deterministic approaches to well correlation and extrapolation, where reservoir continuity/width is greater than the available well spacing, may be utilized; geological judgement is used to generate a single, most realistic interpretation. However, where reservoir bodies are narrower than well spacing, or where marked heterogeneity exists, then stochastic (or probabilistic) methods are employed to complete the interwell volume. Stochastic models are being widely used to describe the distribution of sand bodies, shale layers and reservoir properties. Fundamental to the quality of these models is appropriate quantitative analogue data, which is most commonly acquired from outcrops, but also from mines, modern environments and analogous subsurface datasets. The need for reservoir analogues is giving a major incentive to detailed outcrop studies of many well-known sedimentary successions for reservoir characterization purposes, particularly for clastic deposits. The variability of carbonate reservoirs in terms of pore-space characteristics makes stochastic modelling more difficult and outcrop analogues more restricted in their application. It is essential to acquire analogue data that not only reflect the appropriate depositional environment but also are from successions of comparable accommodation-space history.

13.3 Environmental aspects

Increasing concerns with the environment are prompting research in such areas as environmental change, sea-level fluctuations and climatic variations. Studies are expanding on groundwater aquifers, waste disposal in the subsurface, in shallow seas, ocean deeps and on land, and the disposal of large structures such as oil production and storage platforms. Problems of pollution are increasing in rivers, estuaries, lakes and shallow, enclosed or semi-enclosed seas. Increasing consumption of water, particularly through irrigation, has exacerbated the problem of balancing water input with its extraction. Although society demands an increasing level of extraction of gravels, sand, clays and limestones, it also requires that these are kept to a level that minimizes harm to the environment and has no side-effects. Sedimentologists need to improve the reliability of predictions in all these areas.

The sedimentary record is our principal key to the immediate future. It can indicate the frequency and effects of low frequency events, e.g. major earthquake shocks in areas of low seismicity and major marine landslides as off the Hawaiian islands and Tenerife. It can act as a check on predicted climate models, especially their effects on environmental changes. It is also a benchmark for environmental change prior to human interference. Sedimentologists can thus contribute to the prediction of environmental catastrophes. They can predict the effects of major river floods, glacial surges, and the catastrophic impact on coastal regions of hurricanes, typhoons and tsunamis. In addition, by studying the erosional and depositional effects, sedimentologists can contribute to the work of seismologists and volcanologists in predicting the timing and effects of earthquakes and volcanic explosions.

The upsurge in environmental studies and the necessity to predict the consequences of extraction and amelioration programmes offers a similar spin-off to that of petroleum studies for detailed sedimentary data acquired from a wide range of sedimentary environments and successions. Emphasis will be on Quaternary to Recent deposits. The high resolution data acquired to investigate particular environmental problems will provide additional insight into the evolution of sedimentary rocks, including their depositional, diagenetic and fluid characteristics.

Conversely, those working on ancient sediments can contribute substantially to modern environmental studies through their perspective of longer time scales and knowledge of sequence analysis.

13.4 Future studies

One of the main benefits of sequence stratigraphy has been a significant improvement in our understanding of the interrelationship between shelf and coastal environments, particularly the impact of changes in relative sea level. Partly as a consequence, two previously little studied sand body types, lowstand shoreface and incised valley-fills/estuarine deposits, have received a resurgence of interest, including major revisions of some previously long held models (e.g. the 'offshore bar' model). Many of these new interpretations have been achieved through an improved understanding of the nature and significance of laterally extensive stratal surfaces that define the external geometry of coastal sand bodies and their correlative deposits basinwards. They also allow correlation between shelf and coastal environments during different periods of sea-level history. Interestingly, the mapping and ultimate recognition of these surfaces has resulted from the application of seismic stratigraphic techniques to closely-spaced, high resolution well data (mainly wireline logs). In outcrop the regional significance of many of these surfaces may not be appreciated, but their regional correlation is a prerequisite to correct interpretation.

The development of new models and classification schemes for estuaries and incised valleys has resulted in a dramatic increase in the recognition of their deposits in the stratigraphic record. Many tidal deposits, previously of uncertain origin (e.g. shelf, shoreline or inshore), are being reinterpreted as estuarine on the basis of the nature and geometry of the bounding stratal surfaces.

Sequence stratigraphic concepts have provided added confidence to the widely held view that many ancient sandy shelves are mainly the result of rapidly fluctuating sea levels, with transgression over shallower water sandy sediments the most efficient process for emplacing sands on to ancient shelves. However, there is little doubt that ancient shelf seas at different stratigraphic periods were highly variable and that modern shelves provide only a partial analogy. There was a substantial difference between the relatively common sand-dominated shallow seas of the late Proterozoic and early Palaeozoic and the mud-dominated shallow seas of the Mesozoic. How different were ancient widespread epeiric/epicontinental seas compared with modern shallow seas? Were storm and tidal conditions dramatically different during different periods of Earth history? The demise of the 'offshore bar' model (at least in the Western Interior of North America) and the resurgence of the incised valley-fill model for many tidal sand bodies has emphasized the difficulty of distinguishing between shelf and coastal sand bodies. Storm and tidal sandridges are forming at the present-day, but when and where did they exist in the past? Not all thick shallow marine sandbodies are necessarily shoreline or inshore deposits. Placing shallow marine sandbodies within a sequence stratigraphic context, particularly the nature and significance of bounding erosion surfaces, appears to be one of the best means of making this distinction when sufficient data allow it. In addition, there is a need for a closer integration of physical sedimentary processes with biological aspects, particularly ichnofacies analysis and palaeoecology, within a high-resolution chronostratigraphic framework.

Models for deep-sea clastic systems initially came from sequence analysis of turbidite outcrops, but emphasis has already moved towards three-dimensional studies in the subsurface of both modern and ancient systems and it is through such studies that future models will be developed. However, outcrop studies are still needed to assist in quantifying sand/shale ratios and in testing our ideas on the processes involved in erosion, transport and deposition of sediment. In addition to the wealth of subsurface data available for study in productive areas, academic sedimentologists should be aware that

there is an enormous data set, almost untouched by petroleum geologists, in the unproductive sand-deficient horizons of petroleum provinces. Studies of these data could enhance our understanding of the much neglected mud-dominated deep-sea and shallow marine environments.

Aeolian systems are now much better understood, thanks to advances in understanding fluid and sedimentary processes on dunes and computer simulations that have, for the first time, allowed detailed interpretation of cross-stratification and a greater understanding of the complexity of bedforms. However, these need to be related to field studies. Although ideas on bounding surfaces were developed very early on in aeolian systems, the application of sequence stratigraphic concepts is not easy and aeolian systems need to be integrated with other continental systems, especially coastal systems, in order to isolate the controlling factors. It is also essential to link aeolian systems, which are global barometers as well as climatic indicators, with glacial systems and see how far their cyclicities are linked.

Although evaporites are today interpreted using a variety of depositional models rather than, as in the past just two, either as deep-water basinal sediments or as sabkha deposits, we still suffer from the absence of modern analogues for 'saline giants'. The origin of these vast evaporite bodies is still very controversial, and criteria for identifying the depositional environments that formed them are hotly disputed. Only belatedly have sequence stratigraphic concepts been applied to evaporites and there is considerable scope for marrying depositional evaporite basins with sequence stratigraphic concepts.

Perhaps the most important future development in carbonate sedimentology is to rediscover fossils. The biota, by virtue of producing the sediment grains, the frameworks and the matrices, and their role in stabilizing and modifying sediments, are the crucial element. Every stage in the stratigraphic record has its own distinct biota and facies style, and constraining all ancient carbonates into a straight-jacket of Quaternary platform type is unrealistic. By virtue of having unique biotas, carbonate productivity must have varied throughout geologic history. In particular, to overemphasize the Quaternary as *the* model is flawed because the classical Bahamian platforms, and coral atolls, represent growth in largely oligotrophic settings strongly affected by the extreme sea-level changes of 1 My or so. These platforms are dominated by photo- or mixotrophic carbonate producers. It seems that for most of the last 570 My, limestones have been formed largely by heterotrophic organisms, which were less light dependent, but more strongly influenced by oceanographic factors. Analysing the fossils will help us decide what produced the sediment, and where it was produced.

Glacial sedimentology, because of its climatic significance, is particularly relevant to global issues today. Improved understanding of modern and ancient glacial processes and deposits, will contribute to modelling of global climatic change on many different time scales.

Some glacial environments, such as ice cliffs, floating ice margins and below ice sheets and ice shelves, are still poorly known, because they are difficult to reach. On glaciated continental margins we need more drilling of the deposits, combined with geophysical studies. As sequence stratigraphic concepts, in particular better recognition of unconformities and discontinuities, are applied to glacial successions we should understand further how climatic change, glaciation and sea-level change affect the stratigraphic architecture of continental margins. This should lead to an improved understanding of glacio-eustasy and glacio-isostasy and their effects on the sedimentary record. Large-scale regional studies, such as a transect from terrestrial through continental shelf and slope to deep marine regions for specific time slices in the Pleistocene, would be illuminating, particularly if linked to the deep marine $\delta^{18}O$ record.

The future study of volcaniclastic rocks must continue to draw researchers from a wide multitude of disciplines and will no doubt continue to be stimulated by unforeseen geologic events, such as the 1992 eruptions of Mount Unzen (Japan) and Mount Pinatubo (Philippines). Subaqueous volcaniclastic rocks, which have the greatest preservation potential, are the least understood. Whereas the ocean-drilling programme has improved our understanding of deep marine environments, there are only a handful of sites close to volcanic arcs. It is also extremely difficult to recover undisturbed cores from coarse volcaniclastic facies, particularly pumiceous gravels. Technological improvements in these areas, combined with increased use of advanced oceanographic techniques, such as side-scan sonar or visual observation using submersibles, will undoubtedly improve our understanding of subaqueous volcanoes.

We still have far to go in linking changes in sedimentation, whether cyclic or random, to the changing rates of tectonic movements, basin subsidence and climatic change. Sedimentologists now appreciate that sediments were not necessarily deposited in passive depressions with a constantly available sediment supply but that sedimentation and deformation frequently go hand in hand. We have many examples of local sedimentary tectonic patterns illustrating the variety of tecto-facies in extensional rifted basins, in strike-slip basins, around volcanic arcs and in foreland basins. We need to look at the more subtle changes that occur both over quite minor lines and across long-lived fundamental lineaments. Although outcrop mapping can still contribute substantially to the study of sedimentation and tectonics, in many cases it is only possible to prove the relationship by subsurface data.

Although we have abundant models, many derived from well-studied examples, most of these models only show what we would expect to find under idealized changes of uplift, basin subsidence and climatic change (e.g. the simple sinusoidal curves used to explain sequence stratigraphic models). Primarily, this is because our chronostratigraphy is very poorly constrained. In the rare cases where coasts have been studied for the past few thousand years, we can detect minor, but rapid, sea-level oscillations and we know that this apparently smooth and simple curve is not a true representation of the actual sea-level changes either relative or eustatic. We need to develop our chrono-stratigraphic framework. Biostratigraphy, in particular, has been neglected by sedimentologists, though whether it will ever give the resolutions needed here must remain an open question.

Prediction is one of the most important aspects of sedimentary geology. Sedimentary models have helped us extend predictions from the system with which we are concerned into areas for which we have little or no direct data. Some such predictions may have major implications for investment. However, models are commonly of limited value in exact prediction and it is important to try to express predictions in terms of risk or uncertainty, employing the multiple working hypotheses that facies analysis commonly demands. In some cases a prediction may have a large probabilistic element, but even where prediction is deterministic, it is still valuable to express the unknown as a range about a most likely case. This allows much better treatment of the inherent uncertainty in sedimentological prediction. Modern seismic attribute analysis and three-dimensional visualization techniques now allow a much more exact and dynamic discussion of the problems.

Many believe we can relate Milankovitch cycles to sedimentary sequences. Yet the criteria by which we diagnose such relationships are imperfect. Do we fully understand the links between changes in insolation at the top of the Earth's atmosphere and depositional sequences? We need further studies of glacial cyclicity and its linkage with cyclicity in lakes, deserts and oceanic sediments. Ancient lake deposits are likely to become increasingly important in this respect. A combination of high sedimentation rates and accumulation under often anoxic conditions makes lake basins particularly favourable sites for the preservation of high-resolution records of (cyclic?) climatic changes.

Since every sedimentary environment is influenced, if not directly controlled, by the nature, volume and rate of sediment supplied to it, no environment can be fully understood without reference to the supplying system, known or inferred. Where possible we must determine the nature of the source area surface material, whether it is hard rock or loose sediments, the nature and extent of vegetation, the processes that can erode and transport that sediment, the climate and tectonics. Sedimentary petrography and heavy mineral analysis should differentiate between the various processes which have contributed to the mineral population. We must judge whether it is the result of the composition of the source area, transporting processes, depositional processes or diagenesis either in the source area or after deposition. Fortunately, recent success in the use of neodymium isotopic ratios as indicators of provenance promises to provide a tracer for sediment source areas that is independent of diagenesis.

Too little has still been done on the linkage of systems. This is particularly so for alluvial systems where we need to measure the rates and volumes of denudation in their source areas and understand the response of drainage basins to changes in climate and rates of uplift by working more closely with geomorphologists. We must also develop further the range of models based on specific types of source area along the lines of the grain size and feeder system models that have been developed in recent years for alluvial, coastal and deep sea clastic systems. Because of the inherent difficulties of studying them, as well as their relative lack of economic importance, except as organic-rich source rocks, mudstones have been neglected compared with sandstones and carbonates. Yet they make up a very significant proportion of the World's depositional history.

Sedimentologists have traditionally researched on depositional facies and neglected breaks in sedimentation. Sequence stratigraphy has now taught us the importance of flooding surfaces and unconformities. We must therefore renew studies of the smaller erosion surfaces, non-sequences and unconformities. We need to understand the whole biological and chemical diagenetic history both of the non-depositional facies such as palaeosols and hardgrounds and of deeper subsurface alterations using more precise, mineralogical and geochemical techniques which can indicate palaeotemperatures and groundwater chemistries. Diagenetic studies cannot only indicate the depositional environment but also they can be a tool for understanding burial history, overburden pressures and temperatures, and groundwater movement.

Knowledge is expanding at an unprecedented rate. It is becoming increasingly available, not only in the volumes of data, but in the intellectual understanding of the fundamentals of science. This immensity of knowledge precludes us from achieving expertise in many areas essential to pursuing our own research. However much we try to keep abreast of what we ought to know, our relative ignorance is increasing and each of us is becoming less capable of furthering research outside an ever narrowing field. It is therefore essential that studies, if they are to be successful, are multidisciplinary. We must integrate our studies with those of engineers, petrophysicists, seismologists, geomorphologists, palaeontologists, volcanologists, physicists and chemists, as well as other sedimentologists.

One final word of caution. The present is not a master key to all past environments although it may open the door to a few. The majority of past environments differ in some respect from modern environments. We must be prepared, and have the courage, to develop non-uniformitarian models unlike any that exist today and to extend them into sequence stratigraphic syntheses.

References

Abbots F.V. (1989) Sedimentology of Jurassic syn-rift resedimented carbonate sandbodies, 403 pp. Unpublished PhD thesis, University of Bristol, UK. *9.4.4, Fig. 9.44.*

Abdullatif O.M. (1989) Channel-fill and sheet-flood facies sequences in the ephemeral terminal River Gash, Kassala, Sudan. *Sediment. Geol.*, 63, 171–184. *3.3.8.*

Aber J.S., Croot D.G. & Fenton M.M. (1989) *Glaciotectonic Landforms and Structures*, 200 pp. Kluwer Academic, Dordrecht. *11.3.5.*

Aby S.B. (1994) Relation of bank-margin fractures to sea-level change, Exuma Islands, Bahamas. *Geology*, 22, 1063–1066. *9.5.3.*

Achauer C.W. (1982) Sabkha anhydrite: The supratidal facies of cyclic deposition in the Upper Minnelusa Formation (Permian) Rozet field area, Powder River Basin, Wyoming, In: *Deposition and Diagenetic Spectra of Evaporites* (Ed. by C.R. Handford, R.G. Loucks and G.R. Davies), pp. 193–209. *Soc. econ. Paleont. Miner. Core Workshop*, 3, Tulsa 1982. *8.7, Fig. 8.40.*

Acker L.K. & Risk M.J. (1985) Substrate destruction and sediment production by the sponge *Cliona caribbaea* on Grand Cayman Island. *J. sedim. Petrol.*, 55, 705–711. *9.4.5.*

Adams C.E.J., Wells J.T. & Coleman J.M. (1982) Sediment transport on the central Louisiana continental shelf: implications for the developing Atchafalaya River delta. *Contrib. Mar. Sci.*, 25, 133–148. *7.2.2.*

Adams S.S. (1969) Bromine in the Salado Formation, Carlsbad Potash District, New Mexico. *New Mexico State Bur. Mines Miner. Resour. Bull.*, 93, 122 pp. *8.3.2.*

Adey W.H. (1978) Coral reef morphogenesis: a multidimensional model. *Science*, 202, 831–837. *9.4.5.*

Adjas A., Masse J.-P. & Montaggioni L.F. (1990) Fine grained carbonates in nearly closed reef environments: Mataiva and Takapoto atolls, central Pacific Ocean. *Sedim. Geol.*, 67, 115–132. *9.1.3.*

Aharon P., Kolodny Y. & Sass E. (1977) Recent hot brine dolomitization in the 'Solar Lake', Gulf of Elat: Isotopic, chemical and mineralogical study. *J. Geol.*, 85, 27–48. *8.2.3.*

Ahlbrandt T.S. & Andrews S. (1978) Distinctive sedimentary features of cold-climate eolian deposits, North Park, Colorado. *Palaeogeogr. Palaeoclimatol. Palaeoecol.*, 25, 327–351. *11.4.7.*

Ahlbrandt T.S. & Fryberger S.G. (1981) Sedimentary features and significance of interdune deposits. In: *Recent and Ancient Nonmarine Depositional Environments: Models for Exploration* (Ed. by F.G. Ethridge and R.M. Flores), pp. 293–314. *Spec. Publ. Soc. econ. Paleont. Miner.*, 31, Tulsa. *5.4.5.*

Ahr W.M. (1973) The carbonate ramp: an alternative to the shelf model. *Trans. Gulf Coast Ass. Geol.*, 23, 221–225. *9.1.1, 9.3.4.*

Ahr W.M. (1989) Sedimentary and tectonic controls on the development of an early Mississippian carbonate ramp, Sacramento Mountains area, New Mexico. In: *Controls on Carbonate Platform and Basin Development* (Ed. by P.D. Crevello et al.), pp. 203–212. *Spec. Publ. Soc. econ. Paleont. Miner.*, 44, Tulsa. *9.3.4.*

Aigner T. (1982) Calcareous tempestites: storm dominated stratification in Upper Mischelkalk limestones (Middle Trias, SW-Germany). *Cyclic and Event Stratification* (Ed. by G. Einsele and A. Seilacher), pp. 180–198. Springer-Verlag, Berlin. *7.7.2.*

Aigner T. (1984) Dynamic stratigraphy of epicontinental carbonates,

Upper Muschelkalk (M. Triassic), South German Basin. *Neues Jahrb. Geol. Palaeontol. Abh.*, **169**, 127–159. *9.3.4, 9.4.3, 9.5.2.*

Aigner T. (1985) *Storm Depositional Systems*, 174 pp. Lecture Notes in Earth Sciences, Springer-Verlag, Berlin. *7.7.2, 9.4.2, 9.4.3.*

Aigner T. & Reineck H.-E. (1982) Proximality trends in modern storm sands from the Helegoland Bight (North Sea) and their implications for basin analysis. *Senckenbergiana marit.*, **14**, 183–215. *7.2.2, 7.4.3. Fig. 7.23.*

Ainsworth R.B. & Pattison S.A.J. (1994) Where have all the lowstands gone? Evidence for attached lowstand systems tracts in the Western Interior of North America. *Geology*, **22**, 415–418. *6.7.7.*

Aksu A.E. & Hiscott R.N. (1992) Shingled Quaternary debris flow lenses on the north-east Newfoundland Slope. *Sedimentology*, **39**, 193–206. *10.3.5, 11.4.6.*

Alexander C.R., DeMaster D.J. & Nittrouer C.A. (1986) High-resolution seismic stratigraphy and its sedimentological interpretation on the Amazon continental shelf. *Continental Shelf Res.*, **6**, 337–357. *7.6.*

Alexander C.R., DeMaster D.J. & Nittrouer C.A. (1991) Sediment accumulation in a modern epicontinental-shelf setting: The Yellow Sea. *Mar. Geol.*, **98**, 51–72. *7.6.*

Alexander C.R., Nittrouer C.A., DeMaster D.J., Park Y.A. & Park S.C. (1991) Macrotidal mud flats of the southwestern Korean coast: a model for interpretation of intertidal deposits. *J. sedim. Petrol.*, **61**, 805–824. *6.7.3.*

Alexander J. (1992) Nature and origin of a laterally extensive alluvial sandstone body in the Middle Jurassic Scalby Formation. *J. geol. Soc. Lond.*, **149**, 431–441. *3.6.5.*

Alexander J. & Leeder M.R. (1987) Active tectonic control on alluvial architecture. In: *Recent Developments in Fluvial Sedimentology* (Ed. by F.G. Ethridge, R.M. Flores and M.D. Harvey), pp. 243–252. *Spec. Publ. Soc. econ. Paleont. Miner.*, **39**, Tulsa. *3.6.6, 3.6.7.*

Algeo T.J. & Wilkinson B.H. (1988) Periodicity of mesoscale Phanerozoic sedimentary cycles and the role of Milankovitch orbital modulation. *J. Geol.*, **26**, 313–322. *2.1.5, 9.2.2, 9.4.1.*

Allen G.P. (1991) Sedimentary processes and facies in the Gironde estuary: a recent model for macrotidal estuarine systems. In: *Clastic Tidal Sedimentology* (Ed. by D.G. Smith, G.E. Reinson, B.A. Zaitlin and R.A. Rahmani), pp. 29–40. *Mem. Can. Soc. petrol. Geol.*, **16**, Calgary. *6.7.5.*

Allen G.P. & Posamentier H.W. (1993) Sequence stratigraphy and facies model of an incised valley fill: the Gironde estuary, France. *J. sedim. Petrol.*, **63**, 378–391. *3.6.9, 6.7.5, 6.7.6.*

Allen G.P., Laurier D. & Thouvenin J. (1979) Étude sédimentologique du delta de la Mahakam. *Notes et Mémoires*, **15**, Compagnie Française des Pétroles, Paris. *6.1, 6.6.1, Fig. 6.32.*

Allen J.R.L. (1960) The Mam Tor Sandstones: A 'turbidite' facies of the Namurian deltas of Derbyshire, England. *J. sedim. Petrol.*, **30**, 193–208. *6.6.6.*

Allen J.R.L. (1963a) Henry Clifton Sorby and the sedimentary structures of sands and sandstones in relation to flow conditions. *Geol. Mijnbouw.*, **42**, 223–228. *2.2.2, 2.5.*

Allen J.R.L. (1963b) The classification of cross-stratified units, with notes on their origin. *Sedimentology*, **2**, 93–114. *3.5.2, 3.6.5.*

Allen J.R.L. (1964) Studies in fluviatile sedimentation. Six cyclothems from the Lower Old Red Sandstone, Anglo-Welsh

Basin. *Sedimentology*, **3**, 163–198. *3.1, 3.3.4, Fig. 3.18.*

Allen J.R.L. (1965a) Fining upwards cycles in alluvial successions. *Geol. J.*, **4**, 229–246. *3.6.2.*

Allen J.R.L. (1965b) The sedimentation and palaeogeography of the Old Red Sandstone of Anglesey, North Wales. *Proc. Yorks. geol. Soc.*, **35**, 139–185. *Fig. 3.50.*

Allen J.R.L. (1965c) Late Quaternary Niger delta, and adjacent areas: sedimentary environments and lithofacies. *Bull. Am. Ass. petrol. Geol.*, **49**, 547–600. *6.1, 6.6.1, 6.6.5, Fig. 6.33.*

Allen J.R.L. (1968) *Current Ripples*, 433 pp. North-Holland, Amsterdam. *1.1.*

Allen J.R.L. (1970a) Studies in fluviatile sedimentation: A comparison of fining upwards cyclothems with special reference to coarse-member composition and interpretation. *J. sedim. Petrol.*, **40**, 298–323. *3.3.4, 3.6.2.*

Allen J.R.L. (1970b) A quantitative model of grain size and sedimentary structures in lateral deposits. *Geol. J.*, **7**, 129–146. *3.3.4, Fig. 3.18.*

Allen J.R.L. (1970c) *Physical Processes of Sedimentation*, 248 pp. Allen & Unwin, London. *6.2.1.*

Allen J.R.L. (1971) Mixing at turbidity current heads, and its geological implications. *J. sedim. Petrol.*, **41**, 97–113. *10.2.3.*

Allen J.R.L. (1973) Compressional structures (Patterned ground) in Devonian pedogenic limestones. *Nature, Phys. Sci.*, **243**, 84–86. *3.5.4.*

Allen J.R.L. (1974) Studies in fluviatile sedimentation: implications of pedogenic carbonate units, Lower Old Red Sandstone, Anglo-Welsh outcrop. *Geol. J.*, **9**, 181–208. *3.5.4, 3.6.6, Fig. 3.43.*

Allen J.R.L. (1978) Studies in fluviatile sedimentation: An exploratory quantitative model for the architecture of avulsion controlled alluvial suites. *Sedim. Geol.*, **21**, 129–147. *3.6.7.*

Allen J.R.L. (1980) Sand waves: a model of origin and internal structure. *Mar. Geol.*, **26**, 281–328. *7.3.3, 7.7.1, Fig. 7.36.*

Allen J.R.L. (1982a) Mud drapes in sand-wave deposits: a physical model with application to the Folkestone Beds (early Cretaceous, southeast England). *Proc. R. Soc. Lond. Ser. A*, **306**, 291–345. *Fig. 6.14, 7.1.*

Allen J.R.L. (1982b) *Sedimentary Structures: their Character and Physical Basis. Developments in Sedimentology*, **30A & B**. Elsevier, Amsterdam. *Fig. 6.12, 7.4.3, Fig. 7.25.*

Allen J.R.L. (1983) Studies in fluviatile sedimentation: bars, bar complexes and sandstone sheets (low-sinuosity braided streams) in the Brownstones (L. Devonian), Welsh Borders. *Sedim. Geol.*, **33**, 237–293. *2.2.2, 3.1, 3.5.2, 3.6.1, 3.6.2, 3.6.5, 10.3.2.*

Allen J.R.L. (1984) *Sedimentary Structures: their Character and Physical Basis*, 593 pp., 663 pp. *Developments in Sedimentology*, **30A & B**. Elsevier, Amsterdam. *1.1, 1.2, Fig. 12.25.*

Allen J.R.L. (1985) *Principles of Physical Sedimentology*, 272 pp. Allen & Unwin, London. *1.2.*

Allen J.R.L. (1986) Pedogenic calcretes in the Old Red Sandstone facies (Late Silurian–Early Carboniferous) of the Anglo-Welsh area, southern Britain. In: *Paleosols: Their Recognition and Interpretation* (Ed. by V.P. Wright), pp. 58–86. Blackwell Scientific Publications, Oxford. *3.5.4.*

Allen J.R.L. (1991) The Bouma Division A and the possible duration of turbidity currents. *J. sedim. Petrol.*, **61**, 291–295. *10.2.3, Fig. 10.25.*

Allen J.R.L. (1993) An introduction to estuarine lithosomes and their controls. In: *Sedimentology Review* 1 (Ed. by V.P. Wright), pp. 123–138. Blackwell Scientific Publications, Oxford. *6.7.5*.

Allen J.R.L. & Banks N.L. (1972) An interpretation and analysis of recumbent-folded deformed cross-bedding. *Sedimentology*, 19, 257–283. *3.5.2*.

Allen J.R.L. & Collinson J.D. (1974) The superimposition and classification of dunes formed by unidirectional flow. *Sedim. Geol.*, 12, 169–178. *3.2.2*, *3.3.3*.

Allen J.R.L. & Friend P.F. (1968) Deposition of the Catskill Facies, Appalachian region: with notes on some other Old Red Sandstone Basins. In: *Late Paleozoic and Mesozoic Continental Sedimentation, Northeastern North America* (Ed. by G. deV. Klein), pp. 21–74. *Spec. Pap. geol. Soc. Am.*, 106. *3.6.5*.

Allen P.A. (1981) Wave-generated structures in the Devonian lacustrine sediments of south-east Shetland and ancient wave conditions. *Sedimentology*, 28, 369–379. *4.13.1*.

Allen P.A. (1984) Reconstruction of ancient sea conditions with an example from the Swiss Molasse. *Mar. Geol.*, 60, 455–473. *4.13.1*.

Allen, P.A. (1985) Hummocky cross-stratification is not produced purely under progressive gravity waves. *Nature*, 313, 562–564. *7.7.2*.

Allen P.A. & Allen J.R.L. (1990) *Basin Analysis: Principles and Applications*, 451 pp. Blackwell Scientific Publications, Oxford. *1.1*, *2.1.3*.

Allen P.A. & Homewood P. (1984) Evolution and mechanics of a Miocene tidal sandwave. *Sedimentology*, 31, 62–81. *7.7.1*.

Allen R.L. (1988) False pyroclastic textures in altered silicic lavas, with implications for volcanic-associated mineralization. *Econ. Geol.*, 83, 1424–1446. *12.1.1*, *12.14.1*, *Fig. 12.92*.

Allen R.L. (1992) Reconstruction of the tectonic, volcanic, and sedimentary setting of strongly deformed Zn–Cu massive sulphide deposits at Benambra, Victoria. *Econ. Geol.*, 87, 825–854. *12.10*.

Alley R.B., Blankenship D.D., Rooney S.T. & Bentley C.R. (1989) Sedimentation beneath ice shelves; The view from the Ice Stream B. *Mar. Geol.*, 85, 101–120. *11.4.6*.

Alonso B. & Maldonado A. (1990) Late Quaternary sedimentation patterns of the Ebro turbidite systems (northwestern Mediterranean): two styles of deep-sea deposition. In: *The Ebro Continental Margin, Northwestern Mediterranean Sea* (Ed. by C.H. Nelson and A. Maldonado), pp. 353–377. *Mar. Geol.*, 95. *10.3.4*.

Alonso B., Anderson J.B., Diaz J.I. & Bartek L.R. (1992) Pliocene–Pleistocene seismic stratigraphy of the Ross Sea: Evidence for multiple ice sheet grounding episodes. In: *Contributions to Antarctic Research III* (Ed. by D.H. Elliot), pp. 93–103. *Antarctic Res. Ser.*, 57. *11.5.4*.

Alonso B., Canals M., Got H. & Maldonado A. (1991) Sea valleys and related depositional systems in the Gulf of Lion and Ebro continental margins. *Bull. Am. Ass. petrol. Geol.*, 75, 1195–1214. *10.3.4*.

Alonso Zarza A.M., Calvo J.P. & Garcia del Cura M.A. (1993) Palaeogeomorphological controls on the distribution and sedimentary style of alluvial systems, Neogene of the NE of the Madrid Basin (central Spain). In: *Alluvial Sedimentation* (Ed. by M. Marzo and C. Puigdefabregas), pp. 277–292. *Spec. Publ. int Ass. Sediment.*, 17. *3.6.4*.

Alsharhan A.S. & Kendall C.G. St C. (1994) Depositional setting of the Upper Jurassic Hith Anhydrite of the Arabian Gulf: an analog to Holocene evaporites of the United Arab Emirates and Lake MacLeod of Western Australia. *Bull. Am. Ass. petrol. Geol.*, 78, 1075–1096. *8.7*.

Alt J.C., Lonsdale P., Haymon T. & Muehlenbachs K. (1987) Hydrothermal sulfide and oxide deposits on seamounts near 21°N, East Pacific Rise. *Bull. geol. Soc. Am.*, 98, 157–168. *12.11.1*.

Anadón P., Cabrera L. & Julià R. (1988) Anoxic–oxic cyclical lacustrine sedimentation in the Miocene Rubielos de Mora basin, Spain. In: *Lacustrine Petroleum Source Rocks* (Ed. by A.J. Fleet, K. Kelts and M.R. Talbot), pp. 353–367. *Spec. Publ. geol. Soc. Lond.*, 40. *4.16.2*, *Fig. 4.35*.

Anadón P., Cabrera L., Colldeforns B. & Sáez A. (1989) Los sistemas lacustres del Eoceno superior y Oligoceno del sector oriental de la Cuenca del Ebro. In: *Sistemas lacustres Cenozoicos de España* (Ed. by P. Anadón and L. Cabrera), pp. 205–230. *Acta Geol. Hispanica*, 24. *4.10*.

Anadón P., Cabrera L., Julià R., Roca E. & Rosell L. (1989) Lacustrine oil-shale basins in Tertiary grabens from NE Spain (western European rift system). *Palaeogeogr. Palaeoclimatol. Palaeoecol.*, 70, 7–28. *4.10*.

Anadón P., Cabrera L., Julià R. & Marzo M. (1991) Sequential arrangement and asymmetrical fill in the Miocene Rubielos de Mora basin (northeast Spain). In: *Lacustrine Facies Analysis* (Ed. by P. Anadón, L. Cabrera and K. Kelts), pp. 257–275. *Spec. Publ. int. Ass. Sediment.*, 13. *4.11.1*, *4.16.2*, *Fig. 4.36*.

Anati D.A. & Stiller M. (1991) The post-1979 thermohaline structure of the Dead Sea and the role of double-diffusive mixing. *Limnol. Oceanogr.*, 36, 342–354. *8.6.2*.

Anderson E.J. & Goodwin P.W. (1978) Punctuated aggradational cycles (PACS) in Middle Ordovician and Lower Devonian sequences. *Proc. 50th Ann. Mtg New York State geol. Soc.*, 50, 204–224. *7.4.1*. *9.4.1*.

Anderson J.B. (1983) Ancient glacial-marine deposits: Their spatial and temporal distribution. In: *Glacial-Marine Sedimentation* (Ed. by B.F. Molnia), pp. 3–92. Plenum Press, New York. *11.5.5*.

Anderson J.B. (1989) Antarctica's glacial setting. In: *Glacial-Marine Sedimentation* (Ed. by J.B. Anderson and B.F. Molnia), pp. 11–57. 28th International Geological Congress Short Course, American Geological Union, Washington, DC. *11.4.6*.

Anderson J.B. & Bartek L.R. (1992) Cenozoic glacial history of the Ross Sea revealed by intermediate resolution seismic reflection data combined with drill site information. In: *The Antarctic Paleoenvironment: A Perspective on Global Change* (Ed. by J.P. Kennett and D.A. Warnke), pp. 231–263. *Antarctic Res. Ser.*, 56. *11.4.6*.

Anderson J.B. & Thomas M.A. (1991) Marine ice-sheet decoupling as a mechanism for rapid, episodic sea-level change: the record of such events and their influence on sedimentation. *Sedim. Geol.*, 70, 87–104. *11.5.2*, *11.5.3*.

Anderson J.B., Brake C., Domack E., Myers N. & Wright R. (1983) Development of a polar glacial-marine sedimentation model from Antarctic Quaternary deposits and glaciological information. In: *Glacial-Marine Sedimentation* (Ed. by B.F. Molnia), pp. 233–264. Plenum Press, New York. *11.4.6*.

Anderson J.B., Kennedy D.S., Smith M.J. & Domack E.W. (1991)

Sedimentary facies associated with Antarctica's floating ice masses. In: *Glacial Marine Sedimentation; Paleoclimatic Significance* (Ed. by J.B. Anderson and G.M. Ashley), pp. 1–25. *Spec. Pap. geol. Soc. Am.*, **261** Boulder, CO. *11.4.6, Fig. 11.14.*

Anderson J.B., Kurtz D.D. & Weaver F.M. (1979) Sedimentation on the Antarctic continental slope. In: *The Geology of Continental Slopes* (Ed. by O. Pilkey and L. Doyle), pp. 265–283. *Spec. Publ. Soc. econ. Palaeont. Miner.*, **27**, Tulsa. *11.4.6.*

Anderson J.B., Wright R. & Andrews B. (1986) Weddell fan and associated abyssal plain, Antarctica: Morphology, sediment processes, and factors influencing sediment supply. *Geo-Mar. Lett.*, **6**, 121–129. *11.4.6.*

Anderson R.S. (1988) The pattern of grainfall deposition in the lee of aeolian dunes. *Sedimentology*, **35**, 175–188. *5.3.3.*

Anderson R.S. (1990) Eolian ripples as examples of self-organization in geomorphological systems. *Earth Sci. Rev.*, **29**, 77–96. *5.3.1.*

Anderson R.S. & Haff P.K. (1988) Simulation of eolian saltation. *Science*, **241**, 820–823. *5.3.1.*

Anderson R.Y., Dean W.E., Kirkland D.W. & Snider H.I. (1972) Permian Castile varved anhydrite sequence, west Texas and New Mexico. *Bull. geol. Soc. Am.*, **83**, 59–86. *8.9, Fig. 8.48.*

Anderson T. & Flett J.S. (1903) *Report on the eruptions of the Soufrière in St Vincent in 1902, and on a visit to Montagne Pelée in Martinique*, pp. 353–553. *12.1.1.*

Anderton R. (1976) Tidal shelf sedimentation: an example from the Scottish Dalradian. *Sedimentology*, **23**, 429–458. *7.1.2, 7.7.1, 7.8.1, Fig. 7.49.*

Anderton R. (1985) Clastic facies models and facies analysis. In: *Sedimentology: Recent Developments and Applied Aspects* (Ed. by P.J. Brenchley and B.P.J. Williams), pp. 31–47. *Spec. Publ. geol. Soc. Lond.*, **18**. *2.5.*

Andrews J.E., Turner M.S., Nabi G. & Spiro B. (1991) The anatomy of an early Dinantian terraced floodplain: palaeo-environment and early diagenesis. *Sedimentology*, **38**, 271–287. *3.6.9.*

Andrews J.T. (1978) Sea level history of arctic oceans during the Upper Quaternary: dating, sedimentary sequences, and history. *Progr. Phys. Geogr.*, **2**, 375–407. *11.5.2.*

Andrews J.T. & Matsch C.L. (1983) *Glacial Marine Sediments and Sedimentation. An Annotated Bibliography*, 227 pp. Bibliography No. 11 Geo. Abstracts, Norwich. *11.4.6, 11.5.5.*

Andrews P.B. (1970) *Facies and Genesis of a Hurricane Washover Fan, St. Joseph Island, Central Texas Coast*, 147 pp. Bureau of Economic Geology, Report of Investigations, 67, Austin, TX. *6.7.4.*

Anthony E.J. (1989) Chenier plain development in northern Sierra Leone, West Africa. *Mar. Geol.*, **90**, 297–309. *6.7.2.*

Arakel A.V. (1980) Genesis and diagenesis of Holocene evaporitic sediments in Hutt and Leeman lagoons, Western Australia. *J. sedim. Petrol.*, **50**, 1305–1326. *8.1.*

Arakel A.V. & Hongjun T. (1994) Seasonal evaporite sedimentation in desert playa lakes of the Karinga Creek drainage system, central Australia. In: *Sedimentology and Geochemistry of Modern and Ancient Saline Lakes* (Ed. by R.W. Renaut & W.M. Last), pp. 91–100. *Soc. econ. Paleont. Miner., Spec. Publ.* **50**, Tulsa. *4.7.4.*

Aramaki S. (1984) Formation of the Aira caldera, Southern Kyushu, ~22 000 years ago. *J. geophys. Res.*, **86**, B10, 8485–8501. *12.12.4.*

Aramaki S. & Ui T. (1966) The Aira and Ata pyroclastic flows and related caldera depressions in southern Kyushu, Japan. *Bull. Volcanol.*, **29**, 29–47. *12.4.4.*

Arche A. (1983) Coarse grained meander lobe deposits in the Jarama River, Madrid, Spain. In: *Modern and Ancient Fluvial Systems* (Ed. by J.D. Collinson and J. Lewin), pp. 313–332. *Spec. Publ. int. Ass. Sediment.*, **6**. *3.3.4, 3.5.1.*

Arche A. (Ed.) (1989) *Sedimentología*, 1, 541 pp; 2, 526 pp. Neuvas tendencias 11, 12, Consejo Superior de Investigaciones Científica, Madrid. *1.1.*

Archer A.W. & Kvale E.P. (1993) Origin of gray-shale lithofacies ('clastic wedges') in U.S. mid-continental coal measures (Pennsylvanian): An alternative explanation. In: *Modern and Ancient Coal-Forming Environments* (Ed. by J.C. Cobb and C.B. Cecil) *Spec. Pap. geol. Soc. Am.*, **286**, Boulder, CO. *6.7.7.*

Arguden A.T. & Rodolfo K.S. (1990) Sedimentologic and dynamic differences between hot and cold laharic debris flows of Mayon volcano, Philippines. *Bull. geol. Soc. Am.*, **102**, 865–876. *12.5.2.*

Armstrong L.A., Ten Have A. & Johnson H.D. (1987) The geology of the Gannet Fields, central North Sea, UK sector. In: *Petroleum Geology of North-West Europe* (Ed. by J. Brooks and K. Glennie), pp. 533–548. Graham & Trotman, London. *10.3.3, Figs 10.49, 10.50.*

Arnborg L. (1957) Erosion forms and processes on the bottom of the River Ångermanälven; the use of diving in fluvial-morphological investigations. *Geogr. Ann.*, **39**, 32–47. *3.2.1.*

Arnott D. & Southard J. (1990) Exploratory flow-duct experiments on combined-flow bed configurations, and some implications for interpreting storm-event stratification. *J. sedim. Petrol.*, **60**, 211–219. *7.7.2.*

Arthur M.A. & Jenkyns H.C. (1981) Phosphorites and palaeoceanography. In: *Geology of Oceans. Proc. 26th Int. geol. Congr. Paris. 1980. Suppl. Oceanol. Acta*, 83–96. *10.2.1, Fig. 10.11.*

Arthur M.A., Dean W.E. & Stow D.A.V. (1984) Models for the deposition of Mesozoic–Cenozoic fine-grained organic-carbon-rich sediment in the deep sea. In: *Fine-grained Sediments: Deep-water Processes and Facies* (Ed. by D.A.V. Stow and D.J.W. Piper), pp. 527–560. *Spec. Publ. geol. Soc. Lond.*, **15**. *10.2.1, 10.4.1.*

Arthurton R.S. (1973) Experimentally produced halite compared with Triassic layered halite-rock from Cheshire, England. *Sedimentology*, **20**, 145–160. *8.4.1, 8.5.1.*

Arthurton R.S. (1980) Rhythmic sedimentary sequences in the Triassic Keuper Marl (Mercia Mudstone Group) of Cheshire, northwest England. *Geol. J.*, **15**, 43–58. *4.11.3.*

Ashley G.M. (1975) Rhythmic sedimentation in glacial Lake Hitchcock, Massachusetts–Connecticut. In: *Glaciofluvial and Glaciolacustrine Sedimentation* (Ed. A.V. Jopling and B.C. McDonald), pp. 304–320. *Spec. Publ. Soc. econ. Paleont. Miner.*, **23**. *4.8, 11.4.5.*

Ashley G.M. (1979) Sedimentology of a tidal lake, Pitt Lake, British Columbia, Canada. In: *Moraines and Varves* (Ed. by C. Schlüchter), pp. 327–345. *Proc. INQUA Symp. on Genesis and Lithology of Quat. Deposits*, Zürich, 10–20 Sept. 1978. A.A. Balkema, Rotterdam. *4.2, 4.9.*

Ashley G.M. (1988) Proglacial sedimentary environments. In: *Glacial Facies Models: Continental Terrestrial Environments* (Ed. by

J. Shaw and G. Ashley), pp. 79–121. *Geol. Soc. Am. Short Course* Boulder, CO. *11.4.4, 11.4.5.*

Ashley G.M. (1989) Classification of glaciolacustrine sediments. In: *Genetic Classification of Glacigenic Deposits* (Ed. by R.P. Goldthwait and C.L. Matsch), pp. 243–260. Balkema, Rotterdam. *11.4.5.*

Ashley G.M. (1990) Classification of large-scale subaqueous bedforms: a new look at an old problem. *J. sedim. Petrol.*, 60, 160–172. *3.2.2, 7.3.3, 7.7.1, Fig. 7.36.*

Ashley G.M. & Sheridan R.E. (1994) Depositional model for valley fills on a passive continental margin. In: *Incised-valley Systems: Origin and Sedimentary Sequences* (Ed. by R.W. Dalrymple, R. Boyd and B.A. Zaitlin), pp. 285–301. *Spec. Publ. Soc. sedim. Geol.*, 52. *6.7.6.*

Ashmore P. (1991a) Channel morphology and bed load pulses in braided, gravel-bed streams. *Geogr. Ann.*, 73A, 37–52. *3.3.2.*

Ashmore P. (1991b) How do gravel-bed rivers braid? *Can. J. Earth Sci.*, 28, 326–341. *3.3.2.*

Asquith G.B. (1979) *Subsurface Carbonate Depositional Models – A Concise Review.* The Petroleum Corp., Tulsa. *8.9.*

Asquith S.M. (1979) Nature and origin of the lower continental rise hills, off the east coast of the United States. *Mar. Geol.*, 32, 165–190. *10.3.7.*

Assereto R.L. & Kendall C.G. St C. (1977) Nature, origin and classification of peritidal tepee structures and related breccias. *Sedimentology*, 24, 153–210. *9.3.3.*

Atchley S.G. & Loope D.B. (1993) Low-stand aeolian influence on stratigraphic completeness: upper member of the Hermosa Formation (latest Carboniferous), southeast Utah, USA. In: *Aeolian Sediments Ancient and Modern* (Ed. by K. Pye and N. Lancaster), pp. 127–149. *Spec. Publ. int. Ass. Sediment.*, 16. *5.5.2.*

Atkinson C.D. (1986) Tectonic control on alluvial sedimentation as revealed by an ancient catena in the Capella Formation (Eocene) of northern Spain. In: *Paleosols: Their Recognition and Interpretation* (Ed. by V.P. Wright), pp. 193–179. Blackwell Scientific Publications, Oxford. *3.6.6.*

Aubouin J. (1965) *Geosynclines: Developments in Geotectonics*, 335 pp. Elsevier, Amsterdam. *1.1.*

Aubry M.P. (1991) Sequence stratigraphy: Eustasy or tectonic imprint? *J. geophys. Res.*, 96, 6641–6679. *2.4.*

Augustinus P.G.E.F. (1989) Cheniers and chenier plains: a general introduction. *Mar. Geol.*, 90, 219–229. *6.7.2.*

Augustinus P.G.E.F., Hazelhoff L. & Kroon A. (1989) The chenier coast of Suriname; modern and geological development. *Mar. Geol.*, 90, 269–281. *6.7.1.*

Aurell M., Bosence D.W.J. & Waltham D. (1995) Carbonate ramp depositional systems from a late Jurassic epeiric platform (Iberian Basin, Spain): a combined computer modelling and outcrop analysis. *Sedimentology*, 42, 75–94. *9.1.1, 9.4.3.*

Axelsson V. (1967) The Laitaure Delta, a study of deltaic morphology and processes. *Geogr. Ann.*, 49A, 1–127. *Fig. 4.1, 6.5.*

Ayers W.B. Jr (1986) Lacustrine and fluvial-deltaic depositional systems, Fort Union Formation (Paleocene), Powder River Basin, Wyoming and Montana. *Bull. Am. Ass. petrol. Geol.*, 70, 1651–1673. *4.15, Fig. 4.33.*

Ayers W.B. & Kaiser W.R. (1984) Lacustrine-interdeltaic coal in the Fort Union Formation (Paleocene), Powder River Basin, Wyoming

and Montana, U.S.A. In: *Sedimentology of Coal and Coal-bearing Sequences* (Ed. by R.A. Rahmani and R.M. Flores), pp. 61–84. *Spec. Publ. int. Ass. Sediment.*, 7. *4.15.1.*

Babcock R.S., Burmester R.F., Engebretson D.C., Warnock A. & Clark K.P. (1992) A rifted margin origin for the Crescent basalts and related rocks in the northern Coast Range volcanic province, Washington and British Columbia. *J. geophys. Res. B*, 97, 6799–6821. *12.11.2.*

Bachman S.B., Lewis S.D. & Schweller W.J. (1983) Evolution of a forearc basin, Luzon Central Valley, Philippines. *Bull. Am. Ass. petrol. Geol.*, 7, 1143–1162. *12.12.1.*

Back W., Hanshaw B.B., Herman J.S. & Van Driel J.N. (1986) Differential dissolution of a Pleistocene reef in the groundwater mixing zone of coastal Yucatan, Mexico. *Geology*, 14, 137–140. *9.5.3.*

Bagnold R.A. (1941) *The Physics of Blown Sand and Desert Dunes*, 265 pp. Methuen, London. *5.1, 5.3.1, 5.3.3.*

Bagnold R.A. (1954) Experiments on a gravity-free dispersion of large solid spheres in a Newtonian fluid under shear. *Proc. R. Soc. Lond. Ser. A*, 225, 49–63. *10.2.3.*

Bagnold R.A. (1962) Auto-suspension of transported sediment: turbidity currents. *Proc. R. Soc. Lond. Ser. A*, 265, 315–319. *10.2.3.*

Baker V.R. (1973) Paleohydrology and sedimentology of Lake Missoula flooding of eastern Washington. *Spec. Pap. geol. Soc. Am.*, 144, 79 pp. *3.5.1, 11.4.4.*

Ball M.M. (1967) Carbonate sand bodies of Florida and the Bahamas. *J. sedim. Petrol.*, 37, 556–591. *9.1.1, 9.4.2, Fig. 9.32.*

Ballance P. (1988) The Huriwai braidplain delta of New Zealand: a later Jurassic, coarse-grained, volcanic-fed depositional system in a Gondwana forearc basin. In: *Fan Deltas: Sedimentology and Tectonic Settings* (Ed. by W. Nemec and R.J. Steel), pp. 430–444. Blackie, Glasgow. *10.3.4, 12.12.5.*

Ballance P.F. & Gregory M.R. (1991) Parnell grits – large subaqueous volcaniclastic gravity flows with multiple particle-support mechanisms. In: *Sedimentation in Volcanic Settings* (Ed. by R.V. Fisher and G.A. Smith), pp. 189–200. *Spec. Publ. Soc. econ. Paleont. Miner.*, 45, Tulsa. *12.5.1.*

Baltzer F. (1991) Late Pleistocene and Recent detrital sedimentation in the deep parts of northern Lake Tanganyika (East African rift). In: *Lacustrine Facies Analysis* (Ed. by P. Anadón, L. Cabrera and K. Kelts), pp. 147–173. *Spec. Publ. int. Ass. Sediment.*, 13. *4.6.1.*

Baltzer F., Conchon O., Freytet P. & Purser B.H. (1982) Un complexe fluviodeltaique sûrsalé et son contexte: originalité du Mehran, SE Iran. *Soc. Geol. France Mem.*, 144, 27–42. *8.4.2.*

Baltzer F., Kenig F., Boichard R., Plaziat J.-C. & Purser B.H. (1994) Organic matter distribution, water circulation and dolomitization beneath the Abu Dhabi Sabkha (United Arab Emirates). In: *Dolomites, a Volume in Honour of Dolomieu* (Ed. by B. Purser, M. Tucker and D. Zenger), pp. 409–427. *Spec. Publ. int. Ass. Sediment.*, 21. *8.4.2.*

Banerjee I. (1991) Tidal sand sheets of the late Albian Joli Fou–Kiowa–Skull Creek marine transgression, Western Interior Seaway of North America. In: *Clastic Tidal Sedimentology* (Ed. by D.G. Smith, G.E. Reinson, B.A. Zeitlin and R.A. Rahmani), pp. 335–348. *Mem. Can. Soc. petrol. Geol.*, 16, Calgary. *7.7.1, Fig. 7.37.*

Banerjee I. & McDonald B.C. (1975) Nature of esker sedimentation.

In: *Glaciofluvial and Glaciolacustrine Sedimentation* (Ed. by A.V. Jopling and B.C. McDonald), pp. 132–154. *Spec. Publ. Soc. econ. Paleont. Miner.*, 23, Tulsa. *11.4.1.*

Banks N.L. (1970) Trace fossils from the late Precambrian and Lower Cambrian of Finnmark, Norway. In: *Trace Fossils* (Ed. by T.P. Crimes and J.C. Harper), pp. 19–34. *Geol. J. spec. edn*, 3, *Fig. 7.48.*

Banks N.L. (1973a) The origin and significance of some downcurrent-dipping cross-stratified sets. *J. sedim. Petrol.*, 43, 423–427. *3.3.3, 3.5.2, 3.6.2.*

Banks N.L. (1973b) Falling-stage features of a Precambrian braided stream: criteria for subaerial exposure. *Sedim. Geol.*, 10, 147–154. *3.5.2.*

Banks N.L. (1973c) Tide dominated offshore sedimentation, Lower Cambrian, North Norway. *Sedimentology*, 20, 213–228. *7.1.2, 7.7.2, 7.8.1, Fig. 7.48.*

Banks N.L. (1973d) Innerely Member: Late Precambrian marine shelf deposit, east-Finnmark. *Norges. geol. Unders*, 228, 7–25. *7.8.1, Fig. 7.48.*

Banks N.L., Edwards M.B., Geddes W.P., Hobday D.K. & Reading H.G. (1971) Late Precambrian and Cambro-Ordovician sedimentation in East Finnmark. In: *The Caledonian Geology of Northern Norway* (Ed. by D. Robers and M. Gustauson), pp. 197–236. *Norges. geol. Unders.*, 269. *Fig. 7.48.*

Barber K.E. (1981) *Peat Stratigraphy and Climatic Change*, 219 pp. Balkema, Rotterdam. *3.3.6.*

Barberi F., Santacroe R. & Varet J. (1982) Chemical aspects of rift magmatism. In: *Continental and Oceanic Rifts* (Ed. by G. Palmalson), pp. 223–258. American Geophysical Union, Washington, DC. *12.2.2.*

Barberi F., Cioni R., Rosi M., Santacroce R., Sbrana A. & Vecci R. (1989) Magmatic and phreatomagmatic phases in explosive eruptions of Vesuvius as deduced by grain-size and component analysis of the pyroclastic deposits. *J. Volcanol. Geotherm. Res.*, 38, 287–307. *Fig. 12.1.*

Bardaji T., Dabrio C.J., Goy J.L., Somoza L. & Zazo C. (1987) Sedimentological features related to Pleistocene sea-level changes in the SE Spain. In: *Late Quaternary Sea-level Changes in Spain* (Ed. by C. Zazo), pp. 79–93. *Trab. Neog. Cuat. Museo. Nal. Cienc. Nat.*, 10. *6.5.4.*

Bardaji T., Dabrio C.J., Goy J.L., Somoza L. & Zazo C. (1990) Pleistocene fan deltas in southeastern Iberian peninsula: sedimentary controls and sea-level changes. In: *Coarse-grained Deltas* (Ed. by A. Colella and D.B. Prior), pp. 129–151. *Spec. Publ. int. Ass. Sediment.*, 10. *6.5.2, 6.5.4, 6.5.5, Fig. 6.24.*

Barnes P.M. & Lewis K.B. (1991) Sheet slides and rotational failures on a convergent margin: the Kidnappers Slide, New Zealand. *Sedimentology*, 38, 205–221. *10.2.3.*

Barnes P.W. & Lien R. (1988) Icebergs rework shelf sediments to 500 m off Antarctica. *Geology*, 16, 1130–1133. *11.4.6.*

Barnes P.W. & Reimnitz E. (1974) Sedimentary processes on arctic shelves off the northern coast of Alaska. In: *The Coast and Shelf of the Beaufort Sea* (Ed. by J.C. Reed and J.E. Sater), pp. 439–476. Arctic Institute of North America. *11.4.6.*

Barnes P.W., Rearic D.M. & Reimnitz E. (1984) Ice gouge characteristics and processes. In: *The Alaskan Beaufort Sea – Ecosystems and Environments* (Ed. by P.W. Barnes, D.M. Schell

and E. Reimnitz), pp. 184–212. Academic Press, Orlando. *Fig. 11.15.*

Barnett D.M. & Holdsworth G. (1974) Origin, morphology and chronology of sublacustrine moraines, Generator Lake, Baffin Island, Northwest Territories, Canada. *Can. J. Earth Sci.*, 11, 380–408. *11.4.5.*

Barrell J. (1912) Criteria for the recognition of ancient delta deposits. *Bull. geol. Soc. Am.*, 23, 377–446. *6.1.*

Barrell J. (1914) The Upper Devonian delta of the Appalachian geosyncline. *Am. J. Sci.*, 37, 225–253. *6.1.*

Barrett T.J. (1982) Stratigraphy and sedimentology of Jurassic bedded chert overlying ophiolites in the North Apennines, Italy. *Sedimentology*, 29, 353–373. *12.11.1.*

Barrett T.J. & Fralick P.W. (1989) Turbidites and iron formations, Beardmore–Geraldton, Ontario: application of a combined ramp/fan model to Archaean clastic and chemical sedimentation. *Sedimentology*, 36, 221–234. *10.3.4, Fig. 10.62.*

Barrie J.V. & Collins M.B. (1989) Sediment transport on the shelf margin of the Grand Banks of Newfoundland. *Atlantic Geology*, 25, 173–179. *7.5.*

Barron E.J., Sloan J.L. & Harrison C.G.A. (1980) Potential significance of land–sea distribution and surface albedo variations as a climatic forcing factor; 180 m.y. to present. *Palaeogeogr. Palaeoclimatol. Palaeoecol.*, 30, 17–40. *5.2.2.*

Barwis J.H. (1978) Sedimentology of some South Carolina tidal creek point bars, and a comparison with their fluvial counterparts. In: *Fluvial Sedimentology* (Ed. by A.D. Miall), pp. 129–160. *Mem. Can. Soc. petrol. Geol.*, 5, Calgary. *6.7.3.*

Barwis J.H. & Makurath J.H. (1978) Recognition of ancient tidal inlet sequences: an example from the Upper Silurian Keyser Limestone in Virginia. *Sedimentology*, 25, 61–82. *6.7.7.*

Bassett H.G. & Stout J.G. (1967) Devonian of western Canada. *Proc. Inst. Symp. Devonian System, Canada*, 1, 717–725. *9.4.5.*

Bates C.C. (1953) Rational theory of delta formation. *Bull. Am. Ass. petrol. Geol.*, 37, 2119–2162. *4.4.2, 6.2.2, Fig. 6.5.*

Bathurst R.G.C. (1967) Oolitic films on low energy carbonate sand grains, Bimini Lagoon, Bahamas. *Mar. Geol.*, 5, 89–109. *9.3.2.*

Bathurst R.G.C. (1971, 1975) *Carbonate Sediments and their Diagenesis*, 658 pp. Elsevier, Amsterdam. *1.1, 9.1.1, 9.3.2, 9.3.3.*

Bay T.A. Jr (1977) Lower Cretaceous stratigraphic models from Texas and Mexico. In: *Cretaceous Carbonates of Texas and Mexico: Applications to Subsurface Exploration* (Ed. by D.G. Bebout and R.G. Loucks), pp. 12–30. Bureau of Economic Geology, Report of Investigations, 89, Austin, TX. *9.4.5.*

Beach D.K. & Ginsburg R.N. (1980) Facies succession, Plio-Pleistocene carbonates, north western Great Bahama Bank. *Bull. Am. Ass. petrol. Geol.*, 64, 1634–1642. *9.3.2.*

Beach D.K. & Ginsburg R.N. (1982) Facies succession, Plio-Pleistocene carbonates; north western Great Bahama Bank: reply. *Bull. Am. Ass. petrol. Geol.*, 66, 106–108. *9.4.5.*

Beadle L.C. (1981) *The Inland Waters of Tropical Africa*, 365 pp. Longman, London. *4.3, 4.4, 4.4.3, 4.5, 4.9, Fig. 4.3.*

Beaty C.B. (1990) Anatomy of a White Mountains' debris-flow – The making of an alluvial fan. In: *Alluvial Fans: A Field Approach* (Ed. by A.H. Rachocki and M. Church), pp. 69–89. John Wiley, Chichester. *3.3.8.*

Beerbower J.R. (1964) Cyclothems and cyclic depositional

References

....

mechanisms in alluvial plain sedimentation. In: *Symposium on Cyclic Sedimentation* (Ed. by D.F. Merriam), pp. 31–42. *Bull. Kansas geol. Surv.*, **169.** *2.1.6.*

Behrensmeyer A.K. & Tauxe L. (1982) Isochronous fluvial systems in Miocene deposits of northern Pakistan. *Sedimentology*, **29**, 331–352. *3.6.6.*

Beiersdorf H., Bach W., Duncan R., Erzinger J. & Weiss W. (1995) New evidence for the production of Em-type ocean island basalts and large volumes of volcaniclastites during the early history of the Manihiki Plateau. *Mar. Geol.*, **122**, 181–205. *12.11.3.*

Béin A. & Land L.S. (1982) *San Andres Carbonates in the Texas Panhandle: Sedimentation and Diagenesis Associated with Magnesium–Calcium-chloride Brines*, 48 pp. Bureau of Economic Geology, Report of Investigations, **121**, Austin, TX. *8.8.*

Béin A. & Wéiler Y. (1976) The Cretaceous Talme Yafe Formation: a contour current shaped sedimentary prism of calcareous detritus at the continental margin of the Arabian Craton. *Sedimentology*, **23**, 511–532. *10.3.7.*

Belderson R.H. & Stride R.H. (1966) Tidal current fashioning of a basal bed. *Mar. Geol.*, **4**, 237–257. *7.3.3.*

Belderson R.H., Johnson M.A. & Kenyon N.H. (1982) Bedforms. In: *Offshore Tidal Sands, Process and Deposits* (Ed. by A.H. Stride) pp. 27–57. Chapman & Hall, London. *7.3.3, 7.3.4, Fig. 7.10.*

Bellaiche G., Droz L., Gaullier V. & Pautot G. (1994) Small submarine fans on the eastern margin of Corsica: Sedimentary significance and tectonic implications. *Mar. Geol.*, **117**, 177–185. *10.3.4.*

Belt E.S., Sakimoto S.E.H. & Rockwell B.W. (1992) A drainage diversion hypothesis for the origin of widespread coal beds in the Willaston Basin: Examples from the Paleocene strata of Eastern Montana. In: *Coal Geology of Montana* (Ed. by M.A. Scholes and S.M. Vuke-Foster), pp. 21–60. *Spec. Publ. Montana Bur. Mines Geol.*, **102.** *3.6.8.*

Benson L.V., Currey D.R, Dorn R.I. *et al.* (1990) Chronology of expansion and contraction of four Great Basin lake systems during the past 35 000 years. *Palaeogeogr. Palaeoclimatol. Palaeoecol.*, **78**, 241–286. *4.9.*

Bentor Y.K. (1980) Phosphorites – the unsolved problems. In: *Marine Phosphorites* (Ed. by Y.K. Bentor), pp. 3–18. *Spec. Publ. Soc. econ. Paleont. Miner.*, **29**, Tulsa. *10.2.1.*

Berg O.R. (1982) Seismic detection and evaluation of delta and turbidite sequences: their application to exploration for the subtle trap. *Bull. Am. Ass. petrol. Geol.*, **66**, 1271–1288. *6.6.6.*

Berg R.R. (1975) Depositional environment of Upper Cretaceous Sussex Sandstone, House Creek Field, Wyoming. *Bull. Am. Ass. petrol. Geol.*, **59**, 2099–2110. *7.8.4.*

Berger A. (1988) Milankovitch theory and climate. *Rev. Geophys.*, **26**, 624–657. *2.1.5, 10.2.1.*

Berger A. & Loutre M.R. (1994) Astronomical forcing through geological time. In: *Orbital Forcing and Cyclic Sequences* (Ed. by P.L. de Boer and D.G. Smith), pp. 15–24. *Spec. Publ. int. Ass. Sediment.*, **19.** *2.1.5.*

Berger W.H. (1970) Biogenous deep-sea sediments: fractionation by deep-sea circulation. *Bull. geol. Soc. Am.*, **81**, 1385–1402. *10.4.1.*

Berger W.H. (1974) Deep-sea sedimentation. In: *The Geology of*

Continental Margins (Ed. by C.A. Burk and C.L. Drake), pp. 213–241. Springer-Verlag, New York. *10.2.1, 10.4.1, Table 10.1.*

Berger W.H. & Winterer E.L. (1974) Plate stratigraphy and the fluctuating carbonate line. In: *Pelagic Sediments: on Land and Under the Sea* (Ed. by K.J. Hsü and H.C. Jenkyns), pp. 11–48. *Spec. Publ. int. Ass. Sediment.*, **1.** *2.1.4.*

Berger W.H., Bonneau M.C. & Parker F.L. (1982) Foraminifera on the deep-sea floor: lysocline and dissolution rate. *Oceanol. Acta*, **5**, 249–257. *10.2.1.*

Bergh S.D. & Sigvaldason G.E. (1991) Pleistocene mass-flow deposits of basaltic hyaloclastite on a shallow submarine shelf, South Iceland. *Bull. Volcanol.*, **53**, 597–611. *12.4.5, Fig. 12.30.*

Bergman K.M. (1994) Shannon Sandstone in Hartzog Draw–Heldt Draw Fields (Cretaceous, Wyoming, USA) reinterpreted as lowstand shoreface deposits. *J. sedim. Petrol.*, **64**, 184–201. *7.8.4, Fig. 7.54.*

Bergman K.M. & Walker R.G. (1987) The importance of sea-level fluctuations in the formation of linear conglomerate bodies; Carrot Creek Member of the Cardium Formation, Cretaceous Western Interior Seaway, Alberta, Canada. *J. sedim. Petrol.*, **57**, 651–665. *7.8.4, Fig. 7.56.*

Bergman K.M. & Walker R.G. (1988) Formation of Cardium erosion surface E5, and associated deposition of conglomerate: Carrot Creek Field, Cretaceous Western Interior Seaway, Alberta. In: *Sequences, Stratigraphy, Sedimentology; Surface and Subsurface* (Ed. by D.P. James and D.A. Leckie), pp. 15–24. *Mem. Can. Soc. petrol. Geol.*, **15**, Calgary. *7.8.4.*

Bernard H.A. & Major C.F. Jr (1963) Recent meander belt deposits of the Brazos River: an alluvial 'sand' model. *Bull. Am. Assoc. petrol. Geol.*, **47**, 350. *2.2.2, 3.1, 3.3.4.*

Bernard H.A., LeBlanc R.J. & Major C.F. Jr (1962) Recent and Pleistocene geology of southwest Texas. *Geology of the Gulf Coast and Central Texas and Guidebook of Excursion*, pp. 175–225. Houston Geol. Soc. *3.3.4, 6.7.6, Fig. 6.84.*

Berné S., Auffret J.-P. & Walker P. (1988) Internal structures of subtidal sandwaves revealed by high-resolution seismic reflection. *Sedimentology*, **35**, 5–20. *7.3.3, Fig. 7.12.*

Berné S., Durrand J. & Weber O. (1991) Architecture of modern subtidal dunes (sand waves), Bay of Bourgneuf, France. In: *The Three-dimensional Facies Architecture of Terrigenous Clastic Sediments, and its Implications for Hydrocarbon Discovery and Recovery* (Ed. by A.D. Miall and N. Tyler), pp. 244–261. Soc. econ. Paleont. Miner. (Soc. for Sedim. Geol.) Concepts and Models Series, **3**, Tulsa. *7.3.3, Fig. 7.13.*

Bernoulli D. & Jenkyns H.C. (1970) A Jurassic basin: The Glasenbach Gorge, Salzburg, Austria. *Verh. geol. Bundesanst. Wien*, **1970**, 504–531. *10.4.2, Fig. 10.79.*

Bernoulli D. & Jenkyns H.C. (1974) Alpine, Mediterranean, and Central Atlantic Mesozoic facies in relation to the early evolution of the Tethys. In: *Modern and Ancient Geosynclinal Sedimentation* (Ed. by R.H. Dott and R.H. Shaver), pp. 129–160. *Spec. Publ. Soc. econ. Paleont. Miner.*, **19**, Tulsa. *9.3.5, 10.4.2, Fig. 10.80.*

Berry W.B.N. & Wilde P. (1978) Progressive ventilation of the oceans – an explanation for the distribution of the Lower Paleozoic black shales. *Am. J. Sci.*, **278**, 257–275. *7.7.3.*

Bersier A. (1959) Séquence détritiques et divagations fluviales. *Eclog. geol. Helv.*, **51**, 854–893. *3.1.*

Besly B.M. & Collinson J.D. (1991) Volcanic and tectonic controls of lacustrine and alluvial sedimentation in the Stephanian coal-bearing sequence of the Malpas–Sort Basin, Catalonian Pyrenees. *Sedimentology*, 38, 3–26. *3.5.2, 3.5.4, 3.6.7, 12.6.1, Fig. 3.58.*

Besly B.M. & Fielding C.R. (1989) Palaeosols in Westphalian coal-bearing and red-bed sequences, Central and Northern England. *Palaeogeogr. Palaeoclimatol. Palaeoecol.*, 70, 303–330. *3.5.4.*

Besly B.M. & Turner P. (1983) Origin of red beds in a moist tropical climate (Etruria Formation, Upper Carboniferous, UK). In: *Residual Deposits* (Ed. by R.C.L. Wilson), pp. 131–147. *Spec. Publ. geol. Soc. Lond.*, 11. *3.2.3, Fig. 3.44.*

Best J.L. (1987) Flow dynamics at river channel confluences: implications for sediment transport and bed morphology. In: *Recent Developments in Fluvial Sedimentology* (Ed. by F.G. Ethridge, R.M. Flores and M.D. Harvey), pp. 27–35. *Spec. Publ. Soc. econ. Paleont. Miner.*, 39, Tulsa. *3.3.3.*

Best J.L. (1992) Sedimentology and event timing of a catastrophic volcaniclastic mass flow, Volcan Hudson, Southern Chile. *Bull. Volcanol.*, 54, 299–318. *12.5.2.*

Beuf S., Biju-Duval B., Charpal O., Rognon P., Gariel O. & Bennacef A. (1971) *Les Gres du Paléozoique Inferieur au Sahara*, 464 pp. Editions Technip., Paris. *11.4.7, 11.5.5.*

Beverage J.P. & Culbertson J.K. (1964) Hyperconcentrations of suspended sediment. *Proc. Am. Soc. Civil Eng. J. Hydraul. Div.*, 90, 19–46. *12.5.2.*

Beyth M. (1980) Recent evolution and present stage of Dead Sea brines. In: *Hypersaline Brines and Evaporitic Environments*, pp. 155–166. Elsevier, Amsterdam. *8.6.1.*

Bhattacharya J. (1993) The expression and interpretation of marine flooding surfaces and erosional surfaces in core: examples from the Upper Cretaceous Dunvegan Formation, Alberta foreland basin, Canada. In: *Sequence Stratigraphy and Facies Associations* (Ed. by H.W. Posamentier, C.P. Summerhayes, B.U. Haq and G.P. Allen), pp. 125–160. *Spec. Publ. int. Ass. Sedim.*, 18. *6.6.6.*

Bhattacharya J. & Walker R.G. (1991) River- and wave-dominated depositional systems of the Upper Cretaceous Dunvegan Formation, northwestern Alberta. *Bull. Can. petrol. Geol.*, 39, 165–191. *6.6.6.*

Bhattacharya J. & Walker R.G. (1992) Deltas. In: *Facies Models: Response to Sea-level Change* (Ed. by R.G. Walker and N.P. James), pp. 157–177. Geol. Ass. Canada, Waterloo, Ontario. *6.6.6.*

Bice D.M. & Stewart K.G. (1990) The formation and drowning of isolated carbonate seamounts: tectonic and ecological controls in the northern Apennines. In: *Carbonate Platforms* (Ed. by M.E. Tucker, J.L. Wilson, P.D. Crevello, J.R. Sarg and J.F. Read), pp. 145–168. *Spec. Pub. int. Ass. Sediment.*, 9. *9.3.5.*

Biddle K.T. (1980) The basal Cipit boulders: indicators of Middle to Upper Triassic buildup margins, Dolomite Alps, Italy. *Riv. Ital. Paleontol. Stratigr.*, 86, 779–794. *9.3.2.*

Biju-Duval B., Deynoux M. & Rognon P. (1981) Late Ordovician tillites of the Central Sahara. In: *Earth's Pre-Pleistocene Glacial Record* (Ed. by M.J. Hambrey and W.B. Harland), pp. 99–107. Cambridge University Press, Cambridge. *11.5.5.*

Bird E.C.F. (1967) Depositional features in estuaries and lagoons on the South Coast of New South Wales. *Austr. geog. Stud.*, 5, 113–124. *6.7.5.*

Birkeland P.W. (1984) *Soils and Geomorphology*, 372 pp. Oxford University Press, New York. *3.2.3.*

Bjerrum C.J. & Dorsey R.J. (1995) Tectonic controls on deposition of Middle Jurassic strata in a retroarc foreland basin, Utah–Idaho trough, Western Interior, USA. *Tectonics*, in press. *5.5.1.*

Black R.F. (1976) Periglacial features indicative of permafrost: ice and soil wedges. *Quatery Res.*, 6, 3–26. *11.4.7, Fig. 11.16.*

Blackburn E.A., Wilson L. & Sparks R.S.J. (1976) Mechanisms and dynamics of strombolian activity. *J. geol. Soc. Lond.*, 132, 429–440. *12.4.1.*

Blair T.C. (1987a) Sedimentary processes, vertical stratification sequences, and geomorphology of the Roaring River alluvial fan, Rocky Mountain National Park, Colorado. *J. sedim. Petrol.*, 57, 1–18. *3.3.8, Fig. 3.29.*

Blair T.C. (1987b) Tectonic and hydrologic controls on cyclic alluvial fan, fluvial, and lacustrine rift-basin sedimentation, Jurassic–lowermost Cretaceous Todos Santos Formation, Chiapas, Mexico. *J. sedim. Petrol.*, 57, 845–862. *3.6.3.*

Blair T.C. & McPherson J.G. (1994) Alluvial fans and their natural distinction from rivers based on morphology, hydraulic processes, sedimentary processes and facies assemblages. *J. sedim. Petrol.*, A64, 450–489. *3.3.8.*

Blakey R.C. (1988) Basin tectonics and erg response. In: *Late Paleozoic and Mesozoic Eolian Deposits of the Western Interior of the United States* (Ed. by G. Kocurek), pp. 127–151. *Sedim. Geol.*, 56. *5.5.1.*

Blakey R.C., Peterson F. & Kocurek G. (1988) Synthesis of late Paleozoic and Mesozoic eolian deposits of the Western Interior of the United States. In: *Late Paleozoic and Mesozoic Eolian Deposits of the Western Interior of the United States* (Ed. by G. Kocurek), pp. 3–125. *Sedim. Geol.*, 56. *5.5.1.*

Blanchon P. & Shaw J. (1995) Reef drowning during the last deglaciation: Evidence for catastrophic sea-level rise and ice-sheet collapse. *Geology*, 23, 4–8. *11.5.2.*

Blaszczyk J.K. (1981) Paleomorphology of Weissliegendes top as the control on facies variability on ore-bearing series of Lubin Copper field, southwestern Poland. *Geol. Sudetica*, 16, 195–217. *5.5.3.*

Blatt H., Middleton G.V. & Murray R.C. (1972, 1980) *Origin of Sedimentary Rocks*, 634 pp. (2nd edn. 766 pp). Prentice-Hall, New Jersey. *1.1, 2.2.2.*

Blendinger W. & Blendinger E. (1989) Windward–leeward effects on Triassic carbonate bank margin facies of the Dolomites, northern Italy. *Sedim. Geol.*, 64, 143–166. *9.4.2.*

Blodgett R.H. (1985) Calcareous paleosols in the Triassic Dolores Formation, south-western Colorado. In: *Paleosols and Weathering Through Geologic Time: Principles and Applications* (Ed. by J. Reinhardt and W.R. Sigleo), pp. 103–121. *Spec. Pap. geol. Soc. Am.*, 216. *3.5.4.*

Blodgett R.H. & Stanley K.O. (1980) Stratification, bedforms and discharge relationships of the Platte braided river system, Nebraska. *J. sedim. Petrol.*, 50, 139–148. *3.3.3, Fig. 3.12.*

Blount G. & Lancaster N. (1990) Development of the Gran Desierto sand sea, northwestern Mexico. *Geology*, 18, 724–728. *5.4.2, Fig. 5.18.*

Bloxsom W.E. (1972) A Lower Cretaceous (Comanchean)

prograding shelf and associated environments of desposition, northern Coahuila, Mexico. Unpublished MSc thesis, University of Texas, Austin. *9.4.5.*

Bluck B.J. (1967a) Deposition of some Upper Old Red Sandstone conglomerates in the Clyde area: A study in the significance of bedding. *Scott. J. Geol.*, 7, 93–138. *3.5.1.*

Bluck B.J. (1967b) Sedimentation of beach gravels: examples from south Wales. *J. sedim. Petrol.*, 37, 128–156. *6.5.4.*

Bluck B.J. (1971) Sedimentation in the meandering River Endrick. *Scott. J. Geol.*, 7, 93–138. *3.3.4.*

Bluck B.J. (1974) Structure and directional properties of some valley sandur deposits in Southern Iceland. *Sedimentology*, 21, 533–554. *3.3.2, Figs 3.8, 3.9.*

Bluck B.J. (1976) Sedimentation in some Scottish rivers of low sinuosity. *Trans. R. Soc. Edinburgh*, 69, 425–456. *3.3.2.*

Bluck B.J. (1982a) Texture of gravel bars in braided streams. In: *Gravel-bed Rivers* (Ed. by R.D. Hey, J.C. Bathurst and C.R. Thorne), pp. 339–355. John Wiley, New York. *3.6.4.*

Bluck B.J. (1982b) Hyalotuff deltaic deposits in the Ballantrae ophiolite of SW Scotland: evidence for crustal position of the lava sequence. *Trans. R. Soc. Edinburgh*, 72, 217–228. *12.11.3.*

Bluck B.J. & Kelling G. (1963) Channels from Upper Carboniferous Coal Measures in south Wales. *Sedimentology*, 2, 29–53. *3.6.2.*

Boardman E.L. (1989) Coal measures (Namurian and Westphalian) Blackband Iron Formations: fossil bog iron ores. *Sedimentology*, 36, 621–633. *3.5.3, 3.6.8.*

Boardman M.R., Neumann A.C., Baker P.A. *et al.* (1986) Banktop responses to Quaternary fluctuations in sea level recorded in periplatform sediments. *Geology*, 14, 28–31. *9.5.1.*

Boersma J.R. & Terwindt J.H.J. (1981) Neap–spring tide sequences of intertidal shoal deposits in a mesotidal estuary. *Sedimentology*, 28, 151–170. *6.6.1.*

Bogen J. (1983) Morphology and sedimentology of deltas in fjord and fjord valley lakes. *Sedim. Geol.*, 36, 245–267. *6.5.*

Bonatti E. (1975) Metallogenesis at oceanic spreading centres. *Ann. Rev. Earth and Planet. Sci.*, 3, 401–431. *Fig. 12.58.*

Bonatti E. & Harrison C.G.A. (1988) Eruption styles of basalt in oceanic spreading ridges and seamounts: effects of magma temperature and viscosity. *J. geophys. Res.*, 93, 2967–2980. *12.4.5.*

Bond G., Broeker W., Johnsen S. *et al.* (1993) Correlations between climate records from North Atlantic sediments and Greenland ice. *Nature*, 365, 143–147. *11.5.2.*

Bond G., Heinrich H. *et al.* (1992) Evidence for massive discharges of icebergs into the North Atlantic ocean during the last glacial period. *Nature*, 360, 245–249. *11.4.6.*

Bonen R.N. (1988) Recognition of storm impact on the reef sediment record. *Proc. 6th Int. Coral Reef Symp. Austral.*, 1988, 2, 475–478. *9.4.5.*

Boote D.R.D. & Gustav S.H. (1987) Evolving depositional systems within an active rift, Witch Ground Graben, North Sea. In: *Petroleum Geology of North West Europe* (Ed. by J. Brooks and K.W. Glennie), pp. 819–833. Graham & Trotman, London. *10.3.4, Fig. 10.60.*

Boothroyd J.C. (1972) *Coarse-grained Sedimentation on a Braided Outwash Fan, Northeast Gulf of Alaska*, 127 pp. Tech. Rep. No. 6

Coastal Research Division, University of South Carolina, Columbia. *3.3.2, 3.3.3, 3.3.8, 3.6.4, Fig. 3.31.*

Boothroyd J.C. & Ashley G.M. (1975) Process, bar morphology and sedimentary structures on braided outwash fans, Northeastern Gulf of Alaska. In: *Glaciofluvial and Glaciolacustrine Sedimentation* (Ed. by A.V. Jopling and B.C. McDonald), pp. 193–222. *Spec. Publ. Soc. econ. Paleont. Miner.*, 23, Tulsa. *3.3.2, 3.3.3, 3.3.8, Fig. 6.20, 11.4.4.*

Boothroyd J.C. & Nummedal D. (1978) Proglacial braided outwash: A model for humid alluvial-fan deposits. In: *Fluvial Sedimentology* (Ed. by A.D. Miall), pp. 641–668. *Mem. Can. Soc. petrol. Geol.*, 5, Calgary. *Fig. 6.20, 11.4.4.*

Borchert H. & Muir R.O. (1964) *Salt Deposits: the Origin, Metamorphism and Deformation of Evaporites*, 338 pp. Van Nostrand-Reinhold, Princeton, N.J. *8.2.2, Figs 8.2, 8.3.*

Boreen T.D. & James N.P. (1993) Holocene sediment dynamics on a cool-water carbonate shelf: Otway south eastern Australia. *J. sedim. petrol.*, 63, 574–588. *9.1.1.*

Boreen T.D. & James N.P. (1995) Stratigraphic sedimentology of Tertiary cool-water limestones, SE Australia. *J. sedim. Res.*, B65, 142–159. *9.3.4, Fig. 9.16.*

Borgia A., Ferrari L. & Pasquare G. (1992) Importance of gravitational spreading in the tectonic and volcanic evolution of Mouth Etna. *Nature*, 357, 231–235. *12.8.*

Born S.M. (1972) *Late Quaternary History, Deltaic Sedimentation and Mudlump Formation at Pyramid Lake, Nevada*, 97 pp. Center for Water Resources Research, Desert Research Institute, University of Nevada, Reno. *4.6.2.*

Bornhold B. & Prior D.B. (1990) Morphology and sedimentary processes on the subaqueous Noeick River delta, British Columbia, Canada. In: *Coarse-grained Deltas* (Ed. by A. Colella and D.B. Prior), pp. 169–181. *Spec. Publ. int. Ass. Sediment.*, 10. *6.5.5.*

Bosellini A. (1984) Progradation geometries of carbonate platforms: examples from the Triassic of the Dolomites, northern Italy. *Sedimentology*, 31, 1–24. *9.1.1, 9.3.2.*

Bosellini A. (1989) Dynamics of Tethyan carbonate platforms. In: *Controls on Carbonate Platform and Basin Development* (Ed. by P.D. Crevello, J.F. Sarg, J.F. Read and J.L. Wilson), pp. 3–13. *Spec. Publ. Soc. econ. Paleont. Miner.*, 44, Tulsa. *9.3.5.*

Bosellini A. (1991) *Geology of the Dolomites – An Introduction*, 43 pp. Dolomieu Conference on Carbonate Platforms and Dolomitization. Tourist office, Ortisei, Italy. September 1991. *9.3.2, Fig. 9.11.*

Bosellini A. & Hardie L.A. (1973) Depositional theme of a marginal marine evaporite. *Sedimentology*, 20, 5–27. *8.7.*

Bosellini A., Masetti D. & Sarti M. (1981) A Jurassic 'Tongue of the Ocean' filled with oolitic sands: the Belluno Trough, Venetian Alps, Italy. *Mar. Geol.*, 44, 59–95. *9.4.4, Fig. 9.44.*

Bosence D.W.J. (1985) Preservation of coralline algal frameworks. *Proc. 5th Int. Coral Reef Congress, Tahiti*, 2, 39–45. *9.4.5.*

Bosence D.W.J. (1989) Biogenic carbonate production in Florida Bay. *Bull. mar. Sci.*, 44, 419–433. *9.4.5.*

Bosence D.W.J. (1995) Anatomy of Recent biodetrital mud-mound, Florida Bay, USA. In: *Carbonate Mud Mounds* (Ed. by C.L.V. Monty, D.W.J. Bosence, P.H. Bridges and B.R. Pratt), pp. 475–493. *Spec. Publ. int. Ass. Sediment.*, 23. *9.4.5.*

Bosence D.W.J., Rowlands R.J. & Quine M.L. (1985) Sedimentology and budget of a Recent carbonate mound, Florida Keys. *Sedimentology*, 32, 317–343. *9.3.3, 9.4.5.*

Bosscher H. & Schlager W. (1992) Computer simulation of reef growth. *Sedimentology*, 39, 503–512. *9.2, 9.2.1, 9.4.5, Fig. 9.1.*

Boström K. (1973) The origin and fate of ferromanganoan active ridge sediments. *Stockholm Contrib. Geol.*, 27, 149–253. *12.11.1.*

Bottjer D.J. & Droser M.L. (1991) Ichnofabric and basin analysis. *Palaios*, 6, 199–205. *7.7.3.*

Boudon G., Camus G., Gourgaud A. & Lajoie J. (1993) The 1984 nuée-ardente deposits of Merapi volcano, Central Java, Indonesia: stratigraphy, textural characteristics, and transport mechanisms. *Bull. Volcanol.*, 55, 327–342. *12.4.4.*

Boulton G.S. (1970) The deposition of subglacial and melt-out tills at the margins of certain Svalbard glaciers. *J. Glaciol.*, 9, 231–245. *Fig. 11.7.*

Boulton G.S. (1971) Till genesis and fabric in Svalbard, Spitsbergen. In: *Till: A Symposium* (Ed. by R.P. Goldthwait), pp. 41–72. Ohio State University Press, Columbus *11.3.3.*

Boulton G.S. (1972a) Modern Arctic glaciers as depositional models for former ice sheets. *J. geol. Soc. Lond.*, 128, 361–393. *11.2.2, 11.3.2, 11.3.3, 11.4.3.*

Boulton G.S. (1972b) The role of thermal régime in glacial sedimentation. In: *Polar Geomorphology* (Ed. by R.J. Price and D.E. Sugden), pp. 1–19. *Institute of British Geographers Spec. Publ.*, 4. *Fig. 11.8.*

Boulton G.S. (1974) Processes and patterns of glacial erosion. In: *Glacial Geomorphology* (Ed. by D.R. Coates), pp. 41–87. George Allen & Unwin, London. *11.3.1, 11.3.2.*

Boulton G.S. (1975) Processes and patterns of subglacial sedimentation: a theoretical approach. In: *Ice Ages: Ancient and Modern* (Ed. by A.E. Wright and F. Moseley), pp. 7–42. *Geol. J. Spec. issue*, 6, Seel House Press, Liverpool. *11.3.4.*

Boulton G.S. (1978) Boulder shapes and grain-size distributions of debris as indicators of transport paths through a glacier and till genesis. *Sedimentology*, 25, 773–779. *11.3.3, Fig. 11.5.*

Boulton G.S. (1979) Processes of glacier erosion on different substrata. *J. Glaciol.*, 23, 15–37. *11.3.1, 11.4.1.*

Boulton G.S. (1987a) A theory of drumlin formation by subglacial sediment deformation. In: *Drumlin Symposium* (Ed. by J. Menzies and J. Rose), pp. 25–80. A.A. Balkema, Rotterdam. *11.4.1.*

Boulton G.S. (1987b) Progress in glacial geology during the last fifty years. *J. Glaciol. Spec. Issue*, 25–32. *11.6.*

Boulton G.S. (1990) Sedimentary and sea level changes during glacial cycles and their control on glacimarine facies architecture. In: *Glacimarine Environments: Processes and Sediments* (Ed. by J.A. Dowdeswell and J.D. Scourse), pp. 15–52. *Spec. Publ. geol. Soc. Lond.*, 53. *11.5.4, Figs 11.20, 11.21.*

Boulton G.S. & Dent D.L. (1974) The nature and rates of post-depositional changes in recently deposited till from south-east Iceland. *Geogr. Ann.*, 56A (3–4), 121–134. *11.4.7.*

Boulton G.S. & Deynoux M. (1981) Sedimentation in glacial environments and the identification of tills and tillites in ancient sedimentary sequences. *Precambrian Res.*, 15, 397–422. *11.5.3.*

Boulton G.S. & Vivian R.A. (1973) Underneath the glaciers. *Geogr., Mag.*, 45, (4), 311–316. *11.4.1.*

Boulton G.S., Baldwin C.T. *et al.* (1982) A glacio-isostatic facies model and amino acid stratigraphy for late Quaternary events in Spitsbergen and the Arctic. *Nature*, 298, 437–441. *11.5.3.*

Boulton G.S., Thors K. & Jarvis J. (1988) Dispersal of glacially derived sediment over part of the continental shelf of south Iceland and the geometry of the resultant sediment bodies. *Mar. Geol.*, 83, 193–223. *11.5.4.*

Bouma A.H. (1962) *Sedimentology of Some Flysch Deposits: A Graphic Approach to Facies Interpretation*, 168 pp. Elsevier, Amsterdam. *10.1, 10.2.3.*

Bouma A.H., Coleman J.M., Stetling C.E. & Kohl B. (1989) Influence of relative sea level changes on the construction of the Mississippi Fan. *Geo-Mar. Lett.*, 9, 161–170. *10.3.3.*

Bouma A.H., Normark W.R. & Barnes N.E. (Eds) (1985) *Submarine Fans and Related Turbidite Systems*, 351 pp. Springer-Verlag, New York. *10.3.3.*

Bouma A.H., Stetling C.E. & Coleman J.M. (1985) Mississippi Fan, Gulf of Mexico. In: *Submarine Fans and Related Turbidite Systems* (Ed. by A.H. Bouma, W.R. Normark and N.E. Barnes), pp. 143–150. Springer-Verlag, New York. *10.3.2, 10.3.3, Fig. 10.41.*

Bourgeois J. (1980) A transgressive shelf sequence exhibiting hummocky cross stratification: The Cape Sebastian Sandstone (Upper Cretaceous), southwestern Oregon. *J. sedim. Petrol.*, 50, 681–702. *6.7.7, Fig. 6.97.*

Bourgeois J. & Leithold E.L. (1984) Wave-worked conglomerates – depositional processes and criteria for recognition. In: *Sedimentology of Gravels and Conglomerates* (Ed. by E.H. Koster and R.J. Steel), pp. 331–343. *Mem. Can. Soc. petrol. Geol.*, 10, Calgary. *6.2.3, 6.5.4, 7.7.2.*

Bourgeois J., Pautot G. *et al.* (1988) Seabeam and seismic reflection imaging of the neotectonic regime of the Andean continental margin off Peru (4°S to 10°S). *Earth Planet. Sci. Lett.*, 87, 111–126. *12.12.2, 12.12.5.*

Bouroullec J.L., Rehault J.P., Rolet J., Tiercelin J.J. & Mondeguer A. (1991) Quaternary sedimentary processes and dynamics in the northern part of the Lake Tanganyika trough, East African rift system. Evidence of lacustrine eustatism? *Bull. Centre Rech. Pau-SNPA*, 15, 343–368. *4.6.*

Bouroullec J.L., Thouin C., Tiercelin J.J., Mondeguer A., Rolet J. & Rehault J.P. (1992) Séquences sismiques haute résolution du fossé nord-Tanganyika, rift est-africain. Implications climatiques, tectoniques et hydrothermales. *C. R. Acad. Sci. Paris*, 315, 601–608. *4.6.3.*

Bourque P.-A. & Boulvain F. (1993) A model for the origin and petrogenesis of the red stromatactis limestone of Paleozoic carbonate mounds. *J. sedim. Petrol.*, 63, 607–619. *9.4.5.*

Bourque P.-A., Madi A. & Mamet B.L. (1995) Waulsortian-type bioherm development and response to sea-level fluctuations: Upper Viséan of Béchar Basin, western Algeria. *J. sedim. Res.*, B65, 80–95. *9.4.5.*

Bowler J.M. (1973) Clay dunes: their occurrence, formation and environmental significance. *Earth-Sci. Rev.*, 9, 315–338. *4.6.1.*

Bowman D. (1990) Climatically triggered Gilbert-type lacustrine fan deltas, the Dead Sea area, Israel. In: *Coarse-grained Deltas* (Ed. by A. Colella and D.B. Prior), pp. 273–280. *Spec. Publ. int. Ass. Sediment.*, 10. *6.5.5.*

Bown, T.M. & Kraus, M.J. (1987) Integration of channel and floodplain suites, I. Developmental sequences and lateral relations of alluvial paleosols. *J. sedim. Petrol.*, 57, 587–601. *3.5.4, 3.6.6.*

Boyd R., Dalrymple R.W. & Zaitlin B.A. (1992) Classification of clastic coastal depositional environments. *Sedim. Geol.*, 80, 139–150. *6.3, Figs 6.1, 6.16.*

Boyer B.W. (1982) Green River laminites: does the playa-lake model really invalidate the stratified-lake model? *Geology*, 10, 321–324. *4.14.1.*

Brackett R.S. & Bush D.M. (1986) Sedimentary characteristics of modern storm-generated sequences: North Insular shelf, Puerto Rico. In: *Modern and Ancient Shelf Clastics: A Core Workshop* (Ed. by T.F. Moslow and E.G. Rhodes), pp. 125–168. *Soc. econ. Paleont. Miner. Core Workshop No. 9. 7.4.3.*

Bradley W.H. (1929) The varves and climate of the Green River epoch. *Prof. Pap. U.S. geol. Surv.*, 158, 87–110. *4.16.1.*

Bradley W.H. & Eugster H.P. (1969) Geochemistry and palaeolimnology of the trona deposits and associated authigenic minerals of the Green River Formation of Wyoming. *Prof. Pap. U.S. geol. Surv.*, 469-B, 71 pp. *4.14.1.*

Braitsch O. (1971) *Salt Deposits – Their Origin and Composition*, 297 pp. Springer-Verlag, Berlin. *8.2.1, Table 8.1.*

Brandner R., Flügel E. & Senowbari-Daryan B. (1991) Biotic and microfacies criteria of carbonate slope boulders: implications for the reconstruction of source areas (Middle Triassic: Mahlknecht Cliff, Dolomites). *Facies*, 25, 279–296. *9.3.2.*

Branney M.J. (1991) Eruption and depositional facies of the Whorneyside Tuff Formation, English Lake District: an exceptionally large-magnitude phreatoplinian eruption. *Bull. geol. Soc. Am.*, 103, 886–897. *12.12.4.*

Branney M.J. & Kokelaar P. (1992) A reappraisal of ignimbrite emplacement: progressive agradation and changes from particulate to non-particulate flow during emplacement of high-grade ignimbrite. *Bull. Volcanol.*, 54, 504–520. *12.4.1, 12.4.4.*

Branney M.J. & Kokelaar P. (1994) Volcanotectonic faulting, soft-state deformation and rheomorphism of tuffs during development of a piecemeal caldera, English Lake District. *Bull. geol. Soc. Am.*, 106, 507–530. *12.4.6, 12.12.4, Fig. 12.77.*

Branney M.J. & Sparks R.S.J. (1990) Fiamme formed by diagenesis and burial compaction in soils and subaqueous sediments. *J. geol. Soc. Lond.*, 147, 919–922. *12.14.1.*

Branney M.J. & Suthren R.J. (1988) High-level peperitic sills in the English Lake District: distinction from block lavas and implications for Borrowdale Volcanic Group stratigraphy. *Geol. J.*, 23, 177–187. *12.14.1.*

Branney M.J., Kokelaar B.P. & McConnell B.J. (1992) The Bad Step Tuff: a lava-like rheomorphic ignimbrite in a calc-alkaline piecemeal caldera, English Lake District. *Bull. Volcanol.*, 54, 187–199. *12.4.4, Fig. 12.26.*

Brantley S.L. & Donovan B. (1990) Marine evaporites, bittern seepage, and the genesis of subsurface brines. *Chem. Geol.*, 84, 187–189. *8.2.3.*

Brayshaw A.C. (1984) Characteristics and origin of cluster bedforms in coarse-grained alluvial channels. In: *Sedimentology of Gravels and Conglomerates* (Ed. by E.H. Koster and R.J. Steel), pp. 77–85. *Mem. Can. Soc. petrol. Geol.*, 5, Calgary. *12.5.2.*

Breed C.S. & Grow T. (1979) Morphology and distribution of dunes in sand seas observed by remote sensing. In: *Global Sand Seas* (Ed.

by E.D. Mckee), pp. 253–302. *Prof. Pap. U.S. geol. Surv.*, 1052. *5.4.3.*

Breed C.S., Fryberger S.C., Andrews S., McCauley C., Lennartz F., Gebel D. & Horstman K. (1979) Regional studies of sand seas using Landsat (ERTS) imagery. In: *Global Sand Seas* (Ed. by E.D. McKee), pp. 305–397. *Prof. Pap. U.S. geol. Surv.*, 1052. *5.1, 5.4.2.*

Breed C.S., McCauley J.F. & Davis P.A. (1987) Sand sheets of the eastern Sahara and ripple blankets on Mars. In: *Desert Sediments: Ancient and Modern* (Ed. by L. Frostick and I. Reid), pp. 337–359. *Spec. Publ. geol. Soc. Lond.*, 35. *5.4.6.*

Breed C.S., McCauley J.F. & Whitney M.I. (1989) Wind erosion forms. In: *Arid Zone Geomorphology* (Ed. by D.S.G. Thomas), pp. 284–307. Belhaven Press, London. *5.2.1.*

Breitkreuz C. (1991) Fluvio-lacustrine sedimentation and volcanism in a Late Carboniferous tensional intra-arc basin, northern Chile. *Sedim. Geol.*, 74, 173–187. *12.12.4.*

Bremner J.M. & Willis J.P. (1993) Mineralogy and geochemistry of the clay fraction of sediments from the Namibian continental margin and the adjacent hinterland. *Mar. Geol.*, 115, 85–116. *7.6.*

Brenchley P.J. (1985) Storm influenced sandstone beds. *Mod. Geol.*, 9, 369–396. *7.1.2, 7.7.2.*

Brenchley P.J. & Williams B.P.J. (Eds) (1985) *Sedimentology: Recent Developments and Applied Aspects*, 342 pp. *Spec. Publ. geol. Soc. Lond.*, 18. *1.1.*

Brenchley P.J., Newall G. & Stanistreet J.G. (1979) A storm surge origin for sandstone beds in an epicontinental platform sequence, Ordovician, Norway. *Sediment. Geol.*, 22, 185–217. *7.7.2, 7.8.1.*

Brenchley P.J., Pickerill R.K. & Stromberg J. (1993) The role of wave reworking on the architecture of storm sandstone facies, Bell Island Group (Lower Ordovician), eastern Newfoundland. *Sedimentology*, 40, 359–382. *7.8.1.*

Brenchley P.J., Romano M. & Guiterrez Marco J.C. (1986) Proximal and distal hummocky cross-stratified facies on a wide Ordovician shelf in Iberia. In: *Shelf Sands and Sandstones* (Ed. by R.J. Knight and J.R. McLean), pp. 241–256. *Mem. Can. Soc. petrol. Geol.*, 11. *7.8.1.*

Brennand T.A. (1994) Macroforms, large bedforms and rhythmic sedimentary sequences in subglacial eskers, south-central Ontario: implications for esker genesis and meltwater regime. *Sedim. Geol.*, 91, 9–55. *11.4.1.*

Brenner R.L. & Martinsen O.J. (1990) The Fossil Sandstone — a shallow marine sand wave complex in the Namurian of Cumbria and North Yorkshire, England. *Proc. Yorks. geol. Soc.*, 48, 149–162. *6.6.6.*

Brice J.C. (1964) Channel patterns and terraces of the Loup Rivers of Nebraska. *Prof. Pap. U.S. geol. Surv.*, 422-D, 41 pp. *3.3.1, 3.3.3.*

Bridge J.S. (1975) Computer simulation of sedimentation in meandering streams. *Sedimentology*, 22, 3–44. *3.3.4.*

Bridge J.S. (1977) Flow, bed topography and sedimentary structure in open-channel bends: a three-dimensional model. *Earth Surf. Proc.*, 2, 401–416. *3.3.4.*

Bridge J.S. (1984) Large scale facies sequences in alluvial overbank environments. *J. sedim. Petrol.*, 54, 583–588. *3.6.6.*

Bridge J.S. (1985) Paleochannel channel patterns inferred from alluvial deposits: a critical evaluation. *J. sedim. Petrol.*, 55, 579–589. *3.6.2.*

Bridge J.S. (1993) Description and interpretation of fluvial deposits:

a critical perspective. *Sedimentology*, 40, 801–810. *3.6.1.*

Bridge J.S. & Diemer J.A. (1983) Quantitative interpretation of an evolving ancient river system. *Sedimentology*, 30, 599–623. *3.6.5.*

Bridge J.S. & Jarvis J. (1976) Flow and sedimentary processes in the meandering River South Esk, Glen Clova, Scotland. *Earth Surf. Proc.*, 2, 281–294. *3.3.4.*

Bridge J.S. & Jarvis J. (1982) The dynamics of a river bend; a study in flow and sedimentary processes. *Sedimentology*, 29, 499–541. *3.3.4.*

Bridge J.S. & Leeder M.R. (1979) A simulation model of alluvial stratigraphy. *Sedimentology*, 26, 499–541. *3.6.6, 3.6.7.*

Bridge J.S., Smith N.D., Trent F., Gabel S.L. & Bernstein P. (1986) Sedimentology and morphology of a low-sinuosity river: Calamus River, Nebraska Sand Hills. *Sedimentology*, 33, 851–870. *3.3.3, Fig. 3.15.*

Bridges E.M. (1970) *World Soils*, 89 pp. Cambridge University Press, Cambridge. *3.2.3.*

Bridges P.H. (1972) The significance of toolmarks on a Silurian erosional furrow. *Geol. Mag.*, 109, 405–410. *7.7.2.*

Bridges P.H. (1975) The transgression of a hard substrate shelf: the Llandovery (Lower Silurian) of the Welsh Borderland. *J. sedim. Petrol.*, 45, 79–94. *6.4.*

Bridges P.H. (1982) Ancient offshore tidal deposits. In: *Offshore Tidal Sands: Processes and Deposits* (Ed. by A.H. Stride), pp. 172–192. Chapman & Hall, London. *7.7.1.*

Bridges P.H. & Chapman A.J. (1988) The anatomy of a deep water mud-mound complex to the north of the Dinantian platform in Derbyshire, UK. *Sedimentology*, 35, 139–162. *9.4.5.*

Bridges P.H. & Leeder M.R. (1976) Sedimentary model for intertidal mudflat channels with examples from the Solway Firth, Scotland. *Sedimentology*, 23, 533–552. *6.7.3.*

Brierley G.J. (1989) River planform facies models: the sedimentology of braided, wandering and meandering reaches of the Squamish River, British Columbia. *Sedim. Geol.*, 61, 17–35. *3.3.3.*

Brierley G.J. (1991) Bar sedimentology of the Squamish River, British Columbia: definition and application of morphostratigraphic units. *J. sedim. Petrol.*, 61, 211–225. *3.3.3.*

Briggs L.I. (1957) Quantitative aspects of evaporite deposition. *Pap. Michigan Acad. Sci., Arts Lett.*, 42, 115–123. *8.1.*

Bristow C.R. (1988) Controls on the sedimentation of the Rough Rock Group (Namurian) from the Pennine Basin of northern England. In: *Sedimentation in a Synorogenic Basin Complex: the Upper Carboniferous of NW Europe* (Ed. by B.M. Besly and G. Kelling), pp. 114–131. Blackie, Glasgow. *6.6.6.*

Bristow C.S. (1987) Brahmaputra River: Channel migration and deposition. In: *Recent Developments in Fluvial Sedimentology* (Ed. by F.G. Ethridge, R.M. Flores and M.D. Harvey), pp. 63–74. *Spec. Publ. Soc. econ. Paleont. Miner.*, 39, Tulsa. *3.3.3, 3.6.5, Fig. 3.13.*

Bristow C.S., Best J.L. & Roy A.G. (1993) Morphology and facies models of channel confluences. In: *Alluvial Sedimentation* (Ed. by M. Marzo and C. Puigdefabregas), pp. 91–100. *Spec. Publ. int. Ass. Sediment.*, 17. *3.3.3.*

Brodylo L.A. & Spencer R.J. (1987) Depositional environment of the Middle Devonian Telegraph salts, Alberta, Canada. *Bull. Can. petrol. Geol.*, 35, 186–196. *8.5.1.*

Brodzikowski K. & van Loon A.J. (1991) Glacigenic Sediments, 674 pp. *Developments in Sedimentology*, 49, Elsevier, Amsterdam.

11.3.6, 11.4.2, 11.4.5, 11.4.7.

Bromley M.H. (1991) Variations in fluvial style as revealed by architectural elements, Kayenta Formation, Mesa Creek, Colorado, USA: Evidence for both ephemeral and perennial fluvial processes. In: *The Three-dimensional Facies Architecture of Terrigenous Clastic Sediments and its Implications for Hydrocarbon Discovery and Recovery* (Ed. by A.D. Miall and N. Tyler), pp. 94–102. *Soc. econ. Paleont. Miner. Concepts in Sedimentology and Paleontology*, 3, Tulsa. *3.6.2.*

Bromley R.G. & Ekdale A.A. (1984) Trace fossil preservation in flint in the European chalk. *J. Paleont.*, 58, 289–311. *10.4.3.*

Bromley R.G. & Ekdale A.A. (1986) Composite ichnofabrics and tiering of burrows. *Geol. Mag.*, 123, 59–65. *Fig. 10.81.*

Bromley R.G. & Ekdale A.A. (1987) Mass transport in European Cretaceous chalk: fabric criteria for its recognition. *Sedimentology*, 34, 1079–1092. *10.2.3, 10.4.3.*

Brooker M.R., Houghton B.F., Wilson C.J.N. & Gamble J.A. (1993) Pyroclastic phases of a rhyolitic dome-building eruption: Puketarata tuff ring, Taupo volcanic zone, New Zealand. *Bull. Volcanol.*, 55, 395–406. *12.10.*

Brookfield M.E. (1977) The origin of bounding surfaces in ancient aeolian sandstones. *Sedimentology*, 24, 303–332. *3.6.1, 5.3.6, 10.3.2.*

Brooks J. & Fleet A.J. (Eds) (1987) *Marine Petroleum Source Rocks*, 444 pp. *Spec. Publ. geol. Soc. Lond.*, 26, *10.2.1.*

Brooks G.R. & Holmes C.W. (1989) Recent carbonate slope sediments and sedimentary processes bordering a non-rimmed platform, south west Florida continental margin. In: *Controls on Carbonate Platform and Basin Development* (Ed. by P.D. Crevello, J.F. Sarg, J.F. Read and J.L. Wilson), pp. 259–272. *Spec. Publ. Soc. econ. Paleont. Miner.*, 44, Tulsa. *9.4.4.*

Brooks G.R. & Holmes C.W. (1990) Modern configuration of the southwest Florida carbonate slope: development of shelf margin progradation. *Mar. Geol.*, 94, 301–315. *9.4.4.*

Brown A.A. & Loucks R.G. (1993) Influence of sediment type and depositional processes on stratal patterns in the Permian basin-margin Lamar Limestone, McKittrick Canyon, Texas. In: *Carbonate Sequence Stratigraphy* (Ed. by R.G. Loucks and J.F. Sarg), pp. 133–156. *Mem. Am. Ass. petrol. Geol.*, 57, Tulsa. *9.5.5.*

Brown K.M. & Westbrook G.K. (1988) Mud diapirism and subcretion in the Barbados Ridge accretionary prism: the role of fluids in accretionary processes. *Tectonics*, 7, 613–640. *12.12.5.*

Brown L.F. Jr & Fisher W.L. (1977) Seismic-stratigraphic interpretation of depositional systems: examples from Brazilian rift and pull-apart basins. In: *Seismic Stratigraphy – Applications to Hydrocarbon Exploration* (Ed. by C.E. Payton), pp. 213–248. *Mem. Am. Ass. petrol. Geol.*, 26, Tulsa. *2.2.2, 2.3.2, 2.4.*

Bruhn R.L. & Dalziel I.W.D. (1977) Destruction of the early Cretaceous marginal basin in the Andes of Tierra del Fuego. In: *Island Arcs, Deep Sea Trenches and Back-Arc Basins* (Ed. by M. Talwani and W.C.I. Pitman), pp. 395–405. *Am. geophys. Union, Maurice Ewing Series 1*, Washington, D.C. *Fig. 12.71.*

Bruun P. (1962) Sea-level rise as a cause of shore erosion. *Proc. Am. Soc. Civ. Eng., J. Water Harbors Div.*, 88, 117–130. *6.7, Fig. 6.11.*

Bryant I.D. & Smith M.P. (1990) A composite tectonic–eustatic origin for shelf sandstones at the Cambrian–Ordovician boundary in North Greenland. *J. Geol. Soc.*, **147**, 795–809. *7.7.2.*

Budd D.A. & Harris P.M. (1990) Carbonate–siliciclastic mixtures. *Spec. Publ. Soc. econ. Paleont. Miner.*, **14**, Tulsa. *9.5.5.*

Buddemeier R.W. & Kinzie R.A. (1976) Coral growth. *Oceanograph. mar. Biol. Ann. Rev.*, **14**, 183–225. *9.4.5.*

Budding M.C. & Inglin H. (1981) A reservoir geological model of the Brent Sands in southern Cormorant. In: *Petroleum Geology of the Continental Shelf of North-west Europe* (Ed. by L.V. Illing and G.D. Hobson), pp. 326–334. Heyden, London. *6.6.6.*

Buesch D.C. (1991) Changes in depositional environments resulting from emplacement of a large-volume ignimbrite. In: *Sedimentation in Volcanic Settings* (Ed. by R.V. Fisher and G.A. Smith), pp. 139–154. *Spec. Publ. Soc. econ. Paleont. Miner.*, **45**, Tulsa. *12.6.1, 12.13.*

Bunting B.T. (1967) *The Geography of Soils*, 2nd edn, 213 pp. Hutchinson, London. *3.2.3.*

Burchell M.T., Stephani M. & Masetti D. (1990) Cyclic sedimentation in the southern Alpine Rhaetic: the importance of climate and eustasy in controlling platform-basin interactions. *Sedimentology*, **37**, 795–815. *9.4.3.*

Burchette T.P. (1981) European Devonian reefs: a review of current concepts and models. In: *European Fossil Reef Models* (Ed. by D.F. Toomey), pp. 85–142. *Spec. Publ. Soc. econ. Paleont. Miner.*, **30**, Tulsa. *9.3.4, 9.5.2.*

Burchette T.P. (1987) Carbonate-barrier shorelines during the basal Carboniferous transgression: the Lower Limestone Shale Group, South Wales and western England. In: *European Dinantian Environments* (Ed. by J. Miller, A.E. Adams and V.P. Wright), pp. 239–263. Wiley, Chichester. *9.3.4, 9.5.2.*

Burchette T.P. (1988) Tectonic control on carbonate platform facies distribution and sequence development: Miocene, Gulf of Suez. *Sedim. Geol.*, **59**, 179–204. *9.5.3.*

Burchette T.P. (1993) Mishrif Formation (Cenomanian–Turonian), Southern Arabian Gulf: carbonate platform growth along a cratonic basin margin. In: *Cretaceous Carbonate Platforms* (Ed. by J.A. Simo, R.W. Scott and J.-P. Masse), pp. 185–199. *Mem. Am. Ass. petrol. Geol.*, **56**, Tulsa. *9.3.1.*

Burchette T.P. & Wright V.P. (1992) Carbonate ramp depositional systems. *Sedim. Geol.*, **79**, 3–57. *9.3.4, 9.4.5, 9.5.1, 9.5.2, Figs 9.8, 9.14, 9.58.*

Burchette T.P., Wright V.P. & Faulkner T.J. (1990) Oolitic sandbody depositional models and geometries, Mississippian of southwest Britain: implications for petroleum exploration in carbonate ramp settings. *Sedim. Geol.*, **68**, 87–115. *9.3.4, 9.4.2, 9.5.2, Figs 9.19, 9.30.*

Burke C.A. & Drake C.L. (Eds) (1974) *The Geology of Continental Margins*, 1009 pp. Springer-Verlag, New York. *1.1.*

Burke K. (1967) The Yallahs Basin: a sedimentary basin southeast of Kingston, Jamaica. *Mar. Geol.*, **5**, 45–60. *Fig. 6.21.*

Burke K. (1972) Longshore drift, submarine canyons, and submarine fans in development of Niger Delta. *Bull. Am. Ass. petrol. Geol.*, **56**, 1975–1983. *10.3.1, 10.3.3, Fig. 10.52.*

Burkinshaw J., Illenberger W. & Rust I.C. (1993) Wind-speed profiles over a reversing transverse dune. In: *The Dynamics and Environmental Context of Aeolian Sedimentary Systems* (Ed. by K. Pye), pp. 25–36. *Spec. Publ. geol. Soc. London*, **72**. *5.3.2.*

Burkhart B. & Self S. (1985) Extension and rotation of crustal blocks in northern Central America and effect on the volcanic arc. *Geology*, **13**, 22–26. *12.12.2, Fig. 12.68.*

Burne R.V. & Moore L.S. (1987) Microbialites: organosedimentary deposits of benthic microbial communities. *Palaios*, **2**, 241–254. *4.7.1.*

Burne R.V., Bauld J. & de Deckker P. (1980) Saline lake charophytes and their geological significance. *J. sedim. Petrol.*, **50**, 281–293. *4.7.1.*

Burnett W.C. (1977) Geochemistry and origin of phosphorite deposits from off Peru and Chile. *Bull. geol. Soc. Am.*, **88**, 813–823. *10.2.1.*

Burnham C.W. (1979) The importance of volatile constituents. In: *The Evolution of Igneous Rocks* (Ed. by H.S. Yoder), pp. 382–439. Princeton University Press, Princeton, NJ. *12.3.1.*

Bursik M., Sparks R.S.J., Gilbert J. & Carey S. (1992) Sedimentation of tephra from volcanic plumes I: theory and its comparison with a study of the Fogo A plinian deposit, Sao Miguel (Azores). *Bull. Volcanol.*, **54**, 329–344. *12.4.3.*

Busby-Spera C.J. (1985) A sand-rich submarine fan in the lower Mesozoic Mineral King caldera complex, Sierra Nevada, California. *J. sedim. Petrol.*, **55**, 376–391. *12.13.*

Busby-Spera C.J. (1988a) Evolution of a Middle Jurassic back-arc basin, Cedros Island, Baja California: evidence from a marine volcaniclastic apron. *Bull. geol. Soc. Am.*, **100**, 218–233. *12.12.5, Fig. 12.89.*

Busby-Spera C.J. (1988b) Development of fan-deltoid slope aprons in a convergent-margin tectonic setting: Mesozoic, Baja California, Mexico. In: *Fan Deltas: Sedimentology and Tectonic Setting* (Ed. by W. Nemec and R.J. Steel), pp. 419–429. Blackie, Glasgow. *12.12.5, Fig. 12.86.*

Busby-Spera C.J. (1988c) Speculative tectonic model for the early Mesozoic arc of the southwest Cordilleran United States. *Geology*, **16**, 1121–1125. *12.12.4, Fig. 12.75.*

Busby-Spera C.J. & White J.D.L. (1987) Variation in peperite textures associated with differing host-sediment properties. *Bull. Volcanol.*, **49**, 765–775. *12.4.1.*

Busch D.A. (1975) Influence of growth faulting on sedimentation and prospect evaluation. *Bull. Am. Ass. petrol. Geol.*, **59**, 217–230. *6.6.4.*

Bush P.R. (1970) Chloride rich brines from sabkha sediments and their possible role in ore formation. *Trans. Inst. Miner. Metall. Sect. B*, **79**, 137–144. *8.1.*

Bush P.R. (1973) Some aspects of the diagenetic history of the sabkha in Abu Dhabi, Persian Gulf. In: *The Persian Gulf* (Ed. by B.H. Purser), pp. 395–407. Springer-Verlag, New York. *8.2.2, 8.4.2.*

Butler G.P. (1969) Modern evaporite deposition and geochemistry of coexisting brines, the sabkha Trucial Coast, Arabian Gulf. *J. sedim. Petrol.*, **39**, 70–89. *8.1.*

Butler G.P. (1970) Holocene gypsum and anhydrite of the Abu Dhabi Sabkha, Trucial Coast: an alternative explanation of origin. *Third Symposium on Salt. N. Ohio Geol. Soc.*, **1**, 120–152. *8.4.1, 8.4.2.*

Butler G.P., Harris P.M. & Kendall C.G. St. C. (1982) Recent evaporites from the Abu Dhabi coastal flats. In: *Deposition and Diagenetic Spectra of Evaporites* (Ed. by C.R. Handford, R.G. Loucks and G.R. Davies), pp. 33–64. *Soc. econ. Paleont. Miner. Core Workshop 3*, Tulsa 1982. *8.4.1, 8.4.2, Figs 8.13, 8.16, 8.17, 8.18.*

Buurman P. (1980) Palaeosols in the Reading Beds (Paleocene) of Alum Bay, Isle of Wight, U.K. *Sedimentology*, **27**, 593–606. *3.5.4*.

Buxton W.M.N. & Pedley H.M. (1989) A standardized model for Tethyan Tertiary carbonate ramps. *J. geol. Soc. Lond.*, **146**, 746–748. *9.4.3*.

Byrne J.V., LeRoy D.O. & Riley C.M. (1959) The chenier plain and its stratigraphy, southwestern Louisiana. *Trans. Gulf Coast Ass. geol. Soc.*, **9**, 237–260. *6.7.2*.

Bøe R. (1988) Alluvial channel bank-collapse phenomena in the Old Red Sandstone Hitra Group, Western Central Norway. *Geol. Mag.*, **125**, 51–56. *3.5.2*.

Cabrera L. & Saez A. (1987) Coal deposition in carbonate-rich shallow lacustrine systems: the Calaf and Mequinenza sequences (Oligocene, eastern Ebro Basin, NE Spain). *J. geol. Soc. Lond.*, **144**, 451–461. *4.15.2, Fig. 4.34*.

Caldwell R.H. (1976) Holocene gypsum deposits of the Bullara Sunkland, Camarvon Basin, Western Australia. Unpublished PhD thesis, University of Western Australia, 123 pp. *8.4.2*.

Calkin P.E. & Rutford R.H. (1974) The sand dunes of Victoria Valley, Antarctica. *Geograph. Rev.*, **64**, 189–216. *11.4.7*.

Calvert S.E. (1966) Accumulation of diatomaceous silica in the sediments of the Gulf of California. *Bull. geol. Soc. Am.*, **77**, 569–596. *10.4.1*.

Calvet F. & Tucker M.E. (1988) Outer ramp cycles in the Upper Muschelkalk of the Catalan Basin, northeast Spain. *Sedim. Geol.*, **57**, 185–198. *9.3.4, 9.4.3*.

Calvet F., Tucker M.E. & Henton J.M. (1990) Middle Triassic carbonate ramp systems in the Catalan Basin, northeast Spain: facies, systems tracts, sequences and controls. In: *Carbonate Platforms* (Ed. by M.E. Tucker, J.L. Wilson, P.D. Crevello, J.R. Sarg and J.F. Read), pp. 79–108. *Spec. Publ. int. Ass. Sediment.*, **9**, 79–108. *9.4.3, 9.5.2*.

Cameron G.I.F. (1992) Analysis of dipmeter data for sedimentary orientation. In: *Geological Applications of Wireline Logs II* (Ed. by A. Hurst, C.M. Griffiths and P.E. Worthington), pp. 141–154. *Spec. Publ. geol. Soc. Lond.*, **65**. *2.3.4*.

Camoin G.F. & Montaggioni L.F. (1994) High energy coralgal–stromatolite frameworks from Holocene reefs (Tahiti, French Polynesia). *Sedimentology*, **41**, 655–676. *9.4.5*.

Campbell C.S. (1989) Self-lubrication for long runout landslides. *J. Geol.*, **97**, 653–666. *12.5.1*.

Campbell C.V. (1973) Offshore equivalents of Upper Cretaceous Gallup beach sandstones, northwestern New Mexico. In: *Cretaceous and Tertiary rocks of the southern Colorado Plateau*, pp. 78–84. Cretaceous-Tertiary Memoir, Four Corners Geol. Soc., Durango, Colorado. *7.1.2, 7.8.4*.

Campbell C.V. (1976) Reservoir geometry of a fluvial sheet sandstone. *Bull. Am. Ass. petrol. Geol.*, **60**, 1009–1020. *3.6.1*.

Campbell J.E. & Hendry H.E. (1987) Anatomy of a gravelly meander lobe in the Saskatchewan River, near Nipawin, Canada. In: *Recent Developments in Fluvial Sedimentology* (Ed. by F.G. Ethridge, R.M. Flores and M.D. Harvey), pp. 179–189. *Spec. Publ. Soc. econ. Paleont. Miner.*, **39**, Tulsa. *3.6.4*.

Cannon S.J.C., Giles M.R., Whitaker M.F., Please P.M. & Martin S.V. (1992) A regional reassessment of the Brent Group, UK sector, North Sea. In: *Geology of the Brent Group* (Ed. by A.C. Morton, R.S. Haszeldine, M.R. Giles and S. Brown), pp. 81–107. *Spec. Publ. geol. Soc. Lond.*, **61**. *6.6.6*.

Cant D.J. (1978) Development of a facies model for sandy braided river sedimentation: Comparison of the South Saskatchewan River and the Battery Point Formation. In: *Fluvial Sedimentology* (Ed. by A.D. Miall), pp. 627–639. *Mem. Can. Soc. petrol. Geol.* 5, Calgary. *3.6.2*.

Cant D.J. & Walker R.G. (1976) Development of a braided-fluvial facies model for the Devonian Battery Point Sandstone, Quebec. *Can. J. Earth Sci.*, **13**, 102–119. *3.6.2, 3.6.5, Fig. 3.45*.

Cant D.J. & Walker R.G. (1978) Fluvial processes and facies sequences in the sandy braided South Saskatchewan River. *Sedimentology*, **26**, 625–648. *3.3.3, Fig. 3.16*.

Caputo M.V. (1993) Eolian structures and textures in oolitic-skeletal calcarenites from the Quaternary of San Salvador Island, Bahamas: a new perspective on eolian limestones. In: *Mississippian Oolite and Modern Analogs* (Ed. by B.D. Keith and C.W. Zuppan), pp. 243–259. *Am. Ass. petrol. Geol., Studies in Geology*, **35**, Tulsa. *9.4.2*.

Caputo M.V. & Crowell J.C. (1985) Migration of glacial centers across Gondwana during Paleozoic Era. *Bull. geol. Soc. Am.*, **96**, 1020–1036. *11.6*.

Carey S.N. (1991) Transport and deposition by pyroclastic flows and surges. In: *Sedimentation in Volcanic Settings* (Ed. by R.V. Fisher and G.A. Smith), pp. 39–57. *Spec. Publ. Soc. econ. Paleont. Miner.*, **45**, Tulsa. *12.4.4*.

Carey S.N. & Sigurdsson H. (1982) Influence of particle aggregation on deposition of distal tephra from the May 18, 1980, eruption of Mount St. Helens volcano. *J. geophys. Res.*, **87**, 7061–7072. *12.4.3*.

Carey S.N. & Sigurdsson H. (1984) A model of volcanogenic sedimentation in marginal basins. In: *Marginal Basin Geology: Volcanic and Associated Sedimentary and Tectonic Processes in Modern and Ancient Marginal Basins* (Ed. by B.P. Kokelaar and M.F. Howells), pp. 37–58. *Spec. Publ. geol. Soc. Lond.*, **16**, *12.12.5, Fig. 12.88*.

Carey S.N. & Sigurdsson H. (1989) The intensity of Plinian eruptions. *Bull. Vulcanol.*, **51**, 28–40. *12.4.3*.

Carey S.N. & Sparks R.S.J. (1986) Quantitative models of the fallout and dispersal of tephra from volcanic eruption columns. *Bull. Volcanol.*, **48**, 109–125. *12.4.3*.

Carey S.W. (1958) The isostrat: a new technique for the analysis of structure of the Tasmanian dolerite. In: *Dolerite: A Symposium*, pp. 130–164. University of Hobart, Tasmania. *12.9.4*.

Carlisle D. (1963) Pillow breccias and their aquagene tuffs, Quadra Island, British Columbia. *J. Geol.*, **71**, 48–71, *12.4.1*.

Carlson T.N. & Benjamin S.G. (1980) Radiative heating rates for Saharan dust. *J. Geophys. Res.*, **37**, 193–213. *5.2.2*.

Carman G.J. & Young R. (1981) Reservoir Geology of the Forties Oilfield. In: *Petroleum Geology of the Continental Shelf of North-West Europe* (Ed. by L.V. Illing and G.D. Hobson), pp. 371–379. Heyden, London. *10.3.4*.

Carter R.M., Abbott S.T., Fulthorpe C.S., Haywick D.W. & Henderson R.A. (1991) Application of global sea-level and sequence-stratigraphic models in Southern Hemisphere Neogene Strata from New Zealand. In: *Sedimentation, Tectonics and Eustasy: Sea-level Changes at Active Margins* (Ed. by D.I.M. Macdonald), pp. 41–65. *Spec. Publ. int. Ass. Sediment.*, **12**. *2.4*.

Carter R.W.G. & Orford J.D. (1984) Coarse clastic barrier beaches: a discussion of the distinctive dynamic and morphosedimentary characteristics. *Mar. Geol.*, 60, 377–389. *6.5.4.*

Cas R.A.F. (1978) Silicic lavas in Palaeozoic flysch-like deposits in New South Wales, Australian: behaviour of deep silicic lava flows. *Bull. geol. Soc. Am.*, 89, 1708–1714. *12.4.5.*

Cas R.A.F. (1983) Submarine 'crystal tuffs': their origin using a Lower Devonian example from southeastern Australia. *Geol. Mag.*, 120, 471–486. *12.4.4.*

Cas R.A.F. (1992) Submarine volcanism: eruption styles, products, and relevance to understanding the host-rock successions to volcanic-hosted massive sulphide deposits. *Econ. Geol.*, 87, 511–541. *Fig. 12.55.*

Cas R.A.F. & Wright J.V. (1987) *Volcanic Successions: Modern and Ancient*, 528 pp. Allen & Unwin, London. *12.7, 12.14.2, Figs 12.10, 12.12, 12.14, 12.29.*

Cas R.A.F. & Wright J.V. (1991) Subaqueous pyroclastic flows and ignimbrites: an assessment. *Bull. Volcanol.*, 53, 357–380. *Fig. 12.7.*

Cas R.A.F., Allen R.L., Bull S.W., Clifford B.A. & Wright J.V. (1990) Subaqueous, rhyolitic dome-top tuff cones: a model based on the Devonian Bunga Beds, southeastern Australia and a modern analogue. *Bull. Volcanol.*, 52, 159–174. *12.10, Fig. 12.54.*

Cas R.A.F., Landis C.A. & Fordyce R.E. (1989) A monogenetic, Surtla-type, surtseyan volcano from the Eocene–Oligocene Waiarea–Deborah volcanics, Otago, New Zealand: a model. *Bull. Volcanol.*, 51, 281–298. *12.9.3, Fig. 12.52.*

Cas R.A.F., Powell C.M. & Crook K.A.W. (1980) Ordovician paleogeography of the Lachlan Fold Belt: a modern analog and tectonic constraints. *J. geol. Soc. Aust.*, 28, 271–288. *12.12.5.*

Cas R.A.F., Powell C.M., Fergusson C.L., Jones J.G., Roots W.D. & Fergusson J. (1981) The Kowmung volcaniclastics: a deep-water sequence of mass-flow origin. *J. geol. Soc. Aust.*, 28, 271–288. *12.12.5.*

Casanova J. (1986) La saisonnalité climatique au cours des optimums lacustres de l'est africain et la rhythmicité des constructions stromatolitiques: exemple du bassin Natron-Magadi au Pleistocene terminal. In: *INQUA Symposium 'Changements globaux en Afrique'* (Ed. by H. Faure), pp. 57–60. *4.7.1.*

Casanova J. (1991) Biosedimentology of Quaternary stromatolites in intertropical Africa. *J. Afr. Earth Sci.*, 12, 409–415. *4.7.1.*

Casanova J. & Hillaire-Marcel C. (1992) Late Holocene hydrological history of Lake Tanganyika, East Africa, from isotopic data on fossil stromatolites. *Palaeogeogr. Palaeoclimatol. Palaeoecol.*, 91, 35–48. *4.7.1.*

Casas E. & Lowenstein T.K. (1989) Diagenesis of saline pan halite: comparison of petrographic features of modern, Quaternary, and Permian halites. *J. sedim. Petrol.*, 59, 724–739. *8.5.1.*

Casas E., Lowenstein T.K., Spencer R.J. & Pengxi Z. (1992) Carnallite mineralization in the nonmarine, Qidam Basin, China: evidence for the early diagenetic origin of potash evaporites. *J. sedim. Petrol.*, 62, 881–898. *4.7.4, 8.5.1.*

Cashman K.V. & Fiske R.S. (1991) Fallout of pyroclastic debris from submarine volcanic eruptions. *Science*, 253, 275–280. *12.4.2, 12.4.3.*

Casshyap S.M. & Aslam M. (1992) Deltaic and shoreline sedimentation in Saurashtra Basin, Western India: an example of

infilling in an early Cretaceous Failed Rift. *J. sedim. Petrol.*, 62, 972–991. *6.6.6.*

Caston G.F. (1976) The floor of the North Channel, Irish Sea: a side-scan sonar survey. *Inst. Geol. Sci. Rep.*, 76/7. *7.1.2.*

Caston V.N.D. (1972) Linear sand banks in the southern North Sea. *Sedimentology*, 18, 63–78. *Fig. 7.15.*

Cather S.M. & Folk R.L. (1991) Pre-diagenetic sedimentary fractionation of andesitic detritus in a semi-arid climate: an example from the Eocene Datil Group, New Mexico. In: *Sedimentation in Volcanic Settings* (Ed. by R.V. Fisher and G.A. Smith), pp. 211–226. *Spec. Publ. Soc. econ. Paleont. Miner.*, 45, Tulsa. *12.14.1.*

Cazzola C., Mutti E. & Vigna B. (1985) Cengio turbidite system, Italy. In: *Submarine Fans and Related Turbidite Systems* (Ed. by A.H. Bouma, W.R. Normark and N.E. Barnes), pp. 179–183. Springer-Verlag, New York. *10.3.3.*

Cecil C.B. (1990) Paleoclimate controls on stratigraphic repetition of chemical and siliciclastic rocks. *Geology*, 18, 533–536. *2.1.2.*

Cerling T.E. (1986) A mass–balance approach to basin sedimentation: constraints on the recent history of the Turkana basin. *Palaeogeogr. Palaeoclimatol. Palaeoecol.*, 54, 63–86. *4.6.3, 4.8.*

Chafetz H.S., Rush P.F. & Utech N.M. (1991) Microenvironmental controls on mineralogy and habit of $CaCO_3$ precipitates: an example from an active travertine system. *Sedimentology*, 38, 107–126. *4.7.1.*

Chakraborty T. (1991) Sedimentology of a Proterozoic erg: the Venkatpur Sandstone, Pranhita – Godavari Valley, south India. *Sedimentology*, 38, 301–322. *5.5.3.*

Chakraborty C. (1993) Morphology, internal structure and mechanics of small longitudinal (seif) dunes in an aeolian horizon of the Proterozoic Dhandraul Quartzite, India. *Sedimentology*, 40, 79–85. *5.5.4.*

Chakraborty T. & Chaudhuri A.K. (1993) Fluvial–aeolian interactions in a Proterozoic alluvial plain: example from the Mancheral Quartzite, Sullavai Group, Pranhita–Godavari Valley, India. In: *The Dynamics and Environmental Context of Aeolian Sedimentary Systems* (Ed. by K. Pye), pp. 127–141. *Spec. Publ. geol. Soc. Lond.*, 72, *5.5.3.*

Chamley H. (1989) *Clay Sedimentology*, 623 pp. Springer-Verlag, Berlin. *10.4.1.*

Chan M.A. & Dott R.H. Jr (1983) Shelf and deep-sea sedimentation in Eocene forearc basin, western Oregon – fan or non-fan? *Bull. Am. Ass. petrol. Geol.*, 67, 2100–2116. *10.1, 10.3.4, 12.12.5.*

Chan M.A. & Kocurek G. (1988) Complexities in eolian and marine interactions: processes and eustatic controls on erg development. In: *Late Paleozoic and Mesozoic Eolian Deposits of the Western Interior of the United States* (Ed. by G. Kocurek), pp. 283–300. *Sedim. Geol.*, 56. *5.4.2, Fig. 5.16.*

Chandler M., Rind D. & Ruedy R. (1992) Pangaean climate during the Early Jurassic: GMC simulations and the sedimentary record of paleoclimate. *Bull. geol. Soc. Am.*, 104, 543–559. *5.5.1.*

Chapin M.A., Davies P., Gibson J.L. & Pettingill H.S. (1994) Reservoir architecture of turbidite sheet sandstones in laterally extensive outcrops, Ross Formation, western Ireland. In: *Submarine Fans and Turbidite Systems* (Ed. by P. Weimer, A.H.

Bouma and R.F. Perkins), pp. 53–68. Gulf Coast. Sect. Soc. econ. Paleont. Miner., Earth Enterprises, Austin, TX. *10.3.6.*

Chappell J. & Grindrod J. (1984) Chenier plain formation in northern Australia. In: *Coastal Geomorphology in Australia* (Ed. by B.G. Thom), pp. 197–231. Academic Press, Sydney. *6.7.2.*

Chappell J. & Thom B.G. (1986) Coastal morphodynamics in north Australia: review and prospect. *Aust. Geogr. Stud.*, **24**, 110–127. *6.7.2.*

Cheel R.J. (1991) Grain fabric in hummocky cross-stratified storm beds: genetic implications. *J. sedim. Petrol.*, **61**, 102–110. *7.7.2, Fig. 7.42.*

Cheel R.J. & Leckie D.A. (1993) Hummocky cross-stratification. In: *Sedimentology Review* 1 (Ed. by V.P. Wright), pp. 103–122. Blackwell Scientific Publications, Oxford. *6.7.7, 7.7.2, Fig. 7.40.*

Cheel R.J. & Rust B.R. (1982) Coarse-grained facies of glacio-marine deposits near Ottawa, Canada. In: *Research in Glacial, Glacio-fluvial and Glacio-lacustrine Systems* (Ed. by R. Davidson-Arnott, W. Nickling and B.D. Fahey), pp. 279–295. *6th Guelph Symposium on Geomorphology.* Geo Books, Norwich *11.4.6.*

Cheney R.E., Marsh J.G. & Beckley B.D. (1983) Global mesoscale variability from collinear tracks of SEASAT altimeter data. *J. geophys. Res.*, **88**, 4343–4354. *10.2.2.*

Chepil W.S. & Woodruff N.P. (1963) The physics of wind erosion and its controls. *Adv. Agron.*, **15**, 211–302. *5.1.*

Cherry S.T.J. (1993) The interaction of structure and sedimentary process controlling deposition of the Upper Jurassic Brae Formation Conglomerate, Block 16/17, North Sea. In: *Petroleum Geology of Northwest Europe: Proceedings of the 4th Conference* (Ed. by J.R. Parker), pp. 387–400. Geol. Soc. Lond., Bath. *10.3.4.*

Chevron Standard Ltd. Exploration Staff (1979) The geology, geophysics and significance of the Nisku reef discoveries, West Pembina area, Alberta, Canada. *Bull. Can. Petrol. Geol.*, **27**, 326–359. *9.3.4.*

Chien N. (1961) The braided stream of the Lower Yellow River. *Sci. Sin.*, **10**, 734–754. *3.3.3.*

Chiocci F.L. & Clifton H.E. (1991) Gravel-filled gutter casts in nearshore facies: indicators of ancient shoreline trend. In: *From Shoreline to Abyss* (Ed. by R.H. Osborne), pp. 67–76. *Spec. Publ. Soc. econ. Paleont. Miner.*, **46.** *7.7.2.*

Chivas A.R., Andrew A.S., Lyons W.B., Bird M.I. & Donnelly T.H. (1991) Isotopic constraints on the origin of salts in Australian playas. 1. Sulphur. *Palaeogeogr. Palaeoclimatol. Palaeoecol.*, **84**, 309–332. *4.5.*

Chorley R.J., Schumm S.A. & Sugden D.E. (1984) *Geomorphology*, 605 pp. Methuen, London. *2.1.2, Table 2.1.*

Chough S.K. & Sohn Y.K. (1990) Depositional mechanics and sequences of base surges, Songaksan tuff ring. *Sedimentology*, **37**, 1115–1136. *12.4.4, Figs 12.24, 12.50.*

Chough S.K., Choe M.Y. & Hwang I.G. (1994) Fan deltas in the Pohang Basin (Miocene) SE Korea. *3rd Int. Workshop on Fan Deltas, Field Excursion Guidebook*, 149 pp. Woongjin, Seoul. *10.3.5.*

Chough S.K., Hwang I.G. & Choe M.Y. (1990) The Miocene Doumsan fan-delta, southeast Korea: a composite fan-delta system in back-arc margin. *J. sedim. Petrol*, **60**, 445–455. *10.3.5.*

Chrintz T. & Clemmensen L.B. (1993) Draa reconstruction, the

Permian Yellow Sands, northeast England. In: *Aeolian Sediments Ancient and Modern* (Ed. by K. Pye and N. Lancaster), pp. 151–161. *Spec. Publ. int. Ass. Sediment.*, **16.** *5.5.4.*

Church M. (1972) Baffin Island sandurs: A study of Arctic fluvial processes. *Bull. Geol. Surv. Can.*, **216**, 208 pp. *3.3.2.*

Church M. (1983) Patterns of instability in a wandering gravel bed channel. In: *Modern and Ancient Fluvial Systems* (Ed. by J.D. Collinson and J. Lewin), pp. 169–180. *Spec. Publ. int. Ass. Sediment.*, **6.** *3.3.2.*

Churchill J.H., Levine E.R., Connors D.N. & Cornillon P.C. (1993) Mixing of shelf, slope and Gulf Stream water over the continental slope of the Middle Atlantic Bight. *Deep-Sea Res.*, **40**, 1063–1085. *7.5.*

Cita M.B., Camerlenghi A., Kastens K.A. & McCoy F.W. (1984) New findings of Bronze Age homogenites in the Ionian Sea: geodynamic implications for the Mediterranean. *Mar. Geol.*, **55**, 47–62. *10.2.3.*

Clark D.L., Whitman R.R., Morgan K.A. & Mackey S.D. (1980) *Stratigraphy and Glacial-marine Sediments of the Amerasian Basin, Central Arctic Ocean*, 57 pp. *Spec. Pap. geol. Soc. Am.*, **181.** *11.4.6.*

Clark J.D. & Pickering K.T. (1996) Architectural Elements and Growth Patterns of Submarine Channels: Application to Hydrocarbon Exploration. *Bull. Am. Ass. petrol. Geol.*, **80**, 194–221. *10.3.2.*

Clayton K.M., McCave I.N. & Vincent C.E. (1983) The establishment of a sand budget for the East Anglian coast and its implication for coastal stability. *Shoreline Protection*, pp. 91–96, 111–118. Telford, Proc. ICE Conference, Southampton. *7.3.4.*

Clemmensen L.B. & Blakey R.C. (1989) Erg deposits in the Lower Jurassic Wingate Sandstones, northeastern Arizona: oblique dune sedimentation. *Sedimentology*, **36**, 449–470. *5.5.2, 5.5.4, Fig. 5.29.*

Clemmensen L.B., Olsen H. & Blakey R.C. (1989) Erg-margin deposits in the Lower Jurassic Moenave Formation and Wingate Sandstone, southern Utah. *Bull. geol. Soc. Am.*, **101**, 759–773. *5.5.3.*

Clemmensen L.B., Oxnevad I.E.I. & de Boer P.L. (1994) Climatic controls on ancient desert sedimentation: some late Paleozoic and Mesozoic examples from NW Europe and the Western Interior of the USA. In: *Orbital Forcing and Cyclic Sequences* (Ed. P.L. de Boer and D. Smith), pp. 439–457. *Spec. Publ. int. Ass. Sediment.*, **19.** *5.1, 5.5.2.*

Clement J.H. (1985) Depositional sequences and characteristics of Ordovician Red River reservoirs, Pennel Field, Williston Basin, Montana. In: *Carbonate Petroleum Reservoirs* (Ed. by P.O. Roehl and R.W. Choquette), pp. 74–84. Springer-Verlag, New York. *8.9.*

Clifton H.E. (1973) Pebble segregation and bed lenticularity in wave-worked versus alluvial gravel. *Sedimentology*, **20**, 173–187. *6.5.4.*

Clifton H.E. (1976) Wave-formed sedimentary structures—a conceptual model. In: *Beach and Nearshore Sedimentation* (Ed. by R.A. Davis Jr and R.L. Ethington), pp. 126–148. *Spec. Publ. Soc. econ. Paleont. Miner.*, **24**, Tulsa. *6.2.4, 6.5.4.*

Clifton H.E. (1981) Progradational sequences in Miocene shoreline deposits, southeastern Caliente range, California. *J. sedim. Petrol.*, **51**, 165–184. *6.7.7, Fig. 6.96.*

Clifton H.E. (1982) Estuarine deposits. In: *Sandstone Depositional*

Environments (Ed. by P.A. Scholle and D. Spearing), pp. 179–189. Springer-Verlag, New York. *6.7.5*.

Clifton H.E., Hunter R.E. & Phillips R.L. (1971) Depositional structures and processes in the non-barred, high-energy nearshore. *J. sedim. Petrol.*, 41, 651–670. *6.7.6*, *Figs 6.85, 6.94*.

Cloetingh S. (1986) Intraplate stresses: a new mechanism for relative fluctuations of sea level. *Geology*, 14, 617–620. *2.4, 9.4.1*.

Cloetingh S. (1988a) Intraplate stresses: a new element in basin analysis. In: *New Perspectives in Basin Analysis* (Ed. by K.L. Kleinspehn and C. Paola), pp. 205–230. Springer-Verlag, New York. *2.4*.

Cloetingh S. (1988b) Intraplate stresses: tectonic cause of third-order cycles in apparent sea-level. In: *Sea-level Changes. An Integrated Approach* (Ed. by C.K. Wilgus, B.S. Hastings, H. Posamentier, J. Van Wagoner, C.A. Ross and C.G. St C. Kendall), pp. 19–29. *Spec. Publ. Soc. econ. Paleont. Miner.*, 42, Tulsa, *2.4, 9.2.2*.

Cloetingh S., McQueen H. & Lambeck K. (1985) On a tectonic mechanism for regional sea level variations. *Earth planet. Sci. Lett.*, 75, 157–166. *2.1.4, Table 2.2*.

Clough C.T.H., Maufe H.B. & Bailey E.B. (1909) The caudron subsidence of Glen Coe, and the associated igneous phenomena. *Q. J. geol. Soc. Lond.*, 65, 611–678. *12.1.1*.

Cloyd K.C., Demicco R.V. & Spencer R.J. (1990) Tidal channel, levee, and crevasse-splay deposits from a Cambrian tidal channel system: a new mechanism to produce shallowing upwards sequences. *J. sedim. Petrol.*, 60, 73–83. *9.4.1*.

Cohen A.S. (1989) The taphonomy of gastropod shell accumulations in large lakes: an example from Lake Tanganyika, Africa. *Paleobiology*, 15, 26–45. *4.7.1*.

Cohen A.S. (1990) Tectono-stratigraphic model for sedimentation in Lake Tanganyika, Africa. In: *Lacustrine Basin Exploration – Case Studies and Modern Analogs* (Ed. by B.J. Katz), pp. 137–150. *Mem. Am. Ass. petrol. Geol.*, 50, Tulsa. *4.6.1, 4.6.3, 4.7.1*.

Cohen A.S. & Thouin C. (1987) Nearshore carbonate deposits in Lake Tanganyika. *Geology*, 15, 414–418. *4.7.1, Fig. 4.8*.

Cole J.W. (1990) Structural control and origin of volcanism in the Taupo volcanic zone, New Zealand. *Bull. Volcanol.*, 52, 445–459. *12.12.1, Fig. 12.67*.

Colella A. (1988) Pliocene–Holocene fan deltas and braid deltas in the Crati Basin, southern Italy: a consequence of varying tectonic conditions. In: *Fan Deltas: Sedimentology and Tectonic Settings* (Ed. by W. Nemec and R.J. Steel), pp. 50–74. Blackie, London. *6.5, 6.5.2, 6.5.5, Figs 6.22, 6.23*.

Colella A. & Prior D.B. (Eds) (1990) *Coarse-Grained Deltas*, 357 pp. *Spec. Publ. int. Ass. Sedim.*, 10. *6.1, 10.3.3*.

Coleman A.P. (1926) *Ice Ages Recent and Ancient*, 296 pp. MacMillan, London. *11.1*.

Coleman J.M. (1966) Ecological changes in a massive freshwater clay sequence. *Trans. Gulf Coast. Ass. geol. Soc.*, 16, 159–174. *3.3.6*.

Coleman J.M. (1969) Brahmaputra River; Channel processes and sedimentation. *Sedim. Geol.*, 3, 129–239. *3.3.3, 3.3.6, Fig. 3.14*.

Coleman J.M. (1981) *Deltas: Processes and Models of Deposition for Exploration*, 124 pp. Burgess, CEPCO Division, Minneapolis. *6.6.1, 6.6.4, 6.6.5, Fig. 6.38*.

Coleman J.M. (1988) Dynamic changes and processes in the Mississippi River delta. *Bull. geol. Soc. Am.*, 100, 999–1015. *6.6.4*.

Coleman J.M. & Gagliano S.M. (1964) Cyclic sedimentation in the Mississippi river deltaic plain. *Trans. Gulf Coast Ass. geol. Socs.*, 14, 67–80. *6.6.1, 6.6.5, Fig. 6.31*.

Coleman J.M. & Gagliano S.M. (1965) Sedimentary structures: Mississippi River Deltaic plain. In: *Sedimentary Structures and Their Hydrodynamic Interpretation* (Ed. by G.V. Middleton), pp. 133–148. *Spec. Publ. Soc. econ. Paleont. Miner.*, 12, Tulsa. *Fig. 6.34*.

Coleman J.M. & Garrison L.E. (1977) Geological aspects of marine slope instability, northwestern Gulf of Mexico. *Mar. Geotech.*, 2, 9–44. *6.6.4*.

Coleman J.M. & Wright L.D. (1975) Modern river deltas: variability of processes and sand bodies. In: *Deltas, Models for Exploration* (Ed. by M.L. Broussard), pp. 99–149. Houston Geol. Soc., Houston, TX. *6.1, 6.6.5, 6.7.1, Fig. 6.44*.

Coleman J.M., Gagliano S.M. & Webb J.E. (1964) Minor sedimentary structures in a prograding distributary. *Mar. Geol.*, 1, 240–258. *6.6.1*.

Coleman J.M., Prior D.B. & Lindsay J.F. (1983) Deltaic influences on shelfedge instability processes. In: *The Shelfbreak: Critical Interface on Continental Margins* (Ed. by D.J. Stanley & G.T. Moore), pp. 121–137. *Spec. Publ. Soc. econ. Paleont. Miner.*, 33, Tulsa. *6.6.4, Fig. 6.38*.

Coleman J.M., Suhayda J.N., Whelan T. & Wright L.D. (1974) Mass movement of Mississippi river delta sediments. *Trans. Gulf Coast Ass. geol. Socs.*, 24, 49–68. *6.6.2, 6.6.4, Fig. 6.39*.

Coliacicchi R., Passeri L. & Pialli G. (1975) Evidence of tidal environment deposition in the Calcave Massiccio Formation (Central Apennines, Lower Lias). In: *Tidal Deposits* (Ed. R.N. Ginsburg), pp. 345–353. Springer-Verlag, Berlin. *9.4.1, Fig. 9.25*.

Collier R.E.L. (1990) Eustatic and tectonic controls upon Quaternary coastal sedimentation in the Corinth Basin, Greece. *J. geol. Soc. Lond.*, 301–314. *6.4*.

Collins B.D. & Dunne T. (1986) Erosion of tephra from the 1980 eruption of Mount St Helens. *Bull. geol. Soc. Am.*, 97, 896–905. *12.6.1*.

Collinson J.D. (1968) Deltaic sedimentation units in the Upper Carboniferous of northern England. *Sedimentology*, 10, 233–254. *3.5.2, 6.6.6*.

Collinson J.D. (1969) The sedimentology of the Grindslow Shales and the Kinderscout Grit: a deltaic complex in the Namurian of northern England. *J. sedim. Petrol.*, 39, 194–221. *2.2.2, 3.5.2, 6.6.6*.

Collinson J.D. (1970a) Bedforms of the Tana River, Norway. *Geogr. Ann.*, 52A, 31–56. *3.3.3, Figs 3.10, 3.11*.

Collinson J.D. (1970b) Deep channels, massive beds and turbidity current genesis in the Central Pennine Basin. *Proc. Yorks. geol. Soc.*, 37, 495–520. *6.6.6*.

Collinson J.D. (1971) Current vector dispersion in a river of fluctuating discharge. *Geol. Mijnbouw.*, 50, 671–678. *3.3.2*.

Collinson J.D. (1978) Vertical sequence and sand body shape in alluvial sequences. In: *Fluvial Sedimentology* (Ed. by A.D. Miall), pp. 577–586. *Mem. Can. Soc. petrol. Geol.*, 5, Calgary. *3.6.5*.

Collinson J.D. (1983) Sedimentology of unconformities within a fluvio-lacustrine sequence: Middle Proterozoic of eastern North Greenland. *Sedim. Geol.*, 34, 145–166. *4.10*.

Collinson J.D. (1988) Controls on Namurian sedimentation in the Central Province basins of northern England. In: *Sedimentation in a Synorogenic Basin Complex: the Upper Carboniferous of NW Europe* (Ed. by B.M. Besly and G. Kelling), pp. 85–101. Blackie, Glasgow. *6.6.6, 10.3.4.*

Collinson J.D. & Banks N.L. (1975) The Haslingden Flags (Namurian, G_1) of south-east Lancashire: bar finger sands in the Pennine Basin. *Proc. Yorks. geol. Soc.*, 40, 431–458. *6.6.6.*

Collinson J.D. & Thompson D.B. (1982) *Sedimentary Structures*, 194 pp. Allen & Unwin, London. *1.1, 1.2.*

Collinson J.D., Bevins R.E. & Clemmensen L.B. (1989) Post-glacial mass flow and associated deposits preserved in palaeovalleys: the Late Precambrian Moræneso Formation, North Greenland. *Medd. Grønland; Geosci.*, 21, 3–26. *3.5.1, 3.5.2, 3.6.10, 11.5.5.*

Collinson J.D., Holdsworth B.K., Jones C.M. & Martinsen O.J. (1992) Discussion of: 'Millstone Grit (Namurian) of the Southern Pennines viewed in the light of eustatically controlled sequence stratigraphy' by W.A. Read. *Geol. J.*, 27, 173–180. *6.6.6.*

Collinson J.D., Martinsen O.J., Bakken B. & Kloster A. (1991) Early fill of the Western Irish Namurian basin: A complex relationship between turbidites and deltas. *Basin Res.*, 3, 223–242. *6.6.6, Fig. 6.49, 10.3.6.*

Colmenero J.R., Agueda J.A., Fernández L.P., Salvador C.I., Bahamonde J.R. & Barba P. (1988) Fan-delta systems related to the Carboniferous evolution of the Canatabrian Zone, northwestern Spain. In: *Fan Deltas: Sedimentology and Tectonic Settings* (Ed. by W. Nemec and R.J. Steel), pp. 267–285. Blackie, London. *10.3.4.*

Colter V.S. & Reed G.E. (1980) Zechstein 2 Forden Evaporites of the Atwick No. 1 borehole, surrounding areas of N.E. England and the adjacent southern North Sea. In: *The Zechstein Basin with Emphasis on Carbonate Sequences* (Ed. by H. Fuchtbauer and T. Peryt), pp. 115–129. *Contrib. Sedimentol.*, 9. *8.9, 8.10.2, Fig. 8.51.*

Conaghan P.J. & Jones J.G. (1975) The Hawkesbury Sandstone and the Brahmaputra: A depositional model for continental sandstones. *J. geol. Soc. Aust.*, 22, 275–283. *3.5.2, 3.6.2.*

Congxian I., Gang C., Ming Y. & Ping W. (1991) The influences of suspended load on the sedimentation in the coastal zone and continental shelves of China. *Mar. Geol.*, 96, 341–352. *7.6.*

Coniglio M. & Dix G.R. (1992) Carbonate slopes. In: *Facies Models: Response to Sea-level Change* (Ed. by R.G. Walker and N.P. James), pp. 349–373. Geol. Ass. Can., Waterloo, Ontario. *9.4.4.*

Coniglio M. & James N.P. (1990) Origin of fine-grained carbonate and siliciclastic sediments in an Early Palaeozoic slope sequence, Cow Head Group, western Newfoundland. *Sedimentology*, 37, 215–230. *10.2.1, 10.2.3.*

Cook H.E. & Mullins H.T. (1983) Basin margin environment. In: *Carbonate Depositional Environments* (Ed. by P.A. Scholle, D.G. Bebout and C.H. Moore), pp. 540–617. *Mem. Am. Ass. petrol. Geol.*, 33, Tulsa. *9.4.4.*

Cook H.E. & Taylor M.E. (1991) Carbonate slope failures as indicators or sea-level lowerings (Abs.). *Bull. Ass. petrol. Geol.*, 75, 556. *9.5.1, 9.5.3.*

Cook P.J. & Mayo W. (1977) Sedimentology and Holocene history of a tropical estuary (Broad Sound, Queensland). *Bull. Aust. Bur. Min. Res. Geol. Geophys.*, 170, 206 pp. *6.7.5.*

Cooke R., Warren A. & Goudie A. (1993) *Desert Geomorphology*,

526 pp. University College London Press, London. *5.2.1, 5.2.2.*

Cooper A.K., Barrett P.J., Hinz K., Traube V., Leitchenkov G. & Stagg H.M.J. (1991) Cenozoic prograding sequences of the Antarctic continental margin: a record of glacio-eustatic and tectonic events. *Mar. Geol.*, 102, 175–213. *11.5.3, 11.5.4.*

Cooper J.A.G. (1993) Sedimentation in a river dominated estuary. *Sedimentology*, 40, 979–1017. *6.7.5.*

Cooper P. (1988) Ecological succession in Phanerozoic reef ecosystems: is it real? *Palaios*, 3, 136–151. *9.4.5.*

Coque R. (1962) *La Tunisie Presaharienne*, 476 pp. Armand Colin, Paris. *5.4.1.*

Corbett I. (1993) The modern and ancient pattern of sandflow through the southern Namib deflation basin. In: *Aeolian Sediments Ancient and Modern* (Ed. by K. Pye and N. Lancaster), pp. 45–60. *Spec. Publ. int. Ass. Sediment.*, 16. *5.4.1.*

Cotter, E. (1978) The evolution of fluvial style with special reference to the Central Appalachian Paleozoic. In: *Fluvial Sedimentology* (Ed. by A.D. Miall), pp. 361–383. *Mem. Can. Soc. petrol. Geol.*, 5, Calgary. *3.4.*

Cotter E. (1983) Early Silurian fluvial, coastal, and shelf environments and sea-level fluctuations in the origin of the Tuscarora Formation of Pennsylvania. *J. sedim. Petrol.*, 53, 25–49. *7.8.1.*

Cotter E. & Graham J.R. (1991) Coastal plain sedimentation in the late Devonian of southern Ireland; hummocky cross-stratification in fluvial deposits? *Sedim. Geol.*, 72, 201–224. *3.5.2.*

Cowan E.A. & Powell R.D. (1990) Suspended sediment transport and deposition of cyclically interlaminated sediment in a temperate glacial fjord, Alaska, USA. In: *Glacimarine Environments: Processes and Sediments* (Ed. by J.A. Dowdeswell and J.D. Scourse), pp. 75–89. *Spec. Pub. geol. Soc.*, 53. *11.4.6.*

Cowan E.J. (1991) Reservoir geometry of a fluvial sheet sandstone: an architectural reinterpretation of the Jurassic Westwater Canyon Member, Morrison Formation, U.S.A. In: *The Three-dimensional Facies Architecture of Terrigenous Clastic Sediments and its Implications for Hydrocarbon Discovery and Recovery* (Ed. by A.D. Miall and N. Tyler), pp. 80–93. Soc. econ. Paleont. Miner. *Concepts in Sedimentology and Paleontology*, 3, Tulsa. *3.6.2, 3.6.5, Fig. 3.56.*

Cox K.G., Bell J.D. & Pankhurst R.J. (1979) *The Interpretation of Igneous Rocks*, 450 pp. Allen & Unwin, London. *Fig. 12.5.*

Cox K.G. (1978) Flood basalts, subduction and the break-up of Gondwanaland. *Nature*, 274, 47–49. *12.2.2.*

Crabaugh M. & Kocurek G. (1993) Entrada Sandstone – example of a wet aeolian system. In: *The Dynamics and Environmental Context of Aeolian Sedimentary Systems* (Ed. by K. Pye), pp. 103–126. *Spec. Publ. geol. Soc. Lond.*, 72. *5.1, 5.5.2, 5.5.3, 5.5.4, Fig. 5.26.*

Crabaugh M. & Kocurek G. (1996) Continental sequence stratigraphy of a wet eolian system – a key to relative sea level change. In: *Relative Role of Eustasy, Climate, and Tectonism in Continental Rocks* (Ed. by K.W. Stanley and P.J. McCabe). *Spec. Publ. Soc. econ. Paleont. Miner.*, Tulsa, in press. *5.5.2.*

Crans W., Mandl G. & Haremboure J. (1980) On the theory of growth faulting: a geomechanical delta model based on gravity sliding. *J. petrol. Geol.*, 2, 265–307. *10.2.3.*

Creager J.S. & Sternberg R.W. (1972) Some specific problems in understanding bottom sediment distribution and dispersal on the continental shelf. *Shelf Sediment Transport* (Ed. by D.J.P. Swift, D.B. Duane and O.H. Pilkey), pp. 347–362. Dowden, Hutchinson & Ross, Stroudsburg, Penn. *7.1.2.*

Cremer M., Orsolini P. & Ravenne, C. (1985) Cap-Ferret Fan, Atlantic Ocean. In: *Submarine Fans and Related Turbidite Systems* (Ed. by A.H. Bouma, W.R. Normark and N.E. Barnes), pp. 113–120. Springer-Verlag, New York. *10.3.4.*

Crevello P.D. & Schlager W. (1980) Carbonate debris sheets and turbidites, Exuma Sound, Bahamas. *J. sedim. Petrol.*, 30, 1121–1148. *9.4.4.*

Crevello P.D., Wilson J.L., Sarg J.F. & Read J.F. (Eds) (1989) *Controls on Carbonate Platform and Basin Development*, 405 pp. *Spec. Publ. Soc. econ. Paleont. Miner.*, 44, Tulsa. *9.5.*

Croft A.R. (1962) Some sedimentation phenomena along the Wasatch Mountain Front. *J. geophys. Res.*, 67, 1511–1524. *3.6.3.*

Cronan D.S. (1980) *Underwater Minerals*, 362 pp. Academic Press, London. *10.4.1.*

Cross T.A. & Pilger R.A. (1982) Controls of subduction geometry, location of magmatic arcs and tectonics of arc and back-arc regions. *Bull geol. Soc. Am.*, 93, 545–562. *12.2.1.*

Crossley R. (1984) Controls of sedimentation in the Malawi rift valley, central Africa. *Sedim. Geol.*, 40, 33–50. *6.5.5.*

Crough S.T., Morgan J.W. & Hargraves R.B. (1980) Kimberlites: their relation to mantle hotspots. *Earth Planet. Sci. Lett.*, 50, 260–274. *12.2.2.*

Crowe B.M. & Fisher R.V. (1973) Sedimentary structures in base-surge deposits with special reference to cross-bedding, Ubehebe craters, Death Valley, California. *Bull. geol. Soc. Am.*, 84, 663–682. *12.4.4.*

Crowell J.C. (1957) Origin of pebbly mudstones. *Bull. geol. Soc. Am.*, 68, 993–1010. *11.1.*

Crowell J.C. (1974a) Sedimentation along the San Andreas Fault, California. In: *Modern and Ancient Geosynclinal Sedimentation* (Ed. by R.H. Dott Jr and R.H. Shaver), pp. 292–303. *Spec. Publ. Soc. econ. Paleont. Miner.*, 19, Tulsa. *4.10, 4.11.1.*

Crowell J.C. (1974b) Origin of late Cenozoic basins in southern California. In: *Tectonics and Sedimentation* (Ed. by W.R. Dickinson), pp. 190–204. *Spec. Publ. Soc. econ. Paleont. Miner.*, 22, Tulsa. *4.10, 4.11.1.*

Crowell J.C. (1978) Gondwanan glaciation, cyclothems, continental positioning, and climate change. *Am. J. Sci.*, 278, 1345–1372. *11.6.*

Crowell J.C. & Link M.H. (Eds) (1982) *Geologic History of Ridge Basin, southern California. Pacific Section Soc. econ. Paleont. Miner.*, Tulsa, 304 pp. *4.10, 4.11.1.*

Crowley K.D. (1983) Large-scale bed configurations (macroforms), Platte River Basin, Colorado and Nebraska: Primary structures and formative processes. *Bull. geol. Soc. Am.*, 94, 117–133. *3.6.2.*

Csanady G.T. (1972) The coastal boundary layer in Lake Ontario. *J. Phys. Oceanogr.*, 2, 41–53, 168–176. *4.4.*

Csanady G.T. (1978) Water circulation and dispersal mechanisms, In: *Lakes: Chemistry, Geology, Physics* (Ed. by A. Lerman), pp. 21–64, Springer-Verlag, Berlin. *4.4.*

Cucci M.A. & Clarke M.H. (1993) Sequence stratigraphy of a Miocene carbonate buildup, Java Sea. In: *Carbonate Sequence Stratigraphy* (Ed. by R.G. Loucks and J.F. Sarg), pp. 281–304. *Mem. Am. Ass. petrol. Geol.*, 57, Tulsa. *9.5.4.*

Cudzil M.R. & Driese S.G. (1987) Fluvial, tidal and storm sedimentation in the Chilhowee Group (Lower Cambrian), northeastern Tennessee, U.S.A. *Sedimentology*, 34, 861–883. *7.8.1.*

Curray J.R. (1964) Transgressions and regressions. In: *Papers in Marine Geology* (Ed. by R.L. Miller), pp. 175–203. Macmillan, New York. *6.1, 6.3, Fig. 6.16, 7.2.2.*

Curray J.R. (1965) Late Quaternary history, continental shelves of the United States. In: *The Quaternary of the United States* (Ed. by H.E. Wright and D.G. Frey), pp. 723–735. Princeton University Press, New Jersey. *7.2.2.*

Curray J.R., Emmel F.J. & Crampton P.J.S. (1969) Holocene history of a strand plain, lagoonal coast, Nayarit, Mexico. In: *Coastal Lagoons – a Symposium* (Ed. by A.A. Castanares and F.B. Phleger), pp. 63–100. Universidad Nacional Autónoma, Mexico. *6.6.2, 6.7.1, Fig. 6.68, Fig. 7.2.2.*

Curray J.R., Moore D.G. *et al.* (1982) *Initial Reports Deep Sea Drilling Project*, 64. US Government Printing Office, Washington, DC. *10.4.1.*

Curtis R., Evans G., Kinsman D.J.J. & Shearman D.J. (1963) Association of dolomite and anhydrite in the recent sediments of the Persian Gulf. *Nature*, 197, 6779–6800. *8.1, 8.4.2.*

Czapowski G. (1987) Sedimentary facies in the oldest rock salt (Na1) of the Leda Elevation (northern Poland). In: *The Zechstein Facies in Europe* (Ed. by T.M. Peryt), pp. 207–224. *Lecture Notes in Earth Sciences*, 10. Springer-Verlag, Berlin. *8.6.1, 8.9, 8.10.2.*

Dabrio C.J. (1990) Fan-delta facies associations in late Neogene and Quaternary basins of southeastern Spain. In: *Coarse-grained Deltas* (Ed. by A. Colella and D.B. Prior), pp. 91–111. *Spec. Publ. int. Ass. Sediment.*, 10. *6.5.2, 6.5.4, 10.3.3, 10.3.4.*

Dabrio C.J. & Polo M.D. (1985) Interpretación sedimentaria de las calizas de crinoides del Carixiense Subbético. *Mediterránea Ser. Geol.*, 4, 55–77. *9.4.2.*

Dabrio C.J. & Polo M.D. (1988) Late Neogene fan deltas and associated coral reefs in the Almanzora Basin, Almeria Province, southeastern Spain. In: *Fan Deltas: Sedimentology and Tectonic Settings* (Ed. by W. Nemec and R.J. Steel), pp. 354–367. Blackie, London. *6.5.4.*

Dabrio C.J., Esteban M. & Martin J.M. (1981) The coral reef of Nijar, Messinian (uppermost Miocene), Almeria Province, SE Spain. *J. sedim. Petrol.*, 51, 521–539. *9.4.5.*

Dabrio C.J., Zazo C. & Goy J.L. (Eds) (1991) *The Dynamics of Coarse-grained Deltas*, 405 pp. *Cuadernos de Geología Ibérica*, 15, 11–14. Madrid. *6.1.*

Dakyns J.R. & Greenly E. (1905) On the probable Peléan origin of the felsitic slates of Snowdon. *Geol. Mag.*, 2, 541–549. *12.1.1.*

D'Alessandro A., Ekdale A.A. & Sonnino M. (1986) Sedimentologic significance of turbidite ichnofacies in the Saraceno Formation (Eocene), southern Italy. *J. sedim. Petrol.*, 56, 294–306. *10.2.3.*

Dalrymple R.W. (1992) Tidal depositional systems. In: *Facies Models: Response to Sea-level Change* (Ed. by R.G. Walker and N.P. James), pp. 195–218. Geol. Ass. Can., Waterloo, Ontario *Fig. 7.8.*

Dalrymple R.W., Boyd R. & Zaitlin B.A. (1994) History of research, valley types and internal organisation of incised-valley systems: Introduction to the volume. In: *Incised Valley Systems: Origin and Sedimentary Sequences* (Ed. by R.W. Dalrymple, R. Boyd & B.A. Zaitlin), pp. 3–10. *Spec. Publ. Soc. Sedim. Geol.*, 51, Tulsa. *3.6.9, 6.7.6.*

Dalrymple R.W., Knight R.J., Zaitlin B.A. & Middleton G.V. (1990) Dynamics and facies model of a macrotidal sandbar complex, Cobequid Bay – Salmon River estuary (Bay of Fundy). *Sedimentology*, 37, 577–612. *6.7.5, 7.3.1*.

Dalrymple R.W., Makino Y. & Zaitlin B.A. (1991) Temporal and spatial patterns of rhythmite deposition on mudflats in the macrotidal, Cobequid Bay–Salmon River estuary, Bay of Fundy. In: *Clastic Tidal Sedimentology* (Ed. by D.G. Smith, G.E. Reinson, B.A. Zaitlin and R.A. Rahmani), pp. 137–160. *Mem. Can. Soc. petrol. Geol.*, 16, Calgary. *6.2.7, 6.7.3*.

Dalrymple R.W., Narbonne G.M. & Smith L. (1985) Eolian action and distribution of Cambrian shales in North America. *Geology*, 13, 607–610. *5.2.2*.

Dalrymple R.W., Zaitlin B.A. & Boyd R. (1992) Estuarine facies models: conceptual basis and stratigraphic implications. *J. sedim. Petrol.*, 62, 1130–1146. *6.1, 6.3, 6.6, 6.7.5, 6.7.6, Figs 6.17, 6.18, 6.79, 6.81, 6.82, 6.91, 6.92*.

Daly R.A. (1936) Origin of submarine 'canyons'. *Am. J. Sci.*, 31, 401–420. *1.1, 10.1*.

Dalziel I.W.D. & Dott R.H. Jr (1970) Geology of the Baraboo District, Wisconsin. *Wisconsin geol. natural History Surv., Inform. Circ.*, 14, 164 pp. *6.4*.

Dam G. & Surlyk F. (1992) Forced regressions in a large wave- and storm-dominated anoxic lake, Rhaetian–Sinemurian Kap Stewart Formation, East Greenland. *Geology*, 20, 749–752. *4.11.2*.

Dam G. & Surlyk F. (1993) Cyclic sedimentation in a large wave- and storm-dominated anoxic lake; Kap Stewart Formation (Rhaetian–Sinemurian), Jameson Land, East Greenland. In: *Sequence Stratigraphy and Facies Associations* (Ed. by H.W. Posamentier, C.P. Summerhayes, B.U. Haq and G.P. Allen), pp. 419–448. *Spec. Publ. int. Ass. Sediment.*, 18. *4.11.2, Fig. 4.20*.

Damnati B., Taieb M. & Williamson D. (1992) Laminated deposits from Lake Magadi (Kenya). Climatic contrast effect during the maximum wet period between 12 000–10 000 b.p. *Bull. Soc. geol. France.*, 163, 407–414. *Fig. 4.15*.

Damuth J.E. (1977) Late Quaternary sedimentation in the western equatorial Atlantic. *Bull. geol. Soc. Am.*, 88, 695–710. *12.12.5*.

Damuth J.E. & Hayes D.E. (1977) Echo character of the east Brazilian continental margin and its relationship to sedimentary processes. *Mar. Geol.*, 24, 73–95. *10.3.7*.

Damuth J.E., Flood R.D., Kowsmann R.O., Belderson R.H. & Gorini M.A. (1988) Anatomy and growth pattern of Amazon deep-sea fan revealed by long-range side-scan sonar (GLORIA) and high-resolution seismic studies. *Bull. Am. Ass. petrol. Geol.*, 72, 885–911. *10.3.3, Fig. 10.38*.

Darwin C. (1842) *Structure and Distribution of Coral Reefs*, 214 pp. Reprinted 1962 by University of California Press with forward by H.W. Menard. *12.11.2*.

Davidson-Arnott R.G.D. & Greenwood B. (1974) Bedforms and structures associated with bar topography in the shallow water wave environment, Kouchibouguac Bay, New Brunswick, Canada. *J. sedim. Petrol.*, 44, 698–704. *6.7.6, Fig. 6.85*.

Davidson-Arnott R.G.D. & Greenwood B. (1976) Facies relationships on a barred coast, Kouchibouguac Bay, New Brunswick, Canada. In: *Beach and Nearshore Sedimentation* (Ed. by R.A. Davis Jr and R.L. Ethington), pp. 149–168. *Spec. Publ.*

Soc. econ. Paleont. Miner., 24, Tulsa. *6.7.6*.

Davies G.R. (1970a) Algal laminated sediments, Gladstone embayment, Shark Bay, Western Australia. *Mem. Am. Ass. petrol. Geol.*, 13, 169–203. *8.4.2, 9.4.1*.

Davies G.R. (1970b) Carbonate bank sedimentation, eastern Shark Bay, Western Australia. *Mem. Am. Ass. petrol. Geol.*, 13, 85–168. *9.4.1*.

Davies G.R. (1977) Carbonate–anhydrite facies relation in Otto Fiord Formation (Mississippian–Pennsylvanian) Canadian Arctic Archipelago. *Bull. Am. Ass. petrol. Geol.*, 61, 1929–1949. *8.9, Figs 8.46, 8.47*.

Davies J.L. (1964) A morphogenic approach to world shorelines. *Z. Geomorph.*, 8 (Sp. No.), 127–142. *6.1*.

Davies P.J., Symonds P.A., Feary D.A. & Pigram C.J. (1989) The evolution of the carbonate platforms of Northeast Australia. In: *Controls on Carbonate Platform and Basin Development* (Ed. by P.D. Crevello, J.F. Sarg, J.F. Read and J.L. Wilson), pp. 233–258. *Spec. Publ. Soc. econ. Paleont. Miner.*, 44, Tulsa. *9.2, 9.4.5, 9.5.5, Fig. 9.69*.

Davies S.J. & Elliott T. (1996) Spectral gamma ray characterization of high resolution sequence stratigraphy: examples from Upper Carboniferous fluvio-deltaic systems, County Clare, Ireland. In: *High Resolution Sequence Stratigraphy: Innovations and Applications* (Ed. by J.A. Howell and J.F. Aitken), pp. 25–35. *Spec. Publ. geol. Soc. Lond.*, 104. *6.6.6, Fig. 6.55*.

Davis H.R. & Byers C.W. (1989) Shelf sandstones in the Mowry Shale: evidence for deposition during Cretaceous sea level falls. *J. sedim. Petrol.*, 59, 548–560. *7.8.4*.

Davies T.A. & Gorsline D.S. (1976) Oceanic sediments and sedimentary processes. In: *Chemical Oceanography*, 2nd edn (Ed. by J.P. Riley and R. Chester), 5, pp. 1–80. Academic Press, London. *Figs 10.6, 12.56*.

Davis R.A. Jr & Clifton H.E. (1987) Sea-level change and the preservation of wave-dominated and tide-dominated coastal sequences. In: *Sea-level Fluctuations and Coastal Evolution* (Ed. by D. Nummedal, D.H. Pilkey and J.D. Howard), pp. 167–178. *Spec. Publ. Soc. econ. Paleont. Miner.*, 41, Tulsa. *6.7.6, Fig. 6.83*.

Davis R.A. Jr. & Balson P.S. (1992) Stratigraphy of a North Sea tidal sand ridge. *J. sedim. Petrol.*, 62, 116–121. *7.3.2*.

Davis R.A. Jr & Hayes M.O. (1984) What is a wave-dominated coast? *Mar. Geol.*, 60, 313–330. *6.2.4*.

Dawson M.R. & Bryant I.E. (1987) Three-dimensional facies geometries in Pleistocene outwash sediments, Worcestershire, U.K. In: *Recent Developments in Fluvial Sedimentology* (Ed. by F.G. Ethridge, R.M. Flores and M.D. Harvey), pp. 191–196. *Spec. Publ. Soc. econ. Paleont. Miner.*, 39, Tulsa. *3.6.4*.

Dean W.E. (1978) Trace and minor elements in evaporites, In *Marine Evaporites*, (Ed. by W.E. Dean and B.C. Schreiber) pp. 86–104. SEPM Short Course No. 4. *8.9*.

Dean W.E. (1981) Carbonate minerals and organic matter in sediments of modern north temperate hardwater lakes. In: *Recent and Ancient Non-marine Environments: Models for Exploration* (Ed. by F.G. Ethridge and R.M. Flores), 213–231. *Spec. Publ. Soc. econ. Paleont. Miner.*, 31, Tulsa. *4.7.1*.

Dean W.E. & Anders D.E. (1991) Effects of source, depositional environment and diagenesis on characteristics of organic matter in oil shale from the Green River Formation, Wyoming, Utah, and

Colorado. *Bull. U.S. geol. Surv.*, **1973**, F1–F16. *4.14.1.*

Dean W.E. & Anderson R.Y. (1978) Salinity cycles: evidence for subaqueous deposition of Castile Formation and lower part of Salado Formation, Delaware Basin, Texas and New Mexico. In: *Geology and Mineral Deposits of Ochoan Rocks in Delaware Basin and Adjacent Areas* (Compiled by G.S. Austin), pp. 15–20. *New Mexico Bur. Mines Min. Res. Circ.*, **159**, *8.9.*

Dean W.E. & Gardner J.V. (1986) Milankovitch cycles in Neogene deep-sea sediments. *Paleoceanography*, **1**, 539–553. *10.4.2.*

Dean W.E., Davies G.R. & Anderson R.Y. (1975) Sedimentological significance of nodular and laminated anhydrite. *Geology*, **3**, 367–372. *8.9.*

Dean W.E., Gardner J.V., Jansa L.F., Cepek P. & Seibold E. (1978) Cyclic sedimentation along the continental margin of northwest Africa. In: *Initial Reports of the Deep Sea Drilling Project*, **41** (Ed. by Y. Lancelot, E. Seibold *et al.*), pp. 965–989 US Government Printing Office, Washington, DC. *10.2.1.*

Dean W.E., Leinen M. & Stow D.A.V. (1985) Classification of deep-sea, fine-grained sediments. *J. sedim. Petrol.*, **55**, 250–256. *10.2.1.*

De'Ath N.G. & Schuyleman S.F. (1981) The geology of the Magnus oilfield. In: *Petroleum Geology of the Continental Shelf of North-West Europe* (Ed. by L.V. Illing and G.D. Hobson), pp. 342–351. Heyden, London. *10.3.3.*

De Boer P.L. (1982a) Some remarks about the stable isotope composition of cyclic pelagic sediments from the Cretaceous in the Apennines. In: *Nature and Origin of Cretaceous Carbon-rich Facies* (Ed. by S.O. Schlanger and M.B. Cita), pp. 129–143. Academic Press, London. *10.2.1.*

De Boer P.L. (1982b) Cyclicity and the storage of organic matter in Middle Cretaceous pelagic sediments. In: *Cyclic and Event Stratification* (Ed. by G. Einsele and A. Seilacher), pp. 456–475. Springer-Verlag, Berlin. *10.2.1.*

De Boer P.L. (1983) *Aspects of Middle Cretaceous pelagic sedimentation in S. Europe.* 112 pp. *Geol. Ultraiectina*, **31**. *Fig. 2.3.*

De Boer P.L. (1991) Pelagic black shale–carbonate rhythms: orbital forcing and oceanographic response. In: *Cycles and Events in Stratigraphy* (Ed. by G. Einsele, W. Ricken and A. Seilacher), pp. 63–78. Springer-Verlag, Berlin. *10.4.1.*

De Boer P.L. & Smith D.G. (Eds) (1994a) *Orbital Forcing and Cyclic Sequences*, 559 pp. *Spec. Publ. int. Ass. Sediment.*, **19**. *2.1.5, 10.2.1.*

De Boer P.L. & Smith D.G. (1994b) Orbital forcing and cyclic sequences. In: *Orbital Forcing and Cyclic Sequences* (Ed. by P.L. de Boer and D.G. Smith), pp. 1–14. *Spec. Publ. int. Ass. Sediment.*, **19**. *2.1.5, Fig. 2.3.*

DeCelles P.G., Gray M.B., Ridgway K.D., Cole R.B., Pivnik D.A., Pequera N. & Srivastava P. (1991) Controls on synorogenic alluvial-fan architecture, Beartooth Conglomerate (Paleocene), Wyoming and Montana. *Sedimentology*, **38**, 567–590. *3.6.3, Fig. 3.47.*

Decima A. & Wezel F. (1973) Late Miocene evaporites of the Central Sicilian Basin, Italy. In: *Initial Reports of the Deep Sea Drilling Project*, **XIII** (Ed. by W.B.F. Ryan, K.J. Hsü *et al.*), pp. 1234–1240. US Government Printing Office, Washington, DC. *8.9.*

Decker R.W. (1987) Dynamics of Hawaiian Volcanoes: an overview. *Prof. Pap. U.S. geol. Surv.*, **1350**, 997–1018. *12.11.2.*

De Deckker P. (1988) Biological and sedimentary facies of Australian salt lakes. *Palaeogeogr. Palaeoclimatol. Palaeoecol.*, **62**, 237–270. *4.7.1.*

De Deckker P. & Last W.M. (1988) Modern dolomite deposition in continental, saline lakes, western Victoria, Australia. *Geology*, **16**, 29–32. *4.7.1.*

Defant M.J., DeBoer J.Z. & Oles D. (1988) The western central Luzon volcanic arc, the Philippines: two arcs divided by rifting? *Tectonophysics*, **145**, 305–317. *12.12.1.*

De Geer G. (1912) A geochronology of the last 12 000 years. *Proceedings of the 11th International Geological Congress, Stockholm 1910*, 241–253. *4.8.*

Degens E.T. & Kulbicki G. (1973) Hydrothermal origin of metals in some East African rift lakes. *Miner. Deposita*, **8**, 388–404. *4.5.*

Degens E.T., Von Herzen R.P., Wong H.K., Deuser W.G. & Jannasch H.W. (1973) Lake Kivu: structure, chemistry and biology of an East African rift lake. *Geol. Rundsch.*, **62**, 245–277. *4.3.*

De Groot K. (1973) Geochemistry of tidal flat brines at Umm Said, S.E. Qatar, Persian Gulf. In: *The Persian Gulf* (Ed. by B.H. Purser), pp. 377–394. Springer-Verlag, Berlin. *8.4.3.*

Delaney P.T. (1982) Rapid intrusion of magma into wet rock: groundwater flow due to pore pressure increases. *J. geophys. Res.*, **87**, 7739–7756. *12.4.1.*

Demaison G.J. & Moore G.T. (1980) Anoxic environments and oil source bed genesis. *Bull. Am. Ass. petrol. Geol.*, **64**, 1179–1209. *10.2.1.*

Demarest J.M. II & Kraft J.C. (1987) Stratigraphic record of Quaternary sea levels: implications for more ancient strata. In: *Sea-level Fluctuations and Coastal Evolution* (Ed. by D. Nummedal, D.H. Pilkey and J.D. Howard), pp. 223–239. *Spec. Publ. Soc. econ. Paleont. Miner.*, **41**, Tulsa. *6.7.6.*

Demirpolat S. (1991) Surface and near-surface sediments from the continental shelf off the Russian River, northern California. *Mar. Geol.*, **99**, 163–173. *7.6.*

De Mowbray T. (1983) The genesis of lateral accretion deposits in recent intertidal mudflat channels, Solway Firth, Scotland. *Sedimentology*, **30**, 425–435. *6.7.3.*

De Mowbray T. & Visser M.J. (1984) Reactivation surfaces in subtidal channel deposits, Oosterschelde, southwest Netherlands. *J. sedim. Petrol.*, **54**, 811–824. *6.2.7.*

Dengo G., Bohnenberger O. & Bonis S. (1970) Tectonism and volcanism along the Pacific marginal zone of central America. *Geol. Rundsch.*, **59**, 1215–1232. *12.12.2.*

Denis-Clocchiatti M. (1982) *Sédimentation carbonatée et paléoenvironment dans l'ocean au cénozoïque*, 92 pp. *Mém. Soc. géol. Fr.*, **143**. *10.2.1.*

Denlinger R.P. (1987) A model for generation of ash clouds by pyroclastic flows with application to the 1980 eruptions of Mount St Helens, Washington. *J. Geophys. Res.*, **92**, 10284–10298. *12.4.4.*

Denny C.S. (1967) Fans and pediments. *Am. J. Sci.*, **265**, 81–105. *3.3.8.*

De Raaf J.F.M. & Boersma J.R. (1971) Tidal deposits and their sedimentary structures. *Geol. Mijnbouw*, **50**, 479–503. *6.6.1, 7.1.2.*

De Raaf J.F.M., Boersma J.R. & Van Gelder A. (1977) Wave generated structures and sequences from a shallow marine succession. Lower Carboniferous, County Cork, Ireland. *Sedimentology*, **4**, 1–52. *7.1.2, 7.7.2, Fig. 7.38.*

De Raaf J.F.M., Reading H.G. & Walker R.G. (1965) Cyclic sedimentation in the Lower Westphalian of north Devon, England. *Sedimentology*, **4**, 1–52. *6.6.6.*

De Rosen-Spence A., Provost G., Dimroth E., Gochnauer K. & Owen V. (1980) Archean subaqueous felsic flows, Rouyn–Noranda, Quebec, Canada and their Quarternary equivalents. *Precambrian Res.*, **12**, 43–77. *12.4.5.*

Desborough G.A. (1978) A biogenic–chemical stratified lake model for the origin of oil shale of the Green River Formation: an alternative to the playa-lake model. *Bull. geol. Soc. Am.*, **89**, 961–971. *4.14.1.*

De Silva S.L. & Francis P.W. (1991) *Volcanoes of the Central Andes*, 216 pp. Springer-Verlag, Berlin. *12.12.2.*

Desloges J.R. & Church M. (1987) Channel and floodplain facies in a wandering gravel-bed river. In: *Recent Developments in Fluvial Sedimentology* (Ed. by G.G. Ethridge, R.M. Flores & M.D. Harvey), pp. 99–109. *Spec. Publ. Soc. econ. Paleont. Miner.*, **39**, Tulsa. *3.3.2.*

Dewey J. (1980) Episodicity, sequence and style at convergent plate boundaries. In: *The Continental Crust and its Mineral Deposits* (Ed. by D.W. Strangway), pp. 553–573. *Spec. Pap. geol. Ass. Can.*, **20**, Waterloo, Ontario. *12.12.2.*

Dewey J.F. & Bird J.M. (1970) Mountain belts and the new global tectonics. *J. geophys. Res.*, **75**, 2625–2647. *1.1.*

Deynoux M. (1982) Periglacial polygonal structures and sand wedges in the late Precambrian glacial formations of the Taoudeni Basin in Adrar of Mauretania (West Africa). *Palaeogeogr. Palaeoclimatol. Palaeoecol.*, **39**, 55–70. *11.5.5.*

Deynoux M. (1983) Late Precambrian and Upper Ordovician glaciations in the Taoudeni Basin, West Africa. An introduction to the field excursion of 'Till Mauretania 83' Symposium. In: *Till Mauretania 83* (Ed. by M. Deynoux), pp. 43–86. University of Poitiers, France. *11.5.5, Fig. 11.23.*

Deynoux M. (1985) Terrestrial or waterlain glacial diamictites? Three case studies from the Late Precambrian and Late Ordovician glacial drifts in West Africa. *Palaeogeogr. Palaeoclimatol. Palaeoecol.*, **51**, 97–141. *11.5.5, Figs 11.22, 11.23.*

Deynoux M. & Trompette R. (1981a) Late Precambrian tillites of the Taoudeni Basin, West Africa. In: *Earth's Pre-Pleistocene Glacial Record* (Ed. by M.J. Hambrey and W.B. Harland), pp. 123–131. Cambridge University Press, Cambridge. *11.5.5.*

Deynoux M. & Trompette R. (1981b) Late Ordovician tillites of the Taoudeni Basin, West Africa. In: *Earth's Pre-Pleistocene Glacial Record* (Ed. by M.J. Hambrey and W.B. Harland), pp. 89–96. Cambridge University Press, Cambridge. *11.5.5.*

Deynoux M., Kocurek G. & Proust J.N. (1989) Late Proterozoic periglacial aeolian deposits on the West African platform, Taoudeni Basin, western Mali. *Sedimentology*, **36**, 531–549. *5.5.3, 11.5.5.*

Dickinson W.R. (1968) Sedimentation of volcaniclastic strata of the Pliocene Koroimavua Group in northwest Viti Levu, Fiji. *Am. J. Sci.*, **266**, 440–453. *12.1.1.*

Dickinson W.R. (1971a) Plate tectonic models of geosynclines. *Earth Planet. Sci. Lett.*, **10**, 165–174. *1.1.*

Dickinson W.R. (1971b) Clastic sedimentary sequences deposited in shelf, slope and trough settings between magmatic arcs and associated trenches. *Pacific Geol.*, **3**, 15–20. *12.12.5.*

Dickinson W.R. (1973) Widths of modern arc–trench gaps proportional to past duration of igneous activity in associated magmatic arcs. *J. geophys. Res.*, **78**, 3376–3389. *12.12.5.*

Dickinson W.R. (Ed) (1974a) *Tectonics and Sedimentation*, 204 pp. *Spec. Publ. Soc. econ. Paleont. Miner.*, **22**, Tulsa. *1.1.*

Dickinson W.R. (1974b) Plate tectonics and sedimentation. In: *Tectonics and Sedimentation* (Ed. by W.R. Dickinson), pp. 1–27. *Spec. Publ. Soc. econ. Paleont. Miner.*, **22**. *12.1.1, 12.12.1, 12.12.4.*

Dickinson W.R. (1975) Potash-depth (k–h) relations in continental margins and intra-oceanic magmatic arcs. *Geology*, **3**, 53–56. *12.2.2, 12.12.2, 12.12.5.*

Dickinson W.R. (1995) Forearc basins. In: *Tectonics of Sedimentary Basins* (Ed. by C.J. Busby and R.V. Ingersoll), pp. 221–261 Blackwell Science Inc., Boston *12.12.5, Figs 12.66, 12.83, 12.84.*

Dickinson W.R. & Seely D.R. (1979) Structure and stratigraphy of forearc regions. *Bull. Am. Ass. petrol. Geol.*, **63**, 2–31. *12.12.5.*

Dickman M. (1985) Seasonal succession and microlamina formation in a meromictic lake displaying varved sediments. *Sedimentology*, **32**, 109–118. *4.8.*

Diemer J.A. & Belt E.S. (1991) Sedimentology and paleohydraulics of the meandering river systems of the Fort Union Formation, southeastern Montana. *Sediment. Geol.*, **75**, 85–108. *3.6.5.*

Diester-Haass L. (1991) Rhythmic carbonate content varations in Neogene sediments above the oceanic lysocline. In: *Cycles and Events in Stratigraphy* (Ed. by G. Einsele, W. Ricken and A. Seilacher), pp. 94–109. Springer-Verlag, Berlin. *10.2.1, 10.4.1, 10.4.2.*

Diller J.S. (1916) The volcanic history of Lassen Peak. *Science*, **43**, 727–733. *12.1.1.*

Dingle R.V. (1977) The anatomy of a large submarine slump on a sheared continental margin (SE Africa). *J. geol. Soc. Lond.*, **134**, 293–310. *10.2.3.*

Ditchfield P.W. & Marshall J.D. (1989) Isotopic variation in rhythmically bedded chalks: palaeotemperature variation in the Upper Cretaceous. *Geology*, **17**, 842–845. *10.4.3.*

Dodd J.R. & Stanton R.J. Jr (1981) *Paleoecology: Concepts and Applications*, 559 pp. Wiley-Intersciences, New York. *7.7.3.*

Dodd J.R., Zuppann C.W., Harris C.D., Leonard K.W. & Brown T.W. (1993) Petrologic method for distinguishing eolian and marine grainstones, Ste. Genevieve Limestone (Mississippian) of Indiana. In: *Mississippian Oolites and Modern Analogs* (Ed. by B.D. Keith and C.W. Zuppann), pp. 49–59. *Am. Ass. petrol. Geol.*, *Studies in Geology*, **35**. *9.4.2.*

Dohrenwend J.C., Wells S.G. & Turrin B.D. (1986) Degradation of Quaternary cinder cones in the Cima volcanic field, Mojave Desert, California. *Bull. geol. Soc. Am.*, **97**, 421–427. *12.9.1.*

Domack E.W. (1983) Facies of the Late Pleistocene glacial-marine sediments on Whidbey Island, Washington: An isostatic glaciomarine sequence. In: *Glacial-Marine Sedimentation* (Ed. by B.F. Molnia), pp. 535–570. Plenum Press, New York. *11.5.3.*

Domack E.W. (1988) Biogenic facies in the Antarctic glacimarine environment: Basis for a polar glacimarine summary. *Palaeogeogr. Palaeoclimatol. Palaeoecol.* **63**, 357–372. *11.4.6.*

Domack E.W. & Ishman S. (1993) Oceanographic and physiographic controls on modern sedimentation within Antarctic fjords. *Bull. geol. Soc. Am.*, **105**, 1175–1189. *11.4.6.*

Dominguez J.M.L. & Barbosa L.M. (1994) Controls on quaternary evolution of the São Francisco strandplain: roles of sea-level history, trade winds and climate. *Field trip guide 14th Int. Sedim. Congr.*, 39 pp. *6.6, 6.6.2, 6.7.1.*

Dominguez J.M.L. & Wanless H.R. (1991) Facies architecture of a falling sea-level strandplain, Doce River coast, Brazil. In: *Shelf Sand and Sandstone Bodies: Geometry, Facies and Sequence Stratigraphy* (Ed. by D.J.P. Swift, G.F. Oertel, R.W. Tillman and J.A. Thorne), pp. 259–281. *Spec. Publ. int. Ass. Sediment.*, **14**. *6.6, 6.7, 6.7.1, Fig. 6.11.*

Dominguez J.M.L., Bittencourt A.C.S.P. & Martin L. (1992) Controls on Quaternary coastal evolution of the east-northeastern coast of Brazil: roles of sea-level history, trade winds and climate. *Sedim. Geol.*, **80**, 213–232. *Figs 6.69, 6.70.*

Dominguez J.M.L., Martin L. & Bittencourt A.C.S.P. (1987) Sea-level history and Quaternary evolution of river mouth-associated beach-ridge plains along the east-southeast Brazilian coast: a summary. In: *Sea-level Fluctuations and Coastal Evolution* (Ed. by D. Nummedal, D.H. Pilkey and J.D. Howard), pp. 115–127. *Spec. Publ. Soc. econ. Paleont. Miner.*, **41**. *6.7.1.*

Dominguez L.L., Mullins H.T. & Hine A.C. (1988) Cat Island platform, Bahamas: an incipiently drowned Holocene carbonate shelf. *Sedimentology*, **35**, 805–819. *9.3.5.*

Donaldson A.C., Martin R.H. & Kanes W.H. (1970) Holocene Guadalupe delta of Texas Gulf Coast. In: *Deltaic Sedimentation Modern and Ancient* (Ed. by J.P. Morgan and R.H. Shaver), pp. 107–137. *Spec. Publ. Soc. econ. Paleont. Miner.*, **15**, Tulsa. *6.6.1, 6.7.4.*

Donovan D.T. & Jones E.J.W. (1979) Causes of world-wide changes in sea level. *J. geol. Soc.*, **136**, 187–192. *2.1.4, Table 2.2.*

Donovan R.N. (1975) Devonian lacustrine limestones at the margin of the Orcadian Basin, Scotland. *J. geol. Soc.*, **131**, 489–510. *4.9.1, Fig. 4.27.*

Donovan R.N. (1980) Lacustrine cycles, fish ecology and stratigraphic zonation in the Middle Devonian of Caithness. *Scott. J. Geol.*, **16**, 35–50. *4.13.1.*

Dott R.H. Jr (1961) Squantum 'tillite', Massachusetts – evidence of ancient glaciation or subaqueous mass movements? *Bull. geol. Soc. Am.*, **72**, 1289–1306. *11.1.*

Dott R.H. Jr (1974) Cambrian tropical storm waves in Wisconsin. *Geology*, **2**, 243–246. *6.4.*

Dott R.H. Jr & Bourgeois J. (1982) Hummocky stratification: significance of its variable bedding sequences. *Bull. geol. Soc. Am.*, **93**, 663–680. *7.7.2, 9.4.2.*

Dott R.H. Jr & Byers C.W. (1981) SEPM research conference on modern shelf and ancient cratonic sedimentation – the orthoquartzite-carbonate suite revisited. *J. sedim. Petrol.*, **51**, 329–347. *7.8.1.*

Dott R.H. Jr & Shaver R.H. (Eds) (1974) *Modern and Ancient Geosynclinal Sedimentation*, 380 pp. *Spec. Publ. Soc. econ. Paleont. Miner.*, **19**, Tulsa. *1.1.*

Dott R.H. Jr., Byers C.W., Fielder G.W., Stenzel S.R. & Winfree K.E. (1986) Aeolian to marine transition in Cambro-Ordovician cratonic sheet sandstones of the northern Mississippi Valley, USA. *Sedimentology*, **33**, 345–367. *5.5.3, Fig. 5.28, 7.8.1.*

Dowdeswell J.A. & Sharp M.J. (1986) Characterization of pebble fabrics in modern terrestrial glacigenic sediments. *Sedimentology*, **33**, 699–710. *11.4.1.*

Dowdeswell J.A., Hambrey M.J. & Wu R. (1985) A comparison of clast fabric and shape in Late Precambrian and modern glacigenic sediments. *J. Sedim. Pet.*, **55**, 691–704. *11.4.1, Fig. 11.6.*

Doyle E.J. & Roberts H.H. (Eds) (1988) *Carbonate–Clastic Transitions*, 304 pp. *Developments in Sedimentology*, **42**. Elsevier, Amsterdam. *9.5.5.*

Drake D.E., Hatcher P.G. & Keller G.H. (1978) Suspended particulate matter and mud deposition in Upper Hudson submarine canyon. In: *Sedimentation in Submarine Canyons, Fans, and Trenches* (Ed. by D.J. Stanley and G. Kelling), pp. 33–41. Dowden, Hutchinson & Ross, Stroudsburg, PA. *10.2.1.*

Dravis J.J. (1977) Holocene sedimentary depositional environments of Eleuthera Bank, Bahamas, 288 pp. MS thesis, University of Miami. *9.4.2.*

Dreimanis A. (1989) Tills: their genetic terminology and classification. In: *Genetic Classification of Glacigenic Deposits* (Ed. by R.P. Goldthwait and C.L. Matsch), pp. 17–83. Balkema, Rotterdam. *11.1, 11.4.1.*

Dreimanis A. & Lundqvist J. (1984) What should be called till? In: *Ten Years of Nordic Till Research* (Ed. by L.-K. Konigsson), pp. 5–10. *Striae*, **20**. *11.1.*

Drewry D.J. Ed. (1983) *Antarctica: Glaciological and Geophysical Folio*. Scott Polar Research Institute, University of Cambridge, Cambridge. *Fig. 11.3.*

Drewry D.J. (1986) *Glacial Geological Processes*, 276 pp. Edward Arnold, London *11.3.1, 11.3.3, 11.3.6, 11.4.6, Figs 11.1, 11.13.*

Dreyer T. (1993) Quantified fluvial architecture in ephemeral stream deposits of the Esplugafreda Formation (Palaeocene), Tremp–Graus Basin, northern Spain. In: *Alluvial Sedimentation* (Ed. by M. Marzo and C. Puigdefabregas), pp. 337–362. *Spec. Publ. int. Ass. Sediment.*, **17**. *3.6.5.*

Driese S.G., Byers C.W. & Dott R.H. Jr (1981) Tidal deposition in the basal Upper Cambrian Mt Simon Formation in Wisconsin. *J. sedim. Petrol.*, **51**, 367–381. *7.8.1.*

Driese S.G., Fischer M.W., Easthouse K.A., Marks G.T., Gogola A.R. & Schoner A.E. (1991) Model for genesis of shoreface and shelf sandstone sequences, southern Appalachians: palaeoenvironmental reconstruction of an Early Silurian shelf system. In: *Shelf Sand and Sandstone Bodies: Geometry, Facies and Sequence Stratigraphy* (Ed. by D.J.P. Swift, G.F. Oertel, R.W. Tillman and J.A. Thorne), pp. 309–339. *Spec. Publ. int. Asso. Sediment.*, **14**. *7.8.1.*

Droste H. (1990) Depositional cycles and source rock development in an epeiric intraplatform basin: the Hanifa Formation of the Arabian Peninsula. *Sedim. Geol.*, **69**, 281–296. *9.3.1, 9.3.4, 9.5.2.*

Droxler A.W. & Schlager W. (1985) Glacial versus interglacial sedimentation rates and turbidite frequency in the Bahamas. *Geology*, **13**, 799–802. *9.4.4, 9.5.1.*

Droxler A.W., Schlager W. & Whallon C.C. (1983) Quaternary aragonite cycles and oxygen-isotope record in Bahamian carbonate ooze. *Geology*, **11**, 235–239. *9.4.4.*

Droz L. & Bellaiche G. (1991) Seismic facies and geologic evolution of the central portion of the Indus fan. In: *Seismic Facies and Sedimentary Processes of Submarine Fans and Turbidite Systems*

(Ed. by P. Weimer and M.H. Link), pp. 383–402. Springer-Verlag, New York. *10.3.3.*

Druitt T.H. (1992) Emplacement of the 18 May 1980 lateral blast deposit ENE of Mount St. Helens, Washington. *Bull. Volcanol.*, **54**, 554–572. *12.4.4.*

Druitt T.H. (1995) Settling behaviour of concentrated dispersions and some volcanological applications. *J. Volcanol. geotherm. Res.*, **65**, 27–39. *12.4.4.*

Druitt T.H. & Francaviglia V. (1992) Caldera formation on Santorini and the physiography of the islands in the late Bronze age. *Bull. Volcanol.*, **54**, 484–493. *12.4.6, Fig. 12.1.*

Druitt T.H. & Sparks R.S.J. (1982) A proximal ignimbrite breccia facies on Santorini, Greece. *J. Volcanol. geotherm. Res.*, **13**, 147–171. *12.4.4, 12.4.6.*

Druitt T.H., Mellors R.A., Pyle D.M. & Sparks R.S.J. (1989) Explosive volcanism on Santorini, Greece. *Geol. Mag.*, **126**, 95–126. *12.6, 12.12.3, Fig. 12.1.*

Drummond C.N. & Wilkinson B.H. (1993) Carbonate cycle stacking patterns and hierarchies of orbitally forced eustatic sea-level change. *J. sedim. Petrol.*, **63**, 369–377. *9.4.1.*

Duan T., Gao Z., Zeng Y. & Stow D.A.V. (1993) A fossil carbonate contourite drift on the Lower Ordovician palaeocontinental margin of the middle Yangtze Terrane, Jiuxi, northern Hunan, southern China. In: *Contourites and Bottom Currents* (Ed. by D.A.V. Stow and J.-C. Faugères), pp. 271–284. *Sedim. Geol.*, **82**. *10.3.7.*

Duane D.B., Field M.E., Meisburger E.P., Swift D.J.P. & Williams S.J. (1972) Linear shoals on the Atlantic inner continental shelf, Florida to Long Island. In: *Shelf Sediment Transport: Process and Pattern* (Ed. by D.J.P. Swift, D.B. Duane and O.H. Pilkey), pp. 447–499. Dowden, Hutchinson & Ross, Stroudsburg. *7.4.2.*

Duchaufour P. (1982) *Pedology*, 449 pp. George Allen & Unwin, London. *3.2.3, 3.3.7, 3.5.4, Figs 3.6, 3.25.*

Duff P.McL.D. (Ed.) (1992) *Holmes' Principles of Physical Geology*, 791 pp. Chapman & Hall, London. *Fig. 2.1.*

Duff P.McL.D., Hallam A. & Walton E.K. (1967) *Cyclic Sedimentation*, 280 pp. *Developments in Sedimentology* 10. Elsevier, Amsterdam. *6.6.6.*

Duffield W.A., Stieltjes L. & Varet J. (1982) Huge landslide blocks in the growth of Piton de la Fournaise, La Reunion and Kilauea Volcano, Hawaii. *J. Volcanol. geotherm. Res*, **12**, 147–160. *12.5.1, 12.11.2.*

Duke W.L. (1984) Paleohydraulic analysis of hummocky cross-stratified sands indicates equivalence with wave-formed flat bed: Pleistocene Lake Bonneville deposits, northern Utah. *Bull. Am. Ass. petrol. Geol.*, **68**, 472–473. *4.11.2.*

Duke W.L. (1985) Hummocky cross-stratification, tropical hurricanes, and intense winter storms. *Sedimentology*, **32**, 167–194. *7.1.2, 7.7.5.*

Duke W.L. (1990) Geostrophic circulation or shallow marine turbidity currents? The dilemma of paleoflow patterns in storm-influenced prograding shoreline systems. *J. sedim. Petrol.*, **60**, 870–883. *6.7.7, 7.7.2, Fig. 7.43.*

Duke W.L., Arnott R.W. & Cheel R.J. (1991) Shelf sandstones and hummocky cross stratification; new insights on a stormy debate. *Geology* **19**, 625–628. *7.7.2, Fig. 7.42.*

Duke W.L., Fawcett P.J. & Brusse W.C. (1991) Prograding shoreline deposits in the Lower Silurian Medina Group, Ontario and New York: storm- and tide-influenced sedimentation in a shallow epicontinental sea, and the origin of enigmatic shore-normal channels encapsulated by open shallow marine deposits. In: *Shelf Sand and Sandstone Bodies: Geometry, Facies and Sequence Stratigraphy* (Ed. by D.J.P. Swift, G.F. Oertel, R.W. Tillman and J.A. Thorne), pp. 339–375. *Spec. Publ. int. Ass. Sediment.*, **14**. *6.7.7.*

Dulhunty J.A. (1982) Holocene sedimentary environments in Lake Eyre, South Australia. *J. geol. Soc. Aust.*, **29**, 437–442. *4.6.1, 4.9.*

Dunbar R.B., Anderson J.B., Domack E.W. & Jacobs S.S. (1985) Oceanographic influences on sedimentation along the Antarctic continental shelf. In: *Oceanology of the Antarctic Continental Shelf* (Ed. by S.S. Jacobs), pp. 291–312. *Antarctic Res. Ser.*, **43**. *11.4.6.*

Dunham R.J. (1962) Classification of carbonate rocks according to depositional texture. In: *Classification of Carbonate Rocks* (Ed. by W.E. Hamm), pp. 108–121. *Mem. Am. Ass. petrol. Geol.*, **1**, Tulsa. *9.1.1.*

Dunne L.A. & Hempton M.R. (1984) Deltaic sedimentation in the Lake Hazar pull-apart basin, southeastern Turkey. *Sedimentology*, **31**, 401–412. *6.5.5.*

Dvorak J.J. & Gasparini P. (1991) History of earthquakes and vertical ground movements in Campi Flegrei caldera, southern Italy: comparison of precursory events at the A.D. 1538 eruption of Monte Nuovo and of activity since 1968. *J. Volcanol. geotherm. Res.*, **48**, 199–222. *12.4.6.*

Dyni J.R. (1974) Stratigraphy and nahcolite resources of the saline facies of the Green River Formation in northwest Colorado. In: *Energy Resources of the Piceance Creek Basin, Colorado* (Ed. by D.K. Murray), pp. 111–112. Rocky Mountain Association of Geologists, Denver. *4.14.1.*

Dzulynski S., Ksiazkiewicz M. & Kuenen P.H. (1959) Turbidites in flysch of the Polish Carpathian Mountains. *Bull. geol. Soc. Am.*, **70**, 1089–1118. *10.1.*

Ebanks W.J. & Bubb J.N. (1975) Holocene carbonate sedimentation Matecumbe Keys tidal bank, South Florida. *J. sedim. Petrol.*, **45**, 422–439. *9.4.2.*

Eberli G.P. (1987) Carbonate turbidite sequences deposited in rift-basins of the Jurassic Tethys Ocean (eastern Alps, Switzerland). *Sedimentology*, **34**, 363–388. *9.4.4.*

Eberli G.P. & Ginsburg R.N. (1987) Segmentation and coalescence of Cenozoic carbonate platforms in northwestern Great Bahama Bank. *Geology*, **15**, 75–79. *9.3.2.*

Eberli G.P. & Ginsburg R.N. (1989) Cenozoic progradation of northwestern Great Bahama Bank, a record of lateral platform growth and sea-level fluctuation. In: *Controls on Carbonate Platform and Basin Development* (Ed. by P.D. Crevello, J.F. Sarg, J.F. Read and J.L. Wilson), pp. 339–351. *Spec. Publ. Soc. econ. Paleont. Miner.*, **44**, Tulsa. *9.1.1, 9.3.1, 9.3.2, 9.5.1, 9.5.4, Figs 9.10, 9.57.*

Edwards D.A., Leeder M.R., Best J.L. & Pantin H.M. (1994) On experimental reflected density currents and the interpretation of certain turbidites. *Sedimentology*, **41**, 437–461. *10.2.3.*

Edwards M.B. (1975) Glacial retreat sedimentation in the Smalfjord Formation, Late Precambrian, North Norway. *Sedimentology*, **22**, 75–94. *11.5.5, Fig. 11.24.*

References

Edwards M.B. (1976) Growth faults in Upper Triassic deltaic sediments, Svalbard. *Bull. Am. Assoc. petrol. Geol.*, 60, 341–355. *6.6.6, Fig. 6.66.*

Edwards M.B. (1979) Late Precambrian glacial loessites from North Norway and Svalbard. *J. sedim. Petrol.*, 49, 85–92. *11.5.5.*

Edwards M.B. (1984) Sedimentology of the Upper Proterozoic glacial record, Vestertana Group, Finmark, North Norway, 76 pp. *Nor. geol. Unders. Bull.*, 394. *11.5.5, Fig. 11.26.*

Edwards M.B. (1986) Glacial environments. In: *Sedimentary Environments and Facies*, 2nd Edn, (Ed. by H.G. Reading), pp. 445–470. Blackwell Scientific Publications, Oxford. *11.5.5.*

Edwards M.B. & Føyn S. (1981) Late Precambrian tillites in Finnmark, North Norway. In: *Earth's Pre-Pleistocene Glacial Record* (Ed. by M.J. Hambrey and W.B. Harland), pp. 606–610. Cambridge University Press, Cambridge. *11.5.5.*

Edwards M.B., Eriksson K.A. & Kier R.S. (1983) Paleochannel geometry and flow patterns determined from exhumed Permian point bars in North-Central Texas. *J. sedim. Petrol.*, 53, 1261–1270. *3.6.5.*

Eggert D.L. (1984) The Leslie Cemetery and Fransisco distributary fluvial channels in the Petersburg Formation (Pennsylvanian) of Gibson County, Indiana, U.S.A. In: *Sedimentology of Coal and Coal-bearing Sequences* (Ed. by R.A. Rahmani and R.M. Flores), pp. 309–315. *Spec. Publ. int. Ass. Sediment.*, 7. *3.6.8.*

Ehlers J. (1981) Some aspects of glacial erosion and deposition in north Germany. *Ann. Glaciol.*, 2, 143–146. *11.4.1.*

Ehrlich R.N., Barrett S.F. & Guo B.J. (1990) Seismic and geologic characteristics of drowning events on carbonate platforms. *Bull. Am. Ass. petrol. Geol.*, 74, 1523–1537. *9.3.5, 9.5.4.*

Ehrlich R.N., Longo A.P. & Hyare S. (1993) Response of carbonate platform margins to drowning: evidence of environmental collapse. In: *Carbonate Sequence Stratigraphy: Recent Developments and Applications* (Ed. by R.G. Loucks and J.F. Sarg), pp. 241–266. *Mem. Am. Ass. petrol. Geol.*, 57, Tulsa. *9.5.1, 9.5.3, 9.5.4.*

Eicher D.L. & Diner R. (1991) Environmental factors controlling cretaceous limestone–marlstone rhythms. In: *Cycles and Events in Stratigraphy* (Ed. by G. Einsele, W. Ricken and A. Seilacher), pp. 79–93. Springer-Verlag, Berlin. *10.4.1.*

Einsele G. (1991) Submarine mass flow deposits and turbidites. In: *Cycles and Events in Stratigraphy* (Ed. by G. Einsele, W. Ricken and A. Seilacher), pp. 313–339. Springer-Verlag, Berlin. *10.2.3.*

Einsele G. & Ricken W. (1991) Limestone–marl alternation – an overview. In: *Cycles and Events in Stratigraphy* (Ed. by G. Einsele, W. Ricken and A. Seilacher), pp. 23–47. Springer-Verlag, Berlin. *9.4.3, 10.2.1, 10.4.1, 10.4.2.*

Einsele G., Ricken W. & Seilacher A. (Eds) (1991) *Cycles and Events in Stratigraphy*, 955 pp. Springer-Verlag, Berlin. *7.7.3, 10.2.1.*

Eittreim S. & Ewing M. (1972) Suspended particulate matter in the deep waters of the North American Basin. In: *Studies in Physical Oceanography*, vol. 2 (Ed. by A.L. Gordon), pp. 123–168. Gordon & Breach, New York. *Fig. 10.14.*

Ekdale A.A. & Bromley R.G. (1984) Comparative ichnology of shelf-sea and deep-sea chalk. *J. Paleont.*, 58, 322–332. *10.4.3.*

Ekdale A.A., Bromley R.G. & Pemberton S.G. (1984) Ichnology – the use of trace fossils in sedimentology and stratigraphy. *Soc. econ. Paleont. Miner. Short Course*, 15, 88–96. *7.7.2, 7.7.3, Fig. 7.41.*

Ekes C. (1993) Bedload-transported pedogenic mud and aggregates in the Lower Old Red Sandstone of southwest Wales. *J. geol. Soc. Lond.*, 150, 469–471. *3.2.2, 3.5.3.*

El-Ashry M.T. & Wanless H.R. (1965) Birth and early growth of a tidal delta. *J. Geol.*, 73, 404–406. *6.7.4.*

Elliott T. (1974a) Abandonment facies of high-constructive lobate deltas, with an example from the Yoredale Series. *Proc. Geol. Ass.*, 85, 359–365. *6.6.6.*

Elliott T. (1974b) Interdistributary bay sequences and their genesis. *Sedimentology*, 21, 611–622. *6.6.5, Fig. 6.43.*

Elliott T. (1975) The sedimentary history of a delta lobe from a Yoredale (Carboniferous) cyclothem. *Proc. Yorks. geol. Soc.*, 40, 505–536. *6.6.6.*

Elliott T. (1976a) Upper Carboniferous sedimentary cycles produced by river-dominated, elongate deltas. *J. geol. Soc.*, 132, 199–208. *6.6.6.*

Elliott T. (1976b) The morphology, magnitude and regime of a Carboniferous fluvial-distributary channel. *J. sedim. Petrol.*, 46, 70–76. *6.6.6.*

Elliott T. (1976c) Sedimentary sequences from the upper Limestone Group of Northumberland. *Scott. J. Geol.*, 12, 115–124. *6.6.6.*

Elliott T.L. (1982) Carbonate facies, depositional cycles, and the development of secondary porosity during burial diagenesis, Mission Canyon Formation, Haas Field, North Dakota. In: *Proc. 4th Int. Williston Basin Symp., Regina* (Ed. by J.E. Christopher and J. Kaldi), pp. 131–151. *8.7.*

Ellis J.P. & Milliman J.D. (1986) Calcium carbonate suspended in Arabian Gulf and Red Sea waters: biogenic and detrital, not 'chemogenic'. *J. sedim. Petrol.*, 55, 805–808. *9.3.4.*

Elmore R.D., Pilkey O.H., Cleary W.J. & Curran H.A. (1979) Black Shell turbidite, Hatteras Abyssal Plain, western Atlantic Ocean. *Bull. geol. Soc. Am.*, 90, 1165–1176. *10.3.6.*

Elrick M. & Read J.F. (1991) Cyclic ramp-to-basin carbonate deposits, Lower Mississippian, Wyoming and Montana: a combined field and computer modelling study. *J. sedim. Petrol.*, 61, 1194–1224. *9.4.3, 9.5.2.*

Elrick M., Read J.F. & Coruh C. (1991) Short term paleoclimatic fluctuations expressed in Lower Mississippian ramp-slope deposits, southwestern Montana. *Geology*, 19, 799–802. *9.4.3.*

Elson J.A. (1989) Comment on glacitectonite, deformation till, and comminution till. In: *Genetic Classification of Glacigenic Deposits* (Ed. by R.P. Goldthwait and C.L. Matsch), pp. 85–88. Balkema, Rotterdam. *11.4.1.*

Elverhøi A., Lønne O. & Seland R. (1983) Glaciomarine sedimentation in a modern fjord environment, Spitsbergen. *Polar Res.*, 1, 127–149. *11.4.6.*

Embleton C. & King C.A.M. (1975) *Glacial Geomorphology*, 583 pp. Edward Arnold, London. *Fig. 11.4.*

Embley R.W., Hooje P.J., Lonsdale P., Mayer L. & Tucholke B.E. (1980) Furrowed mud-waves on the western Bermuda rise. *Bull. geol. Soc. Am.*, 91, 731–740. *10.3.7.*

Emery K.O. (1952) Continental shelf sediments off southern California. *Bull. geol. Soc. Am.*, 63, 1105–1108. *7.1.2.*

Emery K.O. (1968) Positions of empty pelecypod valves on the continental shelf. *J. sedim. Petrol.*, 38, 1264–1269. *7.1.2.*

Endo K., Sumita M., Machida M. & Furuichi M. (1989) The 1984 collapse and debris avalanche deposits of Ontake volcano, central

Japan. In: *Volcanic Hazards* (Ed. by J.H. Latter), pp. 210–229. *IAVCEI Proc. Volcanol.*, **1**. *12.5.1.*

Enos P. (1977) Holocene sediment accumulation of the South Florida shelf margin. In: *Quaternary Sedimentation in South Florida* (Ed. by P. Enos and R.D. Perkins), pp. 1–130. *Mem. geol. Soc. Am.*, **147**, Boulder, CO. *9.3.3, 9.4.1, Fig. 9.12.*

Enos P. & Perkins R.D. (1977) Quaternary sedimentation in South Florida. *Mem. geol. Soc. Am.*, **147**, 1–130. *9.1.1.*

Enos P. & Perkins R.D. (1979) Evolution of Florida Bay from island stratigraphy. *Bull. geol. Soc. Am.*, **90**, 59–83. *9.4.5.*

Epting M. (1980) Sedimentology of Miocene carbonate buildups, Central Luconia, offshore Sarawak. 1. *Bull. geol. Soc. Malaysia*, **12**, 17–30. *9.5.5, 9.5.6, Fig. 9.72.*

Ericson D.B., Ewing M. & Heezen B.C. (1951) Deep-sea sands and submarine canyons. *Bull. geol. Soc. Am.*, **62**, 961–965. *10.1.*

Eschard R., Ravenne C., Houel P. & Knox R. (1991) Three-dimensional reservoir architecture of a valley-fill sequence and a deltaic aggradation sequence: Influences of minor relative sea-level variations (Scalby Formation, England). In: *The Three-dimensional Facies Architecture of Terrigenous Clastic Sediments and its Implications for Hydrocarbon Discovery and Recovery* (Ed. by A.D. Miall and N. Tyler), pp. 133–147. *Soc. econ. Paleont. Miner.*, Concepts in Sedimentology and Paleontology, **3**, Tulsa. *3.6.5.*

Eschner T.B. & Kocurek G. (1986) Marine destruction of eolian sand seas: origin of mass flows. *J. sedim. Petrol.*, **56**, 401–411. *5.5.3.*

Eschner T.B. & Kocurek G. (1988) Origins of relief along contacts between eolian sandstones and overlying marine strata. *Bull. Am. Ass. petrol. Geol.*, **72**, 932–943. *5.5.3.*

Ethridge F.G. & Wescott W.A. (1984) Tectonic setting, recognition and hydrocarbon reservoir potential of fan-delta deposits. In: *Sedimentology of Gravels and Conglomerates* (Ed. by E.H. Koster and R.J. Steel), pp. 217–235. *Mem. Can. Soc. petrol. Geol*, **10**, Calgary. *6.1, 6.3, 6.5, 6.5.4.*

Ethridge F.G., Jackson T.J. & Youngbergh A.D. (1981) Floodbasin sequence of a fine-grained meander belt subsystem; The coal-bearing Lower Wasatch and Upper Fort Union Formations, southern Powder River Basin, Wyoming. In: *Recent and Ancient Nonmarine Depositional Environments* (Ed. by F.G. Ethridge and R.M. Flores), pp. 191–209. *Spec. Publ. Soc. econ. Paleont. Miner.*, **31**, Tulsa. *3.6.8.*

Ethridge F.G., Tyler N. & Burns L.K. (1984) Sedimentology of a Precambrian quartz-pebble conglomerate, southwest Colorado. In: *Sedimentology of Gravels and Conglomerates* (Ed. by E.H. Koster and R.J. Steel), pp. 165–174. *Mem. Can. Soc. petrol. Geol.*, **10**, Calgary. *3.6.3.*

Eugster H.P. (1969) Inorganic bedded cherts from the Magadi area, Kenya. *Contrib. Miner. Petrol.*, **22**, 1–31. *4.7.4, 4.14.1.*

Eugster H.P. (1970) Chemistry and origin of brines of Lake Magadi, Kenya. In: *Mineralogy and Geochemistry of Non-Marine Evaporites* (Ed. by B.A. Morgan), pp. 215–235. *Spec. Pap. miner. Soc. Am.*, **3**. *4.5. 4.14.1.*

Eugster H.P. (1980) Lake Magadi, Kenya, and its precursors. In: *Hypersaline Brine and Evaporitic Environments* (Ed. by A. Nissenbaum), 195–232. Elsevier, Amsterdam. *4.7.4.*

Eugster H.P. & Hardie L.A. (1978) Saline lakes. In: *Lakes:*

Chemistry, Geology, Physics (Ed. by A. Lerman), pp. 237–293. Springer-Verlag, Berlin. *4.7.4, Fig. 4.12.*

Eugster H.P. & Kelts K. (1983) Lacustrine chemical sediments. In: *Chemical Sediments and Geomorphology* (Ed. by A.S. Goudie and K. Pye), pp. 321–368. Academic Press, London. *4.7.1.*

Eugster H.P. & Surdam R.C. (1973) Depositional environment of the Green River Formation of Wyoming: A preliminary report. *Bull. geol. Soc. Am.*, **84**, 1115–1120. *4.14.1.*

Evamy B.D., Haremboure J., Kamerling P., Knaap W.A., Molloy F.A. & Rowlands P.H. (1978) Hydrocarbon habitat of Tertiary Niger delta. *Bull. Am. Ass. petrol. Geol.*, **62**, 1–39. *6.6.4, Fig. 6.42.*

Evans G. (1965) Intertidal flat sediments and their environments of deposition in the Wash. *Q. J. geol. Soc. Lond.*, **121**, 209–245. *6.7.3.*

Evans G., Schmidt V., Bush P. & Nelson H. (1969) Stratigraphy and geologic history of the sabkha, Abu Dhabi, Persian Gulf. *Sedimentology*, **12**, 145–159. *9.4.1.*

Evans G., Murray J.W., Biggs H.E.J., Bate R. & Bush P.R. (1973) The oceanography, ecology, sedimentology and geomorphology of parts of the Trucial Coast barrier island complex, Persian Gulf. In: *The Persian Gulf* (Ed. by B.H. Purser), pp. 233–277, Springer-Verlag, Berlin. *9.3.4.*

Evenari M., Gutterman Y. & Gavish E. (1985) Botanical studies on coastal salinas and sabkhas of the Sinai. In: *Hypersaline Ecosystems, The Gavish Sabkha* (Ed. by G.M. Friedman and W.E. Krumbein), pp. 145–182. Springer-Verlag, Berlin. *8.2.2.*

Ewart A. (1979) A review of the mineralogy and chemistry of Tertiary–Recent dacitic, latitic, rhyolitic and realted salic volcanic rocks. In: *Trondhjemites, Dacites and Related Rocks* (Ed. by F. Baker), pp. 113–121. Elsevier, Amsterdam. *12.3.1.*

Eybergen F.A. (1987) Glacier snout dynamics and contemporary push moraine formation at the Turtmannglacier, Wallis, Switzerland. In: *Tills and Glaciotectonics* (Ed. by J.J.M. Van der Meer), pp. 217–231. Balkema, Rotterdam. *11.4.3, Fig. 11.9.*

Eyles C.H. (1987) Glacially influenced submarine-channel sedimentation in the Yakataga Formation, Middleton Island. *J. sedim. Petrol.*, **57**, 1004–1017. *10.3.5, 11.4.5, 11.5.5, Fig. 11.29.*

Eyles C.H. (1988a) Glacially- and tidally-influenced shallow marine sedimentation of the late Precambrian Port Askaig Formation, Scotland. *Palaeogeogr. Palaeoclimatol. Palaeoecol.*, **68**, 1–25. *7.8.1.*

Eyles C.H. (1988b) A model for striated boulder pavement formation on glaciated, shallow-marine shelves: An example from the Yakataga Formation, Alaska. *J. sedim. Petrol.*, **58**, 62–71. *11.5.5.*

Eyles C.H. & Eyles N. (1989) The Late Cenozoic White River 'tillites' of southern Alaska: subaerial slope and fan-delta deposits in a strike-slip setting. *Bull. geol. Soc. Am.*, **101**, 1091–1102. *12.6.2.*

Eyles C.H. & Lagoe M.B. (1990) Sedimentation patterns and facies geometries on a temperate glacially-influenced continental shelf: the Yakataga Formation, Middleton Island, Alaska. In: *Glacimarine Environments: Processes and Sediments* (Ed. by J.A. Dowdeswell and J.D. Scourse), pp. 363–386. *Spec. Pub. geol. Soc.*, **53**. *11.5.5.*

Eyles C.H., Eyles N. & Lagoe M.B. (1991) The Yakataga Formation: A late Miocene to Pleistocene record of temperate glacial marine sedimentation in the Gulf of Alaska. In: *Glacial Marine*

Sedimentation; Paleoclimatic Significance (Ed. by J.B. Anderson and G.M. Ashley), pp. 159–180. *Spec. Pap. geol. Soc. Am.*, **261**. *11.5.5.*

Eyles N. (1979) Facies of supraglacial sedimentation on Icelandic and Alpine temperate glaciers. *Can. J. Earth Sci.*, **16**, 1341–1361. *11.4.2.*

Eyles N. (1993) Earth's glacial record and its tectonic setting. *Earth-Sci. Rev.*, **35**, 1–248. *11.5.5.*

Eyles N. & Clark B.M. (1986) Significance of hummocky and swaley cross-stratification in late Pleistocene lacustrine sediments of the Ontario basin. *Geology*, **14**, 679–682. *4.11.2.*

Eyles N. & Clark B.M. (1988) Storm-influenced deltas and ice scouring in a late Pleistocene glacial lake. *Bull. geol. Soc. Am.*, **100**, 793–809. *11.4.5.*

Eyles N., Clark B.M. & Clague J.J. (1987) Coarse-grained gravity flow facies in a large supraglacial lake. *Sedimentology*, **34**, 193–216. *4.6.3.*

Eyles N., Eyles C.H. & Miall A.D. (1983) Lithofacies types and vertical profile models; an alternative approach to the description and environmental interpretation of glacial diamict and diamictite sequences. *Sedimentology*, **30**, 393–410. *11.1, 11.5.1.*

Eyles N., Sladen J.A. & Gilroy S. (1982) A depositional model for stratigraphic complexes and facies superimposition in lodgement tills. *Boreas*, **11**, 317–333. *11.4.1.*

Eynon G. (1981) Basin development and sedimentation in the Middle Jurassic of the northern North Sea. In: *Petroleum Geology of the Continental Shelf of North-west Europe* (Ed. by L.V. Illing and G.D. Hobson), pp. 196–204. Heyden, London. *6.6.6, Fig. 6.58.*

Fabricius F.H. (1977) *Origin of Marine Ooids and Grapestones*, 130 pp. *Contrib. Sedimentol.*, **7**. *9.1.3.*

Fagerstrom J.A. (1987) *The Evolution of Reef Communities*, 600 pp. Wiley-Interscience, New York. *9.4.5.*

Fairchild I.J. (1993) Balmy shores and icy wastes; the paradox of carbonates associated with glacial deposits in Neo proterozoic times. In: *Sedimentology Review 1* (Ed. by V.P. Wright), pp. 1–16. Blackwell Scientific Publications, Oxford. *11.5.5.*

Fairchild I.J. & Hambrey M.J. (1984) The Vendian succession of northwestern Spitsbergen: Petrogenesis of a dolomite–tillite association: *Precambrian Res.*, **26**, 111–167. *11.5.5.*

Falt L.M. & Steel R.J. (1990) A new palaeogeographic reconstruction for the Middle Jurassic of the northern North Sea: a discussion. *J. geol. Soc. Lond.*, **147**, 1085–1090. *6.6.6.*

Farrell K.M. (1987) Sedimentology and facies architecture of overbank deposits of the Mississippi River, False River Region. Louisiana. In: *Recent Developments in Fluvial Sedimentology* (Ed. by F.G. Ethridge, R.M. Flores and M.D. Harvey), pp. 113–120. *Spec. Publ. Soc. econ. Paleont. Miner.*, **39**, Tulsa. *3.3.6.*

Faugères J.-C. & Stow D.A.V. (1993) Bottom-current-controlled sedimentation: a synthesis of the contourite problem. *Sedim. Geol.*, **82**, 287–297. *10.1, 10.2.2, 10.3.7.*

Faugères J.-C., Mézerais M.L. & Stow D.A.V. (1993) Contourite drift types and their distribution in the North and South Atlantic Ocean basins. In: *Contourites and Bottom Currents* (Ed. by D.A.V. Stow and J.-C. Faugères), pp. 189–203. *Sedim. Geol.*, **82**. *10.3.7, 10.4.1, Fig. 10.78.*

Faulkner T.J. (1988) The Shipway Limestone of Gower: sedimentation on a storm-dominated early Carboniferous ramp.

Geol. J., **23**, 85–100. *9.3.4, 9.4.3, 9.5.2.*

Fedo C.M. & Cooper J.D. (1990) Braided fluvial to marine transition: The basal Lower Cambrian Wood Canyon Formation, Southern Marble Mountains, Mojave Desert, California. *J. sedim. Petrol.*, **60**, 220–234. *7.8.1.*

Feldhausen P.H., Stanley D.J., Knight R.J. & Maldonado A. (1981) Homogenization of gravity-emplaced muds and unifites: models from the Hellenic Trench. In: *Sedimentary Basins of Mediterranean Margins* (Ed. by F.C. Wezel), pp. 203–226. CNR Italian Project of Oceanography, Tecnoprint, Bologna. *10.2.3.*

Fenster M.S., Fitzgerald D.M., Bohlen W.F., Lewis R.S. & Baldwin C.T. (1990) Stability of giant sand waves in eastern Long Island Sound, USA. *Mar. Geol.*, **91**, 207–225. *7.3.*

Ferentinos G., Papatheodorou G. & Collins M.B. (1988) Submarine transport processes on an active submarine fault escarpment: Gulf of Corinth, Greece. *Mar. Geol.*, **83**, 43–61. *10.3.3.*

Fernández J., Bluck B.J. & Viseras C. (1993) The effects of fluctuating base level on the structure of alluvial fan and associated fan delta deposits: an example from the Tertiary of the Betic Cordillera, Spain. *Sedimentology*, **40**, 879–893. *6.5.6.*

Fernández J., Soria J. & Viseras C. (1996) Stratigraphic architecture of the Neogene basins in the central sector of the Betic Cordillera (Spain); tectonic control and base-level changes. In: *Tertiary Basins of Spain: the Stratigraphic Record of Crustal Kinematics* (Ed. by P.F. Friend and C.J. Dabrio), pp. 353–365. Cambridge University Press, Cambridge. *6.5.6.*

Feth J.H. (1964) Review and annotated bibliography of ancient lake deposits (Precambrian to Pleistocene) in the Western States. *Bull. U.S. geol. Surv.*, **1080**, 119 pp. *4.10.1.*

Field M.E. (1980) Sand bodies on coastal plain shelves: Holocene record of the U.S. Atlantic shelf off Maryland. *J. sedim. Petrol.*, **50**, 505–528. *Fig. 7.20.*

Field M.E., Gardner J.V., Jennings A.E. & Edwards B.D. (1982) Earthquake induced sediment failures on a 0.25° slope, Klamath River delta, California. *Geology*, **10**, 542–546. *6.6.4.*

Fielding C.R. (1984a) Upper delta plain lacustrine and fluviolacustrine facies from the Westphalian of the Durham coalfield, NE England. *Sedimentology*, **31**, 547–567. *3.5.3, 3.6.8.*

Fielding C.R. (1984b) A coal depositional model for the Durham Coal Measures of NE England. *J. geol. Soc. Lond.*, **141**, 919–932. *3.5.3, 3.6.8.*

Fielding C.R. (1986) Fluvial channel and overbank deposits from the Westphalian of the Durham coalfield, NE England. *Sedimentology*, **33**, 119–140. *3.5.3, 3.6.8.*

Fielding C.R. & Crane R.C. (1987) An application of statistical modelling to the prediction of hydrocarbon recovery factors in fluvial reservoir sequences. In: *Recent Developments in Fluvial Sedimentology* (Ed. by F.G. Ethridge, R.M. Flores and M.D. Harvey), pp. 321–327. *Spec. Publ. Soc. econ. Paleont. Miner.*, **39**, Tulsa. *3.6.5.*

Fierstein J. & Hildreth W. (1992) The plinian eruptions of 1912 at Novarupta, Katmai National Park, Alaska. *Bull. Volcanol.*, **54**, 646–684. *12.4.2, 12.4.3, Figs 12.1, 12.21.*

Figueiredo A.G.J., Sanders J.E. & Swift D.J.P. (1982) Storm graded layers on inner continental shelves: examples from south Brazil and the Atlantic coast of the central United States. *Sedim. Geol.*, **31**, 171–190. *7.4.3.*

Fink J.H. & Manley C.R. (1987) Origin of pumiceous and glassy textures in rhyolite flows and domes. In: *The Emplacement of Silicic Domes and Lava Flows* (Ed. by J.H. Fink), pp. 77–88. *Spec. Pap. geol. Soc. Am.*, 212, Boulder, CO. *12.4.5.*

Fink J.H. & Manley C.R. (1989) Explosive volcanic activity generated from within advancing silicic lava flows. In: *Volcanic Hazards* (Ed. by J.H. Latter), pp. 169–179. *IAVCEI Proc. Volcanol.*, 1. *12.4.5.*

Fink J.H., Anderson S.W. & Manley C.R. (1992) Textural constraints on effusive silicic volcanism: beyond the permeable foam model. *J. geophys. Res*, 97, 9073–9083. *12.4.*

Finley R.J. (1978) Ebb-tidal delta morphology and sediment supply in relation to seasonal wave energy flux, North Inlet, South Carolina. *J. sedim. Petrol.*, 48, 227–238. *6.7.4.*

Fischer A.G. (1961) Stratigraphic record of transgressing seas in the light of sedimentation on Atlantic coast of New Jersey. *Bull. Am. Ass. petrol. Geol.*, 45, 1656–1666. *6.7.6, Fig. 6.86.*

Fischer A.G. (1981) Climatic oscillations in the biosphere. In: *Biotic Crises in Ecological and Evolutionary Time* (Ed. by M. Nitecki), pp. 103–131. Academic Press, New York. *10.2.1.*

Fischer A.G. (1991) Orbital cyclicity in Mesozoic strata. In: *Cycles and Events in Stratigraphy* (Ed. by G. Einsele, W. Ricken and A. Seilacher), pp. 48–62. Springer-Verlag, Berlin. *10.4.1.*

Fischer A.G. & Arthur M. (1977) Secular variations in the pelagic realm. In: *Deep-water Carbonate Environments* (Ed. by H.E. Cook and P. Enos), pp. 19–50. *Spec. Publ. Soc. econ. Paleont. Miner.*, 25, Tulsa. *10.2.1.*

Fischer A.G. & Roberts L.T. (1991) Cyclicity in the Green River Formation (lacustrine, Eocene) of Wyoming. *J. sedim. Petrol.*, 61, 1146–1154. *4.16.1.*

Fischer A.G., de Boer P.L. & Premoli Silva I. (1990) Cyclostratigraphy. In: *Cretaceous Resources, Events and Rhythms,* (Ed. by R.N. Ginsburg and B. Beaudoin), pp. 139–172. Kluwer Academic, Dordrecht. *10.2.1.*

Fisher R.V. (1961) Proposed classification of volcaniclastic sediments and rocks. *Bull. geol. Soc. Am.*, 72, 1395–1408. *12.1, 12.7, Fig. 12.9.*

Fisher R.V. (1966a) Mechanism of deposition from pyroclastic flows. *Am. J. Sci.*, 264, 350–363. *12.1.1, 12.4.4.*

Fisher R.V. (1966b) Textural comparisons of the John Day volcanic siltstone with loess and volcanic ash. *J. sedim. Petrol.*, 36, 706–718. *12.4.3, 12.4.4.*

Fisher R.V. (1968) Puu Hou littoral cones, Hawaii. *Geol. Rundsch.*, 57, 837–864. *12.4.1, 12.9.3.*

Fisher R.V. (1979) Models for pyroclastic surges and pyroclastic flows. *J. Volcanol. geotherm. Res.*, 6, 305–318. *12.4.2, Fig. 12.28.*

Fisher R.V. (1983) Flow transformations in sediment gravity flows. *Geology*, 11, 273–274. *Fig. 12.22, Fig. 12.28.*

Fisher R.V. (1984) Submarine volcaniclastic rocks. In: *Marginal Basin Geology: Volcanic and Associated Sedimentary and Tectonic Processes in Modern and Ancient Marginal Basins* (Ed. by B.P. Kokelaar and M.F. Howells), pp. 5–27. *Spec. Publ. geol. Soc. Lond.*, 16. *12.12.5.*

Fisher R.V. (1995) Decoupling of pyroclastic currents: hazard assessments. *J. Volcanol. geotherm. Res.*, *12.4.4, Fig. 12.23.*

Fisher R.V. & Heiken G. (1982) Mt. Pelée, Martinique: May 8 and 20, 1902, pyroclastic flows and surges. *J. Volcanol. geotherm. Res.*, 13, 339–371. *12.4.4, Figs 12.1, 12.23, 12.26.*

Fisher R.V. & Schmincke H.U. (1984) *Pyroclastic Rocks*, 472 pp. Springer-Verlag, Berlin. *12.1, 12.6, 12.7, Fig. 12.38.*

Fisher R.V. & Waters A.C. (1970) Base surge bedforms in maar volcanoes. *Am. J. Sci.*, 268, 157–180. *12.1.1, 12.4.4.*

Fisher R.V., Glicken H.X. & Hoblitt R.H. (1987) May 18, 1980 Mount St. Helens deposits in South Coldwater Creek, Washington. *J. geophys. Res.*, 92, 102647–10283. *12.4.4.*

Fisher W.L. (1969) Facies characterisation of Gulf Coast Basin delta systems with some Holocene analogues. *Trans. Gulf Coast. Ass. geol. Soc.*, 19, 239–261. *6.6.6.*

Fisher W.L. & McGowen J.H. (1969) Depositional systems in the Wilcox Group (Eocene) of Texas and their relationship to occurrence of oil and gas. *Bull. Am. Ass. petrol. Geol.*, 53, 30–54. *6.6.5.*

Fisher W.L., Brown L.F., Scott A.J. & McGowen J.H. (1969) *Delta Systems in the Exploration for Oil and Gas.* Bur. econ. Geol., Univ. Texas, Austin, 78 pp. *6.1, 6.3, Fig. 6.34.*

Fisk H.N. (1944) *Geological Investigation of the Alluvial Valley of the Lower Mississippi River*, 78 pp. Mississippi River Commission, Vicksburg. *3.3.4, 3.6.9, 6.1, 6.7.6.*

Fisk H.N. (1947) *Fine Grained Alluvial Deposits and their Effects on Mississippi River Activity*, 82 pp. Mississippi River Commission Vicksburg. *3.3.4, Fig. 3.21, 6.1.*

Fisk H.N. (1952) Mississippi River Valley geology: relation to river regime. *Trans. Am. Soc. Civ. Eng.*, 117, 667–682. *3.3.4.*

Fisk H.N. (1955) Sand facies of recent Mississippi delta deposits. *Wld. Petrol. Cong.*, Rome, 377–398. *6.1.*

Fisk H.N. (1959) Padre Island and the Laguna Madre flats, coastal south Texas. *National Academy of Science–National Research Council, Second Coastal Geography Conference*, pp. 103–151. *6.7.4.*

Fisk H.N. (1961) Bar finger sands of the Mississippi delta. In: *Geometry of Sandstone Bodies – a Symposium* (Ed. by J.A. Peterson and J.C. Osmond), pp. 29–52. Am. Ass. petrol. Geol., Tulsa. *6.1, 6.6.4.*

Fisk H.N., McFarlan E. Jr, Kolb C.R. & Wilbert L.J. Jr (1954) Sedimentary framework of the modern Mississippi delta. *J. sedim. Petrol.*, 24, 76–99. *6.1, 6.6.5.*

Fiske R.S. & Jackson E.D. (1973) Orientation and growth of Hawaiian volcanic rifts: the effect of regional structure and gravitational stresses. *Proc. R. Soc. Lond.*, A329, 299–326. *12.11.2.*

Fiske R.S. & Matsuda T. (1964) Submarine equivalents of ash flows in the Tokiwa Formation, Japan. *Am. J. Sci.*, 262, 76–106. *12.4.4, Fig. 12.28.*

Fitzgerald D.M., Baldwin C.T., Ibrahim N.A. & Humpries S.M. (1992) Sedimentological and morphological evolution of a beach-ridge barrier along an indented coast: Buzzards Bay, Massachusetts. In: *Quaternary Coasts of the United States: Marine and Lacustrine Systems* (Ed. by C.H. Fletcher III and J.F. Wehmiller), pp. 65–75. *Spec. Publ. Soc. econ. Paleont. Miner.*, 48, Tulsa. *6.7.6.*

FitzPatrick E.A. (1980) *Soils: Their Formation, Classification and Distribution*, 353 pp. Longman, London. *3.2.3, 3.3.7.*

Fleet A.J., Kelts K. & Talbot M.R. (Eds) (1988) *Lacustrine Petroleum Source Rocks*, 391 pp. Spec. Publ. Geol. Soc. Lond., 40. *4.17.*

References

Flemming B.W. (1980) Sand transport and bedform patterns on the continental shelf between Durban and Port Elizabeth, southeast African continental margin. *Sedim. Geol.*, **26**, 179–205. *7.1.2, 7.5, Figs 7.26, 7.27, 7.28.*

Flemming B.W. (1981) Factors controlling shelf sediment dispersal along the South-east African continental margin. In: *Sedimentary Dynamics of Continental Shelves* (Ed. by C.A. Nittrouwer), pp. 259–277. *Mar. Geol.*, **42**. *7.1.2, 7.5, Figs 7.26, 7.28.*

Flint R.F. (1945) Glacial map of North America. *Spec. Pap. geol. Soc. Am.*, **60**, Pt. 1 Glacial Map; Pt. 2 Explanatory Notes, 37 pp. *11.5.3.*

Flint R.F. (1959) *Glacial Map of the United States East of the Rocky Mountains: Scale 1: 750 000.* Geological Society of America, 2 sheets. *11.5.3.*

Flint R.F. (1971) *Glacial and Quaternary Geology*, 892 pp. Wiley, New York. *4.9, 11.1, 11.2, 11.2.3.*

Flint R.F. (1975) Features other than diamicts as evidence of ancient glaciations. In: *Ice Ages: Ancient and Modern* (Ed. by A.E. Wright and F. Moseley), pp. 121–136. Geol. J. Spec. issue, **6**. Seel House Press, Liverpool. *11.1.*

Flint R.F., Sanders J.E. & Rodgers J. (1960a) Symmictite: a name for nonsorted terrigenous sedimentary rocks that contain a wide range of particle sizes. *Bull. geol. Soc. Am.*, **71**, 507–510. *11.1.*

Flint R.F., Sanders J.E. & Rodgers J. (1960b) Diamictite: a substitute term for symmictite. *Bull. geol. Soc. Am.*, **71**, 1809–1810. *11.1.*

Flood R.D. & Shor A.N. (1988) Mud waves in the Argentine Basin and their relationship to regional bottom circulation patterns. *Deep Sea Res.*, **35**, 943–971. *10.3.7.*

Flood R.D., Manley P.L., Kowsmann R.O., Appi C.J. & Pirmez C. (1991) Seismic facies and late Quaternary growth of Amazon submarine fan. In: *Seismic Facies and Sedimentary Processes of Submarine Fans and Turbidite Systems* (Ed. by P. Weimer and M.L. Link), pp. 415–433. Springer-Verlag, New York. *10.3.3, Fig. 10.38.*

Flores R.M. (1975) Short-headed stream delta: model for Pennsylvanian Haymond Formation, west Texas. *Bull. Am. Ass. petrol. Geol.*, **59**, 2288–2301. *6.5.4.*

Flores R.M. (1981) Coal deposition in fluvial paleoenvironments of the Paleocene Tongue River Member of the Fort Union Formation, Powder River area, Powder River Basin, Wyoming and Montana. In: *Recent and Ancient Non-Marine Depositional Environments: Models for Exploration* (Ed. by F.G. Ethridge and R.M. Flores), pp. 169–190. *Spec. Publ. Soc. econ. Paleont. Miner.*, **31**, Tulsa. *3.6.8.*

Flores R.M. (1990) Transverse and longitudinal Gilbert-type deltas, Tertiary Coalmont Formation, North Park Basin, Colorado. USA. In: *Coarse-grained Deltas* (Ed. by A. Colella and D.B. Prior), pp. 223–233. *Spec. Publ. int. Ass. Sediment.*, **10**. *6.5.6.*

Folk R.L. (1959) Practical petrographic classification of limestones. *Bull. Am. Ass. petrol. Geol.*, **43**, 1–38. *9.1.1.*

Folk R.L. (1973) Carbonate petrography in the post-Sorbian age. In: *Evolving Concepts in Sedimentology* (Ed. by R.N. Ginsburg), pp. 118–158. Johns Hopkins University Press, Baltimore. *9.1.1, 9.4.1.*

Forbes D.L. (1983) Morphology and sedimentology of a sinuous gravel-bed channel system: lower Babbage River, Yukon coastal plain, Canada. In: *Modern and Ancient Fluvial Systems* (Ed. by J.D. Collinson and J. Lewin), pp. 195–206. *Spec. Publ. int. Ass. Sed.*, **6**. *3.3.4.*

Forel F.A. (1892) *Le Léman: Monographie Limnologique, Vol. 1 Géographie, Hydrographie, Géologie, Climatologie, Hydrologie*, 543 pp. F. Rouge, Lausanne. *4.4.2, 4.6.2.*

Forristall G.S., Hamilton R.C. & Cardone V.J. (1977) Continental shelf currents in Tropical Storm Delia: observation and theory. *J. phys. Oceanogr.*, **7**, 532–546. *7.4.4.*

Förstner U., Müller G. & Reineck H.E. (1968) Sedimente und Sedimentgefüge des Rheindeltas im Bodensee. *Neues Jahrb. Miner. Abh.*, **109**, 33–62. *4.6.2.*

Fortuin A.R., Rope T.B., Sumosusastro P.A., Weering T.C.E.v. & Werff W.v.d. (1992) Slumping and sliding in Miocene and Recent developing arc basins, onshore and offshore Sumba (Indonesia). *Mar. Geol.*, **108**, 345–363. *12.5.1.*

Fouqué F. (1879) *Santorin et ses Eruptions*, 39+440 pp. Masson, Paris. *12.1.1.*

Fournari D.J., Garcia M.O., Tyce R.C. & Gallo D.G. (1988) Morphology and structure of Loihi seamount based on Seabeam sonar mapping. *J. geophys. Res.*, **93**, 15227–15238. *12.11.2.*

Fracasso M.A. & Hovorka S.D. (1986) Cyclicity in the middle Permian San Andres Formation, Palo Duro Basin, Texas Panhandle. *Bur. Econ. Geol. Rept. Invest.* University of Texas, Austin 48 pp. *8.8, Fig. 8.43.*

Frakes L.A. (1978) Diamictite. In: *The Encyclopedia of Sedimentology* (Ed. by R.W. Fairbridge and J. Bourgeois), pp. 262–263. Dowden, Hutchinson & Ross, Inc., Stroudsburg. *11.1.*

Frakes L.A. (1979) *Climates Throughout Geologic Time*, 310 pp. Elsevier, Amsterdam. *11.6.*

Frakes L.A., Francis J.E. & Syktus J.I. (1992) *Climate Modes of the Phanerozoic*, 274 pp. Cambridge University Press, Cambridge. *11.6.*

Francis P. (1993) *Volcanoes: a Planetary Perspective.* Oxford University Press, Oxford, 443 pp. *12.8, Fig. 12.23.*

Francis P.W. & Self S. (1983) The eruption of Krakatau. *Sci. Am.*, **249**, 172–187. *Fig. 12.1.*

Francis P.W. & Wells G.L. (1988) Landsat thematic mapper observations of debris avalanche deposits in the Central Andes. *Bull. Volcanol.*, **50**, 258–278. *12.1.1, 12.5.1.*

Frank A. & Kocurek G. (1994) Effects of atmospheric conditions on wind profiles and eolian sand transport with an example from White Sands National Monument. *Earth Surf. Proc. Landforms*, **19**, 735–745. *5.3.1.*

Frank A. & Kocurek G. (1996) Airflow over dunes – limitations to current understanding. *Geomorphology*, in press. *5.3.2.*

Franseen E.K., Fekete T.E. & Pray L.C. (1989) Evolution and destruction of a carbonate bank at the shelf margin: Grayburg Formation (Permian), western escarpment, Guadalupe Mountains, Texas. In: *Controls on Carbonate Platform and Basin Development* (Ed. by P.D. Crevello, J.F. Sarg, J.F. Read and J.L. Wilson), pp. 289–304. *Spec. Publ. Soc. econ. Paleont. Miner.*, **44**, Tulsa. *9.4.4, Fig. 9.37.*

Fraser G.S. & DeCelles P.G. (1992) Geomorphic controls on sediment accumulation at margins of foreland basins. *Basin Res.*, **4**, 233–252. *3.6.3.*

Frazier D.E. (1967) Recent deltaic deposits of the Mississippi delta:

their development and chronology. *Trans. Gulf Coast. Ass. geol. Soc.*, 17, 287–315. *6.6.5*, **Fig. 6.46**.

Frazier D.E. (1974) *Depositional Episodes: their Relationship to the Quaternary Stratigraphic Framework in the Northwestern Portion of the Gulf basin*. Bur. econ. Geol. Univ. Texas, Austin, **74-1**, 28 pp. *2.2*.

Frazier D.E. & Osanik A. (1961) Point-bar deposits; Old River Locksite, Louisiana. *Trans. Gulf Coast. Ass. geol. Socs.*, 11, 121–137. *3.3.4*.

Freeman-Lynde R.P. & Ryan W.B.F. (1985) Erosional modification of Bahama Escarpment. *Bull. geol. Soc. Am.*, 96, 481–494. *9.4.4*.

Freundt A. & Schmincke H.-U. (1986) Emplacement of small-volume pyroclastic flows at Laacher See (East-Eifel, Germany). *Bull. Volcanol.*, 48, 39–59. *12.4.4*.

Freundt A. & Schmincke H.U.-v. (1995) Eruption and emplacement of a basaltic welded ignimbrite during caldera formation on Gran Canaria. *Bull. Volcanol.*, 56, 640–659. *12.4.4*.

Frey R.W. & Howard J.D. (1969) A profile of biogenic sedimentary structures in a Holocene barrier island–salt marsh complex, Georgia. *Trans. Gulf Coast. Ass. geol. Soc.*, 19, 427–444. *6.7.4*.

Frey R.W. & Pemberton S.G. (1984) Trace fossil facies models. In: *Facies Models* (Ed. by R.G. Walker), pp. 801–828. Geol. Ass. Can., Waterloo, Ontario. *7.7.2*.

Frey R.W., Howard J.D., Han S.J. & Park B.K. (1989) Sediments and sedimentary sequences on a modern macrotidal flat, Inchon, Korea. *J. sedim. Petrol.*, 59, 28–44. *6.7.3*.

Frey R.W., Pemberton S.G. & Saunders T.D.A. (1990) Ichnofacies and bathymetry, a passive relationship. *J. Paleontol.*, 64, 155–158. *10.2.3*.

Freytet P. & Plaziat J.C. (1982) Continental carbonate sedimentation and pedogenesis – Late Cretaceous and Early Tertiary of Southern France. *Contrib. Sedim.*, 12, 213. *4.12.1*, **Fig. 4.23**.

Friedman G.M. (1965) Origin of aragonite in the Dead Sea. *Israel J. Earth-Sci.*, 14, 79–85. *8.6.2*.

Friedman G.M. (1973) Petrologic data and comments on the depositional environment of the marine sulfates and dolomites at sites 124, 132 and 134, western Mediterranean Sea. In: *Initial Reports of the Deep Sea Drilling Project*, **XIII** (Ed. by W.B.F. Ryan and K.J. Hsü *et al.*), pp. 695–707. US Government Printing Office, Washington D.C. *8.9*.

Friedman G.M. (1985) The problems of submarine cement in classifying reefrock: an experience in frustation. In: *Carbonate Cements* (Ed. by N. Schneiderman and P.M. Harris), pp. 121–177. *Spec. Publ. Soc. econ. Paleont. Miner.*, 36, Tulsa. *9.4.5*.

Friedman G.M., Sneh A. & Owen R.W. (1985) The Ras Muhammed Pool: Implications for the Gavish Sabkha. In: *Hypersaline Ecosystems: The Gavish Sabkha* (Ed. by G.M. Friedman and W.E. Krumbein), pp. 218–237. Springer-Verlag, New York. *8.2.3*, *8.5.2*.

Friend P.F. (1983) Towards the field classification of alluvial architecture or sequence. In: *Modern and Ancient Fluvial Systems* (Ed. by J.D. Collinson and J. Lewin), pp. 345–354. *Spec. Publ. int. Ass. Sediment.*, 6. *3.6.5*, *3.6.7*.

Friend P.F., Hirst J.P.P. & Nichols G.J. (1986) Sandstone-body structure and river process in the Ebro Basin of Aragon, Spain. *Cuad. Geol. Iberica*, 10, 9–30. *3.6.5*.

Friend P.F., Marzo M., Nijman W. & Puigdefabregas C. (1981) Fluvial sedimentology in Tertiary South Pyrenean and Ebro Basins, Spain. In: *Field Guide to Modern and Ancient Fluvial Systems in Britain and Spain* (Ed. by T. Elliott), pp. 4.1–4.50. University of Keele. *3.6.5*.

Fritz W.J., Howells M.F., Reedman A.J. & Campbell S.D.G. (1990) Volcaniclastic sedimentation in and around an Ordovician subaqueous caldera, Lower Rhyolitic Tuff Formation, North Wales. *Bull. geol. Soc. Am.*, 102, 1246–1256. *12.13*.

Frostick L.E. & Reid I. (1989) Climatic versus tectonic controls on fan sequences: lessons from the Dead Sea, Israel. *J. geol. Soc. Lond.*, 146, 527–538. *6.5.5*.

Fryberger S.G. (1986) Stratigraphic traps for petroleum in wind-laid rocks. *Bull. Am. Ass. petrol. Geol.*, 70, 1765–1776. *5.5.3*.

Fryberger S.G. (1991) Unusual sedimentary structures in the Oregon coastal dunes. *J. arid Environ.*, 21, 131–150. *5.4.4*.

Fryberger S.G. & Ahlbrandt T.S. (1979) Mechanisms for the formation of eolian sand seas. *Z. Geomorph. N.F.*, 23, 440–460. *5.3.1*.

Fryberger S.G., Ahlbrandt T.S. & Andrews S. (1979) Origin, sedimentary features, and significance of low-angle eolian 'sand sheet' deposits, Great Sand Dunes National Monument and vicinity, Colorado. *J. sedim. Petrol.*, 49, 733–746. *5.4.6*.

Fryberger S.G., Al-Sari A.M. & Clisham T.J. (1983) Eolian dune, interdune, sand sheet and siliciclastic sabkha sediments of an offshore prograding sand sea, Dhahran area, Saudi Arabia. *Bull. Am. Ass. petrol. Geol.*, 63, 280–312. *5.4.1*, *5.4.5*.

Fryberger S.G., Al-Sari A.M., Clisham T.J., Rizvi S.A.R. & Al-Hiani K. (1984) Wind sedimentation in the Jafurah sand sea, Saudi Arabia. *Sedimentology*, 31, 413–431. *5.4.5*.

Fryberger S.G., Hesp P. & Hastings K. (1992) Aeolian granule ripple deposits, Namibia. *Sedimentology*, 39, 319–331. *5.4.6*.

Fryberger S.G., Krystinik L.F. & Schenk C.J. (1990) Tidally flooded back-barrier dunefield, Guerrero Negro area, Baja California, Mexico. *Sedimentology*, 37, 23–43. *5.4.1*, *5.4.2*, **Fig. 5.17**.

Fryberger S.G., Schenk C.J. & Krystinik L.F. (1988) Stokes surfaces and the effects of near-surface groundwater-table on aeolian deposition. *Sedimentology*, 35, 21–41. *5.4.1*, *5.4.5*.

Füchtbauer H. & Peryt T. (1980) Introduction. In: *The Zechstein Basin* (Ed. by H. Füchtbauer and T. Peryt) pp. 1–2. E. Schweizerbart'sche Verlagsbuchhandlung, Stuttgart. **Fig. 8.49**.

Füchtbauer H., von der Brelie *et al.* (1977) Tertiary lake sediments of the Ries, research borehole Nordlingen 1973 – a summary. *Geol. Bavarica*, 75, 13–19. *4.12.3*.

Fürsich F.T. & Wendt J. (1977) Biostratinomy and palaeoecology of the Cassian Formation (Triassic) of the southern Alps. *Palaeogeogr. Palaeoclimatol. Palaeoecol.*, 22, 257–323. *9.3.2*.

Gadow S. & Reineck H.-E. (1969) Ablandiger sand transport bei sturmfluten. *Senckenberg. Marit.*, 1, 63–78. *7.4.3*.

Gagan M.K., Chivas A.R. & Herczeg A.L. (1990) Shelf-wide erosion, deposition and suspended sediment transport during Cyclone Winifred, central Great Barrier Reef, Australia. *J. sedim. Petrol.*, 60, 456–470. *9.4.3*.

Gagliano S.M. & van Beek J.L. (1970) Geologic and geomorphic aspects of deltaic processes, Mississippi delta system. In: *Hydrologic and Geologic Studies of Coastal Louisiana*, Report No. 1, 140 pp. Centre for Wetland Resources, Louisiana State University. *6.6.1*.

Galat D.L. & Jacobsen R.L. (1985) Recurrent aragonite precipitation in saline-alkaline Pyramid Lake, Nevada. *Arch. Hydrobiol.*, 105, 137–159. *4.7.1.*

Gale A.S. (1989) A Milankovitch scale for Cenomanian time. *Terra Nova*, 1, 420–425. *10.4.3.*

Galloway W.E. (1975) Process framework for describing the morphologic and stratigraphic evolution of deltaic depositional systems. In: *Deltas, Models for Exploration* (Ed. by M.L. Broussard), pp. 87–98. Houston Geol. Soc. Houston, TX. *6.1, 6.3, 6.7.1, Figs 6.2, 6.17.*

Galloway W.E. (1976) Sediments and stratigraphic framework of the Copper River fan-delta, Alaska. *J. sedim. Petrol.*, 46, 726–737. *6.5, Fig. 6.20.*

Galloway W.E. (1981) Depositional architecture of Cenozoic Gulf Coastal Plain fluvial systems. In: *Recent and Ancient Nonmarine Depositional Systems* (Ed. by F.G. Ethridge and R.M. Flores), pp. 127–156. *Soc. econ. Paleont. Miner.*, 31, Tulsa. *3.6.7.*

Galloway W.E. (1989a) Genetic stratigraphic sequences in basin analysis. 1: Architecture and genesis of flooding-surface bounded depositional units. *Bull. Am. Ass. petrol. Geol.*, 73, 125–142. *2.2, 2.2.2, 2.4, Fig. 2.10, 9.5.5.*

Galloway W.E. (1989b) Genetic stratigraphic sequences in basin analysis II: application to northwest Gulf of Mexico Cenozoic basin. *Bull. Am. Ass. petrol. Geol.*, 73, 143–154. *2.4.*

Galloway W.E. & Hobday D.K. (1983) *Terrigenous Clastic Depositional Systems*, 423 pp. Springer-Verlag, New York. *1.1.*

Gamboa L.A.P., Buffler R.T. & Barker P.F. (1983) Seismic stratigraphy and geologic history of the Rio Grande Gap and southern Brazil basin. *Init. Rep. DSDP*, 72, 481–498. *10.3.7.*

Garber R.A., Levy Y. & Friedman G.M. (1987) The sedimentology of the Dead Sea. *Carbon. Evap.*, 2, 43–57. *8.6.2, Fig. 8.35.*

García M.O. & Hull D.M. (1994) Turbidites from giant Hawaiian landslides: results from Ocean Drilling Program Site 842. *Geology*, 22, 159–162. *12.5.1.*

García-Mondéjar J. (1990) Sequence analysis of a marine Gilbert-type delta, La Miel, Albian Lunada Formation of northern Spain. In: *Coarse-grained Deltas* (Ed. by A. Colella and D.B. Prior), pp. 255–269. *Spec. Publ. int. Ass. Sediment.*, 10. *6.5.4, 6.5.6, Fig. 6.30.*

García-Mondéjar J. & Fernández-Mendiola P.A. (1993) Sequence stratigraphy and systems tracts of a mixed carbonate and siliciclastic platform-basin setting: the Albian of Lunada and Soba, northern Spain. *Am. Ass. petrol. Geol. Bull.*, 77, 245–275. *9.4.4, Fig. 9.62.*

Gardner J.V. (1975) Late Pleistocene carbonate dissolution cycles in the eastern equatorial Atlantic. In: *Dissolution of Deep-sea Carbonates* (Ed. by W.V. Sliter, A.H.H. Bé and W.H. Berger), pp. 129–141. *Cushman Found. foraminiferal Res. Spec. Publ.*, 13. *10.2.1.*

Gardner W.D. & Sullivan L.G. (1981) Benthic storms: temporal variability in a deep ocean nepheloid layer. *Science*, 213, 329–331. *10.2.2.*

Garland C.R. (1993) Miller Field: reservoir stratigraphy and its impact on development. In: *Petroleum Geology of Northwest Europe: Proceedings of the 4th Conference* (Ed. by J.R. Parker), pp. 401–414. Geol. Soc. Lond., Bath. *10.3.3.*

Garrison R.E. (1973) Space–time relations of pelagic limestones and volcanic rocks, Olympia Peninsula, Washington. *Bull. geol. Soc.*

Am., 84, 583–594. *12.11.2.*

Garrison R.E. (1974) Radiolarian cherts, pelagic limestones and igneous rocks in eugeosynclinal assemblages. *Pelagic sediments: on land and under the sea* (Ed. by K.J. Hsü and H.C. Jenkyns), pp. 367–399. *Spec. Publ. int. Ass. Sedim.*, 1. *Fig. 12.56.*

Garrison R.E., Douglas R.G., Pisciotto K.E., Isaacs C.M. & Ingle J.D. (Eds) (1981) *The Monterey Formation and Related Siliceous Rocks of California*, 327 pp. *Spec. Publ. Pacific Sect. Soc. econ. Paleont. Miner.*, Los Angeles. *10.4.1.*

Garrison R.E., Schreiber B.C., Bernoulli D., Fabricius F.H., Kidd R.B. & Meliere F. (1978) Sedimentary petrology and structures of Messinian evaporitic sediments in the Mediterranean Sea. Leg 42A, Deep Sea Drilling Project. In: *Initial Reports of the Deep Sea Drilling Project*, 42 (Ed. by K.J. Hsü, L. Montadet *et al.*) pp. 571–611. *8.9.*

Gavish E. (1974) Geochemistry and mineralogy (*sic*) of a recent sabkha along the coast of Sinai, Gulf of Suez. *Sedimentology*, 21, 397–414. *8.4.3.*

Gavish E. (1980) Recent sabkhas marginal to the southern coasts of Sinai, Red Sea. In: *Hypersaline Brines and Evaporitic Environments* (Ed. by N. Nissenbaum), pp. 233–251. *8.2.3, 8.4.2, 8.4.3, 8.5.2.*

Gavish E., Krumbein W.E. & Halevy J. (1985) Geomorphology, mineralogy and groundwater geochemistry as factors of the hydrodynamic system of Gavish Sabkha. In: *Hypersaline Ecosystems: The Gavish Sabkha* (Ed. by G.M. Friedman and W.E. Krumbein), pp. 186–217. Springer-Verlag, New York. *8.5.2.*

Gawthorpe R.L. (1986) Sedimentation during carbonate ramp-to-slope evolution in a tectonically active area: Bowland Basin (Dinantian), N. England. *Sedimentology*, 33, 185–206. *9.3.4.*

Gawthorpe R.L. & Colella A. (1990) Tectonic controls on coarse-grained delta depositional systems in rift basins. In: *Coarse-grained Deltas* (Ed. by A. Colella and D.B. Prior), pp. 113–127. *Spec. Publ. int. Ass. Sediment.*, 10. *6.1, 6.5.4, 6.5.5, Fig. 6.26.*

Gaynor G.C. & Swift D.J.P. (1988) Shannon Sandstone depositional model; sand ridge formation on the Campanian western interior shelf. *J. sedim. Petrol.*, 58, 868–880. *7.8.4, Fig. 7.54.*

Gebelein C.D. (1974) *Guidebook of Modern Bahamian Platform Environments*, 93 pp. Geol. Soc. Am. Ann. Mtd. Fieldtrip Guidebook, Boulder, CO. *Fig. 9.9.*

Gebelein C.D. (1977) *Dynamics of Recent Carbonate Sedimentation and Ecology: Cape Sable, Florida*. 120 pp. E.J. Brill, Leiden. *9.4.1.*

Gebelein C.D., Steinen R.P., Garrett P., Hoffman E.J., Queen J.M. & Plummer L.N. (1980) Subsurface dolomitization beneath the tidal flats of Central West Andros Island, Bahamas. In: *Concepts and Models of Dolomitization* (Ed. by D.H. Zenger, J.B. Danham and R.L. Ethington), pp. 31–49. *Spec. Publ. Soc. econ. Paleont. Miner.*, 28. *9.4.1, Fig. 9.24.*

Geisler D. (1982) De la mer au sel: les facies superficiels des marais salants de Salin-de-Giraud (Sud de la France). In: *Données hydrochimiques, biologiques, sedimentologiques et diagénétiques sur les marais de Salin-de-Giraud, Sud de la France. Geol. Méditerranéene*, IX, 521–549. *8.1.*

Geist E.L., Childs J.R. & Scholl D.W. (1988) The origin of summit basins of the Aleutian Ridge: implications for block rotation of an arc massif. *Tectonics*, 7, 327–341. *12.12.2.*

Geldof H.J. & de Vriend H.J. (1983) Distribution of main flow velocity in alternating river bends. In: *Modern and Ancient Fluvial*

Systems (Ed. by J.D. Collinson and J. Lewin), pp. 85–95. *Spec. Publ. int. Ass. Sediment.*, 6. *3.3.4.*

Geldsetzer H.H.J., James N.P. & Tebbutt G.E. (1988) *Reefs, Canada and Adjacent Areas*, 775 pp. *Mem. Can. Soc. petrol. Geol.*, 13, Calgary. *9.4.5.*

Gersib G.A. & McCabe P.J. (1981) Continental coal-bearing sediments of the Port Hood Formation (Carboniferous), Cape Linzee, Nova Scotia, Canada. In: *Recent and Ancient Non-marine Depositional Environments: Models for Exploration* (Ed. by F.G. Ethridge and R.M. Flores), pp. 95–108. *Spec. Publ. Soc. econ. Paleont. Miner.*, 31, Tulsa. *3.5.2, 3.5.3, 3.6.8.*

Ghibaudo G., Mutti E. & Rosell J. (1974) Le spiagge fossili delle Arenarie di Aren (Cretacico superiore) nella valle Noguera-Ribagorzana (Pirenei centro-meridionali, Province di Lerida e Huesca, Spagna). *Mem. Soc. geol. Ital.*, 13, 497–537. *6.7.7, Fig. 6.93.*

Ghosh B. & Lowe D.R. (1993) The architecture of deep-water channel complexes, Cretaceous Venado Sandstone Member, Sacramento Valley, California. In: *Advances in the Sedimentary Geology of the Great Valley Group, Sacramento Valley, California* (Ed. by S.A. Graham and D.R. Lowe), pp. 51–65. Field Trip Guidebook, Soc. econ. Paleont. Miner., Tulsa, Pacific Section. *10.3.2, Figs 10.32, 10.33; Table 10.2.*

Ghosh B. & Lowe D.R. (1996) Architectural Element Analysis of deep-water sedimentary sequences: Cretaceous Venado Sandstone member, Sacramento Valley, California. *J. sedim. Res.*, *Fig. 10.32, 10.3.2, Table 10.2.*

Gibbard P.L. & Dreimanis A. (1978) Trace fossils from late Pleistocene glacial lake sediments in southwestern Ontario, Canada. *Can. J. Earth Sci.*, 15, 1967–1976. *11.4.5.*

Giberti G., Jaupart C. & Sartoris G. (1992) Steady-state operation of Stromboli volcano, Italy: constrainst on the feeding system. *Bull. Volcanol.*, 54, 535–541. *12.3.2, Fig. 12.1.*

Gibling M.R. & Rust B.R. (1984) Channel margins in a Pennsylvanian braided fluvial deposit: The Morien Group near Sydney, Nova Scotia, Canada. *J. sedim. Petrol.*, 54, 773–782. *3.5.2.*

Gibling M.R. & Rust B.R. (1987) Evolution of a mud-rich meander belt in the Carboniferous Morien Group, Nova Scotia, Canada. *Bull. Can. petrol. Geol.*, 35, 24–33. *3.6.5.*

Gibling M.R. & Rust B.R. (1990) Ribbon sandstones in the Pennsylvanian Waddens Cove Formation, Sydney Basin, Atlantic Canada: the influence of siliceous duricrusts on channel-body geometry. *Sedimentology*, 37, 45–65. *3.6.9, Fig. 3.62.*

Gibling M.R. & Rust B.R. (1993) Alluvial ridge-and-swale topography: a case study from the Morien Group of Atlantic Canada. In: *Alluvial Sedimentation* (Ed. by M. Marzo and C. Puigdefabregas), pp. 133–150. *Spec. Publ. int. Ass. Sediment.*, 17. *3.6.5.*

Gilbert G.K. (1885) The topographic features of lake shores. *Ann. Rep. U.S. geol. Surv.*, 5, 75–123. *4.6.2, 6.1, 6.5, Fig. 6.22.*

Gilbert G.K. (1890) Lake Bonneville. *Mon. U.S. geol. Surv.*, 1, 438 pp. *6.1, Fig. 6.22.*

Gilbert G.K. (1894/1895) Sedimentary measurement of Cretaceous time. *J. Geol.*, 3, 121–127. *2.1.5.*

Gill A.E. (1973) Circulation and bottom water production in the Weddell Sea. *Deep-Sea Res.*, 20, 111–140. *10.2.2.*

Gill J.B. (1981) *Orogenic Andesites and Plate Tectonics*, 358 pp. Springer-Verlag, Berlin. *12.2.2.*

Ginsburg R.N. (1953) Beach rock in South Florida. *J. sedim. Petrol.*, 23, 85–92. *9.1.1.*

Ginsburg R.N. (1956) Environmental relationships of grain size and constituent particles in some south Florida carbonate sediments. *Bull. Am. Ass. petrol. Geol.*, 40, 2384–2427. *9.1.1, Fig. 9.12.*

Ginsburg R.N. (1957) Early diagenesis and lithification of shallow water carbonate sediments in South Florida. In: *Regional Aspects of Carbonate Deposition* (Ed. by R.J. LeBlanc and J.C. Breeding), pp. 80–99. *Spec. Publ. Soc. econ. Paleont. Miner.*, 5, Tulsa. *9.4.1.*

Ginsburg R.N. (1971) Landward movement of carbonate mud: new model for regressive cycles in carbonates (abstract). *Am. Ass. petrol. Geol. Ann. Meet., Abstr. Programs*, 55, 340. *9.4.1.*

Ginsburg R.N. (1975) *Tidal Deposits: A Casebook of Recent Examples and Fossil Counterparts*, 428 pp. Springer-Verlag, New York. *9.1.1, 9.4.1.*

Ginsburg R.N. & James N.P. (1974) Holocene carbonate sediments of continental margins. In: *The Geology of Continental Margins* (Ed. by C.A. Burke and C.L. Drake), pp. 137–155. Springer-Verlag, New York. *9.3.4, Figs 9.12, 9.18.*

Ginsburg R.N., Hardie L.A., Bricker O.P., Garrett P. & Wanless H.R. (1977) Exposure index: a quantitative approach to defining position within the tidal zone. In: *Sedimentation on the Modern Tidal Flats of Northwest Andros Island, Bahamas* (Ed. by L.A. Hardie), pp. 7–11. Johns Hopkins University Press, Baltimore. *9.4.1, Fig. 9.24.*

Ginsburg R.N., Harris P.M., Erbli P. & Swart P.K. (1991) The growth potential of a by-pass margin, Great Bahama Bank. *J. sedim. Petrol.*, 61, 976–987. *9.4.5.*

Giovanoli F. (1990) Horizontal transport and sedimentation by interflows and turbidity currents in Lake Geneva. In: *Large Lakes: Ecological Structure and Function* (Ed. by M.M. Tilzer and C. Serruya) pp. 175–195. Springer-Verlag, Berlin. *4.4.2, 4.6.2, 4.6.3, Fig. 4.5.*

Gipp M.R. (1994) Architectural styles of glacially influenced marine deposits on tectonically active and passive continental margins. In: *Earth's Glacial Record* (Ed. by M. Deynoux, J.M.G. Miller, E.W. Domack, N. Eyles, I.J. Fairchild and G.M. Young), pp. 109–120. Cambridge University Press, Cambridge. *11.5.4.*

Glaser K.S. & Droxler A.W. (1991) High production and highstand shedding from deeply submerged carbonate banks, northern Nicaragua Rise. *J. sedim. Petrol.*, 61, 128–142. *9.3.5, 9.4.4.*

Glennie K.W. (1970) *Desert Sedimentary Environments, Developments in Sedimentology*, 14, 222 pp. Elsevier, Amsterdam. *5.1, 5.4.5.*

Glennie K.W. (1983) Lower Permian Rotliegend desert sedimentation in the North Sea. In: *Eolian Sediments and Processes* (Ed. by M.E. Brookfield and T.S. Ahlbrandt), pp. 521–541. *Developments in Sedimentology*, 38, Elsevier, Amsterdam. *5.5.1.*

Glennie K.W. (1986) Early Permian Rotliegend. In: *Introduction to the Petroleum Geology of the North Sea* (Ed. by K.W. Glennie), pp. 63–85. Blackwell, Oxford. *5.5.1, 5.5.3.*

Glennie K.W. & Buller A.T. (1983) The Permian Weissliegendes of N.W. Europe: the partial deformation of eolian dune sands caused by the Zechstein transgression. *Sedim. Geol.*, 35, 43–81. *5.5.3.*

Glennie K.W. & Provan D.M.J. (1990) Lower Permian Rotliegend reservoir of the southern North Sea gas province. In: *Classic Petroleum Provinces* (Ed. by J. Brooks), pp. 399–416. *Spec. Publ.*

geol. Soc. Lond., 50. *5.1.*

Glicken H. (1991) Facies architecture of large volcanic-debris avalanches. In: *Sedimentation in Volcanic Settings* (Ed. by R.V. Fisher and G.A. Smith), pp. 99–108. *Spec. Publ. Soc. econ. Paleont. Miner.*, 45, Tulsa. *12.5.1*, *Fig. 12.32.*

Glover B.W., Powell J.H. & Waters C.N. (1993) Etruria Formation (Westphalian C) palaeoenvironments and volcanicity on the southern margin of the Pennine Basin, South Staffordshire, England. *J. geol. Soc. Lond.*, 150, 737–750. *3.5.4.*

Godchaux M.M., Bonnischen B. & Jenks M.D. (1992) Types of phreatomagmatic volcanoes in the western Snake River Plain, Idaho, USA. *J. Volcanol. geotherm. Res.*, 52, 1–25. *12.9.3.*

Godin P.D. (1991) Fining-upward cycles in the sandy braided-river deposits of the Westwater Canyon Member (Upper Jurassic), Morrison Formation, New Mexico. *Sedim. Geol.*, 70, 61–82. *3.6.5*, *Figs 3.55*, *3.56.*

Goggin D.J., Chandler M.A., Kocurek G. & Lake L.W. (1992) Permeability transects in eolian sands and their use in generating random permeability fields. *Engin. Form. Eval.*, 7, 7–16. *5.1.*

Gohain K. & Parkash B. (1990) Morphology of the Kosi Megafan. In: *Alluvial Fans: A Field Approach* (Ed. by A.H. Rachocki and M. Church), pp. 151–178. John Wiley, Chichester. *3.3.8.*

Goldbery R. (1982) Palaeosols of the Lower Jurassic Mishhor and Ardon Formations ('Laterite derivative facies'), Makhtesh Ramon, Israel. *Sedimentology*, 29, 669–690. *3.5.4.*

Goldhammer R.K. & Harris M.T. (1989) Eustatic controls on the stratigraphy and geometry of the Latemar buildup (Middle Triassic), the Dolomites of northern Italy. In: *Controls on Carbonate Platform and Basin Development* (Ed. by P.D. Crevello, J.F. Sarg, J.F. Read, and J.L. Wilson), pp. 323–338. *Spec. Publ. Soc. econ. Paleont. Miner.*, 44, Tulsa. *9.3.2*, *Fig. 9.11.*

Goldhammer R.K., Dunn P.A. & Hardie L.A. (1987) High frequency glacio-eustatic sea level oscillations with Milankovitch characteristics recorded in Middle Triassic platform carbonates of northern Italy. *Am. J. Sci.*, 287, 853–892. *9.4.1.*

Goldhammer R.K., Dunn P.A. & Hardie L.A. (1990) Depositional cycles, composite sea level changes, cycle stacking patterns, and the hierarchy of stratigraphic forcing: examples from platform carbonates of the Alpine Triassic. *Bull. geol. Soc. Am.*, 102, 535–562. *9.3.2*, *9.4.1*, *9.5.3.*

Goldhammer R.K., Harris M.T., Dunn P.A. & Hardie L.A. (1993) Sequence stratigraphy and systems tract development of the Latemar Platform, Middle Triassic of the Dolomites (northern Italy): outcrop patterns keyed by cycle stacking patterns. In: *Carbonate Sequence Stratigraphy* (Ed. by R.G. Loucks and J.F. Sarg), pp. 353–388. *Mem. Am. Ass. petrol. Geol.*, 57, Tulsa. *9.5.1*, *9.5.3.*

Goldhammer R.K., Lehmann P.J. & Dunn P.A. (1993) The origin of high-frequency platform cycles and third-order sequences (Lower Ordovician El Paso Group, west Texas): constraints from outcrop data and stratigraphic modelling. *J. sed. Petrol.*, 63, 318–359. *9.4.1.*

Goldhammer R.K., Oswald E.J. & Dunn P.A. (1991) The hierarchy of stratigraphic forcing, an example from the Middle Pennsylvanian shelf carbonates of the Paradox Basin. In: *Sedimentary Modeling: Computer Simulations and Methods for Improved Parameter Definition* (Ed. by E.K. Franseen *et al.*), pp. 361–414. *Bull. Kansas Geol. Surv.*, 233. *9.1.1*, *9.4.1*, *9.5.3.*

Goldhammer R.K., Oswald E.J. & Dunn P.A. (1994) High frequency, glacio-eustatic cyclicity in the Middle Pennsylvanian of the Paradox Basin: an evaluation of Milankovitch forcing. In: *Orbital Forcing and Cyclic Sequences* (Ed. by P.L. de Boer and D.G. Smith), pp. 243–283. *Spec Publ. int. Assoc. Sediment.*, 19. *Figs 8.52*, *8.53*, *8.55.*

Goldring R. (1965) Sediments into rock. *New Scientist*, 26, 863–865. *2.1.12.*

Goldring R. & Bridges P. (1973) Sublittoral sheet sandstones. *J. sedim. Petrol.*, 43, 736–747. *7.1.2.*

Gole C.V. & Chitale S.V. (1966) Inland delta building activity of Kosi River. *Proc. Am. Soc. Civ. Engr., J. Hydraul. Div.*, 92, 111–126. *3.3.8*, *Fig. 3.32.*

Gonthier E.G., Faugères J.-C. & Stow D.A.V. (1984) Contourite facies of the Faro Drift, Gulf of Cadiz. In: *Fine-grained Sediments: Deep-water Processes and Facies* (Ed. by D.A.V. Stow and D.J.W. Piper), pp. 275–292. *Spec. Publ. geol. Soc. Lond.*, 15. *10.3.7*, *Fig. 10.16.*

Good T.R. & Bryant I.D. (1985) Fluvio-aeolian sedimentation – an example from Banks Island, NWT, Canada. *Geogr. Ann.*, 67A (1–2), 33–46. *11.4.7.*

Goodwin R.H. & Prior D.B. (1988) Geometry and depositional sequences of the Mississippi Canyon, Gulf of Mexico. *J. sedim. Petrol.*, 59, 318–329. *10.3.3.*

Gornitz V.M. & Schreiber B.C. (1981) Displacive halite hoppers from the Dead Sea: some implications for ancient evaporite deposits. *J. sedim. Petrol.*, 51, 787–794. *8.5.1.*

Gorshkov G.S. (1959) Gigantic eruption of the volcano Bezymianny. *Bull. Volcanol.*, 20, 77–109. *12.1.1.*

Gorsline D.S. (1978) Anatomy of margin basins. *J. sedim. Petrol.*, 48, 1055–1068. *10.3.2*, *10.3.6.*

Gorsline D.S. (1984) A review of fine-grained sediment origins, characteristics, transport and deposition. In: *Fine-Grained Sediments: Deep-Water Processes and Facies* (Ed. by D.A.V. Stow and D.J.W. Piper), pp. 17–34. *Spec. Publ. geol. Soc. Lond.*, 15, Bristol. *7.6.*

Gorsline D.S. & Emery K.O. (1959) Turbidity-current deposits in San Pedro and Santa Monica basins off southern California. *Bull. geol. Soc. Am.*, 70, 279–290. *10.1*, *10.3.5.*

Goudie A. (1973) *Duricrusts in Tropical and Sub-tropical Landscapes*, 174 pp. Clarendon Press, Oxford. *3.2.3*, *3.3.7.*

Gould H.R. (1970) The Mississippi delta complex. In: *Deltaic Sedimentation Modern and Ancient* (Ed. by J.P. Morgan and R.H. Shaver), pp. 3–30. *Spec. Publ. Soc. econ. Paleont. Miner.*, 15, Tulsa. *6.6.1.*

Gould H.R. (1960) Comprehensive survey of sedimentation in Lake Mead, 1948–49. Q. Turbidity currents. *Prof. Pap. U.S. geol. Surv.*, 295, 201–207. *4.6.3.*

Gould H.R. & McFarlan E. (1959) Geological history of the chenier plain, southwestern Louisiana. *Trans. Gulf Coast. Ass. geol. Soc.*, 9, 261–270. *6.7.2*, *Fig. 6.71.*

Gradzinski R. (1992) Deep blowout depressions in the aeolian Tumlin Sandstone (Lower Triassic) of the Holy Cross Mountains, central Poland. *Sedim. Geol.*, 81, 231–242. *5.5.4.*

Grammar G.M. & Ginsburg R.N. (1992) Highstand vs lowstand deposition on carbonate platform margins: insight from Quaternary fore-slopes in the Bahamas. *Mar. Geol.*, 103, 125–136. *Fig. 9.59.*

Grammar G.M., Ginsburg R.N. & Harris P.M. (1993) Timing of deposition, diagenesis, and failure of steep carbonate slopes in response to a high-amplitude/high-frequency fluctuation in sea level, Tongue of the Ocean, Bahamas. In: *Carbonate Sequence Stratigraphy* (Ed. by R.G. Loucks and J.F. Sarg), pp. 107–132. *Mem. Am. Ass. petrol. Geol.*, 57, Tulsa. *9.5.3.*

Grande L. (1980) Paleontology of the Green River Formation, with a review of the fish fauna. *Bull. Univ. geol. Surv. Wyoming*, 63, 333 pp. *4.14.1.*

Grant C.W., Goggin D.J. & Harris P.M. (1994) Outcrop analogue for cyclic-shelf reservoirs, San Andrew Formation of Permian Basin: stratigraphic framework, permeability distribution, geostatistics, and fluid flow modelling. *Bull. Am. Ass. petrol. Geol.*, 78, 23–54. *9.5.*

Graué E., Helland-Hansen W., Johnson J., Lømo L., Nottvedt A., Ronning K., Ryseth K. & Steel T.J. (1987) Advance and retreat of Brent Delta system, Norwegian North Sea. In: *Petroleum Geology of North West Europe* (Ed. by J. Brooks and K.W. Glennie), pp. 915–937. Graham & Trotman, London. *6.6.6, Fig. 6.59.*

Graus R.R. & Macintyre I.G. (1989) The zonation patterns of Caribbean coral reefs as controlled by wave and light energy input, bathymetric setting and reef morphology. *Coral Reefs*, 8, 9–18. *9.4.5.*

Graus R.R., Macintyre I.G. & Herghendroder B.E. (1984) Computer simulation of reef zonation at Discovery Bay, Jamaica – hurricane disruption and long term physical oceanographic controls. *Coral Reefs*, 3, 59–68. *9.4.5.*

Gravenor C.P. (1982) Chattermarked garnets in Pleistocene glacial sediments. *Bull. geol. Soc. Am.*, 93, 751–758. *Table 11.2.*

Greeley R. and Iversen J.D. (1985) *Wind as a Geological Process*, 333 pp. Cambridge University Press, Cambridge. *5.1.*

Greenwood B. & Mittler P.R. (1985) Vertical sequence and lateral transitions in the facies of a barred nearshore environment. *J. sedim. Petrol.*, 55, 366–375. *6.7.6.*

Gressly A. (1838) Observations géologiques sur le Jura Soleurois. *Neue Denkschr. allg. schweiz, Ges. ges. Naturw.*, 2, 1–112. *2.2.1.*

Griffiths R.W. & Fink J.H. (1992) Solidification and morphology of submarine lavas: a dependence on extrusion rate. *J. geophys. Res.*, 97, 19729–19737. *12.4.5.*

Grötsch J., Wu G. & Berger W.H. (1991) Carbonate cycles in the Pacific: reconstruction of saturation fluctuations. In: *Cycles and Events in Stratigraphy* (Ed. by G. Einsele, W. Ricken and A. Seilacher), pp. 110–125. Springer-Verlag, Berlin. *10.4.1.*

Grousset F.E., Biscaye P.E., Revel M., Petit J.-R., Pye K., Joussaume S. & Jouzel J. (1992) Antarctic (Dome C) ice-core dust at 18 ky. B.P.: isotopic constraints on origins. *Earth Planet. Sci. Lett.*, 111, 175–182. *5.1.*

Grover N.C. & Howard C.S. (1938) The passage of turbid water through Lake Mead. *Trans. Am. Soc. Civ. Eng.*, 103, 720–790. *4.6.3.*

Gruszczynski M., Rudowski S., Semil J., Skominski J. & Zrobek J. (1993) Rip currents as a geological tool. *Sedimentology*, 40, 217–236. *6.2.5.*

Guardado L.R., Gamboa L.A.P. & Lucchesi C.F. (1989) Petroleum geology of the Campos Basin, Brazil, a model for a producing Atlantic type basin. In: *Divergent/Passive Margin* Basins (Ed. by J.D. Edwards and P.A. Santogrossi), pp. 3–79. *Mem. Am. Ass.*

petrol. Geol., 48, Tulsa. *10.3.4, Fig. 10.61.*

Guccione M.J. (1993) Grain-size distribution of overbank sediment and its use to locate channel positions. In: *Alluvial Sedimentation* (Ed. by M. Marzo and C. Puigdefabregas), pp. 185–194. *Spec. Publ. int Ass. Sediment.*, 17. *3.3.6.*

Guffanti M., Clynne M.A., Smith J.G., Muffler L.J.P. & Bullen T.D. (1990) Late Cenozoic volcanism, subduction and extension in the Lassen Region of California, southern Cascade range. *J. geophys. Res.*, 95, 19453–19464. *12.12.2.*

Guilcher A. (1988) *Coral Reef Morphology*, 248 pp. John Wiley, Chichester. *9.3.2, 9.4.5.*

Gulbrandsen A. (1987) Agat field. In: *Geology of the Norwegian Oil and Gas Fields* (Ed. by A.M. Spencer), pp. 363–370. Graham & Trotman, London. *10.3.4.*

Gunatilaka A., Saleh A., Al-Temeeni A. & Nassar N. (1987) Calcium-poor dolomite from the sabkhas of Kuwait. *Sedimentology*, 34, 999–1006. *8.4.3.*

Gustavson T.C. (1975) Sedimentation and physical limnology in proglacial Malaspina Lake, southeastern Alaska. In: *Glaciofluvial and Glaciolacustrine Sedimentation* (Ed. by A.V. Jopling and B.C. McDonald), pp. 249–263. *Spec. Publ. Soc. econ. Paleont. Miner.*, 23, Tulsa. *4.2, 4.3, 11.4.5.*

Gustavson T.C. (1978) Bedforms and stratification types of modern gravel meander lobes, Nueces River, Texas. *Sedimentology*, 25, 401–426. *3.3.4, 3.5.1.*

Gustavson T.C. (1991) Buried vertisols in lacustrine facies in the Pliocene Fort Hancock Formation, Hueco Bolson, West Texas and Chihuahua, Mexico. *Bull. geol. Soc. Am.*, 103, 448–460. *3.2.3.*

Gustavson T.C., Ashley G.M. & Boothroyd J.C. (1975) Depositional sequences in glaciolacustrine deltas. In: *Glaciofluvial and Glacio-lacustrine Sedimentation* (Ed. by A.V. Jopling and B.C. McDonald), pp. 264–280. *Spec. Publ. Soc. econ. Paleont. Miner.*, 23, Tulsa. *11.4.5.*

Györke O. (1973) Hydraulic model study of sediment movement and changes in the bed configuration of a shallow lake. In: *Proc. Helsinki Symp. Hydrol Lakes*, pp. 410–416. IAHS-AISH Publ. No. 109. *4.4.1.*

Haak A.B. & Schlager W. (1989) Compositional variations in calciturbidites due to sea-level fluctuations, Late Quaternary, Bahamas. *Geol. Rundsch.*, 78, 477–486. *9.4.4, 9.5.1.*

Haberyan K.A. (1985) The role of copepod fecal pellets in the deposition of diatoms in Lake Tanganyika. *Limnol. Oceanogr.*, 30, 1010–1023. *4.6.3.*

Haberyan K.A. & Hecky R.E. (1987) The Late Pleistocene and Holocene stratigraphy and paleolimnology of Lakes Kivu and Tanganyika. *Palaeogeogr., Palaeoclimatol., Palaeoecol.*, 61, 169–197. *4.4.3, 4.9.*

Hackett W.R. & Houghton B.F. (1989) A facies model for a Quaternary andesitic composite volcano: Ruapehu, New Zealand. *Bull. Volcanol.*, 51, 51–68. *12.6.2.*

Hacquebard P.A. & Donaldson J.R. (1969) Carboniferous coal deposition associated with flood-plain and limnic environments in Nova Scotia. In: *Environments of Coal Deposition* (Ed. by E.C. Dapples and M.E. Hopkins), pp. 143–191. *Spec. Pap. geol. Soc. Am.*, 114, Boulder, CO. *3.5.5.*

Hadley D.F. & Elliott T. (1993) The sequence-stratigraphic

significance of erosive based shoreface sequences in the Cretaceous Mesaverde Group of northwestern Colorado. In: *Sequence Stratigraphy and Facies Associations* (Ed. by H.W. Posamentier, C.P. Summerhayes, B.U. Haq and G.P. Allen), pp. 521–535. *Spec. Publ. int. Ass. Sediment.*, **18**. *6.6.6.*

Hallam A. & Sellwood B.W. (1976) Middle Mesozoic sedimentation in relation to tectonics in the British area. *J. Geol.*, **84**, 301–321. *7.7.3.*

Halley R.B. (1977) Ooid fabric and fracture in the Great Salt Lake and the geologic record. *J. sedim. Petrol.*, **47**, 1099–1120. *4.7.1.*

Hallock P. & Schlager W. (1986) Nutrient excess and the demise of coral reefs and carbonate platforms. *Palaios*, **1**, 389–398. *9.2, 9.3.5, 9.4.5.*

Ham W.E. (Ed.) (1962) *Classification of Carbonate Rocks—a Symposium*, 279 pp. Mem. Am. Ass. petrol. Geol., **1**, Tulsa. *1.1.*

Hamblin A.P. & Walker R.G. (1979) Storm-dominated shallow marine deposits: the Fernie–Kootenay (Jurassic) transition, southern Rocky Mountains. *Can. J. Earth Sci.*, **16**, 1673–1690. *7.7.2, Fig. 7.43.*

Hambrey M. & Alean J. (1992) *Glaciers*, 208 pp. Cambridge University Press, Cambridge. *11.5.3.*

Hambrey M.J. (1994) *Glacial Environments*; 296 pp. UCL Press, London. *11.5.1.*

Hambrey M.J. & Harland W.B. (Eds) (1981) *Earth's Pre-Pleistocene Glacial Record*, 1004 pp. Cambridge University Press, Cambridge. *11.5.1, 11.6.*

Hambrey M.J., Barrett P.J., Ehrmann W.V. & Larsen B. (1992) Cenozoic sedimentary processes on the Antarctic continental margin and the record from deep drilling. *Z. Geomorph. N.F., Suppl.-Bd.*, **86**, 77–103. *11.4.6, 11.5.3, 11.5.4.*

Hamilton D. & Smith A.J. (1972) The origin and sedimentary history of the Hurd Deep, English Channel, with additional notes on deeps in the western English Channel. *Mem. Bur. Rech. Minieres*, **79**, 59–78. *7.3.3.*

Hamilton R.F.M. & Trewin N.H. (1988) Environmental controls on fish faunas of the Middle Devonian Orcadian Basin. In: *Devonian of the World* (Ed. by N.J. McMillan, A.F. Embry & D.J. Glass), 589–600. Mem. Can. Soc. petrol. Geol., **14–3**, Calgary. *4.13.1.*

Hamilton W.R. (1988) Plate tectonics and island arcs. *Bull. geol. Soc. Am.*, **100**, 1503–1527. *12.1.1.*

Hampton M.A. (1979) Buoyancy in debris flows. *J. sedim. Petrol.*, **49**, 753–758. *10.2.3.*

Hand B.M., Wessel J.M. & Hayes M.O. (1969) Antidunes in the Mount Toby Conglomerate (Triassic), Massachusetts. *J. sedim. Petrol.*, **39**, 1310–1316. *3.6.10.*

Handford C.R. (1986) Facies and bedding sequences in shelf-storm deposited carbonates. Fayettville Shale and Pitkin Limestone (Mississippian), Arkansas. *J. sedim. Petrol.*, **56**, 123–137. *9.4.3.*

Handford C.R. (1988a) Depositional interaction of siliciclastics and marginal marine evaporites. In: *Evaporites and Hydrocarbons* (Ed. by B.C. Schreiber), pp. 139–181. Columbia University Press, New York. *8.4.2.*

Handford C.R. (1988b) Review of carbonate sand-belt deposition of ooid grainstones and the application to Mississippian reservoir, Damme Field, southwestern Kansas. *Bull. Am. Ass. petrol. Geol.*, **72**, 1184–1199. *9.4.2, Fig. 9.32.*

Handford C.R. (1990) Halite depositional facies in a solar salt pond – a key to interpreting physical energy and water depth in ancient systems? *Geology*, **18**, 691–694. *8.1.*

Handford C.R. (1991) Marginal marine halite: sabkhas and salinas. In: *Evaporites, Petroleum and Mineral Resources* (Ed. by J.L. Melvin), pp. 1–66. Elsevier, Amsterdam. *8.4.1, 8.4.2, Figs 8.10, 8.22, 8.28.*

Handford C.R. & Loucks R.G. (1993) Carbonate depositional sequences and systems tracts – responses of carbonate platforms to relative sea-level changes. In: *Carbonate Sequence Stratigraphy* (Ed. by R.G. Loucks and J.S. Sarg), pp. 3–41. *Mem. Am. Ass. petrol. Geol.*, **57**, Tulsa. *2.4, 8.10, 9.5.1, 9.5.3.*

Handford C.R., Kendall A.C., Prezbindowski D.R., Dunham J.B. & Logan B.W. (1984) Salina margin tepees, pisoliths and aragonite cements, Lake MacLeod, Western Australia: their significance in interpreting ancient analogs. *Geology*, **12**, 523–527. *8.5.2.*

Haner B.E. (1971) Morphology and sediments of Redondo Submarine Fan, southern California. *Bull. geol. Soc. Am.*, **82**, 2413–2432. *10.1, 10.3.3.*

Hanson R.E. (1991) Quenching and hydroclastic disruption of andesitic to rhyolitic intrusions in a submarine island-arc sequence, northern Sierra Nevada, California. *Bull. geol. Soc. Am.*, **103**, 804–816. *12.4.1.*

Haq B.U. (1991) Sequence stratigraphy, sea-level change, and significance for the deep sea. In: *Sedimentation, Tectonics and Eustasy: Sea-level Changes at Active Margins* (Ed. by D.I.M. Macdonald), pp. 3–39. *Spec. Publ. int. Ass. Sediment.*, **12**. *2.4, Fig. 2.15.*

Haq B.U., Hardenbol J. & Vail P.R. (1988) Mesozoic and Cenozoic chronostratigraphy and cycles of sea-level change. In: *Sea-level Changes – An Integrated Approach* (Ed. by C.K. Wilgus, B.S. Hastings, C.G. St C. Kendall, H.W. Posamentier, C.A. Ross and J.C. Van Wagoner), pp. 71–108. *Spec. Publ. Soc. econ. Paleont. Miner.*, **42**, Tulsa. *2.4. Figs 2.11, 2.12, 2.13, 2.14, 2.15.*

Hardie L.A. (Ed.) (1977) *Sedimentation on the Modern Tidal Flats of Northwest Andros Island, Bahamas*. Johns Hopkins University Press, Baltimore. *9.1.1, Fig. 9.24.*

Hardie L.A. (1984) Evaporites: marine or non-marine? *Am. J. Sci.*, **284**, 193–249. *8.1.*

Hardie L.A. (1986a) Ancient carbonate tidal-flat deposits. *Q. J. Colorado Sch. Mines*, **81**, 37–57. *7.4.1, 8.4.1, 8.5.2, 8.7, 9.4.1, Figs 8.28, 8.37.*

Hardie L.A. (1986b) Stratigraphic models for carbonate tidal flat deposition. *Q. J. Colorado Sch. Mines.*, **81**, 59–74. *9.4.1.*

Hardie L.A. (1990) The roles of rifting and hydrothermal $CaCl_2$ brines in the origin of potash evaporites: a hypothesis. *Am. J. Sci.*, **290**, 43–106. *8.1, 8.2.1.*

Hardie L.A. & Eugster H.P. (1971) The depositional environment of marine evaporites, a case for shallow, clastic accumulation. *Sedimentology*, **16**, 187–220. *8.9.*

Hardie L.A. & Ginsburg R.N. (1977) Layering: the origin and environmental significance of lamination and thin bedding. In: *Sedimentation on the Modern Carbonate Tidal Flats of Northwest Andros Island, Bahamas* (Ed. by L.A. Hardie), pp. 50–123. *Johns Hopkins Studies in Geology*, **22**. Johns Hopkins University Press, Baltimore. *9.4.1.*

Hardie L.A., Smoot J.P. & Eugster H.P. (1978) Saline lakes and their deposits. In: *Modern and Ancient Lake Sediments* (Ed. by A.

Matter and M.E. Tucker), *Spec. Publ. Int. Ass. Sediment.*, 2, 7–42. *4.6.1, 8.4.1.*

Hardie L.A., Wilson E.N. & Goldhammer R.K. (1991) *Cyclostratigraphy and Dolomitization of the Middle Triassic Latemar Buildup, the Dolomites, Northern Italy*, 56 pp. Guidebook Excursion F. Tourist office, Ortisei, Italy, Sept. 1991. *9.3.2.*

Harland W.B. (1983) The Proterozoic glacial record. In: *Proterozoic Geology: Selected Papers from an International Proterozoic Symposium* (Ed. by L.G. Medaris, Jr, C.W. Byers, D.M. Mickelson and W.C. Shanks), pp. 279–288. *Mem. Am. geol. Soc.*, 161, Boulder, CO. *11.6.*

Harland W.B., Herod K.N. & Krinsley D.H. (1966) The definition and identification of tills and tillites. *Earth Sci. Rev.*, 2, 225–256. *11.1, 11.5.1.*

Harms J.C. (1975) Stratification produced by migrating bed forms. In: *Depositional Environments as Interpreted from Primary Sedimentary Structures and Stratification Sequences*, pp. 45–61. *Soc. econ. Paleont. Miner. Short Course*, 2, Dallas. *7.1.2.*

Harms J.C. & Fahnestock R.K. (1965) Stratification, bed forms, and flow phenomena (with an example from the Rio Grande). In: *Primary Sedimentary Structures and their Hydrodynamic Interpretation* (Ed. by G.V. Middleton), pp. 84–115. *Spec. Publ. Soc. econ. Paleont. Miner.*, 12, Tulsa. *10.1.*

Harms J.C., McKenzie D.B. & McCubbin D.G. (1963) Stratification in modern sands of the Red River, Louisiana. *J. Geol.*, 71, 566–580. *3.3.4.*

Harms J.C., Southard J.B. & Walker R.G. (1982) *Structure and Sequence in Clastic Rocks. Lecture Notes: Soc. econ. Paleont. Miner. Short Course* No. 9, Calgary. *7.7.2.*

Harms J.C., Southard J.B., Spearing D.R. & Walker R.G. (1975) *Depositional Environments as Interpreted from Primary Sedimentary Structures and Stratification Sequences*, 161 pp. Lecture Notes: Soc. econ. Paleont. Miner. Course Notes, 2, Dallas. *Fig. 3.3.*

Harms J.C., Tackenberg P., Pollock R.E. & Pickles E. (1981) The Brae Field area. In: *Petroleum Geology of the Continental Shelf of Northwest Europe* (Ed. by L.V. Illing and G.D. Hobson), pp. 352–357. Heyden, London. *3.4, 6.5.4.*

Harper C.W. (1984) Improved methods of facies sequence analysis. In: *Facies Models*, 2nd edn (Ed. by R.G. Walker), pp. 11–13. *Geosci. Can. Reprint Ser.*, *Fig. 12.24.*

Harris J.P. & Fowler R.M. (1987) Enhanced prospectivity of the Mid–Late Jurassic sediments of the South Viking Graben, northern North Sea. In: *Petroleum Geology of North West Europe* (Ed. by J. Brooks and K. Glennie), pp. 879–898. Graham & Trotman, London. *10.3.4, Fig. 10.63.*

Harris M.T. (1993) Reef fabrics, biotic crusts and syndepositional cements of the Latemar reef margin. *Sedimentology*, 40, 383–401. *9.3.2.*

Harris M.T. (1994) The foreslope and toe-of-slope facies of the Middle Triassic Latemar buildup (Dolomites, northern Italy). *J. sedim. Res.*, B64, 132–145. *9.3.2, 9.4.4.*

Harris P.M. (1979) *Facies Anatomy and Diagenesis of a Bahamian Oolite Shoal*, 163 pp. *Sedimenta*, 7. Comparative Sedim. Lab., Miami. *9.1.1, 9.4.2.*

Harris P.T. & Collins M.B. (1985) Bedform distribution and sediment transport paths in the Bristol Channel and Severn Estuary, U.K. *Mar. Geol.*, 62, 153–166. *6.7.5.*

Harrison C.G.A., Brass G.W., Saltzman E. & Sloan J. II (1981) Sea level variations, global sedimentation rates and the hypsographic curve. *Earth planet. Sci. Lett.*, 54, 1–16. *2.1.4.*

Hart B.S. & Plint A.G. (1989) Gravelly shoreface deposits: a comparison of modern and ancient facies sequences. *Sedimentology*, 36, 551–557. *6.5.4.*

Hart M.B. (1987) Orbitally-induced cycles in the Chalk facies of the United Kingdom. *Cretaceous. Res.*, 8, 335–348. *10.4.3.*

Hartley A.J., Flint S., Turner P. & Jolley E.J. (1992) Tectonic controls on the development of a semi-arid alluvial basin as reflected in the stratigraphy of the Purilactis Group (Upper Cretaceous–Eocene), northern Chile. *J. S. Am. Earth Sci.*, 5, 275–296. *12.12.4.*

Hartshorn J.H. (1958) Flowtill in southeastern Massachusetts. *Bull. geol. Soc. Am.*, 69, 477–482. *11.4.3.*

Harvey A.M. (1987) Ancient fan dissection: relationships between morphology and sedimentation. In: *Desert Sediments: Ancient and Modern* (Ed. by L.E. Frostick and I. Reid), pp. 87–103. *Spec. Publ. geol. Soc. Lond.*, 35. *3.3.8.*

Harvie C.E., Weare J.H., Hardie L.A. & Eugster H.P. (1980) Evaporation of seawater: calculated mineral sequences. *Science*, 208, 498–500. *8.2.1.*

Harwood G.M. & Towers P.A. (1988) Seismic sedimentologic interpretation of a carbonate slope, north margin of Little Bahama Bank. *Init. Rep. ODP*, 101, 263–277. *9.4.4.*

Haszeldine R.S. (1983) Descending tabular cross-bed sets and bounding surfaces from a fluvial channel in the Upper Carboniferous coalfield of north-east England. In: *Modern and Ancient Fluvial Systems* (Ed. by J.D. Collinson and J. Lewin), pp. 449–456. *Spec. Publ. int. Ass. Sediment.*, 6. *3.3.3, 3.5.2, 3.6.2, Fig. 3.40.*

Haszeldine R.S. (1984) Muddy deltas in freshwater lakes, and tectonism in the Upper Carboniferous Coalfield of NE England. *Sedimentology*, 31, 811–822. *3.5.3, 3.6.8, Fig. 3.59.*

Hatton I.R. (1986) Geometry of allochthonous Chalk Group members Central Trough, North Sea. *Mar. petrol. Geol.*, 3, 79–98. *10.4.3.*

Haughton P.D.W. (1993) Simultaneous dispersal of volcaniclastic and non-volcanic sediment in fluvial basins: examples from the Lower Old Red Sandstone, east-central Scotland. In: *Alluvial Sedimentation* (Ed. by M. Marzo and C. Puigdefabregas), pp. 451–471. *Spec. Publ. int. Ass. Sediment.*, 17. *3.5.2.*

Havholm K.G. & Kocurek G. (1994) Factors controlling eolian sequence stratigraphy: clues from super bounding surface features in the Middle Jurassic Page Sandstone. *Sedimentology*, 41, 913–934. *5.5.2, 5.5.3, Fig. 5.25.*

Havholm K.G., Blakey R.C., Capps M., Jones L.S., King D.D. & Kocurek G. (1993) Eolian genetic stratigraphy: an example from the Middle Jurassic Page Sandstone, Colorado Plateau. In: *Aeolian Sediments Ancient and Modern* (Ed. by K. Pye and N. Lancaster), pp. 87–111. *Spec. Publ. int. Ass. Sediment.*, 16. *5.5.2, Fig. 5.25.*

Hawkins J., Parson L., Allan J. *et al.* (1994) *Proc. Ocean Drilling Program Sci. Results*, 135. College Station, TX. *10.4.1.*

Hay R.L. (1952) The terminology of fine-grained detrital volcanic rocks. *J. sedim. Petrol.*, 22, 119–120. *12.7.*

Hay R.L. (1968) Chert and its sodium silicate precursors in sodium carbonate lakes of East Africa. *Contrib. Miner. Petrol.*, 17, 255–274. *4.14.1.*

Hay R.L. (1986) Role of tephra in the preservation of fossils in Cenozoic deposits of East Africa. In: *Sedimentation in the African Rifts* (Ed. by L.E. Frostick, R.W. Renaut, I. Reid and J.-J. Tiercelin), pp. 339–344. *Spec. Publ. geol. Soc. Lond.*, 25. *2.1.10.*

Hay W.W., Sibuet J.-C. *et al.* (1984) *Initial Reports of the Deep Sea Drilling Project*, 75, 1303 pp. US Government Printing Office, Washington, DC. *10.2.1, Fig. 10.7.*

Hayes B.J.R. (1988) Incision of a Cadotte Member paleovalley-system at Noel, British Columbia – evidence of a Late Albian sea-level fall. In: *Sequences, Stratigraphy, Sedimentology; Surface and Subsurface* (Ed. by D.P. James and D.A. Leckie), pp. 97–105. *Mem. Can. Soc. petrol. Geol.*, 15, Calgary.

Hayes M.O. (1967) *Hurricanes as Geological Agents: Case Studies of Hurricanes Carla, 1961, and Cindy, 1963.* 54 pp. Bur. econ. Geol., Rep. Invest. 61, Austin, TX. *6.7.4, 7.4.3, 7.4.4, Fig. 7.22.*

Hayes M.O. (1975) Morphology of sand accumulation in estuaries: an introduction to the symposium. In: *Estuarine Research*, Vol. II *Geology and Engineering* (Ed. by L.E. Cronin), pp. 3–22. Academic Press, London. *6.1, Fig. 6.3.*

Hayes M.O. (1979) Barrier island morphology as a function of tidal and wave regime. In: *Barrier Islands – from the Gulf of St Lawrence to the Gulf of Mexico* (Ed. by S.P. Leatherman), pp. 1–27. Academic Press, New York. *6.1, 6.3, Figs 6.3, 6.19, 6.75.*

Hayes M.O. & Kana T.W. (1976) *Terrigenous Clastic Depositional Environments – Some Modern Examples*, pp. I-131, II-184, *Tech. Rept, 11–CRD*, Coastal Res. Div., Univ. South Carolina. *6.7.4.*

Haymon R.M. (1983) Growth history of hydrothermal black smoker chimneys. *Nature*, 301, 695–698. *12.11.1.*

Haymon R.M., Koski R.A. & Sinclair C. (1984) Fossils of hydrothermal vent worms from Cretaceous sulfide ores of the Samail Ophiolite, Oman. *Science*, 223, 1407–1409. *12.11.1.*

Heckel P.H. (1972) Recognition of ancient shallow marine environments. In: *Recognition of Ancient Sedimentary Environments* (Ed. by J.K. Rigby and W.K. Hamblin), pp. 226–286. *Spec. Publ. econ. Paleont. Miner.*, 16, Tulsa. *4.10.1.*

Heckel P.H. (1974) Carbonate buildups in the geologic record: a review. In: *Reefs in Time and Space* (Ed. by L.F. Laporte), pp. 90–154. *Spec. Publ. Soc. econ. Paleont. Miner.*, 18. *9.1.1, 9.2.1, 9.4.5.*

Hedenquist J.W. & Henley R.W. (1985) Hydrothermal eruptions in the Waiotapu Geothermal system, New Zealand: their origin, associated brecciation, and relation to precious metal mineralization. *Econ. Geol.*, 80, 1640–1680. *12.4.1.*

Hedenquist J.W. & Lowenstern J.B. (1994) The role of magmas in the formation of hydrothermal ore deposits. *Nature*, 370, 519–527. *Fig. 12.72.*

Heezen B.C. & Ewing M. (1952) Turbidity currents and submarine slumps, and the 1929 Grand Banks earthquake. *Am. J. Sci.*, 250, 849–873. *10.2.3.*

Heezen B.C. & Hollister C.D. (1971) *The Face of the Deep*, 659 pp. Oxford University Press, New York. *10.2.3, 10.3.7.*

Heezen B.C., Hollister C.D. & Ruddiman W.F. (1966) Shaping of the continental rise by deep geostrophic contour currents. *Science*, 152, 502–508. *10.1.*

Heiken G.H. (1971) Tuff rings: examples from the Fort Rock–Christmas Lake valley basin, South-central Oregon. *J. geophys. Res.*, 76, 5615–5626. *12.9.2.*

Heiken G.H. & Wohletz K.H. (1987) Tephra deposits associated with silicic domes and lava flows. *Spec. Pap. geol. Soc. Am.*, 212, 55–76. *12.10, Fig. 12.53.*

Heiken G.H. & Wohletz K.H. (1991) Fragmentation processes in explosive volcanic eruptions. In: *Sedimentation in Volcanic Settings* (Ed. by R.V. Fisher and G.A. Smith), pp. 19–26. *Spec. Publ. Soc. econ. Paleont. Miner.*, 45, Tulsa. *12.4.1.*

Hein F.J. (1984) Deep-sea and fluvial braided channel conglomerates: a comparison of two case studies. In: *Sedimentology of Gravels and Conglomerates* (Ed. by E.H. Koster and R.J. Steel), pp. 33–49. *Mem. Can. Soc. petrol. Geol.*, 10, Calgary. *3.4, 6.5.4.*

Helland-Hansen W., Ashton M., Lømo L. & Steel R. (1992) Advance and retreat of the Brent delta: recent contributions to the depositional model. In: *Geology of the Brent Group* (Ed. by A.C. Morton, R.S. Haszeldine, M.R. Giles and S. Brown), pp. 109–127. *Spec. Publ. geol. Soc. Lond.*, 61. *6.6.6, Fig. 6.60.*

Helland-Hansen W., Steel R., Nakayama K. & Kendall C.G. St C. (1989) Review and computer modelling of the Brent Group stratigraphy. In: *Deltas: Sites and Traps for Fossil Fuels* (Ed. by M.K.G. Whateley and K.T. Pickering), pp. 237–252. *Spec. Publ. geol. Soc. Lond.*, 41. *6.6.6.*

Heller P.L. & Dickinson W.R. (1985) Submarine ramp facies model for delta-fed, sand-rich turbidite systems. *Bull. Am. Ass. petrol. Geol.*, 69, 960–976. *10.1, 10.3.4, 12.12.5.*

Heller P.L. & Paola C. (1992) The large-scale dynamics of grain-size variation in alluvial basins, 2: Application to syntectonic conglomerate. *Basin Res.*, 4, 91–102. *3.6.3.*

Hendry H.E. & Stauffer M.R. (1977) Penecontemporaneous folds in cross-bedding: inversion of facing criteria and mimicry of tectonic folds. *Bull. geol. Soc. Am.*, 88, 809–812. *3.5.2.*

Henry C.D. & Wolff J.A. (1992) Distinguishing strongly rheomorphic tuffs from extensive silicic lavas. *Bull. Volcanol.*, 54, 171–186. *12.4.4, 12.4.5.*

Hequette A. & Hill P.R. (1993) Storm-generated currents and offshore sediment transport on a sandy shoreface, Tibjak Beach, Canadian Beaufort Sea. *Mar. Geol.*, 113, 283–304. *7.4.3.*

Hereford R. (1977) Deposition of the Tapeats Sandstone (Cambrian) in central Arizona. *Bull. geol. Soc. Am.*, 88, 199–211. *7.8.1.*

Héritier F.E., Lossel P. & Wathne E. (1979) Frigg Field – large submarine-fan trap in Lower Eocene rocks of North Sea Viking Graben. *Bull. Am. Ass. petrol. Geol.*, 63, 1999–2020. *10.3.3.*

Hesse R. (1975) Turbiditic and non-turbiditic mudstone of Cretaceous flysch sections of the East Alps and other basins. *Sedimentology*, 22, 387–416. *10.2.1.*

Hesse R. (1992) Continental slope sedimentation adjacent to an ice margin. In: Seismic facies of Labrador slope. *Geo. Mar. Letts*, 12, 189–199. *10.3.5.*

Heward A.P. (1978) Alluvial fan sequence and mega sequence models: with examples from Westphalian D–Stephanian B coalfields, Northern Spain. In: *Fluvial Sedimentology* (Ed. by A.D. Miall), pp. 669–702. *Mem. Can. Soc. petrol. Geol.*, 5, Calgary. *3.6.3, 6.5.6, Fig. 6.27.*

Heward A.P. (1981) A review of wave-dominated clastic shoreline deposits. *Earth-Sci. Rev.*, 17, 223–276. *6.7.6*, *Figs 6.1, 6.53,* *Fig. 6.87.*

Hickin E.J. (1974) The development of meanders in natural river channels. *Am. J. Sci.*, 274, 414–442. *3.3.4.*

Hicock S.R. (1991) On subglacial stone pavements in till. *J. Geol.*, 99, 607–619. *11.4.1.*

Higgins G.M., Ahmad M. & Brinkman R. (1973) The Thal interfluve, Pakistan; Geomorphology and depositional history. *Geol. Mijnbouw.*, 52, 147–155. *3.2.2.*

Higgs J. (1990) Is there evidence for geostrophic currents preserved in the sedimentary record of inner to middle-shelf deposits? Discussion. *J. sedim. Petrol.*, 60, 630–632. *7.7.2.*

Hildebrand R.S. & Bowring S.A. (1984) Continental intra-arc depressions: a nonextensional model for their origin with a Proterozoic example from Wopmay orogen. *Geology*, 12, 73–77. *12.12.1.*

Hildreth W. (1981) Gradients in silicic magma chambers: implications for lithospheric magmatism. *J. geophys. Res.*, 86, 10153–10192. *12.3.2, Fig. 12.7.*

Hildreth W. & Mahood G. (1985) Correlation of ash flow tuffs. *Bull. geol. Soc. Am.*, 96, 968–974. *12.14.1.*

Hill D.P., Bailey R.A. & Ryall A.S. (1985) Active tectonic and magmatic processes beneath Long Valley Caldera, Eastern California: an overview. *J. geophys. Res.*, 90, 11111–11120. *Fig. 12.31.*

Hill P.R. (1984) Sedimentary facies of the Nova Scotian upper and middle continental slope, offshore eastern Canada. *Sedimentology*, 31, 293–309. *10.3.5.*

Hillaire-Marcel C. & Casanova J. (1987) Isotopic hydrology and paleohydrology of the Magadi (Kenya) – Natron (Tanzania) basin during the Late Quaternary. *Palaeogeogr. Palaeoclimatol. Palaeoecol.*, 58, 155–181. *4.7.1.*

Hine A.C. (1975) Bedform distribution and migration patterns on tidal deltas in the Chatham Harbor estuary, Cape Cod, Massachusetts. In: *Estuarine Research*, Vol. II *Geology and Engineering* (Ed. by L.E. Cronin), pp. 235–252. Academic Press, London. *6.7.4.*

Hine A.C. (1977) Lily Bank, Bahamas: history of an active oolite sand shoal. *J. sedim. Petrol.*, 47, 1554–1581. *9.1.1, 9.4.2, Fig. 9.32.*

Hine A.C. (1979) Mechanisms of berm development and resulting beach growth along a barrier spit complex. *Sedimentology*, 26, 333–351. *Fig. 6.77.*

Hine A.C. & Neumann A.C. (1977) Shallow carbonate bank margin growth and structure, Little Bahama Bank, Bahamas. *Bull. Am. Ass. petrol. Geol.*, 61, 376–406. *9.3.2.*

Hine A.C., Harris M.W., Locker S.D. *et al.* (1994) Sedimentary infilling of an open seaway: Bawihka Channel, Nicaraguan Rise. *J. sedim. Res.*, B64, 2–25. *9.3.1.*

Hine A.C., Wilber R.J. & Neumann A.C. (1981) Carbonate sand-bodies along contrasting shallow-bank margins facing open seaways, northern Bahamas. *Bull. Am. Ass. petrol. Geol.*, 65, 261–290. *9.3.2, 9.4.2.*

Hirst J.P.P. (1991) Variations in alluvial architecture across the Oligo-Miocene Huesca fluvial system, Ebro Basin, Spain. In: *The Three-dimensional Facies Architecture of Terrigenous Clastic Sediments and its Implications for Hydrocarbon Discovery and Recovery* (Ed. by A.D. Miall and N. Tyler), pp. 111–121. Soc. econ. Paleont.

Miner. Concepts in Sedimentology and Paleontology, 3, Tulsa. *3.6.5, Fig. 3.49.*

Hiscott R.N. & James N.P. (1985) Carbonate debris flows. Cow Head Group, Western Newfoundland. *J. sedim. Petrol.*, 55, 735–745. *9.4.4.*

Hiscott R.N., James N.P. & Pemberton S.G. (1984) Sedimentology and ichnology of the Lower Cambrian Bradore Formation, coastal Labrador: fluvial to shallow-marine transgressive sequence. *Bull. Can petrol. Geol.*, 32, 11–26. *7.8.1.*

Hite R.J. (1970) Shelf carbonate sedimentation controlled by salinity in the Paradox Basin, southeast Utah. In: *Third International Symposium on Salt*. N. Ohio Geol. Soc., 1, 48–66, Cleveland. *8.10.2, Fig. 8.54.*

Hjulström F. (1952) The geomorphology of the alluvial outwash plains (Sandurs) of Iceland and the mechanism of braided rivers. *8th Gen. Assembly and Proc. 17th Internat. Congress. Internat. Geograph. Union*, 337–342. *3.3.2.*

Ho K.F. (1978) Stratigraphic framework for oil exploration in Sarawak. *Bull. geol. Soc. Malaysia*, 1–13. *6.6.6, 9.5.5.*

Hobday D.K. & Reading H.G. (1972) Fair weather versus storm processes in shallow marine sand bar sequences in the late Pre-Cambrian of Finnmark, north Norway. *J. sedim. Petrol.*, 42, 318–324. *7.1.2.*

Hobday D.K. & Tankard A.J. (1978) Transgressive-barrier and shallow-shelf interpretation of the lower Paleozoic Peninsula Formation, South Africa. *Bull. geol. Soc. Am.*, 89, 1733–1734. *6.7.7, 7.8.1.*

Ho Kiam Fui (1978) Stratigraphic framework for oil exploration in Sarawak. *Bull. geol. Soc. Malaysia*, 10, 1–13. *6.6.6.*

Holdsworth G. (1973) Ice calving into the proglacial Generator Lake, Baffin Island, NWT, Canada. *J. Glaciol.*, 12, 235–250. *11.4.5.*

Hollister C.D. & Heezen B.C. (1967) Contour current evidence from abyssal sediments. *Trans. Am. geophys. Union*, 48, 142. *10.1.*

Hollister C.D. & Heezen B.C. (1972) Geological effects of ocean bottom currents: western North Atlantic. In: *Studies in Physical Oceanography*, vol. 2 (Ed. by A.L. Gordon), pp. 37–66. Gordon & Breach, London. *10.1.*

Hollister C.D. & McCave I.N. (1984) Sedimentation under deep-sea storms. *Nature*, 309, 220–225. *10.2.2, 10.3.7.*

Hollister C.D., Flood R.D., Johnson D.A., Lonsdale P.F. & Southard J.B. (1974) Abyssal furrows and hyperbolic echo-traces on the Bahama Outer Ridge. *Geology*, 2, 395–400. *10.3.7.*

Holmes A. (1965) *Principles of Physical Geology*, 2nd edn, 1288 pp. Nelson, London. *6.5.*

Holser W.T. (1979) Mineralogy of evaporites, trace elements and isotopes in evaporites. In: *Marine Minerals* (Ed. by R.G. Burns), pp. 211–346. Mineralogical Society of America. *8.2.1.*

Hon K., Kauahikaua J., Denlinger R. & McKay K. (1994) Emplacement and inflation of pahoehoe sheet flows – observations and measurements of active flows on Kilauea volcano, Hawaii. *Bull. geol. Soc. Am.*, *12.4.5.*

Honnorez J. & Kirst P. (1975) Submarine basaltic volcanism: morphometric parameters for discriminating hyaloclastites from hyalotuffs. *Bull. Volcanol.*, 39, 441–465. *12.4.1.*

Hooke R. Le B. (1967) Processes on arid-region alluvial fans. *J. Geol.*, 75, 438–460. *3.3.8, Fig. 3.28.*

References

....

Hopson C.A. & Melson W.G. (1990) Compositional trends and eruptive cycles at Mount St Helens. *Geosci. Can.*, **17**, 131–141. *12.6.*

Horowitz A. (1979) *The Quaternary of Israel*, 394 pp. Academic Press, New York. *4.9.*

Hottinger L. (1989) Conditions for generating carbonate platforms. *Mem. Soc. geol. Ital.*, **40**, 265–271. *9.2.2.*

Houbolt J.H.C. (1968) Recent sediments in the southern bight of the North Sea. *Geol. Mijnbouw*, **47**, 254–273. *7.3.3, 7.3.4, 7.7.1.*

Houbolt J.H.C. (1982) A comparison of recent shallow marine sand ridges with Miocene sand ridges in Belgium. In: *The Ocean Floor* (Ed. by R.A. Scrutton and M. Talwani), pp. 69–80. Wiley, New York. *7.3.4, 7.7.1.*

Houghton B.F. & Hackett W.R. (1984) Strombolian and phreatomagmatic deposits of Ohakune Craters, Ruapehu, New Zealand: a complex interaction between external water and rising basaltic magma. *J. Volcanol. geotherm. Res.*, **21**, 207–231. *12.9.2.*

Houghton B.F. & Landis C.A. (1989) Sedimentation and volcanism in a Permian arc-related basin, southern New Zealand. *Bull. Volcanol.*, **51**, 433–450. *12.12.1, 12.12.5.*

Houghton B.F. & Nairn I.A. (1991) The 1976–1982 Strombolian and phreatomagmatic eruptions of White Island, New Zealand: eruptive and depositional mechanisms at a 'wet' volcano. *Bull. Volcanol.*, **54**, 25–49. *12.4.1.*

Houghton B.F. & Schmincke H.-U. (1989) Rothenberg scoria cone, East Eiffel: a complex Strombolian and phreatomagmatic volcano. *Bull. Volcanol.*, **52**, 28–48. *12.4.1, 12.9.1, 12.9.2.*

Houghton B.F. & Smith R.T. (1993) Recycling of magmatic clasts during explosive eruptions: estimating the true juvenile content of phreatomagmatic volcanic deposits. *Bull. Volcanol.*, **55**, 414–420. *12.4.1.*

Houghton B.F. & Wilson C.J.N. (1989) A vesicularity index for pyroclastic deposits. *Bull. Volcanol.*, **51**, 451–462. *12.4.1.*

Houghton B.F., Wilson C.J.N., McWilliams M.O., Lanphere M.A., Weaver S.D., Briggs R.M. & Pringle M.S. (1995) Chronology and dynamics of a large silicic magmatic system: central Taupo Volcanic Zone, New Zealand. *Geology*, *12.13.*

House M.R. & Gale A.S. (Eds) (1995) *Orbital Forcing Timescales and Cyclostratigraphy*, 210 pp. *Spec. Publ. geol. Soc. Lond.*, **85**. *10.2.1.*

Houthuys R. & Gullentops F. (1988) The Vlierzele Sands (Eocene, Belgium): a tidal ridge system. In: *Tide-Influenced Sedimentary Environments and Facies* (Ed. by P.L. de Boer, A. van Gelder and S.-D. Nio), pp. 139–152. Reidel, Dordrecht. *7.7.1.*

Hovorka S. (1987) Depositional environments of marine-dominated bedded halite, Permian San Andres Formation, Texas. *Sedimentology*, **34**, 1029–1054. *8.8, 8.10.1, Fig. 8.42.*

Hovorka S. (1992) Halite pseudomorphs after gypsum in bedded anhydrite – clue to gypsum–anhydrite relationships. *J. sedim. Petrol.*, **62**, 1098–1111. *8.5.1, 8.8.*

Howard J.D. & Reineck H.-E. (1981) Depositional facies of high-energy beach-to-offshore sequence, comparison with low energy sequence. *Bull. Am. Ass. petrol. Geol.*, **65**, 807–830. *6.2.3, 6.7.6, Fig. 6.85, Fig. 7.22.*

Howard J.D., Frey R.W. & Reineck H.-E. (1972) Introduction. *Senckenberg. Mar.*, **4**, 3–14. *6.7.6, Fig. 6.85.*

Howe J.A., Stoker M.S. & Stow D.A.V. (1994) Late Cenozoic sediment drift complex, northeast Rockall Trough, North Atlantic. *Paleoceanography*, **9**, 989–999. *10.3.7.*

Howells M.F., Reedman A.J. & Campbell S.D.G. (1986) The submarine eruption and emplacement of the Lower Rhyolitic Tuff Formation (Ordovician), N Wales. *J. geol. Soc. Lond.*, **143**, 411–423. *12.4.6, Fig. 12.91.*

Howells M.F., Reedman A.J. & Campbell S.D.G. (1991) *Ordovician (Caradoc) Marginal Basin Volcanism in Snowdonia (North-west Wales)*, 191 pp. HMSO for the British Geological Survey, London. *12.13.*

Hoyt J.H. (1967) Barrier island formation. *Bull. geol. Soc. Am.*, **78**, 1125–1135. *6.6.5.*

Hoyt J.H. (1969) Chenier versus barrier: genetic and stratigraphic distinction. *Bull. Am. Ass. petrol. Geol.*, **53**, 299–306. *6.7.2, Fig. 6.72.*

Hoyt J.H. & Henry V.J. (1967) Influence of island migration on barrier island sedimentation. *Bull. geol. Soc. Am.*, **78**, 77–86. *6.7.4, Fig. 6.78.*

Hsü K.J. (1972) Origin of saline giants: a critical review after the discovery of the Mediterranean evaporite. *Earth Sci. Rev.*, **8**, 371–396. *8.1.*

Hsü K.J. (1975) Catastrophic debris streams (sturzstroms) generated by rockfalls. *Bull. geol. Soc. Am.*, **86**, 129–140. *12.5.1.*

Hsü K.J. & Kelts K. (1978) Late Neogene chemical sedimentation in the Black Sea. In: *Modern and Ancient Lake Sediments* (Ed. by A. Matter and M.E. Tucker), pp. 129–145. *Spec. Publ. int. Ass. Sediment.*, *2.4.1, 4.12.*

Hsü K.J. & Siegenthaler C. (1980) Preliminary experiments on hydrodynamic movement induced by evaporation and the bearing on the dolomite problem. *Sedimentology*, **12**, 11–25. *8.4.2.*

Hsü K.J., Kelts K. & Valentine J.W. (1980) Resedimented facies in Ventura Basin, California, and model of longitudinal transport of turbidity currents. *Bull. Am. Ass. petrol. Geol.*, **64**, 1034–1051. *10.3.6.*

Hsueh Y., Wang J. & Chern C.-S. (1992) The intrusion of the Kuroshio across the continental shelf northeast of Taiwan. *J. geophys. Research*, **97**, 14323–14330. *7.6.*

Huang Z., Boyd R. & O'Connell S. (1992) Upper Cretaceous cyclic sediments from Hole 762C, Exmouth Plateau, NW Australia. In: *Proceedings of the Ocean Drilling Program Sci. Results, 122* (Ed. by U. Von Rad, B.U. Haq *et al.*), pp. 259–278. College Station, Texas. *10.4.2.*

Hubbard D.K., Burke R.B. & Gill I.P. (1986) Styles of reef accretion along a steep, shelf edge reef, St Croix, US Virgin Islands. *J. sedim. Petrol.*, **56**, 848–861. *9.4.5.*

Hubbard D.K., Miller A.I. & Scaturo D. (1990) Production and cycling of calcium carbonate in a shelf edge reef system (St Croix, US Virgin Islands): applications to the nature of reef systems in the fossil record. *J. sedim. Petrol.*, **60**, 335–360. *9.4.5.*

Hubbard D.K., Oertel G. & Nummedal D. (1979) The role of waves and tidal currents in the development of tidal-inlet sedimentary structures and sand body geometry: examples from North Carolina, South Carolina and Georgia. *J. sedim. Petrol.*, **49**, 1073–1092. *6.7.4, Fig. 6.76.*

Hubbard R.J. (1988) Age and significance of sequence boundaries on Jurassic and early Cretaceous rifted continental margins. *Bull. Am. Ass. petrol. Geol.*, **72**, 49–72. *2.4.*

Hubert J.F. & Filipov A.J. (1989) Debris-flow deposits in alluvial fans on the west flank of the White Mountains, Owens Valley, California, U.S.A. *Sedim. Geol.*, 61, 177–205. *3.3.8.*

Hubert J.F. & Hyde M.G. (1983) Sheet flow deposits of graded beds and mudstones on an alluvial sandflat-playa system: Upper Triassic Blomidon redbeds, St Mary's Bay, Nova Scotia. *Sedimentology*, 29, 457–474. *3.5.3, 3.6.10.*

Huc A.Y., Le Fournier J., Vandenbroucke M. & Bessereau G. (1990) Northern Lake Tanganyika – an example of organic sedimentation in an anoxic rift lake. In: *Lacustrine Basin Exploration – Case Studies and Modern Analogs* (Ed. by B.J. Katz), 169–185. *Mem. Am. Ass. petrol. Geol.*, 50, Tulsa. *4.7.5.*

Huddle J.W. & Patterson S.H. (1961) Origin of Pennsylvanian underclay and related seat rocks. *Geol. Soc. Am. Bull.*, 72, 1643–1660. *3.5.4.*

Hughes D.A. & Lewin J. (1982) A small-scale flood plain. *Sedimentology*, 29, 891–895. *3.3.6.*

Hughes Clarke J.E., Shor A.N., Piper D.J.W. & Mayer L.A. (1990) Large-scale current-induced erosion and deposition in the path of the 1929 Grand Banks turbidity current. *Sedimentology*, 37, 613–629. *10.2.3.*

Hulen J.B., Nielson D.L. & Little T.M. (1991) Evolution of the Western Valles caldera complex, New Mexico: evidence from intracaldera sandstones, breccias, and surge deposits. *J. geophys. Res.*, 96, 8127–8142. *12.4.6, 12.12.4.*

Hummel G. & Kocurek G. (1984) Interdune areas of the back-island dune field, north Padre Island, Texas. *Sedim. Geol.*, 39, 1–26. *5.4.5.*

Hunt C.B. (1972) *Geology of Soils*, 344 pp. Freeman, San Francisco. *3.2.3.*

Hunt D. & Tucker M.E. (1993) The sequence stratigraphy of carbonate shelves with an example from the Middle Cretaceous of SE France. In: *Sequence Stratigraphy and Facies Associations* (Ed. by H. Posamentier, C.P. Summerhayes, B.U. Haq and G.P. Allen), pp. 307–341. *Spec. Publ. int. Ass. Sediment.*, 18. *9.5.1.*

Hunt J.W. & Hobday D.K. (1984) Petrographic composition and sulphur content of coals associated with alluvial fans in the Permian Sydney and Gunnedah basins, eastern Australia. In: *Sedimentology of Coal and Coal-bearing Sequences* (Ed. by R.A. Rahmani and R.M. Flores), pp. 43–60. *Spec. Publ. int. Ass. Sediment.*, 7. *3.5.5.*

Hunt R.E., Swift D.J.P. & Palmer H. (1977) Constructional shelf topography, Diamond Shoals, North Carolina. *Bull. geol. Soc. Am.*, 88, 299–311. *7.4.2.*

Hunter R.E. (1973) Pseudo-cross lamination formed by climbing adhesion ripples. *J. sedim. Petrol.*, 43, 1125–1127. *5.4.5.*

Hunter R.E. (1977a) Basic types of stratification in small eolian dunes. *Sedimentology*, 24, 361–387. *5.3.3, 5.3.4, 5.4.4, Fig. 5.3.*

Hunter R.E. (1977b) Terminology of cross-stratified sedimentary layers and climbing-ripple structures. *J. sedim. Petrol.*, 47, 697–706. *5.3.4.*

Hunter R.E. (1981) Stratification styles in eolian sandstones: some Pennsylvanian to Jurassic examples from the Western Interior U.S.A. In: *Recent and Ancient Nonmarine Depositional Environments: Models for Exploration* (Ed. by F.G. Ethridge and R.M. Flores), pp. 315–329. *Spec. Publ. Soc. econ. Paleont. Miner.*, 31, Tulsa. *5.3.3, 5.5.4.*

Hunter R.E. (1985a) A kinematic model for the structure of lee-side deposits. *Sedimentology*, 32, 409–422. *5.3.3.*

Hunter R.E. (1985b) Subaqueous sand-flow cross-strata. *J. sedim. Petrol.*, 55, 886–894. *5.3.3.*

Hunter R.E. (1993) An eolian facies in the Ste. Genevieve Limestones of southern Indiana. In: *Mississippian Oolites and Modern Analogs* (Ed. by B.D. Keith and C.W. Zuppann), pp. 31–48. *Am. Ass. petrol. Geol. Studies Geol.*, 35, Tulsa. *5.5.4, 9.4.2.*

Hunter R.E. & Clifton H.E. (1982) Cyclic deposits and hummocky-cross-stratification of probable storm origin in Upper Cretaceous rocks of Cape Sebastian area, southwestern Oregon. *J. sedim. Petrol.*, 52, 127–143. *7.7.2.*

Hunter R.E. & Kocurek G. (1986) An experimental study of subaqueous slipface deposition. *J. sedim. Petrol.*, 56, 387–394. *5.3.3.*

Hunter R.E. & Richmond B.M. (1988) Daily cycles in coastal dunes. *Sedim. Geol.*, 55, 43–67. *5.4.4.*

Hunter R.E. & Rubin D.M. (1983) Interpreting cyclic crossbedding, with an example from the Navajo Sandstone. In: *Eolian Sediments and Processes* (Ed. by M.E. Brookfield and T.S. Ahlbrandt), pp. 429–454. *Developments in Sedimentology*, 38. Elsevier, Amsterdam. *5.5.4.*

Hunter R.E., Clifton H.E. & Phillips R.L. (1979) Depositional processes, sedimentary structures and predicted vertical sequences in barred nearshore systems, southern Oregon coast. *J. sedim. Petrol.*, 49, 711–726. *6.7.6, Fig. 6.85.*

Hunter R.E., Richmond B.M. & Alpha T.R. (1983) Storm-controlled oblique dunes of the Oregon coast. *Bull. geol. Soc. Am.*, 94, 1450–1465. *5.3.3, 5.4.3, 5.4.4, 5.4.5, Figs 5.19, 5.21.*

Huntoon J.E. & Chan M.A. (1987) Marine origin of paleotopographic relief on eolian White Rim Sandstone (Permian), Elaterite Basin, Utah. *Bull. Am. petrol. Geol.*, 71, 1035–1045. *5.5.3.*

Hurst A., Griffiths C.M. & Worthington P.F. (Eds) (1992) *Geological Applications of Wireline Logs II*, 406 pp. *Spec. Publ. geol. Soc. Lond.*, 65. *2.3.4.*

Hurst A., Lovell M.A. & Morton A. (Eds) (1990) *Geological Applications of Wireline Logs*, 357 pp. *Spec. Publ. geol. Soc. Lond.*, 48. *2.3.4.*

Hurst J.M. & Surlyk F. (1984) Tectonic control of Silurian carbonate-shelf margin morphology and facies, North Greenland. *Bull. Am. Ass. petrol. Geol.*, 68, 1–17. *9.5.3.*

Hussain M. & Warren J.K. (1989) Dolomitization in a sulfate-rich environment. Modern example from Salt Flat Sabkha (dried playa lake) in west Texas–New Mexico. *Carbon. Evap.*, 3, 165–173. *8.10.1.*

Hutchinson D.R., Golmshtok A.J., Zonenshain L.P., Moore T.C., Scholz C.A. & Klitgord K.D. (1992) Depositional and tectonic framework of the rift basins of Lake Baikal from multichannel seismic data. *Geology*, 20, 589–592. *4.2.*

Hutchinson G.E. (1957) *A Treatise on Limnology, Vol. 1: Geography, Physics and Chemistry*, pp. 1015. Wiley, New York. *4.4, 4.4.1.*

Hwang I.G. & Chough S.K. (1990) The Miocene Chunbuk Formation, southeastern Korea: marine Gilbert-type fan-delta system. In: *Coarse-grained Deltas* (Ed. by A. Colella and D.B. Prior), pp. 235–254. *Spec. Publ. int. Ass. Sediment.*, 10. *6.5.6.*

Hyne N.J., Cooper W.A. & Dickey P.A. (1979) Stratigraphy of inter-montane, lacustrine delta, Catatumbo River, Lake

Maracaibo, Venezuela. *Bull. Am. Ass. petrol. Geol.*, 63, 2042–2057. *4.6.1.*

Ikebe N. & Yokoyama T. (1976) General explanation of the Kobiwako Group – ancient lake deposits of Lake Biwa. In: *Palaeoclimatology of Lake Biwa and the Japanese Pleistocene* (Ed. by Shoji Horie), pp. 31–51. Kyoto University, Shiga-Kev. *4.2.*

Ikehara K. (1988) Ocean current generated sedimentary facies in the Osumi Strait, south of Kyushu, Japan. *Prog. Oceanogr.*, 21, 515–524. *7.6.*

Illenberger W.K. & Rust I.C. (1988) A sand budget for the Alexandria coastal dunefield, South Africa. *Sedimentology*, 35, 513–521. *5.4.1.*

Illing L.V. (1954) Bahamian calcareous sands. *Bull. Am. Ass. petrol. Geol.*, 38, 1–95. *9.1.1.*

Imbrie J. & Imbrie J.Z. (1980) Modeling the climatic response to orbital variations. *Science*, 207, 943–953. *10.2.1.*

Imbrie J., Hays J.D. *et al.* (1984) The orbital theory of Pleistocene climate; support from a revised chronology of the marine $\delta^{18}O$ record. In: *Milankovitch and Climate Pt. 1* (Ed. by A. Berger, J. Imbrie, J. Hays, G. Kukla and B. Saltmann), pp. 269–305. Reidel, Dordrecht. *11.5.2.*

Inden R.F. & Moore C.H. (1983) Beach. In: *Carbonate Depositional Environments* (Ed. by P.A. Scholle, D.G. Bebout and C.H. Moore), pp. 211–265. *Mem. Am. Ass. petrol. Geol.*, 33. *9.4.2.*

Ineson J.R. (1989) Coarse-grained submarine fan and slope apron deposits in a Cretaceous back-arc basin, Antarctica. *Sedimentology*, 36, 793–819. *10.3.4.*

Ingersoll R.V. & Busby C.J. (1995) Tectonics of sedimentary basins. In: *Tectonics of Sedimentary Basins* (Ed. by C.J. Busby and R.V. Ingersoll), pp. 1–51. Blackwell Science, Oxford. *12.2.1, Fig. 12.70.*

Ingle J.C. Jr (1981) Origin of Neogene diatomites around the North Pacific Rim. In: *The Monterey Formation and Related Siliceous Rocks of California* (Ed. by R.E. Garrison, R.G. Douglas *et al.*), pp. 159–179. *Spec. Publ. Pacific Sect. Soc. econ. Paleont. Miner.*, Los Angeles. *10.4.1.*

Ingels J.J.C. (1963) Geometry, paleontology and petrography of Thornton Reef Complex, Silurian of northeastern Illinois. *Bull. Am. Ass. petrol. Geol.*, 47, 405–440. *9.4.5, Fig. 9.53.*

Isaacs C.M., Pisciotto K.A. & Garrison R.E. (1983) Facies and diagenesis of the Miocene Monterey Formation, California: a summary. In: *Siliceous Deposits in the Pacific Region* (Ed. by A. Iijima, J.R. Hein and R. Siever), pp. 247–282. *Developments in Sedimentology*, 36. Elsevier, Amsterdam. *10.4.1.*

Ito M. (1995) Volcanic-ash layers facilitate high-resolution sequence stratigraphy at convergent plate margins: an example from the Plio-Pleistocene forearc basin fill in the Boso Peninsula, Japan. *Sedim. Geol.*, 95, 187–206. *12.12.5.*

Ito M. & Masuda F. (1986) Evolution of clastic piles in an arc–arc collision zone: late Cenozoic depositional history around the Tanzawa Mountains, central Honshu, Japan. *Sedim. Geol.*, 49, 223–259. *12.12.5.*

Jacka A.D., Beck R.H., St. Germain L.C. & Harrison S.G. (1968) Permian deep-sea fans of the Delaware Mountain Group (Guadalupian), Delaware Basin. In: *Guadalupian Facies, Apache Mountain Area, West Texas* (Ed. by B.A. Silver), pp. 49–90. *Soc. econ. Paleont. Miner. Permian Basin Section Publication*, 68-11, 49–90. *10.1.*

Jackson R.G. II (1975) Velocity–bed-form–texture patterns of meander bends in the Lower Wabash River of Illinois and Indiana. *Bull. geol. Soc. Am.*, 86, 1511–1522. *3.3.4.*

Jackson R.G. II (1976) Depositional model of point bars in the Lower Wabash River. *J. sedim. Petrol.*, 46, 579–594. *3.3.4.*

Jackson R.G. II (1978) Preliminary evaluation of lithofacies models for meandering alluvial streams. In: *Fluvial Sedimentology* (Ed. by A.D. Miall), pp. 543–576. *Mem. Can. Soc. petrol. Geol.*, 5, Calgary. *3.6.2.*

Jackson R.G. II (1981) Sedimentology of muddy fine-grained channel deposits in meandering streams of the American Middle West. *J. sedim. Petrol.*, 51, 1169–1192. *3.3.4.*

Jacobs S.S., Gordon A.L. & Ardai J.J. (1979) Circulation and melting beneath Ross Ice Shelf. *Science*, 203, 439–443. *11.4.6.*

Jacquin T., Arnaud-Vanneau A., Arnaud H., Ravenne C. & Vail P.R. (1991) Systems tracts and depositional sequences in a carbonate setting: a study of continuous outcrops from platform to basin at the scale of seismic lines. *Mar. petrol. Geol.*, 8, 122–139. *9.5.2.*

James D.M.D. (1984) *The Geology and Hydrocarbon Resources of Negara Brunei Darussalam*, 169 pp. Muzium Brunei. *6.6.6., Figs 9.23, 9.49.*

James N.P. (1983) Reef Environment. In: *Carbonate Depositional Environments* (Ed. P.A. Scholle, D.G. Bebont and C.H. Moore), pp. 345–440. *Mem. Am. Ass. petrol. Geol.*, 33. *9.1.1, 9.2.1, 9.4.5.*

James N.P. & Bone Y. (1991) Origin of a cool water, Oligo-Miocene deep shelf limestone, Eucla Platform, southern Australia. *Sedimentology*, 38, 323–341. *9.2, 9.2.1, 9.3.4, Fig. 9.2.*

James N.P. & Bourque P.A. (1992) Reefs and mounds. In: *Facies Models: Response to Sea Level Change* (Ed. by R.G. Walker and N.P. James), pp. 323–347. Geological Ass. Canada, St Johns, Newfoundland. *9.4.3, 9.4.5.*

James N.P. & Ginsburg R.N. (1979) *The Seaward Margin of the Belize Barrier and Atoll Reefs*, 191 pp. *Spec. Publ. int. Ass. Sediment.*, 3. *9.4.5, 9.5.1.*

James N.P. & Gravestock D. (1990) Lower Cambrian shelf and shelf margin buildups, Flinders Ranges, South Australia. *Sedimentology*, 37, 455–480. *9.4.5.*

James N.P. & Kendall A.C. (1992) Carbonate and evaporite models. In: *Facies Models: Response to Sea-level Change* (Ed. by R.G. Walker and N.P. James), pp. 375–409. Geol. Ass. Can., Waterloo, Ontario. *8.2.3, 8.3.1, 8.10.*

James N.P., Bone Y., van der Borch C.C. & Gostin V.A. (1992) Modern carbonate and terrigenous clastic sediments on a cool water, high energy, mid-latitude shelf: Lacepede, southern Australia. *Sedimentology*, 39, 877–903. *9.3.4, Fig. 9.16.*

James N.P., Stephens R.K., Barnes C.R. & Knight I. (1989) Evolution of a Palaeozoic continental margin carbonate platform, northern Canadian Appalachians. In: *Controls on Carbonate Platform and Basin Development* (Ed. by P.D. Crevello, J.F. Sarg, J.F. Read and J.L. Wilson), pp. 123–146. *Spec. Publ. Soc. econ. Paleont. Miner.*, 44, Tulsa. *9.4.4.*

Janaway T.M. & Parnell J. (1989) Carbonate production within the Orcadian basin, northern Scotland: a petrographic and geochemical study. *Palaeogeogr. Palaeoclimatol. Palaeoecol.*, 70, 89–105. *4.13.1.*

Jarrard R.D. (1986) Relations among subduction zone parameters. *Rev. Geophys.*, 24, 217–284. *12.2.1.*

Jaupart C. & Allegre C.J. (1991) Gas content, eruption rate, and instabilities of eruption regime in silicic volcanoes. *Earth Planet. Sci. Lett.*, 102, 413–429. *12.3.1.*

Jaupart C. & Vergniolle S. (1988) Laboratory models of Hawaiian and Strombolian activity. *Nature*, 300, 427–429. *12.3.1.*

Jenkyns H.C. (1976) Sediments and sedimentary history of the Mannihiki Plateau, South Pacific Ocean. In: *Initial Reports of the Deep Sea Drilling Project* (Ed. by S.O. Schlanger, E.D. Jackson *et. al.*), pp. 873–890. US Government Printing Office, 33, Washington, DC. *12.11.3.*

Jenkyns H.C. (1978) Pelagic environments. In: *Sedimentary Environments and Facies*, 1st edn (Ed. by H.G. Reading), pp. 314–371. Blackwell Scientific Publications, Oxford. *10.2.1.*

Jenkyns H.C. (1980) Cretaceous anoxic events: from continents to oceans. *J. geol. Soc. Lond.*, 137, 171–188. *10.2.1.*

Jenkyns H.C. (1986) Pelagic environments. In: *Sedimentary Environments and Facies*, 2nd edn (Ed. by H.G. Reading), pp. 343–397. Blackwell Scientific Publications, Oxford. *10.4.2, Fig. 10.4, 12.11.1.*

Jenkyns H.C. & Hsü K.J. (1974) Pelagic sediments: on land and under the sea – an introduction. In: *Pelagic Sediments: on Land and Under the Sea* (Ed. by K.J. Hsü and H.C. Jenkyns), pp. 1–10. *Spec. Publ. int. Ass. Sediment.*, *1.10.1.*

Jenkyns H.C. & Winterer E.L. (1982) Palaeoceanography of Mesozoic ribbon radiolarites. *Earth Planet. Sci. Lett.*, 60, 351–375. *10.4.1.*

Jennette D.C. & Pryor W.A. (1993) Cyclic alternation of proximal and distal storm facies: Kope and Fairview Formations (Upper Ordovician), Ohio and Kentucky. *J. sedim. Petrol.*, 63, 183–203. *9.4.3.*

Jervey M.T. (1988) Quantitative Geological Modeling of Siliciclastic Rock Sequences and their Seismic Expression. In: *Sea-level Changes: an Integrated Approach* (Ed. by C.K. Wilgus, B.S. Hastings, C.G.St.C. Kendall, H.W. Posamentier, C.A. Ross & J.C. Van Wagoner), pp. 47–69. *Spec. Publ. Soc. econ. Paleont. Miner.*, 42, Tulsa. *2.4.*

Jerzykiewicz T. & Wojewode J. (1986) The Radkow and Szczeliniec Sandstones: an example of giant foresets on a tectonically controlled shelf of the Bohemian Cretaceous basin (Central Europe). In: *Shelf Sands and Sandstones* (Ed. by R.J. Knight and J.R. McLean), pp. 1–15. *Mem. Can. Soc. petrol. Geol.*, 11, Calgary. *7.8.3, Fig. 7.53.*

Johansen S.J. (1988) Origins of upper Paleozoic quartzose sandstones, American southwest. In: *Late Paleozoic and Mesozoic Eolian Deposits of the Western Interior of the United States* (Ed. by G. Kocurek), pp. 153–166. *Sedim. Geol.*, 56. *5.5.1.*

John B.S. (1979) Ice ages: A search for reasons. In: *The Winters of the World* (Ed. by B.S. John), pp. 29–57. Halsted Press Book, John Wiley and sons, New York. *11.1.*

Johnson A.M. (1970) *Physical Processes in Geology*, 577 pp. Freeman, Cooper & Co., San Fransisco. *10.2.3.*

Johnson D. (1938) The origin of submarine canyons. *J. Geomorph.*, 1, 230–243. *10.1.*

Johnson D.L., Keller E.A. & Rockwell T.K. (1990) Dynamic pedogenesis: new views on some key soil concepts, and a model for interpreting Quaternary soils. *Quaternary Res.*, 33, 306–319. *3.3.7.*

Johnson D.W. (1919) *Shore Processes and Shoreline Development*, 584 pp. John Wiley, New York. *6.3, 7.1.2.*

Johnson G.A.L. (1967) Basement control of Carboniferous sedimentation in northern England. *Proc. Yorks. geol. Soc.*, 36, 175–194. *6.6.6.*

Johnson H.D. (1975) Tide- and wave-dominated inshore and shoreline sequences from the late Precambrian, Finnmark, north Norway. *Sedimentology*, 22, 45–73. *6.7.7, 7.8..1, Fig. 7.48.*

Johnson H.D. (1977a) Shallow marine sand bar sequences: an example from the late Precambrian of North Norway. *Sedimentology*, 24, 245–270. *7.1.2. 7.7.1, 7.8.1, Figs 7.35, 7.48.*

Johnson H.D. (1977b) Sedimentation and water escape structures in some late Precambrian shallow marine sandstones from Finnmark, North Norway. *Sedimentology*, 24, 389–411. *7.8.1, Fig. 7.48.*

Johnson H.D. & Baldwin C.T. (1986) Shallow siliciclastic seas. In: *Sedimentary Environments and Facies*, 2nd edn (Ed. by H.G. Reading), pp. 229–282. Blackwell Scientific Publications, Oxford. *7.7.*

Johnson H.D. & Levell B.K. (1980) Sedimentology of a Cretaceous sub-tidal sand complex, Woburn Sands, southern England. *Bull. Am. Ass. petrol. Geol.*, 64, 728–729. *7.7.1.*

Johnson H.D. & Levell B.K. (1995) Sedimentology of a transgressive, estuarine sand complex: the Lower Cretaceous Woburn Sands (Lower Greensand), southern England. In: *Sedimentary Facies Analysis* (Ed. by A.G. Plint) pp. 17–46. *Spec. Publ. int. Ass. Sedim.*, 22. *6.7.7, Fig. 6.99, 7.7.1.*

Johnson H.D. & Stewart D.J. (1985) Role of clastic sedimentology in the exploration and production of oil and gas in the North Sea. In: *Sedimentology: Recent Developments and Applied Aspects* (Ed. by P.J. Brenchley & B.J.P. Williams), pp. 249–310. *Spec. Publ. Geol. Soc. Lond.*, 18. *3.6.5, 3.6.7, 6.6.6.*

Johnson H.D., Kuud T. & Dundang A. (1989) Sedimentology and reservoir geology of the Betty field, Baram Delta Province, offshore Sarawak, NW Borneo. *Bull. geol. Soc. Malaysia*, 25, 119–161. *6.6.6.*

Johnson H.D., Levell B.K. and Siedlecki S. (1978) Late Precambrian sedimentary rocks in east Finnmark, North Norway and their relationship to the Trollfjord–Komagelv Fault. *J. geol. Soc.*, 135, 517–534. *Fig. 7.48.*

Johnson M.A. & Belderson R.H. (1969) The tidal origin of some vertical sedimentary changes in epicontinental seas. *J. Geol.*, 77, 353–357. *7.1.2.*

Johnson M.A., Kenyon N.H., Belderson R.H. & Stride A.H. (1982) Sand transport. In: *Offshore Tidal Sands: Processes and Deposits* (Ed. by A.H. Stride), pp. 58–94. Chapman & Hall, London. *Figs 7.9, 7.11.*

Johnson M.E. (1988) Why are ancient rocky shores so uncommon? *J. Geol.*, 96, 469–480. *6.4.*

Johnson M.E. (1992) Studies on ancient rocky shores: a brief history and annotated bibliography. *J. coastal Res.*, 8, 797–812. *6.1, 6.4.*

Johnson T.C. (1980) Sediment redistribution by waves in lakes, reservoirs and embayments. *Symposium on Surface Water Impoundment*. Am. Soc. Civ. Eng., Minneapolis, 1307–1317. *4.4.1.*

Johnson T.C. (1984) Sedimentation in large lakes. *Ann. Rev. Earth Planet. Sci.*, 12, 179–204. *4.4.1, 4.4.3, 4.6.3.*

Johnson T.C. & Ng'ang'a P. (1990) Reflections on a rift lake. In:

Lacustrine Basin Exploration – Case Studies and Modern Analogs (Ed. by B.J. Katz), pp. 113–135. *Mem. Am. Assoc. petrol. Geol.*, 50, Tulsa. *4.6.3.*

Johnson T.C., Carlson T.W. & Evans J.E. (1980) Contourites in Lake Superior. *Geology*, 9, 437–441. *4.6.3.*

Johnston W.A. (1921) Sedimentation of the Fraser River delta. *Mem. Can. geol. Surv.*, 125, 46. *6.1.*

Jones B. & Desrochers A. (1992) Shallow platform carbonates. In: *Facies Models: Response to Sea Level Change* (Ed. by R.G. Walker and N.P. James), pp. 277–301. Geological Ass. Canada, St John's, Newfoundland. *Fig. 9.3.*

Jones B. & Pemberton S.G. (1988) *Lithofaga* borings and their influence on the diagenesis of corals in the Pleistocene Ironshore Formation of Grand Cayman Island. *Palaeios*, 3, 3–21. *9.4.5.*

Jones B. & Renaut R.W. (1994) Crystal fabrics and microbiota in large pisoliths from Laguna Pastos Grandes, Bolivia. *Sedimentology*, 41, 1171–1202. *4.7.1.*

Jones B.F. (1965) The hydrology and mineralogy of Deep Springs Lake, Inyo Country, California. *Prof. Pap. U.S. geol. Surv.*, 502-A, 56 pp. *4.7.4.*

Jones B.G. & Rust B.R. (1983) Massive sandstone facies in the Hawkesbury Sandstone, a Triassic fluvial deposit near Sydney, Australia. *J. sedim. Petrol.*, 53, 1249–1259. *3.5.2.*

Jones C.M. (1977) The effects of varying discharge regimes on bed form sedimentary structures in modern rivers. *Geology*, 5, 567–570. *3.3.3.*

Jones C.M. (1980) Deltaic sedimentation in the Roaches Grit and associated sediments (Namurian R$_2$b) in the south-west Pennines. *Proc. Yorks. geol. Soc.*, 43, 39–67. *6.6.6.*

Jones F.G. & Wilkinson B.H. (1978) Structure and growth of lacustrine pisoliths from recent Michigan marl lakes. *J. sedim. Petrol.*, 48, 1103–1110. *4.12.2.*

Jones J.G. & Nelson P.H.H. (1970) The flow of basalt lava from air into water – its structural expression and stratigraphic significance. *Geol. Mag.*, 107, 13–21. *12.11.2.*

Jordan D.W. & Pryor W.A. (1992) Hierarchical levels of heterogeneity in a Mississippi River meander belt and application to reservoir systems. *Bull. Am. Assoc. petrol. Geol.*, 76, 1601–1624. *3.3.4, 3.3.6, Figs 3.19, 3.22, 3.24.*

Jordan T.E. (1995) Retroarc foreland and related basins. In: *Tectonics of Sedimentary Basins* (Ed. by C.J. Busby and R.V. Ingersoll), pp. 331–362. Blackwell Science, Oxford. *12.12.4.*

Kamata H. (1989) Volcanic and structural history of the Hohi volcanic zone, central Kyushu, Japan. *Bull. Volcanol.*, 51, 315–332. *12.12.4.*

Kamb B. & La Chapelle (1964) Direct observations on the mechanism of glacier sliding over bedrock. *J. Glaciol.*, 5, 159–172. *11.4.1.*

Kanamori H. (1986) Rupture process of subduction-zone earthquakes. *Ann. Rev. Earth Planet. Sci.*, 14, 293–322. *12.2.1.*

Kanes W.H. (1970) Facies and development of the Colorado river delta in Texas. In: *Deltaic Sedimentation, Modern and Ancient* (Ed. by J.P. Morgan and R.H. Shaver), pp. 78–106. *Spec. Publ. Soc. econ. Paleont. Miner.*, 15, Tulsa. *6.6.1.*

Kano K., Orton G.J. & Kano T. (1994) A hot Miocene subaqueous scoria-flow deposit in the Shimane Peninsula, SW Japan. *J. Volcanol. Geotherm. Res.*, 60, 1–14. *12.4.4.*

Kano K., Takeuchi K., Yamamoto T. & Hoshizumi H. (1991) Subaqueous rhyolite block lavas in the Miocene Ushikiri Formation, Shimane Peninsula, SW Japan. *J. Volcanol. Geotherm. Res.*, 46, 241–253. *12.4.5.*

Kano K., Yamamoto T. & Takeuchi K. (1993) A Miocene island-arc volcanic seamount: the Takashibiyama Formation, Shimane Peninsula, SW Japan. *J. Volcanol. Geotherm. Res.*, 59, 110–119. *12.12.5.*

Karcz I. & Zak I. (1987) Bed forms in salt deposits of the Dead Sea brines. *J. sedim. Petrol.*, 57, 723–735. *8.3.2.*

Karig D.E. (1971) Origin and development of marginal basins in the western Pacific. *J. geophys. Res.*, 76, 2452–2561. *12.12.1.*

Karig D.E. (1972) Remnant arcs. *Bull. geol. Soc. Am.*, 83, 1057–1068. *12.12.1.*

Karig D.E. & Ingle J.C. Jr et al. (1975) *Initial Reports of the Deep Sea Drilling Project*, 31, 927 pp. US Government Printing Office, Washington. *10.4.1.*

Karig D.E. & Moore G.F. (1975) Tectonically controlled sedimentation in marginal basins. *Earth and Planet. Sci. Letts.*, 26, 233–238. *12.12.5.*

Karig D.E. & Sharman G.F. (1975) Subduction and accretion in trenches. *Bull. geol. Soc. Am.*, 86, 377–389. *12.12.5.*

Karson J.A. (1991) Accommodation zones and transfer faults: integral components of mid-Atlantic ridge extensional systems. In: *Ophiolite Genesis and Evolution of the Oceanic Lithosphere* (Ed. by T. Peters, A. Nicolas and R.G. Coleman), pp. 21–38. Kluwer Academic, Dordrecht. *12.11.1.*

Katz B.J. (1995) A survey of rift basin source rocks. In: *Hydrocarbon Habitat in Rift Basins* (Ed. by J.J. Lambiase), pp. 213–242. *Spec. Publ. geol. Soc. Lond.*, 80. *4.17.*

Kauffman E.G. (1974) Cretaceous assemblages, communities, and associations: Western Interior United States and Caribbean Islands. In: *Principles of Benthic Community Analysis* (Ed. by K.R.W. A.M. Zeigler, E.J. Anderson, E.G. Kauffman, R.N. Ginsburg and N.P. James), pp. 12.1–12.25. University of Miami. *7.7.3.*

Kay M. (1951) *North American Geosynclines*, 143 pp. *Mem. geol. Soc. Am.*, 48, Boulder, CO. *1.1.*

Keen C.E. (1985) The dynamics of rifting: deformation of the lithosphere by active and passive driving forces. *Geophys. J. R. Astron. Soc.*, 80, 95–120. *12.2.1.*

Keith B.D. & Zuppann C.W. (Eds) (1993) *Mississippian Oolites and Modern Analogs*, 265 pp. Am. Ass. petrol. Geol., Studies in Geology, 35, Tulsa. *9.4.2.*

Kelleher G.T. & Smosna R. (1993) Oolitic tidal-bar reservoirs in the Mississippian Greenbrier Group of West Virginia. In: *Mississippian Oolites and Modern Analogs* (Ed. by B.D. Keith and C.W. Zuppann), pp. 162–173. Am. Ass. petrol. Geol., Studies in Geology, 35, Tulsa. *9.4.2, Fig. 9.33.*

Kelling G. & George G.T. (1971) Upper Carboniferous sedimentation in the Pembrokeshire coalfield. In: *Geological Excursions in South Wales and the Forest of Dean* (Ed. by D.A. Bassett and M.G. Bassett), pp. 240–259. Geol. Ass. South Wales Group, Cardiff. *6.6.6.*

Kellogg T.B. & Kellogg D.E. (1988) Antarctic cryogenic sediments; biotic and inorganic facies of ice shelf and marine-based ice sheet environments. *Palaeogeogr. Palaeoclimatol. Palaeoecol.*, 67, 51–74. *11.4.6.*

Kelly S. & Olsen H. (1993) Terminal fans – with reference to

Devonian examples. In: *Current Research in Fluvial Sedimentology* (Ed. by C.R. Fielding), pp. 339–374. *Sedim. Geol.*, 85. *3.3.8, 3.6.10, Fig. 3.64.*

Kelts K. (1988) Environments of deposition of lacustrine petroleum source rocks: an introduction. In: *Lacustrine Petroleum Source Rocks* (Ed. by A.J. Fleet, K. Kelts and M.R. Talbot), pp. 3–26. *Spec. Publ. Geol. Soc. Lond.*, 40. *4.5.*

Kelts K. & Arthur M.A. (1981) Turbidites after ten years of deep-sea drilling – wringing out the mop? In: *The Deep Sea Drilling Project: A Decade of Progress* (Ed. by J.E. Warme, R.G. Douglas and E.L. Winterer), pp. 91–127. *Spec. Publ. Soc. econ. Paleont. Miner.*, 32, Tulsa. *10.2.3.*

Kelts K. & Hsü K.J. (1978) Freshwater carbonate sedimentation. In: *Lakes: Chemistry, Geology, Physics* (Ed. by A. Lerman), pp. 295–323. Springer-Verlag, Berlin. *4.7.1, 4.8, Fig. 4.15.*

Kelts K. & Hsü K.J. (1980) Resedimented facies of the 1875 Horgen slumps in Lake Zürich and a process model of longitudinal transport of turbidity currents. *Eclog. geol. Helv.*, 73, 271–281. *4.6.3.*

Kelts K. & Shahrabi M. (1986) Holocene sedimentology of hypersaline Lake Urmia, northwestern Iran. *Palaeogeogr. Palaeoclimatol. Palaeoecol.*, 54, 105–130. *4.7.1.*

Kemp G.J. & Wilson B.L. (1990) The seismic expression of Middle to Upper Devonian reef complexes, Canning Basin. *J. Austral. petrol. Explor. Ass.*, 30, 280–289. *9.5.3.*

Kempe S., Kazmierczak J., Landmann G., Konuk T., Reimer A. & Lipp A. (1991) Largest known microbialites discovered in Lake Van, Turkey. *Nature*, 349, 605–608. *4.7.1.*

Kendall A.C. (1976) *The Ordovician Carbonate Succession (Bighorn Group) of Southeastern Saskatchewan*, 185 pp. Saskatchewan Dept Mineral Resources Report, 180. *8.9.*

Kendall A.C. (1977) Patterned carbonate – a diagenetic study, by James Dixon (discussion). *Bull. Can. petrol. Geol.*, 25, 695–697. *8.7.*

Kendall A.C. (1979) Continental and supratidal (sabkha) evaporites. No. 13, In: *Facies Models* (Ed. by R.G. Walker), pp. 145–158. *Geosci. Can. Reprint Ser, 1.8.7, Figs 8.38, 8.44.*

Kendall A.C. (1984) Evaporites. In: *Facies Models* (Ed. by R.G. Walker), pp. 259–296. Geol. Ass. Can., Toronto. *4.14, Fig. 4.28.*

Kendall A.C. (1988) Aspects of evaporite basin stratigraphy. In: *Evaporites and Hydrocarbons* (Ed. by B.C. Schreiber) pp. 11–65. Columbia University Press, New York. *8.2.3, 8.9, 8.10, 8.10.2, Figs 8.8, 8.45, 8.55.*

Kendall A.C. (1992) Evaporites. In: *Facies Models: Response to Sea-level Change* (Ed. by R.G. Walker and N.P. James), pp. 375–409. Geol. Ass. Can., Waterloo, Ontario. *8.1, 8.2.3, 8.3.1, 8.9, 8.10.2, Fig. 8.7.*

Kendall A.C. & Harwood G.M. (1989) Shallow-water gypsum in the Castile Formation – significance and implications. In: *Subsurface and Outcrop Examination of the Capitan Shelf Margin, Northern Delaware Basin* (Ed. by P.M. Harris and G.A. Grover), pp. 441–450. Soc. econ. Paleont. Miner. Core Workshop, 13, Tulsa. *8.5.1, 8.9.*

Kendall C.G. St C. & Schlager W. (1981) Carbonates and relative changes in sea-level. *Mar. Geol.*, 44, 181–212. *9.2.1, 9.2.2, 9.5, 9.5.3.*

Kendall C.G. St C. & Skipwith P.A.D.'E. (1968) Recent algal mats of a Persian Gulf Lagoon. *J. sedim. Petrol.*, 38, 1040–1058. *8.4.2.*

Kendall C.G. St C. & Skipwith P.A.D'E. (1969) Holocene shallow-water carbonate and evaporite sediments of Khor al Bazam, Abu Dhabi, southwest Persian Gulf. *Bull. Am. Ass. petrol. Geol.*, 53, 841–869. *8.4.2, 9.3.4.*

Kennedy W.J. (1987) Late Cretaceous and Early Palaeocene Chalk Group sedimentation in the Greater Ekofisk area, North Sea Central Graben. *Bull. Centres Rech. Explor.-Prod. Elf-Aquitaine*, 11 (1), 91–126. *10.4.3, Fig. 10.83.*

Kennedy W.J. & Garrison R.E. (1975) Morphology and genesis of nodular chalk and hardgrounds in the Upper Cretaceous of southern England. *Sedimentology*, 22, 311–386. *10.4.3, Fig. 10.82.*

Kennett J.P. & Watkins N.D. (1975) Deep-sea erosion and manganese nodule development in the south-east Indian Ocean. *Science*, 188, 1011–1013. *10.4.1.*

Kenter J.A.M. (1990) Carbonate platform flanks: slope angle and sediment fabric. *Sedimentology*, 37, 777–794. *9.4.4, 9.5.6, Fig. 9.35.*

Kenter J.A.M. & Schlager W. (1989) A comparison of shear strength in calcareous and siliciclastic sediments. *Mar. Geol.*, 88, 145–152. *9.4.4, 9.5.6.*

Kenyon N.H. (1970) Sand ribbons of European tidal seas. *Mar. Geol.*, 9, 25–39. *7.3.3.*

Kenyon N.H. & Stride A.H. (1970) The tide-swept continental shelf sediments between the Shetland Isles and France. *Sedimentology*, 14, 159–173. *Fig. 7.9.*

Kenyon N.H., Belderson R.H., Stride A.H. & Johnson M.A. (1981) Offshore tidal sand banks as indicators of net sand transport and as potential deposits. In: *Holocene Marine Sedimentation in the North Sea Basin* (Ed. by S-D. Nio, R.T.E. Shüttenhelm and Tj. C.E. van Weering), pp. 257–268. *Spec. Publ. int. Ass. Sediment.*, 5. *7.3.4.*

Kerans C., Fitchen W.M., Gardner M.H. & Wardlow B.R. (1993) *New Mexico Geological Society Guidebook*, pp. 175–184. 44th Field Conference, Carlsbad Region, New Mexico and West Texas. *Fig. 9.13.*

Kerans C., Lucia F.J. & Senger R.K. (1994) Integrated characterization of carbonate ramp reservoirs using Permian San Andres Formation outcrop analogues. *Bull. Am. Ass. petrol. Geol.*, 78, 181–216. *9.5.*

Kidd R.B. & Hill P.R. (1986) Sedimentation on mid-ocean sediment drifts. In: *North Alantic Palaeoceanography* (Ed. by C.P. Summerhayes and N.J. Shackleton), pp. 87–102. *Spec. Publ. geol. Soc. Lond.*, 21. *10.2.2.*

Kidd R.B., Ruddiman W.F., *et al.* (1986) *Init. Repts DSDP 94*. US Government Printing Office, Washington, DC. *10.4.*

Kieffer G. (1971) Aperçu sur la morphologie des régions volcaniques du Massif Centrale. In: *Symposium Jean Jung: Géologie, Géomorphologie et Structure Profonde du Massif Central Francais*, pp. 479–510. Clermont-Ferrand. *12.9.1, Fig. 12.49.*

Kieffer S.W. (1981) Blast dynamics at Mount St Helens on 18 May 1980. *Nature*, 291, 568–570. *12.4.2, Fig. 12.17.*

Kiene W.E. (1988) A model for bioerosion on the Great Barrier Reef. *Proc. 6th Intern. Coral Reef Symp., Australia*, 3, 449–454. *9.4.5.*

Kier J.S. & Pilkey O.H. (1971) The influence of sea-level changes on sediment carbonate mineralogy, Tongue of the Ocean, Bahamas. *Mar. Geol.*, 11, 189–200. *9.4.4.*

References

Kikuchi Y., Tono S. & Funayama M. (1991) Petroleum resources in the Japanese island-arc setting. *Episodes*, **14**, 236–241. *12.12.5*.

Killworth P.D. (1973) A two dimensional model for the formation of Antarctic bottom water. *Deep-Sea Res.*, **20**, 941–971. *10.2.2*.

Kindinger J.L. (1988) Seismic stratigraphy of the Mississippi-Alabama shelf and upper continental slope. *Mar. Geol.*, **83**, 74–94. *6.6.5*.

King L.H. (1993) Till in the marine environment. *J. Quatern. Sci.*, **8**, 347–358. *11.4.6*.

King L.H., Rokoengen K., Fader G.B.J. & Gunleiksrud T. (1991) Till-tongue stratigraphy. *Bull. geol. Soc. Am.*, **103**, 637–659. *11.4.6*.

King R.H. (1947) Sedimentation in Permian Castile Sea. *Bull. Am. Ass. petrol. Geol.*, **26**, 535–563. *8.1*.

Kinsman D.J.J. (1966) Gypsum and anhydrite of Recent age, Trucial Coast, Persian Gulf. In: *Second Symposium on Salt*, I (Ed. by J.L. Rau), pp. 302–326. *8.1*, *8.2.2*, *8.4.2*.

Kinsman D.J.J. (1969) Modes of formation, sedimentary associations, and diagnostic features of shallow-water supratidal evaporites. *Bull. Am. Ass. petrol. Geol.*, **53**, 830–840. *8.1*, *8.4.2*.

Kinsman D.J.J. (1976) Evaporites: relative humidity control on primary mineral facies. *J. sedim. Petrol.*, **46**, 273–279. *8.2.2*.

Kinsman D.J.J. & Park R.K. (1976) Algal belt and coastal sabkha evolution, Trucial Coast, Persian Gulf. In: *Stromatolites* (Ed. by M.R. Walter), pp. 421–433. Elsevier, Amsterdam. *9.4.1*, *Fig. 9.25*.

Kirk M. (1983) Bar developments in a fluvial sandstone (Westphalian 'A'), Scotland. *Sedimentology*, **30**, 727–742. *3.6.2*.

Kirschbaum M.A. & McCabe P.J. (1992) Controls on the accumulation of coal and on the development of anastomosed fluvial systems in the Cretaceous Dakota Formation of southern Utah. *Sedimentology*, **39**, 581–598. *3.6.5*.

Kite G.W. (1972) An engineering study of crustal movement around the Great Lakes. *Inland Waters Directorate, Dept Environ., Tech. Bull.*, **63**, Ottawa, 57 pp. *4.9*.

Klein G.D. (1985a) Intertidal flats and intertidal sand bodies. In: *Coastal Sedimentary Environments*, 2nd edn (Ed. by R.A. Davis), pp. 187–224. Springer-Verlag, New York. *6.7.3*.

Klein G.D. (1985b) The control of depositional depth, tectonic uplift, and volcanism on sedimentation processes in the back-arc basins of the western Pacific Ocean. *J. Geol.*, **93**, 1–26. *12.12.5*.

Kleinspehn K.L., Steel R.J., Johannessen E. & Netland A. (1984) Carboniferous fan-delta sequences, late Carboniferous – early Permian, western Spitsbergen. In: *Sedimentology of Gravels and Conglomerates* (Ed. by E.H. Koster and R.J. Steel), pp. 279–294. *Mem. Can. Soc. petrol. Geol.*, **10**, Calgary. *6.5.6*.

Kleverlaan K. (1989) Three distinctive feeder-lobe systems within one time slice of the Tortonian Tabernas fan, SE Spain. *Sedimentology*, **36**, 24–45. *10.3.3*.

Kling G.W., Clark M.A., Compton H.R. Devine J.D., Evans W.C., Humphrey A.M., Koenigsberg E.J., Lockwood J.P., Tuttle M.L. & Wagner G.N. (1987) The 1986 Lake Nyos gas disaster in Cameroon, West Africa. *Science*, **236**, 169–174. *4.5*, *12.9.2*.

Klug C. & Cashman K.V. (1992) Submarine pumice formation. Abstracts. Volume 2 of 3. *29th International Geological Congress, Kyoto, Japan, 24 August–3 September*, 499 pp. *12.4.1*.

Kneller B.C. & Branney M.J. (1995) Sustained high-density turbidity currents and the deposition of thick, massive beds. *Sedimentology*, **42**, 607–616. *10.2.3*, *12.4.4*.

Kneller B.C., Edwards D.A., McCaffrey W.D. & Moore R. (1991) Oblique reflection of turbidity currents. *Geology*, **19**, 250–252. *10.2.3*.

Knight M.J. (1975) Recent crevassing of the Erap River, Papua New Guinea. *Aust. geol. Stud.*, **13**, 77–84. *3.3.8*.

Knight R.J. (1980) Linear sand bar development and tidal current flow in Cobequid Bay, Bay of Fundy, Nova Scotia. In: *The Coastline of Canada* (Ed. by S.B. McCann), pp. 123–152. Geol. Surv. Can., Ottawa. *7.3.1*.

Kocurek G. (1981) Significance of interdune deposits and bounding surfaces in aeolian dune sands. *Sedimentology*, **28**, 753–780. *5.4.5*.

Kocurek G. (1988) First-order and super bounding surfaces in eolian sequences – bounding surfaces revisited. In: *Late Paleozoic and Mesozoic Eolian Deposits of the Western Interior of the United States* (Ed. by G. Kocurek), pp. 193–206. Sedim. Geol., **56**. *5.3.7*.

Kocurek G. (1991) Interpretation of ancient eolian sand dunes. *Ann. Rev. Earth Planet. Sci.*, **19**, 43–75. *Figs 5.2*, *5.4*, *5.19*.

Kocurek G. & Dott R.H. (1981) Distinctions and uses of stratification types in the interpretation of eolian sands. *J. sedim. Petrol.*, **51**, 579–595. *5.5.4*.

Kocurek G. & Fielder G. (1982) Adhesion structures. *J. sedim. Petrol.*, **52**, 1229–1241. *5.4.5*.

Kocurek G. & Havholm K.G. (1993) Eolian sequence stratigraphy – a conceptual framework. In: *Recent Advances in and Applications of Siliciclastic Sequence Stratigraphy* (Ed. by P. Weimer and H. Posamentier), pp. 393–409. *Mem. Am. Ass. petrol. Geol.*, **58**, Tulsa. *5.3.1*, *5.3.7*, *5.3.8*, *Figs 5.5*, *5.9*, *5.11*, *5.12*, *5.13*.

Kocurek G. & Hunter R.E. (1986) Origin of polygonal fractures in sand, uppermost Navajo and Page Sandstones, Page, Arizona. *J. sedim. Petrol.*, **56**, 895–904. *5.5.2*.

Kocurek G. & Nielson J. (1986) Conditions favourable for the formation of warm-climate aeolian sand sheets. *Sedimentology*, **33**, 795–816. *5.4.6*.

Kocurek G., Deynoux M., Havholm K. & Blakey R. (1991) Amalgamated accumulations resulting from climatic and eustatic changes, Akchar Erg, Mauritania. *Sedimentology*, **38**, 751–772. *5.4.1*, *5.4.3*, *Fig. 5.24*.

Kocurek G., Knight J. & Havholm K. (1991) Outcrop and semi-regional three-dimensional architecture and reconstruction of a portion of the eolian Page Sandstone (Jurassic). In: *The Three-Dimensional Facies Architecture of Terrigenous Clastic Sediments* (Ed. by A.D. Miall and N. Tyler), pp. 25–43. Soc. econ. Paleont. Miner. Concepts Geol., **3**, Tulsa. *5.5.4*.

Kocurek G., Townsley M., Yeh E., Havholm K. & Sweet M.L. (1992) Dune and dune-field development on Padre Island, Texas, with implications for interdune deposition and water-table-controlled accumulation. *J. sedim. Petrol.*, **62**, 622–635. *5.3.1*, *5.4.1*.

Kohm J.A. & Louden R.O. (1978) Ordovician Red River of eastern Montana and western North Dakota: relationships between lithofacies and production. In: *The Economic Geology of the Williston Basin: Montana, North Dakota, South Dakota, Saskatchewan, Manitoba* (Ed. by D. Rehig) pp. 99–117. Williston Basin Symposium, Montana Geol. Soc. 24th Annual Conference, Billings, Montana. *8.9*.

Kokelaar B.P. (1982) The fluidization of wet sediments during the emplacement and cooling of various igneous bodies. *J. geol. Soc., Lond.*, 139, 21–33. *12.4.1.*

Kokelaar B.P. (1983) The mechanism of Surtseyan volcanism. *J. geol. Soc., Lond.*, 140, 939–944. *12.9.3, Fig. 12.1.*

Kokelaar B.P. (1986) Magma–water interactions in subaqueous and emergent basaltic volcanism. *Bull. Volcanol.*, 48, 275–289. *12.4.1.*

Kokelaar B.P. (1988) Tectonic controls of Ordovician arc and marginal basin volcanism in Wales. *J. geol. Soc., Lond.*, 145, 759–775. *12.13.*

Kokelaar B.P. (1993) Ordovician marine volcanic and sedimentary record of rifting and volcanotectonism: Snowdon, Wales, UK. *Bull. geol. Soc. Am.*, 104, 1443–1455. *12.13.*

Kokelaar B.P. & Busby C. (1992) Subaqueous explosive eruptions and welding of pyroclastic deposits. *Science*, 257, 196–201. *12.4.2, 12.4.4, Fig. 12.20.*

Kokelaar B.P. & Romagnoli C. (1995) Sector collapse, sedimentation and clast population evolution at an active island-arc volcano: Stromboli, Italy. *Bull. Volcanol.*, 57, 240–262. *12.12.3, Fig. 12.1.*

Kolla V. (1993) Lowstand deep-water siliciclastic depositional systems: characteristics and terminologies in sequence stratigraphy and sedimentology. *Bull. Centres Rech. Explor.-Prod. Elf-Aquitaine*, 17, 1; 67–78. *10.3.3.*

Kolla V. & Coumes F. (1985) Indus Fan, Indian Ocean. In: *Submarine Fans and Related Turbidite Systems* (Ed. by A.H. Bouma, W.R. Normark and N.E. Barnes), pp. 129–136. Springer-Verlag, New York. *10.3.3.*

Kolla V. & Coumes F. (1987) Morphology, internal structure, seismic stratigraphy, and sedimentation of Indus Fan. *Bull. Am. Ass. petrol. Geol.*, 71, 650–677. *10.3.3, Fig. 10.39.*

Kolla V. & Perlmutter M.A. (1993) Timing of turbidite sedimentation on the Mississippi Fan. *Bull. Am. Ass. petrol. Geol.*, 77, 1129–1141. *10.3.1, 10.3.3, Figs 10.44, 10.45.*

Kolla V., Eittreim S., Sullivan L., Kostecki J.A. & Burckle L.H. (1980) Current-controlled, abyssal microtopography and sedimentation in Mozambique basin, Southwest Indian Ocean. *Mar. Geol.*, 34, 171–206. *10.3.7.*

Komar P.D. (1976) *Beach Processes and Sedimentation*, 429 pp. Prentice-Hall, Englewood Cliffs, NJ. *6.1, 6.2.3, Figs 6.9, 6.10.*

Komar P.D. (1985) The hydraulic interpretation of turbidites from their grain sizes and sedimentary structures. *Sedimentology*, 32, 395–407. *10.2.3.*

Koschel R. & Raidt H. (1988) Morphologische Merkmale der Phacotus-Hüllen in Hartwasserseen der Mecklenburger seenplatte. *Limnologica*, 19, 13–25. *4.7.1.*

Koschel R., Benndorf J., Proft G. & Recknagel F. (1983) Calcite precipitation as a natural control mechanism of eutrophication. *Arch. Hydrobiol.*, 98, 380–408. *4.7.1.*

Kostaschuk R.A. (1985) River mouth processes in a fjord delta, B.C. Canada. *Mar. Geol.*, 69, 1–23. *Fig. 6.5.*

Koster E.A. (1988) Ancient and modern cold-climate aeolian sand deposition: a review. *J. Quatern. Sci.*, 3, 69–83. *5.1, 5.5.3.*

Koster E.H. & Steel R.J. (Eds) (1984) *Sedimentology of Gravels and Conglomerates*, 441 pp. *Mem. Can Soc. petrol. Geol.*, 10, Calgary. *6.1.*

Kosters E.C. & Suter J.R. (1993) Facies relationships and systems tracts in the late Holocene Mississippi delta plain. *J. sedim. Petrol.*, 63, 727–733. *6.6.5.*

Kosters E.C., Chmura G.L. & Bailey A. (1987) Sedimentary and botanical factors influencing peat accumulation in the Mississippi Delta. *J. geol. Soc. Lond.*, 144, 423–434. *6.6.1.*

Kranck K. (1984) Grain-size characteristics of turbidites. In: *Fine-grained Sediments: Deep-water Processes and Facies* (Ed. by D.A.V. Stow and D.J.W. Piper), pp. 83–92. *Spec. Publ. geol. Soc. Lond.*, 15. *10.2.3.*

Kraus M.J. (1987) Integration of channel and floodplain suites, II. Vertical relations of alluvial paleosols. *J. sedim. Petrol.*, 57, 602–612. *3.6.6, Fig. 3.57.*

Kraus M.J. & Aslan A. (1993) Eocene hydromorphic paleosols: significance for interpreting ancient floodplain processes. *J. sedim. Petrol.*, 63, 453–463. *3.5.4, 3.6.6.*

Kraus M.J. & Bown T.M. (1988) Pedofacies analysis: A new approach to reconstructing ancient fluvial sequences. In: *Paleosols and Weathering Through Geologic Time* (Ed. by J. Reinhardt and W.R. Sigleo), pp. 143–152. *Spec. Paper. geol. Soc. Am.*, 216, Boulder, CO. *3.6.6.*

Krause F.F. & Nelson D.A. (1991) Evolution of an Upper Cretaceous (Turonian) shelf sandstone ridge: analysis of the Crossfield-Cardium pool, Alberta, Canada. In: *Shelf Sand and Sandstone Bodies: Geometry, Facies and Sequence Stratigraphy* (Ed. by D.J.P. Swift, G.F. Oertel, R.W. Tillman and J.A. Thorne), pp. 427–456. *Spec. Publ. int. Ass. Sediment.*, 14. *7.8.4.*

Krause F.F. & Oldershaw A.E. (1979) Submarine carbonate breccia beds – a depositional model for two-layer sediment gravity flows from the Sekwi Formation (Lower Cambrian). *Can. J. earth Sci.*, 16, 189–199. *9.4.4, Fig. 9.41.*

Kreisa R.D. (1981) Storm-generated sedimentary structures in subtidal marine facies with examples from the Middle and Upper Ordivician of southwestern Virginia. *J. sedim. Petrol.*, 51, 823–848. *7.7.2.*

Kreisa R.D., Moiola R.J. & Nøttvedt A. (1986) Tidal sand wave facies, Rancho Rojo Sandstone (Permian), Arizona. In: *Shelf Sands and Sandstones* (Ed. by R.J. Knight and J.R. McLean), pp. 277–291. *Mem. Can. Soc. petrol. Geol.*, 11, Calgary. *7.7.1.*

Krigström A. (1962) Geomorphological studies of sandur plains and their braided streams in Iceland. *Geogr. Ann.*, 44, 328–346. *3.3.2.*

Krinsley D.H. & Doornkamp J.C. (1973) *Atlas of Quartz Sand Surface Textures*, 91 pp. Cambridge University Press, Cambridge. *Table 11.2.*

Kruit C. (1955) Sediments of the Rhône delta. Grain size and microfauna. *Ned. geol. Mijnb. Genoot. Verh. Geol. Ser.*, 15, 357–514. *Fig. 6.36.*

Krumbein W.C. & Sloss L.L. (1963) *Stratigraphy and Sedimentation*, 660 pp. W.H. Freeman, San Francisco. *2.2.2, 8.3.1, Fig. 8.9.*

Kuehl S.A., DeMaster D.J. & Nittrouer C.A. (1986) Nature of sediment accumulation on the Amazon continental shelf. *Continental Shelf Res.*, 6, 209–225. *7.6, Fig. 7.30.*

Kuehl S.A., Nittrouer C.A. & DeMaster D.J. (1982) Modern sediment accumulation and strata formation on the Amazon continental shelf. *Mar. Geol.*, 49, 279–300. *7.6.*

Kuenen Ph.H. (1937) Experiments in connection with Daly's hypothesis on the formation of submarine canyons. *Leids geol.*

Meded., 8, 327–335. *1.1, 10.1.*

Kuenen Ph.H. (1950) Turbidity currents of high density. *18th Intl. geol. Congr. London, 1948*, Rept, pt 8, 44–52. *1.1, 10.1.*

Kuenen Ph.H. & Migliorini C.I. (1950) Turbidity currents as a cause of graded bedding. *J. Geol.*, 58, 91–127. *1.1, 2.5, 10.1.*

Kuenzi W.D., Horst O.H. & McGhee R.V. (1979) Effect of volcanic activity on fluvio-deltaic sedimentation in a modern arc–trench gap, SW Guatemala. *Bull. geol. Soc. Am.*, 90, 827–838. *12.6.1, Fig. 12.41.*

Kuhn G., Melles M., Ehrmann W.V., Hambrey M.J. & Schmiedl G. (1993) Character of clasts in glaciomarine sediments as an indicator of transport and depositional processes, Weddell and Lazarev Seas, Antarctica. *J. Sedim. Petrol.*, 63, 477–487. *11.3.3.*

Kulm L.D., Roush R.C., Harlett J.C., Neudeck R.H., Chambers D.M. & Runge E.J. (1975) Oregon continental shelf sedimentation: interrelationships of facies distribution and sedimentary processes. *J. Geol.*, 83, 145–176. *7.6, Fig. 7.31.*

Kumar N. & Sanders J.E. (1974) Inlet sequence: a vertical succession of sedimentary structures and textures created by the lateral migration of tidal inlets. *Sedimentology*, 21, 491–532. *6.7.4.*

Kumar N. & Sanders J.E. (1976) Characteristics of shoreface storm deposits: modern and ancient examples. *J. sedim. Petrol.*, 46, 145–162. *Fig. 7.22.*

Kumar N. & Slatt R.M. (1984) Submarine-fan and slope facies of Tonkawa (Missourian–Virgilian) Sandstone in Deep Anadarko Basin. *Bull. Am. Ass. petrol. Geol.*, 68, 1839–1856. *10.3.5.*

Kuno H. (1959) Origin of Cenozoic petrographic provinces of Japan and surrounding areas. *Bull. Volcanol.*, 20, 37–76. *12.2.2.*

Kushnir J. (1981) Formation and early diagenesis of varved evaporitic sediments in a coastal hypersaline pool. *J. sedim. Petrol.*, 51, 1193–1203. *8.5.2.*

Kutzbach J.E. & Gallimore R.G. (1989) Pangaean climates: megamonsoons of the megacontinent. *J. geophys. Res.*, 94, 3341–3357. *5.5.1.*

Kutzbach J.E. & Street-Perrott F.A. (1985) Milankovitch forcing of fluctuations in the level of tropical lakes from 18 to 0 kyr BP. *Nature*, 317, 130–134. *4.9, 5.4.7.*

Kvale E.P. & Archer A.W. (1990) Tidal deposits associated with low-sulfur coals, Brazil Fm. (Lower Pennsylvanian), Indiana. *J. sedim. Petrol.*, 60, 563–574. *6.7.7.*

Kvale E.P. & Barnhill M.L. (1994) Evolution of Lower Pennsylvanian estuarine facies within two adjacent paleovalleys, Illinois Basin, Indiana. In: *Incised-valley Systems: Origin and Sedimentary Sequences* (Ed. by R.W. Dalrymple, R. Boyd and B.A. Zaitlin), pp. 191–207. *Spec. Publ. econ. Paleont. Miner., Soc. sedim. Geol.*, 52, Tulsa. *6.7.7.*

Lackschewitz K.S., Dehn J. & Wallrabe-Adams H.-J. (1994) Volcaniclastic sediments from mid-oceanic Kolbeinsey Ridge, north of Iceland: evidence for submarine volcanic fragmentation processes. *Geology*, 22, 975–978. *12.4.1.*

Lacroix A. (1904) *La Montagne Pelée et Ses Eruptions*, 662 pp. Masson et Cie, Paris. *12.1.1.*

La Fon N.A. (1981) Offshore bar deposits of Semilla Sandstone Member of Mancos Shale (Upper Cretaceous), San Juan basin, New Mexico. *Bull. Am. Assoc. petrol. Geol.*, 65, 706–721. *7.8.4.*

Lagaaij R. & Kopstein F.P.H.W. (1964) Typical features of a fluviomarine offlap sequence. In: *Deltaic and Shallow Marine Deposits* (Ed. by L.M.J.U. van Straaten), pp. 216–226. Elsevier, Amsterdam. *6.6.5.*

Lajoie J., Boudon G. & Bourdier J.L. (1989) Depositional mechanics of the 1902 pyroclastic nuée-ardente deposits of Mt Pelé, Martinique. *J. Volcanol. Geotherm. Res.*, 38, 131–142. *12.4.4, Fig. 12.1.*

Lambeck K. (1990) Late Pleistocene, Holocene and present sea-levels: constraints on future change *Palaeogeogr. Palaeoclimatol. Palaeoecol. (Global and Planetary Change Section)*, 89, 205–217. *11.5.2.*

Lambeck K., Cloetingh S. & McQueen H. (1987) Intraplate stresses and apparent changes in sea level: the basins of northwestern Europe. In: *Sedimentary Basins and Basin-Forming Mechanisms* (Ed. by C. Beaumont and A.J. Tankard) pp. 259–268. *Mem. Can. Soc. Petrol. Geol.*, 12. *2.1.4.*

Lambert A. & Giovanoli F. (1988) Records of riverborne turbidity currents and indications of slope failures in the Rhone delta of Lake Geneva. *Limnol. Oceanogr.*, 33, 458–468. *4.6.2, 4.6.3.*

Lambert A.M., Kelts K.R. & Marshall N.F. (1976) Measurements of density underflows from Walensee, Switzerland. *Sedimentology*, 23, 87–105. *4.4.2.*

Lambrick H.T. (1967) The Indus flood-plain and the 'Indus' civilisation. *Geogr. J.*, 133, 483–495. *3.2.2, 3.3.6, 3.5.2.*

Laming D.J.C. (1966) Imbrication, palaeocurrents and other sedimentary features in the Lower New Red Sandstone, Devonshire, England. *J. sedim. Petrol.*, 36, 940–959. *3.5.1.*

Lancaster N. (1988) The development of large aeolian bedforms. *Sedim. Geol.*, 55, 69–89. *5.2.2, 5.4.3.*

Lancaster N. (1989a) *The Namib Sand Sea*, 180 pp. A.A. Balkema, Rotterdam. *5.4.1.*

Lancaster N. (1989b) The dynamics of star dunes: an example from the Gran Desierto, Mexico. *Sedimentology*, 36, 273–289. *5.4.4, Fig. 5.23.*

Lancaster N. (1990) Palaeoclimatic evidence from sand seas. *Palaeogeogr. Palaeoclimatol. Palaeoecol.*, 76, 279–290. *5.1.*

Lancaster N. (1992) Relations between dune generations in the Gran Desierto of Mexico. *Sedimentology*, 39, 631–644. *5.4.2, Fig. 5.18.*

Lancaster N. (1993) Origins and sedimentary features of supersurfaces in the northwestern Gran Desierto sand sea. In: *Aeolian Sediments: Ancient and Modern* (Ed. by K. Pye and N. Lancaster), pp. 71–83. *Spec. Publ. int. Ass. Sediment.*, 16. *5.4.2.*

Land L.S. & MacPherson G.L. (1992) Origin of saline formation waters, Cenozoic section, Gulf of Mexico sedimentary basin. *Bull. Am. Ass. petrol. Geol.*, 76, 1344–1362. *8.2.3.*

Land L.S. & Moore C.H. (1977) Deep forereef and upper island slope, north Jamaica. *Am. Ass. petrol. Geol., Studies in Geology*, 4, 53–65. *9.4.4.*

Landmesser C.W., Johnson T.C. & Wold R.J. (1982) Seismic reflection study of recessional moraines beneath Lake Superior and their relationship to regional deglaciation. *Quatern. Res.*, 17, 173–190. *11.5.3.*

Langbein R. (1987) The Zechstein sulphates: the state of the art. In: *The Zechstein Facies in Europe* (Ed. by T.M. Peryt), pp. 143–188. *Lecture Notes in Earth Sciences*, 10. Springer-Verlag, Berlin. *8.9.*

Langford R.P. (1989) Fluvial–aeolian interactions: part I, modern systems. *Sedimentology*, 36, 1023–1035. *5.4.2.*

Langford R.P. & Chan M.A. (1989) Fluvial–aeolian interactions: part II, ancient systems. *Sedimentology*, 36, 1037–1051. *5.5.3, Fig. 5.27.*

Langford R.P. & Chan M.A. (1993) Downwind changes within an ancient dune sea, Permian Cedar Mesa Sandstone, southeast Utah. In: *Aeolian Sediments: Ancient and Modern* (Ed. by K. Pye and N. Lancaster), pp. 109–126. *Spec. Publ. int. Ass. Sediment.*, 16. *5.5.3.*

Larter R.D. & Barker P.F. (1989) Seismic stratigraphy of the Antarctic Peninsular Pacific margin: A record of Pliocene–Pleistocene ice volume and paleoclimate. *Geology*, 17, 731–734. *11.5.4.*

Larter R.D. & Cunningham A.P. (1993) The depositional pattern and distribution of glacial–interglacial sequences on the Antarctic Peninsular Pacific margin. *Mar. Geol.*, 109, 203–219. *11.5.4.*

Larsen V. & Steel R.J. (1978) The sedimentary history of a debris flow-dominated, Devonian alluvial fan – a study of textural inversion. *Sedimentology*, 25, 37–59. *3.5.1.*

Larue D.K., Smith A.L. & Schellekens J.H. (1991) Oceanic island arc stratigraphy in the Caribbean region: don't take it for granite. *Sedim. Geol.*, 74, 289–308. *Fig. 12.79.*

Lattman L.H. & Lauffenberger S.K. (1974) Proposed role of gypsum in the formation of caliche. *Z. Geomorph. Suppl. Bd*, 20, 140–149. *3.2.3.*

Laury R.L. (1971) Stream bank failure and rotational slumping: preservation and significance in the geological record. *Bull. geol. Soc. Am.*, 82, 1251–1266. *3.2.1, 3.5.2.*

Lavoie D. (1995) A late Ordovician high-energy temperate-water carbonate ramp, southern Quebec, Canada: implications for Late Ordovician oceanography. *Sedimentology*, 42, 95–116. *9.1.1.*

Lawrence D.A. & Williams B.J.P. (1987) Evolution of drainage systems in response to Acadian deformation: the Devonian Battery Point Formation, Eastern Canada. In: *Recent Developments in Fluvial Sedimentology* (Ed. by F.G. Ethridge, R.M. Flores and M.D. Harvey), pp. 287–300. *Spec. Publ. Soc. econ. Paleont. Miner.*, 39, Tulsa. *3.6.2.*

Lawrence M.J.F. & Hendy C.H. (1989) Carbonate deposition and Ross Sea ice advance, Fryxell basin, Taylor Valley, Antarctica. *N. Z. J. Geol. Geophys.*, 32, 267–277. *4.7.1.*

Lawson D.E. (1979) Sedimentological analysis of the western terminus region of the Matanuska Glacier, Alaska. *Cold Regions Research and Engineering Laboratory Report 79–97*, 112 pp. US Army Corps of Engineers. *11.3.3, 11.3.6, 11.4.3, Fig. 11.6.*

Lawson D.E. (1981) Distinguishing characteristics of diamictons at the margin of the Matanuska Glacier, Alaska. *Ann. Glaciol.*, 2, 78–84. *11.4.1.*

Lawson D.E. (1982) Mobilization, movement and deposition of active subaerial sediment flows, Matanuska Glacier, Alaska. *J. Geol.*, 90, 279–300. *11.4.3.*

Lea P.D. (1990) Pleistocene periglacial eolian deposits in southwestern Alaska: sedimentary facies and depositional processes. *J. sedim. Petrol.*, 60, 582–591. *11.4.7.*

Leatherman S.P., Williams A.T. & Fisher J.S. (1977) Overwash sedimentation associated with a large-scale northeaster. *Mar. Geol.*, 24, 109–121. *6.7.4.*

Leckie D.A. (1988) Wave formed, coarse-grained ripples and their relationship to hummocky cross-stratification. *J. sedim. Petrol.*, 58, 607–622. *7.7.2.*

Leckie D.A. (1994) Canterbury Plains, New Zealand – implications for sequence stratigraphic models. *Bull. Am. Ass. petrol. Geol.*, 78, 1240–1256. *7.4.2, Fig. 7.21.*

Leckie D.A. & Krystinick L.F. (1989) Is there evidence for geostrophic currents preserved in the sedimentary record of inner to middle-shelf deposits? *J. sedim. Petrol.*, 59, 862–870. *6.7.7, 7.7.2, Fig. 7.39.*

Leckie D.A. & Singh C. (1991) Estuarine deposits of the Albian Paddy Member (Peace River Formation) and lowermost Shaftesbury Formation, Alberta, Canada. *J. sedim. Petrol.*, 61, 825–849. *6.7.7, Fig. 6.100.*

Leckie D.A. & Walker R.G. (1982) Storm- and tide-dominated shorelines in Cretaceous Moosebar–Lower Gates interval-outcrop equivalents of deep basin gas trap in western Canada. *Bull. Am. Assoc. petrol. Geol.*, 66, 138–157. *6.7.7, 7.1.2, 7.7.2.*

Leckie D., Fox C. & Tarnocai C. (1989) Multiple paleosols of the late Albian Boulder Creek Formation, British Columbia, Canada. *Sedimentology*, 36, 307–323. *3.6.6.*

Ledbetter M.T. & Sparks R.S.J. (1979) Duration of large-magnitude explosive eruptions deduced from graded bedding in deep sea ash layers. *Geology*, 7, 240–244. *12.4.3.*

Lee C.B., Park Y.A. & Koh C.H. (1985) Sedimentology and geochemical properties of intertidal surface sediments of the Banweol area in the southern part of Kyeonggi Bay, Korea. *J. Oceanogr. Soc. Korea*, 20, 20–29. *6.7.3.*

Leeder M.R. (1973) Fluviatile fining-upward cycles and the magnitude of palaeochannels. *Geol. Mag.*, 110, 265–276. *3.6.5.*

Leeder M.R. (1974) Lower Border Group (Tournaisian) fluvio-deltaic sedimentation and palaeogeography of the Northumberland Basin. *Proc. Yorks. geol. Soc.*, 40, 129–180. *3.5.3.*

Leeder M.R. (1975) Pedogenic carbonate and floodplain accretion rates: a quantitative model of alluvial, arid-zone lithofacies. *Geol. Mag.*, 112, 257–270. *3.5.4.*

Leeder M.R. (1978) A quantitative stratigraphic model for alluvium with special reference to channel deposit density and interconnectedness. In: *Fluvial Sedimentology* (Ed. by A.D. Miall), pp. 587–596. *Mem. Can. Soc. petrol. Geol.*, 5, Calgary. *3.6.7.*

Leeder M.R. & Zeidan R. (1977) Giant Late Jurassic sabkhas of Arabian Tethys. *Nature*, 268, 42–44. *8.7.*

Leeder M.R., Ord D.M. & Collier R.E.L. (1988) Development of alluvial fans and fan deltas in neotectonic settings: implications for the interpretation of basin fills. In: *Fan Deltas: Sedimentology and Tectonic Settings* (Ed. by W. Nemec and R.J. Steel), 173–185. Blackie, London. *6.1, 6.5.5.*

Lees A. (1975) Possible influences of salinity and temperature on modern shelf carbonate sedimentation. *Mar. Geol.*, 19, 159–198. *9.1.1, 9.2.*

Lees A. & Miller J. (1985) Facies variations in Waulsortian buildups, Part 2; Mid-Dinantian buildups from Europe and North America. *Geol. J.*, 20, 159–180. *9.3.4, 9.4.5.*

Lees A. & Miller J. (1995) Waulsortian banks. In: *Carbonate Mud Mounds* (Ed. by C.L.V. Monty, D.W.J. Bosence, P.H. Bridges and B.R. Pratt), pp. 191–271. *Spec. Publ. int. Ass. Sediment.*, 23. Blackwell Science, Oxford. *9.4.5.*

Leggett J.K., McKerrow W.S. & Casey D.M. (1982) The anatomy of a Lower Palaeozoic accretionary forearc: the Southern Uplands of Scotland. In: *Trench–Forearc Geology: Sedimentation and Tectonics on Modern and Ancient Active Plate Margins* (Ed. by J.K. Leggett), pp. 495–520. *Spec. Publ. geol. Soc. Lond.*, 10. *10.2.1, 10.4.1.*

Leggett J.K., McKerrow W.S. & Eales M.H. (1979) The Southern Uplands of Scotland: a Lower Palaeozoic accretionary prism. *J. geol. Soc. Lond.*, 136, 755–770. *Fig. 12.81.*

Leggett J.K., McKerrow W.S., Cocks L.R.M. & Rickards R.B. (1981) Periodicity in the Early Palaeozoic marine realm. *J. geol. Soc. Lond.*, 138, 167–176. *10.2.1.*

Leighton M.W. & Pendexter C. (1962) Carbonate rock types. In: *Classification of Carbonate Rocks* (Ed. by W.E. Ham), pp. 33–61. *Mem. Am. Ass. petrol. Geol.*, 1. *9.1.1.*

Leinfelder R.R., Nose M., Schmid D.U. & Werner W. (1993) Microbial crusts of the Late Jurassic: composition, palaeoecological significance and importance in reef construction. *Facies*, 29, 195–230. *9.4.5.*

Lemoalle J. & Dupont B. (1976) Iron-bearing oolites and the present conditions of iron sedimentation in Lake Chad. In: *Ores in Sediments* (Ed. by G.C. Amstutz and A.J. Bernard), pp. 167–178. Int. Union geol. Sci., A3, Springer-Verlag, Berlin. *4.7.3.*

Leopold L.B. & Wolman M.G. (1957) River channel.patterns: braided, meandering and straight. *Prof. Pap. U.S. geol. Surv.*, 282-B, 39–85. *3.1, 3.3.1.*

Le Pichon X., Ewing M. & Truchan M. (1971) Sediment transport and distribution in the Argentine Basin, 2. Antarctic Bottom Current passage into the Brazil Basin. *Phys. Chem. Earth*, 8, 31–48. *10.3.7.*

Levell B.K. (1980b) A late Precambrian tidal shelf deposit, the Lower Sanjfjord Formation, Finnmark, North Norway. *Sedimentology*, 27, 539–557. *7.7.2, 7.8.1.*

Levell B.K. (1980a) Evidence for currents associated with waves in Late Precambrian shelf deposits from Finnmark, North Norway. *Sedimentology*, 27, 153–166. *7.7.1, 7.8.1, Fig. 7.50.*

Levell B.K., Braakman J.H. & Rutten K.W. (1988) Oil-bearing sediments of Gondwana glaciation in Oman. *Bull. Am. Ass. petrol. Geol.*, 72, 775–796. *11.5.5.*

Levey R.A. (1978) Bed-form distribution and internal stratification of coarse-grained point bars, Upper Congaree River, S.C. In: *Fluvial Sedimentology* (Ed. by A.D. Miall), pp. 105–127. *Mem. Can. Soc. petrol. Geol.*, 5, Calgary. *3.3.4.*

Levy Y. (1977) Description and mode of formation of the supratidal evaporite facies in northern Sinai coastal plain. *J. sedim. Petrol.*, 47, 463–484. *8.4.3.*

Levy Y. (1980) Evaporitic sediments in Northern Sinai. In: *Hypersaline Brines and Evaporitic Environments* (Ed. by A. Nissenbaum), pp. 131–143. Elsevier, Amsterdam. *8.4.3.*

Lewis K.B. (1971) Slumping on a continental slope inclined at 1°–4°. *Sedimentology*, 16, 97–110. *10.2.3.*

Lezine A.-M. & Casanova J. (1989) Pollen and hydrologic evidence for the interpretation of past climates in tropical west Africa during the Holocene. *Quatern. Sci. Rev.*, 8, 45–55. *5.4.7.*

Lindquist S.J. (1988) Practical characterization of eolian reservoirs for development: Nugget Sandstone, Utah–Wyoming thrust belt. In: *Late Paleozoic and Mesozoic Eolian Deposits of the Western Interior of the United States* (Ed. by G. Kocurek), pp. 315–339. *Sedim. Geol.*, 56. *5.1.*

Lindsay J.F. (1989) Depositional controls on glacial facies associations in a basinal setting, Late Proterozoic Amadeus Basin, Central Australia. *Palaeogeogr. Palaeoclimatol. Palaeoecol.*, 73,

205–232. *11.5.5.*

Lindsay J.F., Prior D.B. & Coleman J.M. (1984) Distributary-mouth bar development and role of submarine landslides in delta growth, South Pass, Mississippi Delta. *Bull. Am. Ass. petrol. Geol.*, 68, 1732–1743. *6.6.4.*

Lindsay R.F. (1985) Rival, North and Black Slough, Foothills and Lignite Oil Fields: their depositional facies, diagenesis and reservoir characteristics, Burke County, North Dakota. In: *Rocky Mountain Reservoirs – a Core Workshop* (Ed. by M.W. Longman, K.W. Shanley, R.F. Lindsay and D.E. Eby), pp. 217–263. Soc. econ. Paleont. Miner., Core Workshop, 7, Tulsa. *Fig. 8.39.*

Lindsay R.F. & Roth M.S. (1982) Carbonate and evaporite facies, dolomitization and reservoir distribution of the Mission Canyon Formation, Little Knife Field, North Dakota. In: *Proc. 4th Intern. Williston Basin Symp., Regina* (Ed. by J.E. Christopher and J. Kaldi), pp. 153–179. *8.7.*

Link M.H. & Nilsen T.H. (1980) The Rocks Sandstone, an Eocene Sand-Rich Deep Sea Fan Deposit, Northern Santa Lucia Range, California. *J. sedim. Petrol.*, 50, 583–601. *10.3.3.*

Link M.H. & Osborne R.H. (1978) Lacustrine facies in the Pliocene Ridge Basin, California. In: *Modern and Ancient Lake Sediments* (Ed. by A. Matter and M.E. Tucker), pp. 167–187. *Spec. Publ. int. Ass. Sediment.*, 2. *Fig. 4.19.*

Link M.H. & Welton J.E. (1982) Sedimentology and reservoir potential of Matilija Sandstone: an Eocene sand-rich deep-sea fan and shallow marine complex, California. *Bull. Am. Ass. petrol. Geol.*, 66, 1514–1534. *10.3.4.*

Link M.H., Osborne R.H. & Awramik S. (1978) Lacustrine stromatolites and associated sediments of the Pliocene Ridge Route Formation, Ridge Basin, California. *J. sedim. Petrol.*, 48, 143–158. *4.11.1.*

Lipman P.W. (1976) Caldera-collapse breccias in the western San Juan Mountains, Colorado. *Bull. geol. Soc. Am.*, 87, 1397–1410. *12.4.6.*

Lipman P.W. (1984) Roots of ash-flow calderas in western North America: windows into the tops of granitic batholiths. *J. geophys. Res.*, 89, 8801–8841. *12.4.6, 12.12.4, Fig. 12.76.*

Lipman P.W. & Mullineaux D.R. (1981) The 1980 eruptions of Mt. St. Helens, Washington. *US Geol. Surv. Prof. Pap.*, 1250, 843 pp. *Fig. 12.1.*

Lipman P.W., Clague D.A., Moore J.G. & Holcomb R.T. (1989) South Arch volcanic field – newly identified young lava flows on the sea floor south of the Hawaiian ridge. *Geology*, 17, 611–614. *12.4.5.*

Lipman P.W., Normark W.R., Moore J.G., Wilson J.B. & Gutmacher C.E. (1988) The giant submarine Alika debris slide, Mauna Loa, Hawaii. *J. geophys. Res.*, 93, 4279–4299. *12.5.1.*

Lirer L., Pescatore T., Booth B. & Walker G.P.L. (1973) Two Plinian pumice-fall deposits from Somma–Vesuvius, Italy. *Bull. geol. Soc. Am.*, 84, 759–772. *12.4.3.*

Lisitzin A.P. (1971) Distribution of siliceous microfossils in suspension and in bottom sediments. In: *The Micropalaeontology of Oceans* (Ed. by B.M. Funnell and W.R. Riedel), pp. 173–195. *Cambridge University Press, Cambridge.* *10.4.1.*

Lisitzin A.P. (1985) The silica cycle during the last ice age. *Palaeogeogr. Palaeoclimatol. Palaeoecol.*, 50, 241–270. *10.4.1.*

Liu Y. & Gastaldo R.A. (1992) Characteristics and provenance of log-transported gravels in a Carboniferous channel deposit. *J. sedim. Petrol.*, **62**, 1072–1083. *3.5.2.*

Livera S.E. (1989) Facies associations and sand body geometries in the Ness Formation of the Brent Group, Brent Field. In: *Deltas: Sites and Traps for Fossil Fuels* (Ed. by M.K.G. Whateley and K.T. Pickering), pp. 269–286. *Spec. Publ. geol. Soc. Lond.*, **41**. *6.6.6, Fig. 6.63.*

Livingstone D.A. & Melack J.M. (1984) Some lakes of subsaharan Africa. In: *Lakes and Reservoirs* (Ed. by F.B. Taub), pp. 467–497. Elsevier, Amsterdam. *4.5, 4.6.1, 4.9.*

Lloyd R.M., Perkins R.D. & Kerr S.D. (1987) Beach and shoreface ooid deposition on shallow interior banks, Turks and Caicos Island, British West Indies. *J. sedim. petrol.*, **57**, 976–982. *9.4.2.*

Lockley M.G. (1990) How volcanism affects the biostratigraphic record. In: *Volcanism and Fossil Biotas* (Ed. by M.G. Lockley and A. Rice), pp. 1–13. *Spec. Pap. geol. Soc. Am.*, **244**, Boulder, CO. *12.4.1.*

Lof P. (Compiler) (1987) *Soils of the World* (Ed. by J. Van Baren). Wallchart compiled in conjunction with ISRIC. Elsevier, Amsterdam. *3.3.7.*

Logan B.W. (1974) Inventory of diagenesis in Holocene–Recent carbonate sediments, Shark Bay, Western Australia. *Mem. Am. Ass. petrol. Geol.*, **22**, 195–249. *9.4.1, Fig. 9.21.*

Logan B.W. (1987) *The MacLeod Evaporite Basin, Western Australia*, 140 pp. *Mem. Am. Ass. petrol. Geol.*, **44**, Tulsa. *8.1, 8.2.3, 8.5.1, 8.5.2, Figs 8.7, 8.31.*

Logan B.W., Davies G.R., Read J.F. & Cebulski D.E. (Eds) (1970) *Carbonate Sedimentation and Environments, Shark Bay, Western Australia*, 223 pp. *Mem. Am. Ass. petrol. Geol.*, **13**, Tulsa. *9.1.1.*

Logan B.W., Harding J.L., Ahr W.M., Williams J.D. & Snead R.G. (1969) *Carbonate Sediments and Reefs, Yucatan Shelf, Mexico*, 198 pp. *Mem. Am. Ass. petrol. Geol.*, **11**, Tulsa. *9.3.4, Fig. 9.17.*

Logan B.W., Hoffman P. & Gebelein C.D. (1974) Algal mats, crypt-algal fabrics and structures, Hamelin Pool, Western Australia. *Mem. Am. Ass. petrol. Geol.*, **22**, 140–193. *9.4.1.*

Logan B.W., Read J.F., Hagan G.M., Hoffman P., Brown R.G., Woods P.J. & Gebelein C.D. (Eds) (1974) *Evolution and Diagenesis of Quaternary Carbonate Sequences, Shark Bay, W. Australia*, 258 pp. *Mem. Am. Ass. petrol. Geol.*, **22**, Tulsa. *9.1.1.*

Longman M.W. (1981) A process approach to recognizing facies of reef complexes. In: *European Fossil Reef Models* (Ed. by D.F. Toomey), pp. 9–40. *Spec. Publ. Soc. econ. Paleont. Miner.*, **30**, 9–40. *9.4.5.*

Longman M.W., Fertal T.G. & Glennie J.S. (1983) Origin and geometry of Red River dolomite reservoirs, western Williston Basin. *Bull. Am. Ass. petrol. Geol.*, **67**, 744–771. *8.9.*

Lonsdale P. (1981) Drifts and ponds of reworked pelagic sediment in part of the southwest Pacific. *Mar. Geol.*, **43**, 153–193. *10.4.1.*

Lonsdale P. & Batiza R. (1980) Hyaloclastite and lava flows on young seamounts examined with a submersible. *Bull. geol. Soc. Am.*, **91**, 545–554. *Fig. 12.60.*

Lonsdale P. & Hollister C.D. (1979) A near-bottom traverse of Rockall Trough: hydrographic and geologic inferences. *Oceanol. Acta*, **31**, 91–105. *10.3.7.*

Lonsdale P.F., Normark W.R. & Newman W.A. (1972) Sedimentation and erosion on Horizon Guyot. *Bull. geol. Soc. Am.*, **83**, 289–316. *12.11.2.*

Loope D.B. (1984) Eolian origin of upper Paleozoic sandstones, southeastern Utah. *J. sedim. Petrol.*, **54**, 563–580. *5.5.4.*

Loope D.B. (1985) Episodic deposition and preservation of aeolian sands: a late Paleozoic example from southeastern Utah. *Geology*, **13**, 73–76. *5.5.1, 5.5.3.*

Lorenz V. (1974) Vesiculated tuffs and associated features. *Sedimentology*, **21**, 273–291. *12.4.1.*

Loucks R.G. & Longman M.W. (1982) Lower Cretaceous Ferry Lake Anhydrite, Fairway field, east Texas; product of shallow-subtidal deposition. In: *Deposition and Diagenetic Spectra of Evaporites* (Ed. by C.R. Handford, R.G. Loucks and G.R. Davies), pp. 130–173. *Soc. econ. Paleont. Miner.*, *Core Workshop*, **3**, Calgary 1982. *8.5.1, 8.8, Fig. 8.41.*

Loucks R.G. & Sarg J.F. (Eds) (1993) *Carbonate Sequence Stratigraphy*, 545 pp. *Mem. Am. Ass. petrol. Geol.*, **47**, Tulsa. *9.5.*

Loutit T.S., Hardenbol J., Vail P.R. & Baum G.R. (1988) Condensed sections: the key to age determination and correlation of continental margin sequences. In: *Sea-level Changes: An Integrated Approach* (Ed. by C.K. Wilgus, B.S. Hastings, C.G.St.C. Kendall, H.W. Posamentier, C.A. Ross and J.C. Van Wagoner), pp. 183–213. *Spec. Publ. Soc. econ. Paleont. Miner.*, **42**, Tulsa. *6.7.6.*

Lovell J.P.B. & Stow D.A.V. (1981) Identification of ancient sandy contourites. *Geology*, **9**, 347–349. *10.2.2.*

Lowe D.R. (1975) Water escape structures in coarse-grained sediments. *Sedimentology*, **22**, 157–204. *10.2.3.*

Lowe D.R. (1976) Subaqueous liquified and fluidized sediment flows and their deposits. *Sedimentology*, **23**, 285–308. *3.5.1.*

Lowe D.R. (1979) Sediment gravity flows: their classification and some problems of application to natural flows and deposits. In: *Geology of Continental Slopes* (Ed. by L.J. Doyle and O.H. Pilkey), pp. 75–82. *Spec. Publ. Soc. econ. Paleont. Miner.*, **27**, Tulsa. *10.2.3, Fig. 10.17.*

Lowe D.R. (1982) Sediment gravity flows: II. Depositional models with special reference to the deposits of high-density turbidity currents. *J. sedim. Petrol.*, **52**, 279–297. *10.1, 10.2.3, Figs 10.23, 10.24, 10.27, 12.28.*

Lowe D.R. (1988) Suspended-load fallout rate as an independent variable in the analysis of current structures. *Sedimentology*, **35**, 765–776. *12.4.3, Fig. 12.24.*

Lowe D.R. & LoPiccolo R.D. (1974) The characteristics and origins of dish and pillar structures. *J. sedim. Petrol.*, **44**, 484–501. *10.2.3.*

Lowenstein T.K. (1982) Primary features in a potash evaporite deposit, the Permian Salado Formation of west Texas and New Mexico. In: *Deposition and Diagenetic Spectra of Evaporites* (Ed. by C.R. Handford, R.G. Loucks and G.R. Davies), pp. 276–301. *SEPM Core Workshop*, **3**, Calgary 1982. *8.5.1.*

Lowenstein T.K. & Hardie L.A. (1985) Criteria for the recognition of salt pan evaporites. *Sedimentology*, **32**, 627–644. *4.7.4, Fig. 4.13, 8.5.2.*

Lowenstein T.K. & Spencer R.J. (1990) Syndepositional origin of potash evaporites: petrographic and fluid inclusion evidence. *Am. J. Sci.*, **290**, 1–42. *8.1, 8.5.1.*

Lucia F.J. (1972) Recognition of evaporite–carbonate shoreline sedimentation. In: *Recognition of Ancient Sedimentary Environments* (Ed. by J.D. Rigby and W.K. Hamblin), pp. 160–191. *Spec. Publ. Soc. econ. Paleont. Miner.*, **16**, Tulsa. *8.2.3, Fig. 8.5.*

Lundegard P.D., Samuels N.D. & Pryor W.A. (1985) Upper

Devonian turbidite sequence, central and southern Appalachian basin. In: *The Catskill Delta* (Ed. by D.L. Woodrow and W.D. Sevon), pp. 107–121. *Spec. Pap. geol. Soc. Am.*, **201**. *10.3.4.*

Lustig L.K. (1965) Clastic sedimentation in Deep Springs Valley, California. *Prof. Pap. U.S. geol. Surv.*, 352F, 131. *3.6.3.*

Luyten J.R. (1977) Scales of motion in the deep Gulf Stream and across the Continental Rise. *J. mar. Res.*, **35**, 49–74. *10.2.2.*

MacAyeal D.R. (1993) Binge–purge oscillations of the Laurentide ice sheet as a cause of the North Atlantic's Heinrich events. *Paleoceanography*, **8**, 775–784. *11.5.2.*

MacDonald A.C. & Halland E.K. (1993) Sedimentology and shale modelling of a sandstone-rich fluvial reservoir: Upper Statfjord Formation, Statfjord Field, northern North Sea. *Bull. Am. Assoc. petrol. Geol.*, **77**, 1016–1040. *3.6.7.*

Macdonald D.I.M. (1986) Proximal to distal sedimentological variation in a linear trough: implications for the fan model. *Sedimentology*, **33**, 243–259. *10.3.6, 12.12.5.*

Macdonald D.I.M., Moncrieff A.C.M. & Butterworth P.J. (1993) Giant slide deposits from a Mesozoic fore-arc basin, Alexander Island, Antarctica. *Geology*, **21**, 1047–1050. *12.5.1.*

MacDonald K.C. (1982) Mid-ocean ridges: fine scale tectonic, volcanic, and hydrothermal processes within the plate boundary. *Ann. Rev. Earth Planet. Sci.*, **10**, 155–190. *12.11.1.*

MacDonald K.C. & Luyendyk B.P. (1981) The crest of the East Pacific Rise. *Sci. Am.*, **244/5**, 86–99. *Fig. 12.57.*

MacEachern J.A. & Pemberton S.G. (1994) Ichnological aspects of incised-valley fill systems from the Viking Formation of the western Canada sedimentary basin, Alberta, Canada. In: *Incised-valley Systems: Origin and Sedimentary Sequences* (Ed. by R.W. Dalrymple, R. Boyd & B.A. Zaitlin), pp. 129–157. *Spec. Publ. Soc. sedim. Geol.*, **51**. *6.7.7.*

Macintyre I. (1984) Preburial and shallow subsurface alteration of modern scleractinian corals. *Palaeontogr. Am.*, **54**, 229–244. *9.4.5.*

Machette M.N. (1978) Dating Quaternary faults in the southwestern United States by using buried calcic paleosols. *U.S. geol. Surv., J. Res.*, **6**, 369–381. *3.3.7, Fig. 3.27.*

Machette M.N. (1985) Calcic soils of the southwestern United States. *Spec. Pap. geol. Soc. Am.*, **203**, 1–21. *3.3.7, 3.5.4.*

Mack G.H., James W.C. & Monger H.C. (1993) Classification of paleosols. *Bull. geol. Soc. Am.*, **105**, 129–136. *3.5.4.*

Mackiewicz N.E., Powell R.D., Carlson P.R. & Molnia B.F. (1984) Interlaminated ice-proximal glacimarine sediments in Muir Inlet, Alaska. *Mar. Geol.*, **57**, 113–147. *11.4.6.*

MacLeod C.J. & Rothery D.A. (1992) Ridge axial segmentation in the Oman ophiolite: evidence from along strike variations in the sheeted dyke complex. In: *Ophiolites and their Modern Analogues* (Ed. by L.M. Parson, B.J. Murton and P. Browning), pp. 39–63. *Spec. Pub. geol. Soc. Lond.*, **60**. *12.11.1.*

MacPherson B.A. (1978) Sedimentation and trapping mechanism in Upper Miocene Stevens and older turbidite fans of southeastern San Joaquin Valley, California. *Bull. Am. Ass. petrol. Geol.*, **62**, 2243–2274. *10.3.3.*

Magee J.W., Bowler J.M., Miller G.H. & Williams D.L.G. (1995) Stratigraphy, sedimentology, chronology and palaeohydrology of Quaternary lacustrine deposits at Madigan Gulf, Lake Eyre, South Australia. *Palaeogeogr. Palaeoclimatol. Palaeoecol.*, **113**, 3–42. *4.2, 4.6.1.*

Maglione G.F. (1980) An example of recent continental evaporitic sedimentation: the Chadian Basin (Africa). In: *Evaporite Deposits*. Chambres syndicale de la recherche et de la production du pétrole et du gaz naturel, Edtns. Technip Paris, 5–9. *4.5.*

Maguregui J. & Tyler N. (1991) Evolution of Middle Eocene tide-dominated deltaic sandstones, Lagunillas Field, Maracaibo Basin, Western Venezuela. In: *The Three-dimensional Facies Architecture of Terrigenous Clastic Sediments, and its Implications for Hydrocarbon Discovery and Recovery* (Ed. by A.D. Miall and N. Tyler), pp. 233–244. Concepts in Sedimentology and Paleontology, 3. SEPM (Soc. for Sedim. Geol.), Tulsa. *6.6.6, Fig. 6.64, 6.65, 7.7.1.*

Maiklem W.R. (1971) Evaporative draw-down – a mechanism for water-level lowering and diagenesis in the Elk Point Basin. *Bull. Can. petrol. Geol.*, **19**, 487–501. *8.1, 8.2.3, Fig. 8.6.*

Mainguet M. (1991) *Desertification*, 306 pp. Springer-Verlag, New York. *5.1.*

Mainguet M. & Chemin M.-C. (1983) Sand seas of the Sahara and Sahel: an explanation of their thickness and sand dune type by the sand budget principle. In: *Eolian Sediments and Processes* (Ed. by M.E. Brookfield and T.S. Ahlbrandt), pp. 353–363. *Developments in Sedimentology*, **38**. Elsevier, Amsterdam. *5.3.7, 5.4.1, Fig. 5.14.*

Maizels J.K. (1983) Proglacial channel systems: change and thresholds for change over long, intermediate and short time-scales. In: *Modern and Ancient Fluvial Systems* (Ed. by J.D. Collinson and J. Lewin), pp. 251–266. *Spec. Publ. int. Ass. Sediment.*, **6**. *11.4.4.*

Maizels J.K. (1989) Sedimentology, paleoflow dynamics and flood history of jökulhlaup deposits: paleohydrology of Holocene sediment sequences in southern Iceland sandur in deposits. *J. sedim. Petrol.*, **59**, 204–223. *11.4.4, 12.5.2.*

Major J.J. & Newhall C.G. (1989) Snow and ice perturbation during historical eruptions and the formation of lahars and floods. *Bull. Volcanol.*, **52**, 1–27. *12.5.2, 12.6.2.*

Malahoff A. (1987) Geology of the summit of Loihi submarine volcano. *Prof. U.S. geol. Surv.*, **1350**, 133–144. *12.11.2.*

Maldonado A. (1975) Sedimentation, stratigraphy and development of the Ebro delta, Spain. In: *Deltas, Models for Exploration* (Ed. by M.L. Broussard), pp. 311–338. Houston geol. Soc., Houston, TX. *6.6.5.*

Maldonado A., Got H., Monaco A., O'Connell S. & Mirabile L. (1985) Valencia Fan (northwestern Mediterranean): distal deposition fan variant. *Mar. Geol.*, **62**, 295–319. *10.3.4.*

Mancini E.A., Mink R.M., Bearden B.L. & Wilkerson R.P. (1985) Norphlet Formation (Upper Jurassic) of southwestern and offshore Alabama: environments of deposition and petroleum geology. *Bull. Am. Ass. petrol. Geol.*, **69**, 881–898. *5.1.*

Mandeville C.W., Carey S., Sigurdsson H. & King J. (1994) Paleomagnetic evidence for high-temperature emplacement of the 1883 subaqueous pyroclastic flows from Krakatau Volcano, Indonesia. *J. Geophys. Res.*, **99**, 9487–9504. *Fig. 12.1.*

Manspeizer W. (1985) The Dead Sea Rift: Impact of climate and tectonism on Pleistocene and Holocene sedimentation. In: *Strike-Slip Deformation, Basin Formation and Sedimentation* (Ed. by K.T. Biddle and N. Christie-Blick), pp. 143–158. *Spec. Publ. Soc. econ. Paleont. Miner.*, **37**, Tulsa. *6.5.5.*

Manz P.A. (1978) Bedforms produced by fine cohesionless granular and flakey sediments under subcritical water flows. *Sedimentology*, 25, 83–103. *3.5.2.*

Marani M., Argnani A., Roveri M. & Trincardi F. (1993) Sediment drifts and erosional surfaces in the central Mediterranean: seismic evidence of bottom-current activity. *Sedim. Geol.*, 82, 207–220. *10.3.7.*

Marriott S.B. & Wright V.P. (1993) Palaeosols as indicators of geomorphic stability in two Old Red Sandstone alluvial suites, South Wales. *J. geol. Soc. Lond.*, 150, 1109–1120. *3.5.4, 3.6.6, Fig. 3.26.*

Marsaglia K.M. (1995) Interarc and backarc basins. In: *Tectonics of Sedimentary Basins* (Ed. by C.J. Busby and R.V. Ingersoll), pp. 299–329. Blackwell Science, Oxford. *12.12.5.*

Marsaglia K.M. & Klein G. de V. (1983) The paleogeography of Paleozoic and Mesozoic storm depositional systems. *J. Geol.*, 91, 117–142. *7.7.2.*

Marsh B.D. (1984) On the mechanics of caldera resurgence. *J. geophys. Res.*, 89, 8245–8251. *12.4.6.*

Marshall J.F. & Davies P.J. (1982) Internal structure and Holocene evolution of One Tree Reef, southern Great Barrier Reef. *Coral Reefs*, 1, 21–28. *9.4.5.*

Marshall J.F. & Davies P.J. (1984) Facies variation of Holocene reef growth in the southern Great Barrier Reef. In: *Coastal Geomorphology of Australia* (Ed. by B.G. Thom), pp. 123–134. Academic Press, Sydney. *9.4.5.*

Marshall P. (1935) Acid rocks of the Taupo–Rotorua volcanic district. *Trans. R. Soc. NZ.*, 64, 323–366. *12.1.1.*

Martí J. (1991) Caldera-like structures related to Permo–Carboniferous volcanism of the Catalan Pyrenees (NE Spain). *J. Volcanol. Geotherm. Res.*, 45, 173–186. *12.4.6.*

Martin-Chivelet J. & Gimenez R. (1992) Palaeosols in microtidal carbonate sequences, Sierra de Utiel Formation, Upper Cretaceous, SE Spain. *Sedim. Geol.*, 81, 125–145. *9.4.1.*

Martinsen O.J. (1989) Styles of soft-sediment deformation on a Namurian (Carboniferous) delta slope, western Irish Namurian Basin, Ireland. In: *Deltas: Sites and Traps for Fossil Fuels* (Ed. by M.K.G. Whateley and K.T. Pickering), pp. 167–177. *Spec. Publ. geol. Soc. Lond.*, 41. *6.6.6, 10.2.3, 10.3.5, Figs 10.20, 10.21.*

Martinsen O.J. (1990) Fluvial, inertia-dominated deltaic deposition in the Namurian (Carboniferous) or northern England. *Sedimentology*, 37, 1099–1113. *6.6.6.*

Martinsen O.J. (1993) Namurian (late Carboniferous) depositional systems of the Craven-Askrigg area, northern England: implications for sequence-stratigraphic models. In: *Sequence Stratigraphy and Facies Associations* (Ed. by H.W. Posamentier, C.P. Summerhayes, B.U. Haq and G.P. Allen), pp. 247–281. *Spec. Publ. int. Ass. Sediment.*, 18. *2.4.*

Martinsen O.J. & Bakken B. (1990) Extensional and compressional zones in slumps and slides in the Namurian of County Clare, Ireland. *J. geol. Soc. Lond.*, 147, 153–164. *6.6.6, 10.2.3.*

Marzo M. & Anadón P. (1988) Anatomy of a conglomeratic fan-delta complex: the Eocene Montserrat Conglomerate, Ebro Basin, northeastern Spain. In: *Fan Deltas: Sedimentology and Tectonic Settings* (Ed. by W. Nemec and R.J. Steel), pp. 318–340. Blackie, London. *6.5.4, 6.5.6, Fig. 6.29.*

Marzolf J.E. (1988) Controls on late Paleozoic and early Mesozoic eolian deposition of the western United States. In: *Late Paleozoic and Mesozoic Eolian Deposits of the Western Interior of the United States* (Ed. by G. Kocurek), pp. 167–191. *Sedim. Geol.*, 56. *5.5.1.*

Massari F. (1983) Tabular cross-bedding in Messinian fluvial channel conglomerates, southern Alps, Italy. In: *Modern and Ancient Fluvial Systems* (Ed. by J.D. Collinson & J. Lewin), pp. 287–300. *Spec. Publ. int. Ass. Sediment.*, 6. *3.5.1, 3.6.4.*

Massari F. & Colella A. (1988) Evolution and types of fan-delta systems in some major tectonic settings. In: *Fan Deltas: Sedimentology and Tectonic Settings* (Ed. by W. Nemec and R.J. Steel), pp. 103–122. Blackie, London. *6.5.*

Massari F. & Parea G.C. (1988) Progradational gravel beach sequences in a moderate- to high-energy, microtidal environment. *Sedimentology*, 35, 881–913. *6.5.4.*

Massari F. & Parea G.C. (1990) Wave-dominated Gilbert-type gravel deltas in the hinterland of the Gulf of Taranto (Pleistocene, southern Italy). In: *Coarse-grained Deltas* (Ed. by A. Colella and D.B. Prior), pp. 311–331. *Spec. Publ. int. Ass. Sediment.*, 10. *6.5.2, 6.5.4.*

Massé L., Faugères J.-C., Mézerais M.-L. & Stow D.A.V. (1991) Contourites quaternaires dans le bassin sud-brésilien: enregistrement sédimentaire des paléocirculations de fond antarctiques (AABW). *3eme Congr. Fr. Sedimentol.*, Brest, Résumés, p. 209. *10.3.7.*

Masson D.G. (1994) Late Quaternary turbidity current pathways to the Madeira Abyssal Plain and some constraints on turbidity current mechanisms. *Basin Res.*, 6, 17–33. *10.3.6, Fig. 10.75.*

Masson D.G., Kidd R.B., Gardner J.V., Huggett Q.J. & Weaver P.P.E. (1992) Saharan Continental Rise: facies distribution and sediment slides. In: *Geologic Evolution of Atlantic Continental Rises* (Ed. by C.W. Poag and P.C. de Graciansky), pp. 327–343. Van Nostrand Reinhold, New York. *10.3.5, Fig. 10.66.*

Mastin L.G. (1991) The roles of magma and groundwater in the phreatic eruptions at Inyo Craters, Long Walley Caldera, California. *Bull. Volcanol.*, 53, 579–596. *12.4.1.*

Matalucci R.V., Shelton J.W. & Abdel-Hady M. (1969) Grain orientation in Vicksburg loess. *J. sedim. Petrol.*, 39, 969–979. *11.4.7.*

Mathews M.D. & Perlmutter M.A. (1994) Global cyclostratigraphy: an application to the Eocene Green River Basin. In: *Orbital Forcing and Cyclic Sequences* (Ed. by P.L. de Boer and D.G. Smith), pp. 459–481. *Spec. Publ. int. Ass. Sediment.*, 19. *2.1.2, Fig. 2.1.*

Mathews W.H. & Shepard F.P. (1962) Sedimentation of Fraser River delta: British Columbia. *Bull. Am. Ass. petrol. Geol.*, 46, 1416–1443. *6.6.4.*

Mathisen M.E. & McPherson J.G. (1991) Volcaniclastic deposits: implications for hydrocarbon exploration. In: *Sedimentation in Volcanic Settings* (Ed. by R.V. Fisher and G.A. Smith), pp. 27–38. *Spec. Publ. Soc. econ. Paleont. Miner.*, 45, Tulsa. *12.14.1.*

Mathisen M.E. & Vondra C.F. (1983) The fluvial and pyroclastic deposits of the Cagayan Basin, Northern Luzon – an example of non-marine volcaniclastic sedimentation in an interarc basin. *Sedimentology*, 30, 369–392. *12.12.4.*

Matthews R.K. (1984) Oxygen isotope record of ice-volume history: 100 million years of glacio-eustatic sea-level fluctuation. In: *Interregional Unconformities and Hydrocarbon Accumulation* (Ed. by J.S. Schlee), pp. 97–107. *Mem. Am. Assoc. petrol. Geol.*, 36, Tulsa. *11.5.2.*

References

....

Maxwell T.A. & Haynes C:V. (1989) Large-scale, low-amplitude bedforms (chevrons) in the Selima Sand Sheet, Egypt. *Science*, 243, 1179–1182. *5.4.6.*

Maxwell W.H.G. (1968) *Atlas of Great Barrier Reef*, 258 pp. Elsevier, Amsterdam. *9.4.5.*

May S.R., Ehman K.D., Gray G.G. & Crowell J.C. (1993) A new angle on the tectonic evolution of the Ridge basin, a 'strike-slip' basin in southern California. *Bull. geol. Soc. Am.*, 105, 1357–1372. *4.11.1.*

Mazzullo J., Alexander A., Tieh T. & Mengun D. (1992) The effects of wind transport on the shapes of quartz silt grains. *J. sedim. Petrol.*, 62, 961–971. *11.4.7.*

Mazzullo S.J. & Reid A.M. (1989) Lower Permian platform and basin depositional systems, northern Midland Basin, Texas. In: *Controls on Carbonate Platform and Basin Development* (Ed. by P.D. Crevello, J.F. Sarg, J.F. Read and J.L. Wilson), pp. 305–320. *Spec. Publ. Soc. econ. Paleont. Miner.*, 44, Tulsa. *9.3.3.*

McBride E.F. (1962) Flysch and associated beds of the Martinsburg Formation (Ordovician), central Appalachians. *J. sedim. Petrol.*, 32, 39–91. *10.1.*

McBride E.F. (1970) Flysch sedimentation in the Marathon region, Texas. In: *Flysch Sedimentation in North America* (Ed. by J. Lajoie), pp. 67–83. *Spec. Pap. geol. Ass. Can.*, Waterloo, Ontario. *7. 6.5.4.*

McCabe A.M., Bowen D.Q. & Penney D.N. (1993) Glaciomarine facies from the western sector of the last British ice sheet, Malin Beg, Country Donegal, Ireland. *Quatern. Sci. Rev.*, 12, 35–45. *11.5.3.*

McCabe P.J. (1977) Deep distributary channels and giant bedforms in the Upper Carboniferous of the Central Pennines, northern England. *Sedimentology*, 24, 271–290. *3.5.2, 3.6.2, Fig. 3.38.*

McCabe P.J. (1978) The Kinderscoutian delta (Carboniferous) of northern England: a slope influenced by density currents. In: *Sedimentation in Submarine Canyons, Fans and Trenches* (Ed. by D.J. Stanley and G. Kelling), pp. 116–126. Dowden, Hutchinson & Ross, Stroudsburg. *6.6.6.*

McCabe P.J. (1984) Depositional environments of coal and coal-bearing strata. In: *Sedimentology of Coal and Coal-bearing Sequences* (Ed. by R.A. Rahmani and R.M. Flores), pp. 13–42. *Spec. Publ. int. Ass. Sediment.*, 7. *3.3.6, 3.5.5, 3.6.8, Fig. 3.60.*

McCabe P.J. & Jones C.M. (1977) The formation of reactivation surfaces within superimposed deltas and bedforms. *J. sedim. Petrol.*, 47, 707–715. *3.3.3, 3.5.2.*

McCaffrey M.A., Lazar B. & Holland H.D. (1987) The evaporation path of seawater and the coprecipitation of Br⁻ and K⁺ with halite. *J. sedim. Petrol.*, 57, 928–937. *8.2.1.*

McCauley J.F., Schaber G.G., Breed C.S., Grolier M.J., Haynes C.V., Issawi B., Elachi E.C. & Blom R. (1982) Subsurface valleys and geoarcheology of the eastern Sahara revealed by shuttle radar. *Science*, 218, 1004–1020. *5.4.7.*

McCave I.N. (1970) Deposition of fine-grained suspended sediment from tidal currents. *J. geophys. Res.*, 75, 4151–4159. *7.7.1.*

McCave I.N. (1971a) Wave effectiveness at the sea-bed and its relationship to bed-forms and deposition of mud. *J. sedim. Petrol.*, 41, 89–96. *6.7.3, 7.3.2.*

McCave I.N. (1971b) Sand waves in the North Sea off the coast of Holland. *Mar. Geol.*, 10, 199–225. *7.3.3.*

McCave I.N. (1971c) Mud in the North Sea. In: *North Sea Science, NATO North Sea Science Conference* (Ed. by E.D. Goldberg), pp. 75–100. *7.3.3.*

McCave I.N. (1972) Sediment transport and escape of fine-grained sediment in shelf areas. In: *Shelf Sediment Transport, Process and Patterns* (Ed. by D.J.P. Swift, D.B. Duane and O.H. Pilkey), pp. 215–248. Dowden, Hutchinson & Ross. Stroudsburg, PA. *7.3.3, 10.2.1.*

McCave I.N. (1979) Suspended material over the central Oregon continental shelf in May 1974: I, concentrations of organic and inorganic components. *J. sedim. Petrol.*, 49, 1181–1194. *7.6.*

McCave I.N. (1984) Erosion, transport and deposition of fine-grained marine sediments. In: *Fine Grained Sediments: Deep-water Processes and Facies* (Ed. by D.A.V. Stow and D.J.W. Piper), pp. 35–69. *Spec. Publ. geol. Soc. Lond.. 7.6.*

McCave I.N. (1985) Recent shelf clastic sediments. In: *Sedimentology: Recent Developments and Applied Aspects* (Ed. by P.J. Brenchley and B.P.J. Williams), pp. 49–65. *Spec. Publ. geol. Soc. Lond.*, 18. *7.3, 7.7.1.*

McCave I.N. & Jones P.N. (1988) Deposition of ungraded muds from high-density non-turbulent turbidity currents. *Nature*, 333, 250–252. *10.2.3.*

McCave I.N. & Tucholke B.E. (1986) Deep current-controlled sedimentation in the western North Atlantic. In: *The Geology of North America*, vol. M. *The Western North Atlantic Region, Decade of North America Geology* (Ed. by P.R. Vogt and B.E. Tucholke), pp. 451–468. Geol. Soc. Am., Boulder, CO. *10.2.2, 10.3.7.*

McCave I.N., Lonsdale P.F., Hollister C.D. & Gardner W.D. (1980) Sediment transport over the Hatton and Gardar contourite drifts. *J. sedim. Petrol.*, 50, 1049–1062. *10.2.2.*

McClelland E.A. & Druitt T.H. (1989) Palaeomagnetic estimates of emplacement temperatures of pyroclastic deposits on Santorini, Greece. *Bull. Volcanol.*, 51, 16–27. *12.4.4.*

McConnaughey T. (1991) Calcification in *Chara corallina*: CO₂ hydroxylation generates protons for bicarbonate assimilation. *Limnol. Oceanogr.*, 36, 619–628. *4.7.1.*

McCubbin D.G. (1982) Barrier island and strand plain facies. In: *Sandstone Depositional Environments* (Ed. by P.A. Scholle and D. Spearing), pp. 247–279. *Am. Ass. petrol. Geol.*, 31. Tulsa. *6.7.7, Figs 6.84, 6.95.*

McDonald B.C. & Shilts W.W. (1975) Interpretation of faults in glacio fluvial sediments. In: *Glaciofluvial and Glaciolacustrine Sedimentation* (Ed. by A.V. Jopling & B.C. McDonald), pp. 123–331. *Spec. Publ. Soc. econ. Paleont. Miner.*, 23. *11.4.1, 11.4.4.*

McEwan I.K. & Willetts B.B. (1993) Sand transport by wind: a review of the current conceptual model. In: *The Dynamics and Environmental Context of Aeolian Sedimentary Systems* (Ed. by K. Pye), pp. 7–16. *Spec. Publ. geol. Soc. Lond.*, 72. *5.3.1.*

McGetchin T.R., Settle M. & Chouet B.H. (1974) Cinder cone growth modelled after Northeast crater, Mt. Etna, Sicily. *J. geophys. Res.*, 79, 3257–3272. *Fig. 12.49.*

McGovney J.E. & Radovich B.J. (1985) Seismic stratigraphy and facies of the Frigg fan complex. In: *Seismic Stratigraphy II, An Integrated Approach* (Ed. by O.R. Berg and D.G. Woolverton), pp. 139–156. *Mem. Am. Ass. petrol. Geol.*, 39. *10.3.3.*

McGowen J.H. & Garner L.E. (1970) Physiographic features and stratification types of coarse-grained point bars: modern and ancient examples. *Sedimentology*, 14, 77–111. *3.3.4*.

McGowen J.H. & Groat C.G. (1971) *Van Horn Sandstone, West Texas: an Alluvial Fan Model for Mineral Exploration*, 57 pp. Bur. econ. Geol., Report of Investigations, 72, Austin, TX. *3.6.4*.

McGowen J.H. & Scott A.J. (1975) Hurricanes as geologic agents on the Texas Coast. In: *Estuarine Research*, Vol. II *Geology and Engineering* (Ed. by L.E. Cronin), pp. 23–46. Academic Press. London. *6.7.4*, *Fig. 6.73*.

McHargue T.R. (1991) Seismic Facies, Processes, and Evolution of Miocene Inner Fan Channels, Indus Submarine Fan. In: *Seismic Facies and Sedimentary Processes of Submarine Fans and Turbidite Systems* (Ed. by P. Weimer and M.H. Link), pp. 403–413. Springer-Verlag, New York. *10.3.3*, *Fig. 10.40*.

McHargue T.R. & Webb J.E. (1986) Internal geometry, seismic facies, and petroleum potential of canyons and inner fan channels of the Indus Submarine fan. *Bull. Am. Ass. petrol. Geol.*, 70, 161–180. *10.3.3*.

McKee E.D. (1966) Structure of dunes at White Sands National Monument, New Mexico (and a comparison with structures of dunes from other selected areas). *Sedimentology*, 7, 1–69. *5.1*, *5.3.5*.

McKee E.D. (1979) An introduction to the study of global sand seas. In: *Global Sand Seas* (Ed. by E.D. McKee), pp. 1–19. *Prof. Pap. U.S. geol. Surv.*, 1052. *5.4.3*.

McKee E.D. & Moiola J. (1975) Geometry and growth of the White Sands dune field, New Mexico. *J. Res. U.S. geol. Surv.*, 3, 59–66. *5.1*.

McKee E.D. & Ward W.C. (1983) Eolian environment. In: *Carbonate Depositional Environments* (Ed. by P.A. Scholle, D.G. Bebout and C.H. Moore), pp. 132–170. *Mem. Am. Ass. petrol. Geol.*, 33, Tulsa. *9.4.2*.

McKee E.D., Crosby E.J. & Berryhill H.L. Jr (1967) Flood deposits, Bijou Creek, Colorado, June 1965. *J. sedim. Petrol.*, 37, 829–851. *3.5.3*, *3.6.2*.

McKee E.D., Douglass J.R. & Rittenhouse S. (1971) Deformation of lee-side laminae in eolian dunes. *Bull. geol. Soc. Am.*, 82, 359–378. *5.1*, *5.3.3*.

McKenna G.T. & Luternauer (1987) First documented large failure at the Fraser River delta front, British Columbia. In: *Current Research. Geol. Surv. Can. Pap.*, 87-1A, 919–924. *6.6.4*.

McKenzie J.A., Hsü K.J. & Schneider J.F. (1980) Movement of subsurface waters under the sabkha, Abu Dhabi, UAE, and its relation to evaporative dolomite genesis. In: *Concepts and Models of Dolomitization* (Ed. by D.H. Zenger, J.B. Danham and R.L. Ethington), pp. 11–30. *Spec. Publ. Soc. econ. Paleont. Miner.*, 28, Tulsa. *8.4.2*.

McKie T. (1990a) Tidal sandbank evolution in the Lower Cambrian Salterella Grit. *Scott. J. Geol.*, 26, 77–88. *7.7.1*, *7.8.1*.

McKie T. (1990b) Tidal and storm influenced sedimentation from a Cambrian transgressive passive margin sequence. *J. Geol. Soc.*, 147, 785–794. *7.7.1*, *7.8.1*.

McLean H. & Howell D.G. (1985) Blanca turbidite system, California. In: *Submarine Fans and Related Turbidite Systems* (Ed. by A.H. Bouma, W.R. Normark and N.E. Barnes), pp. 167–172. Springer-Verlag, New York. *10.3.4*.

McManus D.A. (1975) Modern versus relict sediment on the continental shelf. *Bull. geol. Soc. Am.*, 86, 1154–1160. *7.1.2*.

McPherson J.G., Shanmugam G. & Moiola R.J. (1987) Fan-deltas and braid deltas: varieties of coarse-grained deltas. *Bull. geol. Soc. Am.*, 99, 331–340. *6.3*, *6.5*.

McPherson J.G., Shanmugam G. & Moiola R.J. (1988) Fan deltas and braid deltas: conceptual problems. In: *Fan Deltas: Sedimentology and Tectonic Settings* (Ed. by W. Nemec and R.J. Steel), pp. 14–22. Blackie, London. *6.1*.

McPhie J. (1987) Andean analogue for Late Carboniferous volcanic arc and arc flank environments of the western New England, Oregon, New South Wales, Australia. *Tectonophysics*, 138, 269–288. *12.12.4*.

McPhie J. (1995) A Pliocene shoaling basaltic seamount: Ba volcanic Group at Rakiraki, Fiji. *J. Volcanol. Geotherm. Res.*, 64, 193–210. *12.7*, *12.12.5*.

McPhie J., Doyle M. & Allen R. (1993) *Volcanic Textures: A Guide to the Interpretation of Textures in Volcanic Rocks*, 198 pp. Centre for Ore Deposit and Exploration Studies, University of Tasmania. *12.7*.

Meade R.H., Sachs P.L., Manheim F.T., Hathaway J.C. & Spencer D.W. (1975) Sources of suspended matter in waters of the Middle Atlantic Bight. *J. sedim. Petrol.*, 45, 171–188. *7.6*.

Mearns E.W. (1992) Samarium–neodymium isotopic contraints on the provenance of the Brent Group. In: *Geology of the Brent Group* (Ed. by A.C. Morton, R.S. Haszeldine, M.R. Giles and S. Brown), pp. 213–225. *Spec. Publ. geol. Soc. Lond.*, 61. *6.6.6*.

Meckel L.D. (1975) Holocene sand bodies in the Colorado delta area, northern Gulf of California. In: *Deltas, Models for Exploration* (Ed. by M.L. Broussard), pp. 239–265. Houston geol. Soc., Houston, TX. *6.6.1*, *7.3*.

Melim L.A. & Scholle P.A. (1995) The forereef facies of the Permian Capitan Formation: the role of sediment supply versus sea-level changes. *J. sedim. Res.*, B65, 107–118. *9.3.3*, *9.4.4*, *9.5.5*.

Mellors R.A., Waitt R.B. & Swanson D.A. (1988) Generation of pyroclastic flows and surges by hot-rock avalanches from the dome of mount St Helens volcano, USA. *Bull. Volcanol.*, 50, 14–25. *12.4.1*.

Melosh H.J. (1987) The mechanics of large rock avalanches. *Geol. Soc. Am. Rev. Eng. Geol.*, 7, 41–49. *12.5.1*.

Melvin J.L. (Ed.) (1991) *Evaporites, Petroleum and Mineral Resources*, 556 pp. Amsterdam, Elsevier. *8.1*.

Menard H.W. (1960) Possible pre-Pleistocene deep sea fans off central California. *Bull. geol. Soc. Am.*, 71, 1271–1278. *10.1*.

Menning M., Katzung G. & Lutzner (1988) Magnetostratigraphic investigations in the Rotliegendes (300–252 Ma) of central Europe. *Z. geol. Wiss.*, 16, 1045–1063. *8.9*.

Merz M.U.E. (1992) The biology of carbonate precipitation by cyanobacteria. *Facies*, 26, 81–102. *4.7.1*.

Mesolella K.J., Robinson J.D., Mccormick L.M. & Ormiston A.R. (1974) Cyclic deposition of Silurian carbonates and evaporites in the Michigan Basin. *Bull. Am. Ass. petrol. Geol.*, 58, 34–62. *9.3.4*.

Mesolella K.J., Sealy H.A. & Matthews R.K. (1970) Facies geometries within Pleistocene reef, Barbados, West Indies. *Bull. Am. Ass. petrol. Geol.*, 54, 1899–1917. *9.4.5*.

Messing C.G., Neumann A.C. & Lang J.C. (1990) Biozonation of deep-water lithoherms and associated hardgrounds in the northeastern Straits of Florida. *Palaios*, 5, 15–34. *9.4.5*.

Mézerais M.-L., Faugères J.-C., Figueiredo A.G. Jr & Massé L.

References

....

(1993) Contour current accumulation off the Vema Channel mouth, southern Brazil Basin: pattern of a 'contourite' fan. *Sedim. Geol.*, 82, 173–187. *10.4.1.*

Miall A.D. (1977) A review of the braided-river depositional environment. *Earth Sci. Rev.*, 13, 1–62. *3.5, 3.6.2, 11.4.4, 11.5.1.*

Miall A.D. (1978) Lithofacies types and vertical profile models of braided river deposits, a summary. In: *Fluvial Sedimentology* (Ed. by A.D. Miall), pp. 597–604. *Mem. Can. Soc. petrol. Geol.*, 5, Calgary. *2.2.1, 3.6.2, 6.5.4, Table 6.1.*

Miall A.D. (1983) Glaciofluvial transport and deposition. In: *Glacial Geology: An Introduction for Engineers and Earth Scientists* (Ed. by N. Eyles), pp. 168–183. Pergamon Press, Oxford. *11.4.4.*

Miall A.D. (1984) *Principles of Sedimentary Basin Analysis*, 490 pp. Springer-Verlag, New York. *1.1, 11.5.2.*

Miall A.D. (1985a) Architectural-element analysis: a new method of facies analysis applied to fluvial deposits. *Earth Sci. Rev.*, 22, 261–308. *2.2.2, 3.1, 3.5.2, 3.6.1, 3.6.2, 3.6.4, 10.3.2.*

Miall A.D. (1985b) Sedimentation on an early Proterozoic continental margin under glacial influence: the Gowganda Formation (Huronian), Elliot Lake area, Ontario, Canada. *Sedimentology*, 32, 763–788. *10.3.5.*

Miall A.D. (1986) Eustatic sea-level changes interpreted from seismic stratigraphy: a critique of the methodology with particular reference to the North Sea Jurassic record. *Bull. Am. Ass. petrol. Geol.*, 70, 131–137. *2.4.*

Miall A.D. (1988) Reservoir heterogeneities in fluvial sandstones: Lessons from outcrop studies. *Bull. Am. Ass. petrol. Geol.*, 72, 682–697. *3.5.2, 3.6.1, 3.6.2, 3.6.5, Fig. 3.46, Table 3.1.*

Miall A.D. (1990) *Principles of Sedimentary Basin Analysis*, 2nd edn, 668 pp. Springer-Verlag, New York. *2.1.3.*

Miall A.D. (1991) Stratigraphic sequences and their chronostratigraphic correlation. *J. sedim. Petrol.*, 61, 497–505. *2.4.*

Miall A.D. (1992) Exxon global cycle chart: an event for every occasion.*Geology*, 20, 787–790. *2.4.*

Miall A.D. (1994) Reconstructing fluvial macroform architecture from two-dimensional outcrops: examples from the Castlegate Sandstone, Book Cliffs, Utah. *J. sedim. Res.*, B64, 146–158. *3.6.2, 3.6.5.*

Middleton G.V. (Ed.) (1965) *Primary Sedimentary Structures and their Hydrodynamic Interpretation*, 265 pp. Spec. Publ. Soc. econ. Paleont. Miner., 12, Tulsa. *1.1.*

Middleton G.V. (1966) Experiments on density and turbidity currents. I. Motion of the head. *Can. J. Earth Sci.*, 4, 523–546. *10.2.3.*

Middleton G.V. (1973) Johannes Walther's law of correlation of facies. *Bull. geol. Soc. Am.*, 84, 979–988. *2.2.1, 2.2.2.*

Middleton G.V. & Hampton M.A. (1976) Subaqueous sediment transport and deposition by sediment gravity flows. In: *Marine Sediment Transport and Environmental Management* (Ed. by D.J. Stanley and D.J.P. Swift), pp. 197–218. Wiley, New York. *10.2.3, Fig. 10.18.*

Middleton G.V. & Southard J.B. (1984) Mechanics of sediment movement. *Soc. econ. Paleont. Miner., Short Course*, 3, Tulsa. *5.3.4.*

Middleton L.T. & Blakey R.C. (1983) Processes and controls on the intertonguing of the Kayenta and Navajo Formations, northern Arizona: eolian–fluvial interactions. In: *Eolian Sediments and Processes* (Ed. by M.E. Brookfield and T.S. Ahlbrandt), pp. 613–

634. *Developments in Sedimentology*, 38. Elsevier, Amsterdam. *5.5.3.*

Miller G.H. (1982) Quaternary depositional episodes, western Spitsbergen, Norway: Aminostratigraphy and glacial history. *Arctic Alpine Res.*, 14, 321–340. *11.5.3.*

Miller J. (1986) Facies relationships and diagenesis in Waulsortian mud mounds from the Lower Carboniferous of Ireland and N. England. In: *Reef Diagenesis* (Ed. by J.H. Schoeder and B.H. Purser), pp. 311–335. Springer-Verlag, Berlin. *9.4.5.*

Miller J.A. (1975) Facies characteristics of Laguna Madre wind-tidal flats. In: *Tidal Deposits: A Casebook of Recent Examples and Fossil Counterparts* (Ed. by R.N. Ginsburg), pp. 93–101. Springer-Verlag, Berlin. *6.7.4, 8.4.2.*

Miller J.M.G. (1989) Glacial advance and retreat sequences in a Permo-Carboniferous section, central Transantarctic Mountains: *Sedimentology*, 36, 419–430. *11.5.5, Fig. 11.25.*

Miller K.B. (1991) An undulating discontinuity surface on a gently inclined muddy ramp: possible evidence of secondary helical flow from the geological record. *Sedimentology*, 38, 1097–1112. *9.3.4, 9.4.3.*

Miller M.M. (1989) Intra-arc sedimentation and tectonism: Late Paleozoic evolution of the eastern Klamath terrane, California. *Bull. geol. Soc. Am.*, 101, 170–187. *12.12.5, Fig. 12.87.*

Milliman J.D. & Meade R.H. (1983) World-wide delivery of river sediment to the oceans. *J. Geol.*, 91, 1–21. *6.2.1, 7.6.*

Milliman J.D. & Syvitski J.P.M. (1992) Geomorphic/tectonic control of sediment discharge to the ocean; the importance of small mountainous rivers. *J. Geol.*, 100, 525–544. *7.6.*

Milliman J.D., Butenko J., Barbot J.P. & Hedberg J. (1982) Deposition patterns of modern Orinoco–Amazon muds on the northern Venezuelan shelf. *J. Mar. Res.*, 40 (3), 643–657. *7.6.*

Milliman J.D., Freile D., Steinen R.P. & Baker P.A. (1993) Great Bahama Bank aragonite muds: mostly inorganically precipitated, mostly exported. *J. sedim. Petrol.*, 63, 589–595. *9.1.3.*

Milliman J.D., Qin Y. & Park Y.-H. (1989) Sediments and Sedimentary Processes in the Yellow and East China Seas. In: *Sedimentary Facies in the Active Plate Margin* (Ed. by A. Taira and F. Masuda), pp. 233–249. Terra Scientific Publishing (TERRAPUB), Tokyo. *7.6, Fig. 7.29.*

Mills A.A. (1984) Pillow lavas and the Leidonfrost effect. *J. geol. Soc., Lond.*, 141, 183–186. *12.4.1.*

Milne G. (1935) Some suggested units for classification and mapping, particularly for East African soils. *Soil Res.*, 4, 183–198. *3.3.7.*

Mitchell A.H.G. (1970) Facies of an early Miocene volcanic arc, Malekula Island, New Hebrides. *Sedimentology*, 14, 201–243. *12.1.1, 12.5.2, 12.12.5.*

Mitchell A.H.G. & Reading H.G. (1969) Continental margins, geosynclines and ocean floor spreading. *J. Geol.*, 77, 629–646. *1.1.*

Mitchell A.H.G. & Reading H.G. (1971) Evolution of island arcs. *J. Geol.*, 79, 253–284. *12.1.1, 12.12.5.*

Mitchell R.W. (1985) Comparative sedimentology of shelf carbonates of the Middle Ordovician St Paul Group, central Appalachians. *Sedim. Geol.*, 43, 1–41. *9.4.1.*

Mitchener B.C., Lawrence D.A., Partington M.A., Bowman M.B.J. & Gluyas J. (1992) Brent Group: sequence stratigraphy and regional implications. In: *Geology of the Brent Group* (Ed. by A.C.

Morton, R.S. Haszeldine, M.R. Giles and S. Brown), pp. 45–80. *Spec. Publ. geol. Soc. Lond.*, **61**. *6.6.6.*

Mitcham R.M. & Uliana M.A. (1985) Seismic stratigraphy of carbonate depositional sequences, Upper Jurassic-Lower Cretaceous, Neuquén Basin, Argentina. In: *Seismic stratigraphy II: an integrated approach to hydrocarbon exploration* (Ed. by O.R. Berg and D.G. Woolverton), pp. 255–274. *Mem. Am. Ass. petrol. Geol.*, **39**, Tulsa. *Fig. 9.73.*

Mitchum R.M. Jr, Vail P.R. & Sangree J.B. (1977) Seismic stratigraphy and global changes of sea level, Part 6: stratigraphic interpretation of seismic reflection patterns in depositional sequences. In: *Seismic Stratigraphy – Applications to Hydrocarbon Exploration* (Ed. by C.E. Payton), pp. 117–133. *Mem. Am. Ass. Petrol. Geol.*, **26**, Tulsa. *2.2.2, 2.3.1, Figs 2.4, 2.5, 2.6, 2.7.*

Mitchum R.M. Jr, Vail P.R. & Thompson III S. (1977) Seismic stratigraphy and global changes of sea level, Part 2: The depositional sequence as a basic unit for stratigraphic analysis. In: *Seismic Stratigraphy – Applications to Hydrocarbon Exploration* (Ed. by C.E. Payton), pp. 53–62. *Mem. Am. Ass. petrol. Geol.*, **26**, Tulsa. *2.3.2.*

Mitchum R.M. Jr, Sangree J.B., Vail P.R. & Wornardt W.W. (1990) Sequence stratigraphy in Late Cenozoic expanded sections, Gulf of Mexico. In: *Ann. Res. Conf. Prog. and Abs. Gulf Coast Sect. Soc. econ. Paleont. Miner.*, 1990. *Fig. 2.19.*

Mjøs R., Walderhaug O. & Prestholm E. (1993) Crevasse splay sandstone geometries in the Middle Jurassic Ravenscar Group of Yorkshire, UK. In: *Alluvial Sedimentation* (Ed. by M. Marzo and C. Puigdefabregas), pp. 167–184. *Spec. Publ. int. Ass. Sediment.*, **17**. *3.5.3, 3.6.5.*

Mohamed I. & Meng O.C. (1992) Sequence stratigraphy of Tertiary sediments offshore Sarawak (Balingian and Luconia provinces). *Symposium on Tectonic Framework and Energy Resources of the Western Margin of the Pacific Basin*, Kuala Lumpur, Malaysia, December 1992. Paper **42**, p. 67. (Abs.) *9.5.5.*

Mohindra R., Parkash B. & Prasad J. (1992) Historical geomorphology and pedology of the Gandak Megafan, Middle Gangetic Plains, India. *Earth Surf. Proc. Landf.*, **17**, 643–662. *3.3.8.*

Molina J.M.C. (1987) Analisis de Facies del Mesozoico en el Subbetico Externo Province de Córdoba y Sur de Jaén, 517 pp. Unpubl. PhD thesis, Universidad de Granada. *9.4.2.*

Molnia B.F. (1983) Processes on a glacier-dominated coast, Alaska. *Z. Geomorph. Suppl.*, **57**, 141–153. *7.6.*

Moncrieff A.C.M. (1989) Classification of poorly-sorted sedimentary rocks. *Sedim. Geol.*, **65**, 191–194. *11.5.1.*

Mondeguer A., Tiercelin J.J., Hoffert M., Larque P., Le Fournier J. & Tucholka P. (1986) Sedimentation actuelle et recents dans un petit bassin en contexte extensif et decrochant: la baie de Burton, fosse nord-Tanganyika, rift est-Africain. *Bull. Centres de Rech. Pau-SNPA*, **10**, 229–247. *4.6.3.*

Monty C.L.V. & Hardie L.A. (1976) The geological significance of the freshwater blue-green algal calcareous marsh. In: *Stromatolites* (Ed. by M.R. Walter), pp. 447–477. Elsevier, Amsterdam. *9.4.1.*

Monty C.L.V., Bosence D.W.J., Bridges P.H. & Pratt B.R. (Eds) (1995) *Carbonate Mud-mounds: Their Origin and Evolution*, 537 pp. *Spec. Publ. int. Ass. Sediment.*, **23**. *9.4.5.*

Mooers C.N.K. (1976) Introduction to the physical oceanography

and fluid dynamics of continental margins. In: *Marine Sediment Transport and Environmental Management* (Ed. by D.J. Stanley and D.J.P. Swift), pp. 7–21. Wiley, New York. *7.2.1.*

Moore D. (1959) Role of deltas in the formation of some British Lower Carboniferous cyclothems. *J. Geol.*, **67**, 522–539. *2.2.2, 6.6.6.*

Moore G.F., Curray J.R. & Emmel F.J. (1982) Sedimentation in the Sunda Tench and forearc region. In: *Trench–Forearc Geology* (Ed. by J.K. Legget), pp. 245–258. *Spec. Publ. geol. Soc. Lond.*, **10**, *12.12.5.*

Moore J.G. (1967) Base surge in recent volcanic eruptions. *Bull. Volcanol.*, **30**, 337–363. *12.4.4.*

Moore J.G. & Schilling J.G. (1973) Vesicles, water, and sulfur in Reykjanes Ridge basalts. *Contrib. Mineral. Petrol.*, **41**, 105–118. *12.4.1.*

Moore J.G., Bryan W.B. & Ludwig K.R. (1994) Chaotic deposition by a giant wave, Molokai, Hawaii. *Bull. geol. Soc. Am.*, **106**, 962–967. *12.5.1.*

Moore J.G., Nakamura K. & Alcarez A. (1966) The 1965 eruption of Taal volcano. *Science*, **151**, 955–960. *Fig. 12.1.*

Moore J.G., Bryan W.B., Beeson M.H. & Normark W.R. (1995) Giant blocks in the South Kona landslide. *Geology*, **23**, 125–128. *12.5.1.*

Moore J.G., Normark W.R. & Holcomb R.T. (1994) Giant Hawaiian landslides. *Ann. Rev. Earth Planet. Sci.*, **22**, 119–144. *12.5.1.*

Moore J.G., Phillips R.L., Grigg R.W., Peterson D.W. & Swanson D.A. (1973) Flow of lava into the sea, 1969–1971, Kilauea volcano, Hawaii. *Bull. geol. Soc. Am.*, **84**, 537–546. *12.4.1, 12.11.2.*

Moore R.C. (1959) Geological understanding of cyclic sedimentation represented by Pennsylvanian and Permian rocks of northern Mid-Continent region. In: *Kansas Field Conference, Guidebook for Association American State Geologists* (Ed. by R.C. Moore and D.F. Merriam) pp. 46–55. Kansas Geol. Surv., Lawrence, Kansas. *6.1.*

Moran S.R. (1971) Glaciotectonic structures in drift. In: *Till – A Symposium* (Ed. by R.P. Goldthwait), pp. 127–148. Ohio State University Press, Columbus. *11.3.5.*

Morgan J.P. (1961) Mudlumps at the mouth of the Mississippi River. In: *Genesis and Paleontology of the Mississippi River Mudlumps*, 116 pp. *Louisiana Dept Conservation Geol. Bull.*, **35**(1). *6.6.4.*

Morgan J.P., Coleman J.M. & Gagliano S.M. (1963) Mudlumps at the mouth of South Pass, Mississippi River; sedimentology, paleontology, structure, origin and relation to deltaic processes. *Louisiana State Univ. Coastal Studies Ser.*, **10**, 190 pp. *6.6.4.*

Morgan J.P., Coleman J.M. & Gagliano S.M. (1968) Mudlumps: diapiric structures in Mississippi delta sediments. In: *Diapirism and Diapirs* (Ed. by J. Braunstein and G.D. O'Brien), pp. 145–161. *Mem. Am. Ass. petrol. Geol.*, **8**, Tulsa. *6.6.4, Fig. 6.40.*

Mörner N.-A. (1976) Eustasy and geoid changes. *J. Geol.*, **84**, 123–151. *2.1.5.*

Mörner N.-A. (1980) The Fennoscandian uplift: Geological data and their geodynamical implication. In: *Earth Rheology, Isostasy and Eustasy* (Ed. by N.-A. Mörner), pp. 251–284. Wiley and sons, Chichester. *11.5.2.*

Mörner N.-A. (1994) Internal response to orbital forcing and external cyclic sedimentary sequences. In: *Orbital Forcing and*

Cyclic Sequences (Ed. by P.L. de Boer and D.G. Smith), pp. 25–33. Spec. Publ. int. Ass. Sediment., **19**. *2.1.4, 2.1.5, Table 2.2.*

Morris W.R. & Busby-Spera C.J. (1988) Sedimentologic evolution of a submarine canyon in a forearc basin, Upper Cretaceous Rosario Formation, San Carlos, Mexico. *Bull. Am. Ass. petrol. Geol.*, **72**, 717–737. *12.12.5.*

Morris W.R. & Busby-Spera C. (1990) A submarine-fan valley–levee complex in the Upper Cretaceous Rosario Formation: implication for turbidite facies models. *Bull. geol. Soc. Am.*, **102**, 900–914. *10.3.2.*

Morrison R.B. (1978) Quaternary soil stratigraphy: concepts, methods and problems. In: *Quaternary Soils* (Ed. by W.C. Mahaney), pp. 77–108. Geo-Abstracts, Norwich. *3.3.7, 3.5.4.*

Morton A.C. (1985) A new approach to provenance studies: electron microprobe analysis of detrital garnets from Middle Jurassic sandstones of the northern North Sea. *Sedimentology*, **32**, 553–566. *6.6.6.*

Morton A.C. (1992) Provenance of Brent Group sandstones: heavy mineral contraints. In: *Geology of the Brent Group* (Ed. by A.C. Morton, R.S. Haszeldine, M.R. Giles and S. Brown), pp. 227–244. *Spec. Publ. geol. Soc. Lond.*, **61**. *6.6.6.*

Morton A.C., Haszeldine R.S., Giles M.R. & Brown S. (Eds) (1992) *Geology of the Brent Group*, 506 pp. Spec. Publ. geol. Soc. Lond., **61**. *6.6.6.*

Morton R.A. (1981) Formation of storm deposits by wind-forced currents in the Gulf of Mexico and the North Sea. In: *Holocene Marine Sedimentation in the North Sea Basin* (Ed. by S.D. Nio, R.T.E. Shüttenhelm and Tj.C.E. van Weering), pp. 385–396. *Spec. Publ. int. Ass. Sediment.*, **5**. *7.4.1, 7.4.3, 7.4.4.*

Morton R.A. & Donaldson A.C. (1973) Sediment distribution and evolution of tidal deltas along a tide-dominated shoreline, Washapreague, Virginia. *Sedim. Geol.*, **10**, 285–299. *6.6.1, 6.7.4.*

Morton R.A. & Donaldson A.C. (1978) Hydrology, morphology, and sedimentology of the Guadalupe fluvial-deltaic system. *Bull. geol. Soc. Am.*, **89**, 1030–1036. *6.6.1.*

Moslow T.F. & Heron S.D. (1978) Relict inlets: preservation and occurrence in the Holocene stratigraphy of southern Core Banks, North Carolina. *J. sedim. Petrol.*, **48**, 1275–1286. *6.7.4.*

Moslow T.F. & Tye R.S. (1985) Recognition and characterization of Holocene tidal inlet sequences. *Mar. Geol.*, **63**, 129–151. *6.7.4.*

Mossop G.D. (1974) The Evaporites of the Baumann Fiord Formation, Ellesmere Island, Arctic Canada. *Bull. geol. Surv. Can.*, **298**, 52 pp. *8.7.*

Mossop G.D. & Flach P.D. (1983) Deep channel sedimentation in the Lower Cretaceous McMurray Formation, Athabasca Oil Sands, Alberta. *Sedimentology*, **30**, 493–509. *3.6.5.*

Motti M.J. & Seyfred W.E.J. (1980) Sub sea-floor hydrothermal systems. In: *Seafloor spreading centers: hydrothermal systems* (Ed. by P.A. Rona and R.P. Lowell), pp. 66–82. Dowden, Hutchinson and Ross, Stroudsberg. *Fig. 12.58.*

Mount J.F. & Kidder D. (1993) Combined flow origin of edgewise intraclast conglomerates: Sellick Hill Formation (Lower Cambrian). South Australia. *Sedimentology*, **40**, 315–329. *9.4.3.*

Mueller W. & White J.D.L. (1992) Felsic fire-fountaining beneath Archean Seas: pyroclastic deposits of the 2730 Ma Hunter Mine Group, Quebec, Canada. *J. Volcanol. Geotherm. Res.*, **54**, 117–134. *12.4.5, Fig. 12.30.*

Muffler L.J.P., White D.E. & Truesdell A.H. (1971) Hydrothermal explosion craters in Yellowstone National Park. *Bull. geol. Soc. Am.*, **82**, 723–740. *12.4.1.*

Mukherji A.B. (1976) Terminal fans of inland streams in Sutlej-Yamuna plain. *Geomorphology*, **20**, 190–204. *3.3.8.*

Mulder T. & Syvitski J.P.M. (1995) Turbidity currents generated at river mouths during exceptional discharges to the World Oceans. *J. Geol.*, **103**, 285–299. *10.2.3.*

Müller G. (1966) The new Rhine delta in Lake Constance. In: *Deltas in their Geologic Framework* (Ed. by L. Shirley), pp. 108–124. Houston Geol. Soc., Houston, TX. *4.6.2.*

Müller G. & Oti M. (1981) The occurrence of calcified planktonic green algae in freshwater carbonates. *Sedimentology*, **28**, 897–902. *4.7.1.*

Müller G., Irion G. & Förstner U. (1972) Formation and diagenesis of inorganic Ca–Mg carbonates in the lacustrine environment. *Naturwissenchaften*, **59**, 158–164. *4.7.1.*

Mullins H.T. (1983) Modern carbonate slopes and basins of the Bahamas. In: *Platform Margin and Deepwater Carbonates* (Ed. by H.E. Cook, A.C. Hine and H.T. Mullins), pp. 4.1–4.138. *Soc. econ. Paleont. Miner. Short Course*, **12**, Tulsa. *9.4.4, Fig. 9.42.*

Mullins H.T. & Cook H.E. (1986) Carbonate apron models: alternatives to the submarine fan model for paleoenvironmental analysis and hydrocarbon exploration. *Sedim. Geol.*, **48**, 37–79. *9.1.1, 9.4.4, Fig. 9.43.*

Mullins H.T. & Neumann A.C. (1979) Deep carbonate bank margin structure and sedimentation in the northern Bahamas. In: *Geology of Continental Slopes* (Ed. by L.J. Doyle and O.H. Pilkey), pp. 165–192. *Spec. Publ. Soc. econ. Paleont. Miner.*, **27**. *9.4.4, Fig. 9.42.*

Mullins H.T., Gardulski A.F. & Hinchey E.J. (1988) The modern carbonate ramp slope of central west Florida. *J. sedim. Petrol.*, **58**, 273–290. *9.3.4, Fig. 9.18.*

Mullins H.T., Gardulski A.F. & Hine A.C. (1986) Catostrophic collapse of the west Florida carbonate platform margin. *Geology*, **14**, 167–170. *9.4.4.*

Mullins H.T., Heath K.C., van Buren H.M. & Newton C.R. (1984) Anatomy of modern deep-ocean carbonate slope: Northern Little Bahama Bank. *Sedimentology*, **31**, 141–168. *9.1.1, 9.4.4, Fig. 9.38.*

Mullins H.T., Hine A.C. & Wilber R.J. (1982) Facies succession of Plio-Pleistocene carbonates, north-western Great Bahama Bank: discussion. *Bull. Am. Ass. petrol. Geol.*, **66**, 103–105. *9.3.2, 9.4.5.*

Mullins H.T., Neumann A.C., Wilbur R.J. & Boardman M.R. (1980) Nodular carbonate sediment on Bahamian slopes: possible precursors to nodular limestones. *J. sedim. Petrol.*, **50**, 117–131. *9.4.4, 10.4.2, Fig. 9.42.*

Mullins H.T., Newton C.R., Heath K. & van Buren H.M. (1981) Modern deep-water coral mounds north of Little Bahama Bank: criteria for recognition of deep-water coral bioherms in the rock record. *J. sedim. Petrol.*, **51**, 999–1013. *9.4.5.*

Murphy D.H. & Wilkinson B.H. (1980) Carbonate deposition and facies distribution in a central Michigan marl lake. *Sedimentology*, **27**, 123–135. *4.4.1, 4.7.1, Fig. 4.10.*

Murray J. & Renard A.F. (1891) Report on deep-sea deposits based on specimens collected during the voyage of HMS Challenger in the years 1873–1876. In: *'Challenger' Reports*, 525 pp. HMSO, London. *10.1.*

Murray J.W. (1976) A method of determining proximity of marginal seas to an ocean. *Mar. Geol.*, 22, 103–119. *7.4.4.*

Muto T. (1988) Stratigraphical patterns of coastal-fan sedimentation adjacent to high-gradient slopes affected by sea-level changes. In: *Fan Deltas: Sedimentology and Tectonic Settings* (Ed. by W. Nemec and R.J. Steel), pp. 84–90. Blackie, London. *6.5.5.*

Mutti E. (1979) Turbidites et cônes sous-marins profonds. In: *Sédimentation Détrique (fluviatile, littorale et marine)* (Ed. by P. Homewood), pp. 353–419. Institut Géologie Université de Fribourg, Fribourg. *10.1, 10.3.2.*

Mutti E. (1985) Turbidite systems and their relations to depositional sequences. In: *Provenance of Arenites* (Ed. by G.G. Zuffa), pp. 65–93. Reidel, Amsterdam. *10.3.1, 10.3.2, 10.3.6, Fig. 10.31.*

Mutti E. (1992) *Turbidite Sandstones*, 275 pp. Agip Instituto di Geologia Università di Parma, Milano. *10.3.2, 10.3.7.*

Mutti E. & Johns D.R. (1978) The role of sedimentary bypassing in the genesis of fan fringe and basin plain turbidites in the Hecho Group System (South-Central Pyrenees). *Mem. Soc. geol. Ital.*, 18, 15–22. *10.1.*

Mutti E. & Normark W.R. (1987) Comparing examples of modern and ancient turbidite systems: Problems and concepts. In: *Marine Clastic Sedimentology: Concepts and Case Studies* (Ed. by J.K. Leggett and G.G. Zuffa), pp. 1–38. Graham & Trotman, London. *2.2.2, 10.3.2, Fig. 10.32, Table 10.2.*

Mutti E. & Normark W.R. (1991) An integrated approach to the study of turbidite systems. In: *Seismic Facies and Sedimentary Processes of Submarine Fans and Turbidite Systems* (Ed. by P. Weimer and M.H. Link), pp. 75–106. Springer-Verlag, New York. *2.2.2, 10.3.2, Fig. 10.32, Table 10.2.*

Mutti E. & Ricci Lucchi F. (1972) Le torbiditi dell' Appennino Settentrionale: introduzione all'analisi di facies. *Mem. Soc. geol. Ital.*, 11, 161–199. *2.2.1, 2.2.2, 10.1, 10.2, 10.2.3, 10.2.4, 10.3.2, Fig. 10.29, Fig. 10.35.*

Mutti E. & Ricci Lucchi F. (1975) *Turbidite Facies and Facies Associations. Field Trip Guidebook*, A-11, pp. 21–36. 9th Int. Sedimentology Congr., Nice, France. *10.2.4.*

Mutti E., Rosell J., Allen J.P., Fonnesu F. & Sgavetti M. (1985) The Eocene Baronia tide dominated delta–shelf system in the Ager Basin. In: *Excursion Guidebook: 6th European Regional Meeting, International Association of Sedimentologists, Lérida, Spain* (Ed. by M.D. Mila and J. Rosell), pp. 579–600. *7.7.1, Fig. 7.34.*

Nadon G.C. (1994) The genesis and recognition of anastomosed fluvial deposits: data from the St Mary's River Formation, southwestern Alberta, Canada. *J. sedim. Res.*, B64, 451–463. *3.3.5, 3.6.4, 3.6.5.*

Nakada S., Miyake Y., Sato H., Oshima O. & Fujinawa A. (1995) Endogenous growth of dacite dome at Unzen volcano (Japan). *Geology*, 23, 157–160. *Fig. 12.1.*

Nakamura K. (1977) Volcanoes as possible indicators of tectonic stress orientation: principal and proposal. *Journal of Volcanology and Geothermal Research*, 2, 1–16. *12.3.2, 12.8.*

Nami M. (1976) An exhumed Jurassic meander belt from Yorkshire, England. *Geol. Mag.*, 113, 47–52. *3.6.5.*

Nami M. & Leeder M.R. (1978) Changing channel morphology and magnitude in the Scalby Formation (M. Jurassic) of Yorkshire, England. In: *Fluvial Sedimentology* (Ed. by A.D. Miall), pp. 431–440. *Mem. Can. Soc. petrol. Geol.*, 5, Calgary. *3.6.5, Fig. 3.53.*

Nanson G.C. & Page K. (1983) Lateral accretion of fine-grained concave benches in meandering streams. In: *Modern and Ancient Fluvial Systems* (Ed. by J.D. Collinson and J. Lewin) pp. 133–143. *Spec. Publ. int. Ass. Sediment.*, 6. *3.6.5.*

Nappi G., Renzulli A. & Santi P. (1991) Evidence of incremental growth in the Vulsinian calderas (central Italy). *J. Volcanol. Geotherm. Res.*, 47, 13–31. *12.4.6.*

Narayan J. (1971) Sedimentary structures in the Lower Greensand of the Weald, England, and Bas-Boulonnais, France. *Sedim. Geol.*, 6, 73–109. *7.7.1.*

Nardin T.R. (1983) Late Quaternary depositional systems and sea level changes – Santa Monica and San Pedro Basins, California Continental Borderland. *Bull. Am. Ass. petrol. Geol.*, 67, 1104–1124. *10.3.1, Fig. 10.1.*

Nardin T.R., Hein F.J., Gorsline D.S. & Edwards B.D. (1979) A review of mass movement processes, sediment and acoustic characteristics, and contrasts in slope and base-of-slope systems versus canyon–fan–basin floor systems. In: *Geology of Continental Slopes* (Ed. by L.J. Doyle and O.H. Pilkey), pp. 61–73. *Spec. Publ. Soc. econ. Paleont.*, 27, Tulsa. *10.2.3, 10.3.5.*

Nassichuk W.W. & Davies G.R. (1980) Stratigraphy and sedimentation of the Otto Fiord Formation – a major Mississippian–Pennsylvanian evaporite of subaqueous origin in the Canadian Arctic Archipelago, 87 pp. *Bull geol. Surv. Can.*, 286. *8.9.*

Natland M.L. & Kuenen Ph.H. (1951) Sedimentary history of the Ventura Basin, California, and the action of turbidity currents. In: *Turbidity Currents and the Transportation of Coarse Sediment into Deep Water* (Ed. by J.L. Hough), pp. 76–107. *Spec. Publ. Soc. econ. Paleont. Miner.*, 2, Tulsa. *10.1.*

NEDECO (1959) *River Studies and Recommendations on Improvement of Niger and Benue*, 1000 pp. North Holland, Amsterdam. *3.3.3.*

Nedell S.S., Andersen D.W., Squyres S.W. & Love F.G. (1987) Sedimentation in ice-covered Lake Hoare, Antarctica. *Sedimentology*, 34, 1093–1106. *11.4.7.*

Neev D. & Emery K.O. (1967) The Dead Sea, depositional processes and environments of evaporites, 147 pp. *Bull. geol. Surv. Israel*, 41. *8.6.2.*

Neill C.R. (1969) Bed forms of the Lower Red Deer River, Alberta. *J. Hydrol.*, 7, 58–85. *3.2.2, 3.3.3.*

Nelson A.R. (1981) Quaternary glacial and marine stratigraphy of the Qivitu Peninsula, northern Cumberland Peninsula, Baffin Island, Canada. *Bull. geol. Soc. Am., Part II*, 92, 1143–1261. *11.5.3.*

Nelson C.H. (1982) Modern shallow water graded sand layers from storm surges, Bering Shelf: a mimic of Bouma sequences and turbidite systems. *J. sedim. Petrol.*, 52, 537–545. *7.2.2, 7.4.3, Fig. 7.22.*

Nelson C.H. & Maldonado A. (1988) Factors controlling depositional patterns of Ebro turbidite systems, Mediterranean Sea. *Bull. Am. Ass. petrol. Geol.*, 72, 698–716. *10.3.4, Fig. 10.57.*

Nelson C.H. & Maldonado A. (1990) Factors controlling late Cenozoic continental margin growth from the Ebro Delta to the western Mediterranean deep sea. *Mar. Geol.*, 95, 419–440. *6.6.5.*

Nelson C.H., Bacon C.R., Robinson S.W., Adam D.P., Bradbury D.P., Barber J.H., Schwartz D. & Vagenas G. (1994) The volcanic, sedimentologic, and paleolimnologic history of the Crater Lake caldera floor, Oregon: evidence for small caldera evolution. *Bull. geol. Soc. Am.*, 106, 684–704. *12.4.6, 12.12.4.*

Nelson C.S. (Ed.) (1988) *Non-tropical Shelf Carbonates – Modern and Ancient*, 367 pp. Sedim. Geol., 60. *9.1.1.*

Nemec W. (1990a) Deltas – remarks on terminology and classification. In: *Coarse-grained Deltas* (Ed. by A. Colella and D.B. Prior), pp. 3–12. *Spec. Publ. int. Ass. Sediment.*, 10. *6.3, 6.5.*

Nemec W. (1990b) Aspects of sediment movement on steep delta slopes. In: *Coarse-grained Deltas* (Ed. by A. Colella and D.B. Prior), pp. 29–73. *Spec. Publ. int. Ass. Sediment.*, 10. *6.5, 6.5.3, 6.5.4, 12.5.1.*

Nemec W. (1992) Depositional controls on plant growth and peat accumulation in a braidplain delta environment: Helvetiafjellet Formation (Barremian–Aptian), Svalbard. In: *Controls on the Distribution and Quality of Cretaceous Coals* (Ed. by P.J. McCabe and J.T. Parrish), pp. 209–226. *Spec. Pap. geol. Soc. Am.*, 267. *3.6.5, 3.6.8, Fig. 3.54.*

Nemec W. & Postma G. (1993) Quaternary alluvial fans in southwestern Crete: sedimentation processes and geomorphic evolution. In: *Alluvial Sedimentation* (Ed. by M. Marzo and C. Puigdefabregas), pp. 235–276. *Spec. Publ. int. Ass. Sediment.*, 17. *3.3.8, 3.5.1, 3.6.3, 3.6.4, Fig. 3.35.*

Nemec W. & Steel R.J. (1984) Alluvial and coastal conglomerates: Their significant features and some comments on gravelly mass-flow deposits. In: *Sedimentology of Gravels and Conglomerates* (Ed. by E.H. Koster and R.J. Steel), pp. 1–31. *Mem. Can. Soc. petrol. Geol.*, 10, Calgary. *3.2.2, 3.4, 3.5.1, Figs 3.2, 3.34, 6.5.4, 10.2.3, Fig. 10.19.*

Nemec W. & Steel R.J. (Eds) (1988a) *Fan Deltas: Sedimentology and Tectonic Settings*, 444 pp. Blackie, London. *6.1, 10.3.3.*

Nemec W. & Steel R.J. (1988b) What is a fan delta and how do we recognize it? In: *Fan Deltas: Sedimentology and Tectonic Settings* (Ed. by W. Nemec and R.J. Steel), pp. 3–13. Blackie, London. *6.1, 6.5.*

Nemec W., Steel R.J., Porebski S.J. & Spinnanger Å. (1984) Domba Conglomerate, Devonian, Norway: process and lateral variability in a mass flow-dominated, lacustrine fan-delta. In: *Sedimentology of Gravels and Conglomerates* (Ed. by E.H. Koster and R.J. Steel), pp. 295–320. *Mem. Can. Soc. petrol. Geol.*, 10. *6.5.6.*

Nemec W., Steel R.J., Gjelberg J., Collinson J.D., Prestholm E. & Øxnevad I.E. (1988) Anatomy of collapsed and re-established delta front in Lower Cretaceous of eastern Spitsbergen: gravitational sliding and sedimentation processes. *Bull. Am. Ass. petrol. Geol.*, 72, 454–476. *6.6.6, Fig. 6.67.*

Neuman G. (1968) *Ocean Currents*, 352 pp. Elsevier, Amsterdam. *10.2.2.*

Neumann A.C. & Land L.S. (1975) Lime mud deposition and calcareous algae in the Bight of Abaco, Bahamas: a budget. *J. sedim. Petrol.*, 45, 763–786. *9.3.2.*

Neumann A.C. & Macintyre I.G. (1985) Response to sea-level rise: keep-up, catch-up or give-up. *Proc. 5th Coral Reef Congr., Tahiti*, 3, 105–110. *9.2.2, 9.4.5.*

Neumann A.C., Gebelein C.D. & Scoffin T.P. (1970) The composition, structure and erodibility of subtidal mats, Abaco, Bahamas. *J. sedim. Petrol.*, 40, 274–297. *9.3.2.*

Neumann van Padang M. (1929) De Noordelijke doorbraak in den Papandajan kraterwand. *De Mijningeniur*, 10(3), 55–57. *12.1.1.*

Newell N.D., Imbrie J., Purdy E.G. & Thurber D.L. (1959) Organism communities and bottom facies, Great Bahama Bank. *Bull. Am. Mus. Nat. Hist.*, 117, 177–228. *9.1.1.*

Newell N.D., Rigby J.K., Fischer A.G., Whiteman A.J., Hickox J.E. & Bradley J.S. (1953) *The Permian Reef Complex of the Guadalupe Mountains Region, Texas and New Mexico*, 236 pp. W.H. Freeman, San Francisco. *9.1.1.*

Newhall C.G. & Dzurisin D. (1988) *Historical Unrest at Large Calderas of the World*, 598 pp. Bull. US geol. Surv., 1855. *12.4.6, 12.13.*

Newhall C.G. & Self S. (1982) The Volcanic Explosivity Index (VEI): an estimate of explosive magnitude for historical volcanism. *J. geophys. Res.*, 87, 1231–1238. *12.6.2.*

Newman M. St J., Reeder M.L., Woodruff A.H.W. & Hatton I.R. (1993) The geology of the Gryphon Oil Field. In: *Petroleum Geology of Northwest Europe: Proceedings of the 4th Conference* (Ed. by J.R. Parker), pp. 123–133. Geol. Soc. Lond., Bath. *10.3.3.*

Newton R.S., Seibold E. & Werner F. (1973) Facies distribution patterns on the Spanish Sahara continental shelf mapped with side-scan sonar. *Meteor. Forsch. Engl.*, C15, 55–77. *7.5.*

Nichol S.L. (1991) Zonation and sedimentology of estuarine facies in an incised-valley, wave-dominated, microtidal setting, New South Wales, Australia. In: *Clastic Tidal Sedimentology* (Ed. by D.G. Smith, G.E. Reinson, B.A. Zaitlin and R.A. Rahmani), pp. 41–58. *Mem. Can. Soc. petrol. Geol.*, 16, Calgary. *6.7.5.*

Nichol S.L., Boyd R. & Penland S. (1994) Stratigraphic response of wave-dominated estuaries to different relative sea-level and sediment supply histories: Quaternary case studies from Nova Scotia, Louisiana and eastern Australia. In: *Incised-valley Systems: Origin and Sedimentary Sequences* (Ed. by R.W. Dalrymple, R. Boyd & B.A. Zaitlin), pp. 265–283. *Spec. Publ. Soc. sedim. Geol.*, 51. *6.7.6.*

Nichols M.M. & Biggs R.B. (1985) Estuaries. In: *Coastal Sedimentary Environments* (Ed. by R.A. Davis) pp. 77–186. Springer-Verlag, New York. *6.7.5.*

Nichols M.M., Johnson G.H. & Peebles P.C. (1991) Modern sediments and facies models for a microtidal coastal plain estuary, the James Estuary, Virginia. *J. sedim. Petrol.*, 61, 883–899. *6.7.5.*

Nickling W.G. & Gillies J.A. (1993) Dust emission and transport in Mali, West Africa. *Sedimentology*, 40, 859–868. *5.1.*

Nielson J. & Kocurek G. (1986) Climbing zibars of the Algodones. *Sedim. Geol.*, 48, 1–15. *5.4.6.*

Nielson J. & Kocurek G. (1987) Surface processes, deposits, and development of star dunes. *Bull. geol. Soc. Am.*, 99, 177–186. *5.4.3, 5.4.4.*

Nieuwenhuyse A. & Kroonenberg S.B. (1994) Volcanic origin of Holocene beach ridges along the Caribbean coast of Cost Rica. *Mar. Geol.*, 120, 12–26. *12.5.3.*

Nilsen T.H. & Zuffa G.G. (1982) The Chugach Terrane, a Cretaceous trench-fill deposit, southern Alaska. In: *Trench–Forearc Geology* (Ed. by J.K. Legget), pp. 213–227. *Spec. Publ. geol. Soc. Lond.*, 10. *12.12.5.*

Nio S.-D. (1976) Marine transgressions as a factor in the formation of sand wave complexes. *Geol. Mijnbouw*, 55, 18–40. *7.7.1.*

Nio S.-D. & Yang C.-S. (1991) Diagnostic attributes of clastic tidal deposits: a review. In: *Clastic Tidal Sedimentology* (Ed. by D.G. Smith, G.E. Reinson, B.A. Zaitlin and R.A. Rahmani), pp. 3–28. *Mem. Can. Soc. petrol. Geol.*, 16, Calgary. *6.2.7, 7.3.2, 7.7.1.*

Nio S.-D., Siegenthaler C. & Yang C.-S. (1983) Megaripple cross-bedding as a tool for the reconstruction of the palaeohydraulics in a Holocene subtidal environment, SW Netherlands. *Geol. Mijnbouw*, 62, 499–510. *6.2.7.*

Nio S.-D., van den Berg J.H., Goesten J.H. & Smulders F. (1980) Dynamics and sequential analysis of a mesotidal shoal and intershoal channel complex in the Eastern Scheldt (southwestern Netherlands). *Sedim. Geol.*, 26, 263–279. *7.3.2.*

Nishimura A., Marsaglia K.M. *et al.* (1991) Pliocene–Quaternary submarine pumice deposits in the sumisu rift area, Izu–Bonin. In: *Sedimentation in Volcanic Settings* (Ed. by R.V. Fisher and G.A. Smith), pp. 201–210. *Spec. Publ. Soc. econ. Paleont. Miner.*, 45, Tulsa. *12.5.5.*

Nittrouer C.A. & DeMaster D.J. (1986) Sedimentary processes on the Amazon continental shelf. *Continental Shelf Res.*, 6(1–2), 361 pp. *7.6.*

Nittrouer C.A. & Sternberg R.W. (1981) The formation of sedimentary strata in an allochthonous shelf environment: the Washington continental shelf. *Mar. Geol.*, 42, 201–232. *7.2.2.*

Nittrouer C.A. & Wright L.D. (1994) Transport of particles across continental shelves. *Rev. Geophys.*, 32, 85–113. *Fig. 7.17.*

Nittrouer C.A., Kuehl S.A., DeMaster D.J. & Kowsmann R.O. (1986) The deltaic nature of Amazon shelf sedimentation. *Bull. geol. Soc. Am.*, 97, 444–458. *6.6, 6.6.4.*

Normark W.R. (1970) Growth patterns of deep sea fans. *Bull. Am. Ass. petrol. Geol.*, 54, 2170–2195. *10.1, 10.3.3.*

Normark W.R. (1974) Submarine canyons and fan valleys: Factors affecting growth patterns of deep-sea fans. In: *Modern and Ancient Geosynclinal Sedimentation* (Ed. by R.H. Dott, Jr and R.H. Shaver), pp. 56–68. *Spec. Publ. Soc. econ. Paleont. Miner.*, 19, Tulsa. *10.1.*

Normark W.R. (1978) Fan valleys, channels, and depositional lobes on modern submarine fans: Characters for recognition of sandy turbidite environments. *Bull. Am. Ass. petrol. Geol.*, 62, 912–931. *10.1.*

Normark W.R. & Piper D.J.W. (1985) Navy Fan, Pacific Ocean. In: *Submarine Fans and Related Turbidite Systems* (Ed. by A.H. Bouma, W.R. Normark and N.E. Barnes), pp. 87–94. Springer-Verlag, New York. *10.3.3, Fig. 10.47.*

Normark W.R. & Piper D.J.W. (1991) Initiation processes and flow evolution of turbidity currents: implications for the depositional record. In: *From Shoreline to Abyss; Contributions in Marine Geology in Honor of Francis Parker Shepard* (Ed. by R.H. Osborne), pp. 207–230. *Spec. Publ. econ. Paleont. Miner.*, 46, Tulsa. *10.2.3.*

Nota D.J.G. (1958) *Reports of the Orinoco Shelf Expedition*, 2, 98 pp. H. Veenman en Zönen, Wageningen. *6.6.4.*

Nøttvedt A. & Kreisa R.D. (1987) Model for the combined flow origin of hummocky cross bedding. *Geology*, 15, 357–361. *7.7.2.*

Nowell A.R.M. & Hollister C.D. (Eds) (1985) *Deep Ocean Sediment Transport: Preliminary Results of the High Energy Benthic Boundary Layer Experiment*, 409 pp. *Mar. Geol.*, 66. *10.2.2, 10.3.7.*

Nummedal D. (1991) Shallow marine storm sedimentation – the oceanographic perspective. In: *Cycles and Events in Stratigraphy* (Ed. by G. Einsele, W. Ricken and A. Seilacher), pp. 565–571. Springer-Verlag, Berlin. *7.4.1.*

Nummedal D. & Swift D.J.P. (1987) Transgressive stratigraphy at sequence-bounding unconformities: some principles derived from Holocene and Cretaceous examples. In: *Sea-level Fluctuations and Coastal Evolution* (Ed. by D. Nummedal, D.H. Pilkey and J.D. Howard), pp. 241–260. *Spec.. Publ. Soc. econ. Paleont. Miner.*, 41, Tulsa. *6.7.6.*

Nurmi R.D. & Friedman G.M. (1977) Sedimentology and depositional environments of basin-central evaporites, Lower Salina Group (Upper Silurian), Michigan Basin. In: *Reefs and Evaporites – Concepts and Depositional Models* (Ed. by J.H. Fisher), pp. 23–52. *Am. Ass. petrol. Geol. Studies in Geology*, 5, Tulsa. *8.6.1.*

Nygaard E., Lieberkind K. & Frykman P. (1983) Sedimentology and reservoir parameters of the Chalk Group in the Danish Central Graben. *Geol. Mijnbouw*, 62, 77–190. *10.4.3.*

O'Brien N.R., Nakazawa K. & Tokuhashi S. (1980) Use of clay fabric to distinguish turbidite and hemipelagic siltstones and silts. *Sedimentology*, 27, 47–61. *10.2.3.*

O'Brien P.E. & Wells A.T. (1986) A small, alluvial crevasse splay. *J. sedim. Petrol.*, 56, 876–879. *3.3.6.*

Ocamb R.D. (1961) Growth faults of south Louisiana. *Trans. Gulf Coast. Ass. geol. Soc.*, 11, 139–175. *6.6.4.*

Ochsenius K. (1877) *Die Bildung der Steinsalzlager und ihrer Mutter langensalze*, 172 pp. CEM Pfeffer Verlag, Halle. *8.1.*

O'Connell S., Ryan W.B.F. & Normark W.R. (1991) Evolution of a fan channel on the surface of the Outer Mississippi Fan: evidence from side-looking sonar. In: *Seismic Facies and Sedimentary Processes of Submarine Fans and Turbidite Systems* (Ed. by P. Weimer and M.H. Link), pp. 365–381. Springer-Verlag, New York. *10.3.3.*

Odin G.S. & Matter A. (1981) De glauconarium origine. *Sedimentology*, 28, 611–641. *10.2.1.*

O'Driscoll D., Hindle A.D. & Long D.C. (1990) The structural controls on Upper Jurassic and Lower Cretaceous reservoir sandstones in the Witch Ground Graben, UK North Sea. In: *Tectonic Events Responsible for Britain's Oil and Gas Reserves* (Ed. by R.F.P. Hardman and J. Brooks), pp. 299–323. *Spec. Publ. geol. Soc. Lond.*, 55. *10.3.4.*

Oertel G.F. (1985) The barrier island system. *Mar. Geol.*, 63, 1–18. *6.7.4.*

Oertel G.F., Henry V.J. & Foyle A.M. (1991) Implications of tide-dominated lagoonal processes on the preservation of buried channels on a sediment-starved continental shelf. In: *Shelf Sand and Sandstone Bodies: Geometry, Facies and Sequence Stratigraphy* (ed. by D.J.P. Swift, G.F. Oertel, R.W. Tillman and J.A. Thorne), pp. 379–393. *Spec. Publ. int. Ass. Sediment.*, 14. *6.7.6.*

Off T. (1963) Rhythmic linear sand bodies caused by tidal currents. *Bull. Am. Ass. petrol. Geol.*, 47, 324–341. *7.3.4.*

Ogg J.G., Haggerty J., Sarti M. & Von Rad U. (1987) Lower Cretaceous pelagic sediments of Deep Sea Drilling Project Site 603, western North Atlantic: a synthesis. In: *Initial Reports of the Deep Sea Drilling Project*, 93 (Ed. by J.E. Van Hinte, S.W. Wise *et al.*), pp. 1305–1331. US Government Printing Office, Washington, DC. *10.4.2.*

Okolo S.A. (1983) Fluvial distributary channels in the Fletcher Bank Grit (Namurian R$_2$), at Ramsbottom, Lancashire, England. In: *Modern and Ancient Fluvial Systems*, (Ed. by J.D. Collinson and J. Lewin), pp. 421–433. *Spec. Publ. int. Ass. Sediment.*, 6. *3.6.9.*

Ollier C.D. & Brown M.J.F. (1971) Erosion of a young volcano in New Guinea. *Z. Geomorph.*, 15, 12–28. *12.9.1.*

Olsen H. (1987) Ancient ephemeral stream deposits: a local terminal fan model from the Bunter Sandstone Formation (L. Triassic) in the Tronder-3, -4 and -5 wells, Denmark. In: *Desert Sediments: Ancient and Modern* (Ed. by L.E. Frostick and I. Reid), pp. 69–86. *Spec. Publ. geol. Soc. Lond.*, 35. *3.6.10, Fig. 3.63.*

Olsen H. (1989) Sandstone-body structures and ephemeral stream processes in the Dinosaur Canyon Member, Moenave Formation (Lower Jurassic), Utah, U.S.A. *Sedim. Geol.*, 61, 207–221. *3.6.10.*

Olsen H. & Larsen P.-H. (1993) Structural and climatic controls on fluvial depositional systems – Devonian, North-East Greenland. In: *Alluvial Sedimentation* (Ed. by M. Marzo and C. Puigdefabregas), pp. 401–423. *Spec. Publ. int. Ass. Sediment.*, 17. *3.6.7.*

Olsen P.E. (1984) Periodicity of lake-level cycles in the Late Triassic Lockatong Formation of the Newark Basin (Newark Supergroup, New Jersey and Pennsylvania). In: *Milankovitch and Climate, Part 1* (Ed. by A.L. Berger), pp. 129–146. A.L. Reidel, Dordrecht. *4.16.3, Fig. 4.37.*

Olsen P.E. (1986) A 40-million-year lake record of early Mesozoic orbital climatic forcing. *Science*, 234, 842–848. *4.10, 4.16.3, Fig. 4.38.*

Olsen P.E. (1990) Tectonic, climatic, and biotic modulation of lacustrine ecosystems – examples from Newark Supergroup of eastern North America. In: *Lacustrine Basin Exploration – Case Studies and Modern Analogs* (Ed. by B.J. Katz), pp. 209–224. *Am. Ass. petrol. Geol. Memoir*, 50, Tulsa. *4.10, 4.16.3.*

Oomkens E. (1967) Depositional sequences and sand distribution in a deltaic complex. *Geol. Mijnbouw*, 46, 265–278. *6.6.5.*

Oomkens E. (1970) Depositional sequences and sand distribution in the post-glacial Rhône delta complex. In: *Deltaic Sedimentation – Modern and Ancient* (Ed. by J.P. Morgan and R.H. Shaver), pp. 198–212. *Spec. Publ. Soc. econ. Paleont. Miner.*, 15, Tulsa. *6.1, 6.6.1, 6.6.5.*

Oomkens E. (1974) Lithofacies relations in the Late Quaternary Niger delta complex. *Sedimentology*, 21, 195–222. *6.1, 6.6.1, 6.6.5, Fig. 6.45.*

Oomkens E. & Terwindt J.H.J. (1960) Inshore estuarine sediments in the Haringvliet, Netherlands. *Geol. Mijnbouw*, 39, 701–710. *6.6.1.*

Open University Course Team (1989a) *Waves, Tides and Shallow-water Processes*, 187 pp. Open University, Milton Keynes/ Pergamon Press, Oxford. *7.3.1, Fig. 7.7.*

Open University Course Team (1989b) *Ocean Circulation*, 238 pp. Open University, Milton Keynes/Pergamon Press, Oxford. *7.5.*

Ore H.T. (1963) Some criteria for recognition of braided stream deposits. *Contrib. Geol. Wyoming Univ. Dept Geol.*, 3, 1–14. *3.1.*

Orford J.D. (1975) Discrimination of particle zonation on a pebble beach. *Sedimentology*, 22, 441–463. *6.5.4.*

Orford J.D. & Carter R.W.G. (1982) Crestal overtop and washover sedimentation on a fringing sandy gravel barrier coast, Carnsmore Point, southwest Ireland. *J. sedim. Petrol.*, 52, 265–278. *6.5.4.*

Orheim O. & Elverhøi A. (1981) Model for submarine glacial deposition. *Ann. Glaciol.*, 2, 123–128. *11.4.6.*

Ori G.G. & Roveri M. (1987) Geometries of Gilbert-type deltas and large channels in the Meteora Conglomerate, Meso-Hellenic basin (Oligo-Miocene), central Greece. *Sedimentology*, 34, 845–859. *3.5.1, 6.5.6.*

Orti Cabo F., Pueyo Mur J.J., Geisler-Cussey D. & Dulau N. (1984) Evaporitic sedimentation in the coastal salinas of Santa Pola (Alicante, Spain). *Instit. Invest. Geolog. Diput. Prov. Univ. Barcelona*, 38/39, 169–220. *8.1.*

Orton G.J. (1988) A spectrum of Middle Ordovician fan deltas and braid-plain deltas North Wales: a consequence of varying fluvial clastic input. In: *Fan Deltas: Sedimentology and Tectonic Settings* (Ed. by W. Nemec and R.J. Steel), pp. 23–49. Blackie, London. *6.1, 6.3, 6.5, 12.12.5.*

Orton G.J. (1991) Emergence of subaqueous depositional environments in advance of a major ignimbrite eruption, Capel Curig Volcanic Formation, North Wales: an example of regional volcanotectonic uplift? *Sedim. Geol.*, 74, 251–288. *12.13.*

Orton G.J. (1995) Facies models in volcanic terrains: time's arrow versus time's cycle. In: *Sedimentary Facies Analysis* (Ed. by A.G. Plint), pp. 157–193. Blackwell Science, Oxford. *12.4.6, 12.13, 12.14.2, Fig. 12.42.*

Orton G.J. & Reading H.G. (1993) Variability of deltaic processes in terms of sediment supply, with particular emphasis on grain size. *Sedimentology*, 40, 475–512. *3.3.1, Figs 3.7, 6.1, 6.3, 6.6, Tables 6.1, 6.2, Figs 6.2, 6.5, 6.8, 6.19.*

Osborne R.H., Licari G.R. & Link M.H. (1982) Modern lacustrine stromatolites, Walker Lake, Nevada. *Sedim. Geol.*, 32, 39–61. *4.7.1.*

Oschmann W. (1991) Anaerobic–poikiloaerobic–aerobic: a new facies zonation for Modern and Ancient neritic facies. In: *Cycles and Events in Stratigraphy* (Ed. by G. Einsele, W. Ricken and A. Seilacher), pp. 565–571. Springer-Verlag, Berlin. *Fig. 7.46.*

Osleger D.A. & Read J.F. (1991) Relation of eustasy to stacking patterns of meter-scale carbonate cycles, Late Cambrian, U.S.A. *J. sedim. Petrol.*, 61, 1225–1252. *9.1.1, 9.4.1, 9.5.3.*

Otvos E.G. & Price W.A. (1979) Problems of chenier genesis and terminology – an overview. *Mar. Geol.*, 31, 251–263. *6.7.2.*

Oudin E. & Constantinou G. (1984) Black smoker chimney fragments in Cyprus sulphide deposits. *Nature*, 308, 349–353. *12.11.1.*

Ovenshine A.T. (1970) Observations of iceberg rafting in Glacier Bay, Alaska, and the identification of ancient ice-rafted deposits. *Bull. geol. Soc. Am.*, 81, 891–894. *11.4.5, Table 11.2.*

Oviatt C.G. & Nash, W.P. (1989) Late Pleistocene basaltic ash and volcanic eruptions in the Bonneville basin, Utah. *Bull. geol. Soc. Am.*, 101, 292–303. *12.9.3.*

Owen R.B. & Crossley R. (1992) Spatial and temporal distribution of diatoms in sediments of Lake Malawi, central Africa, and ecological implications. *J. Paleolimnol.*, 7, 55–71. *4.7.2.*

Owen R.B., Crossley R., Johnson T.C., Tweddle D., Kornfield L., Davison S., Eccles D.H. & Engstrom D.E. (1990) Major low levels of Lake Malawi and implications for speciation rates in cichlid fishes. *Proc. R.Soc. Lond., Ser. B*, 240–519. *4.9.*

Page K. & Nanson G. (1982) Concave-bank benches and associated flood plain formation. *Earth Surf. Process. Landf.*, 7, 529–543. *3.3.5.*

Pages F. & Gili J.-M. (1992) Influence of Agulhas waters on the population structure of planktonic cnidarians in the southern Benguela region. *Sci. Mar., Barcelona*, 56, 109–123. *7.5.*

Palmer B.A. & Neall V.E. (1991) Contrasting lithofacies architecture in ring-plain deposits related to edifice construction and destruction, the Quaternary Stratford and Opunake Formations, Egmont Volcano, New Zealand. *Sedim. Geol.*, 74, 71–88. *12.12.3, Fig. 12.36.*

Palmer B.A. & Walton A.W. (1990) Accumulation of volcaniclastic aprons in the Mount Dutton Formation (Oligocene–Miocene), Marysvale volcanic field, Utah. *Bull. Geol. Soc. Am.*, 102, 734–748. *12.6.2, 12.12.4, 12.14.1.*

Palmer B.A., Alloway B.V. & Neall V.E. (1991) Volcanic-debris-avalanche deposits in New Zealand – lithofacies organization in unconfined wet avalanche flows. In: *Sedimentation in Volcanic Settings* (Ed. by R.V. Fisher and G.A. Smith), pp. 89–98. *Spec. Publ. Soc. econ. Paleont. Miner.*, 45, Tulsa. *12.5.1, Fig. 12.33.*

Palmer B.A., Purves A.M. & Donoghue S.L. (1993) Controls on accumulation of a volcaniclastic fan, Ruapehu composite volcano, New Zealand. *Bull. Volcanol.*, 55, 176–189. *12.6.2.*

Palmer T.J. (1979) The Hampen Marly and White Limestone formations: Florida-type carbonate lagoons in the Jurassic of central England. *Palaeontology*, 22, 189–228. *9.4.2, Fig. 9.23.*

Pantin H.M. (1979) Interaction between velocity and effective density in turbidity flow: phase-plane analysis, with criteria for autosuspension. *Mar. Geol.*, 31, 59–99. *10.2.3.*

Paola C. & Borgman L. (1991) Reconstructing random topography from preserved stratification. *Sedimentology*, 38, 553–565. *5.3.4, Fig. 5.6.*

Paola C., Heller P.L. & Angevine C.L. (1992) The large-scale dynamics of grain-size variation in alluvial basins, 1: Theory. *Basin Res.*, 4, 73–90. *3.6.3.*

Parea G.C. & Ricci Lucchi F. (1972) Resedimented evaporites in the Periadriatic trough (Upper Miocene, Italy). *Israel J. Earth Sci.*, 21, 125–141. *8.9.*

Parfitt E.A., Wilson L. & Head J.W. (1993) Basaltic magma reservoirs: factors controlling their rupture characteristics and evolution. *J. Volcanol. Geotherm. Res.*, 55, 1–14. *12.3.2.*

Park R.K. (1976) A note on the significance of lamination in stromatolites. *Sedimentology*, 23, 379–393. *8.4.2.*

Park R.K. (1977) The preservation potential of some stromatolites. *Sedimentology*, 24, 485–506. *9.4.1.*

Park Y.-H. (1986) Water characteristics and movements of the Yellow Sea Warm Current in summer. *Prog. Oceanogr.*, 17, 243–254. *7.6.*

Parkash B. & Middleton G.V. (1970) Downcurrent textural changes in Ordovician turbidite greywackes. *Sedimentology*, 14, 259–293. *10.2.3.*

Parkash B., Awasthi A.K. & Gohain K. (1983) Lithofacies of the Merkanda terminal fan, Kurukshetra district, Haryana, India. In: *Modern and Ancient Fluvial Systems* (Ed. by J.D. Collinson and J. Lewin), pp. 337–344. *Spec. Publ. int. Ass. Sediment.*, 6. *3.3.8.*

Parker B., Simmons G., Love F.G., Wharton R.A. & Seaburg K. (1991) Modern stromatolites in Antarctic dry valley lakes. *BioScience*, 31, 656–661. *4.7.1.*

Parker J.R. (1975) Lower Tertiary sand development in the Central

North Sea. In: *Petroleum and the Continental Shelf of North-West Europe* (Ed. by A.W. Woodland), pp. 447–451. Applied Science, London. *10.3.4.*

Parnell J. (1986) Devonian Magadi-type cherts in the Orcadian Basin, Scotland. *J. sedim. Petrol.*, 56, 595–600. *4.13.1.*

Parrish J.T. & Peterson F. (1988) Wind directions predicted from global circulation models and wind directions determined from eolian sandstones of the western United States – a comparison. In: *Late Paleozoic and Mesozoic Eolian Deposits of the Western Interior of the United States* (Ed. by G. Kocurek), pp. 261–282. *Sedim. Geol.*, 56. *5.1, 5.5.1, 5.5.4.*

Parry C.C., Whitley P.K.J. & Simpson R.D.H. (1981) Integration of palynological and sedimentological methods in facies analysis of the Brent Formation. In: *Petroleum Geology of the Continental Shelf of North-west Europe* (Ed. by L.V. Illing and G.D. Hobson), pp. 205–215. Heyden, London. *6.6.6.*

Parson L.M., Walker C.L. & Dixon D.R. (1995) Hydrothermal vents and processes. *Spec. Publ. geol. Soc. Lond.*, 87, 416 pp. *12.11.1.*

Parsons T. & Thompson G.A. (1993) Does magmatism influence low-angle normal faulting? *Geology*, 21, 247–250. *12.4.6.*

Paterson W.S.B. (1969) *The Physics of Glaciers*, 250 pp. Pergamon Press, Oxford. *11.2.1, Fig. 11.1.*

Patterson C.J. (1994) Tunnel-valley fans of the St Croix moraine, east-central Minnesota, USA. In: *Formation and Deformation of Glacial Deposits* (Ed. by W.P. Warren and D.G. Croot), pp. 69–87. Balkema, Rotterdam. *11.4.1.*

Patterson R.J. & Kinsman D.J.J. (1981) Hydrologic framework of a sabkha along the Arabian Gulf. *Bull. Am. Ass. petrol. Geol.*, 65, 1457–1475. *8.4, 8.4.2.*

Paul M.A. & Eyles N. (1990) Constraints on the preservation of diamict facies (melt-out tills) at the margins of stagnant glaciers. *Quatern. Sci. Rev.*, 9, 51–69. *11.4. 1.*

Pautot G. & Melguen M. (1976) Deep bottom currents, sedimentary hiatuses and polymetallic nodules. In: *Marine Geological Investigations in the Southwest Pacific and Adjacent Areas* (Ed. by G.P. Glasby and H.R. Katz), pp. 54–61. *Tech. Bull. Comm. Co-ord. Joint Prospect., Econ. Soc. Comm. Asia and Pacific (UN)*, 2. *10.4.1.*

Payton C.E. (Ed.) (1977) *Seismic Stratigraphy—Applications to Hydrocarbon Exploration*, 516 pp. *Mem. Am. Ass. petrol. Geol.*, 26, Tulsa. *1.1, 2.4.*

Pearce J.A., Lippard S.J. & Roberts S. (1984) Characteristics and tectonic significance of supra-subduction zone ophiolites. In: *Marginal Basin Geology* (Ed. by B.P. Kokelaar and M.F. Howells), pp. 77–94. *Spec. Publ. geol. Soc., Lond.*, 16. *12.11.1.*

Pearce T.J. & Jarvis I. (1992) Application of geochemical data to modelling sediment dispersal patterns in distal turbidites: Late Quaternary of the Madeira Abyssal Plain. *J. sedim. Petrol.*, 62, 1112–1129. *10.3.6.*

Pedersen G.K. & Surlyk F. (1977) Dish structures in Eocene volcanic ash layers, Denmark. *Sedimentology*, 24, 581–590. *12.4.3.*

Pedley M. (1992) Bio-retexturing: early diagenetic fabric modifications in outer-ramp settings – a case-study from the Oligo-Miocene of the Central Mediterranean. *Sedim. Geol.*, 79, 173–188. *9.4.3.*

Pedley M., Gugno C. & Grasso M. (1992) Gravity slide and

resedimentation processes in a Miocene carbonate ramp, Hyblean Plateau, southeastern Sicily. *Sedim. Geol.*, 79, 189–202. *9.5.2.*

Pelet R. (1987) A model of organic sedimentation on present-day continental margins. In: *Marine Petroleum Source Rocks* (Ed. by J. Brooks and A.J. Fleet), pp. 167–180. *Spec. Publ. geol. Soc. Lond.*, 26. *Fig. 10.5.*

Penland S. & Suter J.R. (1989) The geomorphology of the Mississippi River chenier plain. *Mar. Geol.*, 90, 231–258. *6.7.2.*

Penland S., Boyd R. & Suter J.R. (1988) Transgressive depositional systems of the Mississippi delta plain: a model for barrier shoreline and shelf sand development. *J. sedim. Petrol.*, 58, 932–949. *6.6.5, Figs 6.46, 6.47, 6.48.*

Penland S., Suter J.R. & Boyd R. (1985). Barrier island arcs along abandoned Mississippi river deltas. *Mar. Geol.*, 63, 197–233. *7.4.2.*

Penland S., Suter J.R. & Moslow T.F. (1986) Inner-shelf shoal sedimentary facies and sequences: Ship Shoal, northern Gulf of Mexico. In: *Modern and Ancient Shelf Clastics: A Core Workshop* (Ed. by T.F. Moslow and E.G. Rhodes), pp. 73–123. Soc. econ. Paleont. Miner. Core Workshop No. 9. *7.4.2.*

Pentecost A. & Riding R. (1986) Calcification in cyanobacteria. In: *Biomineralization in Lower Plants and Animals* (Ed. by B.S.C. Leadbeater and R. Riding), pp. 73–90. Clarendon Press, Oxford. *4.7.1.*

Percival C.J. (1986) Palaeosols containing an albic horizon: examples from the Upper Carboniferous of Northern England. In: *Palaeosols: their Recognition and Interpretation* (Ed. by V.P. Wright), pp. 87–111. Blackwell Scientific Publications, Oxford. *3.5.4.*

Percival C.J. (1992) The Harthorpe Gannister – a transgressive barrier island to shallow-marine sand-ridge from the Namurian of Northern England. *J. sedim. Petrol.*, 62, 442–454. *6.6.6.*

Perkins R.D., Dwyer G.S., Rosoff D.B., Fuller J., Baker P.A., Lloyd R.M. (1994) Salina sedimentation and diagenesis: West Caicos Island, British West Indies. In: *Dolomites*, a *volume in Honour of Dolomieu* (Ed. by B. Purser, M. Tucker and D. Zenger), pp. 37–54. *Spec. Publ. int. Ass. Sediment.*, 21. *8.5.2, Figs 8.29, 8.30.*

Perlmutter M.A. & Mathews M.D. (1989) Global cyclostratigraphy – a model. In: *Quantitative Dynamic Stratigraphy* (Ed. by T.A. Cross), pp. 233–260. Prentice Hall, Englewood Cliffs, NJ. *2.1.2.*

Perthuisot J.P. (1974) Les depots salins de la sebkha El Melah de Zarzis: conditions et modalites de la sedimentation evaporitique. *Rev. Geogr. Phys. Geol. Dynam.*, 16, 177–188. *8.5.2.*

Perthuisot J.P. (1975) *La Sebkha El Melah de Zarzis. Génèse et evolution d'un bassin salin paralique*, 252 pp. Ecol. Norm. Super., Paris. Trav. Lab. Géol., 9. *8.5.2.*

Perthuisot J.P. (1980) Sites et processe de la formation d'evaporites dans la nature actuelle. *Bull. Cent. Rech. Explor. Elf-Aquitaine*, 4, 207–233. *8.2.3.*

Perthuisot J.P. (1989) Recent evaporites. In: *Brines and Evaporites* (Ed. by P. Sonnenfeld and J.P. Perthuisot), pp. 65–126. *28th Int. Geol. Congress Short Course in Geology*, 3. American Geophysical Union, Washington, DC. *8.5.2.*

Perthuisot J.P., Floridia S. & Jauzein A. (1972) Un modèle récent de bassin côtier à sedimentation saline: La sebkha El Melah (Zarzis, Tunisie). *Rev. Geogr. Phys. Géol. Dynam.*, 14, 67–84. *8.5.2.*

Peryt T.M. (1994) The anatomy of a sulphate platform and adjacent basin system in the Leba sub-basin of the Lower Werra Anhydrite (Zechstein, Upper Permian), northern Poland. *Sedimentology*, 41, 83–113. *8.9.*

Peryt T.M., Orti F. & Rosell L. (1993) Sulfate platform–basin transition of the Lower Werra Anhydrite (Zechstein, Upper Permian), western Poland: facies and petrography. *J. sedim. Petrol.*, 63, 646–658. *8.9.*

Peters J.M., Troelstra S.R. & van Harten D. (1985) Late Neogene and Quaternary vertical movements in eastern Crete and their regional significance. *J. geol. Soc. Lond.*, 142, 501–513. *6.4.*

Peterson D.W. & Moore J.G. (1987) Geologic history and evolution of geologic concepts, Island of Hawaii. In: *Volcanism in Hawaii* (Ed. by R.W. Decker, T.L. Wright and P.H. Stauffer), pp. 149–189. *Prof. Pap. US geol. Surv.*, 1350. *Fig. 12.62.*

Peterson F. (1979) Sedimentary and tectonic controls of uranium mineralization in Morrison Formation (Upper Jurassic) of south-central Utah. *Bull. Am. Ass. petrol. Geol.*, 63, 837. *4.17.*

Peterson F. (1986) Jurassic paleotectonics in the west-central part of the Colorado Plateau, Utah and Arizona. In: *Paleotectonics and Sedimentation, Rocky Mountain Region, United States* (Ed. by J.A. Peterson), pp. 563–596. *Mem. Am. Ass. petrol. Geol.*, 41, Tulsa. *5.5.1.*

Peterson F. (1988) Pennsylvanian to Jurassic eolian transportation systems in the western United States. In: *Late Paleozoic and Mesozoic Eolian Deposits of the Western Interior of the United States* (Ed. by G. Kocurek), pp. 207–260. *Sedim. Geol.*, 56. *5.5.1.*

Peterson J.A. & Hite R.J. (1969) Pennsylvanian evaporite–carbonate cycles and their relation to petroleum occurrence, southern Rocky Mountains. *Bull. Am. Ass. petrol. Geol.*, 53, 884–908. *8.10.2, Figs 8.52, 8.53.*

Petterson M.G., Beddoe-Stephens B., Millward D. & Johnson E.W. (1992) A pre-caldera plateau–andesite field in the Borrowdale Volcanic Group of the English Lake District. *J. geol. Soc., Lond.*, 149, 889–906. *12.12.4.*

Petit-Maire N. (1989) Interglacial environments in presently hyperarid Sahara: palaeoclimatic implications. In: *Paleoclimatology and Paleometeorology: Modern and Past Patterns of Global Atmospheric Transport* (Ed. by M. Leinen and M. Sarnthein), pp. 637–661. Kluwer Academic, Boston. *5.1, 5.4.7.*

Pettijohn F.J. (1957) *Sedimentary Rocks*, 718 pp. Harper, New York. *7.8.1.*

Pharo C.H. & Carmack E.C. (1979) Sedimentation processes in a short residence time intermontane lake, Kamloops Lake, British Columbia. *Sedimentology*, 26, 523–541. *4.4.2.*

Phillips F.M., Campbell A.R., Kruger C., Johnson P., Roberts R. & Keyes E. (1992) *A Reconstruction of the Response of the Water Balance in Western United States Lake Basins to Climatic Change*. New Mexico Water Resources Research Institute, Technical Completion Report, 167 pp. *4.9.*

Phleger F.B. (1965) Sedimentology of Guerrero Negro Lagoon, Baja California, Mexico. In: *Geology and Geophysics, Colston Research Society 17th Symposium, Colston Paper*, 17, pp. 205–327. Butterworth, London. *6.7.4.*

Picard M.D. & High L.R. Jr (1972) Criteria for recognizing lacustrine rocks. In: *Recognition of Ancient Sedimentary Environments* (Ed. by J.K. Rigby and W.K. Hamblin), pp. 108–145. *Spec. Publ. Soc. econ. Paleont. Miner.*, 16, Tulsa. *4.10.1.*

Picha F. (1978) Depositional and diagenetic history of Pleistocene and Holocene oolitic sediments and sabkhas in Kuwait, Persian Gulf. *Sedimentology*, 25, 427–449. *8.4.3.*

Pickering K.T. (1979) Possible retrogressive flow slide deposits from the Kongsfjord Formation: a Precambrian submarine fan, Finnmark, N. Norway. *Sedimentology*, 26, 295–305. *10.2.3.*

Pickering K.T. (1981) The Kongsfjord Formation – a late Precambrian submarine fan in north-east Finnmark, North Norway. *Norge geol. Unders.*, 367, 77–104. *10.3.3.*

Pickering K.T. (1983) Transitional submarine fan deposits from the late Precambrian Kongsfjord Formation submarine fan, NE Finnmark, N. Norway. *Sedimentology*, 30, 181–199. *10.3.4.*

Pickering K.T., Hiscott R.N. & Hein F.J. (1989) *Deep-marine Environments: Clastic Sedimentation and Tectonics*, 416 pp. Unwin Hyman, London. *10.3.2, Fig. 10.12.*

Pickering K.T., Stow D.A.V., Watson M. & Hiscott R.N. (1986) Deep-water facies, processes and models: a review and classification scheme for modern and ancient sediments. *Earth Sci. Rev.*, 22, 75–174. *10.2.3, 10.2.4, Figs 10.3, 10.29.*

Pickford M. (1986) Sedimentation and fossil preservation in the Nyanza Rift System, Kenya. In: *Sedimentation in the African Rifts* (Ed. by L.E. Frostick, R.W. Renaut, I. Reid and J.-J. Tiercelin), pp. 345–362. *Spec. Publ. geol. Soc. Lond.*, 25. *2.1.10.*

Pierson T.C. (1981) Dominant particle support mechanisms in debris flows at Mt Thomas, New Zealand, and implications for flow mobility. *Sedimentology*, 28, 49–60. *3.2.2.*

Pierson T.C. (1995) Flow characteristics of large eruption-triggered debris flows at snow-clad volcanoes: constraints for debris-flow models. *J. Volcanol. Geotherm. Res.*, 66, 283–294. *12.5.2.*

Pierson T.C. & Costa J.C. (1987) A rheologic classification of subaerial sediment–water flows. In: *Debris Flows/Avalanches: Process, Recognition and Mitigation* (Ed. by J.E. Costa and G.F. Wieczorek), pp. 1–12. *Geol. Soc. Am. Rev. Eng. Geol.*, 7. *12.5.2.*

Pierson T.C. & Janda R.J. (1994) Volcanic mixed avalanches: a distinct eruption-triggered mass-flow process at snow-clad volcanoes. *Bull. geol. Soc. Am.*, 106, 1351–1358. *12.5.2.*

Pierson T.C. & Scott K.M. (1985) Downstream dilution of a lahar: transition from debris flow to hyperconcentrated streamflow. *Water Resour. Res.*, 21, 1511–1524. *12.5.2.*

Pierson T.C., Janda R.J., Thouret J.-C. & Borrero C.A. (1990) Perturbation and melting of snow and ice by the 13 November eruption of Nevado del Ruiz, Columbia, and consequent mobilization, flow and deposition of lahars. *J. Volcanol. Geotherm. Res.*, 41, 17–66. *12.5.2, Fig. 12.1.*

Pierson T.C., Janda R.J., Umbal J.V. & Daag A.S. (1992) Immediate and long-term hazards from lahars and excess sedimentation in rivers draining Mt Pinatubo, Philippines. *U.S. Geol. Surv. Water-resour. Investig. Rep.*, 92-4039, 1–35. *12.5.2, 12.6.1, Figs 12.1, 12.40.*

Pilkey O.H. (1987) Sedimentology of basin plains. In: *Geology and Geochemistry of Abyssal Plains* (Ed. by P.P.E. Weaver and J. Thomson), pp. 1–12. *Spec. Publ. geol. Soc. Lond.*, 31. *10.3.6.*

Pilkey O.H. & Hokanson C. (1991) A proposed classification of basin plains. In: *From Shoreline to Abyss: Contributions in Marine Geology in Honor of Francis Parker Shepard* (Ed. by R.H. Osborne), pp. 249–257. *Spec. Publ. Soc. econ. Paleont. Miner.*, 46, Tulsa. *10.3.6.*

Pilkey O.H., Locker S.D. & Cleary W.J. (1980) Comparison of sand-layer geometry on flat floors of 10 modern depositional basins. *Bull. Am. Ass. petrol. Geol.*, 64, 841–856. *10.3.1, 10.3.6.*

Pilskaln C.H. & Johnson T.C. (1991) Seasonal signals in Lake Malawi sediments. *Limnol. Oceanogr.*, 36, 544–557. *Fig. 4.15.*

Pinet P.R. & Popence P. (1985) A scenario of Mesozoic–Cenozoic ocean circulation over the Blake Plateau and its environs. *Bull. geol. Soc. Am.*, 96, 618–626. *9.5.3.*

Piper D.J.W. (1978) Turbidite muds and silts on deepsea fans and abyssal plains. In: *Sedimentation in Submarine Canyons, Fans and Trenches* (Ed. by D.J. Stanley and G. Kelling), pp. 163–176. Dowden, Hutchinson & Ross, Stroudsburg, PA. *10.2.3.*

Piper D.J.W. & Normark W.R. (1983) Turbidite depositional patterns and flow characteristics, Navy Submarine Fan, California Borderland. *Sedimentology*, 30, 681–694. *10.3.3, 10.3.6, Fig. 10.47.*

Piper D.J.W. & Stow D.A.V. (1991) Fine-grained turbidites. In: *Cycles and Events in Stratigraphy* (Ed. by G. Einsele, W. Ricken and A. Seilacher), pp. 360–376. Springer-Verlag, Berlin. *10.2.3.*

Piper D.J.W., Kontopoulos N., Anagnostou C., Chronis G. & Panagos A.G. (1990) Modern fan deltas in the western Gulf of Corinth, Greece. *Geo-Mar. Lett.*, 10, 5–12. *10.3.3.*

Piper D.J.W., Normark W.R. & Ingle J.C. Jr (1976) The Rio Dell Formation: a Plio-Pleistocene basin slope deposit in Northern California. *Sedimentology*, 23, 309–328. *10.3.5.*

Piper D.J.W., Shor A.N. & Hughes Clarke J.E. (1988) The 1929 Grand Banks earthquake, slump and turbidity current. *Spec. Pap. geol. Soc. Am.*, 229, 77–92. *10.2.3.*

Pisciotto K.A. & Garrison R.E. (1981) *Lithofacies and Depositional Environments of the Monterey Formation, California* (Ed. by R.E. Garrison and R.G. Douglas *et al.*), pp. 97–122. *Spec. Publ. Pacific Sect. Soc. econ. Paleont. Miner.*, Los Angeles. *10.4.1.*

Pisciotto K.A., Ingle J.C. Jr. *et al.* (1992) Proc. Ocean Drilling Program, Sci. Results, 127/128 pt. 1: College Station Texas. *10.4.1.*

Pittmann J.G. (1985) Correlation of beds within the Ferry Lake Anhydrite of the Gulf Coastal Plain. *Trans. Gulf Coast Ass. geol. Soc.*, 35, 251–260. *8.8, Fig. 8.41.*

Pitman III W.C. & Golovchenko X. (1983) The effect of sea level change on the shelf edge and slope of passive margins. In: *The Shelfbreak: Critical Interface on Continental Margins* (Ed. by D.J. Stanley and G.T. Moore), pp. 41–58. *Spec. Publ. Soc. econ. Paleont. Miner.*, 33, Tulsa. *2.1.4, Table 2.2.*

Pizzuto J.E. (1987) Sediment diffusion during overbank flows. *Sedimentology*, 34, 301–317. *3.3.6.*

Platt J.P., Leggett J.K., Young J., Raza H. & Alam S. (1985) Large-scale sediment underplating in the Makran accretionary prism, southwest Pakistan. *Geology*, 13, 507–511. *12.12.5.*

Platt N.H. & Keller B. (1992) Distal alluvial deposits in a foreland basin setting – the Lower Freshwater Molasse (Lower Miocene), Switzerland: sedimentology, architecture and palaeosols. *Sedimentology*, 39, 545–565. *3.6.6.*

Platt N.H. & Wright V.P. (1991) Lacustrine carbonates: facies models, facies distributions and hydrocarbon aspects. In: *Lacustrine Facies Analysis* (Ed. by P. Anadón, L. Cabrera and K. Kelts), pp. 57–74. *Spec. Publ. int. Ass. Sediment.*, 13. *4.12, Fig. 4.22.*

Platt N.H. & Wright V.P. (1992) Palustrine carbonates and the

Florida Everglades: towards an exposure index for the fresh-water environment. *J. sedim. Petrol.*, 62, 1058–1071. *4.7.1*, *4.12.1*, *Fig. 4.11*, *Fig. 4.24*.

Playford P.E. & Cockbain A.E. (1976) Modern algal stromatolites at Hamelin Pool, a hypersaline barred basin in Shark Bay. In: *Stromatolites* (Ed. by M.R. Walter), pp. 389–411. Elsevier, Amsterdam. *9.4.1*.

Playford P.E., Hurley N.F., Kerans C. & Middleton M.F. (1989) Reefal platform development, Devonian of the Canning Basin, Western Australia. In: *Controls on Platform Development* (Ed. by P.D. Crevello *et al.*), pp. 187–202. *Spec. Publ. Soc. econ. Paleont. Miner.*, 44. *9.4.5*, *9.5.6*, *Fig. 9.74*.

Plint A.G. (1983) Facies, environments and sedimentary cycles in the Middle Eocene, Bracklesham Formation of the Hampshire Basin: evidence for global sea-level changes? *Sedimentology*, 30, 625–653. *6.7.7*.

Plint A.G. (1986) Slump blocks, intraformational conglomerates and associated erosional structures in Pennsylvanian fluvial strata of eastern Canada. *Sedimentology*, 33, 387–399. *3.5.2*.

Plint A.G. (1988a) Global eustasy and the Eocene sequence in the Hampshire Basin, England. *Basin Res.*, 1, 11–22. *6.7.7*, *Fig. 6.98*.

Plint A.G. (1988b) Sharp-based shoreface sequences and 'offshore bars' in the Cardium Formation of Alberta: their relationship to relative changes in sea level. In: *Sea Level Changes – an Integrated Approach* (Ed. by C.K. Wilgus, B.S. Hastings, C.G. St C. Kendall, H.W. Posamentier, C.A. Ross and J.C. Van Wagoner), pp. 357–370. *Spec. Publ. Soc. econ. Paleont. Miner.*, 42, Tulsa. *7.1.2*, *7.8.4*, *Fig. 7.57*.

Plint A.G. (1991) High frequency relative sea level oscillations in Upper Cretaceous shelf clastics of the Alberta foreland basin; possible evidence for a glacio-eustatic control? In: *Sedimentation, Tectonics and Eustacy* (Ed. by D.I.M. Macdonald), pp. 409–428. *Spec. Publ. int. Ass. Sediment.*, 12. *7.7.2*.

Plint A.G. & Walker R.G. (1987) Morphology and origin of an erosional surface cut into the Bad Heart Formation during major sea-level change, Santonian of west-central Alberta, Canada. *J. sedim. Petrol.*, 57, 639–650. *7.8.4*.

Plint A.G. & Walker R.G. (1992) Wave- and storm-dominated shallow marine systems. In: *Facies Models: Response to Sea Level Change* (Ed. by R.G. Walker and N.P. James), pp. 219–238. Geol. Ass. Can., Waterloo, Ontario. *7.8.4*, *Fig. 7.57*.

Plint A.G., Walker R.G. & Bergman K.M. (1986) Cardium Formation 6. Stratigraphic framework of the Cardium in the subsurface. *Bull. Can. petrol. Geol.*, 34, 213–225. *7.7.3*, *Fig. 7.55*.

Pomar L. (1991) Reef geometries, erosion surfaces and high frequency sea-level changes, upper Miocene Reef Complex, Mallorca, Spain. *Sedimentology*, 38, 243–269. *9.4.5*, *Fig. 9.55*.

Pomar L. (1993) High resolution sequence stratigraphy in prograding Miocene carbonate: application to seismic interpretation. In: *Carbonate Sequence Stratigraphy: Recent Developments and Applications* (Ed. by R.G. Loucks and J.F. Sarg), pp. 389–407. *Mem. Am. Ass. petrol. Geol.*, 57, Tulsa. *9.4.5*.

Pomar L., Fornos J.J. & Rodriguez-Perea A. (1985) Reef and shallow carbonate facies of the Upper Miocene of Mallorca. Excursion II. In: *6th I.A.S. European Regional Meeting, Lleida, Guidelbook* (Ed. by M.D. Mila and J. Rosell), pp. 493–518. Inst. d'Estudis Ilerendos, Lleida. *9.4.5*, *Fig. 9.49*.

Popp B.N. & Wilkinson B.H. (1983) Holocene lacustrine ooids from Pyramid Lake, Nevada. In: *Coated Grains* (Ed. by T.M. Peryt), pp. 142–153. Springer-Verlag, Berlin. *4.7.1*.

Porebski S.J. & Gradzinski R. (1990) Lava-fed Gilbert-type delta in the Plonez Cove Formation (Lower Oligocene), King George Island, West Antarctica. In: *Coarse-grained Deltas* (Ed. by A. Colella and D.B. Prior), pp. 335–351. *Spec. Publ. int. Ass. Sediment.*, 10. *12.11.2*, *Fig. 12.64*.

Porebski S.J., Meischner D. & Gorlich K. (1991) Quaternary mud turbidites from the South Shetland Trench (West Antarctica): recognition and implications for turbidite facies modelling. *Sedimentology*, 38, 691–715. *10.2.3*.

Posamentier H.W. & James D.P. (1993) An overview of sequence-stratigraphic concepts: uses and abuses. In: *Sequence Stratigraphy and Facies Associations* (Ed. by H.W. Posamentier, C.P. Summerhayes, B.U. Haq and G.P. Allen), pp. 3–18. *Spec. Publ. int. Ass. Sediment.*, 18. *2.1.4*, *2.4*.

Posamentier H.W. & Vail P.R. (1988) Eustatic controls on clastic deposition II – sequence and systems tract models. In: *Sea-level Changes – an Integrated Approach* (Ed. by C.K. Wilgus, B.S. Hastings, C.G. St C. Kendall, H.W. Posamentier, C.A. Ross and J.C. Van Wagoner), pp. 125–154. *Spec. Publ. Soc. econ. Paleont. Miner.*, 42, Tulsa. *2.2*, *6.1*, *10.3.3*, *Fig. 10.45*.

Posamentier H.W., Allen G.P., James D.P. & Tesson M. (1992) Forced regressions in a sequence stratigraphic framework: concepts, examples, and exploration significance. *Bull. Am. Ass. petrol. Geol.*, 76, 1687–1709. *2.4*, *6.1*, *6.3*, *7.8.4*.

Posamentier H.W., Jervey M.T. & Vail P.R. (1988) Eustatic controls on clastic deposition I – conceptual framework. In: *Sea-level Changes – an Integrated Approach* (Ed. by C.K. Wilgus, B.S. Hastings, C.G. St C. Kendall, H.W. Posamentier, C.A. Ross and J.C. Van Wagoner), pp. 109–124. *Spec. Publ. Soc. econ. Paleont. Miner.*, 42, Tulsa. *2.1.4*, *2.4*.

Postma G. (1984) Mass-flow conglomerates in a submarine canyon: Abrioja Fan-delta, Pliocene, southeastern Spain. In: *Sedimentology of Gravels and Conglomerates* (Ed. by E.H. Koster and R.J. Steel), pp. 237–258. *Mem. Can. Soc. petrol. Geol.*, 10, Calgary. *6.5.6*, *Fig. 6.28*.

Postma G. (1990) Depositional architecture and facies of river and fan deltas: a synthesis. In: *Coarse-grained Deltas* (Ed. by A. Colella and D.B. Prior), pp. 13–27. *Spec. Publ. int. Ass. Sediment.*, 10. *6.1*, *6.3*, *6.5*, *6.5.1*, *6.5.3*, *6.5.4*, *Fig. 6.15*.

Postma G. & Roep T.B. (1985) Resedimented conglomerates in the bottomset of a Gilbert-type gravel delta. *J. sedim. Petrol.*, 55, 874–885. *6.5.6*, *Fig. 6.22*.

Postma G., Nemec W. & Kleinspehn K.L. (1988) Large floating clasts in turbidites: a mechanism for their emplacement. *Sedim. Geol.*, 58, 47–61. *12.5.2*.

Potter P.E. & Pettijohn F.J. (1963) *Paleocurrents and Basin Analysis*, 296 pp. Springer-Verlag, Berlin. *1.1*.

Powell R.D. (1981) A model for sedimentation by tidewater glaciers. *Ann. Glaciol.*, 2, 129–134. *11.4.6*.

Powell R.D. (1988) *Processes and Facies of Temperate and Sub-polar Glaciers with Tidewater Fronts*, 114 pp. Short Course Notes, Geol. Soc. Am. Centennial Annual Meeting, Denver, CO. *11.4.6*, *Fig. 11.12*.

Powell R.D. (1990) Glacimarine processes at grounding-line fans and their growth to ice-contact deltas. In: *Glacimarine Environments: Processes and Sediments* (Ed. by J.A. Dowdeswell and J.D. Scourse), pp. 53–73. *Spec. Publ. geol. Soc.*, 53. *11.4.6.*

Powell R.D. & Molnia B.F. (1989) Glacimarine sedimentary processes, facies and morphology of the south-southeast Alaska shelf and fjords. *Mar. Geol.*, 85, 359–390. *11.4.6.*

Powers D.W. & Hassinger B.W. (1985) Synsedimentary dissolution pits in halite of the Permian Salado Formation, southeastern New Mexico. *J. Sedim. Petrol.*, 55, 769–773. *8.10.1.*

Pozzobon J.G. & Walker R.G. (1990) Viking Formation (Albian) at Eureka, Saskatchewan: a transgressed and degraded shelf sand ridge. *Bull. Am. Ass. petrol. Geol.*, 74, 1212–1227. *7.7.1.*

Pratt B.R. & James N.P. (1986) The St George Group (Lower Ordovician) of western Newfoundland: tidal flat island model for carbonate sedimentation in shallow epeiric seas. *Sedimentology*, 33, 313–343. *9.3.1, 9.4.1.*

Pratt B.R., James N.P. & Cowan C.A. (1992) Peritidal carbonates. In: *Facies, Models and Response to Sea Level Change* (Ed. by R.G. Walker and N.P. James), pp. 303–322. Geosci. Can., St John's, Newfoundland. *9.4.1.*

Pratt R.M. & Heezen B.C. (1964) Topography of the Blake Plateau. *Deep-Sea Res.*, 11, 721–728. *10.4.2.*

Prell W.L., Niitsuma N. *et al.* (1989) *Proc. Ocean Drilling Program, Init. Rep.*, 117, 1236 pp. College Station, TX. *10.4.2.*

Prell W.L., Niitsuma N. *et al.* (1990) Neogene tectonics and sedimentation of the SE Oman continental margin: results from ODP Leg 117. In: *The Geology and Tectonics of the Oman Region* (Ed. by A.H.F. Robertson, M.P. Searle and A.C. Ries), pp. 745–758. *Spec. Publ. geol. Soc. Lond.*, 49, Bristol. *10.4.2.*

Prest V.K., Grant D.R. & Rampton V.N. (1968) *Glacial Map of Canada. Scale 1:5 000 000*, 30 pp. Geol. Surv. Can., Ottawa, Map 1253A. *11.5.3.*

Pretious E.S. & Blench T. (1951) *Final Report on Special Observations on Bed Movement in the Lower Fraser River at Ladner Reach during 1959 Freshet*. Nat. Res. Coun. Can., Vancouver, 12 pp. *3.3.3, Fig. 3.4.*

Price R.J. (1971) The development and destruction of a sandur, Breidamerkurjökull, Iceland. *Arctic Alpine Res.*, 3, 225–237. *11.4.4.*

Prior D.B. & Bornhold B.D. (1988) Submarine morphology and processes of fjord fan deltas and related high-gradient systems: modern examples from British Columbia. In: *Fan Deltas: Sedimentology and Tectonic Settings* (Ed. by W. Nemec and R.J. Steel), pp. 125–143. Blackie, London. *10.3.3, 10.3.4, Fig. 10.53.*

Prior D.B. & Bornhold B.D. (1989) Submarine sedimentation on a developing Holocene fan delta. *Sedimentology*, 36, 1053–1076. *6.5.3, 10.3.3, 10.3.4, Fig. 10.54.*

Prior D.B. & Bornhold B.D. (1990) The underwater development of Holocene fan deltas. In: *Coarse-grained Deltas* (Ed. by A. Colella and D.B. Prior), pp. 75–90. *Spec. Publ. int. Ass. Sediment.*, 10. *6.5, 6.5.3, 6.5.4, 10.3.3, Fig. 10.53.*

Prior D.B. & Coleman J.M. (1978) Disintegrating retrogressive landslides on very low-angle subaqueous slopes, Mississippi delta. *Mar. Geotech.*, 3, 37–60. *6.6.4.*

Prior D.B., Bornhold B.D., Coleman J.M. & Bryant W.R. (1982) Morphology of a submarine slide, Kitimat arm, British Columbia. *Geology*, 10, 588–592. *6.6.4.*

Prior D.B., Yang Z.-S., Bornhold B.D., Keller G.H., Lu N., Wiseman W.J., Wright L.D. & Zhang J. (1986) Active slope failure, sediment collapse, and silt flows on the modern subaqueous Huanghe (Yellow River) Delta. *Geomarine Lett.*, 6, 77–84. *6.6.4.*

Pritchard D.W. (1967) What is an estuary? Physical viewpoint. In: *Estuaries* (Ed. by G.D. Lauff), pp. 3–5. Am. Assoc. Adv. Sci., 38, Washington, DC. *6.7.5, Fig. 6.79.*

Proust J.N. & Deynoux M. (1994) Marine to non-marine sequence architecture of an intracratonic glacially related basin: Late Proterozoic of the West African platform in western Mali. In: *Earth's Glacial Record* (Ed. by M. Deynoux, J.M.G. Miller, E.W. Domack, N. Eyles, I.J. Fairchild & G.M. Young), pp. 121–145. Cambridge University Press, Cambridge. *11.5.5.*

Psuty N.P. (1967) The geomorphology of beach ridges in Tabasco, Mexico. *Louisiana State Univ. Coast. Stud. Ser.*, 18, 51 pp. *6.6.2, Fig. 6.37.*

Pudsey C.J. (1984) Fluvial to marine transition in the Ordovician of Ireland – a humid-region fan-delta? *Geol. J.*, 19, 143–172. *6.5.6.*

Pudsey C.J. & Reading H.G. (1982) Sedimentology and structure of the Scotland Group, Barbados. In: *Trench Forearc Geology: Sedimentation and Tectonics on Modern and Ancient Active Plate Margins* (Ed. by J.K. Leggett), pp. 291–308. *Spec. Publ. geol. Soc. Lond.*, 10. *10.4.1.*

Puigdefabregas C. (1973) Miocene point bar deposits in the Ebro Basin, Northern Spain. *Sedimentology* 20, 133–144. *3.6.5.*

Puigdefabregas C. & Van Vliet A. (1978) Meandering stream deposits from the Tertiary of the Southern Pyrenees. In: *Fluvial Sedimentology* (Ed. by A.D. Miall), pp. 469–485. *Mem. Can. Soc. petrol. Geol.*, 5, Calgary. *3.6.5, Fig. 3.51.*

Pujos M. & Javelaud O. (1991) Depositional facies of a mud shelf between the Sinu River and the Darien Gulf (Caribbean coast of Colombia): environmental factors that control its sedimentation and origin of deposit. *Continental Shelf Res.*, 11, 601–623. *7.6, Fig. 7.32.*

Pulham A.J. (1989) Controls on internal structure and architecture of sandstone bodies within Upper Carboniferous fluvial-dominated deltas, County Clare, western Ireland. In: *Deltas: Sites and Traps for Fossil Fuels* (Ed. by M.K.G. Whateley and K.T. Pickering), pp. 179–203. *Spec. Publ. geol. Soc. Lond.*, 41. *6.6.6, Figs 6.51, 6.54.*

Purdy E.G. (1961) Bahamian oolite shoals. In: *Geometry of Sandstone Bodies* (Ed. by J.A. Peterson and J.C. Osmond), pp. 53–62. Am. Ass. Petrol. Geol., Tulsa. *9.1.1.*

Purdy E.G. (1963a) Recent calcium carbonate facies of the Great Bahama Bank, I. Petrography and reaction groups. *J. Geol.*, 71, 334–355. *9.1.1.*

Purdy E.G. (1963b) Recent carbonate facies of the Great Bahama Bank, II. Sedimentary facies. *J. Geol.*, 71, 472–497. *9.1.1.*

Purdy E.G., Pusey W.C. & Wantland K.F. (1975) Continental Shelf of Belize – regional shelf attributes. In: *Belize Shelf – Carbonate Sediments, Clastic Sediments, and Ecology* (Ed. by K.F. Wantland and W.C. Pusey), pp. 1–39. *Am. Ass. petrol. Geol. Studies in Geology*, 2, Tulsa. *9.4.1.*

Purser B.H. (Ed.) (1973) *The Persian Gulf: Holocene Carbonate*

Sedimentation and Diagenesis in a Shallow Epicontinental Sea, 471 pp. Springer-Verlag, Berlin. *9.3.4, 9.4.2, Fig. 9.15*.

Purser B.H. (1985) Coastal evaporite systems. In: *Hypersaline Ecosystems: the Gavish Sabkha* (Ed. by G.M. Friedman and W.E. Krumbein), pp. 72–102. Springer-Verlag, New York. *8.4.2*.

Purser B.H. & Evans G. (1973) Regional sedimentation along the Trucial Coast, SE Persian Gulf. In: *The Persian Gulf: Holocene Carbonate Sedimentation and Diagenesis in a Shallow Epicontinental Sea* (Ed. by B.H. Purser), pp. 211–231. Springer-Verlag, Berlin. *9.5.1, Fig. 9.23*.

Purser B.H., Azzawi Al, Hassini N. Al, Baltzer F. *et al.* (1982) Sedimentation et évolution du complexe deltaique Tigre-Euphrate. *Mém. Soc. géol. France*, **144**, 207–216. *8.4.2*.

Pusey W.C. (1975) Holocene carbonate sedimentation on Northern Belize shelf. In: *Belize Shelf-Carbonate Sediments, Clastic Sediments, and Ecology* (Ed. by K.F. Wantland and W.C. Pusey), pp. 131–233. *Am. Ass. Petrol. Geol. Studies in Geology*, **2**. *9.4.1*.

Pye K. (1983) Dune formation on the humid tropical sector of the north Queensland coast, Australia. *Earth Surf. Proc. Landf.*, **8**, 371–381. *5.2.2*.

Pye K. (1987) *Aeolian Dust and Dust Deposits*, 334 pp. Academic Press, London. *5.2.1, 10.4.1*.

Pye K. (Ed.) (1994) *Sediment Transport and Depositional Processes*, 397 pp. Blackwell Scientific Publications, Oxford. *1.2*.

Pye K. (1995) The nature, origin and accumulation of loess. *Quat. Sci. Rev.*, **14**, 653–668. *5.2.1, 5.4.2*.

Pye K. & Tsoar H. (1987) The mechanics and geological implications of dust transport and deposition in deserts with particular reference to loess formation and dune sand diagenesis in the northern Negev, Israel. In: *Desert Sediments: Ancient and Modern* (Ed. by L.E. Frostick and I. Reid), pp. 139–156. *Spec. Publ. geol. Soc. Lond.*, **35**. *5.3.1, 5.3.7*.

Pye K. & Tsoar H. (1990) *Aeolian Sand and Sand Dunes*, 396 pp. Unwin Hyman, London. *5.1*.

Pyle D.M. (1989) The thickness, volume, and grain size of tephra fall deposits. *Bull. Volcanol.*, **51**, 1–15. *12.4.3, Fig. 12.21*.

Quine M.L. & Bosence D.W.J. (1991) Stratal geometries, facies and sea floor erosion in Upper Cretaceous Chalk, Normandy, France. *Sedimentology*, **38**, 1113–1152. *9.4.3*.

Radwanski A. (1970) Dependence of rock-borers and burrowers on the environmental conditions within the Tortonian littoral zone of southern Poland. In: *Trace Fossils* (Ed. by T.P. Crimes and J.C. Harper), pp. 371–390. Letterpress, Liverpool. *6.4*.

Ragan D.H. & Sheridan M.F. (1972) Compaction of the Bishop Tuff, California. *Bull. geol. Soc. Am.*, **83**, 95–106. *12.4.4*.

Ragotzkie R.A. (1978) Heat budgets of lakes. In: *Lakes: Chemistry, Geology, Physics* (Ed. by A. Lerman), pp. 1–20. Springer-Verlag, Berlin. *4.3, Fig. 4.4*.

Rahmani R.A. (1988) Estuarine tidal channel and near shore sedimentation of a Late Cretaceous epicontinental sea, Drumheller, Alberta, Canada. in: *Tide-influenced Sedimentary Environments and Facies* (Ed. by P.L. de Boer, A. van Gelder and S.D. Nio), pp. 433–474. Reidel, Dordrecht. *6.2.7, 6.7.6, 6.7.7*.

Rahmani R.A. & Smith D.G. (1988) The Cadotte Member of northwestern Alberta: a high energy barred shoreline. In: *Sequences, Stratigraphy, Sedimentology: Surface and Subsurface* (Ed. by D.P. James and D.A. Leckie), pp. 431–437. *Mem. Can.*

Soc. petrol. Geol., **15**, Calgary. *6.7.7*.

Rahn P.H. (1967) Sheetfloods, streamfloods and the formation of sediments. *Ann. Ass. am. Geogr.*, **57**, 593–604. *3.3.8*.

Raidt H. & Koschel R. (1988) Morphology of calcite crystals in hardwater lakes. *Limnologica*, **19**, 3–12. *4.7.1*.

Rains B., Shaw J., Skoye R., Sjogren D. & Kvill D. (1993) Late Wisconsin subglacial megaflood paths in Alberta. *Geology*, **21**, 323–326. *11.4.1*.

Rains R.B., Selby M.S. & Smith C.J.R. (1980) Polar desert sandar, Antarctica. *NZ J. Geol. Geophy.*, **23**, 595–604. *11.4.4*.

Raistrick A. & Marshall C.E. (1939) *The Nature and Origin of Coal and Coal Seams*, 282 pp. English University Press, London. *3.5.4*.

Raiswell R. (1988) Chemical model for the origin of minor limestone–shale cycles by anaerobic methane oxidation. *Geology*, **16**, 641–644. *9.4.3*.

Ramos A. & Sopeña A. (1983) Gravel bars in low-sinuosity streams (Permian and Triassic, central Spain). In: *Modern and Ancient Fluvial Systems* (Ed. by J.D. Collinson and J. Lewin), pp. 301–312. *Spec. Publ. int. Ass. Sediment.*, **6**. *3.5, 3.5.1, 3.6.4, Fig. 3.33*.

Ramos A., Sopeña A. & Perez-Arlucea M. (1986) Evolution of Buntsandstein fluvial sedimentation in the Northwest Iberian Ranges (Central Spain). *J. sedim. Petrol.*, **56**, 862–875. *3.6.4, 3.6.5, Figs 3.37, 3.48, 6.5.4*.

Rampino M.R. & Sanders J.E. (1980) Holocene transgression in south-central Long Island, New York. *J. sedim. Petrol.*, **50**, 1063–1080. *Fig. 6.86*.

Rampino M.R. & Stothers R.B. (1988) Flood basalt volcanism during the past 250 million years. *Science*, **241**, 663–668. *12.9.4*.

Ramsbottom W.H.C. (1977) Major cycles of transgression and regression (mesothems) in the Namurian. *Proc. Yorks. geol. Soc.*, **41**, 261–291. *6.6.6*.

Ramsbottom W.H.C. (1979) Rates of transgression and regression in the Carboniferous of NW Europe. *J. geol. Soc. Lond.*, **136**, 147–154. *6.6.6*.

Rao C.P. (1981) Criteria for recognition of cold-water carbonate sedimentation: Berriedale Limestone (Lower Permian), Tasmania, Australia. *J. sedim. Petrol.*, **51**, 491–506. *11.5.5*.

Rasmussen K.R., Sorensen M. & Willetts B.B. (1985) Measurement of saltation and wind strength on beaches. In: *Proceedings of International Workshop on the Physics of Blown Sand* (Ed. by O.E. Barndorff-Nielsen), pp. 301–325. Department Theoretical Statistics, Aarhus University, Mem., **8**. *5.3.1*.

Rast N. (1958) Subaerial volcanicity in Snowdonia. *Nature*, **181**, 508. *12.1.1*.

Raup O.B. (1982) Gypsum precipitation by mixing seawater brines. *Bull. Am. Ass. petrol. Geol.*, **66**, 363–367. *8.6.1*.

Raup O.B. & Hite R.J. (1992) Lithology of evaporite cycles and cycle boundaries in the upper part of the Paradox Formation of the Hermosa Group of Pennsylvanian age in the Paradox Basin, Utah and Colorado. *Bull. U.S. geol. Surv.*, **2000-B**, 37 pp. *8.6.1, 8.10.2*.

Read J.F. (1974) Carbonate bank and wave-built platform sedimentation, Edel Province, Shark Bay, Western Australia. *Mem. Am. Ass. petrol. Geol.*, **12**, 250–282. *9.4.1*.

Read J.F. (1982) Carbonate platforms of passive (extensional) continental margins: types, characteristics and evolution. *Tectonophysics*, **81**, 195–212. *9.3.4, 9.5.6*.

Read J.F. (1985) Carbonate platform facies models. *Bull. Am. Ass. petrol. Geol.*, 66, 860–879. *9.3.4, 9.5.2, 9.5.6.*

Read J.F., Osleger D. & Elrick M. (1991) Two-dimensional modelling of carbonate cycles. In: *Sedimentary Modeling: Computer Simulations and Methods for Improved Parameter Definition* (Ed. by E.K. Franseen *et al.*), pp. 473–488. *Bull. Kansas geol. Surv.*, 233. *9.5.2.*

Reading H.G. (Ed.) (1978) *Sedimentary Environments and Facies*, 1st edn, 557 pp. Blackwell Scientific Publications, Oxford. *1.1.*

Reading H.G. (1987) Fashions and models in sedimentology: a personal perspective. *Sedimentology* 34, 3–9. *2.2.2, 2.5.*

Reading H.G. (1991) The classification of deep-sea depositional systems by sediment calibre and feeder systems. *J. geol. Soc. Lond.*, 148, 427–430. *10.3.2.*

Reading H.G. & Orton G.J. (1991) Sediment calibre: a control on facies models with special reference to deep-sea depositional systems. In: *Controversies in Modern Geology* (Ed. by D.W. Müller, J.A. McKenzie and H. Weissert), pp. 85–111. Academic Press, London. *2.1.1, 6.1, 6.3, 10.3.2.*

Reading H.G. & Richards M. (1994) Turbidite systems in deep-water basin margins classified by grain size and feeder system. *Bull. Am. Ass. petrol. Geol.*, 78, 792–822. *9.4.4, 10.1, 10.3.2, 10.3.4, Figs 10.37, 10.46, 10.51, 10.53, 10.55, 10.56, 10.59, 10.64, 10.67, 10.69, 10.70, Table 10.3.*

Reading H.G. & Walker R.G. (1966) Sedimentation of Eocambrian tillites and associated sediments in Finnmark, northern Norway. *Palaeogeogr. Palaeoclimatol. Palaeoecol.*, 2, 177–212. *11.5.4.*

Reddering J.S.V. (1983) An inlet sequence produced by migration of a small microtidal inlet against longshore drift; the Keurbooms inlet, South Africa. *Sedimentology*, 30, 201–218. *6.7.4.*

Reedman A.J., Howells M.F., Orton G.J. & Campbell S.D.G. (1987) The Pitts Head Tuff Formation: a subaerial to submarine welded ash flow tuff of Ordovician age, North Wales. *Geol. Mag.*, 124, 427–439. *12.4.4.*

Rees B.A., Detrick R.S. & Coakley B.J. (1993) Seismic stratigraphy of the Hawaiian flexural moat. *Bull. Geol. Soc. Am.*, 105, 189–205. *12.11.2.*

Reeves C.C. Jr (1970) Origin, classification and geologic history of caliche on the southern High Plains, Texas and eastern New Mexico. *J. Geol.*, 78, 352–362. *3.2.3.*

Reid I. & Frostick L.E. (1985) Beach orientation, bar morphology and the concentration of metalliferous placer deposits: a case study, Lake Turkana, Kenya. *J. geol. Soc. Lond.*, 142, 837–848. *4.6.1.*

Reid R.P. (1987) Non-skeletal peloidal precipitates in Upper Triassic reefs, Yukon Territory, Canada. *J. sedim. Petrol.*, 57, 893–900. *9.4.5.*

Reid R.P., Macintyre I.G. & Post P.E. (1992) Micritized skeletal grains in northern Belize lagoon: a major source of Mg-calcite mud. *J. sedim. petrol.*, 62, 145–156. *9.1.3, 9.4.1.*

Reidel S.P. & Hooper P.R. (1989) Volcanism and tectonism in the Columbia River flood-basalt province. *Spec. Pap. geol. Soc. Am.*, 239, Boulder, CO. *12.9.4.*

Reijmer J.J.G., Ten kate W.G.H.Z., Sprenger A. & Schlager W. (1991) Calciturbidite composition related to the exposure and flooding of a carbonate platform (Triassic, Eastern Alps). *Sedimentology*, 38, 1049–1074. *9.4.4.*

Reineck H.E. (1967) Layered sediments of tidal flats, beaches and shelf bottoms of the North Sea. In: *Estuaries* (Ed. by G.D. Lauff), pp. 191–206. Am. Assoc. Adv. Sci., Washington, DC. *6.7.3.*

Reineck H.E. & Singh I.B. (1971) Der Golf von Gaeta (Tyrrhenisches Meer) III. Die Gefuge von Vorstrand und Schelfsedimenten. *Senckenberg. Mar.*, 3, 185–201. *6.7.6, Fig. 6.85.*

Reineck H.-E. & Singh I.B. (1972) Genesis of laminated sand graded rhythmites in storm-sand layers of shelf mud. *Sedimentology*, 18, 123–128. *7.4.3.*

Reineck H.E. & Singh I.B. (1973) *Depositional Sedimentary Environments—with Reference to Terrigenous Clastics*, 439 pp. Springer-Verlag, Berlin. *1.1, 6.7.6, Fig. 6.85.*

Reineck H.E. & Wunderlich F. (1968) Classification and origin of flaser and lenticular bedding. *Sedimentology*, 11, 99–104. *6.7.3.*

Reinson G.E. (1992) Transgressive barrier island and estuarine systems. In: *Facies Models: Response to Sea-level Change* (Ed. by R.G. Walker and N.P. James), pp. 179–194. Geol. Ass. Can., Waterloo, Ontario. *6.7.5, 6.7.6, Fig. 6.80, 9.4.2.*

Reinson G.E., Clark J.E. & Foscolos A.E. (1988) Reservoir geology of Crystal Viking field, Lower Cretaceous tidal channel-bay complex, south-central Alberta. *Bull. Am. Ass. petrol. Geol.*, 72, 1270–1294. *6.7.7.*

Ren M.-E. & Shi Y.-L. (1986) Sediment discharge of the Yellow River (China) and its effect on the sedimentation of the Bohai and Yellow Sea. *Continental Shelf Res.*, 6, 785–810. *7.6.*

Renaut R.W. (1993) Zeolitic diagenesis of late Quaternary fluviolacustrine sediments and associated calcrete formation in the Lake Bogoria basin, Kenya Rift Valley. *Sedimentology*, 40, 271–302. *4.14.1.*

Renaut R.W. & Owen R.B. (1988) Opaline cherts associated with sublacustrine hydrothermal springs at Lake Bogoria, Kenya rift valley. *Geology*, 16, 699–702. *4.7.2.*

Renaut R.W. & Owen R.B. (1991) Shore-zone sedimentation and facies in a closed rift lake: the Holocene beach deposits of Lake Bogoria, Kenya. In: *Lacustrine Facies Analysis* (Ed. by P. Anadón, L. Cabrera and K. Kelts), pp. 175–195. *Spec. Publ. int. Ass. Sediment.*, 13. *4.6.1.*

Renaut R.W. & Tiercelin J.J. (1994) Lake Bogoria, Kenya rift valley—a sedimentological overview. In: *Sedimentology and Geochemistry of Modern and Ancient Saline Lakes* (Ed. by. R.W. Renaut and W.M. Last), pp. 101–124. *Spec. Publ. Soc. econ. Paleont. Miner.*, 50. *4.6.1, 4.7.4, 4.11.1, 4.14.1.*

Renberg I. (1981) Formation, structure and visual appearance of iron-rich, varved lake sediments. *Verh. Internat. Verein. Limnol.*, 21, 94–101. *Fig. 4.15.*

Retallack G.J. (1981) Fossil soils: indicators of ancient terrestrial environments. In: *Paleobotany, Paleoecology and Evolution*, Vol. 1 (Ed. by K.J. Niklas), pp. 55–102. Praeger, New York, *3.2.3.*

Retallack G.J. (1986a) The fossil record of soils. In: *Palaeosols: Their Recognition and Interpretation* (Ed. by V.P. Wright), pp. 1–57. Blackwell Scientific Publications, Oxford. *3.2.3, 3.4.*

Retallack G.J. (1986b) Fossil soils as grounds for interpreting long-term controls on ancient rivers. *J. sedim. Petrol.*, 56, 1–18. *3.6.6.*

Retallack G.J. (1988) Field recognition of paleosols. In: *Paleosols and Weathering Through Geologic Time* (Ed. by J. Reinhardt and W.R. Sigleo), pp. 1–20. *Spec. Pap. geol. Soc. Am.*, 216, Boulder, CO. *3.5.4.*

References

....

Retallack G.J. (1990) *Soils of the Past: An Introduction to Paleopedology*, 520 pp. Unwin Hyman, Boston. *3.2.3, 3.4, 11.4.7.*

Reynolds A.D. (1994) Sequence stratigraphy from core and wireline log data: the Viking Formation, Albian, south central Alberta, Canada. *Mar. petrol. Geol.*, 11, 258–282. *7.7.1.*

Reynolds M.A., Best J.G. & Johnson R.W. (1980) 1953–1957 eruption of Tuluman Volcano: rhyolitic volcanic activity in the northern Bismarck Sea. *Mem. geol. Surv. Papua New Guinea*, 7, 44. *12.10.*

Rhodes E.E. (1982) Depositional model for a chenier plain, Gulf of Carpentaria, Australia. *Sedimentology*, 29, 201–221. *6.7.2.*

Rhoads D.C. & Morse J.W. (1971) Evolutionary and ecologic significance of oxygen-deficient basins. *Lethaia*, 4, 413–428. *7.7.3.*

Rhoads D.C., Mulsow S.G., Gutschick R., Baldwin C.T. & Stoltz J.F. (1991) The dysaerobic zone revisited: a magnetic facies. In: *Modern and Ancient Continental Shelf Anoxia* (Ed. by R.V. Tyson and T.H. Pearson), pp. 187–199. *Spec. Publ. geol. Soc. Lond.*, 58. *7.7.3.*

Ricci Lucchi F. (1975) Sediment dispersal in turbidite basins: examples from the Miocene of northern Apennines. *9th Int. Congr. Sedimentol., Nice 1975*, Theme 5(2), 347–352. *10.3.2.*

Ricci Lucchi F. (1985) Marnosa–Arenacea turbidite system, Italy. In: *Submarine Fans and Related Turbidite Systems* (Ed. by A.H. Bouma, W.R. Normark and N.E. Barnes), pp. 209–216. Springer-Verlag, New York. *10.3.6.*

Ricci Lucchi F. & Valmori E. (1980) Basinwide turbidites in a Miocene, oversupplied deep-sea plain: a geometrical analysis. *Sedimentology*, 27, 241–270. *10.3.6, Fig. 10.76.*

Rice J.A. & Loope D.B. (1991) Wind-reworked carbonates, Permo-Pennsylvanian of Arizona and Nevada. *Bull. geol. Soc. Am.*, 103, 254–267. *9.4.2.*

Richards M.A., Duncan R.A. & Courtillot V.E. (1989) Flood basalts and hotspot tracks: plume heads and tails. *Science*, 246, 103–107. *12.9.4.*

Richards M.T. (1986) Tidal bed form migration in shallow marine environments: evidence from the Lower Triassic, Western Alps, France. In: *Shelf Sands and Sandstones* (Ed. by R.J. Knight and J.R. McLean), pp. 257–276. *Mem. Can. Soc. petrol. Geol.*, 11, Calgary. *7.7.1.*

Richards P.C. (1991) An estuarine facies model for the Middle Jurassic Sleipner Formation: Beryl Embayment, North Sea. *J. geol. Soc. Lond.*, 148, 459–471. *6.6.6.*

Richards P.C. (1992) An introduction to the Brent Group: a literature review. In: *Geology of the Brent Group* (Ed. by A.C. Morton, R.S. Haszeldine, M.R. Giles and S. Brown), pp. 15–26. *Spec. Publ. geol. Soc. Lond.*, 61. *6.6.6, Figs 6.58, 6.59.*

Richards P.C., Brown S., Dean J.M. & Anderton R. (1988) A new palaeogeographic reconstruction for the Middle Jurassic of the northern North Sea. *J. geol. Soc. Lond.*, 145, 883–886. *6.6.6.*

Richardson M.J., Wimbush M. & Mayer L. (1981) Exceptionally strong near-bottom flows on the continental rise off Nova Scotia. *Science*, 213, 887–888. *10.2.2.*

Richardson M.J., Biscaye P.E., Gardner W.D. & Hogg N.G. (1987). Suspended particulate matter transport through the Vema Channel. *Mar. Geol.*, 77, 171–184. *10.3.7.*

Richardson P.L. (1983) Eddy kinetic energy in the North Atlantic from surface drifters. *J. geophys. Res.*, 88, 4355–4367. *10.2.2.*

Richter-Bernburg G. (1985) *Zechstein–Anhydrit–Facies und Genese*, 82 pp. *Geol. Jahrb., Reihe A*, 85. *8.9.*

Ricketts B.D., Ballance P.F., Hayward B.W. & Mayer W. (1989) Basal Waitemata Group lithofacies: rapid subsidence in an early Miocene interarc basin, New Zealand. *Sedimentology*, 36, 559–580. *6.1, 6.4.*

Rider M.H. (1974) The Namurian of West County Clare. *Proc. R. Irish Acad.*, 74B(9), 125–143. *6.6.6.*

Rider M.H. (1978) Growth faults in the Carboniferous of western Ireland. *Bull. Am. Assoc. petrol. Geol.*, 62, 2191–2213. *6.6.6.*

Rider M.H. & Laurier D. (1979) Sedimentology using a computer treatment of well logs. *Trans. Soc. Professional Well Log Analysts*, 12 pp. 6th European Symp., London. *2.3.4.*

Riding R. (1979) Origin and diagenesis of lacustrine algal bioherms at the margin of the Ries Crater, Upper Miocene, southern Germany, *Sedimentology*, 26, 645–680. *4.12.3.*

Riding R. (1981) Composition, structure and environmental setting of Silurian bioherms and biostromes in northern Europe. In: *European Reef Models* (Ed. by D.F. Toomey), pp. 9–40. *Spec. Publ. Soc. econ. Paleont. Miner.*, 30, Tulsa. *9.4.5.*

Riehle J.R. (1973) Calculated compaction profiles of rhyolitic ash-flow tuffs. *Bull. Geol. Soc. Am.*, 84, 2193–2216. *12.4.1.*

Riggs N.R. & Blakey R.C. (1993) Early and Middle Jurassic paleogeography and volcanology of Arizona and adjacent areas. In: *Mesozoic Paleogeography of the Western United States II* (Ed. by G. Dunn and K. McDougall), pp. 347–375. *Pacific Sect. Soc. econ. Paleont. Miner.*, 71, Los Angeles. *5.5.1.*

Riggs N.R. & Busby-Spera C.J. (1990) Evolution of a multivent volcanic complex within a subsiding arc graben depression: Mt. Wrightson Formation, Arizona. *Bull. geol. Soc. Am.*, 102, 1114–1135. *12.12.4, Fig. 12.73.*

Rine J.M. & Ginsburg R.N. (1985) Depositional facies of a mud shoreface in Suriname, South America. A mud analogue to sandy, shallow-marine deposits. *J. sedim. Petrol.*, 55, 633–652. *6.6.2, 6.7.2, 7.1.2.*

Rine J.M., Tillman R.W., Culver S.J. & Swift D.J.P. (1991) Generation of late Holocene sand ridges on the middle continental shelf of New Jersey, USA — evidence for formation in a mid-shelf setting based on comparisons with a nearshore ridge. In: *Shelf Sand and Sandstone Bodies: Geometry, Facies and Sequence Stratigraphy* (Ed. by D.J.P. Swift, G.F. Oertel, R.W. Tillman and J.A. Thorne), pp. 395–423. *Spec. Publ. int. Ass. Sediment.*, 14. *7.7.2.*

Ripepe M., Roberts L.T. & Fischer A.G. (1991) ENSO and sunspot cycles in varved Eocene oil shales from image analysis. *J. sedim. Petrol.*, 61, 1155–1163. *4.16.1.*

Risacher F. & Eugster H.P. (1979) Holocene pisoliths and encrustations associated with spring-fed surface pools, Pastos Grandes, Bolivia. *Sedimentology*, 26, 253–270. *4.7.1.*

Roberts H.H. (1980) Sediment characteristics of Mississippi River delta-front mudflow deposits. *Trans. Gulf Coast Ass. geol. Soc.*, 30, 485–496. *6.6.4.*

Roberts H.H. & Macintyre I.G. (Eds) (1988) *Halimeda. Coral Reefs*, 6, 121–271. *9.4.5.*

Roberts H.H. & Phipps C.V. (1988) Proposed oceanographic controls on modern Indonesian reefs: a turn-off/turn-on mechanism in a monsoonal setting. *Proc. 6th Int. Coral Reef Symp., Townsville, 1988*, 3, 529–534. *9.2.*

Robertson A.H.F. (1994) Role of tectonic facies concept in orogenic analysis and its application to Tethys in the Eastern Mediterranean region. *Earth Sci. Rev.*, 37, 139–213. *12.11.1.*

Robertson A.H.F. & Degnan P. (1994) The Dras arc complex: lithofacies and reconstruction of a late Cretaceous oceanic volcanic arc in the Indus Suture Zone, Ladakh Himalaya. *Sedim. Geol.*, 92, 117–145. *12.12.5.*

Robertson A.H.F. & Hudson J.D. (1974) Pelagic sediments in the Cretaceous and Tertiary history of the Troodod Massif, Cyprus. In: *Pelagic Sediments: On Land and under the Sea* (Ed. by K.J. Hsü and H.C. Jenkyns), pp. 403–436. *Spec. Publ. int. Ass. Sediment.*, 1. *12.11.1, Fig. 12.59.*

Robin C., Eissen J.-P. & Monzier M. (1994) Ignimbrites of basaltic andesite and andesite compositions from Tanna, New Hebrides Arc. *Bull. Volcanol.*, 56, 10–22. *12.4.4.*

Robin G. de Q. (1979) Formation, flow, and disintegration of ice shelves. *J. Glaciol.*, 24, 259–271. *11.4.6.*

Rocha-Campos A.C., Ernesto M. & Sundaram D. (1981) Geological, palynological and paleomagnetic investigations on Late Paleozoic varvites from the Parana Basin, Brazil. *Atlas 3° Simp. Geol. Reg.*, *Soc. Bras. Geol.*, Núcleo São Paulo, 2, 162–175. *11.5.5.*

Rodolfo K.S. (1989) Origin and early evolution of lahar channel at Mabinit, Mayon Volcano, Philippines. *Bull. geol. Soc. Am.*, 101, 414–426. *1.1.3, 12.5.2, 12.12.3.*

Rodolfo K.S. & Arguden A.T. (1991) Rain-lahar generation and sedimentary delivery systems at Mayon volcano. In: *Sedimentation in Volcanic Settings* (Ed. by R.V. Fisher and G.A. Smith), pp. 71–88. *Spec. Publ. Soc. econ. Paleont. Miner.*, 45, Tulsa. *12.5.2.*

Rogers D.A. & Astin T.R. (1991) Ephemeral lakes, mud pellet dunes and wind-blown sand and silt; reinterpretations of Devonian lacustrine cycles in north Scotland, In: *Lacustrine Facies Analysis* (Ed. by P. Anadón, L. Cabrera and K. Kelts), pp. 199–221. *Spec. Publ. int. Ass. Sedim.*, 13. *4.13.1.*

Roobol M.J., Smith A.L. & Wright J.V. (1985) Dispersal and characteristics of pyroclastic fall deposits from Mt Misery volcano, West Indies. *Geol. Rundsch.*, 74, 321–335. *12.4.3.*

Rose W.I. (1987) Volcanic activity at Santiaguito volcano, 1976–1984. *Spec. Pap. geol. Soc. Am.*, 212, 17–28. *12.10.*

Rosen M.R. (1991) Sedimentologic and geochemical constraints on the evolution of Bristol Dry Lake basin, California, USA. *Palaeogeogr. Palaeoclimatol. Palaeoecol.* 84, 229–257. *4.6.1, 4.7.4.*

Rosendahl B.R. (1987) Architecture of continental rifts with special reference to East Africa. *Ann. Rev. Earth Planet. Sci.*, 15, 445–503. *4.2.*

Rosendahl B.R., Reynolds D.J., Lorber P.M., Burgess C.F., McGill J., Scott D., Lambiase J.J. & Derksen S.J. (1986) Structural expressions of rifting: lessons from Lake Tanganyika, Africa. In: *Sedimentation in the African Rifts* (Ed. by L.E. Frostick, R.W. Renaut, I. Reid and J.J. Tiercelin), pp. 29–44. *Spec. Publ. geol. Soc. Lond.*, 25. *6.5.5.*

Rosenthal L. (1988) Wave-dominated shorelines and incised channel trends: Lower Cretaceous Glauconite Formation, west-central Alberta. In: *Sequences, Stratigraphy, Sedimentology: Surface and Subsurface* (Ed. by D.P. James and D.A. Leckie), pp. 207–219. *Mem. Can. Soc. petrol. Geol.*, 15. *6.6.6.*

Rosenthal L.R.P. & Walker R.G. (1987) Lateral and vertical facies sequences in the Upper Cretaceous Chungo Member, Wapiabi Formation, southern Alberta. *Can. J. sedim. Petrol.*, 24, 771–783. *7.7.2.*

Rosi M. (1992) A model for the formation of vesiculated tuff by the coalescence of accretionary lapilli. *Bull. Volcanol.*, 54, 429–434. *12.4.1.*

Ross D.J. & Skelton P.W. (1993) Rudist formation of the Cretaceous: a palaeoecological, sedimentological and stratigraphical review. In: *Sedimentology Review* (Ed. by V.P. Wright), pp. 73–91. Blackwell Scientific, Oxford. *9.2.1, 9.4.5.*

Ross G.M. (1986) Eruptive style and constructions of shallow marine mafic tuff cones in the Narakay volcanic complex (Proterozoic, Hornby Bay Group, Northwest territories, Canada). *J. Volcanol. Geotherm. Res.*, 27, 265–297. *12.9.3.*

Rothwell R.G., Pearce T.J. & Weaver P.P.E. (1992) Late Quaternary evolution of the Madeira Abyssal Plain, Canary Basin, NE Atlantic. *Basin Res.*, 4, 103–131. *10.3.6, Fig. 10.66.*

Rothwell R.G., Weaver P.P.E. *et al.* (1994) Clayey nanofossil ooze turbidites and hemipelagites at Sites 834 and 835 (Lau Basin, SW Pacific). In: *Proc. ODP Sci. Results*, 135 (Ed. by J. Hawkins, L. Parsons *et al.*) pp. 101–130. *10.2.3, 10.4.1.*

Rouchy J.M. (1976) Sur la genèse de deux principaux types de gypse (finement lité et en chevrons) du Miocène terminal de Sicille et d'Espagne méridionale. *Rev. Géog. Phys. et Géol. Dynam.*, (2), 18, 347–364. *8.5.1.*

Rouchy J.M. (1980) La genese des evaporites messiniennes de Méditerranée: un bilan. *Bull. Cent. Rech. Pau*, 4, 511–545. *Fig. 8.1.*

Rouse J.E. & Sherif N. (1980) Major evaporite deposition from remobilized salts. *Nature*, 285, 470–472. *8.4, 8.4.3.*

Rowland S.K. & Walker G.P.L. (1990) Pahoehoe and aa in Hawaii: volumetric flow rate controls the lava structure. *Bull. Volcanol.*, 52, 625–628. *12.4.5, Fig. 12.1.*

Rowland S.K., Munro D.C. & Perez-Oviedo V. (1994) Volcán Ecuador, Galápagos Islands: erosion as a possible mechanism for the generation of steep-sided basaltic volcanoes. *Bull. Volcanol.*, 56, 271–283. *12.11.2.*

Roy P.S. (1984) New South Wales estuaries: their origin and evolution. In: *Coastal Geomorphology in Australia* (Ed. by B.G. Thom), pp. 99–121. Academic Press, Australia. *6.7.5.*

Roy P.S. (1994) Holocene estuary evolution-stratigraphic studies from southeastern Australia. In: *Incised-valley Systems: Origin and Sedimentary Sequences* (Ed. by R.W. Dalrymple, R. Boyd & B.A. Zaitlin), pp. 241–263. *Spec. Publ. Soc. sedim. Geol.*, 61. *6.7.5.*

Roy P.S., Cowell P.J., Ferland M.A., & Thom B.G. (1994) Wave-dominated Coasts. In: *Coastal Evolution: Late Quaternary Shoreline Morphodynamics* (Ed. by R.W.G. Carter & C.D. Woodroffe), pp. 121–186. Cambridge University Press, Cambridge. *6.2.4.*

Rubin D.M. (1987a) *Cross-bedding, Bedforms, and Paleocurrents*, 187 pp. *Soc. econ. Paleont. Miner., Concepts in Geology*, 1, Tulsa. *5.3.5, 5.3.6, 5.5.4, Fig. 5.7.*

Rubin D.M. (1987b) Formation of scalloped cross-bedding without unsteady flow. *J. sedim. Petrol.*, 57, 39–45. *5.3.6.*

Rubin D.M. & Hunter R.E. (1982) Bedform climbing in theory and nature. *Sedimentology*, 29, 121–138. *5.3.4, Fig. 5.8.*

Rubin D.M. & Hunter R.E. (1983) Reconstructing bedform assemblages from compound crossbedding. In: *Eolian Sediments and Processes* (Ed. by M.E. Brookfield and T.S. Ahlbrandt), pp. 407–427. *Developments in Sedimentology*, 38. Elsevier, Amsterdam. *5.3.6, 5.5.4.*

Rubin D.M. & Hunter R.E. (1987) Bedform alignment in directionally varying flows. *Science*, 237, 276–278. *5.4.3, Fig. 5.20.*

Rubin D.M. & Ikeda H. (1990) Flume experiments on the alignment of transverse, oblique, and longitudinal dunes in directionally varying flows. *Sedimentology*, 37, 673–684. *5.4.3, Fig. 5.20.*

Rudolph K.W. & Lehmann P.J. (1989) Platform evolution and sequence stratigraphy of the Natuna platform, South China Sea. In: *Controls on Carbonate Platform and Basin Development* (Ed. by P.D. Crevello, J.L. Wilson, J.F. Sarg and J.F. Read), pp. 353–361. *Spec. Publ. Soc. econ. Paleont. Miner.*, 44, Tulsa. *9.5.1, 9.5.4, Fig. 9.68.*

Russell A.J. (1993) Obstacle marks produced by flow around stranded ice blocks during a glacier outburst flood (jökulhlaup) in west Greenland. *Sedimentology*, 40, 1091–1111. *11.4.4.*

Russell R.J. & Russell R.D. (1939) Mississippi River delta sedimentation. In: *Recent Marine Sediments* (Ed. by P.D. Trask), pp. 153–177. Am. Ass. petrol. Geol., Tulsa. *6.1.*

Rusnak G.A. (1960) Sediments of Laguna Madre, Texas. In: *Recent Sediments, Northwest Gulf of Mexico* (Ed. by F.P. Shepard, F.B. Phleger and Tj.H. van Andel), pp. 153–196. Am. Ass. petrol. Geol., Tulsa. *6.7.4.*

Rust B.R. (1972) Structure and process in a braided river. *Sedimentology*, 18, 221–245. *3.3.2.*

Rust B.R. (1977) Mass flow deposits in a Quaternary succession near Ottawa, Canada: Diagnostic criteria for subaqueous outwash. *Can. J. Earth Sci.*, 14, 175–184. *11.4.5.*

Rust B.R. (1978) A classification of alluvial channel systems. In: *Fluvial Sedimentology* (Ed. by A.D. Miall), pp. 187–198. *Mem. Can. Soc. petrol. Geol.*, 5, Calgary. *3.3.1.*

Rust B.R. (1981) Sedimentation in an arid-zone anastomosing fluvial system: Cooper's Creek, Central Australia. *J. sedim. Petrol.*, 51, 745–755. *3.3.5.*

Rust B.R. (1984) Proximal braidplain deposits in the Middle Devonian Malbaie Formation of Eastern Gaspé, Quebec, Canada. *Sedimentology*, 31, 675–695. *3.6.4.*

Rust B.R. & Gibling M.R. (1990) Three-dimensional antidunes as HCS mimics in a fluvial sandstone: The Pennsylvanian South Bank Formation near Sydney, Nova Scotia. *J. sedim. Petrol.*, 60, 59–72. *3.5.2.*

Rust B.R. & Jones B.G. (1987) The Hawkesbury Sandstone south of Sydney, Australia: Triassic analogue for the deposit of a large, braided river. *J. sedim. Petrol.*, 57, 222–233. *3.5.2, 3.6.2.*

Rust B.R. & Legun A.S. (1983) Modern anastomosing-fluvial deposits in arid Central Australia and a Carboniferous analogue in New Brunswick, Canada. In: *Modern and Ancient Fluvial Systems* (Ed. by J.D. Collinson & J. Lewin), pp. 385–392. *Spec. Publ. int. Ass. Sediment.*, 6. *3.3.5, 3.6.5.*

Rust B.R. & Nanson G.C. (1989) Bedload transport of mud as pedogenic aggregates in modern and ancient rivers. *Sedimentology*, 36, 291–306. *3.2.2, 3.3.6.*

Rust B.R. & Romanelli R. (1975) Late Quaternary subaqueous outwash deposits near Ottawa, Canada. In: *Glaciofluvial and Glaciolacustrine Sedimentation* (Ed. by A.V. Jopling and B.C. MacDonald), pp. 177–192. *Spec. Publ. Soc. econ. Paleont. Miner.*, 23, Tulsa. *11.4.5.*

Ruttner F. (1952) *Fundamentals of Limnology*, 3rd edn, 307 pp. University of Toronto Press. *4.3.*

Rutzler K. & Macintyre I.G. (1982) The habitat, distribution and community structure of the barrier reef complex at Carrie Bow Cay, Belize. In: *The Atlantic Barrier Reef Ecosystem at Carrie Bow Cay, Belize I, Structure and Communities* (Ed. by K. Rutzler and I.G. Macintyre), pp. 9–45. *Smithsonian Contrib. mar. Sci.*, 12. *9.4.1.*

Ruzyla K. & Friedman G.M. (1985) Factors controlling porosity in dolomite reservoirs of the Ordovician Red River Formation, Cabin Creek Field, Montana. In: *Carbonate Petroleum Reservoirs* (Ed. by P.O. Roehl and P.W. Choquette), pp. 41–58. Springer-Verlag, New York. *8.9.*

Ryan M.P. (1987) Neutral buoyancy and the mechanical evolution of magmatic systems. In: *Magmatic Processes: Physiochemical Principles* (Ed. by M.O. Mysen), pp. 259–287. *Spec. Publ. geochem. Soc.*, 1, University Park, PA. *12.3.2.*

Ryan M.P. (1988) The mechanics and three-dimensional internal structure of active magmatic systems: Kilauea volcano, Hawaii. *J. geophys. Res.*, 93, 4213–4248. *12.3.2.*

Ryan W.B.F. & Hsü K.J. *et al.* (1973) *Initial Reports of the Deep Sea Drilling Project*, 13, 1447 pp. US Government Printing Office, Washington, DC. *8.1, 8.9.*

Ryseth A. (1989) Correlation of depositional patterns in the Ness Formation, Oseberg area. In: *Correlation in Hydrocarbon Exploration* (Ed. by J.D. Collinson), pp. 313–326. Graham & Trotman, London. *6.6.6.*

Rzoska J. (1974) The Upper Nile Swamps, a tropical wetland study. *Freshwat. Biol.*, 4, 1–30. *3.3.6.*

Røe S.-L. (1987) Cross-strata and bedforms of probable transitional dune to upper-stage plane-bed origin from a Late Precambrian fluvial sandstone, northern Norway. *Sedimentology*, 34, 89–101. *3.5.2, Fig. 3.39.*

Røe S.-L. & Hermansen M. (1993) Processes and products of large, Late Precambrian sandy rivers in northern Norway. In: *Alluvial Sedimentation* (Ed. by M. Marzo and C. Puigdefabregas), pp. 151–166. *Spec. Publ. int. Ass. Sediment.*, 17. *3.5.2, 3.6.1, 3.6.2.*

Sageman B.B., Wignall P.B. & Kauffman E.G. (1991) Biofacies models for oxygen-deficient facies in epicontinental seas; tool for palaeoenvironmental analysis. In: *Cycles and Events in Stratigraphy* (Ed. by G. Einsele, W. Ricken and A. Seilacher), pp. 542–564. Springer-Verlag, Berlin. *7.7.3.*

Saller A.H. (1996) Differential compaction and basinward tilting of the prograding Capitan Reef Complex, Permian, west Texas and southeast New Mexico, USA. *Sedim. geol.*, 101, 21–30. *9.4.4.*

Saller A., Armin R., Ichram L.O. & Glenn-Sullivan C. (1993) Sequence stratigraphy of aggrading and backstepping carbonate shelves, Oligocene, Central Kalimantan, Indonesia. *Mem. Am. Ass. petrol. Geol.*, 57, 267–290. *9.5.3.*

Saller A., Barton J.W. & Barton R.E. (1989) Slope sedimentation associated with a vertically building shelf, Bone Spring Formation, Mescalero Escarpe Field, Southeastern New Mexico. *Controls on Carbonate Platform and Basin Development* (Ed. by P.D. Crevello, J.L. Wilson, J.F. Sarg and J.F. Read), pp. 275–288. *Spec. Publ. Soc.*

econ. Paleont. Miner., **44**, Tulsa. *9.5.5, Fig. 9.71.*

Saller A., Dickson J.A.D. & Boyd S.A. (1994) Cycle stratigraphy and porosity in Pennsylvanian and Lower Permian shelf limestones, Eastern Central Basin Platform, Texas. *Bull. Am. Ass. petrol. Geol.*, **78**, 1820–1842. *9.5.3.*

Salvany J.M. & Ortí F. (1994) Miocene glauberite deposits of Alcandre, Ebro Basin, Spain: sedimentary and diagenetic processes. In: *Sedimentology and Geochemistry of Modern and Ancient Saline Lakes* (Ed. by R.W. Renaut and W.M. Last), 203–216. *Soc. econ. Paleont. Miner. Spec. Publ.*, **50**, Tulsa. *4.14.2, Fig. 4.31, Fig. 4.32.*

Salvany J.M., Muñoz A. & Pérez A. (1994) Nonmarine evaporitic sedimentation and associated diagenetic processes of the southwestern margin of the Ebro basin (Lower Miocene), Spain, *J. sediment. Petrol.*, **64**, 190–203. *4.5, 4.14.2, Fig. 4.32.*

Sami T. & Desrochers A. (1992) Episodic sedimentation on an early Silurian, storm-dominated carbonate ramp, Becscie and Merrimack formations, Anticosti Island, Canada. *Sedimentology*, **39**, 355–381. *9.4.3.*

Sandford W.E. & Wood W.W. (1991) Brine evolution and mineral deposition in hydrologically open evaporite basins. *Am. J. Sci.*, **291**, 687–710. *8.2.3.*

Saner S. & Abdulghani W.M. (1995) Lithostratigraphy and depositional environments of the Upper Jurassic Arab-C Carbonate and associated evaporites in the Abqaiq Field, eastern Saudi Arabia. *Bull. Am. Ass. petrol. Geol.*, **79**, 394–409. *8.7.*

Santisteban C. & Taberner C. (1983) Shallow marine and continental conglomerates derived from coral reef complexes after dessication of a deep marine basin: the Tortonian–Messinian deposits of the Fortuna Basin, south east Spain. *J. geol. Soc. Lond.*, **140**, 401–411. *9.4.5.*

Sanz M.E., Rodríguez-Aranda J.P., Calvo J.P. & Ordóñez S. (1994) Tertiary detrital gypsum in the Madrid Basin, Spain: criteria for interpreting detrital gypsum in continental evaporitic sequences. In: *Sedimentology and Geochemistry of Modern and Ancient Saline Lakes* (Ed. by R.W. Renaut and W.M. Last), pp. 217–228. *Soc. econ. Paleont. Miner. Spec. Publ.*, Tulsa, **50**. *4.5, 4.14.2.*

Sarewitz D.R. & Lewis S.D. (1991) The Marinduque intra-arc basin, Philippines: basin genesis and *in situ* ophiolite development in a strike-slip setting. *Bull. geol. Soc. Am.*, **103**, 597–614. *12.11.1.*

Sarg J.F. (1977) Sedimentology of the carbonate–evaporite facies transition of the Seven Rivers Formation (Guadalupian, Permian) in southeast New Mexico. In: *Upper Guadalupian Facies, Permian Reef Complex, Guadalupe Mountains, New Mexico and W. Texas* (Ed. by M.E. Hileman and S.J. Mazzullo), pp. 451–478. *1977 Field Conf. Guidebook.* Soc. econ. Paleont. Miner. Permian Basin Section, Tulsa, Publ. 77-16. *9.3.3.*

Sarg J.F. (1988) Carbonate sequence stratigraphy. In: *Sea-level Changes – an Integrated Approach* (Ed. by C.K. Wilgus, B.S. Hastings, C.G. St. C. Kendall, H. Posamentier, C.A. Ross and J. van Wagoner), pp. 155–181. *Spec. Publ. Soc. econ. Paleont. Miner.*, **42**, Tulsa. *2.4, Fig. 2.17, 8.3.1, 8.10, 9.5, 9.5.1.*

Sarna-Wojcicki A.M., Morrison S.D., Meyer C.E. & Hillhouse J.W. (1987) Correlation of upper Cenozoic tephra layers between sediments of the western United States and eastern Pacific Ocean and comparison with biostratigraphic and magnetostratigraphic ice age data. *Bull. geol. Soc. Am.*, **98**, 207–223. *12.14.1.*

Sarna-Wojcicki A.M., Shipley S., Waitt R.B., Dzurisin D. & Wood S.H. (1981) Areal distribution, thickness, mass, volume, and grain size of air-fall ash from the six major eruptions of 1980. In: *The 1980 Eruptions of Mount St Helens, Washington* (Ed. by P.W. Lipman and D.R. Mulineaux), pp. 577–600. *Prof. Paper. US Geol. Surv.*, 1240. *Fig. 12.19.*

Sarnthein M. (1978) Sand deserts during glacial maximum and climatic minimum. *Nature*, **272**, 43–46. *5.4.1, 5.4.7.*

Sarnthein M. & Diester-Haass L. (1977) Eolian-sand turbidites. *J. sedim. Petrol.*, **47**, 868–890. *5.2.2.*

Sarnthein M. & Koopmann B. (1980) Late Quaternary deep-sea record on northwest African dust supply and wind circulation. In: *Palaeoecology of Africa and Surrounding Islands* (Ed. by M. Sarnthein, E. Seibold and P. Rognon), pp. 239–253. *5.4.7.*

Sato H. & Amano K. (1991) Relationship between tectonics, volcanism, sedimentation and basin development, Late Cenozoic, central part of Northern Honshu, Japan. *Sedim. Geol.*, **74**, 323–343. *12.12.2, 12.12.5, Fig. 12.70.*

Savoye B., Piper D.J.W. & Droz L. (1993) Plio-Pleistocene evolution of the Var deep-sea fan off the French Riviera. *Mar. petrol. Geol.*, **10**, 550–571. *10.3.3.*

Savrda C.E. & Bottjer D.J. (1991) Oxygen-related biofacies in marine strata: an overview and update. In: *Modern and Ancient Continental Shelf Anoxia* (Ed. by R.V. Tyson and T.H. Pearson), pp. 201–219. *Spec. Publ. geol. Soc. Lond.*, **58**. *Figs 7.44, 7.45.*

Savrda C.E., Bottjer D.J. & Seilacher A. (1991) Redox-related benthic events. In: *Cycles and Events in Stratigraphy* (Ed. by G. Einsele, W. Ricken and A. Seilacher), pp. 524–541. Springer-Verlag, Berlin. *7.7.3, Fig. 7.47.*

Scasso R.A., Olivero E.B. & Buatois L.A. (1991) Lithofacies, biofacies, and ichnoassemblage evolution of a shallow submarine volcaniclastic fan–shelf depositional system (Upper Cretaceous, James Ross Island, Antarctica). *J. S. Am. Earth Sci.*, **4**, 239–260. *12.12.5.*

Schaal S. & Ziegler W. (Eds) (1988) *Messel – Ein Schaufenster in die Geschichte der Erde und des Lebens.* Frankfurt am Main, Verlag Waldemar Kramer, 1–315. *4.15.2.*

Schäfer W. (1972) *Ecology and Palaeoecology of Marine Environments* (Trans. by I. Oertel; Ed. by G.Y. Craig), 568 pp. Oliver & Boyd, Edinburgh. *1.1.*

Schäfer A. & Stapf K.R.G. (1978) Permian Saar–Nahe Basin and Recent Lake Constance (Germany): two environments of lacustrine algal carbonates. In: *Modern and Ancient Lake Sediments* (Ed. by A. Matter and M.E. Tucker), pp. 81–106. *Spec. Publ. int. Ass. Sediment.*, **2**. *4.7.1.*

Scheiling M.H. & Gaynor G.C. (1991) The shelf sand plume model: a critique. *Sedimentology*, **38**, 433–444. *7.8.4.*

Schenk C.J., Gautier D.L., Olhoeft G.R. & Lucius J.E. (1993) Internal structure of an aeolian dune using ground-penetrating radar. In: *Aeolian Sediments: Ancient and Modern* (Ed. by K. Pye and N. Lancaster), pp. 59–69. *Spec. Publ. int. Ass. Sediment.*, **16**. *5.3.5.*

Schermerhorn L.J.G. (1974) Late Precambrian mixtites: Glacial and/or nonglacial? *Am. J. Sci.*, **274**, 673–824. *11.1, 11.5.1.*

Schlager W. (1981) The paradox of drowned reefs and carbonate platforms. *Bull. geol. Soc. Am.*, **92**, 197–211. *2.4, 9.2.2, 9.3.5, 9.4.5, 9.5, Fig. 9.2.*

Schlager W. (1989) Drowning unconformities on carbonate platforms. In: *Controls on Carbonate Platform and Basin Development* (Ed. by P.D. Crevello, J.F. Sarg, J.F. Read and J.L. Wilson), pp. 15–25. *Spec. Publ. Soc. econ. Paleont. Miner.*, 44, Tulsa. *2.4, 9.3.5, 9.5.3.*

Schlager W. (1991) Depositional bias and environmental change – important factors in sequence stratigraphy. *Sedim. Geol.*, 70, 109–130. *9.2.2, 9.3.5, 9.4.4, 9.5.1, Fig. 9.39.*

Schlager W. (1992) *Sedimentology and Sequence Stratigraphy of Reefs and Carbonate Platforms*, 71 pp. *Am. Ass. petrol. Geol. Course Notes Ser.*, 34, Tulsa. *9.2.2, Fig. 9.5.*

Schlager W. & Bolz H. (1977) Clastic accumulation of sulphate evaporites in deep water. *J. sedim. Petrol.*, 47, 600–609. *8.6.1.*

Schlager W. & Camber O. (1986) Submarine slope angles, drowning unconformities, and self-erosion of limestone escarpments. *Geology*, 14, 762–765. *9.4.4, 9.5.3.*

Schlager W. & Chermak A. (1979) Sediment facies of platform–basin transition, Tongue of the Ocean, Bahamas. In: *Geology of Continental Slopes* (Ed. by L.J. Doyle and O.H. Pilkey), pp. 193–207. *Spec. Publ. Soc. econ. Paleont. Miner.*, 27, Tulsa. *9.4.4.*

Schlager W. & Ginsburg R.N. (1981) Bahamas carbonate platforms – the deep and the past. *Mar. Geol.*, 44, 1–24. *9.4.4, Fig. 9.36.*

Schlager W., Reijmer J.J.G. & Droxler A. (1994) Highstand shedding of carbonate platforms. *J. sedim. Res.*, B64, 270–281. *9.4.4, 9.5.1, 9.5.3.*

Schlanger S.O. & Cita M.B. (1992) *Nature and Origin of Cretaceous Carbon-rich Facies.* Academic Press, London. *7.7.3.*

Schlanger S.O. & Jenkyns H.C. (1976) Cretaceous oceanic anoxic events: causes and consequences. *Geol. Mijnbouw.*, 55, 179–184. *7.7.3, 10.2.1, Fig. 10.10.*

Schlanger S.O., Arthur M.A., Jenkyns H.C. & Scholle P.A. (1987) The Cenomanian–Turonian Oceanic Anoxic Event, I. Stratigraphy and distribution of organic carbon-rich beds and the marine $\delta^{13}C$ excursion. In: *Marine Petroleum Source Rocks* (Ed. by J. Brooks and A.J. Fleet), pp. 371–399. *Spec. Publ. geol. Soc. Lond.*, 26. *10.2.1.*

Schlanger S.O., Jenkyns H.C. & Premoli Silva I. (1981) Volcanism and vertical tectonics in the Pacific basin related to global Cretaceous transgressions. *Earth planet. Sci. Lett.*, 52, 435–449. *2.1.4.*

Schmalz R.F. (1969) Deep-water evaporite deposition: a genetic model. *Bull. Am. Ass. petrol. Geol.*, 53, 798–823. *8.1, 8.9.*

Schmidt V. (1977) Inorganic and organic reef growth and subsequent diagenesis in Permian Capitan Reef Complex, Guadalupe Mountains, Texas, New Mexico. In: *Upper Guadalupian Facies, Permian Reef Complex, Guadalupe Mountains, New Mexico and West Texas* (Ed. by M.E. Hileman and S.J. Mazzullo), pp. 93–132. *1977 Field Conf. Guidebook.* Soc. econ. Paleont. Miner., Permian Basin Section, Tulsa, Publ., 77-16. *9.3.3.*

Schmincke H.-U. (1967) Graded lahars in the type sections of the Ellensburg Formation, South-central Washington. *J. sedim. Petrol.*, 37, 438–448. *12.5.2.*

Schmincke H.U. & Bogaard P. van den (1991) Tephra layers and tephra events. In: *Cycles and Events in Stratigraphy* (Ed. by G. Einsele, W. Ricken and A. Seilacher), pp. 392–429. Springer-Verlag, Berlin. *Figs 12.38, 12.61.*

Schmincke H.-U., Fisher R.V. & Waters A.C. (1973) Antidune chute

and pool structures in the base surge deposits of the Laacher See area, Germany. *Sedimentology*, 20, 553–574. *12.4.4.*

Schmincke H.-U., Robinson P.T., Ohmacht W. & Flower M.F.J. (1978) Basaltic hyaloclastites from Hole 396B, DSDP leg 46. In: *Initial Reports, Deep Sea Drilling Project*, 46 (Ed. by L. Dmitriev, J. Heimler *et al.*), pp. 341–355. *12.11.1, Fig. 12.12.*

Schoell M., Tietze K. & Schoberth S.M. (1988) Origin of methane in Lake Kivu (East-Central Africa). *Chem. Geol.*, 71, 257–265. *4.5.*

Scholl D.W. (1960) Pleistocene algal pinnacles at Searles Lake, California. *J. sedim. Petrol.*, 30, 414–431. *4.7.1.*

Scholl D.W. & Creager J.S. (1973) Geological synthesis of Leg 19 (DSDP) results: far north Pacific and Aleutian Ridge, and Bering Sea. In: *Initial Reports of the Deep Sea Drilling Project*, 19 (Ed. by J.S. Creager, D.W. Scholl *et al.*), pp. 897–913. US Government Printing Office, Washington, DC. *10.4.1.*

Scholle P.A. & Arthur M.A. (1980) Carbon-isotope fluctuations in Cretaceous pelagic limestones: potential stratigraphic and petroleum exploration tool. *Bull. Am. Ass. petrol. Geol.*, 64, 67–87. *10.2.1.*

Scholle P.A. & Spearing D. (Eds) (1982) *Sandstone Depositional Environments*, 410 pp. *Mem. Am. Ass. petrol. Geol.*, 31, Tulsa. *1.1.*

Scholle P.A., Bebout D.G. & Moore C.H. (Eds) (1983) *Carbonate Depositional Environments*, 708 pp. *Mem. Am. Ass. petrol. Geol.*, 33, Tulsa. *1.1.*

Scholz C.A. & Rosendahl B.R. (1990) Coarse-clastic facies and stratigraphic sequence models from Lakes Malawi and Tanganyika, East Africa. In: *Lacustrine Basin Exploration – Case Studies and Modern Analogs* (Ed. by B.J. Katz), pp. 151–168. *Am. Ass. petrol. Geol., Mem.* 50, Tulsa. *4.6.3, 4.9.*

Scholz C.A., Johnson T.C. & McGill J.W. (1993) Deltaic sedimentation in a rift valley lake: new seismic reflection data from Lake Malawi (Nyasa), East Africa. *Geology*, 21, 395–399. *4.6.2, 4.6.3.*

Scholz C.A., Rosendahl B.R. & Scott D.L. (1990) Development of coarse-grained facies in lacustrine rift basins: examples from East Africa. *Geology*, 18, 140–144. *4.9, Fig. 4.17.*

Schopf J.M. (1980) *Paleoceanography*, 341 pp. Harvard University Press, Cambridge, MA. *8.1.*

Schreiber B.C. (1986) Arid shorelines and evaporites. In: *Sedimentary Environments and Facies*, 2nd edn (Ed. by H.G. Reading), pp. 189–228. Blackwell Scientific Publications, Oxford. *8.1. Fig. 9.28.*

Schreiber B.C. (Ed.) (1988) *Evaporites and Hydrocarbons*, 475 pp. Columbia University Press, New York. *8.1, Fig. 8.21.*

Schreiber B.C. & Hsü K.J. (1980) Evaporites. In: *Developments in Petroleum Geology* (Ed. by G.D. Hobson), pp. 87–138. *8.2.1, Table 8.3.*

Schreiber B.C. & Walker D. (1992) Halite pseudomorphs after gypsum: a suggested mechanism. *J. sedim. Petrol.*, 62, 61–70. *8.5.1.*

Schreiber B.C., Catalano R. & Schreiber E. (1977) An evaporitic lithofacies continuum: latest Miocene (Messinian) deposits of Saleni Basin (Sicily) and a modern analog. In: *Reefs and Evaporites – Concepts and Models* (Ed. by J.H. Fisher), pp. 169–180. *Am. Ass. petrol. Geol., Studies in Geology*, 5, Tulsa. *8.9.*

Schreiber B.C., Friedman G.H., Decima A. & Schreiber E. (1976)

Depositional environments of Upper Miocene (Messinian) evaporite deposits of the Sicilian Basin. *Sedimentology*, 23, 729–760. *8.3.2, 8.6.1, 8.9, Fig. 8.11.*

Schreiber B.C., McKenzie J. & Decima A. (1981) Evaporative limestone: its genesis and diagenesis. *Abs. Am. Ass. petrol. Geol.*, 65, 988. *8.9.*

Schreiber B.C., Roth M.S. & Helman M.L. (1982) Recognition of primary facies characteristics of evaporites and the differentiation of these forms from diagenetic overprints. In: *Depositional and Diagenetic Spectra of Evaporites – a Core Workshop* (Ed. by C.R. Handford, R.D. Loucks and G.R. Davies), pp. 1–32. *Soc. econ. Paleont. Miner. Core Workshop*, 3, Tulsa. *8.7.*

Schroeder J.H. & Purser B.H. (Eds) (1986) *Reef Diagenesis*, 455 pp. Springer-Verlag, Berlin. *9.4.5.*

Schumacher R. & Schmincke H.U. (1995) Models for the origin of accretionary lapilli. *Bull. Volcanol.*, 56, 626–639. *12.4.1.*

Schumm S.A. (1972) Fluvial palaeochannels. In: *Recognition of Ancient Sedimentary Environments* (Ed. by J.K. Rigby and W.K. Hamblin), pp. 98–107. *Spec. Publ. Soc. econ. Paleont. Miner.*, 16, Tulsa. *3.3.1.*

Schumm S.A. (1973) Geomorphic thresholds and complex response of drainage systems. In: *Fluvial Geomorphology* (Ed. by M. Morisawa), pp. 299–310. *Publ. in Geomorphology.* SUNY, Binghampton. *3.6.3.*

Schumm S.A. (1981) Evolution and response of the fluvial system; sedimentologic implications. In: *Recent and Ancient Nonmarine Depositional Environments* (Ed. by F.G. Ethridge and R.M. Flores), pp. 19–29. *Spec. Publ. Soc. econ. Paleont. Miner.*, 31, Tulsa. *3.3.1.*

Schumm S.A. (1993) River response to baselevel change: Implications for sequence stratigraphy. *J. Geol.*, 101, 279–294. *3.6.9, 6.7.6.*

Schwab W.C., Lee H.J. & Twichell D.C. (1993) (Eds) Submarine Landslides: selected studies of the US Exclusive Economic Zone. *USGS Bull.*, 2002, US Govt. Print. Office. *10.2.3.*

Schwartz R.K. (1975) Nature and genesis of some storm washover deposits. *U.S. Army Corps. Engin. Coastal Eng. Res. Centre Tech. Mem.*, 61, 69 pp. *6.7.4, Fig. 6.74.*

Schwartz R.K. (1982) Bedform and stratification characteristics of some modern small-scale washover sand bodies. *Sedimentology*, 29, 835–849. *6.7.4, Fig. 6.74.*

Schwarz H.U. (1982) Subaqueous slope failures – experiments and modern occurrences. In: *Contributions to Sedimentology* (Ed. by H. Füchtbauer, A.P. Lisitzin, J.D. Milliman and E. Seibold), 116 pp. Schweizerbart'sche Verlagsbuchhandlung (Nagele u. Obermüller), Stuttgart. *9.5.3.*

Schwarzacher W. & Fischer A.G. (1982) Limestone–shale bedding and perturbations of the Earth's orbit. In: *Cyclic and Event Stratigraphy* (Ed. by G. Einsele and A. Seilacher), pp. 72–95, Springer-Verlag, Berlin. *10.2.1.*

Scoffin T.P. (1970) The trapping and binding of subtidal carbonate sediments by marine vegetation in Bimini Lagoon, Bahamas. *J. sedim. Petrol.*, 40, 249–273. *9.3.2.*

Scoffin T.P. (1987) *An introduction to carbonate sediments and rocks.* 274 pp. Blackie and Son Ltd, Glasgow. *Fig. 9.52.*

Scoffin T.P. (1992) Taphonomy of coral reefs, a review. *Coral Reefs*, 11, 57–77. *9.4.5.*

Scoffin T.P. & Garrett P. (1974) Processes in the formation and preservation of internal structure in Bermuda patch reefs. *Proc. 2nd Int. Coral Reef Symp., Brisbane*, 2, 429–448. *9.4.5.*

Scoffin T.P. & Tudhope A.W. (1985) Sedimentary environments of the central region of the Great Barrier Reef of Australia. *Coral Reefs*, 4, 81–93. *9.4.1.*

Scoffin T.P. & Tudhope A.W. (1988) Shallowing-upwards sequences in reef lagoon sediments: examples from the Holocene of the Great Barrier Reef of Australia and the Silurian of Much Wenlock, Shropshire, England. *Proc. 6th Intern. Coral Reef Symp., Australia, 1988*, 3, 479–484. *9.4.5, Fig. 9.52.*

Scoffin T.P., Stoddart D.R., Mclean R.F. & Flood P.G. (1978) The recent development of the reefs in the Northern Province of the Great Barrier Reef. *Philos. Trans. R. Soc. Lond.*, B284, 129–139. *9.4.5.*

Scoffin T.P., Alexandersson E.T., Bowes G.E., Clokie J.J., Farrow G.E. & Milliman J.D. (1980) Recent, temperate sub-photic, carbonate sedimentation: Rockall Bank, northeast Atlantic. *J. sedim. Petrol.*, 50, 331–356. *9.4.5.*

Scott A.C. (1978) Sedimentological and ecological control of Westphalian B plant assemblages from West Yorkshire. *Proc. Yorks. geol. Soc.*, 41, 461–508. *3.5.3, 3.5.5.*

Scott E.S. (1992) The palaeoenvironments and dynamics of the Rannock–Etive nearshore and coastal succession, Brent Group, northern North Sea. In: *Geology of the Brent Group* (Ed. by A.C. Morton, R.S. Haszeldine, M.R. Giles and S. Brown), pp. 129–147. *Spec. Publ. geol. Soc. Lond.*, 61. *6.6.6, Figs 6.61, 6.62, 7.7.2.*

Scott K.M. (1967) Intra-bed palaeocurrent variations in a Silurian flysch sequence, Kirkcudbrightshire, Southern Uplands of Scotland. *Scott. J. Geol.*, 3, 268–281. *10.2.3.*

Scott K.M. (1988a) Origin, behaviour, and sedimentology of lahars and lahar-runout flows in the Toutle–Cowlitz River system. *Prof. Pap. US geol. Surv.*, 1447-A, 1–74. *12.5.2, Fig. 12.35.*

Scott K.M. (1988b) Magnitude and frequency of lahars and lahar-runout flows in the Toutle–Cowlitz River system. *Prof. Pap. US geol. Surv.*, 1447-B, 1–33. *12.5.2.*

Scott R.M. & Tillman R.W. (1981) Stevens Sandstone (Miocene), San Joaquin Basin, California. In: *Deep-Water Clastic Sediments: A Core Workshop* (Ed. by C.T. Siemers, R.W. Tillman, and C.R. Williamson), pp. 116–248. *Soc. econ. Paleont. Miner., Core Workshop*, 2, San Francisco. *10.3.3, Fig. 10.48.*

Scruton P.C. (1953) Deposition of evaporites. *Bull. Am. Ass. petrol. Geol.*, 37, 2498–2512. *8.1.*

Scruton P.C. (1960) Delta building and the deltaic sequence. In: *Recent Sediments, Northwest Gulf of Mexico* (Ed. by F.P. Shepard, F.B. Phleger and Tj.H. van Andel), pp. 82–102. *Am. Ass. petrol. Geol.*, Tulsa. *2.2.2.*

Sebrier M. & Soler P. (1992) Tectonics and magmatism in the Peruvian Andes from late Oligocene time to the present. In: *Andean Magmatism and its Tectonic Setting* (Ed. by R.S. Harmon and C.W. Rapela), pp. 259–278. *Spec. Pap. geol. Soc. Am.*, 265, Boulder, CO. *12.12.2.*

Seeling A. (1978) The Shannon Sandstone, a further look at the environment of deposition at Heldt Draw field, Wyoming. *Mount. Geol.*, 15, 133–144. *7.8.4.*

Segerstrom K. (1950) Erosion studies at Paricutin, State of Michoacan, Mexico. *Bull. US geol. Surv.*, 965-A, 164. *12.6.1, 12.9.1.*

Sekiya S. & Kikuji Y. (1889) The eruption of Bandai-san. *Tokyo Imperial Univ. Coll. Sci.*, 3(2), 91–172. *12.1.1.*

Self S. & Sparks R.S.J. (1978) Characteristics of widespread pyroclastic deposits formed by the interaction of silicic magma and water. *Bull. Volcanol.*, 41, 196–212. *12.4.1.*

Self S. & Walker G.P.L. (1994) Ash clouds: characteristics of eruption columns. *Bull. US geol. Surv.*, 2047, 65–74. *Fig. 12.13.*

Self S., Goff F., Gardner J.N., Wright J.V. & Kite W. (1986) Explosive rhyolitic volcanism in the Jemez Mountains: vent locations, caldera development and relation to regional structure. *J. geophys. Res.*, 91, 1779–1798. *12.12.3.*

Self S., Wilson L. & Nairn I.A. (1979) Vulcanian eruption mechanisms. *Nature*, 277, 440–443. *12.4.1.*

Sellwood B.W. (1986) Shallow-marine carbonate environments. In: *Sedimentary Environments and Facies*, 2nd edn (Ed. by H.G. Reading), pp. 283–342. Blackwell Scientific Publications, Oxford. *Figs 9.12, 9.18, 9.25.*

Sellwood B.W. & Netherwood R.E. (1984) Facies evolution in the Gulf of Suez area: sedimentation history as an indicator of rift initiation and development. *Mod. Geol.*, 9, 43–69. *6.5.4.*

Sepkoski J.J. Jr (1982) Flat-pebble conglomerates, storm deposits, and the Cambrian bottom fauna. In: *Cyclic and Event Stratigraphy* (Ed. by G. Einsele and A. Seilacher), pp. 371–385. Springer-Verlag, Berlin. *9.4.3.*

Sepkoski J.J., Bambach R.K. & Droser M.L. (1991) Secular changes in Phanerozoic event bedding and biological overprint. In: *Cycles and Events in Stratigraphy* (Ed. by G. Einsele, W. Ricken and A. Seilacher), pp. 298–312. Springer-Verlag, Berlin. *7.4.3.*

Serra O. & Sulpice L. (1975) Sedimentological analysis of shale-sand series from well logs. *Trans. Soc. Prof. Well Log Analysts*, W1–W23. *16th Annual Logging Symp., New Orleans. Fig. 2.9.*

Servant M. & Servant S. (1970) Les formations lacustres et les diatomées du quaternaire recent du fond de la cuvette tschadienne. *Rev. Géogr. phys. Géol. Dynam.*, 13, 63–76. *4.9.*

Shanley K.W. & McCabe P.J. (1991) Predicting facies architecture through sequence stratigraphy: An example from the Kaiparowits Plateau, Utah. *Geology*, 19, 742–745. *3.6.9.*

Shanley K.W. & McCabe P.J. (1993) Alluvial architecture in a sequence stratigraphic framework: a case history from the Upper Cretaceous of southern Utah. In: *The Geological Modelling of Hydrocarbon Reservoirs and Outcrop Analogues* (Ed. by S.S. Flint and I.D. Bryant), pp. 21–56. *Spec. Publ. int. Ass. Sediment.*, 15. *Fig. 3.61.*

Shanley K.W., McCabe P.J. & Hettinger R.D. (1992) Tidal influence in Cretaceous fluvial strata from Utah, USA: a key to sequence stratigraphic interpretation. *Sedimentology*, 39, 905–930. *3.6.5, 3.6.9, 6.7.6.*

Shanmugam G. (1980) Rhythms in deep sea, fine-grained turbidite and debris-flow sequences, Middle Ordovician, eastern Tennessee. *Sedimentology*, 27, 419–432. *10.3.2.*

Shanmugam G. & Moiola R.J. (1988) Submarine fans: characteristics, models, classification, and reservoir potential. *Earth Sci. Rev.*, 24, 383–428. *10.3.2.*

Shanmugam G. & Moiola R.J. (1991) Types of submarine fan lobes: models and implications. *Bull. Am. Ass. petrol. Geol.*, 75, 156–179. *10.3.2.*

Shanmugam G., Spalding T.D. & Rofheart D.H. (1993) Process sedimentology and reservoir quality of deep-marine bottom-current reworked sands (sandy contourites): an example from the Gulf of Mexico. *Bull. Am. Ass. petrol. Geol.*, 77, 1241–1259. *10.1, 10.3.7.*

Sharma G.D. (1979) *The Alaskan Shelf: Hydrographic, Sedimentary and Geochemical Environments*, 498 pp. Springer-Verlag, New York. *7.6.*

Sharp M.J. (1982) A comparison of the landforms and sedimentary sequences produced by surging and non-surging glaciers in Iceland. Unpublished PhD thesis, University of Aberdeen, 380 pp. *Fig. 11.6.*

Sharp R.P. & Nobles L.H. (1953) Mudflow of 1941 at Wrightwood, southern California. *Bull. geol. Soc. Am.*, 64, 547–560. *3.3.8.*

Shaver R.H. (1977) Silurian reef geometry – new dimensions to explore. *J. sedim. Petrol.*, 47, 1409–1424. *9.4.5, Fig. 9.53.*

Shaw A.B. (1977) A review of some aspects of evaporite deposition. *Mount. Geol.*, 14, 1–16. *8.2.3, Fig. 8.8.*

Shaw J. (1982) Melt-out till in the Edmonton area, Alberta, Canada. *Can. J. Earth Sci.*, 19, 1548–1569. *11.4.1.*

Shaw J. (1985) Subglacial and ice marginal environments. In: *Glacial Sedimentary Environments* (Ed. by G.M. Ashley, J. Shaw and N.D. Smith), 7–84. *Soc. econ. paleont. Miner. Short Course*, 16, Tulsa. *11.3.4, 11.3.6, 11.4.1, 11.4.2, 11.4.3, Fig. 11.4.*

Shaw J. (1989) Sublimation till. In: *Genetic Classification of Glacigenic Deposits* (Ed. by R.P. Goldthwait and C.C. Matsch), pp. 141–142. Balkema, Rotterdam. *11.4.2.*

Shaw J., Kvill D. & Rains B. (1989) Drumlins and catastrophic subglacial floods. *Sedim. Geol.*, 62, 177–202. *11.4.1.*

Shearman D.J. (1966) Origin of marine evaporites by diagenesis. *Trans. Inst. Min. Metall., Ser. B*, 75, 208–215. *2.2.2, 8.1, 8.4.2, 8.7, Fig. 8.36.*

Shearman D.J. (1970) Recent halite rock, Baja California, Mexico. *Trans. Inst. Min. Metall., Ser. B*, 79, 155–162. *8.4.1, 8.5.1, 8.5.2, Fig. 8.23.*

Shearman D.J., McGugan A., Stein C. & Smith A.J. (1989) Ikaite, $CaCO_3.6H_2O$, precursor of thinolites in the Quaternary tufas and tufa mounds of the Lahontan and Mono Lake basins, western United States. *Bull. geol. Soc. Am.*, 101, 913–917. *4.7.1.*

Shepard F.P. (1932) Sediments on continental shelves. *Bull. geol. Soc. Am.*, 43, 1017–1034. *7.1.2.*

Shepard F.P. (1955) Delta front valleys bordering Mississippi distributaries. *Bull. geol. Soc. Am.*, 66, 1489–1498. *6.6.4.*

Shepard F.P. (1973a) Sea floor off Magdalena delta and Santa Marta area, Colombia. *Bull. geol. Soc. Am.*, 84, 1955–1972. *6.6.4.*

Shepard F.P. (1973b) *Submarine Geology*, 348 pp. Harper & Row, New York. *7.1.2.*

Shepard F.P. & Inman D.L. (1950) Nearshore water circulation related to bottom topography and wave refraction. *Trans. Am. geophys. Union*, 31, 196–212. *Fig. 6.9.*

Shepard F.P. & Moore D.G. (1960) Bays of Central Texas Coast. In: *Recent Sediments, northwest Gulf of Mexico* (Ed. by F.P. Shepard, F.B. Phleger & Tj. H. van Andel), pp. 117–152. *Am. Ass. petrol. Geol.*, Tulsa. *6.7.4.*

Shepard F.P., Dill R.F. & Heezen B.C. (1968) Diapiric intrusions in foreset slope sediments off Magdalena delta, Colombia. *Bull. Am. Ass. petrol. Geol.*, 52, 2197–2207. *6.6.4.*

Sheridan M.F. (1970) Fumarolic mounds and ridge of the Bishop Tuff, California. *Bull. geol. Soc. Am.*, 81, 851–868. *12.4.1.*

Sheridan M.F. (1979) Emplacement of pyroclastic flows: a review. In: *Ash Flow Tuffs* (Ed. by C.E. Chapin and W.E. Elston), pp. 125–136. *Spec. Pap. geol. Soc. Am.*, 180. *Fig. 12.26.*

Sheridan M.F. & Wohletz K.H. (1983) Hydrovolcanism: basic considerations and review. *J. Volcanol. Geotherm. Res.*, 17, 1–29. *12.4.1.*

Sheridan M.F., Barberi F., Rose M. & Santacroce R. (1981) A model for plinian eruptions of Vesuvius. *Nature*, 289, 282–285. *12.4.1.*

Sheridan M.F., Frazetta G. & La Volpe L. (1987) Eruptive histories of Lipari and Vulcano, Italy during the past 22 000 years. *Spec. Pap. geol. Soc. Am.*, 212, 29–34. *12.10.*

Sheridan R.E. & Enos P. (1979) Stratigraphic evolution of the Blake Plateau after a decade of scientific drilling. In: *Deep Drilling Results in the Atlantic Ocean: Continental Margins and Paleoenvironment* (Ed. by M. Talwani, W. Hay and W.B.F. Ryan), pp. 109–122. *Maurice Ewing Ser.*, 3. Am. Geophys. Union, Washington. *10.4.2.*

Shimoyama T. (1984) Sulphur concentration in the Japanese Palaeogene coal. In: *Sedimentology of Coal and Coal-bearing Sequences* (Ed. by R.A. Rahmani and R.M. Flores), pp. 361–372. *Spec. Publ. int. Ass. Sediment.*, 7. *3.5.5.*

Shinn E.A. (1973) Carbonate coastal accretion in an area of longshore transport, NE Qatar, Persian Gulf. In: *The Persian Gulf* (Ed. by B.H. Purser), pp. 179–191. Springer-Verlag, Berlin. *8.4.3.*

Shinn E.A. (1980) Geologic history of Grecian rocks, Key Largo Coral Reef Marine Sanctuary. *Bull. mar. Sci.*, 30, 646–656. *9.4.5.*

Shinn E.A., Hudson J.H., Robbin D.M. & Lidz B.H. (1981) Spurs and grooves revisited: construction versus erosion, Looe Key Reef, Florida. *Proc. 4th Int. Coral Reef Symp., Manila*, 1, 475–483. *9.4.5.*

Shinn E.A., Lloyd R.M. & Ginsburg R.N. (1969) Anatomy of a modern carbonate tidal flat, Andros Island, Bahamas. *J. sedim. Petrol.*, 39, 1202–1228. *9.1.1.*

Shinn E.A., Steinen R.P., Lidz B.H. & Swart R.K. (1989) Whitings, a sedimentological dilemma. *J. sedim. Petrol.*, 59, 147–161. *9.1.3.*

Shinn E.A., Hudson J.H., Halley R.B., Lidz B., Robbin D.M. & Macintyre I.G. (1982) Geology and sediment accumulation rates at Barrie Bow Cay, Belize. In: *The Atlantic Barrier Reef Ecosystem at Barrier Bow Cay, Belize, 1, Structure and Communities* (Ed. by K. Rutzler and I.G. Macintyre), pp. 63–75. *Smithsonian Contr. mar. Sci.*, 12. Smithsonian Institute Press, Washington, DC. *9.4.5.*

Shor A., Lonsdale P., Hollister C.D. & Spencer D. (1980) Charlie–Gibbs fracture zone: bottom-water transport and its geologic effects. *Deep-Sea Res.*, 27A, 325–345. *10.2.2.*

Short A.D. (1984) Beach and nearshore facies: SE Australia. *Mar. Geol.*, 60, 261–282. *6.2.4, 6.7.6.*

Short A.D. (1989) Chenier research on the Australian coast. *Mar. Geol.*, 90, 345–351. *6.7.2.*

Shreve R.L. (1972) Movement of water in glaciers. *J. Glaciol.*, 11 (62), 205–214. *11.3.6.*

Shuisky Y.D. (1989) Approaches to study of cheniers along the coastline of the Soviet Union. *Mar. Geol.*, 90, 289–296. *6.7.2.*

Shultz A.W. (1984) Subaerial debris-flow deposition in the Upper Paleozoic Cutler Formation, Western Colorado. *J. sedim. Petrol.*, 54, 759–772. *3.5.1.*

Siebe C., Komorowski J.-C. & Sheridan M.F. (1992) Morphology and emplacement of an unusual debris-avalanche deposit at Jocotilán volcano, Central Mexico. *Bull. Volcanol.*, 54, 573–589. *12.5.1.*

Siebert L. (1984) Large volcanic debris avalanches: characteristics of source areas, deposits, and associated eruptions. *J. Volcanol. Geotherm. Res.*, 22, 163–197. *12.5.1.*

Siegenthaler C. (1982) Tidal cross-strata and the sediment transport rate problem: a geologist's approach. *Mar. Geol.*, 45, 227–240. *6.2.7.*

Siemers C.T. & Ristow J.H. (1986) Marine-shelf bar sand/channelized sand shingled couplet, Terry Sandstone Member of Pierre Shale, Denver Basin, Colorado. In: *Modern and Ancient Shelf Clastics: A Core Workshop* (Ed. by T.F. Moslow and E.G. Rhodes), pp. 269–323. Soc. econ. Paleont. Miner. Core Workshop No. 9. *7.8.4.*

Sigurdsson H. & Carey S.N. (1989) Plinian and co-ignimbrite tephra fall from the 1815 eruption of Tambora Volcano. *Bull. Volcanol.*, 51, 243–270. *12.4.2, Fig. 12.1.*

Sigurdsson H., Carey S.N. & Fisher R.V. (1987) The 1982 eruptions of El Chichón Volcano, Mexico (3): physical properties of pyroclastic surges. *Bull. Volcanol.*, 49, 467–488. *12.4.4.*

Sigurdsson H., Carey S., Cornell W. & Pescatore T. (1985) The eruption of Vesuvius in AD 79. *Nat. Geog. Res.*, 1, 332–387. *Fig. 12.1.*

Sigurdsson H., Sparks R.S.J., Carey S.N. & Huang T.C. (1980) Volcanogenic sedimentation in the Lesser Antilles arc. *J. Geol.*, 88, 523–540. *10.3.6, 12.12.5, Fig. 12.78.*

Silver B.A. & Todd R.G. (1969) Permian cyclic strata, northern Midland and Delaware Basins, west Texas and southeastern New Mexico. *Bull. Am. Ass. petrol. Geol.*, 53, 2223–2251. *9.3.3.*

Simkin T. (1993) Terrestrial volcanism in space and time. *Ann. Rev. Earth planet. Sci.*, 427–452. *Fig. 12.43.*

Simkin T. & Fiske R.S. (1983) *Krakatau 1883: the volcanic eruption and its effects.* Smithsonian Institution, Washington DC, 464 pp. *Fig. 12.1.*

Simkin T., Siebert L., McClelland L., Bridge D., Newhall C. & Latter J.H. (1981) *Volcanoes of the World*, 232 pp. Smithsonian Institution, Hutchinson Ross, Stroudsberg, PA. *12.6, Fig. 12.46.*

Simo J.A., Scott R.W. & Masse J.P. (Eds) (1993) *Cretaceous Carbonate Platforms*, 479 pp. Mem. Am. Ass. petrol. Geol., 56, Tulsa. *9.5.*

Simons D.B. & Richardson E.V. (1961) Forms of bed roughness in alluvial channels. *Trans. Am. Soc. Civ. Eng.*, 128, 284–302. *3.1.*

Simons T.J. & Jordan D.E. (1972) Computed water circulation of Lake Ontario for observed winds 20 April–14 May 1971. *Canada Centre Inland Waters Publ.*, Burlington, 17 pp. *4.4.2.*

Simpson E.L. & Erikson K.A. (1991) Depositional facies and controls on parasequence development in siliciclastic tidal deposits from the Lower Proterozoic, upper Mount Guide Quartzite, Mount Isa Inlier, Australia. In: *Clastic Tidal Sedimentology* (Ed. by D.G. Smith, G.E. Reinson, B.A. Zeitlin and R.A. Rahmani), pp. 371–387. *Mem. Can. Soc. petrol. Geol.*, 16, Calgary. *7.7.1. 7.8.1.*

Simpson F. (1975) Marine lithofacies and biofacies of the Colorado

References

···

Group (middle Albian to Santonian) in Saskatchewan. *Spec. Pap. geol. Ass. Can.*, **13**, 553–587. *7.7.3.*

Singer B.S. & Myers J.D. (1990) Intra-arc extension and magmatic evolution in the central Aleutian arc, Alaska. *Geology*, **18**, 1050–1053. *12.12.2.*

Singh A. & Bhardwaj B.D. (1991) Fluvial facies model of the Ganga River sediments, India. *Sedim. Geol.*, **72**, 135–146. *3.3.3.*

Singh A.B. (1972) On the bedding in the natural-levee and point bar deposits of the Gomti River, Uttar Pradesh, India. *Sedim. Geol.*, **7**, 309–317. *3.3.6.*

Siringan F.P. & Anderson J.B. (1993) Seismic facies, architecture, and evolution of the Bolivar Roads tidal inlet/delta complex, east Texas Gulf Coast. *J. sedim. Petrol.*, **63**, 794–808. *6.7.4.*

Skulski T., Francis D. & Ludden J. (1992) Volcanism in an arc-transform zone: the stratigraphy of the St. Clare volcanic field, Wrangell volcanic belt. *Can. J. Earth Sci.*, **29**, 446–461. *12.2.2.*

Slatt R.M. (1984) Continental shelf topography: key to understanding distribution of shelf sand-ridge deposits from Cretaceous Western Interior Seaway. *Bull. Am. Ass. petrol. Geol.*, **68**, 1107–1120. *7.8.4.*

Slingerland R. (1986) Numerical computation of co-oscillating palaeotides in the Catskill epeiric Sea of eastern North America. *Sedimentology*, **33**, 487–497. *9.3.1.*

Sliter W.V., Bé A.H.H. & Berger W.H. (Eds) (1975) *Dissolution of Deep-Sea Carbonates*, 159 pp. Spec. Publ. Cushman Found. Foraminiferal Res., 13. Washington, DC. *10.2.1.*

Sloss L.L. (1950) Paleozoic stratigraphy in the Montana area. *Bull. Am. Ass. petrol. Geol.*, **34**, 423–451. *1.1, 2.2.*

Sloss L.L. (1953) The significance of evaporites. *J. sedim. Petrol.*, **23**, 143–161. *8.1.*

Sloss L.L. (1963) Sequences in the cratonic interior of North America. *Bull. geol. Soc. Am.*, **74**, 93–114. *1.1, 2.2, 7.8.1.*

Sloss L.L. (1969) Evaporite deposition from layered solutions. *Bull. Am. Ass. petrol. Geol.*, **53**, 776–789. *8.1.*

Sly P.G. & Lewis C.F.M. (1972) The Great Lakes of Canada – Quaternary geology and limnology. *Guide Book Trip A43: 24th Internat. Geol. Congress*, Montreal, 92 pp. *4.9.*

Smalley P.C., Higgins A.C., Howarth R.J., *et al.* (1994) Seawater Sr isotope variations through time: a procedure for constructing a reference curve to date and correlate marine sedimentary rocks. *Geology*, **22**, 431–434. *4.10.1.*

Smedes H.W. & Prostka H.J. (1972) Stratigraphic framework of the Absaroka Volcanic Supergroup in the Yellowstone National Park Region. *Prof. Pap. US Geol. Surv.*, 729-C, C1–C33. *12.12.4.*

Smith D.B. (1973) The origin of the Permian Middle and Upper Potash deposits of Yorkshire: an alternative hypothesis. *Proc. Yorkshire geol. Soc.*, **39**, 327–346. *8.9.*

Smith D.B. (1980) The evolution of the English Zechstein basin. In: *The Zechstein Basin* (Ed. by H. Füchtbauer and T. Peryt), pp. 7–34. E. Schweizerbart'sche Verlagsbuchandlung, Stuttgart. *Fig. 8.49.*

Smith D.B. (1989) The late Permian palaeogeography of north-east England. *Proc. Yorkshire geol. Soc.*, **47**, 285–312. *8.9.*

Smith D.G. (1983) Anastomosed fluvial deposits: examples from Western Canada. In: *Modern and Ancient Fluvial Systems* (Ed. by J.D. Collinson and J. Lewin), pp. 155–168. Spec. Publ. int. Ass. Sediment., 6. *3.3.5, 3.6.5.*

Smith D.G. (1986) Anastomosing river deposits, sedimentation rates and basin subsidence, Magdelena River, Northwestern Colombia, South America. *Sedim. Geol.*, **46**, 177–196. *3.3.5.*

Smith D.G. (1987) Meandering river point bar lithofacies models: Modern and ancient examples compared. In: *Recent Developments in Fluvial Sedimentology* (Ed. by F.G. Ethridge, R.M. Flores and M.D. Harvey), pp. 83–91. Spec. Publ. Soc. econ. Paleont. Miner., 39, Tulsa. *3.3.4, 3.6.5.*

Smith D.G. (1991) Lacustrine deltas. *Can. Geogr.*, **35**, 311–316. *4.6.2.*

Smith D.G. (1994) Cyclicity or chaos? Orbital forcing versus non-linear dynamics. In: *Orbital Forcing and Cyclic Sequences* (Ed. by P.L. de Boer and D.G. Smith), pp. 531–544. Spec. Publ. int. Ass. Sediment., 19. *2.1.5.*

Smith D.G. & Jol H.M. (1992a) Ground-penetrating radar investigation of a Lake Bonneville delta, Provo level, Brigham City, Utah. *Geology*, **20**, 1083–1086. *4.6.2, Fig. 4.6.*

Smith D.G. & Jol H.M. (1992b) GPR results used to infer depositional processes of coastal spits in large lakes. In: *Fourth International Conference on Ground Penetrating Radar* (Ed. by P. Hänninen and S. Autio), pp. 169–177. Geol. Surv. Finland, Spec. Pap., 16. *4.6.2.*

Smith D.G. & Smith N.D. (1980) Sedimentation in anastomosing river systems: Examples from alluvial valleys near Banff, Alberta. *J. sedim. Petrol.*, **50**, 157–164. *3.3.5, Fig. 3.23.*

Smith D.J. & Hopkins T.S. (1972) Sediment transport on the continental shelf of Washington and Oregon in light of recent current measurements. In: *Shelf Sediment Transport: Process and Pattern* (Ed. by D.J.P. Swift, D.B. Duane and O.H. Pilkey), pp. 143–180. Dowden, Hutchinson & Ross, Stroudsbourg. *7.4.1.*

Smith D.K. & Cann J.R. (1992) The role of seamount volcanism in crustal construction at the Mid-Atlantic Ridge. *J. geophys. Res.*, **97**, 1645–1658. *12.11.1.*

Smith G.A. (1986) Coarse-grained nonmarine volcaniclastic sediment: terminology and depositional process. *Bull. geol. Soc. Am.*, **97**, 1–10. *12.5.2.*

Smith G.A. (1987a) The influence of explosive volcanism on fluvial sedimentation: the Deschutes Formation (Neogene) in Central Oregon. *J. sedim. Petrol.*, **57**, 613–629. *3.5.2, 12.5.2, 12.12.4.*

Smith G.A. (1987b) Sedimentology of volcanism-induced aggradation in fluvial basins: Examples from the Pacific Northwest U.S.A. In: *Recent Developments in Fluvial Sedimentology* (Ed. by F.G. Ethridge, R.M. Flores and M.D. Harvey), pp. 215–228. Spec. Publ. Soc. econ. Paleont. Miner., 39, Tulsa. *3.5.2, 12.12.4.*

Smith G.A. (1988) Sedimentology of proximal to distal volcaniclastics dispersed across an active foldbelt: Ellensberg Formation (late Miocene), central Washington. *Sedimentology*, **35**, 953–977. *12.5.2, 12.6.2, 12.12.4, Figs 12.73, 12.74.*

Smith G.A. (1991) Facies sequences and geometries in continental volcaniclastic sequences. In: *Sedimentation in Volcanic Settings* (Ed. by R.V. Fisher and G.A. Smith), pp. 109–122. Spec. Publ. Soc. econ. Paleont. Miner., 45, Tulsa. *12.6.1, 12.6.2, Fig. 12.42.*

Smith G.A. & Landis C.A. (1995) Intra-arc basins. In: *Tectonics of Sedimentary Basins* (Ed. by C.J. Busby and R.V. Ingersoll), pp. 263–298. Blackwell Science Inc., Boston. *Figs 12.65, 12.69.*

Smith G.A. & Katzman D. (1991) Discrimination of eolian and pyroclastic-surge processes in the generation of cross-bedded tuffs,

Jemez Mountains volcanic field, New Mexico. *Geology*, 19, 465–468. *12.4.4*.

Smith G.A. & Landis C.A. (1995) Intra-Arc Basins. In: *Tectonics of Sedimentary Basins* (Ed. by C.J. Busby and R.V. Ingersoll), pp. 263–298. Blackwell Science Inc., Boston. *12.12.1, Figs 12.65, 12.69, 12.87*.

Smith G.A. & Lotosky J.E. (1994) What factors control the composition of andesitic sand? *J. sedim. Res.*, A65, 91–98. *12.14.1*.

Smith G.A. & Lowe D.R. (1991) Lahars: volcano-hydrologic events and deposition in the debris flow–hyperconcentrated flow continuum. In: *Sedimentation in Volcanic Settings* (Ed. by R.V. Fisher and G.A. Smith), pp. 59–70. *Spec. Publ. Soc. econ. Paleontol. Miner.*, 45, Tulsa. *12.5.2*.

Smith G.I. & Street-Perrott F.A. (1983) Pluvial lakes of the western United States. In: *Late Quaternary Environments of the United States. Vol. 1, The Late Pleistocene* (Ed. by S.C. Porter), pp. 190–212. Longman, London. *4.9*.

Smith M.A. (1990) Lacustrine oil shale in the geologic record. In: *Lacustrine Basin Exploration – Case Studies and Modern Analogs* (Ed. by B.J. Katz), pp. 43–60. *Mem. Am. Ass. petrol. Geol.*, 50, Tulsa. *4.17*.

Smith N.D. (1970) The braided stream depositional environment: Comparison of the Platte River with some Silurian clastic rocks, North-Central Appalachians. *Bull. geol. Soc. Am.*, 81, 2993–3014. *3.3.3*.

Smith N.D. (1971) Transverse bars and braiding in the Lower Platte River, Nebraska. *Bull. geol. Soc. Am.*, 82, 3407–3420. *3.3.3*.

Smith N.D. (1974) Sedimentology and bar formation in the Upper Kicking Horse River, a braided outwash stream. *J. Geol.*, 82, 205–224. *3.3.2, 3.5.1*.

Smith N.D. (1985) Proglacial fluvial environment. In: *Glacial Sedimentary Environments* (Ed. by G.M. Ashley, J. Shaw and N.D. Smith), pp. 85–134. *Soc. econ. Paleont. Miner., Short course*, 16, Tulsa. *11.4.4*.

Smith N.D. & Ashley G.M. (1985) Proglacial lacustrine environment. In: *Glacial Sedimentary Environments* (Ed. by G.M. Ashley, J. Shaw & N.D. Smith), pp. 135–216. *Soc. econ. Paleont. miner., Short Course*, 16, Tulsa. *Fig. 4.15, 11.4.5*.

Smith N.D. & Smith D.G. (1984) William River: An outstanding example of channel widening and braiding caused by bed-load addition. *Geology*, 12, 78–82. *3.3.3*.

Smith N.D. & Syvitski J.P.M. (1982) Sedimentation in a glacier-fed lake: The role of pelletization on deposition of fine-grained suspensates. *J. sedim. Petrol.*, 52, 503–513. *11.4.5*.

Smith N.D., Cross T.A., Dufficy J.P. & Clough S.R. (1989) Anatomy of an avulsion. *Sedimentology*, 36, 1–23. *3.3.4, 3.3.6, 3.5.3, 3.6.6, Fig. 3.17*.

Smith R.B. & Braile L.W. (1994) The Yellowstone hotspot. *J. Volcanol. Geotherm. Res.*, 61, 121–187. *12.2.2, 12.13, Fig. 12.2*.

Smith R.C.M. (1991) Landscape response to a major ignimbrite eruption, Taupo volcanic center, New Zealand. In: *Sedimentation in Volcanic Settings* (Ed. by R.V. Fisher and G.A. Smith), pp. 123–138. *Spec. Publ. Soc. econ. Paleont. Miner.*, 45, Tulsa. *12.13*.

Smith R.L. (1960a) Ash flows. *Bull. geol. Soc. Am.*, 71, 795–842. *12.1.1*.

Smith R.L. (1960b) Zones and zonal variation in welded ash flows. *Prof. Pap. US Geol. Surv.*, 354-F, 149–159. *12.1.1, 12.4.1,*

Fig. 12.27.

Smith R.L. (1979) Ash flow magmatism. *Spec. Pap. geol. Soc. Am.*, 180, 5–27. *12.3.2*.

Smith R.L. & Bailey R.A. (1968) Resurgent cauldrons. *Mem. geol. Soc. Am.*, 116, 613–662. *12.4.6, 12.12.4*.

Smith R.M.H. (1986) Sedimentation and paleoenvironment of Late Cretaceous crater-lake deposits in Bushmanland, South Africa. *Sedimentology*, 33, 369–386. *12.9.2*.

Smith R.M.H. (1987) Morphology and depositional history of exhumed Permian point bars in the southwestern Karoo Basin, South Africa. *J. sedim. Petrol.*, 57, 19–29. *3.6.5*.

Smith R.M.H. (1990) Alluvial paleosols and pedofacies sequences in the Permian Lower Beaufort of the southwestern Karoo Basin, South Africa. *J. sedim. Petrol.*, 60, 258–267. *3.6.6*.

Smith S.A. (1990) The sedimentology and accretionary style of an ancient gravel-bed stream: The Buddleigh Salterton Pebble Beds (Lower Triassic), southwest England. *Sedim. Geol.*, 67, 199–219. *3.6.4*.

Smith T.L. & Batiza R. (1989) New field and laboratory evidence for the origin of hyaloclastite flows on seamount summits. *Bull. Volcanol.*, 51, 96–114. *12.11.2*.

Smoot J.P. (1983) Depositional subenvironments in an arid closed basin; the Wilkins Peak Member of the Green River Formation (Eocene), Wyoming, USA. *Sedimentology*, 30, 801–827. *4.14.1*.

Smoot J.P. (1991) Sedimentary facies and depositional environments of Early Mesozoic Newark Supergroup basins, eastern North America. *Palaeogeogr. Palaeoclimatol. Palaeoecol.*, 84, 369–423. *4.16.3, 8.4.1*.

Smoot J.P. & Castens-Seidell B. (1994) Sedimentary features produced by efflorescent salt crusts, Saline Valley and Death Valley, California. In: *Sedimentology and Geochemistry of Modern and Ancient Saline Lakes* (Ed. by R.W. Renaut and W.M. Last), pp. 73–90. *Spec. Publ. Soc. econ. Paleont. Miner.*, 50, Tulsa. *4.6.1, 8.4.1*.

Smoot J.P. & Lowenstein T.K. (1991) Depositional environments of non-marine evaporites. In: *Evaporites, Petroleum and Mineral Resources* (Ed. by J.L. Melvin), pp. 189–347. Elsevier, Amsterdam. *4.7.4, 8.1, 8.4.1*.

Smoot J.P. & Olsen P.E. (1988) Massive mudstones in basin analysis and paleoclimatic interpretation of the Newark Supergroup. In: *Triassic–Jurassic Rifting: Continental Breakup and the Origin of the Atlantic Ocean and Passive Margins* (Ed. by W. Manspeizer), pp. 249–274. Elsevier, Amsterdam. *4.6.1*.

Smosna R. & Koehler B. (1993) Tidal origin of a Mississippian oolite on the West Virginia Dome. In: *Mississippian Oolites and Modern Analogs* (Ed. by B.D. Keith and C.W. Zuppann), pp. 149–162. *Am. Ass. petrol. Geol., Studies in Geology*, 35, Tulsa. *9.4.2*.

Snedden J.W. & Nummedal D. (1991) Origin and geometry of storm-deposited sand beds in modern sediments of the Texas continental shelf. In: *Shelf Sand and Sandstone Bodies: Geometry, Facies and Sequence Stratigraphy* (Ed. by D.J.P. Swift, G.F. Oertel, R.W. Tillman and J.A. Thorne), pp. 283–308. *Spec. Publ. int. Ass. Sediment.*, 14. *7.4.1, 7.4.3, 7.4.4, Fig. 7.24*.

Soh W., Tanaka T. & Taira A. (1995) Geomorphology and sedimentary processes of a modern slope-type fan delta (Fujikawa fan delta), Suruga Trough, Japan. *Sedim. Geol.*, 98, 79–95. *12.12.5*.

Soh W., Pickering K.T., Taira A. & Tokuyama H. (1991) Basin evolution in the arc–arc Izu Collision zone, Mio-Pliocene Miura Group, central Japan. *J. geol. Soc., Lond.*, 148, 317–330. *12.12.5.*

Sohn Y.K. (1995) Geology of Tok Island, Korea: eruptive and depositional processes of a shoaling to emergent island volcano. *Bull. Volcanol.*, 56, 660–674. *12.12.5.*

Sohn Y.K. & Chough S.K. (1989) Depositional processes of the Suwolbong tuff ring, Cheju Island (Korea). *Sedimentology*, 36, 837–855. *12.9.2, Fig. 12.51.*

Sohn Y.K. & Chough S.K. (1992) The Ilchulbong tuff cone, Cheju Island, South Korea: depositional processes and evolution of an emergent Surtseyan-type tuff cone. *Sedimentology*, 39, 523–544. *12.9.3.*

Soil Survey Staff (1975) *Soil Taxonomy*. Agriculture Handbook 436, Soil Conservation Service, Washington, DC. *3.3.7.*

Somoza L., Zazo C., Bardaji T., Goy J.L. & Dabrio C.J. (1987) Recent Quaternary sea-level changes and tectonic movements in SE Spanish Coast. In: *Late Quaternary Sea-level Changes in Spain* (Ed. by C. Zazo), pp. 49–78. *Trab. Neog. Cuat. Museo Nal. Cienc. Nat.*, 10. *6.5.4.*

Sonnenfeld M.D. & Cross T.A. (1993) Volumetric partitioning and facies differentiation within the Permian Upper San Andres Formation of Last Chance Canyon, Guadalupe Mountains, New Mexico. In: *Carbonate Sequence Stratigraphy* (Ed. by R.G. Loucks and J.F. Sarg), pp. 435–474. *Mem. Am. Ass. petrol. Geol.*, 57, Tulsa. *9.5.5.*

Sonnenfeld P. (1984) *Brines and Evaporites*, 613 pp. Academic Press, Orlando. *8.1, Fig. 8.4.*

Sonu C.J. & van Beek J.L. (1971) Systematic beach changes in the Outer Banks, North Carolina. *J. Geol.*, 74, 416–425. *6.2.6.*

Sorby H.C. (1859) On the structures produced by the currents present during the deposition of stratified rocks. *Geologist*, 2, 137–147. *1.1, 12.1, 12.12.4.*

Sorby H.C. (1879) Anniversary address of the President: structure and origin of limestones. *Proc. geol. Soc. Lond.*, 35, 56–95. *1.1.*

Sorby H.C. (1908) On the application of quantitative methods to the study of rocks. *Q. J. geol. Soc.*, 64, 171–233. *12.1, 12.12.4.*

Southard J.B. & Mackintosh M.E. (1981) Experimental test of autosuspension. *Earth Surf. Proc. Landf.*, 6, 103–111. *10.2.3.*

Southard J.B., Lambie J.M., Federico D.C., Pile H.T. & Weidman C.R. (1990) Experiments on bed configurations in fine sand under bidirectional purely oscillatory flow, and the origin of hummocky cross stratification. *J. sedim. Petrol.*, 60, 1–17. *7.7.2.*

Southard J.B., Smith N.D. & Kuhnle R.A. (1984) Chutes and lobes: newly identified elements of braiding in shallow gravelly streams. In: *Sedimentology of Gravels and Conglomerates* (Ed. by E.H. Koster and R.J. Steel), pp. 51–59. *Can. Soc. petrol. Geol.*, 10, Calgary. *3.3.2.*

Southgate P.N., Kennard J.M., Jackson M.J., O'Brien P.E. & Sexton M.J. (1993) Reciprocal lowstand clastic and highstand carbonate sedimentation, subsurface Devonian Reef complex, Canning Basin, Western Australia. In: *Carbonate Sequence Stratigraphy* (Ed. by R.G. Loucks and J.F. Sarg), pp. 157–179. *Mem. Am. Ass. petrol. Geol.*, 57, Tulsa. *9.5.5, Fig. 9.70.*

Sparks R.S.J. (1976) Grain size variations in ignimbrites and implications for the transport of pyroclastic flows. *Sedimentology*, 23, 147–188. *12.4.4.*

Sparks R.S.J. (1978) The dynamics of bubble formation and growth in magmas: a review and analysis. *J. Volcanol. Geotherm. Res.*, 3, 1–37. *12.3.1, 12.4.1.*

Sparks R.S.J. (1986) The dimensions and dynamics of volcanic eruption columns. *Bull. Volcanol.*, 48, 3–15. *12.4.2.*

Sparks R.S.J. & Walker G.P.L. (1977) The significance of vitric-enriched air-fall ashes associated with crystal-enriched ignimbrites. *J. Volcanol. Geotherm. Res.*, 2, 329–341. *12.4.4.*

Sparks R.S.J., Bonnecaze R.T., Huppert H.E., Lister J.R., Hallworth M.A., Mader H. & Phillips J. (1993) Sediment-laden gravity currents with reversing buoyancy. *Earth Planet Sci. Lett.*, 114, 243–257. *10.2.3.*

Sparks R.S.J., Bursik M.I., Ablay G.J., Thomas R.M.E. & Carey S.N. (1992) Sedimentation of tephra by volcanic plumes. Part 2: Controls on thickness and grain-size variations of tephra fall deposits. *Bull. Volcanol.*, 54, 685–695. *12.4.3.*

Sparks R.S.J., Carey S.N. & Sigurdsson H. (1991) Sedimentation from gravity currents generated by turbulent plumes. *Sedimentology*, 38, 839–857. *12.4.2, 12.4.3.*

Sparks R.S.J., Hubbert H.E. & Turner J.S. (1984) The fluid dynamics of evolving magma chambers. *Philos. Trans. R. Soc. Lond.*, A310, 511–534. *12.3.2.*

Sparks R.S.J., Moore J.G. & Rice C.J. (1986) The initial giant umbrella cloud of the May 18, 1980 explosive eruption of Mount St Helens. *J. Volcanol. Geotherm. Res.*, 28, 257–274. *12.4.2, Fig. 12.18.*

Sparks R.S.J., Self S. & Walker G.P.L. (1973) Products of ignimbrite eruptions. *Geology*. *12.1.1, 12.4.4, Fig. 12.26.*

Sparks R.S.J., Wilson L. & Hulme G. (1978) Theoretical modelling of the generation, movement, and emplacement of pyroclastic flows by column collapse. *J. geophys. Res.*, 83, 1727–1739. *12.4.2.*

Spearing D.R. (1975) Shallow marine sands. In: *Depositional Environments as Interpreted from Primary Sedimentary Structures and Stratification Sequences*, pp. 103–132. *Short Course Soc. econ. Paleont. Miner.*, 2, Tulsa. *Fig. 7.54.*

Spearing D.R. (1976) Upper Cretaceous Shannon Sandstone: an offshore shallow marine sand body. *Wyoming Geol. Ass. Guidebook, 28th Field Conf.*, pp. 65–72. *7.1.2, 7.8.4, Fig. 7.54.*

Speed R.C. & Larue D.K. (1982) Barbados: architecture and implications for accretion. *J. geophys. Res.*, 87, 3633–3643. *10.4.1.*

Speed R.C., Torrini R. & Smith P.L. (1989) Tectonic evolution of the Tobago Trough forearc basin. *J. Geophys. Res.*, 94, 2913–2936. *12.12.5.*

Speight J.G. (1965) Flow and channel characteristics of the Angabunga River, Papua. *J. Hydrol.*, 3, 16–36. *3.3.4.*

Spencer R.J., Baedecker M.J., Eugster H.P. *et al.* (1984) Great Salt Lake, and precursors, Utah: the last 30 000 years. *Contrib. Mineral. Petrol.*, 86, 321–334. *4.7.1, 4.9.*

Spera F.J. (1984) Some numerical experiments on the withdrawal of magma from crustal reservoirs. *J. geophys. Res.*, 89, 8219–8842. *12.3.2.*

Spera F.J. & Crisp J.A. (1981) Eruption volume, periodicity, and caldera area: relationships and inferences on development of compositional zonation in silicic magma chambers. *J. Volcanol. Geotherm. Res.*, 11, 169–187. *12.4.6.*

Spotl C. & Wright V.P. (1992) Groundwater dolocretes from the Upper Triassic of the Paris Basin, France: a case study of an arid, continental diagenetic facies. *Sedimentology*, 39, 1119–1136. *3.5.4.*

Stabel H.H. (1986) Calcite precipitation in Lake Constance: chemical equilibrium, sedimentation and nucleation by algae. *Limnol. Oceanogr.*, 31, 1081–1093. *4.7.1.*

Stanley D.J. (1988a) Deep-sea current flow in the late Cretaceous Caribbean. *Mar. Geol.*, 79, 127–133. *10.3.7.*

Stanley D.J. (1988b) *Turbidites Reworked by Bottom Currents: Upper Cretaceous Examples from St Croix, US, Virgin Islands*, 79 pp. *Smithsonian Contrib. Mar. Sci.*, 22. *10.3.7.*

Stanley D.J. (1993) Model for turbidite-to-contourite continuum and multiple process transport in deep marine settings: examples in the rock record. In: *Contourites and Bottom Currents* (Ed. by D.A.V. Stow and J.-C. Faugères), pp. 241–255. *Sedim. Geol.*, 82. *10.2.2, 10.3.7.*

Stanley D.J. & Swift D.J.P. (Eds) (1976) *Marine Sediment Transport and Environmental Management*, 602 pp. Wiley, New York. *7.1.2.*

Stanley D.J., Addy S.K. & Beherens E.W. (1983) The mudline: variability of its position relative to the shelf break. In: *The Shelfbreak: Critical Interface on Continental Margins* (Ed. by D.J. Stanley and G.T. Moore), pp. 279–298. *Spec. Publ. Soc. econ. Paleont. Miner.*, 33, Tulsa. *7.6.*

Stanley D.J., Palmer H.D. & Dill R.F. (1978) Coarse sediment transport by mass flow and turbidity current processes and downslope transformations in Annot Sandstone canyon–fan valley systems. In: *Sedimentation in Submarine Canyons, Fans and Trenches* (Ed. by D.J. Stanley and G. Kelling), pp. 85–115. Dowden, Hutchinson & Ross, Stroudsburg, PA. *10.3.3.*

Staudigel H. & Schmincke H.-U. (1984) The Pliocene seamount series of La Palma (Canary Islands). *J. geophys. Res.*, 89. *12.4.1, 12.12.5, Fig. 12.63.*

Stauffer P.H. (1967) Grain flow deposits and their implications, Santa Ynez Mountains, California. *J. sedim. Petrol.*, 37, 487–508. *10.1.*

Stear W.M. (1978) Sedimentary structures related to fluctuating hydrodynamic conditions in flood plain deposits of the Beaufort Group near Beaufort West, Cape. *Trans. geol. Soc. S. Afr.*, 81, 393–399. *3.5.2.*

Stear W.M. (1985) Comparison of the bedform distribution and dynamics of modern and ancient sandy ephemeral flood deposits in the southwestern Karoo region, South Africa. *Sedim. Geol.*, 45, 209–230. *3.5.2, 3.5.3, 3.6.10.*

Stearn C.W. (1982) The shapes of Paleozoic and modern reef builders: a critical review. *Paleobiology*, 8, 228–241. *9.4.5.*

Steel R.J. (1974) New Red Sandstone floodplain and piedmont sedimentation in the Hebridean Province. *J. sedim. Petrol.*, 44, 336–357. *3.5.4, 3.6.3.*

Steel R.J. (1976) Devonian basins of Western Norway – sedimentary response to tectonism and to varying tectonic context. *Tectonophysics*, 36, 207–224. *3.6.3.*

Steel R.J. & Aasheim S.M. (1978) Alluvial sand deposition in a rapidly subsiding basin (Devonian, Norway). In: *Fluvial Sedimentology* (Ed. by A.D. Miall), pp. 385–412. *Mem. Can. Soc. petrol. Geol.*, 5, Calgary. *3.5.3, Fig. 3.42.*

Steel R.J. & Thompson D.B. (1983) Structures and textures in Triassic braided stream conglomerates ('Bunter' Pebble Beds) in the Sherwood Sandstone Group, North Staffordshire, England.

Sedimentology, 30, 341–367. *3.5.1, 3.6.4, Fig. 3.36.*

Steele R.P. (1983) Longitudinal draa in the Permian Yellow Sands of north-east England. In: *Eolian Sediments and Processes* (Ed. by M.E. Brookfield and T.S. Ahlbrandt), pp. 543–550. *Developments in Sedimentology*, 38. Elsevier, Amsterdam. *5.5.3, 5.5.4.*

Steinhorn I. (1985) The disappearance of the long term meromictic stratification of the Dead Sea. *Limnol. Oceanogr.*, 30, 451–472. *8.6.2.*

Steinmann G. (1905) Geologische Beobachtungen in den Alpen. II. Die Schardtsche Überfaltungs-theorie und die geologische Bedeutung der Tiefseeabasätze und der ophiolithischen Massengesteine. *Ber. naturf. Ges. Freiburg*, 16, 18–67. *10.1.*

Steinmann G. (1925) Gibt es fossile Tiefseeablagerungen von erdgeschichtlichen Bedeutung? *Geol. Rundsch.*, 16, 435–468. *10.1.*

Stemmerik L. (1991) Reservoir evaluation of the Upper Permian buildups in the Jameson Land Basin, east Greenland. *Grønl. Geol. Unders. Rep.*, 149, 23 pp. *9.4.5.*

Stephenson M.A. (1991) The North Brae Field, Block 16/7a, UK North Sea. In: *United Kingdom Oil and Gas Fields, 25 Years Commemorative Volume. Mem. geol. Soc. Lond.*, 14, 43–48. *10.3.4.*

Stern R.J., Bloomer S.H., Lin P.-N. & Smoot N.C. (1989) Submarine arc volcanism in the southern Mariana Arc as an ophiolite analogue. *Tectonophysics*, 168, 151–170. *12.11.1.*

Stevens C.H. (1977) Was development of brackish oceans a factor in Permian extinctions? *Bull. geol. Soc. Am.*, 88, 133–138. *8.1.*

Stewart D.J. (1983) Possible suspended-load channel deposits from the Wealden Group (Lower Cretaceous) of Southern England. In: *Modern and Ancient Fluvial Systems* (Ed. by J.D. Collinson and J. Lewin), pp. 369–384. *Spec. Publ. int. Ass. Sediment.*, 6. *3.6.5.*

Stix J. (1991) Subaqueous, intermediate to silicic-composition explosive volcanism: a review. *Earth-Science Rev.*, 31, 21–53. *12.4.4, 12.14.1.*

Stoakes F.A. (1980) Nature and control on shale basin fill and its effect on reef growth and termination: Upper Devonian Duvernay and Ireton Formations of Alberta, Canada. *Bull. Can. petrol. Geol.*, 28, 234–410. *9.5.5.*

Stoffers P. & Singer A. (1979) Clay minerals in Lake Mobutu Sese Seko (Lake Albert) – their diagenetic changes as an indicator of the paleoclimate. *Geol. Rundsch.*, 68, 1009–1024. *4.7.1.*

Stoker M.S., Leslie A.B., Scott W.D. *et al.* (1994) A record of late Cenozoic stratigraphy, sedimentation and climate change from the Hebrides Slope, NE Atlantic Ocean. *J. geol. Soc. Lond.*, 151, 235–249. *10.3.5.*

Storey B.C. & Macdonald D.I.M. (1984) Processes of formation and filling of a Mesozoic back-arc basin on the island of South Georgia. In: *Marginal Basin Geology: Volcanic and Associated Sedimentary and Tectonic Processes in Modern and Ancient Marginal Basins* (Ed. by B.P. Kokelaar and M.F. Howells), pp. 207–218. *Spec. Publ. geol. Soc. Lond.*, 16. *12.12.5.*

Stow D.A.V. (1979) Distinguishing between fine-grained turbidites and contourites on the Nova Scotian deep water margin. *Sedimentology*, 26, 371–387. *10.2.3.*

Stow D.A.V. (1981) Laurentian Fan: morphology, sediments, processes and growth pattern. *Bull. Am. Ass. petrol. Geol.*, 65, 375–393. *10.3.5, Fig. 10.65.*

Stow D.A.V. (1982) Bottom currents and contourites in the North Atlantic. *Bull. Inst. Geol. Bassin d'Aquitaine*, 31, 151–166. *10.1, 10.2.2, 10.3.7.*

Stow D.A.V. (1985) Deep-sea clastics: where are we and where are we going? In: *Sedimentology: Recent Developments and Applied Aspects* (Ed. by P.J. Brenchley and B.P.J. Williams), pp. 67–93. *Spec. Publ. geol. Soc. Lond.*, 18. *10.1, 10.2, 10.2.1, 10.2.4, 10.3.2, 10.3.4, 10.3.6, Figs 10.9, 10.29, 10.36.*

Stow D.A.V. (1986) Deep clastic seas. In: *Sedimentary Environments and Facies* (Ed. by H.G. Reading), pp. 399–444. Blackwell Scientific Publications, Oxford. *10.1, 10.2.1, 10.2.4, Fig. 10.8.*

Stow D.A.V. (1987) South Atlantic organic-rich sediments: facies, processes and environments of deposition. In: *Marine Petroleum Source Rocks* (Ed. by J. Brooks and A.J. Fleet), pp. 287–299. *Spec. Publ. geol. Soc. Lond.*, 26. *10.2.1.*

Stow D.A.V. (1994) Deep sea processes of sediment transport and deposition. In: *Sediment Transport and Depositional Processes* (Ed. by K. Pye), pp. 257–291. Blackwell Scientific Publications, Oxford. *10.3.7, Figs 10.2, 10.13, 10.16, 10.77.*

Stow D.A.V. & Atkin B. (1987) Sediment facies and geochemistry of Upper Jurassic mudrocks in the central North Sea area. In: *Petroleum Geology of NW Europe* (Ed. by J. Brooks and K.W. Glennie), pp. 797–808. Graham & Trotman, London. *10.2.1.*

Stow D.A.V. & Bowen A.J. (1980) A physical model for the transport and sorting of fine-grained sediments by turbidity currents. *Sedimentology*, 27, 31–46. *10.2.3.*

Stow D.A.V. & Holbrook J.A. (1984) North Atlantic contourites: an overview. In: *Fine-grained Sediments: Deep-water Processes and Facies* (Ed. by D.A.V. Stow and D.J.W. Piper), pp. 245–256. *Spec. Publ. geol. Soc. Lond.*, 15. *6.5.4, 10.2.2, 10.3.7.*

Stow D.A.V. & Lovell J.P.B. (1979) Contourites; their recognition in modern and ancient sediments. *Earth Sci. Rev.*, 14, 251–291. *10.1, 10.2.2.*

Stow D.A.V. & Shanmugam G. (1980) Sequence of structures in fine-grained turbidites; comparison of recent deep-sea and ancient flysch sediments. *Sedim. Geol.*, 25, 23–42. *10.1, 10.2.3.*

Stow D.A.V. & Wetzel A. (1990) Hemiturbidite: a new type of deep-water sediment. In: *Proc. ODP Sci. Results*, 116 (Ed. by J.R. Cochran, D.A.V. Stow et al.), 25–34. *10.2.3, Fig. 10.26.*

Stow D.A.V., Bishop C.D. & Mills S.J. (1982) Sedimentology of the Brae oilfield, North Sea: Fan models and controls. *J. petrol. Geol.*, 5, 129–148. *3.4, 6.5.4, 10.2.4.*

Stow D.A.V., Braakenburg N.E. & Johansson M. (1994) Atlas of deepwater, massive sands. Unpublished Report, Southampton University. *10.2.3, 10.2.4, Fig. 10.22.*

Stow D.A.V., Faugères J.-C. & Gonthier E.G. (1986) Facies distribution and textural variation in Faro Drift contourites: velocity fluctuation and drift growth. *Mar. Geol.*, 72, 71–100. *10.1, 10.2.2.*

Stow D.A.V., Wezel F.C., Savelli D., Rainey S.C.R. & Angell G. (1984) Depositional model for calcilutites: Scaglia Rossa Limestones, Umbro–Marchean Apennines. In: *Fine-grained Sediments: Deep-water Processes and Facies* (Ed. by D.A.V. Stow and D.J.W. Piper), pp. 223–241. *Spec. Publ. geol. Soc. Lond.*, 15. *10.2.3, 10.4.2.*

Straccia F.G., Wilkinson B.H. & Smith G.R. (1990) Miocene lacustrine algal reefs – southwestern Snake River Plain, Idaho.

Sedim. Geol., 67, 7–23. *4.12.3, Fig. 4.26.*

Strasser A. (1991) Lagoonal–peritidal sequences in carbonate environments: autocyclic and allocyclic processes. In: *Cyclic and Event Stratification* (Ed. by G. Einsele, W. Ricken and A. Seilacher), pp. 709–721. Springer-Verlag, Berlin. *9.4.1.*

Street-Perrott F.A. & Harrison S.P. (1984) Temporal variations in lake levels since 30000 yr BP – an index of the global hydrological cycle. *Geophys. Monogr.*, 29, 118–129. *4.1, 4.9.*

Street-Perrott F.A. & Harrison S.P. (1985) Lake levels and climate reconstruction. In: *Paleoclimate Analysis and Modeling* (Ed. by A.D. Hecht), pp. 291–340. Wiley, New York. *4.9.*

Stride A.H. (1963) Current swept floors near the southern half of Great Britain. *Q. J. geol. Soc. Lond.*, 119, 175–199. *7.1.2, 7.3.3, Fig. 7.9.*

Stride A.H. (1970) Shape and size trends for sand waves in a depositional zone of the North Sea. *Geol. Mag.*, 107, 469–477. *7.3.3.*

Stride A.H. (1974) Indications of long term tidal control of net sand loss or gain by European coasts. *Estuar. Coast. Mar. Sci.*, 2, 27–36. *7.3.4.*

Stride A.H. (Ed.) (1982) *Offshore Tidal Sands: Process and Deposits*, 222 pp. Chapman & Hall, London. *7.1.2, 7.2.2, 7.3, 7.7.1, Fig. 7.9.*

Stride A.H. (1988) Preservation of marine sand wave structures. In: *Tide-influenced Sedimentary Environments and Facies* (Ed. by P.L. de Boer, A. van Gelder and S.D. Nio), pp. 13–22. Reidel, Dordrecht. *7.3.3.*

Stride A.H., Belderson R.H., Kenyon N.H. & Johnson M.A. (1982) Offshore tidal deposits: sand sheet and sand bank facies. In: *Offshore Tidal Sands, Process and Deposits* (Ed. by A.H. Stride), pp. 95–125. Chapman and Hall, Edinburgh. *6.7.6, Figs 7.14, 7.16.*

Sturm M. (1979) Origin and composition of clastic varves. In: *Moraines and Varves* (Ed. by C. Schlüchter), pp. 281–285. Balkema, Rotterdam. *4.8, Fig. 4.14.*

Sturm M. & Matter A. (1978) Turbidites and varves in Lake Brienz (Switzerland): deposition of clastic detritus by density currents. In: *Modern and Ancient Lake Sediments* (Ed. by A. Matter and M.E. Tucker), pp. 147–168. *Spec. Publ. int. Ass. Sediment.*, 2. *4.6.3, 4.8, Fig. 4.5, 11.4.5.*

Styan W.B. & Bustin R.M. (1984) Sedimentology of Fraser River delta peat deposits: a modern analogue for some deltaic coals. In: *Sedimentology of Coal and Coal-bearing Sequences* (Ed. by R.A. Rahmani and R.M. Flores), pp. 241–271. *Spec. Publ. int. Ass. Sediment.*, 7. *6.6.1.*

Sugden D.E. & John B.S. (1988) *Glaciers and Landscape: A Geomorphological Approach*, 376 pp. Edward Arnold, London. *11.3.2, 11.3.3, 11.3.4, 11.4.3, 11.5.3. Figs 11.7, 11.17.*

Summerfield M.A. (1991) *Global Geomorphology*, 537 pp. Longman, Harlow. *2.1.2, Fig. 2.2, Table 2.1, 6.1.*

Sun S.Q. & Wright V.P. (1989) Peloidal fabrics in Upper Jurassic reefal limestones, Weald Basin, southern England. *Sedim. Geol.*, 65, 165–181. *9.4.5.*

Sundborg Å. (1956) The River Klarälven: a study of fluvial processes. *Geogr. Ann.*, 38, 127–316. *3.1, 3.3.4, Fig. 3.20.*

Surdam R.C. & Boles J.R. (1979) Diagenesis of volcanic sandstones. In: *Aspects of Diagenesis* (Ed. by P.A. Scholle and P.R. Schluger), pp. 227–242. *Spec. Publ. Soc. econ. Paleont. Miner.*, **26**, Tulsa. *12.14.1.*

Surdam R.C. & Wolfbauer C.A. (1975) Green River Formation, Wyoming: A playa-lake complex. *Bull. geol. Soc. Am.*, **86**, 335–345. *Fig. 4.29, Fig. 4.30.*

Surlyk F. (1978) Submarine fan sedimentation along fault scarps on tilted fault blocks (Jurassic–Cretaceous boundary), East Greenland. *Bull. Grønlands geol. Unders.*, **128**, 1–108. *10.3.5.*

Surlyk F. (1984) Fan-delta to submarine fan conglomerates of the Volgian–Valanginian Wollaston Forland Group, East Greenland. In: *Sedimentology of Gravels and Conglomerates* (Ed. by E.H. Koster and R.J. Steel), pp. 359–382. *Mem. Can. Soc. petrol. Geol.*, **10**, Calgary. *10.3.5, 12.12.5.*

Surlyk F. (1987) Slope and deep shelf gully sandstones, Upper Jurassic, East Greenland. *Bull. Am. Ass. petrol. Geol.*, **71**, 464–475. *10.3.5.*

Surlyk F. (1989) Mid-Mesozoic syn-rift turbidite systems, controls and predictions. In: *Correlation in Hydrocarbon Exploration* (Ed. by J.D. Collinson), pp. 231–241. Norwegian Petrol. Soc., Graham & Trotman, London. *6.5.5, 10.3.5, Figs 10.68, 10.71, 10.72.*

Surlyk F. & Christensen W.K. (1974) Epifaunal zonation on an Upper Cretaceous rocky coast. *Geology*, **2**, 529–534. *6.1, 6.4.*

Surlyk F. & Noe-Nygaard (1991) Sand bank and dune facies architecture of a wide intracratonic seaway: late Jurassic–early Cretaceous Raukelv Formation, Jameson Land, East Greenland. In: *The Three-dimensional Facies Architecture of Terrigenous Clastic Sediments, and its Implications for Hydrocarbon Discovery and Recovery* (Ed. by A.D. Miall and N. Tyler), pp. 261–276. *Concepts in Sedimentology and Paleontology, 3.* SEPM (Soc. Sedim. Geol.), Tulsa. *7.7.1, 7.8.2, Figs 7.51, 7.52.*

Suter J.R. & Berryhill H.L. Jr. (1985) Late Quaternary shelf-margin deltas, northwest Gulf of Mexico. *Bull. Am. Ass. petrol. Geol.*, **69**, 77–91. *6.6.5.*

Svela K.E. (1988) Sedimentology of Lower and Middle Coal Measures (Westphalian A and B) in the Broadhaven–Little Haven Coalfield, Pembrokeshire, SW Wales. Unpublished Cand. Scient. Thesis, University of Bergen. *Fig. 6.50.*

Swagor N.S., Oliver T.A. & Johnson B.A. (1976) Carrot Creek field, central Alberta. In: *The Sedimentology of Selected Clastic Oil and Gas Reservoirs in Alberta* (Ed. by M.M. Lerand), pp. 78–95. Can. Soc. Petrol. Geol., Calgary. *7.8.4.*

Swanson D.A. (1990) A decade of dome growth at Mount St. Helens 1980–90. *Geosci. Can.*, **17**, 154–157. *12.10.*

Swanson D.A., Wright T.L. & Helz R.T. (1975) Linear vent systems and estimated rates of magma production and eruption for the Yokima basalt of the Columbia Plateau. *Am. J. Sci.*, **275**, 877–905. *12.9.4.*

Sweet M.L. (1992) Lee-face airflow, surface processes, and stratification types: their significance for refining the use of eolian cross-strata as paleocurrent indicators. *Bull. geol. Soc. Am.*, **104**, 1528–1538. *5.3.2, 5.4.4, Fig. 5.2.*

Sweet M.L. & Kocurek G. (1990) An empirical model of aeolian dune lee-face airflow. *Sedimentology*, **37**, 1023–1038. *5.3.2, 5.4.4, Fig. 5.2.*

Sweet M.L., Nielson J., Havholm K. & Farrelley J. (1988) Algodones dune field of southeastern California: case study of a migrating modern dune field. *Sedimentology*, **35**, 939–952. *5.4.6.*

Swift, D.J.P. (1969a) Inner shelf sedimentation: process and products. In: *The New Concepts of Continental Margin Sedimentation: Application to the Geological Record* (Ed. by D.J. Stanley), pp. DS-5-1–DS-5-26. Am. geol. Inst., Washington, DC. *7.2.2.*

Swift D.J.P. (1969b) Outer shelf sedimentation: process and products. In: *The New Concepts of Continental Margin Sedimentation: Application to the Geological Record.* (Ed. by D.J. Stanley), pp. DS-4-1–DS-4-46. Am. geol. Inst., Washington, DC. *7.2.2.*

Swift D.J.P. (1970) Quaternary shelves and the return to grade. *Mar. Geol.*, **8**, 5–30. *7.2.2.*

Swift D.J.P. (1974) Continental shelf sedimentation. In: *The Geology of Continental Margins* (Ed. by C.A. Burk and C.L. Drake), pp. 117–135. Springer-Verlag, Berlin. *7.1.2.*

Swift D.J.P. (1975a) Barrier island genesis: evidence from the Middle Atlantic Shelf of North America. *Sedim. Geol.*, **14**, 1–43. *Fig. 6.86, 7.4.2.*

Swift D.J.P. (1975b) Tidal sand ridges and shoal-retreat massifs. *Mar. Geol.*, **18**, 105–134. *7.3.4, 7.4.2.*

Swift D.J.P. (1976) Coastal sedimentation. In: *Marine Sediment Transport and Environmental Management* (Ed. by D.J. Stanley and D.J.P. Swift), pp. 255–310. Wiley, New York. *6.7.*

Swift D.J.P. & Field M.E. (1981) Evolution of a classic sand ridge field; Maryland sector, North American inner shelf. *Sedimentology*, **28**, 461–482. *7.4.2.*

Swift D.J.P. & Nummedal D. (1987) Hummocky cross-stratification, tropical hurricanes and intense winter storms: a discussion. *Sedimentology*, **34**, 388–344. *7.7.2.*

Swift D.J.P. & Thorne J.A. (1991) Sedimentation on continental margins, I: a general model for shelf sedimentation. In: *Shelf Sand and Sandstone Bodies: Geometry, Facies and Sequence Stratigraphy* (Ed. by D.J.P. Swift, G.F. Oertel, R.W. Tillman and J.A. Thorne), pp. 3–31. *Spec. Publ. int. Ass. Sediment.*, **14**. *7.1.2, 7.2.2, 7.3.3, 7.4.1, 7.7.3, Figs 6.4, 6.7, 7.3, 7.5, 7.6.*

Swift D.J.P., Duane D.B. & McKinney T.F. (1973) Ridge and swale topography of the Middle Atlantic Bight, North America: secular response to the Holocene hydraulic regime. *Mar. Geol.*, **15**, 227–247. *Fig. 7.19.*

Swift D.J.P., Duane D.B. & Pilkey O.H. (Eds) (1972) *Shelf Sediment Transport: Process and Pattern*, 656 pp. Dowden, Hutchinson & Ross, Stroudsberg. *7.1.2.*

Swift D.J.P., Stanley D.J. & Curray J.R. (1971) Relict sediment on continental shelves: a reconsideration. *J. Geol.*, **79**, 322–346. *7.2.2.*

Swift D.J.P., Figueiredo A.G., Freeland G.L. & Oertel G.F. (1983) Hummocky cross-stratification and megaripples: a geological double standard. *J. sedim. Petrol.*, **53**, 1295–1317. *7.4.1, 7.7.2.*

Swift D.J.P., Han G. & Vincent C.E. (1986) Fluid processes and sea-floor responses on a modern storm-dominated shelf: middle Atlantic shelf of North America. Part I. The Storm Current Regime. In: *Shelf Sands and Sandstones* (Ed. R.J. Knight & J.R. McLean), pp. 99–119. Mem. Can. Soc. petrol. Geol., **11**, Calgary. *7.2.1, 7.2.2.*

Swift D.J.P., Hudelson P.M., Brenner R.L. & Thompson P. (1987) Shelf construction in a foreland basin: storm beds, shelf sandbodies and shelf–slope depositional sequences in the Upper Cretaceous Mesaverde Group, Brook Cliffs, Utah. *Sedimentology*, **34**, 423–457. *7.8.4, Fig. 7.5.*

Swift D.J.P., Phillips S. & Thorne J.A. (1991a) Sedimentation on continental margins, IV: lithofacies and depositional systems. In: *Shelf Sand and Sandstone Bodies: Geometry, Facies and Sequence Stratigraphy* (Ed. by D.J.P. Swift, G.F. Oertel, R.W. Tillman and J.A. Thorne), pp. 89–152. *Spec. Publ. int. Ass. Sediment.*, **14**. *7.3.2, Fig. 6.4, Table 7.1.*

Swift D.J.P., Phillips S. & Thorne J.A. (1991b) Sedimentation on continental margins, V: parasequences. In: *Shelf Sand and Sandstone Bodies: Geometry, Facies and Sequence Stratigraphy* (Ed. by D.J.P. Swift, G.F. Oertel, R.W. Tillman and J.A. Thorne), pp. 153–187. *Spec. Publ. int. Ass. Sediment.*, **14**. *6.1, 7.2.2, Fig. 7.18.*

Swift D.J.P., Thorne J.A. & Oertel G.F. (1986) Fluid process and sea floor responses on a storm dominated shelf: middle Atlantic shelf of North America. Part II. Response of the shelf floor. In: *Shelf Sands and Sandstones* (Ed. by R.J. Knight and J.R. McLean), pp. 191–211. *Mem. Can. Soc. Petrol. Geol.*, **11**, Calgary. *7.2.2.*

Swirydczuk K., Wilkinson B.H. & Smith G.R. (1979) The Pliocene Glenns Ferry Oolite: lake-margin carbonate deposition in the southwestern Snake River plain. *J. sedim. Petrol.*, **49**, 995–1004. *4.12.2, Fig. 4.25.*

Swirydczuk K., Wilkinson B.H. & Smith G.R. (1980) The Pliocene Glenns Ferry Oolite – II: sedimentology of oolitic lacustrine terrace deposits. *J. sedim. Petrol.*, **50**, 1237–1248. *4.12.2.*

Sydow J. & Roberts H.H. (1994) Stratigraphic framework of a Late Pleistocene shelf-edge delta, northeast Gulf of Mexico. *Bull. Am. Ass. petrol. Geol.*, **78**, 1276–1312. *6.6.*

Sykes G.G. (1937) *The Colorado Delta*, 193 pp. Carnegie Inst. Washington Publ. **460**. *Spec. Publ. Am. geogr. Soc.*, **19**. *6.1.*

Sykes T. & Kidd R.B. (1994) Volcanogenic sediment distributions in the Indian Ocean through the Cretaceous and Cenozoic, and their paleoenvironmental implications. *Mar. Geol.*, **116**, 267–291. *12.11.3.*

Syvitski J.P.M. & Farrow G.E. (1989) Fjord sedimentation as an analogue for small hydrocarbon-bearing fan deltas. In: *Deltas: Sites and Traps for Fossil Fuels* (Ed. by M.K.G. Whateley and K.T. Pickering), pp. 21–43. *Spec. Publ. geol. Soc. Lond.*, **41**. *6.5.3, 6.6.4, 10.3.3.*

Syvitski J.P.M., LeBlanc K.W.G. & Cranston R.E. (1990) The flux and preservation of organic carbon in Baffin Island fjords. In: *Glacimarine Environments: Processes and Sediments* (Ed. by J.A. Dowdeswell & J.D. Scourse), pp. 177–199. *Spec. Publ. geol. Soc.*, **53**. *11.4.6.*

Taira A. (1985) Sedimentary evolution of the Shikoku subduction zone: the Shimanto Belt and Nankai Trough. In: *Formation of Active Margins* (Ed. by N. Nasu *et al.*), pp. 835–851. Terrapub., Tokyo. *Fig. 12.85.*

Taira A. & Ogawa Y. (1988) *The Shimanto Belt, Southwest Japan – Studies on the Evolution of an Accretionary Prism. Modern Geol.*, **12**, 542 pp. *12.12.5.*

Taira A., Okada H., Whitaker J.H. McD. & Smith A.J. (1982) The Shimanto Belt of Japan: Cretaceous–lower Miocene active-margin sedimentation. In: *Trench–Forearc Geology: Sedimentation and Tectonics on Modern and Ancient Active Plate Margins* (Ed. by J.K. Legget), pp. 5–26. *Spec. Publ. geol. Soc. Lond.*, **10**. *10.4.1, 12.12.5.*

Takada A. (1994a) Accumulation of magma in space and time by crack interaction. In: *Magmatic Systems* (Ed. by M. P. Ryan), pp. 241–257. Academic Press, San Diego. *12.3.2.*

Takada A. (1994b) The influence of regional stress and magmatic input on styles of monogenetic and polygenetic volcanism. *J. geophys. Res.*, **99**, 13563–13573. *12.8, Fig. 12.47.*

Talbot M.R. (1980) Environmental responses to climatic change in the west African Sahel over the past 20,000 years. In: *The Sahara and the Nile* (Ed. by M.A.J. Williams and H. Faure), pp. 37–62. A.A. Balkema, Rotterdam. *5.4.7.*

Talbot M.R. (1984) Late Pleistocene rainfall and dune building in the Sahel. In: *Palaeoecology of Africa and Surrounding Islands* (Ed. by J.A. Coetzee), pp. 203–214. A.A. Balkema, Rotterdam. *5.4.7.*

Talbot M.R. (1985) Major bounding surfaces in aeolian sandstones – a climatic model. *Sedimentology*, **32**, 257–265. *5.3.7, 5.4.7.*

Talbot M.R. (1988) The origins of lacustrine oil source rocks: evidence from the lakes of tropical Africa. In: *Lacustrine Petroleum Source Rocks* (Ed. by A.J. Fleet, K. Kelts and M.R. Talbot), pp. 29–43. *Spec. Publ. geol. Soc. Lond.*, **40**, Bristol. *4.1, 4.7.5, Fig. 4.16.*

Talbot M.R. (1994) Paleohydrology of the late Miocene Ridge basin lake, California. *Bull. geol. Soc. Am.*, **106**, 1121–1129. *4.11.1.*

Talbot M.R. & Kelts K. (1990) Paleolimnological signatures from carbon and oxygen isotopic ratios in carbonates from organic carbon-rich lacustrine sediments. In: *Lacustrine Basin Exploration – Case Studies and Modern Analogs* (Ed. by. B.J. Katz), pp. 99–112. *Mem. Am. Ass. petrol. Geol.*, **50**, Tulsa. *4.14.1.*

Talbot M.R. & Livingstone D.A. (1989) Hydrogen index and carbon isotopes of lacustrine organic matter as lake level indicators. *Palaeogeogr. Palaeoclimatol. Palaeoecol.*, **70**, 121–137. *4.9.*

Talbot M.R., Holm K. & Williams M.A.J. (1994) Sedimentation in low-gradient desert margin systems: A comparison of the Late Triassic of northwest Somerset (England) and the late Quaternary of east-central Australia. In: *Paleoclimate and Basin Evolution of Playa Systems* (Ed. by M.R. Rosen), pp. 97–117. *Spec. Pap. geol. Soc. Am.*, **289**, Boulder, CO. *4.11.3, Fig. 4.21.*

Tamaki K. & Honza E. (1991) Global tectonics and formation of marginal basins: role of the western Pacific. *Episodes*, **14**, 224–230. *12.12.5.*

Tamaki K., Suyehiro K. *et al.* (1992) Proc. Ocean Drilling Program, Sci. Results, 127/128 Pt. 2. College Station, Texas. *10.4.1.*

Tamura Y., Koyama M. & Fiske R.S. (1991) Paleomagnetic evidence for hot pyroclastic debris in the shallow submarine Shirahama Group (Upper Miocene–Pliocene), Japan. *J. geophys. Res.*, **96**, 21779–21787. *12.4.4.*

Tandon S.K. & Friend P.F. (1989) Near-surface shrinkage and carbonate replacement processes, Arran Cornstone Formation, Scotland. *Sedimentology*, **36**, 1113–1126. *3.5.4.*

TANGANYDRO (1992) Sublacustrine hydrothermal seeps in northern Lake Tanganyika, East African rift: 1991 TANGANYDRO Expedition. *Bull. Cent. Recherch. Explor. Product. Elf-Aquitaine*, **16**, 55–81. *4.5.*

Tankard A.J. & Hobday D.K. (1977) Tide-dominated back-barrier sedimentation, early Ordivician Cape Basin, Cape Peninsula, South Africa. *Sedim. Geol.*, **18**, 135–159. *7.8.1.*

Tanner L.H. & Hubert J.F. (1991) Basalt breccias and conglomerates in the Lower Jurassic McCoy Brook Formation, Fundy Basin, Nova Scotia: Differentiation of talus and debris-flow deposits. *J. sedim. Petrol.*, **61**, 15–27. *3.5.1.*

Tassé N., Lajoie J. & Dimroth E. (1978) The anatomy and interpretation of an Archean volcaniclastic sequence, Noranda region. *Can. J. Earth Sci.*, 15, 874–888. *12.4.4.*

Taylor B. & Karner G.D. (1983) On the evolution of marginal basins. *Rev. Geophys. Space Phys.*, 21, 1727–1741. *12.12.1, 12.12.5.*

Taylor B., Brown G., Fryer P. *et al.* (1990) ALVIN-Sea beam studies of the Sumisu Rift, Izu–Bonin arc. *Earth Planet. Sci. Lett.*, 100, 127–147. *12.12.5.*

Taylor B., Klaus A., Brown G.R. & Moore G.F. (1991) Structural development of Sumisu Rift, Izu–Bonin arc. *J. geophys. Res.*, 96, 16113–16129. *12.12.5.*

Taylor G. & Woodyer K.D. (1978) Bank deposition in suspended-load streams. In: *Fluvial Sedimentology* (Ed. by A.D. Miall), pp. 257–275. *Mem. Can. Soc. petrol. Geol.*, 5, Calgary. *3.3.5, 3.6.5.*

Taylor J.C.M. (1980) Origin of the Werraanhydrit in the U.K. Southern North Sea – a reappraisal. In: *The Zechstein Basin with Emphasis on Carbonate Sequences.* (Ed. by H. Fuchtbauer and T. Peryt), pp. 91–113. *Contrib. Sedimentology*, 9. *8.9, Fig. 8.51.*

Taylor J.C.M. (1990) Late Permain–Zechstein. In: *Introduction to the Petroleum Geology of the North Sea* (Ed. by K.W. Glennie), pp. 153–190. Blackwell Scientific Publications, Oxford. *8.9, Fig. 8.50.*

Taylor J.C.M. & Colter V.S. (1975) Zechstein of the English sector of the southern North Sea Basin. In: *Petroleum and the Continental Shelf of Northwest Europe*, 1, *Geology*, pp. 249–263. Applied Science, Barking. *8.9.*

Ten Brink U.S. & Brocher T.M. (1987) Multichannel seismic evidence for a subcrustal intrusive complex under Oahu and a model for Hawaiian volcanism. *J. geophys. Res.*, 92, 13687–13707. *12.11.2.*

Terwindt J.H.J. (1971a) Lithofacies of inshore estuarine and tidal inlet deposits. *Geol. Mijnbouw*, 50, 515–526. *6.6.1.*

Terwindt J.H.J. (1971b) Sand waves in the Southern Bight of the North Sea. *Mar. Geol.*, 10, 51–67. *7.3.3.*

Terwindt J.H.J. & Brouwer M.J.N. (1986) The behaviour of intertidal sandwaves during neap–spring tide cycles and the relevance for palaeoflow reconstructions. *Sedimentology*, 33, 1–31. *7.7.1.*

Tessier B., Monfort Y., Gigot P. & Larsonneur C. (1988) Enregistrement vertical des cyclicités tidales; adaptation d'un outil de traitement mathématique, exemples en Baie du Mont Saint-Michel et dans la molasse marine Miocène de Digne. Journée L. Dangeard, Dynamique des Milieux Tidaux, Soc. Geol. de France, 69–70. *6.2.7, 6.7.3.*

Tesson M., Allen G.P. & Ravenne C. (1993) Late Pleistocene shelf-perched lowstand wedges on the Rhône continental shelf. In: *Sequence Stratigraphy and Facies Associations* (Ed. by H.W. Posamentier, C.P. Summerhayes, B.U. Haq and G.P. Allen), pp. 183–196. *Spec. Publ. int. Ass. Sediment.*, 18. *6.6, 6.6.5.*

Tewes D.W. & Loope D.B. (1992) Palaeo-yardangs: wind-scoured desert landforms at the Permo-Triassic unconformity. *Sedimentology*, 39, 251–261. *5.5.3.*

Teyssen T.A.L. (1984) Sedimentology of the Minette oolitic ironstones of Luxembourg and Lorraine: a Jurassic subtidal sandwave complex. *Sedimentology*, 31, 195–211. *7.7.1.*

Thickpenny A. & Leggett J.K. (1987) Stratigraphic distribution and palaeoceanographic significance of European early Palaeozoic organic-rich sediments. In: *Marine Petroleum Source Rocks* (Ed. by J. Brooks and A.J. Fleet), pp. 231–247. *Spec. Publ. geol. Soc. Lond.*, 26. *10.2.1.*

Thomas D.S.G. & Allison R.J. (1993) *Landscape Sensitivity*, 347 pp. John Wiley, New York. *5.2.2.*

Thomas D.S.G. & Tsoar H. (1990) The geomorphological role of vegetation in desert dune systems. In: *Vegetation and Erosion* (Ed. by J.B. Thornes), pp. 471–489. Wiley, Chichester. *5.2.2.*

Thomas G.S.P. & Connell R.J. (1985) Iceberg drop, dump, and grounding structures from Pleistocene glacio-lacustrine sediments, Scotland. *J. sedim. Petrol.*, 55, 243–249. *11.4.5, Fig. 11.11.*

Thomas R.G., Smith D.G., Wood J.M., Visser J., Calverley-Range E.A. & Koster E.H. (1987) Inclined heterolithic stratification — terminology, description, interpretation and significance. *Sedim. Geol.*, 53, 123–179. *6.7.3.*

Thomas R.H. (1979) The dynamics of marine ice sheets. *J. Glaciol.*, 24, 167–177. *11.5.3, Fig. 11.19.*

Thompson D.B. (1970) Sedimentation of the Triassic (Scythian) Red Pebbly Sandstone in the Cheshire Basin and its margins. *Geol. J.*, 7, 183–216. *3.5.2.*

Thompson J.B. & Ferris F.G.I. (1990) Geomicrobiology and sedimentology of the mixolimnion and chemocline in Fayetteville Green Lake, NY. *Palaios*, 5, 52–75. *4.7.1, Fig. 4.9.*

Thompson R.W. (1968) Tidal flat sedimentation on the Colorado River delta, northwestern Gulf of California. *Mem. geol. Soc. Am.*, 107, 1–133. *6.7.2, 6.7.3, 8.5.2.*

Thornburg T.M. & Kulm L.D. (1987) Sedimentation in the Chile Trench: depositional morphologies, lithofacies, and stratigraphy. *Bull. Geol. Soc. Am.*, 98, 33–52. *12.12.5.*

Thorne J.A. & Swift D.J.P. (1991a) Sedimentation on continental margins, II: application of the regime concept. In: *Shelf Sand and Sandstone Bodies: Geometry, Facies and Sequence Stratigraphy* (Ed. by D.J.P. Swift, G.F. Oertel, R.W. Tillman and J.A. Thorne), pp. 33–58. *Spec. Publ. int. Ass. Sediment.*, 14. *7.1.2.*

Thorne J.A. & Swift D.J.P. (1991b) Sedimentation on continental margins, VI: a regime model for depositional sequences, their component system tracts, and bounding surfaces. In: *Shelf Sand and Sandstone Bodies: Geometry, Facies and Sequence Stratigraphy* (Ed. by D.J.P. Swift, G.F. Oertel, R.W. Tillman and J.A. Thorne), pp. 189–255. *Spec. Publ. int. Ass. Sediment.*, 14. *7.1.2.*

Tiercelin J.-J. (1991) Natural resources in the lacustrine facies of the Cenozoic rift basins of east Africa. In: *Lacustrine Facies Analysis* (Ed. by P. Anadón, L. Cabrera and K. Kelts), pp. 3–37. *Spec. Publ. int. Ass. Sediment.*, 13. *4.1, 4.17.*

Tiercelin J.-J. & Mondeguer A. (1991) The geology of the Tanganyika Trough. In: *Lake Tanganyika and its Life* (Ed. by G.W. Coulter), pp. 76–89. Oxford University Press, Oxford. *4.4.2, Fig. 4.18.*

Tiercelin J.J., Mondeguer A., Gasse F., *et al.* (1988) 25 000 ans d'histoire hydrologique et sédimentaire du lac Tanganyika, rift est-africain. *C.R. Acad. Sci. Paris*, 307 série II, 1375–1382. *4.6.3.*

Tiercelin J.-J., Soreghan M., Cohen A.S., Lezzar K.E. & Bouroullec J.L. (1992) Sedimentation in large rift lakes: example from the Middle Pleistocene – Modern deposits of the Tanganyika trough,

East African rift system. *Bull. Cent. Recherch. Explor. Product. Elf-Aquitaine*, 16, 83–111. *4.4.2, 4.6.3, 4.7.1, 4.7.2, 4.11.1, Fig. 4.7.*

Tiercelin J.-J., Vincennes A. *et al.* (1987) Le demi-graben de Baringo-Bogoria, Rift Valley, Kenya. 30 000 years of hydrological and sedimentary history. *Bull. Centre Rech. Explor. -Prod Elf-Aquitaine*, 11–2, 249–544. *3.6.4.*

Tillman R.W. (1985) A spectrum of shelf sands and sandstones. In: *Shelf Sands and Sandstone Reservoirs* (Ed. by D.J.P. Swift, R.G. Walker and R.W. Tillman), pp. 1–46. *Soc. econ. Paleontol. Miner. Short Course*, 13, Tulsa. *7.1.2, 7.8.4.*

Tillman R.W. & Martinsen R.S. (1984) The Shannon shelf-ridge sandstone complex, Salt Creek anticline area, Powder River Basin, Wyoming. In: *Siliciclastic Shelf Sediments* (Ed. R.W. Tillman and C.T. Seimers), pp. 85–142. *Spec. Publ. Soc. econ. Paleont. Miner.*, 34, Tulsa. *7.8.4, Fig. 7.54.*

Tillman R.W. & Martinsen R.S. (1987) Sedimentological model and production characteristics of Hartzog Draw Field, Wyoming. In: *Reservoir Sedimentology* (Ed. by R.W. Tillman and K.J. Weber), pp. 15–112. *Spec. Publ. Soc. econ. Paleont. Miner.*, 40, Tulsa. *7.8.4, Fig. 7.54.*

Tilzer M.M. & Serruya C. (Eds) (1990) *Large Lakes: Ecological Structure and Function*, 691 pp. Springer-Verlag, Berlin. *4.4.*

Todd S.P. (1989) Stream-driven, high-density gravelly traction carpets: possible deposits in the Trabeg Conglomerate Formation, SW Ireland and some theoretical considerations of their origin. *Sedimentology*, 36, 513–530. *3.5.1.*

Todd S.P. & Went D.J. (1991) Lateral migration of sand-bed rivers: examples from the Devonian Glashabeg Formation, SW Ireland and the Cambrian Alderney Sandstone Formation, Channel Islands. *Sedimentology*, 38, 997–1020. *3.6.5.*

Törnqvist T.E. (1993) Holocene alternation of meandering and anastomosing fluvial systems in the Rhine–Meuse delta (Central Netherlands) controlled by sea-level rise and subsoil erodibility. *J. sedim. Petrol.*, 63, 683–693. *3.6.9, 6.7.6.*

Torres R.C., Self S. & Martinez M. (1996) Secondary pyroclastic flows from the June 15, 1991 ignimbrite of Mount Pinatubo. In: *Mount Pinatubo and the 1991 eruption* (Ed. by R.S. Punongbayan and C.G. Newhall), PHILVOLCS, University of Washington Press, in press. *12.4.4, Fig. 12.1.*

Treese K.L. & Wilkinson B.H. (1982) Peat-marl deposition in a Holocene paludal–lacustrine basin – Sucker Lake, Michigan. *Sedimentology*, 29, 375–390. *4.7.1, 4.7.5.*

Trenhaille A.S. (1987) *The Geomorphology of Rock Coasts*, 384 pp. Clarendon Press, Oxford. *6.4.*

Trewin N.H. (1986) Palaeoecology and sedimentology of the Achanarras fish bed of the Middle Old Red Sandstone, Scotland. *Trans. Roy. Soc. Edin.: Earth Sci.*, 77, 21–46. *4.13.1, 4.15.*

Treworgy C.G. & Jacobson R.J. (1985) Paleoenvironments and distribution of low-sulfur coal in Illinois. *Compte Rendu, Neuvième Congrès International de Stratigraphie et de Géologie du Carbonifère, Washington and Champaign-Urbana*, 1979, 4, 349–359. *3.6.8.*

Tricart J. & Cailleux A. (1972) *Introduction to Climatic Geomorphology* (translated by C.J. Kiewiet de Jonge), 295 pp. Longman, London. *2.1.2, Fig. 2.2.*

Triffleman N.J., Hallock P. & Hine A.C. (1992) Morphology,

sediments, and depositional environments of a small carbonate platform: Serranilla Bank, Nicaraguan Rise, southwest Caribbean Sea. *J. sedim. Petrol.*, 62, 591–606. *9.3.5.*

Trowbridge A.C. (1930) Building of Mississippi delta. *Bull. Am. Ass. petrol. Geol.*, 14, 867–901. *6.1.*

Truc G. (1978) Lacustrine sedimentation in an evaporitic environment: the Ludian (Palaeogene) of the Mormoiron Basin, southwestern France. In: *Modern and Ancient Lake Sediments* (Ed. by A. Matter and M.E. Tucker), pp. 189–203. *Spec. Publ. int. Ass. Sediment.*, 2. *4.12.1.*

Trudell L.G., Beard T.N. & Smith J.W. (1974) Stratigraphic framework of Green River Formation oil shales in the Piceance Creek Basin, Colorado. In: *Energy Resources of the Piceance Creek Basin, Colorado* (Ed. by D.K. Murray), pp. 65–69. *Rocky Mountain Ass. Geol.*, Denver. *4.14.1.*

Tsoar H. (1982) Internal structure and surface geometry of longitudinal (seif) dunes. *J. sedim. Petrol.*, 52, 823–831. *5.4.4.*

Tsoar H. (1983) Dynamic processes acting on a longitudinal (seif) sand dune. *Sedimentology*, 30, 567–578. *5.3.2, 5.4.4, Fig. 5.22.*

Tsoar H. & Pye K. (1987) Dust transport and the question of desert loess formation. *Sedimentology*, 34, 139–153. *5.3.7, 5.4.2, 5.4.7.*

Tucker M.E. (1978) Triassic lacustrine sediments from South Wales: shore-zone clastics, evaporites and carbonates. In: *Modern and Ancient Lake Sediments* (Ed. by A. Matter and M.E. Tucker), pp. 205–224. *Spec. Publ. int. Ass. Sediment.*, 2. *4.11.3.*

Tucker M.E. (1985) Shallow marine carbonate facies and facies models. In: *Sedimentology: Recent Developments and Applied Aspects* (Ed. by P.J.. Brenchley and B.P.J. Williams), pp. 139–161. *Spec. Publ. geol. Soc. Lond.*, 18. *9.4.2.*

Tucker M.E. (1986) Formerly aragonitic limestones associated with tillites in the late Proterozoic of Death Valley, California. *J. sedim. Petrol.*, 56, 818–830. *11.5.5.*

Tucker M.E. (1991) Sequence stratigraphy of carbonate–evaporite basins: models and application to the Upper Permian (Zechstein) of northeast England and adjoining North Sea. *J. geol. Soc. Lond.*, 148, 1019–1036. *8.2.3, 8.3.1, 8.10, 8.10.2, 9.5.1.*

Tucker M.E. (1993) Carbonate diagenesis and sequence stratigraphy. In: *Sedimentology Review*, 1 (Ed. by V.P. Wright), pp. 51–72. Blackwell Scientific Publications, Oxford. *Figs. 9.60, 9.61, 9.65.*

Tucker M.E. & Wright V.P. (1990) *Carbonate Sedimentology*, 496 pp. Blackwell Scientific Publications, Oxford. *9.1.3, 9.3.2, 9.3.3, 9.4.1, 9.4.2, 9.4.4, 9.4.5, Figs 9.9, 9.31, 9.41, 9.46, 9.49.*

Tucker M.E., Wilson J.L., Crevello P.D. Sarg J.R. & Read J.F. (Eds) (1990) *Carbonate Platforms*, 328 pp. *Spec. Publ. int. Ass. Sediment.*, 9. *2.4, 9.5.*

Tudhope A.W. (1989) Shallowing-upwards sedimentation in a coral reef lagoon, Great Barrier Reef of Australia. *J. sedim. Petrol.*, 59, 1036–1051. *9.4.5, Fig. 9.52.*

Tudhope A.W. & Risk M.J. (1985) Rate of dissolution of carbonate sediments by microboring organisms, Davies Reef, Australia. *J. sedim. Petrol.*, 55, 440–447. *9.4.5.*

Tudhope A.W. & Scoffin T.P. (1984) The effects of *Callianasa* bioturbation on the preservation of carbonate grains in Davies Reef lagoon, Great Barrier Reef, Australia. *J. sedim. Petrol.*, 54, 1091–1096. *9.4.1, 9.4.5.*

Tunbridge I.P. (1981) Sandy high-energy flood sediments – some criteria for their recognition, with an example from the Devonian of S.W. England. *Sedim. Geol.*, 28, 79–95. *3.5.3, 3.6.10.*

Tunbridge I.P.. (1984) Facies model for a sandy ephemeral stream and clay playa complex; the Middle Devonian Trentishoe Formation of North Devon, U.K. *Sedimentology*, 31, 697–715. *3.5.3.*

Turbeville, B.N. (1991) The influence of ephemeral processes on pyroclastic sedimentation in a rift-basin, volcaniclastic–alluvial sequence, Española basin, New Mexico. *Sedim. Geol.*, 74, 139–155. *12.6.2.*

Turmel R.J. & Swanson R.G. (1976) The development of Rodriguez Bank, a Holocene mudbank in the Florida reef tract. *J. sedim. Petrol.*, 46, 497–518. *9.4.5.*

Turnbull W.J., Krinitzsky E.L. & Weaver F.S. (1966) Bank erosion in soils of the Lower Mississippi valley. *Soil Mech. Found. Proc. Am. Soc. Civ. Eng.*, 92, 121–136. *3.2.1, Fig. 3.1.*

Turner B.R. & Munro M. (1987) Channel formation and migration by mass-flow processes in the Lower Carboniferous fluviatile Fell Sandstone Group, northeast England. *Sedimentology*, 34, 1107–1122. *3.5.2, 3.6.2, Fig. 3.41.*

Turner C.C., Cohen J.M., Connell E.R. & Cooper D.M. (1987) A depositional model for the South Brae oilfield. In: *Petroleum Geology of North-West Europe* (Ed. by J. Brooks and K. Glennie), pp. 853–864. Graham & Trotman, London. *6.5.4, 10.3.4.*

Turner J.P. (1992) Evolving alluvial stratigraphy and thrust front development in the West Jaca piggyback basin, Spanish Pyrenees. *J. geol. Soc., Lond.*, 149, 51–63. *3.6.5.*

Turner J.S. & Campbell I.H.. (1986) Convection and mixing in magma chambers. *Earth Sci. Rev.* 23, 255–352. *12.3.2.*

Turner P. (1980) Continental red beds. *Developments in Sedimentology*, 29, 562 pp. Elsevier, Amsterdam. *3.2.3.*

Turner-Peterson C.E. (1979) Lacustrine-humate model – sedimentologic and geochemical model for tabular uranium deposits. *Bull. Am. Ass. petrol. Geol.*, 63, 843. *4.17.*

Tuttle M.L. (1991) Introduction to geochemical, biogeochemical and sedimentological studies of the Green River Formation, Wyoming, Utah and Colorado. *Bull. U.S. geol. Surv.*, 1973-A-G, A1–A11. *4.14.1.*

Twichell D.C., Kenyon N.H., Parson L.M. & McGregor B.A. (1991) Depositional patterns of the Mississippi Fan surface: Evidence from GLORIA II and high-resolution seismic profiles. In: *Seismic Facies and Sedimentary Processes of Submarine Fans and Turbidite Systems* (Ed. by P. Weimer and M.H. Link), pp. 349–363. Springer-Verlag, New York. *10.3.3.*

Tye R.S. & Coleman J.M. (1989a) Depositional processes and stratigraphy of fluvially dominated lacustrine deltas: Mississippi delta plain. *J. sedim. Petrol.*, 59, 973–996. *6.6.1, 6.6.5.*

Tye R.S. & Coleman J.M. (1989b) Evolution of Atchafalaya lacustrine deltas, South-Central Louisiana. *Sedim. Geol.*, 65, 95–112. *6.6.1.*

Tye R.S. & Kosters E.C. (1986) Styles of interdistributary basin sedimentation: south central Louisiana. *Trans. Gulf Coast Ass. geol. Soc.*, 36, 575–588. *6.6.1.*

Tye R.S., Ranganathan V. & Ebanks W.J. Jr. (1986) Facies analysis and reservoir zonation of a Cretaceous shelf sand ridge: Hartzog Draw Field, Wyoming. In: *Modern and Ancient Shelf Clastics: A Core Workshop* (Ed. by T.F. Moslow and E.G. Rhodes), pp. 169–216. Soc. econ. Paleont. Miner. Core Workshop No. 9. *7.8.4.*

Tyson R.V. & Pearson T.H. (Eds) (1991) *Modern and Ancient Continental Shelf Anoxia*, 470 pp. *Spec. Publ. geol. Soc. Lond. 58. 7.7.3.*

Ui T. (1983) Volcanic dry avalanche deposits – identification and comparison with nonvolcanic debris stream deposits. *J. Volcanol. Geotherm. Res.*, 18, 135–150. *12.5.1.*

Underhill J.R. (1991) Controls on Late Jurassic seismic sequences, Inner Moray Firth, UK North Sea: a critical test of a key segment of Exxon's original global cycle chart. *Basin Res.*, 3, 79–98. *2.4.*

Underwood M.B. (1986) Transverse filling of the central Aleutian Trench by unconfined turbidity currents. *Geo-Mar. Lett.*, 6, 7–13. *12.12.5.*

Underwood M.B. & Bachman S.B. (1982) Sedimentary facies associations within subduction complexes. In: *Trench–Forearc Geology: Sedimentation and Tectonics on Modern and Ancient Plate Margins* (Ed. by J.K. Legget), pp. 537–550. *Spec. Publ. geol. Soc. Lond.*, 10. *12.12.5, Fig. 12.80.*

Underwood M.B. & Moore G.F. (1995) Trenches and trench-slope basins. In: *Tectonics of Sedimentary Basins* (Ed. by C.J. Busby and R.V. Ingersoll), pp. Blackwell Science Inc., Boston. *12.12.5, Fig. 12.82.*

Urabe T. & Marumo K. (1991) A new model for Kuroko-type deposits of Japan. *Episodes*, 14, 246–251, *12.10, 12.11.1.*

Usiglio J. (1849) Analyse de l'eau de la Méditerranée sur les cotes de France. *Ann. Chem.* 27, 92–107; 172–191. *8.2.1.*

Usui A. & Nishimura A. (1992) Submersible observations of hydrothermal manganese deposits on the Kaikata Seamount, Izu–Ogawasara (Bonin) arc. *Mar. Geol.*, 106, 203–216. *12.12.5.*

Uyeda S. (1982) Subduction zones: an introduction to comparative subductology. *Tectonophysics*, 81, 133–159. *12.2.1.*

Uyeda S. (1991) The Japanese island arc and the subduction process. *Episodes*, 14, 190–198. *12.12.5.*

Vai G.B. & Ricci Lucchi F. (1978) Algal crusts, autochthonous and clastic gypsum in a cannibalistic evaporite basin: a case history from the Messinian of Northern Apennines. *Sedimentology*, 24, 211–244. *8.10.1.*

Vail P.R. (1987) Seismic stratigraphy interpretation procedure. In: *Seismic Stratigraphic Atlas* (Ed. by B. Balley), pp. 1–10. *Am. Ass. petrol. Geol., Studies in Geology*, 27, Tulsa. *2.4.*

Vail P.R. & Todd R.G. (1981) Northern North Sea Jurassic unconformities, chronostratigraphy and sea-level changes from seismic stratigraphy. In: *Petroleum Geology of the Continental Shelf of North-West Europe; Proceedings of the Second Conference* (Ed. by L.V. Illing and G.D. Hobson), pp. 216–235. Heyden, London. *2.4, 6.6.6, 7.7.3.*

Vail P.R. & Wornardt W.W. (1990) Well log-seismic sequence stratigraphy: an integrated tool for the 90's. In: *Sequence Stratigraphy as an Exploration Tool concepts and practices in the Gulf Coast. Gulf Coast Section SEPM 11th annual research conference, program and abstracts, Dec. 2nd 1990, pp. 379–397. Fig. 2.20.*

Vail P.R., Mitchum R.M. & Thompson III, S. (1977a) Seismic stratigraphy and global changes in sea level, Part 3: Relative changes of sea level from coastal onlap. In: *Seismic Stratigraphy – Applications to Hydrocarbon Exploration* (Ed. by C.W. Payton), pp. 63–81. *Mem. Am. Ass. petrol. Geol.*, 26, Tulsa. *9.5.*

Vail P.R., Mitchum R.M. & Thompson III, S. (1977b) Seismic

stratigraphy and global changes of sea level, Part 4: Global cycles of relative changes of sea level. In: *Seismic Stratigraphy – Applications to Hydrocarbon Exploration* (Ed. by C.E. Payton), pp. 83–97. *Mem. Am. Ass. petrol. Geol.*, 26, Tulsa. *2.4.*

Valentine G.A. (1987) Stratified flow in pyroclastic surges. *Bull. Volcanol.*, 49, 616–630. *12.4.4, Fig. 12.22.*

Valentine G.S. & Fisher R.V. (1986) Origin of layer 1 deposits in ignimbrites. *Geology*, 14, 146–148. *12.4.4.*

Valyashko M.G. (1972) Playa lakes: necessary stage in the development of a salt bearing basin. In: *Geology of Saline Deposits* (Ed. by G. Richter-Bernburg) pp. 41–51. *Earth Sci. Series*, 7, UNESCO, Paris. *8.9.*

Van Andel T.H. & Postma H. (1954) Recent sediments of the Gulf Paria. *Verh. K. Akad. Wet., Afd. Natuurkd.*, 20, 1–245. *7.0.0.*

Van Andel Tj. H. & Curray J.R. (1960) Regional aspects of modern sedimentation in northern Gulf of Mexico and similar basins, and palaeogeographic significance. In: *Recent Sediments, Northwest Gulf of Mexico* (Ed. by F.P. Shepard, F.B. Phleger and Tj.H. van Andel), pp. 345–364. *Am. Ass. petrol. Geol.*, Tulsa. *6.1, 6.6, Fig. 6.36.*

Van Andel Tj. H. & Komar P.D. (1969) Ponded sediments of the Mid-Atlantic Ridge between 20° and 23° north latitude. *Bull. geol. Soc. Am.*, 80, 1163–1190. *10.3.6, 12.11.1.*

Van Bendegom L. (1947) Eenige beschouwingen over riviermorphologie en rivierbetering. *De Ingenieur*, 59 (4), 1–11. *3.3.4.*

Van den Berg J.H. (1982) Migration of large-scale bedforms and preservation of crossbedded sets in highly accretional parts of tidal channels in the Oosterschelde, S.W. Netherlands. *Geol. Mijnbouw*, 61, 253–263. *7.3.2.*

Van den Bogaard P., Schmincke H.-U., Freundt A. & Park C. (1990) Evolution of complex Plinian eruptions: the Late Quaternary Laacher See case history. In: *Thera and the Aegean World 3. Volume two. Earth Sciences* (Ed. by D.A. Hardy, J. Keller, V.P. Galanopoulos, N.C. Fleming and T.H. Druitt), pp. 463–483. The Thera Foundation, London. *12.9.2.*

Van der Weil A.M., Van den Bergh G.D. & Hebeda E.H. (1992) Uplift, subsidence, and volcanism in the southern Neiva Basin, Columbia, Part 2: influence on fluvial deposition in the Miocene Gigante Formation. *J. S. Am. Earth Sci.*, 5, 175–196. *12.12.4.*

Van Heerden I.L. & Roberts H.H. (1988) Facies development of Atchafalaya Delta, Louisiana: a modern bayhead delta. *Bull. Am. Ass. petrol. Geol.*, 72, 439–453. *6.6.1.*

Van Heerden I.L., Wells J.T. & Roberts H.H. (1981) Evolution and morphology of sedimentary environments, Atchafalaya delta, Louisiana. *Trans. Gulf Coast. Ass. geol. Soc.*, 31, 399–408. *6.6.5.*

Van Houten F.B. (1964) Cyclic lacustrine sedimentation, Upper Triassic Lockatong Formation, central New Jersey and adjacent Pennsylvania. In: *Symposium on Cyclic Sedimentation* (Ed. by D.F. Merriam), pp. 495–531. *Bull. geol. Surv. Kansas*, 169. *4.10, 4.16.3.*

Van Houten F.B. (1972) Iron and clay in tropical savanna alluvium: A contribution to the origin of red beds. *Bull. geol. Soc. Am.*, 83, 2761–2772. *3.2.3.*

Van Houten F.B. (1976) Late Cenozoic volcaniclastic deposits, Andean foredeep, Colombia. *Bull. geol. Soc. Am.*, 87, 481–495. *12.12.4.*

Van Straaten L.M.J.U. (1954) Composition and structure of Recent marine sediments in the Netherlands. *Leidse. geol. Meded.*, 19, 1–110. *6.7.3.*

Van Straaten L.M.J.U. (1961) Sedimentation in tidal flat areas. *J. Alberta Soc. petrol. Geol.*, 9, 203–226. *6.7.3.*

Van Steenwinkel M. (1990) Sequence stratigraphy from 'spot' outcrops – example from a carbonate-dominated setting: Devonian–Carboniferous transition, Dinant synclinorium (Belgium). *Sedim. Geol.*, 69, 259–280. *9.5.2.*

Van Veen J. (1935) Sand waves in the southern North Sea. *Int. Hydrograph. Rev.*, 12, 21–29. *7.1.2.*

Van Veen J. (1936) *Onderzoekingen in de Hoofden*, 252 pp. Algemene Landsrukkerij, The Hague. *7.1.2.*

Van Wagoner J.C. (1985) Reservoir facies distribution as controlled by sea-level change. *Soc. econ. Paleont. Miner. Ann. Midyear Meeting*. Golden Colorado, 2, 91–92. *2.4.*

Van Wagoner J.C., Posamentier H.W., Mitchum H.W., Vail P.R., Sarg J.F., Loutit T.S. & Hardenbol J. (1988) An overview of sequence stratigraphy and key definitions. In: *Sea-level Changes – an Integrated Approach* (Ed. by C.K. Wilgus, B.S. Hastings, C.G. St C Kendall, H. Posamentier, C.A. Ross & J. Van Wagoner) ,pp. 39–45. *Spec. Publ. Soc. econ. Paleont. Miner.*, 42, Tulsa. *2.2, 2.2.2, 2.4, Figs 2.16, 2.18, 6.7.6, 9.5.5.*

Van Wagoner J.C., Mitchum R.M., Campion K.M. & Rahmanian V.D. (1990) Siliciclastic Sequence Stratigraphy in Well Logs, Cores, and Outcrops: Concepts for High-Resolution Correlation of Time and Facies. 55 pp. AAPG Methods in Exploration Series No. 7. *7.2.2.*

Veevers J.J. & Powell C.McA. (1987) Late Paleozoic glacial episodes in Gondwanaland reflected in transgressive–regressive depositional sequences in Euramerica. *Bull. geol. Soc. Am.*, 98, 475–487. *11.5.2.*

Vera J.-A., Ruiz-Ortiz P.A., Garcia-Hernandez M. & Molina J.M. (1988) Paleokarst and related pelagic sediments in the Jurassic of the Subbetic Zone, southern Spain. In: *Paleokarst* (Ed. by N.P. James and P.W. Choquette), pp. 364–384. Springer-Verlag, New York. *Fig. 9.20.*

Verbeek R.D.M. (1886) *Krakatau*, 546 pp. Batavia. *12.1.1, 12.6.2.*

Vernon J.E.N. & Hudson R.C.L. (1978) Ribbon reefs of the Northern Region. *Philos. Trans. R. Soc. Lond.*, B284, 3–21. *9.4.5.*

Vessel R.K. & Davies D.K.. (1981) Non-marine sedimentation in an active fore-arc basin. In: *Recent and Ancient Non-marine Depositional Environments: Models for Exploration* (Ed. by F.G. Ethridge and R.M. Flores), pp. 31–45. *Spec. Publ. Soc. econ. Paleont. Mineral.*, 31, Tulsa. *3.3.8, 12.6.1, 12.6.2, 12.12.3.*

Vianna M.L., Solewicq R., Cabral A.P. & Testa V. (1991) Sandstream on the northeast Brazilian shelf. *Continental Shelf Res.*, 11, 509–524. *7.4.3.*

Viau C. (1983) Depositional sequences, facies and evolution of the Upper Devonian Swan Hills reef buildup, central Alberta, Canada. In: *Carbonate Buildups – a Core Workshop* (Ed. by P.M. Harris), pp. 112–143. *Soc. econ. Paleont. Miner., Core Workshop*, 4, Tulsa. *9.4.5, Fig. 9.51.*

Visher G.S. (1965) Fluvial processes as interpreted from ancient and recent fluvial deposits. In: *Primary Sedimentary Structures and their Hydrodynamic Interpretation* (Ed. by G.V. Middleton), pp. 116–132. *Spec. Publ. Soc. econ. Paleont. Miner.*, 12, Tulsa. *2.2.2.*

Visser J.N.J. (1986) Lateral lithofacies relationships in the glacigene Dwyka Formation in the western and central parts of the Karoo Basin. *Trans. geol. Soc. S. Afr.*, 89, 373–383. *11.5.1.*

Visser J.N.J. (1991) The paleoclimatic setting of the late Paleozoic marine ice sheet in the Karoo Basin of Southern Africa. In: *Glacial Marine Sedimentation: Paleoclimatic Significance* (Ed. by J.B. Anderson and G.M. Ashley), pp. 181–189. *Spec. Pap. geol. Soc. Am.*, 261, Boulder, CO. *11.5.5, Figs 11.27, 11.28.*

Visser J.N.J. & Kingsley C.S. (1982) Upper Carboniferous glacial valley sedimentation in the Karoo Basin, Orange Free State. *Trans. geol. Soc. S. Afr.*, 85, 71–79. *11.5.5.*

Visser M.J. (1980) Neap–spring cycles reflected in Holocene subtidal large-scale bedform deposits: a preliminary note. *Geology*, 8, 543–546. *6.2.7, 7.7.1, Figs 6.13, 7.12.*

Von Brunn V. & Gold D.J.C. (1993) Diamictite in the Archean Pongola Sequence of southern Africa. *J. Afr. Earth Sci.*, 16, 367–374. *11.6.*

Von der Borch C., Bolton B. & Warren J.K. (1977) Environmental setting and microstructure of subfossil lithified stromatolites associated with evaporites, Marion Lake, Australia. *Sedimentology*, 24, 693–708. *8.5.2.*

Vorren T.O., Hald M. & Thomsen E. (1984) Quaternary sediments and environments on the continental shelf off northern Norway. *Mar. Geol.*, 57, 229–257. *11.4.6.*

Vorren T.O., Hald M., Edvardsen M. & Lind-Hansen O. (1983) Glacigenic sediments and sedimentary environments on continental shelves: General principles with a case study from the Norwegian shelf. In: *Glacial Deposits in North-West Europe* (Ed. by J. Ehlers), pp. 61–73. Balkema, Rotterdam. *11.4.6.*

Vorren T.O., Lebesbye E., Andreassen K. & Larsen K.-B. (1989) Glacigenic sediments on a passive continental margin as exemplified by the Barents Sea. *Mar. Geol.*, 85, 251–272. *11.5.4.*

Vossler S.M. & Pemberton S.G. (1988) Ichnology of the Cardium Formation (Pembina Oilfield): implications for depositional and sequence stratigraphic interpretations. *Sequences, Stratigraphy, Sedimentology; Surface and Subsurface* (Ed. by D.P. James and D.A. Leckie), pp. 237–253. *Mem. Can. Soc. petrol. Geol.*, 15, Calgary. *7.7.2.*

Wadge G., Francis P.W. & Ramirez C.F. (1995) The Socompa collapse and avalanche event. *J. Volcanol. Geotherm. Res.*, 66, 309–336. *12.5.1.*

Wagner C.W. & Van der Togt C. (1973) Holocene sediment types and their distribution in the southern Persian Gulf. In: *The Persian Gulf* (Ed. by B.H.. Purser), pp. 123–156. Springer-Verlag, Berlin. *9.3.4.*

Waite L.E. (1993) Upper Pennsylvanian seismic sequences and facies of the eastern Southern Horseshoe Atoll, Midland Basin, West Texas. In: *Carbonate Sequence Stratigraphy. Recent Development and Applications.* (Ed. by R.G. Loucks and J.F. Sarg), pp. 213–240. *Mem. Am. Ass. petrol. Geol.*, 57, Tulsa. *9.5.4.*

Walder J.S. (1982) Stability of sheet flow of water beneath temperate glaciers and implications for glacier surging. *J. Glaciol.*, 28, 273–293. *11.3.6.*

Walker G.P.L. (1973a) Explosive volcanic eruptions – a new classification scheme. *Geol. Rundsch.*, 62, 431–446. *12.1.1, 12.4.1, 12.4.3. Fig. 12.10.*

Walker G.P.L. (1973b) Lengths of lava flows. *Philos. Trans. R. Soc.*

Lond., 274, 107–118. *Fig. 12.29.*

Walker G.P.L. (1981) Plinian eruptions and their products. *Bull. Vulcanol.*, 44, 223–240. *12.4.1.*

Walker G.P.L. (1984) Downsag calderas, ring faults, caldera sizes, and incremental caldera growth. *J. geophys. Res.*, 89, 8407–8416. *12.4.6.*

Walker G.P.L. (1988) Three Hawaiian calderas: an origin through loading by shallow intrusions? *J. geophys. Res.*, 93, 14773–14784. *12.4.6.*

Walker G.P.L. (1992) Coherent intrusion complexes in large basaltic volcanoes – a structural model. *J. Volcanol. Geotherm. Res.*, 50, 41–54. *12.11.2.*

Walker G.P.L. (1993) Basaltic-volcano systems. In: *Magmatic Processes and Plate Tectonics* (Ed. by H.M. Prichard, T. Alabaster, N.B.W. Harris and C.R. Neary), pp. 3–38. *Spec. Publ. geol. Soc. Lond.*, 76. *12.4.5, 12.8, 12.14.2.*

Walker G.P.L., Hayashi J.N. & Self S. (1995) Travel of pyroclastic flows as transient waves: implications for the energy line concept and particle-concentration modelling. *J. Volcanol. Geotherm. Res.*, *12.4.4.*

Walker G.P.L., Wright J.V., Clough B.J. & Booth B. (1981) Pyroclastic geology of the rhyolitic volcano of La Primavera, Mexico. *Geol. Rundsch.*, 70, 1100–1118. *Fig. 12.90.*

Walker K.R. & Alberstadt L.P. (1975) Ecological succession as an aspect of structure in fossil communities. *Paleobiology*, 3, 238–257. *9.4.5.*

Walker R.G. (1965) The origin and significance of the internal sedimentary structures of turbidites. *Proc. Yorks. geol. Soc.*, 35, 1–32. *10.1, 10.2.3.*

Walker R.G. (1966a) Shale Grit and Grindslow Shales: transition from turbidite to shallow water sediments in the Upper Carboniferous of northern England. *J. sedim. Petrol.*, 36, 90–114. *6.6.6, 10.1, 10.3.4.*

Walker R.G. (1966b) Deep channels in turbidite-bearing formations. *Bull. Am. Ass. petrol. Geol.*, 50, 1899–1917. *10.3.4.*

Walker R.G. (1967) Turbidite sedimentary structures and their relationship to proximal and distal depositional environments. *J. sedim. Petrol.*, 37, 25–43. *10.2.3.*

Walker R.G. (1973) Mopping up the turbidite mess. In: *Evolving Concepts in Sedimentology* (Ed. by R.N. Ginsburg), pp. 1–37. Johns Hopkins University Press, Baltimore. *10.1.*

Walker R.G. (1978) Deep-water sandstone facies and ancient submarine fans: models for exploration for stratigraphic traps. *Bull. Am. Ass. petrol. Geol.*, 62, 932–966. *10.1, 10.3.3, Fig. 10.3.*

Walker R.G. (Ed.) (1979, 1984) *Facies Models. Geosci. Can. Reprint Ser.*, 1. Geol. Soc. Can., Waterloo, Ontario. *1.1.*

Walker R.G. (1984) Shelf and shallow marine sands. In: *Facies Models* (Ed. by R.G. Walker), pp. 141–170. *Geosci. Can. Reprint Ser.*, 1. Geol. Soc. Can., Waterloo, Ontario. *7.7.2, 7.8.4.*

Walker R.G. (1990) Facies modeling and sequence stratigraphy. *J. sedim. Petrol.*, 60, 777–786. *2.2.2, 2.4, 2.5.*

Walker R.G. & Bergman K.M. (1993) Shannon Sandstone in Wyoming: a shelf-ridge complex reinterpreted as lowstand shoreface deposits. *J. sedim. Petrol.*, 63, 839–851. *7.8.4, Fig. 7.54.*

Walker R.G. & James N.P. (1992) Facies, facies models, and modern stratigraphic concepts. In: *Facies Models: Response to Sea-level Change* (Ed. by R.G. Walker and N.P. James), pp. 1–14. Geol. Ass.

Can., Waterloo, Ontario. *1.1.*

Walker R.G. & Plint A.G. (1992) Wave- and storm-dominated shallow marine systems. In: *Facies Models: Response to Sea-level Change* (Ed. by R.G. Walker and N.P. James), pp. 219–238. Geol. Ass. Can., Waterloo, Ontario. *6.7.5, 7.1.2.*

Walker T.R. (1967) Formation of red reds in ancient and modern deserts. *Bull. geol. Soc. Am.*, 78, 353–368. *3.2.3.*

Walker T.R., Waugh B. & Crone A.J. (1978) Diagenesis of first-cycle desert alluvium of Cenozoic age, southwestern United States and northwestern Mexico. *Bull. geol. Soc. Am.*, 89, 19–32. *3.2.3, 3.3.8.*

Walter M.R. & Bauld J. (1983) The association of sulphate evaporites, stromatolitic carbonates and glacial sediments: Examples from the Proterozoic of Australia and the Cainozoic of Antarctica. *Precambrian Res.*, 21, 129–148. *11.4.7.*

Walther J. (1894) *Einleitung in die Geologie als Historische Wissenschaft*, **Bd. 3.** Lithogenesis der Gegenwart, pp. 535–1055. Fischer Verlag, Jena. *2.2.2, 13.1.1.*

Walton A.W. (1986) Effect of Oligocene volcanism on sedimentation in the Trans-Pecos volcanic field of Texas. *Bull. geol. Soc. Am.*, 97, 1192–1207. *12.13.*

Walton A.W. & Palmer B.A. (1988) Lahar facies of the Mount Dutton Formation (Oligocene–Miocene) in the Marysvale Volcanic Field, southwestern Utah. *Bull. geol. Soc. Am.*, 100, 1078–1091. *12.5.2.*

Walton E.K. (1967) The sequence of internal structures in turbidites. *Scott. J. Geol.*, 3, 306–317. *10.2.3.*

Wanless H.R. (1950) Late Paleozoic cycles of sedimentation in the United State. *Intern. Geol. Congr., 18th, London, 1948, Rep.*, 4, 17–28. *6.6.6.*

Wanless H.R. (1991) Observational foundation for sequence modeling. In: *Sedimentary Modeling: Computer Simulations and Methods for Improved Parameter Definition* (Ed. by E.K. Franseen *et al.*), pp. 42–62. *Bull. Kansas geol. Survey*, 233. *9.4.1, 9.4.5.*

Wanless H.R. & Dravis J.J. (1989) *Carbonate Environments and Sequences of Caicos Platform*, 75 pp. Field Trip Guidebook T374: 28th Int. Geol. Congr. Washington Geophysical Union. *9.4.1.*

Wanless H.R. & Tagett M.G. (1989) Origin, growth and evolution of carbonate mudbanks in Florida Bay. *Bull. mar. Sci.*, 44, 454–489. *9.3.3, 9.4.5.*

Wanless H.R. & Tedesco L.P. (1993) Comparison of oolitic sand bodies generated by tidal vs. wind-wave agitation. In: *Mississippian Oolites and Modern Analogs* (Ed. by B.D. Keith and C.W. Zuppan), pp. 199–225. *Am. Ass. petrol. Geol., Studies in Geology*, 35, Tulsa. *9.4.2.*

Wanless H.R., Cottrell D.J., Tagett M.G., Tedesco L.P. & Warzeski Jr E.R. (1995) Origin and growth of carbonate banks in south Florida. In: *Carbonate Mud-Mounds: Their Origin and Evolution* (Ed. by C.L.V. Monty, D.W.J. Bosence, P.H. Bridges and B.R. Pratt), pp. 439–473. *Spec. Publ. int. Ass. Sediment.*, 23. *9.4.5.*

Wanless H.R., Tedesco L.P. & Tyrrell K.M. (1988) Production of subtidal tabular and surficial tempestites by hurricane Kate, Caicos Platform, British West Indies. *J. sedim. Petrol.*, 58, 739–750. *7.4.3, 9.4.1, 9.4.3.*

Wanless H.R., Tyrell K.M., Tedesco L.P. & Dravis J.J. (1988) Tidal-flat sedimentation from Hurricane Kate, Caicos Platform, British West Indies. *J. sedim. Petrol.*, 58, 724–738. *9.4.1.*

Wantland K.F. & Pusey W.C. (eds.) (1975) *Belize Shelf-carbonate Sediments, Clastic Sediments, and Ecology*, 599 pp. Am. Ass. petrol. Geol., Studies in Geology, 2, Tulsa. *9.1.1.*

Ward W.C. & Brady M.J. (1979) Strandline sedimentation of carbonate grainstones, Upper Pleistocene, Yucatan Peninsula, Mexico. *Bull. Am. Ass. petrol. Geol.*, 63, 362–369. *9.4.2, Fig. 9.29.*

Ward W.C., Kendall C.G. St C. & Harris P.M. (1986) Upper Permian (Guadalupian) facies and their association with hydrocarbons – Permian Basin, West Texas and New Mexico. *Bull. Am. Ass. petrol. Geol.*, 70, 239–262. *9.3.3.*

Ward W.C., Weidie A.E. & Back W. (1985) *Geology and Hydrology of the Yucatan*, 160 pp. New Orleans Geol. Soc. *9.4.2.*

Wardlaw N.C. (1972) Synsedimentary folds and associated structures in Cretaceous salt deposits of Sergip, Brazil. *J. sedim. Petrol.*, 42, 572–577. *8.5.1.*

Wardlaw N.C. & Christie D.L. (1975) Sulphates of submarine origin in Pennsylvanian Otto Fiord Formation of Canadian Arctic. *Bull. Can. petrol. Geol.*, 23, 149–171. *8.9.*

Waresback D.B. & Turbeville B.N. (1990) Evolution of a Plio-Pleistocene volcanogenic alluvial fan: the Puye Formation, Jemez Mountains, New Mexico. *Bull geol. Soc. Am.*, 102, 298–314. *12.6.2.*

Warren A. & Kay S. (1987) Dune networks. In: *Desert Sediments: Ancient and Modern* (Ed. by L.E. Frostick and I. Reid), pp. 205–212. *Spec. Publ. geol. Soc. Lond.*, 35. *5.4.3.*

Warren J.K. (1982) The hydrological setting, occurrence and significance of gypsum in late Quaternary salt lakes in South Australia. *Sedimentology*, 29, 609–638. *8.3.2, 8.5.2, Figs 8.26, 8.27.*

Warren J.K. (1985) On the significance of evaporite lamination. In: *Proc. 6th Intern. Symp. on Salt*, Toronto, 1982 (Ed. by B.C. Schrieber), pp. 161–170. *8.5.1, 8.5.2.*

Warren J.K. (1989) *Evaporite Sedimentology: Importance in Hydrocarbon Accumulation*, 285 pp. Prentice Hall, NJ. *8.1.*

Warren J.K. (1991) Sulphate dominated sea-marginal and platform evaporitic settings: sabkhas and salinas, mudflats and salterns. In: *Evaporites, Petroleum and Mineral Resources* (Ed. by J.L. Melvin), pp. 69–187. Elsevier, Amsterdam. *8.4.2, 8.4.3, 8.8.*

Warren J.K. (1992) Sulfate dominated sea-marginal and platform evaporitic settings: sabkhas and salinas, mudflats and salterns In: *Evaporites, Petroleum and Mineral Resources* (Ed. by J.L. Melvin), pp. 69–187 Elsevier, Amsterdam. *8.4.2.*

Warren J.K. & Kendall C.G. St. C. (1985) Comparison of sequences formed in marine sabkha (subaerial) and salina (subaqueous) settings – modern and ancient. *Bull. Am. Ass. petrol. Geol.*, 69, 1013–1023. *8.4.2, 8.5.2, 8.7, 8.10.1, Fig. 8.25.*

Warren W.P. & Ashley G.M. (1994) Origins of the ice-contact stratified ridges (eskers) of Ireland. *J. sedim. Res.*, A64, 433–449. *11.4.1.*

Washburn A.L. (1980) *Geocryology: A Survey of Periglacial Processes and Environments*, 406 pp. Haltstead Press, John Wiley and sons, New York. *11.4.7.*

Wasson R.J. (1974) Intersection point deposition on alluvial fans: an Australian example. *Geogr. Ann.*, 56A, 83–92. *3.3.8.*

Wasson R.J. (1984) Late Quaternary palaeoenvironments in the desert dunefields of Australia. In: *Palaeoclimates of the Southern*

Hemisphere (Ed. by J.C. Vogel), pp. 419–432. A.A. Balkema, Rotterdam. *5.2.2.*

Waters A.C. & Fisher R.V. (1971) Base surges and their deposits: Capelinhos and Taal volcanoes. *J. Geophys. Res.*, **76**, 5596–5614. *Fig. 12.1.*

Watts N.R. (1981) Sedimentology and diagenesis of the Hogklint reefs and their associated sediments. Lower Silurian, Gotland, Sweden. Unpubl. PhD thesis, University of Wales, Cardiff. *Fig. 9.47.*

Watts N.R. (1988) The role of carbonate diagenesis in exploration and production from Devonian pinnacle reefs, Alberta, Canada. *Bull. geol. Soc. Malaysia*, **22**, 1–22. *9.5.2.*

Weaver P.P.E. (1994) Determination of turbidity current erosional characteristics from reworked coccolith assemblages, Canary Basin, north-east Atlantic. *Sedimentology*, **41**, 1025–1038. *10.3.6.*

Weaver P.P.E. & Rothwell R.G. (1987) Sedimentation on the Madeira Abyssal Plain over the last 300,000 years. In: *Geology and Geochemistry of Abyssal Plains* (Ed. by P.P.E. Weaver and J. Thomson), pp. 71–86. *Spec. Publ. geol. Soc. Lond.*, **31**. *10.3.6.*

Weaver P.P.E. & Thomson J. (Eds) (1987) *Geology and Geochemistry of Abyssal Plains*, 246 pp. *Spec. Publ. geol. Soc. Lond.*, **31**. *10.3.6.*

Weaver P.P.E., Thomson J. & Hunter P.M. (1987) Introduction. In: *Geology and Geochemistry of Abyssal Plains* (Ed. by P.P.E. Weaver and J. Thomson), vii–xii. *Spec. Publ. geol. Soc. Lond.*, **31**. *10.3.6*, *Fig. 10.73.*

Weaver P.P.E., Rothwell R.G., Ebbing J., Gunn D. & Hunter P.M. (1992) Correlation, frequency of emplacement and source directions of megaturbidites on the Madeira Abyssal Plain. *Mar. Geol.*, **109**, 1–20. *Fig. 10.74.*

Webb P.N., Ronan T.E., Lipps J.H. & DeLaca T.E. (1979) Miocene glaciomarine sediments from beneath the Southern Ross Ice Shelf, Antarctica. *Science*, **203**, 435–437. *11.4.6.*

Weber H.P. (1981) Sedimentologische und Geochemische Untersuchungen in Greifensee (Kanton ZH/Schweiz). Unpublished PhD dissertation, ETH-Zürich, Nr 6811. *4.7.1.*

Weber K.J. (1971) Sedimentological aspects of oilfields of the Niger delta. *Geol. Mijnbouw*, **50**, 559–576. *6.1, 6.6.4, 6.6.6.*

Weber K.J. & Daukoru E. (1975) Petroleum geology of the Niger delta. *Proc. 9th World Petrol. Conf.*, 209–221. *6.6.4, Fig. 6.41.*

Weedon G.P. (1991) The spectral analysis of stratigraphic time series. In: *Cycles and Events in Stratigraphy* (Ed. by G. Einsele, W. Ricken and A. Seilacher), pp. 840–854. Springer-Verlag, Berlin. *2.1.5.*

Weedon G.P. (1993) The recognition and stratigraphic implications of orbital-forcing of climate and sedimentary cycles. In: *Sedimentology Review*, *1* (Ed. by V.P. Wright), pp. 31–50. Blackwell Scientific Publications, Oxford. *2.1.5.*

Weedon G.P. & Jenkyns H.C. (1990) Regular and irregular climatic cycles and the Belemnite Marls (Pliensbachian, Lower Jurassic, Wessex Basin). *J. geol. Soc. Lond.*, **147**, 915–918. *9.4.3.*

Weibe P.H. (1982) Rings of the Gulf Stream. *Sci. Am.*, **246**, 60–79. *7.5.*

Weiler Y., Sass E. & Zak I. (1974) Halite oolites and ripples in the Dead Sea, Israel. *Sedimentology*, **21**, 623–632. *8.5.1.*

Weimer P. (1990) Sequence stratigraphy, facies geometries, and depositional history of the Mississippi Fan, Gulf of Mexico. *Bull. Am. Ass. petrol. Geol.*, **74**, 425–453. *10.3.3, Figs 10.42, 10.43, 10.45.*

Weimer P. (1991) Seismic facies, characteristics, and variations in channel evolution, Mississippi Fan (Plio-Pleistocene), Gulf of Mexico. In: *Seismic Facies and Sedimentary Processes of Submarine Fans and Turbidite Systems* (Ed. by P. Weimer and M.H. Link), pp. 323–347. Springer-Verlag, New York. *10.3.3.*

Weimer P., Bouma A.H. & Perkins R.F. (Eds) (1994) *Submarine Fans and Turbidite Systems*, 440 pp. Gulf Coast Sect., Soc. econ. Paleont. Miner. Earth Enterprises, Austin. *10.3.6.*

Weimer R.J., Howard J.D. & Lindsay D.R. (1982) Tidal flats. In: *Sandstone Depositional Environments* (Ed. by P.A. Scholle and D. Spearing), pp. 191–245. *Mem. Am. Ass. petrol. Geol.*, **31**, Tulsa. *6.7.3.*

Weirich F.H. (1986) A study of the nature and incidence of density currents in a shallow glacial lake. *Ann. Ass. Am. Geogr.*, **76**, 396–413. *4.4.2.*

Weise B.R. (1980) *Wave-dominated Delta Systems of the Upper Cretaceous San Miguel Formation, Maverick Basin, South Texas*, 33 pp. Bureau of Economic Geology, Report of Investigations, 107, Austin, TX. *6.6.6, Figs 6.56, 6.57.*

Weller J.M. (1956) Argument for diastrophic control of Late Paleozoic cyclothems. *Bull. Am. Ass. petrol. Geol.*, **40**, 17–50. *6.6.6.*

Wells J.T. & Coleman J.M. (1981) Physical processes and fine-grained sediment dynamics, coast of Surinam, South America. *J. sedim. Petrol.*, **51**, 1053–1068. *7.6.*

Wells J.T., Prior D.B. & Coleman J.M. (1980) Flowslides in muds on extremely low angle tidal flats, northeastern South America. *Geology*, **8**, 272–275. *6.7.3.*

Wells N.A. & Dorr Jr J.A. (1987) A reconnaissance of sedimentation on the Kosi alluvial fan of India. In: *Recent Developments in Fluvial Sedimentology* (Ed. by F.G. Ethridge, R.M. Flores and M.D. Harvey), pp. 51–61. *Spec. Publ. Soc. econ. Paleont. Miner.*, **39**, Tulsa. *3.3.8.*

Wells S.G. & Harvey A.M. (1987) Sedimentological and geomorphic variations in storm-generated alluvial fans, Howgill Fells, northwest England. *Bull. geol. Soc. Am.*, **95**, 182–198. *Fig. 12.34.*

Wendte J. (1992a) Cyclicity of Devonian strata in the Western Canada sedimentary basin. In: *Devonian–Early Mississippian Carbonates of the Western Canada Sedimentary Basin: A Sequence Stratigraphic Framework* (Ed. by J. Wendte, F.A. Stoakes and C.V. Campbell), pp. 25–39. Soc. econ. Paleont. Miner., Short Course, 28, Tulsa. *9.5.2, 9.5.6.*

Wendte J. (1992b) Platform evolution and its control on reef inception and localisation. In: *Devonian–Early Mississippian Carbonates of the Western Canada Sedimentary Basin: A Sequence Stratigraphic Framework* (Ed. by J. Wendte, F.A. Stoakes and C.V. Campbell), pp. 41–87. Soc. econ. Paleont. Miner., Short Course, 28, Tulsa. *9.5.4.*

Werner F. & Wetzel A. (1982) Interpretation of biogenic structures in oceanic sediments. *Bull. Inst. Geol. Bassin d'Aquitaine*, **31**, 275–288. *10.2.3.*

Wescott W.A. & Ethridge F.G. (1980) Fan-delta sedimentology and tectonic setting – Yallahs fan-delta, southeast Jamaica. *Bull. Am. Ass. petrol. Geol.*, **64**, 374–399. *6.5, Fig. 6.21.*

Wescott W.A. & Ethridge F.G. (1983) Eocene fan delta–submarine fan deposition in the Wagwater Trough, east-central Jamaica. *Sedimentology*, **30**, 235–247. *10.3.4.*

References

Wescott W.A. & Ethridge F.G. (1990) Fan deltas – alluvial fans in coastal settings. In: *Alluvial Fans: a Field Approach* (Ed. by A.H. Rachocki and M. Church), pp. 195–211. Wiley, Chichester. *6.1, 6.3, 6.5, 6.5.4, 6.5.5, Figs 6.20, 6.21, 6.22, 6.25.*

West I.M. (1964) Evaporite diagenesis in the Lower Purbeck Beds of Dorset. *Proc. Yorkshire geol. Soc.*, 34, 315–326. *8.7.*

West I.M., Ali Y.A. & Hilmy M.E. (1979) Primary gypsum nodules in a modern sabkha on the Mediterranean coast of Egypt. *Geology*, 7, 354–358. *8.4.3, Fig. 8.19.*

Westbrook G.K. (1982) The Barbados Ridge complex: tectonics of a mature forearc system. In: *Trench–Forearc Geology* (Ed. by J.K. Leggett), pp. 275–290. *Spec. Publ. geol. Soc. Lond.*, 10. *12.12.5.*

Wetzel A. (1991a) Ecologic interpretation of deep-sea trace fossil communities. *Palaeogeogr. Palaeoclimatol. Palaeoecol.*, 85, 47–69. *10.2.3.*

Wetzel A. (1991b) Stratification in black shales: depositional models and timing – an overview. In: *Cycles and Events in Stratigraphy* (Ed. by G. Einsele, W. Ricken and A. Seilacher), pp. 508–523. Springer-Verlag, Berlin. *7.6.*

Wetzel A. (1993) The transfer of river load to deep-sea fans: a quantitative approach. *Bull. Am. Ass. petrol. Geol.*, 77, 1679–1692. *10.3.3.*

Wetzel R.G. (1983) *Limnology*, 767 pp. Saunders, Philadelphia. *4.7.1, Fig. 4.2.*

Wezel F.C., Savelli D., Bellagamba M., Tramontana M. & Bartole R. (1981) Plio-Quaternary depositional style of sedimentary basins along insular Tyrrhenian margins. In: *Sedimentary Basins of Mediterranean Margins* (Ed. by F.C. Wezel), pp. 239–269. CNR Italian Project of Oceanography. *10.3.5.*

Whitaker J.H.M. (1965) Primary sedimentary structures from the Silurian and lower Devonian of the Oslo region, Norway. *Nature*, 207, 709–711. *7.7.2.*

White G.W., Totten S.M. & Gross D.L. (1969) Pleistocene Stratigraphy of Northwestern Pennsylvania. *Pennsylvania Geol. Surv. Bull.*, G55, 88 pp. *11.4.7, 11.5.3.*

White J.D.L. (1991) The depositional record of small, monogenetic volcanoes within terrestrial basins. In: *Sedimentation in Volcanic Settings* (Ed. by R.V. Fisher and G.A. Smith), pp. 155–174. *Spec. Publ. Soc. econ. Paleont. Miner.*, 45, Tulsa. *12.9, 12.9.1, 12.9.2.*

White J.D.L. (1992) Pliocene subaqueous fans and Gilbert-type deltas in maar crater lakes, Hopi Battes, Navajo Nation (Arizona), USA. *Sedimentology*, 39, 931–946. *12.9.2.*

White J.D.L. & Busby-Spera C.J. (1987) Deep marine arc apron deposits and syndepositional magmatism in the Alistos group at Punta Cono, Baja California, Mexico. *Sedimentology*, 34, 911–927. *10.3.5, 12.12.5.*

White J.D.L. & Robinson P.T. (1992) Intra-arc sedimentation in a low-lying marginal arc, Eocene Clarno Formation, central Oregon. *Sedim. Geol.*, 80, 89–114. *12.12.4.*

White R. & McKenzie D. (1989) Magmatism at rift zones: the generation of volcanic continental margins and flood basalts. *J. geophys. Res.*, 94, 7685–7729. *12.2.1.*

Whitham A. (1989) The behaviour of subaerially produced pyroclastic flows in a subaqueous environment: evidence from the Roseau eruption, Dominica, West Indies. *Mar. Geol.*, 86, 27–40. *12.4.4.*

Whitham A.G. (1993) Facies and depositional processes in an Upper Jurassic to Lower Cretaceous pelagic sedimentary sequence, Antarctica. *Sedimentology*, 40, 331–349. *10.2.3.*

Whittard W.F. (1932) The stratigraphy of the Valentian rocks of Shropshire: The Longmynd–Shelve and Breidden outcrops. *Q. J. geol. Soc. Lond.*, 88, 859–902. *6.1, 6.4.*

Whyatt M., Bowen J.M. & Rhodes D.N. (1991) Nelson – successful application of a development geoseismic model in North Sea exploration. *First Break*, 9, 265–280. *10.3.4, Fig. 10.58.*

Wilber R.J., Milliman J.D. & Halley R.B. (1990) Accumulation of bank-top sediment on the western slope of Great Bahama Bank: rapid progradation of a carbonate megabank. *Geology*, 18, 970–974. *9.4.4.*

Wilde P. & Berry W.B.N. (1992) Progressive ventilation of the oceans – potential for return to anoxic conditions in the post-Paleozoic. In: *Nature and Origin of Cretaceous Carbon-rich Facies* (Ed. by S.O. Schlanger and M.B. Cita), pp. 209–224. Academic Press, London. *7.7.3.*

Wilgus C.K., Hastings B.S., Kendall C.G. St C., Posamentier H.W., Ross C.A. & Van Wagoner J.C. (Eds) (1988) *Sea-level Changes – an Integrated Approach*, 404 pp. *Spec. Publ. Soc. econ. Paleont. Miner.*, 42, Tulsa. *1.1, 7.1.2.*

Wilkinson B.H., Opdyke B.N. & Algeo T.J. (1991) Time partitioning in cratonic carbonate rocks. *Geology*, 19, 1093–1096. *9.2.1.*

Wilkinson B.H., Pope B.N. & Owen R.M. (1980) Nearshore ooid formation in a modern temperate region marl lake. *J. Geol.*, 88, 697–704. *4.7.1.*

Williams G.E. (1969) Piedmont sedimentation and late Quaternary chronology in the Biskra region of the northern Sahara. *Z. Geomorphol., Suppl.*, 10, 40–63. *3.3.7.*

Williams G.E. (1979) Sedimentology, stable-isotope geochemistry and palaeoenvironment of dolostones capping late Precambrian glacial sequences in Australia. *J. geol. Soc. Aust.*, 26, 377–386. *11.5.5.*

Williams G.E. (1989) Late Precambrian tidal rhythmites in South Australia and the history of the Earth's rotation. *J. geol. Soc. Lond.*, 146, 97–111. *6.2.7.*

Williams P.F. & Rust B.R. (1969) Sedimentology of a braided river. *J. sedim. Petrol.*, 39, 649–679. *3.3.2.*

Williams T.M. & Owen R.B. (1992) Geochemistry and origins of lacustrine ferromanganese nodules from the Malawi Rift, central Africa. *Geochim. Cosmochim. Acta*, 56, 2703–2712. *4.7.3.*

Williams-Stroud S. (1994) The evolution of an inland sea of marine origin to a non-marine saline lake: the Pennsylvanian Paradox salt. In: *Sedimentology and Geochemistry of Modern and Ancient Saline Lakes* (Ed. by R.W. Renaut and W.M. Last), pp. 293–306. *Spec. Publ. Soc. econ. Paleont. Miner.*, 50, Tulsa. *8.10.2.*

Williamson C.R. (1977) Deep-sea channels of the Bell Canyon Formation (Guadalupian) Delaware Basin, Texas–New Mexico. In: *Upper Guadalupian Facies, Permian Reef Complex, Guadalupe Mountains, New Mexico and West Texas* (Ed. by M.E. Hileman and S.J. Mazzullo), pp. 409–432. *Soc. econ. Paleont. Miner., Permian Basin Section*, Tulsa, Publ. *77–16. 9.3.3.*

Willis B.J. & Behrensmeyer A.K. (1994) Architecture of Miocene overbank deposits in northern Pakistan. *J. sedim. Res.*, B64, 60–67. *3.6.6.*

Willman H.B., Glass H.D. & Frye J.C. (1966) Mineralogy of glacial tills and their weathering profiles in Illinois: Part II. Weathering profiles. *Illinois State Geol. Surv. Circ.*, 400, 76 pp. *11.4.7.*

Wilson C.J.N. (1980) The role of fluidization in the emplacement of pyroclastic flows: an experimental approach. *J. Volcanol. Geotherm. Res*, 8, 231–249. *12.4.4.*

Wilson C.J.N. (1985) The Taupo eruption, New Zealand II. The Taupo ignimbrite. *Phil. Trans. R. Soc. London*, A314, 229–310. *Fig. 12.1.*

Wilson C.J.N. (1994) Ash-fall deposits from large-scale phreatomagmatic volcanism: limitations of available eruption-column models. *Bull US geol. Surv.*, 2047, 93–99. *12.4.1.*

Wilson C.J.N. & Walker G.P.L. (1982) Ignimbrite depositional facies: the anatomy of a pyroclastic flow. *J. geol. Soc. Lond.*, 139, 581–592. *12.4.4.*

Wilson C.J.N., Houghton B.F., McWilliams M.O., Lanphere M.A., Weaver S.D. & Briggs R.M. (1995) Volcanic and structural evolution of Taupo Volcanic Zone, New Zealand: a review. *J. Volcanol. Geotherm. Res.*, 68, 1–28. *12.13.*

Wilson C.J.N., Rogan A.M., Smith I.E.M., Northey D.J., Nairn I.A. & Houghton, B.F. (1984) Caldera volcanoes of the Taupo Volcanic zone, New Zealand. 89 B10, 8463–8484. *12.4.6.*

Wilson I.G. (1971) Desert sandflow basins and a model for the development of ergs. *Geogr. J.*, 137, 180–199. *5.1.*

Wilson I.G. (1973) Ergs. *Sedim. Geol.*, 10, 77–106. *5.1, 5.2.2, 5.4.1.*

Wilson J.L. (1975) *Carbonate Facies in Geologic History*, 471 pp. Springer-Verlag, Berlin. *1.1, 2.2.2, 9.1.1, 9.4.1.*

Wilson J.T. (1963) A possible origin for the Hawaiian islands. *Can. J. Phys.*, 41, 863–870. *12.2.1.*

Wilson J.T. (1966) Did the Atlantic close and then re-open? *Nature*, 211, 676–681. *12.12.5.*

Wilson L. (1972) Explosive volcanic eruptions – 2. The atmospheric trajectories of pyroclasts. *Geophys. J. R. Astron. Soc.*, 30, 381–392. *12.4.3.*

Wilson L. (1980) Relationships between pressure, volatile content and ejecta velocity in three types of volcanic explosion. *J. Volcanol. Geotherm. Res.*, 8, 297–313. *12.4.1.*

Wilson L. & Head J.W. (1981) Ascent and eruption of basaltic magma on the earth and the moon. *J. geophy. Res.*, 86, 2971–3001. *12.4.1.*

Wilson L., Sparks R.S.J. & Walker G.P.L. (1980) Explosive volcanic eruptions – 4. The control of magma properties and conduit geometry on eruption column behaviour. *Geophys. J. R. Astron. Soc.*, 63, 117–148. *12.4.2, Fig. 12.15.*

Wilson M. (1989) *Igneous Petrogenesis*, 466 pp. Unwin Hyman, London. *12.2.2, Figs 12.6, 12.8.*

Wilson M.V.H. (1977) Paleoecology of Eocene lacustrine varves at Horsefly, British Columbia. *Can. J. Earth Sci.*, 14, 953–962. *4.15.2.*

Wilson M.V.H. (1980) Eocene lake environments: depth and distance-from-shore variation in fish insect, and plant assemblages. *Palaeogeogr. Palaeoclimatol. Palaeoecol.*, 32, 21–44. *4.15.*

Wilson P.A. & Roberts H.H. (1995) Density cascading: off-shelf sediment transport, evidence and implications, Bahama Banks. *J. sedim. Res.*, A65, 45–56. *9.4.4.*

Windom H.L. (1975) Eolian contributions to marine sediments. *J. sedim. Petrol.*, 45, 520–529. *10.4.1.*

Winker C.D. & Edwards M.B. (1983) Unstable progradational clastic shelf margins. In: *The Shelfbreak: Critical Interface on Continental Margins* (Ed. by D.J. Stanley and G.T. Moore), pp. 139–157. *Spec. Publ. Soc. econ. Paleont. Miner.*, 33, Tulsa. *6.6, 6.6.5, Fig. 6.42.*

Winn R.D. & Dott R.H. (1978) Submarine-Fan Turbidites and Resedimented Conglomerates in a Mesozoic Arc-Rear Marginal Basin in Southern South America. In: *Sedimentation in submarine canyons, fans and trenches* (Ed. by D.J. Stanley and G. Kelling), pp. 362–373. Dowden, Hutchinson and Ross. Stroudsburg, Pennsylvania. *12.13.5.*

Winterer E.L. & Bosellini A. (1981) Subsidence and sedimentation on Jurassic passive continental margin, southern Alps, Italy. *Bull. Am. Ass. petrol. Geol.*, 65, 394–421. *9.3.5.*

Winterer E.L. & Sarti M. (1994) Neptunian dykes and associated features in southern Spain: mechanisms of formation and tectonic implications. *Sedimentology*, 41, 1109–1132. *Fig. 9.20.*

Wintle A.G. (1993) Luminescence dating of aeolian sands: an overview. In: *The Dynamics and Environmental Context of Aeolian Sedimentary Systems* (Ed. by K. Pye), pp. 49–58. *Spec. Publ. geol. Soc. Lond.*, 72. *5.4.7.*

Wiseman W.J., Fan Y.-B. & Bornhold B.D. (1986) Suspended sediment advection by tidal currents off the Huanghe (Yellow River) delta. *Geo-Mar. Lett.*, 6, 107–113. *7.6.*

Wohletz, K.H. (1986) Explosive magma–water interactions: thermodynamics, explosion mechanisms, and field studies. *Bull. Volcanol.*, 48, 245–264. *12.4.1.*

Wohletz K.H. & Sheridan M.F. (1979) A model of pyroclastic surge. *Spec. Pap. geol. Soc. Am.*, 180, 177–194. *12.4.4.*

Wohletz K.H. & Sheridan M.F. (1983) Hydrovolcanic explosions II: evolution of basaltic tuff rings and tuff cones. *Am. J. Sci.*, 283, 384–413. *12.9.2, Fig. 12.12.*

Wohletz K.H., McGetchin T.R., Sandford M.T. & Jones E.M. (1984) Hydrodynamic aspects of caldera-forming eruptions: numerical models. *J. geophy. Res.*, 89, 8269–8285. *12.4.2, 12.4.4.*

Wohletz K.H., Heiken G., Ander M., Goff F., Vuataz F.-D. & Wadge G. (1986) Qualibou caldera, St. Lucia, West Indies. *J. Volcanol. Geotherm. Res.*, 27, 77–155. *12.12.3.*

Woldstedt P. (1970) *International Quaternary Map of Europe, Sheet 6, København.* Bundesanstalt für Bodenforschung, Hannover. *11.5.3.*

Woldstedt P. (1971) *International Quaternary Map of Europe, Sheet 7, Moskva.* Bundesanstalt für Bodenforschung, Hannover. *11.5.3.*

Wolf T. (1878) *Der Cotopaxi und seine letzte Eruption am 26 Juni 1877*, pp. 113–167. *12.1.1.*

Wonham J.P. & Elliott T. (1996) High-resolution sequence stratigraphy of a mid-Cretaceous estuarine complex: the Woburn Sands of the Leighton Buzzard area, southern England. In: *Sequence Stratigraphy in British Geology* (Ed. by S.P. Hesselbo and D.N. Parkinson), pp. 9–30. *Spec. Publ. geol. Soc. Lond.*, 103, Bristol. *6.7.7, 7.7.1.*

Wood A. & Smith A.J. (1959) The sedimentation and sedimentary history of the Aberystwyth Grits (Upper Llandoverian). *Q. J. geol. Soc. Lond.*, 114, 163–195. *10.1.*

Wood C.A. (1980a) Morphometric evolution of cinder cones. *J. Volcanol. Geotherm. Res.*, 7, 387–413. *12.9.1.*

Wood C.A. (1980b) Morphometric evolution of cinder cone

degradation. *J. Volcanol. Geotherm. Res.*, 8, 137–160. *12.6.1, 12.9.1.*

Wood G.V. & Wolfe M.J. (1969) Sabkha cycles in the Arab/Darb Formation of the Trucial Coast of Arabia. *Sedimentology*, 12, 165–191. *8.7.*

Wood R. (1993) Nutrients, predation and the history of reef-building. *Palaios*, 8, 526–543. *9.2, 9.4.5.*

Woodcock N.H. (1979) The use of slump structures as palaeoslope orientation estimators. *Sedimentology*, 26, 83–99. *10.2.3.*

Woodland A.W. (1970) The buried tunnel-valleys of East Anglia. *Proc. Yorkshire geol. Soc.*, 37(4), 521–578. *11.4.1.*

Woodroffe C.D., Curtis R.J. & McLean R.F. (1983) Development of a chenier plain, Firth of Thames, New Zealand. *Mar. Geol.*, 53, 1–22. *6.7.2.*

Woodroffe C.D., Chappell J., Thom B.G. & Wallensky E. (1989) Depositional model of a macrotidal estuary and floodplain, South Alligator River, Northern Australia. *Sedimentology*, 36, 737–756. *6.7.3, 6.7.5.*

Woodrow D.L. & Isley A.M. (1983) Facies, topography, and sedimentary processes in the Catskill Sea (Devonian), New York and Pennsylvania. *Bull. geol. Soc. Am.*, 94, 459–470. *10.3.4.*

Woodyer K.D., Taylor G. & Crook K.A.W. (1979) Depositional processes along a very low-gradient, suspended-load stream: the Barwon River, New South Wales. *Sedim. Geol.*, 22, 97–120. *3.3.5.*

Wright A.E. & Bowes D.R. (1963) Classification of volcanic breccias: a discussion. *Bull. geol. Soc. Am. Bull.*, 74, 79–86. *12.7.*

Wright I.C. (1992) Shallow structure and active tectonism of an offshore continental back-arc spreading system: the Taupo Volcanic zone, New Zealand. *Mar. Geol.*, 103, 287–309. *12.12.1.*

Wright J.V. & Walker G.P.L. (1977) The ignimbrite source problem: significance of a co-ignimbrite lag fall deposit. *Geology*, 5, 729–732. *12.4.4.*

Wright J.V. & Walker G.P.L. (1981) Eruption, transport and deposition of ignimbrite: a case study from Mexico. *J. Volcanol. Geotherm. Res.*, 9, 111–131. *12.4.4.*

Wright J.V., Smith A.L. & Self S. (1980) A working terminology of pyroclastic deposits. *J. Volcanol. Geotherm. Res.*, 8, 315–336. *Fig. 12.14.*

Wright L.D. (1977) Sediment transport and deposition at river mouths: a synthesis. *Bull. geol. Soc. Am.*, 88, 857–868. *6.2.2, 6.6.2, Figs 6.5, 6.35.*

Wright L.D. & Coleman J.M. (1973) Variations in morphology of major river deltas as functions of ocean wave and river discharge regimes. *Bull. Am. Ass. petrol. Geol.*, 57, 370–398. *6.1, 6.2.4.*

Wright L.D. & Coleman J.M. (1974) Mississippi River mouth processes: effluent dynamics and morphologic development. *J. Geol.*, 82, 751–778. *6.6.2, Fig. 6.34.*

Wright L.D. & Short A.D. (1984) Morphodynamic variability of surf zones and beaches: a synthesis. *Mar. Geol.*, 56, 93–118. *6.2.4, Fig. 12.37.*

Wright L.D., Chappell J., Thom B.G., Bradshaw M.P. & Cowell P. (1979) Morphodynamics of reflective and dissipative beach and inshore systems, Southeastern Australia. *Mar. Geol.*, 32, 105–140. *6.2.4.*

Wright L.D., Coleman M. & Thom B.G. (1973) Processes of channel development in a high tide range environment: Cambridge

Gulf–Ord River delta, western Australia. *J. Geol.*, 81, 15–41. *6.7.5.*

Wright L.D., Yang Z.-S., Bornhold B.D. *et al.* (1986) Hypepycnal plumes and plume fronts over the Huanghe (Yellow River) delta front. *Geo-Mar. Lett.*, 6, 97–105. *7.6.*

Wright R. & Anderson J.B. (1982) The importance of sediment gravity flow to sediment transport and sorting in a glacial marine environment: Eastern Weddell Sea, Antarctica. *Bull. geol. Soc. Am.*, 93, 951–963. *11.4.6.*

Wright R., Anderson J.B. & Frisco P.P. (1983) Distribution and association of sediment gravity flow deposits and glacial/glacial-marine sediments around the continental margin of Antarctica. In: *Glacial-Marine Sedimentation* (Ed. by B.F. Molnia), pp. 265–300. Plenum Press, New York. *11.4.6.*

Wright R.F. & Nydegger P. (1980) Sedimentation of detrital particulate matter in lakes: influence of currents produced by inflowing rivers. *Water Resour. Res.*, 16, 597–601. *4.4.2.*

Wright V.P. (1984) Peritidal carbonate facies models: a review. *Geol. J.*, 19, 309–325. *9.4.1, Figs 9.23, 9.24, 9.25.*

Wright V.P. (1986) Facies sequences on a carbonate ramp: the Carboniferous Limestone of South Wales. *Sedimentology*, 33, 221–241. *9.5.2, Fig. 9.23.*

Wright V.P. (1992a) Paleosol recognition: A guide to early diagenesis in terrestrial settings. In: *Diagenesis, III* (Ed. by K.H. Wolf and G.V. Chillingarian), pp. 591–619. *Developments in Sedimentology*, 47. Elsevier, Amsterdam. *3.5.4.*

Wright V.P. (1992b) Paleopedology: stratigraphic relationships and empirical models. In: *Weathering, Soils and Paleosols* (Ed. by I.P. Martini and W. Chesworth), pp. 475–499. Elsevier, Amsterdam. *4.11.3.*

Wright V.P. (1994a) Early Carboniferous carbonate systems: an alternative to the Cainozoic paradigm. *Sedim. Geol.*, 93, 1–5. *9.2.*

Wright V.P. (1994b) Paleosols in shallow marine carbonate sequences. *Earth Sci. Rev.*, 35, 367–395. *9.4.1.*

Wright V.P. & Alonso Zarza A.M. (1990) Pedostratigraphic models for alluvial fan deposits: a tool for interpreting ancient sequences. *J. geol. Soc. Lond.*, 147, 8–10. *3.3.8, Fig. 3.30.*

Wright V.P. & Faulkner T.J. (1990) Sediment dynamics of early Carboniferous ramps: a proposal. *Geol. J.*, 25, 139–144. *9.3.4.*

Wright V.P. & Robinson D. (1988) Early Carboniferous floodplain deposits from South Wales: a case study of the controls on palaeosol development. *J. geol. Soc. Lond.*, 145, 847–857. *3.5.4, 3.6.6.*

Wright V.P. & Wilson R.C.L. (1984) A carbonate submarine-fan sequence from the Jurassic of Portual. *J. sedim. Petrol.*, 54, 394–412. *9.4.4.*

Wright V.P., Sloan R.J., Valero Garces B. & Garvie L.A.J. (1992) Groundwater ferricretes from the Silurian of Ireland and Permian of the Spanish Pyrenees. *Sedim. Geol.*, 77, 37–49. *3.5.4.*

Xue C. (1993) Historical changes in the Yellow River delta. *Mar. Geol.*, 113, 321–329. *6.6.5.*

Yaalon D.H. & Dan J. (1974) Accumulation and distribution of loess derived deposits in the semi-desert and desert fringe areas of Israel. *Z. Geomorph., N.F. Suppl. Bd*, 20, 91–105. *3.2.3.*

Yamada E. (1984) Subaqueous pyroclastic flows: their development and deposits. In: *Marginal Basin Geology* (Ed. by B.P. Kokelaar and M.F. Howells), pp. 29–35. *Spec. Publ. geol. Soc., Lond.*, **16**. *12.4.4, Fig. 12.28.*

Yamada E. (1988) Geological development of the Onikobe caldera, Northeast Japan, with special reference to its hydrothermal system. In: *Research in the Kurikoma Geothermal Area* (Ed. by E. Yamada, H. Hase and K. Ogawa), pp. 61–162. *Geol. Surv. Jpn*, **268**. *12.12.4.*

Yamagishi H. (1991) Morphological and sedimentological characteristics of the Neogene submarine coherent lavas and hyaloclastites of southwest Hokkaido, Japan. *Sedim. Geol.*, **74**, 5–23. *12.4.5.*

Yamamoto T., Takarada S. & Suto S. (1993) Pyroclastic flows from the 1991 eruption of Unzen volcano, Japan. *Bull. Volcanol.*, **55**, 166–175. *12.4.4, Fig. 12.1.*

Yang C.-S. (1989) Active, moribund and buried tidal sand ridges in the East China Sea and southern Yellow Sea. *Mar. Geol.*, **88**, 97–116. *7.3.4.*

Yang C.-S. & Nio S.-D. (1985) The estimation of palaeohydrodynamic processes from subtidal deposits using time series analysis methods. *Sedimentology*, **32**, 41–57. *6.2.7, Fig. 6.14.*

Yang C.-S. & Nio S.D. (1989) An ebb-tide delta depositional model — a comparison between the modern Eastern Scheldt tidal basin (southwest Netherlands) and the lower Eocene Roda Sandstone in the southern Pyrenees (Spain). *Sedim. Geol.*, **64**, 175–196.

Yang C.-S. & Sun J.-S. (1988) Tidal sand ridges on the East China Sea shelf. In: *Tide-influenced Sedimentary Environments and Facies* (Ed. by P.L. de Boer, A. van Gelder and S.D. Nio), pp. 23–38. Reidel, Dordrecht. *6.7.6.*

Yemane K. (1993) Contribution of Late Permian palaeogeography in maintaining a temperate climate in Gondwana. *Nature*, **361**, 51–54. *4.10.*

Young G.M. (1991) The geologic record of glaciation: Relevance to the climatic history of Earth. *Geosci. Can.*, **18**(3), 100–108. *11.6.*

Young G.M. & Gostin V.A. (1989) An exceptionally thick upper Proterozoic (Sturtian) glacial succession in the Mount Painter area, South Australia. *Bull. geol. Soc. Am.*, **101**, 834–845. *11.5.5.*

Young G.M. & Gostin V.A. (1991) Late Proterozoic (Sturtian) succession of the North Flinders Basin, South Australia: An example of temperate glaciation in an active rift setting. In: *Glacial marine sedimentation; Paleoclimatic Significance* (Ed. by J.B. Anderson and G.M. Ashley), pp. 207–222. *Spec. Pap. geol. Soc. Am.*, **261**, Boulder, CO. *11.5.5.*

Young R.W. & Bryant E.A. (1992) Catastrophic wave erosion on the southeastern coast of Australia: impact of the Lanai tsunamis ca.

150 ka. *Geology*, **20**, 199–202. *12.5.1.*

Yuretich R.F. (1979) Modern sediments and sedimentary processes in Lake Rudolf (Lake Turkana), eastern Rift Valley, Kenya. *Sedimentology*, **26**, 313–332. *4.4.2.*

Yuretich R.F. (1989) Paleocene lakes of the central Rocky Mountains, western United States. *Palaeogeogr. Palaeoclimatol. Palaeoecol.*, **70**, 53–63. *4.15.1.*

Yurewicz D.A. (1977) The origin of massive facies of the Lower and Middle Capitan Limestone (Permian) Guadalupe Mountains, New Mexico and West Texas. In: *Upper Guadalupian Facies, Permian Reef Complex, Guadalupe Mountains, New Mexico and West Texas* (Ed. by M.E. Hileman and S.J. Mazzullo), pp. 45–92. *Soc. econ. Paleont. Miner., Permian Basin Section*, Tulsa, Publ. 77–16. *9.3.3.*

Zaitlin B.A. & Schultz B.C. (1990) Wave-influenced estuarine sand body, Senlac heavy oil pool, Saskatchewan, Canada. In: *Sandstone Petroleum Reservoirs* (Ed. by J.H. Barwis, J.G. McPherson and J.R.J. Studlick), pp. 363–387. Springer-Verlag, New York. *6.7.5.*

Zaitlin B.A., Dalrymple R.W. & Boyd R. (1994) The stratigraphic organisation of incised-valley systems associated with relative sea-level change. In: *Incised Valley Systems: Origin and Sedimentary Sequences* (Ed. by R.W. Dalrymple, R. Boyd and B.A. Zaitlin), pp. 45–60. *Spec. Publ. Soc. sedim. Geol.*, **51**, Tulsa. *3.6.9, 6.7.6, Figs 6.88, 6.89, 6.90.*

Zempolich W.G. (1993) The drowning succession in Jurassic carbonates of the Venetian Alps, Italy: a record of supercontinent break-up, gradual eustatic rise, and eutrophication of shallow water environments. In: *Carbonate Sequence Stratigraphy: Recent Developments and Applications* (Ed. by R.G. Loucks and J.F. Sarg), pp. 63–105. *Mem. Am. Ass. petrol. Geol.*, **57**, Tulsa. *9.3.5.*

Zhuang W.Y. & Chappell J. (1991) Effects of seagrass on tidal flat sedimentation, Corner Inlet, southeast Australia. In: *Clastic Tidal Sedimentology* (Ed. by D.G. Smith, G.E. Reinson, B.A. Zaitlin & R.A. Rahmani) pp. 291–300. *Mem. Can. Soc. petrol. Geol.*, **16**. *6.7.3.*

Ziegler M.A. (1989) North German Zechstein facies patterns in relation to their substrate. *Geol. Rundsh.*, **78**, 105–127. *8.9.*

Ziegler P.A. (1982) *Geological Atlas of Western and Central Europe*, 130 pp. Shell Internationale Petrol. Maatschappij B.V., The Hague. *7.7.3.*

Ziegler P.A. (1990) *Geological Atlas of Western and Central Europe*, 239 pp. Shell Internationale Petrol. Maatschappij B.V., The Hague. *4.11.3. 7.7.3.*

Index

Numbers refer to pages, whether in text, figures or tables. Numbers in **bold** indicate either the principal references, or where a definition may be found.